INTERSTELLAR PROCESSES

ASTROPHYSICS AND SPACE SCIENCE LIBRARY

A SERIES OF BOOKS ON THE RECENT DEVELOPMENTS

OF SPACE SCIENCE AND OF GENERAL GEOPHYSICS AND ASTROPHYSICS

PUBLISHED IN CONNECTION WITH THE JOURNAL

SPACE SCIENCE REVIEWS

VOLUME 134

PROCEEDINGS

INTERSTELLAR PROCESSES

PROCEEDINGS OF THE SYMPOSIUM ON INTERSTELLAR PROCESSES,
HELD IN GRAND TETON NATIONAL PARK, JULY 1986

Edited by

DAVID J. HOLLENBACH

NASA Ames Research Center,
Moffett Field, California, U.S.A.

and

HARLEY A. THRONSON, Jr.

Department of Physics and Astronomy,
The University of Wyoming, Laramie, Wyoming, U.S.A.

D. REIDEL PUBLISHING COMPANY

A MEMBER OF THE KLUWER  ACADEMIC PUBLISHERS GROUP

DORDRECHT / BOSTON / LANCASTER / TOKYO

Library of Congress Cataloging in Publication Data

Interstellar processes.

(Astrophysics and space science library; v. 134)
Includes index.
1. Astrophysics—Congresses. 2. Interstellar matter—Congresses.
I. Hollenbach, D. J. (David J.) II. Thronson, Harley A. III. Summer School
on Interstellar Processes (1986: Grand Teton National Park, Wyo.) IV. Series.
QB460.I59 1987 523.1′135 87-4880
ISBN 90-277-2482-2 (hbk.)
ISBN 90-277-2485-7 (pbk.)

Published by D. Reidel Publishing Company,
P.O. Box 17, 3300 AA Dordrecht, Holland.

Sold and distributed in the U.S.A. and Canada
by Kluwer Academic Publishers,
101 Philip Drive, Assinippi Park, Norwell, MA 02061, U.S.A.

In all other countries, sold and distributed
by Kluwer Academic Publishers Group,
P.O. Box 322, 3300 AH Dordrecht, Holland.

TABLE OF CONTENTS

PART 2

SECTION I: INTERSTELLAR DUST GRAINS

SECTION III: HEATING, COOLING AND RADIATIVE PROCESSES

Preface

The idea for an international symposium on the interstellar medium
was first discussed at the University of Wyoming during the summer of
1984. It was obvious that the outstanding natural beauty of the Teton
mountain range in northwestern Wyoming must be matched by a meeting with
the broadest appeal to the astronomical community. If the meeting was to
produce a book, it must likewise be an important contribution to the
astronomical literature. It was for these reasons that early in the
discussions, it was decided that the University should host a "school",
with the invited speakers presenting tutorials on a broad range of topics
involving the interstellar medium. The symposium proceedings would then
be a compilation of the written versions of these presentations. It has
been nearly a decade since Lyman Spitzer published his classic text on
the interstellar medium and we felt the need for a school and book that
would focus on the recent developments in our understanding of the interstellar medium. Thus, we view this two-volume set as an adjunct text
to Spitzer's book.

From the first, it was important to the symposium organizers that
younger scientists -- graduate students and post-docs, in particular --
play an important role in the summer school. Not only is it important
to include these junior scientists in major scientific conferences for
their personal benefit, but it is our belief that the larger the number
of young people in attendance, the greater the influence the meeting will
have long after individuals have departed to their home institutions.
Two organizing committees were assembled, composed of scientists with
broad experience in studying the interstellar medium. All the individuals
who worked on the two committees agreed from the start that junior
scientists should play a major role in the meeting, in the audience and
as speakers.

Two institutions strongly supported our goals for the meeting with
financial support: the University of Wyoming and the National Aeronautics
and Space Administration. In the first case, the University of Wyoming
supported the meeting with organizational assistance and substantial
financial help. NASA underwrote the publication of an inexpensive
copy of this text, specifically designed for students. In addition,
NASA has published the extended abstracts from the contributed papers
in their series of technical memoranda.

As usual, numerous individuals whose primary goal was a successful
meeting, rather than personal recognition, helped enormously. We are,
of course, grateful to every member of the two organizing committees
and the long discussions that they put up with. Mike Shull and Nick
Scoville, especially, went out of their way to advise the editors on
the components of a successful book. At the University of Wyoming,
Drs. Joan Wadlow and Walter Eggers (Deans of the College of Arts &
Sciences) patiently listened to every new reason for additional money
and, in all cases, gave in. The University's Vice-President for

Research, Dr. Ralph DeVries, needed only short phone calls to be prompted to contribute to our budget. In the Department of Physics & Astronomy, our secretaries --- Evelyn Haskell, Judy Spaulding, and Michael Mann --- typed, stapled, copied, and mailed the enormous amount of paper that was required to keep the momentum of a large international meeting going. Evelyn Haskell also handled the duties of registrar at the meeting site. Not to forget Marce Mitchum, who paid the bills, without asking too many questions about where the money was going to come from. We are especially grateful to our department chairman, Dr. Glen Rebka, who found the personnel, money, and patience for the summer school.

At NASA, we appreciate the support from Louis Haughney at the NASA-Ames Medium Altitude Missions Branch, for his work in support of the Kuiper Airborne Observatory and the Summer School on Interstellar Processes. Ed Weiler, at NASA Headquarters, also readily contributed to the support of an inexpensive student's edition of the symposium proceedings. Janice Varney, secretary for the Astrophysical Experiments Branch at NASA Ames, worked diligently to prepare the final manuscript.

At nearly the last minute, with the book due at the publisher's, Eric Herbst and Paul Goldsmith agreed to write major sections. We appreciate their agreeing to this intrusion on their time.

This book consists of contributions from nearly 30 authors. Each of them took our request to produce a tutorial on a major topic very seriously and this book reflects their work. We also appreciate the prompt return of each section, so that these volumes could be produced in a timely manner. Our goal was to produce a pair of books that could be used as an adjunct text in a graduate course on the interstellar medium. We believe that the authors of the sections succeeded and, for that, we are most grateful.

David Hollenbach
NASA-Ames Research Center

Harley A. Thronson, Jr.
Wyoming Infrared Observatory

December 15, 1986

The Summer School on Interstellar Processes
Grand Teton National Park
Wyoming, United States
July 1 - 7, 1986

Supporting Institutions
University of Wyoming
National Aeronautics and Space Administration

Sponsoring Institutions
American Astronomical Society
Canadian Astronomical Society
Inst. de Astronomia (UNAM)

Scientific Organizing Committee

T. de Jong
Univ. of Amsterdam

D. Flower
Univ. of Durham

J. Hackwell
Univ. of Wyoming

D. Hollenbach
NASA-Ames Research Center

C. Lada
Univ. of Arizona

P. Martin
Univ. of Toronto

L. Rodriguez
UNAM

N. Scoville
Caltech

J. M. Shull
Univ. of Colorado

H. Thronson (Chairman)
Univ. of Wyoming

Local Organizing Committee
(University of Wyoming)

M. Greenhouse

T. Hayward

P. Johnson (Chairman)

H. Thronson

Photographs by B. Stormy Burns & Harley Thronson

Editorial Supervision by Gert Kiers, D. Reidel Publishing Company

PART 1

The plane of the Milky Way at infrared wavelengths. IRAS data at 12, 60, and 100 μm was used to produce this image of the emission from the first galactic quadrant ($\ell \cong$ 0 – 90°). Each strip covers 45° in longitude and ± 5° in latitude. A constant logarithmic stretch in intensity was used throughout. Emission from relatively cool dust (25 – 30 K) associated with molecular clouds appears red in color, while the higher-temperature emission (> 100 K) associated with HII regions and late-type stars registers blue. The center of the galaxy appears in the upper right; the extreme right border of the upper panel is at ℓ = -5°. The nearby Cygnus complex of molecular clouds and HII regions is in the center of the lowest panel. The figure was prepared at the Infrared Processing and Analysis Center by G. Laughlin, J. Good, and N. Scoville.

Section I: The Milky Way as a Galaxy

Jay Lockman and Larry Marschall

THE MILKY WAY AS A GALAXY

M. Jura
Department of Astronomy
UCLA
Los Angeles CA 90024

ABSTRACT.

The physical conditions in the interstellar medium of the Milky Way are briefly reviewed. It is noted that the local interstellar medium shows relatively uniform abundances and dust/gas ratios; in fact recent measurements of the $^{12}C/^{13}C$ ratio indicate that the gas is homogeneous to better than 15%. Since we appear to live in an orderly system, observations of external galaxies similar to the Milky Way can provide useful insight into the nature of the interstellar medium of the Galaxy.

1. INTRODUCTION

The goals of research on the interstellar medium include both providing a detailed physical description of the environment and developing insight into the processes that govern the observed conditions. The past 20 years have seen the application of submillimeter, millimeter, infrared, ultraviolet, X-ray and γ ray observational techniques to studying the interstellar medium, in addition to the more traditional optical, radio and cosmic ray methods, and we now have quite sophisticated ways of investigating the interstellar medium. Compared to 20 or even 10 years ago, we have a much better picture of interstellar conditions ranging from the giant molecular clouds to the regions of hot ($T \sim 10^6$ °K gas).

It should be recognized that substantial observational improvements can still be made. For example, much of the entire energy budget of the Galaxy is measured in the infrared because dust grains absorb light initially emitted by stars. Although the IRAS survey has given us a vastly improved picture of the dust emission from the Galaxy, the lowest energy band was centered at 100 μm. Perhaps half of the infrared from the Milky Way is radiated at wavelengths greater than 100 μm, and further studies in the submillimeter range are required.

Some of the more salient structures in the interstellar medium are listed in Table 1; they are discussed in much more detail in other papers in this conference. The interstellar medium contains both regions well separated from stars, such as diffuse clouds, and structures that are intimately associated with stars such as supernova remnants.

D. J. Hollenbach and H. A. Thronson, Jr. (eds.), Interstellar Processes, 3–17.
© *1987 by D. Reidel Publishing Company.*

Table 1   --   Some Structures in the Interstellar Medium

Name	Density (cm^{-3})	Temperature $(°K)$	Main Diagnostics
Hot, intercloud gas	$3 \ 10^{-3}$	10^6	Soft X-ray emission, UV Absorption lines
Warm, intercloud gas	0.1	10^4	If ionized -- optical emission lines, pulsar dispersion measures If neutral -- 21 cm observations
Diffuse Cloud	>10	100	Optical, UV absorption lines, 21 cm line
Dark cloud	$>10^3$	10	Dark patches, IR emission, molecules
H II Regions	>10	10^4	Optical, IR, UV emission lines
Supernova Remnants	>1	10^4 to 10^7	Radio to X-rays

The interstellar medium is very far from thermodynamic equilibrium. While it is still possible to use a temperature parameter to approximate the microscopic velocity distribution of the atoms and molecules as a Maxwellian, the gas temperature generally only refers to the mean kinetic energy. The state of ionization and the molecular composition are all controlled by non-equilibrium processes. However, in many but not all circumstances, a steady state does prevail.

Evidently, the interstellar medium is very complex. However, the goal of this review is to indicate that there are some striking regularities in all this diversity and that therefore one can sensibly study either the Milky Way or other galaxies as a whole to gain insight into the interstellar medium. In Section II, I discuss the energy density of the interstellar medium while in Section III, I consider the chemical homogeneity of the Galaxy. Finally, because the Galaxy does appear to be a predictable system, in Section IV, I describe some consequences that can be derived from looking at the Milky Way as a whole.

II. ENERGY DENSITY OR PRESSURE

A fundamental feature of the interstellar medium is that even though there is a wide range of temperatures and pressures, the typical energy density of important constituents is on the order of 1 eV cm^{-3}. That is, although the gas may be as cool as 10 °K or as hot as 10^6 °K, the typical gas pressure, nT, is $\sim 3 \ 10^3$ cm^{-3} °K (Jenkins, Jura and

Loewenstein 1983). A number of years ago, it was suggested that the gas was largely at constant pressure (Field, Goldsmith and Habing 1969). However, more sophisticated observations and models show that while the matter is driven towards uniform pressure, there are large and important excursions from the mean behind shock waves produced by supernovae and winds from hot stars, H II regions, and self-gravitating clouds where there is considerable compression of the gas.

An important consequence of the current analysis of the interstellar medium is that much if not most of the volume of the region is filled with very hot gas at $\sim 10^6$ °K, but with a density of only about 0.003 cm^{-3} (see Cox and Smith 1974, McKee and Ostriker 1977, Cox and McCammon 1986). Most of the mass of the interstellar medium is contained within clouds with densities larger than 10 cm^{-3} and temperatures lower than 100 °K.

Not only does the gas have an energy density of about 1 eV cm^{-3}, but so do other important constituents as well. For example, Mathis, Mezger and Panagia (1983) have described in some detail the energy density of starlight in the solar neighborhood. They find that between 912 Å, the cutoff provided by the opacity of neutral hydrogen, and 8 μm, the long wavelength cutoff beyond which most stars do not radiate, the mean energy density is 0.5 eV cm^{-3}. The radiation field near the Sun seems to be representative of most of the volume of the local Milky Way, but there are locations either near stars where the radiation energy density is particularly high or inside dust clouds where it is unusually low (Jura 1974).

The energy density of cosmic rays is also close to 1 eV cm^{-3} (see, for example, Leger, Jura and Omont 1985). This conclusion is reached both from measurements of the cosmic ray flux at the earth and, indirectly, by studying the density of ionized molecules (see Spitzer and Jenkins 1975) in the interstellar matter and the production of γ rays that result from interactions between cosmic rays and cold interstellar matter (Cesarsky and Volk 1978). The energy density of cosmic rays seems to be relatively constant in the solar neighborhood.

Finally, the magnetic field strength is characteristically on the order of 3 10^{-6} Gauss (Heiles 1976, Troland and Heiles 1982) which corresponds to an energy density, $B^2/8\pi$, of close to 1 eV cm^{-3}. Measurement of the magnetic field in the Milky Way is difficult, but it does seem to be close to this value at least in partially ionized regions where pulsar dispersion measures can be obtained and in some clouds where Zeeman splitting of the 21 cm line of hydrogen is observed.

It should also be recognized that while the energy densities of the magnetic field, cosmic rays, starlight and the gas thermal pressure are all roughly comparable, the theory to explain this rough equality has not yet been described. It may even be a coincidence, but this seems unlikely.

III.   ABUNDANCES AND THEIR FLUCTUATIONS

For a number of years, the zero order approximation has been to assume
that the interstellar abundances are characteristic of Population I in
general and of the Sun in particular. This hypothesis of "cosmic
abundances" has been successful in understanding the relatively crudely
determined abundances in stars, H II regions, diffuse interstellar
clouds and the sun. However, sufficient progress has been made in the
precision of measurements of abundances that it is now possible to ask
questions about deviations from "cosmic abundances" (see Grevesse 1984).
It has even been suggested that the solar system does not have cosmic
abundances (Olive and Schramm 1982). Different elements and isotopes
are formed in different sorts of stars, and it is at least possible that
interstellar clouds are not well mixed. Alternatively, if the
interstellar medium is homogeneous, this may serve as an important
constraint on the physical processes in the interstellar medium.
        The discussion about abundances in this review is restricted to
cold clouds within 1 kpc of the sun since it is difficult to measure
accurate abundances beyond this distance. Also, 1 kpc is characteristic
of the extent of the epicyclic orbits for motions around the Galactic
Center, and we might expect that Galactic abundance gradients are not
important within this region.
        Roughly half of all the elements in the interstellar medium heavier
than helium are contained within solid grains (Aannestad and Purcell
1973, Spitzer 1978). As a result, one way to measure crudely the
fluctuations of the local abundances is to measure the dust to gas
ratio. By using Copernicus to determine H and H_2 (but not H^+), Bohlin,
Savage and Drake (1978) found on the average that $N_H/E(B-V) = W = 5.8$
10^{-22} mag cm^{-2} and that for $E(B-V) > 0.10$, almost all the lines of sight
agreed with this value to within a factor of two. The most conspicuous
deviation from the mean value of W is the measurement toward ρ Oph. In
this direction, it is possible that grain coagulation has occured so
that the grains are unusual in the sense that $E(B-V)$ does not serve as a
measure of the amount of mass in dust (Jura 1980). For stars with
$E(B-V) < 0.10$, there are also some marked deviations from the mean value
of W. At the moment it is not clear whether these deviations arise
because of the difficulties of determining small amounts of reddening or
because the fractional fluctuations about the mean are greater in the
smaller clouds.
        Burstein and Heiles (1978) and Heiles, Stark and Kulkarni (1981)
have argued from 21 cm measurements that at high galactic latitudes
there is gas without dust. The column densities are sufficiently small
that this does not obviously contradict the Copernicus results, but
there is certainly some dust at high latitudes as is shown by the
existence of optical reflection nebulae (Sandage 1976), interstellar
polarization (Appenzeller 1975, Markannen 1979) and emission by infrared
cirrus (Low et al. 1984). Also, Knude (1978) has argued on the basis of
very careful photometry that there is high latitude dust. The dust to
gas ratio in directions where there is little gas remains somewhat
uncertain.

Typically, more than 99% of refractory elements such as calcium, iron and aluminum are contained within the grains (Field 1974, Spitzer and Jenkins 1975). However, some elements such as oxygen are largely undepleted (York et al. 1983). By measuring the fluctuations of the abundances of these volatile elements, it is possible to estimate fluctuations in the true local abundances. At the moment, there is still some controversy about the amount of depletion of even the volatiles onto grains (see Jenkins, Savage and Spitzer 1986 and references therein), and it has proven difficult to use total gas phase abundances to measure the inhomogeneity of the interstellar medium.

It is usually thought that measurements of the abundances of different isotopes may not be sensitive to differential depletion onto grains, and isotopes may serve to measure true variations of interstellar abundances. The most commonly studied isotope ratio is that of $^{12}C/^{13}C$. Most interstellar observations of this ratio have been obtained from radio frequency emission lines of various carbon-bearing molecules in dark and giant molecular clouds throughout the Galactic plane (see Wannier 1980). However, from all these observations, it is difficult to derive highly precise values of $^{12}C/^{13}C$ due to the effects of large optical depths, photon trapping, and differential excitation and photodissociation of the ^{12}C and ^{13}C molecular species (Glassgold, Huggins and Langer 1985).

In contrast to other methods, optical observations of CH^+ are straightforward to interpret. Watson, Anicich and Huntress (1976) have shown that CH^+ is unaffected by chemical fractionation in diffuse clouds. The CH^+ molecule is believed to form in high-temperature post shock regions (Elitzur and Watson 1980) or some other zone of high temperature (White 1984) as supported by its higher velocity line width when compared to other molecules in the cloud (Willson 1981). This high temperature gas does not constitute a propitious environment for CH^+ fractionation to occur. Consequently, the accuracy of the carbon isotope ratio derived from the CH^+ line strengths is limited only by the quality of the observations, the uncertainty in measuring the equivalent widths and the errors introduced when the small optical depth effects are taken into account.

Until recently, there has been a large range (between 39 and 125) in the value of the interstellar CH^+ isotope ratio measured toward ζ Oph (Augason and Herbig 1967, Bortolot and Thaddeus 1969, Hobbs 1972, Vanden Bout 1972, Snell et al. 1977, Vanden Bout and Snell 1980). In view of the recent progress made in linear detectors for high resolution ground-based spectroscopy, it has proven worthwhile to remeasure the interstellar CH^+ ratio by studying the lines at 4232 Å. Hawkins, Jura and Meyer (1985) report that toward ζ Oph, $^{12}C/^{13}C = 43 \pm 6$ (1σ). The sum of the equivalent widths for the $^{12}CH^+$ and $^{13}CH^+$ lines toward ζ Oph was 22.3 ± 0.2 mÅ, consistent with the values apparently measured for the $^{12}CH^+$ and $^{13}CH^+$ lines blended together, obtained with much lower signal to noise, of 23.2 ± 0.7 mÅ and 22.9 ± 0.7 mÅ by Lambert and Danks (1986) and Crutcher and Chu (1985), respectively. Modern observations appear to be consistent with each other.

Table 2 --$^{12}C/^{13}C$ RATIOS FROM RECENT RADIO AND INFRARED
OBSERVATIONS OF OBJECTS AT THE SOLAR GALACTO-CENTRIC RADIUS

Site Observed	Molecular Ratio	$^{12}C/^{13}C$	Reference
dust cloud (L1529)	$^{12}C^{18}O/^{13}C/^{16}O$	64, 77	1
NGC 2024	$H_2CO/H_2^{13}CO$	72 ± 11	2
DR 21	$H_2CO/H_2^{13}CO$	73 ± 11	2
W 49 A	"	53 ± 8	
4 local clouds	$^{12}C^{18}O/^{13}C^{18}O$	55 −65	3
4 local clouds	$^{12}C/^{18}O/^{13}C/^{16}O$	83 ± 4	4
Ori A	$HCO^+/H^{13}CO^+$	~ 40	5
Ori A	$HCN/H^{13}CN$	~ 30	6
3 local clouds	$^{13}C^{18}O/^{13}C^{18}O$	75 ± 8	7
ζ Oph	$^{12}CO/^{13}CO$	> 75	8
NGC 2024	$H_2CO/H_2^{13}CO$	68 ± 11	9
9 local clouds	$H_2CO/H_2^{13}CO$	80 ± 7	9
OMC-1	$^{12}CO/^{13}CO$	96 ± 5	10
3 local clouds	$^{12}C^{18}O/^{13}C^{18}O$	100 ± 14	11
Ori A	$CH_3OH/^{13}CH_3OH$ $OCS/O^{13}CS$ $HC_3N/H^{13}CCCN$	~ 40	12
Ori A	$^{12}CN/^{13}CN$	~35	13
OMC-1	$H_2CO/H_2^{13}CO$	~ 80	14

All hybrid ratios are multiplied by the appropriate solar ratio for the
oxygen isotopes from Wannier (1980). The references in the above Table
are: 1. Combes et al. (1980), 2. Henkel et al. (1980), 3. Langer
et al. (1980), 4. McCutcheon et al. (1980), 5. Stark (1981), 6. Rydbeck
et al. (1981), 7. Wilson et al. (1981), 8. Crutcher and Watson (1981),
9. Henkel et al. (1982), 10. Scoville et al. (1983), 11. Penzias (1983)
12. Johansson et al. (1984), 13. Gerin et al. (1984), 14. Bastien et al
(1985).

Many of these measurements do not agree very well with the study of
ζ Oph by Hawkins, Jura and Meyer (1985). At the moment, this
discrepancy is not understood. In any case, it seems that the $^{12}C/^{13}C$
ratio is known to at least a factor of 2.

With further optical observations of weak CH^+ absorption lines,
Hawkins and Jura (1986) have obtained the ratios listed in Table 3.

TABLE 3 -- WEIGHTED MEAN $^{12}CH^+/^{13}CH^+$ ISOTOPE RATIOS

ζ Oph	43 ± 6
ξ Per	46 ± 11
P Cyg	44 ± 11
20 Tau	41 ± 9
23 Tau	41 ± 8

Unless there is some serious systematic error in the observational
procedure, these results strongly indicate that the interstellar $^{12}C/^{13}C$
ratio is about 43 ± 4 and that the interstellar medium is very
homogeneous.

IV. INTERSTELLAR MATTER -- SOURCES AND SINKS

The interstellar medium appears to have reached a state of
quasi-equipartition of energy densities and to be locally chemically
homogeneous. In order to understand these results, it is essential to
describe the sources and sinks of interstellar matter and the dynamic
processes in the medium. Here, I concentrate on the origin and fate of
interstellar matter.

The major sink of interstellar matter is usually thought to be star
formation. While the rate for this process is somewhat uncertain
because, for example, the rate of formation of stars with $M < 0.4 \ M_\odot$ is
not yet directly measurable, it has been possible to make plausible
estimates from observations of the numbers of stars in the local
neighborhood and estimates for the lifetimes of these objects. It seems
that the minimum rate at which interstellar matter is transformed into
stars is $3 \ 10^{-3} \ M_\odot \ kpc^{-2} \ yr^{-1}$ (Miller and Scalo 1982). This rate
applies locally to the solar neighborhood for stars in the mass range
$0.4 \ M_\odot < M < 50 \ M_\odot$. By extrapolating to all masses, Tinsley (1980)
estimated that within a factor of 2, the actual star formation rate in
the solar neighborhood is $10 \ 10^{-3} \ M_\odot \ kpc^{-2} \ yr^{-1}$.

There is a very uncertain extrapolation from the local star
formation rate to that over the entire Galaxy. Following Knapp and
Morris (1985), we assume an area for the disc of the Milky Way of 880
kpc^2; this then implies a total star formation rate in the Galaxy of
about $9 \ M_\odot \ yr^{-1}$.

The two main sources of new interstellar matter are mass-loss from stars and infall of extragalactic material. The rate of infall of material into the Milky Way Galaxy is not well known. While it has been suggested that the high velocity clouds observed at 21 cm at high galactic latitudes are infalling material (Oort 1970), this interpretation of the data is quite uncertain. Oort's suggested rate of $2 \ 10^{17}$ H atom cm^{-2} (10^6 year)$^{-1}$ corresponds to 1.4 $M_\odot$ yr^{-1} for the entire Milky Way. At the moment, there are no firm measurements of the rate of infall of material into the Galaxy; this quantity could even be essentially zero (Songaila et al. 1985).

It is possible to be much more quantitative about the rate for mass loss from evolved stars. Observations during the past decade have greatly increased our knowledge of this process, and we are in a much better position to make quantitative descriptions of mass loss.

The most important source of interstellar matter is mass loss from red giants. It is known that post asymptotic giant branch (AGB) stars evolve into planetary nebulae (see, for example, Iben and Renzini 1983), and the traditional way to calculate the mass loss from stars that have just evolved off the AGB is to examine planetary nebulae. However, the distance scale to planetaries is uncertain as are estimates of the total amount of both ionized and neutral mass in these objects. While it has been difficult to make precise estimates of the mass injection rate from the stars that produce planetary nebulae into the interstellar medium, a standard value is 1 $M_\odot$ yr^{-1} (Salpeter 1976), or, correspondingly, about $1 \ 10^{-3}$ $M_\odot$ kpc^{-2} yr^{-1}. This result is consistent with a white dwarf birth rate of $\sim$ 1 yr^{-1}; each such birth being associated with the injection of $\sim$ 1 $M_\odot$ into the interstellar medium (Weidemann and Koester 1983).

Although planetary nebulae have the traditional advantage that they are bright sources of optical radiation, more precise estimates of mass loss rates can be derived from radio and infrared measurements of AGB stars. In particular, although the hydrogen is usually not detected in these objects, measurements of the strength of the radio CO emission can be used to infer mass loss rates if one makes a reasonable estimate for the carbon and oxygen abundances in these stars (see Morris 1985, Knapp and Morris 1985). The basic idea is that the stronger CO lines result from larger amounts of mass loss.

One advantage of using red giants to estimate the mass injection rate into the interstellar medium is that the inferred quantity is independent of the quite uncertain distance to individual red giants. That is, if a star's distance is mis-estimated, then so is its mass loss rate so that the calculated rate for the injection of mass per kpc^{-2} is independent of assumed distance.

Knapp and Morris (1985) estimate a lower bound of 0.4 10^{-3} $M_\odot$ kpc^{-2} yr^{-1} of mass loss from red giants into the interstellar medium. However, this result was obtained only by directly counting mass loss from stars with an incomplete selection criteria. There is mass loss from stars that were simply unobserved by Knapp and Morris. It would be realistic to increase their result by a factor of three although the exact quantitative amount remains to be determined (see Zuckerman and Dyck 1986, Zuckerman, Dyck and Claussen 1986). This increase in the mass loss rate from red giants derived by Knapp and Morris would correspond to about 1 $M_\odot$ yr^{-1} in the Galaxy in agreement with Salpeter's (1976) estimate from the numbers and lifetimes of planetary nebulae.

Winds from hot stars are another source of interstellar matter, and it is estimated that they contribute from about 0.09 10^{-3} (Abbott 1983) to 0.6 10^{-3} $M_\odot$ kpc^{-2} yr^{-1} (Van Buren 1985) to the interstellar medium. The discrepancy between these two results is largely due to different values for the luminosity function of very massive (M > 30 $M_\odot$) stars. Additionally, supernovae probably eject about 0.03 $M_\odot$ yr^{-1} into the interstellar medium (Trimble 1983, Van Buren 1985). Finally, if there are 40 novae per year in the Galaxy and each ejects 10^{-4} $M_\odot$ (Clayton 1984), then the mass ejection from these stars is about 0.003 $M_\odot$ yr^{-1}.

In Table 4, we summarize the sources and sinks of interstellar matter in the Milky Way. Below, we now consider two consequences of using these rates.

TABLE 4 — SOURCES AND SINKS OF INTERSTELLAR MATTER

Process	Rate ($M_\odot$ yr^{-1})
Star formation	3 -10
AGB stars/planetary nebulae	0.3 - 1
Early type stars	0.08 - 0.5
Supernovae	0.03
Novae	0.003
Infall	<1.4

A.   INTERSTELLAR DEPLETION OF ELEMENTS HEAVIER THAN HELIUM

The discovery of depletion of interstellar matter raises the question of whether the depletion occurs at the sources of the interstellar matter -- mainly stars -- or whether it occurs within the interstellar medium itself. Field (1974) noted that there is a strong correlation between the condensation temperature of an element and the amount of its depletion. Therefore, Field proposed that the observed interstellar depletions of the most refractory species occur mainly within the envelopes of mass losing red giants. This idea is attractive because red giants are the main source for the replenishment of the interstellar medium and these stars are known to contain large amounts of dust. A traditional view of interstellar grains is that of refractory cores surrounded by mantles composed of relatively volatile material. In Field's picture, the cores are synthesized in the outflows from red giants, while these cores then accrete mantles within interstellar clouds. However, quantitative evaluation of this picture shows that it is not correct. While condensation of elements certainly does occur in

the outflows from red giants, this cannot explain the full pattern of
depletions observed in the interstellar medium.

Although there are some uncertainites in the analysis, it seems
that the red giants generally have dust to gas ratios by mass of about
1% (Sopka et al. 1985, Knapp 1985, Jura 1986a), a value similar to that
which is found in the general interstellar medium (Spitzer 1978).
Therefore, it is conceivable that interstellar and circumstellar grain
cores are identical. However, mass loss from early type stars
apparently accounts for at least about 10% of all the matter ejected
into the interstellar medium (see Table 4), and there is no reason to
think that grains form in the outflows from these stars. Therefore, any
element where more than 90% of the material is contained within grains,
such as iron, calcium or titanium, or any of the other particularly
refractory species that was thought to be mainly depleted in the
atmospheres of red giants, must actually stick to grains in the
interstellar medium itself. Additional evidence that grain processing
occurs in the interstellar medium is given by:
i) Gas phase abundances of normally depleted elements like calcium are
much higher in high velocity clouds. This presumably occurs because of
the removal of these elements from grains by sputtering and grain-grain
collisions (see, for example, Seab and Shull 1983).
ii) The depletion of markedly refractory elements such as iron is
highly correlated with cloud density (Savage and Bohlin 1979, Jenkins,
Savage and Spitzer 1986).
iii) Phosphorus and iron have nearly the same condensation temperature
(Grossman and Olsen 1974, Wai and Wasson 1977) yet phosphorus displays a
much lower depletion than does iron (Jura and York 1978). It seems
that thermodynamic equilibrium does not govern the depletion for these
two elements.

B.   GLOBAL EMISSION OF INFRARED RADIATION

The IRAS observations have shown that in spiral galaxies, much of the
energy emitted by stars is absorbed by grains and re-emitted in the
infrared (de Jong et al. 1984) as would be expected on the basis of
straightforward models for these objects (Jura 1982). Most of the
observed 60 μm and 100 μm fluxes are produced by interstellar dust, the
question arises as to how galaxies produce their integrated 12 μm
emission that is often detected by IRAS.

There are at least four conceivable sources of the observed 12 μm
emission. In our own Galaxy, the strongest 12 μm point sources are
either evolved mass losing red giants or very young pre-main sequence
stars (Kleinmann, Gillett and Joyce 1981). Also, the diffuse infrared
cirrus emits at 12 μm, a likely explanation of this result is the
presence of very small grains, perhaps polycyclic aromatic hydrocarbons
in the interstellar medium (Leger and Puget 1984). Finally, at least in
some external galaxies, there may be powerful nonthermal sources of
infrared radiation.

First we consider the contribution to the 12 μm emission from
mass-losing red giants. Faber and Gallagher (1979) estimated that the
absolute magnitude of the Milky Way galaxy in B is -20.3 mag which

corresponds to a B band luminosity of 2×10^{10} $L_\odot$. If the total mass loss rate from red giants in the Milky Way is 1 $M_\odot$ yr^{-1} (see Table 4), we then expect that $\dot{M} = 5 \times 10^{-11}$ L_B ($M_\odot$ yr^{-1}) where L_B is the Sun's luminosity in the B band ($M_{B,\odot} = 5.48$ mag, Allen 1973). In contrast, for the bulge of M 31 which has a somewhat different stellar population, Soifer et al. (1986) estimate $\dot{M} = 1.5 \times 10^{-11}$ L_B ($M_\odot$ yr^{-1}).

In the red giants in the Milky Way, there is a good correlation between mass loss rate and infrared colors. Specifically, Jura (1986b) has found that $L_\nu(60\ \mu m) = 7.0 \times 10^{27}$ erg s^{-1} Hz^{-1} for AGB stars with "standard" properties of a luminosity of 10^4 $L_\odot$ and an outflow velocity of 15 km s^{-1}. Extrapolating to 12 μm from the results derived for the fluxes at 60 μm, one can show that a mass loss rate of 1 $M_\odot$ yr^{-1} corresponds to a luminosity at 12 μm of 8.0×10^{28} erg Hz^{-1} s^{-1}. Alternatively, one can write for the flux from an ensemble of mass-losing stars that:

$$F_\nu(12\ \mu m) = 7 \times 10^7\ \dot{M}\ r_{kpc}^{-2} \qquad\qquad Jy$$

where r_{kpc} is the distance in units of kpc and $\dot{M}$ is the total mass loss rate ($M_\odot$ yr^{-1}). Soifer et al. (1986) use a similar result with a numerical coefficient of 8×10^7 instead of our value of 7×10^7.

We now consider all the Sb galaxies in the revised Shapley Ames Catalogue (Sandage and Tammann 1981) which have sufficiently small angular sizes that they come close to fitting into the IRAS apertures and which were actually detected by IRAS. In this way, we consider all galaxies with angular diameters given by de Vaucouleurs, de Vaucouleurs and Corwin (1976) that are $\leqslant 3.0'$. As is estimated above for the Milky Way, we assume for all the galaxies that $\dot{M} = 5 \times 10^{-11}$ L_B ($M_\odot$ yr^{-1}). From this estimate of the total mass loss rate, we can predict the 12 μm flux from AGB stars. The galaxies, their distances derived on the basis of H $= 50$ km s^{-1} Mpc^{-1}, and the predicted and observed IRAS 12 μm fluxes are given in Table 5.

TABLE 5 —— IRAS 12 μm FLUXES AND UPPER LIMITS TO MASS LOSS RATES FOR SHAPLEY-AMES Sb GALAXIES WITH $D_{25} \leqslant 3.0'$

Galaxy	M_B	D (Mpc)	$F_\nu(12\ \mu m)$ Predicted Jy	$F_\nu(12\ \mu m)$ Observed Jy
NGC 23	-22.86	96.7	0.081	0.47
NGC 473	-20.71	46.9	0.048	<0.25
NGC 670	-22.34	80.5	0.073	<0.27
NGC 955	-20.74	32.8	0.10	<0.25
NGC 1140	-20.36	31.2	0.078	<0.25
NGC 1309	-21.61	42.9	0.13	0.29
NGC 1417	-22.57	82.8	0.085	<0.25
NGC 2347	-22.32	93.8	0.052	<0.25
NGC 2764	-20.29	52.7	0.026	<0.25
NGC 2889	-22.19	63.9	0.10	<0.25

Galaxy	M_B	D (Mpc)	$F_\nu(12\ \mu m)$ Predicted Jy	$F_\nu(12\ \mu m)$ Observed Jy
NGC 3067	−20.52	28.6	0.11	0.68
NGC 3177	−19.49	24.4	0.057	0.55
NGC 3241	−20.80	51.7	0.043	<0.26
NGC 3259	−20.90	40.1	0.078	<0.25
NGC 3504	−21.11	29.6	0.17	1.05
NGC 3583	−21.65	43.6	0.13	0.53
NGC 4032	−19.38	23.6	0.055	<0.33
NGC 4102	−19.80	19.0	0.13	1.46
NGC 4457	−20.58	14.2	0.46	<0.29
NGC 4679	−22.55	90.2	0.07	<0.60
NGC 4750	−21.10	36.6	0.11	0.38
NGC 4800	−19.32	16.2	0.11	0.36
NGC 4868	−22.15	95.0	0.044	<0.61
NGC 5134	−19.96	20.5	0.12	<0.75
NGC 5150	−21.86	82.5	0.044	<0.25
NGC 5362	−20.90	46.4	0.058	<0.25
NGC 5395	−22.69	71.7	0.13	<0.33
NGC 5600	−20.55	45.9	0.043	0.40
NGC 5612	−21.65	50.1	0.10	0.36
NGC 5691	−20.34	35.4	0.059	<0.25
NGC 5740	−20.51	29.8	0.098	<0.25
NGC 5806	−20.23	25.4	0.10	<0.25
NGC 6753	−22.64	60.0	0.17	0.58
NGC 6769	−22.57	74.1	0.11	<0.36
NGC 7171	−21.46	55.2	0.069	<0.25
NGC 7232	−19.79	33.4	0.040	<0.60
NGC 7782	−22.88	112	0.062	<0.25

The results displayed in Table 5 show that the mass loss from evolved stars can lead to some 12 μm emission from galaxies, but that the dominant contributor to the 12 μm flux from Sb galaxies often has some other source. That is, with the exception of NGC 4457 which is the closest galaxy in this particular sample, the observed 12 μm fluxes, or their upper limits, are all above what one would expect on the basis of a simple extrapolation from the somewhat uncertain mass loss rate for the Milky Way. However, since in most cases, IRAS provides only an upper limit to the 12 μm flux, it is not obvious how much of this infrared emission comes from sources other than evolved stars. For those galaxies where there is a detected 12 μm flux, it seems that mass-losing evolved stars contribute on the order of 10% to 40% of the observed flux. It would be interesting to determine if these galaxies with particularly strong 12 μm emission have either particularly high star formation rates or particularly strong emission from PAH's. In either case, we would expect somewhat higher ultraviolet fluxes, unless, of course, there is an unusually large amount of interstellar extinction.

V. CONCLUSIONS

The main point of this review is that despite its very wide diversity, the interstellar medium in some ways is an orderly system. That is, the energy density (or pressure) seems similar for a number of different constituents as if some sort of quasi-equipartition has been reached. Similarly, the chemical composition of the interstellar medium in the solar neighborhood exhibits a striking uniformity.

Since the interstellar medium seems to be so well organized, it should be possible to develop comprehensive models of the system. Here, we have illustrated two arguments that can be derived from global parameters. First, much of the depletion of the most refractory elements such as calcium and iron must occur within the interstellar medium, itself, rather than in the sources of interstellar matter such as red giants. Also, in other spiral galaxies and presumably in the Milky Way, mass-losing red giants contribute a significant though not necessarily dominant portion of the total 12 μm luminosity.

This work has been partly supported by NASA and the NSF. I thank Mark Morris and Ben Zuckerman for their comments.

VI.   REFERENCES

Aannestad, P. A., and Purcell, E. M. 1973, Ann. Rev. Astr. Ap., 13, 133.

Abbott, D. C. 1982, Ap. J., 263, 723.

Allen, C. W. 1973, Astrophysical Quantities, 3rd Ed. (Athlone Press: London).

Appenzeller, I. 1975, Astr. Ap., 38, 313.

Augason, G. C., and Herbig, G. H. 1967, Ap. J., 150, 729.

Bastien, P., Batrla, W., Henkel, C., Pauls, T., Walmsley, C. M., and Wilson, T. L. 1985, Astr. Ap., 146, 86.

Bohlin, R. C., Savage, B. D., and Drake, J. F. 1978, Ap. J., 224, 132.

Bortolot, V. J., and Thaddeus, P. 1969, Ap. J. (Letters), 155, L17.

Burstein, D., and Heiles, C. 1978, Ap. J., 225, 40.

Cesarsky, C. J., and Volk, H. J. 1978, Astr. Ap., 70, 367.

Clayton, D. D. 1984, Ap. J., 280, 144.

Combes, F., Falgarone E., Guibert, J., and Nguyen-Q-Rieu 1980, Astr. Ap., 90, 88.

Cox, D. P., and McCammon, D. 1986, Ap. J., 304, 657.

Cox, D. P., and Smith, B. W. 1974, Ap. J. (Letters), 189, L105.

Crutcher, R. M., and Chu, Y.-H. 1985, Ap. J., 290, 251.

Crutcher, R. M., and Watson, W. D. 1981, Ap. J., 244, 855.

de Jong, T. et al. 1984, Ap. J. (Letters), 278, L67.

de Vaucouleurs, G., de Vaucouleurs, A., and Corwin, H. G. 1976, Second Reference Catalogue of Bright Galaxies (University of Texas: Austin).

Elitzur, M., and Watson, W. D. 1980, Ap. J., 236, 172.

Faber, S. M., and Gallagher, J. S. 1979, Ann. Rev. Astr. Ap., 17, 135.

Field, G. B. 1974, Ap. J., 187, 453.

Field, G. B., Goldsmith, D. W., and Habing, H. J. 1969, Ap. J. (Letters), 155, L149.

Gerin, M., Combes, F., Encrenaz, P., Linke, R., Destombes, J. L., and Demuynck, C. 1984, Astr. Ap., 136, L17.

Glassgold, A. E., Huggins, P. J., and Langer, W. D. 1985, Ap. J., 290, 615.

Grevesse, N. 1984, Phys. Scripta, T8, 49.

Grossman, L., and Olsen, E. 1974, Geochim. Cosmochim. Acta, 38, 173.

Hawkins, I., Jura, M., and Meyer, D. M. 1985, Ap. J. (Letters), 294, L131.

Hawkins, I., and Jura, M. 1986, Ap. J. Suppl., submitted.

Heiles, C. 1976, Ann. Rev. Astr. Ap., 14, 1.

Heiles, C., Stark, A. A., and Kulkarni, S. 1981, Ap. J. (Letters), 247, L73.

Henkel, C., Walmsley, C. M., and Wilson, T. L. 1980, Astr. Ap., 82, 41.

Henkel, C., Wilson, T. L., and Bieging, J. 1982, Astr. Ap., 109, 344.

Hobbs, L. M. 1972, Ap. J. (Letters), 175, L39.

Iben, I., and Renzini, A. 1983, Ann. Rev. Astr. Ap., 21, 271.

Jenkins, E. B., Jura, M., and Loewenstein, M. 1983, Ap. J., 270, 88.

Jenkins, E. B., Savage, B. D., and Spitzer, L. 1986, Ap. J., 301, 355.

Johansson, L. E. B., Andersson, C., Elldér, J., Friberg, P., Hjalmarson, A., Hoglund, B., Irvine, W. M., Olofsson, H., and Rydbeck, G. 1984, Astr. Ap., 130, 227.

Jura, M. 1974, Ap. J., 191, 375.

Jura, M. 1980, Ap. J., 235, 63.

Jura, M. 1982, Ap. J., 254, 70.

Jura, M. 1986a, Ap. J., 303, 327.

Jura, M. 1986b, Ap. J., submitted.

Jura, M., and York, D. G. 1978, Ap. J., 219, 861.

Kleinmann, S. G., Gillett, F. C., and Joyce, R. R. 1981, Ann. Rev. Astr. Ap., 19, 411.

Knapp, G. R. 1985, Ap. J., 293, 273.

Knapp, G. R., and Morris, M. 1985, Ap. J., 292, 640.

Knude, J. 1978, Astronomical Papers Dedicated to Bengt Stromgren (ed. A. Reiz and T. Anderson).

Lambert, D. L., and Danks, A. C. 1986, Ap. J., 303, 401.

Langer, W. D., Goldsmith, P. F., Carlson, E. R., and Wilson, R. W. 1980, Ap. J. (Letters), 235, L39.

Leger, A., Jura, M., and Omont, A. 1985, Astr. Ap., 144, 147.

Leger, A., and Puget, J.-L. 1984, Astr. Ap., 137, L5.

Low, F. J. et al. 1984, Ap. J. (Letters), 278, L19.

Markannen, T. 1979, Astr. Ap., 74, 201.

Mathis, J., Mezger, P., and Panagia, N. 1983, Astr. Ap., 128, 212.

McCutcheon, W. H., Dickman, R. L., Shuter, W. L. H., and Roger, R. S. 1980, Ap. J., 237, 9.

McKee, C. F., and Ostriker, J. P. 1977, Ap. J., 218, 148.
Miller, G. E., and Scalo, J. M. 1979, Ap. J. Suppl., 41, 513.
Morris, M. 1985, in Mass Loss from Red Giants, eds. M. Morris and B.
    Zuckerman, (Dordrecht: Reidel), p. 129.
Olive, K., and Schramm, D. N. 1982, Ap. J., 257, 276.
Oort, J. H. 1970, Astr. Ap., 7, 381.
Penzias, A. A. 1983, Ap. J., 273, 195.
Rydbeck, O. E. H., Hjalmarson, A., Rydbeck, G., Ellder, J., Olofsson,
    H., and Sume, A. 1981, Ap. J. (Letters), 243, L41.
Salpeter, E. E. 1976, Ap. J., 206, 673.
Sandage, A. 1976, Astr. J., 81, 954.
Sandage, A., and Tammann, G. A. 1981, A Revised Shapley-Ames Catalogue
    of Bright Galaxies (Washington: Carnegie Institute of Washington).
Savage, B. D., and Bohlin, R. C. 1979, Ap. J., 229, 136.
Scoville, N., Kleinmann, S. G., Hall, D. N. B., and Ridgway, S. T. 1983,
    Ap. J., 275, 201.
Seab, C. G., and Shull, J. M. 1983, Ap. J., 275, 652.
Snell, R., Tull, R., Vanden Bout, P., and Vogt, S. 1977, in CNO Isotopes
    in Astrophysics, ed. J. Audouze (Dordrecht: Reidel), p. 85.
Soifer, B. T., Rice, W. L., Mould, J. R., Gillett, F. C., Rowan
    Robinson, M., and Habing, H. J. 1986, Ap. J., 304, 651.
Songaila, A., York, D. G., Cowie, L. L., and Blades, J. C. 1985, Ap. J.
    (Letters), 293, L15.
Sopka, R. J., Hildebrand, R., Jaffe, D. T., Gatley, I., Roelligh, T.,
    Werner, M., Jura, M., and Zuckerman, B. 1985, Ap. J., 294, 242.
Spitzer, L. 1978, Physical Processes in the Interstellar Medium (J.
    Wiley: New York).
Spitzer, L., and Jenkins, E. B. 1975, Ann. Rev. Astr. Ap., 13, 133.
Stark, A. A. 1981, Ap. J., 245, 99.
Tinsley, B. M. 1980, Fund. of Cosmic Physics, 5, 287.
Trimble, V. 1983, Rev. Mod. Phys., 55, 511.
Troland, T. H., and Heiles, C. 1982, Ap. J., 252, 179.
Van Buren, D. 1985, Ap. J., 294, 567.
Vanden Bout, P. A. 1972, Ap. J. (Letters), 176, L127.
Vanden Bout, P. A., and Snell, R. L. 1980, Ap. J., 236, 460; erratum
    1981, Ap. J., 246, 1045.
Wai, C. M., and Wasson, J. T. 1977, Earth Planet Sci. Letters, 36, 115,
Wannier, P. G. 1980, Ann. Rev. Astr. Ap., 18, 399.
Watson, W. D., Anicich, V. G., and Huntress, W. T. 1976, Ap. J.
    (Letters), 205, L165.
Weidemann, V., and Koester, D. 1983, Astr. Ap., 121, 77.
White, R. E. 1984, Ap. J., 284, 695.
Willson, R. F. 1981, Ap. J., 247, 116.
Wilson, R. W., Langer, W. D., and Goldsmith, P. F. 1981, Ap. J.
    (Letters), 243, L47.
York, D. G., Spitzer, L., Bohlin, R. C., Hill, J., Jenkins, E. B.,
    Savage, B. D., and Snow, T. P. 1983, Ap. J. (Letters), 266, L55.
Zuckerman, B., and Dyck, H. M. 1986, Ap. J., 304, 394.
Zuckerman, B., Dyck, H. M., and Claussen, M. 1986, Ap. J., 304, 401.

Section II: Observations of Components of the Interstellar Medium

Ron Snell and Bel Campbell

H_2 IN THE GALAXY

N. Z. Scoville and D. B. Sanders
Physics, Mathematics, and Astronomy 105-24
California Institute of Technology
Pasadena, CA 91125

ABSTRACT. Molecular clouds are the active, star forming component of the interstellar medium. In this article, we review the theoretical and empirical basis for using the 2.6 mm CO emission line as a tracer of H_2 and we summarize the galactic distribution and properties of the molecular clouds. [All quantitative estimates are reduced to the recent IAU adopted value of R_o=8.5 kpc.]

1. INTRODUCTION

Over the last decade, extensive observations of carbon monoxide emission in the disk of our galaxy have shown that molecular (H_2) gas rather than atomic (HI) hydrogen is the major active component of the interstellar medium. In the Galaxy, virtually all known regions of star formation activity are associated with molecular clouds.

The relationship of the molecular gas clouds to other phases of the interstellar medium is rather different than what might have been imagined two decades ago. The giant molecular clouds (GMCs) are found to be self-gravitating with an effective internal pressure approximately an order of magnitude higher than the pressure in other phases of the ISM. Thus, these clouds are not in pressure equilibrium with the hotter, more diffuse phases; yet the young stars formed within them ultimately provide the massive stars and super-novae which energize the rest of the ISM.

In the next section, the foundation for CO as a tracer of the H_2 gas is reviewed. Despite the high opacity of the 2.6 mm (J=1-0) line, the emission, proves to be an effective proportional tracer of the mass of molecular gas in the giant molecular clouds. In Section 3, the large scale, galactic distribution of H_2 is presented and compared with that of the atomic gas derived from 21 cm (HI) line studies. The large-scale distribution of molecules in the Milky Way is then compared with recent determinations of the molecular gas in nearby external galaxies (§ 4). The properties of the individual GMCs are described in Section 5 followed by a discussion of the mechanisms for initiating star formation in these clouds (§ 6).

D. J. Hollenbach and H. A. Thronson, Jr. (eds.), Interstellar Processes, 21–50.
© 1987 by D. Reidel Publishing Company.

2. CO AS A TRACER FOR H_2

The 2.6 mm (J=1-0) rotational transition of CO has become the most widely
used probe of the interstellar molecular gas. Since molecular hydrogen
itself has a small moment of inertia and no permanent electric dipole
moment (due to the fact that it is made up of two identical nuclei
whose center of mass coincides with the center of charge distribution),
it has no permitted radio frequency transitions. The first pure
rotational transition (J=2-0) occuring at λ=28 μm, is relatively weak
and occurs in a spectral region of poor atmospheric transmission and
detector performance. The J=4-2 transition at λ=12 μm (Beck et al 1982)
and the rotation-vibration band at λ=2 μm (cf. Beckwith et al 1978;
Scoville et al 1982) have been detected in 1000-2000 K shock heated
gas but these transitions have negligible emissivities in the low
excitation gas outside the active, star forming, cloud cores. Electronic
transitions in the Lyman and Werner bands of H_2 were studied using the
Copernicus satellite from low opacity clouds along the line of sight to
nearby bright ultraviolet stars (cf. Spitzer and Jenkins 1975). Neither
the near infrared nor the ultraviolet transitions serve as general
probes of the cold molecular gas at large distances in the galaxy.

2.1 Theoretical Basis

The J=1-0 CO fundamental transition at 115.2712 GHz, corresponding to
$h\nu/k$=5.5 K, is excited readily by collisions with H_2, even in clouds at
very low kinetic temperature. In Figure 1, the low lying rotational
energy states of CO are shown. For the first transition, the spontaneous
decay rate is A_{1-0}=6x10^{-8} s^{-1}. Since the collisional cross-section
for rotational excitation of CO by H_2 is approximately 2x10^{-15} cm^2
(Green and Chapman 1978), the critical density for significant excitation
of the J=1 state, given by

$$n_{H_2} = A_{1-0}/\langle \sigma v \rangle \quad , \tag{1}$$

is $n_{H_2} \geq 3000$ cm^{-3}. This estimate is appropriate if the level excitation
is determined solely by a balance between H_2-CO collisions and spontaneous
decay.
       On the other hand, measurements of the rare ^{13}CO isotope indicate
that the CO emission is optically thick. [^{13}CO is typically seen at an
intensity approximately 0.1-0.5 of CO, yet the terrestial abundance
ratio ^{13}C/C=1/89.] For optically thick transitions, the excitation
analysis must also include induced radiative processes since the
radiation energy density in the lines builds up due to trapping of
spontaneous decay photons. The escape probability formalism, first
applied to molecular clouds by Scoville and Solomon (1974) and Goldreich
and Kwan (1974), was originally developed for circumstellar envelopes
with radial velocity gradients large compared to random thermal
motions. The relevant result is that the spontaneous decay rate (A) in
equation (1) can simply be multiplied by a factor equal to the effective
probability for escape of line photons from the emission region. For
a spherically symmetric cloud with radial velocity field ($V \propto r$), the

CO ROTATIONAL LEVELS

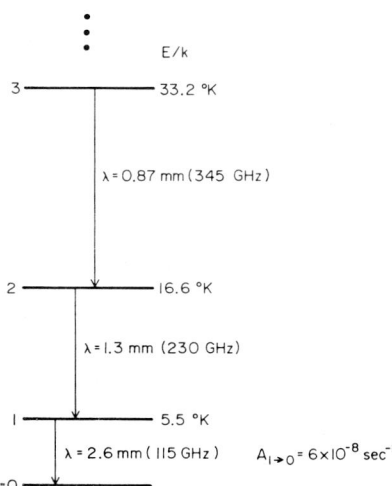

Figure 1. The lowest rotational states of CO are shown. The 2.6 mm transition arises from spontaneous decay from J=1-0 with a decay rate of 6×10^{-8} sec^{-1}.

escape probability, β, is given by

$$\beta = \frac{1}{\tau}(1 - e^{-\tau})$$ (2)

where

$$\tau = \frac{A_{ul}\lambda^3}{8\pi} \frac{g_u}{dv/dr} \left\{ \frac{n_1}{g_1} - \frac{n_u}{g_u} \right\}$$

For $\tau \ll 1$, $\beta = 1$; while for $\tau \gg 1$, $\beta = \frac{1}{\tau}$.

Thus the critical density for excitation of the CO J=1 level is reduced by a factor of τ in the optically thick regime. For most GMCs, $\tau_{CO} \gtrsim 10$ and the critical density for excitation is reduced to <300 cm^{-3}.
Based on the above, we expect that the J=1-0 CO transition will be thermalized ($T_x = T_k$) at densities of a few hundred per cc. Since the transition is optically thick, the observed brightness temperature will therefore be equal to the gas kinetic temperature (once corrections are made for the cosmic background radiation and telescope inefficiencies, cf. Kutner and Ulich 1981).
If I_{CO} ($\equiv \int T \, dv$) is the CO brightness temperature integrated over the line profile, then we can define the CO luminosity of a cloud by

$$L_{CO} = d^2 \int I_{CO} \, d\Omega$$ (3a)

where d is the distance.
    Thus,

$$L_{CO} \equiv T_{CO} \; \Delta V \; \pi R^2 \qquad\qquad\qquad (3b)$$

where T_{CO} is the peak brightness temperature in the CO transition, ΔV
is the line width, and R is the cloud radius. For clouds in virial
equilibrium, $\Delta V = (GM/R)^{\frac{1}{2}}$ and

$$L_{CO} = (\frac{3\pi G}{4\rho})^{\frac{1}{2}} \; T_{CO} \; M \qquad\qquad\qquad (4)$$

    In general, the cloud volume (rather than the density) is the
dominant factor varying between GMCs of different mass. To the extent
that most of the clouds have similar density (ρ) and temperature (T_{CO},
see below), the CO luminosity will therefore be a proportional measure
of the total mass of molecular clouds contained in the telescope beam.
The reason that this proportionality holds for optically thick lines is
that for a collection of clouds with varying mass but fairly constant
density, the virial line-width increases linearly with cloud radius
(ie. linearly witn the column density of gas). Thus, an increasing
mass of gas along the line of sight is registered in a linear increase
in the width, ΔV, of the emission line.
    For clouds with differing temperature and mean density, the constant
of proportionality between the CO luminosity and the molecular gas
mass will scale like $T_{CO}/\rho^{\frac{1}{2}}$. Since hotter clouds, eg. in galactic
nuclei, may also tend to be denser, the temperature and density
dependences will to some extent compensate each other.

2.2 Empirical Basis

In recent years, there have been several observational attempts to
evaluate the constant proportionality between the CO emission and H_2
column density and to test the linearity of the relation.
    Four general techniques, all pseudo-empirical, have been applied:
measurement of ^{13}CO emission which is presumably less saturated than
CO; correlation of visual extinction in nearby dark clouds with CO
intensity; estimation of virial masses using the CO and ^{13}CO line widths
and cloud sizes; and estimates of the total column density of nucleons
derived from gamma ray observations. The derived ratios of N_{H_2}/I_{CO} are
presented in Table 1; the total range for the inferred constant of
proporLionality is $1.8-4.8 \times 10^{20}$ cm^{-2} (K km s^{-1})$^{-1}$. The relatively good
agreement of these totally independent methods suggests that global mass
estimates for H_2 are probably correct to a factor of two, although
there will certainly be larger uncertainties when the relation is applied
to individual clouds.
    It should be recognized that several of these determinations were
based on different areas of the galaxy, suggesting that global gradients
in the metallicity and the conditions of the gas clouds do not significantly affect the CO emissivity per unit mass of H_2. In particular the
virial mass estimates were derived for clouds within the molecular ring
at galactic radii 3-7 kpc while the extinction analysis was based on

clouds in the solar neighborhood, whithin 1 kpc of the sun. Also
noteworthy are the gamma ray analyses which yield nearly identical
calibration constants for the Orion molecular cloud and the inner
galaxy. [The constancy of the γ-ray results is in fact rather disturbing
since one might anticipate that the cosmic rays which give rise to the
γ-rays would vary somewhat with galactic radius, and it is also
surprising that the cosmic ray penetration of the clouds is so constant
from one cloud to the next. It should be pointed out that the γ-ray
data, made with 1° resolution, are extremely biased (by a factor ∼10)
to the gas on the near side of the galaxy relative to the far side.]

Table 1

Empirical Measures of the Ratio $N(H_2)/\int T(CO)dv$

Source	Value
CO vs. Av (Young and Scoville 1982)	4×10^{20}
^{13}CO vs. Av in Taurus (Frerking, Langer and Wilson 1982)	4.8×10^{20}
^{13}CO vs. Av for 5 dark clouds (Sanders, Solomon, and Scoville 1984)	2.9×10^{20}
^{13}CO vs. Av in ρ Oph (Frerking, Langer, and Wilson 1982)	1.8×10^{20}
Virial masses (Sanders, Scoville and Solomon 1985, Scoville et al 1986)	$2.5-4.5 \times 10^{20}$
γ-Rays inner Galaxy (Lebrun et al 1983)	$\leq 3 \times 10^{20}$
γ-Rays Orion (Bloemen et al 1984)	2.6×10^{20}

Figure 2 illustrates the correlation found between the CO line
luminosity from clouds and their virial masses. The excellent correlation
found for GMCs of mass $10^5-2\times10^6$ M_O is equivalent to

$$N_{H_2}/I_{CO} = 3.6\times10^{20} \text{ cm}^{-2} \text{ (K km s}^{-1})^{-1} \quad . \tag{5}$$

and we adopt this CO to H_2 conversion ratio consistently throughout
the remainder of this contribution.

Although most observational analyses employ a constant ratio for
N_{H_2}/I_{CO}, it is clear that in some instances the dependence on the mean
gas temperature and density will show up [see Equation (4)]. Thus, in
regions with abundant high-mass star formation where the gas is hotter,
the H_2 mass implied by a given I_{CO} could be somewhat less than is
obtained from Equation (5). This is also probably the case in the high

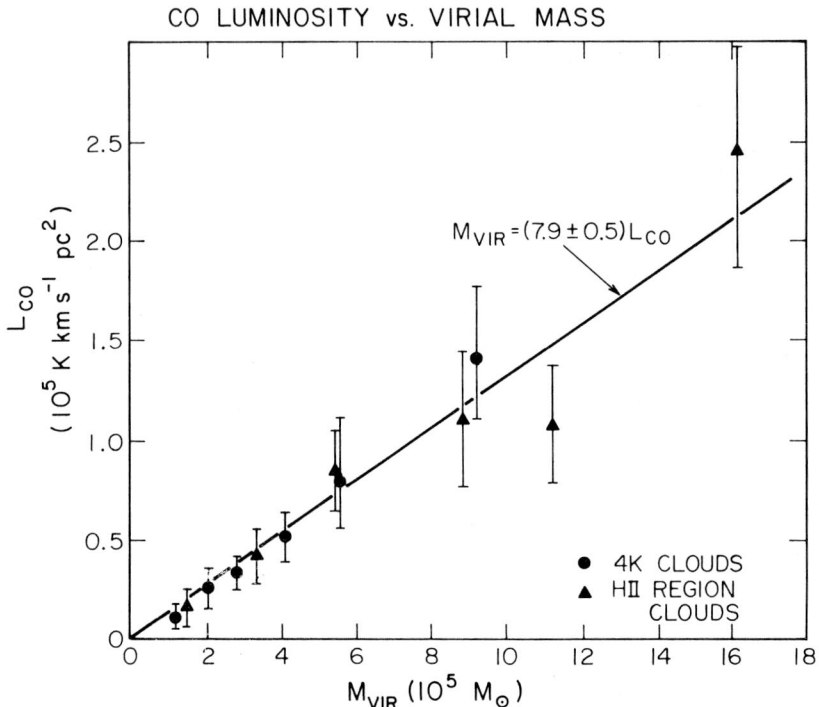

Figure 2. The CO luminosity is closely correlated with the virial masses of the clouds both with and without massive OB star formation (Scoville et al 1986). This linear proportionality justifies the use of CO as a tracer of the galactic H_2 mass. The best fit is equivalent to a constant of proportionality of 3.6×10^{20} H_2 cm^{-2} (K km s^{-1})$^{-1}$.

luminosity infrared galaxies such as Arp 220. The fact that larger variations in the constant of proportionality do not occur is probably due in part to a compensating increase in the mean density of the hotter clouds [see discussion following Equation (4)].

3. THE LARGE SCALE DISTRIBUTION OF H_2

The first large scale survey of carbon monoxide in the inner galaxy was done by Scoville and Solomon (1975) and shortly thereafter, a similar survey was done by Burton et al (1975). The principal result of both surveys was the discovery that the molecular emission was concentrated in a ring at 3-7 kpc radius. Scoville and Solomon (1975) also noted that most of the molecular material was confined to discrete clouds. Both characteristics are entirely different from those found previously for HI, for which the distribution is rather flat in radius with little tendency to be clumped (cf. Kulkarni 1987). Table 2 lists a number of

Table 2

Recent Studies of H_2 in the Galaxy

MAJOR RECENT SURVEYS

	Coverage	Resolution	Completion
CO: Massachusetts/Stony Brook	North	0.8	1986 (1)
Goddard	North & South	8'	1987 (2)
^{13}CO: NRAO	North	1.1	1981 (3)
Bordeaux	North	4'	1983 (4)
Bell Labs	North and Gal. Center	2'	1987+ (5)

LARGE SCALE H_2 DISTRIBUTION

   Axisymmetric: Sanders, Solomon and Scoville (1984)

                 Bronfman et al (1987)

   Non-Axisymmetric: Clemens, Sanders, and Scoville (1986)

CLOUD PROPERTIES

   Stark (1979)
   Liszt, Xiang, and Burton (1981)
   Sanders, Scoville, and Solomon (1985)
   Scoville et al (1986)

NOTES

   (1) Sanders et al (1986), (2) Bronfman et al (1987), (3) Liszt, Xiang, and Burton (1981), (4) Despois and Baudry (1983), (5) Stark et al (1987)

the most recent studies of the Galactic H_2 distribution and cloud
properties.

The general characteristics of the CO emission in the galactic
plane are illustrated in the sample CO and ^{13}CO spectra shown in
Figure 3. Figure 4 shows schematically how emission regions along a
typical line-of-sight in the inner galactic plane map into radial
velocities of features in a spectrum. In the first quadrant, gas
in circular orbits will be seen at positive (or negative) radial velocities
depending upon whether it is inside (or outside) of the solar circle.
The maximum positive velocity (the "terminal velocity") corresponds to
the subcentral point where the line-of-sight passes closest to the
galactic center. Below the terminal velocity, there exists a distance
ambiguity with two locations, equidistant from the subcentral point,
possibly contributing to the emission at a given velocity. Both locations
are at the same galactic radius but different azimuthal positions.

The strongest CO emission is seen in the galactic center ($|1|<4°$)
where the integrated CO emisison is a factor of 10 greater than that
seen at higher longitudes and the velocity spread is approximately
500 km s^{-1}. At longitudes between 5 and 20°, the emission is reduced
by a factor ~15 and the velocity range is much smaller, with little or
no emission seen near the terminal velocity, corresponding to the point
of closest approach of the line-of-sight to the galactic center. In this
longitude range, the line-of-sight is cutting across the inside of the
molecular cloud ring. At $1=20-40°$, the emission concentrates towards
the terminal velocity indicating that significant CO emission exists
near the subcentral point; here, the line of sight passes through the
densest part of the molecular cloud ring. Beyond $1=50°$, the emission
falls off with increasing galactocentric radius. Finally, at $1=75-85°$,
strong emission is seen from the nearby Cygnus arm. Figure 3 also
illustrates the generally close correspondence between CO and ^{13}CO
emission features, supporting the view that the optically thick CO line
is indeed a proportional indicator of the H_2 mass.

The longitude-velocity distribution for the CO emission, including
both northern and southern hemisphere data (Robinson et al 1984) is
shown in Figure 5 with the lines of constant galactic radius superposed.
Outside of the galactic center region, most of the strong CO emission in
both hemispheres is concentrated between the lines corresponding to 0.45
R_0 and 0.85 R_0. The molecular gas content is approximately symmetric
about $1=0°$ with the total CO emissivity (weighted by area at the
different radii) being about 20% greater in the south.

3.1   The H_2 Distribution in R and Z

Based on the observed longitudes and velocities of the CO emission, the
galactocentric radii of the emitting clouds may be derived from the
Galactic rotation curve. Although the galactocentric radii of emission
features in the first quadrant are unambiguously determined from the
longitude and velocity, derivation of the mass density in the inner
galactic plane requires knowledge of the distance to the emitting region.
Clemens, Sanders, and Scoville (1986) have recently done a full non-
axisymmetric analysis of the Massachusetts-Stony Brook CO survey making

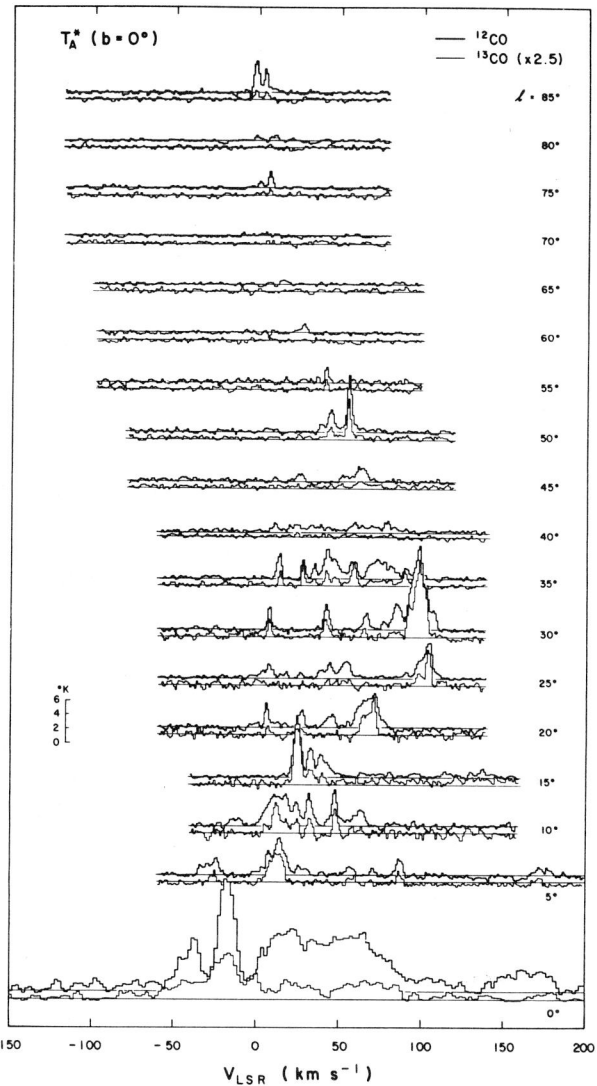

Figure 3. CO and ^{13}CO spectra of emission from the galactic plane at 5° intervals in longitude between l=0° and 85° (Sanders, Solomon, and Scoville 1984). The intensities are corrected for atmospheric attenuation but not for antenna efficiency. ^{13}CO intensities are scaled up by a factor of 2.5. Note that the strongest CO emission outside the galactic center arises from l=20-50°, corresponding to the molecular cloud ring.

use of the latitude dependence of emission features to resolve the distance ambiguities.

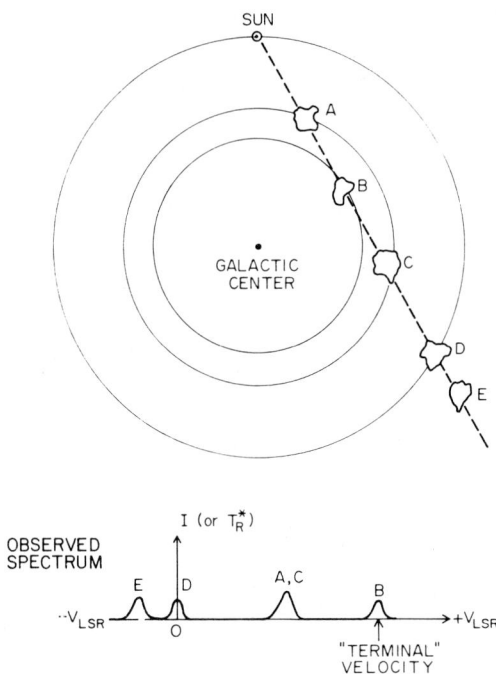

Figure 4. The mapping of emission regions into radial velocities along a line-of-sight in the inner galaxy is shown schematically. In the first galactic quadrant, gas interior to the solar circle (moving in circular orbits) will be at positive radial velocities while gas outside the solar circle will appear at negative velocities. The maximum positive velocity corresponds to emission at the subcentral point along the line-of-sight.

In Figure 6, the radial distribution of H_2 at the midplane of the Z distribution is compared with those for atomic hydrogen and giant radio HII regions. As noted earlier, the dominant features of the molecular distribution are a strong concentration within 1 kpc of the galactic center and the molecular cloud ring at 3-7 kpc radius. The ring feature is also prominent in the HII region distribution. In contrast, HI is relatively constant at most radii and exhibits a depression in the central region of the galaxy.

The Z centroid and ΔZ (FWHM) for the H_2 are shown in Figure 7. In the first galactic quadrant, the midplane of the molecular cloud ring is displaced approximately -25 pc below the galactic equator, but at larger radii outside the solar circle, the midplane rises to over +80 pc. The thickness of the molecular gas increases monotonically with radius from approximately 80 pc at 3 kpc radius to 140 pc at the solar circle (as originally noted by Cohen and Thaddeus 1977). The thickness in Z obtained from the molecular gas agrees with that derived for OB associations but is about 50% less than that of HI clouds and nearly a factor of 2 smaller

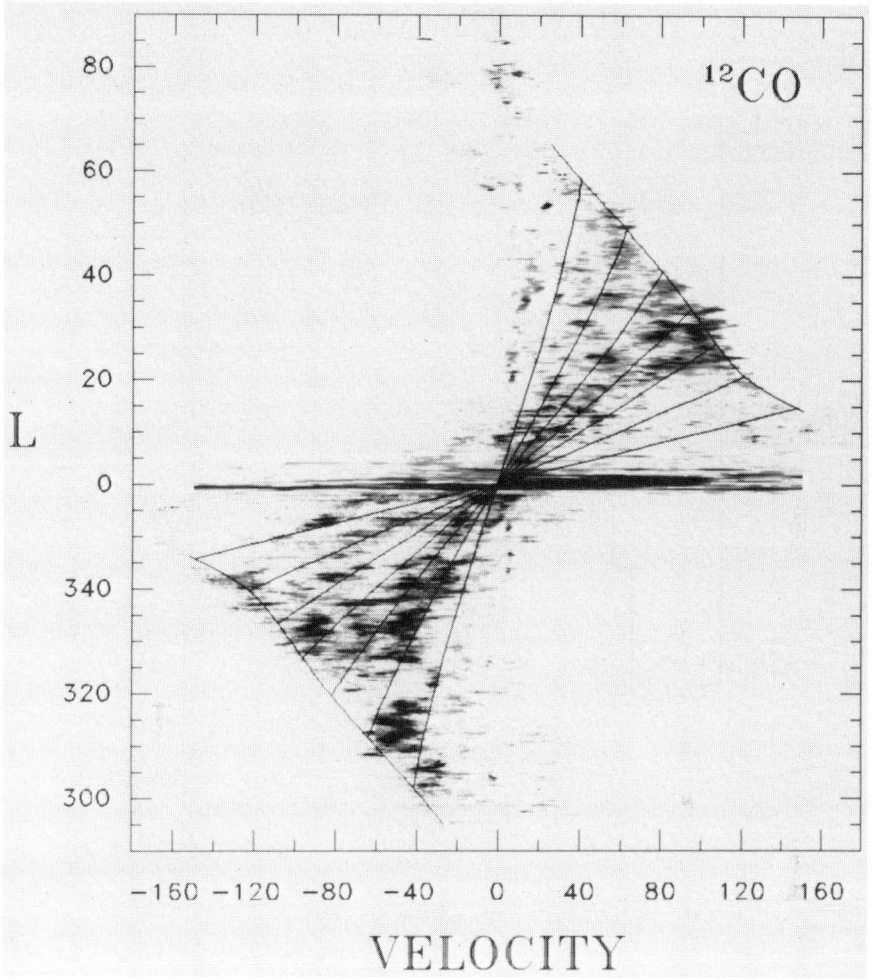

Figure 5. The longitude-velocity distribution of CO emission observed using the FCRAO and CSIRO telescopes (Robinson <u>et al</u> 1984). The data was sampled every 3' and smoothed to an effective resolution of 9'; the effective velocity resolution is 1 km s⁻¹. Superposed lines of constant galactocentric radius at 0.1 R_0 intervals between 0.25 R_0 and 1.05 R_0 were computed using the galactic rotation curve.

than that of the HI intercloud medium.[1]

---

1. Throughout this paper we have adopted the new IAU solar constants R_0=8.5 km s⁻¹ and θ_0=220 km s⁻¹. The rotation curve recently determined by Clemens (1985) from CO and HI data uses these new values. When making use of previous results for the gas distribution, we have scaled volume densities $n \propto R_0^{-1}$, scale heights $\Delta Z \propto R_0$, surface densities $N \propto R_0^0$,

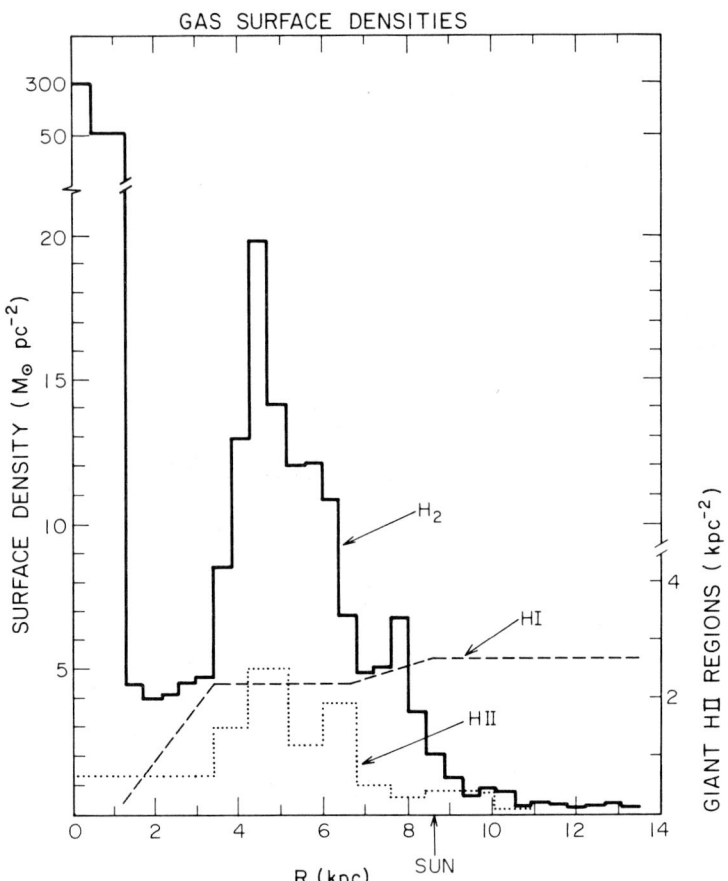

Figure 6. Comparison of gas surface densities in the Milky Way disk. Values for H_2 (Clemens, Sanders, and Scoville 1986) and HI (Burton and Gordon 1978) include a 1.36 correction factor for He and they have been scaled to R_0=8.5 kpc.

and total masses $M \propto R_0^2$. These scalings are used for both H_2 and HI.

In the galactic center, the thickness of the molecular layer is reduced to about 65 pc and here, one also sees a 7° tilt with respect to the galactic plane in the sense that at negative longitudes, the molecular gas rises above the galactic plane and at positive longitudes, it falls below (Sanders, Solomon, and Scoville 1984). A tilted disk has previously been postulated for the HI distribution in the galactic center (Liszt and Burton 1980), but the HI model involves a tilt of 29°.

In Table 3, parameters for the galactic H_2 distribution are summarized (assuming R_0=8.5 kpc). The mass estimates for the molecular gas at the galactic center are particularly uncertain due to the possi-

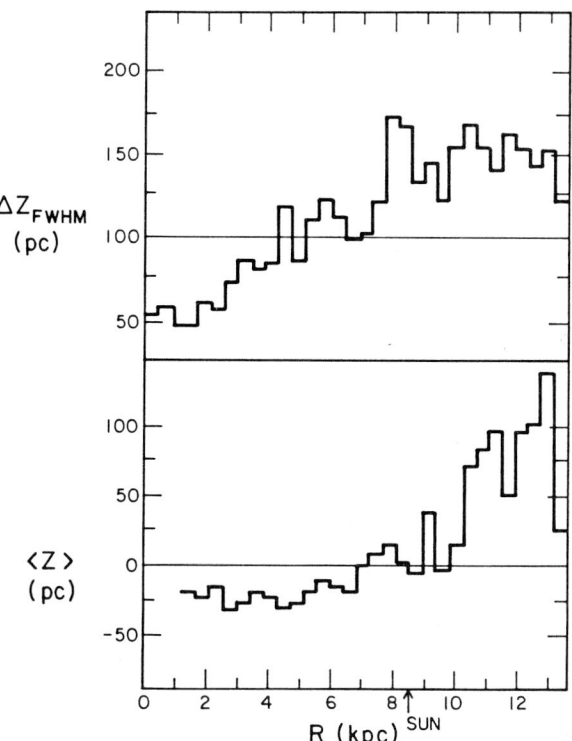

Figure 7. The scale height (full-width at half maximum density), and midplane displacement ⟨Z⟩, of CO emission is shown as a function of galactocentric radius. These determinations are based upon a non-axisymmetric analysis of CO emission in the first galactic quadrant (Clemens, Sanders, and Scoville 1986).

bility that the molecular gas in that region is, on average, both hotter and denser (see discussion in §2). Overall in the galaxy, the HI and H₂ masses are equal within the uncertainties of their determinations. On the other hand, the radial distributions of the mass in the two phases is entirely different. Approximately 90% of the total H₂ mass occurs inside the solar circle whereas, only 33% of the HI mass is in the same region.

3.2 Spiral Structure

The question of whether or not all of the observed CO emission from the galactic disk is confined to a regular pattern of a few spiral arms is important for understanding the origin and evolution of molecular clouds. If molecular clouds are short-lived objects which form out of

Table 3

Axisymmetric Distribution of H_2

Galactic Center (R≤400 pc)

Total H_2 Mass	∿2×10^8 $M_\odot$
FWHM in Z	65 pc
$\overline{n}_{H_2}$	60-120 cm^{-3}
H_2/HI Mass Ratio	>20

Molecular Cloud Ring (R=3-7 kpc)

Total H_2 Mass	1.8×10^9 $M_\odot$
FWHM in Z	90 pc
Midplane in Z	-25 pc
$\overline{n}_{H_2}$	2.7 cm^{-3}
Surface Density	12 $M_\odot$ pc^{-2}
H_2/HI Mass Ratio	3
Volume Filling Factor for Clouds	0.8-1.5%

Entire Galactic Disk (R<14 kpc)

Total H_2 Mass	2.3×10^9 $M_\odot$
Total HI Mass	2.2×10^9 $M_\odot$
H_2/HI Mass Ratio	1.0

Notes
    H_2 estimates from Sanders, Solomon, and Scoville
(1984) and Clemens, Sanders, and Scoville (1986) were
corrected to R_0=8.5 kpc. HI estimates from Burton and
Gordon (1978) have been increased by 20% to compensate
for optical depth in the 21-cm line (Dickey and Benson
1982) and corrected to R_0=8.5 kpc. All mass estimates
exclude the contribution of He (a factor of 1.36).

HI in spiral arm shocks and are destroyed upon leaving the arms, their spatial distribution should mimic the spiral potential and result in an high contrast between arm peaks and troughs, similar to what is inferred for the distribution of giant HII regions (eg. Georgelin and Georgelin 1976; Lockman 1979). On the other hand, if clouds are formed by some other mechanism and survive through several spiral arm passages, they should be widely distributed throughout the disk and show a similar low contrast in and out of arms as is observed for HI (eg. Kerr 1969; Yuan 1969).

Prior to the recent large-scale CO surveys, there were several attempts to determine the degree of confinement of molecular clouds to a spiral pattern which relied on relatively simple interpretations of the appearance of CO emission in the 1-V plane. Lizst (1983) provides a summary of the conflicting results of these searches for CO spiral structure. More detailed studies have compared the observed CO 1-V distribution (see Figure 5) with models simulating the two extreme cases of short-lived clouds (Leisawitz and Bash 1982) and those with infinitely long lived clouds in the presence of a spiral potential (Levinson and Roberts 1981; Liszt and Burton 1981). Although models with long lived clouds can account for the widespread distribution of molecules, a simple two or three arm spiral pattern cannot account for all of the clustering of features observed in the CO distribution in areas outside of the arms. In particular, the filling of CO emission along the tangent point at l=35-45° (see Figure 4) suggests a more chaotic distribution of clouds with azimuth (Burton and Gordon 1978), or at the very least, many arm segments rather than a simple grand design.

Studies of the properties of individual molecular clouds throughout the galactic disk now suggest that spiral structure as delineated by giant HII regions is a phenomenon that happens to preexisting molecular clouds. Stark (1979) first pointed out that classic arm and interarm zones in the inner galaxy contained equal numbers of small clouds (<30 pc), but the largest clouds appeared to exist only in arm regions. From a survey of ∿300 giant molecular clouds, Sanders (1981) found that the distribution of hot molecular clouds (T_K>11 K) more closely resembled the distribution of giant HII regions while cooler clouds which represented nearly 60% of the total emission were more uniformily distributed throughout the disk. These results have been confirmed in a much larger survey of over 2000 cloud cores by Solomon, Sanders, and Rivolo (1985) and Scoville et al (1986).

In Figure 8, the longitude-velocity distributions for different types of clouds are compared. The galactic distribution of the general cloud population (clouds with CO emission exceeding T_K=5 K) is much less tightly confined in the 1-V space than that of either the hot cloud cores or the clouds associated with HII regions. The strong similarity of the 1-V distributions for hot cloud cores and HII region clouds is probably due to the fact that they both represent CO emission regions which are presently undergoing high mass star formation and the large power output of those stars raises the temperature of the surrounding molecular gas (Goldreich and Kwan 1974). The two zones with highest abundance of hot molecular gas running from l=0°, V=0 km s^{-1} to l=32 and 50° at the terminal velocity correspond to Scutum and Sagittarius

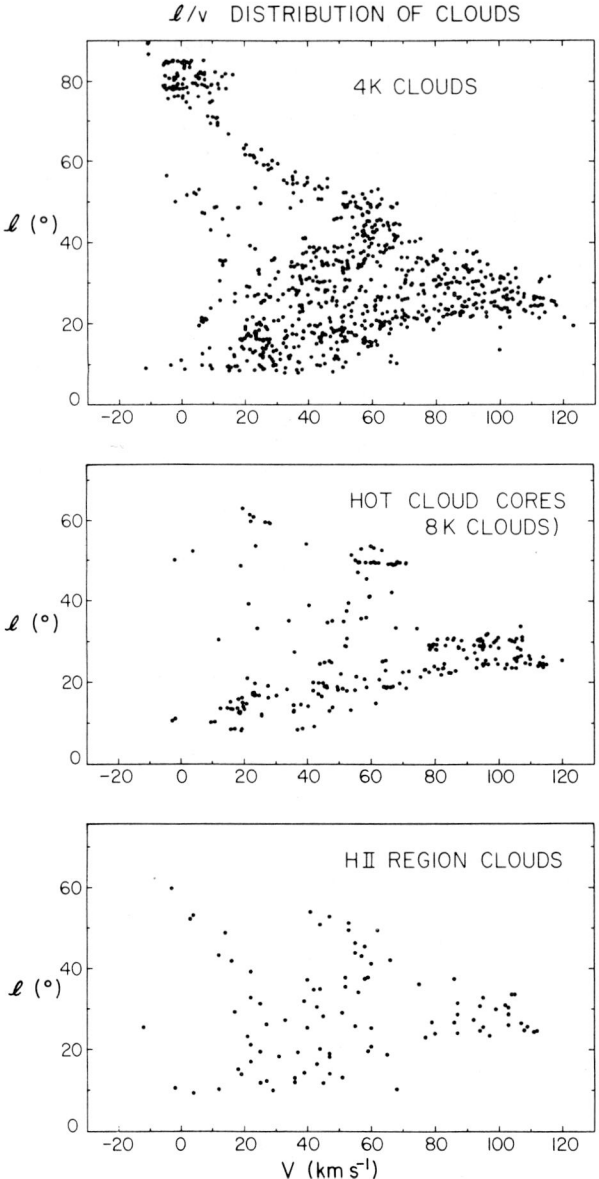

Figure 8. The longitude-velocity distribution of the general cloud population in the first galactic quadrant is contrasted with the distributions of hot cloud cores and clouds associated with HII regions (Scoville et al 1986). The hot cloud cores and HII region clouds exhibit a much greater confinement to the loops in the 1-V plane corresponding to the Scutum and Sagittarius spiral arms.

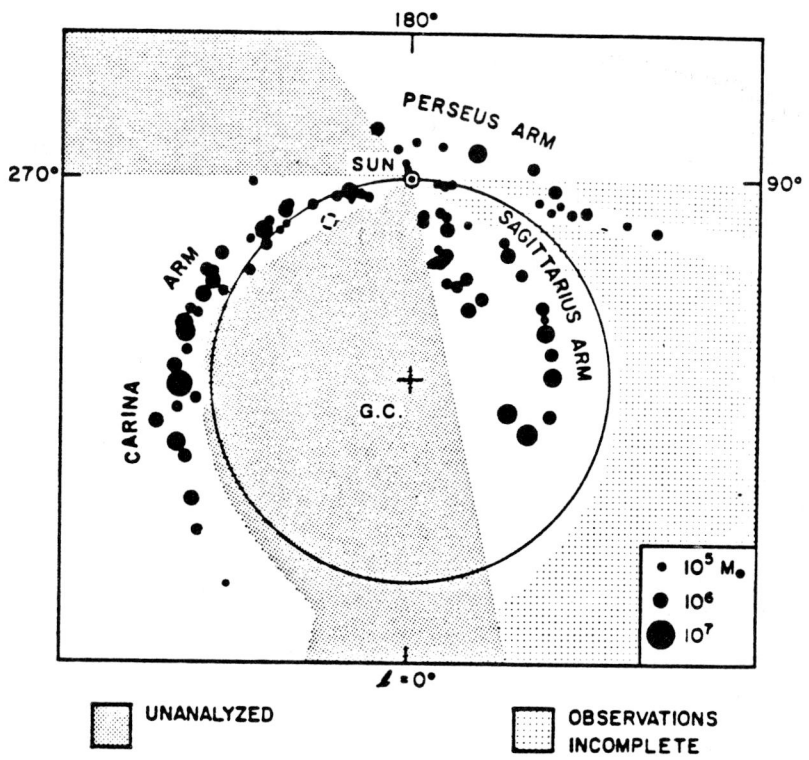

Figure 9. Face-on view of large GMCs along the Carina, Perseus, and Sagittarius arms from Grabelsky (1985). The stippled areas were left unanalyzed.

spiral arms.

Face-on pictures of the molecular gas distribution have been derived from the large scale GISS and Massachusetts-Stony Brook CO surveys. Cohen et al (1980), Dame et al (1986) and Grabelsky et al (1986) identified selected discrete cloud complexes associated with the Perseus arm (at $l\approx100°-160°$), the Sagittarius arm (between $l\approx35°-60°$ at $R\approx0.7~R_0$), and the Carina arm (as shown in Figure 9).

The first comprehensive image of all of the CO emission is shown in Figure 10 (Clemens, Sanders, and Scoville 1986). The inner galaxy is dominated by the molecular ring at $R\approx3-7$ kpc. Two lesser, extended features are also present outside the ring. One feature crosses the solar circle near $l=60°$ and may be associated with an outer Galaxy HI arm. The other feature resides at a Galactic radius of about 7 kpc and has been previously identified as the Sagittarius spiral arm. However, it is clear from the distribution in Figure 10, that the CO emission is abundant between the arms. The mean of the arm-interarm density contrast for the three major arms is 3.6:1.

The existence of molecular clouds in interarm regions has also been

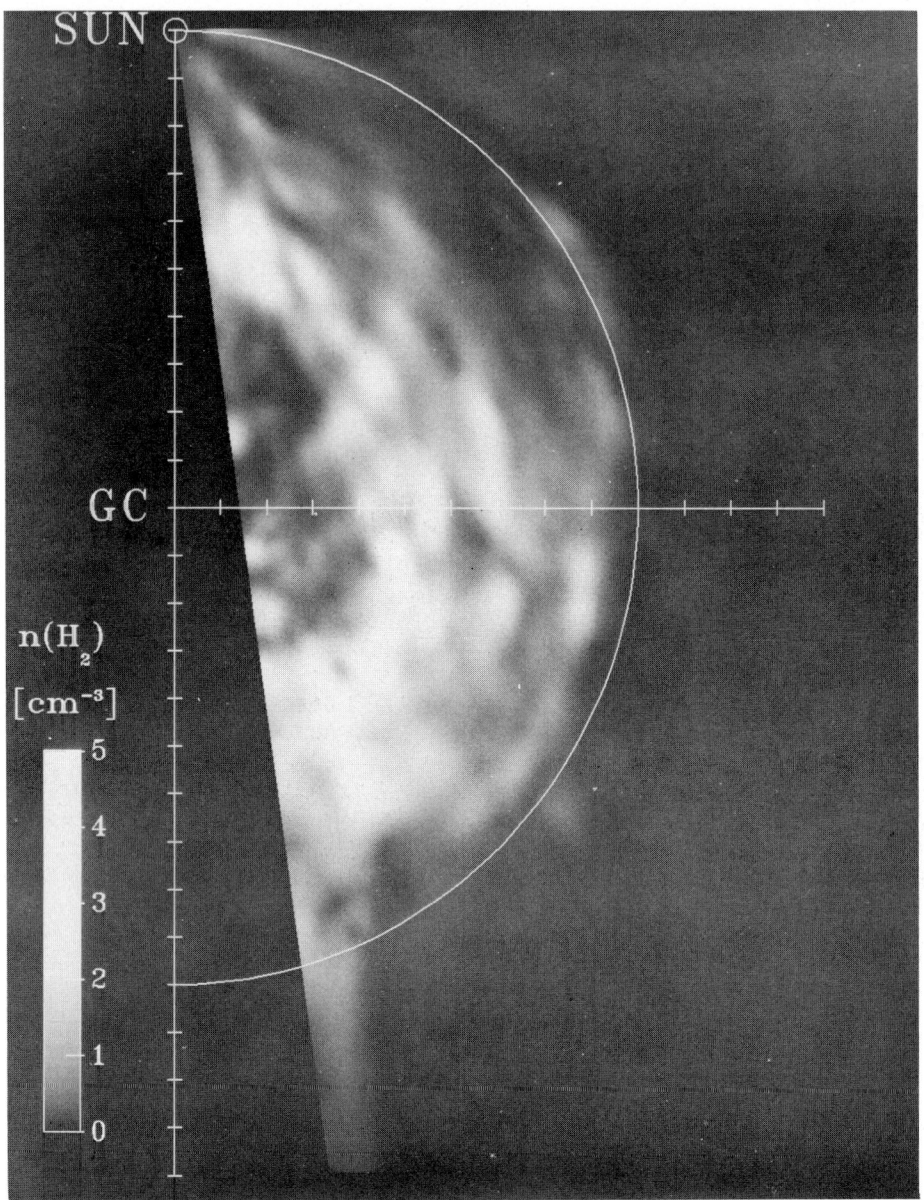

Figure 10.   Face-on image of the mid-plane molecular hydrogen density
in the Milky Way derived from a non-axisymmetric analysis of CO emission
by Clemens, Sanders, and Scoville (1986).   The dominant feature is the
molecular ring at 3-7 kpc.   The "spur" at $l=8°$ extending to 13 kpc at
the bottom of the figure is due to incorrect "mapping" of the 3 kpc
expanding arm into the outer galaxy by virtue of its non-circular
(negative velocity) motion.   [In this figure, the value $R_0=10$ kpc
was not changed to 8.5 kpc.]

clearly demonstrated in recent high resolution CO mapping of external
galaxies. In M51, the CO is peaked along the spiral arm loci, yet
the intensity of the emission from the interarm regions is well over
50% of that seen in the arms (Rydbeck et al 1985). Interarm molecular
clouds have also been inferred from VLA maps of 21 cm emission. In M83,
Allen, Atherton, and Tilanus (1986) find that the HI emission peaks
downstream from the dust lanes; they point out that the absence of HI
emission from either the dust lanes or the upstream side of the spiral
arms implies that the pre-arm gas must be largely molecular.

4. COMPARISON WITH OTHER SPIRAL GALAXIES

At present there exist single dish CO measurements in over 100 spiral
galaxies and well sampled radial distributions exist for 23 Sc and 19
Sb/Sbc galaxies (cf. Young 1986). In Figure 11 the distributions of
H$_2$ for these two classes of galaxies are compared and the Milky Way
distribution is included as a dashed line.

In virtually all of the external spiral galaxies now mapped in CO,
the emission exhibits a monotonic fall-off with increasing radius. In
many of the Sc spirals, the radial distributions of CO can be modeled
as an exponential in r with scale lengths similar to that of the blue
light in the disks (Young and Scoville 1982, Scoville and Young 1983).
It is noteworthy that none of the Sc spirals and only three of the earlier
type spirals (NGC 2841, NGC 7331, and M31) exhibit a ring-like distribution
similar to the Milky Way.

Figure 11 illustrates the large range in the H$_2$ surface densities
from one galaxy to another--over two orders of magnitude within galaxies
of similar morphological type. In contrast the HI varies only factors
of 2-4 between galaxies of similar type and the HI radial distributions
are relatively flat, like that in the Milky Way. A general characteristic
of the overall gas distributions (HI and H$_2$) in late-type spiral galaxies
is that the HI surface density is approximately constant at 10^{21} cm^{-2}
while the H$_2$ gas typically exhibits a monotonic increase at smaller
radii. That is, increases in the total gas density above 10^{21} cm^{-2} go
into molecular rather than diffuse atomic gas.

The large masses of molecular gas found in the above sample of
galaxies should not be taken as indicative that the molecular mass
always exceeds the atomic hydrogen mass in late-type spirals since the
galaxies chosen for CO mapping have to some extent, been selected observationally for their strong CO emission.

5. MOLECULAR CLOUD PROPERTIES

As is clear from the spectra shown in Figure 3, the bulk of the CO
emission arises from discrete clouds (2-8 along lines of sight through
the molecular midplane of the Galaxy), each of which has a small velocity
dispersion relative to the full range of the galactic velocity field.
The best sampling of the properties for these emission regions is
provided by the Massachusetts-Stony Brook Galactic CO Survey, comprising
over 40,000 CO spectra at l=8-90° sampled on a 3-6' grid with a 45"

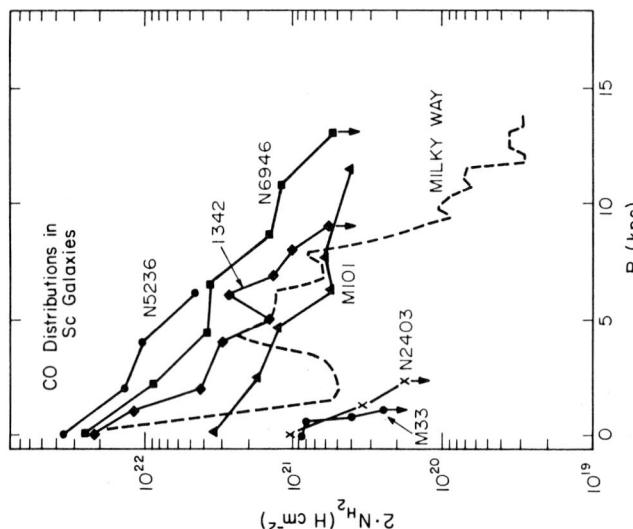

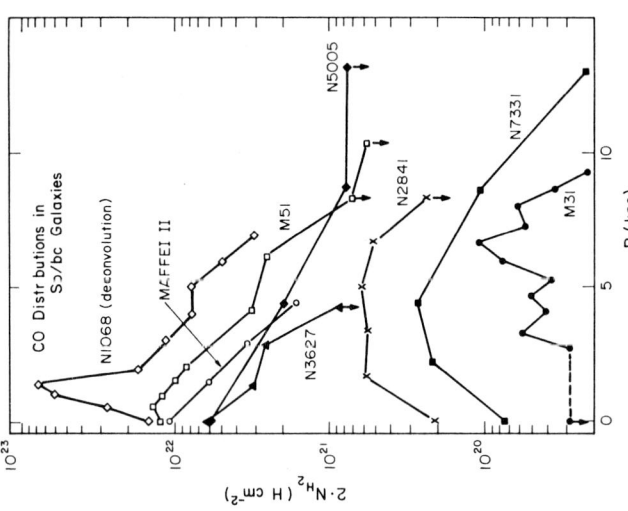

Figure 11. Radial distributions of molecular hydrogen surface density, corrected to inclination 0° are shown for 8 Sb-Sbc galaxies and 6 Sc galaxies for comparison with the Milky Way distribution (dashed line; Clemens, Sanders, and Scoville 1986). The distribution for M31 has been scaled down by a factor of 5 for clarity. This figure was adapted from Young (1986). The extragalactic H_2 densities were scaled from the CO emission using a conversion factor of 3.6×10^{20} cm^{-2} (K km s^{-1})$^{-1}$. References for the extragalactic data are N1068 (Scoville, Young, and Lucy 1983), M51 (Scoville and Young 1983), Maffei II (Rickard et al 1977), N5005 (Young 1986), N3627 (Young et al 1983), N2841 and N7331 (Young and Scoville 1982), M31 (Stark 1979), N5236 (Combes et al 1978), N6946 and IC342 (Young and Scoville 1982), and M101 (Solomon et al 1983).

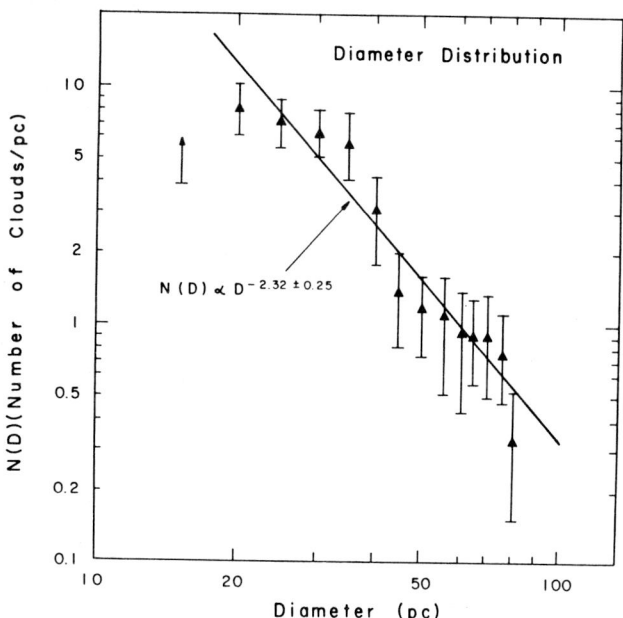

Figure 12. The number distribution of cloud diameters for clouds with D>10 pc within the sample volume studied by Sanders, Scoville, and Solomon (1984). The distribution is adequately fit by a power law: $N(D) \propto D^{-2.32}$.

beam (Sanders et al. 1986). A complete cataloging of the emission regions (Scoville et al 1986) yields 1,427 clouds with CO temperatures exceeding 5 K (corresponding to $T_K \geq 10.5$ K), 255 hot cloud cores with CO temperatures exceeding 9 K ($T_K \geq 16.3$ K), and 95 clouds associated with 171 radio HII regions. These cloud samples and earlier high resolution but more limited observations obtained by Sanders, Scoville and Solomon (1984) provide a basis for study of the distribution of cloud properties: their diameters, velocity dispersions, temperatures, and masses. Liszt, Xiang, and Burton (1981) and Stark (1979) have also measured samples of GMCs in ^{13}CO.

5.1 Size and Mass Spectrum of GMCs.

In Figure 12, we show the distribution of cloud diameters obtained by Sanders, Scoville, and Solomon (1985). A reasonable power law fit to the distribution is

$$N(D) \propto D^{-2.32 \pm 0.25} \tag{6}$$

The observed internal velocity dispersion (one-dimensional) also shows a systematic correlation with the cloud diameter (D in pc), given by

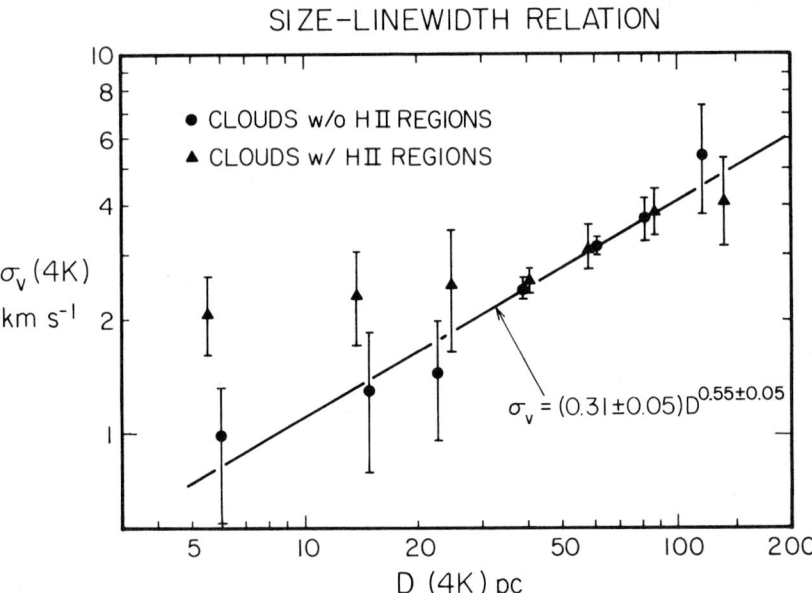

Figure 13. The diameter-velocity dispersion relation is shown for clouds with and without giant HII regions (Scoville et al 1986).

$$\sigma_v = 0.5 \ D^{0.55} \ km \ s^{-1} \tag{7}$$

where D is measured at a CO intensity threshold of 4 K (see Scoville et al 1986). This relation (shown in Figure 13) holds for GMCs with and without HII regions for D>30 pc but for smaller diameter clouds, the HII region GMCs have a constant dispersion of ∿3-4 km s⁻¹. The larger velocity dispersion for the small clouds with HII regions could be attributed to the disruptive effects of the HII regions or related to the suggestion that the OB stars are formed in GMCs which have recently undergone a cloud-cloud collision (see §6).

For a cloud with a modestly peaked density distribution ($\rho = \rho_R R/r$ where R=D/2), the three dimensional velocity dispersion averaged over the cloud in virial equilibrium is

$$\sigma_v(3-d) = (\frac{\pi}{4} \ G\rho_R)^{\frac{1}{2}} \ D \tag{8}$$

If σ_v is the observed one-dimensional velocity dispersion, then the mean density is

$$\langle \rho \rangle = \frac{3}{2} \ \rho_R = \frac{18}{\pi G}(\frac{\sigma_v}{D})^2 \tag{9}$$

and the virial mass is

$$M_{vir} = \frac{3}{G} \ D \ \sigma_v^2 \tag{10}$$

For D and σ_v measured in pc and km s^{-1}, equation (10) becomes

$$M_{vir} = 698 \, D \, \sigma_v^2 \, M_O \quad . \tag{11}$$

[For a uniform density cloud, the derived virial mass is 25% less than that given by equation (11)].

Combining equations (7) and (9), we find a mean density

$$\langle \rho \rangle = 8.0 \times 10^{-22}(\frac{D}{40 \text{ pc}})^{-0.9} \quad \text{gr cm}^{-3} \quad . \tag{12a}$$

and

$$\langle n_{H_2} \rangle = 180 \, (\frac{D}{40 \text{ pc}})^{-0.9} \quad \text{cm}^{-3} \quad . \tag{12b}$$

[To go from equation (12a) to (12b), the mass density was divided by a factor 1.36 to remove the He content.] Equation (12b) predicts that clouds of diameter 10 to 100 pc have mean H$_2$ number densities of 625 to 80 cm^{-3}. The latter value is approximately a factor of 5 higher than the mean density required to be stable against the Galactic tidal force and thus it is unlikely that even the lowest density giant molecular clouds are significantly affected by tidal disruption (except possibly at R<1 kpc).

Combining the cloud size spectrum given by equation (6) with equation (12) yields a mass spectrum

$$N(M) \propto M^{-1.61} \quad . \tag{13}$$

This mass spectrum is similar to that expected if the larger GMCs are built up by a agglomeration of smaller clouds (Kwan and Valdes 1982). The time required for the GMCs in the molecular ring to double their mass by sweeping up smaller GMCs is approximately 10^8 years (Scoville and Hersh 1979) for the parameters given in Table 3.

For the mass spectrum given in equation (13), 50% of the total galactic H$_2$ is contained in 1,000 clouds with D>50 pc (M>10^6 M$_O$) and 90% is contained in 5,000 clouds with D>20 pc and M>10^{5}M$_O$. In Table 4 the properties are given of a "typical" GMC of radius 40 pc. This size cloud represents the midpoint in the GMC mass function; that is, 50% of the total H$_2$ mass is contained in clouds larger than this size and 50% contained in clouds smaller than this size.

Although the thermal pressure cf the gas (P$_{TH}$/k$\sim$2000 cm^{-3} K) is comparable to that of the diffuse intercloud medium, the internal velocity dispersion is far in excess of that expected for 10 K gas. Thus, the _effective_ internal pressure in the GMCs is at least an order of magnitude higher than that of the intercloud medium. Since the virial masses estimated from the cbserved velocity dispersions are consistent with mass estimates based on molecular column density measurements, the GMCs <u>are bound by self-gravity rather than being in pressure equilibrium with other phases of the external ISM.</u>

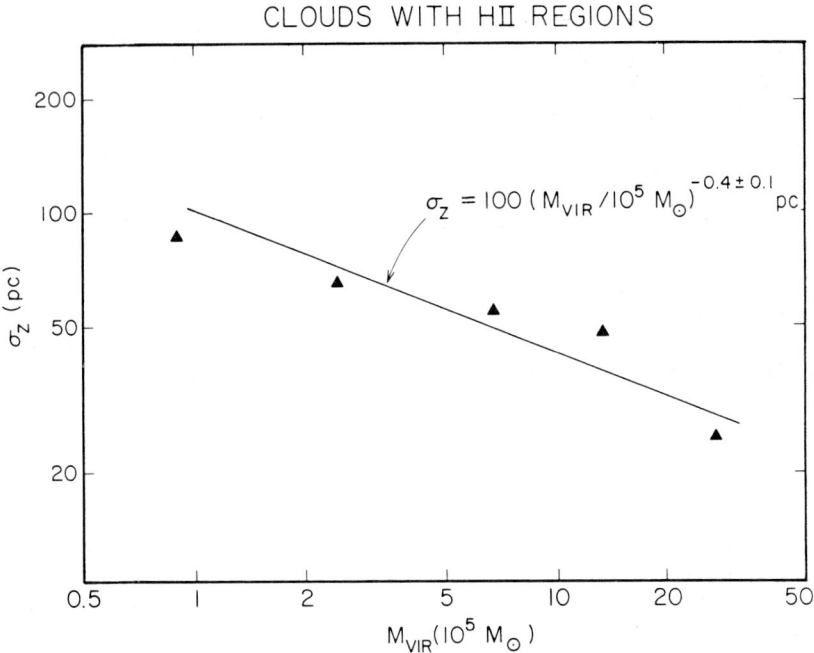

Figure 14. The dispersion in Z is shown as a function of cloud mass for GMCs containing HII regions (Scoville et al 1986).

Table 4

Properties of a "Typical" GMC

Diameter	40 pc
M_{H_2+He}	$4 \times 10^5 \, M_\odot$
$\bar{n}_{H_2}$	$180 \, cm^{-3}$
T_K	$10 \, K$
P_{TH}/k	$2000 \, cm^{-3} \, K$
σ_V	$3.8 \, km \, s^{-1}$

5.2 GMC Lifetimes

The presence of a substantial number of GMCs in interarm regions (§3.2), strongly suggests that the lifetimes of the GMCs are longer than the time between spiral arm passages (ie. $>10^8$ years). Further evidence

for such longevity comes from the observation that the Z scale-height
decreases with increasing cloud mass ($\sigma_Z \propto M^{-0.4}$ for $M > 10^5$ M_O) as
shown in Figure 14. A similar result was found by Stark (1982)
for a limited sample of "spiral arm" clouds. Assuming that the
Z motion of the clouds is determined solely by the gravitational
potential of the Galactic disk, this relation implies that the cloud-cloud
velocity dispersion decreases approximately as $M_{VIR}^{-\frac{1}{2}}$ as expected
for equipartition of the GMC kinetic energy. In order for the random
kinetic energy of clouds to be in equipartition, they must survive
several cloud-cloud collision times or longer than 10^8 years. (A
more precise evaluation of this lower limit on the cloud lifetime
requires a detailed modeling of the cloud collisions taking account
of gravitational focussing and the actual space density of clouds
in and outside of arms.) This lower limit to the cloud lifetime
is $\sim$30 times the typical free-fall collapse time for the GMCs.

6. STAR FORMATION IN GMCs

Given the fact that the GMCs are self-gravitating, it is remarkable
that they have such low efficiency of star formation (Scoville and
Solomon 1975). At the typical density of 200 cm^{-3}, the gravitational
free-fall time is approximately 3×10^6 years and the star formation
rate for 2.3×10^9 M_O of H_2 could be as high as 10^3 M_O yr^{-1}. This
maximum rate is more than 100 times greater than estimates for the
global star formation rate based on HII region studies (5 M_O yr^{-1},
eg. Mezger and Smith 1977). The molecular clouds must therefore
be generally stable against gravitational collapse.

What supports the clouds against collapse? and what are the
dominant mechanisms for initiating star formation within the stable
clouds?. Suggested mechanisms for initiating collapse include the
compression due to expansion of HII regions and supernova remnants
(eg. Elmegreen and Lada 1977, Herbst and Assousa 1977) and external
triggers such as shock waves associated with galactic spiral arms.

There now exists considerable evidence that star formation
is a bimodal process -- with high and low mass star formation occurring
independently (cf. Larson 1986). This is suggested by both the
discontinuous slope of the initial mass function and observations
of nearby dark clouds with a high abundance of low mass, pre-main
sequence stars, but virtually no high mass stars. We note also
that there are approximately 4,000 giant molecular clouds with mass
greater than 10^5 M_O, yet only 100 giant HII regions (Downes et al
1980), and often, several of these HII regions are clustered within
a single cloud. Thus, high mass stars form only in select clouds
with extraordinary characteristics (eg. larger mass or density)
or via an external stimulus (associated with the spiral arms).
And when the conditions are right, several clusters of OB stars
may form, almost simultaneously, in the same cloud.

The formation of low mass stars appears to occur at a rate
proportional to the mass density of H_2 since the blue light from
the disks of late type spiral galaxies exhibits a linear correlation

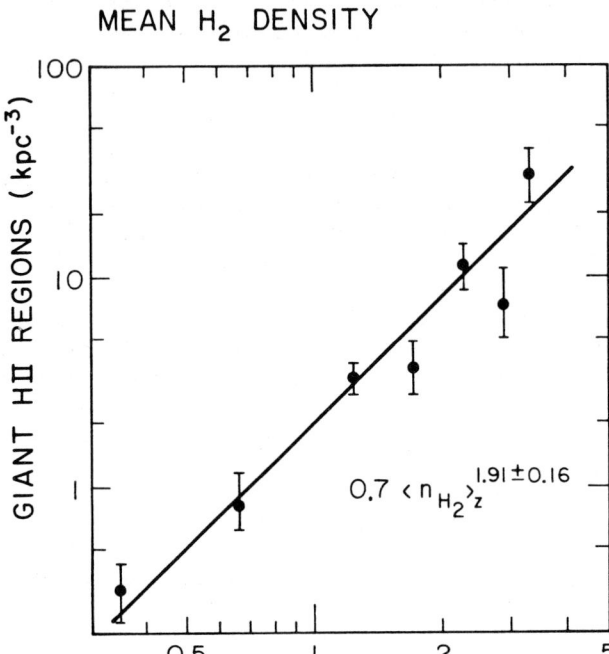

Figure 15. The number density of HII regions is shown as a function of the local mean H_2 density shown in Figure 9 (averaged over length scales of $\sim$300 pc in the galactic plane and $\sim$100 pc perpendicular to the plane). The HII region number density shows a quadratic dependence on the density of molecular gas, suggesting that massive star formation is initiated by collisions of GMCs. This figure is from Clemens, Sanders, and Scoville (1986).

with the CO emission, both as a function of radius and from one galaxy to another (Young and Scoville 1982, Scoville and Young 1983). Thus, the low mass stars apparently form at a low efficiency (per free-fall time) in the molecular clouds more or less independent of the external stimuli such as spiral arms. Substantial low mass star formation must therefore occur in interarm clouds.

Just the opposite situation pertains to the formation of high mass stars. Figure 15 shows the correlation between the concentration of giant HII regions in the Milky Way (averaged over length scales of $\sim$300 pc) and the mean H_2 density (averaged over the same area, Clemens, Sanders and Scoville 1986). The <u>quadratic</u> dependence of the high mass star formation rate on the local mean molecular hydrogen density strongly suggests that the OB stars form as a result of a collision process such as cloud-cloud collisions (Scoville, Sanders, and Clemens 1986).

   Since the 4 km s^{-1} rms velocity dispersion of giant molecular
clouds (Clemens 1985) is comparable to or less than the typical
internal velocities in the clouds, it is clear that most collisions
between GMCs will result in a bound complex rather than disruption
of the clouds. During a cloud-cloud collision, the interface gas
will remain molecular but be heated to peak temperatures of 10^3-$2x10^3$ K.
Since the Jeans length at the highest temperatures is much greater
than the thickness of the hottest zone in the interface, stars would
not be expected to form until the post-shock gas has cooled to approximately
100 K. At that point, the highest mass stars will form first in
the cooling gas, thus favoring massive stars in the initial mass
function.

   The mechanism of cloud-cloud collisions to form OB stars is
consistent with a number of previous observations of OB star formation.
Most important is the fact that one may now understand the concentration
of HII regions along the spiral arms as resulting from the convergence
of cloud orbits in the spiral potential minimum associated with
the density wave. With a modest 5% spiral perturbation, Kwan and
Valdes (1983) found that the number density of clouds increased
by a factor of approximately three in the spiral arms which would
result in an increase of the collision rate by a factor of nine.
This corresponds well to the observed contrast between arm and interarm
HII regions--Mezger and Smith (1977) find 15% of the giant HII regions
in classic interarm areas of the 1-V plane. In addition, it has
often been noted that OB star formation occurs with a low duty cycle
in molecular clouds, that is, the time during which massive star
formation occurs is relatively short compared to the spread of ages
for lower mass young stars within the clouds. In the cloud collision
model, it is natural to expect a low duty cycle as a result of the
fact that the cloud crossing time (this is, the interaction time)
is a factor of 5-10 less than the mean time between cloud collisions.

7. SUMMARY

The 2.6 mm CO transition has become widely adopted as a tracer of
the large scale distribution and properties of molecular clouds
in the disk of our galaxy. The theoretical and empirical foundations
for using this trace molecule as a probe of the H_2 suggests that
the CO emissivity is a good proportional tracer of the mass of H_2,
probably accurate to a factor of two when averaged over a large
number of emission regions. The total mass of molecular gas at
R<14 kpc is $2.3x10^9$ M_0 which is approximately equal to the mass
of atomic gas within the same region.

   The large-scale distribution of H_2, derived from CO observations,
exhibits a sharp peak within the central 400 pc of the galaxy and
a broad ring-like maximum at radii of 3-7 kpc. This is markedly
different from the atomic gas which has an approximately constant
surface density outside 3 kpc and a minimum in the central region
of the galaxy. The mean thickness of the molecular disk in the
vicinity of the molecular cloud ring is 90 pc, increasing monotonically

at larger radii to over 150 pc outside the solar circle. The mass
of molecular hydrogen within the molecular cloud ring is approximately
1.8×10^9 M_o and that in the galactic center at R<400 pc is 2×10^8
M_o. The mean mass ratio for H_2/HI is 3 at R<7 kpc.

In contrast to the atomic gas, most of the molecular gas is
contained in discrete clouds with typical diameter ∿40 pc and mass
∿4×10^5 M_o. The high internal velocity dispersions within the giant
molecular clouds imply an effective internal pressure approximately
10 times that of the intercloud medium. The GMCs are self-gravitating
and not in pressure equilibrium with the other phases of the interstellar
medium.

The variation of the Z scale-height as a function of cloud
mass suggests an equipartition of kinetic energy amoung GMCs with
M>10^5 M_o, implying that their ages must exceed the typical cloud-cloud
collision time of 10^8 years. Both the cloud mass spectrum and the
lower limit to the lifetime suggests that build-up of the GMCs occurs
by collisional agglomeration of smaller clouds.

The GMCs are also generally stable against collapse to form
stars. If they were in free-fall gravitational collapse, the global
star formation rate would be over 100 times current estimates for
the galactic rate (5 M_o yr^{-1}). There now exists strong evidence
that the formation of high and low mass stars occurs by fundamentally
different mechanisms. The general correlation of the blue light
and the mean molecular emission from external spiral galaxies suggests
that low-mass star formation occurs at a rate linearly dependent
on the mass of molecular gas. On the other hand, the density of
HII regions exhibits a higher order of dependence on the density
of molecular gas, possibly indicating that massive star formation
is triggered by the collision of GMCs.

Approximately 50% of the molecular gas in the inner galaxy
is not in spiral arms, but the spiral arm GMCs are significantly
hotter than those in the general cloud population, presumably due
to the large heat input of the massive stars. In the outer galaxy,
the arm/interarm contrast is much higher with most of the molecular
gas in the second galactic quadrant confined to the Perseus arm.

ACKNOWLEDGMENTS

This research was supported in part by NSF Grant 84-12473 and an
IRAS General Investigator Grant. It is a pleasure to acknowledge
the assistance of Leona Kershaw in preparation of this manuscript.

8. REFERENCES

Allen, R.J., Atherton, P.D., and Tilanus, R.P.J. 1986, _Nature_, **319**, 296.

Beck, S.C., Bloemhof, G.E., Serabyn, E., Townes, C.H., Tokunaga, A.T., Lacy, J.H., and Smith, H.A. 1982, _Ap. J. (Letters)_, **253**, L83.

Beckwith, S., Persson, S.E., Neugebauer, G., and Becklin, E.E. 1978, _Ap. J._, **223**, 464.

Bloemen, J.B.G.M., Caraveo, P.A., Hermsen, W., Lebrun, F., Maddalena, R.J., Strong, A.W., and Thaddeus, P. 1984, _Astr. Ap._, **139**, 37.

Bronfman, L., Cohen, R.S., Alvarez, H., May, J., and Thaddeus, P. 1987, preprint.

Burton, W.B. and Gordon, M.A. 1978, _Astr. Ap._, **63**, 7.

Burton, W.B., Gordon, M.A., Bania, T.M., and Lockman, F.J. 1975, _Ap. J._, **202**, 30.

Clemens, D.P. 1985, _Ap. J._, **271**, 604.

Clemens, D.P., Sanders, D.B., and Scoville, N.Z. 1986, _Ap. J._ (submitted).

Cohen, R.S., Cong, H., Dame, T.M., and Thaddeus, P. 1980, _Ap. J._ ((Letters), **239**, L53.

Cohen, R.S., and Thaddeus, P. 1977, _Ap. J. (Letters)_, **217**, L155.

Combes, F., Encrenaz, P.J., Lucas, R., and Weliachew, L. 1978, _Astr. Ap._, **67**, L13.

Dame, T.M., Elmegreen, B.G., Cohen, R.S., and Thaddeus, P. 1986, _Ap. J._, **305**, 892.

Despois, D. and Baudry, A. 1983 in "Surveys of the Southern Galaxy", ed. W.B. Burton and F.P. Israel, (Dordrecht:Reidel), p. 173.

Dickey, J.M. and Benson, J.M. 1982, _A. J._, **87**, 278.

Downes, D., Wilson, T.L., Bieging, J., and Wink, J. 1980, _Astr. Ap. Suppl._, **40**, 379.

Elmegreen, B.G. and Lada, C.J. 1977, _Ap. J._, **214**, 725.

Frerking, M.A., Langer, W.D., and Wilson, R.W. 1982, _Ap. J._, **262**, 590.

Georgelin, Y.M. and Georgelin, Y.P. 1976, _Astr. Ap._, **49**, 57.

Goldreich, P. and Kwan, J. 1974, _Ap. J._, **189**, 441.

Grabelsky, D.A. 1985, Ph.D. Thesis, Columbia University.

Grabelsky, D.A., Cohen, R.S., May, J., Bronfman, L., and Thaddeus, P. 1986, _Ap. J._ (submitted).

Green, S. and Chapman, S. 1978, _Ap. J. Suppl._, **37**, 167.

Herbst, W. and Assousa, G.E. 1977, _Ap. J._, **217**, 473.

Kerr, F.J. 1979, _Ann. Rev. Astr. Ap._, **7**, 39.

Kulkarni, S.R. 1987, this volume.

Kutner, M.L. and Ulich, B.L. 1981, _Ap. J._, **250**, 341.

Kwan, J. and Valdes, F. 1983, _Ap. J._, **271**, 604.

Larson, R.B. 1986, _M.N.R.A.S._, **218**, 409.

Lebrun, _et al_. 1983, _Ap. J._, **274**, 231.

Leisawitz, D. and Bash, F. 1982, _Ap. J._, **259**, 133.

Levinson, F.H. and Roberts, W.W. 1981, _Ap. J._, **245**, 465.

Liszt, H.S. 1983, in "The Milky Way Galaxy", IAU Symp. 106, eds. H. van Woerden, R.J. Allen, and W.B. Burton (Dordrecht:Reidel), p. 283.

Liszt, H.S. and Burton, W.B. 1980, _Ap. J._, **236**, 779.

Liszt, H.S. and Burton, W.B. 1981, Ap. J., **243**, 778.
Liszt, H.S., Xiang, D., and Burton, W.B. 1981, Ap. J., **249**, 532.
Lockman, F.J. 1979, Ap. J., **232**, 761.
Mezger, P.G. and Smith, L.F. 1977, "Star Formation", IAU Symp. No. 75,
    ed. T. de Jong and A. Maeder, (Dordrecht:Reidel), p. 133.
Rickard, L.J., Palmer, P., Morris, M., Zuckerman, B., and Turner,
    B. 1977, Ap. J., **213**, 673.
Robinson, B.J., Manchester, R.N., Whiteoak, J.B., Sanders, D.B.,
    Scoville, N.Z., Clemens, D.P., McCutcheon, W.H., and Solomon,
    P.M. 1984, Ap. J. (Letters), **283**, L31.
Rydbeck, G., Hjalmarson, A., and Rydbeck, O.E.H. 1985, Astr. Ap.,
    144, 282.
Sanders, D.B. 1981, Ph.D. Thesis, State University of New
    York at Stony Brook.
Sanders, D.B., Clemens, D.B., Scoville, N.Z., and Solomon, P.M.
    1986, Ap. J. Suppl., **60**, 1.
Sanders, D.B., Scoville, N.Z., and Solomon, P.M. 1985, Ap. J.,
    276, 182.
Sanders, D.B., Solomon, P.M., and Scoville, N.Z. 1984, Ap. J.,
    276, 182.
Scoville, N.Z., Hall, D.N.B., Kleinmann, S.G., and Ridgway,
    S.T. 1982, Ap. J., **253**, 136.
Scoville, N.Z. and Hersh, K. 1979, Ap. J., **229**, 578.
Scoville, N.Z., Sanders, D.B., and Clemens, D.P. 1986, Ap. J.
    (Letters), November 15.
Scoville, N.Z. and Solomon, P.M. 1974, Ap. J. (Letters), **187**, L67.
Scoville, N.Z. and Solomon, P.M. 1975, Ap. J. (Letters), **199**, L105.
Scoville, N.Z. and Young, J.S. 1983, Ap. J., **265**, 148.
Scoville, N.Z., Young, J.S., and Lucy, L.B. 1983, Ap. J., **270**, 443.
Scoville, N.Z., Yun, M.S., Clemens, D.P., Sanders, D.B., and Waller,
    W.H. 1986, Ap. J. Suppl. (submitted)
Solomon, P.M., Barrett, J., Sanders, D.B., and de Zafra, R.
    1983, Ap. J. (Letters), **266**, L103.
Solomon, P.M., Sanders, D.B., and Rivolo, R. 1985, Ap. J.
    (Letters), **292**, L19.
Spitzer, L. and Jenkins, E.B. 1975, Ann. Rev. Astr. Ap., **13**, 133.
Stark, A.A. 1979, Ph.D. Thesis, Princeton University.
Stark, A.A. 1982, "Kinematics, Dynamics, and Structure of the Milky
    Way", ed. W.L.H. Shuter (Dordrecht:Reidel), p. 127.
Stark, A.A., Bally, J., Knapp, G.R., Krahnert, A., Penzias,
    A.A., and Wilson, R.W. 1987, Bell Laboratories CO Survey,
    in progress.
Young, J.S. 1986 in "Star-Forming Regions", IAU Symp. 115. eds. M.
    Piembert and J. Junyaku (Dordrecht:Reidel).
Young, J.S. and Scoville, N.Z. 1982, Ap. J., **258**, 467.
Young, J.S. and Scoville, N.Z. 1982, Ap. J. (Letters), **260**, L11.
Young, J.S., Tacconi, L., and Scoville, N.Z. 1983, Ap. J., **269**, 136.
Yuan, C. 1969, Ap. J., **158**, 889.

MOLECULAR CLOUDS: AN OVERVIEW

Paul F. Goldsmith
FCRAO, Department of Physics and Astronomy
University of Massachusetts
Amherst, MA 01003

ABSTRACT. Molecular material is found in differing regions in our
galaxy. We attempt here to give a summary of the observationally
determined properties of molecular clouds. Some of the most important
methods of obtaining information on physical conditions within these
regions are reviewed, and a summary of different types of molecular
regions is presented.

1. INTRODUCTION

Molecular clouds share the obvious characteristic that they consist
largely of molecular material. However, molecular clouds may well have
envelopes with a relatively large abundance of atomic species including
hydrogen (Wannier, Lichten, and Morris 1983) and (at least
theoretically) carbon (Langer 1976; Tielens and Hollenbach 1985). From
an observational point of view, molecular clouds do have edges. Very
sensitive observations have delineated the drop in the abundance of
tracers such as $C^{18}O$ at the edges of clouds where the visual extinction
is less than 3 mag. (Goldsmith et al. 1980; Frerking, Langer, and
Wilson (1982). It is generally possible to define the extent of a
molecular cloud by a cutoff in the antenna temperature of the tracer
being employed. Due to the nature of the photodissociation and
chemical processes affecting the abundance of carbon monoxide in the
relatively unshielded cloud edges, the size of molecular clouds is not,
in general, *critically* sensitive to whether the ^{12}C or the ^{13}C isotopic
species of carbon monoxide isotope is observed (e.g. Kutner et al.
1977; Young et al. 1982; compare maps by Schloerb et al. **1987** and Bally
et al. **1987**). Thus, the carbon monoxide istotopes are the most often-
used tracers of cloud boundaries.

It is also the case that even the interiors of "molecular" clouds
are not necessarily purely molecular. There is evidence for an
appreciable percentage of atomic hydrogen present (Knapp 1974). In the
past few years it has also become apparent that a substantial fraction
of the carbon in these regions remains in atomic form (Phillips and
Huggins 1981; Keene et al. 1985; Zmuidzinas, Betz, and Goldhaber 1986),

D. J. Hollenbach and H. A. Thronson, Jr. (eds.), Interstellar Processes, 51–70.

with [CI]/[CO] $\approx$ 0.1 being generally accepted as a representative
value.

The structure of the gas within the boundaries which define the
molecular clouds is poorly understood. Progress in unravelling the
complexities of these regions has largely been through empirically
developed methods of study. Determination of key properties such as
gas temperature, density, and the velocity field, has been made with
increasingly refined techniques, which have become available as the
instrumentation at millimeter and centimeter wavelengths has gradually
improved. The extension of these studies to clouds in other galaxies
is in its infancy, being hampered by angular resolution and sensitivity
limitations. Consequently, the discussion will be restricted
essentially to the Milky Way. Circumstellar envelopes are really
outside the usual definition of the interstellar medium; these regions
do have a rich and distinctive molecular chemistry which is discussed
by, e.g., Olofsson et al. (1982) and Omont (1986). The methods of
determining most physical properties – as well as the results – are
sufficiently distinctive that we will not refer further to these
regions except in summary in Section 3.

In the following section we discuss some of the techniques which
have been developed to study important parameters of molecular clouds.
We attempt to give representative results for different types of
regions that have been studied. Finally, we give a classification of
the different types of molecular regions found in our galaxy. Articles
in these proceedings that cover material particularly relevant to the
present topic include those by Black (heating and cooling processes);
Irvine, Goldsmith, and Hjalmarson; Herbst and Prasad (chemistry); Scalo
(fragmentation and turbulence); Myers (turbulence in clouds); Heiles
(magnetic field); and Scoville and Sanders (distribution and properties
of giant molecular cloud complexes).

2.   DETERMINATION OF PHYSICAL CONDITIONS IN MOLECULAR CLOUDS

Understanding the structure of molecular clouds in the galaxy, as well
as their role in such processes as star formation, requires knowledge
of the basic physical conditions in these regions. In the
approximately 20 years that molecular observations have been widely
used as diagnostic probes, a number of relatively "standard" techniques
have been developed for determining certain parameters. It must,
however, be recognized that these are subject to many types of errors,
and that there is room for significant improvement. It is fair to
state that the techniques presently in use often give apparently
contradictory information, and that reconciling the results obtained
from different cloud diagnostics should give considerably improved
insight into the detailed structure of these regions.

2.1   Temperature

The molecular transitions which can presently be observed have a wide
variation in spontaneous decay rate, from $<10^{-8}$ s^{-1} to $>10^{-4}$ s^{-1},
which, combined with the peculiar energy level schemes of certain

species, has the result that an equilibrium population distribution cannot be assumed. Consequently, the excitation temperature (defined by the relative population of the upper and lower states involved) of any given transition is not obviously related to the kinetic temperature of the gas in which H_2 is the dominant molecular species. Observation of certain transitions of particular molecules can give information about the gas kinetic temperature in a relatively direct fashion.

Most important of these are the rotational transitions of ^{12}CO. Since shortly after initial detection of this species in interstellar clouds, it has been recognized that its low-J transitions are in general optically thick; the opacity is determined observationally by the relative intensity of the corresponding ^{13}CO transition. In the low opacity limit this is ≈40-90 times weaker, but is typically found to be a factor of only 3-10 less intense than the common isotope, indicative of substantial opacity for the latter. Together with the relatively low spontaneous decay rate of the J=1-0 transition, 6×10^{-8} s^{-1}, which requires a correspondingly low density for thermalization, it is apparent that the emission from a cloud observed in this transition will be that of a blackbody at a temperature equal to the gas kinetic temperature, and that the peak antenna temperature directly yields the kinetic temperature of the H_2 (Penzias et al. 1972).

Most molecular material not associated with high luminosity heating sources has temperatures between 10 K and 20 K. This range includes not only the classical "dark clouds" which exhibit no evidence of star formation, but also many regions associated with low mass young stars (as indicated by infrared emission), as well as the bulk of the volume of clouds in which massive stars have formed. The thermal balance of the molecular gas involves heating by cosmic rays, by photoelectrons (cloud edges only) and by collisions with warm dust. The cooling is primarily by molecular line emission, with carbon monoxide the dominant coolant. Temperatures at the low end of the range given above are consistent with cosmic ray heating being the primary energy input, while additional heating at cloud edges and in the vicinity of embedded sources produces a moderate increase in temperature, limited by the cooling rate which varies as temperature to the 1.5 to 3.0 power for typical molecular cloud densities (Goldsmith and Langer 1977). Several studies of the edges of dark clouds (Snell 1981, Young et al. 1982, Dickman and Clemens 1983) require temperatures of 20 to 50 K to reproduce the observations, and although consistent with the elevated temperatures expected from photoelectric heating for Av ≲ few, these high temperatures have not been directly measured. It has also been suggested that the edges of molecular clouds in the galaxy M82 are heated by an enhanced radiation field (Young and Scoville 1984).

Only in a small number of clouds with very large heating rates such as M17 and Orion does the measured kinetic temperature exceed 40 K over a substantial region. This is undoubtedly a result of observational selection in part resulting from limited angular resolution, and in part from requiring a favorable geometry to expose the heated material to our view.

The major problem associated with this method of determining the gas kinetic temperature is determining *to what gas* in the cloud the measurement applies. If the velocity field is microturbulent, we are measuring the temperature at optical depth approximately unity from the near side; if there are large scale motions present, the measurement may apply to gas at quite a different location. If there are temperature gradients on the relevant scale, the different absorption coefficients of various CO transitions can result in differing temperatures being measured (Phillips et al. 1981). Sometimes there are indications of this in terms of self-absorbed lines, but often this is not the case. In extreme situations with sufficiently large column densities, embedded heating sources and presumably very hot gas in their vicinity are not detected at all in ^{12}CO (e.g. Goldsmith and Mao 1983). At present, uncertainty about the velocity field within molecular clouds, together with our desire to know the kinetic temperature as a function of position within regions expected to have large gradients (as produced by embedded heating sources) encourages us to find other methods of temperature determination which may be able to sample the cloud volume less problematically.

A number of low opacity cloud thermometers have been proposed, analyzed, and used to some extent. The most widely accepted are techniques involving the observation of several transitions whose *relative* intensity is determined by the temperature of the gas. The most often used of these is the ammonia molecule (NH_3). Observations of the inversion transitions from different K ladders can yield the gas kinetic temperature (e.g. Stutzki and Winnewisser 1985a). Problems with this technique include the limited angular resolution that can be achieved with single dish antennas ($\approx 1'$) and the relatively low sensitivity of synthesis arrays at this frequency. It has also become apparent that ammonia emission is not always optically thin, and this can complicate analysis of data, although interesting information about cloud structure can also result (Stutzki and Winnewisser 1985b). On the whole, kinetic temperatures derived from NH_3 agree fairly well with those from CO, for regions where extreme variations along the line of sight are not thought to be present. In the case of molecular outflows, Takano (1986) has compared results from the two species, and finds good agreement for the fairly low temperature gas associated with outflows of moderate mechanical (or source infrared) luminosity. For high-luminosity outflows, there are greater differences, with the ammonia giving systematically higher temperatures; this can most likely be ascribed to optical depth effects in the CO as discussed above.

Symmetric top molecules without inversion transitions can also be used as temperature probes, because transitions between different K-ladders are essentially forbidden. The relative intensity of transitions of the type (J,K) − (J−1,K) for different K will, to first order, be determined by the gas kinetic temperature. The species which have been utilized for this purpose include methyl cyanide (CH_3CN) and methyl acetylene (CH_3C_2H). The latter molecule is preferable in the sense that its much smaller dipole moment (.75 D compared to 3.75 D for methyl cyanide) means that the spontaneous emission rates are much

smaller, and the level populations can be brought into thermal
equilibrium at a lower hydrogen density (Askne et al. 1984).

   Comparison of the temperatures derived from observations of
different probes yields interesting results about the structure and
chemistry of molecular clouds. For the case of the Orion molecular
cloud, the three lowest CO transitions (which have been studied in some
detail) all yield kinetic temperatures $\gtrsim$100 K in the vicinity of the KL
nebula, and >60 K over a region several parsecs in extent. The
symmetric top molecules yield values approximately half of this
(Churchwell and Hollis 1983; Kuiper et al. 1984; see comparison in
Goldsmith et al. 1986). This is consistent with the cloud being heated
from the near side (as we see it) by the HII region. The decreasing
temperature as a function of distance into the cloud results in a lower
temperature for optically thin transitions such as those of the
symmetric tops discussed here, compared to that of the CO which becomes
optically thick in a relatively thin, hot layer of material.

   Multiple transitions from any optically thin species can be used
to determine the degree of molecular excitation (e.g. Irvine,
Goldsmith, and Hjalmarson 1986). If the level populations so-determined
conform to a single "rotation" temperature, one may be tempted to
equate this with the kinetic temperature. There are problems with
doing this which center around the relative degree of thermalization of
different rotational transitions and the effects of gradients within
the clouds. Statistical equilibrium calculations (e.g. Linke et al.
1982; Loren and Mundy 1984) and realistic cloud models are required to
analyze this data accurately. Fortunately, calculations of cross
sections for symmetric tops are becoming available (e.g. Green, 1986)
and further progress can be hoped for in this area.

   Vibrationally excited emission is a particularly good tracer of
hot material which is localized within an extended cloud of cooler gas,
since the foreground material cannot absorb the excited state
emission. The Sgr B2 molecular cloud is a good example of this
situation. The vibrationally excited cyanoacetylene (HC_3N) emission is
confined to the immediate vicinity of the Sgr B2 (North) cluster of
radio continuum sources (Goldsmith et al. 1987); this latter emission
presumably indicates the presence of a cluster of massive stars there,
but the foreground column density is so large that this source is
visible in the continuum only at wavelengths $\gtrsim$ 100 microns. Gas
temperatures in excess of 200 K are indicated for the vicinity of this
source which considerably exceeds the $\approx$50 K found in the bulk of the
cloud by Gusten et al. (1985) and Churchwell and Hollis (1983) using a
number of the thermometers discussed above.

2.2 Density

Techniques for determining the density of molecular clouds are of
several different varieties, which average along the line of sight in
different ways and which are susceptible to different types of errors.
At one extreme are the methods based on a determination of cloud mass
(such as discussed by Scoville and Sanders 1986), and derivation of the
density from an assumed volume for the object under study. The mass

may be obtained, for example, from application of the virial theorem.
The resulting density is clearly an average value, and may not really
apply to any portion of the cloud.

Somewhat related is the use of the column density along the line
of sight and an assumed linear dimension to obtain the space density of
material. The column density of molecular hydrogen is rarely measured
directly, but some surrogate quantity such as visual extinction
(determined by the dust grain column density, cf. Bok, Cordwell, and
Cromwell 1970), or the ^{13}CO integrated intensity (Dickman 1978) is
obtained, and the H_2 column density obtained by using what we hope is a
well-calibrated relationship. Other optically thin tracers may be
used, but the above are the most widely employed methods. The major
uncertainties in this approach involve the constancy of the conversion
factor from observed quantity to total column density. For example,
the ^{13}CO integrated intensity (together with the assumption of LTE and
^{12}CO antenna temperature to obtain the cloud kinetic temperature) has
been shown to correlate well with visual extinction for A_V > 2 mag
(Encrenaz, Falgarone, and Lucas 1975; Dickman 1978; Bachiller and
Cernicharo 1986). Frerking, Langer, and Wilson (1982) have compared
column densities of various rare isotopes of carbon monoxide with
visual extinctions derived from infrared absorption. This technique
has enabled study of the relationship at larger column densities than
was previously possible. The result is again consistent with a fairly
constant ratio for extinctions > 2-4 mag. It thus appears fairly well
established that ^{13}CO is a column density tracer except (perhaps) at
cloud edges. For the cores of large clouds where the visual extinction
cannot be determined, the situation is necessarily less certain. There
is some evidence that the ratio of rare carbon monoxide isotopes to
total column density may drop by a factor of a few for Av ≳ 100 mag,
but not nearly as drastically as has been suggested for H_2CO by
Wootten, Snell, and Evans (1980).

The ability to readily measure column density permits the
determination of the space density, given a reasonable degree of
symmetry and an assumed density profile for the region under study.
This technique has been employed by Arquilla and Goldsmith (1985) for
analyzing the density profile of a number of relatively isolated dark
clouds without clear evidence for star formation; the resulting central
densities are in the range 5-10x10^3 cm^{-3}, and the column densities are
well reproduced by space densities varying as (radius)$^{-2}$. In the case
of CRL437, a moderately luminous IR source has formed, but a high
degree of symmetry has been retained. The space density exhibits a
relatively rapid r^{-3} dependence (Arquilla and Goldsmith 1984). A
somewhat different, but related technique has been used by Dickman and
Clemens (1983) to analyze an isolated dark cloud; the observed column
density distribution was compared to different pressure-bounded
polytropic models, and one having a modest temperature increase at the
edge, with density varying as r$^{-1.5}$ was found to give the best
agreement.

A more direct technique of inferring space densities in molecular
clouds is the measurement of the relative intensities of two or more
transitions which are not thermalized. In this case the level

populations and hence the emitted intensity in the optically thin case
are functions of the collision rate. A minimum of two transitions is
required because the total column density of the species being observed
must also be determined. In order to derive densities of molecular
hydrogen, collision rates at the kinetic temperature of the cloud must
be known. These are only obtained from detailed quantum mechanical
calculations. At the present time such information is available for a
number of the more common linear molecules, e.g. CO, CS, OCS, and HC_3N
(Green and Chapman 1978), as well as for a few symmetric and asymmetric
top molecules.

Given the availability of the relevant collision rates, the space
density is obtained by fitting the relative intensities of the two or
more observed transitions to the output of a statistical equilibrium
calculation. With two transitions of a linear molecule, there is a
monotonic relationship between the ratio of a higher to a lower
transition and the hydrogen density. A definite value of $n(H_2)$ is
obtained if the ratio falls within certain limits (Goldsmith 1984).
For molecules with more complex structures, this is not necessarily the
case. If more than two transitions are observed there is no guarantee
of a solution, and in this case results are usually obtained by
carrying out a least-squares fit to the intensities of the various
transitions observed (e.g. Snell et al. 1984). It must be recognized
that particular transitions, or combinations of transitions, are
sensitive only to a *limited range* of density. Thus, for example, the
lowest two transitions of carbon monoxide function as a good
densitometer only for $10^{2.5} {\leq} n(H_2) {\leq} 10^4$ cm^{-3}, while the J=7-6 and J=4-3
transitions of CS together are diagnostics for the range between $10^{5.5}$
and $10^{7.5}$ cm^{-3} (Snell et al. 1986). This means that different
transitions can give apparently inconsistent results if a wide range of
densities occurs along the line of sight (Evans 1980). An obvious test
for this situation, and a critical ingredient for constructing a more
realistic model, is to employ transitions which sample a wide range of
densities; this has been the trend in recent studies of this type.

Serious deviations of the intensity of a single transition from
the best fit model can be merely an indication of an error in
calibration (which is a particular challenge for this type of work),
but may also be an important clue that the model is not adequate. This
could be a result of optical depth effects such as absorption by
(typically) lower excitation foreground material, or even more
fundamentally to the indadequacy of a single density to describe the
excitation of the molecules along the line of sight being observed. If
there is a good fit for a large number of transitions, our confidence
in the results may be enhanced. This can at best, however, be
construed in the sense of a single density characterizing a particular
molecular species. Chemical selectivity can result in emission from
different species effectively sampling very different regions. This
effect must be considered along with that discussed above in assessing
the correctness of a given density model.

Massive molecular clouds appear to have a complex density
structure. The bulk of observations have been in directions where star
formation is taking place, and thus it might at first appear that the

density inhomogeneities are restricted to regions where young stars are
forming. Indeed, these regions, essentially by definition, do possess
much higher densities than the surrounding molecular cloud material.
One important issue is the scale size over which appreciable, small
scale density variations are found. A number of interferometric
studies have revealed the presence of small fragments at distances of
several arcminutes ($\approx$0.5 pc) from the BN-KL star forming region in the
Orion molecular cloud (Harris et al. 1983; Mundy et al. 1986). The
condensations studied by Mundy et al. in CS have masses $\approx$ 50 M_0, and
densities 10^6-10^7 cm^{-3}, about an order of magnitude larger than those
observed by Harris et al. in NH$_3$. They form an intermediate phase
between the higher densities present immediately around the infrared
source source IRc2, which is thought to be the dominant luminosity
source in this region, and those of the ambient cloud material. There
is very clearly an increase of the effective space density of material
as one approaches IRc2. It is plausible that this is caused by an
increase in the volume filling factor of clumps (such as those
discussed above) in the lower density interclump medium.

     Multitransition studies of CS in three molecular clouds, excluding
Orion, by Snell et al. (1984), also suggest the existence of high ($\approx 10^6$
cm^{-3}) density clumps embedded in a much lower density interclump
medium. This model is not uniquely determined by the CS data alone,
but appears to be reasonable, when CO data demanding much lower
densities along the same lines of sight is taken into account. The
clump filling factor drops as one moves to greater distances from the
centers of star-forming activity, but the characteristic clump density
derived from the CS observations remains approximately constant,
reflecting essentially uniform conditions in the high density regions
responsible for the emission in this species. Observations relevant to
small-scale structure and clumping in molecular clouds are discussed by
Wilson (1986). It appears that a highly inhomogeneous structure
characterizes the portions of massive molecular clouds which have been
intensively studied. The density contrast may well be sustained by the
energy input from the star formation process, and this may be an
indication of the propagation of this process through an extended
region.

2.3 Column Density and Mass

The only molecular species which directly trace column density along a
particular line of sight are the rare isotopes of carbon monoxide. As
discussed in the preceeding section, a number of studies have verified
the correlation of the emission in these species with visual
extinction. The good correlation of C^{18}O emission with A$_v$, as well as
with 100 micron flux observed from IRAS in the Heiles Cloud 2 region,
has been demonstrated by Cernicharo and Guelin (1986). The value of
the carbon monoxide isotopes as tracers of total column density arises
first from the chemistry of this species. It is not easy to put much
more than the canonical 10 percent of available carbon in this form,
nor is it straightforward either to destroy a significant fraction of
this molecular species or to deplete it significantly, in the bulk of

molecular cloud material. A second factor is the ease of excitation of
the lower rotational transitions and the consequently relatively
unbiased tracing of various regions, as long as the optical depth is
not too large.

Most other molecular species have abundance variations that,
although from a chemical point of view are relatively small (e.g.
Irvine, Goldsmith and Hjalmarson **1987**), render them useless as general
tracers of total column density. Emission from particular molecular
species can be used to identify certain morphological features or
structures in molecular clouds (e.g. ammonia and other relatively rare
species trace dense cores within large, lower density regions as
discussed by Myers and Benson 1983, and Benson and Myers 1983), but
cannot be used to determine column densities with confidence comparable
to that for carbon monoxide isotope observations. Masses of molecular
regions are thus widely determined by integrating the column densities
determined from ^{13}CO or $C^{18}O$ observations as discussed in section 2.2.

A quite different method of mass determination involves the highly
saturated emission of ^{12}CO. As discussed by Scoville and Sanders
(**1987**), and by Dickman, Snell, and Schloerb (**1987**), the assumption of
a virialized cloud (or ensemble thereof), together with constant gas
temperature, allows one to obtain the mass of molecular material in the
telescope beam. This technique is particularly valuable for studies of
extragalactic material, and of the large scale distribution of material
in the Milky Way.

2.4 Chemical Composition

Since the molecular content of molecular clouds has been described in
some detail in recent reviews (e.g. Irvine et al. 1985; Irvine,
Goldsmith, and Hjalmarson **1987**), this topic need not be covered in
detail here. The issue of important atomic species has been discussed
in Section 1 above. Our knowledge about one aspect of cloud
composition appears to have regressed recently; measurements of
dissociative recombination coefficients for important molecular ions
(particularly H_3^+) at low temperatures indicate that previous
calculations of electron abundances were seriously in error, and that,
in fact, we do not have any useful upper limit to the fractional
ionization in molecular clouds (Smith and Adams 1984; Dalgarno and Lepp
1984). Thus, this critical parameter for the dynamics of clouds
remains quite unknown.

2.5 Cloud Morphology

Any discussion of cloud morphology has to deal with the significant
observational selection effect that most studies of very extended
objects have been made with relatively low angular resolution, and vice-
versa. Fortunately, the improvement in receiver technology has
increased mapping speed to the point that high angular resolution
telescopes can cover large areas of the sky. As shown clearly in the
maps of the Orion region by Bally et al. (**1987**) and Schloerb et al.
(**1987**), the complex small-scale structure such as found in a star-

forming center of activity does not wash out on larger scales
characteristic of an entire molecular cloud or cloud complex.

The highly inhomogeneous distribution of molecular material on a
wide range of scale sizes has been visible in data available for some
time. The early, undersampled studies of the galactic plane indicated
that the emission from carbon monoxide is quite irregular. More
extensive surveys have confirmed this basic point; the subject of
molecular cloud distribution on the largest scale is the topic of a
number of recent reviews, e.g. Solomon and Sanders (1985); Scoville and
Sanders (1986). As an example of the irregular distribution on the
scale of a molecular cloud complex, we show first a map of an
approximately 5 degree by 12 degree region in Orion in ^{12}CO, J=1-0
obtained by Kutner et al. (1977) with the GISS telescope having 8'
(equivalent to 1 pc) resolution. The K-L Nebula in the L1640 region,
and NGC2024 in the L1630 region to the north stand out clearly, but
other structure down to the angular resolution limit of the instrument
is apparent. Following this, we show a map of the central portion of
this region in ^{13}CO J=2-1, taken at the MWO, with 1.75' resolution
(Goldsmith et al. 1982). The figure gives the integrated intensity in
a 1 km/s wide interval at 11 km/s, but is a good indication of the
distribution of emission. Again, structure down to the (.25 pc)
resolution of the telescope used can be seen. Finally, we show a map
of the central part of the molecular peak near the Kleinmann-Low
nebula, obtained in C^{18}O by Wilson et al. 1986. The resolution of the
IRAM telescope is 21" (which corresponds to a linear dimension of .05
pc) and still, obvious structure at the resolution limit can be seen.
Some of the interferometric work, which has yet higher resolution, has
been mentioned in Section 2.2. The number of fields mapped to date is
limited, but there certainly are cases where structure on a
significantly smaller scale is visible.

The same type of structure on different size scales is found in
M17. Structures ranging from 250x30 pc for the entire cloud complex,
to 15 pc for the I-N cloud, to 3 pc for a core within this cloud, to
0.7 pc for condensations and finally to <0.3 pc (the limit of
resolution available to date) for clumps are discussed by Snell et al.
(1986).

A very large range of size scales can be found in dark clouds,
where the large energies associated with high mass star formation are
not present. The map of ^{13}CO in the Perseus region obtained by
Bachiller and Cernicharo (1986) with 5' resolution shows a high degree
of clumpiness in the string of molecular material extending over at
least 5 degrees from NGC1333. Towards the end is the fairly isolated
cloud B5, observed in a number of different species by Young et al.
(1982). In a nearly fully-sampled study of C^{18}O with a 1.6' beam,
Goldsmith, Langer, and Wilson (1986) find a number of condensations,
some of which are apparently unresolved. The infrared sources, which
appear associated with molecular outflows, have definitely perturbed
the distribution of material, but they are probably not the sole agents
responsible for the clumping observed.

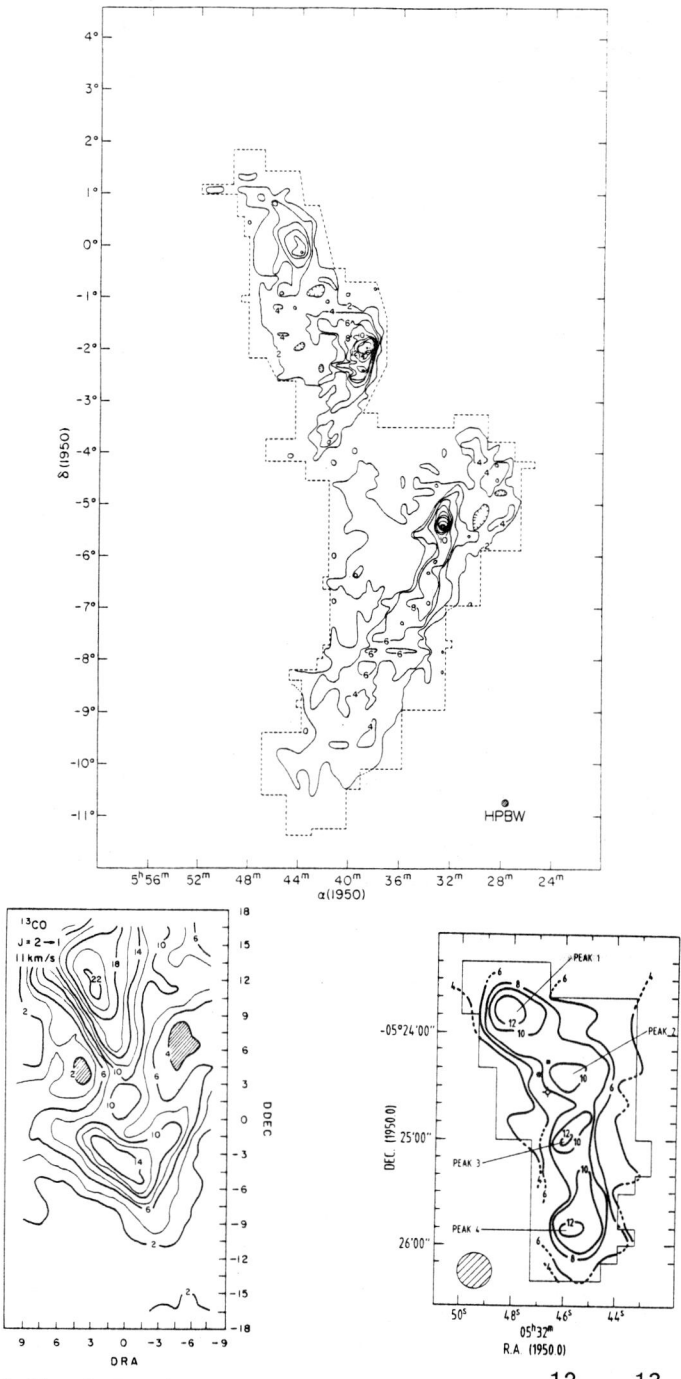

Maps of the Orion Molecular Cloud Complex in ^{12}CO, ^{13}CO, and C^{18}O. The central position for the 13CO map (which is J=2-1; the others are J=1-0) is 5^{h}32^{m}47^s, -5°24'30", and the beamsize is 1.75'.

Beyond the general observation that molecular clouds exhibit structure on a wide range of scale sizes, what can be said about the apparent *form* of these regions? Perhaps the most striking feature is the prevalence of filamentary structures in the molecular material, both within giant cloud complexes such as Orion, and dark cloud complexes including Taurus and Perseus (e.g. Clark et al. 1977). Filamentary structures in the visual extinction of relatively isolated regions have also been catalogued by Schneider and Elemegreen (1979), and one such filament has been studied polarimetrically, as well as in molecular emission, by McCutcheon et al. (1986). Elongated "elephant-trunk" globules are generally seen optically against the background of an HII region. Schneps, Ho, and Barrett (1980) studied some of these regions in the Rosette nebula and found that conditions are similar to those in dense isolated dark clouds. They conclude that the expansion of the HII region has produced the distorted shape observed.

The highly clumped nature of material in molecular clouds in general is emphasized by the observations of Perault, Falgarone, and Puget (1985). A hydrostatic model for cloud substructures has been developed by Falgarone and Puget (1985) and the possible role of the magnetic field is discussed by Myers (**1987**). A wide variety of physical processes may contribute to the morphology of molecular clouds; the subject is reviewed in some detail by Scalo (1985).

2.6 Association

The widespread distribution of molecular material in the galaxy makes it eminently plausible that molecular clouds are associated with a wide variety of other types of objects. Perhaps the most widely studied association is with different phases of stellar birth and death. Walmsley (1982), Solomon and Sanders (1985), and Scoville, Sanders, and Clemens (1986) discuss molecular clouds and star formation; Harris (1980) reviews HII regions and molecular clouds; Huang and Thaddeus (1986) discuss molecular clouds and supernova remnants, and Olofsson (1986) reviews return of molecular material to the interstellar medium by evolved stars. Many detailed studies have been published; the recent conference proceedings (e.g. Roger and Dewdney 1982; Black and Matthews 1985; Shaver and Kjar 1986) represent a good starting point for gaining an overview of the complex relationships between molecular clouds and the other phases of the interstellar medium.

2.7 Velocity Field

The velocity field which characterizes the material in interstellar clouds remains highly problematic. It has gradually become clear that simple, large scale motions are, in general, not the dominant form of mass motion. In some clouds, rotation is present (Schloerb and Snell 1984; Goldsmith and Arquilla 1985), but rarely contributes systematic velocity shifts comparable to the characteristic linewidth. Modelling the velocity field in terms of a turbulent medium has been actively pursued (Dickman 1985; Scalo 1985). This approach has received observational impetus from observed correlations between the linewidth and size of extensive samples of molecular clouds (Larson 1981; Dame et al. 1986). While not uniquely specifying any model, the observed correlation is consistent with clouds being in virial equilibrium. The scale size of the turbulent motions that must be present is not well-determined, but may be an appreciable fraction of the cloud size (cf. Dickman and Kleiner 1986).

Models with empirically chosen probabilistic functions for density and velocity fluctuations have been developed by Kwan and Sanders (1986). These reasonably successfully address the long-standing apparent contradiction that one does "see" to the center of some molecular clouds, despite very high opacity in ^{12}CO. The issues of dissipation of turbulence and replenishment of turbulent energy are serious ones, and may be resolved by having motions constrained in part by the magnetic field (e.g. Myers 1987). Theory, modelling, and observation definitely have a good deal of connecting to do in this area!

2.8 Magnetic Field

The magnetic field in interstellar clouds is often assumed to play a major role in their dynamics and evolution, despite the fact that very little relevant quantitative information is available. It has been proposed that the magnetic field affects the material both in terms of wavelike motions that could play a role in producing the observed linewidths (Arons and Max 1975), and in terms of affecting rotational motion through magnetic braking (cf. Mouschovias and Paleologou 1979). The magnetic field may also affect the translational motion of an entire molecular cloud complex (Elmegreen 1981). The possible effect of the magnetic field, especially on the internal dynamics of a molecular cloud, depends both on the magnetic field strength, and the degree to which it is coupled to the mass in the cloud. Support by the magnetic field (cf. Mestel and Spitzer 1956) will be compromised if the coupling between ions and neutrals drops below some critical value. The coupling is set primarily by the fractional ionization within the cloud, which is why our lack of knowledge of this parameter (discussed in Section 2.4) is particularly unfortunate.

Observational results on the magnetic field in molecular clouds are reviewed by Heiles (1987). Optical and infrared observations of polarization (e.g. Vrba, Strom, and Strom 1976; McDavid 1984; McCutcheon et al. 1986) measure the alignment of interstellar grains, but do not give precise values of the field strength. Molecular Zeeman effect measurements are difficult, and only a few detections of of the effect in OH are presently available (cf. Crutcher, Troland, and Heiles 1981; Troland, Crutcher, and Kazes 1986). These data indicate magnetic field strengths in the 15-125 microgauss range; values which are consistent with the magnetic field increasing as a moderately rapid (0.3-0.4) power of the density. Observations of the Zeeman effect in OH masers suggest magnetic field strengths of 5 to 7 milligauss (Reid et al. 1980). The density in these regions is not well determined but could be on the order of $10^6 - 10^7$ cm^{-3} (Lucas 1980), which would be consistent with extrapolating the B(n) relationship given above.

3. CLOUD CLASSIFICATION

It is certainly not clear to what extent classification of molecular clouds is a rewarding undertaking. Most classification schemes are parallel-a star is a B star or a G star. Molecular regions in the galaxy end up being identified to a large extent in a heirarchical sense-a region may be identified as a cloud core which is part of a dark cloud, which is... There is also, at the present time, a serious lack of uniformity in nomeclature. In particular, one scientist's giant molecular cloud can be another's giant molecular cloud complex. The general confusion is, of course, intimately connected with the wide range of size scales over which molecular gas is found in the Milky Way. Any attempt at a classification scheme is really only a summary of how properties vary as a function of dimension. Nevertheless, by constructing a table of molecular cloud types, we can gain a perspective on different molecular regions. In the accompanying Table we attempt to give representative characteristics of different categories of molecular clouds. There has been no attempt to verify complete consistency with all objects that have at one time or another been put in a particular slot. It is felt, however, that the characteristics given are representative of the objects of each type. The Table is organized vertically in order of decreasing size (from complexes to clumps), and horizontally into three major categories of molecular regions: giant molecular clouds (associated with high-mass star formation), dark clouds (sometimes associated with low-mass star formation), and circumstellar molecular clouds (associated with evolved stars).

    Some references may aid those interested in pursuing the topic of cloud classification. For giant molecular cloud complexes, see e.g. Solomon and Sanders (1985); the values within the range for each parameter are not tied together. Myers (**1987**), among many others, uses the appellation *"giant molecular cloud"* for what we (along with others) denote *"giant molecular cloud complex"*. For dark cloud complexes and cores see Myers (1985, **1987**) and list of representative objects in Goldsmith and Arquilla (1985). For dark clouds see Leung (1985). For circumstellar clouds, the physical conditions are strongly radially dependent and the range given excludes the extremes of the clouds; information is based heavily on observations (Olofsson et al. 1982) and theoretical models (Kwan and Hill 1977; Kwan and Linke 1982) of IRC+10216, which is by far the best-studied circumstellar envelope. The entry for "GIANT MOLECULAR CLOUD" certainly overlaps those for both cores and complexes, but appears to correspond to reasonably well-defined structures within complexes, but ones which are not dominated by proximity to an embedded heating source, as are the GMC cores. It is also true that we may expect to find structures that would go into the "DARK CLOUD CLUMP" category, once instrumentation with adequate sensitivity and angular resolution becomes available, but until now, regions with the appropriate distinctive physical conditions have not been identified.

MOLECULAR REGIONS IN THE INTERSTELLAR MEDIUM

		GIANT MOLECULAR CLOUD	DARK CLOUD	CIRCUM-STELLAR CLOUD
C O M P L E X	Size (pc)	20–80	6–20	
	Density (cm^{-3})	100–300	100–1000	
	Mass (M_O)	$8 \times 10^4 - 2 \times 10^6$	$10^3 - 10^4$	
	Linewidth (km/s)	6–15	1–3	
	Temperature (K)	7–15	≈10	
	Examples	W51, M17, W3	Taurus, Sco–Oph	
	Size (pc)	3–20	0.2–4	≈0.2
	Density (cm^{-3})	$10^3 - 10^4$	$10^2 - 10^4$	$10^2 - 10^7$
	Mass (M_O)	$10^3 - 10^5$	5–500	≈10^{-2}
	Linewidth (km/s)	4–12	0.5–1.5	20–40
	Temperature (K)	15–40	8–15	10–100
	Examples	Orion OMC1, W33 W3 A	B227, B335 (ISO) B5, B18 (COMPL)	IRC +10216
C O R E	Size (pc)	0.5–3	0.1–0.4	
	Density (cm^{-3})	$10^4 - 10^6$	$10^4 - 10^5$	
	Mass (M_O)	$10 - 10^3$	0.3–10	
	Linewidth (km/s)	1–3	0.2–0.4	
	Temperature (K)	30–100	≈10	
	Examples	Orion (Ridge)	B335, L1535	
C L U M P	Size (pc)	<0.5		
	Density (cm^{-3})	>10^6		
	Mass (M_O)	$30 - 10^3$		
	Linewidth (km/s)	4–15		
	Temperature (K)	30–200		
	Examples	M17 (Kleinmann–Wright), Orion (Hot Core) W3(OH)		

4. CONCLUSIONS

This contribution has focused on our knowledge of physical conditions in interstellar molecular clouds. This information is of particular value for two reasons. The first is for developing our understanding of these regions, which contain a large fraction of interstellar material in the Milky Way and presumably in other galaxies. The study of molecular clouds has yielded significant new information in many varied areas such as astrochemistry, and utilizing these regions as laboratories requires knowledge of the conditions within them. Second, molecular clouds appear to be intimately related to many important astrophysical processes, particularly those associated with star formation. Understanding molecular clouds is clearly of crucial importance for unravelling issues such as the initial mass function, rotation in stars, and multiple star systems.

The properties of molecular regions cover an enormous range in size, density, and mass, and a large range in temperature. A number of the key properties such as ionization state and magnetic field remain, in general, quite unknown. The classification of these objects is still in a primitive and confused state. This may be simply a result of the relative newness of the detailed study of these objects. It may, however, be a reflection of the large number of processes that bear on determining the conditions in these regions; the inherent variation of these inputs from point to point in the galaxy, coupled with the complex response of interstellar molecular material may result in molecular clouds being inherently irregular and relatively unstructured. New instrumentation with higher angular resolution and with access to submillimeter wavelengths will surely help in unravelling these questions. The combination of the wide range of physical and chemical processes at work, together with the large variations in the conditions which are present in molecular clouds guarantee that they will be a rich field for study in the forseeable future.

5. ACKNOWLEDGEMENTS

I thank Ron Snell for many valuable discussions about all aspects of interstellar molecular clouds, and Dave Hollenbach, Harley Thronson, and Sheryl Reiss for suggestions which did much to improve this manuscript. This research was supported in part by the National Science Foundation.

REFERENCES

Arons, J., and Max, C. 1975, *Ap. J. (Letters)*, 196, L77.
Arquilla, R., and Goldsmith, P.F. 1984, *Ap. J.*, 279, 664.
Arquilla, R., and Goldsmith, P.F. 1985, *Ap. J.*, 297, 436.
Askne, J., Höglund, B., Hjalmarson, A., and Irvine, W.M. 1984,
    Astron.Astrophys., 130, 311.
Bachiller, R., and Cernicharo, J. 1986, *Astron. Astrophys.*,166, 283.
Bally, J., Dragovan, M., Langer, W.D., Stark, A.A., and Wilson,
    R.W., **1987**, these proceedings.
Benson, P.J., and Myers, P.C. 1983, *Ap. J.*, 270, 589.
Black, D.C., and Matthews, M.S. (ed) 1985, *Protostars and Planets II*
    (Tucson: University of Arizona).
Black, J.H. **1987**, these proceedings.
Bok, B.J., Cordwell, C.S., and Cromwell, R.H. 1970, *Dark Nebulae,
    Globules, and Protostars*, ed, B.T. Lynds
    (Tucson: University of Arizona), p.33
Burton, W.B., Liszt, H.S., and Baker, P.L. 1978,
    Ap. J.(Letters), 219, L67.
Cernicharo, J., and Guelin, M. 1986, *Astron. Astrophys.*, in press.
Churchwell, E., and Hollis, J.M. 1983, *Ap. J.*, 272, 591.
Clark, F.O., Giguere, P.T., and Crutcher, R.M. 1977, *Ap. J.*, 215,
    511.
Crutcher, R.M., Troland, T.H., and Heiles, C. 1981, *Ap. J.*,
    249, 134.
Dalgarno, A., and Lepp, S. 1984, *Ap. J. (Letters)*, 287, L47.
Dame, T., Elmegreen, B.G., Cohen, R.S., and Thaddeus, P.
    1986, *Ap. J.*, 305, 892.
Dickman, R.L. 1978, *Ap. J. Suppl.*, 37, 407.
Dickman, R.L. 1985, *Protostars and Planets II*, ed D.C. Black and
    M.S. Matthews (Tucson: University of Arizona), p. 150.
Dickman, R.L., and Clemens, D.P. 1983, *Ap. J.*, 271, 143.
Dickman, R.L., and Kleiner, S. 1986, *Ap. J.*, in press.
Dickman, R.L., Snell, and Schloerb, F.P. 1986, *Ap. J.*, 309, 326.
Elmegreen, B. 1981, *Ap. J.*, 243, 512.
Encrenaz, P.J., Falgarone, E., and Lucas, R. 1975,
    Astron.Astrophys., 44, 73.
Evans, N.J.II 1980, IAU Symp. 87, *Interstellar Molecules*,
    ed B.H. Andrew (Dordrecht: Reidel), p. 1.
Falgarone, E., and Puget, J.L. 1985, *Astron. Astrophys.*, 142, 157.
Frerking, M.A., Langer, W.D., and Wilson, R.W. 1982, *Ap. J.*,
    262, 590.
Goldsmith, P.F. 1984, *Galactic and Extragalactic Infrared
    Spectroscopy*, ed M.F. Kessler and J.P. Phillips (Dordrecht:
    Reidel), p. 233.
Goldsmith, P.F., and Langer, W.D. 1978, *Ap. J.*, 222, 881.
Goldsmith, P.F., Langer, W.D., Carlson, E.R., and Wilson,
    R.W. 1980, Proc. IAU Symp. 87: *Interstellar Molecules*,
    ed B.H. Andrew (Dordrecht: Reidel), p. 417.
Goldsmith, P.F., Arquilla, R.A., Schloerb, F.P., and Scoville,

N.Z. 1982, *Regions of Recent Star Formation*, ed R.S. Roger and
P.E. Dewdney (Dordrecht: Reidel), p. 295.

Goldsmith, P.F., and Mao, X.-J. 1983, *Ap. J.*, 265, 791.

Goldsmith, P.F., and Arquilla, R.A. 1985, *Protostars and Planets II*.
ed D.C. Black and M.S. Matthews (Tucson: University of
Arizona), p. 137.

Goldsmith, P.F., Langer, W.D., and Wilson, R.W. 1986,
Ap. J. (Letters), 303, L11.

Goldsmith, P.F., Irvine, W.M., Hjalmarson, A., and Ellder, J. 1986,
Ap. J., 310, 383.

Goldsmith, P.F., Snell, R.L., Hasegawa, T., and Ukita, N.
1987, *Ap. J.*, in press.

Green, S. 1986, *Ap. J.*, 309, 331.

Green, S., and Chapman, S. 1978 *Ap. J. Suppl.*, 37, 169.

Gusten, R., Walmsley, C.M., Ungerechts, H., and Churchwell,
E., 1985, *Astron. Astrophys.*, 142, 381.

Harris, A., Townes, C.H., Matsakis, D.N., and Palmer, P.
1983, *Ap. J.(Letters)*, 265, L63.

Harris, S. 1980, *Giant Molecular Clouds in the Galaxy*, ed P.M.
Solomon and M.G. Edmunds (Oxford: Pergamon), p. 201.

Heiles, C. 1987, these proceedings.

Hemeon-Heyer, M.C. 1986, PhD. Thesis, Univ. of Massachusetts.

Herbst, E., and Prasad. S.S. 1987, these proceedings.

Huang, Y.L., and Thaddeus, P. 1986, *Ap. J.*, 309, 804.

Irvine, W.M., Schloerb, F.P., Hjalmarson, A., and Herbst,
E., 1985, *Protostars and Planets II*, ed D.C. Black and
M.S. Matthews (Tucson: University of Arizona), p. 579.

Irvine, W.M., Goldmsith, P.F., and Hjalmarson, Ake, 1987,
these proceedings.

Keene, J., Blake, G.A., Phillips, T.G., Huggins, P.J., and
Beichman,C.A. 1985, *Ap. J.*, 299, 967.

Knapp, G.R. 1974, *A. J.*, 79, 527, and *A. J.*, 79, 541.

Kuiper, T.B.H., Rodriguez Kuiper, E.N., Dickinson, D.F.,
Turner, B.E., and Zuckerman, B. 1984, *Ap. J.*, 276, 211.

Kutner, M.L., Tucker, K.D., Chin, G., and Thaddeus, P. 1977,
Ap. J., 215, 521.

Kwan, J., and Hill, F. 1977, *Ap. J.*, 215, 781.

Kwan, J., and Linke, R.A. 1982, *Ap. J.*, 254, 587.

Kwan, J. and Sanders, D.B. 1986, *Ap. J.*, 309, 783.

Langer, W.D. 1976, *Ap. J.* 206, 699.

Larson, R. 1981, *M.N.R.A.S.*, 194, 809.

Leung, C.M. 1985, *Protostars and Planets II*, ed, D.C. Black and M.S.
Matthews (Tucson: University of Arizona), p. 579.

Linke, R.A., Cummins, S.E., Green, S., and Thaddeus, P.
1982, in *Regions of Recent Star Formation*, ed R.S.
Roger and P.E. Dewdney (Dordrecht: Reidel), p.391.

Loren, R.B., and Mundy, L.G. 1984, *Ap. J.*, 286, 232.

Lucas, R. 1980, IAU Symp. 87, *Interstellar Molecules*,
ed B.H. Andrew (Dordrecht: Reidel), p. 581.

McCutcheon, W.H., Vrba, F.J., Dickman, R.L., and Clemens,
D.P. 1986, *Ap. J.*, 309, 619.

McDavid, D. 1984, *Ap. J.*, 284, 141.

Mestel, L., and Spitzer, L., Jr. 1956, *M.N.R.A.S.*, 116, 583.
Mouschovias, T. Ch. and Paleologou, E.V. 1979, *Ap. J.*, 230, 204.
Mundy, L.G., Scoville, N.Z., Baath, L.B., Masson, C.R.,
    and Woody, D.P. 1986, *Ap.J. (Letters)*, 304, L51.
Myers, P.C. 1985, *Protostars and Planets II*, ed D.C. Black
    and M.S. Matthews (Tucson: University of Arizona), p. 81.
Myers, P.C. **1987, these proceedings.**
Myers, P.C., and Benson, P.J. 1983, *Ap. J.*, 266, 309.
Olofsson, H., Johansson,L.E.B., Hjalmarson, A, and Rieu,
    N.-Q. 1982, *Astron. Astrophys.*, 107, 128.
Olofsson, H. 1986, *ESO-Iram-Onsala Workshop on (Sub)-Millimeter
    Astronomy*, ed P.A. Shaver and K. Kjar (Munchen: ESO), p.535.
Omont, A. 1986, *Space-Borne Sub-Millimetre Astronomy Mission*,
    ed N. Longdon (Paris: European Space Agency), p. 129.
Penzias, A.A., Solomon, P.M., Jefferts, K.B., and Wilson,
    R.W. 1972, *Ap. J. (Letters)*, 174, L43.
Perault, M., Falgarone, W., and Puget, J.L. 1985,
    Astron. Astrophys., 152, 371.
Phillips, T.G., and Huggins, P.J. 1981, *Ap. J.*, 251, 533.
Phillips, T.G., Knapp, G..R., Huggins, P.J., Werner, M.W., Wannier,
    P.G., Neugebauer, G., and Ennis, G. 1981, *Ap. J.*, 245, 512.
Reid, M.J., et al. 1980, *Ap. J.*, 239, 89.
Roger, R.S. and Dewdney, P. (ed) 1982, *Regions of Recent
    Star Formation* (Dordrecht: Reidel).
Scalo, J.M. 1985, *Protostars and Planets II*, ed D.C. Black
    and M.S. Matthews (Tucson: University of Arizona), p.201.
Scalo, J.M. **1987, these proceedings.**
Schloerb, F.P., and Snell, R.L. 1984, *Ap. J.*, 283, 129.
Schloerb, F.P., Snell, R.L., Goldsmith, P.F., and Morgan,
    J.A. **1987, these proceedings.**
Schneider, S., and Elmegreen, B.G. 1979, *Ap. J. Suppl.*,41, 87.
Schneps, M.H., Ho, P.T.P., and Barrett, A.H. 1980, *Ap. J.*, 240, 84.
Scoville, N.Z., and Sanders, D.B. **1987, these proceedings.**
Scoville, N.Z., Sanders, D.B., and Clemens, D.P. 1986,
    Ap. J. (Letters), 310, L77.
Shaver, P.A. and Kjar, K. (ed) 1986, *ESO-IRAM-Onsala
    Workshop on (Sub)Millimeter Astronomy* (Munchen: ESO).
Smith, D., and Adams, N.G. 1984, *Ap. J. (Letters)*, 284, L13.
Snell, R.L. 1981, *Ap. J. Suppl.*, 45, 121.
Snell, R.L., et al. 1984, *Ap. J.*, 276, 625.
Snell, R.L., et al. 1986, *Ap. J.*, 304, 780.
Stutzki, J. and Winnewisser, G. 1985a, *Astron. Astrophys.*, 148, 254.
Stutzki, J. and Winnewisser, G. 1985b, *Astron. Astrophys.*, 144, 1.
Takano, T. 1986, *Ap. J.*, 303, 349.
Tielens, A.G.G.M., and Hollenbach, D. 1985, *Ap. J.*, 291, 722.
Troland, T.H., Crutcher, R.M., and Kazes, I. 1986, *Ap. J.*,
    (Letters), 304, L57.
Vrba, F.J., Strom, S.E., and Strom, K.M. 1976, *A.J.*, 81, 958.
Walmsley, C.M. 1982, *Irish Astron. J.*, 15, 161.
Wannier, P.G., Lichten, S.M., and Morris, M. 1983, *Ap. J.*,
    268, 727.
Wilson, T.L. 1986, *ESO-IRAM-Onsala Workshop on (Sub)-Millimeter*

Astronomy, ed P.A. Shaver and K. Kjar (Munchen: ESO), p. 401.
Wilson, T.L., Serabyn, E., Henkel, C., and Walmsley, C.M.
    1986, *Astron. Astrophys.*, <u>158</u>, L1.
Wootten, A., Snell, R., and Evans, N.J. II, 1980, *Ap. J.*, <u>240</u>, 532.
Young, J.S., Goldsmith, P.F., Langer, W.D., Wilson, R.W.,
    and Carlson, E.R. 1982, *Ap. J.*, <u>261</u>, 513.
Young, J.S., and Scoville, N.Z. 1984, *Ap. J.*, <u>287</u>, 153.
Zmuidzinas, J., Betz, A.L., and Goldhaber, D.M. 1986,
    Ap.J.(Letters), <u>307</u>, L75.

Philip Myers and Paul Goldsmith

OBSERVATIONS OF MOLECULAR CLOUD STRUCTURE AND INTERNAL MOTIONS

Philip C. Myers
Harvard-Smithsonian Center for Astrophysics
60 Garden Street
Cambridge, MA 02138
USA

ABSTRACT. Recent observations of high-latitude clouds, giant molecular clouds, dark cloud complexes, and dark cloud cores are summarized in terms of cloud structure, motions, and stellar content. Data from large numbers of clouds are compared with predictions of equilibrium models. For ~ 100 self-gravitating clouds, power-law relations between velocity dispersion and size, and between density and size can be understood if the observed nonthermal motions arise from magnetic fields 5–150 μG.

1. INTRODUCTION

This paper summarizes recent observational results on molecular cloud structure and motions, and discusses empirical relationships among cloud velocity dispersion σ, size R, and density n in terms of underlying physical principles. The approach emphasizes properties common to large numbers of clouds. The cloud types considered are high-latitude clouds, dark cloud cores, dark cloud complexes, and giant clouds. These are meant to be a representative selection, but not a comprehensive survey, of all cloud types. Reviews of related subjects by Scoville (giant clouds), Scalo (turbulence), and Heiles and Zweibel (magnetic fields) can be found elsewhere in this volume.

It is useful to divide the clouds into *gravitationally unbound clouds* whose mass M is insufficient to bind their internal motions ($M \ll R\sigma^2 G^{-1}$), and *gravitationally bound clouds* ($M \sim R\sigma^2 G^{-1}$). Table I summarizes characteristic properties of each cloud type.

D. J. Hollenbach and H. A. Thronson, Jr. (eds.), Interstellar Processes, 71–86.

TABLE I. Molecular Cloud Properties

Cloud Type	Tracer Line	Radius (pc)	Mass ($M_\odot$)	Density (cm^{-3})	Line FWHM ($km\ s^{-1}$)	Stars
High-Latitude (unbound)	^{12}CO $1 \to 0$	0.3–3	0.5–100	30–500	0.7–1.5	T Tauri (?); rare
Giant (bound)	^{12}CO $1 \to 0$	20–100	10^{5-6}	10–300	5–15	OB (and lower-mass?); common
Dark Cloud Complex (bound)	^{12}CO $1 \to 0$	3–10	10^{3-4}	100–10^3	1–3	T Tauri; common
Dark Cloud Core (bound)	NH_3 1,1	0.05–0.2	0.3–10	10^{4-5}	0.2–0.4	Obscured T Tauri; common

2. GRAVITATIONALLY UNBOUND CLOUDS

In the last few years, it has become clear that large numbers of very tenuous clouds of dust and gas are present within a few hundred pc of the Sun. The IRAS satellite revealed numerous clouds of "infrared cirrus" at 60 and 100 μm, with irregular and complex shape, distributed over the entire sky (Low et al. 1984). In some cases, individual features can be traced for several degrees, suggesting that these clouds are local and lie within $\sim$ 100 pc. Galaxy counts and H I observations indicate visual extinction 0.1–0.2 mag in prominent cirrus regions (Burstein and Heiles 1982). Such clouds are probably present throughout the Galaxy, but more distant ones cannot be detected as easily as near ones due to confusion and sensitivity limitations. The color temperature of the dust is typically $\sim$ 25 K, significantly hotter than in visually opaque dark clouds.

Clouds similar to the infrared cirrus in angular extent and distribution on the sky are also seen in the 2.6-mm line of CO. Blitz, Magnani, and Mundy (1984) and Magnani, Blitz, and Mundy (1985; MBM) identified 493 patches of faint visual extinction ($A_V \lesssim 1$ mag) at high galactic latitude ($b \geq 20°$) by inspection of Palomar Sky Survey prints. They surveyed 448 and detected CO emission in 35 complexes with $b \geq 25°$. A study with similar results was done at Southern declinations by Keto and Myers (1986; KM). Some of these "high-latitude clouds" coincide with, but are less extended than, regions of infrared cirrus (Weiland et al. 1986), while other prominent cirrus regions have no detectable CO emission. There is substantial correlation between far-infrared IRAS emission and HI emission (Burton, Deul, and Walker 1986), and some cirrus clouds appear to be primarily atomic in particular regions and primarily molecular in others (de Vries, Heithausen, and Thaddeus 1986).

The high-latitude clouds show no strong systematic trends in velocity, and their random velocity structure can be described in terms of the dispersion in CO velocity over a complex that contains several clumps (MBM), and the dispersion within one clump (KM). In either case, the dispersion is typically 0.3–1 km s^{-1}, greater by an order of magnitude than can be bound by the gas mass and cloud size deduced from the CO observations. This discrepancy is not due to anomalously low CO abundance, as H_2 column densities deduced from the CO data and from star counts, made with the same angular resolution, agree within a factor 2 (KM). This insufficiency in

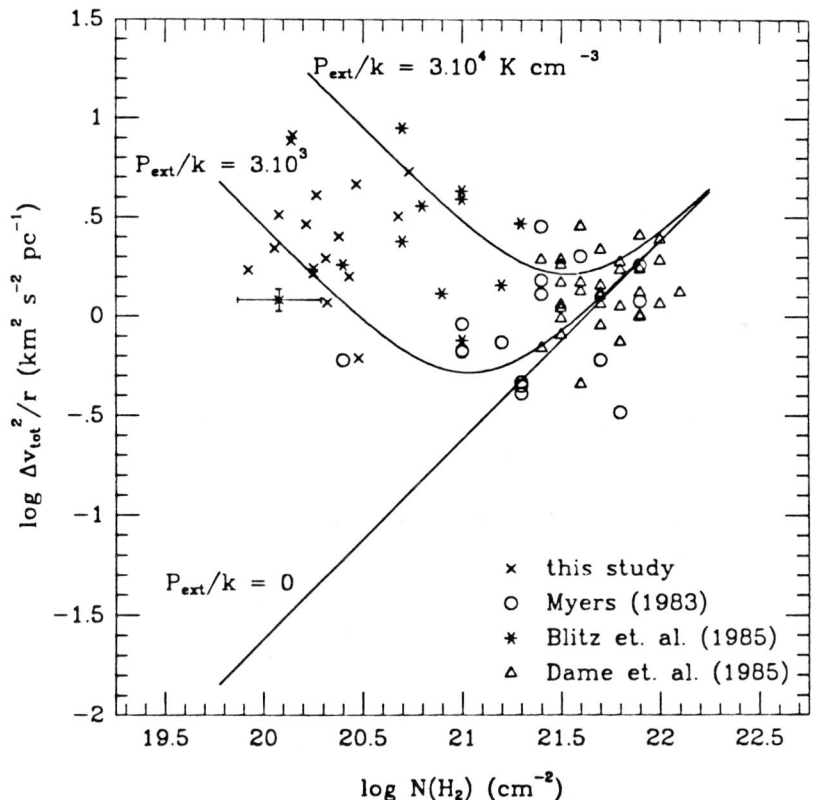

Figure 1. Plot of line width squared over radius vs. column density, for high-latitude clouds ($\times$ and $*$), giant clouds ($\triangle$), and dark cloud cores ($\bigcirc$). *Solid lines:* virial equilibrium for external pressure $P_{ext} = 0$ *(straight)*, and for $P_{ext}/k = 3 \times 10^3$ and 3×10^4 K cm^{-3} *(curved)*. From Keto and Myers (1986).

binding mass may imply that the typical high-latitude cloud or cloud complex is expanding, with age $\sim 10^6$ yr deduced from velocity dispersion and cloud size (MBM). On the other hand, four MBM clouds and two KM clouds have nebulosity indicating associated stars, and the star in MBM #20 (Lynds 1642) is known to be a T Tauri star. The typical stellar age might thus be $\sim 10^6$ yr, characteristic of T Tauri stars, suggesting that the parent cloud must be substantially older. If so, the cloud must be expanding more slowly than indicated by the CO velocity dispersion, and some agent in addition to gravity is needed to oppose or prevent the expansion.

KM modeled an isolated cloud (not a "complex") surrounded by a medium of constant pressure, P_{ext}. They found that the velocity dispersion, cloud size, and column density of most high-latitude clouds are inconsistent with virial equilibrium when $P_{ext} = 0$, but consistent when 3×10^3 K cm^{-3} $\lesssim P_{ext}/k \lesssim 3 \times 10^4$ K cm^{-3}, where k is Boltzmann's constant. This range of pressures coincides with estimates of the gas pressure in the warm, low-density intercloud medium. This coincidence suggests that at least some high-latitude clouds not bound by gravity may be confined by the intercloud medium. Figure 1 illustrates this point for high-latitude clouds. It also shows that clouds of higher column density – dark clouds and giant clouds – have no significant deficit between their internal motions and the mass needed to bind them gravitationally.

3. GRAVITATIONALLY BOUND CLOUDS

3.1. Giant Clouds

Over the last several years, CO surveys of the galactic plane have revealed new details about the physical properties and stellar content of giant molecular clouds. Many cloud maps are now available with sampling and resolution much finer than the characteristic size 20–100 pc (Solomon and Sanders 1985; Solomon et al. 1986; Dame et al. 1986; Myers et al. 1986a). Figure 2 shows maps of CO intensity, integrated over velocity windows 20 km s^{-1} wide, over galactic longitude 20° to 32° and latitude $-1°$ to 1°. Also shown are all H II regions with 5 GHz flux density > 1 Jy and with recombination line velocity within 10 km s^{-1} of the CO velocity. Nearly all the objects in this figure are 5–7 kpc distant. These clumpy structures are ~ 100 pc in size and blend together along the galactic plane. Higher-resolution maps (Solomon et al. 1986) show more detail but similar large-scale structure. Filaments are common, as they are in the well-studied Orion cloud. There, a striking CO "hole" is also evident (Schloerb et al. 1986; Bally et al. 1986). In Figure 2, many of the local maxima coincide with giant H II regions, but others do not, and one large complex, labelled 29,80, has no H II regions. Some half dozen

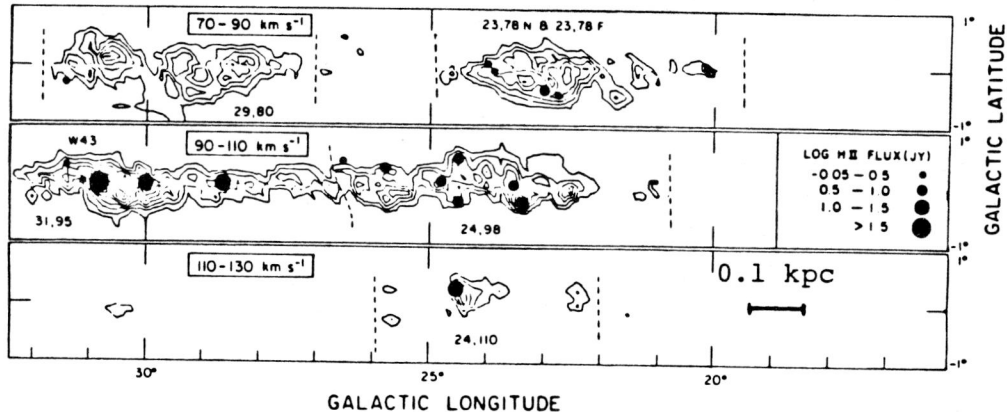

Figure 2. Contour maps of ^{12}CO line intensity integrated over velocity
windows 20 km s^{-1} wide, for longitude 20–32 deg and latitude −1 to 1 deg.
Filled circles: H II regions with 5-GHz flux density greater than 1 Jy,
coincident with the CO emission in position and velocity. The CO observations
have FWHM beam size 8 arcmin. From Myers *et al.* (1986a).

large clouds in the inner galaxy are similarly poor in OB stars. Evidently giant
clouds form OB stars with efficiency that varies widely from cloud to cloud.

Dame *et al.* (1986) analyzed the line width, density, and size of giant
clouds and found relationships similar to those presented by Larson (1981)
for a wide range of cloud size, and by Leung, Kutner, and Mead (1982) and
Myers (1983) for small globules and dense cores. Specifically Dame *et al.* found
that the line widths ΔV are close to the virial values $(8 \ln 2GM/5R)^{1/2}$; that
$\Delta V \sim R^{0.50 \pm 0.05}$ and that $n \sim R^{-1.3 \pm 0.1}$. These findings extend Larson's
results to larger cloud sizes and are based on a more homogeneous sample.
The line width-size correlation of Dame *et al.* has an exponent (0.5) steeper
than that of Larson (0.3) but equal within uncertainty to that found by
Sanders, Scoville, and Solomon (1985), also for CO observations of inner-galaxy
giant clouds. Possible origins of the line width-size and density-size relations
are discussed in Section 3.4.

3.2. Dark Cloud Complexes

The structure and stellar content of nearby dark cloud complexes, including
Taurus-Auriga, Ophiuchus, Lupus, and Chamaeleon, is becoming known in
detail, due to CO line observations of their low-density $(n \lesssim 300 \text{ cm}^{-3})$ gas;
NH$_3$ and other line observations of their dense cores, with $n \gtrsim 10^4 \text{ cm}^{-3}$; IRAS
observations of their highly obscured stars; and optical observations of their
visible T Tauri stars. The best studied complex is in Perseus–Taurus–Auriga,

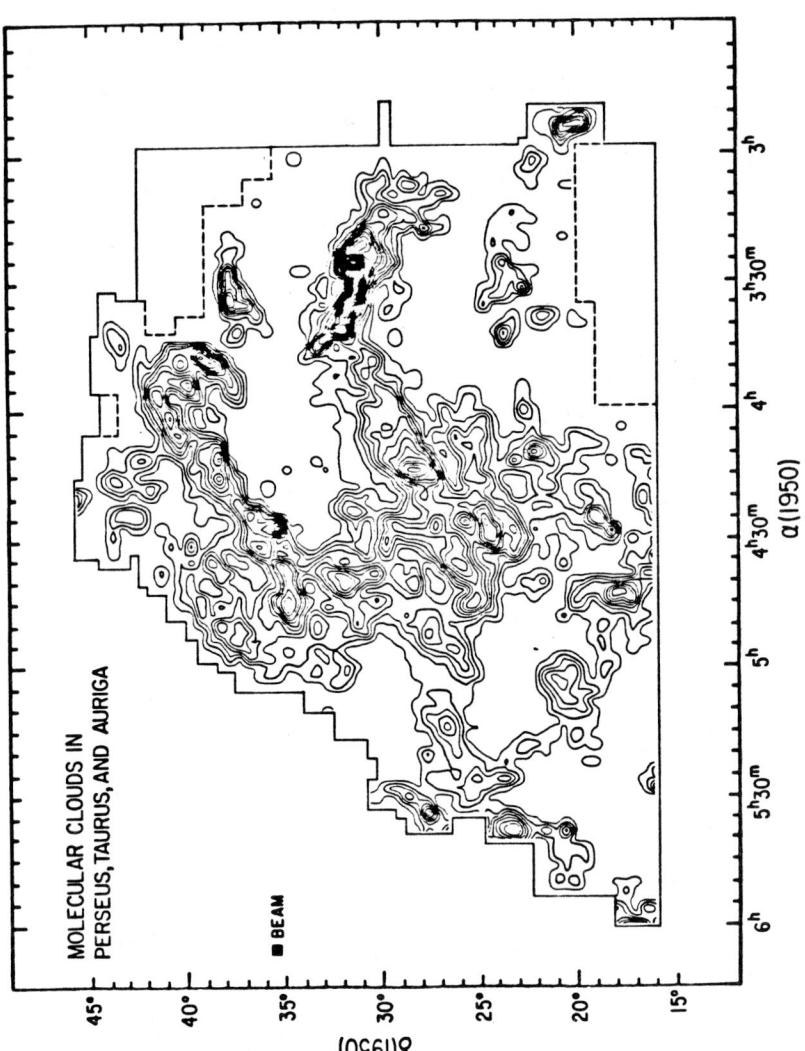

Figure 3. Contour map of velocity-integrated ^{12}CO line intensity in Perseus-Taurus-Auriga. *Straight lines:* boundary of region mapped. FWHM beam size, 0.5 deg. From Ungerechts and Thaddeus (1986).

where Ungerechts and Thaddeus (1986) mapped CO emission over 750 square degrees at 0.5 degree resolution and identified 34 distinct subregions. Figure 3 shows their map of integrated CO intensity, and Figure 4 shows selected contours from their map superposed on the distribution of NH$_3$ cores, obscured IRAS sources, and T Tauri stars (Myers 1986). The CO emission is most intense where the extinction and T Tauri stars are concentrated; but weak

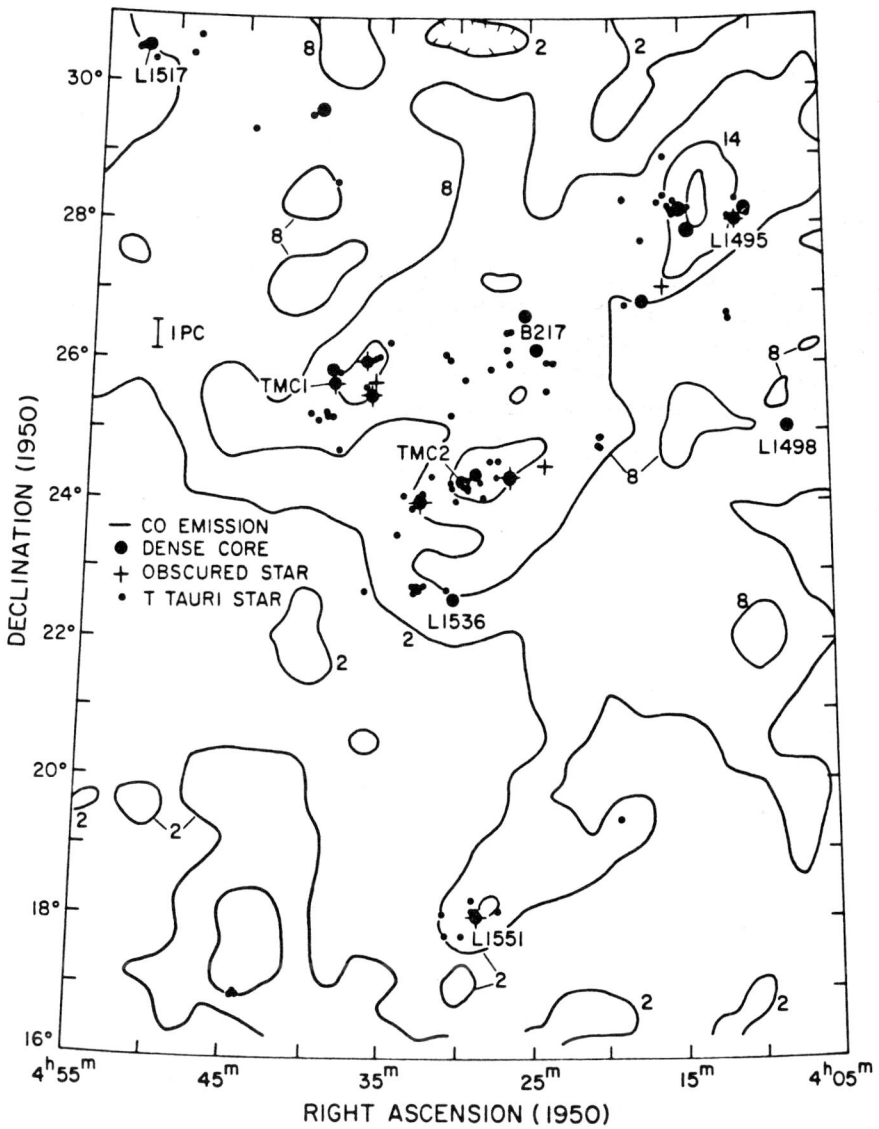

Figure 4. Summary of the distributions of low-density molecular gas (CO emission, Ungerechts and Thaddeus 1986; contours in K km s^{-1}); dense cores (NH$_3$ emission, Myers and Benson 1983 plus recent unpublished results; large filled circles); highly obscured stars (Beichman et al. 1986, Myers et al. 1986c; crosses); and T Tauri stars (Cohen and Kuhi 1979, Jones and Herbig 1979; small filled circles) in Taurus-Auriga. Names refer to prominent dark clouds. From Myers (1986).

CO emission is evident far beyond these narrow lanes, perhaps consistent with the finding that the entire Taurus–Auriga complex has a minimum extinction of about 1 mag (Cernicharo and Bachiller 1984).

Most of the CO emission regions in Taurus–Auriga are close to virial equilibrium; but Ungerechts and Thaddeus note that smaller ($M \sim 100\ M_\odot$), more isolated regions in their maps have mass less than virial by a factor 10–30, as for the high-latitude clouds discussed in Section 2. Sections of the Taurus–Auriga complex have been mapped in enough detail to also reveal systematic motions. A gradient of ~ 0.2 km s^{-1} pc^{-1} along the narrow lane containing TMC–2 was first noted by Kutner (1973). Schloerb and Snell (1984) interpret their observations of Heiles Cloud 2 as due to a rotating ring of gas. Murphy and Myers (1985) and Heyer (1986) found a region ~ 2 pc in diameter in Barnard 18, which resembles an expanding shell – perhaps driven by winds from the two dozen associated T Tauri stars.

3.3. Dark Cloud Cores

The dense cores indicated in Figure 4 are evident in photographs of dark cloud complexes as spots of opaque extinction a few arcmin across. Their physical properties are derived from maps in spectral lines of NH_3, CS, HC_3N, and other molecules whose transitions require gas density $\gtrsim 10^4$ cm^{-3} for collisional excitation. They are not evident as spots of enhanced emission in the ^{12}CO line, which is excited by gas of much lower density.* Their properties are discussed in more detail by Myers (1986, and references therein).

The internal structure of cores is poorly known because the typical core dimension (0.1 pc = 2.5 arcmin in Taurus) is poorly resolved by most single dish telescopes, and because their line brightness temperature contrast on scales $\lesssim 10''$ is at most a few K, making them hard to detect with present-day interferometers. A few dark cloud cores have been detected in the 1.3-cm $(J, K) = (1, 1)$ line of NH_3 at the VLA – B335 (P. Palmer, private communication), B5 and L1172 (Myers et al. 1986b). Each of these has an obscured IRAS source and shows a single region of line emission within $\sim 10''$ of the star, and extended by $\sim 30''$. Each of these VLA maps contains a small fraction of the flux in the corresponding single-dish map. This suggests that the core consists not of numerous independent clumps, but rather a relatively smooth density and temperature structure between 2×10^{16} and 3×10^{17} cm. The small region detected by the VLA probably stands out against the smooth background mapped by single antennas because of gas heating by the star and by its warm circumstellar dust.

Fewer than half of the available core maps show systematic velocity gradients, usually about 1 km s^{-1} pc^{-1}; such gradients are too small to

---

* Unless they contain a star with CO outflow.

dominate the core mechanical energy. In a few cases, however, substantially larger gradients are seen: L1551, 4–6 km s^{-1} pc^{-1} (Menten and Walmsley 1985), ρ Oph cloud, ~ 40 km s^{-1} pc^{-1} (Wadiak et al. 1985). The random velocity dispersions of cores are among the narrowest known. For 13 cores without IRAS sources in Myers and Benson (1983), the median nonthermal portion of the dispersion derived from the line width is only 0.085 km s^{-1}.

The most interesting property of dark cloud cores is their strong correlation with optically obscured stars visible to IRAS. Beichman et al. (1986) compared the positions of steep-spectrum IRAS sources with those of dense cores and found that 47 of 95 cores have an IRAS source within 6 arcmin. Similarly, 28 of 56 NH$_3$ cores have an IRAS source within one NH$_3$ map FWHM (Myers 1986). These IRAS detection rates are far higher than expected from chance coincidence of cores and galaxies sufficiently bright at 60 and 100 μm. The spatial relation of NH$_3$ cores and IRAS sources is well illustrated in Figure 5, which shows contours of integrated (1,1) line intensity and three nearby IRAS sources in the Perseus dark cloud L1448 (Bachiller and Cernicharo 1986). The most luminous of the IRAS sources has the steepest spectral slope and coincides with the maximum of the NH$_3$ map.

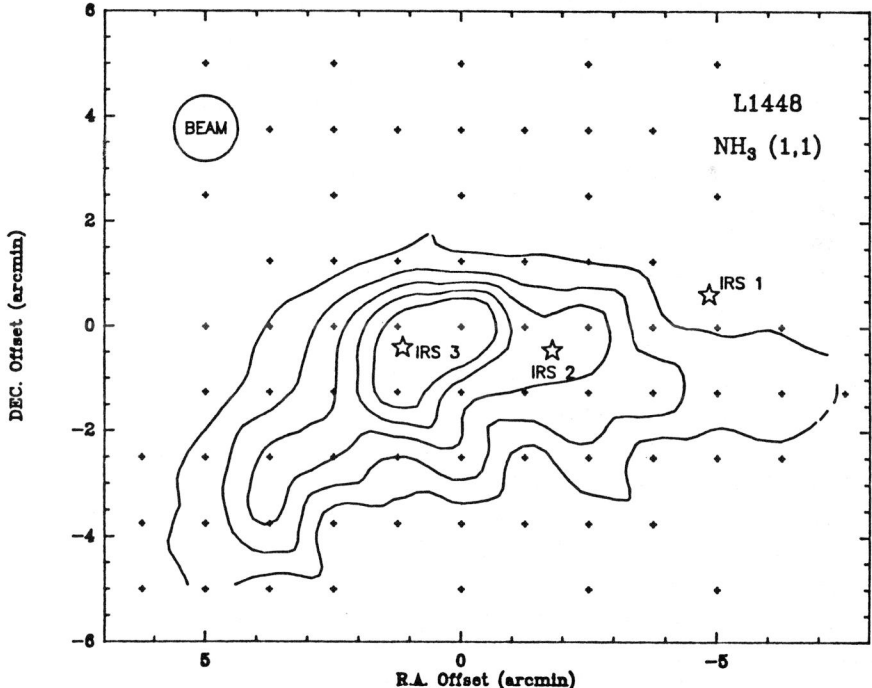

Figure 5. Map of velocity-integrated emission in the $(J, K) = (1,1)$ line of NH$_3$ at 1.3 cm in the Perseus dark cloud L1448. *Stars:* IRAS sources associated with the dense core. From Bachiller and Cernicharo (1986).

3.4. Linewidth-Size-Density Relations and Magnetic Support

All of the gravitationally bound cloud types discussed here obey approximately the relations of virial balance, $\sigma^2 \sim GM/R$; and power-laws between linewidth and size, $\sigma \sim R^{1/2}$, and density and size $n \sim R^{-1}$ first noted by Larson (1981). All of these clouds also have supersonic velocity dispersion. The origins of these power-law relations and of supersonic motions have long been subjects of discussion. Here we point out that if virial clouds are magnetically supported, each of these phenomena has a simple explanation, in good accord with available magnetic field measurements.

The idea that some clouds are supported against collapse by magnetic energy in addition to their thermal motions has been discussed by many authors (e.g., Elmegreen 1978). As a possible formulation of this idea, we assume rough equality between the magnetic energy density and the nonthermal part of the kinetic energy density

$$\frac{B^2}{8\pi} = f\rho\sigma_N^2 \ . \tag{1}$$

Here ρ is the mass density and σ_N the one-dimensional nonthermal velocity dispersion. The dimensionless constant f depends upon the cloud geometry and the degree of flux-freezing; we assume $f = 3$, close to the virial-theorem value for a uniform, spherical, flux-frozen cloud. Equation (1) implies that σ_N equals one-sixth of the Alfvén speed $v_A \equiv B/\sqrt{4\pi\rho}$; and if the nonthermal velocities have a Gaussian distribution, then the FWHM Δv_N of the distribution obeys

$$\Delta v_N \approx V_A \ . \tag{2}$$

Thus a cloud whose energy of nonthermal motions is comparable to its magnetic energy has nonthermal velocity width near the Alfvén speed. If so, the nonthermal motions are likely to arise from the passage of Alfvén waves (Arons and Max 1975; Zweibel and Josafatsson 1983, hereafter ZJ; Falgarone and Puget 1986). Most Alfvén wave modes have decay times shorter than cloud or core lifetimes (ZJ), and we assume that disturbances of intensity and frequency sufficient to maintain Alfvén waves are present. If so, line widths in molecular clouds have a simple explanation: they are supersonic but sub-Alfvénic "turbulence" driven by the propagation of Alfvén waves (Shu, Lizano, and Adams 1986). An alternate form of the foregoing argument is to assume eq. (2), i.e., that the velocity dispersion of nonthermal motions is comparable to the Alfvén speed. Then the equality of energy densities, eq. (1), follows.

We further model the cloud described by eq. (1) as a uniform sphere of radius R, mass M, temperature T, and mean mass per particle m; and assume that it obeys the well-established empirical law of virial balance discussed earlier,

$$\frac{GM}{5R} = \sigma_N^2 + \frac{kT}{m} \ .$$
(3)

The thermal term in eq. (3) can usually be neglected; then eqs. (1) and (3) yield

$$\sigma_N = \left(\frac{G}{90}\right)^{1/4} B^{1/2} R^{1/2}$$
(4)

and

$$n = \left(\frac{5}{2G}\right)^{1/2} \frac{B}{4\pi m} R^{-1} \ ,$$
(5)

where the number density is $n \equiv \rho/m$. These two equations are remarkably similar in form to the empirical relations between line width and size, and between density and size discussed earlier. This similarity was also noted by Scalo (this volume) and Shu (1986).

We have compared eqs. (4) and (5) with data from giant clouds (Dame et al. 1986), dark cloud cores (Myers 1983), and a wide range of cloud types, including dark cloud complexes (Larson 1981). For each data set, we found two values of magnetic field B that make each relation (4) and (5) just bound the data from above and below. The results are shown in Figure 6 as two curves superposed on each of the six plots of data (three of velocity dispersion vs. radius, three of density vs. radius). In each case, the theoretical curves closely match the observed data. For example, the plot of $\log n$ vs. $\log R$ for giant cloud data points is well fit by eq. (5) for B between 15 to 50 μG, and the plot of $\log \Delta v$ (given by $\Delta v^2 = 8 \ln 2 \, (\sigma_N^2 + kT/m)$) is well fit by the appropriate modification of eq. (4) for B within the same range. The "best-fit" or typical value of B is the geometric mean of the upper and lower bounds, or about 30 μG. Similarly, the values of B that approximately bound Larson's data are 15 and 150 μG, with typical $B \sim 50$ μG; and for Myers' data, 5 and 50 μG, with typical value $B \sim 15$ μG. For the latter dense core data, the thermal term in eq. (3) is not negligible for $B = 5\mu$G. When this term is included, the corresponding curves in Figure 6 bend down (e) or up (f) as R decreases; but the curves still largely "bound" the data.

These comparisons show that magnetically supported clouds in virial equilibrium have relationships among line width, size, and density in good agreement with available data for ~ 100 clouds, if B lies in the range 5–150 μG and if B is typically 15–50 μG. These required values of B also agree well with the ~ 10 measurements of B derived from observations of Zeeman splitting of H I and OH absorption lines (Troland and Heiles 1986; Kazés and Crutcher 1986).

The evidence presented here for magnetic support of virial clouds appears more detailed and convincing than that of other possible explanations

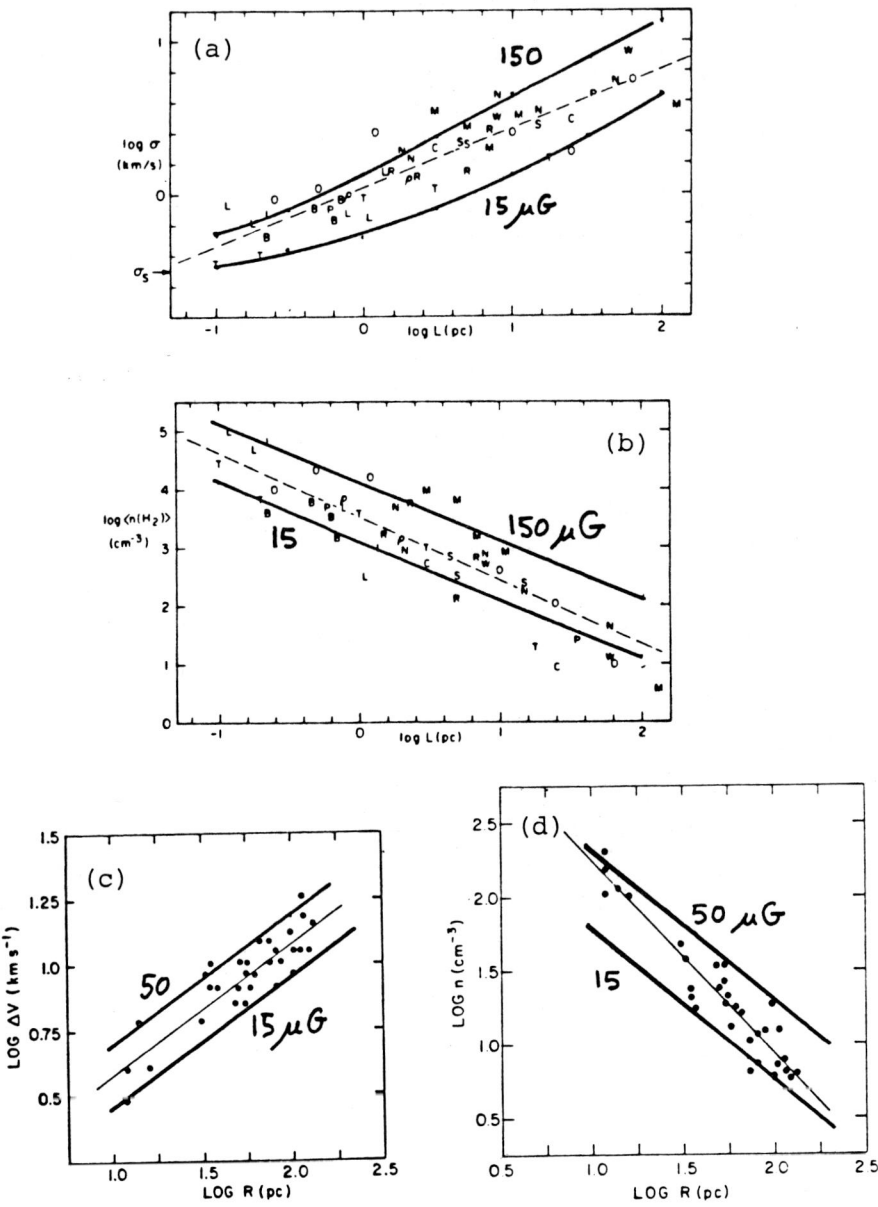

Figure 6(a)–(d).

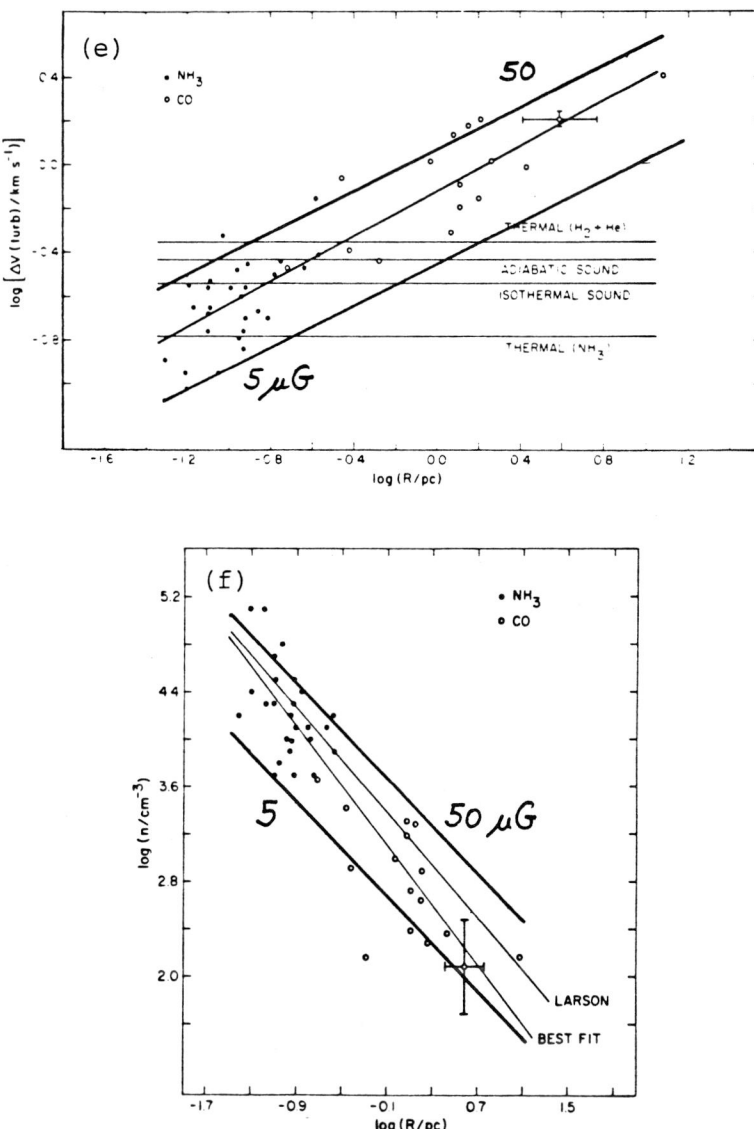

Figure 6. Plots of cloud velocity dispersion vs. radius. (a) and (b), from Larson (1981); (c) and (d), from Dame *et al.* (1986); (e) and (f), from Myers (1983). *Heavy solid lines:* predictions of magnetic support model [eqs. (4) and (5)] for the indicated values of magnetic field.

of the density-size-linewidth relations, summarized in Myers (1985). For
example, when Larson (1981) found that $\sigma \sim R^{0.3}$, he noted the possibility of a
process resembling Kolmogorov turbulence, which has the same dependence
of velocity dispersion on size scale for incompressible subsonic motions.
This interpretation has been criticized on several grounds (e.g., Scalo 1984).
However, in the present model, Larson's exponent of 0.3 is readily explained,
since he plotted the total dispersion

$$\sigma \equiv \left[3\sigma_N^2 + \frac{3kT}{m} \right]^{1/2} . \tag{6}$$

According to eq. (4), this dispersion varies as $R^{1/2}$ for large R when the
thermal term in eq. (6) is negligible, and as R^0 for small R when the thermal
term dominates. This behavior is evident in the curves of σ vs. R in Figure 6.
Therefore, Larson's exponent of 0.3 is consistent with magnetic support, as are
the exponents 0.5 in the data sets of Dame et al. and Myers.

An important implication of the magnetic support model is that the
"flux-freezing" relation $B \sim n^\kappa$, $\kappa \approx \frac{1}{2}$, may not apply to self-gravitating clouds
as widely as was previously believed. According to eq. (1), $B \sim n^{1/2}\sigma_N$, but
σ_N is not constant from cloud to cloud, so magnetic support and flux-freezing
are inconsistent. For example, the best-fit values of B in Figure 6 for giant
clouds is ~ 30 μG, for typical density 30 cm^{-3}. For dark cloud cores with
density 3×10^4 cm^{-3}, flux-freezing would therefore require $B \sim 900$ μG for
consistency whereas the best-fit B in Figure 6 is only ~ 15 μG. Also, if flux-
freezing in its usual formulation applied to dense cores, the magnetic energy
density would exceed the typical gravitational energy density by one to two
orders of magnitude and the core would not be bound (Goodman 1986). This
discrepancy may be interpreted to mean that dense cores do not conserve their
magnetic flux as they condense, but lose much of their flux through a process
such as ambipolar diffusion – as suggested by Shu, Lizano, and Adams (1986).

The foregoing picture of magnetic support is based on a highly idealized
model and on statistics of many different clouds; it is therefore likely to be
incorrect in several details. To critically test the basic idea, self-consistent
measurements are needed of B, n, R, and σ in dense, self-gravitating clouds.
More realistic models are also needed, to better understand excitation and
propagation of Alfvén waves in nonuniform media, where the magnetic flux
is not fully frozen to the cloud.

4. ACKNOWLEDGMENTS

I thank the University of Wyoming for suporting my attendance at this
meeting; Alyssa Goodman for assistance and discussions; Frank Shu and Lee

Rickard for discussions; and Paul Ho, Phil Solomon, and Hans Ungerechts for communicating data in advance of publication.

5. REFERENCES

Arons, J., and Max, C. 1975, *Ap. J. (Lett.)*, **196**, L77.
Bachiller, R., and Cernicharo, J. 1986, *Astron. Ap.*, in press.
Bally, J., Dragovan, M., Langer, W., Stark, A., and Wilson, R. 1987, contributed paper at Summer School on Interstellar Processes, Grand Tetons National Park.
Beichman, C., Myers, P., Emerson, J., Harris, S., Mathieu, R., Benson, P., and Jennings, R. 1986, *Ap. J.*, in press.
Blitz, L., Magnani, L., and Mundy, L. 1984, *Ap. J. (Lett.)*, **282**, L9.
Burstein, D., and Heiles, C. 1982, *A. J.*, **87**, 1165.
Burton, W.B., Deul, E., and Walker, H. 1986, in *First International Symposium on IRAS Results*, F. Israel, ed. (Dordrecht: Reidel).
Cernicharo, J., and Bachiller, R. 1984, *Astron. Ap. (Suppl.)*, **58**, 327.
Cohen, M., and Kuhi, L. 1979, *Ap. J. (Suppl.)*, **41**, 743.
Dame, T., Elmegreen, Cohen, R., and Thaddeus, P. 1986, *Ap. J.*, **305**, 892.
de Vries, H., Heithausen, A., and Thaddeus, P. 1986, in preparation.
Elmegreen, B. 1978, *Ap. J. (Lett.)*, **225**, L85.
Falgarone, E., and Puget, J. 1986, *Astron. Ap.*, in press.
Goodman, A. 1987, contributed paper at Summer School on Interstellar Processes, Grand Tetons National Park.
Heyer, M. 1986, Ph.D. Thesis, U. of Mass. Dept. of Physics and Astronomy.
Jones, B., and Herbig, G. 1979, *A. J.*, **84**, 1872.
Kazés, I., and Crutcher, R. 1986, *Astron. Ap.*, in press.
Keto, E., and Myers, P. 1986, *Ap. J.*, **304**, 466.
Kutner, M. 1973, in *Molecules in the Galactic Environment*, M. Gordon and L. Snyder, eds. (New York: Wiley), p. 199.
Larson, R. 1981, *M. N. R. A. S.*, **194**, 809.
Leung, C., Kutner, M., and Mead, K. 1982, *Ap. J.*, **262**, 583.
Low, F. *et al.* 1984, *Ap. J. (Lett.)*, **278**, L19.
Magnani, L., Blitz, L., and Mundy, L. 1985, *Ap. J.*, **295**, 402.
Menten, K., and Walmsley, C. 1985, *Astron. Ap.*, **146**, 369.
Murphy, D., and Myers, P. 1985, *Ap. J.*, **298**, 818.
Myers, P. 1983, *Ap. J.*, **270**, 105.
Myers, P. 1985, in *Protostars and Planets II.*, D. Black and M. Mathews, eds. (Tucson: University of Arizona Press), p. 81.
Myers, P. 1986, in *Star Forming Regions* (IAU Symposium No. 115), M. Peimbert and J. Jugaku, eds. (Dordrecht: Reidel).
Myers, P., and Benson, P. 1983, *Ap. J.*, **266**, 309.

Myers, P., Dame, T., Thaddeus, P., Cohen, R., Silverberg, R., Dwek, E., and
    Hauser, M. 1986, *Ap. J.*, **301**, 398. (Myers *et al.* 1986a)
Myers, P., Fuller, G., Mathieu, R., Beichman, C., Benson, P., and Schild, R.
    1986, submitted to *Ap. J.* (Myers *et al.*1986c)
Myers, P., Reid, M., Goodman, A., and Benson, P. 1986, in preparation.
    (Myers *et al.* 1986b)
Sanders, D., Scoville, N., and Solomon, P. 1985, *Ap. J.*, **289**, 373.
Scalo, J. 1984, *Ap. J.*, **277**, 556.
Schloerb, F., and Snell, R. 1984, *Ap. J.*, **283**, 129.
Schloerb, F., Snell, R., Goldsmith, P., and Morgan, J. 1987, contributed paper
    at Summer School on Interstellar Processes, Grand Tetons National Park.
Shu, F. 1986, paper presented at 98th Meeting of the A.S.P., Boulder.
Shu, F., Lizano, S., and Adams, F. 1986, in *Star Forming Regions* (IAU
    Symposium No. 115), M. Peimbert and J. Jugaku, eds. (Dordrecht:
    Reidel).
Solomon, P., and Sanders, D. 1985, in *Protostars and Planets II.*, D. Black and
    M. Mathews, eds. (Tucson: University of Arizona Press), p. 59.
Solomon, P., Rivolo, A., Barrett, J., and Yahil, A. 1986, submitted to *Ap. J.*
Troland, T., and Heiles, C. 1986, *Ap. J.*, **301**, 339.
Ungerechts, H., and Thaddeus, P. 1986, submitted to *Ap. J.*
Wadiak, E., Wilson, T., Rood, R., and Johnston, K. 1985, *Ap. J. (Lett.)*, **295**,
    L43.
Weiland, J., Blitz, L., Dwek, E., Hauser, M., Magnani, L., and Rickard, L.
    1986, *Ap. J. (Lett.)*, in press.
Zweibel, E., and Josafatsson, K. 1983, *Ap. J.*, **270**, 511.

John Bally, Ron Snell and Bruce Elmegreen

THE ATOMIC COMPONENT

Shrinivas R. Kulkarni[†] & Carl Heiles[*]

[†]Department of Astronomy, 105-24
Caltech, Pasadena, CA 91125, U. S. A.

[*]Department of Astronomy
Univ. of California, Berkeley, CA 94720, U. S. A.

ABSTRACT. We review the physical conditions and the distribution of the three phases which consitute the atomic component: the cold clouds, the warm, neutral medium and the warm, ionized medium. Together, these three phases occupy about $\gtrsim 40\%$ of the interstellar volume and contain half the interstellar mass. The size of the H I disk is comparable or exceeds that of the stellar disk. In the outer Galaxy, a spiral pattern is clearly discernable. We discuss extensively the distribution of the three phases near the sun as well as the physics and the limitations of various probes that have been used to study the atomic component. The ionization of the phases is studied in depth and we conclude that most of the interstellar electrons reside in the warm, ionized medium. We review various determinations of interstellar pressure and its scale height. Finally, we derive the filling factors of the atomic phases as a function of z. [Unless otherwise stated, $R_0 = 10$ kpc.]

Introduction

The interstellar medium (ISM) is composed of three main components: the molecular, the atomic and the hot component. These are acted upon by cosmic rays, electromagnetic radiation, gravity and magnetic field. The atomic component is composed of neutral atomic hydrogen (H I) and the diffuse, ionized hydrogen (H II). The hot component, while certainly 'atomic', follows such a different cycle as compared to H I and the diffuse H II that it is considered to be an entirely different component.

The Galaxy contains about $4.8 \times 10^9 \ M_\odot$ of H I (Henderson, Jackson and Kerr 1982). Estimates for the total amount of the H_2 range from one nearly equal to the H I estimate (Scoville and Sanders, this volume), to one that is only 25% of the H I estimate (Bloemen, this volume). The surface density distribution of H I is roughly constant from about 4 kpc to 20 kpc; however H I dominates H_2 in mass beyond Galactocentric radius 8 kpc (Blitz, Fich and Kulkarni 1983). Not much is known about the radial distribution and the mass of the diffuse, ionized hydrogen. Based on extrapolation of local data, we estimate at least $\sim 10^9 \ M_\odot$ for the total mass of the diffuse ionized gas. Estimates of the fraction of Galactic interstellar space occupied by the atomic component, range from 40% to 80%. All these factors establish the preeminence of the atomic component in the dynamics and evolution of the ISM.

D. J. Hollenbach and H. A. Thronson, Jr. (eds.), Interstellar Processes, 87–122.

The mass estimates of the interstellar components depend upon the assumed rotation curve and R_0, the distance of the Sun to the Galactic center. The above estimates assume: $R_0 = 10$ kpc, the standard Schmidt rotation curve for the inner Galaxy and a flat rotation curve for the outer Galaxy. Recent analyses suggest $R_0 \sim 8.5$ kpc. To first order, the mass estimates will have to be decreased by a factor $\sim (8.5/10)^2 = 0.72$. After correcting for differences in R_0, the atomic component constitutes 5% of the mass of the visible matter (Bahcall, Schmidt and Soneira 1983).

This review is not an exhaustive review. Much of the material presented here is adapted from a review on a similar topic that we wrote (Kulkarni and Heiles 1986). The reader is advised to consult the aforementioned review for a more detailed treatment and Spitzer (1978) for standard derivations not presented here.

The atomic component is thought to consist of three phases: (1) cold ($T \sim 80$ K), dense, neutral H I (cold medium or CM), (2) warm ($T \sim 8000$ K), neutral H I (the warm, neutral medium or WNM) and (3) a warm ($T \sim 8000$ K), ionized hydrogen (the warm, ionized medium or WIM). The CM is supposed to be distributed in the form of discrete structures, the so-called 'diffuse clouds'. In contrast, the WNM is widely distributed since it can be seen along most lines of sight. The cold clouds, the warm neutral and the ionized phases and the hot component (see Savage, this issue) are supposed to be in rough pressure equilibrium. Whether the interstellar space is largely occupied by the hot component or by the warm, atomic phases is uncertain and controversial. So is the relation of the WNM to the CM. Finally, the topology of all the phases is unknown.

1. Probes of the Atomic Component.

We now discuss the probes employed to study the atomic phases. Our knowledge of the properties of the CM and the WNM have mainly resulted from the study of the 21-cm line of H I (§1.1). The CM and the WNM are distinguished by H I absorption observations. The limitations of such observations are discussed in §1.2. The WIM can be studied using many probes: Hα emission, pulsar dispersion measures and the optical metastable lines of sulfur, oxygen, nitrogen *etc.* and these are briefly discussed in §1.3.

1.1 The 21-cm Line of H I.

In the H I atom, the magnetic moment of the proton interacts with the combined magnetic field generated by the orbiting electron and the magnetic moment of the electron; this interaction leads to the 'hyperfine' splitting of all the energy levels. The 21-cm line is the result of the hyperfine splitting of the ground state of H I. This hyperfine transition, being a magnetic dipole transition, is a forbidden transition. This, together with the low frequency of the 21-cm line, makes the Einstein A-coeffecient exceedingly small: 2.85×10^{-15} s^{-1}.

Consider an H I cloud in the ISM. Let n_1 and n_2 be the density of the H I atoms in the lower and upper hyperfine states, respectively. We define, T_x, the so-called 'excitation temperature' by

$$\frac{n_2/g_2}{n_1/g_1} = e^{-E/(kT_x)}. \tag{1.1-1}$$

For the 21-cm line, the statistical weights g_1 and g_2 are 1 and 3, respectively. The energy separation E/k is tiny: a mere 70 mK. In keeping with general convention for the 21-cm line, we will henceforth refer to the excitation temperature as the spin temperature (T_s).

The population ratio or T_s is determined by the excitation of H I atoms by collisions with particles, the 2.75 K cosmic background radiation and indirectly by Ly α pumping (see Field 1959 and references therein). If the cosmic background radiation is the only source of excitation, then, clearly, $T_s = 2.75$ K. The other two mechanisms drive T_s towards T_k, the kinetic temperature of the gas. Neglecting atomic collisions with electrons, which is justified when $x_e \lesssim 0.05$, collisions dominate over radiative excitation if

$$n_{HI} > n_{crit} = 4.7 \times 10^{-3} (T_k/1000K)^{0.77} \text{cm}^{-3}. \tag{1.1-2}$$

This density is sufficiently low that for all Galactic situations, one can safely assume $T_s = T_k$.

1.1.1 Radiative Transfer of the 21-cm line.

At cm wavelengths, the Rayleigh-Jeans approximation is valid and it is conventional to use T_B, the brightness temperature instead of I, the specific intensity. T_B is linearly proportional to I via the Rayleigh-Jeans relation: $I = 2\nu^2 kT_B c^{-2}$. It is convenient to use T_B since T_B, unlike I, is independent of ν. The equation of radiative transfer in the H I line can then be written as

$$dT_B(\nu)/d\tau(\nu) = T_s - T_B(\nu), \tag{1.1-3}$$

where τ is the optical depth along the line of sight. This equation shows yet another advantage of using T_B: in the limit of large τ, $T_B = T_s$ i.e. the brightness temperature of an optically thick H I cloud is equal to its kinetic temperature, as it should.

Though equation (1.1-3) appears simple, the solution to the general case of both T_s and τ varying from position to position is non-trivial. For the simple case of an isolated, single, isothermal H I cloud, equation (1.1-3) can be solved to yield

$$T_B(v) = T_{bg}e^{-\tau(v)} + T_s[1 - e^{-\tau(v)}], \tag{1.1-4}$$

where T_{bg} is the brightness temperature of the radiation incident on the far side of the cloud. In an actual measurement, we measure the spectrum with respect to the continuum i.e. we measure $\Delta T_B(v) = T_B(v) - T_{bg}$:

$$\Delta T_B(v) = (T_s - T_{bg})[1 - e^{\tau(v)}]. \tag{1.1-5}$$

$\tau(v)$ is related to $N(v)$, the number of atoms with a velocity v in a cylinder of base 1 cm^{-2} (i.e. the column density) by

$$\tau(v) = \frac{N(v)}{C \times T_s}. \tag{1.1-6}$$

Here v is assumed to be in km s^{-1} and the constant C is 1.83×10^{18} cm^{-2} K^{-1} (km s^{-1})$^{-1}$.

In order to gain a physical feeling for these equations we consider two extreme limits of the single cloud with no background radiation (i.e. $T_{bg} << T_s$):

Optically thin case ($\tau(v) << 1$). In this case, $T_B(v) = T_s\tau(v) = N(v)/C$ and the measured brightness temperature is proportional to the column density per unit velocity. Physically what this means is that almost all the spontaneously emitted 21-cm photons escape the cloud without being absorbed. The emission rate is practically independent of

T_s since E/kT_s is exceedingly small for any reasonable T_s. Thus the number of photons leaving the cloud is a direct measure of the H I column density.

Optically thick case $(\tau(v) \gg 1)$. In this case, $T_B = T_s$, i.e. the brightness temperature is simply the spin temperature. Since $\tau \gg 1$, the 21-cm photons emitted within the interior of the cloud get absorbed within the cloud. Only the photons emitted by gas within $\tau \lesssim 1$ of the front surface manage to escape from the cloud. Thus the observed brightness temperature is independent of the column density and depends only on the temperature of the cloud.

1.2 H I Emission and Absorption Observations.

From equation (1.1-6) it should be apparent that H I is seen in 'emission' or in 'absorption' depending whether T_s is bigger or smaller than T_{bg}, respectively. In the former case, the emission from the foreground H I more than makes up for the attenuation of the background radiation and vice-versa for the latter case. For the special situation of a cloud having $T_s = T_{bg}$, the emission from the foreground is exactly compensated by the absorption of the background and there is no net emission. For example, an H I cloud at the same temperature as the 2.75 K cosmic background radiation would not be detectable. However this is not a problem for Galactic H I since T_s is never so small at any Galactic site.

Emission data have been extensively used to map the distribution of Galactic H I. Absorption data is needed to separate cold H I from warm H I and to measure the temperature of H I. For this reason, absorption data are crucial to our understanding of the thermodynamics and the structure of diffuse clouds as well as the warm medium.

1.2.1 H I Absorption Techniques.

Both single dishes and interferometers can be employed to obtain H I absorption spectra. We discuss both these techniques and show that in general, interferometric measurements are more reliable than single dish techniques.

Consider an isothermal H I cloud and a radio source of flux density S behind it. In the single dish technique, we assume that the beamwidth of the telescope is much smaller than the angular size of the cloud but larger than that of the radio source. Then (equation 1.1-5), the antenna temperature measured by the telescope in the direction towards the source (the 'on-source' spectrum) is

$$\Delta T_{A,on}(v) = (T_s - T_{src})[1 - e^{-\tau(v)}]. \tag{1.2-1}$$

Here we implicitly assume that the beam efficiency of the telescope is unity so that, for an extended source such as the H I cloud, the antenna temperature is equal to the brightness temperature. T_{src} is the antenna temperature of the compact source and is given by $S \times G$ where G is the gain of the telescope. G for a single 25-m antenna such as the ones used at the Very Large Array (VLA) is ~ 0.12 K Jy^{-1} and ~ 8 K Jy^{-1} for the giant Arecibo reflector. We have neglected background contributions such as the 2.75 K cosmic background radiation since such contributions are present in both the 'on' and the 'off' spectra and do not affect the determination of the optical depth spectrum.

The single dish technique consists of obtaining an 'off-source' emission spectrum, $\Delta T_{A,off}(v)$, in a direction displaced by $\sim$ beamwidth so as to obtain an independent measure of T_s. Thus, an important assumption is being made: the cloud has little structure over one beamwidth. Under this assumption, the absorption spectrum is readily obtained as:

$$e^{-\tau(v)} = 1 + \frac{\Delta T_{A,on}(v) - \Delta T_{A,off}(v)}{T_{src}}. \tag{1.2-2}$$

In practice, the off-source spectrum is formed from several off-spectra taken around the source in a pattern such as a cross or hexagon. *Linear gradients* in $T_B(v)$ are cancelled out by the use of such symmetrical patterns. However, structure on the scale of the beamwidth is not, which is a fundamental limitation on single-dish work. For example, even for the giant Arecibo dish, a fluctuation of 10% in adjacent emission spectra is typical for $|b| > 10°$. This variation increases at low latitudes, towards intermediate and high-velocity clouds, and of course is worse for larger beamwidths and weaker background sources. For this reason, essentially all single dish H I absorption data towards weak or low-latitude sources are unreliable.

Intereferometers surmount these difficulties because of their ability to act as *spatial filters*. An interferometer with a baseline B responds only to structures in the sky with spatial frequency $\sim B/\lambda$, where λ is the wavelength. Most of the H I features in the sky do not have much structure on angular scales below about an arc minute (see Crovisier, Dickey and Kazes 1985), whereas most background radio sources are generally smaller than 1 arc minute. Hence an interferometer with a fringe spacing smaller than 1 arc minute but bigger than the size of the background source resolves the foreground H I emission and responds only to the background sources. Thus interferometers measure $e^{-\tau(v)}$ directly. The ability of interferometers to form images (aperture synthesis) is not critical for H I absorption measurements. Thus the sensitive Arecibo interferometer is as useful as the multi-element Very Large Array (VLA). In fact, the most extensive low-latitude survey was done at the VLA in the so-called 'phased array' mode in which the VLA is essentially reduced to a 3-element interferometer (Dickey *et al.* 1983).

Interferometric surveys are always more reliable than single dish surveys. However, the Arecibo single dish surveys provide reliable absorption data because of the very narrow Arecibo beam (4 arc minute) and the associated high antenna gain. Kulkarni *et al.* (1985) compare Arecibo single dish data with Arecibo interferometric data and conclude, that, at least for $|b| > 10°$ and strong sources ($S > 3$ Jy), the Arecibo single dish data are reliable.

1.2.2 Limitations of the Derived Spin Temperatures.

Our discussion of the excitation temperature of the 21-cm line was pursued in the context of a single, isothermal H I cloud. In practice, this simple situation is rarely encountered. When more than one H I cloud is present along a given line of sight, the measured spin temperatures is related in a complicated way to the true spin temperatures.

Consider an isothermal H I cloud with a spin temperature T_s and a column density $N_H(v)$. The velocity dependence of N_H is a consequence of both thermal and macroscopic motions. Observationally we can measure $\tau(v)$ and $T_B(v)$. In the spirit of equation (1.1-5), and assuming $T_{bg} \ll T_s$, we define $T_n(v)$, the *naively-derived spin temperature*, in terms of measured quantities as:

$$T_n(v) = T_B(v)/[1 - e^{-\tau(v)}]. \tag{1.2-3}$$

For the simple case of a single, isothermal cloud that we considered in §1.1.1, $T_n(v)$ is indeed equal to T_s.

Complications arise when there is more than one parcel of H I with overlapping velocities along the same line of sight. For example, consider the case of two isothermal clouds with the same velocity distribution, but with two different spin temperatures. Application of equation (1.2-3) yields

$$T_n(v) = \frac{T_{s,1}[1 - e^{-\tau_1}] + T_{s,2}[1 - e^{-\tau_2}]e^{-\tau_1}}{[1 - e^{-\tau_1 - \tau_2}]} \qquad (1.2\text{-}4)$$

where the subscript '1' refers to the cloud nearer to the observer (the 'foreground' cloud) and the velocity dependence of τ has been dropped out for clarity. The derived spin temperature, T_n is now a function of v and *from $T_{s,1}$ through $T_{s,2}$*. Thus, literally interpreting $T_n(v)$ as spin temperature would lead us to the false conclusion that H I at temperatures intermediate to the T_1 and T_2 exists.

Consider three illustrative cases to appreciate the pitfalls and the limitations of $T_n(v)$:

(1) *Optically thick foreground.* Then $T_n(v) \simeq T_{s,1}$ and we have no information about the background cloud.

(2) *Optically thick background and optically thin foreground.* Then $T_n(v) \simeq T_{s,1}\tau_1(v) + T_{s,2}$ and the naively derived spin temperature of the background cloud is increased by the foreground contribution. This is the usual case with the WNM being the foreground and a cold background cloud.

(3) *Optically thin background and foreground.* Then, the naively deduced temperature, $T_n(v) = [N_1(v) + N_2(v)]/[(N_1(v)/T_{s,1}) + (N_2(v)/T_{s,2})]$ *i.e.* $T_n(v)$ is simply the column density weighted *harmonic mean* temperature. This leads to a severe bias when we try to measure the temperature of the WNM *e.g.* a tiny cold ($T \sim 80K$) cloud with only one-tenth the column density of the WNM ($T \sim 8000K$) will decrease the naively derived spin temperature of the WNM from 8000 K to ~ 800 K!

1.3 Probes of the Ionized Gas.

Ionized gas in the ISM is revealed by a variety of observations, each sensitive to a combination of the electron density and temperature:

(a) *Dispersion of Pulsar Signals.* The group velocity of an electromagnetic signal of frequency ν travelling through ionized gas is $V_g = c(1 - \nu_p^2/\nu^2)^{1/2}$, where $\nu_p = 8.97 n_e^{1/2}$ kHz is the plasma frequency. Owing to this dispersion in the ISM, a pulse at higher radio frequency arrives earlier than that at lower frequencies. The observed rate of change of delay with frequency provides a measurement of $DM \equiv \int n_e dl$, the integrated column density of electrons to the pulsar. The conventional units of DM are cm^{-3} pc.

(b) *Optical Recombination Emission.* In the ISM, a free electron eventually recombines with a proton and in the process emits a host of recombination line photons. Nearly all useful work has employed a wide beam (5' to 50') Fabry-Perot spectrometer to observe the Hα $\lambda6563$ Å recombination line (see Reynolds 1984). The observed Hα intensity, $I = 0.36 \int n_e^2 T_4^{-0.9} dl$ Rayleighs (R), where T_4 is the temperature in units of 10^4 K and dl is the path length in parsecs. One Rayleigh is the isotropic intensity emitted by 10^6 recombinations per second. In steady state, I is a measure of the ionization rate. For an assumed constant temperature, the emission measure, $EM \equiv \int n_e^2 dl$, can be obtained. The usual units of EM are cm^{-6} pc.

(c) Low-frequency Radio Absorption. In ionized gas, the encounters of free electrons with ions leads to the emission of radiation by the 'Bremsstrahlung' mechanism. Free-free absorption is the inverse of this process. The optical depth $\tau_{ff} \propto g_{ff} \int n_e^2 T^{-3/2} \nu^{-2} dl$. At radio frequencies, after taking account of the variation of the Gaunt factor, g_{ff}, we find $\tau_{ff} \propto \nu^{-2.1} T^{-1.35} \times EM$. In order to see significant absorption by the Galactic ISM one has to go down to frequencies as low as a few MHz! Tasmania is one of the few locations where, occasionally, the ionosphere becomes transparent to let radio astronomers peek at the low-frequency heavens. Almost all the ground-based observations in this field has been done in Tasmania by the pioneering radio astronomer Grote Reber and his colleagues. Observations below 2 MHz have been obtained from spacecraft; the space observations have almost no angular information due to the small size of the space antennas.

2. Distribution of the Atomic Component.

Locally, the H I layer and the diffuse H II layer can be approximately described by a stratified horizontal atmosphere model, *i.e.* $N_H(b) = N_{H,\perp}/sin(|b|)$, where $N_H(b)$ is the integrated column density towards galactic latitude b and $N_{H,\perp} \equiv \langle N_H(b)sin(|b|)\rangle$ is the mean column density projected to the pole; the total vertical column is thus $2N_{H,\perp}$. Hereafter, we will use the subscript "$\perp$" to indicate any mean column density projected to the Galactic pole.

The distribution of the neutral atomic component has been traced solely from H I emission data. H I emission data are a measure of the column density and hence do not discriminate between the cold medium (CM) and the warm, neutral medium (WNM). Thus the discussion of the large-scale (§2.1), the H I halo (§2.2) and intermediate-scale (§2.3) refer to both the neutral phases. The warm, ionized medium (WIM) has been mostly studied using optical recombination and metastable lines. Obscuration in the galactic plane has thus far hindered the determination of the large-scale distribution of this medium, although some meager information has been obtained using pulsars (§2.4).

Before we discuss the Galactic distribution of the various atomic phases, a brief summary of the local gas (distance < 1 kpc) is in order. Extensive H I absorption surveys have been used to derive the distribution of cold clouds. Detailed emission line studies of H I kinematics have been successfully used to infer the physical conditions and the distribution of the WNM. The CM appears to be organized into discrete structures in contrast to the WNM which appears to be widely distributed. The scale height of the CM layer is about 100 pc, half that of the WNM layer. The cloud-cloud velocity dispersion of cold clouds is about 6.9 km s^{-1} and that of the WNM is slightly larger: 9 km s^{-1}. Typical cloud temperatures are around 80 K whereas various observations suggest a temperature of $\sim$ 8000 K for the WNM. A detailed discussion of the physical conditions and the distribution of the local H I (the CM and the WNM) can be found in §5 and §6, respectively.

From emission data, Heiles (1976) finds $N_{H,\perp} \sim 3.7 \times 10^{20}$ cm^{-2} which translates to a mass surface density of 8 $M_\odot$ pc^{-2} more or less evenly divided between the CM and the WNM. A small caveat: it appears to be the practice of H I observers *not* to include He when estimating the mass surface density. In this article all the quoted *mass* surface densities have been increased by 1.36 to include contribution from He. From pulsar dispersion data, $N_{e,\perp} \sim 0.9 \times 10^{20}$ cm^{-2} (see §2.4.1). The surface density of the WIM is a non-neglible 2 $M_\odot$ pc^{-2}. Thus, altogether, the surface mass density of the atomic phases is 10 $M_\odot$ pc^{-2}, which may be compared to the total disk surface-density of 75 $M_\odot$ pc^{-2} (Bahcall, Schmidt and Soneira 1983) and the molecular mass surface density of $\sim 3\ M_\odot\ pc^{-2}$ (Scoville and Sanders, this volume). [The molecular material is exponentially decreasing around $R = R_0$. Consequently it is difficult to accurately estimate the local mass surface density.] Note that $N_{H,\perp}$ and $N_{e,\perp}$, are based on *local* measurements and hence are *unaffected* by changes in R_0.

2.1 Large-scale distribution of H I.

The large-scale distribution of H I is derived in a straightforward manner: one first obtains H I emission spectra which are a measure of the column density of H I per unit velocity interval; with the help of a rotation curve, the volume density as a function of velocity and therefore of distance can be derived. Various quantities of interest such as the surface density and the scale height are easily derived from the volume density distribution. The derived distribution thus depends upon the choice of the rotation curve including the value of R_0, the distance to the Galactic center. For example, changing R_0 by a factor α changes the derived densities by α^{-1}, the scale height by α, the surface density remains unchanged and the mass by α^2 (the additional dependence on l, the Galactic longitude, has been ignored here).

Distribution of H I in the Inner Galaxy ($R < R_0$). In the inner Galaxy, the so-called 'distance ambiguity' problem prevents a unique relation between velocity and distance. However, the radial (*i.e.* galactocentric) distribution of H I can be determined since the differential velocity depends only on the Galactocentric radius, R. Studies of scale height are best done at the 'tangential points'. These are locations for which the line of sight is tangential to a circle centered at the Galactic center. Tangential points are desirable in two ways: (a) these are easily identified as extreme velocity gas and (b) the distance to tangential points is independent of the assumed rotation curve and determined strictly from geometry; it is simply $R_0 cos(l)$ where l is the Galactic longitude.

The surface density of the H I layer in the inner Galaxy is constant for $R > 0.4R_0$ and decreases rapidly for $R < 0.4R_0$. The precise value of the surface density depends upon the optical depth correction. Assuming the 21-cm line is optically thin, the mean surface density of H I is about $4.5 \times 10^{20}\ cm^{-2}$ (Lockman 1984). No rigorous estimate of the correction due to the optical depth in the 21-cm line has been made. Using Kulkarni's (1983) *estimate* of a correction factor of 1.25, the mass surface density is $\sim 6\ M_\odot\ pc^{-2}$ – apparently somewhat smaller than the local value. Given the uncertainties, this discrepancy may not be significant.

The vertical structure of the H I layer appears to be independent of R for $0.4R_0 < R < R_0$ (Lockman 1984). $\langle z \rangle$, the mean of the H I layer is close to 0 pc throughout the inner Galaxy. The FWHM (full width at half maximum) of the H I layer is about 365 pc, independent of R; this value is comparable to the local value. The constancy of the shape and the *width* of the H I layer is a great puzzle. Ignoring the effect of cosmic rays and magnetic fields,

$$\sigma_z \propto \sigma_v \rho_*(0)^{-1/2} \tag{2-1}$$

(Spitzer 1978); here σ_z is the rms scale height of the gas layer, σ_v is the velocity dispersion of the gas and $\rho_*(0)$ is the stellar mass density at $z = 0$; this expression is valid when $\sigma_* \gg \sigma_z$ where σ_* is the scale height of the stellar disk. From studies of H I in our own Galaxy and in external galaxies, σ_v appears to be independent of R. Assuming an exponential stellar disk with a scale length of 4.3 kpc (Bahcall, Schmidt and Soneira 1983), we find $\rho_*(4kpc)/\rho_*(10kpc) = 4$. So naively we expect the H I scale height at $R = 4$ kpc to be twice as small as locally. Perhaps, pressure from magnetic fields and cosmic rays is higher in the inner Galaxy and conspire to keep the scale height constant. It remains to be demonstrated whether the mild gradients in magnetic field (Heiles, this volume) and cosmic rays (Bloemen, this volume) can actually account for the constancy of the H I z-distribution. In contrast, the width of the molecular layer in the inner Galaxy is smaller than the local value (Scoville and Sanders, this volume).

Distribution of H I in the Outer Galaxy $(R > R_0)$. The structure of H I in the outer Galaxy has been successfully determined since there is no distance ambiguity problem. Recently, there have been two comprehensive analyses: one assuming a flat rotation curve (Henderson, Jackson and Kerr 1982) and the other using a rising rotation curve (Kulkarni, Blitz and Heiles 1982).

In both these analyses, three major *coherent* features are seen very clearly. These features spiral out as one follows them along their length and for this reason we refer to them as spiral arms. The three major arms have been named as Perseus, Cygnus and Carina after the constellations where the major portion of their length lie. There is a hint of a fourth arm across the Galaxy towards $l = 0$. Locally the sun appears to be situated in a minor feature popularly referred to as the Orion arm. These three (and perhaps four) major arms appear to have constant surface density and are also confined to the main H I disk. High-density molecular clouds and their associated H II regions dot these features (Blitz *et al.* 1983), just like the spiral arms in external galaxies.

It is not clear whether the large-scale features we see in the H I map arise due to density variations or systematic non-circular velocity fields or both. In fact, if spiral arms are due to non-axisymmetric potential perturbation then we expect systematic non-circular velocity fields. Incidentally, the non-circular velocities distort the H I distribution when viewed in the R, θ plane; this is one reason (and perhaps the only one!) why the H I structure is best viewed in 'l-v' diagrams. Thus the derived parameters such as pitch angle and the arm-interarm contrast are uncertain. The other parameters such as the surface density and the size of the H I disk depend upon the choice of the rotation curve.

For the sake of 'hard' numbers we briefly summarize the results of one analysis (Kulkarni *et al.* 1982). In this analysis, the spiral arms have a pitch angle of $\sim 25°$, a length of about 20 kpc, an arm-interarm surface-density contrast of about 4 and are confined to $R < 20$ kpc. The H I surface density is roughly constant to $R = 20$ kpc and falls exponentially beyond that.

Verschuur (1975) has suggested that there may be more distant and fainter arms. This intriguing suggestion has not been investigated further.

The H I disk in the outer Galaxy is warped. Large sections of the H I disk in the first two quadrants $(0 < l < 180°)$ are systematically above and that in the other two quadrants systematically below the plane defined by the the H I layer in the inner Galaxy. The large-scale warp is fairly small to a distance of ~ 18 kpc from the center and then rises very sharply. At $R = 20$ kpc the peak-to-peak displacement of the warp is about 3 kpc, which amounts to more than $6°$ in galactic latitude! The outer edge of the H I disk oscillates *i.e.* the outer edge is scalloped. The wave number of this oscillation, $m \sim 10$ and the amplitude is about 1 kpc. If one views the large-scale warp as an $m = 2$ oscillation then perhaps it is not unreasonable to expect additional oscillations with values of m between 2 and 10. Such oscillations have not been carefully searched for.

In the outer Galaxy, unlike in the inner Galaxy, the thickness of the H I layer increases linearly with R. In view of equation (2-1) this is understandable since the stellar density is rapidly decreasing and σ_v is unchanged. Since an exponential disk is inconsistent with either a flat or a rising rotation curve, a spherical massive 'dark halo' has been postulated. The postulated dark halo does not affect the vertical structure of the H I disk because of its small z-gradient.

2.2 The H I Halo.

More than a decade ago, W. W. Shane pointed out the existence of H I at high-z. Recently, the distribution of this H I (hereafter referred to as the 'H I halo') has been studied by Lockman (1984) by observing the vertical structure of H I at the tangential points. Lockman found (a) about 13% of H I exists at $|z| > 500$ kpc and (b) the high-z H I appears to be corotating with the disk. The H I halo disappears for $R \lesssim 0.3R_0$.

Lockman found a three component fit with the following parameters provides an adequate description of the H I layer ($0.3R_0 < R \lesssim 0.9R_0$): (1) a Gaussian with an rms scale height, $\sigma_z \sim 100$ pc, (2) a Gaussian with $\sigma_z \sim 250$ pc, and (3) an exponential with an exponential scale height, $H_z \sim 500$ pc; the total vertical column densities without any corrections are 1.2×10^{20} cm^{-2}, 1.7×10^{20} cm^{-2} and 1.6×10^{20} cm^{-2} respectively. We identify component 1 with the CM from the similarity of the scale heights. Thus, it is not unreasonable to suggest that components 2 and 3 together make up the WNM (see also §4.1 for further justification). The existence of more than one component implies that H I as a whole cannot be characterized by one single velocity dispersion (cf. equation 2-1). This brings us to an important point viz. are the three components merely mathematical decompositions or are they physical components?

There are two biases which affect the parameters for the components quoted above. One is that there was no attempt to apply optical depth correction. Component 1 is most affected by this bias. This bias results in an underestimation of the measured width and the column density of component 1; crudely, we estimate that the measured column density of component 1 has to be increased by a factor of 1.3. The second bias is that in order to convert the observed brightness temperature to volume density of H I in the tangential region, an estimate of Δr, the 'length' of the tangential region must be made. Clearly Δr increases with the assumed velocity dispersion of the gas. However, as discussed above there is more than one velocity dispersion. In fact, it is the higher dispersion gas that reaches the higher $|z|$. In short, Δr is itself a function of z. This bias results in an overestimation of the measured width and column density of component 3. Clearly, careful modelling of the tangential point data is needed to reduce these biases.

What is the evidence for a local H I halo? Lockman, Shull and Hobbs (1985) have compared the amount of H I inferred from Lyα absorption towards high-latitude OB stars with that inferred from 21-cm observations. Such comparisions, though meager when compared to the 21-cm data, confirm the existence of a local halo similar to that in the inner Galaxy.

There is no study of the halo in the outer Galaxy. Studying the halo in the outer Galaxy will require careful modelling since there are no convenient regions like the tangential points. H I from many regions will contribute to the brightness temperature at a given velocity. Additionally, the modelling is complicated by the existence of more than one velocity dispersion and the flaring of the H I disk. Ignoring these complications, Kulkarni *et al.* (1982) found that the *shape* of H I layer in the outer Galaxy as characterized by the ratio of the 50-percentile width to the rms width appears to be invariant with R. Clearly more work is needed here.

2.3 Distribution of H I on Intermediate Scales.

In this section we discuss the distribution of H I on scales below about 1 kpc. This discussion is brief and cursory. The reader is referred to the review by Kulkarni and Heiles (1986) for a complete discussion. The structure of H I has been traced solely from H I emission data. The emission data come mainly from the following surveys: in the north at the Hat Creek Observatory (Heiles and Habing 1974) and Weaver and Williams (1973), in the south at the Instituto Argentino de Radioastronomia, Argentina (Colomb, Poppel and Heiles 1980) and at Parkes, Australia (Cleary, Heiles and Haslam 1979).

In the solar neighbourhood, there are three major regions containing young stellar objects (OB associations and H II regions): Ophiuchus, Perseus/Taurus and Orion. Every one of these regions is enveloped by a large H I concentration ('H I complexes') with high column density. Approximate typical physical properties of these complexes are: $N_H \sim 1.0 \times 10^{21}$ cm^{-2}, $\langle n_H \rangle \sim 2.5$ cm^{-3} (volume average; actual value probably higher due to clumping), linear diameter ~ 120 pc, and mass $\sim 1.0 \times 10^5 M_\odot$.

These parameters vary widely from one regions to another. Orion, for example, is enveloped by about $7 \times 10^4 M_\odot$ of H I with a linear diameter of about 125 pc (Gordon 1971), while the associations I Mon and II Mon are enveloped by H I masses of about $1.5 \times 10^5 M_\odot$ and $2 \times 10^4 M_\odot$, respectively (Raimond 1966). At least in some cases, an equivalent mass is contained in molecular clouds; in Orion, for example, Thaddeus (1982) reports a total of about $3.3 \times 10^5 M_\odot$ in H$_2$ (a somewhat uncertain figure, because it is derived from CO observations using a controversial CO/H$_2$ conversion figure). In the early days of 21-cm astronomy, Deiter(1960), using a small telescope (and hence a large beam), reported the presence of such complexes in 31 out of 40 OB associations investigated by her. We now know that star forming regions are intimately associated with the giant, molecular cloud complexes (see Scoville and Sanders, this volume). The time is now ripe for a systematic, high angular-resolution study of H I around *all* molecular complexes.

Apart from these complexes, the H I sky abounds in filaments on all angular scales. The largest of these filaments appear to be parts of expanding shells. The nearest such objects include the North Polar Spur, the Eridanus Loop, a shell encircling the North Celestial Pole and another running close to Radio Loop II. Some of these shells such as the North Polar Spur and the Eridanus Loop appear to be filled with hot gas emitting X-rays and some are associated with intense radio continuum emission *e.g.* North Polar Spur and Radio Loop I. Less spectacular expanding shells have been inferred from high resolution observations of H I around H II regions. In the latter case, the H I is the H$_2$-dissociation front.

Heiles (1979 and 1984) searched for distant expanding shells such as the ones found locally and discovered quite a large number of them. Most suprising was the discovery of a population of extremely large shells, 'supershells', ranging up to 2 kpc in diameter. The Sun may be located just inside the boundary of a supershell (Lindblad *et al.* 1973). These supershells are truly remarkable. If produced by the instantaneous release of explosive energy, the required energies range from 4×10^{53} erg – equivalent to 400 supernovae. Energy released from a large number of stars in the form of stellar winds and supernovae explosions may, in fact, be the energy source for most or all supershells. However, the large energies and the fact that usually only one hemisphere of an expanding supershell is visible may imply another mechanism in some cases – specifically, the collision of infalling gas with the gas in the Galactic plane.

About half of the H I in the solar vicinity appears to be falling towards the Galactic plane (see Kulkarni and Fich 1985). Most of this H I lies at LSR velocities within ± 40 km s^{-1}. However, some of the gas lies at higher negative velocities: 'intermediate velocity gas (IVG)', extending to about -90 km s^{-1} and 'high velocity gas (HVG)', extending to more extreme velocities (Giovanelli 1980). Both the IVG and the HVG were originally discovered by Dutch radio astronomers. The Dutch work on IVG (Wesselius and Fejes 1973) remains the best and the most comprehensive.

Space and other considerations restrict our discussion of the IVG and HVG. The principal result of the Dutch study is that IVG is located primarily at high positive Galactic latitudes and coincides with a 'hole' in the low velocity gas (LVG) distribution. The anticorrelation between LVG and the IVG is very striking. In some regions, the low velocity gas and the IVG fit together like pieces of a jigsaw puzzle. The IVG column density column density is close to what the LVG column density would be if the LVG 'hole' did not exist. Thus, Wesselius and Fejes argue that the LVG has been 'displaced' to form the 'IVG' – that some agent affected the LVG and changed its velocity without changing the total H I column density.

The physical cause for the origin of the IVG and the HVG is not clearly established. Heiles (1984) has presented photographs of the IVG distribution which makes it look like a portion of an expanding circular shell. Cohen (1981) has shown that the HVG lies precisely on top of the LVG in the direction of Radio Loop II. Nichols-Bohlin and Fesen (1986) suggest that the star HD 50896 is in the middle of an expanding, intermediate velocity shell. The circumstantial evidence i.e. expanding shells and a cluster in the middle of the shell (as in the shell surrounding HD 50896) suggest a supernova origin. Heiles (1984) showed that the IVG in the Galactic anticenter region is organized in the form of a large (30°) one-sided shell. Kulkarni (1983) has investigated this structure and favors an expanding/moving ring model. Mirabel (1982) has argued that this structure has resulted from interaction of high velocity clouds with the LVG. Kulkarni and Mathieu (1986) have placed a lower limit on the distance (2.5 kpc) and favor Mirabel's suggestion. This structure deserves to be investigated further since it is probably the best example of a structure formed by the interaction of HVG with LVG.

2.4 Distribution of the Ionized Gas.

In §7 we show that most of the interstellar electrons come from the warm, ionized medium (WIM). Thus the distribution of the WIM is the same as the distribution of the electrons. The various probes of interstellar electrons are reviewed in §1.3. Locally, the electrons appear to be wide spread as evidenced by diffuse Hα emission along most lines of sight. The Hα line and optical metastable lines have provided us much insight about the physical conditions of the interstellar electrons. Unfortunately, owing to obscuration, we cannot use these optical probes to study the distribution of the interstellar electrons on Galactic scales. At the present moment our knowledge of the electrons on non-local scales comes from analysis of pulsar dispersion measure (DM) data.

2.4.1 Local Distribution

The vertical column density distribution of the electrons can be estimated from the pulsar DM data. H I absorption spectra towards 40 low-latitude pulsars have been used to establish the distance-DM calibration for pulsars (Lyne, Manchester and Taylor 1985). $\langle |z| \rangle$ for these 40 pulsars is ~ 200 pc and hence the exact form of the vertical profile of $\langle n_e \rangle$ cannot be directly inferred from observations. Using a carefully selected sub-sample of pulsars and superimposing the requirement that the spatial distribution of pulsars deduced from the dispersion measures should be consistent with cylindrical symmetry around the galactic center, Vivekanand and Narayan (1982) tested a number of simple models for the distribution of interstellar electrons. Their principal conclusions are: (i) $\langle n_e \rangle = 0.037^{+0.020}_{-0.012}$ cm^{-3} and (ii) the exponential scale height of electrons is greater than 300 pc and probably 1 kpc or larger. Electrons from the diffuse medium as well as discrete H II regions contribute to $\langle n_e \rangle$. After allowing for the contribution from the H II regions we get $\langle n_e(0) \rangle \sim 0.030$ cm^{-3}. The lower limit on the scale height of electrons is comparable to the scale height of pulsars (400 pc). Thus $\langle DM \sin(b) \rangle$ obtained from pulsar data is a *lower* limit to the electron vertical column density. Manchester and Taylor (1977) have attempted to correct this by obtaining the vertical distribution of pulsars and electrons self-consistently and find $DM_\perp \simeq 30$ cm^{-3} pc. If the scale height of interstellar electrons is significantly larger than that of the pulsars then even this determination should be considered a lower limit. If we assume an exponential distribution: $\langle n_e(z) \rangle = \langle n_e(0) \rangle e^{-|z|/H_e}$ then we have $H_e \sim 1000$ pc to reproduce the observed $DM_\perp$.

Almost all the observations in the Hα $\lambda 6563 \overset{\circ}{A}$ line have been done by wide-beam (5 arcmin to 50 arcmin) Fabry-Perot spectrometers and mainly by Reynolds (a good reference is Reynolds 1984). The Hα emission appears to be widely spread with a disk-like distribution. The intensity projected to the pole *i.e.* $I_\perp$ is found to be between 0.5 R and 1.7 R. Adopting a mean value of 1 R, this translates to an $EM_\perp \sim 2.8$ cm^{-6} pc for $T_4 = 1$ (§1.3). From latitude scans in the Perseus arms, Reynolds (1986) finds that the $|z|$-distribution of EM is well represented by an exponential with a scale height of 300 pc. If we assume that the local scale height of the local WIM is also 300 pc, then the observed $EM_\perp$ requires $\langle n_e^2 \rangle \approx 0.009 e^{-|z|/300}$ cm^{-6}.

The observational situation of the low-frequency absorption is unfortunately uncertain. From observations of extragalactic sources at 10 MHz, Bridle and Venugopal (1969) find $\tau_\perp(10 \text{ MHz}) = 0.1 \pm 0.02$. From observations of the diffuse background, Ellis (1982) derived a value five times smaller towards the south Galactic pole; this is approximately consistent with space-based observations extending to frequencies below 1 MHz. We favor Ellis's result because the absorption effects are more pronounced at the lower frequencies. The optical depth per kpc in the Galactic plane, κ_0, can be estimated by measuring τ towards well-known Galactic sources such as the Crab Nebula *etc.* Comparision of $\tau_\perp$ with κ_0 yields a scale height of about 1 kpc (Bridle and Venugopal 1969).

Irregularities in the interstellar electrons give rise to the phenomenon of interstellar scintillation (ISS). ISS dilates images of compact extragalactic sources, broadens pulsar pulse profiles and cause pulsar signals to vary in time and radio frequency. Since ISS depends upon irregularities (*i.e.* δn_e) and not on n_e it is not clear whether the irregularities arise in the WIM or in the hot component. Either medium has more than enough electrons to account for the deduced δn_e. However, quite surprisingly, Readhead and Duffet-Smith (1975) obtain a scale height of about 1 kpc from measurements of ISS scatter-broadening of compact extragalactic sources. This scale height is similar to that inferred from pulsar DM

data. Thus, there is some circumstantial evidence suggesting that the irregularities indeed arise in the WIM. We favor this interpretation but at the same time agree that more data is needed to firmly prove this point.

2.4.2 Distribution on Non-Local Scales.

Our knowledge of the distribution of the interstellar electrons on non-local scales is meager and mainly comes from pulsar dispersion measure (DM) data. H I absorption data have been used to obtain kinematic distances for about 40 pulsars. These data, in conjunction with the DM data, have been used to establish the Galactic mean electron density distribution (Lyne, Manchester and Taylor 1985):

$$\langle n_e \rangle = [0.025 + 0.015e^{-|z|/70}][\frac{2}{1 + R/R_0}]\text{cm}^{-3}. \tag{2.4-1}$$

The first term describes the extended component with a scale height of about 1000 pc and discussed above. The second term describes, in a statistical way, the contribution by discrete, bright H II regions and hence is of not great interest in this review. [In equation 2.4-1, R_0 has been assumed to be 10 kpc.]

Equation (2.4-1) is valid only within 2 kpc of the sun since most of the pulsars are nearby. Thus there appears to be an increase in the interstellar electron density in the inner Galaxy. This is only to be expected since the density of young stars which are probably responsible for most of the ionization in the ISM (§7.1) increases dramatically in the inner Galaxy. Recently, Clifton and Lyne (1986) have discovered 40 new pulsars, all within 30° of the Galactic center. Determination of distances to these pulsars is crucial and will enable us to quantitatively determine the increase of $\langle n_e \rangle$ in the inner Galaxy. Indeed, if the WIM is powered by radiation from hot stars (§7.1), it is possible that *the WIM could become as important as H I in the inner Galaxy!* It is crucial to test this reasonable suggestion. This interplay between star formation and interstellar medium should also be studied by observations of diffuse $H\alpha$ and H I in external galaxies. However, the detection and measurement of the WIM in external galaxies are challenging observations.

3. Pressure of the ISM.

In steady state models of the ISM in which gas is heated by interstellar radiation field or cosmic rays and cooled by collisional deexcitation, interstellar gas can exist either as cold clouds or warm, neutral hydrogen. This is essentially the heart of all steady state models (Field, Goldsmith and Habing 1969). However, theoretically these two phases can *coexist* over a narrow range of pressures. In the supernova-dominated model of McKee and Ostriker (1977), interstellar pressure is predicted, given input parameters such as the supernovae rate *etc.* Clearly, interstellar pressure is a critical parameter for any model of the ISM.

A note on nomenclature: pressure is usually defined as $P = knT$ in standard physics textbooks; here, it is more convenient to drop k, the Boltzman constant and redefine $P = nT$. Thus the units of pressure will be cm^{-3} K. Note that n is the total particle density and thus for a completely ionized medium $P = 2n_eT$.

There are three methods by which pressure has been determined:

a. Excitation of the C I fine-structure lines.

The ground electronic state of C I is split into three fine-structure levels; the transitions between these levels lie in the far-infrared. Only recently has one of these transitions been detected (see Phillips, this volume). Jenkins, Jura and Lowenstein (1983) measured the fractional population of these levels by observing the ultraviolet transitions between the ground and some excited electronic states. The level-population is determined by a balance between collisional excitation and radiative decay; the collisional rate depends both on temperature and density. From the fractional level-populations, Jenkins et $al.$ obtained an estimate of interstellar pressure. Most of the gaseous carbon in the ISM is found as C II since the ionization potential of C I is less than 13.6 eV. Thus, this pressure 'probe' is biased towards high density regions where the fractional abundance of C I is higher than in lower density regions. In short, the pressure determinations from the C I lines are insensitive to the WNM and the WIM and apply $mainly$ to the $cold$ $clumps$ in the CM (see §5.3 for discussion about the clumps). Jenkins et $al.$ found that the pressure in most clumps is between 10^3 cm^{-3} K and 10^4 cm^{-3} K. About 6% of the gas is at pressures greater 10^4 cm^{-3} K and one third is below 10^3 cm^{-3} K. We adopt a representative median pressure of 4000 cm^{-3} K.

b. Excitation of the C II fine-structure line.

The ground electronic state of the C II ion is split into two fine-structure levels with a separation (in units of temperature) of 92 K or a wavelength of 157.7μm. Just like the C I lines, this transition is a nice probe of pressure. Additionally C II, unlike C I, is a majority species and hence is found in all the phases of the atomic ISM. Thus it is a better probe of the $average$ pressure of the ISM.

The C II 157.7μm line is the primary cooling line of diffuse clouds. Not surprisingly, the fraction of excited C II ions is usually stated in terms of the cooling rate per unit hydrogen nucleus: $l_C = f_{21} x_C A_{21} (h\nu)$ erg s^{-1} nucleon^{-1} where f_{21} is the fraction of C II in the excited state, x_C is the fractional abundance of gaseous carbon, $A_{21} = 2.36 \times 10^{-6}$ s^{-1} is the Einstein A-coefficient for the transition and ν is the frequency of the C II line. The cosmic abundance of carbon is thought to be 4×10^{-4} (Spitzer 1978). We will assume that about half of this is depleted onto grains and thus $x_C \sim 2 \times 10^{-4}$.

In the diffuse ISM, C II is excited by collisions with electrons and hydrogen atoms and cools by radiative decay. In the WIM, collisions with electrons dominate and $f_{21} \sim 10^{-2} (n_e/0.33)(T/10^4)^{-1/2}$. For our nominal parameters of the WIM ($n_e \sim 0.25$ cm^{-3}, $T \sim 8000$ cm^{-3} K; see §7), we find $f_{21} \sim 8 \times 10^{-3}$ or $l_C \sim 0.5 \times 10^{-25}$ erg s^{-1} nucleon^{-1}. In the CM, both electronic and atomic collisions are important. A proper calculation of f_{21} for the C II in the CM has to take into account of the existence of CM material over a substantial range of temperature (§5.3). Assuming a cosmic ray ionization rate, $\zeta \sim 10^{-16}$ s^{-1} (§6.1) an estimate of l_C as a function of temperature for various values of pressure is shown in Figure (3-1). This figure shows two important features of the C II line: (i) for a given cosmic ray ionization, ζ_{CR}, $l_C \propto P$, the gas pressure, and (ii) at any given P, l_C varies by less than a factor of 2 for $40 < T < 400$ K, the temperature range spanned by the CM. Thus the C II line is a robust estimator of pressure. Surprisingly, for our nominal median pressure (4000 cm^{-3} K), l_C for the CM is nearly the same as that for the WIM. We do not expect much C II emission from the WNM since both n_H and n_e are small.

There are two measurements of f_{21}: one by direct observations of the far-infrared line itself (Stacey et $al.$ 1985) and the other from ultraviolet absorption observations of the excited C II ions (Pottasch, Wesselius and van Duinen 1979). Stacey et $al.$ detected C II

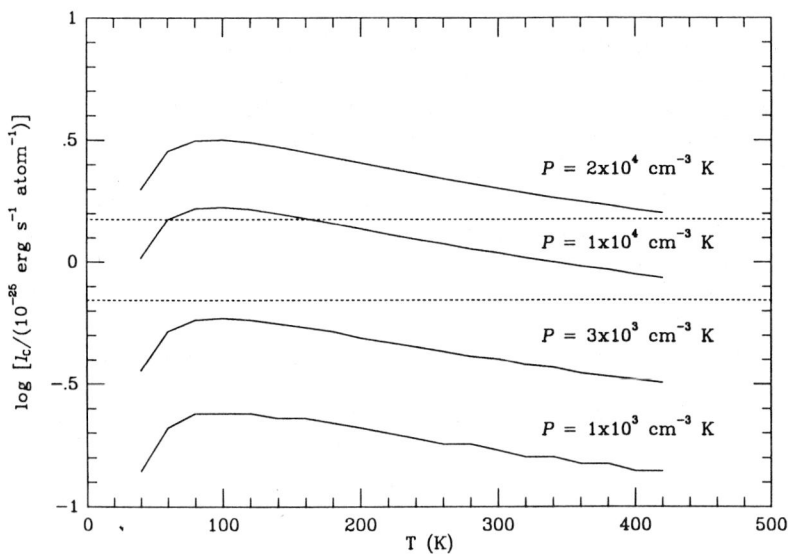

Figure 3-1. Plot of the C II cooling rate per nucleon in cold clouds as a function of temperature for four different values of gas pressure (P). Carbon is assumed to be depleted by 50% and a cosmic ray ionization rate of 10^{-16} s^{-1} is assumed.

emission from the warm interfaces of giant molecular clouds and thus their results are not relevant to our discussion here.

Pottasch *et al.* found the surprising result that $l_C \sim 10^{-25}$ erg s^{-1} nucleon^{-1} along eight different lines of sight; the variation between the lines of sight was less than a factor of 2. Some of these lines of sight are nearby (~ 100 pc) and are dominated by the WIM whereas others are quite distant with many cloud components. $l_C \sim 10^{-25}$ erg s^{-1} nucleon^{-1} corresponds to $P \sim 8000$ cm^{-3} K for both the CM and WIM. This is in excess by a factor of 2 over the mean pressure of 4000 cm^{-3} K inferred from the C I measurements. The discrepancy gets worse when a correction is made for the low C II emission from the WNM; the precise factor depends upon the z-distance of the star and is between 1.4 and 2.

The mean pressure deduced from the C II measurements appears to be larger than the median pressure obtained from C I data by a factor between 2 and 4. Agreement can be obtained (barely) by assuming that $x_C \sim 4 \times 10^{-4}$ which essentially means that there is very little depletion of carbon in the diffuse ISM. The existence of grains can be used to show that the CNO elements (in some combination) must be depleted by at least 25% (see Jenkins, this volume). Independent estimates of the depletion of carbon are needed to resolve this discrepancy. However, the C II UV lines are optically thick and there is some doubt about the accuracy with which the column density of the excited C II ions can be measured (see pitfall #1 in Jenkin's contribution; also Pottasch *et al.* do caution us about this problem). All this discussion shows that the best way to probe the pressure in the ISM is by observations of the 157.7μm line directly in the far infrared.

c. Comparison of Emission Measure and Dispersion Measure.

We postpone the detailed discussion of this method to §8. The basic idea is that the emission measure per unit volume is equal to ϕn_e^2 whereas the dispersion measure per unit volume is equal to ϕn_e; here ϕ is the filling factor of the WIM and n_e is the true volume density of the WIM. From the observed emission and dispersion measures, one obtains $n_e \sim 0.25$ cm^{-3}. Using the nominal temperature of 8000 K for the WIM and assuming it is fully ionized (see §7) we derive a mean WIM pressure ~ 4000 cm^{-3} K.

To summarize, the median pressure in the CM material appears to be about 4000 cm^{-3} K – about the same as the mean pressure in the WIM. There is no direct measurement of the pressure in the WNM.

4. The Warm, Neutral Medium (WNM).

Broad (velocity dispersion, $\sigma_v \sim 9$ km s^{-1}) H I emission is present along all lines of sight at intermediate and high latitudes; in contrast, narrow emission ($\sigma_v \lesssim 5$ km s^{-1}) features are seen in only one third of the directions. Only these narrow features show detectable absorption. Since the optical depth is inversely related to temperature, the narrow component can be identified with the cold medium (CM) and the broad component with the warm, neutral medium (WNM). The existence of the WNM is beautifully illustrated in Figure (4-1) which shows a set of high latitude emission and absorption spectra. The narrow absorption features arise in cold H I 'clouds' *i.e.* the CM.

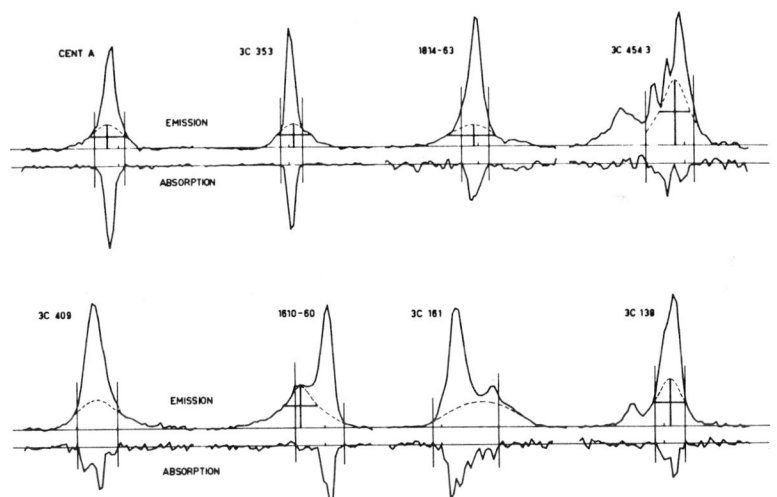

Figure 4-1. A set of high latitude emission and absorption spectra from the Parkes survey (Radhakrishnan *et al.* 1972).

In Figure (4-1), pairs of vertical lines mark velocity ranges within which detectable absorption ($\tau \gtrsim 0.1$) is present. There is clearly considerable H I outside this velocity range which does not show absorption. The broad dashed line is a fit to this optically thin component; this is the operational definition of the WNM. Measurement of the temperature of this broad component is difficult because the opacity is so low. Signal-to-noise ratios better than 10^3 are required to measure the optical depth of the WNM and has been

achieved only towards a handful of the strongest background sources; these results are
discussed in §4.2.

Most of our information about the WNM have come from H I emission studies. The
WNM can be best probed by high-resolution UV absorption lines of warm, neutral species
such as H I, O I, N I, *etc.* and by far infrared cooling lines in emission (C II, Si II, Fe II
etc.) – probes which, unfortunately, are not accessible from ground-based facilities. Owing
to the low density and low ionization in the WNM, even the far infrared cooling lines will
be faint and detectable only by extremely sensitive far infrared telescopes.

Locally, the WNM is widely distributed and constitute about 40% of interstellar
hydrogen. The local WNM has been extensively studied from H I emission data, the detailed
distribution of which is discussed in §4.1. On Galactic scales, not much is known about the
distribution of the WNM; the meager data we have suggests that the WNM is a significant
constituent everywhere in the Galaxy outside ~ 8 kpc Galactocentric radius (Heiles 1980).
Inside $R = 8$ kpc, there are no data.

4.1 Dispersion and Scale Height of the Local WNM.

Mebold (1972) decomposed about 1200 emission spectra at $b \sim 30°$ and $0° < l < 360°$
into narrow ($\sigma_v < 5$ km s^{-1}) and wide ($5 < \sigma_v < 17$ km s^{-1}) Gaussian components
and found: (1) the wide-σ_v H I has a scale height of about 200 pc and (2) an intrinsic
velocity dispersion of about 8.8 km s^{-1}, corresponding to an upper limit of 9600 K for the
temperature of the WNM. These are the 'classical' parameters for the WNM. However,
these values are at variance with those found by Lockman (1984) in the inner Galaxy
(§2.2). To recapitulate, Lockman found a *three* component fit is necessary to adequately
describe the vertical structure of the H I layer. Lockman's component 1 has the same scale
height as the CM and component 2 corresponds to the 'classical' WNM layer. Analyses
of the kinematics of the H I emission spectra, such as by Mebold (1972), are sensitive to
perturbations of the local velocity field and hence the failure to detect the long exponential
tail (*i.e.* component 3) is, in hindsight, understandable. Since there is no evidence for
a long exponential tail in the various H I absorption surveys we have to conclude that
the temperature of Lockman's component 3 is significantly higher than that of the clouds.
The few temperature measurements of the WNM that are available (see §4.2, below) do
not distinguish between H I belonging to component 2 and component 3. Hence, it is not
unreasonable to assume that the temperatures of the two components are the same *i.e.*
$T \gtrsim 6000$ K (§4.2).

It is not clear whether Lockman's three components are merely mathematical artifacts
or physically meaningful components. Perhaps, the WNM is made of gas with a range of
velocity dispersions and hence a range of scale heights. From an analysis of local H I in
emission, Kulkarni and Fich (1985) found that $N_H(v) \propto v^{-2}$ where $N_H(v)$ is the column
density of H I with velocity v (note that since they used H I data at high latitudes, there
is negligible contribution to v from Galactic rotation). Curiously, there appears to be an
equipartition of kinetic energy per unit velocity interval, *i.e.* $\frac{3}{2}[dN_H(v)/dv]v^2$ is a constant!

From equation (2-1) we expect that the higher dispersion gas should have the larger
scale height. In Figure (4-2) we have plotted the rms angular scale height, $\langle b^2 \rangle^{1/2}$ of H I
at a tangential point. Note the clear presence of some H I well in excess of the tangential
velocity. The velocity in excess of the tangential velocity is simply a reflection of the

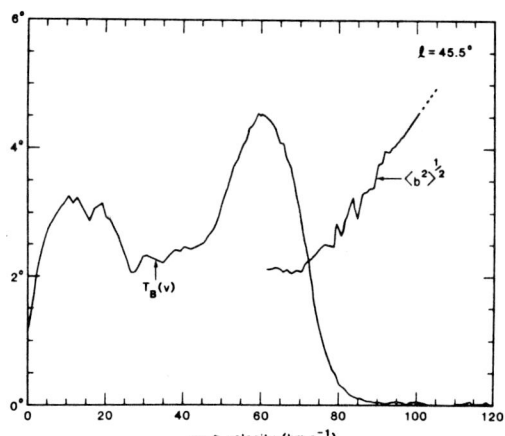

Figure 4-2. A plot of the angular scale height of H I, $\langle b^2 \rangle^{1/2}$ as a function of velocity along a line of sight in the inner Galaxy. The H I emission profile in that direction is superimposed as an aid in locating the tangential velocity.

velocity dispersion of the gas. Figure (4-2) clearly demonstrates that $\langle b^2 \rangle^{1/2}$ increases with σ_v, implying a continuum of scale heights.

4.2 Temperature of the WNM.

At present the temperature of the WNM can be estimated using two different techniques: 21-cm emission and absorption studies and UV-line absorption studies of neutral species. So far, the 21-cm technique has contributed most of the information. The basic method is to measure $\tau(v)$, the optical depth of the WNM and use equation (1.2-3) to estimate the temperature. However, the measurement is difficult because $\tau \propto T^{-1}$, the temperature. This difficulty is compounded by two biases: (i) the presence of a tiny amount of cold H I will bias the derived spin temperature, T_n to smaller values (§1.2.2) and (ii) stray radiation will increase T_n (Kalberla, Mebold and Reich 1980). The latter problem is only a technical hinderance and is surmountable. Since nothing can be done about the former bias, the measured T_n should always be considered as a lower limit.

Many attempts have been made to measure the optical depth of the WNM. However, in most cases no absorption has been detected so that only upper limits to τ, and thus many lower limits to T_n are available. Many of these lower limits lie in the neighbourhood of 3000 K or below (e.g. Mebold et al. 1982). The highest measured temperature of the WNM is $\sim$ 6000 K towards the strong, low-latitude source Cygnus A (Kalberla et al. 1980). The highest 3σ lower limit is 10^4 K, towards 3C 123 (Kulkarni, Dickey and Heiles 1985); however this needs to be confirmed by independent observations before it can be accepted.

From a statistical analysis of the sensitive Arecibo survey Payne, Salpeter and Terzian (1983) find $T_n \sim 5000$ K for gas not explicitly associated with absorbing gas. Even the Arecibo survey did not have enough sensitivity to directly detect the WNM in absorption along any one single line of sight. So Payne et al. averaged all the velocity channels which could not be explicitly associated with an absorption feature. Payne et al. found that this gas amounted to 50% of the local gas and had a mean spin temperature, $T_n \sim 5000$ K. We would like to remind the reader, again, that T_n (§1.2.2) is the harmonic mean temperature and hence is a lower limit to the true temperature. Thus the straightforward conclusion of the analysis of Payne *et al.* is that $\sim 50\%$ of the local gas is the WNM and the other 50% is the CM. The temperature of the WNM is probably not greater than 10^4 K for two reasons: (a) a good fraction of the WNM has a dipersion of ~ 9 km s^{-1}, corresponding to a maximum temperature of $\lesssim 10^4$ K and (b) cooling by hydrogen increases dramatically for $T \gtrsim 10^4$ K (see discussion of the cooling curve in Black's contribution).

The study of UV absorption lines of various ions of the WNM can, in principle, yield the temperature. Several papers presented at the IAU Colloquium on the Local Interstellar Medium (*e.g.* York and Frisch 1984) find $T \gtrsim 6000$ K for a handful of nearby (distance $\lesssim 200$ pc) stars. This technique appears to be reliable and needs to be applied in a large number of directions.

To summarize, the WNM cannot be adequately described by a single component (*i.e.* the 'classical' model). Instead, two components with differing scale heights and dispersion appear to provide a reasonable fit. It is not clear whether these two components are mathematical fits or physically meaningful components. Indeed, one may argue that the WNM is best described by a continuum of scale heights and dispersion. Several observations suggest that the temperature of the WNM is $\gtrsim 6000$ K. Theoretically the temperature cannot exceed 10^4 K. However, temperature data are sparse and we definitely need more measurements.

5. The Cold Medium (CM).

The cold medium (CM) is easily seen in absorption spectra. The narrow-σ_v features seen in absorption in Figure (4-1) have large optical depths and hence are cold. Each absorption dip is assumed to arise in an 'H I cloud' or a 'diffuse cloud'. The nomenclature probably arose from the similarity of the H I absorption spectrum with the optical absorption spectrum such as that of Na I and Ca II.

The nomenclature is justified since the emission features corresponding to the narrow absorption features appear to be confined to discrete structures in the sky; this should be contrasted with the nearly featureless distribution of the broad-σ_v, low-optical depth features which constitute the warm, neutral medium. In one of the earliest models of the ISM (Field, Goldsmith and Habing 1969), the cold medium was distributed into clouds with the WNM filling the space between the clouds. For this reason the WNM is sometimes referred to as the 'intercloud medium'.

At high latitudes, where the emission spectra are simple, absorption features can be typically associated with the corresponding emission features. The absorption features are found to be well represented by Gaussian functions in $\tau(v)$, the optical depth. The decomposition of emission features is somewhat subjective; some observers fit Gaussian profiles whereas others find Gaussian fits inadequate. The distribution function of $\sigma_v(\tau)$, the velocity dispersion of the absorption features, peaks at about 0.75 km s^{-1}; $\langle \sigma_v(\tau) \rangle \sim 1.7$ km s^{-1} and is significantly larger than the peak value because the distribution function has a long tail. In contrast, the emission features are twice as broad as the corresponding absorption features; they peak at ~ 2.2 km s^{-1} (Crovisier 1981). No cloud has been observed with $\sigma_v(\tau) < 0.4$ km s^{-1} and very few with $\sigma_v(\tau) > 4$ km s^{-1}. The lower limit is probably unaffected by observational biases.

For every 'emission-absorption' pair we can obtain the naively-derived spin temperature as a function of velocity (see equation 1.2-3). In almost all cases, $T_n(v)$ is not a constant as would be expected for an isothermal cloud; instead, T_n changes increases on either side of the velocity at which maximum absorption occurs. This is a simple consequence of the differing line widths of the emission and the absorption features. The simplest and in fact the correct interpretation of the velocity dependence of $T_n(v)$ is that the *H I clouds are not isothermal blobs*. The implications of this result are discussed in §5.3. Traditionally, the smallest $T_n(v)$, *i.e.* T_n at the velocity of maximum absorption, is called the spin temperature of the cloud. We will denote this temperature by $T_{n,min}$ and the peak absorption by τ_{max}.

The radial velocity of diffuse clouds can be measured to great precision because of the narrowness of the absorption features. The observed radial velocity of clouds is due to Galactic differential motion and cloud-cloud velocity dispersion. The differential motion depends upon the distance to the cloud as well as the direction. With sufficient data, the mean height, $\langle |z| \rangle$ of the clouds can be extracted from the absorption data. Belfort and Crovisier (1984) find $\langle |z| \rangle \sim 100$ pc. The scale height appears to be a function of the optical depth: $\langle |z| \rangle \sim 88 \pm 13$ pc for $\tau_{max} > 0.1$, while that for $\tau_{max} < 0.1$, $\langle |z| \rangle \sim 229 \pm 48$ pc. The T-τ relation enables us to conclude that *the clouds get warmer at higher* $|z|$ (see §5.3 for an alternative explanation).

5.1 The 'T-τ' Relation.

Lazareff (1975) first noted that $T_{n,min}$'s are inversely related to τ_{max}. This the famous 'T-τ' relation. A fit to the sensitive Arecibo data yields:

$$T_{n,min} = T_0(1 - e^{-\tau_{max}})^{-\alpha} \qquad (5\text{-}1)$$

with $T_0 = 55 \pm 7$ K and $\alpha = 0.34 \pm 0.05$ (Payne, Salpeter and Terzian 1983). Usually $T_{n,min}$ and τ_{max} are plotted on a log-log plot and one does not realize that the scatter in this relation is quite large. Values of $T_{n,min}$ range from 20 K to about 250 K and those of τ_{max} from 0.01 to ~ 2. The lowest observed τ_{max} is limited by sensitivity. There is probably a real cutoff since there is a dearth of emission features with widths between the WNM and the cold clouds.

5.2 Statistics of Diffuse Clouds.

The procedure to derive the statistics of absorption features has been nicely presented by Crovisier (1981) and applied to Arecibo data by Payne, Salpeter and Terzian (1983). The probability of finding an absorption peak with $\tau_{max} > \tau$ along a line of sight, reduced to $|b| = 90°$, is well represented by $P(\tau > \tau_{max}) = 0.3\tau_{max}^{-0.4}$. At $b = 0°$, this translates to

$$N(\tau > \tau_{max}) = 3.0\tau_{max}^{-0.4}, \qquad (5\text{-}2)$$

where $N(\tau > \tau_{max})$ is the number of absorption features (*i.e.* clouds) per kpc with the maximum optical greater than τ_{max}. In order to convert the optical depths to temperature and column density, two assumptions are made *viz.* (i) the T-τ relation is used to yield the temperature and (ii) the emission line width is assumed to be 1.3 times the thermal width of the absorption feature. With these assumptions, the corresponding probabilities for $T_{n,min}$ and N_H are: $P(T < T_{n,min}) \propto T_{n,min}^{-1.2}$ and $P(N > N_{cloud}) \propto N_{cloud}^{-0.8}$. Both of these derived probabilites are consistent with the ones derived directly by Payne *et al.* (1983). At $b = 0°$, the latter relation translates to the number of clouds per kpc with $N_H > N_{cloud}$, $P(N > N_{cloud}) = 4.7 N_{cloud}^{-0.8}$. Here N_{cloud} is the column density expressed in units of 10^{20} cm^{-2}. These probability relations are valid over the range $\tau_{max} \simeq 0.02$ to 1.0; this corresponds to ranges of temperature and column density of 210 K to 84 K and 0.32×10^{20} cm^{-2} to 2.2×10^{20} cm^{-2}, respectively. The median cloud has $\tau_{max} = 0.11$, $T_{n,min} = 115$ K and $N_H = 0.8 \times 10^{20}$ cm^{-2}.

The distribution of column densities of diffuse clouds has been obtained using two other methods: interstellar reddening and optical absorption lines. These methods are ill-suited for cloud statistics. The observed reddening is simply the *integrated* column density of dust. For this reason, the analysis is necessarily crude: the spectrum of diffuse clouds is approximated by 'standard' and 'large' clouds (Spitzer 1978). However, in view of the large range of $N_{H\ I}$ that is actually observed, we seriously question the value of this crude approximation. For example, the *median* H I cloud has $N_{H\ I} = 0.8 \times 10^{20}$ cm^{-2}, nearly four times smaller than a 'standard' cloud. *Thus the 'standard' cloud is not even representative of the cloud population!* The reddening of a median H I cloud is so small ($E_{B-V} = 0.017$ mag.) that reddening data cannot be effectively used to study diffuse clouds.

Hobbs (1974) has attempted to use K I absorption lines to measure the column density distribution function. There are two problems in using K I lines for this purpose: (1) K I lines are weak and hence are tracers of large column density clouds; with typical detectors, a median cloud is not detectable in the K I line. (2) K I absorption lines do not measure $N_{H\ I}$ directly. Hobbs uses an empirical quadratic relation between $N_{H\ I}$ (obtained from Ly α measurements) and $N_{K\ I}$ to determine $N_{H\ I}$. We question this scheme since the Ly α-derived $N_{H\ I}$ refers to the total column density of H I, which may include more than one diffuse cloud and definitely includes a substantial contribution from the WNM. In order to obtain $N_{H\ I}$ from K I data properly, a knowledge of depletion of metals and electron density in clouds is needed.

It has been popular among theorists and observers alike to derive the cloud 'size-spectrum' by using the observed column density distribution function together with the assumptions of (i) constant volume density and (ii) spherical clouds. Assuming pressure equilibrium, the observed spectrum of temperatures translates to a range of densities – making assumption (i) highly questionable. In §5.4 we show that assumption (ii) cannot be justified observationally. Thus one should reject results based on these false assumptions.

5.3 Internal Structure of Clouds.

In §5.1 we noted that emission widths are typically twice as large as the corresponding absorption widths. A straightforward consequence of this is that the naively-derived temperature, $T_n(v)$, will vary across the absorption feature. Two explanations are possible: (i) different portions of the cloud are being viewed in emission compared to absorption or (ii) H I clouds are not isothermal and that the variation in $T_n(v)$ is real. The first hypothesis has been tested quite thoroughly and rejected (see Liszt 1983 for a summary). So we have to conclude that H I clouds are indeed not isothermal structures (Payne, Salpeter and Terzian 1983; Liszt 1983). This is very puzzling since theoretically we expect H I to be either warm ($T \sim 8000$ K) or cool ($T \sim 40$ K; see Draine 1978 or Field, Goldsmith and Habing 1969). The sound-crossing time across 1 pc of a diffuse cloud is only 10^6 y. Unless diffuse clouds are disturbed more frequently than every million years, we expect pressure equilibrium in diffuse clouds. Hence, temperature variations should give rise to density variations.

A number of observers have used double-lobed radio sources as background sources to study the internal structure in diffuse clouds and find very little H I structure on scale lengths below 1.5 pc (see Crovisier, Dickey and Kazes 1985). Internal structure can also be studied by high-resolution mapping of H I in emission. This is best accomplished by aperture synthesis techniques, although the process is laborious and difficult. Kalberla, Schwarz and Goss (1985) have mapped a field containing the bright radio source 3C 147. Even at the small angular scales of their map, extended filaments and/or sheets dominate the distribution of H I. Occasionally, there are sharp edges, probably indicative of shocks in the gas. Also seen are small clumps within the filaments and/or sheets.

The presence of 3C 147 allowed Kalberla *et al.* to derive temperatures independently for the various H I components. They conclude:

(1) The clumps, which are imbedded in the filaments and/or sheets, are primarily responsible for H I absorption. They have densities $T_{cl} \sim 20$ to 50 cm^{-3} and spin temperatures between 30 K and 80 K, averaging 40 K.

(2) The filaments and/or sheets are associated with lukewarm (say, $T \gtrsim 500$ K) H I envelopes; the envelopes account for 80% of the H I emission.

The 3C 147 field is non-ideal in some respects. It is at low latitude ($b \sim 10°$) and in the anti-center direction. Thus the distance estimate is highly uncertain. We urgently need more high resolution studies of H I in emission to understand the topology of the internal structure of diffuse clouds.

The non-isothermal and inhomogeneous nature of diffuse clouds has profound effect on the interpretation of not only 21-cm data but other data as well. Consider two examples:

(1) The distribution of majority species (such as C II) and minority species (such as C I) will be different. In particular, C I will be biased towards the colder regions. Thus the ratio, $n_{C\ I}/n_{C\ II}$ will vary throughout the cloud, complicating the interpretation of the *observed* ratio. Optical and UV absorption line studies should pay particular attention to this problem.

(2) The good agreement between the rotation temperature of H_2, derived from UV studies (Spitzer 1978), with the apparent spin temperatures, $T_{n,min}$ should be considered fortuitous since we now know that the temperature of the cold clumps, $T_{cl} \sim 40$ K. This implies that substantial amounts of H_2 exists outside the clumps.

Given the fact that clouds have temperature structure, how should the T-τ relation be interpreted? Liszt (1983) has simulated various cloud models and concludes that basically the T-τ relation does not constrain any specific cloud model. It is fair to say that the T-τ relation is basically satisfied by any class of reasonable, *inhomogeneous* cloud models. One such model, possibly the simplest, is the 'clump-envelope' model advocated by Payne *et al.* (1983). In this model, the clumps are assumed to be at a fixed *spin* temperature ($T_{s,cl}$) and variable optical depth (τ_{max}) and surrounded by a lukewarm envelope of constant column density, contribution a fixed *brightness* temperature ($T_{B,env}$). This model fits the observed data as good as the T-τ relation does! The best fit parameters are $T_{s,cl} = 55$ K and $T_{B,env} = 4$ K. Incidentally, in this simple model, the parameter $T_{B,env}$ determines the exponent α of the T-τ relation (equation 5-1).

In §5 we learned that the mean height of the cloud layer, $\langle |z| \rangle$ is a function of τ_{max}. In view of the 'clump-envelope' model, two equally plausible interpretations are (i) the column density of the envelope increases with $|z|$ or (ii) the temperature of the clumps increases with $|z|$ (or both!).

All this discussion emphasizes that we need an unambiguous verification of the 'clump- envelope' model. One way is through high-resolution H I maps. The best way, which must await proper spaceborne instrumentation, would employ mapping of the far-IR cooling lines.

If we believe in thermal equilibrium models (Field, Goldsmith and Habing 1969) then the existence of lukewarm H I is a mystery. Thus either we have to give up steady state models altogether or most of H I is in a transient state. In our opinion, this challenging set of problems needs immediate theoretical and observational attention.

5.4 Shapes of H I clouds.

A survey of the literature shows that most workers assume that clouds are spherical. However, even a cursory look at the 21-cm sky will convince the reader that this is patently untrue. Filamentary or sheet-like features abound in the 21-cm sky. Since the evaporation rate depends upon the topology and the surface area of the clouds, the topology of clouds is an important parameter in global models of the ISM. The standard assumption of spherical clouds results in artificially decreasing the evaporation rate since the evaporation rate of spheres is smaller than that of flat objects such as sheets.

Given the complexity of the 21-cm sky, a *quantitative* study of H I cloud shapes is not easy. There are only two quantitative studies – one by Heiles (1967) and the other by Schwarz and van Woerden (1974). Heiles stuided a $40° \times 4°$ region of sky. The gas in this region could be resolved into three physical components: a diffuse, smooth background and two large sheets with a rift in between. The velocity of each sheet is highly ordered over many tens of degrees on the sky. Small concentrations of gas exist within these sheets; however, the density within these concentration is only twice as much as in the sheets.

In another study, Schwarz and van Woerden (1974) mapped a $20° \times 10°$ region of sky with a beam of 36 arcmin and one degree spacing. The emission profiles were decomposed into Gaussian components. A cloud was then defined to be a concentration in (l, b, v, σ) space; here V and σ are the mean velocity and the dispersion of the Gaussian components. The above procedure furnished a total of 88 clouds, of which 10 appeared to belong to large-scale features including the intercloud medium. The most striking characteristics of the clouds are their filamentary shapes (see Figure 1 of their paper). The axis ratio varied from 2 to $\gtrsim 30$; in fact, in a substantial number of cases, the minor axes of the filaments were unresolved.

Clearly, both these studies are subjective to some extent: the first one to visual impressions and the second one to the dangers associated with Gaussian analysis. While we do not offer any specific panacea, we do wish to emphasize that the detailed studies of specific regions of the sky have not yielded the postulated spherical clouds. Diffuse clouds are not self-gravitating objects. Thus the observed filamentary shapes suggest that magnetic fields (see Heiles, this volume) and/or evaporation play an active role in shaping the cloud.

6. Ionization in the Atomic ISM.

The presence of free electrons in the ISM is clearly revealed by the dispersion of pulsar signals and by wide spread diffuse Hα emission. The ionizing power needed to balance the observed recombination rate is comparable to the total Galactic supernovae power! Clearly, the ionizing agent has a great impact on the ISM. In this section, we discuss the ionization in the cold medium (CM) and the warm, neutral medium (WNM). We come to the very important conclusion that most of the electrons reside *not* in the CM or WNM but *in* the warm, ionized medium (WIM). The energetics and distribution of WIM are discussed in the next section (§7).

If interstellar electrons predominantly arise in the WIM then why should we study the ionization in the CM and the WNM? One good reason is that the cooling rate depends sensitively on x_e, the fraction of electrons: in the CM, cooling by electronic excitations dominate over atomic collisions for $x_e \gtrsim 10^{-3}$. Also quantitative measures of x_e are needed to interpret the abundances of ions. In the past, it was thought that the atomic ISM was heated and ionized by a single agent (*e.g.* Field, Goldsmith and Habing 1969) and hence the origin of interstellar electrons was a very important issue. Increasingly, there is much evidence, albeit indirect, that the neutral phases are probably heated by dust via the photoelectric process.

First we discuss the two traditional ionizing agents: energetic particles (*i.e.* cosmic rays; §6.1) and energetic radiation below 912 Å, the Lyman limit (*i.e.* soft X-rays; §6.2). The intensities of soft X-rays and the low energy cosmic rays are either difficult or even impossible to measure directly. Consequently, it is important to clearly understand the basis on which *estimates* of the ionization rates are made; these issues are discussed in §6.3 (the CM) and §6.4 (the WNM). A brief note on notation: traditionally, the symbols ζ_{CR} and ζ_{XR} are used to denote the ionization rate per H I atom. The number of ionizations per unit time per unit volume is simply $\zeta n_{H\ I}$.

6.1 Cosmic Rays.

The cross-section for ionization of hydrogen by cosmic rays varies inversely as the square of the velocity of the ionizing nucleon. Hence the ionization rate is dominated by non-relativistic protons and heavy ions ($E \ll 1$ GeV/nucleon). Unfortunately the solar wind, with its trapped magnetic field, tends to sweep the lower-energy particles back out into interstellar space, necessitating a large upward extrapolation to the observed intensity of cosmic rays incident at earth. Spitzer and Tomasko (1968) made the *ad hoc* assumption that the interstellar cosmic ray intensity is twice that measured at Mars (the solar wind is weaker at Mars as compared to Earth and hence a more desirable location for such observations). Additionally, the interstellar cosmic ray spectrum was assumed to decrease sharply for energies smaller than 40 MeV/nucleon – the lower energy limit of the detector on the space probe at Mars. With these assumptions, they derived a "minimum" ionization rate $\sim 10^{-17}$ s^{-1}.

Historically, $\zeta_{CR} \sim 10^{-15}$ s^{-1} was invoked to explain the heating of diffuse clouds (see Spitzer and Tomasko 1968). This rate can be obtained by extending the observed slope (*i.e.* the relativistic portion) down to a lower limit of 40 MeV/nucleon. The cosmic ray pressure would then be 2×10^7 cm^{-3} K, nearly 10^4 the mean ISM pressure! In order to avoid this embarassing situation, very low energy (2 MeV/nucleon) cosmic rays were invoked (see Spitzer and Tomasko 1968) which would contribute negligible pressure but provide sufficient ionization. The very low energy cosmic rays were supposed to originate in the expanding shells of fast moving Type I supernova shells. However, there is no firm observational basis for expecting such a high ζ_{CR} and the justification for a high ζ_{CR} was weak, even when viewed in the historical context. Additionally, the short lifetime of such low energy particles limits their range (Spitzer and Jenkins 1975) – making low energy cosmic rays useless for ionizing large regions of the ISM. From energetic arguments alone, ζ_{CR} cannot be much higher than 10^{-15} s^{-1} if cosmic rays are produced by supernovae related processes.

Cosmic ray ionization is the chief source of protons in the interiors of diffuse clouds. Hence by measuring ζ_{CR} the abundance of certain molecules whose formation critically depends upon the proton concentration can be estimated. Values of ζ_{CR} ranging from 10^{-17} s^{-1} to 10^{-15} s^{-1}, implying a typical value of 10^{-16} s^{-1} are inferred from the UV-derived abundance ratio [HD/H$_2$] in diffuse clouds (Barsuhn and Walmsley 1977). An upper limit of 3.5×10^{-16} s^{-1} has been derived from the radio-derived abundance ratio [DCO$^+$/HCO$^+$] in three molecular clouds (Guelin, Langer and Wilson 1982). In a comprehensive modelling of four diffuse clouds, van Dishoeck and Black (1986) had to invoke $\zeta_{CR} \sim 7 \times 10^{-17}$ s^{-1} to explain the observed OH abundance. From the absence of a low-frequency recombination line of hydrogen towards 3C 123, Payne, Salpeter and Terzian (1984) obtain an upper limit of 4×10^{-17} s^{-1} for gas associated with the Taurus dust cloud and an upper limit of 1.6×10^{-16} s^{-1} for the diffuse clouds in that direction.

These astrophysical estimates apply to either *cores of diffuse clouds* or *molecular* clouds. Many authors have questioned whether low energy cosmic rays can penetrate clouds. Cesarsky and Volk (1978) have shown that cosmic rays with energies as low as tens of MeV/nucleon are not hindered by instabilites in penetrating dense molecular clouds. Indeed, cosmic ray ionization forms the cornerstone of gas-ion chemistry in dense molecular clouds and a value approximately equal to the 'minimum' ionization rate is invoked to explain the observed abundances of some molecules (see Watson 1978).

To summarize, the minimum value of ζ_{CR} based on a conservative (and somewhat *ad hoc*) correction of the observed cosmic ray spectrum and an arbitrary low energy cutoff is $\sim 10^{-17}$ s^{-1}. Indirect evidence suggests that an ionization rate close to this minimum is present even in the cores of dense molecular clouds. ζ_{CR} in diffuse clouds appears to be higher than this minimum by about an order of magnitude. There is no firm estimate of ζ_{CR} in the WNM; however ζ_{CR} cannot be larger than 10^{-15} s^{-1} if cosmic rays are produced by supernova related processes.

6.2 Soft X-rays.

The cross-section for photoionization of H I atoms is $6 \times 10^{-18}(\lambda/\lambda_0)^3$ cm^{-2} where $\lambda_0 = 912$ Å, the Lyman edge; thus low energy X-rays ('soft' X-rays) dominate ionization. A column density of 1.6×10^{17} cm^{-2} will attenuate photons at the Lyman edge significantly. Given the mean density of ISM of ~ 1 cm^{-3}, this column density translates to a distance of only 0.056 pc! No wonder our knowledge of the soft X-ray radiation field is small.

In the three-phase model (McKee and Ostriker 1977), most of the interstellar volume is occupied by the hot, ionized medium. This medium, by virtue of its high temperature ($T \sim 10^6$ K), radiates X-rays (see Savage, this volume). Draine (1978) estimates, at $z = 0$ pc, a soft X-ray ionization rate of 5×10^{-16} s^{-1} at $z = 0$ pc with a mean photon energy of ~ 74 eV. This rate has been calculated assuming a filling factor of nearly unity for the hot component. Hence, it is a maximum value since in §8 we show that the filling factor of the atomic phases (*i.e.* the WNM, the WIM, and the CM) is $\sim$50%. Also note that soft X-ray flux will decrease with increasing $|z|$; the scale height of ζ_{XR} will depend upon the scale height of the hot component.

6.3. Ionization within Diffuse clouds.

There are no reliable measurements of x_e, the fractional density of electrons in diffuse clouds. *Traditionally*, analyses of optical and UV absorption line data simply assume that the electrons come from starlight photoionization of elements with ionization potential less than 13.6 eV. The main contribution is from carbon and hence $x_e \sim 4 \times 10^{-4} \delta_C$ where δ_C is the depletion of carbon. Some analyses ignore the depletion of carbon which is clearly incorrect since we do observe grains in the ISM (see Jenkins, this volume).

Including cosmic ray ionization of hydrogen and helium, the density of electrons is

$$n_e = n_i + \frac{n_i}{2}[(1 + \frac{4\zeta_{CR} n_{H\ I}}{\alpha n_i^2})^{1/2} - 1] \qquad (6.3\text{-}1)$$

where α is the recombination coeffecient for hydrogen and is 4×10^{-13} $(T/6000\ K)^{-1/2}$ cm³ s⁻¹ and n_i is the number density of metal-ions (*viz.* gaseous carbon; $n_i/n_H \sim 2 \times 10^{-4}$ assuming 50% depletion of carbon). For diffuse cloud conditions: $n_H \sim 40$ cm⁻³, $T \sim 80$ K we estimate $n_e \sim 0.015$ cm⁻³ for $\zeta \sim 10^{-17}$ s⁻¹ with about equal contribution from photoionization of carbon and cosmic ray ionization of hydrogen. If $\zeta \sim 10^{-16}$ s⁻¹ as suggested by van Dishoeck and Black (1986) then cosmic ray ionization dominates and $n_e \sim 0.038$ cm⁻³. x_e is independent of n_H for photoionization of carbon whereas $x_e \sim n_H^{-1/2}$ for cosmic ray ionization of hydrogen. In §5.3 we learned that diffuse clouds are inhomogeneous with a considerable fraction of hydrogen at temperatures greater than 40 K. Assuming pressure equality we see from equation (6.3-1) that cosmic ray ionization plays an increasingly dominant role in the warmer regions of the clouds. Thus, contrary to the standard assumptions made in UV and optical absorption studies, cosmic rays probably dominate the ionization in clouds, even if ζ_R is as low as the minimum ionization rate.

The ionization in clouds is so small that they hardly contribute to the observed emission measure, the dispersion measure or the free-free absorption opacity (§2.4) – this is left as an exercise to the reader.

6.4. Ionization in the Warm, Neutral Medium.

In principal, the WNM can be ionized by soft X-rays and cosmic rays. Whether soft X-rays can actually be effective will depend upon the topology of the WNM. The soft X-ray flux with a mean energy of 74 eV (§6.2) cannot penetrate a column density more than 3×10^{19} cm⁻² and will be effective only if the WNM is organized into blobs smaller than ~ 20 pc. The filling factor of the WNM is estimated to be between 30% and 60% and hence it is not unreasonable to assume that the WNM is distributed more like a pervasive 'intercloud' medium. Theoretically such a morphology also appears to be favored (see Cowie, this volume). To summarize, if the WNM is distributed like an intercloud medium, then the effective ionization rate is equal to ζ_{CR} and $x_e \sim 0.007$ for typical WNM conditions. If, the WNM is distributed into small blobs then soft X-rays dominate and the effective ionization rate is $\sim \zeta_{XR}$ and $x_e \sim 0.05$.

Regardless of this uncertainty two points should be clear. First, as argued in §7.1, on general energetic grounds, any ionizing agent which is a byproduct of supernovae cannot be responsible for the ionization of gas seen in Hα emission. Thus neither soft X-rays nor cosmic rays can account for the observed $EM_\perp$ (§2.4). Second, the WNM cannot be contributing more than 10% of the interstellar electrons. By equating the recombination rate to the ionization rate we find $x_e = 0.05[\zeta_{-15}/n_{H\ I}]^{1/2}$ where ζ_{-15} is the ionization rate in units of 10^{-15} s⁻¹. For an assumed pressure of 4000 cm⁻³ K, $n_{H\ I} \sim 0.5$ cm⁻³ and $x_e \sim 5\%$ for $\zeta_{-15} = 0.5$, the maximum possible ionization rate. Integrating this over z, the expected maximum $N_{e,\perp} \sim 1.5 \times 10^{19}$; this estimate includes a factor of 2 to account for decreased pressure of the WNM at high z. This value of $N_{e,\perp}$ is only 15% of the observed vertical column density (§2.4). For these two reasons, a model in which the Hα emission is produced by regions of high n_e and the pulsar dispersion measure by electrons in the low density WNM is not viable. In addition, the estimated $EM_\perp$ and the free-free absorption opacity are orders of magnitude below the observed values (§2.4); this is left as an exercise to the the reader). Thus, we conclude that neither the CM nor the WNM can account for the observed DM, EM or free-free absorption. On energetic grounds, we argued that neither the CM nor the WNM can account for the observed Hα emission. So the conclusion is that we need another component – a warm, ionized medium and this medium is discussed below.

7. The Warm, Ionized Medium (WIM).

Two lines of reasonings lead us to postulate yet another atomic phase – the warm, ionized medium (WIM). First, a pervasive, largely ionized medium is required to explain the wide spread, diffuse Hα emission. Several optical metastable emission lines such as N II, S II *etc* are also observed from the same gas as the diffuse Hα emission. From the optical line data, we infer that this medium is warm ($T \lesssim 8000$ K) and substantially ionized (see Reynolds 1985). Second, in §6.3 and §6.4 we concluded that neither the warm, neutral medium (WNM) nor the cold medium (CM) can satisfactorily account for the various observations of interstellar electrons. Thus we are forced to invoke yet another phase to account for the interstellar electrons.

The simplest model is one in which most of the interstellar electrons arise in a single phase. This phase must warm since line widths of the metastable lines suggest temperatures ∼ 8000 K; this is consistent with the detection of N II, the excitation of which requires T_e, the temperature of the electrons to be > 3000 K. The ratios [N I/N II] and [O I/O II] are tied to the [H I/H II] ratio. The lack of detection of N I suggests that the medium is substantitally ionized: $x_e > 0.75$ (Reynolds, Roesler and Scherb 1977). We now proceed to examine if other observational measures of interstellar electrons *viz.* the dispersion measure data [$DM_\perp$], the emission measure data [$EM_\perp$], and the low-frequency free-free opacity [$\tau_\perp(10 \text{ MHz})$], can be consistently explained in this model. The subscript '⊥' refers to a projection of an integrated quantity such as the electron column density (for $DM_\perp$) onto the Galactic pole; the physics of these probes is discussed in §1.3 and the observational data may be found in §2.4.

The Hα intensity and the free-free opacity are proportional to the emission measure but have different temperature dependences (§1.3). From the ratio $EM_\perp/\tau_\perp(10 \text{ MHz})$ we derive $T \sim 4400$ K – close enough, given the observational uncertainties, to the 8000 K obtained from the optical observations of the metastable lines. If the hypothesized medium is in pressure equilibrium with the rest of the atomic phases then $P = 2n_eT \sim 4000$ cm^{-3} K (§3). We have assumed that the hypothesized medium is completely ionized to account for the [N I]/[N II] ratio and hence the factor of 2 in the pressure equation. For $T = 7500$ K, we get $n_e \sim 0.26$ cm^{-3}. The filling factor is simply $\langle n_e \rangle / n_e$ where $\langle n_e \rangle$ is the volume-averaged mean density. From pulsar observations(§2.4.1), $\langle n_e \rangle \sim 0.03$ cm^{-3} and thus the filling factor is about 11%. Thus, we find that a warm ($T \sim 8000$ K), fully ionized medium occupying a substantial volume of interstellar space can adequately account for all the present data of interstellar electrons. This *important conclusion* appears not to be well appreciated in the the interstellar medium literature. Henceforth, we will refer to this new phase as the warm, ionized medium (WIM).

The solar system appears to be immersed in the WIM. Studies of scattering of Solar Ly α (*e.g.* Bertaux *et al.* 1985) have established a temperature of 8000 ± 1000 K and a pressure of about 3000 cm^{-3} K – similar to the parameters deduced for the WIM, above.

The local mass surface density of the WIM is estimate to be $\sim 2 M_\odot$ pc^{-2} (§2 or §2.4). Assuming an effective radius of 12 kpc and assuming no increase in the surface density of the WIM in the inner Galaxy, we estimate a total mass of $10^9 M_\odot$. Clearly, this is a conservative value. This mass estimate is comparable to the mass of the molecular material obtained from observations of the diffuse γ-ray radiation (see Bloemen, this volume). This comparision is meant to illustrate the importance of the WIM. [A small note of worry: the γ-ray analyses do not include the contribution of the WIM. Thus, the γ-ray estimate of the molecular material is truly an upper limit and in fact including our estimate for the total mass of the WIM may reduce the γ-ray estimate rather drastically. We suggest a careful investigation of this point to resolve this important issue.]

7.1 Energy Source of the WIM.

The power required to ionized the WIM is very high. Locally, the number of recombinations in a column perpendicular to the Galactic plane is $\sim 4 \times 10^6$ cm^2 s^{-1}. To maintain this over the whole Galaxy requires a *minimum* power of 10^{42} erg s^{-1} – comparable to or exceeding all the power injected by Galactic supernovae (Reynolds 1984)! Clearly, supernova-related processes are inadequate. There are two other possibilites: shocks and stellar photoionization. Shocks are unattractive because the various ratios of metastable lines are sensitive to shock speeds whereas, observationally, the ratios appear to be rather constant across the sky. Also, we have to address the issue of the agent which created the shocks in the first place. Stellar photoionization is an attractive possibility. Mathis (1986) has a model of the WIM with steady state equilibrium between recombination and photoionization by a diffuse Lyman continuum. The model explains the optical data quite well.

However, the principal unanswered question is where does the diffuse Lyman continuum come from? The ionizing power of the O stars alone is more than five times that required to explain the observed Hα emission (Reynolds 1984). However, O stars are rare and hence it is not clear how their Lyman continuum radiation can reach every nook and corner of the Galaxy to produce the widespread WIM. Mathis (1986) estimates that about 10% of the sky as seen from a typical O star must have a column density of H I $< 10^{17}$ cm^{-2} so that the ionizing radiation can escape into the general ISM. There are two problems with this requirement: (1) there is not a single direction from the Sun where the column density of H I is so low and (2) a theoretical study by Elmegreen (1976) showed that O stars tend to destroy clouds in their vicinity, the debris then forms an ionization-bounded H II region. Indeed there appears to be some observational data supporting this theoretical point (Elmegreen 1975; McKee, van Buren and Lazareff 1984). Thus we conclude that O stars are ineffective within the *cloud layer i.e.* $|z| < 100$ pc.

Outside the cloud layer, O stars can indeed ionize large regions of the interstellar volume as evidenced by observations of diffuse H II regions around high-z O stars (Reynolds 1982). A fraction of O stars, the so-called runaway O stars, are found above $|z| > 100$ pc. Accurate statistics of these stars are urgently needed to confirm whether there are sufficient number of runaway O stars to ionize the bulk of the WIM. In this scenario, the more numerous B stars, which are confined to $|z| < 100$ pc, are the principle ionizing agents within the cloud layer.

Another possibility is that the WIM is ionized by a variety of hot stars such as planetary nebula nuclei, hot white dwarfs, B stars (operative only in the cloud layer) and QSO light (operative at the edges of the WIM layer). The collective ionizing energy from these sources is just about enough to account for the observed recombination rate. In this scenario, these stars simply ionize the H I nearest to them, which on the average is the WNM. Unlike the O stars, these stars are numerous and hence the WIM would appear widespread and smooth – just as the observations indicate. High-sensitivity Hα maps of regions around potential sources of ionizations are needed to confirm this scenario.

8. Filling factor of the Atomic Component.

The filling factors of the various phases of the ISM are *very* important numbers. The crucial question is whether most of the interstellar space is occupied by the hot component or the atomic component *viz.* the warm, neutral medium (WNM) and the warm, ionized medium (WIM). If the atomic component dominates then the ISM is best described by the 'two-phase' model (Field, Goldsmith and Habing 1969). If the hot component dominates then the ISM is best described by the 'three-phase' model (McKee and Ostriker 1977). [The reader should be aware that there are really *four* phases in the McKee and Ostriker model - the CM, the WNM, the WIM and the hot component. The model of Field, Goldsmith and Habing in its purest form has only two components – the CM and the WNM; including a sufficiently intense, diffuse Lyman continuum will produce the WIM.]

First we derive the filling factor and its z-dependence $\phi(z)$ of the WIM. The basic ingredients are the z-distribution functions of $\langle n_e \rangle$, from DM observations, and $\langle n_e^2 \rangle$, from EM observations. The exponential scale heights are 1000 and 300 pc, respectively (§2.4). The *true* electron density, $n_e(z)$, is related to the *volume-averaged* electron density, $\langle n_e \rangle$, by $\langle n_e(z) \rangle = \phi(z) n_e(z)$; similarly, $\langle n_e^2(z) \rangle = \phi(z) n_e(z)^2$. A little algebra yields:

$$n_e(z) \sim 0.27 e^{-|z|/428} \, \text{cm}^{-3}, \tag{8-1}$$

and

$$\phi(z) \sim 0.11 e^{|z|/748}. \tag{8-2}$$

In §7 we concluded that the WIM is fully ionized and has a temperature of ~ 7500 K. Thus the run of the pressure of the WIM with z is

$$P(z) \sim 4000 e^{-|z|/428} \, \text{cm}^{-3} \, \text{K}. \tag{8-3}$$

It is important to note that the above equations are only statistical in nature and should not be interpreted too literally. At any given $|z|$, the pressure in the ISM probably varies by at least a factor of 2.

Given our assumption of pressure balance among all gas components in the ISM, equation (8-2) should also describe the run of ambient pressure of the ISM. It is, then, a very important equation and hence needs to be inspected critically. It is remarkably consistent with our meagre, independent knowledge about the interstellar pressure. It predicts the mean pressure within the cloud layer ($|z| \lesssim 100$ pc) to be ~ 3200 cm^{-3} K – in excellent agreement with the observational determination of the mean pressure in the CM from UV observations of C I (§3). This agreement is remarkable because equation (8-2) was derived with very little or no input physics but only some basic algebra.

We now use the assumption of gas pressure equality among the various phases, together with equation (8-2), to derive the filling factors of the CM and WNM. The method is straightforward: knowing the pressures and temperatures of these components, we derive the true densities and then, from the observed volume-averaged densities, the filling factors.

For the CM, the determination of the filling factor is complicated by the inhomogeneous structure of clouds (§5.3). For example, if we assume that the CM contains equal amounts of H I at 40 K and 400 K, then the filling factor of the former is a negligible $\sim 5 \times 10^{-3}$ and of the latter a non-negligible 0.05. Since the mean ISM pressure hardly varies over the width of the cloud layer ($|z| < 100$ pc), these filling factors are independent of z.

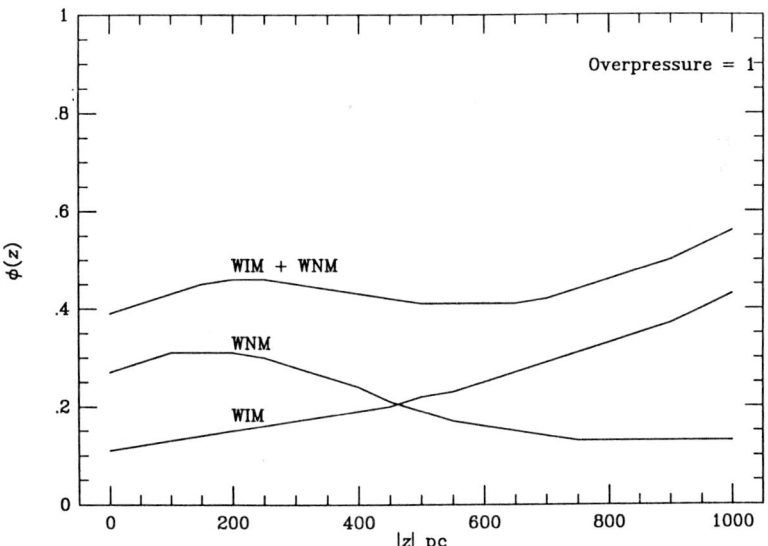

Figure 8-1 Filling factors of the WNM and the WIM as a function of $|z|$. The mean ISM pressure is assumed to be 4000 cm^{-3} K at $z = 0$ pc. Pressure equality between the WIM and WNM is assumed.

For the WNM, instead of the three component fit discussed in §4.1, we use a simplified two component fit: a Gaussian of rms scale height 250 pc and an exponential of scale height 480 pc. Under our assumption of pressure equality, the true density of the WNM is twice that of the WIM since the temperatures are nearly equal. The sum of the filling factors of the WNM and the WIM at $z = 0$ pc is ~ 0.4 and gradually rises to about 0.6 at $z = 1$ kpc (Figure 8-1).

Thus, under the assumption of gas pressure equality, we find $\sim 50\%$ of the interstellar volume to be occupied by the WNM and the WIM. If these numbers were absolutely reliable, we would conclude that the hot component occupied the other 50% and thus the ISM would be borderline between the Field, Goldsmith and Ostriker model and the McKee-Ostriker model. However, the gas pressure in the WIM could be systematically different compared to the WNM. In §7, one of the scenarios discussed for the formation of the WIM is one of ionization of the WNM by stray stellar Lyman continuua. If this scenario is correct then

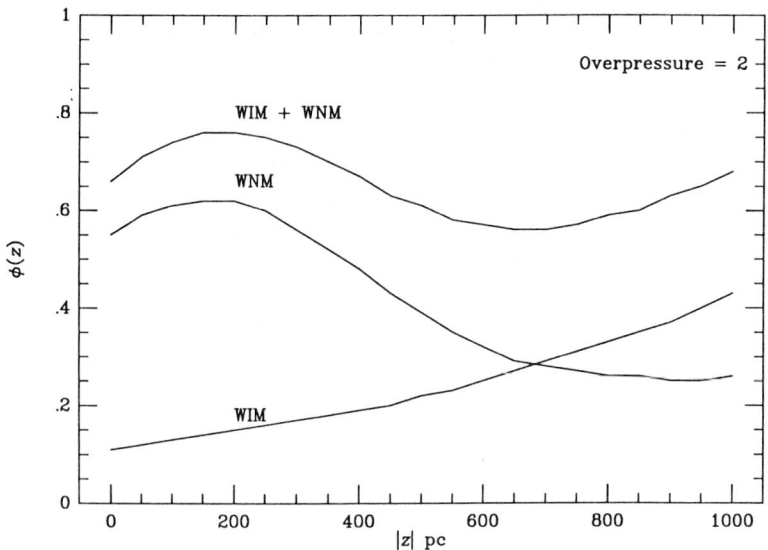

Figure 8-2 The mean ISM pressure is assumed to be 2000 cm^{-3} K at $z = 0$ pc. The pressure in the WIM is assumed to be twice the mean ISM pressure at all z.

the WIM is a dynamically changing medium and could be overpressured by a factor of 4 with respect to WNM (Elmegreen 1976). The C I measurements (§3) suggest that the mean ISM pressure is unlikely to be as low as 1000 cm^{-3} K. However, a mean ISM pressure of 2000 cm^{-3} K corresponding to an overpressure in the WIM by a factor of 2 is well within the uncertainties of observations. In this case, the filling factor of the WNM would double and the atomic component would occupy $\sim$ 80% of the interstellar volume (see Figure 8-2).

9. Summary and Future Work.

In this contribution, we have reviewed the phases which constitute the atomic component of the ISM: the cool diffuse clouds , the warm neutral medium and the warm ionized medium. Together, these three phases occupy a sizable fraction of the interstellar volume and contain at least half the interstellar mass. The emphasis has been on synthesizing relevant observations, from MHz radio measurements to γ-ray measurements, in order to arrive at a comprehensive picture of the atomic component. Rather than simply summarize what we discussed in the previous sections, we will highlight some recent advances, reiterate major questions of current interest and discuss some future goals.

The warm, ionized medium (WIM) has been a neglected phase despite the fact that, locally, it has a filling factor $\sim$ 0.12 and a surface mass density $\sim$ 50% of the local surface mass density of the molecular material. Conservatively we estimate a total mass of $\gtrsim 10^9$ $M_\odot$. Surely with all these attributes this phase deserves more of our attention! The most important recent result is the recognition that most of the interstellar electrons reside in this medium. This result removes a great obstacle hampering all steady state models *viz.* the agent(s) which heat the atomic gas need not necessarily ionize the gas. The radial distribution of the WIM is unknown; this will be useful in constraining possible ionizing agents and in obtaining a firmer total mass estimate. From pulsar DM data, we infer a vertical scale height of ~ 1 kpc. However, this is strictly a lower limit since the scale height of the pulsars is only 400 pc, significantly smaller than 1 kpc. Clearly, we need other probes of interstellar electrons. Observations of interstellar scattering of extragalactic sources appear promising; such observations may also establish in which medium do the irregularities reside: the WIM or the hot component? The power needed to sustain the WIM exceeds the power injected by supernovae. Thus the conclusion that the WIM is powered by radiation from hot stars seems unavoidable. How do the ionizing stellar photons leak into the diffuse ISM, given the high opacity of the atomic component towards the ionizing stellar photons? High resolution Hα observations are needed to determine *which* class of hot stars are the dominant sources of ionization. Finally, the existence of this widely distributed phase is indicative of the great influence of hot stars on the structure of the ISM. Thus global models of ISM should not only include the influence of supernovae on the ISM but also that of hot stars.

The warm, neutral medium (WNM) is an enigma. The presently available data, though sparse, suggests that the temperature of the WIM is indeed ~ 8000 K, in accordance with our theoretical expectation. Space-based ultra violet and far infrared observations are urgently needed. Ground based observations of metastable lines of neutral species such as N I, O I *etc.* would be very valuable; unfortunately, the signals are expected to be low. The filling factor of this medium is estimated to be $\gtrsim 0.3$. If the pressure in the WNM is 2000 cm^{-3} K then the filling factor would rise to 0.6. Clearly, this phase must play an important role in the evolution of the ISM. Yet, we do not understand why about 50% of the local H I is in the WNM phase. What determines this fraction? Does this fraction change with R?

The diffuse clouds which constitute the cold medium (CM) appear to be more complicated than predicted by theory. One might think that diffuse clouds, in some sense, are the simplest structures in the ISM. We have good measurements of the radiation field incident on these clouds and the atomic physics of the cooling collisional processes are well understood. So supposedly we know the inputs and the ouptuts. DESPITE this, observationally we find that substantial portions of the clouds are at temperatures significantly larger than the theoretical expectation of 40 K. Further more there is a great deal of internal structure *viz* clouds are not isothermal. Do we understand diffuse clouds at all? Are clouds best described by an onion model with a cold core and successively warmer skins? What determines the internal structure? The inhomegeneity of diffuse clouds complicates interpretation of *all* absorption line studies, including the 21-cm line and especially saturated optical and UV lines. Is the WNM wrapped around each cloud or does it exist independently of the clouds? Is the large scale filamentary structure of clouds due to cooling instabilites in the presence of magnetic fields and/or evaporation?

The heating agents of the WNM and the CM have still not been identified. There is plenty of energy in the stellar radiation field to heat the diffuse medium. The principal obstacle has been identifying a mechanism to tap this vast reservoir. So far, the most promising mechanism appears to be photoelectric heating by dust grains. However, even with the most optimistic parameters, substantial amounts of lukewarm ($T > 100$ K) H I cannot be produced by this mechanism. At this meeting there were some suggestions that PAHs could be the principal heating agents via the photoelectric process. Clearly, it is worth pursuing this suggestion. From time to time, non-radiative heating methods such as mechanical heating, Alfven wave dissipation *etc.* have been suggested. Such suggestions are difficult to model theoretically and equally difficult to verify observationally.

The pressure scale height is a crucial parameter for models of the ISM. If the ISM is dominated by the hot component, then the pressure scale height should be large ($\gtrsim 2$ kpc). Our determination of the local pressure scale height of 500 pc is significantly below this prediction. In fact, it is comparable to the width of the WNM layer. What is the significance of this? Anyway, an independent determination of the pressure scale height is urgently needed. What determines the scale height of the CM? Cloud-cloud velocity dispersion or some other parameter? Do magnetic fields keep clouds tied together and thus somehow restrict the scale height? Are there really three distinct components of H I as suggested by Lockman's work? Or should we consider H I to have a continuum of velocity dispersions with corresponding scale heights? In either case, can we theoretically explain the observed vertical structure? In the inner Galaxy, why is the shape and the *thickness* of the H I layer independent of R in contrast to the molecular and the stellar layer?

The topology of the various phases are not understood. In the supernova-dominated McKee-Ostriker model, the hot component pervades most of the interstellar space with the atomic component occupying the remaining volume. In the quiescent Field, Goldsmith and Habing model, the WNM occupies most of the interstellar space. Which picture describes the ISM? The applicable model almost certainly changes with R and it is not clear where in the Galaxy the transition occurs. Clearly, the time is ripe for a systematic study of the variation of the different *phases* of the ISM as a function of R and z – including the study of external galaxies.

Locally, both the WNM and the hot component appear to have similar filling factor. Is the hot component distributed in discrete structures? Is the WNM organized into small blobs or more like an 'intercloud' medium? What is the relation between the WNM and the WIM? Is the WIM simply the ionized edges of the WNM, or distributed completely independently of the WNM? Moving to smaller scales, we find that diffuse clouds are rarely spherical. What determines their filamentary shapes? Magnetic fields or evaporation? In a hot component-dominated ISM, the topology of the CM governs the total evaporation rate. For this reason, it is important to determine whether diffuse clouds are filaments or sheets and whether the WNM is distributed like an 'intercloud' medium.

This work was supported in part by an NSF grant to CH and a Millikan Fellowship to SRK. We gratefully acknowledge help of S. Vogel (CIT) for constructive criticisms.

References

Bahcall, J. N., Schmidt, M. and Soneira, R. M. 1983, *Ap. J.* **265**, 730.

Barsuhn, J. and Walmsley, C. M. 1977, *Astr. Ap.* **54**, 345.

Belfort, P. and Crovisier, J. 1984, *Astr. Ap.* **136**, 368.

Bertaux, J. L., Lallement, R., Kurt, V. G. and Mironova, E. N. 1985, *Astr. Ap.* **150**, 1.

Blitz, L., Fich, M. F. and Kulkarni, S. R. 1983, *Science* **220**, 1233.

Bridle, A. H. and Venugopal, V. R. 1980, *Nature* **224**, 545.

Cesarsky, C. J. and Volk, H. J. 1978, *Astr. Ap.* **70**, 367.

Cleary, M. N., Heiles, C. and Haslam, C. G. T. 1979, *Astr. Ap. Suppl.* **36**, 95.

Clifton, T. and Lyne, A. G. 1986, *Nature* **320**, 43.

Cohen, R. J. 1981, *M. N. R. A. S.* **196**, 835.

Colomb, F. R., Poppel, W. G. L. and Heiles, C. 1980, *Astr. Ap. Suppl.* **40**, 47.

Crovisier, J. 1981, *Astr. Ap.* **94**, 162.

Crovisier, J., Dickey. J. M. and Kazes, I. 1985, *Astr. Ap.* **146**, 223.

Deiter, N. H. 1960, *Ap. J.* **132**, 49.

Dickey, J. M., Kulkarni, S. R., van Gorkom, J. H. and Heiles, C. E. 1983, *Ap. J.* **53**, 591.

Draine, B. T. 1978, *Ap. J. Suppl. Ser.* **36**, 595.

Ellis, G. R. A. 1982, *Austr. J. Phys.* **35**, 91.

Elmegreen, B. G. 1975, *Ap. J.* **198**, L31.

Elmegreen, B. G. 1976, *Ap. J.* **205**, 405.

Giovanelli, R. 1980, *A. J.* **85**, 1155.

Gordon, C. P. 1971, *A. J.* **74**, 914.

Guelin, M., Langer, W. D. and Wilson, R. W. 1982, *Astr. Ap.* **107**, 107.

Field, G. B. 1959, *Ap. J.* **129**, 551.

Field, G. B., Goldsmith, D. W. and Habing, H. J. 1969, *Ap. J.* **155**, L149.

Heiles, C. 1967, *Ap. J. Suppl. Ser.* **15**, 97.

Heiles, C. and Habing, H. J. 1974, *Astr. Ap. Suppl.* **14**, 1.

Heiles, C. 1976, *Ap. J.* **204**, 379.

Heiles, C. 1979, *Ap. J.* **229**, 533.

Heiles, C. 1980, *Ap. J.* **235**, 833.

Heiles, C. 1984, *Ap. J.* **55**, 585.

Henderson, A. P., Jackson, P. D. and Kerr, F. J. 1982, *Ap. J.* **263**, 116.

Hobbs, L. 1974, *Ap. J.* **191**, 395.

Jenkins, E. B., Jura, M., and Loewenstein, M. 1983, *Ap. J.* **270**, 88.

Kalberla, P. M. W., Mebold, U. and Reich, W. 1980, *Astr. Ap.* **82**, 275.

Kalberla, P. M. W., Schwarz, U. J. and Goss, W. M. 1985, *Astr. Ap.* **144**, 27.

Kulkarni, S. R., Blitz, L., and Heiles, C. 1982, *Ap. J.* **259**, L63.

Kulkarni, S. R. 1983, *Ph. D. Thesis*, U. C. Berkeley.

Kulkarni, S. R., Turner, K. C., Heiles, C. and Dickey, J. M. 1985, *Ap. J. Suppl.* **57**, 631.

Kulkarni, S. R. and Fich, M. 1985, *Ap. J.* **289**, 792.

Kulkarni, S. R., Dickey, J. M. and Heiles, C. 1985, *Ap. J.* **281**, 796.

Kulkarni, S. R. and Mathieu, R. 1986, *Space Astr.* **118**, 531.

Kulkarni, S. R. and Heiles, C. 1986, Ch. 3 in *Galactic and Extragalactic Radio Astronomy* (Eds. K. I. Kellerman and G. L. Verschuur), *submitted.*

Lazareff, B. 1975, *Astr. Ap.* **42**, 225.

Lindblad, P. O., Grape, K., Sandquist, A. and Scheler, J. 1973, *Ap. Astr.* **24**, 309.

Liszt, H. 1983, *Ap. J.* **275**, 163.

Lockman, F. J. 1984, *Ap. J.* **283**, 90..

Lockman, F. J., Hobbs, L. M. and Shull, M. J. 1986, *Ap. J.* **301**, 380.

Lyne, A. G., Manchester, R. N. and Taylor, J. H, 1985, *M. N. R. A. S.* **213**, 613.

Manchester, R. N. and Taylor, J. H. 1977, *Pulsars* (Freeman:San Francisco).

Mathis, J. S. 1986, *Ap. J.* **301**, 423.

McKee, C. F. and Ostriker, J. P. 1977, *Ap. J.* **196**, 565.

McKee, C. F., van Buren, D. and Lazareff, B. 1984, *Ap. J.* **278**, L115.

Mebold, U. 1972, *Astr. Ap.* **19**, 13.

Mebold, U., Winnberg, A., Kalberla, P. M. W. and Goss, W. M. 1982, *Astr. Ap.* **115**, 223.

Mirabel, I. F. 1982, *Ap. J.* **256**, 112.

Nichols-Bohlin, J. and Fesen, R. A. 1986, *A. J.* **92**, 642.

Pottasch, S. R., Wesselius, P. R., and van Duinen, R. J. 1979, *Astr. Ap.* **74**, L15.
Radhakrishnan, V., Murray, J. D., Lockhart, P. and Whittle, R. P. J. 1972, *Ap. J. Suppl. Ser.* **24**, 15.
Raimond, E. 1966, *Bull. Astr. Inst. Netherlands* **18**, 191.
Readhead, A. C. S. and Duffet-Smith, P. J. 1975, *Astr. Ap.* **42**, 151.
Reynolds, R. J., Roesler, F. L. and Scherb, F. 1977, *Ap. J.* **211**, 115.
Reynolds, R. J. 1982, *A. J.* **87**, 306.
Reynolds, R. J. 1984, *Ap. J.* **282**, 191.
Reynolds, R. J. 1985, *Ap. J.* **285**, 256.
Reynolds, R. J. 1986, *preprint.*
Schwarz, U. J. and van Woerden, H. 1974, in *IAU Symposium on Galactic Radio Astronomy,* Eds. F. J. Kerr and S. C. Simonson (Reidel:Dordrecht).
Spitzer, L. Jr. and Tomasko, 1968, *Ap. J.* **152**, 971.
Spitzer, L. and Jenkins, E. B. 1975, *Ann. Rev. Astr. Ap.* **13**, 133.
Spitzer, L. Jr. 1978, *Physical Processes in the Interstellar Medium* (John Wiley:New York).
Stacey, G. J., Viscuso, P. J., Fuller, C. E., and Kurtz, N. T. 1985, *Ap. J.* **289**, 803.
Thaddeus, P. 1982, *Ann. New York Acad. Sci.* **395**, 9.
van Dischoeck E. F. and Black, J. H. 1986, *preprint.*
Verschuur, G. L. 1975, *Ann. Rev. Astr. Astrophys.* **13**, 257.
Vivekanand, M. and Narayan, R. 1982, *J. Ap. Astr.* **3**, 399.
Watson, W. D. 1978, *Ann. Rev. Astr. Ap.* **16**, 585.
Weaver, H. F. and Williams, D. R. W. 1973, *Astr. Ap. Suppl.* **8**, 1.
Wesselius, P. R. and Fejes, I 1973, *Astr. Ap.* **24**, 15.
York, D. G. and Frisch, P. 1984, *I.A.U. Colloq.* **81**, 3 (NASA Conf. Publ. 2345).

Symposium Barbeque

HOT INTERSTELLAR GAS IN THE GALACTIC DISK AND HALO

Blair D. Savage
Departmant of Astronomy
University of Wisconsin
Madison, Wisconsin 53706 USA

1. INTRODUCTION

Our view of the physical nature and distribution of the interstellar gas has fundamentally changed in the past 15 years (see Table 1). From far ultraviolet absorption line studies and diffuse X-ray emission studes we now know of the existence of a hot (T $\approx$ 0.3 to 2×10^6 K) low density (n $\approx 10^{-3}$ cm^{-3}) phase of the interstellar medium which *may* occupy much of the volume of interstellar space. This hot or "coronal" gas is probably heated by high velocity shocks produced by supernova explosions and the high speed winds of associations of hot young stars. The coronal gas may be the fundamental substrate in which the other phases of the interstellar medium are embedded. The various phases probably exist in rough pressure equilibrium and may have the approximate properties listed in Table 1, which is adapted from Spitzer's (1986) review of interstellar clouds.

TABLE 1
PHASES OF THE INTERSTELLAR MEDIUM

phase	state of H	approximate T(K)	approximate n(cm^{-3})
molecular clouds	H_2	10	>300
cold H I clouds	H I	80	40
warm H I	H I	8000	0.4
H II regions	H II	8000	>1
coronal gas	H II	10^6	0.003

The possible existence of the hot phase of the ISM was first predicted by Spitzer (1956). He recognized that the presence of neutral clouds at large distances above the galactic plane required the support of some kind of exterior medium to keep the clouds from expanding and disappearing. Spitzer proposed that the exterior medium was hot gas (T$\approx 10^6$K) in approximate pressure equilibrium with the halo clouds. The direct detection of this hot gas had to wait 17 years for the technological advances of the space age since the gas is only easily seen at far ultraviolet and X-ray wavelengths.

D. J. Hollenbach and H. A. Thronson, Jr. (eds.), Interstellar Processes, 123–141.
© *1987 by D. Reidel Publishing Company.*

Hot (10^6K) gas in the galactic disk is buoyant. Its thermal scale height is about 7 kpc. The gas will, therefore, tend to flow outward away from the galactic plane, into the halo where it may cool an return to the disk as falling clouds of matter. This process is best described as "galactic convection". If the flow of hot gas into the halo occurs as a consequence of a recent supernova explosion that heats and elevates the pressure in a large volume of disk gas, the region of elevated pressure may breakout and flow into the halo and eventually cool and return to the disk in a flow pattern resembling a "galactic fountain" (Shapiro and Field 1976). In either case the result is the injection of gas into the lower halo of the galaxy.

The organization of this paper which emphasizes the observational results is as follows: The diagnostic methods employed to study the hot ISM are discussed in §2 The observational results for hot gas in the galactic disk and halo are found in §3. A brief overview of the theory of the hot ISM as it might apply to disk and halo gas is found in §4.

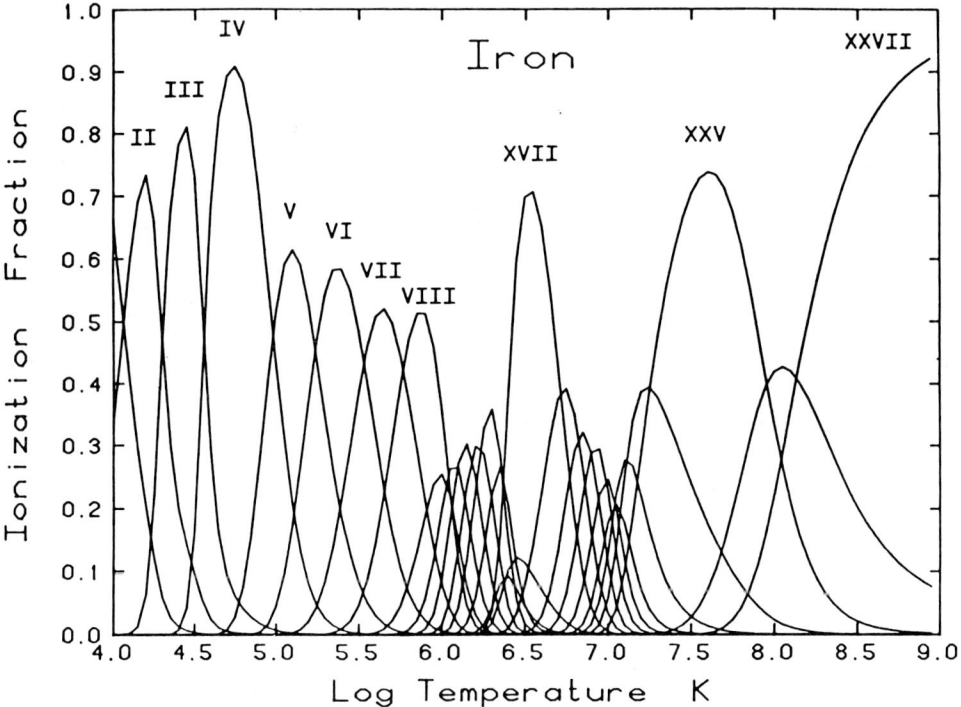

Figure 1. Fractional amounts of iron in different stages of ionization asuming collisional ionization equilibrium (i.e. "coronal ionization") from calculations by Shull and VanSteenberg(1982). The figure is from McCray (1986).

2.DIAGNOSTIC METHODS

The hot interstellar medium is not in thermal dynamical equilibrium; the temperatures quoted only refer to particle kinetic temperatures. In this situation the expected ionization equilibrium can be calculated by assuming time equilibrium and equating the rates of creation and destruction of an ion. The processes usually included in these calculations are electron collisional ionization, radiative recombination and dielectronic recombination. For results for many atoms see Shull and VanSteenberg (1982). Since the rates of ion production and destruction are both proportional to the electron density, the resultant ionization is independent of the density and only depends on the kinetic temperature and atom type. The ionization equilibrium described here is often called "coronal ionization" or simply "collisional ionization" . Sample calculations for the element Fe are shown in Figure 1. Different ion stages of Fe peak in abundance at different temperatures. If the conditions assumed in the coronal ionization calculations are valid then the simple detection of an ion such as Fe X would imply the existence of gas near logT = 6.1. Better temperature estimates could be obtained if measures of the abundance of two adjacent ion stages were available. Although we will generally discuss the hot ISM assuming conditions of coronal ionization are valid, it is important to note that in some situations the very assumption of time equilibrium is in doubt. In this case, the observed abundances of particular ion types will depend in a complex way on the past thermal history of the gas. Another obvious problem is that curves like those shown in Figure 1 are only as good as the quality of the input atomic physics. Reliable experimental and/or theoretical cross-sections for a large number of atomic processes are required to perform the coronal ionization calculations. It would not be surprising to occasionally encounter large errors in the published results.

Hot interstellar gas can be detected either through absorption or emission measurements. The absorption measurements can be made at any wavelength containing a strong resonance line of a highly ionized and abundant atom that might exist in the gas. The emission from hot gas is in the form of continuum and line emission. The emission is most easily seen at those wavelengths where it peaks which is the EUV and X-ray region of the spectrum. The techniques of absorption and emission spectroscopy are described below.

2.1.Absorption Line Spectroscopy

The study of the hot phase of the ISM through absorption line spectroscopy involves searching for expected resonance lines in the spectra of objects with smooth continua. Hot stars with large rotational velocities are often the object of choice. The stars must be hot to produce an adequate flux at the wavelengths where the lines occur. The stellar line broadening produced by a large rotational velocity helps when trying to separate the effects of stellar and interstellar absorption. However, if the rotational broadening is too large the star may begin to exhibit relatively narrow absorption features associated with ejected circumstellar matter. The luminosity class of the star is also an important consideration, particularly in the far ultraviolet spectral region. The star's luminosity influences the appearance of stellar wind features, with the hot luminous stars having pronounced P-Cygni stellar lines indicating mass outflow and the lower luminosity stars sometimes only having photospheric absorption. The P-Cygni features are often so broad (velocities to 3000 km/s), they may provide quite a smooth continuum to measure a narrow interstellar line against.

Table 2 lists information about the resonance absorption lines of abundant ions that exist in collisionally ionized gas with temperatures greater than 0.6×10^5 K. Only those lines with wavelengths greater than the interstellar H I cutoff at 912 A are listed. With two exceptions ([Fe X] ,$\lambda 6374.51$ and [Fe XIV], $\lambda 5302.89$),the lines fall in the ultraviolet region of the spectrum. In collisionally ionized gas, Fe X and Fe XIV peak in abundance at 1.2 and 2×10^6 K, respectively (see Figure 1). These lines have very small f values and are expected to be very weak. They have not yet been detected in absorption (Hobbs and Albert 1984; Hobbs 1985; Pettini and D'Odorico 1986). However, if the techniques of exceedingly high signal to noise spectroscopy continue to improve, the observational study of hot interstellar gas may begin to open to the ground based optical astronomers.

TABLE 2
VARIOUS HIGHLY IONIZED INTERSTELLAR SPECIES

species	$\lambda(A)$	f	IP_{X-1}	IP_X	$n(total)/n(H)$ [a]	$T_{max}(K)$ [b]	sensitivity [c]
Si IV	1402.77	0.262	33.5	45.1	3.5×10^{-5}	0.6×10^5	2.4×10^{-6}
Si IV	1393.76	0.528					4.8×10^{-6}
C IV	1550.76	0.097	47.9	64.5	4.7×10^{-4}	1.0×10^5	1.6×10^{-5}
C IV	1548.19	0.194					3.2×10^{-5}
S VI	944.52	0.210	72.5	88.0	1.6×10^{-5}	2.0×10^5	4.0×10^{-7}
S VI	933.38	0.426					8.2×10^{-7}
N V	1242.80	0.0757	77.5	97.9	9.8×10^{-5}	1.8×10^5	1.6×10^{-6}
N V	1238.81	0.152					3.3×10^{-6}
O VI	1037.63	0.0648	113.9	138.1	8.3×10^{-4}	3.0×10^5	1.2×10^{-5}
O VI	1031.95	0.130					2.4×10^{-5}
Fe X	6374.51	2.1×10^{-7}	235	262	2.5×10^{-5}	1.2×10^6	1.4×10^{-12} [d]
Fe XIV	5302.89	5.1×10^{-7}	355	390	2.5×10^{-5}	2.0×10^6	3.2×10^{-12} [d]

a. Solar CNO abundances are from Lambert (1978), Si , S, and Fe are from Withbroe(1971).

b. T_{max} lists the temperature at which a particular ion reaches maximum abundance assuming conditions of coronal ionization according to Shull and Van Steenberg (1982). For Si IV the effects of charge exchange are included according to Baliunas and Butler(1980).

c. The sensitivity refers to the product f [n(tot)/n(H)] $[n(ion)/n(tot)]_{T(max)}$. This is a measure of the sensitivity of a particular line to collisionally ionized gas assuming solar abundances and a uniform distribution of gas temperatures.

d. Since the cooling rate for gas at $T > 10^6$ K is much less than the rate for cooler gas, the abundance of 10^6 K gas is expected to be much higher than assumed in the sensitivity calculation listed here.

The ultraviolet satellites that have been capable of studying the species listed in Table 2 include the Copernicus (OAO-3) satellite and the International Ultraviolet Explorer (IUE). Copernicus was launched in 1972 and operated until 1980 . IUE was launched in 1978 and is still operating in June 1986. Hopefully in the near future the launch of the Hubble Space Telescope will provide ultraviolet spectroscopists with the High Resolution Spectrograph (HRS). The various ultraviolet satellites mentioned have different characteristics and limitations. For example, the Copernicus satellite could obtain high signal to noise spectra of bright stars at a resolution of 13 km s^{-1} between 912 and about 1400 A . The IUE with a resolution of 25 km s^{-1} , is restricted to the wavelength interval 1150 to 3300 A and can only with great effort achieve a signal to noise of 20. However, the IUE has obtained high dispersion spectra of objects 500 times fainter than the faintest objects observed with Copernicus. The HRS like IUE will also be limited to wavelengths longward of 1150 A, but will be capable of obtaining high resolution (13 and 4 km s^{-1}) *and* high signal to noise spectra of very faint objects. Reference to Table 2 shows that the Copernicus satellite was the only satellite capable of recording interstellar O VI while the IUE and HRS are restricted to N V, C IV , and Si IV.

The ionization potentials, temperatures of maximum abundance, and detection sensitivities listed in Table 2 reveal the importance of O VI to the study of the hot ISM. The production of the ions Si IV, C IV, N V and O VI require approximately 34, 48, 78 and 113 eV, respectively. With increasing energy it is becomes less likely that photoionization in warm (T $\approx 10^4$K) gas is the source of ionization. Photoionization by hot young stars, hot white dwarf stars and the extragalactic EUV background may create C IV and Si IV. This is less likely for N V and exceedingly unlikely for O VI (see Savage 1984 and references therein).

Ultraviolet measures of the highly ionized ions listed in Table 2 have been obtained with the Copernicus and IUE satellites for many lines of sight in the Milky Way. From an ultraviolet absorption spectrum it is possible to estimate the line of sight column density through the interstellar gas , N(ion) = $\int n_x$(ion)dx. To do this it is first necessary to locate the continuum i.e. determine how the spectrum would appear if there were no interstellar absorption. It is very common to have cases where the stellar spectrum has significant curvature and the continuum placement is uncertain (see Figure 2). Once the continuum has been determined,the observed spectrum, I(λ) , can be converted to a normalized line profile , r(λ) = $I(\lambda)/I_c(\lambda)$ = exp[$-\sigma_\lambda$N(ion)]. Here $I_c(\lambda)$ refers to the continuum level and σ_λ is the absorption cross-section which depends on the atomic absorption oscillator strength and the line broadening. If the instrumental resolution has degraded the appearance of the line ,the expression for the observed line profile must be written as r(λ) = exp[$-\sigma_\lambda$N(ion)] x P_λ , where P_λ is the instrumental spectral smearing function.

The wavelength dependent absorption cross-section has the form σ_λ = ($\pi e^2 f \lambda^2/m_e c^2$)φ_λ , where f is the absorption oscillator strength and φ_λ is a normalized line broadening function ($\int \varphi_\lambda$dλ = 1). If the effects of the instrumental smearing are indeed minor, i.e. if the observed line is well resolved, then we can ignore P_λ in the equation for r_λ above and we see ln r_λ = $- \sigma_\lambda$N(ion) , and N(ion) = $- (m_e c^2/\pi e^2 f \lambda^2) \int$ lnr_λ dλ. Therefore ,a simple integration of the line profile provides the column density, N(ion). This procedure breaks down if the line is very strong and approaches zero central

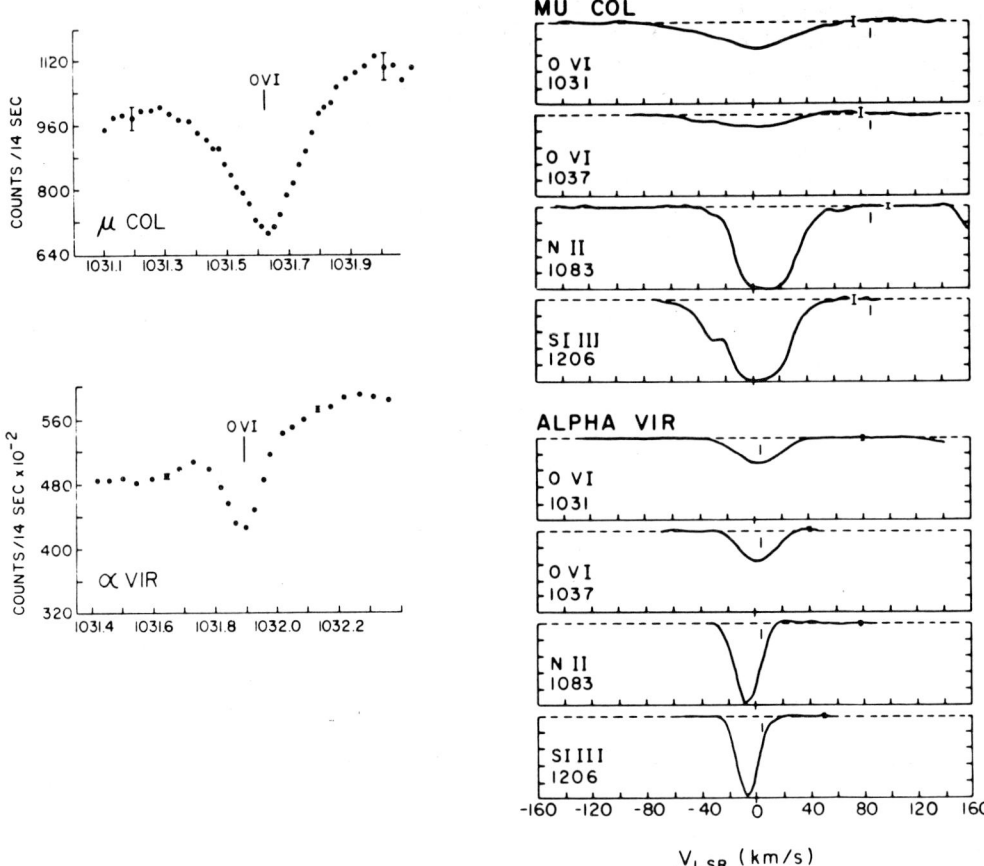

Figure 2. Sample raw and normalized O VI line profiles for the well observed
stars μ Col and α Vir . Normalized profiles for various ions of lower
ionization state are also illustrated. The zero levels of the raw data are off
the bottom of the plotted figure. For the normalized profiles the lower axis is
the zero level. The reliability of a normalized line profile is often lower than
a simple assessment of the noise level would suggest because the accuracy of
the continuum placement influences the accuracy of the final line profile. In
most published papers, it is difficult to independently access this problem
because only normalized profiles are shown. In the raw data illustrated, a
continuum placed higher would have produced broader wings on the normalized
O VI line profiles. Such wings would be consistent with the thermal
broadening occurring in an absorbing medium with a wide range of
temperatures. The raw data illustrated are from York(1974) and the
normalized line profiles are from Cowie et al. (1979).

intensity (very large $-\ln r_\lambda$) because the actual value of the central intensity (i.e. is it 10^{-2} or 10^{-4} or 10^{-6} ?) will be uncertain as a result of uncertain knowledge light scattering in the spectrograph.

For strong lines and in cases where the instrumental resolution degrades the spectrum it is common to use "curve of growth" techniques to derive column densities [see Spitzer (1978) and Jenkins this volume]. In this case the line equivalent width , $W_\lambda = \int (1-r_\lambda)d\lambda$, which is independent of instrumental resolution is measured. The equivalent width depends on N(ion) , atomic parameters, and the line of sight velocity distribution. For very weak lines (r_λ near 1) or for exceedingly strong lines which have broad natural damping wings, the conversion between W_λ and column density only depends on atomic parameters. However, for lines of intermediate strength the conversion requires a knowledge of the line broadening. It is common to assume a simple Maxwellian velocity distribution described by the velocity spread parameter $b = (2kT/m)^{1/2}$ to derive column densities in these intermediate cases. However, the results based on such assumptions are often very uncertain (Nachman and Hobbs 1973) . For the highly ionized species, most absorption multiplets are doublet lines with one component of the doublet having twice the oscillator strength of the other component. In this case, if the ratio of line equivalent widths (i.e. the doublet ratio) lies between 2 and about 1.5 , the simple one component curve of growth as described above will probably yield reliable results. However, for double ratios less than 1.5 it is very risky to try to estimate a column density from a simple one component curve of growth. In this latter case it is probably best to simply quote lower limits to the observed column density. Fortunately many of the measures of O VI, N V , C IV and Si IV involve data with doublet ratios between 2 and 1.5.

In addition to column density information, the ultraviolet measurements provide absorption line velocities and line widths. The line widths and/or shapes can be used to obtain direct information about the temperature of the gas if the broadening is thermal. A doppler broadened line produced by an ion of mass number A at temperature T will have a full velocity width at half maximum intensity, FWHM $= 2(\ln2)^{1/2}(2kT/M)^{1/2}$ or FWHM(km s^{-1}) $= 0.215(T/A)^{1/2}$. In the case of O VI, N V , C IV and Si IV at their temperatures of peak ionic abundance (see Table 2), the individual values of FWHM are 29,26,20 and10 km s^{-1}, respectively. Of the satellites so far placed into orbit, only Copernicus has had the spectral resolution (13 km s^{-1}) to determine reliable temperature information from the observed line profiles. However, this will certainly change with the launch of the HRS (resolution $\approx$ 4 km s^{-1}).

2.2.Emission Spectroscopy

Hot plasmas emit X-ray, EUV and UV radiation. Calculations of the expected soft X-ray (E < 1 keV) emissions for low density optically thin collisionally ionized plasmas with solar abundances are found in Raymond and Smith (1977). Updated versions of these calculations can be obtained directly from Raymond. The radiation produced is a combination of continuum and line emission. The continuum is from Bremsstrahlung, radiative recombination and two photon emission from one and two election atoms. The line emission is from collisionally excited lines and recombination followed by radiative cascades. Examples of the expected emission over the range 0.1 to 0.3 keV are shown in

Figure 3. Calculations for the EUV and UV region are reported by Paresce et al.(1983).

While a few low resolution observations of the line emission from the hot interstellar medium exist (see §3.1.2) , the bulk of our information about emission from interstellar coronal gas has come from broad band measurements obtained from various sounding rocket and satellite programs. In particular, in a series of 10 sounding rocket flights extending over a 7 year period the Wisconsin X-ray group (McCammon et al.1983) mapped the diffuse X-ray sky with 7^0 spatial resolution in 7 broad energy bands using thin-window proportional counters. The energy resolution of the proportional counters varies as $E^{1/2}$ and is E/ΔE $\approx$ 2.5 at 1 keV. Since this provides very little resolution for low energy X-rays, the K-edge absorption of various window materials (eg. boron and carbon) are utilized to isolate various soft X-ray bands. The resulting soft X-ray bands and their respective energy range are: B(0.130-0.188keV) ,C(0.160-0.284keV), M_1(0.440-0.930 keV), and M_2(0.600-1.10 keV). The actual band response functions are found in McCammon et al.(1983). Very recently a counter with a beryllium window has been added which defines a band hereafter referred to as the Be band extending from 0.078 to 0.111 keV (Bloch et al.1986). Important diffuse soft X-ray data have also been obtained

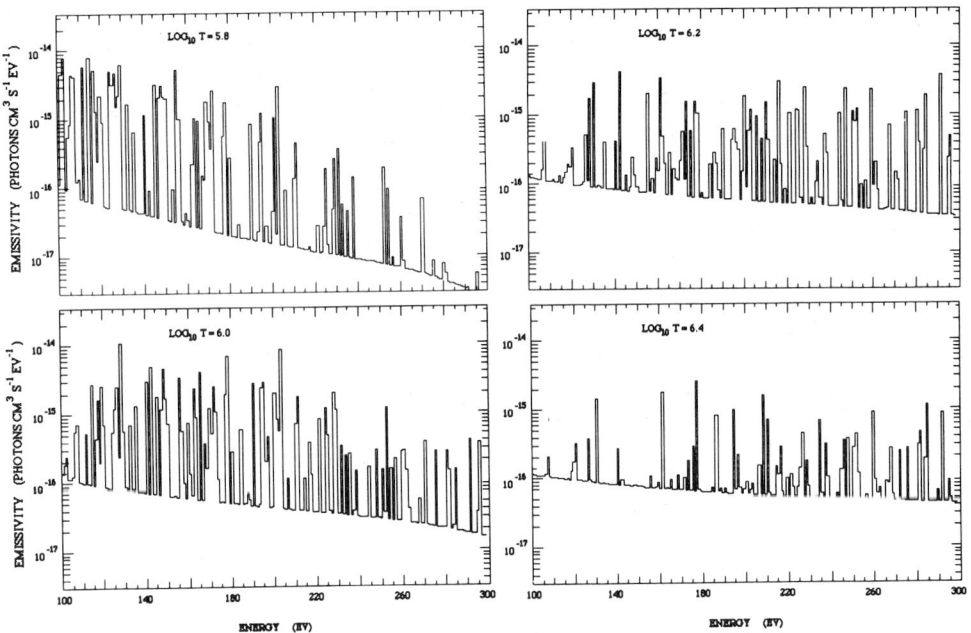

Figure 3. The emissivity ε_{ν}(photons cm^{3}s^{-1}eV^{-1}) of a hot optically thin plasma of solar abundances at logT = 5.8, 6.0, 6.2 and 6.4 according to recent unpublished calculations of Raymond which represent an update to the work of Raymond and Smith (1977). The plasma produces a combination of line and continuum emission. The energy range shown (100 to 300 eV) covers the B(130-188 eV) and C(160-284eV) diffuse X-ray bands. The diagram was kindly provided by S.Snowden.

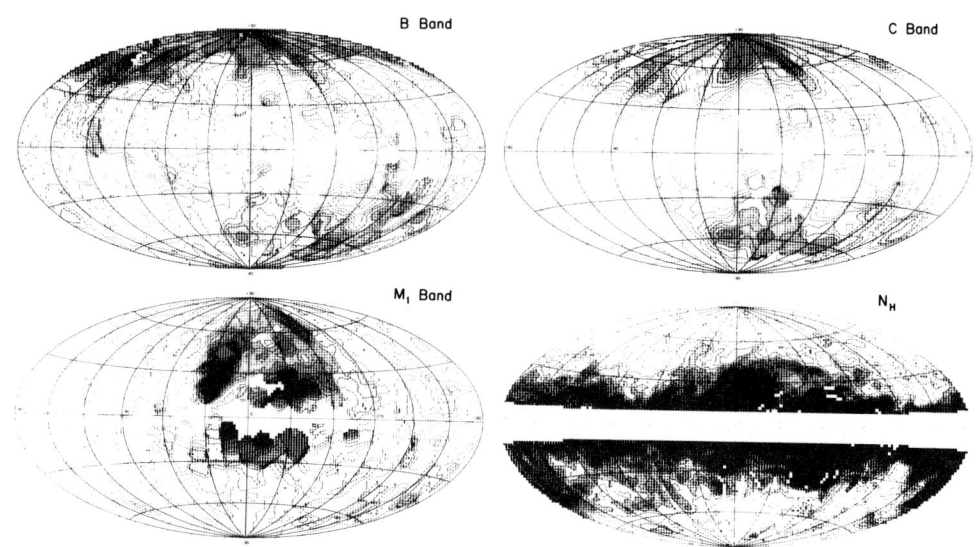

Figure 4. The diffuse B(0.13-0.188 KeV), C(0.16-0.284 KeV) and
M_1(0.440-0.93 KeV) band X-ray maps of McCammon et al.(1983) are shown in an
Aitoff projection centered on l = 0 with l increasing to the left. The degree
of shading is proportional to the observed count rate. A neutral hydrogen
column density map for gas between -90 and +70 km/s is also shown. The
anti-correlation between the B and C band data and N(HI) is very apparent. The
C band map shown here is very similar to the SAS-3 C band map of Marshall and
Clark(1984).

with the third Small Astronomy Satellite (SAS-3). The data from Marshall and
Clark (1984) are illustrated in Figure 4 along with the neutral hydrogen survey
measurements. The SAS-3 measurements shown are for their C band and have
an angular resolution of 2.7^0. This independent all sky survey agrees quite
favorably with the maps published by McCammon et al. (1983).
    The intensity of diffuse X-rays in a given direction is
$$I_{\upsilon} = 1/4\pi \int j_{\upsilon} (x) \exp[- \sigma_{\upsilon} N_H(x)]dx , \quad (1)$$

where j_{υ} (photons cm^{-3} s^{-1}eV^{-1}) is the emission coefficient and $\exp[- \sigma_{\upsilon}$
$N_H(x)]$ allows for line of sight absorption by cooler gas of total hydrogen
column density N_H. The effective absorption cross-section for the absorbing
gas ,σ_{υ} , is mostly due to photoelectric absorption and for E < 0.3 keV about
one third of the absorption is produced by hydrogen and almost all the rest is
produced by helium (see Morrison and McCammon 1983).   The emission
coefficient $j_{\upsilon} = \varepsilon_{\upsilon}(T) n_e^2$, where the emissivity $\varepsilon_{\upsilon}(T)$ (photons cm^3
s^{-1}eV^{-1}) depends on the assumed abundances and the gas temperature (see
Figure 3). Therefore, ignoring the absorption, diffuse X-ray intensity
measurements provide information on the temperature and emission measure,
Em = $\int n_e^2 dx$, of the hot emitting gas.

In actual practice it is necessary to allow for the finite angular and
energy resolutions of the X-ray measurements when applying equation (1) to
real data. We see that the interpretation of the diffuse X-ray emission
requires a good understanding of the line of sight geometry of the hot emitting
gas **and** cooler absorbing gas. Reaching optical depth 1 in the absorbing
medium in the various bands requires larger column densities as the band
energy increases. The approximate values for the Be, B, C and M_1 bands are
$1x10^{19}$, $6x10^{19}$, $1.1x10^{20}$ and $1.1x10^{21}$ atoms cm^{-3}, respectively. Therefore,
the lowest energy X-rays are likely created in the most local gas.

3.OBSERVATIONAL RESULTS

3.1 Hot Gas in the Galactic Disk

3.1.1 *Absorption Line Studies* . The first detection via absorption line
spectroscopy of the hot phase of the ISM came with the O VI measurements of
Rogerson et al. (1973) shortly after the launch of the Copernicus satellite in
1972. Those early measurements were soon followed up with more extensive
work (Jenkins and Meloy 1974 ; York 1974,1977; Jenkins 1978a and b).
Although there have been a number of reviews of this subject (McCray and Snow
1979; Jenkins 1981,1985), many of the fundamental observational results are
contained in the two 1978 papers of Jenkins which report a survey of O VI
absorption toward 72 stars. The derived O VI column densities show a
convincing correlation with path length which points to the interstellar
(rather than circumstellar) origin of the absorption. The large fluctuations in
the correlation between column density and distance provide information about
the irregular distribution or patchyness of the O VI absorption. The data
suggest 6 absorbing regions per kpc each with an O VI column density of about
10^{13} cm^{-3}. This degree of irregularity is very similar to that observed for
the neutral interstellar gas. In about 10 percent of the observed directions
denser volumes of O VI absorbing gas are observed. Ignoring the denser regions,
the average line of sight O VI density derived is about $2x10^{-8}$ cm^{-3}.
The multicomponent O VI absorption line profiles are consistent with gas
components having an internal velocity dispersion of 10 km s^{-1} and moving at
random with an rms dispersion of radial velocity of 26 km s^{-1}. The relatively
narrow width of many O VI profiles (v_{rms} < 14 km s^{-1}) can be used to
constrain the temperature of the absorbing gas. Assuming thermal doppler
broadening with gas at one temperature the result is T< $3.8x10^5$ K. However,
the data are also compatible with a distribution of temperatures, provided the
fall off to higher temperatures of the O VI density is steep enough.
Ions other than O VI can also provide information about the nature of the
absorbing gas. The Copernicus satellite was able to detect N V along a number
of lines of sight (York 1977) with the result< N(N V)/N(O VI)> $\approx$ 0.08.
Savage and Massa (1986) used the IUE satellite to determine the distribution of
Si IV , C IV and N V in the galactic disk and halo. By emphasing observations
of B type stars not located in H II regions, their measurements should refer
to the abundance of these ions in the general interstellar medium rather than
the photoionized regions surrounding hot O-type stars. The observed column
densities were found to roughly correlate with distance yielding average
mid-plane densities of Si IV, C IV, and N V of $2x10^{-9}$, $7x10^{-9}$ and $3x10^{-9}$
cm^{-3}, respectively. The possibility that the gas producing the Si IV , C IV and

N V absorption is warm and photoionized versus hot and collisionally ionized is discussed in §4.

The interrelationship between lines of high and lower ionization may provide clues about the origin of the O VI absorption. In a study of the velocity correspondence between OVI absorption and absorption produced by Si III and N II , Cowie et al. (1979) found that the three species are correlated in velocity space. This correlation led Cowie et al. to interpret their results as demonstrating that the O VI absorption arises in intermediate temperature gas in the interface region between cool clouds and a hot exterior medium. In this situation one would expect a good correspondence between absorption line velocities of cooler cloud material and intermediate temperature gas in the conductive interface region. However, the interface origin for O VI absorption is not firmly established (see §4).

3.1.2. *Diffuse X-ray Emission Studies.* The recognization that a local hot gaseous component of the ISM is producing soft X-ray emission (Williamson et al. 1974) occurred at nearly the same time as the first ultraviolet detections with the Copernicus satellite of O VI absorption. Since those early results, the diffuse soft X-rays from the entire sky have been mapped with angular resolutions ranging from 7 to 3 degrees (McCammon et al. 1983; Marshall and Clark 1984) in the various broad X-ray bands discussed in §2.2. The primary observational results are as follows: With the exception of a northern hemisphere feature known as the north polar spur, the angular distribution of the B (0.13-0.188keV) and C(0.16-0.284 keV) band diffuse emission is very similar with the intensity at high galactic latitudes being 2 to 3 times higher than that in the galactic plane (see Figure 4). The intensity in the plane is approximately constant with galactic longitude. There is a general anti-correlation between X-ray intensity and HI column density (see Figure 4).

The relatively small absorption mean free path for the B and C band X-rays implies that the emission near the galactic plane must originate within a few hundred parsec of the sun. The most viable explanation for the diffuse emission is thermal emission from hot interstellar gas. If the emitting gas is in collisional equilibrium the observed B to C band intensity ratio for the low latitude emission is consistent with gas near 10^6 K with an emission measure of approximately 0.002 cm^{-6}pc. If the emitting gas uniformly fills the volume out to 100 pc, this emission measure implies an electron density $n_e \approx 0.004$ cm^{-3} and a pressure $P/k = 2.1$ n_e $T \approx 8000$ cm^{-3} K.

At a temperature of about 10^6 K, emission from hot gas is mostly in lines of highly ionized atoms (see Figure 3). Although the data have low resolution, there is evidence from spectroscopic observations with solid state detectors for emission from C V, C VI and O VII in the diffuse background (Inoue et al. 1979; Schnopper et al. 1982; Rocchia et al. 1984). These results provide a very convincing case that the broad band soft X-ray measurements do indeed refer to emission from hot gas.

The anticorrelation of the soft X-ray intensity and H I column density as measured from its 21-cm emission and the nearly constant B to C band ratio is very difficult to understand. The most obvious interpretation for the anti-correlation would be that it suggests the existence of an absorbed component to the diffuse background such that the observed background is given by an expression of the form,

$$I_\nu = I_\nu(\text{local}) + I_\nu(\text{distant}) \exp[-\sigma_\nu N(HI)] ,$$

where I_v(local) is a local contribution to the background and I_v(distant) comes from more distant regions of the galaxy and is reduced in intensity by absorption from cool intervening gas. However, such a simple model would predict a different behavior for the B band than for the C band X-rays since the absorption cross-section for the lower energy B band X-rays is larger by a factor of 2. A way around this problem is to clump the absorbing gas into clouds which are optically thick to B and C band X-rays but thin to 21 cm radiation. Clouds having $N(HI) \approx 2 \times 10^{20}$ cm^{-2} would work but the required clumping properties are inconsistent with the angular distribution of the emission observed in the 21 cm line (Jahoda et al. 1985). Another possible explaination for the anti-correlation is to assume that a "displacement " of hot by cool gas occurs and this displacement reduces the hot gas emission measure in those directions where the amount of cool gas is large (McCammon et al 1983). This problem has been a topic of great debate in recent years (eg. see McCammon 1985; Clark 1985; Jakobsen and Kahn 1986). A direct test for the location of the X-ray emitting material would be to look for absorption of the diffuse X-rays against absorbing clouds with known distances. This proceedure has been attempted for a few clouds with distances greater than 150 pc with negative results(McCammon 1985). Measurements in the low energy Be (0.078 -0.111keV) band may eventually help to resolve the issue since optical depth one for these X-rays occurs for $N(H) = 1 \times 10^{19}$ cm^{-2}. The first results for the Be band (Block et al. 1986) show a close correlation between Be and B band data for an arc extending over 120^0 on the sky. This result suggests that most of the X-rays measured in the Be and B bands occurs from regions with intervening absorption by cooler gas which varies by less than 2×10^{18} atoms cm^{-2} over the observed region (Block et al. 1986). It will be very important to obtain additional Be band data to pursue the question of where the X-ray emitting gas is located.

The angular distribution of the diffuse X-ray background at higher energies is completely different than that seen in the B and C bands. The M_1 (0.440-1.10 keV) band emission is bright in the direction of the galactic center and in the region associated with the north polar spur but is nearly isotropic elsewhere . The large mean free path for these X-rays ($N(HI) \approx 1.1 \times 10^{21}$ cm^{-2}) means very distant contributions to the background may be involved. Detailed models for the origin of the nearly isotropic M_1 band emission are given by Sanders et al.(1982) and Nousek et al.(1982). These models balance the contributions from various anisotropic sources (extragalactic background, hot halo gas, dwarf M stars, and a distribution of hot disk regions) to produce an isotropic result. The models are complex and do not rule out a very local contribution from hot gas.

3.2.Hot Gas in the Galactic Halo

3.2.1.*Absorption Line Studies.* The O VI survey of Jenkins (1978a and b) did not reach further than about 1 kpc from the galactic plane because the satellite could not track on stars fainter than about 7'th magnitude. From the limited data available, Jenkins estimated the O VI exponential scale height is 0.3 (+0.2 , -0.15) kpc. However, the estimate was based on only a few stars and is exceedingly uncertain. In searching for systematic motions of gas away from the galactic plane ,Jenkins found for a group of 15 stars with $|z| > 0.2$ kpc that the average motion in the z direction is v_z(O VI) = -8±6 km s^{-1}.

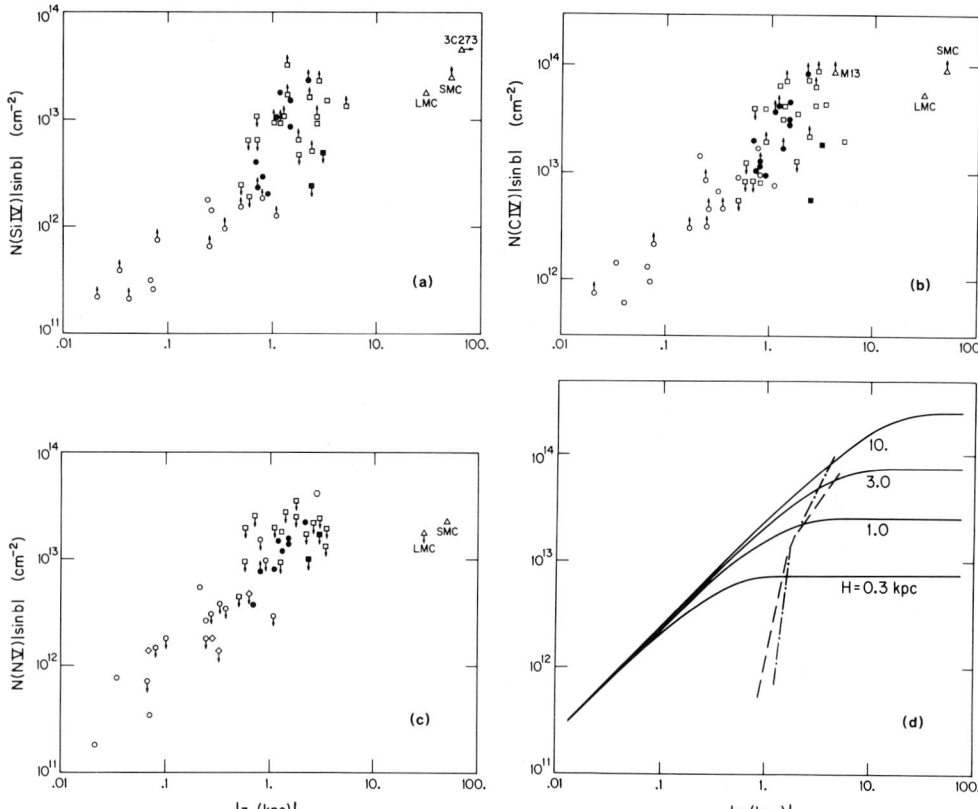

Figure 5. In (a), (b), and (c), N(ion)|sinb| for SiIV, C IV and N V is plotted
against |z(kpc)| for approximately 60 stars located at various distances away
from the galactic plane. The figure is from Savage and Massa(1986) and
includes data for approximately 30 B stars from their survey plotted as open
and closed circles. The results from the Pettini and West(1982) halo gas survey
are shown as squares. Extragalactic measurements from Savage and deBoer
(1981) and York et al. (1983) are shown as triangles. Many of the data points
are plotted as lower limits to N(ion)|sinb| because the doublet ratio measured
was less than 1.5 (see §2). In (d) the solid lines illustrate the expected
behavior of N(ion)|sinb| versus |z| if the gas has a simple exponential density
distribution with scale heights of 0.3, 1.0, 3.0, and 10. kpc. These curves are
most easily compared to the data by making a transparency of the figure and
placing it over the data points. The data points are very roughly described by
an exponential distribution with a scale height of about 3 kpc. However, these
estimates are strongly influenced by the few extragalactic data points (LMC,
SMC, and 3C273). The Si IV and C IV measurements suggest the presence of an
abrupt (factor of 2 to 3) increase in N(ion)|sinb| over the prediction of the
simple exponential curve near |z| ≈ 1 kpc. This increase could be the result
of the photoproduction of Si IV and C IV by the extragalactic EUV radiation.
The dashed and dot-dashed lines in (d) show the model results for the
photoionized halo calculations of Chevalier and Fransson (1984) for Si IV and C
IV, respectively.

Because of the tracking limitations of Copernicus, most of our information about highly ionized halo gas from ultraviolet absorption line studies has come from the IUE satellite. IUE's wavelength coverage is limited to the wavelength region longward of 1150 A which unfortunately does not include the O VI doublet (see Table 2). However, IUE has provided an enormous amount of information on the distribution of Si IV, C IV and N V in the galactic halo (for reviews see York 1982 and Savage 1986). Under conditions of coronal ionization, these species peak in abundance at temperatures of 0.6,1.0 and 2.0×10^5 K, respectively. However, the photon energy required to produce these ions is low enough (see Table 2) that photoionization in cooler gas $(T \approx 10^4 K)$ must be considered as a production process see (§4.2).

Figure 5a, b and c can be used to study the stratification of the highly ionized halo gas away from the galactic plane. The figure shows for about 60 stars $N(ion) |\sin b|$ versus $|z|$ for Si IV , C IV and N V.   Here $N(ion) |\sin b|$ is the projection onto the axis perpendicular to the galactic plane of the column density measured to each star and $|z|$ is the distance of each star away from the galactic plane. The solid lines in Figure 5d show the expected behavior if the gas has a simple exponential density distribution $n(z) = n(0)\exp(-|z|/H)$ with scale heights H = 0.3, 1.0, 3.0 and 10. kpc. The data of Figure 5a ,b and c are roughly consistent with a scale height of approximately 3 kpc. However, it is clear that a simple exponential distribution is only a crude representation of the data. In particular there seems to be evidence for an enhancement in the Si IV and C IV density over the simple exponential curve curve for z near 1 kpc as first noticed by Pettini and West (1982). At a given z distance, the data of Figure 5 suggest a large range for $N(C\ IV) |\sin b|$ and $N(SI\ IV) |\sin b|$. This implies the highly ionized gas of the halo has a patchy spatial distribution similar to that found for O VI in the disk gas. For the three highly ionized atoms accessible to the IUE , the column densities perpendicular to the disk through the halo are $N(Si\ IV) \approx 1.0 \times 10^{13} cm^{-2}$ $N(C\ IV) \approx 1.0 \times 10^{14}$ cm^{-2} , and $N(N\ V) \approx 1.0 \times 10^{13}$ cm^{-2}.

Although the emphasis of this section is the highly ionized and perhaps hot halo gas, it is important to recognize that weakly ionized and neutral gas also extends to large distances away from the galactic plane. In the case of H I, it is known that the z distribution is best described by two exponentials, one with a scale height of about 0.12 kpc (Spitzer 1978) and the other, a more extended distribution with a scale height of about 0.5 kpc (Bohlin, Savage and Drake 1979; Lockman 1984; Lockman, Hobbs and Shull 1986).

3.2.2.*Diffuse X-Ray Emission Studies*. It is difficult to decide if diffuse X-rays from hot gas in the Milky Way halo ($|z|$ > 0.5 kpc) have or have not been detected. The interpretation of the X-ray data depends on the assumptions about the line of sight density distribution of the hot emitting gas and the cool absorbing gas. Different groups have arrived at very different results using similar data (see McCammon 1986 for a brief review). The data which has been interpreted as containing a halo contribution to the diffuse X-ray background is the MI band (0.5 to 1.0 keV) data of Nousek et al. (1981) and the SAS-3 C band data of Marshall and Clark(1984). In both cases, the measurements are compatible with the existence of hot halo gas with an emission measure perpendicular to the galactic plane of about 0.004 cm^{-6} pc and a temperature of about $2-3 \times 10^6$ K. If this gas has a $|z|$ extension of about 3 kpc which is the value inferred from the IUE Si IV, C IV and N V data its density would be about 10^{-3} cm^{-3}. However, it is important to emphasize that both data sets are also be consistent with a more local origin for the emission.

3.2.3.*Diffuse Ultraviolet Emission Studies.* Information about hot halo gas can also be obtained from diffuse ultraviolet emission studies. Feldman, Brune and Henry (1981) have obtained a low resolution spectrum of the diffuse far-ultraviolet background radiation near the north galactic pole. Various emission lines seem to be present in the data but their identities are not certain because of the low resolution of the measurements. An additional complication is that some of the emission might originate in the terrestrial atmosphere above the sounding rocket. Even with all these qualifications, the data can be used to set constraints on the nature of the plasma near 1 to 3×10^5 K in the direction of the north galactic pole (Paresce et al. 1983). However, the measurements contain no information on the distance of the emitting gas from the galactic plane. If data with higher spectral resolution and photometric quality can be obtained, ultraviolet diffuse emission measurements should provide valuable new constraints on the physical state of hot interstellar and halo gas.

4.THEORETICAL CONSIDERATIONS

4.1.Hot Gas in the Galactic Disk

Two quite independent observational techniques have provided information about the existence of hot gas in the galactic disk. The ultraviolet measurements of O VI and other highly ionized atoms are a good probe of gas with temperatures near the values of T_{max} listed in Table 2 (ie. gas near 1 to 3×10^5 K). The diffuse soft X-ray emission measurements mostly probe gas with T near or greater than 10^6 K. When making generalizations about the large scale aspects of the ISM it is important to note that the two techniques sample different volumes of space. The ultraviolet data refer to absorbing gas between the earth and the background source which in the case of the O VI measurements range from 20 to 3200 pc. However, the B and C band diffuse X-ray measurements only provide information on gas out to about one X-ray optical depth which is from less than 50 to more than 300 pc in the galactic plane depending on the viewing direction. Since we know the ISM is very inhomogeneous, we must be careful when generalizing from measurements which may sample only a very small volume of space.

In order to interpret the data pertaining to hot gas in the ISM it is helpful to start with a mental picture for the general distribution of the various components or phases of the ISM. This is particularly important in interpreting the X-ray data since the measurements depend in a complex way on the distribution of hot emitting and warm, cool or cold absorbing gas. In this discussion we will start out by considering the data within the framework of recent detailed models of the ISM and then indicate some possible inconsistencies or limitations of these models in explaining the observations.

The discovery of the hot phase of the ISM motivated the theoretical supernova dominated ISM models of Cox and Smith (1974) and McKee and Ostriker (1977). Cox and Smith first pointed out that if supernova explosions are frequent and energetic enough, their individual remnants will overlap and eventually fill a major portion of the galactic volume with hot gas. In this view, the hot ISM assumed a tunnel like structure quite analogous to the holes in Swiss cheese. In their very detailed model of the ISM, McKee and Ostriker took a more radical view and proposed a medium mostly filled with hot supernova heated gas in which clouds and filaments of cooler neutral gas are embedded. Surrounding the neutral clouds would be halos of warm neutral and warm ionized gas. In this theory, the hot exterior medium produces the diffuse X-ray background while the O VI absorption occurs in the conductive interface region between the cool clouds and the hot exterior medium. Weak evidence supporting the conductive interface idea was provided by the O VI, Si III and N II absorption line velocity correlations of Cowie et al. (1979). The interface theory for the origin of O VI absorption has the important advantage of being able to explain the relatively small rms dispersion (26 km s^{-1}) among individual O VI components. If instead , the O VI is created directly in shock heated gas , large shock speeds of about 150 km s^{-1} are required to heat the gas to the temperature, 3×10^5 K , needed to produce O VI. In this situation one would expect to measure a large velocity dispersion among the different O VI absorbing components.

According to the equilibrium collisional ionization estimates of McKee and Ostriker (1977) each interface between cold gas and a hot exterior medium will produce an O VI column density of about 5×10^{11} cm^{-2} and therefore 120 such interfaces are required per kpc to explain the average observed space density of O VI. However , the statistical analysis of the patchy distribution of interstellar O VI (Jenkins1978b) implies that the individual O VI absorption occurs in units of about 10^{13} cm^{-2} with an estimated 6 such absorbing regions per kpc. Very recently Ballet, Arnaud and Rothenflug(1986) have carried out non-equilibrium ionization calculations of the expected amount of O VI and other ions in the intermediate temperature gas surrounding evaporating clouds and find the column densities of the highly ionized species are enhanced by an order of magnitude compared to the equilibrium ionization results. These authors go on and say that the conductive interface idea for the origin of O VI may have major problems because their calculations predict too much O VI, N V and C IV regardless of the (reasonable) interstellar parameters adopted. Clearly more theoretical work is needed to evaluate the likely contributions to the observed O VI column density from cloud interfaces.

In a new theory of the general ISM, Cox(1986) has proposed a model which provides another explanation for the origin of interstellar O VI. In this model it is concluded that the neutral intercloud phase of the ISM with n ≈ 0.1 atoms cm^{-3} is not unstable to disruption by supernovae into a medium with a high filling factor of coronal gas. The intercloud phase is instead quite stable being merely pushed about by large scale remnants. These remnants produce hot bubbles with T > 4×10^5 K and radius of order 50 pc. In this theory the gas in the cooling bubbles would provide the observed individual O VI absorbing sites. The filling factor of these bubbles is estimated to be about 0.2 although it could be large enough to produce a network of tunnels as originally postulated by Cox and Smith(1974).

On a more local scale it now seems likely from the diffuse X-ray observations that we actually live within a "middle aged" supernova remnant with a radius of perhaps 100 pc containing hot gas with T≈10^6 K and n_e ≈ 0.004 cm^{-3} (Cox and Anderson 1982; Edgar 1986). It is possible that the observed local hot gas survived from as long ago as 2×10^7 years, but it could have been reheated by a supernova occurring as recently as 10^5 years (Cox and Snowden 1986).

4.2. Hot Gas in the Galactic Halo

It appears well established that *highly ionized* gas (eg. Si IV, C IV and N V) exists at large distances (1 to 3 kpc) from the galactic plane (see §3.2). A theory for the origin of this gas must explain what supports it at such large distances from the galactic plane and what ionizes it. Two competing ideas for the support of the gas are the galactic fountain model (Shapiro and Field 1976; Bregman 1980; also see Chevalier and Oegerle 1979 and Habe and Ikeuchi 1980) and the cosmic ray supported halo models of Chevalier and Fransson (1984). The competing ideas for the production of the highly ionized species include electron collisional ionization in 0.6 to 2×10^5 K gas and photoionization by hot white dwarfs (Dupree and Raymond 1983), by normal population I stars and by the extragalactic EUV background. A number of detailed calculations have been performed to evaluate the production of the highly ionized gas by photoionization (Hartquist, Pettini and Tallant 1984; Fransson and Chevalier 1985; Bregman and Harrington 1986). From these calculations it seems possible to understand the observed amounts of Si IV and C IV from photoionization by the EUV background. However, the various calculations are not able to produce enough N V. It is also unlikely that the required amount of N V can be produced by stellar sources. This is because N V requires 77 eV for its production and most hot stars which contain He have strong absorption edges at 54 eV. The only stellar sources that might be capable of converting N IV into N V are the very hot hydrogen white dwarfs (Dupree and Raymond 1983). However, they do not appear to be numerous enough to explain the results since exceedingly hot ones are required (Savage and Massa 1985).

A full explanation for the support and ionization of halo gas will probably require a combination of ideas from the photoionized halo models and the galactic fountain models. In the galactic fountain model hot (10^6K) disk gas which is bouyant (thermal scale height $\approx$ 7 kpc) flows outward away from the galactic disk, cools and returns to the plane as falling clouds. In this situation, the Si IV, C IV and N V ions would be expected to exist in the collisionally ionized cooling gas of the flow. However, such a flow will occur in the presence of sources of radiation (extragalactic EUV and stellar EUV) which are capable of producing Si IV and C IV. Therefore some of these ions may be produced by photoionization. The EUV extragalactic background radiation produces Si IV and C IV in the photoionized halo models of Chevalier and Fransson(1984). These models also predict the $|z|$ distribution for these ions shown in Figure 5d. The theoretical curve for N V is not illustrated; it lies below the curves shown by a factor of 60. With the source of ionizing radiation exterior to the galaxy , the expected run of column density with $|z|$ rises abruptly near $|z| \approx 1$ kpc. There is some indication in the data of Figure 5a and b for a sharp (factor of 2 to 3) rise in $N(ion)|\sin b|$ near 1 kpc as predicted by the photoionized halo models. The complex picture that finally emerges is therefore a blending of the ideas of the galactic fountain (and/or galactic convection) model and the photoionized halo models.

ACKNOWLEDGEMENTS. I acknowledge helpful discussions with many members of the Wisconsin X-ray astrophysics group. This research was in part supported by NASA grant NAG5-186 to the University of Wisconsin.

REFERENCES
Baliunas,S.L., and Butler,S.E. 1980, *Ap.J. (Letters)*,**235**, L45.
Ballet, J.,Arnaud, M.,and Rothenflug, R., 1986,*Astr. and Ap.* ,(in press).
Bloch,J.J.,Jahoda, K.,Juda, M., McCammon, D. and Sanders, W.T. 1986, (preprint).
Bohlin, R.C., Savage, B.D., and Drake, J.F. 1978,*Ap.J.* ,**224**,132.
Bregman,J.N. 1980,*Ap.J.* ,**236**,577.
Bregman, J.N., and Harrington,P.J. 1986,*Ap.J.* , (in press).
Chevalier, R. A., and Fransson, C. 1984,*Ap.J. (Letters)*,**279**,L43.
Chevalier, R. A., and Oegerle, W.R.1979,*Ap.J* , **227**,398.
Clark,G.W. 1985, in*IAU Colloquium No 81:The Local Interstellar Medium* , eds.
    Y.Kondo, F.C.Bruhweiler,and B.D.Savage(Greenbelt:NASA CP 2345), p.204.
Cowie,L.L.Jenkins, E.B.,Songaila,A.,and York, D.B.1979,*Ap.J.* ,**232**,467.
Cox, D.P. 1986,in *Proceedings of the MeudonWorkshop on Model Nebulae* ,
    ed. D. Pequinot (in press).
Cox,D.P., and Anderson, P.R. 1982,*Ap.J.* ,**253**,268.
Cox,D.P.,and Smith,B.W. 1974, *Ap.J.(Letters)* ,**189**,L105.
Cox,D.P.,and Snowden,S.L.1986, in*Advances in Space Research* ,eds.C.Gry and
    W.Wamsteker (Pergaman Press) (in press).
Dupree, A.K., and Raymond, J.C.1983,*Ap.J.(Letters)* ,275,L71.
Edgar,R.J. 1986, Wisconsin Astrophysics No 236.
Feldman,P.D.,Brune,W.H.,Henry,R.C. 1981,*Ap.J.(Letters)* ,**249**,L51.
Fransson, C., and Chevalier, R. A. 1985, *Ap.J.* , **296** ,35.
Habe,A., and Ikeuchi,S. 1980,*Prog.Theor. Phys.* ,**64**,1995.
Hartquist, T.W., Pettini, M., and Tallant, A. 1984, *Ap.J.* , **276**,519.
Hobbs,L.M. 1985,*Ap.J.* , **298**,357.
Hobbs,L.M., and Albert,C.E. 1984,*Ap.J.* , **281**,639.
Inoue,H., Koyama, K.,Matsuoka, M.,Ohashi,T., Tanaka, Y., and Tsunemi, H. 1979,
    Ap.J. (Letters),**227**,L85.
Jahoda,K.,McCammon, D.,Dickey, J.M., and Lockman,F.J. 1985,*Ap.J.* ,**290**,229.
Jakobsen,P. and Kahn,S.M. 1986, (preprint).
Jenkins, E. B. 1978a, *Ap.J.* , **219** ,845.
_____. 1978b, *Ap.J.* , **220**,107.
_____. 1981, in *The Universe at Ultraviolet Wavelengths* , ed. R.D.
    Chapman, (Greenbelt: NASA CP2171), p.541.
_____. 1985, in *IAU Colloquium No 81:The Local Interstellar Medium* ,
    eds.Y.Kondo, F.C.Bruhweiler,and B.D.Savage(Greenbelt:NASA CP 2345),
    p.155.
Jenkins,E B. and Meloy,D.A. 1974,*Ap.J.(Letters)* ,**193**,L121.
Lambert,D.L. 1978,*M.N.R.A.S.* ,**182**,249.
Lockman, F. J. 1984, *Ap.J.* ,**283**, 90.
Lockman, F.J., Hobbs, L. M., and Shull, M. 1986, (preprint).
Marshall,F.J., and Clark,G.W. 1984,*Ap.J.* , **287**, 633.
McCammon,D.,Burrows,D.N.,Sanders,W.T.,and Kraushaar,W.L. 1983,*Ap.J.* ,**269**,
    107.
McCammon,D. 1985, in*IAU Colloquium No 81:The Local Interstellar Medium* ,
    eds.Y.Kondo, F.C.Bruhweiler,and B.D.Savage(Greenbelt:NASA CP2345),p195.
McCammon,D.1986, in *Proceedings of the NRAO Conference on Gaseous Galactic
    Halos* , eds. J.N.Bregman and F.J.Lockman (Greenbank: NRAO SP),in press.
McCray,R.C. 1986 (preprint).
McCray,R.C.,and Snow,T.P. 1979, *Ann. Rev. Astr. Ap.* ,**17**,213.
McKee,C.F.,and Ostriker,J.P. 1977,*Ap.J.* ,**218**,148.

Morrison,R.,and McCammon,D. 1983,*Ap.J.* , **270**,119.
Nachman,P.,and Hobbs,L.M. 1973,*Ap.J.* ,**182**,481.
Nousek,J.A.,Fried,P.M.,Sanders,W.T.,and Kraushaar,W.L. 1982,*Ap.J.* ,**258**,83.
Paresce,F.,Monsignori Fossi,B.C.,and Landini,M. 1983, *Ap.J.(Letters)* , **266**,L107.
Pettini, M.,and West, K.A. 1982, *Ap.J.* , **260** ,561.
Pettini, M., and D'Odorico, S.1986,(preprint).
Raymond,J.C.,and Smith,B.W.1977, *Ap.J. Suppl.*, **35**, 419.
Rocchia,R.,Arnaud,M.,Blondel,C.,Cheron,C.,Christy,J.C.,Rothenflug,R.,Schnopper,H.
        W., and Delvaille,J.P. 1984,*Astron.Ap.* ,**130**,53.
Rogerson,J.B.,York,D.G.,Drake,J.F.,Jenkins,E.B.,Morton,D.C.,and Spitzer,L. 1973,
        Ap.J.(Letters) ,**181**,L110.
Sanders, W.T.,Burrows,D.N.,Kraushaar,W.L.,and McCammon,D. 1982,in*IAU
        Symposium No.101,Supernova Remnants and their X-ray Emission,*
        ed.J.Danziger and P.Gorenstein(Dordrecht:Reidel).
Savage,B. D. 1984, in *Future of UV Astronomy Based on Six Years of IUE
        Research* , eds.J.M.Meade, R.D.Chapman, Y.Kondo (Greenbelt: NASA
        CP 2349), p.3.
_____. 1986, in *Proceedings of the NRAO Conference on Gaseous Galactic
        Halos* , eds. J.N.Bregman and F.J.Lockman (Greenbank: NRAO SP),in press.
Savage, B.D., and deBoer, K.S. 1979, *Ap.J. (Letters)* , **230**, L77.
_____ . 1981, *Ap.J.* , **243**,460.
Savage, B.D., and Massa, D. 1985, *Ap.J. (Letters)* ,**295** , L9.
_____. 1986,*Ap.J.* , (in press).
Schnopper,H.W.,Delvaille,J.P.,Rocchia,R.,Blondel,C.,Cheron,C.,Christy,J.C.,Ducros,
        R.,Koch,L.,and Rothenflug,R. 1982,*Ap.J.* ,**253**,131.
Shapiro, P.R., and Field, G.B. 1976, *Ap.J.* , **205**,762.
Shull, M., and VanSteenberg,M.E. 1985,*Ap.J.* ,**294**,599.
Spitzer, L. 1956, *Ap.J.* , **124**, 20.
_____. 1978, *in Physical Processes in the Interstellar Medium* , (New
        York: John Wiley and Sons).
_____. 1986,(Crafoord Lecture preprint).
Williamson,F.O.,Sanders,W.T.,Kraushaar,W.L.,McCammon,D.,Borken,R.,and Bunner,
        A.N. 1974, *Ap.J. (Letters*),**193**,L133.
Withbroe,G.D.1971,in*The Menzel Symposium* , ed.K.B.Gebbie(Washington:NBS
        SP353).
York, D.G. 1974, *Ap.J. (Letters*),**193**,L127.
_____.1977, *Ap.J.* ,**213**,43.
_____. 1982, *Ann. Rev. Astr. Ap.* , **20**,221.
York, D.G., Wu, C.C., Ratchiff, S., Blades, J.C., Cowie, L.L., and Morton, D.C.
        1983, *Ap.J.* ,**274**, 136.

Peter Pesch and Vern Pankonin

THE HIGH-ENERGY COMPONENT OF THE ISM — COSMIC-RAY PHENOMENA

Hans Bloemen
Astronomy Department
University of California, Berkeley, CA 94720

ABSTRACT.   Salient properties of cosmic rays and their interactions with the other interstellar components are summarized. It is concentrated on observational constraints, both from direct detections of the energetic particles near the earth and from indirect detections by low-frequency radio-continuum and high-energy γ-ray observations, with the emphasis on recent γ-ray results. In addition, γ-ray constraints on the molecular hydrogen content of the Galaxy are described. The role of cosmic rays among the magnetic-field, gaseous, and stellar components in a possible large-scale hydrostatic equilibrium of the Galaxy is discussed.

1.   INTRODUCTION

High-energy astrophysics of the interstellar medium (ISM) is governed by cosmic-ray (CR) particles. As the cosmic rays are tied to the magnetic field, which is confined to the interstellar gas threaded by the field, there is an intimate relation between these energetic particles and the other interstellar components. The energy density of cosmic rays is comparable to that of the interstellar gas (both thermal and from macroscopic motions), the magnetic field, and the interstellar photon field, at least in the solar vicinity (~ 1 eV cm^{-3}), and it is therefore inevitable that cosmic rays play an important role in various interstellar processes and in our understanding of the structure of the Galaxy.

The CR energy density is largely due to protons with energies between 1 and 10 GeV; helium nuclei contribute about 10% to the particle flux arriving near the earth and heavier nuclei and electrons each contribute less than a few percent. Although particles with energies as large as 10^{20} eV have been detected, most of our knowledge about cosmic rays is restricted to energies of 100 MeV – 100 GeV/nucleon, either from direct detections by balloon and satellite experiments or from observations of the high-energy γ-ray and radio synchrotron emission originating from CR interactions with the other interstellar components. Cosmic rays at higher energies, which can be studied mainly through the extensive showers they produce in the earth's atmosphere, are not of direct importance for studies of the ISM and will not be discussed here. Cosmic rays at lower energies ($\lesssim$ 100 MeV) may be important for heating and ionization of the diffuse low-density medium and cloud exteriors, but the information on these particles is scanty. This is due to the increasing impact of solar modulation on the CR spectrum with decreasing energy and to the technical difficulties that have hampered observations of low-energy γ-rays, tracing the low-energy cosmic rays throughout the Galaxy. Particularly the *COMPTEL* γ-ray

D. J. Hollenbach and H. A. Thronson, Jr. (eds.), Interstellar Processes, 143–168.
© *1987 by D. Reidel Publishing Company.*

telescope aboard the *Gamma-Ray Observatory* can be expected to give further insight in the near future. This paper is concentrated on cosmic rays in the 100 MeV – 100 GeV range.

This is not a balanced review of the field. I will give some background information, refer to several review papers where proper references can be found, and concentrate on new developments that may be of interest here. Basic information on CR astrophysics can be found in the monographs of Ginzburg and Syrovatskii (1964), Hayakawa (1969), and Longair (1982). Large collections of up-to-date information are available in the proceedings of the *International Cosmic-Ray Conference*, the most recent ones in Paris (1981), Bangelore (1983), and La Jolla (1985).

2.   ON THE ORIGIN OF COSMIC RAYS

There is no conclusive proof of the origin and trapping region of cosmic rays. The CR electrons have at least to be of galactic origin, because they cannot survive the inverse-Compton losses in the microwave background throughout intergalactic space, but there is even no consensus on this basic point for the CR nuclei — either origin and confinement in the Galaxy or extragalactic origin, with the cosmic rays filling up the volume of a (super)cluster [*cf.* Ginzburg and Syrovatskii (1964) and Ginzburg *et al.* (1980) & Brecher and Burbidge (1972) and Burbidge (1974, 1983)]. A major argument against the extragalactic hypothesis, namely the existence of a strong galacto-centric gradient in the CR density, obtained from high-energy γ-ray observations, has recently essentially vanished (Bloemen *et al.* 1986; §4), but this does obviously not exclude the galactic hypothesis.

A variety of galactic phenomena that could provide the required CR energies has been put forward in the past. Supernova explosions have mostly played a major role in the models (e.g. Colgate and Johnson 1969; Gunn and Ostriker 1969; Kulsrud *et al.* 1972; Scott and Chevalier 1975), but there have always been doubts about the efficiency of the proposed acceleration mechanisms. For CR energies below $\sim 10^5$ GeV, stochastic acceleration processes in the interstellar medium form a promising alternative. In such a scenario, cosmic rays gain their principal energy from (2nd order) Fermi acceleration (Fermi 1949, 195?) or from acceleration by shock waves (based on Fermi's 1st order mechanism) induced mainly by supernova remnants (Krymsky 1977; Axford *et al.* 1977; Bell 1978; Blandford and Ostriker 1978; for a review see Drury 1983). In particular the latter, so called diffusive shock acceleration, has become very popular [see however Ginzburg and Ptuskin 1981], because it produces quite naturally a power-law CR spectrum with a spectral index of 2 (for a strong shock) or somewhat larger, which is consistent with the CR source spectrum inferred from observations (§3.3). In addition, if the interstellar medium can at least to some extent be described by the McKee and Ostriker (1977) model, then diffusive shock acceleration seems hard to avoid and CR acceleration might even be a continuous process of reacceleration of existing cosmic rays (Blandford and Ostriker 1980) [but there are potential problems; §3.3]. Although the non-linear aspects of the shock acceleration mechanism are not fully understood yet (see review by Völk 1984), the mechanism is very attractive and the problem of the origin of cosmic rays ($\lesssim 10^5$ GeV) seems to be shifting towards finding the CR injectors, i.e. the sources that provide the CR material and speed it up to moderate suprathermal energies. On the basis of energetics as well as composition, stellar flares and winds are considered to be likely candidates (§3.1). It has been argued, however, that CR injectors are not needed and that shocks may (or have to) accelerate particles out of a thermal plasma (§3.3; Eichler 1979, 1980; Axford 1981). Obviously several potential sources of energy are available (not yet mentioned here are the possible injection/acceleration mechanisms near rotating compact objects); observations of the elemental and isotopic compositions of cosmic rays are of crucial importance to discriminate among the various alternatives. Observational constraints on the CR origin will be addressed throughout this paper.

3. INFERENCES FROM OBSERVATIONS OF COSMIC RAYS NEAR THE EARTH

Cosmic rays suffer inelastic collisions with interstellar gas nuclei during propagation in interstellar space. Two classes of arriving CR nuclei can be distinguished on this basis. First, a class containing elements which mainly originate directly from the CR source (the so called primary nuclei) and second, a class of nuclei whose largest fraction is produced by nuclear interactions of the primaries during propagation (the secondary nuclei). Although also the composition of the first class is affected in subtle (but important) ways, interstellar propagation has not destroyed the details of the original composition (and thus the injection/acceleration characteristics). The second class forms the basis for CR propagation and confinement models, although in the case of stochastic acceleration in the ISM, the acceleration and propagation processes might occur simultaneously. The observed variations in the secondary-to-primary ratios with energy provide important constraints for the modelling. A complete set of nuclear cross sections, describing the spallation of CR nuclei in collisions with interstellar matter, is required to disentangle the CR source composition and the modifications in interstellar space. A set of semi-empirical spallation cross sections has been developed (Silberberg *et al.* 1985a and references therein), which turns out to be generally in remarkably good agreement with recent direct measurements in accelerator particle beams. Diffusion, convection, and adiabatic deceleration processes, resulting from the solar wind, modify the spectra of all particles reaching the heliosphere; these solar modulation effects are reasonably well understood at energies above ~ 1 GeV/nucleon.

3.1. Primary nuclei & cosmic-ray sources.

A comprehensive review of the CR composition and its energy dependence is given by Simpson (1983). Abundance studies of the CR elements are concentrated around the energy range 100 MeV – 100 GeV/nucleon. At higher energies, the information is still limited. Measurements of the elemental composition of cosmic rays span now the entire periodic table, although the individual nuclear charges of the so called ultra-heavy nuclei $(Z > 28)$ are only partially resolved at the moment. It is concentrated here on the most common elements $(Z \leq 28)$. The abundance distribution of cosmic rays is very similar to that of the solar system and the local region of the Galaxy (Fig. 1), but there are also some significant deviations. The overall similarity indicates that the nuclear species are produced in nucleosynthesis processes. The most obvious deviations between the observed CR composition and the solar-system and local galactic compositions are the overabundances of the spallation products lithium, beryllium, boron, and the sub-iron elements $(20 < Z < 26)$ (Fig. 1). Li, Be, and B are good examples of pure secondary nuclei, because these light elements would have been destroyed during nucleosynthesis. The impact of these secondaries on our understanding of CR propagation and confinement is discussed in §3.3.

There are other deviations, which cannot be attributed to interstellar propagation. Hydrogen and helium are the most obvious underabundant elements in cosmic rays (by roughly an order of magnitude, depending on the normalization), but their fluxes are nevertheless two to three orders of magnitude higher than the fluxes of the most common heavier nuclei. Contributions of secondaries from heavier nuclei to the measured H and He abundances are therefore negligible. Several other (small) deviations between

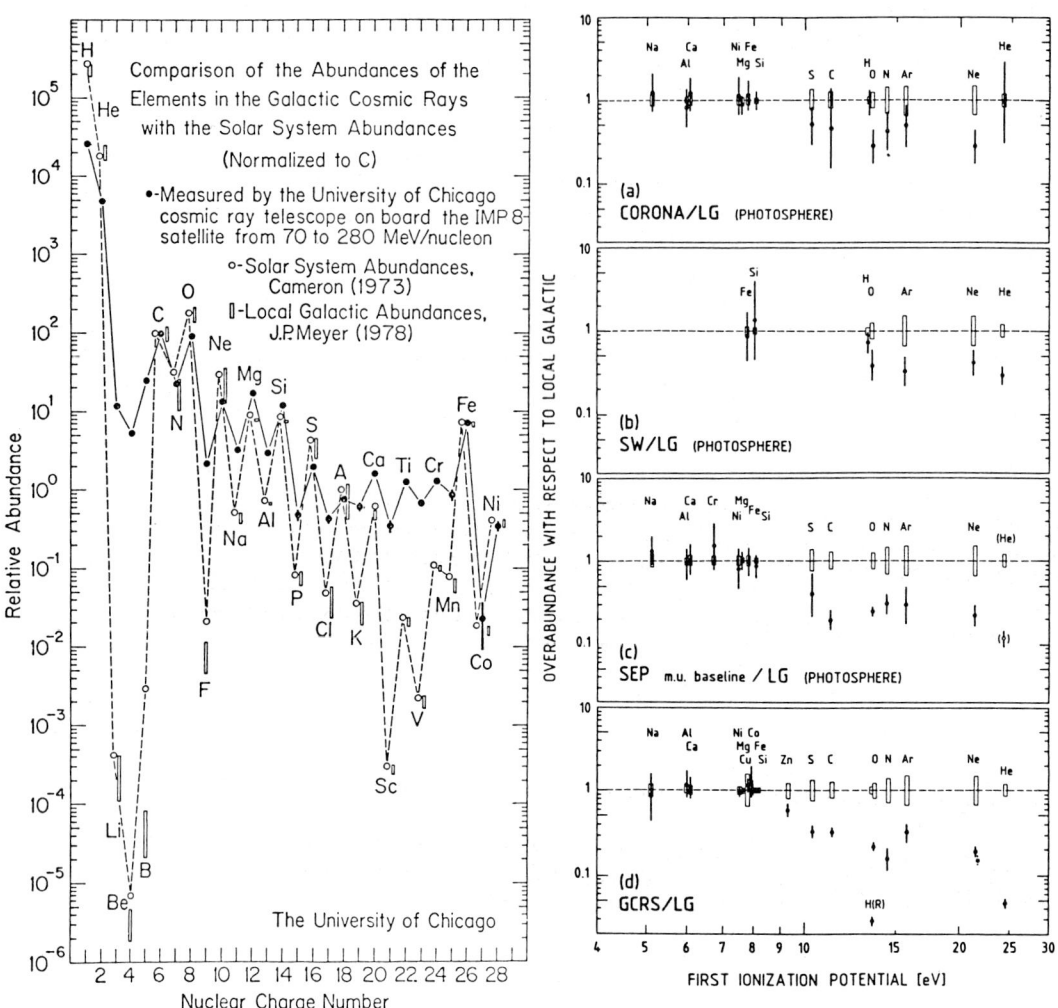

Figure 1 *(left)*: The cosmic-ray abundances for $Z \leq 28$ measured at earth compared to the solar system and local galactic abundances.

Figure 2 *(right)*: Ratios of abundances in (*a*) the solar corona, (*b*) solar wind (SW), (*c*) solar energetic particles (SEP), and (*d*) galactic cosmic ray sources (GCRS) over local galactic (LG) abundances, versus the first ionization potential (black dots and errorbars). The small boxes for each element represent the uncertainties on the local galactic composition. [for details, see Meyer 1985b]

the elemental composition of cosmic rays and that of the the local galactic environment and solar system remain after the fragmentation processes in the ISM have been taken into account. [1] Although these discrepancies seem rather random at first glance, they have been ordered in a very interesting way, namely as a function of the first ionization potential (FIP) of the elements (Fig. 2; see e.g. Cassé and Goret 1978). The low-FIP ($\leq$ 9 eV) CR elements show a clear enhancement compared to the high-FIP elements by a factor of about five. The ultra-heavy elements show the same FIP relation (e.g. Israel 1983). Although not all elements follow smoothly the trend (particularly H and He), it is clear that atomic – rather than nuclear – effects are important during the selection/injection/acceleration process, which implies that the temperature of the primary reservoir has to be moderate ($\sim 10^4$ K). The same FIP relation has been found for solar (flare) energetic particles, the solar corona, and probably the solar wind [2] (Fig. 2; Meyer 1985b and references therein). These remarkable similarities have led to the hypothesis that cosmic rays are extracted from the coronae of solar-like stars [ordinary, unevolved later type stars (F–M)], where the composition might originate from selective filtering of particles rising from the chromosphere (sensitive to whether an element is ionized or not), and then injected into the surrounding ISM by stellar flares and winds, preserving approximately the composition of the corona (Meyer 1985b). Note that the solar wind and solar energetic particles are extracted from the solar corona, which has a temperature of $\sim 10^6$ K, whereas the underlying chromosphere has the required temperature of $\sim 10^4$ K. Diffusive shock acceleration (which is not selective) could boost (part of) the injected particles to relativistic energies, without distorting the elemental abundance distribution. Injection and acceleration have to be sufficiently close in space and time in order to avoid a distortion of the elemental abundance distribution by Z^2-dependent ionization losses (Eichler 1980), but this may be only a weak constraint because of the pick-up of electrons by low-energy particles (Havnès 1982; Meyer 1985b). This model cannot account for the low abundances of H and He in cosmic rays. The behaviour of these light elements may be governed by different processes; they do also not behave like other elements in solar energetic particles.

Whereas in this scenario of selective feeding of the corona the elemental composition of cosmic rays is determined before injection, other stellar models of CR origin proposed in the past decade are concentrated on selective injection (acceleration) from warm stellar plasmas (e.g. Cassé and Goret 1978; Arnaud and Cassé 1985). However, the common FIP relation for cosmic rays, solar energetic particles, corona, and wind is striking and renders the model sketched above favourite (although it has to be further substantiated). In an alternative view, a stellar injection mechanism is not present and particles in the hot phase of the ISM are directly accelerated to CR energies by interstellar shocks (Eichler 1979; Axford 1981). In such a hot plasma of $10^5 - 10^6$ K, shock acceleration favours particles of high mass-to-charge ratio, A/Z^*, where Z^* is the effective charge (Eichler and Hainebach 1981; Ellison et al. 1981). Observations show indeed a global trend, but the predicted (smooth) increase of the CR abundances with A/Z^* cannot account for the observed abundance variations (Cesarsky et al. 1981a). In another non-stellar injection model, cosmic rays are grain-destruction products (Epstein 1980; Tarafdar and Apparao 1981; Bibring and Cesarsky 1981). A clear disadvantage of non-stellar injection models

---

[1] The derivation of these "CR source abundances" is model dependent (§3.2), but the resulting differences have no significant impact on the results presented in this section.

[2] Note the terminology: "solar system abundance" is not the same as "abundance of solar energetic particles, solar corona, etc." Solar system (i.e. photospheric) abundances and local galactic abundances are similar or complementary. [see e.g. Meyer 1985a]

is that the similarity between the CR source composition and the composition of solar energetic particles is purely accidental.

Important information on the nucleosynthesis of cosmic rays can be obtained from abundance measurements of the ultra-heavy nuclei (see review by Israel 1983) and from measurements of the isotopic composition of species that are predominantly of primary origin. These are active areas of CR research, but only a limited amount of data is available at the moment. With regard to the isotopic composition, it has been found that the relative fractions of the most common primary isotopes in cosmic rays are similar to those in the solar system (and solar energetic particles), but there are good indications of weak enhancements of some neutron-rich CR isotopes (^{22}Ne, ^{25,26}Mg, and ^{29,30}Si) [see e.g. review by Wiedenbeck 1983]. An outstanding anomaly is the ^{22}Ne/^{20}Ne isotope ratio, which has an enhancement factor of ~ 4 in cosmic rays. There is no unique explanation of this effect (see review by Cassé 1983); carbon-rich Wolf-Rayet stars appear to be likely candidates and could also explain the overabundance of ^{12}C in cosmic rays compared to solar energetic particles (the most distinctive difference in the elemental compositions).

3.2. Cosmic-ray propagation and confinement models.

The cosmic-ray particles with relatively small rigidities [3] considered here ($R \lesssim 100$ GV) have small gyroradii and must be closely attached to the magnetic field lines. The propagation of these particles in the Galaxy is generally believed to be dominated by a diffusion process. Convection may however play an important role in CR transport away from the galactic disk into the halo (§3.3). Cosmic-ray diffusion along the field lines results from resonant scattering of cosmic rays by hydromagnetic waves on scales of the gyroradius. The rate of scattering perpendicular to the field lines is very low; it has been shown, however, that cosmic rays can be spread around to other field lines by a random walk of field lines in a turbulent magnetic field as well as in irregularities superimposed on a uniform field (see e.g. Jokipii 1973; Ptuskin 1979).

Cosmic rays can generate the resonant hydromagnetic waves themselves when their streaming velocity becomes larger than the Alfvén velocity [$v_A = B/(4\pi\rho)^{1/2}$, where ρ is the density of the medium]. These Alfvén waves in turn scatter the cosmic rays and thus limit the streaming velocity, in principal to the Alfvén velocity (for a review on these effects see Wentzel 1974). Collisions between the charged particles and the neutral particles in the ISM, however, are very effective in dissipating the wave energy (Kulsrud and Pearce 1969); basically the damping rate increases with the density of the neutral particles. The gradient in the gas density perpendicular to the galactic plane may therefore lead to "self-confinement" of cosmic rays (Skilling 1971): beyond some critical height above the plane, where the damping rate is sufficiently low, the cosmic rays can limit their streaming velocity and they will tend to propagate with the local Alfvén velocity whereas below the critical height, they are not scattered and stream freely. The onset of the waves at some distance from the plane acts as a reflecting boundary, preventing free propagation out of the Galaxy. This can be regarded as a basic physical picture of the "leaky-box" model, which has turned out to be very successful in the interpretation of CR observations, as discussed below. If the galactic field is connected to the intergalactic field, as suggested by

---

[3] The rigidity of a particle with charge Ze, mass number A, and momentum p is defined as $R = pc/Ze$ (usually expressed in gigavolts). At relativistic energies ($\gamma \gg 1$), R converges to the ratio between the kinetic energy and the charge of the particle, and is thus proportional to A/Z. For convenience (if $\gamma \gg 1$): $R(\text{GV}) \simeq (A/Z) \cdot E(\text{GeV/nucleon})$, so $R(\text{GV}) \simeq E(\text{GeV})$ for protons and electrons. The gyroradius of a particle in a magnetic field with strength B equals $r_g = R/Bc$, or $r_g \simeq 10^{-6} R(\text{GV})/B(\mu\text{G})$ parsec.

Piddington (1973), the cosmic rays will convect out of the Galaxy along the field lines at essentially the local Alfvén velocity once they have passed the boundary (see however §5 for an alternative picture, suggested by Parker). The growth rate of the waves decreases with decreasing CR density and thus with increasing CR energy. Consequently, high-energy cosmic rays can escape from the Galaxy more readily than those of lower energy (Holmes 1974). Cosmic rays with energies above ~ 100 GeV/nucleon are generally found to be not sufficiently numerous at all to constrain their streaming velocity and to participate in the diffusion process in any phase of the ISM. The mean escape length $\lambda_e(R)$ for cosmic rays of rigidity R (i.e. the average amount of matter traversed before escape from the confinement region; in g cm^{-2}) is $\propto R^{-\delta}$ in this scenario. If the neutral and charged particles have the same distribution perpendicular to the galactic plane, then $\delta \simeq 0.4$; if the scale height of the charged particles is larger, then δ is larger due to the increasing importance of non-linear damping effects in the highly ionized regions beyond some height above the plane (Holmes 1975). In a more inhomogeneous picture of the ISM the situation becomes evidently complex, but it is beyond the scope of the present paper to extend the simple considerations given here [see reviews by Wentzel (1974) and Cesarsky (1980)].

Cosmic-ray diffusion in the Galaxy may also be due to scattering by already existing hydromagnetic waves (see review by Cesarsky 1980), if an appropriate wave spectrum is present [i.e., in particular, a *smooth* spectrum of wavelengths (on scales of the gyroradii) in order to be consistent with the observed smooth CR spectra]. The spectrum of hydromagnetic turbulence in the ISM has to be deduced from a variety of measurements of fluctuations in the density distribution of thermal electrons over a wide range of scale sizes (assuming that magnetic inhomogenities are associated with electron density fluctuations). Armstrong *et al.* (1981) concluded from a compilation of measurements that the power spectrum of fluctuations can most simply be represented by a power-law spectrum $\propto q^{-\alpha}$ with $3.4 \le \alpha \le 3.8$ over $10^{-18} \lesssim q \lesssim 10^{-6}$ m^{-1}, where q is the spatial wave number ($\equiv 2\pi/[\text{scale size}]$). It has been suggested that a hydromagnetic spectrum of this smooth form might result (in the hot phase of the ISM) from a cascade of turbulent energy, from long to successively shorter scales, induced at long scale sizes by supernova explosions and subsequent cloud motions (Cesarsky 1975, 1980). Interpreting the spectrum proposed by Armstrong *et al.* in terms of the mean CR escape length $\lambda_e(R) \propto R^{-\delta}$ gives $0.2 \lesssim \delta \lesssim 0.6$ (see e.g. Cesarsky 1983). For comparison, the Kolmogorov and Kraichnan theories of hydromagnetic turbulence lead to $\delta = 1/3$ and $\delta = 1/2$, respectively. In §3.3 the rigidity dependence deduced from observations of CR secondary abundances is discussed.

Observations of secondaries and the highly isotropic distribution of CR arrival directions form essentially the basis for confinement models. Only the basic ideas of proposed models are summarized here, with some additional elaborations for the illustrative leaky-box model; in the next section, the observational constraints are considered. For a review and references, see Cesarsky (1980).

In the *leaky-box model*, cosmic rays are trapped within a reflecting boundary surrounding the Galaxy, but there is a certain (energy-dependent) probability of escaping into extragalactic space at each encounter (a possible physical background was described above); the CR density is uniform throughout the confinement volume. Above a few GeV/nucleon, Coulomb interactions in the ISM are negligible and the spallation cross sections are energy-independent. In the leaky-box model, the relation between the production rate P_i of primary nuclei with atomic number i and the observed flux F_i of these nuclei is then simply given by $F_i \propto (P_i + \sum_{j>i} S_{j \rightarrow i})/(\lambda_{int,i}^{-1} + \lambda_e^{-1})$, where $S_{j \rightarrow i}$ is the production rate of secondary nuclei i from fragmentation of nuclei j on the interstellar gas nuclei, $\lambda_{int,i}$ is the pathlength for nuclear interactions of nuclei i with interstellar matter, and λ_e is the mean pathlength for escape from the confinement region. The mean escape lifetime

τ_e (or leakage time, confinement time) is proportional to the mean escape length $\lambda_e = \rho v \tau_e$, where v is the particle velocity and ρ is the mean interstellar gas density. The actual escape length (and lifetime) has an exponential distribution, $P(\lambda) = (1/\lambda_e)\exp(-\lambda/\lambda_e)$, around the mean value λ_e in the leaky-box model, which is a direct consequence of the "leaking" boundary. In the *nested leaky-box model* cosmic rays are first trapped near their sources [with a certain (energy-dependent) probability of escaping] before pervading the Galaxy and escaping into extragalactic space. The pathlength distribution in this case can be described by the convolution of two exponential distributions, resulting into a depletion of short pathlengths.

In the *closed model*, cosmic rays cannot escape from the Galaxy and lose all their energy in the ISM. In a modified version, cosmic rays are trapped in spiral arms [with a certain (energy-dependent) probability of escaping], which makes it indistinguishable from the leaky-box model at CR energies $\lesssim 100$ GeV if the sun is in a spiral arm (but this may not be the case). Although CR diffusion forms the physical background of all these models, the term *diffusion models* has been reserved for models that take the diffusion processes explicitly into account, allowing for different diffusion coefficients perpendicular/parallel to the plane, actual source distributions, etc. The mean escape length is expected to have a power-law rigidity dependence, $\lambda_e(R) \propto R^{-\delta}$, in simple diffusion models with $K(R) \propto R^{\delta}$, where K is the diffusion coefficient. In the *dynamical halo model* a galactic wind is present; cosmic rays both diffuse and are convected away from the galactic disk into the halo. There are several models that describe the distribution of cosmic rays on smaller scales, taking into account the structure of the gas distribution in the Galaxy and considering possible phenomena such as CR trapping (or exclusion) in (from) clouds, propagation in interstellar tunnels, etc. Finally, the stochastic acceleration processes discussed in §2 can lead to a totally different picture, depending on the ubiquity of these phenomena throughout the ISM. In the case of shock acceleration by a single shock the models mentioned above may be applicable, but if the cosmic rays are gradually accelerated by multiple shocks, acceleration and propagation are intimately related, thus violating the basic assumptions of the previous models.

3.3. Secondary nuclei *vs.* models.

The amount of interstellar matter traversed by cosmic rays can be estimated from the abundance ratios between secondary nuclei and their primary progenitors, such as (Li+Be+B)/(C+N+O) and (Z=21–25)/Fe. A uniform slab of material cannot explain consistently the ratios for both the light (Li, Be, and B) and heavy secondaries (Z=21–25) (e.g. Shapiro and Silberberg 1970). A pathlength distribution of an exponential form (as in the leaky-box model) was found to be in reasonable agreement with the observations, but there may be a deficiency of short pathlengths (as for instance in the nested leaky-box model), which is discussed below. Around 1 GeV/nucleon, the mean matter pathlength is found to be 7–10 g cm^{-2}.

Above ~ 1 GeV/nucleon, the ratio of secondaries to primaries decreases with increasing energy (rigidity) to the highest energies the spectra have been measured (~ 150 GeV/nucleon) [see Ormes and Freier (1978) and Webber (1983a) and references in these papers]. This suggests a decrease in the mean pathlength with energy, thus reducing the production of secondaries with increasing energy; it should be kept in mind that at energies above a few GeV/nucleon the cross sections of the interactions are essentially independent of energy and once the secondaries have been produced, they are subject to the same loss processes as the primaries. In the leaky-box model (and diffusion models in general) this energy dependence of the secondary-to-primary ratio is interpreted as a power-law decrease of the mean escape length with rigidity, $\lambda_e = \lambda_o R^{-\delta}$. Propagation calculations, based on very accurate abundance measurements obtained recently with the Danish-French experiment aboard the *HEAO*–3 satellite (0.9–15 GeV/nucleon), have led to an accurate estimate of the exponent δ above ~ 3 GeV/nucleon: Koch *et al.* (1983) found $\delta = 0.6$ and Ormes and Protheroe (1983) found $\delta = 0.7 \pm 0.1$; at 5 GeV/nucleon $\lambda_e \simeq 6$ g cm^{-2}. These recent estimates are larger than the values found in the past from balloon data (energy/nucleon was mostly used, giving $\lambda_e \propto E^{-(0.3 \leftrightarrow 0.5)}$), but there seems to be agreement now on the new higher δ-values.

What is the impact of this rigidity dependence of λ_e on the source spectrum of cosmic rays in the leaky-box model? Assume the source spectrum is a power-law spectrum $\propto R^{-\alpha}$, then the observed spectrum for primary nuclei is $\propto R^{-\alpha}/(1/\lambda_{int,i} + R^\delta/\lambda_o)$. For protons, $\lambda_{int} \simeq 60$ g cm$^{-2} \gg \lambda_e$, so the observed spectrum would be $\propto R^{-(\alpha+\delta)}$. The observed proton spectrum (Ryan *et al.* 1972; Gregory *et al.* 1981; Tasaka *et al.* 1982) can indeed be fitted by a power-law spectrum, with $\alpha + \delta = 2.70 \pm 0.05$ for energies between ~ 10 GeV and $\sim 10^5$ GeV. So the protons with energies $\gtrsim 10$ GeV may be produced with a R^{-2} spectrum (probably up to $\sim 10^5$ GeV, because of the absence of structure in the observed proton spectrum, such as e.g. a change in slope), as expected for diffusive shock acceleration in a strong shock. Since certainly $\alpha < 2.2$ for the new high δ-values, the compression ratio of the shock has to be > 3.5 if the source spectrum is produced by shock acceleration (Ormes and Protheroe 1983). Other primary nuclei have also a power-law spectrum above 10–100 GeV/nucleon with a spectral index of ~ 2.7 (Webber 1983a and references therein), which leads to similar production spectra.

In the previous section it was briefly described how the rigidity(energy)-dependent pathlength is introduced in the other confinement models. Apart from the original closed model, all these models seem to be able to account for the observed decrease of the secondary-to-primary ratio with increasing energy (see e.g. Cesarsky 1983). However, this decrease is a potential problem (Hayakawa 1969; Eichler 1980; Cowsik 1980) for models of continuous stochastic acceleration throughout the ISM, as e.g. in the case of acceleration by successive interstellar shocks. These models (with energy-independent escape lifetime) predict a (logarithmic) *increase* in the secondary-to-primary ratio with energy. Lerche and Schlickeiser (1985) argued that this logarithmic increase is mainly due to the common assumption that the primary and secondary particles have the same escape lifetimes. Cowsik (1986) agrees that this explains indeed the logarithmic behaviour. Lerche and Schlickeiser then showed that the observed decrease in the secondary-to-primary ratio with increasing energy can be explained if (*i*) the escape lifetime for secondaries $\tau_{e,s}$ is smaller than for primaries ($\tau_{e,p}$) and (*ii*) both decrease with energy (in an identical way). Point (*i*) they attribute to different galactic scale heights of the primary and secondary "sources": most of the interstellar gas is concentrated in clouds with a relatively small scale height, so the secondary sources (fragmentation of primaries in collisions with the interstellar gas particles) have a narrower spatial distribution than the primary sources. They show that the broader the source distribution, the longer the mean age. Point (*ii*) they attribute to the energy dependence of the diffusion coefficient, as discussed above. Cowsik (1986) showed that $\tau_{e,s} < \tau_{e,p}$, solely, cannot explain the observed decrease in the secondary-to-primary ratio and noted that introducing a decrease in escape lifetime with energy does probably not reproduce the observed power-law CR spectra. In general, the Fermi acceleration mechanism predicts spectra which are cut-off exponentially when the escape lifetime increases with energy. Cowsik concludes that the problem is not solved. Silberberg *et al.* (1983, 1985b) noted that reacceleration could explain several anomalies in the CR composition. Obviously, there is no consensus on the important question concerning the existence of continuous acceleration/reacceleration in interstellar space.

Below a few GeV/nucleon, the interpretation of the CR observations is hampered by uncertainties in the spallation cross sections (which become energy dependent) and by the increasing importance of solar modulation effects (adiabatic energy losses become important). It seems clear from the measured secondary-to-primary ratios, however, that the escape length deviates from the simple power law found at higher energies; λ_e must flatten off or decrease with decreasing energy below a few GeV/nucleon (Garcia-Munoz et al. 1979; Ormes and Protheroe 1983). This cannot be explained by simple diffusion models. Jones (1979) has shown that a relative independence of the mean path length on energy below ~ 1 GeV/nucleon could be explained in the framework of the dynamical halo model proposed by Jokipii (1976) as a transition from diffusion-dominated propagation ($\gtrsim 1$ GeV/nucleon), as considered above, to convection-dominated propagation ($\lesssim 1$ GeV/nucleon). This diffusion/convection picture forms a natural explanation for a flattening of λ_e, but encounters problems with a rapid turnover, which is possibly indicated by the B/C ratio (Garcia-Munoz et al. 1979; Ormes and Protheroe 1983). This brings us to another dispute: there appears to be a discrepancy between the mean escape length values derived from light (B) and heavier (sub-Fe) secondaries that might become stronger with decreasing energy below a few GeV/nucleon (Garcia-Munoz et al. 1984 and references therein; Soutoul et al. 1985). The significance of this discrepancy is a matter of debate; if it is real, it suggests a deficiency of short path lengths, largest at low energies and decreasing with increasing energy. In the framework of the nested leaky-box model this would imply that the low-energy particles are more efficiently trapped near the sources.

A potential problem for all CR propagation and confinement models is the observed high flux of CR antiprotons, particularly at low energies (~ 200 MeV; Buffington et al. 1981). A primary origin for antiprotons requires rather exotic sources (Protheroe 1983 and references therein). If these antiprotons are of secondary origin (originating mainly from inelastic collisions of CR protons with the interstellar gas particles), then the $\bar{p}/p$ ratio is expected to decrease rapidly below a few GeV in the leaky-box model (e.g. Gaisser and Maurer 1973), contrary to the observations, which show barely any decrease. This steep cut-off arises from the kinematics of the $\bar{p}$ production process, which is quite unlike the general fragmentation process of CR nuclei, in combination with the steep CR primary spectrum (Gaisser and Levy 1974). At energies above a few GeV, the observed $\bar{p}$ flux is about four times higher than expected for the leaky-box model with values for λ_e obtained from heavier nuclei, as descibed above. It has been shown that this high-energy excess can be explained in the frame work of nested leaky-box models (enhanced $\bar{p}$ production in a high-density region around the CR source; Cowsik and Gaisser 1981; Cesarsky and Montmerle 1981) or by the (modified) closed model, with antiprotons (and protons) being trapped for longer than nuclei (Protheroe 1981; Stephens 1981). These models can however not account for the high $\bar{p}/p$ ratio at low energies. If the observation by Buffington et al. is confirmed, all CR propagation models have severe problems (or one has to admit that part of the antiprotons is of primary origin). If the antiprotons are indeed of secondary origin, then one might expect excesses in other secondaries of protons and perhaps helium (positrons, helium-3, deuterium). An enhancement by a factor of about two has been found for the ^{3}He/^{4}He ratio at energies of a few GeV (Jordan and Meyer 1984), but there is no evidence for an excess at low energies (e.g. Mewaldt 1985); positron measurements are not a sensitive test for various reasons and ^{2}H observations at high energies are in progress. Morfill et al. (1985) have shown that diffusive shock acceleration in a highly structured ISM, as in the McKee and Ostriker model, could explain the high $\bar{p}/p$ and ^{3}He/^{4}He ratios above ~ 1 GeV/nucleon.

Radioactive secondary nuclei with relatively long half-lives, such as ^{10}Be, ^{26}Al, ^{36}Cl, and ^{54}Mn can provide information on the mean age of cosmic rays. Combined with measurements of the CR pathlength, an estimate of the average gas density of the confinement region in the (nested) leaky-box model can be obtained. Only the abundance of ^{10}Be ($\tau_{1/2} \simeq 2 \times 10^6$ yr) has been measured with sufficient accuracy, suggesting a mean age of about $(1-2) \times 10^7$ yr around ~ 100 to a few hundred MeV/nucleon and a mean gas density of 0.05–0.5 cm^{-3} (Garcia-Munoz et al. 1977, 1981; Wiedenbeck and Greiner 1980; Guzik et al. 1985); for the nested leaky-box model this density range holds for the outer region in which a higher-density source region is nested. This interpretation may be too simple; in diffusion models with a halo the average confinement time in the Galaxy may be larger than this mean age (Ginzburg et al. 1980) and the interpretation would even be more complicated in case of continuous acceleration (Cesarsky et al. 1981b). Anyway, the resulting gas density estimates are lower than the average density in the disk and probably higher than the estimates of the density in a possible extended halo ($\sim 10^{-3}$ cm^{-3}), suggesting that either the cosmic rays are excluded from dense clouds or they fill a flattened halo. The former is unlikely, because theoretical work (e.g. Cesarsky and Völk 1978) and gamma-ray observations (§4) indicate that cosmic rays with energies $\gtrsim 100$ MeV/nucleon can penetrate dense clouds; confinement in a "thick disk" seems most likely.

3.4. The cosmic-ray electron spectrum.

Cosmic-ray electrons are mainly of primary origin; positron measurements indicate that secondary electrons constitute roughly 10%–20% of the observed electron flux (approximately equal numbers of e$^+$ and e$^-$ are expected from CR interactions with the interstellar gas). The observed electron spectrum does not impose strong additional constraints on the propagation and confinement models discussed above, because the primary electron spectrum is strongly modified in interstellar (and interplanetary) space by several processes that cannot be disentangled conclusively. In addition to escape and ionization losses, that are important for the CR nuclei, CR electrons suffer bremsstrahlung, synchrotron, and inverse-Compton losses. The distribution of the CR sources plays therefore an essential role in the modelling. On the other hand, the strong interactions of the electrons with the interstellar gas, magnetic field, and photon field provide very useful possibilities to trace CR electrons throughout the Galaxy at radio and gamma-ray frequencies, as discussed in §4.

Recent measurements of the CR electron spectrum are reviewed by Webber (1983b). Above ~ 10 GeV, the true local interstellar electron spectrum is observed at earth, but below 10 GeV, solar-modulation effects become increasingly important. The differential energy spectrum has a spectral index of ~ 3.0 around 10 GeV and steepens to an index of ~ 3.3 at several hundred GeV. The shape of the local electron spectrum between ~ 100 MeV and ~ 10 GeV is reflected in the spectral shape of the low-frequency (10 MHz – 1 GHz) radio-continuum observations of the synchrotron emission in the directions of the galactic poles and anticenter (e.g. Rockstroh and Webber 1978; Webber et al. 1980) and can be normalized to the observed electron spectrum at 10 GeV. This approach suffered in the past from uncertainties in the electron spectrum around 10 GeV observed at earth, but the measurements have converged recently (see Webber 1983b). At low energies, the shape of the deduced interstellar electron spectrum is similar to that of the CR protons, but above ~ 10 GeV, the electron spectrum is significantly steeper than the proton spectrum (§3.3). A steepening is expected because of the radiation losses mentioned above (particularly synchrotron losses). When coarse estimates of these losses are taken into account, as well as the energy dependence of the escape length derived from the secondary-to-primary ratios (§3.3), the resulting electron source spectrum is consistent with a power-law spectrum that has a spectral index of 2 or somewhat higher, i.e. similar to that found for the proton source spectrum (and primary nuclei in general).

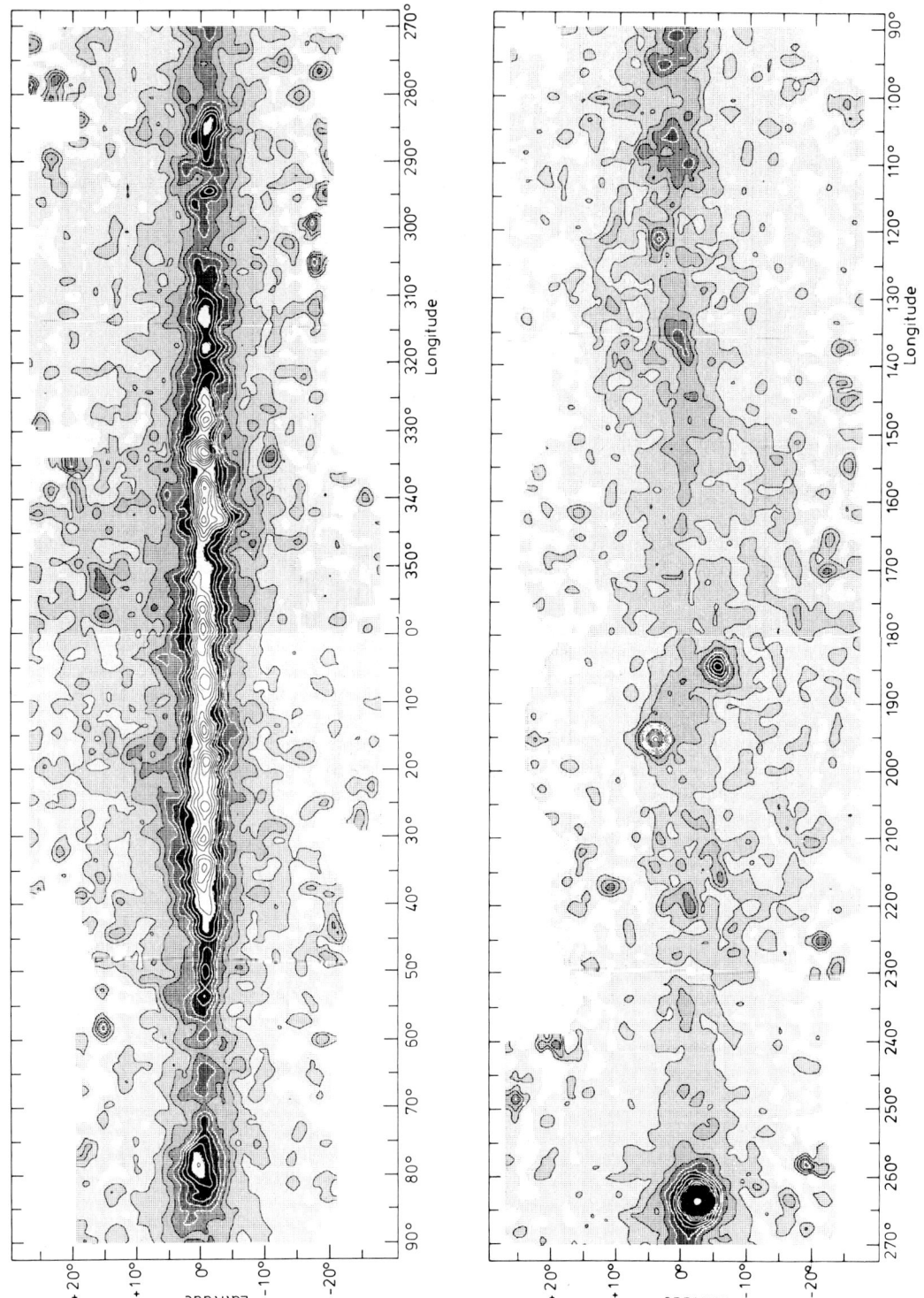

4. GAMMA RAYS AS A TRACER OF COSMIC RAYS (& INTERSTELLAR GAS)

4.1. Towards the current status.

Gamma-ray astronomy provides a powerful tool to study the distribution of CR electrons and protons throughout the Galaxy. The diffuse component of the galactic high-energy ($\sim$ 50 MeV – 5 GeV) γ-rays results mainly from the interaction of CR nuclei (predominantly 1–10 GeV protons) and electrons (<1 GeV) with the nuclei of the interstellar gas (via the decay of π°-mesons and bremsstrahlung, respectively). An additional contribution is obtained from the interaction of CR electrons (>1 GeV) with the interstellar photons (mainly in the optical and infrared range) through the inverse-Compton process. For a review of the various production mechanisms see Stecker (1971).

Gamma radiation is a tracer of the product of the CR density and the interstellar gas density, integrated along the line of sight: $I_\gamma \propto \int \rho_{cr}\rho_{gas}dl$ (γ-rays do not suffer absorption in the ISM). If the CR density can be assumed to be constant, γ-ray measurements provide a probe of the total gas column density. This has been verified for the local ($\lesssim 1$ kpc) region of the Galaxy at intermediate latitudes ($|b| \simeq 10^\circ - 20^\circ$), using 21-cm HI observations and galaxy counts as other gas tracers (e.g. Fichtel et al. 1978; Lebrun and Paul 1983; Strong et al. 1982, 1985). In order to use γ-ray observations as a CR tracer throughout the Galaxy, it is essential to know the distribution of the target interstellar gas particles. The numerous γ-ray studies of the galactic CR distribution performed in the past, using the γ-ray observations obtained by the SAS-2 and COS-B satellites, suffered severely from uncertainties in the galactic distribution of the interstellar molecular hydrogen. The best tracer of the large-scale distribution of H_2 is the carbon-monoxide molecule, but the sky coverage of the CO observations was poor and the relation between the measured CO intensities and H_2 column densities was uncertain. Usually the sparse CO observations and the 21-cm HI observations were converted into (spiral arm) models of the gas distribution. By assuming that the CR density on a large scale is proportional to the gas density, some studies showed that the observed and predicted γ-ray intensities agreed (e.g. Bignami et al. 1975; Kniffen et al. 1977; Fichtel and Kniffen 1984) while others required modifications, such as abundance gradients or a partial exclusion of cosmic rays from molecular clouds (e.g. Cesarsky et al. 1977). Without a priori assumptions on the proportionality between CR and gas density, several studies indicated that the CR density in the inner Galaxy is higher than in the solar vicinity (e.g. Stecker et al. 1975; Higdon 1979; Issa et al. 1981; Harding and Stecker 1985).

The situation seemed to be less confusing for the outer Galaxy, particularly because the H_2 contribution to the gas column densities appears to be small on average. Dodds et al. (1975), analyzing the SAS-2 γ-ray data, first suggested a decrease in the density of cosmic rays with increasing distance beyond the solar circle; other studies of the SAS-2 data confirmed this gradient (e.g. Cesarsky et al. 1977; Higdon 1979) and it has been adopted ever since as the main observational evidence against a universal origin of CR nuclei in the GeV range. However, this strong gradient does barely show up in recent studies of the (final) SAS-2 data ($E_\gamma > 100$ MeV; Strong et al. 1978; Arnaud et al. 1982; Bloemen et al. 1984c). Although Strong et al. and Arnaud et al. ascribed this result to

---

Figure 3: COS-B γ-ray intensity maps for the 150 MeV - 5 GeV range. Although a smoothing algorithm has suppressed fluctuations of angular scale smaller than the point-spread function, not all individual features in the maps are statistically significant, particularly at medium latitudes near the edges of the field of view. Contour values are indicated at multiples of 5×10^{-5} ph cm^{-2} s^{-1} sr^{-1}. The isotropic background (mainly instrumental) is subtracted.

a weaker gradient in the second galactic quadrant (which they studied) compared to the third quadrant, it turned out recently (see Bloemen et al. 1986) that the disappearance of the strong gradient is in fact a direct consequence of an improvement in the calibration of the SAS-2 data around 1978.

More refined analyses have been performed recently, using the COS-B observations which have larger counting statistics than the SAS-2 observations (by a factor of ~25). Moreover, large-scale CO surveys are available now. It is concentrated here on these new developments.

4.2. A method to disentangle the γ-ray observations.

A large-scale mm-wave CO survey obtained with the 1.2m telescopes of Columbia University & Goddard Institute for Space Studies, covering the entire Milky Way up to $|b| \simeq 7° - 10°$, was completed recently; the large sky coverage and 0.5° angular resolution make it very suitable for correlation studies with the γ-ray observations. The COS-B observations have sufficient resolution and sensitivity to constrain the relation between the integrated CO line intensity W_{CO} and the H_2 column density $N(H_2)$ (Lebrun et al. 1983; Bloemen et al. 1984a). These possibilities were exploited by Bloemen et al. (1986) in a large-scale γ-ray study of the CR distribution as well as the H_2 distribution throughout the Galaxy. They used the velocity information of the HI and CO observations as a distance indicator to ascertain the spatial distribution of the interstellar gas. Using this distance information, the galacto-centric distribution of the γ-ray emissivity (the production rate per H atom) was determined for different energy ranges from a correlation study of the γ-ray intensity maps and the HI and CO gas-tracer maps for selected galacto-centric distance intervals. On the assumption that unresolved γ-ray point sources do not contribute significantly to the observed γ-ray emission, the γ-ray emissivity is proportional to the CR density. More specifically, since the spectral shapes of the γ-rays produced by the CR nuclei and electrons are different (the $\pi°$-decay spectrum has a maximum at ~70 MeV, whereas the bremsstrahlung spectrum decreases monotonically with energy), the spatial distribution of the γ-ray emissivity spectrum can be used to trace CR electrons and nuclei separately. At high energies ($E_\gamma \gtrsim 300$ MeV) the diffuse γ-ray emission is dominated by $\pi°$-decay, whereas the emission at lower energies has a large electron bremsstrahlung contribution. In an internally consistent approach they determined also the ratio $X \equiv N(H_2)/W_{CO}$ and studied possible variations of this quantity throughout the Galaxy.

A few details about their approach. Skymaps of HI column densities, $N(HI)_i$, and integrated CO line intensities, $W_{CO,i}$, were constructed in four ($i = 1, 2, 3, 4$) galacto-centric distance ranges (2 kpc $< \Re <$ 8 kpc, 8 kpc $< \Re <$ 10 kpc ($=\Re_\odot$), 10 kpc $< \Re <$ 15 kpc, and $\Re >$ 15 kpc) and convolved with the energy-dependent COS-B point-spread function. These four distance intervals were selected because the angular distributions of the gas in each interval show distinct differences, which are needed to ascertain the contribution of the gas in each interval to the observed γ-ray intensities. Similarly, determining the conversion factor X requires distinct differences between the structures in the HI and CO maps. Due to the still limited CO coverage available at that time, the correlation analysis was restricted to the first and second galactic quadrants and the Carina region ($270° < \ell < 300°$), and $|b| \lesssim 5°$. Using a likelihood analysis for three different energy ranges (70–150 MeV, 150–300 MeV, 300 MeV – 5 GeV), it was investigated which combination of the free parameters (q_i, Y_i, and I_B) best describes the observed γ-ray intensity distributions I_γ for each energy range, assuming a relation of the form:

$$I_\gamma = \left\{ \sum_i \frac{q_i}{4\pi} [N(HI)_i + 2Y_i \cdot W_{CO,i}] \right\} + I_{IC} + I_B.$$

The term enclosed by braces represents the intensities that originate from CR collisions with atomic and molecular hydrogen (q_i is the γ-ray emissivity for annulus i), I_{IC} represents the modelled (small) inverse-Compton contribution (Bloemen 1985), and I_B is the isotropic background, including the (dominant) instrumental background. The parameter Y_i is related to the $N(H_2)/W_{CO}$ ratio X_i for each annulus: $Y_i = \{q_i(H_2)/q_i\} \cdot X_i$. If CR particles are not excluded from, or concentrated in, molecular clouds, then Y_i equals X_i, independent of energy (§4.3). Y was first assumed to be the constant throughout the Galaxy. Starting from this general model with six parameters for each energy range, it was tested (using the likelihood ratio) whether various simpler models with fewer parameters (i.e. constant emissivity distributions as a function of $\Re$, identical values of Y for each energy range and/or identical emissivity distributions for each energy range) give significantly worse fits to the data. The next two sections describe the results and implications.

4.3. Gamma-ray constraints on the CO–H_2 calibration and the mass of H_2 in the Galaxy.

The ratio $X = N(H_2)/W_{CO}$ is frequently used to determine the distribution of H_2 in the Galaxy from CO surveys. The $J=1\rightarrow0$ transition of CO is generally optically thick, so it is not evident that this transition can be used to obtain H_2 column densities. The quantity X is probably not constant for individual clouds, but when averaged over a large number of clouds on a galactic scale it appears to be useful to estimate the mass of H_2. The determination and application of X have been controversial, in part because galactic temperature and abundance gradients can have a significant effect on the conversion. The $COS-B$ γ-ray data provide a way of determining X (§4.2) that is independent of excitation, abundance and optical-depth effects, which have plagued previous determinations.

The likelihood-ratio tests mentioned in §4.2 showed that the hypothesis of an identical Y-value for each energy range gives a satisfactory fit of the data, which indicates that the spectral shapes of the γ-rays from HI and H_2 are roughly the same. The theoretical work of Skilling and Strong (1976) and Cesarsky and Völk (1978) indicates that only very low-energy cosmic rays (< 50 MeV/nucleon) may not penetrate a dense cloud completely. In addition, the γ-ray observations of the Orion molecular complex show conclusively that cosmic rays do penetrate at least the local giant molecular clouds (Bloemen et al. 1984a). If the average CR density inside the molecular clouds is indeed the same as in the HI gas, then the Y-parameter represents X (see §4.2); its value was found to be $X = (2.75\pm0.35) \times 10^{20}$ mol. cm^{-2} K^{-1} km^{-1} s. Taking into account systematic uncertainties (see Bloemen et al. 1986), a more realistic estimate of the uncertainty on X is probably a factor ~2 times the formal error. To investigate possible large-scale variations of X throughout the Galaxy, the Y-values were determined separately for 2 kpc $< \Re <$ 8 kpc and $\Re >$ 8 kpc; they were found to be identical within uncertainties.

A few notes. (i) If a population of unresolved γ-ray point sources distributed like CO exists, Y would be an overestimate of X, so the X-value given here should in fact be regarded as an upper limit. The γ-ray observations do not provide a stringent lower limit on X. However, [a] the good correlation between the observed γ-ray intensities and the estimates from the gas and [b] the agreement between the X-value obtained from this large-scale analysis and the value derived by Bloemen et al. (1984a) from a similar analysis of the Orion region (where the source contribution is probably negligible) suggest that the point-source contribution is not dominant in the γ-ray component that was attributed to H_2. Therefore, X is probably not smaller than about 1.5 $\times$ 10^{20} mol. cm^{-2} K^{-1} km^{-1} s. (ii) The X-value was determined using the Columbia CO survey; systematic differences between the Columbia data and other surveys would change the value given here proportionally when these data would have been used. (iii) The γ-ray

TABLE I.   Principal determinations of $N(H_2)/W_{CO}$.

source	quantity determined	derived $N(H_2)/W_{CO}{}^a$	calibration difference[b]
Gordon & Burton (1976)	$N(H_2)/W_{CO}$	2.3	1.66[h]
Dickman (1978)	$N(H_2)/N(^{13}CO)$	1.5	–
Solomon & Sanders (1980)	$N(H_2)/W_{CO}$	6.0	1.27[i]
Frerking et al. (1982)	$N(H_2)/W_{CO}$	1.8[d]	–
Lebrun et al. (1983)[c]	$N(H_2)/W_{CO}$	1–3	–
Black and Willner (1984)	$N(H_2)/N(CO)$	<3	–
Bloemen et al. (1984a)[c]	$N(H_2)/W_{CO}$	2.6	–
Lebrun & Huang (1984)	$N(H_2)/W_{CO}$	1.1	–
Sanders et al. (1984)	$N(H_2)/W_{CO}$	3.6	1.27[i]
Bhat et al. (1985) [c]	$N(H_2)/W_{CO}$	0.7[e]–1.4[f]	–
Dickman (1985)	$N(H_2)/N(^{13}CO)$	2.2[g]	–
Bloemen et al. (1986) [discussed here]	$N(H_2)/W_{CO}$	2.8	–

Footnotes on next page.

TABLE II.   Mass of HI and H_2 in the Galaxy (Bloemen et al. 1986).

R (kpc)	$\sigma(H_2)/\sigma(HI)$	$M(HI)^a$ $(M_\odot)$	$M(H_2)$ $(M_\odot)$	R (kpc)	$M(HI)^a$ $(M_\odot)$	$M(H_2)$ $(M_\odot)$
< 0.5	–	2.3×10^6	$< 2 \times 10^{7b}$	< 0.5	2.3×10^6	$< 2 \times 10^{7b}$
2–8	1.35	5.5×10^8	7.4×10^8	< 8	5.9×10^8	8.1×10^8
8–10	0.40	3.3×10^8	1.3×10^8	< 10	9.2×10^8	9.4×10^8
10–15	0.20	1.1×10^9	2.3×10^8	< 15	2.1×10^9	1.2×10^9

(a) Based on constant value of $\sigma(HI) = 2.9$ $M_\odot$ pc^{-2}.
(b) Blitz et al. (1985), 1σ upper limit; also from γ-ray data.

TABLE III.   Other determinations of $M(H_2)$ (2–10 kpc).

source	$M(H_2)$ $(M_\odot)$	$M(H_2)/M(HI)^a$
Scoville & Solomon (1975)	$1 - 3 \times 10^9$	1–3
Gordon & Burton (1976)	2.1×10^9	2.4
Solomon & Sanders (1980)	4×10^9	4.5
Blitz & Shu (1980)	1×10^{9b}	1.1
Liszt et al. (1981)	2.3×10^9	2.6
Dame (1983)	0.7×10^9	0.8
Lebrun et al. (1983)	0.9×10^{9b}	1.0
Sanders et al. (1984)	2.6×10^9	3.0
Bhat et al. (1985)	0.6×10^9	0.7
Bloemen et al. (1986) [discussed here]	0.9×10^9	1.0

(a) M(HI) is taken to be 0.9×10^9 $M_\odot$ (Henderson et al. 1982).
(b) This mass is derived from quantities given in the source.

analysis indicates that X is roughly constant on a few-kpc scale throughout the Galaxy, but small-scale variations in X cannot be excluded, so the use of this conversion factor at a resolution finer than the one used to determine X, should take this possibility of variations into account.

The principal previous determinations of X are given in Table I (column 3). Column 4 gives the known calibration differences between the various surveys (Cohen et al. 1986). These differences have to be considered when the γ-ray estimates of H_2-masses are compared to those from other authors (see below). Since the X-value derived from the γ-ray analysis is a reliable upper limit, it favours the low value originally derived by Dickman (1978) and subsequently used by Blitz and Shu (1980) and others.

Table II summarizes the masses of the H_2 gas in the radial bins selected for the γ-ray analysis, using the estimate of X given above (no helium correction is applied). Note that these mass estimates result basically from the different angular distributions of HI and H_2 and are not subject to uncertainties in the calibration of the CO observations. The ratios between the H_2 and HI surface densities, $\sigma(H_2)/\sigma(HI)$, given in the table, show that the radial distributions of HI and H_2 are intrinsically different, as first found by Scoville and Solomon (1975), Gordon and Burton (1976), and Cohen and Thaddeus (1977). Whereas these authors (and others) *assumed* a constant X-value, the γ-ray analysis shows that excitation and abundance effects in the CO do not alter this conclusion. The γ-ray analysis differs, however, with several previous studies in the molecular fraction of the interstellar gas (Table III). These discrepancies can partly be explained by calibration differences and different X-values (Table I). The γ-ray results favour rough equality in the HI and H_2 masses inside the solar circle and indicate that the H_2 mass even in the molecular ring exceeds the HI mass by only $\sim35\%$. This value may be higher or lower at the peak of the ring depending on the true variation of X within the 2–8 kpc annulus. The constant HI surface density of 2.9 $M_\odot$ pc^{-2} assumed here in the comparison with the H_2 estimates accounts only for the HI near the mid plane with a Gaussian scale height of ~135 pc for $\Re \lesssim \Re_\odot$; taking into account the HI at larger distances from the plane recently discovered by Lockman (1984) and Lockman et al. (1986), which appears to have an (exponential) scale height of about 400–500 pc, increases $\sigma(HI)$ by 1–2 $M_\odot$ pc^{-2} (at least for $\Re \lesssim \Re_\odot$). When this high-z HI is taken into account, $\sigma(H_2)/\sigma(HI)$ is also $\lesssim 1$ in the 2–8 kpc ring, because the relative contribution from high-z H_2 to $\sigma(H_2)$ is probably much smaller (e.g. Magnani et al. 1985).

Footnotes TABLE I: (a) Units: 10^{20} mol. cm^{-2} K^{-1} km^{-1} s. Other isotopic ratios (column 2) determined by some authors for the solar vicinity have been converted to $N(H_2)/W_{CO}$ using standard assumptions (e.g. Solomon and Sanders 1980).   (b) This is the ratio between the CO intensities of the surveys used by the authors listed in the first column and those of the Columbia survey used in the γ-ray analysis described in this paper, as derived by Cohen et al. (1986).   (c) Using γ-ray observations.   (d) Determined in the ρ Oph region and recommended for galactic-survey work (no correlation between $N(H_2)$ and W_{CO} was found in the Taurus region).   (e) At $R = 6$ kpc.   (f) Local.   (g) Private communication.   (h) Data are similar to those used by Burton and Gordon (1978), which have been studied by Cohen et al. (1986).   (i) Sanders et al. (1984) contained data both from the NRAO 11m and the FCRAO 14m telescopes; the NRAO data are the same as in Solomon and Sanders (1980), and the FCRAO data were scaled to agree with the NRAO calibration (Sanders 1985; private communication).

4.4. Gamma-ray constraints on the galactic distribution of cosmic rays.

Bloemen *et al.* (1984bc) applied the method described in §4.2 to the *COS-B* observations and HI observations of the second and third galactic quadrants alone (they showed that H_2 can be neglected in the outer Galaxy within the uncertainties of the analysis), and found that the galacto-centric distribution of the γ-ray emissivity beyond the solar circle shows a stronger decrease for low energies than for high energies. In fact, the γ-ray emissivity for the 300 MeV – 5 GeV range shows barely any fall-off. The results from the recalibrated *SAS-2* data mentioned in §4.1 are in agreement with this near constancy. The extension of this analysis (Bloemen *et al.* 1986), including the inner Galaxy as described in §4.2, resulted in a similar picture for the Galaxy as a whole (although for the inner Galaxy the interpretation of the low-energy emission is debatable [4]). On the assumption that the shapes of the CR spectra do not vary strongly throughout the galactic plane, it was argued that the energy dependence of the emissivity distribution suggests that the galacto-centric gradient for CR electrons is stronger than for CR nuclei. It was shown that the data are consistent with radial CR density distributions of an exponential form for $\Re \gtrsim 4$ kpc: $\rho_{cr}(\Re) \propto \exp(-\Re/L)$. There are some small-scale discrepancies, which might be due to genuine point sources or local enhancements of the CR density (or H_2/CO ratio), but all deviations are fairly small. For the CR nuclei, the radial scale length L was found to be $\gtrsim 18$ kpc, for CR electrons $5 - 11$ kpc. When the radial distributions of the electrons and nuclei are forced to be the same, then $L \simeq 15$ kpc. These CR gradients are upperlimits if a population of unresolved galactic γ-ray sources exists with a latitude distribution similar to that of the gas but with a stronger concentration towards the inner parts of the Galaxy.

It has to be stressed that the argument for different distributions of electrons and nuclei is based on the assumption that the CR spectral shape does not change significantly with galacto-centric radius. Work in progress (Bloemen 1986a) indicates that this assumption may be incorrect. The radial CR scale length would remain very large, as discussed here, but it may be energy dependent (and not different for CR electrons). The results are too preliminary to give further details, but the reader should keep in mind that such an alternative interpretation may be necessary.

Inside $\Re \simeq 4$ kpc, the situation is uncertain. There is no conclusive evidence from the *COS-B* data for neither a significant increase nor decrease of the CR density. The γ-ray flux (>300 MeV) from the central few hundred parsecs of the Galaxy is, however, an order of magnitude smaller than the value expected from the large H_2 masses generally estimated to be present near the centre and from the average γ-ray emissivity measured in the disk (Blitz *et al.* 1985). If the cosmic rays do penetrate the clouds (as expected

---

[4] In summary, the problem is the following. There is little doubt that the interpretation of the *COS-B* γ-ray data requires an energy-dependent model, such that the intensity spectrum towards the inner Galaxy is softer than for the remainder of the disk. If this effect is not due to a concentration of unresolved steep-spectrum γ-ray sources in the inner Galaxy, then, in the modelling of Bloemen *et al.* (1986), it has to be ascribed to either a higher Y-value for low energies than for high energies (because most of the H_2 is located inside the solar circle) or to a steeper γ-ray emissivity gradient for low energies. In principle, these two effects are distinguishable in the correlation analysis because of the differences between the angular distributions of HI and H_2. In practice, this method applies less satisfactorily to the observations at low energies than at high energies because the angular resolution of *COS-B* is degrading with decreasing energy. The steeper low-energy emissivity gradient was favoured by likelihood tests, although marginally. An analysis of the whole Galaxy is in progress and may give further insight.

— §4.3), this deficit of γ-rays suggests that either the density of GeV CR nuclei near the centre is anomalously small relative to the galactic disk or that H_2 is nearly an order of magnitude less abundant than estimates made from CO observations.

The exponential scale length of the CR nuclei (and electrons?) is much larger than for the type of objects generally considered as candidates for CR sources or injectors; e.g. supernova remnants, pulsars, early-type stars, and disk stars in general, all have radial scale lengths of typically ~5 kpc for $\Re \gtrsim 4$ kpc (e.g. Kodaira 1974; Lyne et al. 1985; Mathis et al. 1983 and references therein). It seems that the distribution of at least the CR protons in the Galaxy does not reflect the distribution of the objects that have probably injected the particles and/or accelerated them, either directly or indirectly by producing shocks in the ISM. In the case of diffusive shock acceleration, however, the impact of the ambient medium needs further investigation in view of the probably widely different structure of the ISM in the inner and outer regions of the Galaxy (see e.g. Heiles 1986). If the production rate of cosmic rays indeed increases towards the inner Galaxy with a radial scale length of ~5 kpc, it is unlikely that extensive diffusion of the particles into the outer Galaxy can reproduce the near constancy. This is an inevitable consequence of the rapid increase of the volume to be filled as the radius increases (although this problem would be alleviated if the particles propagated predominantly in the disk instead of 3-dimensionally). Remains the interesting alternative that cosmic rays in the inner Galaxy escape easier from the disk than those in the outer Galaxy. This possibility needs further investigation. The data are also consistent with an extragalactic origin (§2); the gradient of the CR nuclei is so weak, if it exists, that on the basis of γ-ray observations it is no longer certain that CR nuclei (with energies of several GeV) are produced in the Galaxy.

4.5 Comparison with synchrotron emission.

The observed diffuse radio-continuum emission of the Galaxy at low frequencies is primarily synchrotron emission from CR electrons spiraling around the magnetic-field lines. Observations around 100–1000 MHz are most appropriate for a large-scale study, because they do not suffer from free-free absorption and they have only a small thermal contribution. In a typical interstellar field of a few μG, 100 MHz observations mainly trace electrons with energies of a few GeV, which is approximately an order of magnitude larger than that of the electrons traced by the γ-ray observations. For CR electrons with a power-law spectrum $kE^{-\alpha}$, the synchrotron emissivity at frequency ν is proportional to $kB_{\perp}^{(1+\alpha)/2}\nu^{(1-\alpha)/2}$, where $B_{\perp}$ is the magnetic-field component perpendicular to the line of sight. Phillipps et al. (1981) and Beuermann et al. (1985) both found, from different unfolding techniques applied to the 408 MHz survey of Haslam et al. (1981ab), that the synchrotron volume emissivity in the galactic plane for 5 kpc $\lesssim \Re \lesssim$ 15 kpc can be represented by an exponential distribution $\propto \exp(-\Re/L_{syn})$ with a radial scale length L_{syn} of 3.9 kpc. This result appears to depend not strongly on the large-scale geometry and fluctuations of the galactic magnetic-field. The synchrotron emissivity distribution is steeper than the CR electron distribution derived from the γ-ray observations, as expected if the magnetic-field strength has a weak galacto-centric gradient; representing the magnetic-field strength also by an exponential distribution, this comparison shows that the field strength has a radial scale length $\gtrsim 10$ kpc, even if the electron distribution would have a radial scale length as large as 15 kpc.

5.   COSMIC RAYS AND GALACTIC STRUCTURE — A STATIC EQUILIBRIUM?

Since the classical work of Parker (1966, 1969 and references therein) it is well-known that cosmic rays and magnetic fields play an important role in a possible (quasi) hydrostatic-equilibrium configuration of the Galaxy. In such an equilibrium state the gravitational attraction towards the galactic plane is balanced by the pressure of the gas, the magnetic field, and the CR particles. Parker considered the case where the lines of force of the magnetic field are parallel to the galactic plane (a horizontal equilibrium) [5], which makes the equilibrium equation simple:

$$\frac{d}{dz}\Big\{P_g(z) + P_{mf}(z) + P_{cr}(z)\Big\} = -\rho(z)g(z) \quad \text{or} \quad P_{tot}(z) = \int_z^\infty \rho(x)g(x)dx,$$

where P_{tot} represents the total internal pressure (given by the term in braces) as a function of distance z to the mid plane ($P_{tot} \to 0$ if $z \to \infty$). P_g is the gas pressure (thermal and due to macroscopic motions), P_{mf} is the magnetic-field pressured ($B^2/8\pi$, where B is the strength of the plane-parallel magnetic field), P_{cr} is the CR pressure (equal to $1/3$ of the CR energy density), g is the z-component of the gravitational acceleration, and ρ is the mean gas density. Parker solved this equation for the case that the gas is isothermal assuming that $P_{mf} = \alpha P_g$ and $P_{cr} = \beta P_g$, where α and β are dimensionless constants, and showed that the resulting equilibrium is unstable on a time scale of the order 10^7 years. This instability arises from any perturbation which depresses the magnetic lines of force at one point and raises them at another. The gas tends to slide into the depression and concentrate into clouds or cloud complexes (Shu 1974; Mouschovias et al. 1974), further depressing the field there, and the CR pressure (which remains uniform along the MF lines) tends to inflate the raised portions of the field. At the disk boundary, the CR inflation of such a bubble may lead to escape from the Galaxy if the inflation is fast. The rate of inflation is probably governed by the random walk of the magnetic-field lines and the CR production rate. If the inflation is slow, most cosmic rays may return to the galactic disk; this can be regarded as an alternative to the physical representation of the boundary conditions in the leaky-box model discussed in §3.2. For equilibrium studies of the final state of the Parker instability see Mouschovias (1974, 1975).

Others in subsequent studies of possible horizontal equilibrium configurations for the present state of the Galaxy (Kellman 1972; Wentzel et al. 1975; Fuchs et al. 1976; Fuchs and Thielheim 1979) also assumed that the gas is isothermal and that the pressure components all decrease in the same way with distance from the mid plane; all these equilibrium models are unstable. Some treated the molecular clouds as a separate dynamical system, not coupled to the magnetic field, but they made the same assumptions for the gas-field system. It is evident now that these assumptions are incompatible with a variety of observations. Badhwar and Stephens (1977) dropped these assumptions and derived horizontal-equilibrium models for the solar vicinity that account for the observed radio synchrotron emission in the direction of the galactic poles. Lachièze-Rey et al. (1980; see below) showed, however, that also the models of Badhwar and Stephens are unstable.

---

[5]   The observed properties of the galactic magnetic field do not preclude a large-scale horizontal equilibrium. Studies of the interstellar polarization of starlight, the polarization of synchrotron emission, and the rotation measures of pulsars and extragalactic radio sources all indicate that there is a systematic magnetic-field component, preferentially aligned parallel to the galactic plane (see e.g. paper by C. Heiles in this volume). The observed random field component of comparable strength, superimposed on the systematic field, has a stabilizing effect (Zweibel and Kulsrud 1975).

Also the hydrostatic-equilibrium models that have been proposed for the gaseous halo (e.g. Weisheit and Collins 1976; Chevalier and Fransson 1984; Fransson and Chevalier 1985) are unstable.

This leads to the obvious question: can the present state of the Galaxy at all be described by a stable horizontal-equilibrium configuration? This problem is discussed in the remainder of this section. For details see Bloemen (1986b). The ISM in the galactic plane is subject to a permanent agitation (supernova explosions – cloud motions), but this may not preclude a large-scale hydrostatic equilibrium, particularly away from the plane. In fact, Zweibel and Kulsrud (1975) showed that a small-scale (much smaller than the scale height) turbulent component of the magnetic field, which they attributed to cloud motions, in a horizontal equilibrium of the Parker type has a stabilizing effect due to magnetic shear stresses. Parker (1975) noted that they ignored the most unstable modes (those with short wavelengths perpendicular to both the gravitation and magnetic field), but Lachièze-Rey et al. (1980) showed that the latter ones are completely stabilized by the turbulence in the field. From their results it can be calculated that the energy density of the turbulent field has to be only a minor fraction of that of the systematic field. In fact, the first effect of the onset of a Parker instability is to generate this small-scale turbulence. The stabilizing effect of the turbulence on the two-dimensional instability, considered by Zweibel and Kulsrud, is only significant if the energy density of the turbulent field is comparable to that of the systematic field. There is good observational evidence for such a turbulent field on scales of tens of parsecs, but let us forget about this for a while and consider the possibility of a stable equilibrium without additional stabilizing effects.

The normal-modes technique applied by Parker cannot be extended easily to more complex configurations (such as for instance those with α and β being functions of z), but Lachièze-Rey et al. (1980) derived a local instability criterion, based on an energy principle, that applies to any horizontal equilibrium state. Using the equilibrium equation, their instability criterion can be written as

$$\frac{1}{P_g} \frac{dP_{tot}}{dz} < \gamma \frac{1}{\rho} \frac{d\rho}{dz},$$

which shows its close resemblence to the well known criterion for convective instability of a gas $(dS/dz < 0;\ S$ is the entropy), which contains P_g instead of P_{tot}. γ is the adiabatic index. A configuration meeting the stability criterion of Lachièze-Rey et al. is also not convectively unstable, assuming that the sum of magnetic-field and CR pressures does not increase with z. Basically, this criterion defines the minimum fraction of the total internal pressure that has to be due to gas pressure in order to have a system that is not unstable. For an equilibrium of the Parker type, the criterion reduces again to $\gamma < 1 + \alpha + \beta$. It is assumed here that cosmic rays distribute very rapidly along a field line, which is the most stringent case. In addition, let us impose another stringent requirement, namely $\gamma \lesssim 1$.

Only for the solar vicinity the information on the stellar, gaseous, cosmic-ray, and magnetic-field components is sufficiently accurate to perform a meaningful analysis of the equilibrium and stability of the galactic disk. Still, some relevant characteristics of the system are not well determined by observations, particularly the density, scale height, and temperature of the gaseous halo and the scale height of the cosmic rays and the magnetic field, but there are some observational constraints.

There is conclusive evidence for highly ionized gas far from the plane, obtained primarily from *IUE* observations of C IV and Si IV absorption lines (see e.g. paper by B. Savage in this volume). The absorbing halo gas appears to be concentrated at about 1 – 3 kpc from the plane and, depending on the ionizing mechanism adopted, characteristic temperatures of $10^4 - 10^5$ K and densities of $10^{-3} - 10^{-2}$ cm^{-3} have been derived. A

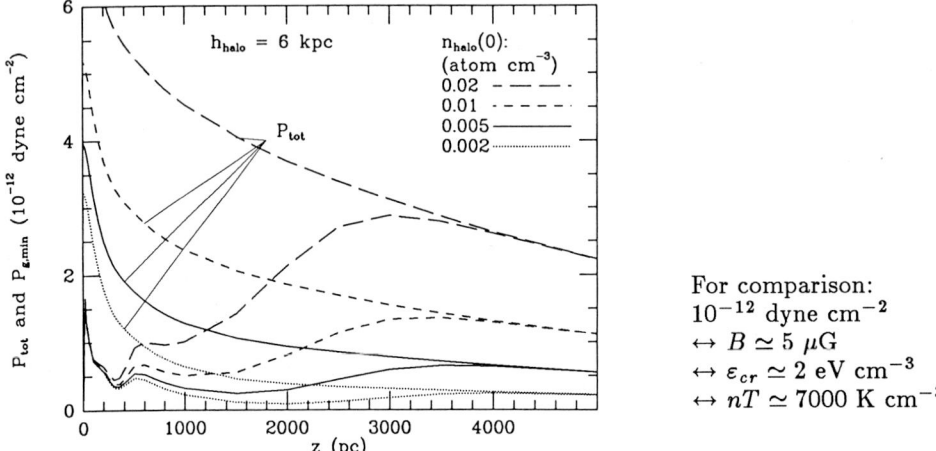

Figure 4: Examples of the distribution of the total internal pressure that is required to balance the gravitational attraction in the solar vicinity and the corresponding distributions of the minimum gas pressure that is required for stability ($\gamma < 1$).

significantly hotter halo component extending to much larger distances from the plane might exist [a 10^6 K halo was suggested by Spitzer (1956)], which would not be observable with *IUE*. There is some evidence from soft X-ray observations that such a halo exists (Nousek *et al.* 1982). Although the halo density is small, it has a major impact on the total pressure that is required to balance the gravitational attraction, because of the large z-integration range. The halo density and scale height have therefore to be taken as free parameters in an equilibrium analysis. Exponential density distributions are considered here: $n_{halo}(z) \propto \exp(-z/h_{halo})$. Figure 4 shows some examples of the total pressure distribution, determined from the equilibrium equation, together with the minimum gas pressure that is required for stability. Near the plane, the gas distribution used is essentially the same as described by S. Kulkarni in this volume [including HI with an exponential scale height of 400–500 pc found by Lockman (1984) and molecular gas — see §4.3]. The g-distribution presented by Bahcall (1984) was used ($z \lesssim 1$ kpc); the dark material is assumed to be distributed as the observed material (stellar and interstellar) and a massive halo, consistent with the rotation curve of the Galaxy, is included $(\rho_{halo}(0)/\rho_{disk}(0) - 0.1)$ [see Bahcall]. At $z \gtrsim 1300$ pc, g was taken to be constant ($g_c = 8.3 \times 10^{-9}$ cm s^{-2}). From the examples shown in figure 4, it is clear that the gas pressure is allowed to be significantly smaller than the total internal pressure up to a few kpc from the mid plane, but at larger distances the gas has to be supported by its own (thermal) pressure, requiring a temperature of $\sim 2 \times 10^5 g_c h_{halo}$ (g_c in units of 10^{-8} cm s^{-2} and h_{halo} in kpc).

The residual pressure $(P_{tot} - P_{g,min})$ increases with halo density and is the maximum that can be attributed to magnetic-field and CR pressure. From a comparison with radio continuum observations towards the galactic poles, one can derive lower limits for the density of the halo as a function of its scale height. It turns out that, essentially independent of further assumptions, this relation is approximately given by ($h_{halo} \lesssim 15$ kpc)

$$\left(\frac{n_{halo}(0)}{0.01 \text{ atom cm}^{-3}}\right) \left(\frac{h_{halo}}{1 \text{ kpc}}\right)^2 > \Gamma,$$

with $\Gamma \simeq 17$ if the magnetic-field pressure and the cosmic rays are assumed to have the same (free) z-distribution. So $n_{halo}(0) \gtrsim 0.005$ atom cm^{-3} for the example in figure 4. It was pointed out already by Badhwar and Stephens (1977) that the presence of a gaseous halo is of essential importance to reconcile equilibrium configurations with radio-continuum observations, but the stability criterion requires that its temperature is at least an order of magnitude higher than the value (10^4 K) assumed by Badhwar and Stephens (see below). This high temperature suggests that the maximum halo density is about twice the minimum density given above and that the scale height is probably $\gtrsim 5$ kpc. This is necessary in order (i) to have rough pressure equilibrium with the other phases of the ISM, (ii) to avoid that radiative cooling dominates the supernova heat input, and (iii) to limit the amount of soft X-rays produced (see Bloemen 1986b).

The temperature of the hot medium in the disk and at large distances from the plane ($z \gtrsim 3$ kpc) is found to be typically $\sim 10^6$ K, whereas around $z = 1 - 3$ kpc the temperature is only $(2 - 3) \times 10^5$ K. Such a layer of gas with relatively low temperature is in good agreement with the IUE observations of C IV and Si IV absorption lines from the halo, which seem to originate at these distances from the plane and which require a temperature less than a few times 10^5 K. The results presented here favour collisional ionisation, rather than photoionisation of a 10^4 K halo as the source of this ionized gas. If the isotropic component of the observed soft X-ray emission in the $0.5 - 1.2$ keV MI band comes from the halo, it has to originate at $z \gtrsim 3$ kpc in this scenario. The resulting scale height of the sum of cosmic-ray and magnetic-field pressures is only weakly dependent on the scale height of the gaseous halo: the half-equivalent width is $2 - 2.5$ kpc for $h_{halo} \gtrsim 5$ kpc. It cannot be excluded that the scale height of the CR nuclei is significantly smaller.

This approach leads to a consistent picture, but it should be kept in mind that the actual parameter values given here are flexible because possible stabilizing effects, as discussed above, have been ignored. The stability criterion is possibly not sufficient, but this exercise shows at least that the simplifying assumptions made in the past may have led to doubtful results.

ACKNOWLEDGEMENTS. I thank the University of Wyoming for financial support to attend the meeting and gratefully acknowledge receipt of a Miller Fellowship. I thank my colleagues and co-authors of the COS-B gamma-ray work summarized in §4 for many helpful discussions.

REFERENCES

Armstrong, J.W., Cordes, J.M., Rickett, B.J. 1981, *Nature*, **291**, 561.
Arnaud, K., *et al.* 1982, *M.N.R.A.S.*, **201**, 745.
Arnaud, M., and Cassé, M. 1985, *Astr. Ap.*, **144**, 64.
Axford, W.I., Leer, E., Skandron, K.G. 1977, *Proc. 15th Int. Cosmic Ray Conf.*, **11**, 132.
Axford, W.I. 1981; in *Origin of Cosmic Rays*, IAU Symp. 94, Reidel, p. 339.
Bahcall, J.N. 1984, *Ap. J.*, **276**, 169.
Badhwar, G.D., and Stephens, S.A. 1977, *Ap. J.*, **212**, 494.
Bell, A.R. 1978, *M.N.R.A.S.*, **182**, 147.
Beuermann, K., Kanbach, G., and Berkhuijsen, E.M. 1985, *Astr. Ap.*, **153**, 17.
Bhat, C.L., *et al.* 1985, *Nature*, **314**, 511.
Bibring, J.P., and Cesarsky, C.J. 1981, *17th Int. Cosmic Ray Conf.*, **2**, 289.
Bignami, G.F., Fichtel, C.E., Kniffen, D.A., and Thompson, D.J. 1975, *Ap. J.*, **199**, 54.
Black, J.H., and Willner, S.P. 1984, *Ap. J.*, **279**, 673.
Blandford, R.D., and Ostriker, J.P. 1978, *Ap. J.*, **221**, L29.
Blandford, R.D., and Ostriker, J.P. 1980, *Ap. J.*, **237**, 793.

Blitz, L., and Shu, F.H. 1980, *Ap. J.*, **238**, 148.
Blitz, L., Bloemen, J.B.G.M., Hermsen, W., and Bania, T.M. 1985, *Astr. Ap.*, **143**, 267.
Bloemen, J.B.G.M., *et al.* 1984a, *Astr. Ap.*, **139**, 37.
Bloemen, J.B.G.M., Blitz, L., and Hermsen, W. 1984b, *Ap. J.*, **279**, 136.
Bloemen, J.B.G.M., *et al.* 1984c, *Astr. Ap.*, **135**, 12.
Bloemen, J.B.G.M. 1985, *Astr. Astr.*, **145**, 391.
Bloemen, J.B.G.M., *et al.* 1986, *Astr. Ap.*, **154**, 25.
Bloemen, J.B.G.M. 1986a, in prep.
Bloemen, J.B.G.M. 1986b, in prep.
Brecher, K., and Burbidge, G.R. 1972, *Ap. J.*, **174**, 253.
Buffington, A., Schindler, S.M., and Pennypacker, C.R. 1981, *Ap. J.*, **248**, 1179.
Burbidge, G.R. 1974, *Phil. Trans. Roy. Soc. London*, **A 277**, 481.
Burbidge, G.R. 1983, in *Comp. and Origin of Cosmic Rays*, ed. M. Shapiro, Reidel, 245.
Cassé, M., and Goret, P. 1978, *Ap. J.*, **221**, 703.
Cassé, M. 1983, in *Comp. and Origin of Cosmic Rays*, ed. M.M. Shapiro, Reidel, p. 193.
Cesarsky, C.J. 1975, *Proc. 14th Int. Cosmic Ray Conf.*, **12**, 4166.
Cesarsky, C.J., Cassé, M., and Paul, J.A. 1977, *Astr. Ap.*, **60**, 139.
Cesarsky, C.J., and Völk, H.J. 1978, *Astr. Ap.*, **70**, 367.
Cesarsky, C.J. 1980, *Ann. Rev. Astr. Ap.*, **18**, 289.
Cesarsky, C.J., and Montmerle, T. 1981, *Proc. 17th Int. Cosmic Ray Conf.*, **9**, 207.
Cesarsky, C.J., Rothenflug, R., and Cassé, M. 1981a, *Proc. 17th Int. CR Conf.*, **2**, 269.
Cesarsky, C.J., Koch, L., and Perron, C. 1981b, *Proc. 17th Int. Cosmic Ray Conf.*, **2**, 22.
Cesarsky, C.J. 1983, in *Comp. and Origin of Cosmic Rays*, ed. M. Shapiro, Reidel, 175.
Chevalier, R.A., and Fransson, C. 1984, *Ap. J. (Letters)*, **279**, L43.
Cohen, R.S., and Thaddeus, P. 1977, *Ap. J. (Letters)*, **217**, L155.
Cohen, R.S., Dame, T.M., and Thaddeus, P. 1986, *Ap. J. Suppl.*, in press.
Colgate, S.A., and Johnson, M.H. 1969, *Phys. Rev. Lett.*, **5**, 235.
Cowsik, R. 1980, *Ap. J.*, **241**, 1195.
Cowsik, R., and Gaisser, T.K. 1981, *Proc. 17th Int. Cosmic Ray Conf.*, **2**, 218.
Cowsik, R. 1986, *Astr. Ap.*, **155**, 344.
Cox, D.P., and Smith, B.W. 1974, *Ap. J. (Letters)*, **189**, L105.
Dame, T.M. 1983, Ph. D. thesis, Columbia University.
Dame, T.M., and Thaddeus, P. 1985, *Ap. J.*, **297**, 751.
Dickman, R.L. 1978, *Ap. J. Suppl.*, **37**, 407.
Dodds, D., Strong, A.W., and Wolfendale, A.W. 1975, *M.N.R.A.S.*, **171**, 569.
Drury, L.O'C. 1983, *Rept. Progr. Phys.*, **46**, 473.
Eichler, D. 1979, *Ap. J.*, **229**, 419.
Eichler, D. 1980, *Ap. J.*, **237**, 809.
Eichler, D., and Hainebach, K.L. 1981, *Phys. Rev. Lett.*, **47**, 1560.
Ellison, D.C., Jones, F.C., and Eichler, D. 1981, *J. Geophys.*, **50**, 113.
Epstein, R.I. 1980, *M.N.R.A.S.*, **193**, 723.
Fermi, E. 1949, *Phys. Rev.*, **75**, 1169.
Fermi, E. 1954, *Ap. J.*, **119**, 1.
Fichtel, C.E., Simpson, G.A., and Thompson, D.J. 1978, *Ap. J.*, **222**, 833.
Fichtel, C.E., and Kniffen, D.A. 1984, *Astr. Ap.*, **134**, 13.
Fransson, C., and Chevalier, R.A 1985, *Ap. J.*, **296**, 35.
Frerking, M.A., Langer, W.D., and Wilson, R.W. 1982, *Ap. J.*, **262**, 590.
Fuchs, B., and Thielheim, K.O. 1979, *Ap. J.*, **227**, 801.
Fuchs, B., Schlickeiser, R., and Thielheim, K.O. 1976, *Ap. J.*, **206**, 589.
Gaisser, T.K., and Maurer, R.H. 1973, *Phys. Rev. Lett.*, **30**, 1264.
Gaisser, T.K., and Levy, E.H. 1974, *Phys. Rev.*, **D10**, 1731.

Garcia-Munoz, M., Mason, G.M., and Simpson, J.A. 1977, *Ap. J.*, **217**, 859.
Garcia-Munoz, M., *et al.* 1979, *Proc. 16th Int. Cosmic Ray Conf.*, **1**, 310.
Garcia-Munoz, M., Simpson, J.A., and Wefel, J.P. 1981, *Proc. 17th Int. CR Conf.*, **2**, 72.
Garcia-Munoz, M., Guzik, T.G., Simpson, J.A., and Wefel, J.P. 1984, *Ap. J.*, **280**, L13.
Ginzburg, V.L., and Syrovatskii, S.I. 1964, The Origin of Cosmic Rays, Pergamon Press.
Ginzburg, V.L., Khazan, Y.M., and Ptuskin, V.S. 1980, *Ap. Sp. Sci.*, **68**, 295.
Ginzburg, V.L., and Ptuskin, V.S. 1981, *Proc. 17th Int. Cosmic Ray Conf*, **2**, 336.
Gordon, M.A., and Burton, W.B. 1976, *Ap. J.*, **208**, 346.
Gregory, J.C., *et al.* 1981, *Proc. 17th Int. Cosmic Ray Conf.*, **9**, 154.
Gunn, J.E., and Ostriker, J.P. 1969, *Phys. Rev. Lett.*, **22**, 728.
Guzik, T.G., *et al.* 1985, *Proc. 19th Int. Cosmic Ray Conf.*, **2**, 76.
Harding, A.K., and Stecker, F.W. 1985, *Ap. J.*, **291**, 471.
Haslam, C.G.T., *et al.* 1981a, *Astr. Ap.*, **100**, 209.
Haslam, C.G.T., *et al.* 1981b, *Astr. Ap. Suppl.*, **47**, 1.
Havnès, O. 1982, *Astr. Ap.*, **110**, 203.
Hayakawa, S. 1969, Cosmic Ray Physics, Wiley-Interscience, New York.
Heiles, C. 1986, *Ap. J.* (in press).
Henderson, A.P., Jackson, P.D., and Kerr, F.J. 1982, *Ap. J.*, **263**, 182.
Higdon, J.C. 1979, *Ap. J.*, **232**, 113.
Holmes, J.A. 1974, *M.N.R.A.S.*, **166**, 155.
Holmes, J.A. 1975, *M.N.R.A.S.*, **170**, 251.
Israel, M.H. 1983, in *Comp. and Origin of Cosmic Rays*, ed. M.M. Shapiro, Reidel, p. 47.
Issa, M.R., Riley, P.A., Strong, A.W., and Wolfendale, A.W. 1981, *J. Phys.*, G7, 973.
Jokipii, J.R. 1973, *Ap. J.*, **183**, 1029.
Jokipii, J.R. 1976, *Ap. J.*, **208**, 900.
Jones, F.C. 1979, *Ap. J.*, **229**, 747.
Jordan, S.P., and Meyer, P. 1984, *Phys. Rev. Lett.*, **53**, 505.
Kellman, S.A. 1972, *Ap. J.*, **175**, 353.
Kniffen, D.A., Fichtel, C.E., and Thompson, D.J. 1977, *Ap. J.*, **215**, 765.
Koch, L. *et al.* 1983, *Proc. 18th Int. Cosmic Ray Conf.*, **9**, 275.
Kodaira, K. 1974, *Publ. Astr. Soc. Japan*, **26**, 255.
Krymsky, G.F. 1977, *Dok. Akad. Nauk. S.S.S.R.*, **234**, 1306.
Kulsrud, R.M., and Pearce, W.P. 1969, *Ap. J.*, **156**, 445.
Kulsrud, R.M., Ostriker, J.P., and Gunn, J.E. 1972, *Phys. Rev. Lett.*, **28**, 636.
Lachièze-Rey, M., Asséo, E., Cesarsky, C.J., and Pellat, R. 1980, *Ap. J.*, **238**, 175.
Lebrun, F., and Paul, J.A. 1983, *Ap. J.*, **266**, 276.
Lebrun, F. *et al.* 1983, *Ap. J.*, **274**, 231.
Lebrun, F., and Huang, Y.-L. 1984, *Ap. J.*, **281**, 634.
Lerche, I., and Schlickeiser, R. 1985, *Astr. Ap.*, **151**, 408.
Liszt, H.S., Xiang, D.-L., and Burton, W.B. 1981, *Ap. J.*, **249**, 532.
Lockman, F.J. 1984, *Ap. J.*, **283**, 90.
Lockman, F.J., Hobbs, L.M., and Shull, J.M. 1986, *Ap. J.* (in press).
Longair, M.S. 1982, High Energy Astrophysics, Cambridge Univ. Press, Cambridge.
Lyne, A.G., Manchester, R.N., and Taylor, J.H. 1985, *M.N.R.A.S.*, **213**, 613.
Magnani, L., Blitz, L., and Mundy, L. 1985, *Ap. J.*, **295**, 402.
Mathis, J.S., Mezger, P.G., and Panagia, N. 1983, *Astr. Ap.*, **128**, 212.
McKee, C.F., and Ostriker, J.P. 1977, *Ap. J.*, **218**, 148.
Meyer, J.-P. 1985a, *Ap. J. Suppl.*, **57**, 151.
Meyer, J.-P. 1985b, *Ap. J. Suppl.*, **57**, 173.
Mewaldt, R.A. 1985, *Proc. 19th Int. Cosmic Ray Conf.*, **2**, 64.
Morfill, G.E., Meyer, P., and Lüst, R. 1985, *Ap. J.*, **296**, 670.

Mouschovias, T.C. 1974, *Ap. J.*, **192**, 37.
Mouschovias, T.C., Shu, F.H., and Woodward, P.R. 1974, *Astr. Ap.*, **33**, 73.
Mouschovias, T.C. 1975, *Astr. Ap.*, **40**, 191.
Nousek, J.A., Fried, P.M., Sanders, W.T., and Kraushaar, W.L. 1982, *Ap. J.*, **258**, 83.
Ormes, J.F., and Freier, P. 1978, *Ap. J.*, **222**, 471.
Ormes, J.F., and Protheroe, R.J. 1983, *Ap. J.*, **272**, 756.
Protheroe, R.J. 1981, *Ap. J.*, **251**, 387.
Protheroe, R.J. 1983, in *Comp. and Origin of Cosmic Rays*, ed. M. Shapiro, Reidel, 119.
Parker, E.N. 1966, *Ap. J.*, **145**, 811.
Parker, E.N. 1969, *Space Sci. Rev.*, **9**, 651.
Parker, E.N. 1975, *Ap. J.*, **201**, 74.
Phillipps, S., *et al.* 1981, *Astr. Ap.*, **98**, 286.
Piddington, J.H. 1973, *M.N.R.A.S.*, **162**, 73.
Ptuskin, V.S. 1979, *Astrophys. Space Sci.*, **61**, 259.
Rockstroh, J.M., and Webber, W.R. 1978, *Ap. J.*, **224**, 677.
Ryan, M.J., Balasubrahmanyan, V.K., and Ormes, J.F. 1972, *Phys. Rev. Lett.*, **28**, 985.
Sanders, D.B., Solomon, P.M., and Scoville, N.Z. 1984, *Ap. J.*, **276**, 182.
Scott, J.S., and Chevalier, R.A. 1975, *Ap. J. (Letters)*, **197**, L5.
Scoville, N.Z., and Solomon, P.M. 1975, *Ap. J. (Letters)*, **199**, L105.
Shu, F.H. 1974, *Astr. Ap.*, **33**, 55.
Shapiro, M.M., and Silberberg, R. 1970, *Ann. Rev. Nuc. Sci.*, **20**, 323.
Silberberg, R., *et al.* 1983, *Phys. Rev. Lett.*, **51**, 1217.
Silberberg, R., Tsao, C.H., and Letaw, J.R. 1985a, *Ap. J. Suppl.*, **58**, 873.
Silberberg, R., *et al.* 1985b, *Proc. 19th Int. Cosmic Ray Conf.*, **3**, 238.
Simpson, J.A. 1983, *Ann. Rev. Nucl. Part. Sci.*, **33**, 323.
Skilling, J. 1971, *Ap. J.*, **170**, 265.
Skilling, J., and Strong, A.W. 1976, *Astr. Ap.*, **53**, 253.
Solomon, P.M., and Sanders, D.B. 1980, in *Giant Molecular Clouds in the Galaxy*, eds.
        P.M. Solomon and M.G. Edmunds, Pergamon Press, Oxford, p. 41.
Soutoul *et al.* 1985, *Proc. 19th Int. Cosmic Ray Conf.*, **2**, 8.
Spitzer, L. 1956, *Ap. J.*, **124**, 20.
Stecker, F.W. 1971, Cosmic Gamma Rays, Mono Book, Corp., Baltimore.
Stecker, F.W., Solomon, P.M., Scoville, N.Z., and Ryter, C.E. 1975, *Ap. J.*, **201**, 90.
Stephens, S.A. 1981, *Nature*, **289**, 267.
Strong, A.W., Wolfendale, A.W., Bennett, K., Wills, R.D. 1978, *M.N.R.A.S.*, **182**, 751.
Strong, A.W., *et al.* 1982, *Astr. Ap.*, **115**, 404.
Strong, A.W., *et al.* 1985, *Proc. 19th Int. Cosmic Ray Conf.*, **1**, 317.
Tarafdar, S.P., and Apparao, K.M.V. 1981, *Ap. Sp. Sci.*, **77**, 521.
Tasaka, S. *et al.* 1982, *Phys. Rev.*, **D25**, 1765.
Völk, H.J. 1984, in *High Energy Astrophysics*, eds. J. Audouze and J. Tran Thanh Van,
        Editions Frontieres, Gif-sur-Yvette, p. 281.
Webber, W.R., Simpson, G.A., and Cane, H.V. 1980, *Ap. J.*, **236**, 448.
Webber, W.R. 1983a, in *Comp. and Origin of Cosmic Rays*, ed. M. Shapiro, Reidel, 25.
Webber, W.R. 1983b, in *Comp. and Origin of Cosmic Rays*, ed. M. Shapiro, Reidel, 83.
Weisheit, J.C., and Collins, L.A. 1976, *Ap. J.*, **210**, 299.
Wentzel, D.G. 1974, *Ann. Rev. Astr. Ap.*, **12**, 71.
Wentzel, D.G., Jackson, P.D., Rose, W.K., and Sinha, R.P. 1975, *Ap. J. (Lett.)*, **201**, L5.
Wiedenbeck, M.E., and Greiner, D.E. 1980, *Ap. J. (Lett.)*, **239**, L139.
Wiedenbeck, M.E. 1983, in *Comp. and Origin of Cosmic Rays*, ed. M. Shapiro, Reidel,
        65.
Zweibel, E.G., and Kulsrud, R.M. 1975, *Ap. J.*, **201**, 63.

Section III: Interstellar Magnetic Properties

Carl Heiles

INTERSTELLAR MAGNETIC FIELDS

CARL HEILES

Astronomy Department
University of California, Berkeley

ABSTRACT. Methods to observe magnetic fields are considered. Next, results for external galaxies are reviewed; I conclude that most results are questionable. Next, the Galactic field is reviewed. The large-scale field decreases slowly with Galactic radius and z, and has a strength ~ 4 μG near the Sun. It is a roughly circular field, which may reverse one or more times inside the Solar circle. The local value of the uniform component is $\gtrsim 1.6$ μG. The Galactic field is not uniform. A few 'magnetic bubbles' of diameter ~ 100 pc stand out quite prominently, and statistical analyses show that the nonuniform component of the field is at least as strong as the uniform component.

Finally, observations of the field on small scales are reviewed. There is no evidence for an increase in field strength with volume density for densities $\lesssim 100$ cm^{-3}, except behind interstellar shocks and near some dark clouds and star-forming regions. For larger densities the data are consistent with the field increasing as a fairly weak power of the volume density. The field is often morphologically related to the interstellar gas—$e.g.$ parallel or perpendicular to filaments, and systematically oriented in large shells. This can also be true on smaller scales; for example, bipolar flows tend to be aligned with the large-scale, ambient magnetic field.

1. METHODS OF MEASUREMENT

Methods fit into two categories: methods to measure the *direction* of the component lying *perpendicular* to the line of sight ($B_\perp$, the projected direction on the plane of the sky), and methods to measure the *strength* of the component of magnetic field lying *parallel* to the line of sight, $B_\parallel$.

1.1. Methods to Measure the Direction of $B_\perp$.

$B_\perp$: **Polarization of Starlight.** Historically, the very existence of the interstellar magnetic field was deduced from the discovery of linear polarization of starlight. The polarization is produced by the alignment of dust grains in the interstellar magnetic field. A grain is continually bombarded by interstellar gas atoms, which makes it spin. For a spinning grain in collisional equilibrium, the energies associated with spin about all three principal axes of inertia are equal. This means that the largest angular momentum lies along the axis of *largest* moment of inertia. For this reason, needle-like grains tend to spin end-over-end, like a well-kicked American football.

An observer situated on a grain spinning in a magnetic field sees a time-variable magnetic field, unless the spin axis is parallel to the magnetic field. If the grain sees a time-variable magnetic field, then two energy-dissipating mechanisms occur: one, the time-variable field induces electric currents in the grain; two, the magnetic susceptibility

D. J. Hollenbach and H. A. Thronson, Jr. (eds.), Interstellar Processes, 171–194.

of any real material has an imaginary component, so that the field inside the grain is not perfectly parallel to the external field. The latter mechanism is usually more important. This dissipation tends to drive the grain toward a configuration in which the grain sees a time-*independent* magnetic field as it spins. Imagine a small insect sitting on a needle-like grain spinning about its short axis. The only way this insect can see a constant field as the grain spins is if the spin vector is parallel to the field—or, for a needle-like grain, if the short axis is parallel to the external magnetic field. Thus, grains tend to line up such that their time-averaged projected lengths are longer in the direction perpendicular to the magnetic field. This 'shorts out' the E-vector in a passing light wave, producing linear polarization of the observed wave that is parallel to $B_\perp$.

The original 'Davis-Greenstein' theory of grain alignment involves dissipation of the *thermal* spin energy of grains by the imaginary component of magnetic susceptibility in the grain (see Jones and Spitzer 1967). However, this theory requires magnetic field strengths that are much higher than those observed in the diffuse interstellar medium. This important point is apparently not generally appreciated, because some authors still derive field strengths using this theory. A modification of the theory allows alignment even by weak fields: the spin energy of grains is much larger than thermal—it is *suprathermal* because of the unbalanced *time average* torque on a grain produced by a small number of sites where H_2 molecules are produced and ejected with large kinetic energy, spinning the grain up like a steam turbine. The theory involves a remarkably wide range of physical principles which are presented in an elegant paper by Purcell (1979).

Circular polarization of starlight has also been observed. There are two mechanisms. In one, the circular polarization is produced by birefringence of aligned grains; this depends on the grain properties alone (Martin 1972). In the other, it is produced a by change in the direction of $B_\perp$ with distance along the line of sight (Martin 1974, 1975; Martin and Campbell 1976). These effects are weak, and few useful results have been obtained yet.

$B_\perp$: **Linear Polarization of Synchrotron Radiation.** Synchrotron radiation is produced by the acceleration of relativistic electrons gyrating in a magnetic field. The direction of linear polarization is parallel to the direction of acceleration. Thus, the polarization is perpendicular to the magnetic field, and the observed polarization is perpendicular to $B_\perp$.

If the field is perfectly uniform, the linear polarization is large, typically $\sim 70\%$ (Ginzburg and Syrovatskii 1969). In 'real astronomy', the magnetic field is unlikely to be perfectly uniform. A better description is a uniform plus a random component. The linear polarization is smaller in this case, and in addition it depends on the angle between the uniform component and the observer's line of sight. This is easy to see for the case in which the uniform component is parallel to the line of sight. In this case, the polarization is zero: the uniform component contributes nothing to the intensity of the emitted radiation, and by definition the polarization of the random component is zero.

Thus, in almost any real astronomical situation, as the field direction changes from being perpendicular to parallel to the observer's line of sight, the direction of linear polarization remains perpendicular to $B_\perp$ but the intensity of the polarization goes to zero. If the observer thinks she knows the angle between the uniform component of the field and the line of sight, then she can derive the ratio of the random and uniform components of the field from the observed degree of polarization.

$B_\perp$: **Linear Polarization of Radio-wavelength Spectral Lines.** It has been predicted (Kylafis 1983 and Deguchi and Watson, 1984), but not yet observed, that molecular spectral lines at radio wavelengths should be linearly polarized parallel to $B_\perp$. The polarization depends on a number of complicating factors, including the characteristics of the local velocity and radiation fields; it is insensitive to the field strength. This may be the only way to probe the field structure in dense molecular clouds and protostars.

1.2. Methods to Measure the Strength of $B_\parallel$.

First, we note a fact that statistically relates $B_\parallel$ to the total field strength B_{tot}. For a random distribution of directions of the magnetic field, it is easy to show that with an *a priori* probability of x^{-1} the total field is $> x B_\parallel$. Thus, for any single measured value of $B_\parallel$, $2B_\parallel$ represents a median value for the total field strength, because the total field strength has an equal probability of lying above or below $2B_\parallel$.

$B_\parallel$: **Faraday Rotation.** The Rotation Measure $RM \propto \int n_e B_\parallel dl$ is derived from the Faraday rotation of the plane of linear polarization of a background radio continuum source. Along the line of sight all of the electrons rotate either clockwise or counterclockwise, depending on the direction of the magnetic field. This causes the index of refraction, and thus the phase velocity, to be different for the two senses of circular polarization. Linear polarization with a specific position angle is produced by combining the two circular polarizations with a specific phase difference. Because the phase difference changes with wavelength, the position angle also changes with wavelength, λ; it is proportional to $RM\lambda^2$ (see Spitzer 1978).

RM's can be measured for ordinary radio sources that exhibit linear polarization. Many pulsars are also linearly polarized. Pulsars provide the additional important piece of information, dispersion measure ($DM \propto \int n_e dl$). The ratio RM/DM provides the magnetic field strength directly, with no ambiguity about the electron density.

There are two important cautions concerning the derivation of RM's from the wavelength dependence of the position angle of linear polarization. One is obvious, concerning the $n\pi$ ambiguity of the angle. This ambiguity can be removed by obtaining measurements at sufficiently closely-spaced wavelengths. The second is more subtle, and involves the complications that arise when the relativistic electrons that produce the polarized emission coexist with the thermal electrons that produce the Faraday rotation. In this case, the position angle rotates with λ^2 only for rotation angles $\lesssim 1$ radian (Burn 1966; Cioffi and Jones 1980), the 'Faraday-thin' regime. Reliable results can be obtained only by restricting the wavelength range to this regime. As emphasized by Vallée (1980), some observers and interpreters of data have fallen into this pitfall.

$B_\parallel$: **Zeeman Splitting.** We first consider the hydrogen atom as the most important example. An external magnetic field splits the upper level of the 21-cm line into three levels. The splitting $\Delta\nu$ between the highest and the lowest levels is $2.8 B_{tot}$ Hz, where B_{tot} is the *total* field strength in μG. The atom radiates two circularly-polarized σ components, separated in frequency by $\Delta\nu$, with maximum amplitude parallel to the magnetic field. The atom radiates three linearly-polarized π components with maximum amplitude perpendicular to the magnetic field.

In all diffuse clouds, the splitting is much smaller than the typical line width. In this limiting case, the components cannot be distinctly separated. To observe the splitting, one observes the difference between the circular (Stokes parameter V) or linear (Stokes parameters Q and U) polarizations. In the case of the σ components, the V spectrum is the first derivative with respect to frequency of the line profile with an amplitude proportional to $(\Delta\nu/\delta\nu)$, where $\delta\nu$ is the line width. In the case of the π components, the Q and U spectra are the *second* frequency derivative of the line profile with an amplitude proportional to $(\Delta\nu/\delta\nu)^2$. With the small $\Delta\nu$'s encountered in practice, the π components are virtually undetectable and the σ components only barely so. Because of the directional dependence of the amplitude of the σ components, for small $(\Delta\nu/\delta\nu)$ the the *observed* amplitude of the V spectrum is proportional only to the *parallel* component of magnetic field, *i.e.* proportional to $(B_{\parallel}/\delta\nu)$. Note that high frequency resolution is not required to observe the Zeeman effect, even though $\Delta\nu$ is small.

Zeeman splitting was first detected in the 21-cm line seen in absorption against the classical strong radio sources (Verschuur 1969a); the early history is an interesting story of the frustrations involved with inadequate instrumentation (Verschuur 1979). Zeeman splitting is now being observed in the H I line seen in emission, a very difficult undertaking because of the instrumental problems involved (see Troland and Heiles 1982a). Zeeman splitting of the H I line in absorption against bright radio sources is also being observed with high angular resolution using aperture synthesis techniques.

Any atom or molecule having an electron with an unpaired electron spin yields Zeeman splitting comparable to the H atom. In practice, the only such molecules having cm-wavelength lines are those exhibiting Λ-doubling (the hydrides). Several such molecules have mm-wavelength lines, but Doppler widths are so large that the ratio $(\Delta\nu/\delta\nu)$ is too small to produce detectable V spectra with system temperatures currently available. Zeeman splitting has been searched for in H recombination lines (Troland and Heiles, 1977) and C recombination lines (Silverglate 1984). In diffuse clouds, it has been measured in the 18-cm OH lines. In OH masers, splitting larger than the line width—and therefore proportional to B_{tot} rather than $B_{\parallel}$—has been observed in several cm-wavelength OH lines.

1.3. Faraday Rotation *vs.* Zeeman Splitting.

These two techniques for measuring $B_{\parallel}$ sample different kinds of region. Zeeman splitting favors high H I column density and narrow line width, so it samples the cold H I clouds—the CM (cold medium). Faraday rotation samples ionized regions, the same as sampled by pulsar DM's. Kulkarni and Heiles (1986) show that pulsar DM's are produced mainly by the WIM. The WIM—'warm ionized medium'—is a nearly fully ionized, rarefied, temperature ~ 8000 K component of the interstellar medium whose source of ionization is probably starlight (see Reynolds 1984 and Mathis 1986).

The fact that Zeeman splitting and Faraday rotation sample different regions is elegantly illustrated by the observational results towards the Crab nebula and pulsar, against which both Zeeman splitting and Faraday rotation have been measured. Zeeman splitting shows a field directed towards the Earth, while Faraday rotation shows the opposite.

There is confusion with regard to signs. Radio astronomers adopt the IEEE convention for circular polarization, not the definition of classical optics. In the IEEE convention, a right-hand circularly (RHC) polarized wave rotates clockwise as it propagates away from the observer. In Zeeman splitting, if the RHC component is observed at a higher frequency then the magnetic field points towards the observer. A field pointing *towards* the observer is a *negative* magnetic field. In Faraday rotation, if the position angle of the plane of polarization increases with increasing wavelength (corresponding to counterclockwise rotation against the plane of the sky: the position angle is measured eastward from north), then the *RM* is positive. This corresponds to a field pointed towards the observer. Thus, *positive RM*'s correspond to *negative* magnetic fields (!).

2. FIELDS IN EXTERNAL GALAXIES

Sofue, Fujimoto, and Wielebinski (1986) have recently reviewed the field, so we omit many details and stress one fundamental point: in my opinion, the results are *highly suspect*. Attempts to determine magnetic field geometry and strength in external galaxies have been based on measurements of linear polarization at optical and radio wavelengths. In most cases optical measurements can be done only of the diffuse light from the disk of a galaxy. These are plagued by contamination from scattered light, and their usefulness for determining information about the magnetic field is very limited.

The radio emission is polarized because it is nonthermal. In the absence of Faraday rotation, the observed polarization is perpendicular to $B_\perp$. At 6-cm wavelength, where a number of galaxies have been measured with reasonable accuracy, Faraday rotation is expected to be small so that, to zeroth order, the observed polarization vectors trace out the direction of $B_\perp$. The results show that, again to zeroth order, the field lines are circular or, perhaps, aligned along the spiral arms. I believe these zeroth-order results. Furthermore, they are in accord with results on our Galaxy.

The problems arise with attempts to go beyond this zeroth order picture. Typically, the assumption is made that the magnetic field is either circular or aligned with the arms, and that departures of the observed polarization angles from those expected under this assumption, $\Delta\chi$, arise from Faraday rotation. Faraday rotation is largest when the field points toward the observer, as it does along the major axis of an inclined spiral galaxy if the field is parallel to the galactic plane. Thus, as we follow $\Delta\chi$ around the ellipses that correspond to constant galactic radii, Faraday rotation should make $\Delta\chi$ largest along the major axes. A least square fit to this expected behavior allows the determination of Faraday rotation from measurements of polarization at a single wavelength! Such determinations are used to infer the direction of $B_\parallel$ and, with suitable assumptions about electron density, the strength of $B_\parallel$. In this way, it has been concluded that a minority of galaxies have circular fields. A majority have 'bisymmetric' fields, in which magnetic field lines lie along the spiral arms, appearing to be 'wound up' by the differential rotation and alternating in direction from one spiral arm to the next. As a natural consequence, the field also reverses several times with galactocentric radius.

The cleanest example of this technique is M31 (Beck 1982). $\triangle\chi$ has a clear, unambiguous sinusoidal variation with azimuthal angle measured from the center of M31 for galactic radii $\gtrsim 7$ kpc; inside 7 kpc the polarized radiation is weak and undetectable. Beck concludes that the field outside 7 kpc is circular, with no reversals. He estimates the strength and other parameters of the magnetic field by adopting values for other parameters that are appropriate to our Galaxy, and derives values that are close to those of our Galaxy. Overall rough circularity of the field is corroborated by observations of optical polarization of globular clusters located on the far side of M31's disk (Martin and Shawl 1982). The whole picture is self-consistent and quite satisfying.

In my opinion, the application of this method to most galaxies, and perhaps even M31, is highly suspect for two reasons. One, measured polarizations are so weak that both noise and systematic errors are problems. $\triangle\chi$ is expected to be largest on the major axes, the very places where the polarized intensity is smallest because the field direction is most nearly parallel to the observer's line of sight. Thus, in many cases the position angles do not show the behavior expected under the model in a clear and unambiguous fashion. In my opinion, M31 is the only galaxy that does not suffer in this way.

Two, there is only one galaxy (M81) for which good polarization measurements exist at more than one wavelength—*i.e.*, only in M81 can the hypothesis that $\triangle\chi$ really results from Faraday rotation be checked. In M81, $\triangle\chi$'s do *not* vary as λ^2 (Beck, Klein, and Krause 1985). These authors conclude that $\triangle\chi$ is due to real structure in the magnetic field, and that the observations cannot distinguish between the circular and the bisymmetric models. In fact, the most direct conclusion is that the observations are not really consistent with either model!

In principle, accurate *RM* measurements would allow much to be learned about magnetic fields in external galaxies, and with today's availability of sensitive aperture synthesis instruments the field is ripe for investigation.

3. THE GALACTIC FIELD ON SCALES ABOVE 100 PC

3.1. Crude Statements about the Galaxy-wide Field Strength.

Rough information on the large-scale behavior of the magnetic field strength can be obtained from models fitting the all-sky brightness distribution of the diffuse Galactic synchrotron emission (Phillips *et al.* 1981a, 1981b, hereafter PKOHS; Beuermann, Kanbach, and Berkhuijsen 1985, hereafter BKB). These models provide estimates of the volume emissivity as a function of Galactocentric radius R_G and height above the plane z. In the Galactic plane, spiral arms can be discerned in the continuum emission, a complication which we neglect here.

The emissivity decreases outwards with R_G. If one fits this with an exponential, the radial exponential scale length is about 3.9 kpc or somewhat less (for an assumed Solar R_G of 10 kpc). Surprisingly, the models can distinguish between a systematic, circular field and a mixture of systematic and random components. They assume that the systematic field lies parallel to the spiral arms. Since synchrotron radiation in a uniform field is directional, with no emission along the field, a perfectly uniform field with no random component leads to no emission tangential to the spiral arms. Instead, observations show enhanced brightness in these directions. A totally random field is consistent with the data inside roughly the Solar circle, but not outside. Equal strengths in the random and systematic components are consistent with the data throughout the Galaxy, and perhaps this should be favored for simplicity.

The emissivity decreases with $|z|$. PKOHS model this in a manner that can be represented roughly with a thick disk plus a long, linearly-decreasing $|z|$ tail. In contrast, BKB model it with two disks, a thin and a thick disk with roughly equal volume emissivities at $z = 0$, but no long tail. In both models, the $|z|$-scale heights of the disks *increase* with R_G, roughly as $\exp(R_G/10 \text{ kpc})$. At the Solar circle, the disk scale height of PKOHS is roughly 0.6 kpc and the long tail extends to ~ 8 kpc; the disk scale heights of BKB are ~ 0.2 and 1.8 kpc.

The strength of the local magnetic field can be obtained from the derived emissivity near the Sun, together with the observed energy spectrum of the electron component of cosmic rays. PKOHS and BKB derive values for the local emissivity that differ by a factor of ~ 3.5. The derived values for the total local field strength are about 4 and 9 μG for PKOHS and BKB, respectively. If we assume an equal split between uniform and random components of magnetic field, each component would be $2^{1/2}$ times smaller than the total. The resulting value for PKOHS is about 3 μG. As we shall see, this is larger than the strength of the uniform component derived from RM's and pulsar DM's, but not by much.

The above emissivities are not well-determined. The reason is that the observations cover long path lengths, and the volume emissivity—which varies along the path length—must be modelled and determined by an unfolding process. The differences between PKOHS and BKB are indicative of the uncertainties. Unless Nature has placed the Sun in an unusually bad location, the basic conclusions should be valid: the emissivity decreases outwards with R_G, extending to some 20 kpc, and decreases outwards with $|z|$, extending to at least ~ 2 kpc.

The synchrotron emissivity depends not only on the magnetic field strength, but also on the energy density of the electron component of the cosmic rays. In the direction of R_G, Bloemen *et al.* (1986) have derived the distribution of the electron cosmic rays from observations of diffuse Galactic gamma rays. Folding their result together with the R_G-dependence of synchrotron emissivity, we have roughly $B \propto \exp(-R_G/20 \text{ kpc})$—roughly independent of R_G.

In the $|z|$ direction, we have no information on the variation of relativistic electrons. Perhaps the most reasonable assumption is that the energy density of cosmic rays is equal to that of the magnetic field. This is close to the 'minimum energy' configuration required to generate the observed radiation. If, in addition, we assume that the energy of the *electron* component of the cosmic rays (which is a negligible fraction of the *total*) is proportional to the magnetic energy density, then the emissivity $\propto B^{7/2}$ (Ginzburg and Syrovatskii 1969). For PKOHS's single disk, the local B scale height would be about 2.4 kpc. For BKB's disks, the thin disk would raise the field strength by only $2^{2/7} \approx 1.2$ at $z = 0$, where the two disks have equal synchrotron emissivities, so that the B scale height would be determined by the thick disk and would be $\sim$ 6 kpc. While these numbers are subject to considerable uncertainty because of the assumptions involved, the synchrotron emissivity *in any case* depends on a high power of B. Thus, the B scale height must be considerably larger than the scale height of synchrotron emissivity.

The numerical details of the R_G and z dependence of magnetic field strength are quite uncertain, but the main conclusion is clear because the synchrotron emissivity depends on a high power of B. The main conclusion is that the field strength varies only slowly with R_G and z, much more slowly than, for example, the volume density and pressure of interstellar gas.

3.2. Crude Statements about the Galaxy-wide Field Direction.

Optical polarization showed long ago that the field lines lie parallel to the Galactic plane; the comprehensive map by Mathewson and Ford (1970) exhibits more details. Towards $l_o \approx 80°$ the polarization vectors tend to 'focus' as we look down the field and see the field lines receding into the distance. Some analyses of Faraday rotation data are in agreement with the result, but others are in disagreement, yielding $l_o \approx 100°$ (see below). To zeroth order, then, the field is circular—just as it is in external galaxies. However, the exact value of l_o is more than just a detail: the local spiral arm has $l_o < 90°$, so if in fact $l_o \approx 100°$ then the field is significantly tilted with respect to the spiral arms.

The above optical and RM data sets refer, in fact, to regions within a few kpc of the Sun. One would like to extrapolate and conclude that the magnetic field is roughly circular everywhere in the Galaxy. However, there is an absence of large RM's toward the Galactic interior. This is important, for it indicates that the Galaxy cannot have a roughly circular uniform field that points in the same direction everywhere. In the Galactic plane near, say, $l = 30°$, such a field would make every volume element along the line of sight contribute in the same sense to the RM of an extragalactic source, so that with the long path lengths of tens of kpc we would expect RM's to be systematically larger than ~ 500 rad m^{-2}. Such large RM's are not observed. In addition, Manchester and Taylor (1977) point out that closely-spaced pulsars, located at different distances, sometimes have different algebraic signs of RM.

These facts indicate the existence of either one or more field reversals or a significant radial field component inside the Solar circle (Heiles 1976a). Alternatively, the uniform component of the field may vanish, leaving only a random component; this is consistent with the model fitting of Philips *et al.* (1981a) described above in section 3.1. As discussed below in section 3.3, analyses of RM's of extragalactic sources probably favor field reversal(s) inside the Solar circle, but the data are not yet good enough to determine this unambiguously, much less the number and location(s) of the field reversals. There is no evidence for field reversals *outside* the Solar circle (Vallée 1983).

3.3. The Uniform Component of the Magnetic Field within a Few Kpc of the Sun.

The data are solely RM's of extragalactic radio sources and of pulsars. For extragalactic radio sources, statistically adequate discussions involving more than 600 sources are given by Inoue and Tabara (1981) (IT) and by Simard-Normandin and Kronberg (1979) (SK).

At negative latitudes, for $b \lesssim -30°$, the signs of the RM's reverse near $l = 0°$. This is consistent with a uniform circular field in the Galaxy. However, at positive latitudes the distribution is more complicated. This is usually attributed to the perturbing influence of the North Polar Spur (Radio Loop I); a clear explanation and pictorial representation of the perturbation is given by Vallée and Kronberg (1973). Near the Galactic plane, for $|b| \lesssim 30°$, the distribution is also more complicated; this is often attributed to a reversal of the Galactic field inside the Solar circle.

Thus, in line with nearly *everyone's* prejudiced expectation of a large-scale roughly circular Galactic field, IT do the logical thing and derive the properties of the 'local Galactic field' from sources having $b < -30°$. This is quite a severe selection—it discards sources occupying 75% of the sky! I know of no other case in which many astronomers consider such a small minority of data to be the true, unbiased representatives of the statistical sample! IT obtain $l_o = 102°$. In contrast, if they restrict the fit to sources near the Galactic plane ($|b| < 30°$), they obtain $l_o = 80°$, which is also in agreement with SK's result for sources near the plane. In the plane, SK find evidence for one or more reversals in the circular field, while IT do not; the two papers use different data in the plane, and include different latitude ranges, which are probably the sources of the discrepancy.

For pulsars the results are more ambiguous, mainly because of inadequate number statistics but also because of differences in analysis. The total available sample of pulsars within ~ 3 kpc distance numbers only 48; this was analyzed by Thomson and Nelson (1980; hereafter TN). Manchester and Taylor (1977) (MT) and IT use a somewhat smaller sample, restricting the distance to 2 kpc. For the straightforward approach to the least-squares fit, involving no extra parameters or 'fancy' statistical techniques, TN and IT agree on $l_0 \approx 110°$, while MT exclude pulsars within 30° of the Galactic center and obtain $l_o = 90°$. All three agree that the uniform component of magnetic field $B_u \approx 1.6$ μG. However, TN go on to invoke a more complicated model, including a $|z|$-scale height in the magnetic field and a reversal inside the Solar circle; they obtain $l_o = 74°$, $B_u = 3.5$ μG, and a scale height of 75 pc. In contrast, IT used RM's of extragalactic sources to find a scale height of ~ 1000 pc. The disagreement between the scale heights of IT and TN is severe. The larger value of IT is, in my opinion, more reliable because it is based on a larger statistical sample and is in much better agreement with the models based on the diffuse Galactic synchrotron emission (section 3.1). TN's small value of 170 pc for the distance to the reversal is much smaller than the value ~ 1.3 kpc found by SK. I favor the larger value, because if only a few reversals occur within the Galaxy, there is only a small *a priori* probability that the Sun lies so close to a reversal, and in particular the outermost reversal.

To summarize: Within a few kpc of the Sun, the uniform component of the field is roughly circular. Its exact direction depends on which part of the sky is analyzed. Its strength is ~ 1.6 μG in the most straightforward fits, but larger for fits that include more parameters. Its z-scale height has been fit but the value is uncertain. The model of a roughly circular field with no reversals inside the Solar circle cannot be valid. Evidence for a reversal is suggestive.

Clearly, there are no widely-accepted conclusions in this field. This is partly because of inadequate statistics and differences in techniques of analysis, but probably equally as important is real structure in the magnetic field on various length scales. Some analyses are weakened by the inclusion of incorrect data and by excess zeal in matching observations to models; no paper in this field can be read without critical care. Further progress requires a quantum jump in the number of RM's, particularly for extragalactic sources in the Galactic plane and for pulsars everywhere.

3.4. The Nonuniform Component of the Magnetic Field within a Few Kpc of the Sun.

There is considerable observational support for a substantial nonuniform component of the magnetic field. SK found four distinctly anomalous regions in the RM sky, and Vallée and his collaborators (see Vallée 1984; Broten, MacLeod, and Vallée 1985) have used Faraday rotation of extragalactic sources to delineate several 'magnetic bubbles' (Vallée 1984).

Magnetic bubbles are structures that show enhanced RM's over angular diameters of 30° to 130°, corresponding to linear diameters of 100 to 200 pc. Detection of these bubbles is difficult, because Faraday rotation integrates over the line of sight and the enhancement in RM is, in some cases, only a small fraction of the total. In my opinion, the reality of the bubble associated with the Monogem Ring $[(l,b) \approx (203°, 11°)]$ is questionable for this reason. In my opinion, the reality of the other three bubbles is statistically sound. These bubbles are associated with the North Polar Spur (Radio Loop I) $[(l,b) \approx (329°, 18°)]$, Radio Loop II $[(l,b) \approx (110°, -32°)]$, and the Gum (Vela) Nebula $[(l,b) \approx (260°, 0°)]$.

There is a serious question about the identification of the Gum Nebula bubble with the Vela supernova remnant. Vallée and Bignell (1983) ascribe an RM of about 130 rad m^{-2} to the Gum Nebula itself, while the Vela pulsar has $RM=34$ rad m^{-2}, four times smaller. If the pulsar lies at the center of the Gum Nebula, its RM should equal the ordinary interstellar contribution plus half the contribution of the whole Gum Nebula. In fact, at the distance of 400 pc, the ordinary interstellar contribution alone amounts to nearly 34 rad m^{-2}. This argues that the Vela pulsar lies on the very near side of the Gum nebula.

The nomenclature 'magnetic' bubble is misleading, for it implies an enhanced magnetic field. However, RM's can also be increased by an enhanced electron density. This is the main cause of the enhanced RM's toward the Gum Nebula (Vallée and Bignell 1983). The enhanced RM's toward Radio Loops I and II are to some extent a result of enhanced and perturbed magnetic fields, because these structures exhibit enhanced diffuse synchrotron emission. However, Radio Loop I encircles the Sco-Oph stellar associations, which contain many OB stars; the UV radiation from these stars certainly enhances the electron density, and thus also the RM's.

IT statistically analyzed the nonuniform magnetic field component by analyzing the angular dependence of the correlation of the residuals from their model fits of the uniform field component to the observed extragalactic RM's. They found a decorrelation angle of about $10°$ and an r.m.s. residual RM of about 20 rad m^{-2}. This residual is large: the RM produced by the uniform field component is essentially equal, 21 rad m^{-2}! Since the residual RM is comparable to the systematic RM, the product $n_e B_\parallel$ for the random component is larger by the ratio of the path length in the uniform component to the scale length of the random component. Judging from the angular decorrelation length, this factor is about five. One doesn't know whether the product $n_e B_\parallel$ in the random component is large because of fluctuations in n_e or in $B_\parallel$. IT assume that n_e in the fluctuations is lower than the average value in the Galaxy, and derive random magnetic field strengths ~ 20 μG. This is unrealistically high: with the observed spectrum of cosmic ray electrons, it would produce far too much Galactic synchrotron radiation. From section 3.1, we would conclude that 3 to 6 μG would be much more reasonable. This means that there are very large positive fluctuations in n_e, ranging up to some 5 times the average value of 0.03 cm^{-3}, over scale lengths of ~ 100 pc.

To conclude: Results on the nonuniform field component are hard to come by because of the inadequate statistical sample of RM's. RM's of extragalactic sources suffer two deficiencies for this purpose: one, extragalactic sources have their own, intrinsic RM's, whose values are not negligible compared to the residuals from the fits to the uniform Galactic field. Two, their RM's integrate over the entire line of sight through the Galaxy, thus merging both inhomogenieties and the uniform component. Pulsar RM's are, in principle, much better because pulsars have no intrinsic RM's and because the pulsar DM's can be used as distance indicators. However, the current pulsar sample is so small that, in my opinion, extragalactic RM's provide a better picture at the present time. Real progress would result from the measurement of RM's for a significant fraction of today's cataloged pulsars.

4. THE MAGNETIC FIELD ON SCALES BELOW 100 PC

4.1. The Typical Field Strength in the Diffuse ISM.

In perusing the literature and the present review article, one finds quoted values for measured field strength. However, these are not representative of the typical field strength in the diffuse interstellar medium. Most of these quoted values are obtained for special regions of interest, such as interstellar shocks or dense clouds. Upper limits are rarely quoted. For example, we have the well-known case of the double H I absorption line of the Perseus spiral arm seen in against Cas A, which gives $B_\parallel = 11$ and 18 μG (see section 4.9 below). However, it is not often pointed out that the H I absorption line of the Orion arm—which, like that of the Perseus arm, is associated with molecular emission and thus a dense cloud—has a very low upper limit of field strength. My own measurements place an upper limit on $B_\parallel$ from this line at considerably smaller than 0.5 μG.

The field strength in the diffuse ISM can only be determined from Zeeman splitting of the H I line, and such measurements are very time-consuming because of the small ratio of splitting to line width. For this reason, one doesn't want to 'waste time' observing randomly selected positions in the diffuse ISM. The only data I know of are my own. I observed a set of positions in 'region A' of Simard-Normandin and Kronberg (1979), an area of ~ 1 steradian in which RM's are consistently large and negative. I had hoped to determine whether the large RM's are a result of a large $B_\parallel$ in the diffuse ISM. This may in fact invalidate the use of these data as a 'random sample'. Nevertheless, it is the most random selection I have observed. Strict upper limits on field strength are typically 7 μG for a selection of some ten positions. These upper limits are usually limited by noise, and could be lowered by using longer integration times. At $(l, b) = (62.2°, -10.6°)$, for example, I integrated much longer and measured $B_\parallel = +3.8$ μG.

However, not all positions yielded upper limits or small fields. Some had substantial field strengths. One of these positions, $(l, b) = (96.8°, -19.6°)$, has $B_\parallel \approx +12$ μG in a narrow component centered roughly at zero LSR velocity. Inspection of the photographs of Colomb, Pöppel, and Heiles (1980) shows that this component comes from an H I filament. Such filaments are probably produced by interstellar shocks (section 4.4).

4.2. The Typical Field Strength in Dust Clouds.

Specifying representative values for $B_\parallel$ in dust clouds is subject to the same diffi-culties as in diffuse clouds: we tend to remember the detections rather than the upper limits. Dust clouds have high gas density, and the H I line is not necessarily suitable for determining field strengths within dust clouds. Instead, Zeeman splitting of a molecular line should be used. The only suitable lines at present are the 18-cm lines of OH.

Troland and Heiles (1986) summarize the results obtained for OH Zeeman splitting. There exist eight upper limits on $B_\parallel < 30$ μG, three of which are < 20 μG. There exist four detections: Cas A (9 μG), W22 (11 μG), Ori B (38 μG), and Ori A (120 μG). The last two, with large $B_\parallel$'s, are near H II regions and probably not typical of ordinary dust clouds. Thus, for all ten 'typical' dust clouds, $B_\parallel < 30$ μG. Volume densities in these molecular clouds are typically of order $\gtrsim 10^3$ cm^{-3}. Below in sections 4.8 and 4.11, we find some suggestion that, roughly, $B \propto n^{0.3}$. The upper limits on $B_\parallel$ of 20 to 30 μG are consistent with this.

4.3. The Overall Situation in the Diffuse Interstellar Medium.

Troland and Heiles (1986) have recently reviewed the overall observational situation of the magnetic field in the diffuse interstellar medium. Because of flux freezing, we generally expect an increase in magnetic field strength with volume density. However, magnetic field strengths show little or no evidence of increase over the density range 0.1 to 100 cm^{-3}. The lower densities are sampled by Faraday rotation, and the higher densities by Zeeman splitting. Thus, there is no significant increase in magnetic field strength between the WIM and CM. This constancy can be rationalized with theory, in that magnetic field enhancement is not expected if density enhancement occurs by relatively quiescent streaming of low-density gas along the field lines. However, such quiescent streaming is probably rare: we have observational evidence for shocks, and for field enhancement behind shocks. Thus we do expect enhancement. But the enhancement is expected to be small, because we will see in section 4.4 that the magnetic field dominates the gas dynamics in such circumstances. The overall situation can probably be understood in these terms.

4.4. Enhancement of the Field Strength in Interstellar Shocks.

H I Zeeman splitting has been used to measure the field strength in shocks associated with expanding interstellar shells (Troland and Heiles 1982b). One of these is the Eridanus shell, expanding at about 23 km/s (Heiles 1976b). The H I behind this shock has temperatures $\sim$ 100 K, typical of interstellar clouds (Heiles, 1982). Since a 23 km/s shock produces post-shock temperatures $\gtrsim 10^4$ K, the cool gas should be well-described by the theory of isothermal shocks. If there were no magnetic field, the H I behind an isothermal shock would rise in density by a factor of several hundred. However, the magnetic field ($B_\parallel \approx 7$ μG) has prevented this large increase. The actual factor is only 3 or so. In addition, the magnetic field is being stretched on very large scales by the expansion of the shell as a whole. This distortion is now becoming sufficiently large to allow the magnetic field to play a role decelerating the overall shell expansion.

Other shells have comparable field strengths. Troland and Heiles (1982b) measured Zeeman splitting in H I emission and found a 7 μG field at one position in a shell that appears to be associated with Radio Loop II. The overall bulk characteristics of this shell are similar to those of the Eridanus shell, except for its expansion velocity which is probably about 10 km/s. At one position $[(l,b) = (36.5°, 40.2°)]$ in the North Polar Spur (Radio Loop I), Heiles *et al.* (1980) found $B_\parallel$ between 1.2 and 6 μG. I have measured Zeeman splitting of H I in emission at about ten positions within $\sim$ 25° of $(l,b) = (290°, 40°)$. These positions lie on H I filaments associated with the North Polar Spur, but in this region the radio continuum brightness associated with the Spur is small. At most of these positions, $B_\parallel \approx -5$ μG; it is smaller at a few positions.[1] Again, the bulk characteristics are very roughly similar to those of the Eridanus shell. Finally, there is a prominent shell that surrounds the North Celestial pole, about 30° in diameter and easily visible on the photographs of Colomb, Pöppel, and Heiles (1980). I have measured (but not yet published) values of about +9, +8, and +4 μG at three widely-separated positions in the periphery of this shell.

Even though the results for only one shell have been analyzed in detail, the properties of at least two of the other three shells are similar. Field strengths in shells are larger than those in arbitrarily-selected positions, and this makes them fairly easy to measure using Zeeman splitting of H I in emission. The enhancement of magnetic field strength by interstellar shocks, and the effect of the enhanced field in providing an important or even dominant contribution to the total interstellar pressure, is observationally well-established (if not well-published!).

---

[1] We note that $B_\parallel$ as determined from Zeeman splitting points *toward* the Sun in this region, while that determined from Faraday rotation points *away*. Thus the H I and the thermal electrons responsible for Faraday rotation occupy different regions of space.

4.5. Alignment of Field Lines Near Elongated Clouds.

Optical polarization reveals a tendency for magnetic field lines to lie either parallel or perpendicular to filaments or elongated clouds. On large scales this is spectacularly the case for the North Polar Spur H I shell, for which the polarization vectors (and thus the magnetic field) lies parallel to the filaments,[2] and may also be the case for the H I shell located in the vicinity of radio loop II (see Heiles and Jenkins 1976). On smaller scales, Vrba, Strom, and Strom (1976) find parallel alignment in dust filaments in Ophiuchus and R Coronae Australis. The opposite tendency, with perpendicular alignment, is observed to some degree in Taurus (Moneti *et al.*, 1984) and to a spectacular degree in L204, a 4°-long, very thin dust cloud (section 4.6).

In R Coronae Australis (RCA), Vrba, Coyne, and Tapia (1981) have made a comprehensive study of the optical polarization. This dark cloud complex consists of a well-defined filament that terminates in a circular blob; inside and near the periphery of the blob, star formation is occuring. The optical polarization vectors, and thus $B_\perp$, are well-aligned with the filament. In the blob, their directions are not parallel to the filament, but tend to be perpendicular to the filament. Vrba *et al.* derive the magnetic field strengths required to produce the observed dust alignment, assuming that the classical Davis and Greenstein (1951) theory is valid. They obtain total field strengths of ~ 120 μG in the filament. I have tried to make measurements of H I Zeeman splitting in the filament, with conflicting results because of instrumental effects. However, I can state with absolute confidence that the field strengths are smaller than about 7 μG. The natural conclusion is, as discussed in section 1.1, that Purcell's (1979) suprathermally spinning grains are required to produce the observed optical polarization.

The structure of the Ophiuchus dust cloud is similar to that of the RCA cloud. Two well-defined dust filaments lead into two blobs where star formation is occuring (see Minn 1981). The optical polarization vectors are reasonably well aligned with the filaments, and the vectors in the blob are not parallel to the filaments, but tend to be perpendicular (Vrba, Strom, and Strom 1976). I have made H I Zeeman observations at a number of positions within this complex. In the filaments, the splitting is undetectable, yielding upper limits on $B_\parallel$ of a few μG. In the blobs, however, the Zeeman splitting is easy to detect. The H I profiles in the blobs clearly show narrow self-absorption dips, which must be produced by the cold H I within the blobs. It is these dips that exhibit the Zeeman splitting. The circular polarization is quite large, but unfortunately it is not straightforward to derive the magnetic field strength in these complicated profiles. At the moment I can only say that at some positions $B_\parallel$ is at least 10 μG, and possibly much larger.

The Taurus dark cloud complex consists of several filamentary features oriented in different directions, and a large, somewhat elongated ($\sim 1°$ by $2°$) cloud usually called 'cloud 2'. This cloud contains subcondensations with very high molecular density and is also the site of star formation. I have made some observations of H I Zeeman splitting in this region, with no clear detections. The most interesting result is on cloud 2 itself,

---

[2]It is worth recalling that the polarization of the diffuse synchrotron emission from the North Polar Spur implies a field parallel to that obtained optically for $b \gtrsim 40°$, but *perpendicular* for $b \lesssim 40°$ (Spoelstra 1971). The only reasonable conclusion is that the synchrotrom-emitting region does not contain very much dust.

which clearly exhibits self-absorption in the H I line, as does the Ophiuchus cloud. However, contrary to the situation in Ophiuchus, the self-absorption component exhibits no detectable Zeeman splitting. Long ago, Vershuur (1970) established an upper limit of 10 μG; my measurements provide a strict upper limit of 5 μG.

4.6. The Strong, Aligned Magnetic Field near L204.

L204 is a striking 4°-long, very thin dark cloud easily visible on the Palomar Sky Survey (PSS) prints. McCutcheon *et al.* (1986) have mapped $B_\perp$ using optical polarization of starlight, and mapped the cloud dynamics using the lower two mm-wave transitions of CO. A reproduction of the PSS, together with their optical polarization vectors and my H I Zeeman measurements, is shown in Figure 1. They assign a distance of 170 pc, making its vertical length $\sim$ 10 pc. They determine its opacity from star counts; from these, together with the CO data, they calculate its total mass to be $\sim$ 400 $M_\odot$.

This long cloud is not quite straight, but instead has the shape of the letter 'S' greatly stretched in the vertical direction. The radial velocities show a strikingly similar behavior, in that the portions of the cloud displaced to the right on Figure 1 have the most positive radial velocities, and *vice versa*. The radial velocities range from about 2.6 to 4.6 km/s.

To zeroth order, $B_\perp$ tends to be horizontal in Figure 1, and perpendicular to the axis of the elongated cloud. In more detail, however, the pattern departs significantly from perpendicularity. Near the center, in the densest portion of the cloud, there are three vectors that are *parallel* to the local direction of the axis of the cloud. Just above center, the cloud slopes upward and to the left; here the polarization vectors do the same—but in a more exaggerated fashion. Near the top of the cloud, the cloud axis slopes upward and to the right; again, the polarization vectors exaggerate this behavior. At the bottom of Figure 1, the cloud axis is nearly horizontal, but tilted upward and to the right; the polarization vectors are tilted in the opposite direction.

I have measured the Zeeman splitting of the H I line in emission in this region. The results are indicated on Figure 1. These are preliminary results, obtained by a least-squares fit to the whole H I profile. The final details may differ, but the overall picture is accurate. The plus symbols indicate that the field points away from us; the size of the symbols indicates the strength of $B_\parallel$. Where the symbols are small, the errors due to noise are large; thus the lone circle, which indicates a field pointing toward us, may well be noise.

This region is characterized by a magnetic field pointing away from us. $B_\parallel$ is largest at the top, decreases to essentially zero at the middle where the cloud is thickest, and increases again toward the bottom. The field strength at the top right, where $B_\parallel$ is strongest, is 9.6 μG.

This cloud looks like a shock front. However, McCutcheon *et al.* suggest that instead it is a cylinder, because correlation of the left-to-right position on Figure 1 with the radial velocity implies that the total velocity has a significant component across the line of sight. This is a convincing argument. They go on to calculate the virial equilibrium of the cloud. With the total mass of 400 $M_\odot$, gravity completely overwhelms the macroscopic

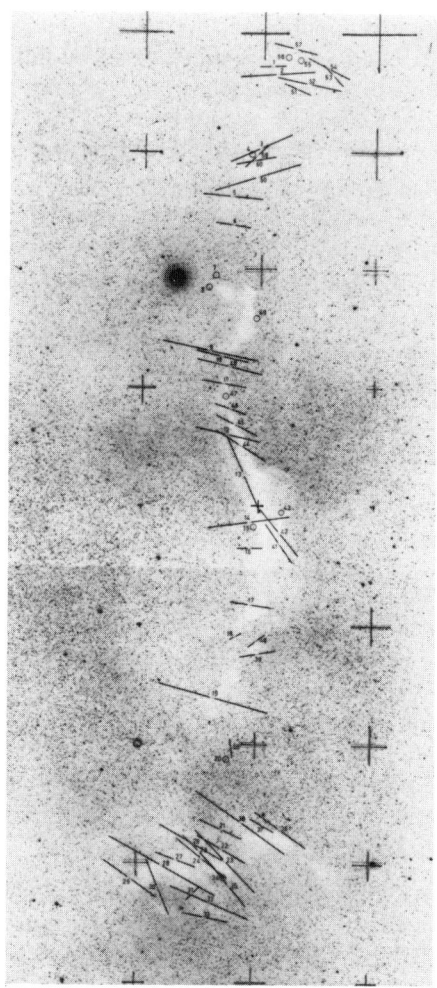

Figure 1. The long, dark cloud L204, together with optical polarization vectors and $B_\parallel$'s from Zeeman splitting of H I in emission. The photograph and optical data are from McCutcheon *et al.* (1986). Circles indicate a field pointing toward us; the size of the symbols is proportional to $B_\parallel$. The largest $B_\parallel$ is about 10 μG.

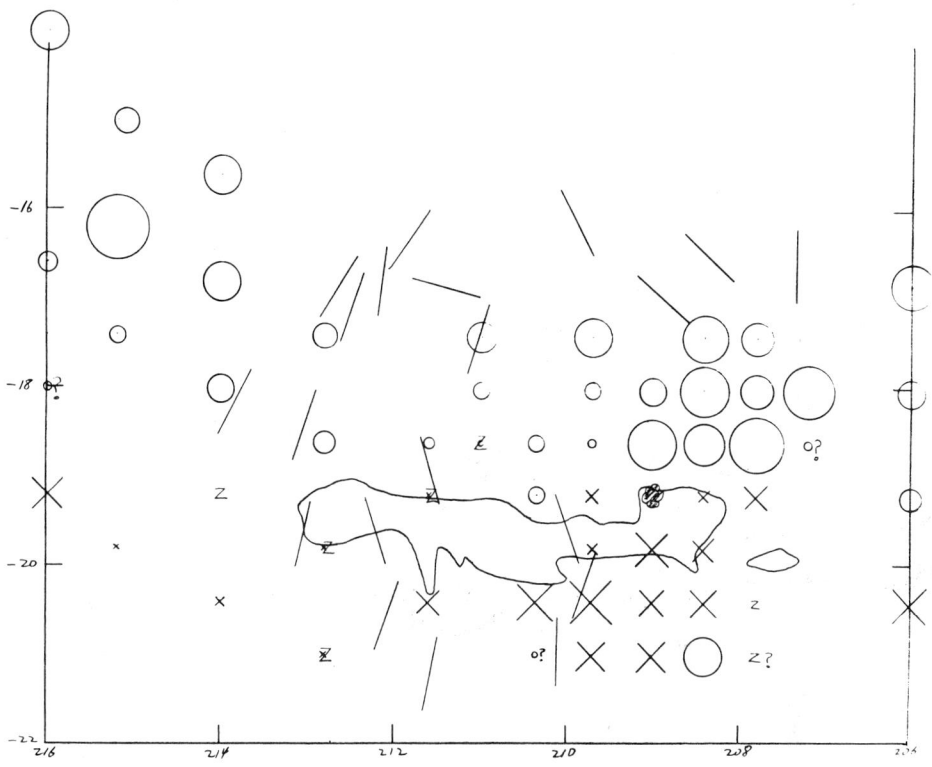

Figure 2. Sketch of the ^{13}CO distribution in the Orion region, together with $B_\parallel$'s from Zeeman splitting of H I in emission. Galactic longitude is 206° at the right and 216° at the left; latitude is −22° at the bottom and −14° at the top. Circles indicate a field pointing toward us; the size of the symbols is proportional to $B_\parallel$. The largest $B_\parallel$ is about 10 μG.

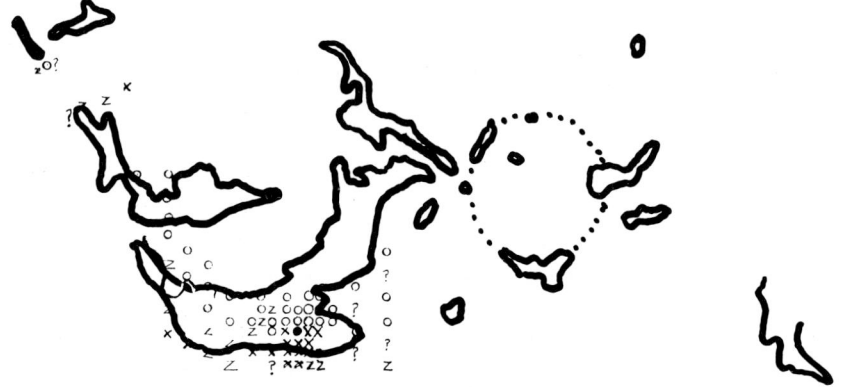

Figure 3. Sketch of the large-scale ^{12}CO distribution, together with the sign of $B_{\parallel}$'s from Zeeman splitting of H I in emission (circles indicate a field pointing toward us).

and thermal motions. A total magnetic field strength of ~ 50 μG is required for virial equilibrium. Measured $B_\parallel$'s shown on Figure 1, outside the cloud in the ambient H I, are much smaller. The decrease in $B_\parallel$ in the H I near the center of Figure 1, where the dark cloud is thickest, suggests that perhaps the cloud has gathered together some of the field lines from the ambient medium, increasing the field strength within the cloud itself. Clearly, this portion of the cloud is an excellent candidate for OH Zeeman observations.

4.7. The Large-Scale Magnetic Field in the Orion Region.

The vicinity of the Orion nebula contains coherent, large-scale structures seen not only in atomic and molecular gas, but also in the magnetic field. Kutner *et al.* (1977) mapped the CO distribution and found two long filaments north and south of the nebula. Heiles and Troland (1982) measured Zeeman splitting near the southern cloud, not only in the H I line near the cloud seen in emission but also from H I inside the cloud seen in self-absorption. I subsequently mapped the Zeeman splitting from the H I near the cloud, seen in emission. The ^{13}CO distribution, together with the measured $B_\parallel$'s, is shown in Figure 2. There is a striking pattern, with $B_\parallel$ pointing *toward* us on 'top' of the cloud and *away* from us on the bottom. Optical polarization vectors, also shown on Figure 2, suggest that $B_\perp$ is roughly perpendicular to the axis of the cloud. Taken in a restricted perspective, the observations are consistent with a helical field 'wrapped around' the cloud.

However, a larger angular scale may reveal a different situation. Maddalena *et al.* (1986) mapped a much larger region in CO, and found that the filamentary cloud sketched in Figure 2 is part of a much larger CO structure, which looks like a circular shell. Figure 3 shows a sketch of this shell, together with a grey-scale picture of the H I distribution and the sign of $B_\parallel$ at those locations where I have measured it. At every position within the shell, $B_\parallel$ points toward us. Furthermore, $B_\parallel$ is strong enough to measure quite easily inside the shell, while outside the shell it is often too weak to measure easily. My interpretation is that the field observed outside the shell may belong to H I that is located at a different distance. Thus the apparent field reversal shown in Figure 2 may simply be a systematically negative field within the shell combined with a weak positive field at some other location along the line of sight. Deciding unambiguously what the actual situation is requires examining the H I kinematics and will be very difficult, if not impossible. This illustrates the difficulties involved with measurements of Zeeman splitting of multicomponent H I lines!

One thing is clear: the large shell of CO and H I is associated with a large negative field pointing toward us. $B_\parallel$ is typically measured to be between 5 and 10 μG.

4.8. The Magnetic Field Under High Resolution Toward Ori A.

The H II region Ori A exhibits a deep H I absorption line that has the largest Zeeman splitting measured in single-dish spectra—some 50 μG (Verschuur 1969b). Troland, Heiles, and Goss (1986) have used the VLA to map the H I Zeeman splitting. They find considerable structure in the field, but no field reversals. Measured $B_\parallel$'s easily exceed the average value and range up to 90 μG. In the 18-cm OH lines, Troland, Crutcher, and Kazès (1986) have measured 120 μG in a Nancay single-dish spectrum. The H_2 volume density in the OH-emitting region exceeds 5000 cm^{-3}. These authors compare the H I and OH observations and find that the field strength increases roughly as $n^{0.3}$. This is in sharp contrast to the situation near Cas A (section 4.9). Instead, it is in accord with theoretical expectations for clouds bound by their own self-gravity. Since Ori A is a star-forming region, this might actually be the case.

4.9. Structure of the Magnetic Field Toward Cas A.

Interstellar matter in the Perseus arm has dense clouds that produce two spectacularly deep H I absorption lines and multicomponent molecular spectral lines. When observed with a single dish that integrates over the whole source, Zeeman splitting of the two H I absorption lines yields $B_\parallel$'s of 11 and 18 μG (Verschuur 1969a). Schwarz *et al.* (1986) used the WSRT to map these fields with an angular resolution of 1 arcmin. The magnetic field is enhanced in H I clumps, with measured values ranging from 20 to 40 μG. Actual $B_\parallel$'s are 1.5 to 2 times larger because the clumps constitute only a fraction of the total H I. The H I clumps are associated with molecular clumps observed in H_2CO cm-wavelength lines. Unfortunately, however, the angular resolution of neither the H I nor the H_2CO observations is not sufficient to determine whether the H I surrounds the molecular clumps or simply lies in proximity.

With the above association of strong H I fields with H_2CO clumps, one would expect the measured *molecular* field strengths to be high. Surprisingly, this is not the case. Heiles and Stevens (1986) used single-dish measurements of Zeeman splitting of the 18-cm OH lines to find that the observed field strengths in the molecular clumps are about only about 9 μG—five times smaller than in the atomic clumps. This result goes against the expected tendency of field strength to increase with gas density. One might argue that the *true* field strength is larger than the observed field strength for two reasons: one, the observed field strength only refers to the line-of-sight component $B_\parallel$; and two, the observed field strength refers only to the uniform component, not the random component. Nevertheless, it is difficult reduce the observed component $B_\parallel$ by a factor of five without invoking improbable geometrical alignment or severe field strength/density relationships within the clouds.

It appears that neither the atomic nor the molecular clumps is in virial equilibrium because self-gravity is negligible, the magnetic field is high, and there is apparently insufficient external pressure. The clumps are almost certainly transient condensations located behind a passing shock wave. Under these transient conditions, one doesn't necessarily expect the magnetic field strength to increase with density.

For example, consider the scenario proposed to me by Frank Shu (private communication). It rests crucially on the fact that the density condensations result from isothermal shocks. First consider the situation before the shock has passed, and suppose that the medium was of roughly constant gas density and had a magnetic field whose direction changed slowly but roughly randomly throughout. In those regions where the shock velocity was parallel to the magnetic field, the increase in gas density was large but the increase in field strength was small. In regions where the shock velocity was perpendicular to the magnetic field, the compression enhanced the field strength, which prevented the gas density from rising very much—just like the situation in the shocks associated with H I shells, discussed above in section 4.4. In the regions of dense gas, molecules formed quickly. This scenario results in the observed low field strengths in the molecular lines. However, the quantitative details and molecular formation rates have not yet been worked out.

4.10. The Magnetic Fields Associated with Bipolar Flows.

A remarkable result has been found by Cohen, Rowland, and Blair (1984), who compiled optical polarization data obtained from stars located behind bipolar flows of protostars. In nine out of ten cases, the bipolar flows—having *arcminute* angular scales—are aligned with the optical polarization vectors—often having *tens of arcminute* or *degree* angular scales. Thus, the *large-scale, diffuse* interstellar field lies parallel to the bipolar flow and is the primary factor that determines its direction! Theorists have put forward two processes that can explain this result. In one, Alfvén waves transfer angular momentum from the collapsing protostellar cloud to the surrounding medium. This process is more efficient for the component of angular momentum that is perpendicular to the field lines in the surrounding medium (Mouschovias and Paleologou 1980), which leads to the observed alignment. In the other, a spherically-symmetric outflow from the protostar blows a cavity whose shape is influenced by the pressure in the external medium (Königl 1982); the pressure of a magnetic field is anisotropic, in the sense that also leads to the observed alignment.

Another result relating to bipolar flows, which is remarkable if it is real, has been found by Simonetti and Cordes (1986). In L1551, they observe what they claim are significant variations in RM that can only be ascribed to Faraday rotation occuring within the flow itself. The implied values of magnetic field in the flow are $\lesssim 300~\mu G$. In Cep A, they find very large variations in RM which, if produced in the bipolar flow, imply very high magnetic fields. In my opinion, there is a chance that the RM variations could occur elsewhere along the line of sight, either in the Galaxy or in the extragalactic sources used as the RM probes.

4.11. Milligauss Magnetic Fields in OH Masers.

The 18-cm OH masers exhibit high circular polarization. Linear polarization is weak, probably because of large Faraday rotation which occurs because of the strong magnetic fields. In many cases, the spectra are too complex to interpret as simple Zeeman splitting. Furthermore, OH masers tend to occur in clusters, so that even when a spectrum can be interpreted as Zeeman splitting, VLBI measurements are required to prove that the oppositely-polarized 'Zeeman' components indeed come from the same maser.

Lo *et al.* (1975) have done this for six masers. The derived field strengths range from 2.5 to 9.0 mG (*milli*gauss). With such strong fields, the splitting exceeds the line width; thus these are *total* field strengths, not just the line-of-sight component. Total gas densities in OH masers are probably of order 10^6 cm^{-3} (Elitzur 1986). In OH masers, then, the expected increase of field strength with volume density is definitively established. A crude fit to a power law implies that B increases roughly as $n^{0.5}$. This exponent is larger than that obtained by Troland, Crutcher, and Kazès (1986) in Ori A, which applies to smaller volume densities (section 4.8).

All of the observations of Zeeman splitting of H I in emission reported here were obtained using the 85-foot telescope of the Hat Creek Observatory. It is a pleasure to thank J.B.G.M. ('Hans') Bloemen for informative discussions. Portions of this manuscript are exerpted from the paper 'H I and the Diffuse Interstellar Medium', chapter 3 of the second edition of *Galactic and Extragalactic Radio Astronomy* (ed. K. I. Kellerman and G.L. Verschuur), by Shrinivas R. Kulkarni and Carl Heiles. Preparation of this paper was supported in part by an NSF grant to the author.

It is a great pleasure to thank Harley Thronson for organizing this meeting, which was both productive and extremely enjoyable; and to acknowledge financial support from the University of Wyoming, which enabled me and my graduate students to attend the meeting.

REFERENCES

Beck, R. 1982, *Astron. Ap.*, **106**, 121.

Beck, R., Klein, U. and Krause, M. 1985, *Astron. Ap.*, **152**, 237.

Bloemen, J.B.G.M., Strong, A.W., Blitz, L., Cohen, R.S., Dame, T.M., Grabelsky, D.A., Hermsen, W., Lebrun, F., Mayer-Hasselwander, H.A., and Thaddeus, P. 1986, *Astron. Ap.*, **154**, 25.

Beuermann, K., Kanbach, G., and Berkhuijsen, E.M. 1985, *Astron. Ap.*, **153**, 17.

Broten, N.W., MacLoed, J.M., and Vallée, J.P. 1985, *Ap. Lett.*, **24**, 165.

Burn, B.J. 1966, *Mon. Not. Roy. Astron. Soc.*, **133**, 67.

Cioffi, D.F. and Jones, T.W. 1980, *Astron. J.*, **85**, 368.

Cohen, R.J., Rowland, P.R., and Blair, M.M. 1984, *M.N.R.A.S.*, **210**, 425.

Colomb, F.R., Pöppel, W.G.L., and Heiles, C. 1980, *Astr. Ap. Suppl.*, **40**, 47.

Davis, L. and Greenstein, J.L. 1951, *Ap. J.*, **114**, 206.

Deguchi, S. and Watson, W.D. 1984, *Ap. J.*, **285**, 126.

Elitzur, M. 1986, private communication.

Ginzburg, V.L. and Syrovatskii, S.I. 1969, *Ann. Rev. Astron. Ap.*, **7**, 375; see also *Ann. Rev. Astron. Ap.*, **3**, 297.

Heiles, C. 1976a, *Ann. Rev. Astron. Ap.*, **14**, 1.

Heiles, C. 1976b, *Ap. J.*, **208**, L137.

Heiles, C. 1982, *Ap. J.*, **262**, 135.

Heiles, C. 1968, in preparation.

Heiles, C., Chu, Y-H., Reynolds, R.J., Yegingil, I., and Troland, T.H. 1980, *Ap. J.*, **242**, 533.

Heiles, C. and Jenkins, E.B. 1976, *Astr. Ap.*, **46**, 333.

Heiles, C. and Stevens, M. 1986, *Ap. J.*, **301**, 331.

Heiles, C. and Troland, T.H. 1982, *Ap. J.*, **260**, L23.

Inoue, M. and Tabara, H. 1981, *Pub. Astr. Soc. Japan.*, **33**, 603.

Jones, R.V. and Spitzer, L. 1967, *Ap. J.*, **147**, 943.

Königl, A. 1982, *Ap. J.*, **261**, 115.

Kulkarni, S.R. and Heiles, C. 1986, chapter 3 of the second edition of *Galactic and Extragalactic Radio Astronomy*, ed. K.I. Kellerman and G.L. Verschuur, in press.

Kutner, M.L., Tucker, K.D., Chin, G. and Thaddeus, P. 1977, *Ap. J.*, **215**, 521.

Kylafis, N.D. 1983, *Ap. J.*, **275**, 135.

Lo, K.Y., Walker, R.C., Burke, B.F., Moran, J.M., Johnston, K.J., and Ewing, M.S. 1975, *Ap. J.*, **202**, 650.

Maddalena, R.J., Morris, M., Moscowitz, J. and Thaddeus, P. 1986, *Ap. J.*, **303**, 375.

Manchester, R. N. and Taylor, J. H. 1977; *Pulsars*, W. H. Freeman and Co., San Francisco

Martin, P.G. 1972, *Mon. Not. Roy. Astron. Soc.*, **159**, 179.

Martin, P.G. 1974, *Ap. J.*, **187**, 461.

Martin, P.G. 1975, *Ap. J.*, **201**, 373.

Martin, P.G. and Campbell, B. 1976, *Ap. J.*, **208**, 727.

Martin, P.G. and Shawl, S.J. 1982, *Ap. J.*, **253**, 86.

Mathewson, D.S. and Ford, V.L. 1970, *Mem. Roy. Astr. Soc.*, **74**, 139.

Mathis, J.S. 1986, *Ap. J.*, **301**, 423.

McCutcheon, W.H., Vrba, F.J., Dickman, R.L., and Clemens, D.P. 1985, *Ap. J.*, **309**, ???.

Minn, Y.K. 1981, *Astron. Ap.*, **103**, 269.

Moneti, A., Pipher, J.L., Helfer, H.L., McMillan, R.S., and Perry, M.L. 1984, *Ap. J.*, **282**, 508.

Mouschovias, T.Ch. and Paleologou, E.V. 1980, *Ap. J.*, **237**, 877.

Phillips, S., Kearsey, S., Osborne, J.L., Haslam, C.G.T., and Stoffel, H. 1981, *Astron. Ap.*, **98**, 286.

Phillips, S., Kearsey, S., Osborne, J.L., Haslam, C.G.T., and Stoffel, H. 1981, *Astron. Ap.*, **103**, 405.

Purcell, E.M. 1979, *Ap. J.*, **231**, 404.

Reynolds, R.J. 1984, *Ap. J.*, **282**, 191.

Schwarz, U.J., Troland, T.H., Albinson, J.S., Bregman, J.D., Goss, W.M., and Heiles, C. 1986, *Ap. J.*, **301**, 320.

Silverglate, P.R. 1984, *Ap. J.*, **279**, 694.

Simard-Normandin, M. and Kronberg, P.P. 1979, *Ap. J.*, **242**, 74.

Simonetti, J.H. and Cordes, J.M., *Ap. J.*, **303**, 130.

Sofue, Y., Fujimoto, M. and Wielebinski, R. 1986, *Ann. Rev. Astron. Ap.*, **24**, ???.

Spitzer, L. Jr. 1978, *Physical Processes in the Interstellar Medium*, John Wiley and Sons (New York).

Spoelstra, T.A.Th. 1971, *Astron Ap.*, **13**, 237.

Thomson, R.C. and Nelson, A.H. 1980, *M.N.R.A.S.*, **191**, 863.

Troland, T.H., Crutcher, R.M. and Kazès, I. 1986, *Ap. J.*, **304**, L57.

Troland, T.H. and Heiles, C. 1977, *Ap. J.*, **214**, 703.

Troland, T.H. and Heiles, C. 1982a, *Ap. J.*, **252**, 179.

Troland, T.H. and Heiles, C. 1982b, *Ap. J.*, **260**, L19.

Troland, T.H. and Heiles, C. 1986, *Ap. J.*, **301**, 339.

Troland, T.H., Heiles, C., and Goss, W.M. 1986, in preparation.

Vallée, J.P. 1980, *Astr. Ap.*, **86**, 251.

Vallée, J.P. 1983, *Astr. Ap.*, **124**, 147.

Vallée, J.P. 1984, *Astr. Ap.*, **136**, 373.

Vallée, J.P. and Bignell, R.C. 1983, *Ap. J.*, **272**, 131.

Vallée, J.P., and Kronberg, P. P. 1973, *Nature Phys. Sci.*, **246**, 49.

Verschuur, G.L. 1969a, *Ap. J.*, **156**, 861.

Verschuur, G.L. 1969b, *Nature*, **223**, 140.

Verschuur, G.L. 1970, *Ap. J.*, **161**, 867.

Verschuur, G.L. 1979, *Fund. Cosmic Phys.*, **5**, 113.

Vrba, F.J., Coyne, G.V., and Tapia, S. 1981, *Ap. J.*, **243**, 489.

Vrba, F.J., Strom, S.E. and Strom, K.M. 1976, *A. J*, **81**, 958.

The Theory of the Galactic Magnetic Field

Ellen G. Zweibel
Department of Astrophysical, Planetary and Atmospheric Sciences
University of Colorado
Boulder, CO 80309

Abstract. We discuss the role of the magnetic field in determining the large scale structure and dynamics of the interstellar medium. We then discuss the origin and maintenance of the galactic field. The two major competing theories are that the field is primordial and connected to an intergalactic field or that the field is removed from and regenerated within the galaxy. We review the theoretical and observational basis for these theories. Finally, we discuss cosmic ray acceleration and confinement in the interstellar medium.

I. Introduction

Magnetic fields appear to be universally present in spiral galaxies, and without a magnetic field our galaxy would be quite a different place. There would be no (or an insignificant number) of cosmic rays with drastic results for the ionization and heating of interstellar clouds. Because both magnetic fields and cosmic rays play an important role in the overall hydrostatic equilibrium of the interstellar medium, and possibly in maintaining the galactic halo and in forming giant molecular cloud complexes, the global structure of the interstellar medium would not be the same. Finally, small scale processes in clouds, such as star formation, would probably proceed rather differently. But, at present, we do not know the overall structure of the galactic magnetic field, how the field originated, or how it is maintained in its present topology and strength.

This article discusses the interstellar magnetic field on global scales. There is little discussion of the internal magnetic fields of clouds, except for a review of some important physical processes. There is little observational background because that is the subject of the paper by Heiles. Literature through mid 1986 is cited and although the reference list is not exhaustive it should enable the reader to pursue any topic further.

Section II is concerned with large scale dynamics. After a review of basic physical processes, the effects of magnetic fields on cloud random motion, on the evolution of cloud angular momentum, and on the equilibrium and stability of the interstellar medium are discussed. In Section III theories of the origin and maintenance of the galactic field, including the creation of primordial flux, the hydromagnetic dynamo, the evolution of a primordial field, and the possible role of magnetic monopoles, are reviewed. Section IV is a brief treatment of cosmic ray propagation and acceleration in the galaxy. In Section V some outstanding theoretical and observational problems are identified.

D. J. Hollenbach and H. A. Thronson, Jr. (eds.), Interstellar Processes, 195–221.

II. Global Dynamics

A. Physical Processes

Large scale phenomena which involve the galactic magnetic field can be described using the magnetohydrodynamic approximation (e.g. Cowling 1976), in which the displacement current is neglected, so that

$$J = \tfrac{c}{4\pi} \nabla \times B \tag{1}$$

and the electric field is determined from a simple scalar form of Ohm's law

$$E = \eta J - \tfrac{V_c}{c} \times B \tag{2}$$

Here, V_c is the bulk velocity of the charged component of the gas. As we will see, it is sometimes necessary to distinguish between V_c and V_n, the bulk velocity of the neutral component.

Equations (1) and (2) can be substituted into Faraday's Law to give an equation for the evolution of the magnetic field:

$$\tfrac{\partial B}{\partial t} = \nabla \times (V_c \times B) + \tfrac{\eta c^2}{4\pi} \nabla^2 B \tag{3}$$

where, in writing the last term, the resistivity η has been assumed uniform in space. The first term on the right hand side represents induction of magnetic field by plasma flow perpendicular to the fieldlines. The second term represents resistive decay and is diffusive in character. If the field changes on a lengthscale L then the dynamical timescale associated with the first term is $\tau_D = L/V_c$ and the diffusive timescale associated with the second term is $\tau_R = 4\pi L^2/\eta c^2$.

Ordinarily resistive decay is negligible. In ionized regions of the interstellar medium resistivity is due to Coulomb encounters between electrons and other charged particles, and $\tau_{Ri} \sim 10^{16} \, T^{3/2} \, L_{pc}^2$ yr. In predominantly neutral regions resistivity is due to collisions between electrons and atoms or molecules, and $\tau_{Rn} \sim 10^{27} \, (n_e/n_n) \, (10^{-9}/ < \sigma \, v >) \, L_{pc}^2$ yr. (In this expression, n_e and n_n are the number densities of electrons and neutral particles, respectively.) We see that even in dense molecular clouds where $(n_e/n_n) \sim 10^{-9}$, resistive times are very long over typical interstellar lengthscales.

There are two types of processes which might enhance the importance of resistivity. One is anomalous resistivity and is produced by microscopic fluctuations in the medium which scatter the electrons (Tang 1978). In the absence of any definite source of large amplitude microturbulence, it is ad hoc to postulate anomalous resistivity. In particular, the electron drift velocity associated with the typical current density in the interstellar medium is far too small to be unstable.

The second mechanism is turbulent diffusion. It is known from simulations of weak magnetic fields interacting kinematically with fluid convection (Moffatt 1978) that the field can be expelled from the eddies and concentrated into thin sheets or ropes. Thus the effective lengthscale which should be used in estimating τ_R can be much smaller than lengthscales in the flow, and ohmic dissipation can be quite rapid. These considerations form the physical basis for the eddy diffusivity that is invoked in the theories of the large scale magnetic field discussed below. We see, though, that L_{pc} must still be extremely small to achieve significant diffusion on timescales of 10^8 yr, and we should remember that these studies of field concentration do not include the effect of magnetic forces on the gas.

We now consider equation (3) for zero resistivity. It is then possible to prove (Spitzer 1962) that the magnetic flux through any surface in the fluid

$$\Phi \equiv \int_S B \cdot n da$$

is constant as the surface moves with the fluid, so the flux is said to be frozen in.

An important example of magnetic induction described by equation (3) is the winding up of the galactic field by differential rotation. Let $V_c = \hat{\theta} r \Omega(r,z)$ and assume that B is axisymmetric. Then equation (3) becomes

$$\frac{\partial B}{\partial t} = \hat{\theta} r B \cdot \nabla \Omega \tag{4}$$

which shows that an azimuthal component of field is produced unless Ω is constant on magnetic fieldlines; the fieldlines are stretched in the azimuthal direction by the differentially rotating fluid.

As another example, suppose that the galactic disk initially contained a straight, uniform magnetic field of the form

$$B(r,\theta,0) = \hat{r} B_o \cos\theta - \hat{\theta} B_o \sin\theta + \hat{z} \lambda B_o \cos\theta \tag{5}$$

The solution of the induction equation (3) with the initial conditions (5) is

$$B(r,\theta,t) = \hat{r} B_o \cos(\theta - \Omega t) - \hat{\theta}\left(B_o \sin(\theta - \Omega t) - r t B_o\left(\lambda \frac{\partial \Omega}{\partial z} + \frac{\partial \Omega}{\partial r}\right)\cos(\theta - \Omega t)\right) +$$

$$+ \hat{z} \lambda B_o \cos(\theta - \Omega t) \tag{6}$$

so we see that differential rotation causes the azimuthal field to increase linearly with time. More problematic is the fact that at large times the field reverses frequently with radius and vertical height. At fixed z, the arguments of the trigonometric functions change by π over a distance Δr given approximately by

$$\Delta r \sim \pi / \left(t \frac{d\Omega}{dr}\right) \approx r\pi/\Omega t$$

In the solar neighborhood, $r \sim 10$ kpc and $\Omega t \sim 300$ so $\Delta r \sim 100$ pc. The reversal of the field with height is more difficult to estimate because the z dependence of Ω is not well known. We return to a discussion of this model in Section IIIC. The fields given in equations (5) and (6) are sketched in Figure 1.

Freezing in of magnetic flux also implies that if material contracts perpendicular to the magnetic field, the strength of the field should increase. Isotropic contraction gives $B \sim \rho^{2/3}$. There is no systematic observed variation in fieldstrength over several orders of magnitude in gas density, however, in the diffuse gas (Heiles, this volume).

We now consider the effect of the field on the gas motions. The Lorentz force on the charged component is $(J \times B)/c$ and is of course zero on the neutral component.

Using equation (1) the Lorentz force can be rewritten as

$$\frac{J \times B}{c} = \frac{(\nabla \times B) \times B}{4\pi} = -\nabla \frac{B^2}{8\pi} + \frac{B \cdot \nabla B}{4\pi} \tag{7}$$

The first term shows that a gradient in magnetic energy produces a force like a pressure

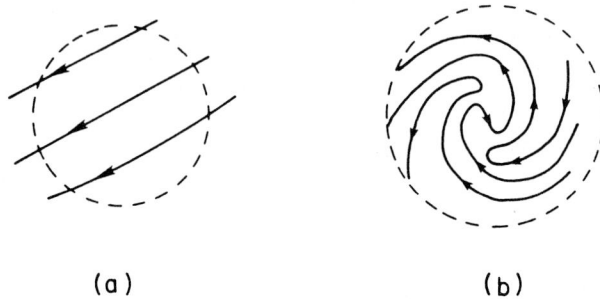

Figure 1. a) Projection of a primordial field onto the galactic disk. b) Winding up of the field into a bisymmetric spiral.

gradient; the second term is a tension term arising from curvature in the magnetic fieldlines.

The charged and neutral components are coupled together by an effective friction force provided by momentum exchanging collisions between the particles. Thus the equations of motion for the two components are

$$\rho_c \frac{dV_c}{dt} = F_c + \frac{J \times B}{c} - \rho_c \nu_{cn}(V_c - V_n) \tag{8a}$$

$$\rho_n \frac{dV_n}{dt} = F_n - \rho_n \nu_{nc}(V_n - V_c) \tag{8b}$$

where

$$F_{c,n} = \rho_{c,n} g - \nabla P_{c,n}$$

and

$$\rho_c \nu_{cn} = \rho_n \nu_{nc} = \frac{m_c m_n n_c n_n}{m_c + m_n} < \sigma v >$$

with $< \sigma v >$ defined acording to Spitzer (1978) p. 25.

Equations (8) should also contain terms which represent the exchange of momentum through ionization and recombination. Shu (1983) has noted that the effect is small in a quasistationary molecular cloud, and we will not consider these terms further here.

When the flow is subsonic, so that the Reynolds stress terms on the left hand sides of equation (8a) and (8b) can be neglected, the equations for the charged and neutral components can be combined to yield

$$\frac{\partial}{\partial t}(V_c - V_n) = \frac{F_c}{\rho_c} - \frac{F_n}{\rho_n} + \frac{J \times B}{\rho_c c} - (\nu_{cn} + \nu_{nc})(V_c - V_n)$$

$$= \frac{-\nabla P_c}{\rho_c} + \frac{\nabla P_n}{\rho_n} + \frac{J \times B}{\rho_c c} - (\nu_{cn} + \nu_{nc})(V_c - V_n) \tag{9}$$

Equation (9) shows that when the acceleration per particle is different for the charged and neutral components, a relative velocity appears between the two. This relative velocity tends to be damped out by momentum exchanging collisions. The pressure gradient terms in equation (9) are not included in any papers in the literature and we will not discuss them here; they might be important if there is a sharp spatial gradient in ionization.

The frictional timescale $(\nu_{cn} + \nu_{nc})^{-1}$ is typically short compared to the dynamical timescale, so we can assume a steady state solution to equation (9). Neglecting the pressure terms we then find

$$V_c - V_n = \frac{J \times B}{\rho_c c(\nu_{cn} + \nu_{nc})} \tag{10}$$

The relative velocity $V_c - V_n$ is called the ambipolar drift V_D. If we approximate J from equation (1) by $cB/4\pi L$ then

$$V_c - V_n \sim \frac{V_{Ac}^2 t_s}{L}$$

where t_s is the ion slowing down time and $V_{AC} \equiv B/\sqrt{4\pi\rho_c}$ is the Alfven speed defined with respect to the charged component. (We have assumed, in deriving this relation, that $\nu_{cn} \gg \nu_{nc}$). Equation (10) can also be derived by assuming that $\rho_c \to 0$ and balancing the Lorentz force with the friction term in equation (8a).

The magnitude of V_D is usually quite small. In a medium with ionization fraction x and fieldstrength $10^{-6} B_\mu$ G, with $< \sigma v >$ taken from Draine, Roberge and Dalgarno (1983) we find

$$V_D \sim \frac{8.3 B_\mu^2}{L_{pc} \mu_r x n_n^2} \text{cm s}^{-1} \tag{11}$$

where

$$\mu_r \equiv \frac{m_c m_n}{(m_c + m_n) m_H}.$$

As an example of the dynamical role of ambipolar drift consider a one dimensional slab of weakly ionized material which contains a straight magnetic field such that the magnetic pressure gradient force (see eq. (7)) is antiparallel to the gravitational acceleration. Such a configuration cannot be in true static equilibrium because although the charged component may be supported against gravity (or actually pushed outward) by the magnetic field, the neutrals feel the effect of the field only by slipping inward relative to the ions. Thus, although the fieldlines remain tied to the ions, the field moves through the neutrals. This process is called ambipolar diffusion, and is thought to be responsible for removing magnetic flux from material which is collapsing to form stars (Mestel and Spitzer 1956, Mouchovias and Paleologou 1981, Nakano 1981, Black and Scott 1982, Shu 1983).

The ion-neutral friction also heats the gas (Scalo 1977, Draine, Dalgarno and Roberge 1983). The heating terms for the ions and neutrals, added together are,

$$H = \rho_c \nu_{cn} V_D^2 = \frac{2.2 \ 10^{-31} B_\mu^4}{L_{pc}^2 \mu_r x n_n^2} \text{erg cm}^{-3} \text{s}^{-1}. \tag{12}$$

It is interesting to note that H is inversely proportional to n_n, in contrast to most other heating mechanisms. This is because the larger the density, the smaller the drift velocity and the lower the available energy for heating. Equation (12) can also be generalized to include heating by waves with wavevector k and amplitude δB. The factor B^4/L^2 should then be replaced by $k^2 B^2 \delta B^2$. Studies of heating by ion-neutral slip have been made by Scalo (1977), Zweibel and Josafatsson (1983), Hartquist and Morfill (1984), Elmegreen (1985), and Falgarone, Puget and Pérault (1986).

B. Magnetic Effects on Cloud Random Motion

We next consider the effects of magnetic fields on small scale cloud motions. Diffuse clouds in the interstellar medium have a velocity dispersion of about 10 km s^{-1} (Spitzer 1978), a value which presumably represents a balance between the formation or re-acceleration of clouds and the deceleration of clouds in collisions. Giant molecular clouds seem to consist of fragments with a velocity dispersion of several km s^{-1} and these fragments also must maintain their kinetic energy despite dissipative collisions (e.g. Scalo and Pumphrey 1982).

A simple way to illustrate some of the relevant processes is illustrated in Figure (2a), which depicts a magnetic fieldline $\hat{x}B_o$ on which several clouds are strung. The space between the clouds is filled with a tenuous medium. If, as shown in Figure (2b), one of the clouds is given a downward displacement then the fieldline will be dragged along. The resulting bend in the fieldline will exert an upward magnetic tension force on the cloud. It is of some interest to calculate what this tension force will do to the cloud before it collides with the next cloud below it.

To obtain an equation of motion for the cloud, we integrate the Lorentz force over the cloud volume

$$F_z = \int d^3 r \frac{(J \times B)_z}{c} \approx \int d^3 r \frac{B_x}{4\pi} \frac{\partial B_z}{\partial x} \tag{13}$$

where we assume that the force is primarily due to magnetic tension. Applying Gauss' theorem we see that the tension force on a cloud of size D is of order $\alpha D^2 < B_x B_z > /4\pi$, where $<>$ denotes the jump from one side of the cloud to the other and α is a geometrical factor which is unity for a cubical cloud.

We now evaluate $< B_x B_z >$ in terms of the cloud displacement. B_x is approximately B_o. If the cloud is at $z = 0$ at $t = 0$ when it begins to move then a disturbance travels out along the fieldlines at speed V_{ai}, the Alfven speed $B_x/\sqrt{4\pi\rho_i}$ in the intercloud medium, and at time t will have gone a distance $V_{ai}t$. If the cloud is displaced a distance z at this time then at the position of the cloud, $|B_z/B_x| \sim z/V_{ai}t$. When the disturbance reaches the next cloud on the fieldline, a distance L away, it will be almost perfectly reflected and from that time on $|B_z/B_x| \sim z/L$. Thus the equation of motion for a cloud of mass M_c is

$$\frac{d^2 z}{dt^2} = -\frac{\alpha B_o^2 D^2 z}{2\pi M_c V_{ai} t} \qquad t < L/V_{ai} \tag{14a}$$

$$\frac{d^2 z}{dt^2} = -\frac{\alpha B_o^2 D^2 z}{2\pi M_c L} \qquad t > L/V_{ai} \tag{14b}$$

The solution of equation (14a) with initial conditions $z = 0$, $dz/dt = V_o$ is

$$z(t) = \frac{V_o}{K} t^{1/2} J_1(2K t^{1/2}) \tag{15}$$

$$K^2 = \frac{\alpha B_o^2 D^2}{2\pi M_c V_{ai}} = 3.2 \ 10^{-14} \frac{B_\mu}{n_c D_{pc}} n_i^{1/2} \alpha \ \text{s}^{-1}$$

where n_i and n_c are the intercloud and cloud number densities, respectively and J_1 is a

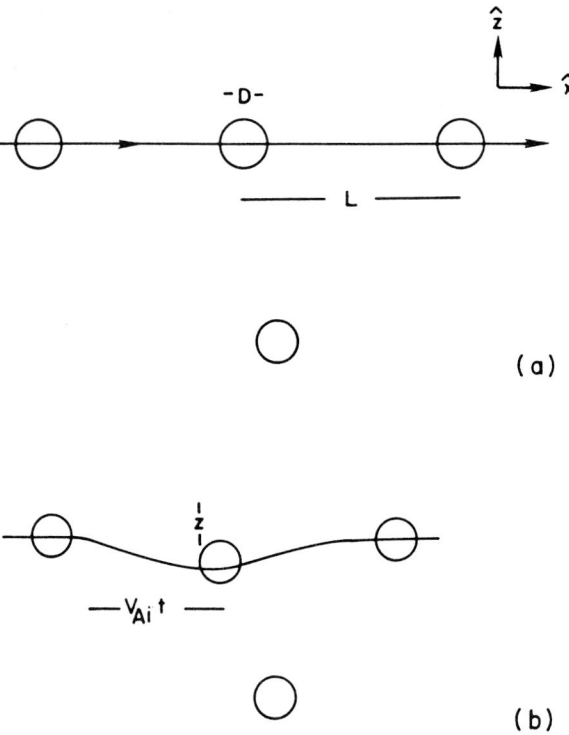

Figure 2. a) Magnetic fieldline with attached clouds. b) Distortion of the fieldline by cloud motions.

Bessel function of order 1. After $t > L/V_{ai}$ the cloud oscillates harmonically. In practice, as long as $V_o/V_{ai} \ll 1$, it is the oscillating solution which is most important because the disturbance travels to the next cloud before the perturbing cloud has moved very far. The amplitude of the oscillation is $V_o\sqrt{L/K^2V_{ai}}$. For $V_o = 5.8$ km s^{-1}, $n_c = 40$, $D_{pc} = 5$, $L = 1/6$ kpc, $B_\mu = 5$, the amplitude is

$$z_{max} \sim 4.5\ 10^{20}\text{cm} \approx L$$

which implies that clouds may be appreciably decelerated by magnetic tension forces before running into each other. We reach a similar conclusion by inserting parameters appropriate to cloudlets in giant molecular complexes.

One can vary this simple model in many ways. The gravitational acceleration perpendicular to the galactic plane can easily be included in the equations of motion (14). The oscillations of an ensemble of clouds can also be worked out.

Cloud collisions are highly dissipative and in the example worked out here, the collision is softened or even prevented. Yet we should not conclude that magnetic fields always reduce the dissipation rate of cloud kinetic energy. Widely separated clouds can interact remotely if the magnetic flux tubes threading the clouds become mutually entwined, as considered by Clifford and Elmegreen (1983). The magnetic disturbances excited by cloud motions are subject to decay, primarily through ion-neutral friction. When we take dissipation into account it appears that the magnetic field may be an energy sink for cloud motions (Elmegreen 1985).

C. Magnetic Transfer of Angular Momentum

Comparison of the angular momentum present in a stellar mass sized chunk of interstellar matter (estimated from galactic rotational shear) and the angular momentum of a star shows that the angular momentum must decrease by several orders of magnitude during the star formation process. One way to accomplish this is by magnetic torques.

Detailed calculations of the magnetic braking process have been carried out by a number of authors under different assumptions (Elmegreen 1981; Gillis, Mestel, and Paris 1979; Königl 1986; Mestel and Paris 1979, 1984; Mouschovias 1981; Mouschovias and Paleologou 1979, 1980), and the basic mechanism is as follows. If a magnetic field connects the rotating cloud to an intercloud medium with density ρ_i then, as the fieldlines are twisted up by the cloud rotation, a signal (similar to a torsional Alfven wave) propagates away from the cloud at the Alfven speed in the intercloud medium, and the intercloud medium is sent into rotation at the angular velocity of the cloud at the time the signal was launched. By requiring that the angular momentum of the cloud plus the intercloud medium be constant we readily find (Spitzer 1978) that the timescale on which a disk of thickness H and density ρ_o loses angular momentum is of order $\alpha H \rho_o / 2\rho_i V_{ai}$, where α is a geometrical factor. If a number of clouds are magnetically connected then angular momentum can be transferred between them. This possibility has recently been studied by Mouschovias and Morton (1985a, b). The ambipolar drift present in a weakly ionized rotating cloud has been included by Königl (1986).

D. Global Equilibrium and Stability

We now consider the effects of magnetic fields on the global structure of the interstellar medium. As reviewed by Heiles in this volume, the magnetic field within 1-2 kpc of the Sun is fairly parallel to the galactic plane, inclined in the direction of galactic rotation, and a few μG in strength. Cosmic rays are confined by the galactic field and contribute a source of pressure which is comparable to the magnetic pressure and to the turbulent pressure associated with cloud motions (the latter is the relevant gas pressure for describing the global dynamics of the gas of clouds).

If we assume that on a large scale the interstellar medium is in hydrostatic equilibrium then the system is described by the set of equations

$$0 = -\nabla P_g + \frac{J_g \times B}{c} + \rho_g g \tag{16a}$$

$$0 = -\nabla P_{cr} + \frac{J_{cr} \times B}{c} \tag{16b}$$

$$\nabla \times B = \frac{4\pi}{c}(J_{cr} + J_g) \tag{16c}$$

$$P_g = \frac{1}{3}\rho_g < V_c^2 > \tag{16d}$$

where we have neglected the cosmic ray mass density, the subscripts g and cr denote gas and cosmic rays, respectively, and $< V_c^2 >$ is the cloud velocity dispersion. Using equations (16b) and (16c) to eliminate J_g in favor of B and P_{cr} in equation (16a) we find

$$0 = -\nabla(P_g + P_{cr}) + \frac{(\nabla \times B) \times B}{4\pi} + \rho_g g \tag{17}$$

From the form of equation (17) the cosmic ray pressure gradient appears to exert a force on the thermal gas. This term does not represent the usual kinetic momentum transfer but rather the effect of the cosmic ray current as a source term for the magnetic field.

The structure of the interstellar medium perpendicular to the galactic plane can be studied with a simple one dimensional model in which B is straight and parallel to the galactic plane and all quantities depend only on the vertical coordinate z. Equation (17) then reduces to its $\hat{z}$ component:

$$0 = -\frac{d}{dz}\left(P_g + P_{cr} + \frac{B^2}{8\pi}\right) - g\rho_g \tag{17'}$$

If g, $< V_c^2 >$, $P_{cr}/P_g \equiv \beta$, $B^2/8\pi P_g \equiv \alpha$ are all constant then equation (17') has the simple solution

$$P_g = P_g(0)e^{-z/H} \tag{18}$$

$$H = < V_c^2 > (1 + \alpha + \beta)/3g \tag{19}$$

Equations (18) and (19) show that the scale height of the gas is increased by magnetic field and cosmic ray pressure. If we take $\alpha = 0.25$ and $\beta = 0.4$ (Spitzer 1978) we see that this inflation is nonnegligible. The scale height given by equation (19) is of order 200 pc, which is in reasonable agreement with the observed HI thickness. Still, this model leaves aside two important properties of the interstellar medium: the presence of hot coronal gas, which has a larger scale height then that implied by equation (19), and the scale height of the galactic synchrotron background, which may be several kpc thick (e.g. Rockstroh and Webber 1978). These facts and others suggest that hot gas, magnetic fields, and cosmic rays extend into a halo.

Nevertheless, the system described by equation (18) and basic variations of it have stimulated much discussion, so we review the model here.

Because the scale height of the interstellar medium in this model is increased by magnetic fields and cosmic rays we might suspect that the system is unstable to something like a Rayleigh-Taylor instability (the instability of a heavy fluid supported by a light one). The instability problem is actually closely akin to the problem of convective instability in stellar interiors. It was shown by Newcomb (1961) that gas in a horizontal magnetic field and a vertical gravitational field is stable if and only if

$$-\frac{d\rho}{dz} > -\frac{d\rho}{dz}\Big|_{\text{adiabatic}} = \frac{\rho^2 g}{\gamma P} \tag{20}$$

where $\gamma \equiv \rho/P \ (dP/d\rho)|_{\text{adiabatic}}$ is the adiabatic index. Equation (20) is just the familiar Schwarzschild criterion for convection. Although the magnetic field does not enter the stability criterion directly it does affect the spatial structure of the unstable modes. The modes which become unstable first have a small wavenumber $k_{\parallel h}$ parallel to the magnetic field and a large wavenumber $k_{\perp h}$ in the horizontal direction perpendicular to the field, a choice which suppresses the stabilizing effects of magnetic tension and magnetic pressure. The unstable motions consist primarily of translation along B with an interchange of position between vertically stacked magnetic flux tubes. If we assume that the cloud gas can be described by an adiabatic index γ (which measures the dependence of $<V_c^2>$ on the number of clouds per unit volume) and set the cosmic ray pressure equal to zero then equation (20) implies instability if and only if

$$\gamma > \alpha + 1$$

Parker (1966) was the first to calculate this instability in the context of the interstellar medium including the effects of cosmic rays, and the instability bears his name. The most stringent criterion for stability occurs when $k_{\perp h} \to \infty$, $k_{\parallel h} \to 0$, and is

$$\gamma > \alpha + \beta + 1$$

which is also what is predicted from equation (20). The fastest growing modes have $k_z = 0$ and $k_{\parallel h} \sim .8H^{-1}$; the growth rates are quite insensitive to $k_{\perp h}$ (Shu 1974). The maximum growth rates are of order $<V_c^2>^{1/2}/H$, implying growth times of a few times 10^7 years.

A physical picture of the instability is as follows. The gas can lower its potential energy by sliding along a rippled fieldline. Pockets of gas form in magnetic troughs, and magnetic crests, partly unburdened of gas, are free to rise (see Figure (3)). If the perturbation is unstable then the energy released by these effects must be greater than the energy required to compress the gas and bend the field. Therefore low compressibility (small γ) and small parallel wavenumber $k_{\parallel h}$ promote instability.

Figure 3. Gas condensations (shaded) and undulating magnetic field produced by Parker's instability.

Parker chose $\gamma = 1$, reasoning that compression of the interstellar gas would bring the cloud centers closer together and increase the rate of cloud-cloud collisions, which dissipate velocity. Thus $\gamma = 1$ is really an upper limit in his model. He also treated the cosmic rays as a $\gamma = 0$ fluid, assuming that cosmic ray pressure would rapidly be made uniform along a fieldline by cosmic ray streaming.

Parker (1979 and references therein) argued that the instability would have two important consequences for the interstellar medium. Clouds would be formed in the gas pockets and the magnetic crests described in linear theory would inflate into giant loops, rising above the galactic disk and carrying cosmic rays with them. Magnetic diffusion by small scale turbulence at the base of the loops would eventually cause the fieldlines to snap and the resulting bubbles would float away into the intergalactic medium. Thus the magnetic field would be destroyed in a few linear instability growth times, or a few 10^8 yr. Cosmic rays would also be removed from the disk.

Parker's picture stimulated extensive further work on the subject. Shu (1974) included galactic rotation and rotational shear. He found that these effects (acting through the Coriolis force) can reduce the growth rate but do not affect the criterion for stability. This can be understood qualitatively by noting that a marginally stable (zero frequency) perturbation is static and hence not acted on by Coriolis forces.

Zweibel and Kulsrud (1975) reevaluated γ by setting up an equilibrium for the gas of clouds in which cloud deceleration and destruction by collisions is balanced by cloud re-acceleration and new sources of cloud material. Most sources of cloud kinetic energy are associated with young massive stars, and since empirical evidence suggests that the star formation rate increases with increasing gas density there will be more cloud energy sources as well as more cloud collisions in compressed regions. Thus γ need not be 1; Zweibel and Kulsrud found that γ is probably between 5/3 and 2, which suppresses the instability for a plausible range of α and β.

Lachièze Rey et al. (1980) reconsidered Parker's original assumption that the cosmic rays behave as a $\gamma = 0$ fluid. The reason for this is that a substantial body of work now suggests (see Cesarsky 1980 and Section IVB of this review) that cosmic ray streaming excites instabilities which scatter the cosmic rays so that they diffuse rather than stream. Thus, cosmic ray pressure gradients along an interstellar fieldline might be maintained for long periods of time. Since work must be done to compress a cosmic ray fluid with nonzero γ_{cr}, the effect of γ_{cr} is stabilizing. For a planar system the critical γ for stability is reduced by an amount $\beta\gamma_{cr}$.

On a microscopic scale the field should be quite tangled by cloud random motions. Observational evidence suggests that the strength of the small scale ($\lesssim 50$ pc) random component is of the same order as the average field (Manchester 1974). Zweibel and Kulsrud (1975) and Lachièze-Rey et al. (1980) considered the effect of magnetic microturbulence on the Parker instability and found it to be stabilizing; such a field adds stiffness and resists being bent.

In 1974 Mouschovias published two dimensional equilibrium states which might represent the nonlinear saturation of the Parker instability. These states can be reached from the exponentially stratified isothermal equilibrium by flux preserving motions and have a horizontal periodicity with magnetic hills and valleys. It was later shown (Asseo et al. 1978, 1980) that these states are unstable to three dimensional perturbations. The instability can be interpreted partially in terms of the more stringent stability criterion in three dimensions than in two, so that some free energy remains untapped by the original perturbation. However the curvature in the equilibrum field is also destabilizing.

Mouschovias, Shu and Woodward (1974) suggested that the Parker instability triggers the formation of the OB associations which delineate the spiral arms of many galaxies. In this picture the gas, magnetic field, and cosmic rays are first compressed by shocked flow through the gravitational potential of a spiral arm. The relative increase of magnetic and cosmic ray energy density within this compressed gas layer favors the growth of the Parker instability. The observed spacing of the OB associations, which is of order 1 kpc, is consistent with the wavelength of the fastest growing Parker modes. Elmegreen (1982a, b) has further investigated this picture by including the self-gravity of the gas. For a sufficiently high mean density ($n \gtrsim 5$ cm^{-3}) the condensation process is dominated by the Jeans instability; at intermediate densities both the Jeans mechanism and the Parker mechanism are important.

Finally we mention the role of a hot phase of the interstellar medium on the Parker instability (Lachièze-Rey et al. 1980). We can include hot materal by modeling the gas with a hot and cold component.

$$P = P_c(0)e^{-z/H_c} + P_h(0)e^{-z/H_h} \tag{21}$$

$$\rho = \frac{P_c(0)}{C_c^2}e^{-z/H_c} + \frac{P_h(0)}{C_h^2}e^{-z/H_h}$$

The stability criterion (20) evaluated at the midplane is

$$\gamma > \frac{H_c g}{C_c^2} \frac{(1+\rho_h(0)/\rho_c(0))^2}{\left(1+\frac{\rho_h H_c}{\rho_c H_h}\right)(1+P_h(0)/P_c(0))} \tag{22}$$

Since $\rho_h(o)/\rho_c(o) \ll 1$ we see that for fixed H_c and C_c the local stability at midplane is improved by the hot gas. (Of course, magnetic fields and cosmic rays must be added to (22)) In the models of Lachièze-Rey et al. the hot gas is destabilizing at large heights.

What is the current status of the Parker instability and what are the most promising future directions in research? The investigations of Parker and others from the mid 1960's to the late 1970's were based essentially on the two phase model of the interstellar medium, in which clouds provide most of the mass, the turbulent velocity of the clouds is approximately equal to the thermal velocity of the intercloud gas, and the system is assumed to be statistically homogeneous over a density scale height. Analysis of such systems shows that a laminar equilibrium with $\gamma = 1$ is unstable with instability growth times of a few times 10^7 years and that magnetic microturbulence and a stiffer equation of state can suppress the instability while galactic rotation slows the growth rates. Possible consequences of the instability are destruction of the galactic field in a few 10^8 yr and the formation of condensations, which become OB associations, in gas which has been shocked by flow through a spiral arm.

The interstellar medium appears to extend into a halo, which contains hot gas, magnetic fields, cosmic rays, and clouds. It is at present not known whether the halo is expanding as a wind or whether it is bound to the galaxy, and the geometry and extent of the halo are uncertain. Although the Parker instability predicts a magnetic halo, the halo may also be formed from energy input by supernovae. It is now also known that hot gas is present in the disk, possibly with a large filling factor.

We suggest on the theoretical side that interstellar medium models which are closer to the McKee and Ostriker (1977) model than to the older two phase models should be investigated. The work of Lachièze-Rey et al. (1980) is an important step in that direction. On the observational side, more information about the magnetic fieldstrength and configuration and the cosmic ray spectrum above the plane is necessary to constrain the theoretical models.

A very important point which is brought out by Parker's instability is that the magnetic field and cosmic ray energy densities are probably bounded by some number of order the gas energy density. Thus the fact that these three energy densities are approximately equal is probably not a coincidence.

III. Origin and Maintenance of the Galactic Field

A discussion of the origin of the galactic magnetic field begins with a cosmological problem:

in the standard Big Bang model, magnetic fields are not present. However, magnetic flux can be created by plasma processes, although the resulting fields are generally weak. We also note that Witten (1985) has discussed the possibility of magnetic field amplification by superconducting strings.

After discussing flux generation, we turn to methods of amplifying and destroying the galactic field. In particular we consider whether the field can be primordial or whether the original seed flux has been completely replaced and reconfigured by dynamo action. Finally we discuss the effect of magnetic monopoles on the galactic field.

A. Creation of Magnetic Flux

Whenever the forces acting on electrons and ions in a plasma are different, there is a tendency for charge separation. Any such charge separation produces an electric field, which tends to couple the two species together. If the electric field is nonpotential, a magnetic field can be produced by Faraday's Law.

Harrison (1970) considered magnetic field induction in rotating, expanding eddies in the early universe. Prior to recombination, the electrons were coupled very strongly to photons by Compton scattering. The angular velocity ω of ions in an eddy of size a scales as a^{-2} while the photon angular velocity ω_γ scales as a^{-1}. The electromagnetic force on the electrons must therefore balance the frictional force between the electron and photons (neglecting ordinary conductivity). Harrison found that a magnetic field $B \sim m_i c \frac{\omega}{e}$ would be produced by the polarization electric field. This is of order 10^{-19} G for ω appropriate to a present galaxy and should be reduced by $(1 + z)^2$ for a protogalactic fragment assuming angular momentum conservation; this gives a field of $B < 10^{-25}$ G for z measured at recombination.

Harrison (1973a, b) later went on to consider amplification of these seed fields by primordial turbulence and concluded that fields as large as 10^{-8} G on scales of 1 Mpc could be produced. Such fields are marginally consistent with observations (Zeldovich, Ruzmaikin, and Sokoloff, 1983). The level of turbulence required is more problematic, because the turbulence must be sustained against damping and must not cause observable fluctuations in the microwave background.

Another method of generating flux was proposed by Mishtustin and Ruzmaikin (1972). In a rotating protogalaxy (after recombination), Compton scattering of cosmic background radiation will exert a drag force on the electrons so that the electron fluid tends to drift inward. In this case the polarization electric field must balance the Compton drag. The resulting magnetic field generation rate is of order

$$\frac{1}{a^2} \frac{\partial}{\partial t} (Ba^2) = \frac{4}{3} \sigma_T U_\gamma \frac{\omega}{e}$$

where a is the cosmic scale factor, U_γ is the photon energy density, σ_T is the Thomson cross-section, and ω is the rotation rate. If galaxies formed at $z \sim 10$ and $\omega \sim 10^{-15}$, then after 10^{10} yr, $B \sim 10^{-21}$ G.

These two estimates, and consideration of other mechanisms (e.g. Biermann 1950), demonstrate the need for field amplification by fluid motions, that is, for a hydromagnetic dynamo.

B. The Hydromagnetic Dynamo

We have already seen (eqs. 4 and 6) that a magnetic field can be amplified by differential rotation. After a time t, differential rotation $\omega(r)$ will have amplified the field by a factor of

order $rtd\omega/dr$, which is about 300 at the solar circle over 10^{10} yr. This factor obviously fails
to bring the seed fields discussed above up to the observed value. It is possible, however,
to make a more favorable estimate using the theory of hydromagnetic dynamos, which has
been applied extensively to stars and planets. Some general references on dynamos and
their application to the galactic magnetic field include the books by Moffatt (1978), Parker
(1979), and Zeldovich, Ruzmaikin, and Sokoloff (1983). Here we only sketch some basic
ideas.

The magnetic field of a three dimensional object consists of poloidal (in meridional
planes) and toroidal (azimuthal) components. Equation (4) shows that differential rotation
generates toroidal flux from poloidal flux. Under certain conditions, toroidal flux can act
as a source of poloidal flux. This mutual reinforcement by the toroidal and poloidal field
leads to exponential instead of algebraic growth of the field with time.

Consider the situation sketched in Figure (4). In Figure (4a) a rising fluid element
lifts a toroidal fieldline into a loop. In Figure (4b) rotation of the fluid element twists the
loop. The mean of a great number of such convection cells with a *mean net twist or helicity*
will thereby generate a poloidal field. This was first pointed out by Parker (1955). In
astrophysical situations this helicity is provided by Coriolis forces and it is necessary to
postulate some asymmetry between rising and sinking motions. The production of poloidal
field from toroidal field by helical turbulence is called the α effect.

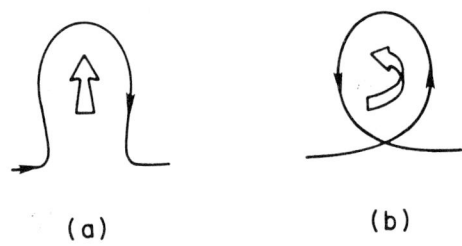

$\qquad\qquad$ (a) $\qquad\qquad\qquad\qquad\qquad$ (b)

Figure 4. a) Formation of a flux loop by upward convection. b) Twisting of
the loop by helical convection.

The α effect is usually computed in the following way (Steenbeck, Krause, and
Rädler, (1966)). In the magnetic induction equation

$$\frac{\partial B}{\partial t} = \nabla \times (V \times B) + \lambda \nabla^2 B \qquad\qquad (23)$$

(where $\lambda = \eta c^2/4\pi$ for ordinary resistive dissipation or, more often is a turbulent diffusivity), the velocity and magnetic fields are assumed to consist of small scale parts with zero
mean and large scale parts with nonzero mean

$$B = B_o + b \qquad\qquad (24)$$

$$V = V_o + v$$

Substituting the forms (24) into equation (23) and averaging over space or time or the turbulent ensemble we derive an equation for the mean field

$$\frac{\partial B_o}{\partial t} = \nabla \times (V_o \times B_o) + \nabla \times <v \times b> + \lambda \nabla^2 B_o \tag{25}$$

Subtracting equation (25) from equation (23) gives an equation for the fluctuating field

$$\frac{\partial b}{\partial t} = \nabla \times [v \times B_o + V_o \times b + v \times b - <v \times b>] + \lambda \nabla^2 b \tag{26}$$

The term $v \times b - <v \times b>$ is usually neglected (first order smoothing) and the velocity field V_o and v are usually prescribed. Thus, the fluid flow is assumed not to be influenced by the field, meaning that the dynamo is only a kinematic dynamo. One then solves for b in terms of v and computes $<v \times b>$ for insertion into equation (25). The $<v \times b>$ term is the mean electric field produced by the small scale motions.

Under certain conditions (homogeneous, isotropic velocity fluctions; long length scale for B_o, c.f. ch. 7 of Moffatt) it turns out that

$$<v \times b> \cong \alpha B_o \tag{27}$$

where α is a pseudoscalar related to the helicity of the velocity field. The parameter α can be calculated in a number of cases. For a small angular rotation over the lifetime of a highly resistive eddy with rotational velocity v_ϕ, Parker (1979, p. 580) estimated that $\alpha \sim 0.4 <v_\phi>$. When resistivity is weak (large magnetic Reynolds number) it has been shown (Childress 1979, Perkins and Zweibel 1986) that α scales as $\lambda^{-1/2}$.

We now consider differential rotation combined with the α effect. To get some idea of the field amplification timescales we consider plane wave magnetic fields of the form

$$B_o = (\hat{x} B_x + \hat{z} B_z) \exp(pt + iky) \tag{28}$$

and take $V_o = \hat{x} V(z)$. Then, using equations (27) and (28), equation (25) becomes

$$pB_x = B_z V' + ik\alpha B_z - \lambda k^2 B_x \tag{29a}$$

$$pB_z = -ik\alpha B_x - \lambda k^2 B_z, \tag{29b}$$

where $V' \equiv dV/dz$ is to be treated as a constant. We can generally neglect the α effect compared to the shear effect on the right hand side of equation (29a). Doing so, we find the dispersion relation

$$p = -\lambda k^2 \pm (1 - i)(|k\alpha V'|/2)^{1/2} \tag{30}$$

so the solutions are either damped or growing oscillations. Short wavelength modes are damped by diffusion. The fastest growing mode occurs for

$$k = (|\alpha V'|/32\lambda^2)^{1/3} \tag{31}$$

and has growth rate

$$p_{max} = \tfrac{3}{4}(\alpha^2 V'^2/16\lambda)^{1/3} \tag{32}$$

If we assume the resistivity is due to eddies with lengthscale L and velocity v then $\lambda \sim 0.2Lv$ while Parker's estimate for α is $\alpha \sim 0.04\Omega L$, where Ω is the galactic rotation rate (which induces helicity in the turbulence through the Coriolis force). Taking $V' \sim \Omega$ we find from equations (31) and (32)

$$p_{max} \sim 4 \times 10^{-17} \text{ s}^{-1}$$

and $2\pi/k(p_{max}) \sim 4$ kpc

Thus the field amplifies exponentially on a timescale of $\sim 8 \times 10^8$ yr with a lengthscale of ~ 4 kpc. This value still falls short of amplifying a 10^{-21} G field in 10^{10} yr but all the factors could probably be adjusted slightly as needed.

Of course these estimates are very crude and one should do much better in modeling galactic structure. Parker (1971, 1979) has modeled the galactic dynamo in a slab geometry with free escape of flux (such as by buoyancy instability) at the upper and lower boundaries and α antisymmetric across the midplane. Stix (1975) has used the same basic approach for an oblate spheriodal model. It is possible to find steady solutions in which the loss of flux (primarily through the boundaries) balances regeneration of flux. The desired solutions have predominantly toroidal fields which are symmetric about the midplane, in agreement with observations. Notice that equations (25) and (27) are linear in B_o and so do not determine the amplitude of the field.

This model depends crucially on the assumed resistivity of the interstellar medium, which in the estimates above is given a value appropriate to large scale eddies. Yet, as we saw in Section (IIA), very thin flux sheets must form if true field dissipation is achieved. (The sheet thickness based on Ohmic diffusion may be broadened if resistive instabilities are important, but scales remain small). The question of resistivity is perhaps not so important if the loss of flux is primarily through buoyant escape, but it is important for computation of the α effect.

Another important approximation is that the feedback of the magnetic forces on the fluid flow is negligible. Energetic arguments (and Section IIB) suggest that this is not the case. Feedback has been shown to be very important in models of the solar dynamo (Gilman 1983). It is interesting to note that attempts to reproduce the solar cycle with hydromagnetic dynamos are now stalled and the suspicion is that flux rope formation on a scale below the numerical resolution of the codes should play an important role.

C. Bisymmetric Spiral Fields

We saw in Section (IIA) that an initially straight magnetic field lying in the plane of a differentially rotating disk would be amplified into a spiral-like configuration in which the field direction changes sign frequently with radius. (Equation (6), Figure (1)) At the solar circle in the present epoch the field should reverse about every 100 pc but this is not seen.

It is claimed, however, that the magnetic fields in many galaxies, including our own, do have the bisymmetric spiral shape, but with much milder winding than we would expect from equation (6) (Sofue, Fujimoto, and Wielebinski, 1986). The bisymmetric spiral was predicted by Piddington (1964, 1978).

Sawa and Fujimoto (1980) made a model of the galactic field which is bisymmetric but not severely twisted. In their model the field in the disk is dominated by the field in a halo which rotates slowly and essentially rigidly. Turbulence in the halo and the disk causes halo field to diffuse into the disk, where it is wound up by differential rotation. But turbulent diffusion soon sends the flux back into the halo, so not much winding up occurs.

This model is in complete contrast to Parker's picture; the halo is a souce as well as a sink for the disk field and no dynamo action is assumed. In both the halo driven and the dynamo driven field models, turbulent diffusion plays an essential role.

Kulsrud (1986) has suggested a model of the field which avoids the winding up problem without recourse to turbulent diffusion. Recall from equation (10) that in a weakly ionized, magnetized medium there is a relative drift between ions and neutrals which allows hydromagnetic forces to act on the neutrals through friction. If the interstellar gas is partially supported above the galactic plane by magnetic fields then there must be a drift of the fieldlines away from the galactic plane, which from equation (11) is of order 0.04 km s^{-1}. This velocity, although small, moves the lines upward by one to two scale heights over 10^{10} yr.

To fully work out the effect of this vertical motion we should solve equation (4) including the ambipolar drift. This would give a nonlinear diffusion-like equation for the magnetic field. We can understand the basic mechanism, however, from kinematic considerations. Let the galactic disk occupy $|z| < D$. Consider a magnetic fieldline at time $t = 0$ that enters the disk at $z = -D$, $r = r_1$, $\theta = \theta_1$, and exits at $z = D$, $r = r_2$, $\theta = \theta_2$. Assuming for simplicity that the line is straight within the disk, the r, θ coordinates of a point on the line are given by

$$r(z,0) = \frac{(r_1 + r_2)}{2} + \frac{(r_2 - r_1)}{2} \frac{z}{D}$$

$$\theta(z,0) = \frac{(\theta_1 + \theta_2)}{2} + \frac{(\theta_2 - \theta_1)}{2} \frac{z}{D}$$

Now let differential rotation and a velocity $\hat{z}V$ (positive above the plane and negative below it) act on the fieldline. After a time Δt the points on the line at $z = \pm D$ are points which were at $z_\pm = \pm(D - V\Delta t)$ at $t = 0$. The angles θ_1 and θ_2 of these two points are

$$\theta_2(D, \Delta t) = \theta_2 - (\theta_2 - \theta_1)\frac{V\Delta t}{D} + \Delta t \Omega(r_1) + \Delta t(r_2 - r_1)(1 - \frac{V\Delta t}{D})\frac{d\Omega}{dr}$$

$$\theta_1(-D, \Delta t) = \theta_1 + (\theta_2 - \theta_1)\frac{V\Delta t}{D} + \Delta t \Omega(r_1) + \Delta t(r_2 - r_1)\frac{V\Delta t}{D}\frac{d\Omega}{dr}$$

and

$$\theta_2(D, \Delta t) - \theta_1(-D, \Delta t) = \left[\theta_2 - \theta_1 + \Delta t(r_2 - r_1)\frac{d\Omega}{dr}\right]\left(1 - \frac{2V\Delta t}{D}\right)$$

so the expansion velocity V reduces the shearing of the lines of force. This occurs because points a given distance z apart were closer together in the past and had values of θ and r which were closer together. Kulsrud estimates that the galactic field need have wound up by only ~ 1 radian over the lifetime of the galaxy.

This configuration is predicted if the field is primordial. One would then have an intergalactic field of $\sim 10^{-9}$ G which is compressed to larger values by collapse of the galaxy and then evolved by rotation. However, the existence of an intergalactic field of this magnitude seems to imply a pregalactic dynamo that amplifies the very weak fields generated by plasma processes.

D. Magnetic Monopoles

Physicists have speculated on the existence of magnetic monopoles for a long time, and such particles are now predicted by grand unified theories (t'Hooft 1974, Polyakov 1974).

The monopoles would have a charge equal to the Dirac charge; $|g| = \hbar c/2e = 3.3 \ 10^{-8}$ esu and a mass m of order $10^{16} \ Gev/c^2$. Cabrera (1982) suggested that monopoles could comprise a significant fraction of the dark matter in the galaxy.

Parker (1970) was the first to point out that the galactic magnetic field might be used to set limits on the properties of the monopole distribution. He treated the monopoles as test charges with rms velocity V_m and number density n_m, which are accelerated magnetostatically by the galactic field, taken to have lengthscale ℓ. Energy is transferred from the field to the monopoles on a timescale

$$\tau = mV_m/4\pi n_m g^2 \ell = V_m/\omega_m^2 \ell$$

where $\omega_m^2 \equiv 4\pi n_m g^2/m$ is to be identified as the square of the monopole plasma frequency. If the mass density of monopoles is 10^{-24} g cm^{-3}, $V_m = 200$ km s^{-1}, and $\ell = 1$ kpc then $\tau = 5.3 \ (m/10^{16} \ Gev/c^2)^2$ yr. Since this time is some 8 orders of magnitude faster than the regeneration time for the galactic field, monopoles cannot be a dynamically significant component of the galaxy. Turner, Parker, and Bogdan (1982) extended these arguments.

Arons and Blandford (1983), Farouki, Shapiro, and Wasserman (1984) and Chernoff, Shapiro, and Wasserman (1985) showed that by including collective effects this limit could be drastically altered. In their picture the galaxy contains equal numbers of monopoles and antimonopoles. This monopole plasma can sustain longitudinal magnetostatic plasma oscillations which are directly analogous to electrostatic plasma oscillations. These oscillations are Landau damped by resonant monopoles, that is, monopoles with velocity equal to the phase velocity of the oscillation. (An unstable distribution of monopoles, e.g. a beam, could in principle destabilize the oscillation, leading to magnetic field growth.) The period of these oscillations is the monopole plasma period $(3.2 \ 10^4(m/10^{16} \ Gev/c^2))$ yr. The Landau damping rates can be low enough that conventional dynamo action can regenerate the field. Of course, if the monopole density is sufficiently low, the test particle treatment is valid. But it does not seem possible to draw any straightforward conclusions about monopoles from the properties of the galactic field at this time. A field which reversed every 10^4 yr might have some interesting consequences for cosmic ray confinement, however.

IV. Cosmic Rays

Cosmic rays are confined and possibly accelerated by the galactic magnetic field. Hence it is appropriate to discuss cosmic rays in this paper. After a very brief review of cosmic ray properties we discuss cosmic ray transport in the galactic magnetic field and acceleration of cosmic rays in the interstellar medium. Various aspects of cosmic ray astrophysics have been reviewed by Hillas (1975), Cesarsky (1980), Axford (1981), Simpson (1983), Eichler (1986) and Israel (1986).

The spectrum of cosmic rays at the Earth can be approximated by a piecewise continuous power law. At energies of about 1 Gev and less, cosmic rays are strongly scattered by magnetic fluctuations in the solar wind. The effect of this solar modulation is to make the interstellar low energy spectrum rather uncertain. Up to about 3×10^6 Gev the differential spectrum is $f(E) \sim E^{-2.6}$. From 3×10^6 Gev to $\sim 10^7$ Gev the spectral index steepens from 2.6 to 3.5, and above 10^8 Gev flattens to about 3. Above $\sim 10^{10}$ Gev the small number of particles makes the spectrum difficult to measure. The integrated interstellar energy density in cosmic rays is about $1.3 \ 10^{-12}$ erg cm^{-3} which is comparable with the energy densities in other major components of the interstellar medium. Cosmic rays of energies $\lesssim 10^6$ Gev are directionally isotropic to within a few parts in 10^4. There is some evidence for increasing anisotropy at the highest energies.

The chemical composition of cosmic rays carries information about both propagation and sources. The cosmic rays are enriched in Li, Be, and B; these nuclei are assumed to be spallation products of collisions between CNO nuclei and hydrogen. With this interpretation, the light element abundance shows that cosmic rays have interacted with an average of 3-5 gm/cm^2 of matter. The grammage appears to decrease with cosmic ray energy as $E^{-0.3-0.6}$, at least up to about 10^2 Gev.

The ^{10}Be isotope has a half-life at rest of 1.5 10^6 yr. From the measured abundance of ^{10}Be in cosmic rays, Garcia-Munoz, Mason, and Simpson (1977) derive a cosmic ray age of $\sim 2 \times 10^7$ yr. When combined with the traversed grammage, this implies that the mean density of the material in which cosmic rays spend most of their time is $\sim .18$cm^{-3}, a value which might be increased by up to a factor of 2. This value is several times lower than the accepted mean density of the interstellar medium, and it has been proposed that cosmic rays are stored in a low density halo (Cesarsky (1980)).

Theories of cosmic ray propagation attempt to explain the residence time of cosmic rays in the galaxy, the interaction of cosmic rays with matter, and the spatial distribution of cosmic rays. The propagation turns out to depend not only on the large scale topology of the field, but also on the microscale field structure.

The Larmor radius of a proton with energy E_{Gev} (in Gev) in a $3\mu G$ field is $r_L = 10^{12}\beta E_{Gev}$cm, where $\beta \equiv v/c$. Thus, cosmic rays with moderate energies have orbits well confined within the galaxy, but at sufficiently high energies ($E_{Gev} \gtrsim 3 \ 10^8$) the field of the galactic disk does not confine the cosmic rays. The propagation and probably the origin of the highest energy cosmic rays is quite different from that of the bulk of galactic cosmic rays, and we focus on the latter in this article.

The observed isotropy of cosmic rays suggests either an extraordinarily dense and homogeneous distribution of sources or an efficient mechanism for isotropizing the particles. The mechanism most commonly invoked is resonant scattering of the cosmic rays by hydromagnetic waves. Consider a wave with frequency ω, wavenumber component $k_\parallel$ along the ambient magnetic field B, and fluctuating electric and magnetic field components $\delta E_\perp$, $\delta B_\perp$ transverse to B. A cosmic ray which moves with velocity $v_\parallel$ along the magnetic field and gyrates at the relativistic gyrofrequency $\Omega = \Omega_o/\gamma$ will see the wave at the Doppler shifted frequency $\omega' = \omega - k_\parallel v_\parallel$. If the resonance conditions $\omega' = \pm\Omega$ is satisfied then the fields $\delta E_\perp$ and $\delta B_\perp$ maintain a fixed phase relative to the gyration of the particle and exert a steady instead of a fluctuating force. In a frame traveling with the wave, this force cannot change the particle energy, but only the direction of the particle velocity (labelled by pitch angle $\mu \equiv v_\parallel/v$). For a hydromagnetic wave with speed $V_A \ll c$, $|\delta E_\perp|/|\delta B_\perp| \sim V_A/c$, so the relative energy change $\Delta E/E$ of the particle is smaller than the relative pitch angle change by a factor $\sim V_A/c$ and is usually ignored. Thus, a particle is scattered in pitch angle by resonant waves. Approximating the resonance condition by $k_\parallel = \pm\Omega_o m/\mu v\gamma$ we see that the wavelength of a resonant wave is of order the particle gyroradius and that for a given wave there is a minimum particle energy that can resonate (corresponding to $|\mu| = 1$). For a random wave distribution, the scattering produces diffusion in pitch angle and isotropizes the cosmic rays. The diffusion coefficient $< (\Delta\theta)^2 > /\Delta t$ is of order $\Omega(\delta B/B)^2$.

Why should such waves be present in the interstellar medium? The waves which resonate with Gev particles are much shorter in wavelength and period than typical lengthscales and timescales in the interstellar medium, but a smooth spectrum extending to short wavelengths might result from nonlinear interactions between waves, as suggested by Lee and Jokipii (1976). Cesarsky and Kulsrud (1973) have considered cosmic ray diffusion due to a spectrum of interstellar turbulence. However, the theoretical and observational description of such turbulence is still quite vague.

Fortunately there is a mechanism by which the waves necessary for isotropizing the cosmic rays can be produced by the cosmic rays themselves (Kulsrud and Pearce 1969). Suppose that the cosmic rays are anisotropic in velocity space. A bulk streaming velocity is naturally produced, for example, by a nonuniform distribution of sources along a fieldline; the cosmic rays then stream down their pressure gradient. Resonant interactions between the cosmic rays and Alfven waves exchange net momentum between the cosmic rays and the waves unless the cosmic rays are isotropic in the wave frame. If the cosmic ray bulk velocity V_{cr} exceeds the wave velocity V_A, momentum is lost by the cosmic rays and the waves are amplified.

The growth rate for waves with wavenumber k can be written (Kulsrud and Cesarsky 1971) as

$$\Gamma_{cr}(k) \sim C\Omega_o \frac{n_{cr}(>p_1(k))}{n_c} \left(\frac{V_{cr}}{V_{ac}} - 1 \right) \text{ s}^{-1} \tag{33}$$

where C is a constant, Ω_o is the proton gyrofrequency, $n_{cr}(> p_1(k))$ is the number density of cosmic rays with momentum greater than $p_1(k) \sim m\Omega_o/k$; the minimum resonant cosmic ray momentum, and n_c is the density of the charged background gas component. The Alfven speed V_{ac} is computed with respect to the density of the charged component (Equation (1) must be modified in molecular clouds; see Zweibel and Shull, 1982). When the cosmic ray spectrum is a power law such that $n_{cr}(> p_1) \sim p_1^{-(\gamma-1)}$ the constant C is $\pi(\gamma - 1)/4\gamma$.

If we evaluate equation (33) using galactic cosmic ray parameters with $\gamma = 2.6$ we find

$$\Gamma_{cr}(k) \sim 1.3 \frac{B_\mu}{n_c} \gamma_1(k)^{-1.6} \left(\frac{V_{cr}}{V_{ac}} - 1 \right) 10^{-12} \text{ s}^{-1} \tag{34}$$

where γ_1 is the Lorentz factor corresponding to p_1. Typical values of B_μ and n_c thus give quite rapid growth rates over interstellar timescales. But the $\gamma^{-1.6}$ dependence shows that the growth rate for waves which can scatter particles of energy E is a decreasing function of E. As long as $\Gamma_{cr} > 0$ the waves continue to grow in amplitude, scattering the cosmic rays more and more efficiently and reducing V_{cr}.

In practice, the growth rate Γ_{cr} is offset by damping mechanisms which increase the value of V_{cr} needed for net growth. In a partially ionized medium, Alfven waves are damped by ion-neutral collisions (the physics is the same as that responsible for ambipolar diffusion). In the regime of weak collisions the damping rate is

$$\Gamma_{cn} = \frac{\nu_{cn}}{2} \sim 10^{-9} n_c \text{ s}^{-1} \tag{35}$$

where ν_{cn} is given below equation (8b). Equating expressions (34) and (35) then gives V_{cr}/V_A at marginal stability

$$\frac{V_{cr}}{V_A} = 1 + 7.6 \frac{10^2 n_c^2}{B_\mu} \gamma_1^{1.6} \tag{36}$$

With $n_c = 3 \times 10^{-2}$ (the mean value obtained from pulsar dispersion measures) and $B_\mu = 3$,

$$\frac{V_{cr}}{V_A} = 1 + 2.3 \ 10^{-1} \gamma_1^{1.6}$$

Cosmic rays of moderate energy are thus constrained to stream near $V = V_A$, but for $\gamma_1 > 10^2$, $V_{cr}/V_A > 10^2$ and the cosmic rays can stream rapidly without exciting waves. Equation (36) suggests that ion-neutral damping very effectively limits wave growth in clouds, but this formula should not be applied to very dense clouds in which collisions are strong (Zweibel and Shull 1982).

Other forms of damping must be considered in highly ionized regions of the interstellar medium. If the sound speed is less than the Alfven speed then Alfven waves are subject to decay into backscattered Alfven waves and readily dissipated sound waves (Chin and Wentzel 1972). The damping dominates wave growth for $E_1 \gtrsim 10^3$ Gev (Wentzel 1974). The decay instability is probably suppressed in the coronal phase of the interstellar gas because the sound speed is too high. It then turns out that Alfven waves decay by nonlinear Landau damping, in which thermal ions resonantly absorb energy from the low frequency envelope that modulates a packet of waves (Lee and Volk 1983). Cesarsky and Kulsrud (Cesarsky 1980) have shown that this damping mechanism is effective enough to destroy cosmic ray confinement above $10^3 - 10^4$ Gev.

In summary, the isotropy of cosmic rays at energies less than $10^2 - 10^4$ Gev can be explained by self-confinement, but the cosmic rays at higher energies apparently are isotropized by waves produced by other sources (e.g. Flewelling and Coroniti 1976).

Whether or not we completely understand the origin of the scattering, the isotropy of the cosmic rays strongly suggests that they are well scattered, and thus locked more or less completely to the gas. What, then, determines the confinement time of cosmic rays in the galaxy? The answer obviously depends strongly on the structure of the galactic field. If the field is primordial and connects fairly smoothly to an intergalactic field then the cosmic rays generated in the disk leak into the intergalactic medium. We have also discussed the possibility in Section (IID) that giant magnetic bubbles are formed as as result of Parker's instability. The bubbles break off and escape, carrying cosmic rays with them. Bubbles might be inflated near the boundary of the disk simply as the result of cosmic ray pressure (Jokipii and Parker 1969). Depending on the ratio of the transit time of cosmic rays through the bubble to the time for bubble ascension the cosmic rays either escape once they enter a bubble (large ratio) or likely return to the disk after spending some time in the halo (small ratio). Jokipii (1976) has modified the halo storage model to allow for a galactic wind. Ipavich (1975) has modeled winds in which the driving force is provided by cosmic rays. The cosmic rays transfer momentum to the waves which scatter them, and the waves transfer momentum to the gas. There are also models in which the cosmic rays remain confined to the galaxy while suffering spallation and degradation in energy (Rasmussen and Peters 1975, Peters and Westergaard 1977).

In principle the isotropic composition, spectra of secondary nuclei, and positron and antiproton spectra allow some constraints on the various escape models. In practice the multiplicity of free parameters that occur in various propagation theories as well as their phenomenological formulation, have made it difficult to discriminate firmly between the various models. The situation is reviewed by Cesarsky (1980).

We now briefly consider cosmic ray acceleration. The extent to which acceleration and propagation are comingled is not known. In some models cosmic rays are accelerated in discrete sources and propagate thereafter with little change in energy. In other models acceleration and reacceleration occur at large in the interstellar medium throughout the cosmic ray lifetime.

Astrophysicists have linked supernovae with the origin of cosmic rays since Baade and Zwicky's original (1934) suggestion. A few percent of the available supernova energy (or $\sim 10^{51}$ few Gev particles per supernova) is required to replenish the cosmic ray pool. However two developments have drawn most theorists away from acceleration mechanisms that operate at close quarters to the supernova explosion itself. First, there is evidence that cosmic rays are not abundant in the r-process and ultra heavy nuclei that characterize supernova ejecta (Israel 1986 and references therein). Second, there is the problem of adiabatic decompression (Kulsrud and Zweibel 1975, Zweibel 1979). A larger number of cosmic rays created in a small supernova remnant or by a supernova explosion will rapidly generate Alfven waves through the streaming anisotropy mechanism. These waves will trap the particles in the expanding remnant so that the particles lose energy adiabatically. If a particle is trapped at radius R_i and freed at radius R_f its energy will decrease by a factor R_f/R_i, which is probably 10-100. Thus, supernova must put most of their energy into cosmic rays, and accelerate individual particles to very high energies.

Several authors (Axford, Leer, and Skadron 1977, Bell 1978, Blandford and Ostriker 1978) independently proposed a mechanism for cosmic ray acceleration that draws its energy from supernovae but avoids r-process enrichment and adiabatic decompression. This mechanism is first order Fermi acceleration in supernova shocks, and the basic process has found wide application to many other astrophysical problems.

In the Fermi process (Fermi 1949) a particle is reflected from moving magnetic fluctuations. A particle with momentum P reflected from a fluctuation with velocity $V \ll c$ changes its energy by the amount $-2P \cdot V$, which is positive for a head on collision and negative for an overtaking collision. If the velocity distribution of scatterers is isotropic, the head on and overtaking collisions nearly cancel, with only a slight statistical energy gain due to the greater frequency of head on collisions. This statistical process is second order Fermi acceleration.

In the frame of a shock wave with speed U_s and compression ratio R, incoming (upstream) fluid has speed U_s and outgoing (downstream) fluid has speed U_s/R. Suppose there are upstream and downstream fluctuations which are advected with the flow. A particle which reflects (head on) from a fluctuation upstream, crosses the shock, and is reflected by a downstream fluctuation (overtaking) will gain a net energy of $2(P \cdot U_s)(1 - 1/R)$. This situation is first order Fermi acceleration.

Bell (1978) gave a simple argument for the energetic particle spectrum resulting from this process (Blandford and Ostriker (1978) made similar physical assumptions but followed a different method). Particles are injected at some energy E_o and bounce back and forth across the shock, always with some probability of escape downstream. Frequent collisions keep the particle distribution isotropic. If the probability that a particle escapes after ℓ loops is $P(\ell)$ and the energy after ℓ loops is $E(\ell)$ (for large ℓ the $E(\ell)$ essentially form a continuum) then the particle distribution function $f(E(\ell))$ is

$$f\left(E\left(\ell\right)\right) = P(\ell) \left(\tfrac{dE(\ell)}{d\ell}\right)^{-1}. \tag{37}$$

Assuming that the particles are nearly isotropic at the shock front and that escaping particles are advected downstream with the fluid leads to an expression for the escape probability

$$P(\ell) = \left(1 - 4U_s/Rv\right)^\ell \approx 1 - \tfrac{4\ell U_s}{Rv} \tag{38}$$

where $v \sim c \gg U_s$ is the particle speed. Again, for an isotropic distribution of particles

$$E(\ell) \sim E_o \exp\left(\tfrac{4}{3}\tfrac{\ell U_s}{c}\left(1 - \tfrac{1}{R}\right)\right) \tag{39}$$

and from equations (38) and (39) it follows that

$$P(\ell) \sim (E(\ell)/E_o)^{-3/(R-1)}$$

Using equation (37) we then have

$$f(E) \sim E^{-(R+2)/(R-1)} \tag{40}$$

The limiting value of R for a strong shock in a $\gamma = 5/3$ gas is $R = 4$, giving an E^{-2} spectrum. To produce an $E^{-2.6}$ spectrum, R must be 2.9.

There have been many refinements and embellishments of this mechanism since the original papers. Various authors have considered the effect of a smooth instead of a discontinuous shock profile, time dependent effects, shock curvature, rigorous calculation of the fluctuations spectrum or diffusion coefficient, and modification of the shock structure itself by cosmic ray pressure. Many of the references can be found in the reviews by Axford (1981) and Eichler (1986), in the volume edited by Koch-Miramond and Lee (1984), and in International Cosmic Ray Conference Proceedings.

Blandford and Ostriker (1980) used the basic shock acceleration model to develop a theory of cosmic ray acceleration and propagation in the galaxy. They assume that the filling factor of low density coronal gas is very large (McKee and Ostriker 1977) so that supernova remnants expand until they overlap. Most of the cosmic ray acceleration is effected by shock waves with a large surface area and cosmic rays are likely to encounter many shocks before leaving the galaxy. A consequence of this model (or any model with protracted acceleration in the interstellar medium) is that the most energetic cosmic rays tend to be the oldest. Thus the fraction of secondaries tends to be flat (or even increasing) with energy. This now appears to be at variance with measured secondary energy spectra (Cesarsky 1980, Jones 1985). Simon, Heinrich, and Mathis (1986) have recently proposed a composite model which attempts to circumvent this difficulty.

V. Conclusions

In this article we have discussed the role of the magnetic field in the large scale structure and dynamics of the interstellar medium and we have reviewed some current ideas about the origin and maintenance of the field.

Probably the most important single question is whether the galactic field is primordial, and connected to an intergalactic field, or whether it is generated wholly within the galaxy from ancient seed flux, independently of any intergalactic field. In the former case we expect a bisymmetric spiral, and cosmic rays can naturally escape the galaxy by diffusing along magnetic fieldlines. In the latter case the field is amplified by a dynamo and destroyed by turbulent resistivity and buoyancy losses. Cosmic rays escape in bubbles of inflated field.

Some major theoretical sub-problems in this area are the following. First, the theory of buoyant escape of the galactic field should be reconsidered for models of the interstellar medium which contain large volumes of hot gas and which include a fat disk or halo. Second, most theories of galactic field evolution invoke a large turbulent resistivity, but there is insufficient theoretical basis for this and the problem needs more work. Third, dynamo theories should be extended from the kinematic to the dynamic regime. This extension has had enormous consequences for the theory of the solar dynamo. Fourth, the windup problem for bisymmetric spiral fields needs more attention, although Kulsrud's model may be the first step. Finally, an efficient dynamo at some stage seems essential to amplify the very weak seed fields created by plasma processes.

Observations in three areas are especially crucial. It is difficult to map the large scale morphology of the field in our galaxy, so it is perhaps more practical to rely on observations of other galaxies. Such observations are needed to determine whether the fields are bisymmetric or primarily azimuthal. In addition to these large scale observations, more data is needed about field and cosmic rays in the halo. This data is especially important for calculations of buoyant escape. Finally, theoretical and computational work suggest that if turbulent resistivity and a turbulent dynamo are operative, the field is concentrated into thin sheets or ropes. The observational limits on this structure for the field would be very interesting to explore.

Aside from questions dealing with the origin of the field, there are still unsolved problems relating to cloud dynamics in the presence of a field, and to the dissipation of cloud energy by dissipation of magnetic disturbances. The heating that accompanies such dissipation should also be taken into account.

Acknowledgements: Support from NASA Grant NAG W-91 to the University of Colorado and comments by Don Cox are gratefully acknowledged.

References

Arons, J. and Blandford, R. D. 1983, *Phys. Rev. Lett.* **50**, 544.

Asseo, E., Cesarsky, C. J., Lachièze-Rey, M., and Pellat, R. 1978, *Ap. J.* **225**, L21.

Asseo, E., Cesarsky, C. J., Lachièze-Rey, M., and Pellat, R. 1980, *Ap. J.* **237**, 752.

Axford, W. I. 1981, *17th International Cosmic Ray Conference Proc.* **v 12**, p. 155.

Axford, W. I., Leer, E., and Skadron, G. 1977, *15th International Cosmic Ray Conference Proc.* **v 11**, p. 132.

Baade, W. and Zwicky, F. 1934, *Phys. Rev.* **46**, 76.

Bell, A. R. 1978, *M.N.R.A.S.* **182**, 147.

Biermann, L. 1950, *Z. Naturforsch* **5a**, 65.

Black, D. C. and Scott, E. H. 1982, *Ap. J.* **263**, 696.

Blandford, R. D. and Ostriker, J. P. 1978, *Ap. J.* **221**, 129.

Blandford, R. D. and Ostriker, J. P. 1980, *Ap. J.* **237**, 793.

Cabrera, B. 1982, *Phys. Rev. Lett.* **48**, 1378.

Cesarsky, C. J. 1980, *Ann. Rev. Astron. Astrophys.* **18**, 289.

Cesarsky, C. J. and Kulsrud, R. M. 1973, *Ap. J.* **185**, 153.

Chernoff, D., Shapiro, S. L. and Wasserman, I. 1985, preprint.

Childress, S. 1979, *Phys. Earth. Planet. Int.* **20**, 172.

Chin, Y. and Wentzel, D. G. 1972, *Astrophys. Sp. Sci.* **16**, 465.

Clifford, P. and Elmegreen, B. G. 1983, *M.N.R.A.S.* **202**, 629.

Cowling, T. G. 1976, *Magnetohydrodynamics* (Hilger, London).

Draine, B. T., Roberge, W. G. and Dalgarno, A. 1983, *Ap. J.* **264**, 485.

Eichler, D. 1986, in *Twelfth Texas Symposium on Relativistic Astrophysics*, ed. M. Livio and G. Shaviv, Ann. N.Y. Acad. Sci., **v 470**, p. 205.

Elmegreen, B. G. 1981, *Ap. J.* **243**, 512.

Elmegreen, B. G. 1982a, *Ap. J.* **253**, 634.

Elmegreen, B. G. 1982b, *Ap. J.* **253**, 655.

Elmegreen, B. G. 1985, *Ap. J.* **299**, 196.

Falgarone, E., Puget, J. L. and Pérault, M. 1986, to be published in *Astronomy and Astrophysics*.

Farouki, R. T., Shaprio, S. L. and Wasserman, I. 1984, *Ap. J.* **284**, 282.

Fermi, E. 1949, *Phys. Rev.* **75**, 1169.

Flewelling, R. F. and Coroniti, F. V. 1976, *Ap. J.* **205**, L135.

Garcia-Munoz, M., Mason, G. M., and Simpson, J. A. 1977, *Ap. J.* **217**, 859.

Gillis, J., Mestel, L., and Paris, R. B. 1979, *M.N.R.A.S.* **187**, 311.

Gilman, P. 1983, *Ap. J. Supp.* **53**, 243.

Harrison, E. R. 1970, *M.N.R.A.S.* **147**, 279.

Harrison, E. R. 1973a, *M.N.R.A.S.* **165**, 185.

Harrison, E. R. 1973b, *Phys. Rev. Lett.* **30**, 188.

Hartquist, T. W. and Morfill, G. 1984, *Ap. J.* **287**, 194.

Hillas, A. M. 1975, *Phys. Rep.* **2**, 59.

Ipavich, F. M. 1975, *Ap. J.* **196**, 107.

Israel, M. H. 1986, in *Twelfth Texas Symposium on Relativistic Astrophysics*, ed. M. Livio and G. Shaviv, Ann. N.Y. Acad. Sci., **V 470**, p. 188.

Jokipii, J. R. 1976, *Ap. J.* **208**, 900.

Jokipii, J. R. and Parker, E. N. 1969, *Ap. J.* **155**, 799.

Jones, M. D. 1985, *Abundances of Heavy Cosmic Rays at Energies up to 1000 Gev/amu*, unpublished PhD thesis, Washington University.

Koch-Miramond, L. and Lee. M. A. 1984, *Advance in Space Res.* **4**, nos. 2, 3.

Konigl, A. 1986, preprint.
Kulsrud, R. M. 1986, preprint.
Kulsrud, R. M. and Cesarsky, C. J. 1971, *Astrophys. Lett.* **8**, 189.
Kulsrud, R. M. and Pearce, W. P. 1969, *Ap. J.* **156**, 445.
Kulsrud, R. M. and Zweibel, E. G., *14th International Cosmic Ray Conference Proc.* **v 2**, 465.
Lachièze-Rey, M., Asseo, E., Cesarsky, C. J., and Pellat, R. 1980, *Ap. J.* **238**, 175.
Lee, L. C. and Jokipii, J. R. 1976, *Ap. J.* **206**, 735.
Lee, M. A. and Völk, H. J. 1973, *Astrophys. Sp. Sci.* **24**, 31.
Manchester, R. N. 1974, *Ap. J.* **188**, 637.
McKee, C. F. and Ostriker, J. P. 1977 *Ap. J.* **218**, 148.
Mestel, L. and Paris, R. B. 1979, *M.N.R.A.S.* **187**, 337.
Mestel, L. and Paris, R. B. 1984, *Astr. Ap.* **136**, 98,.
Mestel, L. and Spitzer, L. 1956, *M.N.R.A.S.* **116**, 503.
Mishustin, I. N. and Ruzmaikin, A. A. 1972, *Sov. Phys. JETP* **34**, 233.
Moffatt, H. K. 1978, *Magnetic Field Generation in Electrically Conducting Fluids* (Cambridge, London).
Mouschovias, T. Ch. 1974, *Ap. J.* **192**, 37.
Mouschovias, T. Ch. 1985, *Astr. Ap.* **142**, 41.
Mouschovias, T. Ch. and Morton, S. A. 1985a, *Ap. J.* **298**, 190.
Mouschovias, T. Ch. and Morton, S. A. 1985b, *Ap. J.* **298**, 205.
Mouschovias, T. Ch. and Paleologou, E. V. 1979, *Ap. J.* **230**, 204.
Mouschovias, T. Ch. and Paleologou. E. V. 1980, *Ap. J.* **237**, 877.
Mouschovias, T. Ch. and Paleologou, E. V. 1981, *Ap. J.* **246**, 48.
Mouschovias, T. Ch., Shu, F. H., and Woodward, P. R. 1974, *Astron. Ap.* **33**, 73.
Nakano, T. 1981, *Prog. Theoret. Phys. Supp.* No. 70, 54.
Newcomb, W. A. 1961, *Phys. Fl.* **4**, 391.
Parker, E. N. 1955, *Ap. J.* **122**, 293.
Parker, E. N. 1966, *Ap. J.* **145**, 811.
Parker, E. N. 1970, *Ap. J.* **160**, 383.
Parker, E. N. 1971, *Ap. J.* **163**, 255.
Parker, E. N. 1979, *Cosmical Magnetic Fields*, (Clarendon, Oxford).
Perkins, F. W. and Zweibel, E. G. 1986, submitted to *Phys. Fl.*.
Peters, B. and Westergaard, N. J. 1977, *Astrophys. Sp. Sci.* **48**, 21.
Piddington, J. H. 1964, *M.N.R.A.S.* **128**, 345.
Piddington, J. H. 1978, *Astr. Sp. Sci.* **59**, 237.
Rasmussen, I. L. and Peters, B. 1975, *Nature* **258**, 412.
Rockstroh, J. M. and Webber, W. R. 1978, *Ap. J.* **244**, 677.
Sawa, T. and Fujimoto, M. 1980, *Prog. Ast. Soc. Jap.* **32**, 551.
Scalo, J. M. 1977, *Ap. J.* **213**, 705.
Scalo, J. M. and Pumphrey, W. A. 1982, *Ap. J.* **258**, 229.
Shu, F. H. 1974, *Astr. Ap.* **33**, 55.
Shu, F. H. 1983, *Ap. J.* **273**, 202.
Simon, M., Heinrich, W. and Mathis, K. D. 1986, *Ap. J.* **300**, 32.
Simpson, J. A. 1983, in *Ann. Rev. Nuc. Part. Sci.* **33**.
Sofue, Y., Fujimoto, M., and Wielebinski, R. 1986 *Ann. Rev. Astr. Astrophys.*, in press.
Spitzer, L. 1962, *Physics of Fully Ionized Gases* (Wiley, New York).
Spitzer, L. 1978, *Physical Processes in the Interstellar Medium* (Wiley, New York).
Steenbeck, M., Krause, F., and Rädler, K. H. 1966, *Z. Naturforsch* **21a**, 369.
Stix, M. 1975, *Astr. Ap.* **42**, 85.

Tang, W. M. 1978, *Nuclear Fusion* **18**, 1089.

Turner, M., Parker, E. N. and Bogdan, T. J. 1982, *Phys. Rev.* **D26**, 1296.

Wentzel, D. G. 1974, *Ann. Rev. Astron. Astrophys.* **12**, 71.

Witten, E. 1985, *Nuc. Phys.* B. **249**, 557.

Zeldovich, Ya. B., Ruzmaikin, A. A., and Sokoloff, D. D. 1983, *Magnetic Fields in Astrophysics*, (Gordon Breach, New York).

Zweibel, E. G. 1979, in *Particle Acceleration Mechanisms in Astrophysics*, ed. J. Arons, C. Max, and C. McKee (New York, AIP), p. 319.

Zweibel, E. G. and Josafatsson, K. 1983, *Ap. J.* **270**, 511.

Zweibel, E. G. and Kulsrud, R. M. 1975, *Ap. J.* **201**, 63.

Zweibel, E. G. and Shull, J. M. 1982, *Ap. J.* **259**, 859.

Ellen Zweibel

Section IV: Interstellar Processes on a Galactic Scale

Richard McCray

PANEL DISCUSSION: PHASES OF THE INTERSTELLAR MEDIUM

J. Michael Shull
Center for Astrophysics and Space Astronomy, and
Joint Institute for Laboratory Astrophysics
University of Colorado and National Bureau of Standards
Boulder, CO 80309 (USA)

ABSTRACT. This article summarizes the panel discussion on "Phases of the Interstellar Medium". The panelists were asked to identify key issues related to the structure of the ISM, and the audience participated in a two-hour discussion. Beginning with a summmary of the physics of thermal phases, I provide some background on the observations and theories of 2-phase and 3-phase models. While the 3-phase model has had many successes, several recent observations disagree with its predictions. The major unresolved issue is whether the model can be "fixed" by tinkering with cloud geometries and supernova rates and by including a galactic fountain. The Milky Way supernova rate may have been overestimated, many supernovae may be less than 10^{51} ergs, and modelers have probably neglected important physics in mass transport, cloud formation and halo input.

1 INTRODUCTION

Since fireworks were forecasted, it was appropriate that this panel discussion occurred on July 4, 1986. The panelists were Leo Blitz (Maryland), Carl Heiles (Berkeley), Dick McCray (Colorado), Chris McKee (Berkeley), and Don York (Chicago), with Mike Shull (Colorado) as moderator. The discussion began with an introduction to the concept of "thermal phases", followed by a 5-10 minute presentation by each of the panelists, and finally a full discussion involving the panelists and the audience.

In this summary, I will attempt to present the issues raised and give some flavor of the discussion. In section 2, I provide some background on the physics behind thermal equilibrium, thermal instability, and phases of the interstellar medium (ISM). In section 3, I describe the major points raised by the panelists. Finally, in section 4, I will attempt a synthesis of the conclusions reached and the questions that still elude a consensus.

D. J. Hollenbach and H. A. Thronson, Jr. (eds.), Interstellar Processes, 225–244.
© *1987 by D. Reidel Publishing Company.*

2 THERMAL PHASES OF INTERSTELLAR MATTER

2.1 Thermal Equilibrium and Thermal Instability

Interstellar gas may be in a variety of phases depending on its local heating, ionization, and, in some cases, its past history. Field, Goldsmith, and Habing (1969), or FGH, showed that a model for the ISM, heated and ionized by cosmic rays, would have two stable phases: cold clouds (10^2 K) and warm intercloud matter (10^4 K). For selected values of the heating or ionization rate, these two phases can coexist in pressure equilibrium. In this subsection, I summarize the physics of thermal equilibrium and stability of interstellar gas, based on the pioneering work of Field (1965).

The basis for the two-phase model can be understood as follows. Let the gas have hydrogen density n, with a cooling rate per volume $n^2 \Lambda(T)$ and a heating rate $n\mathcal{H}$. In the case of cosmic ray heating, $\mathcal{H}$ is proportional to the primary ionization rate ς (s^{-1}). Thermal equilibrium requires that $\Lambda(T) \propto (\mathcal{H}/n) \propto (\varsigma/n)$. The locus of equilibria is thus determined by the shape of the cooling curve $\Lambda(T)$. As n increases at fixed zeta, both $\Lambda(T)$ and T must decrease. Above some critical density, T drops precipitously from ~8000 K (Lyα cooling dominant) to < 300 K (trace metal cooling dominant). Thus, the pressure curve (nT) possesses a kink at the transition from warm to cold phases. Because of the quadratic dependence of the cooling rate, if a small region becomes colder than its neighborhood, it will be compressed and cool more rapidly, leading to a condensational thermal instability.

A detailed analysis of thermal stability was given by Field (1965) and, more recently, by Balbus (1986). Consider a gas in thermal and ionization equilibrium at constant pressure nT, with optically thin heating and cooling, mass density $\rho = \rho_0$, hydrogen density n (cm^{-3}) and temperature $T = T_0$. The equation of thermal balance may be written as $\mathcal{L}(\rho_0, T_0) = 0$, in terms of a generalized loss function $\mathcal{L}(\rho, T)$ equal to the cooling rate minus heating rate per unit mass (ergs g^{-1} s^{-1}). This equation determines the equilibrium temperature T_0, for any value of ρ_0. Stability is tested for by introducing a perturbation in density and temperature such that some thermodynamic variable A (e.g., pressure) is held constant. The gas will be thermally unstable if the change in $\mathcal{L}$ has a sign opposite to the change in entropy S; that is, if

$$\left(\frac{\partial \mathcal{L}}{\partial S}\right)_A < 0. \tag{1}$$

[Note that the sign of the inequality is mistakenly reversed in Field's original paper]. If we assume a perfect gas, the instability criteria for isochoric and isobaric perturbations become:

$$\left(\frac{\partial \mathcal{L}}{\partial T}\right)_\rho < 0 \quad ; \quad \text{isochoric} \tag{2}$$

$$\left(\frac{\partial \mathcal{L}}{\partial T}\right)_P = \left(\frac{\partial \mathcal{L}}{\partial T}\right)_\rho - \left(\frac{\rho_0}{T_0}\right)\left(\frac{\partial \mathcal{L}}{\partial \rho}\right)_T < 0 \quad ; \quad \text{isobaric.} \tag{3}$$

These criteria refer to perturbations away from equilibrium ($\mathcal{L} = 0$). Balbus (1986) has shown that the corresponding criterion for a dynamic thermal instability (*e.g.*, perturbations to a cooling flow or a post-shock flow) is $[\partial(\mathcal{L}/T)/\partial S]_A < 0$.

What temperature regions are prone to instability acoording to these criteria ? This depends on the specific forms of heating and cooling. In certain instances, such as cosmic ray heating and optically thin, low-density radiative cooling, $\mathcal{L}$ may be written in the form $\mathcal{L} = n^2 \Lambda(T) - n\mathcal{H}$. Thermal equilibrium ($\mathcal{L} = 0$) at constant pressure requires that $\Lambda(T)/T = (\mathcal{H}/nT)$. Thus, the equilibrium temperature T is a function of the single parameter $(\mathcal{H}/nT)$, and may be found once the cooling curve $\Lambda(T)$ is known. For a constant heating rate $\mathcal{H}$, the Field instability criterion at constant pressure, $(\partial \mathcal{L}/\partial T)_P < 0$, implies instability for temperature regimes in which the cooling function is not steeply rising, $d(\ln \Lambda)/d(\ln T) < 1$.

Figure 1 shows a schematic radiative cooling curve for interstellar gas of cosmic abundances in coronal equilibrium (Cox, Raymond, and Smith 1976; Gaetz and Salpeter 1983), together with the regions prone to thermal instability ($nT = constant$). The salient features of this cooling function $\Lambda(T)$ are: (1) steeply rising portions below 100 K and between about 10^4 and 10^5 K; (2) a relatively flat portion from 10^2 to 10^4 K; (3) a falling portion for $5.3 < \log T < 7.0$; and (4) a high temperature portion, with $\Lambda \propto T^{1/2}$, due to thermal bremsstrahlung. The shape of $\Lambda(T)$ results from the efficiency of radiative cooling of the dominant species at each temperature: excitation of the [C II] infrared fine structure line at low temperature; excitation of [C II] and optical forbidden lines at moderate temperatures; H I (Lyα) excitation above about 7000 K; and excitation of permitted and semi-forbidden lines of ions of C, O, and Fe above 20,000 K. For temperatures between about 10^6 and 10^8 K, the cooling is dominated by iron ions, which are the only abundant species which retain their bound electrons at such temperatures.

Figure 2 shows the equilibria for a constant heating rate $\mathcal{H}$ and the standard interstellar cooling function $\Lambda(T)$. The function $\Lambda(T)/T$, which equals $(\mathcal{H}/nT)$, plotted versus T represents the locus of equilibria for which heating equals cooling ($\mathcal{L} = 0$). Above the curve, the gas is net heating ($\mathcal{L} < 0$); below the gas is net cooling ($\mathcal{L} > 0$). Equilibrium solutions may be found where the curve $\Lambda(T)/T$ is intersected by a horizontal line corresponding to the ratio $(\mathcal{H}/nT)$. Four solutions are shown. Two are thermally stable according to the Field criterion, $(\partial \mathcal{L}/\partial T)_P > 0$, and we identify them with the cloud and intercloud phases. The other two equilibria, on the downward sloping portions of the curve, are thermally unstable. Small perturbations away from these unstable equilibria tend to heat up to the warm intercloud phase or cool down to the cold (cloud) phase, as indicated in the diagram. Note that the high temperature gas is thermally unstable in this model; even thermal bremsstahlung with $\Lambda \propto T^{1/2}$ does not have a sufficiently steep cooling function to stabilize isobaric perturbations. The hot gas observed in the interstellar medium may be stabilized by other heat loss mechanisms, such as thermal conduction or inverse Compton cooling (in X-ray photoionized nebulae), or it must exist in a transient state, with an age less than the cooling time of the gas.

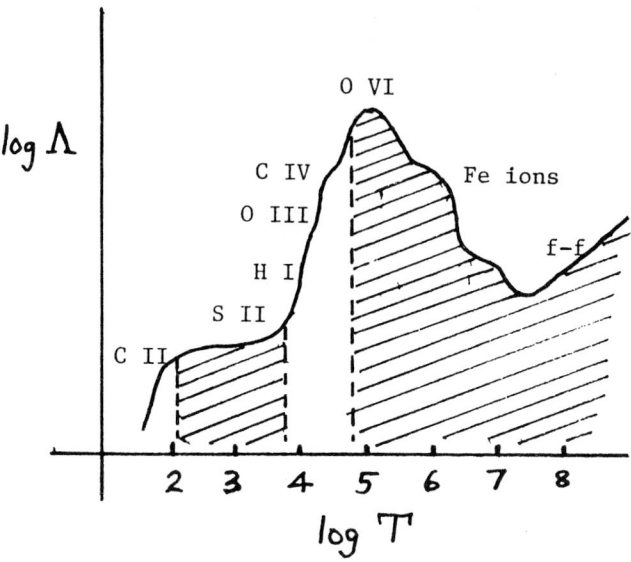

Figure 1: Radiative cooling curve for optically thin, low density interstellar gas. Regions prone to thermal instability are cross-hatched. Dominant coolants are shown.

2.2 The Two-Phase Model

Because much of the current debate centers around the question of phases, clouds, and intercloud medium, it is worthwhile defining some terms. The term "cloud" has come to represent any interstellar absorption (or 21-cm emission) component at fixed velocity. The term "thermal phases" generally refers to stable distributions of interstellar gas at radically different temperatures and densities. For example, the ISM is believed to exist in "cold clouds" $(T < 100K)$ in approximate pressure equilibrium with a lower density warm "intercloud medium" $(T \approx 10^4 K)$ and hot "coronal" gas at temperature $T > 10^6 K$. Each of these components forms a "phase", which is thermally stable or at least dynamically and thermally long-lived. The interstellar pressure, nT, averages between 10^3 and 10^4 cm^{-3} K in each phase.

Observing the true intercloud medium may be a problem in semantics, since any detectable absorption could be interpreted as a cloud. In some instances, the density of intercloud gas may be too low to produce detectable absorption in optical or UV resonance lines. In practice, the ionization state is often used to separate cold clouds from a hotter substrate. For example, the *Copernicus* observations of broad O VI absorption lines were initially identified with hot $(3 \times 10^5$ to $10^6 K)$ "coronal" gas in the disk, although it now appears more likely that the O VI arises in conductive interfaces between clouds and gas at somewhat hotter, X-ray emitting temperatures (Jenkins 1978a,b; McCray and Snow 1979).

Thus the intercloud medium remains somewhat of a theoretical construct. Spitzer (1956) predicted the existence of a hot intercloud medium in the Galactic halo, to confine

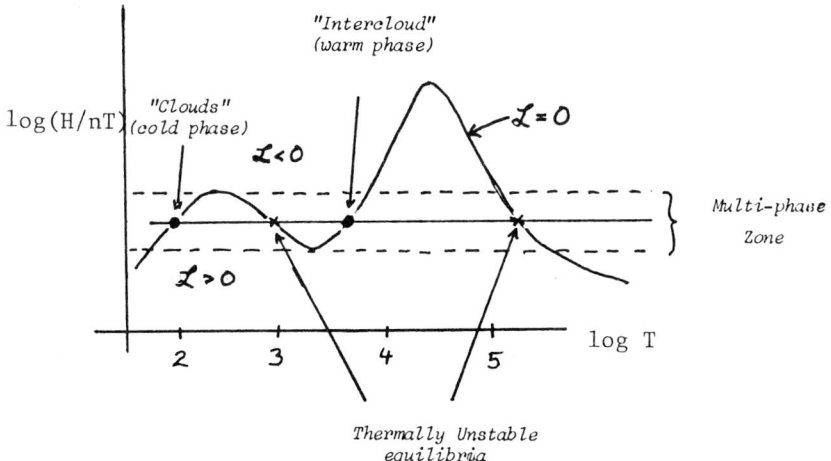

Figure 2: Locus of equilibria for interstellar gas with constant heating rate $\mathcal{H}$ and constant pressure nT. Two of the equilibria are thermally stable and two are unstable.

high-latitude Ca II clouds. A warm, partially ionized intercloud medium at 10^4 K was predicted by the thermal stability analysis of Field, Goldsmith, and Habing (1969). As we have seen, the tendency of interstellar gas to favor discrete ranges $(10^2, 10^4$ and 10^6 K) follows generally from the atomic properties of the coolants. In general, a stable phase is determined by the onset of a new cooling mechanism or the decline of a heating source. The heat source can be quite general, and suggested mechanisms include cosmic rays, impulsive bursts of X-rays, magnetic ion-slip, and photoelectric heating from dust grains. The most current investigations of two-phase models are by Draine (1978) and Shull and Woods (1985), who examined heating by cosmic rays as well as grain photoelectric emission. Shull and Woods (1985) also allowed for changes in metallicity (to treat galactic abundance gradients or gas in the Magellanic Clouds) by including a scaling factor D. This factor D equals 1 for cosmic abundances and standard depletion factors and may be increased or decreased appropriately.

Figure 3 shows the locus of equilibria of cosmic ray heated gas in the pressure-density plane, where $P = nT$ is the gas pressure, n_H is the hydrogen number density, D is the abundance scaling factor, and the cosmic ray ionization rate is fixed at $\varsigma = 10^{-17}$ s^{-1}. The onset of Lyα cooling at 8000 K produces a kink in the equilibrium curves, resulting in two thermally stable phases at about 100 and 7000 K. Gas at temperatures between these phases is thermally unstable. Note also that increasing the metallicity decreases the amplitude of the "kink"; that is, the range in pressure over which both cloud and intercloud phases can coexist is decreased when the metal abundance increases.

Figure 4 shows the cooling rates for various atomic species for one model. The exact temperature of the clouds depends on elemental abundances and the gas-phase depletion into grains of C, O, Si, and Fe, whose fine structure lines dominate the cooling below 1000

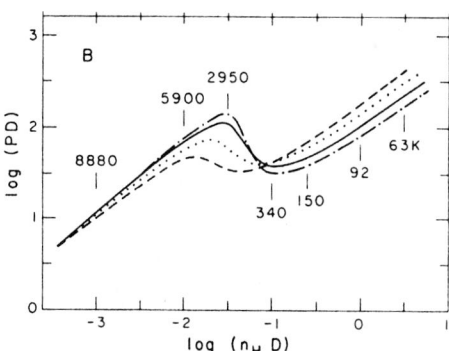

Figure 3: Locus of equilibria for interstellar gas heated and ionized by cosmic rays ($\varsigma = 10^{-17}$ s^{-1}) and photoelectric grain emission. The pressure $P = nT$ (cm^{-3} K), n_H is the hydrogen density (cm^{-3}), and D is the metallicity scaling factor. The curves correspond to $D = 0.5(\cdot - \cdot), D = 1(\text{———}), D = 2.5(\cdots)$, and $D = 5$ (- - -). Labeled temperatures refer to the $D = 1$ model.

K. A simple analysis (Shull and Woods 1985) shows that curves of different cosmic ray ionization rate and metallicity lie along a similar locus, plotted as (PD/ς) versus (nD/ς).

Figure 3 showed two-phase models for gas heated by cosmic rays and grain photoelectric emission. These heating sources have rates linearly proportional to density. However, some sources of heating, such as magnetic ion slip or wave heating, may have an inverse dependence on density. This may alter the temperature range in which gas is thermally unstable in pressure equilibrium, but the general effect remains the same.

2.3   The Violent Interstellar Medium

The FGH model has several problems. First, even with some depletion of the dominant coolants into grains, it is difficult to explain the observed (40 to 100 K) temperatures of diffuse H I clouds with cosmic ray heating (Spitzer 1978). The required cosmic ray ionization rate ($\varsigma \approx 10^{-15}$ s^{-1}) disagrees with *Copernicus* UV absorption measurements of H$_2$ and HD, which imply a rate of about 10^{-17} s^{-1}. Second, the partially ionized intercloud medium predicted by FGH has not been observed, and is inconsistent with $\langle n_e \rangle$ determined from pulsar dispersion measures (Kulkarni and Heiles 1987). Finally, and probably most important, new observations and theories suggest that the structure of the ISM is dominated by the dynamics of supernova remnants (SNRs) and OB-star winds, and regulated by thermal conduction and spatial transport of gaseous shells in the disk and halo (Fig. 5). Thus, the ISM is far from quiescent, and the "violent interstellar medium" has led to consideration of a hot "third phase" at millions of degrees.

Cox and Smith (1974) first pointed out that the hot, low density interiors of SNRs would cool slowly, and suggested that the ISM would take millions of years to fill the

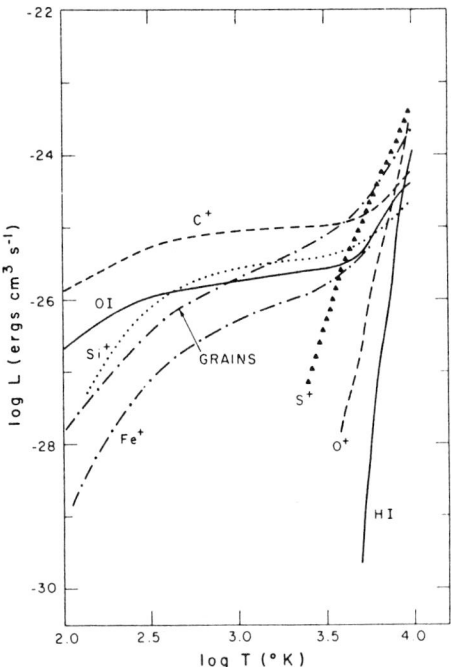

Figure 4: Cooling rate coefficients $\Lambda(T)$ versus temperature T for equilibrium models (Shull and Woods 1985), heated and ionized by cosmic rays ($\varsigma = 10^{-17}$ s^{-1}) and with $D = 5$. The 158 μm fine structure line of [C II] dominates the cooling up to about 4000 K, while optical forbidden [S II] lines and Lyα dominate the warm phase cooling.

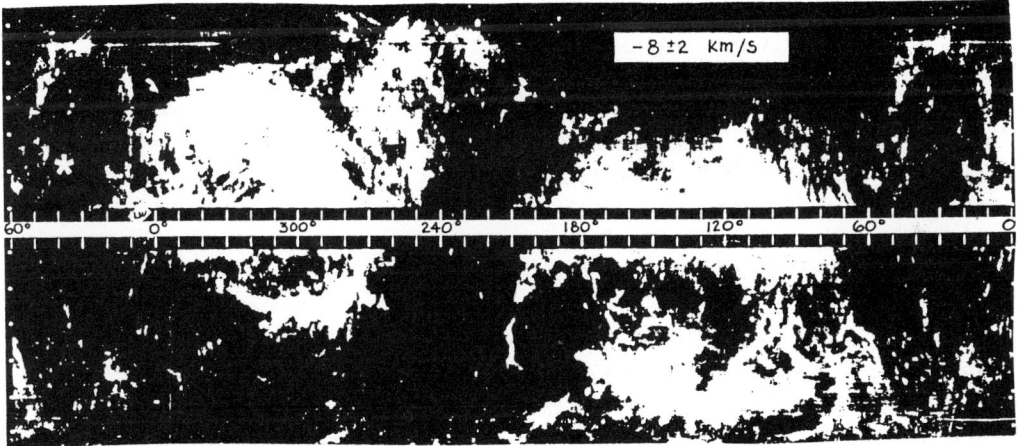

Figure 5: 21-cm survey of high-latitude gas from Colomb, Poppel, and Heiles (1980). Note the supershells of gas near $l = 300^\circ$ and 210°.

cavities. With a sufficiently high SN rate, r, per volume, a long SNR lifetime τ, and a large final remnant volume V_f, these cavities would overlap, connect, and form a tunnel network. They characterized this topology by the "porosity parameter" $Q = rV_f\tau$, which provides a measure of the fraction of total volume occupied by hot cavities. Both V_f and τ depend sensitively on the density of the substrate. If the medium is uniform and if Q remains constant over several growth times, τ/Q, then the ISM should establish a time-averaged fraction of coronal gas, $f = Q/(1+Q)$. However, the porosity theory is strictly valid only for $Q < 1$.

Using the approximate Sedov-Taylor similarity solution for SNR evolution one may show (Shull and Draine 1987, this volume) that radiative shells begin to form at a time, radius, and velocity,

$$t_{sh} = (1.91 \times 10^4 \text{ yr}) E_{51}^{3/14} n_0^{-4/7} \tag{4}$$

$$R_{sh} = (16.2 \text{ pc}) E_{51}^{2/7} n_0^{-3/7} \tag{5}$$

$$V_{sh} = (332 \text{ km s}^{-1}) E_{51}^{1/14} n_0^{1/7}, \tag{6}$$

where E_{51} is the supernova energy in units 10^{51} ergs and n_0 is the hydrogen density of the ambient medium in cm^{-3}. The subsequent evolution of a remnant in the radiative phase approximately follows the pressure modified "snowplow" solution, with $R_s \propto t^{2/7}$ and $V_s \propto t^{-5/7}$. By matching R_s and V_s at $t = t_{sh}$, one can solve for the time and radius at which the remnant slows to about 20 km s^{-1} and merges with the ISM:

$$t_f \approx (10^6 \text{ yr}) E_{51}^{11/35} n_0^{-13/35} \tag{7}$$

$$R_f \approx (50 \text{ pc}) E_{51}^{11/35} n_0^{-13/35}. \tag{8}$$

We can now demonstrate the sensitivity of the porosity parameter $Q = rV_f\tau$ to the density of the substrate in which the remnants evolve. Assume a standard supernova rate $r = (4 \times 10^{-5} \text{ kpc}^{-3} \text{ yr}^{-1}) N_{100}$, characteristic of one supernova per 100 years in a disk of effective area 830 kpc^2 and thickness 300 pc. If the supernovae are distributed randomly, with final volume V_f and lifetime τ corresponding to R_f and t_f above, we find,

$$Q \approx (0.02) N_{100} E_{51}^{44/35} n_0^{-52/35}. \tag{9}$$

Thus, if the substrate had a density $n_0 \approx 1$ cm^{-3} comparable to the mean density of the ISM, the porosity would be small and the remnants would not overlap. If, however, much of the ISM mass is contained in dense clouds occupying a small volume and if the intercloud density is less than 0.1 cm^{-3}, then $Q > 1$ and remnant overlap occurs frequently. The transition from isolated remnants to overlapping cavities happens quickly as the intercloud density drops, and the change has all the characteristics of a phase transition.

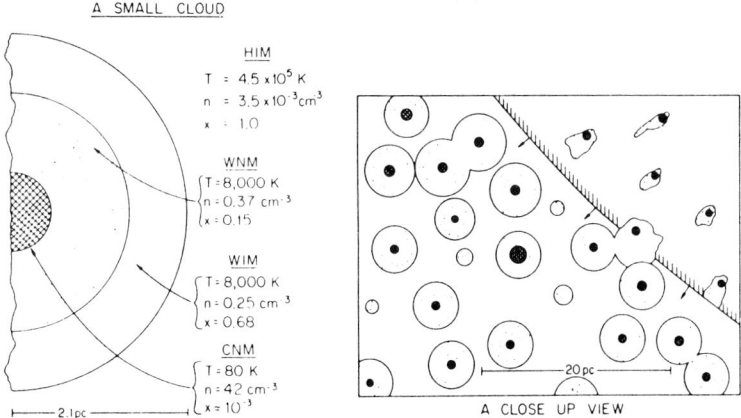

Figure 6: Schematic picture (MO 1977) of the four components of three-phase model plus their densities n, temperatures T, and ionization fractions x. Left panel shows a "standard" spherical cloud core (CNM) and its WNM and WIM envelopes. Right panel shows effect of SNR blast wave on cloud populations.

Increased supernova activity clears out cavities, lowers the substrate density, and results in larger remnant volumes.

McKee and Ostriker (1977, hereafter denoted MO) developed a theory for a SNR-dominated interstellar medium, in which the "third phase" of hot cavities occupies most of the volume. In the MO model, the H I mass is concentrated in clouds with a range of mass. The cores of these clouds are cold and neutral, but the cloud envelopes are warm (~ 8000 K) and partially ionized by conduction from the hot gas and by photoionization by OB stars, white dwarfs, and planetary nebulae. Evaporation is the dominant mass exchange from clouds to the hot gas, while new clouds are produced by the radiative shells of old SNRs. As illustrated in Fig. 6, the gas is distributed among four forms:

1. Hot Ionized Medium (HIM): The hot, low-density SNR cavities.

2. Warm Ionized Medium (WIM): The partially ionized cloud envelopes.

3. Warm Neutral Medium (WNM): The warm neutral cloud envelopes.

4. Cold Neutral Medium (CNM): The cold cloud cores.

The three cloud components (CNM, WNM, WIM) have volume filling factors $f_{CNM} \approx 0.025, f_{WIM} \approx 0.23, f_{WNM} \approx 0.15$, and therefore $f_{HIM} = 1 - f(clouds) \approx 0.6$. [MO chose a disk supernova rate equivalent, in our units, to $r = 10^{-4}$ kpc^{-3} yr^{-1}, or one per 40 years in the disk]. The cloud envelopes occupy more volume than the cloud cores, but contain far less mass (the volume filling factor of the WIM and WNM cloud envelopes is 38%, but they contain only 4% of the mass). Note that the filling factor of hot gas in this model, $f_{HIM} = 1 - f(clouds)$, which is not necessarily equal to the fraction predicted by porosity theory, $Q/(1 + Q)$ when Q becomes large.

Successful predictions of the MO theory include: (1) the interstellar pressure, $nT \approx$ 3700 cm^{-3} K; (2) the electron densities from the WIM cloud envelopes, $\langle n_e \rangle = 0.04$ cm^{-3} and $\langle n_e^2 \rangle^{1/2} = 0.08$ cm^{-3}, which are near those found from pulsar dispersion measures; (3) the observed cloud velocity dispersion, $\langle V_{cl} \rangle \approx 8$ km s^{-1}; and (4) the soft X-ray intensities in the local bubble. The theory also provides a framework for examining (and perhaps understanding) O VI absorption, cosmic ray propagation and acceleration by shocks, grain survival, and cloud size spectrum. In these repects, the three-phase model has been a most successful theory.

However, several aspects of this "three-phase model" have been challenged by observations. Lockman, Hobbs, and Shull (1986) find that ratios of Lyα and 21-cm column densities do not show the expected fluctuations if clouds occupy such a small volume fraction. Kulkarni and Heiles (1987) point out that the fraction of warm, "not strongly absorbing" H I (Payne, Salpeter, and Terzian 1983) is larger than that predicted by the MO model. Heiles (1987) claims that the 21-cm distribution of H I appears too smooth for a three-phase medium with pervasive hot gas. He also notes that if one adopts standard supernova rates for the disk (Pop I, Type II supernovae) and halo (Pop II, Type I supernovae) one finds $Q \gg 1$, violating the assumption of porosity theory. Clearly, before we make a judgment about the validity of the three-phase model, we must understand a few new effects: SNR shell overlap, "venting" of hot SNR interior gas and radiative shells to the galactic halo, and the accuracy of estimates for supernova rates in the disk and halo.

A recent attempt to study some of these new effects was made by Cioffi (1985). His numerical simulations of overlapping SNRs in disk galaxies (Fig. 7) included the effects of spatial correlations (supernovae tend to form in OB associations), transport of matter by SNR shells into pre-existing cavities, and injection of mass to the halo. The first two effects tend to smooth out the H I distribution, and perhaps can mitigate the disagreement with the 21-cm data, particularly if beam dilution by small clouds is important. However, other 21-cm observations (Kulkarni and Heiles 1986) argue for substantial amounts of smoothly distributed warm H I, which may require changes in the volume fractions of the three phases.

3   THE PANEL DISCUSSION

3.1   Summary of Issues by the Panelists

The panel was charged with addressing several "large questions", in addition to technical points. These questions fall in five categories:

1. Problems with the Three-Phase Model

2. Galactic Ecology

3. Structure of Clouds

4. Structure of the Halo

5. Starburst Galaxies and the ISM

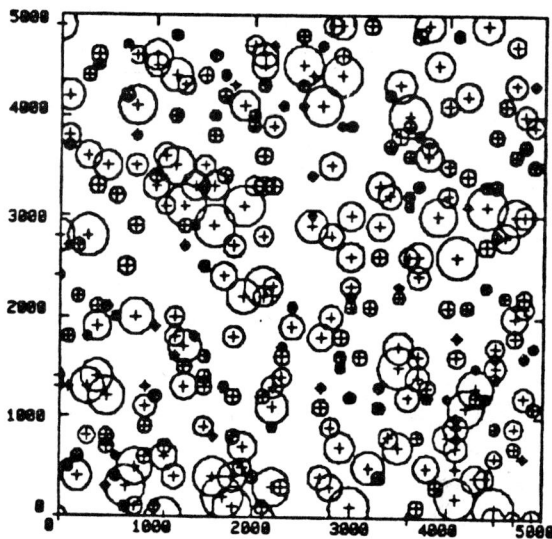

Figure 7: Model of overlapping SNRs for (5 kpc) × (5 kpc) region of galactic disk, based on uncorrelated supernovae in an ISM with variable substrate densities (Cioffi 1985).

The first four questions are obviously interrelated, and most of the discussion revolved around these issues. To provide a focus for some of the discussion that follows, Figure 8 illustrates the processes that exchange mass and energy between stars and interstellar gas.

In addition to these "large issues", each panelist was asked to list several key problems that need to be discussed and resolved in the future. Their main points, paraphrased from notes, are given here:

Leo Blitz

In the cycles shown in Figure 8, probably the least is known observationally or theoretically about the link:

$$(\text{Diffuse Clouds}) \implies (\text{Molecular Clouds}) \implies (\text{Star Formation})$$

We know that gravity ultimately plays a role in the last step, but what happens earlier ? How are clouds assembled out of the low density gas produced by stellar winds, supernovae, and planetary nebulae ? What are the effects of magnetic fields, turbulent velocity fields, and dynamic interactions with spiral density waves and interstellar blast waves ? What is the significance of the observation that many molecular cloud complexes contain a hierarchy of sub-structure which occupies only a fraction of the total volume?

Carl Heiles

The three-phase theory relied on the fact that the supernova rate was sufficiently high that SNRs overlap to form cavities of hot interstellar gas. However, there are two major objections to this model: (1) the distribution of H I in our galaxy and nearby spirals

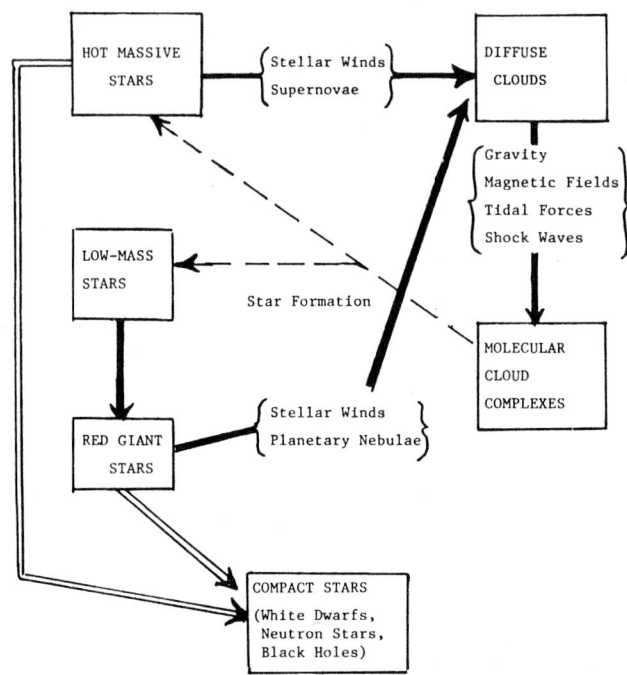

Figure 8: Schematic diagram of "Galactic Ecology", showing interchange of gas between stars and the ISM.

appears much smoother than predicted; and (2) SNR cavities appear to fill much smaller volumes than predicted by estimates of the porosity parameter, $Q = rV_f\tau$.

Heiles (1987) has estimated the supernova rate and porosity in the Milky Way, taking into account the fact that there are two types of supernovae, Type I (Pop II) and Type II (Pop I), distributed with galactocentric radius R and vertical height z. He finds that the inner Milky Way disk ($R < 3$ kpc) should be dominated by old Type I SNRs; the ring of molecular clouds ($4 < R < 7$ kpc) should be dominated by young Type II SNRs; and the outer disk ($R > 12$ kpc) should be a 2-phase medium with few overlapping remnants ($Q \ll 1$). Although the solar vicinity is in the transition region between the two regimes, it is difficult to reconcile the observations with the theoretical models or the observational input parameters (SN rate, interstellar densities, SN energies, etc.)

Donald York

An emerging consensus is that both OB stars and KM-giants are important sources of mass input to the ISM. Jura (1987, this meeting) discussed evidence for the mass input and noted that, because more than 99% of the Fe and Ca atoms are locked up in grains while approximately 10% of the new mass input (from O star winds) comes dust free, some grains must grow *in situ* in the quiescent ISM.

The possibility of several sites of dust formation has significance for depletion studies, grain history in the ISM, and cosmology. If metal formation is accompanied by prompt grain formation, it will make the $z > 3$ universe opaque. If grains grow slowly ($\sim 2 \times 10^9$ yrs) we may still see out to $z > 3$. Given this uncertainty, it is important to be extremely careful in depletion studies; abundance determinations have subtle problems due to line saturation and high velocity components.

Studies of hot gas produced by SNR cavities need a good measurement of T_{max}. The O VI absorption probes $T \approx 3 \times 10^5$ K, the soft X-ray emission comes from $T \approx 10^6$ K, while the SNR interiors may reach 10^7 K. X-ray absorption lines from highly ionized states are just barely detectable with spectrometers planned for AXAF. Equally important is understanding the morphology of H I with respect to O VI. Is the O VI formed in conductive interfaces between cold H I and the hot substrate? What are the filling factors of each phase? Do ionization and pressure equilibrium apply ? Is some O VI associated with warm H I (5000 - 10,000 K) ?

Finally, there needs to be a sociological advance, to merge techniques of observation, each of which has its own problems. For example, 21-cm has no gauge of distance; UV and optical absorption studies suffer from line saturation and component problems; infrared data (IRAS) select hot spots and often measure a convolution of dust and UV flux.

Christopher McKee

The MO three-phase model has had notable successes: predicting the correct interstellar pressure, electron density, cloud velocity dispersion, and soft X-ray intensity. Subsequent applications of the theory have been made to shock acceleration of cosmic rays (Blandford and Ostriker 1980) and grain survival (Seab et al. 1985). A significant observation for the 3-phase theory is that the mean density of the ISM along some lines of sight is less than the minimum density for a cloudy phase in the 2-phase model. Observational uncertainties still remain, such as the O VI distribution, the clumpiness of the cloudy ISM, the filling factor of hot gas, and the variations of the ISM structure with galactocentric radius R, vertical height z, and from galaxy to galaxy. Outstanding theoretical questions are the effects of the Galactic magnetic field, thermal conduction, and cloud evaporation on SNR evolution and multi-phase equilibria.

There is no doubt that the structure of the ISM is affected by SNR evolution. The question raised here is whether the filling factor of hot gas is sufficiently high that SNR evolution is determined by a hot, low-density substrate (HIM) or smoothly distributed warm H I with moderate density, $n \approx 0.1$ cm^{-3}. The current rates and the MO theory suggest $f_{HIM} \approx 0.6$. Arguments for a high $f(hot)$ include: (1) O VI absorption observations require $f(hot) > 0.2$ (Jenkins 1978a,b); (2) cosmic ray shock acceleration theories suggest that $f(hot) > 0.5$ [although the observed ages and "grammage" of cosmic rays mean that $f(hot)$ cannot be as high as 1]; (3) grain survival in a shock dominated ISM suggests $f(hot) > 0.5$; and (4) observations of distributed warm H I suggest $f(hot) \approx 1 - f(warm) \approx 0.5$ (Kulkarni 1987, this volume). Recent observations (Lockman, Hobbs, and Shull 1986) are best explained by two components of H I, to match the low degree of fluctuations in H I columns toward halo B-stars. If the smoothly distributed component has a 500 pc exponential scale height, then $f(warm)n(warm) \approx 0.1$ cm^{-3} in the disk. Since $n(warm) \approx 0.3$ cm^{-3} for pressure equilibrium, the fraction of warm H I is $f(warm) \approx 0.3$, so that $f(hot) \approx 0.7$. There may be more WNM in the halo, since its lower pressure may lie below the critical level to induce a phase transition (see Fig. 3). All of these arguments are consistent with a filling factor $f(hot) \approx 0.4 - 0.7$.

Richard McCray

The first point, made by "guerrilla theatre", is that 21-cm astronomers and UV astronomers rarely speak the same language. The next point is that it seems apparent that, although the MO model has had many successes, it may need to be modified to account for new data from IRAS, IUE, and 21-cm radio observations.

For example, clouds are sheet-like, not round (IRAS data show this vividly). This is probably because shocks can't easily envelop clouds, and because pressure tends to flatten gas distributions, while gravity makes them round. The IUE and 21-cm observations (Kulkarni and Heiles 1987; Lockman, Hobbs, and Shull 1986), which suggest that the volume filling factors $f_{HIM} \approx f_{WNM} \approx 0.5$, are best understood if the H I is smoothly distributed in sheets rather than in spherical clouds.

One may also question supernova injection rates. Type I supernovae have a scale height of 300-400 pc, but 20% - 50% are of Type Ib and occur in the disk spiral arms (Branch 1986). Type II supernovae are correlated in space and time, since they occur in OB associations in groups of $N \approx 20 - 40$. Both of these effects will lower $f(hot)$. But, as Cowie (1987, this volume) suggests, the number of O VI absorption components requires a high rate of supernova energy injection. The solution to both constraints may be a new energy sink: a "galactic fountain" which vents matter and energy to the halo. In other words, the disk interstellar medium is not a closed system.

3.2   Discussion with Audience

The points described here are guaranteed to be incomplete, since they are an amalgam of notes taken during a lively two-hour discussion. I have tried to mention some of the high points, but I apologize if I have omitted important topics.

Verschuur began the discussion by noting that the physical processes in the ISM are complex and certainly non-linear. He questioned whether the system was in equilibrium, and asked whether there had been any work on the ISM as an example of Prigogine's non-linear systems. The answer appears to be no, although a number of authors have attempted to deal with the non-linearities through numerical modeling (Ikeuchi, Habe, and Tanaka 1984; Cioffi 1985).

Zweibel noted that if cosmic rays are accelerated in shocks, there should be a correlation between the age of a cosmic ray, as measured by spallation products, and its energy. This effect arises because to reach higher energy, cosmic rays must spend a longer time in the ISM and experience more shock acceleration. The HEAO-3 observations (Jones 1985) appear to contradict this effect. Another relevant point is that the mean density of the ISM determined by cosmic ray propagation is $\bar{n} \approx 0.1$ cm^{-3}, assuming a total path length ("grammage") of 5 g cm^{-2} and a spallation age of $t \approx 20$ million years (the corresponding column density $N_H = 2 \times 10^{24}$ cm$^{-2} \approx \bar{n}ct$). This probably means that cosmic rays penetrate the denser cold clouds as well as the hot, low density cavities. It also constrains the volume filling factor of hot low density gas. The calculation is uncertain, though, since it is based on the averages of skew distributions of grammages and spallation ages.

Seab raised the issue of grain survival. Current estimates of the input of mass to the ISM suggest that "sinks" outnumber "sources" of grains, and the population should disappear in about 10^9 years. This calculation bears on multi-phase models of the ISM, since shock propagation distances and shock velocities are primarily determined by the density of the substrate and the filling factors of clouds. These estimates are uncertain to a factor of 2, and depend on modeling of grain destruction in shocks (see Seab 1987, this volume). The panel discussed the "robustness" of the gaseous component of the ISM, suggested that several $M_\odot$ yr^{-1} of gas may come from infalling high velocity clouds, and pointed out that the mean ISM density is marginally unstable to collapse triggered by a spiral density wave.

Lockman brought up the "local bubble" – the cavity of low-density gas near the Sun. Where did the gas go, if it was cleared out by a SNR ? Is there structure in the H I? It is well known from Lyα observations that the local ISM has a small H I column density out to 30 - 100 pc in many directions (Frisch and York 1983). However, several dark clouds exist within this cavity, in the directions of Perseus and Scorpius-Ophiuchi. The cavity extends to more than 100 pc in the third quadrant (galactic longitudes $l \approx 200 - 240^o$). The soft X-ray emission is not intense in this quadrant, though, which may mean that the gas has cooled below 10^6 K.

Scoville discussed estimates of the growth time ($\sim 10^8$ yr) for molecular clouds. Gravitational free fall times greatly exceed observed star formation rates, so something (magnetic fields?) must be supporting the clouds against collapse. Age estimates for the giant molecular clouds exceed 5×10^7 years, based on rough calculations of velocity dispersions in equipartition. However, not everyone agrees with this method. Nevertheless, the hierarchical structure observed in molecular cloud complexes may have been there at formation.

Several from the audience brought up the question of infalling high velocity clouds in the context of a galactic fountain. Danly described her IUE observations with Blair Savage of total line widths of interstellar absorption lines, which show significant displacements to negative velocities in directions associated with infalling 21-cm clouds. This is indicative that the gas comes from stars, rather than the infall of pristine intergalactic medium. Buss asked whether the observed galactic metallicity gradients could persist in the presence of a vigorous fountain. The answer is probably yes, since the radial mixing length is not likely to exceed 500 pc.

A large discussion ensued concerning the role of the galactic halo in the dynamics and energy balance of the ISM. Using the halo as a "vent" could reduce the porosity of the hot third phase. One would predict large pressure excursions, as SNRs burst through the gaseous layer. McCray described recent work with Kafatos and Mac Low, which shows that the ISM is most affected by OB associations and "superbubbles" in a stratified disk. Since the H I layer is 100 - 200 pc thick locally, individual Pop I SNRs will be unable to break through. However, the combined effects of 30 to 100 supernovae will produce bubbles 200 - 500 pc in size. When a supershell breaks through the layer, it is Rayleigh-Taylor unstable and "bursts" when $z \approx 1.5$ to 2 times the gaseous scale height. This theory helps explain why the Heiles supershells are most prominent toward the galactic anticenter, since the larger H I scale height confines the shells.

How many supershells should be seen? If there are 10,000 O-type stars in the Milky Way (based on extrapolation of observed O stars within 3 kpc), there are perhaps 40,000 potential Type II supernovae, assuming an initial mass function $N(M)dM \propto M^{-2.35}dM$ extending from $100M_\odot$ to $20M_\odot$ (O stars) and down to the minimum mass, $8M_\odot$ at spectral type B4, believed responsible for Type II supernovae. Since B4 stars live 30 million years, this is a supernova rate of 1 every 750 years. If these supernovae are clustered in groups of $N \approx 40$, then the supershell formation rate is approximately one per 30,000 years. If the supershell lifetime is of order 10^8 years (it could be less, since the driving source fades after 30 million years), then there should be about $N_{SS} \approx 3000$ supershells ("holes") of radius $R_{SS} \approx 200$ pc throughout the Milky Way. The area covering factor ("two-dimensional porosity") of the disk is:

$$C \approx \left(\frac{R_{SS}}{R_{disk}} \right)^2 N_{SS} \approx \left(\frac{200 \text{ pc}}{15,000 \text{ pc}} \right)^2 (3000) \approx 0.5 \qquad (10)$$

Nearly the entire Milky Way disk should be covered with supershells! Since it clearly is not, we have a problem. Perhaps the number of Type II supernovae is overestimated, perhaps the supernova energies are weaker than the canonical 10^{51} ergs, or perhaps we have neglected higher-order clustering. Alternatively, we may have underestimated the substrate density, which determines the final size of shells. The main point is that supershells do more damage to the ISM than individual SNRs, and it is difficult to escape the conclusion that the H I distribution does not agree with predictions based on "standard" supernova rates.

4   SUMMARY AND SYNTHESIS

As is probably clear from the previous sections, the panel and audience did not resolve all the issues. On the question of "Galactic Ecology", most participants seem to agree that the ISM contains three phases: clouds, warm intercloud gas, and a hot coronal phase. The controversy is over the relative volumes and masses.

The McKee-Ostriker model predicts a large volume filling factor, $f_{HIM} \approx 0.6$ in the hot cavities, and a small filling factor for clouds and warm H I. But several facts argue against this. These objections may be summarized as follows:

1. The MO theory assumed random supernovae. But Type II supernovae occur in clusters of 10 - 40 stars, and a substantial fraction of Type I's occur in the disk.

2. The MO theory assumed spherical clouds, whereas most clouds appear sheet-like. Shocks tend to make sheets, rather than spheres.

3. The H I distribution looks smooth and exhibits fewer fluctuations than predicted if the clouds are small and occupy a small volume.

4. Absorption studies at 21-cm show that a substantial fraction (30-40%) of the H I is warm, whereas only $\sim 2\%$ of the H I is warm in the MO model.

5. Theoretical estimates of the porosity parameter yield $Q \approx 1$, unless the substrate has a density $n_H > 0.1$ cm^{-3}. Even with correlated OB stars, there should be ~ 3000 "supershells" covering $\sim 50\%$ of the disk area. Where are they?

6. The number of O VI conductive interfaces requires a large energy injection rate. Is it supernovae? Where does the energy go?

7. Grain destruction in shock waves is so effective that the population would disappear in 10^9 years unless we can identify new grain sources or increase the substrate density.

Taking all of these arguments into account, it seems likely that the MO theory requires some changes. The basic premise must be correct, but several assumptions or input parameters probably must be revised. Stripped down to its simplest ingredients, the 3-phase model depends on four main parameters:

1. The supernova rate $r(\text{kpc}^{-3}\ \text{yr}^{-1})$.

2. The mean density $\bar{n}$ of gas in the disk.

3. The energy per supernova.

4. The ionizing EUV photon production rate per unit volume.

The model also requires input on cloud sizes and geometries, on the temperatures of the cloud and intercloud phases, and on the cloud formation process during shell fragmentation. The characteristics of the ISM (densities, pressures, scale heights, cloud velocity dispersions, and cloud column densities) all follow from this input and the theory of SNR evolution. Ideally, the cloud characteristics would follow too, but it is doubtful whether we understand cloud formation or shell formation well enough to go so far.

Of the four ingredients above, (2) is the only one that is well understood. We know that $\bar{n} \approx 1$ cm^{-3} in the ISM, so that interstellar matter comprises about 10% of the local mass density. But a related ingredient, the surface gravity of the disk, is neglected in the MO model. In a realistic theory, high velocity remnant shells should be allowed to "vent" pressure to the halo. The theory must also allow for the spatial transport of mass between shells when they overlap. Since SNRs push matter into pre-existing cavities, this might destroy some of the "holes" and account for some of the observed smoothness. Ingredient (3) is usually taken to be the canonical 10^{51} ergs per supernova. But this energy is uncertain by at least a factor of 2, and there is no guarantee that many supernovae are not "anemic". For instance, the energy of the Crab Nebula appears to be less than 10^{50} ergs (Chevalier 1981). Ingredient (4) is poorly known, since we lack a reliable estimate of the Lyman continuum fluxes from early-type stars with "wind-blanketed" atmospheres (Hummer 1982), and the effects of runaway OB stars and planetary nebulae are uncertain. Cloud geometries are even more uncertain, although as we have noted, data suggest that clouds are compressed sheets rather than spheres. Major physical processes, such as evaporative mass loss and SNR evolution, are sensitive to the spectrum of cloud sizes and their geometry. Furthermore, observational predictions, such as $\langle n_e \rangle$, $\langle n_e^2 \rangle$, and the distribution of warm H I, depend on how the gas is distributed.

If there is a single pressing need, it is for a new three-phase model that includes sheetlike clouds, and allows for spatial mass transport between shells and into the halo. But even then, the most basic ingredient to the models would be missing – the supernova rate in the Milky Way. Historical supernovae, corrected for selection effects, give a rate of about one every 30 years (Clark and Stephenson 1977), and a comparison with supernova rates in external galaxies (Tammann 1982) gives one every 20-40 years. Clark and Caswell (1976) find a characteristic interval of $\tau = 150$ years, based on the number-radius relation of radio SNRs. However, this method is sensitive to the assumed density of the substrate; Caswell and Lerche (1979) find $\tau = 80$ years when one accounts for the lower densities above the disk, and Higdon and Lingenfelter (1980) suggest $\tau = 10 - 30$ years if many supernovae explode in the low density hot phase. Cioffi (1985) demonstrated that all of these estimates are incorrect, when one considers a distribution of substrate densities. Of course, each of these methods suffers from serious selection effects and uncertainties due to unreliable distance indicators.

Pulsars provide an additional estimate of the supernova interval, but not without its own uncertainties. Although pulsar estimates have yielded intervals as low as $\tau = 6$ years (Taylor and Manchester 1977), depending on the assumed electron densities used in deriving dispersion measure distances, the most recent work (Lyne, Manchester, and Taylor 1985) yields $\tau = 30 - 120$ years. This estimate uses data on 316 pulsars, together with distances derived from a three-component model of the electron density distribution and a pulsar magnetic field decay model for the evolution of the luminosity function. Their minimum rate has $\tau = 230$ years, and application of an uncertain "beaming correction factor" for unseen pulsars gives the stated range in $\tau = 30 - 120$ years.

Disk supernova rates as high as one per 30 years may be too fast by a factor of 10. For such a high rate, calculations of porosity and grain survival give unreasonable predictions. The X-ray SNR counts in the Large Magellanic Cloud also hint that the Milky Way supernova rate has been overestimated. The supernova rate in the LMC is approximately one per 1000 years, based on X-ray remnants with well known distances (Hughes, Helfand, and Kahn 1984). Since the LMC has one-tenth the mass of the Milky Way, yet is undergoing a burst of massive star formation, it is surprising that the "standard rate" for the Milky Way (1 per 40 years) is 25 times faster. New observations, such as the recent discovery that 20-50% of Type I supernovae occur in spiral arms (Branch 1986), challenge our set ideas about the stellar progenitors of supernovae. Lyne, Manchester, and Taylor (1985) also note the difficulties of reconciling high SN rates with standard massive stellar birthrates (Miller and Scalo 1979).

My own feeling is that a new model of the diffuse interstellar medium will be needed to incorporate the ideas developed at this conference. The theory will probably require two- or even three-dimensional modeling of the fluid dynamics of SNR interactions, and must include more refined observational input. In the meantime, it would be useful to explore the "robustness" of the ISM phases with changes in these parameters, since ours is not the only spiral galaxy. Understanding the dynamics of SNR-dominated interstellar media will likely be a key ingredient in studying the "Starburst Galaxy" phenomenon, if not other galactic phenomena such as chemical evolution, the fuel supply for active galactic nuclei, galactic halo structure, and QSO absorption lines.

This work was supported in part by grants to the University of Colorado from NASA (NSG-7128 and NAGW-766). I thank Drs. Rob Fesen, Ellen Zweibel, Chris McKee, and Richard McCray for the opportunity to discuss various ideas put forth in this paper, and I am grateful to all of the panelists for providing such a stimulating discussion.

REFERENCES

Balbus, S.A. 1986, *Ap. J. Letters*, **303**, L79.

Blandford, R.D., and Ostriker, J.P. 1980, *Ap. J.*, **237**, 793.

Branch, D. *Ap. J. Letters*, **300**, L51.

Caswell, J.L., and Lerche, I. 1979, *M.N.R.A.S.*, **187**, 201.

Chevalier, R.A. 1981, *Fund. Cosmic Physics*, **7**, 1.

Cioffi, D.F. 1985, Ph.D. Thesis, University of Colorado.

Clark, D.H., and Caswell, J.L. 1976, *M.N.R.A.S.*, **174**, 267.

Clark, D.H., and Stephenson, F.R. 1977, *M.N.R.A.S.*, **179**, 87p.

Colomb, F.R., Poppel, W.G.L., and Heiles, C. 1980, *Astr. Ap. Suppl.*, **40**, 47.

Cox, D.P., and Smith, B.W. 1974, *Ap. J. Letters*, **89**, L105.

Cox, D.P., Raymond, J.C., and Smith, B.W. 1976, *Ap. J.*, **204**, 290.

Cowie, L.L. 1987, in *Interstellar Processes*, eds. D.J. Hollenbach and H. Thronson, (Dordrecht: Reidel).

Draine, B.D. 1978, *Ap. J. Suppl.*, **36**, 595.

Field, G.B. 1965, *Ap. J.*, **142**, 531.

Field, G.B., Goldsmith, D., and Habing, H. 1969, *Ap. J. Letters*, **15**, L49.

Frisch, P.C., and York, D.G. 1983, *Ap. J.*, **271**, L59.

Gaetz, T.J., and Salpeter, E.E. 1983, *Ap. J. Suppl.*, **52**, 155.

Heiles, C. 1987, *Ap. J.*, submitted.

Higdon, J.C., and Lingenfelter, R.E. 1980, *Ap. J.*, **239**, 867.

Hughes, J.P., Helfand, D.J., and Kahn, S.M. 1984, *Ap. J. Letters*, **281**, L25.

Hummer, D.G. *Ap. J.*, **257**, 724.

Ikeuchi, S., Habe, A., and Tanaka, Y.D. 1984, *M.N.R.A.S.*, **207**, 909.

Jenkins, E.B. 1978a, *Ap. J.*, **219**, 845.

―――― 1978b, *Ap. J.*, **220**, 107.

Jones, M. 1985, Ph.D. Thesis, Washington University.

Jura, M. 1987, in *Interstellar Processes*, eds. D.J. Hollenbach and H. Thronson, (Dordrecht: Reidel).

Kulkarni, S. 1987, in *Interstellar Processes*, eds. D.J. Hollenbach and H. Thronson, (Dordrecht: Reidel).

Kulkarni, S., and Heiles, C. 1987, "H I and the Diffuse Interstellar Medium", Chap. 3, in *Galactic and Extragalactic Radio Astronomy*, eds. K.I. Kellerman and G.L. Verschuur.

Lockman, F.J., Hobbs, L.M.. and Shull, J.M. 1986, *Ap. J.*, **301**, 380.

Lyne, A.G., Manchester, R.N., and Taylor, J.H. 1985, *M.N.R.A.S.*, **213**, 613.

McCray, R., and Snow, T.P. 1979, *Ann. Rev. Astr. Ap.*, **17**, 213.

Miller, G.E., and Scalo, J.M. 1979, *Ap. J. Suppl.*, **41**, 513.

McKee, C.F., and Ostriker, J.P. 1977, *Ap. J.*, **218**, 148.

Payne, H.E., Salpeter, E.E., and Terzian, Y. 1983, *Ap. J.*, **272**, 540.

Seab, C.G. 1987, in *Interstellar Processes*, eds. D.J. Hollenbach and H. Thronson, (Dordrecht: Reidel).

Seab, C.G., Hollenbach, D.J., McKee, C.F., and Tielens, A.G.G.M. 1985, in *Interrelationships among Circumstellar, Interstellar, and Interplanetary Grains*, NASA Conf. Publ. 2403, pA-28.

Shull, J.M., and Draine, B.D. 1987, in *Interstellar Processes*, eds. D.J. Hollenbach and H. Thronson, (Dordrecht: Reidel).

Shull, J.M., and Woods, D.T. 1985, *Ap. J.*, **288**, 50.

Spitzer, L. 1956, *Ap. J.*, **124**, 20.

Spitzer, L. 1978, *Physical Processes in the Interstellar Medium*, (New York: Wiley-Interscience).

Taylor, J.H., and Manchester, R.N. 1977, *Ap. J.*, **215**, 885.

Tammann, G. 1982, in *Supernovae: A Survey of Current Research*, eds. M.J. Rees and R.J. Stoneham, (Dordrecht: Reidel), p371.

Richard McCray, Christopher McKee, Carl Heiles, and Leo Blitz

MASS AND ENERGY BALANCE IN THE INTERSTELLAR MEDIUM:
HOW TO MODEL THE DISK GAS IN OUR GALAXY

Lennox L. Cowie
Institute for Astronomy
2680 Woodlawn Drive
Honolulu, Hawaii   96822
U.S.A.

ABSTRACT. A model for the interstellar medium is given which addresses
the energy balance between cold, warm, and hot components, and which is
consistent with recent observational data on pressures, cloud cooling,
and OVI measurements. It is further demonstrated that the proposed
model is stable under a wide range of parameter space and converges to
a solution approximating conditions which are observed for our Galaxy.
Finally, the consequences of extreme conditions on model solutions are
considered, and the implications for the interstellar medium in other
galaxies are discussed.

1.   INTRODUCTION

For the particular conditions holding in our own galaxy the gas exists
in roughly equal parts in molecular clouds and in more diffuse gas
(Sanders et al. 1984). The diffuse gas itself is split almost equally
between cold (T ~ 100° K) and warm (~5000° K) neutral material (e.g.
Liszt 1983). In addition, the distributed nature of the OVI absorption
leaves little doubt that a significant fraction of the volume lies in
hot gas (T ~ $10^{6°}$ K) (Jenkins 1978b). Recent reviews of the structure
of the diffuse interstellar gas are given in Cowie and Songaila (1986)
and Kulkarni and Heiles (1986). The gas system is very complex and
also at first sight very fragile. Superficially it appears that relatively small changes in the mass or energy input could radically change
its state.
    A model of the ISM should ideally account for the existence of all
of the structure. It should be capable of explaining the energy and
mass balance in the disk and of providing governing mechanisms which
ensure that the system is stable and long lived. It should also allow
us to decide how unique conditions in our Galaxy must be and hopefully
demonstrate that there is at least some substantial range of parameter
space where the ISM would appear similar to the conditions which are
actually observed. With such a model in hand we should hopefully be
able to predict how different conditions might be in other galaxies.

D. J. Hollenbach and H. A. Thronson, Jr. (eds.), Interstellar Processes, 245–258.

Some elements of the correct theoretical description of the ISM are clear. There is little doubt that the presence of the cold and warm phases results from the functional form of the cooling curve at these temperatures, as described in the Field, Goldsmith, and Habing (1969) model. There also seems little doubt that the hot gas is formed by the dynamical interactions of supernova remnants, as postulated by Cox and Smith (1974) and McKee and Ostriker (1977). It is likely that energy balance in the hot phase determines the pressure of the interstellar gas (McKee and Ostriker 1977, Cox 1981) and it is plausible that the OVI absorption forms in evaporating regions on the boundaries between hot and cool gas. However, there is no satisfactory overall model of the ISM at this time. The ambitious and elegant attempt of McKee and Ostriker (1977) is incomplete in a number of respects. It does not provide a heating mechanism for either cold or warm gas or account for the balance between them. In addition, it does not explain why this cooler gas should lie in a relatively anomalous two-phase state when the pressure and energy balance is controlled in the hot gas (e.g. Ikeuchi, Habe, and Tanaka 1984 for the wide range of possible structures that could exist in a McKee and Ostriker model). Finally, the model in at least one form assumes that energy loss terms in the hot gas are primarily through radiation in the disk and this is not consistent with the ultraviolet absorption line results we discuss in the next section.

The latter problem can be solved by allowing the cooling to take place in the halo in a fountain model (Shapiro and Field 1976, McKee and Ostriker 1977). A reasonable "low fountain" model is given by Cox (1981). Cox also pointed out that compression and expansion could heat the warm gas as the pressure fluctuated in the ISM (Cox 1979,1981). Spitzer (1982) and Ikeuchi and Spitzer (1984) discussed the generation of acoustic waves and how these could heat the warm gas. However, as we shall also discuss in the next section, none of these mechanisms can heat the cold gas, nor do any of these models deal in any way with the problems of balance and stability.

I will argue in Section 2 that recent observations leave us very little choice about the basic model of the ISM and a qualitative synopsis of the resulting model is given in Section 3. Most importantly, this model provides all the features we have required above. It can account for the observed properties of the ISM. One can also demonstrate that the model is totally stable and robust and that under a wide variety of circumstances the ISM would evolve to a state very similar to its current appearance (Section 4). Finally, in Section 5 we discuss at what point the model might break down under extreme conditions in a galaxy and what the corresponding consequences might be.

2. INFERENCES FROM CERTAIN OBSERVATIONS

Since the previous rounds of global modeling in the 1970s, a number of important observational results have been obtained that closely constrain any description of the ISM. It is my belief, as I shall outline in this section, that the existing constraints are so tight

that we are left with only one plausible model. Fortunately, as we
shall see in the following section, this model seems able to describe
all the properties of the ISM and is, in addition, extremely stable.
     Perhaps the most important measurement on the ISM is the direct
determination of the cooling rate in the cold diffuse ISM using UV
measurements of the column densities of the excited levels of the fine
structure lines (cf. Penston 1974). The cooling in the cold clouds is
dominated by the CII 158 micron line which is also a significant
coolant in the warm gas (Dalgarno and McCray 1972, Penston 1974, Shull
and Woods 1985). By measuring the column density of the upper fine
structure level [$N^*(CII)$] using Copernicus or IUE UV absorption line
observations, the energy loss along a line of sight is obtained.
Dividing this by the column density of neutral hydrogen and hydrogen
molecules [$N(H) + 2N(H_2)$] which is also determined from the UV
measurements gives the energy loss per hydrogen nucleon as

$$e = \frac{N^*(CII) \ A \ h\nu}{N(H) + 2N(H_2)} \tag{1}$$

where A (= 2.4×10^{-6} s^{-1}) is the transition probability and $h\nu$ the
energy of the CII line. Pottasch et al. (1979) have made measurements
of the energy loss along eight lines of sight using IUE and Copernicus
measurements. Their measured loss averages to 1.2×10^{-25} ergs/s per
nucleon with variations by up to a factor of three from line of sight
to line of sight.
     I have compiled values obtained for a selected sample of stars
where good Copernicus data is available together with the IUE measurements given by Pottasch et al. In this much larger data set there is a
fairly large spread in the values which may be caused at least in part
by problems in measuring the column densities but which may also
reflect real variations. Low column density lines of sight cannot
contain cold clouds and the typical value of slightly less than 10^{-25}
ergs/s nucleon in these regions of must represent loss values in the
warm component (neutral and ionized). [Kulkarni in the meeting argued
that this was dominated by ionized regions.] On the other hand, the
lines of sight to stars such as ζ Oph are dominated by cold gas and
here the typical value of 5×10^{-26} ergs/s nucleon should be representative of the cooling rate in the cold clouds. However, it is possible
that the values in the cold clouds may be particularly underestimated
because of the larger saturation effects in the narrow cold components,
so that this value could be higher.
     If these rates apply throughout the galaxy then the 3×10^9 M$_\odot$ of
diffuse gas (Sanders et al. 1984), which is split roughly equally
between cold and warm gas, emits 2×10^{41} ergs/s in the CII 158 micron
line alone. This value slightly exceeds typical directly measured
values in our own and external galaxies (Stacey et al. 1983, 1985,
Crawford et al. 1985). However, these direct measurements are
sensitive only to high surface brightness regions and are almost
certainly dominated by photodissociation regions on the surfaces of
molecular clouds (e.g. Tielens and Hollenbach 1985). Therefore they

underestimate the total luminosity of the galaxy if this is dominated
by the diffuse gas. As an aside, the total CII 158 micron luminosity
of a normal galaxy (~10^8 $L_\odot$) then translates into an extremely high
luminosity (~10^{11} $L_\odot$) for a gas-rich protogalaxy and suggests that with
new technological advances in submillimeter astronomy searching for
this line may become an excellent method of searching for the protogalaxies in the next few years.

Now the total dynamical energy input into the ISM is at most 6 x
10^{41} ergs/s even for a generous <u>field</u> supernova rate in the galactic
disk of 1/50 year each with an energy of 10^{51} ergs. This dynamical
energy is (marginally) sufficient to balance the losses in the diffuse
gas. What is more difficult is to find a plausible energy source for
the cold component. The reason for this is that the small filling factor of this component makes efficient deposition of energy difficult.
Thus if a typical cold region has a density of 50 cm^{-3} (e.g. Cowie and
Songaila 1986) we need at least 2.5 x 10^{-24} ergs/cm^3 s in heating.
However, if energy is transferred at the external energy density (ε)
across the surface at near the sound speed (c) (or the Alfven velocity)
into a cloud of radius a, we can input at most

$$e \cong \frac{\varepsilon c}{a} \sim 10^{-26} \ \mathrm{ergs/cm}^3 \ s \ \{\frac{\varepsilon}{10^{-12} \ \mathrm{ergs/cm}^3}\}\{\frac{3pc}{a}\}\{\frac{c}{1 \ km/s}\}$$

which fails by around two orders of magnitude. This is <u>not</u> the case
for the warm gas where such terms <u>can</u> supply typical requirements of
3 x 10^{-26} ergs/cm^3 s (e.g. Cox 1979,1981). However, neither compressive heating mechanisms, acoustic waves, nor cosmic rays can adequately
heat the cold clouds. Nor can soft X-rays, which simply do not have
enough energy.

This leaves only the Galactic UV luminosity as an energy source
for heating the cold clouds (de Jong 1980). Predictions of photoelectric heating rates <u>do</u> roughly agree with the measured cooling rates
from the CII* discussed above though with substantial uncertainty
because of our lack of knowledge of yields and threshold energies (e.g.
Draine 1978). It is also likely that the presence of micrograins or
polycyclic aromatic hydrocarbons (e.g. Sellgren 1984, Puget, Leger, and
Boulanger 1985) will substantially increase the heating rate, so that
even if revised CII* measurements show a higher value in the cold
clouds, this will be matched by evolution of the theory. For the
present we shall simply assume that there must be a conversion of UV
photons via some grain mediation process to provide the energy balance
in the cold clouds. This mechanism will provide a comparable heating
rate per nucleon in the warm gas. Depending on the grain parameters
and the warm gas temperature, electron and hydrogen impacts may imply
that the net effect of the grains is to cool the warm regions (Draine
1978). <u>However, if, as we have speculated above, the micrograins</u>
<u>dominate the heating, then photo-electric net heating of the warm gas</u>
<u>will result.</u> We shall assume this in the following discussion.

The CII 158 micron emission may be written as a sum of contributions from cold regions, warm neutral regions, and warm ionized

regions. As Kulkarni stressed in his talk here, the breakdown between warm neutral and warm ionized gas is complex. The cold gas is more straightforward and the energy loss per nucleon can be written as

$$e = D_c \, X_c \, P_c \, \{ \frac{\varepsilon_H(T_c)}{T_c} + f_e \, \frac{\varepsilon_e(T_c)}{T_c} \} \tag{2}$$

where D_c (= 0.5) is the depletion and X_c (= 8 x 10^{-4}) is the gas phase abundance of carbon, P_c and T_c are the cold cloud pressure and temperature, f_e is the electron fraction and ε_e and ε_H are emissivities owing to electron and hydrogen excitation. For T_c in the 50-100° K range, we may rewrite this equation as

$$e = 2 \times 10^{-25} \, \{ \frac{100° \, K}{T_c} \} (1 + 500 \, f_e) \, \exp - (\frac{92}{T_c}) \{ \frac{P_c}{4000° \, K \, cm^{-3}} \} \tag{3}$$

and for T_c in this range and f_e small the value is quite constant at

$$e = 8 \times 10^{-26} \, \{ \frac{P_c}{4000° \, K \, cm^{-3}} \} \{ \frac{D_c}{0.5} \} \tag{4}$$

Thus the measured energy loss rate may be viewed as determining, at least crudely, the carbon depletion times the pressure in the cold clouds. As such it is crudely consistent with measured value of $\langle P \rangle \approx$ 4000° K cm^{-3} using the CI fine structure line (Jenkins, Jura, and Loewenstein 1983) and H_2 (Bohlin et al. 1978) provided that the electron fraction in the cold clouds is small.

I want to turn now to the hot ($10^{6°}$ K) gas and the controversy of its filling factor in the disk (e.g. Cowie and Songaila 1986). The OVI absorption lines as seen by Copernicus (Jenkins 1978a,b) are the crucial diagnostic.

The ionization energy for OVI (113.9 eV) is too high for any plausible photoionization mechanism so these ions must be thermally ionized. However, here the narrowness of these lines ($v_{rms} \approx$ 26 km/s) means that they cannot be produced by shock heating, which would require v >> 100 km/s, or by dynamical heating of a turbulent hot gas, which would give comparable broadening. This leaves little doubt that the OVI must be produced by relatively quiescent heating of a cool gas. The only mechanism suggested for this is thermal conduction from a hot to a cold gas (Weaver et al. 1977, McKee and Cowie 1977, McKee and Ostriker 1977, Cowie, Jenkins, Songaila, and York 1979) and indeed it is hard to see any other way to produce the observed result.

Now the average line of sight distance to an observed OVI producing region is about 160 pc (Jenkins 1978b). The OVI column densities also increase quite smoothly with distance indicating that hot and cold gas must be routinely interspersed. Finally, in order to produce the observed density, we must heat a sufficient amount of material from the cold to the hot phase to produce enough OVI, since this ion is a trace species, is transient, and ultimately ionizes to a higher stage. That is, we must have a rate of transfer from cool to hot gas given by

$$\dot{n} = \frac{n_{OVI}}{X_o D_o t_{coll}} \tag{5}$$

Where t_{coll} is the timescale for collisional ionization of OVI to OVII, and can be written

$$t_{coll} = \frac{1}{n_e C(T)} = \frac{2T}{C(T) P} \tag{6}$$

where $C(T)$ is the collisional ionization rate, and T and P are the effective temperature and pressure at which OVI ionizes to OVII. $T/C(T)$ is relatively independent of the temperature at which ionization takes place (over the expected range) and for $P = 4000°$ K cm^{-3} the timescale is 9×10^{12} s at $4.5 \times 10^{5°}$ K, 2×10^{12} s at $10^{6°}$ K and 2×10^{12} s at $3 \times 10^{6°}$ K (Lotz 1966). For $D_o = 0.6$, $X_o = 8.1 \times 10^{-4}$ (Cowie and Songaila 1986) and $n_{OVI} = 2.8 \times 10^{-8}$ cm^{-3} (Jenkins 1978a) we require

$$\dot{n} = 9 \times 10^{-10} \left\{ \frac{2 \times 10^{12} \text{ s}}{t_{coll}} \right\} \text{ cm}^{-3}/\text{yr} \tag{7}$$

The equivalent argument may be expressed in a more formal and exact (but less general) way for the case where the OVI is formed in thermally evaporating surfaces of cool regions. Now the evaporative mass loss rate for clouds of dimension a_{pc} pc immersed in a hot fluid of asymptotic temperature $10^6 T_6°$ K and density $10^{-2} n_{-2}$ cm^{-3} is

$$\dot{M} = 2.8 \times 10^{19} T_6^{5/2} a_{pc} \phi \text{ g/s} \tag{8}$$

from each cloud (Cowie and McKee 1977), where ϕ measures the reduction owing to magnetic fields. The OVI on the two sides of the cloud is

$$N_{OVI} = 3 \times 10^{12} \left\{ \frac{\beta}{3} \right\} T_6^{-3/2} n_{-2} a_{pc} \text{ cm}^{-2} \tag{9}$$

where β measures the effect of time-dependent ionization (Weaver et al. 1977, Cowie et al. 1979). Now the distance to an OVI surface is

$$\ell = \frac{4/3 \, a_{pc}}{f \phi} \text{ pc} \tag{10}$$

where f is the cloud filling factor and the product

$$\ell N_{OVI} = 1.2 \times 10^{31} \{\frac{\beta}{3}\} \frac{a_{pc}^2}{f\phi \, T_6^{5/2}} (T_6 \, n_{-2}) \text{ cm}^{-1} \qquad (11)$$

Now

$$\dot{n} = \frac{\dot{M}}{\frac{4\pi}{3} a^3} \frac{f}{m} = 1.2 \times 10^{-13} \{\frac{f\phi \, T_6^{5/2}}{a_{pc}^2}\} \text{ cm}^{-3}/s$$

$$= 1.8 \times 10^{-9} \{\frac{160 \text{ pc}}{\ell}\} \{\frac{10^{13} \text{ cm}^{-2}}{N_{OVI}}\} \{\frac{P}{4000° \text{ K cm}^{-3}}\} \{\frac{\beta}{3}\} \text{ cm}^{-3}/yr \qquad (12)$$

Within the uncertainties of t_{coll} and β this is consistent with equation (7). Over the volume of the disk ($\sim6 \times 10^{66}$ cm^3) equation (12) translates to 10 $M_\odot$/yr being heated from the cold to the hot phase.

This is a large mass flow which has to be matched by equivalent cooling, and the problem of obtaining this balance is difficult and tells us a great deal about the structure of the hot gas. Firstly, the cooling time in the hot gas at pressure P can be written as

$$t_{cool} = \frac{3kT}{n_e \, \Lambda(T)} = 4 \times 10^7 \, T_6^{2.6} \{\frac{4000° \text{ K cm}^{-3}}{P}\} \text{ yrs} \qquad (13)$$

where we have approximated $\Lambda(T)$ to the cooling curve of Raymond, Cox, and Smith (1976). If the hot gas is forced to balance the mass input by radiative cooling we require

$$\langle n \rangle_{hot} = \dot{n} \, t_{cool} = \frac{P}{2T} f_{hot} \qquad (14)$$

where f_{hot} is the filling factor of the hot gas. Even adopting the lower n estimate of equation (7) we now have

$$f_{hot} = 18 \, T_6^{3.6} \{\frac{4000° \text{ K cm}^{-3}}{P}\}^2 \qquad (15)$$

Now this would require the unphysical result $f_{hot} > 1$ unless the pressure in the hot gas were nearly an order of magnitude higher than the measured value in the cold gas. This statement of the argument actually underestimates the problem since in most models the heating of the gas will take place at higher pressures than the cooling, accentuating the discrepancy.

It is possible to invoke additional cooling terms—as for example cooling by compression of cold clouds (McKee and Ostriker 1977, Cox 1979) to deal with this problem, but there is an observational test which directly demonstrates that the hot gas cannot balance the mass input to the hot gas by radiatively cooling in the disk. This is the

observation of Cowie and York (1978) that there is very little material
at velocities greater than 50 km/s in the disk with column densities in
excess of 10^{17} cm^{-2} in stars at up to kpc distances. Cowie and York
showed that the average line of sight distance to a component with v >
100 km/s is between 12 and 100 kpc. However, such material should be
omnipresent at these low column densities if the dynamically active hot
gas were locally cooling at the required rates.

The only apparent solution to this problem is to vent material
into the halo. Because the mass flow is so large, it must cool and
fall in so that the disk does not become gas depleted. It cannot be a
high fountain without violating constraints on high-velocity gas. Thus
we are forced to the conclusion that a low halo fountain must exist
(Shapiro and Field 1976, McKee and Ostriker 1977, Cox 1981). From the
discussion of the observations given above the fountain must be circu-
lating about 10 M$_\odot$ of gas/yr.

Turning back again to the question of the filling factor of the
hot gas, we must have good connectivity of the hot gas to allow this
venting. We can go beyond this qualitative point and estimate a minimum filling function for the hot gas in the disk. The minimum time for
gas to escape from the disk [which we take to have a height (H) of 150
pc (e.g. Cowie and Songaila 1986)] is given if it can freely stream
directly from the disk as

$$t_{escape} = \frac{H}{\{\frac{10kT}{3\bar{m}}\}^{1/2}} = 10^6 \ T_6^{-1/2} \ (\frac{H}{150 \ pc}) \ yrs \qquad (16)$$

With the lowest $\dot{n}$ estimate of equation (7) we now have

$$\langle n \rangle_{hot} = 9 \times 10^{-4} \ T_6^{-1/2} \ \{\frac{P}{4000° \ K \ cm^{-3}}\} \ cm^{-3} = f_{hot} \ n_{hot}$$

$$= 2 \times 10^{-3} \ T_6^{-1} \ \{\frac{P}{4000° \ K \ cm^{-3}}\} \ f_{hot} \ cm^{-3}$$

or

$$f_{hot} = 45\% \ T_6^{1/2}$$

while with the higher mass heating estimate the value would be 90%
$T_6^{1/2}$. Thus for $T_6 = 0.5$ the filling factor lies between 30% and
60%. Since at least 20% of the disk gas is warm and neutral gas (Cowie
and Songaila 1986) the upper bound is more realistically about 75%.

The final point that I want to note is that at least a substantial
part of the warm gas must exist in large very uniform regions. The
existence of such regions, possibly with embedded denser cold clouds,
is in no way incompatible with any of the observational data (or
inferences) discussed above and in particular is not inconsistent with

a large filling factor for the hot gas. Indeed referring back to equation (10) we see that the warm gas must have a scalesize of

$$a_{warm} \geqslant 60 \ \phi \ \left\{\frac{f}{0.5}\right\} \ pc$$

to be consistent with the distances to the OVI producing regions. Even for $\phi = 1/3$ as might be expected from a turbulent magnetic field this would imply scalesizes greater than 20 pc for f = 50%. Thus the warm gas exists primarily in large structures rather than in small-scale clouds. There can be very few pieces of embedded warm material in the hot gas regions before the average distance to an OVI producing front becomes too small. Indeed this is the case with the local region. There is very little warm gas within the local hot bubble which is however surrounded, at distances of 50 to 200 pc, by warm neutral gas which seems to exist in the form of fairly well defined walls (Paresce 1984, Cowie and Songaila 1986).

Cold clouds larger than about a parsec have long lifetimes against thermal evaporation ($\sim 3 \times 10^7$ yrs) even in the hot gas and can survive passages through any reasonably sized hot phase regions. Thus the cold cloud population should be uniformly dispersed through both hot and cool regions. Since the average distance to a cold cloud is about 170 pc (e.g. Cowie and York 1978) the mean free path to an OVI producing cold cloud surface is 170 pc/f_{hot}. Thus if these are the dominant OVI-producing regions as suggested by the column densities, we again find $f_{hot} \approx 1$. Most likely some OVI regions arise on the surfaces of large warm regions and some on the surfaces of the cold clouds. Some evidence for this exists in the work of Jenkins (1978b) who found two distinct populations of OVI producers, one with column densities of 10^{13} cm^{-2} (the ℓ = 170 pc component) and one with much larger column densities [which was much more infrequent ($\sim 10\%$)] which might naturally be associated with the warm regions. If this division is correct the typical size of a warm region would be of order 100 pc.

3. A MODEL FOR THE ISM

Based on the considerations of the previous section, I would conclude that there is little choice but to adopt the following model of the diffuse ISM (particular details are postponed to the next section). As noted in the introduction, many of the individual processes closely follow earlier work with elements drawn particularly from Field, Goldsmith, and Habing 1969, Cox and Smith 1974, McKee and Ostriker 1977, Cox 1979, 1981, and de Jong 1980.

3.1. Cold Clouds

The heating mechanism is some type of photoelectric or grain heating mechanism with a rate around e = 5×10^{-26} ergs/s (or larger) per nucleon. Following McKee and Ostriker (1977), lower bounds to the sizes (about 1 pc) are probably determined by evaporation of smaller

clouds on passage through the hot gas and upper bounds by gravitational
stability. The ratio of cold to warm gas probably is governed by the
pressure fluctuation spectrum in the ISM. This drives material between
the two phases and determines the mass balance. The cloud velocity
dispersion is most likely determined by acceleration in the turbulence
of the hot phase balancing cloud-cloud interaction and slowing in the
warm phase. Cold clouds immersed in the hot gas are responsible for
most of the observed OVI.

3.2. Warm Gas

The warm gas may also be heated by the photoelectric mechanism heating
the cold clouds with a similar energy rate per nucleon (de Jong 1980).
In addition, the warm clouds will receive heat from compression and
expansion governed by pressure fluctuations in the hot gas (Cox 1979,
1981) and by dissipation of acoustic waves generated in the hot gas
(Spitzer 1981, Ikeuchi and Spitzer 1984). Little warm gas exists
within the hot phase where it would be rapidly destroyed by evaporation; rather the warm gas exists in large clumps with sizes well in
excess of 20 pc and probably closer to 100 pc. Such regions are disrupted only by supernova explosions occurring in them and the size
scale is a natural consequence of this mechanism (cf. Jones 1975).
A small part of the OVI arises on the warm cloud surfaces.

3.3. Hot Gas

The hot gas is created and energized by the dynamical energy input of
supernovae as postulated by Cox and Smith (1974). Mass input is by
evaporation (McKee and Ostriker 1977) from embedded cold clouds and
warm cloud surfaces. The mass transfer rate from cool to hot gas is
about 10 $M_\odot$/yr. The filling factor of the hot gas is substantial--
about 50% with an uncertainty of 20% either way and it must have good
connectivity. Energy is not dissipated in the plane but instead is
dissipated in the low halo with mass balance achieved through a foun-
taining effect (Shapiro and Field 1976, McKee and Ostriker 1977, Cox
1981). The average pressure is determined in the hot phase by a balance between the supernova energy injection and energy transfer to the
halo and compressive heating of the warm gas (Cox 1981), while the
pressure fluctuation spectrum is determined by the evolution of individual supernova remnants within the disk (McKee and Ostriker 1977).
The density, and consequently the temperature, of the escaping hot gas
are determined by the balance between the evaporated mass injection
rate and the mass transfer rate to the halo.

4. BALANCE AND STABILITY AMONG HOT, WARM, AND COLD GAS

Because the heating of the cold gas is produced by high-energy photons
(hν ~ 10 eV) from short-lived, massive stars which also produce at

least some of the type II supernovae, it is reasonable to set the cold
gas heating rate proportional to the supernova rate S (per unit
volume), which we normalize to the local supernova rate (S_{local}) and
denote s. With this definition we now have the energy balance equation

$$n_c \ \Lambda(T_c) = es \qquad (17)$$

in the cold phase. Here Λ is the usual cooling function. In the warm
phase we have volume heating terms caused by the dynamical energy input
which is also proportional to S and a heating term owing to the photoelectric terms and we may write

$$n_w \ \Lambda(T_w) = es + \frac{\Phi hs}{n_w} \qquad (18)$$

The term $h = S_{local} E_{SN}$ while Φ is a factor of approximately unity and
is ≈ 2 for the particular case of compressive heating (Cox 1981). We
note that grain cooling is included in the cooling function Λ_w. For
$S_{local} = 10^{-13} S_{-13}/pc^3$ yr and $E_{SN} = 10^{51} E_{51}$ then $h = 10^{-25} S_{-13} E_{51}$
ergs/cm^3 s and it can be seen that this term is comparable to the pho-
toelectric term. (Recall that $e \geqslant 5 \times 10^{-26}$ ergs/s nucleon). Finally
the energy balance equation in the hot gas is given by

$$\frac{5k \ T_h n_h}{t} = wh \ s \qquad (19)$$

where t is the escape time for the gas to vent from the disk,
(10^6(H/150 pc) $T_6^{-1/2}$ yrs) and w is the fraction of the energy which is
not lost in compressing cool gas (Cox 1981).

Now assume that we can write the pressure balance equations as

$$P_w = \alpha \ P_c$$
$$P_h = \beta \ P_c$$

where $\beta \leqslant \alpha \leqslant 1$ allows for possible contributions by magnetic fields to
the pressure support and $P_{c,w,h}$ are the dimensionless thermal pressures
in the cold, warm, and hot gas. Then we can rewrite the energy balance
equations as

$$\frac{\Lambda(T_c)}{T_c} = \frac{es}{P_c} \qquad (20)$$

$$\frac{\Lambda(T_w)}{T_w} = \frac{es}{\alpha P_c} + \frac{\Phi h \ s \ T_w}{\alpha^2 \ P_c^2} \qquad (21)$$

and

$$P_c = \frac{2}{5} \frac{wh\ st}{\beta k} \tag{22}$$

The final equation determines the mean interstellar pressure and follows the concepts of McKee and Ostriker (1977) and Cox (1981). Substituting it back into the two remaining equations we obtain

$$\frac{\Lambda(T_c)}{T_c} = \frac{5}{2w} \frac{\beta ke}{ht} \tag{23}$$

and

$$\frac{\Lambda(T_w)}{T_w} = \frac{5}{2w} \frac{\beta ke}{\alpha ht} + \frac{5}{2w} \frac{\Phi\beta\ kT_w}{\alpha^2\ P_c t} \tag{24}$$

$$= \frac{5}{2w} \frac{\beta ke}{\alpha ht} \left\{ 1 + \frac{\Phi\ T_w}{\alpha\ P_c} \cdot \frac{h}{e} \right\}$$

As in the classical two-phase model the functional form of the cooling curve Λ determines whether we can simultaneously have both a warm and cold phase. However, before turning to this we may note some extremely interesting points about these two basic equations. The cold gas equation, which determines T_c, is independent of the supernova rate, except through a weak dependence on the hot gas temperature--the hot gas temperature itself being extremely weakly dependent on S as we shall show subsequently. Furthermore, it does not depend on the gas density or even on the metallicity (since Λ and e both depend linearly on the metal abundance). Thus, T_c is independent of the state of evolution of the disk and virtually independent of the star formation rate. If a solution to the equation exists, it will hold under all perturbed conditions. Thus the cold phase is extremely stable.

The same remarks hold for the first term on the R.H.S. in equation (24). Now, if the pressure in the disk were high and the second term negligible, the existence of the warm phase would not depend on the properties of the disk or the star formation rate, given that photo-electric heating dominates cooling. If the pressure is low, however, the heating rate rises because of the second term--but because the cooling curve rises rapidly in this temperature range even when P_c drops several orders of magnitude below the current value, the warm phase would still exist. Thus except at extremely low pressures the temperature of the warm phase is also determined and does not depend on any special condition holding. The warm phase is also extremely stable.

Turning now to the hot gas we see that

$$P_h \equiv \beta\ P_c = 9000\ w\ (E_{51}\ S_{-13})\ \left[\frac{H}{150\ pc}\ T_6^{-1/2}\right]^\circ\ K\ cm^{-3} \tag{25}$$

Best agreement with the observed values [$P_c \approx 4000°$ K cm^{-3} (Jenkins, Jura, and Loewenstein 1983)] is obtained for w E_{51} S_{-13} $\approx 1/2$ though given the uncertainty in β even a higher value is not implausible. This equation may now be combined with the mass flow rates to obtain densities and temperatures. Semiempirically, this may be written as

$$n = \dot{n}t = 2 \times 10^{-3} \left[\eta \frac{H}{150 \text{ pc}} T_6^{-1/2} \right] \left\{ \frac{P}{4000° \text{ K cm}^{-3}} \right\} \text{ cm}^{-3} \tag{26}$$

where η represents the uncertainty in equations (7) and (12). This in turn gives

$$T_6 = \left\{ \eta \frac{H}{150 \text{ pc}} \right\}^{-2} \tag{27}$$

We may also note that while the pressure and hence the density of the hot gas varies almost linearly with the supernova energy input rate, the hot gas temperature is near invariant. The escape time scales as $T^{-1/2}$ and, as we noted earlier, is quite insensitive to any changes in the star formation or supernova rate.

5. SUMMARY

What section (4) has demonstrated is that, given heating by a photo-electric effect and compressive energy input, the interstellar medium structure is very stable and robust. That is, its character would be the same even if the density, star-formation rate, or metallicity of the galaxy were different. These are very powerful statements but the description above is by no means complete. It does not address how the fraction by mass of warm and cold gas is established (this is likely a result of the local pressure fluctuation spectrum) or of how the star formation process is regulated (this may be a balance process where enough supernovae are formed to stir the gas to gravitational stability). However, a complete theory is clearly possible at this stage.
     Perhaps even more interesting is how a system can depart from the equilibrium situation. Clearly one possibility is that star formation becomes rapid on the evolution timescales of the stars and gas (10^6-10^7 yrs). Such a situation could lead to a runaway where the gas warms up and can no longer be stirred sufficiently to remain gravitationally stable. The system one would then move into runaway star formation. Abnormal cases of this sort, possibly induced by tidal interactions, might constitute a large fraction of the starburst galaxies.

ACKNOWLEDGMENTS

I would like to thank T. de Jong and C. McKee for the interesting discussions on which this talk was partly based.

REFERENCES

Bohlin, R. C., Savage, B. D., and Drake, J. F. 1978, Ap. J., 224, 132.
Cowie, L. L., Jenkins, E. B., Songaila, A., and York, D. G. 1979,
    Ap. J., 232, 467.
Cowie, L. L., and McKee, C. F. 1977, Ap. J., 211, 135.
Cowie, L. L., and Songaila, A. 1986, Ann. Rev. Astr. Ap., 24, 499.
Cowie, L. L., and York, D. G. 1978, Ap. J., 223, 876.
Cox, D. P. 1979, Ap. J., 234, 863.
Cox, D. P. 1981, Ap. J., 245, 534.
Cox, D. P., and Smith, B. W. 1974, Ap. J. (Letters), 189, L105.
Crawford, M. K., Genzel, R., Townes, C. H., and Watson, D. M. 1985,
    Ap. J., 291, 755.
Dalgarno, A., and McCray, R. A. 1972, Ann. Rev. Astr. Ap., 10, 375.
de Jong, T. 1980, Highlights Astron., 5, 301.
Draine, B. T. 1978, Ap. J. Suppl., 36, 595.
Field, G. B., Goldsmith, D. W., and Habing, H. J. 1969, Ap. J.
    (Letters), 155, L149.
Ikeuchi, S., Habe, A., and Tanaka, Y. D. 1984, M.N.R.A.S., 207, 909.
Ikeuchi, S., and Spitzer, L. 1984, Ap. J., 283, 825.
Jenkins, E. B. 1978a, Ap. J., 219, 845.
_____. 1978b, Ap. J., 220, 107.
Jenkins, E. B., Jura, M., Loewenstein, M. 1983, Ap. J., 270, 88.
Jones, E. M. 1975, Ap. J., 201, 377.
Kulkarni, S., and Heiles, C. 1986, preprint.
Liszt, H. 1983, Ap. J., 275, 163.
Lotz, W. 1966, Ap. J. Suppl., 14, 207.
McKee, C. F., and Cowie, L. L. 1977, Ap. J., 215, 213.
McKee, C. F., and Ostriker, J. P. 1977, Ap. J., 218, 148.
Paresce, F. 1984, A. J., 89, 1022.
Penston, M. V. 1974, M.N.R.A.S., 166, 21P.
Pottasch, S. R., Wesselius, P. R., and van Duinen, R. J. 1979, Astr.
    Ap., 74, L15.
Puget, J. L., Leger, A., and Boulanger, F. 1985, Astr. Ap., 142, L19.
Raymond, J. C., Cox, D. P., and Smith, B. W. 1976, Ap. J., 204, 290.
Sanders, D. B., Solomon, P. M., and Scoville, N. Z. 1984, Ap. J., 276,
    182.
Sellgren, K. 1984, Ap. J., 277, 623.
Shapiro, P. R., and Field, G. B. 1976, Ap. J., 205, 762.
Shull, J. M., and Woods, D. T. 1985, Ap. J., 288, 50.
Spitzer, L. 1982, Ap. J., 262, 315.
Stacey, G. J., Smyers, S. D., Kurtz, N. T., and Harwit, M. 1983,
    Ap. J. (Letters), 268, L99.
Stacey, G. J., Viscuso, P. J., Fuller, C. E., and Kurtz, N. T. 1985,
    Ap. J., 289, 803.
Tielens, A. G. G. M., and Hollenbach, D. 1985, Ap. J., 291, 747.
Weaver, R., McCray, R., Castor, J., Shapiro, P., and Moore, R. 1977,
    Ap. J., 218, 377.

CLOUD FORMATION AND DESTRUCTION

Bruce G. Elmegreen
IBM Thomas J. Watson Research Center
P.O. Box 218, Yorktown Heights, N.Y. 10598 USA

ABSTRACT. Cloud formation and destruction by processes related to shock fronts, cloud-cloud collisions and condensation instabilities are reviewed. The importance of magnetic fields in increasing the cohesion and cross section for cloud collisions is emphasized. The hierarchical structure of interstellar clouds is considered as a constraint on their possible formation mechanisms.

1. INTRODUCTION

Density variations in the interstellar medium are often described in terms of clouds embedded in an intercloud medium. This concept is reinforced by optical photographs of sharp-edged dark clouds, by well-defined absorption and emission lines from interstellar gas, and by spatial maps of color excess from dust concentrations. The picture is oversimplified, of course. "Cirrus" clouds observed by IRAS (Low *et al.* 1984) are stringy and highly textured. Most cloud boundaries are not well defined, and can vary with the atom or molecule used to observe them. A cloud's velocity can also depend on the spectral line used to observe it.

Theories of cloud formation and destruction usually concentrate on a particular type of cloud, such as a cool diffuse cloud or a giant molecular cloud. The heterogeneity of the interstellar medium limits this approach, however. Some diffuse clouds are local density structures on the periphery of giant molecular clouds (Federman and Willson 1982) and some giant molecular clouds are only the core regions of larger atomic clouds (Elmegreen and Elmegreen 1983; Grabelsky *et al.* 1986). This heterogeneity, or hierarchy of structures, implies that any particular cloud-like feature may have several formation mechanisms, one to make the density enhancement that is observed, and another to make the associated larger-scale structure. The concept of cloud and intercloud medium is particularly confusing when clouds appear hierarchically clumped (Pérault, Falgarone and Puget 1985; Scalo 1985); structure exists on a variety of scales within each cloud, and even the concept of a cloud boundary becomes ambiguous.

Bearing in mind these obvious difficulties with cloud definition, and the possibility of multiple formation mechanisms, we concentrate here on the three basic processes that give structure to interstellar gas: (1) shock accumulation of ambient gas into shells and filaments; (2) collisional build-up of large clouds from small clouds and (3) instabilities in the ambient medium or inside pre-existing clouds or shells. A discussion in § 5 illustrates how various processes may combine to give the observed cloudy structure.

259

D. J. Hollenbach and H. A. Thronson, Jr. (eds.), Interstellar Processes, 259–280.

2. CLOUD FORMATION BY SHOCK COMPRESSION OF AMBIENT GAS

2.1. Evidence that Some Interstellar Clouds are Swept-up Layers Behind Shock Fronts

Interstellar gas pressures vary by at least two orders of magnitude on time scales of 10^4 to 10^7 years as a result of supernova explosions, large-scale galactic flows (as in a density wave) and winds or HII regions from OB stars. Because only a factor of ~ 2 variation in pressure can drive a shock front, most of the distributed component of interstellar gas should be continuously shocked. The lowest density gas ($n_H < 10^{-3}$ cm^{-3}) may contain such fast and hot shocks that it doesn't have time to cool before it is shocked again (Cox and Smith 1974; McKee and Ostriker 1977). Higher-density gases can cool after being shocked, and they can form dense moving layers behind each front. An interstellar absorption line or cloud-like object can form whenever a shock front accumulates a sufficiently large column density of co-moving material. Because there are many possible sources of pressure, and many length scales over which these pressures operate, shock-related cloud formation can be extremely diverse. Many of the transparent "diffuse" clouds in the solar neighborhood could be swept-up regions behind active or former shock fronts originating in the Sco-Cen-Oph association (Frisch 1981; Crutcher 1982). Even some of the opaque (i.e., dark, molecular) clouds could be swept-up regions, but possibly older and gravitationally collapsed into distinct cores.

The simplest picture of interstellar clouds as standardized spheres began to change when Verschuur (1974), Heiles (1976) and others found numerous filaments and shells in wide-angle maps of neutral hydrogen (§2.2). The alignment of the interstellar magnetic field along these filaments and associated high-velocity gas suggested that they were swept up by shock fronts (§2.2). Other observations also suggest that interstellar clouds can be post-shock layers. Section 2.3 discusses dark clouds as shocked material, including wind-swept globular filaments and cometary globules. Section 2.4 discusses diffuse clouds, including observations of variable depletion of the elements and grain destruction behind fronts (§2.4.1), small line-of-sight cloud thicknesses (§2.4.2), variations of the thermal pressure inside the clouds (§2.4.3), CH$^+$ and rotationally excited H$_2$ behind the fronts (§2.4.4), and high excitation atomic transitions (§2.4.5). Global models of cloud formation behind shock fronts are discussed in §2.5.

2.2. Shells and Supershells

Shell or sheet-like structures in the distribution of interstellar hydrogen were noted by Heiles (1967), Ames and Heiles (1970) and Verschuur (1970). The shells often appear in conjunction with giant Hα loops (Elliot and Meaburn 1970) and radio continuum loops (Berkhuijsen, Haslam and Salter 1971; Haslam, Kahn and Meaburn 1971). Early models of these loops as accumulated gas behind shock fronts were made by Fejes (1971), Spoelstra (1972), Fejes and Verschuur (1973), Verschuur (1973) and Elliot and Meaburn (1973). Wide-angle HI surveys by Verschuur (1974) and Heiles (1975, 1976), and optical surveys by Sivan (1974) and Brand and Zealey (1975), demonstrated the ubiquity of interstellar shells.

Distant shells much larger than those found locally were catalogued by Heiles (1979), who termed them "supershells." A second catalog of large shells was made by Hu (1981), and an additional 5 loops that appear intermediate in size and age were found by Sofue and Nakai (1983). An optical study was made by Meaburn (1983). The kinetic energy of a supershell may be 3×10^{52} ergs or more.

Aside from the swept-up appearance of many interstellar shells and supershells, other evidence for shell formation by shock compression comes from the generally parallel mag-

netic field alignment (Spoelstra 1972; Cleary, Heiles and Haslam 1979) and increased mag-
netic field strengths in these regions (Troland and Heiles 1982; Broten, MacLeod and
Vallée 1985). Parallel field alignments are also observed in some dark clouds and filaments
(Schlosser and Schmidt-Kaler 1974; Appenzeller 1974; Vrba, Strom and Strom 1976;
McDavid 1984), which may be swept-up as well (§2.3). High velocity gas is also associated
with the loops (Bates et al. 1983).

Several explanations for the formation of shells and supershells have been proposed.
Some shells are obviously supernova remnants (as in the catalog by van den Bergh, Marscher
and Terzian 1973). Other shells are so large that they require several supernovae (Fich
1985), or a combination of pressures acting for 10^6 - 10^7 years, as might be expected in the
neighborhood of an OB association (Bruhweiler et al. 1980; Tomisaka, Habe and Ikeuchi
1981; Silich 1985). Giant shells may also result from collisions between high velocity halo
clouds and the galactic disk (Chow 1971; Tenorio-Tagle 1981). Observations of such a
collision, with or without shell formation, were discussed by Cohen (1981), Mirabel (1982),
Kulkarni, Dickey and Heiles (1985) and Mebold et al. (1985). The Cygnus superbubble
(Cash et al. 1980) is as large as a supershell, but it may be the superposition of many smaller
shells along the line of sight (Bochkarev 1984) or the result of a single supernova explosion
in a low-density medium (Higdon 1981). Some of the largest shells could also be driven by
radiation pressure from field stars (Elmegreen and Chiang 1982).

A well-studied example of a local interstellar feature that may be an expanding shell
is Lindblad's ring, which was originally recognized by the systematic recession of HI gas at
high galactic latitudes (Lindblad 1967; 1983). The extension of this feature into the southern hemisphere was discussed by Rickard (1975) and Franco and Pöppel (1978). Early
models of the ring assumed a ballistic expansion of the gas shell (Houghes and Routledge
1972; Lindblad et al. 1973); more recent models include back pressure from the ambient
interstellar medium (Olano 1982; Elmegreen 1985a). The age of the ring was estimated by
Olano (1982) to be 3×10^7 years, which is comparable to the age of the youngest expanding
group of stars in Gould's Belt (Tsioumis and Fricke 1979), and comparable to the age of the
centrally located Cas-Tau association (Blaauw 1956, 1984). Thus the pressure for the ring's
expansion may have come from the Cas-Tau association. The ring may be responsible for
the formation of the Ori OB1, Per OB2 and Sco-Oph molecular cloud complexes by
gravitational instabilities in the compressed gas (Olano 1982; Elmegreen 1982c). In that
case, even some giant molecular clouds could be post-shock gas.

2.3. Dark Clouds as Shocked Gas

Many dark clouds also appear to have been swept up by pressures from nearby stars or
supernova explosions (Schneider and Elmegreen 1979). Some filaments have an obvious
difference in the angular density of background stars on their two sides, as if the more
transparent side were cleared away, and the missing gas is in the filament. Some filaments
also have one edge sharp and the other diffuse, as if they were either folded sheets observed
edge-on, or cylinders partially confined by stellar wind or radiation pressure. The alignment
of a magnetic field perpendicular to the Taurus filaments suggests that they condensed by
gas motions along the field (Moneti et al. 1984). Many filaments also contain dense, round
globules that are more-or-less equally spaced, as in the Taurus complex (Gaida, Ungerechts
and Winnewisser 1984; Kleiner and Dickman 1984; Cernicharo, Bachiller and Duvert
1985). This suggests that the filaments collapsed by some regular instability after they were
swept up. A linewidth-size-density relation, and virialized motions in the embedded
globules, suggests that the instability is driven by self-gravity (Leung, Kutner and Mead
1982).

Cometary globules also show evidence for interactions between dense gas and an external pressure source. In some cases the interaction may only produce the comet tail (which is a reflection nebula) and not the cloud itself (as shown in models by Woodward 1976 or Nittmann 1981). In other cases, the comet head could be formed by direct compression from a wind, and the tail could be peripheral gas streaming away from the wind. Cometary globules are found on the periphery of the Gum nebula and elsewhere (Hawarden and Brand 1976; Sandqvist 1976; Brand *et al.* 1983; Zealey *et al.* 1983; Taylor, Taylor and Viale 1982). Similar clouds were found earlier by Dibai (1970), and one in the direction of M17 was mapped at 21 cm by Simonson (1973); this latter cloud shows a velocity gradient along its tail.

Globules in HII regions may form by Rayleigh-Taylor instabilities in the expanding flow (Spitzer 1954; Frieman 1954; van de Hulst 1955; Schneps, Ho and Barrett 1980; Brand 1980, 1981), or by expansion around density inhomogeneities in the adjacent cloud (Pikel'ner 1973; Pikel'ner and Sorochenko 1974). In either case, their formation is triggered by high pressures and discontinuous flows. Isolated globules may be formed by compression as well (Tomita, Saito and Ohtani 1979; Reipurth 1983).

2.4. Diffuse Clouds as Shocked Gas

2.4.1. *Velocity-Dependent Depletion of the Elements.* The variation with cloud velocity of the ratio of interstellar NaI to CaII abundance has long been taken as suggestive of grain destruction accompanied by release of refractory elements behind interstellar shock fronts (Routley and Spitzer 1952; Siluk and Silk 1974; Jura 1976; Shull, York and Hobbs 1977; Cohen 1977; Cowie and York 1978a; Hobbs 1983). Other elements, such as Fe and Ti show the same variations with cloud velocity (for a review of observations see Spitzer and Jenkins 1975).

Several theories of grain destruction behind shock fronts have been proposed. Some involve grain sputtering in the high temperature region behind the front (Barlow and Silk 1977, Barlow 1978), and others consider grain-grain collisions, with the grains continuously accelerated by the compressed magnetic field (Spitzer 1976; Shull 1977a, 1978; Cowie 1978). Both theory and observations are consistent with the idea that many, if not all, high and intermediate velocity diffuse clouds ($v > 20$km s^{-1}) are the swept-up regions behind shock fronts (Siluk and Silk 1974). Many low velocity diffuse clouds could be older and decelerated versions of the faster clouds, if grain growth is fast enough to deplete refractory elements in the deceleration time.

2.4.2. *Sheet-Like Geometries for Diffuse Clouds.* The line of sight thickness of an interstellar cloud can be determined from the ratio of the column density, N, to the space density, n. Observations of column density have been available for a long time. Observations of space density require density-sensitive atomic or molecular transitions, such as the CI fine structure line or the CO ultraviolet bands, which are collisionally excited. Observations of these and other transitions with the Copernicus satellite indicate that many of the local diffuse clouds have ratios $N/n \simeq 0.1$ pc or less, which is much smaller than the clouds' transverse dimensions, as determined by mapping which stars have that particular absorption line feature.

Morton (1975) determined the electron density in a cloud on the line of sight to ζ Oph from the ionization equilibrium of C, Mg, S and Ca. He assumed that the gas was photoionized, so $n_e/n_H \simeq 7 \times 10^{-5}$, and then derived a density of $n_H \simeq 10^4$cm^{-3}. This was combined with the observed column density to give a cloud thickness of 0.05 parsec. Smith, Krishna Swamy and Stecher (1978) measured a lower density in this same cloud from CO collisional excitation, and derived a larger cloud thickness of 0.5 to 4.6 pc. They concluded that the cloud was a thin layer swept up by a supernova explosion in the Sco OB1 association.

de Jong (1980) suggested that this cloud is a shock front ahead of an HII region. A similar cloud thickness (~0.2 pc) was suggested for a cloud on the line of sight to ζ Per (Snow 1977); the transverse cloud dimension is probably much larger than this, 12 pc, because this velocity feature also appears in the spectrum of o Per, which is 2° away.

Spitzer and Morton (1976) derived cloud thicknesses from observations of H_2 excitation. They found that the absorbing regions are generally very thin, ~0.02 pc, that the uv radiation field in the neighborhood of these regions usually exceeds the interstellar average by a factor of ~20, and that the rotationally excited H_2 is systematically shifted in velocity from the lower rotational states (cf. § 2.4.4). These observations indicate that the absorbing gas is near the background OB star, and is probably shocked gas in the high-pressure environment of the OB cluster. Another survey of cloud thicknesses was made by Giovanelli et al. (1978), who determined N from 21 cm observations and n from H_2 excitation. The 21-cm emitting regions were found to be very thin (~0.03 pc), again resembling compressed layers behind shock fronts.

2.4.3. *Variations in the Thermal Pressure in Diffuse Clouds.* Large variations in the internal pressures of diffuse clouds (Tarafdar 1978; Jenkins and Shaya 1979; Jenkins, Jura and Lowenstein 1983) are consistent with a shock-related origin for some of these objects, or a compression of pre-existing clouds by a passing shock front (McKee and Ostriker 1977; Cox 1979; Ferlet et al. 1980; Tenorio-Tagle and Rozyczka 1986). Pressure variations among clouds in obviously active regions are found in Orion (Cowie, Songaila and York 1979), the Gum nebula (Wallerstein, Silk and Jenkins 1980) and the Vela supernova remnant (Jenkins et al. 1981). High resolution HI maps of the vicinity of supernova remnants (Braun and Strom 1986) show some clouds with dense cores and high-velocity envelopes, as if the envelope were swept back, and other clouds resembling recombined gas comoving with a shock front.

2.4.4. CH^+ *and Rotationally Excited* H_2 *in Diffuse Clouds.* CH^+ and rotationally excited H_2 have been attributed to chemical processing and excitation behind shock fronts (Elitzur and Watson 1978, 1980; Aannestad and Field 1973). This interpretation is consistent with the observed velocity shift between CH and CH^+, as if these two molecules occur in different parts of a post-shock flow (Chaffee 1975; Frisch 1979; Crutcher 1979; Federman 1980; 1982a), and with a correlation between cloud velocity and the abundance of rotationally excited H_2 (Spitzer and Morton 1976; Martin and York 1982). Also consistent is the correlation between CH^+ and rotationally excited H_2 (Frisch and Jura 1980; Lambert and Danks 1985), and the lack of a correlation between CH^+ and ground state H_2 (Federman 1982b).

The theory of shock chemistry and excitation is complicated by several factors. The formation of OH behind the front can cool the gas and limit the amount of CH^+ that forms (Flower et al. 1982). Continuous magnetic shocks instead of hot, jump-type shocks also affect the chemistry (Flower et al. 1985, 1986), as can ultraviolet radiation (Freeman and Williams 1982). Even the most complex reaction schemes computed so far have not yet succeeded in reproducing the observed CH^+ abundances (Mitchell and Watt 1985).

2.4.5. *High Excitation Spectra of Diffuse Clouds.* Observations of highly excited transitions in interstellar clouds are another indication that the gas has been recently shocked. A review of these high temperature features may be found in McCray and Snow (1979) and McKee and Hollenbach (1980). High ratios of SiIII/SiII imply that 50 - 100 km s^{-1} shocks are present in some clouds (Shull 1977b). This diagnostic has been used, for example, by York (1983), who studied 5 absorption line systems on the line of sight to λSco, and found that 2 of them originate in hot post-shock regions.

These observations of shocks or compression in diffuse clouds reveal the most violent aspects of the interstellar medium. The fraction of clouds that form by these processes is not yet known.

2.5. Global Models for Shock-Related Formation and Destruction of Interstellar Clouds

When the steady-state model of cloud formation by thermal instability (Field, Goldsmith and Habing 1969) began to look implausible because of insufficient heating from cosmic rays or X-rays (see §4.1), the time-dependent models of interstellar gas evolution began to look more promising (Dalgarno and McCray 1972; Schwarz, McCray and Stein 1972; Mansfield 1973; Schwarz 1973; Gerola et al. 1973, 1974; Kafatos et al. 1974). These time-dependent models relied on sporadic events of high pressures and temperatures to give the interstellar medium its ionization and temperature structure, and to make some of the diffuse clouds either by thermal instabilities in the cooling gas or by shock compression. Most of the global models of interstellar gas evolution discussed today might be viewed as descendants of the early time-dependent models, with important modifications to account for revised concepts of the intercloud medium. In the late 1960's the intercloud medium was thought to be warm (8000K), of moderate density (0.2 cm^{-3}), and partially ionized ($n_e/n_H \simeq 0.1$); in the early 1970's, following the observation of bare O-type stars (Torres-Peimbert, Lazcano-Araujo and Peimbert 1974) and diffuse Hα emission (Reynolds, Roesler and Scherb 1974), the intercloud medium was thought to be warm (8000K), still of moderate density (0.03 to 0.2 cm^{-3}), and highly ionized ($n_e/n_H \simeq 1$), like an HII region; and since the mid-1970's, following observations of the diffuse soft X-ray background (Burstein et al. 1976) and interstellar absorption lines from OVI (Jenkins and Meloy 1974; York 1974), the intercloud medium has been thought to be hot (10^6 K), at low density ($10^{-2} - 10^{-3}$cm^{-3}), and fully ionized and heated by stellar winds and supernova explosions. Models of pressurized cloud formation were made for each of these concepts of the intercloud medium.

The first shock-related cloud formation models in this historical sequence were those by Hills (1972; 1973), who conjectured that uv stars ionize and heat the intercloud medium, and that neutral clouds are regions between the Strömgren spheres (see also Castle and Hills 1974). Rose and Wentzel (1973) considered the time evolution of uv stars, and calculated the filling factor of the ionized volume. Rose and Wentzel also pointed out that the Strömgren spheres would not get enough time to expand, which implies that the clouds are not formed behind shock fronts; they are only the pockets of residual gas that could not be reached by uv photons. The Hills' model gave a steady rate of cloud formation and destruction and could account for the observed sizes and space densities of diffuse clouds. A problem with the model arose when Mezaros (1973a) noted that the ionization structure observed by the Copernicus satellite was not consistent with uv photon energies.

A modification to the model came from Lyon (1975), who suggested that main sequence OB stars heat and ionize the intercloud medium, and that the expanding Strömgren spheres compress the gas and make new clouds. A detailed computer model by Bania and Lyon (1980) matched the observed cloud column densities, radii, temperatures and internal velocity dispersions, and it gave the observed line-of sight cloud separation, the cloud-cloud velocity dispersion (if additional energy input from supernova was allowed), and the cloud mass spectrum. The model also obtained acceptable temperatures, ionization fractions and filling factors for the warm neutral intercloud medium. None of the input physics could produce a hot intercloud phase however, so the model could not explain the soft X-ray background or the OVI.

A new generation of models with a hot intercloud medium was discussed by Salpeter (1976) in a 1974 talk, and by Cox and Smith (1974), who discovered that random supernova could produce and maintain high temperature cavities that overlap in space. A more com-

prehensive model of a supernova-dominated intercloud medium was mad_ by McKee and
Ostriker (1977). They included enhanced energy losses from radiative interfaces in evaporating cloud envelopes, and solved for the steady state in which energy input from explosions
balances the energy lost at the interfaces. A steady state between the rate of cloud destruction by evaporation and the rate of cloud growth by accretion of cooled shock fronts
was also found. The model explained the cloud velocity dispersion, the interstellar pressure,
and the line-of-sight spacing between diffuse clouds, and it predicted a cool-core/warm-halo
structure for HI condensations.

Several predictions of the McKee-Ostriker model have been confirmed by observations. The minimum cloud size that is allowed by evaporative losses (see also Chièze and
Lazareff 1980) has been observed as a decrease in the cloud size distribution function for
clouds smaller than ~2 parsecs (Knude 1981b; Crovisier, Dickey and Kazés 1985). The
cool-core/warm-halo structure of diffuse HI clouds has been discussed extensively (Dickey,
Salpeter and Terzian 1977; Mebold et al. 1982; Liszt 1983; Payne, Salpeter and Terzian
1983; Kalberla, Schwarz and Goss 1985), and the observed cloud filling factor seems to be
in agreement (Knude 1981a). The occurrence of OVI at the evaporative interface between
a cloud and the hot intercloud medium has also been discussed (Cowie et al. 1979). Two
possible densities for the intercloud medium are allowed by observations of cloud space
densities (Cowie and York 1978b), one exceeds 0.1 cm^{-1}, and the other is less than
7×10^{-3}cm^{-2}, which is sufficiently low that supernova remnants can overlap before they
reach the radiative phase.

Theoretical developments in the McKee-Ostriker model include further studies of
cloud evaporation, energy input from winds, and cloud-supernova interactions. Following
the initial work on evaporation by Cowie and McKee (1977) and McKee and Cowie (1977),
the effect of infrared radiation from hot dust in evaporating clouds was considered by Dwek
(1981), suprathermal evaporation was discussed by Balbus and McKee (1982), the opacity
of the evaporating outflow was discussed by Königl (1984), gradual saturation was studied
by Giuliani (1984), viscosity was included by Draine and Giuliani (1984), geometric effects
were calculated by Balbus (1985), and magnetic fields were considered by Balbus (1986b).
Other models of supernova-dominated interstellar media were discussed by Habe, Ikeuchi
and Tanaka (1981) and Ikeuchi and Tomita (1983). The additional importance of stellar
winds as a source of energy input to the interstellar medium was discussed by Basu (1972),
Abbott (1982), Tarrab (1983), McCray (1983) and van Buren (1985). Acoustic heating
of neutral cloud halos subject to supernova shock waves was considered by Ikeuchi and
Spitzer (1984), heating by weak shocks was considered by Sabano and Tosa (1978), and the
formation of some of this warm HI during cloud-cloud collisions was discussed by Clifford
(1984). The potential importance of magnetic coupling between expanding supernova shells
and ambient diffuse clouds was emphasized by Clifford (1984) and illustrated by Oettl,
Hillebrandt and Müller (1985).

A potential problem with the McKee-Ostriker model is indicated by the lack of detectable X-ray emission from the galaxy M101 (Cox and McCammon 1986). This seems to
imply that remnant evolution with evaporation is impossible unless the supernova rate is
lower than expected, perhaps by a factor of 3 to 10. Another problem is that shocks may
remove the proposed cloud halos so quickly that not much gas can be in this phase
(Heathcote and Brand 1983).

More elaborate models that illustrate the general behavior of interstellar gas subject to
various pressures, energy sources, star formation and so on, were made by Shore (1983),
Franco and Cox (1983), Fujimoto and Ikeuchi (1984), Ikeuchi, Habe and Tanaka (1984),
Chiang and Prendergast (1985), Bodefée and de Loore (1985) and Dopita (1985). These
models illustrate how complex the galactic gas system can be when cloud formation and destruction are parts of a feedback cycle including star formation and supernovae.

3. CLOUD FORMATION BY THE AGGLOMERATION OF SMALLER CLOUDS

3.1. The Physics of Cloud-Cloud Collisions

3.1.1. *Non-Magnetic Collisions*. The outcome of a collision between two non-magnetic clouds depends on their relative sizes, densities and velocities, and on the collision's impact parameter. A cloud collision produces two shocks and a high pressure shocked layer at the interface (Kahn 1955). The high pressure layer is confined in the collision direction by the ram pressure at the shock front, but it is not confined in a direction perpendicular to the collision, so it tends to expand along the shock plane (Stone 1970). This expansion pushes on the non-overlapping parts of the cloud and enlarges the original shock into a wake, which may go all the way through both clouds. The wake transfers momentum from the overlapping part of the collision to the non-overlapping parts. If the total momentum of the overlapping part is small at the beginning of the collision, or if it can be decreased sufficiently because of momentum transfer to the non-overlapping parts of the clouds, then the overlapping part will coalesce with the non-overlapping parts. A criterion for coalescence is that the initial Mach number of the overlapping part relative to the non-overlapping part must be less than approximately 2 (Hausman 1981).

Fast collisions produce fragmentation, with possibly 2 or 3 fragments resulting (Chièze and Lazareff 1980; Hausman 1981; Lattanzio *et al*. 1985). Off center collisions tend to produce fragmentation or disruption also (Gilden 1984a). Two of the resulting fragments are the non-overlapping parts of the clouds and the third fragment is the coalesced overlapping part. This third fragment may take the form of a filament or bridge between the other two parts.

The tendency toward fragmentation rather than coalescence at high collision velocities implies that collisional build-up of (non-magnetic) clouds is unlikely if the component clouds have a velocity comparable to the observed velocity dispersion of diffuse clouds (Hausman 1981). Only very slow collisions can build up large clouds. This is a severe limitation to the collisional build-up model, and it is especially restrictive if the most important collisions are in the shock-like region of a spiral density wave arm (§3.2.2.). Clouds enter a spiral arm at such high velocities that their collisions with other clouds already in the arm are more likely to disrupt each other than coalesce.

Chièze and Lazareff (1980) proposed that clouds increase their mass in part by collisions, and in part by accretion from cooled supernova shock fronts, as in the McKee-Ostriker model (§2.5). They also allowed collisions to break down the clouds into smaller pieces. They obtained reasonable agreement with the observed mass spectrum of clouds.

The formation of molecules in the shocked layer between two clouds was discussed by Smith (1980) and instabilities in the layer were found by Hunter *et al*. (1986).

3.1.2. *Magnetic Collisions*. Magnetic collisions can be more cohesive than non-magnetic collisions because the field lines from one cloud can become entangled in the fields lines of the other cloud (Clifford and Elmegreen 1983; Clifford 1985). Such magnetic entanglement can rapidly transfer all of the momentum from the one cloud to the other. The resulting pair may oscillate around their interconnecting field lines and dissipate their energy slowly during this oscillation, or they may eventually collide and then lose their energy rapidly.

Magnetic fields also increase the cross section for collisions by allowing remote field lines to ensnare a cloud (Clifford and Elmegreen 1983). The collision cross section for magnetic collisions is approximately $(\rho_c v_A / \rho v)^{2/3}$ multiplied by the physical cross section, where ρ_c is the density in the cloud, ρ is the average density of all of the cloudy material spread out uniformly, v is the collision velocity and v_A is the Alfvén velocity in the cloud. For typical $\rho_c / \rho \simeq 50$ and $v / v_A \simeq 10$, the effective cross section equals $\simeq 2.5$ times the geometric cross section. The coalescence time is smaller for magnetic collisions than non-magnetic collisions by this same factor.

Clumps that are observed in giant molecular cloud complexes (Pérault, Falgarone and Puget 1985) could be remnants of the former clouds that coalesced to make the larger structure. If the collisions are magnetic, then the build-up process can be relatively soft and non-destructive. If the collisions are not magnetic, then most of the collisions may just shred or destroy the primordial clouds, in which case the observed clumpy structure may have a different origin. Coalescence by magnetic entanglement appears to be the only way a giant molecular cloud can preserve the cloudy structure of the multi-component gas that went into it (if in fact the observed clumps are remnants of this structure).

3.2. Models of Cloud Formation by Collisional Agglomeration

3.2.1. *Random Collisional Agglomeration.* Collisional build-up of large clouds from small clouds is random when the relative cloud velocities are random and isotropic. Random coagulation models have been discussed by a number of authors (Oort 1954; Field and Saslaw 1965; Field and Hutchings 1968; Penston *et al.* 1969; Taff and Savedoff 1973; Handbury, Simon and Williams 1977; Scoville and Hersh 1979; Kwan 1979; Cowie 1980, 1981; Hausman 1982; Pumphrey and Scalo 1983; Nozakura and Ikeuchi 1984; Scalo and Struck-Marcell 1984, 1986; Struck-Marcell and Scalo 1984). The basic model has small clouds coalesce into a large cloud, and then the large cloud breaks apart and disperses back into small clouds. This process is sometimes assumed to reach a steady state, and the resulting mass and velocity distribution functions are then calculated.

Casoli and Combes (1982) point out that cloud collisions may not reach a steady state between spiral density wave shocks. This is an important modification to the random build-up model because the observed velocity distribution function, i.e., where velocity is nearly independent of cloud mass, can only be obtained without kinetic energy equipartition. Casoli and Combes obtain the observed cloud mass spectrum and velocity distribution function with their model.

A problem with the random agglomeration model for the formation of giant molecular clouds is that (without magnetic entanglement as the primary type of interaction) there are too few molecular clouds to coalesce with each other before each one self-destructs by star formation. This can be understood as follows. A unifying relationship between mass, M, density, n, velocity dispersion, Δv, and radius, R, appears to exist for molecular clouds, such that (Falgarone and Puget 1986)

$$\left(\frac{M}{100\ M_\odot}\right)^1 \simeq \left(\frac{\Delta v}{0.36\text{km s}^{-1}}\right)^4 \simeq \left(\frac{R}{\text{parsecs}}\right)^2 \simeq \left(\frac{n}{570\text{cm}^{-3}}\right)^{-2}. \quad (1)$$

The differential number density of molecular clouds is

$$n(M) \simeq 10^{-8.18} \left(\frac{M}{100\ M_\odot}\right)^{-1.5} \text{pc}^{-3}, \quad (2)$$

where the observed mass spectrum, $n \propto M^{-1.5}$ (Solomon, Sanders and Scoville 1979; Liszt, Xiang and Burton 1981; Dame 1983; Casoli, Combes and Gerin 1984; Bhatt, Rowse and Williams 1984), has been normalized to an average molecular density in the solar neighborhood of 0.2 molecules cm^{-3} (Dame and Thaddeus 1985) for cloud masses between $10^2\ M_\odot$ and $10^6\ M_\odot$. Equation (1) implies that the cloud cross section is $\sigma = \pi R^2 = \pi M_2$, where $M_2 = M/100\ M_\odot$. The average mean free path for collisions between clouds with masses between $10^4\ M_\odot$ and $10^6\ M_\odot$ is therefore

$$\lambda = \left(\int_{10^4}^{10^6} n(M)\sigma(M)dM \right)^{-1} = 2.7 \text{ kpc.} \tag{3}$$

The one-dimensional velocity dispersion for clouds of this mass is 6.6 km s^{-1} (Stark 1984), so the average cloud collision velocity is $v = 6.6 \times 6^{0.5} = 16$ km s^{-1}. Thus the mean collision time is

$$\tau_{coll} = \lambda/v = 1.7 \times 10^8 \text{years} \quad . \tag{4}$$

for 10^5 M$_\odot$ clouds. This assumes that the most important collisions are between clouds of equal mass, which is approximately true for random build-up models. The time to build a giant molecular cloud by accreting smaller clouds is also between 10^8 and 10^9 years, because the collision time for a single diffuse cloud and a giant molecular cloud is between 10^6 and 10^7 years, and at least 100 diffuse clouds must coalesce to make a 10^5 M$_\odot$ cloud.

These build-up times are too long to be reasonable. Giant cloud complexes probably dissipate their turbulent energy, form stars and disperse in only 5 to 10 internal crossing times, which is between 10^7 and 10^8 years (Blitz and Shu 1980; Scalo and Pumphrey 1982; Elmegreen 1985b). The internal crossing time for the above scaling laws is

$$\tau_{cross} = (G\rho_c)^{-0.5} = 2.6 \times 10^6 M_2^{0.25} \text{ years .} \tag{5}$$

for cloud density $\rho_c = n\mu$ and mean molecular weight $\mu = 2.3m_H$. Thus the internal crossing time in a 10^5 M$_\odot$ cloud is 1.5×10^7 years, and

$$\tau_{coll}/\tau_{cross} \simeq 11 \quad . \tag{6}$$

This implies that collisions between 10^5 M$_\odot$ clouds occur after an average of 11 internal crossing times, and therefore that giant molecular may evolve and disperse before they get a chance to collide with each other. The mean collision velocity of ~16 km s^{-1} may also be too large to promote collisional build-up; most collisions may only disrupt non-magnetic clouds.

Magnetic interactions (§3.1.2) change this conclusion because they soften the collisions and shorten the collisional build-up time enough that random coalescence can occur in less than an internal evolution time. The ratio of the effective magnetic collision cross section to the geometric cross section is $(\rho_c v_A/\rho v)^{2/3}$, as given above. For the scaling laws of molecular clouds, this ratio equals

$$\frac{\sigma_{eff}}{\sigma} \simeq \left(\frac{\rho_c/\rho}{v/v_A} \right)^{2/3} = \left(\frac{2850M_2^{-0.5}}{143M_2^{-0.32}} \right)^{2/3} = 7.4M_2^{-0.12} \quad . \tag{7}$$

Here we have assumed that the average density, ρ, equals 0.2μ gm cm^{-3}, that the internal Alfvén velocity equals $1/2$ of the internal cloud velocity dispersion, Δv, and that the cloud-cloud collision velocity equals $6^{1/2}$ times the one-dimensional velocity dispersion of $10.5M_2^{-0.07}$, which was evaluated to give 9 km s^{-1} at 10^3 M$_\odot$ and 6.6 km s^{-1} at 10^5 M$_\odot$ (Stark 1984). For a 10^5 M$_\odot$ cloud, the effective cross section equals 3.2 times the geometric cross section. The number of internal crossing times in a magnetic collision time for such a cloud is therefore

$$\frac{\tau_{coll}}{(\sigma_{eff}/\sigma)\tau_{cross}} = 3.4 \quad , \tag{8}$$

which implies that some internal evolution takes place between collisions of 10^5 M$_\odot$ clouds, but perhaps not enough to exclude the collisional build-up model.

Because σ_{eff}/σ is large for most cloud masses, molecular cloud collisions should be magnetic; only a fraction of collisions equal to the inverse of this ratio will be physical impacts. The resulting cloud complex should therefore be very loose, with well-separated molecular cores that were once individual clouds. Presumably these cores will begin to coalesce physically in a few internal crossing times, so the internal structure will have changed by the time stars form (§5).

3.2.2. Forced Agglomeration in Converging Flows. The converging flow in a spiral density wave can force the clouds to move closer together, and this will decrease their collision time because of the higher cloud density (Norman and Silk 1980; Casoli and Combes 1982; Yuan and Wang 1982; Kwan and Valdes 1983, Stark 1983; Levinson and Roberts 1981; Roberts and Hausman 1984; Hausman and Roberts 1984; Combes and Gerin 1985; Tomisaka 1984, 1986; Carlberg and Freedman 1985). If the clouds are magnetic they should coalesce rapidly, without severe fragmentation and disruption. Magnetic agglomeration of giant cloud complexes seems inevitable when density waves are present. It may account for the appearance of the largest regions of star formation in the spiral arms of galaxies.

Forced agglomeration also occurs when pre-existing clouds are swept up into a giant shell by a large-scale pressure disturbance; thus a supershell is probably a collection of individual diffuse clouds. Large scale instabilities also force the agglomeration of individual clouds, because they bring together the ambient gas that is composed of these clouds.

4. CLOUD FORMATION BY SPONTANEOUS INSTABILITIES IN THE AMBIENT MEDIUM

4.1. The Thermal Instability and Related Instabilities

The original concept that a thermal instability (Field 1965; Field, Goldsmith and Habing 1969) produces the two-phase structure of interstellar matter observed in HI (Clark 1965) was found to be unacceptable after the Copernicus satellite failed to observe the degree of ionization that was expected from cosmic ray and X-ray heating (Mezaros 1973b; O'Donnell and Watson 1974; Hobbs 1974). More recent work on the instability (Shull and Woods 1985; Lepp *et al.* 1985) includes elemental depletion and X-ray heating to a third, "coronal" phase. The three dominant phases of interstellar gas, i.e., cold atomic or molecular clouds, warm HI or HII, and a hot intercloud medium, could each be thermally stable phases, and transitions between these phases could form some of the observed cloud structures.

The thermal instability also has other applications. Bregman (1980) discussed the formation of galactic halo clouds by thermal instabilities in the hot halo gas. Several authors applied this instability to the formation of condensations behind shock fronts (Avedisova 1974; McCray, Stein and Kafatos 1975; Mufson 1974, 1975; Smith 1980; Hong and Koo 1984; Kritsuk 1985). The application to density wave shocks was made by Shu *et al.* (1972), Biermann *et al.* (1972) and Marochnik *et al.* (1983). A slightly different application is to molecular clouds, where the thermal instability has been combined with molecule formation (Sabano 1972; Yoneyama 1973; Reddish 1975; Glassgold and Langer 1976; Sabano and Kannari 1978; Yoshii and Sabano 1980; Sofue and Sabano 1980; Kannari 1981; Ibañez 1981; Ibañez and Parravano 1983; Gilden 1984b). Self-gravity has been included by Kannari, Sabano and Tosa (1979) and Balbus (1986a). Thermal instabilities, molecule-forming instabilities (Giaretta 1979), and acoustic instabilities (Flannery and Press 1979) apparently induce some of the fine structure inside clouds, and they may also form clouds directly.

4.2. The Parker Instability

The Parker (1966) instability is a magnetic Rayleigh-Taylor instability in the galactic gas layer, with cosmic rays and magnetic buoyancy taking the place of the light fluid, and the interstellar gas representing the heavy fluid. The characteristic length for the instability is approximately $2\pi H$ for scale height H, so the important condensations that form are presumably the largest cloud complexes that are observed in galaxies (Mouschovias, Shu and Woodward 1974; Mouschovias 1974; Levy 1978; Blitz and Shu 1980; Elmegreen 1982b). The growth time of the instability is approximately the free fall time over one scale height, which is $\sim 2 \times 10^7$ years. The growth time can be considerably longer (by a factor of at least 2) when rotation is included (Zweibel and Kulsrud 1975), but the instability is still present with rotation (Lerch 1967a; Shu 1974).

The instability grows rapidly on short length scales in a direction perpendicular to the field and in the plane of the galaxy (Parker 1967; Lerch 1967b; Asseo *et al.* 1978; Lachièze-Rey *et al.* 1980; Cesarsky 1981). This transverse growth may imply that only chaotic or turbulent structures can be formed, and not coherent cloud complexes. The instability is also driven entirely by pressures from cosmic rays, magnetism and thermal or turbulent motions, so if a distinct condensation forms, it cannot have a pressure much larger than a factor of ~ 3 more than the ambient pressure. This implies that other forces are necessary to produce giant molecular cloud complexes, which have internal turbulent pressures much larger than the ambient pressure (by a factor of 10 - §5).

The inclusion of self-gravity may solve both of these problems (i.e., chaotic growth and low internal pressures), because it binds the condensing gas in all directions, even transverse to the mean field direction, and because it compresses the condensation to high densities (Elmegreen 1982a). Self-gravity also makes the instability grow faster. The relative importance of cosmic ray pressure, P_{CR}, magnetic field pressure, $P_B = B^2/8\pi$ for field strength B, thermal pressure, $P = \rho c^2$ for density ρ and sound speed c, and self-gravity can be assessed from the relative values of the dimensionless pressure ratios, $\alpha = P_{CR}/P$ and $\beta = P_B/P$, and mid-plane density, $s = 8\pi G\rho H^2/c^2$. For the interstellar medium in the solar neighborhood, $\alpha \simeq 0.2$, $\beta = 0.4$ and $s \simeq 0.9$ if the average density is 0.5 atom cm^{-3}, $c = 7$km s^{-1}, $B = 2.2 \times 10^{-6}$ Gauss, $H = 160$pc and $P_{CR} = 4 \times 10^{-13}$dyn (Elmegreen 1982a). The relatively large value of s implies that self-gravity is equally important as a driving force for the instability at ambient densities, and that self-gravity may dominate at slightly higher densities.

Realistic equilibrium models for the interstellar gas consider the cloud and intercloud media as distinct components (Fuchs and Thielheim 1979). These authors also account for height variations in the gravitational acceleration and relative proportion of cloud and intercloud gas. The equilibrium of a three-phase medium was discussed by Schlickeiser and Lerch (1985). The importance of an isotropic cosmic ray pressure was emphasized by Nelson (1985), who suggested that anisotropic pressures are likely, and that this will tend to stabilize the gas. Stability may also occur if cosmic ray diffusion is slow (Kuznetsov and Ptuskin 1983). Application of the Parker instability on a small scale, to fragmentation inside a cloud, was made by Baierlein, Schwing and Herbst (1981) and Fleck (1984).

4.4. The Gravitational Instability

Gravitational instabilities in the interstellar medium have a length scale similar to the Parker instability, and the two mechanisms may operate simultaneously to form the largest structures. Self-gravity should be the dominant force, however, especially after the density increases over the ambient value by a factor of 1.5 to 2 (Elmegreen 1982b). The self-gravitational instability operates on the Jeans length scale, $c/(G\rho)^{1/2}$, so it can produce both the largest structures (at low ρ), and also the hierarchy of clumped structures inside of these (at higher ρ). The time scale for the instability is approximately $(G\rho)^{-1/2}$, which is $\sim 2 \times 10^7$ years in the ambient medium.

Large-scale gravitational instabilities in galactic disks should be unavoidable if the magnetic field is relatively continuous over the Jeans length scale (Elmegreen 1987a). The magnetic field removes angular momentum from a growing condensation and counteracts the resistance to collapse imposed by rotation (Chandrasekhar 1954; Lynden-Bell 1966). Without a magnetic field, the gas in a galaxy is normally stabilized by rotation (Toomre 1964); the instability grows either by transient non-axisymmetric instabilities that are sheared away (Goldreich and Lynden-Bell 1965), or it grows after large-scale gas compression, as in a density wave (Elmegreen 1979; Cowie 1981; Balbus and Cowie 1985). When a uniform magnetic field is present, such large scale compression is not necessary to trigger the instability, although it might decrease the growth time. The instability is also likely to grow in swept-up shells and supershells, and it can produce molecular cloud complexes along the rims of these objects (Tomisaka and Ikeuchi 1983; see reviews by Elmegreen 1985a, 1987b).

Because self-gravitational forces are always present in the interstellar gas, the gravitational instability can operate in conjunction with many other cloud formation mechanisms. Applications to the thermal and Parker instabilities have already been discussed. Gravity's contribution to cloud coagulation models is obvious because gravity increases the collision cross section, and because it helps bind the resulting complex together, thereby minimizing fragmentation and disruption during individual impacts. Large scale gravitational instabilities might also drive the coagulation in the first place, as small-scale clouds are brought together during the growth of a large-scale perturbation. Gravitational instabilities also form clouds or clumps in the dense gas behind shock fronts and in expanding shells (Elmegreen and Elmegreen 1978; Nakano and Nakamura 1978; Ghosh and Bhattacharyya 1985).

5. AN OVERVIEW OF CLOUD FORMATION AND DESTRUCTION

The interstellar medium is inhomogeneous, with clouds separated by an intercloud medium on scales larger than approximately 0.1 to 1 parsec. This inhomogeneity often appears hierarchical, in the sense that the largest complexes contain separate clouds, each of which may contain denser and smaller clumps, and each of these clumps may contain even denser and smaller cores (Schneider and Elmegreen 1979; Pérault, Falgarone and Puget 1985; Falgarone and Puget 1985; Scalo 1985). The unifying relationship between mass, M, density, n, velocity dispersion, Δv, and radius, R, for molecular clouds, discussed in §3.2.1, suggests that the largest hierarchies are virialized ($M/R\Delta v^2 =$ constant), and they have approximately the same turbulent pressure ($n\Delta v^2 =$ constant) and column density ($nR =$ constant). The turbulent pressure equals $2.1 \times 10^4 k_B$, which exceeds the average pressure inside diffuse clouds by a factor of ~ 8.

Because of this hierarchical structure, there are essentially two important scales, the smallest scale, below which there are no more hierarchies, and the largest scale, which is apparently limited by available space in the Galaxy (i.e., the Galactic scale height). There may be no need for a distinction between diffuse clouds and molecular clouds. The apparent distinction between these two types of clouds may only be an artifact of different observation techniques. In the solar neighborhood, molecules begin to form at the column density where self-gravity in a cloud becomes important (Elmegreen 1985c). Thus dark clouds tend to be round and at high pressures because of self-gravity, and molecular because of self-shielding. Transparent clouds are more amorphous and transient because self-gravity is not involved in their confinement. The formation mechanisms for these two types of clouds may be the same for a given mass. Perhaps diffuse clouds can be viewed as the low column density portion of the cloud size spectrum.

The smallest scales in the hierarchy are probably in total pressure equilibrium with the surrounding gas; their internal pressures are determined by thermal motions, turbulence or magnetosonic waves, and magnetic fields. These smallest scales might be viewed as the McKee-Ostriker clouds, in addition to any substructure inside these clouds. They are not strongly self-gravitating, they should have a core and multiple-halo structure determined by the various penetration depths of different types of energy (cosmic rays, X-rays, magnetosonic waves, non-ionizing starlight and ionizing starlight), and they should be continuously compressed and pushed around as supernovae perturb their environment. They may grow by accretion from cool supernova shells as in the McKee-Ostriker model, by condensation of warm or hot gas from their surroundings (McKee and Cowie 1977), or by fragmentation in supernova or other shells, possibly as a result of post-shock thermal instabilities (§4.1), gravitational instabilities (§4.3), or other types of instabilities (e.g., Vishniac 1983). They could also form by collision-induced fragmentation of larger clouds (Chièze and Lazareff 1980). They are probably destroyed by evaporation in a hot environment (McKee and Ostriker 1977), by ionization from nearby OB stars or uv stars (Elmegreen 1975), by direct collisions with similar clouds at velocities exceeding ~4 km s^{-1} (Hausman 1981), and by passing shock fronts (Nittmann and Falle 1982; Heathcote and Brand 1983).

The next larger hierarchy is a collection of the smallest clouds, possibly brought together by random motions and then stuck together by magnetic entanglements, or possibly swept up and forced to coalesce by some large-scale pressure disturbance, as in a shell. Such coagulation will preserve most of the original cloud-intercloud structure if magnetic fields are involved in the collisions, but a few direct collisions could destroy some of the components and possibly add to the intercloud or interclump gas in the region. The turbulent pressure of these larger hierarchies probably exceeds the ambient pressure. This turbulence may be sustained by continued collisions (Bash, Hausman and Papaloizou 1981), star formation (Norman and Silk 1980), galactic shear (Fleck 1981) and magnetic cloud interactions (Falgarone and Puget 1986).

Successively larger hierarchies may be brought together by the same magnetic coagulation, but larger-scale gas flows, as in a density wave or giant pressurized shell, or larger-scale forces, such as self-gravity and galactic gravity directed along curved magnetic field lines, are probably involved. In this sense, the largest clouds form by galactic-scale shocks and gravitational and magnetic instabilities, although the detailed process is one of agglomeration of pre-existing small clouds.

Once a strongly self-gravitating object is formed, no matter what its mass, the structure and chemistry of each component cloud should change. The pressure increases with the enhanced self-gravity, and some of the components will coalesce. Atomic clouds brought into a small cloud complex may turn molecular even if their density does not change because they will begin to shield each other from the ambient radiation field. They may also become smaller at first because of the increased pressure, but after several internal crossing times they may appear larger because of coalescence. An old cloud should appear less clumpy than a young cloud (except for the clumps made by stellar winds - Norman and Silk 1980), and the old cloud should have a core-halo structure and R^{-2} density dependence after the clumps scatter off each other and dissipate their kinetic energy. New clumps may form by gravitational or thermal instabilities in the cloud cores.

Giant cloud complexes are presumably destroyed by the high pressures from winds and HII regions created by the young stars they form (Elmegreen and Lada 1977; Whitworth

1979; Bally 1980; Mazurek 1980). They may also be destroyed by collisions with other clouds (Leisawitz and Bash 1982). This destruction should not be complete, however; an entire cloud cannot readily be converted into a uniform low density gas. Instead the cloud should be torn into pieces and pushed around by these starbirth pressures or collisions. The chunks that are torn off may later appear as independent clouds, or they may coalesce with other chunks of former cloud complexes, as in the Oort model (§3.2.1). While the age of an individual cloud complex is probably less than 10^8 years, the ages of some of the pieces from this complex could be older than this. The cycle time for gas into and out of the low density intercloud medium could be 10^9 years or more. In that case, some distinction between cloud lifetimes and molecule lifetimes will be necessary.

Large cloud complexes formed by coagulation and gravitational instabilities in the spiral arms of galaxies could also be destroyed by increased tidal forces in the interarm regions.

6. REFERENCES

Aannestad, P.A. and Field, G.B. 1973, *Ap.J.*(*Letters*), **186**, L29.

Abbott, D.C. 1982, *Ap.J.*, **263**, 723.

Ames, S. and Heiles, C. 1970, *Ap.J.*, **160**, 59.

Appenzeller, I. 1974, *Astr.Ap.*, **36**, 99.

Asseo, E., Cesarsky, C.J., Lachieze-Ray, M. and Pellat, R. 1978, *Ap.J.*(*Letters*), **225**, L21.

Avedisova, V.S. 1974, *A.Zh.*, **51**, 479 (*Sov.A.AJ*, **18**, No.3).

Baierlein, R., Schwing, E. and Herbst, W. 1981, *Icarus*, **48**, 49.

Balbus, S.A. 1985, *Ap.J.*, **291**, 518.

Balbus, S.A. 1986a, *Ap.J.*(*Letters*), **303**, L79.

Balbus, S.A. 1986b, *Ap.J.*, **304**, 787.

Balbus, S.A. and Cowie, L.L. 1985, *Ap.J.*, **297**, 61.

Balbus, S.A. and McKee, C.F. 1982, *Ap.J.*, **252**, 529.

Bally, J. 1980, IAU Symposium No. 87, *Interstellar Molecules*, ed. B.H. Andrew, (Dordrecht: Reidel), p. 151.

Bania, T.M. and Lyon, J.G. 1980, *Ap.J.*, **239**, 173.

Barlow, M.J. 1978, *M.N.R.A.S.*, **183**, 367.

Barlow, M.J. and Silk, J. 1977, *Ap.J.*(*Letters*), **211**, L83.

Bash, F., Hausman, M. and Papaloizou, J. 1981, *Ap.J.*, **245** , 92.

Basu, B. 1972, *Indian J.Phys.*, **46**, 379.

Bates, B., Brown-Kerr, W. Giaretta, D.L. and Keenan, F.P. 1983, *Astr.Ap.*, **122** , 64.

Berkhuijsen, E.M., Haslam, C.G.T. and Salter, C.J. 1971, *Astr.Ap.*, **14** , 252.

Bhatt, H.C., Rowse, D.P. and Williams, I.P. 1984, *M.N.R.A.S.*, **209** , 69.

Biermann, P., Kippenhahn, R., Tscharnuter, W. and Yorke, H. 1972, *Astr.Ap.*, **19** , 113.

Blaauw, A. 1956, *Ap.J.*, **123**, 408.

Blaauw, A. 1984, *Irish A.J.*, **16**, 141.

Blitz, L. and Shu, F.H. 1980, *Ap.J.*, **238**, 148.

Bochkarev, N.G. 1984 *Pis'ma A.Zh.*, **10**, 184 (*Sov.A.Letters*, **10**).

Bodefée, G. and de Loore, C. 1985, *Astr.Ap.*, **142**, 297.

Brand, P.W.J.L. 1980, *Observatory*, **100**, 62.

Brand, P.W.J.L. 1981, *M.N.R.A.S.*, **197**, 217.

Brand, P.W.J.L. and Zealey, W.J. 1975, *Astr.Ap.*, **38**, 363.

Brand, P.W.J.L., Hawarden, T.G., Longmore, A.J., Williams, P.M., Caldwell, J.A.R. 1983, *M.N.R.A.S.*, **203**, 215.

Braun, R. and Strom, R.G. 1986, *Astr.Ap.Suppl.*, **63**, 345.

Bregman, J.N. 1980, *Ap.J.*, **236**, 577.
Broten, N.W., MacLeod, J.M. and Vallée, J.P. 1985, *Ap.Letters*, **24** , 165.
Bruhweiler, F.C., Gull, T.R., Kafatos, M. and Sofia, S. 1980, *Ap.J.(Letters)*, **238** , L27.
Burstein, P., Borken, R.J., Kraushaar, W.L. and Sanders, W.T. 1976, *Ap.J.* **213**, 405.
Carlberg, R.G. and Freedman, W.L. 1985, *Ap.J.*, **298**, 486.
Cash, W., Charles, P., Bowyer, S., Walter, F., Garmire, G. and Riegler, G. 1980,
    Ap.J.(Letters), **238**, L71.
Casoli, F. and Combes, F. 1982, *Astr.Ap.*, **110**, 287.
Casoli, F., Combes, F. and Gerin, M. 1984, *Astr.Ap.*, **133**, 99.
Castle, K.G. and Hills, J.G. 1974, *Astr.Ap.*, **30**, 455.
Cernicharo, J. Bachiller, R. and Duvert, G. 1985, *Astr.Ap.*, **149** , 273.
Cesarsky, C. 1981, in *Plasma Astrophysics*, ed. T.D. Guyenne and G. Lévy, (Paris: European
    Space Agency), p. 51.
Chaffee, F.H., Jr. 1975, *Ap.J.*, **199**, 379.
Chandrasekhar, S. 1954, *Ap.J.*, **119**, 7.
Chiang, W.H. and Prendergast, K. 1985, *Ap.J.*, **297**, 507.
Chièze, J.P. and Lazareff, B. 1980, *Astr.Ap.*, **91**, 290.
Chow, T.L. 1971, Ph.D. Dissertation, University of Rochester, NY.
Clark, B.G. 1965, *Ap.J.*, **142**, 1398.
Cleary, M.N., Heiles, C. and Haslam, C.G.T. 1979, *Astr.Ap.Suppl.*, **36** , 95.
Clifford, P. 1984, *M.N.R.A.S.*, **211**, 125.
Clifford, P. 1985, *M.N.R.A.S.*, **216**, 93.
Clifford, P. and Elmegreen, B.G. 1983, *M.N.R.A.S.*, **202**, 629.
Cohen, J.G. 1977, *Pub.A.S.P.*, **89**, 626.
Cohen, R.J. 1981, *M.N.R.A.S.*, **196**, 835.
Combes, F. and Gerin, M. 1985, *Astr.Ap.*, **150**, 85.
Cowie, L.L. 1978, *Ap.J.*, **225**, 887.
Cowie, L.L. 1980, *Ap.J.*, **236**, 868.
Cowie, L.L. 1981, *Ap.J.*, **245**, 66.
Cowie, L.L. and McKee, C.F. 1977, *Ap.J.*, **211**, 135.
Cowie, L.L. and York, D.G. 1978a, *Ap.J.*, **220**, 129.
Cowie, L.L. and York, D.G. 1978b, *Ap.J.*, **223**, 826.
Cowie, L.L., Songaila, A. and York, D.G. 1979, *Ap.J.*, **230** , 469.
Cowie, L.L., Jenkins, E.B., Songaila, A. and York, D.G. 1979, *Ap.J.*, **232** , 467.
Cox, D.P. 1979, *Ap.J.*, **234**, 863.
Cox, D.P. and Smith, B.W. 1974, *Ap.J.(Letters)*, **189**, L105.
Cox, D.P. and McCammon, D. 1986, *Ap.J.*, **304**, 657.
Crovisier, J., Dickey, J.M. and Kazés 1985, *Astr.Ap.*, **146**, 223.
Crutcher, R.M. 1979, *Ap.J.(Letters)*, **231**, L151.
Crutcher, R.M. 1982, *Ap.J.*, **254**, 82.
Dalgarno, A. and McCray, R. 1972, *Ann.Rev.Astr.Ap.*, **10**, 375.
Dame, T.M. 1983, Ph.D. Dissertation, Columbia University.
Dame, T.M. and Thaddeus, P. 1985, *Ap.J.*, **297**, 751.
de Jong, 1980 in *Les spectres des molécules simples au laboratoire et en astrophysique*, XXI
    International Colloque Liege, (Université de Liège, Institut d'Astrophysique), p. 117.
Dibai, E.A. 1970, *A.Zh.*, **47**, 977 (*Sov.A.AJ*, **14**, No.5).
Dickey, J.M., Salpeter, E.E. and Terzian, Y. 1977, *Ap.J.(Letters)*, **211** , L77.
Dopita, M.A. 1985, *Ap.J.*, **295**, L5.
Draine, B.T. and Giuliani, J.L., Jr. 1984, *Ap.J.*, **281**, 690.
Dwek, E. 1981, *Ap.J.*, **246**, 430.
Elitzur, M. and Watson, W.D. 1978, *Ap.J.(Letters)*, **222**, L141.

Elitzur, M. and Watson, W.D. 1980, *Ap.J.*, **236**, 172.
Elliot, K.H. and Meaburn, J. 1970, *Ap.Sp.Sci.*, **7**, 252.
Elliot, K.H. and Meaburn, J. 1973, *Ap.Sp.Sci.*, **20**, 111.
Elmegreen, B.G. 1975, *Ap.J.*, **205**, 405.
Elmegreen, B.G. 1979, *Ap.J.*, **231**, 372.
Elmegreen, B.G. 1982a, *Ap.J.*, **253**, 634.
Elmegreen, B.G. 1982b, *Ap.J.*, **253**, 655.
Elmegreen, B.G. 1982c, in *Submillimetre Wave Astronomy*, ed. J.E. Beckman and J.P. Phillips, (Cambridge: Cambridge University Press), p. 3.
Elmegreen, B.G. 1985a, in *Birth and Infancy of Stars*, ed. R. Lucas, A. Omont and R. Stora, (Amsterdam: North Holland), p. 215.
Elmegreen, B.G. 1985b, *Ap.J.*, **299**, 196.
Elmegreen, B.G. 1985c, in *Protostars and Planets, II*, ed. D.C. Black and M.S. Matthews, (Tucson: University of Arizona), p. 33.
Elmegreen, B.G. 1987a, *Ap.J.*, **312**, in press.
Elmegreen, B.G. 1987b, in IAU Symposium No. 115, *Star Forming Regions*, ed. M. Peimbert and J. Jugaku, (Dordrecht: Reidel).
Elmegreen, B.G. and Chiang, W.H. 1982, *Ap.J.*, **253**, 666.
Elmegreen, B.G. and Lada, C.J. 1977, *Ap.J.*, **214**, 725.
Elmegreen, B.G. and Elmegreen, D.M. 1978, *Ap.J.*, **220**, 1051.
Elmegreen, B.G. and Elmegreen, D.M. 1983, *M.N.R.A.S.*, **203**, 31.
Falgarone, E. and Puget, J.L. 1985, *Astr.Ap.*, **142**, 157.
Falgarone, E. and Puget, J.L. 1986, *Astr.Ap.*, in press.
Federman, S.R. 1980, *Ap.J.(Letters)*, **241**, L109.
Federman, S.R. 1982a, *Ap.J.*, **253**, 601.
Federman, S.R. 1982b, *Ap.J.*, **257**, 125.
Federman, S.R. and Willson, R.F. 1982, *Ap.J.*, **260**, 124.
Fejes, I. 1971, *Astr.Ap.*, **15**, 419.
Fejes, I. and Verschuur, G.L. 1973, *Astr.Ap.*, **25**, 85.
Ferlet, R., Vidal-Madjar, A., Laurent, C. and York, D.G. 1980, *Ap.J.*, **242**, 576.
Field, G.B. 1965, *Ap.J.*, **142**, 531.
Field, G.B. and Saslaw, W.C. 1965, *Ap.J.*, **142**, 568.
Field, G.B. and Hutchins, J. 1968, *Ap.J.*, **153**, 737.
Field, G.B., Goldsmith, D.W. and Habing, H.J. 1969, *Ap.J.(Letters)*, **155**, L149.
Fich, M. 1985, *Ap.J.*, **303**, 465.
Flannery, B.P. and Press, W.H. 1979, *Ap.J.*, **231**, 688.
Fleck, R.C., Jr. 1981, *Ap.J.(Letters)*, **246**, L151.
Fleck, R.C., Jr. 1984, *A.J.*, **89**, 506.
Flower, D.R., Guilloteau, S. and Hartquist, T.W. 1982, *M.N.R.A.S.*, **200**, 550.
Flower, D.R., Pineau des Forets, G. and Hartquist, T.W. 1985, *M.N.R.A.S.*, **216**, 775.
Flower, D.R., Pineau des Forets, G. and Hartquist, T.W. 1986, *M.N.R.A.S.*, **218**, 729.
Franco, J. and Cox, D.P. 1983, *Ap.J.*, **273**, 243.
Franco, M.L. and Pöppel, W.G.L. 1978, *Ap.Sp.Sci.*, **53**, 91.
Freeman, A. and Williams, D.A. 1982, *Ap.Sp.Sci.*, **83**, 417.
Frieman, E.A. 1954, *Ap.J.*, **120**, 18.
Frisch, P.C. 1979, *Ap.J.*, **227**, 474.
Frisch, P.C. 1981, *Nature*, **293**, 377.
Frisch, P.C. and Jura, M. 1980, *Ap.J.*, **242**, 560.
Fuchs, B. and Thielheim, K.O. 1979, *Ap.J.*, **227**, 801.
Fujimoto, M. and Ikeuchi, S. 1984, *Pub.Astr.Soc.Japan*, **36**, 319.
Gaida, M., Ungerechts, H. and Winnewisser, G. 1984, *Astr.Ap.*, **137**, 17.

Gerola, H., Iglesias, E. and Gamba, Z. 1973, *Astr.Ap.*, **24** , 369.
Gerola, H., Kafatos, M. and McCray, R. 1974, *Ap.J.*, **189**, 55.
Ghosh, K.K. and Bhattacharyya, J.C. 1985, *Ap.Sp.Sci.*, **117**, 89.
Giaretta, D.L. 1979, *Astr.Ap.*, **78**, 328.
Gilden, D.L. 1984a, *Ap.J.*, **279**, 335.
Gilden, D.L. 1984b, *Ap.J.*, **283**, 679.
Giovanelli, R., Haynes, M.P., York, D.G. and Shull, J.M. 1978, *Ap.J.*, **219** , 60.
Giuliani, J.L., Jr. 1984, *Ap.J.*, **277**, 605.
Glassgold, A.E. and Langer, W.D. 1976, *Ap.J.*, **204**, 403.
Goldreich, P. and Lynden-Bell, D. 1965, *M.N.R.A.S.*, **130**, 97.
Grabelsky, D.A., Cohen, R.S., May, J., Bronfman, L. and Thaddeus, P. 1986, *Ap.J.*, submitted.
Habe, A., Ikeuchi, S. and Tanaka, Y.D. 1981, *Pub.Astr.Soc.Japan*, **33** , 23.
Handbury, M.J., Simon, S. and Williams, I.P. 1977, *Astr.Ap.*, **61**, 443.
Haslam, C.G.T., Kahn, F.D. and Meaburn, J. 1971, *Astr.Ap.*, **12** , 388.
Hausman, M.A. 1981, *Ap.J.*, **245**, 72.
Hausman, M.A. 1982, *Ap.J.*, **261**, 532.
Hausman, M.A. and Roberts, W.W., Jr. 1984, *Ap.J.*, **282**, 106.
Hawarden, T.G. and Brand, P.W.J.L. 1976, *M.N.R.A.S.*, **175**, 19p.
Heathcote, S.R. and Brand, P.W.J.L. 1983, *M.N.R.A.S.*, **203**, 67.
Heiles, C. 1967, *Ap.J.Suppl.*, **15**, 97.
Heiles, C. 1975, I.A.U.Symposium No. 60, *Galactic Radio Astronomy*, ed. F. Kerr (Dordrecht: Reidel).
Heiles, C. 1976, *Ap.J.(Letters)*, **208**, L137.
Heiles, C. 1979, *Ap.J.*, **229**, 533.
Higdon, J.C. 1981, *Ap.J.*, **244**, 88.
Hills, J.G. 1972, *Astr.Ap.*, **17**, 155.
Hills, J.G. 1973, *Astr.Ap.*, **26**, 197.
Hobbs, L.M. 1974, *Ap.J.(letts.)*, **188**, L107.
Hobbs, L.M. 1983, *Ap.J.*, **265**, 817.
Hong, S.S. and Koo, B.C. 1984, *J.Korean A.S.*, **17**, 115.
Houghes, V.A. and Routledge, D. 1972, *Astr.Ap.*, **77**, 210.
Hu, E.M. 1981, *Ap.J.*, **248**, 119.
Hunter, J.H., Jr., Sandford, M.T., II., Whitaker, R.W. and Klein, R.I. 1986, *Ap.J.*, **305**, 309.
Ibanez S., M.H. 1981, *Rev.Mex.Astr.Ap.*, **6**, 277.
Ibanez S., M.H. and Parravano B., A. 1983, *Ap.J.*, **275**, 181.
Ikeuchi, S. and Tomita, H. 1983, *Pub.Astr.Soc.Japan*, **35**, 77.
Ikeuchi, S., Habe, A. and Tanaka, Y.D. 1984, *M.N.R.A.S.*, **207**, 909.
Ikeuchi, S. and Spitzer, L., Jr. 1984, *Ap.J.*, **283**, 825.
Jenkins, E.B. and Meloy, D.A. 1974, *Ap.J.(Letters)*, **193**, L121.
Jenkins, E.B. and Shaya, E.J. 1979, *Ap.J.*, **231**, 55.
Jenkins, E.B., Silk, J., Wallerstein, G. and Leep, E.M. 1981, *Ap.J.*, **248**, 977.
Jenkins, E.B., Jura, M. and Lowenstein, M. 1983, *Ap.J.*, **270** , 88.
Jura, M. 1976, *Ap.J.*, **206**, 691.
Kafatos, M., Gerola, H., Hatchett, S. and McCray, R. 1974, *Ap.J.(Letters)*, **187** , L113.
Kahn, F. 1955, in *Gas Dynamics in Cosmic Gas Clouds*, ed. H.C. Van de Hulst and J.M. Burgers, (Amsterdam: North Holland), Ch.12.
Kalberla, P.M.W., Schwarz, U.J. and Goss, W.M. 1985, *Astr.Ap.*, **144** , 27.
Kannari, Y. 1981, *Sci.Rep.Tohoku Univ.*, Ser 8, **2**, 1.
Kannari, Y., Sabano, Y. and Tosa, M. 1979, *Pub.Astr.Soc.Japan*, **31** , 395.
Kleiner, S.C. and Dickman, R.L. 1984, *Ap.J.*, **286**, 255.

Knude, J. 1981a, *Astr.Ap.*, **97**, 380.
Knude, J. 1981b, *Astr.Ap.*, **98**, 74.
Königl, A. 1984, *Ap.J.*, **284**, 303.
Kritsuk, A.G. 1985, *A.Zh.*, **62**, 66 (*Sov.A.AJ*, **29**, No.1).
Kulkarni, S.R., Dickey, J.M. and Heiles, C. 1985, *Ap.J.*, **291**, 716.
Kuznetsov, V.D. and Ptuskin, V.S. 1983, *Ap.Sp.Sci.*, **94**, 5.
Kwan, J. 1979, *Ap.J.*, **229**, 567.
Kwan, J. and Valdes, F. 1983, *Ap.J.*, **271**, 604.
Lachièze-Rey, M., Asséo, E., Cesarsky, C.J. and Pellat, R. 1980, *Ap.J.*, **238**, 175.
Lambert, D.L. and Danks, A.C. 1985, *Ap.J.*, **303**, 401.
Lattanzio, J.C., Monaghan, J.J., Pongracic, H. and Schwarz, M.P. 1985, *M.N.R.A.S.*, **215**, 125.
Leisawitz, D. and Bash, F. 1982, *Ap.J.*, **259**, 133.
Lepp, S., McCray, R., Shull, J.M., Woods, D.T. and Kallman, T. 1985, *Ap.J.*, **288**, 58.
Lerch, I. 1967a, *Ap.J.*, **148**, 415.
Lerch, I. 1967b, *Ap.J.*, **149**, 553.
Leung, C.M., Kutner, M.L. and Mead, K.N. 1982, *Ap.J.*, **262**, 583.
Levinson, F.H. and Roberts, W.W., Jr. 1981, *Ap.J.*, **245**, 465.
Levy, E.H. 1978, IAU Symposium No.77, *Structure and Properties of Nearby Galaxies*, ed. E.M. Berkhuijsen and R. Wielebinski, (Dordrecht: Reidel), p. 57.
Lindblad, P.O. 1967, *B.A.N.*, **19**, 34.
Lindblad, P.O. 1983, in *Kinematics, Dynamics and Structure of the Milky Way*, ed. W.L.H. Shuter, (Dordrecht: Reidel), p. 55.
Lindblad, P.O., Grope, K., Sandqvist, Aa. and Schober, J. 1973, *Astr.Ap.*, **24**, 309.
Liszt, H.S. 1983, *Ap.J.*, **275**, 163.
Liszt, H.S., Xiang, D. and Burton, W.B. 1981, *Ap.J.*, **249**, 532.
Low, F.J., Bientema, D.A., Gautier, T.N., Gillett, F.C., Beichman, C.A., Neugebauer, G., Young, E., Aumann, H.H., Boggess, N., Emerson, J.P., Habing, H.J., Hauser, M.G., Houck, J.R., Rowan-Robinson, M., Soifer, B.T., Walker, R.G., Wesselius, P.R. 1974, *Ap.J.(Letters)*, **278**, L19.
Lynden-Bell, D. 1966, *Observatory*, **86**, 57.
Lyon, J. 1975, *Ap.J.*, **201**, 168.
Mansfield, V.N. 1973, *Ap.J.*, **179**, 815.
Marochnik, L.S., Berman, B.G., Mishurov, Yu.N. and Suchkov, A.A. 1983, *Ap.Sp.Sci.*, **89**, 177.
Martin, E.R. and York, D.G. 1982, *Ap.J.*, **257**, 135.
Mazurek, T.J. 1980, *Astr.Ap.*, **90**, 65.
McCray, R. 1983, *Highlights of Astronomy*, **6**, 565.
McCray, R., Stein, R.F. and Kafatos, M. 1975, *Ap.J.*, **196**, 565.
McCray, R. and Snow, T.P., Jr. 1979, *Ann.Rev.Astr.Ap.*, **17**, 213.
McDavid, D. 1984, *Ap.J.*, **284**, 141.
McKee, C.F. and Cowie, L.L. 1977, *Ap.J.*, **215**, 213.
McKee, C.F. and Ostriker, J.P. 1977, *Ap.J.*, **218**, 148.
McKee, C.F. and Hollenbach, D.J. 1980, *Ann.Rev.Astr.Ap.*, **18**, 219.
Meaburn, J. 1983, *Highlights of Astronomy*, **6**, 665.
Mebold, U., Winnberg, A., Kalberla, P.M.W. and Goss, W.M. 1982, *Astr.Ap.*, **115**, 223.
Mebold, U., Cernicharo, J., Velden, L., Reif, K., Crezelius, C. and Goerigk, W. 1985, *Astr.Ap.*, **151**, 427.
Mezaros, P. 1973a, *Ap.J.*, **183**, 469.
Mezaros, P. 1973b, *Ap.J.(Letters)*, **185**, L41.
Mirabel, I.F. 1982, *Ap.J.*, **256**, 112.

Mitchell, G.F. and Watt, G.D. 1985, *Astr.Ap.*, **151**, 121.

Moneti, A., Pipher, J.L., Helfer, H.L., McMillan, R.S. and Perry, M.L. 1984, *Ap.J.*, **282**, 508.

Morton, D.C. 1975, *Ap.J.*, **197**, 85.

Mouschovias, T.Ch. 1974, *Ap.J.*, **192**, 37.

Mouschovias, T.Ch., Shu, F. and Woodward, P. 1974, *Astr.Ap.*, **33** , 73.

Mufson, S.L. 1974, *Ap.J.*, **193**, 561.

Mufson, S.L. 1975, *Ap.J.*, **202**, 372.

Nakano, T. and Nakamura, T. 1978, *Pub.Astr.Soc.Japan*, **30**, 671.

Nelson, A.H. 1985, *M.N.R.A.S.*, **215**, 161.

Nittmann, J. 1981, *M.N.R.A.S.*, **197**, 699.

Nittmann, J. and Falle, S.A.E.G. 1982, *M.N.R.A.S.*, **201**, 833.

Norman, C. and Silk, J. 1980, IAU Symposium No. 87, *Interstellar Molecules*, ed. B.H. Andrew, (Dordrecht: Reidel), p. 137.

Norman, C. and Silk, J. 1980, *Ap.J.*, **238**, 158.

Nozakura, T. and Ikeuchi, S. 1984, *Ap.J.*, **279**, 40.

O'Donnell, E.J. and Watson, W.D. 1974, *Ap.J.*, **191**, 89.

Oettl, R., Hillebrandt, W. and Müller, E. 1985, *Astr.Ap.*, **151** , 33.

Olano, C.A. 1982, *Astr.Ap.*, **112**, 195.

Oort, J.H. 1954, *B.A.N.*, **12**, 177.

Parker, E.N. 1966, *Ap.J.*, **145**, 811.

Parker, E.N. 1967, *Ap.J.*, **149**, 517.

Payne, H.E., Salpeter, E.E. and Terzian, Y. 1983, *Ap.J.*, **272** , 540.

Penston, M.V., Munday, V.A., Stickland, D.J. and Penston, M.J. 1969, *M.N.R.A.S.*, **142**, 355.

Pérault, M., Falgarone, E. and Puget, J.L. 1985, *Astr.Ap.*, **152** , 371.

Pikel'ner, S.B. 1973, *Comments Ap.*, **6**, 151.

Pikel'ner, S.B. and Sorochenko, R.L. 1974, *Sov.A.A.J.*, **17**, 443.

Pumphrey, W.A. and Scalo, J.M. 1983, *Ap.J.*, **269**, 531.

Reddish, V.C. 1975, *M.N.R.A.S.*, **170**, 261.

Reipurth, B. 1983, *Astr.Ap.*, **117**, 183.

Reynolds, R.J., Roesler, F.L. and Scherb, F. 1974, *Ap.J.(Letters)*, **192** , L53.

Rickard, J.J. 1975, *Astr.Ap.*, **41**, 403.

Roberts, W.W., Jr. and Hausman, M.A. 1984, *Ap.J.*, **277**, 744.

Rose, W.K. and Wentzel, D.G. 1973, *Ap.J.*, **181**, 115.

Routley, P. McR. and Spitzer, L. 1952, *Ap.J.*, **115**, 227.

Sabano, Y. 1972, *Sci.Rep.Tohoku Univ,*, First Ser., **55** , 155.

Sabano, Y. and Kannari, K. 1978, *Pub.A.S.P.*, **30**, 77.

Sabano, Y. and Tosa, M. 1978, *Pub.A.S.P.*, **30**, 67.

Salpeter, E.E. 1976, *Ap.J.*, **206**, 673.

Sandqvist, A., 1976, *M.N.R.A.S.*, **177**, 69p.

Scalo, J.M. 1985, in *Protostars and Planets, II.*, ed. D.C. Black and M.S. Matthews, (Tucson: University of Arizona), p. 201.

Scalo, J.M. and Pumphrey, W.A. 1982, *Ap.J.(Letters)*, **258**, L29.

Scalo, J.M. and Struck-Marcell, C. 1984, *Ap.J.*, **276**, 60.

Scalo, J.M. and Struck-Marcell, C. 1986, *Ap.J.*, **301**, 77.

Schlickeiser, R. and Lerche, I. 1985, *Astr.Ap.*, **151**, 151.

Schlosser, W., Schmidt-Kaler, T. 1974, *Planets, Stars and Nebulae Studied with Photopolarimetry*, ed. T. Gehrels, (Tucson: University of Arizona), p. 972.

Schneider, S. and Elmegreen, B.G. 1979, *Ap.J.Suppl.*, **41**, 87.

Schneps, M.H., Ho, P.T.P. and Barrett, A.H. 1980, *Ap.J.*, **240** , 84.

Schwarz, J. 1973, *Ap.J.*, **182**, 449.
Schwarz, J., McCray, R. and Stein, R.F. 1972, *Ap.J.*, **175**, 673.
Scoville, N. and Hersh, K. 1979, *Ap.J.*, **229**, 578.
Shore, S.N. 1983, *Ap.J.*, **265**, 202.
Shu, F.H. 1974, *Astr.Ap.*, **33**, 55.
Shu, F.H., Milione, V., Gebel, W., Yuan, C., Goldsmith, D.W. and Roberts, W.W., Jr. 1972, *Ap.J.*, **173**, 557.
Shull, J.M. 1977a, *Ap.J.*, **215**, 805.
Shull, J.M. 1977b, *Ap.J.*, **216**, 414.
Shull, J.M. 1978, *Ap.J.*, **226**, 858.
Shull, J.M., York, D.G. and Hobbs, L.M. 1977, *Ap.J.(Letters)*, **211**, L139.
Shull, J.M. and Woods, D.T. 1985, *Ap.J.*, **288**, 50.
Silich, S.A. 1985, *Astrofizika*, **22**, 563 (*Astrophysics*, **22**, No.3).
Siluk, R.S. and Silk, J. 1974, *Ap.J.*, **192**, 51.
Simonson, S.C. III 1973, *Astr.Ap.*, **23**, 19.
Sivan, J.P. 1974, *Astr.Ap.Suppl.*, **16**, 163.
Smith, A.M., Krishna Swamy, K.S. and Stecher, T.P. 1978, *Ap.J.*, **220**, 138.
Smith, J. 1980, *Ap.J.*, **238**, 842.
Snow, T.P., Jr. 1977, *Ap.J.*, **216**, 724.
Sofue, Y. and Nakai, N. 1983, *Astr.Ap.Suppl.*, **53**, 57.
Sofue, Y. and Sabano, Y. 1980, *Pub.Astr.Soc.Japan*, **32**, 623.
Solomon, P.M., Sanders, D.B. and Scoville, N.Z. 1979, in IAU Symposium No. 84, *The Large Scale Characteristics of the Galaxy*, ed. W.B. Burton, (Dordrecht: Reidel)., p. 35.
Spitzer, L. 1954, *Ap.J.*, **120**, 1.
Spitzer, L. 1976, *Comments Ap.*, **6**, 177.
Spitzer, L. and Jenkins, E.B. 1975, *Ann.Rev.Astr.Ap.*, **13**, 133.
Spitzer, L. and Morton, W.A. 1976, *Ap.J.*, **204**, 731.
Spoelstra, T.A.Th. 1972, *Astr.Ap.*, **21**, 61.
Stark, A.A. 1983, in *Kinematics, Dynamics and Structure of the Milky Way*, ed. W.L.H. Shuter, (Dordrecht: Reidel), p. 127.
Stark, A.A. 1984, *Ap.J.*, **281**, 624.
Stone, M.E. 1970, *Ap.J.*, **159**, 293.
Struck-Marcell, C. and Scalo, J.M. 1984, *Ap.J.*, **277**, 132.
Taff, L.G. and Savedoff, M.P. 1973, *M.N.R.A.S.*, **164**, 357.
Tarafdar, S.P. 1978, *Astr.Ap.*, **68**, 165.
Tarrab, I. 1983, *Astr.Ap.*, **125**, 308.
Taylor, M.I., Taylor, K.N.R. and Vaile, R.A. 1982, *Proc.Astr.Soc.Australia*, **4**, 440.
Tenorio-Tagle, G. 1981, *Astr.Ap.*, **94**, 338.
Tenorio-Tagle, G. and Rozyczka, M. 1986, *Astr.Ap.*, **155**, 120.
Tomisaka, K., 1984, *Pub.Astr.Soc.Japan*, **36**, 457.
Tomisaka, K., 1986, *Pub.Astr.Soc.Japan*, **38**, 95.
Tomisaka, K., Habe, A. and Ikeuchi, S. 1981, *Ap.Sp.Sci.*, **78**, 273.
Tomisaka, K., and Ikeuchi, S. 1983, *Pub.Astr.Soc.Japan*, **85**, 187.
Tomita, Y., Saito, T. and Ohtani, H. 1979, *Pub.Astr.Soc.Japan*, **31**, 407.
Toomre, A. 1964, *Ap.J.*, **139**, 1217.
Torres-Peimbert, S., Lazcano-Araujo, A. and Peimbert, M. 1974, *Ap.J.*, **191**, 401.
Troland, T.H. and Heiles, C. 1982, *Ap.J.(Letters)*, **260**, L19.
Tsioumis, A. and Fricke, W. 1979, *Astr.Ap.*, **75**, 1.
van Buren, D. 1985, *Ap.J.*, **294**, 567.
van de Hulst, H.C. 1955, in *Gas Dynamics of Cosmic Clouds*, ed. J.M. Burgers and H.C. van de Hulst, (New York: Interscience), p. 111.

van den Bergh, S., Marscher, A.P. and Terzian, Y. 1973, *Ap.J.Suppl.*, **26**, 19.
Verschuur, G.L. 1970, *A.J.*, **75**, 687.
Verschuur, G.L. 1973, *A.J.*, **78**, 573.
Verschuur, G.L. 1974, *Ap.J.Suppl.*, **27**, 65.
Vishniac, E.T. 1983, *Ap.J.*, **274**, 152.
Vrba, F.J., Strom, S.E. and Strom, K.M. 1976, *A.J.*, **81**, 958.
Wallerstein, G., Silk, J. and Jenkins, E.B. 1980, *Ap.J.*, **240**, 834.
Whitworth, A. 1979, *M.N.R.A.S.*, **186**, 59.
Woodward, P.R. 1976, *Ap.J.*, **207**, 484.
Yoneyama, T. 1973, *Pub.Astr.Soc.Japan*, **25**, 349.
York, D.G. 1974, *Ap.J.(Letters)*, **193**, L127.
York, D.G. 1983, *Ap.J.*, **264**, 172.
Yoshii, Y. and Sabano, Y. 1980, *Pub.Astr.Soc.Japan*, **32**, 229.
Yuan, C. and Wang, C.Y. 1982, *Ap.J.*, **252**, 508.
Zealey, W.J., Ninkov, Z., Rice, E., Hartley, M. and Tritton, S.B. 1983, *Ap.Letters*, **23**, 119.
Zweibel, E.G. and Kulsrud, R.M. 1975, *Ap.J.*, **201**, 63.

Bruce Elmegreen

Section V: Dynamical Processes in Interstellar Clouds

Michael Shull, Katherine Thronson, and John Stocke

THE PHYSICS OF INTERSTELLAR SHOCK WAVES

J. Michael Shull
Center for Astrophysics and Space Astronomy, and
Joint Institute for Laboratory Astrophysics
University of Colorado and National Bureau of Standards
Boulder, CO 80309 (USA)

and

Bruce T. Draine
Princeton University Observatory
Peyton Hall
Princeton, NJ 08544 (USA)

ABSTRACT. This review discusses the observations and theoretical models of interstellar shock waves, in both diffuse cloud and molecular cloud environments. We summarize the relevant gas dynamics, atomic, molecular and grain processes, radiative transfer, and physics of radiative and magnetic precursors in shock models. We then describe the importance of shocks for observations, diagnostics, and global interstellar dynamics. We conclude with current research problems and data needs for atomic, molecular and grain physics.

1 INTRODUCTION

This review of interstellar shock waves is intended both as a pedagogical reference and as an indicator of current and future research in the field. During the past decade, the study of shock waves has become extremely important, attracting a number of observers and theorists. This reflects primarily the realization in the 1970's that the interstellar medium (ISM) is far from quiescent, but rather is dominated by the action of supernovae, stellar winds, and other violent activity (Cox and Smith 1974; McKee and Ostriker 1977; McCray and Snow 1979).

However, the study of shock waves has attracted observers and theorists for more basic reasons. To the theorist, shock models require a wide range of astrophysical tools: gas dynamics, MHD, atomic and molecular physics, spectroscopy, radiative transfer, plasma physics, interstellar chemistry, and even the solid state physics of dust grains. To the observer, shocks are a rich field owing to the ubiquity of astrophysical sites of shock production and the value of shocks as physical diagnostics. Shocks are believed to play an

D. J. Hollenbach and H. A. Thronson, Jr. (eds.), Interstellar Processes, 283–319.

important role in the dynamical evolution, emission processes, and interpretation of the following astronomical objects:

1. Supernovae and supernova remnants (SNRs)

2. Stellar winds (OB, Wolf-Rayet, Red Giants, etc.)

3. High and intermediate velocity diffuse clouds

4. Expanding H II regions

5. Molecular outflows, Herbig-Haro objects, jets

6. Cloud-cloud collisions

7. Mergers of spiral galaxy disks (IRAS galaxies ?)

8. Starburst galaxies (and some LINERs ?)

9. Intergalactic and pre-galactic shock waves

10. Accretion onto protostars and pre-stellar disks.

The environments of these shocks range in density from diffuse clouds, which contain mostly atomic gas $(n_H = 0.1 - 10^3 \text{ cm}^{-3})$ and cool by optical and ultraviolet emission lines of H, He, and atomic ions, up to molecular clouds $(n(H_2) = 10^3 - 10^8 \text{ cm}^{-3})$ which cool by optical, infrared, sub-millimeter, and millimeter wavelength lines of atoms and molecules.

Shock waves heat, compress, accelerate, and irreversibly change the entropy of the interstellar gas ahead of them. They can trigger gravitational (Elmegreen and Lada 1977) and thermal instabilities (McCray, Stein, and Kafatos 1975), provide a coupling of energy and momentum from stars to the interstellar gas (Spitzer 1978), destroy or process interstellar grains (Draine and Salpeter 1979; Dwek and Scalo 1980; Seab and Shull 1983), and produce molecular species, such as CH^+ or H_2O, whose reaction barriers or pathways inhibit formation in cold clouds (Elitzur and Watson 1978a,b; Hollenbach and McKee 1979). In addition, the shock-excited vibrational and rotational lines of H_2 currently provide the only direct measurement of this dominant component of dark molecular clouds (see review by Shull and Beckwith 1982).

Emission lines are widely used by astronomers as diagnostics of density, temperature, abundance, and excitation mechanism (*e.g.*, shock excitation vs. photoionization). The techniques of observation have resulted in a natural division between shocks in diffuse atomic gas and dense molecular clouds. In this review, we follow a similar division, based on the theoretical distinctions which govern diffuse and molecular shock models. In section 2, we discuss single-fluid shocks in diffuse gas, the gas dynamic, atomic, molecular, and grain processes involved in their study, and the relevant observations. In section 3, we discuss multi-fluid MHD shocks with magnetic precursors and their application to shock waves in the Orion molecular cloud, other "bipolar flows" and possible shocks in diffuse molecular clouds. In section 4, we discuss current problems and uncertainties in the theoretical models and conclude with suggestions for future directions of research.

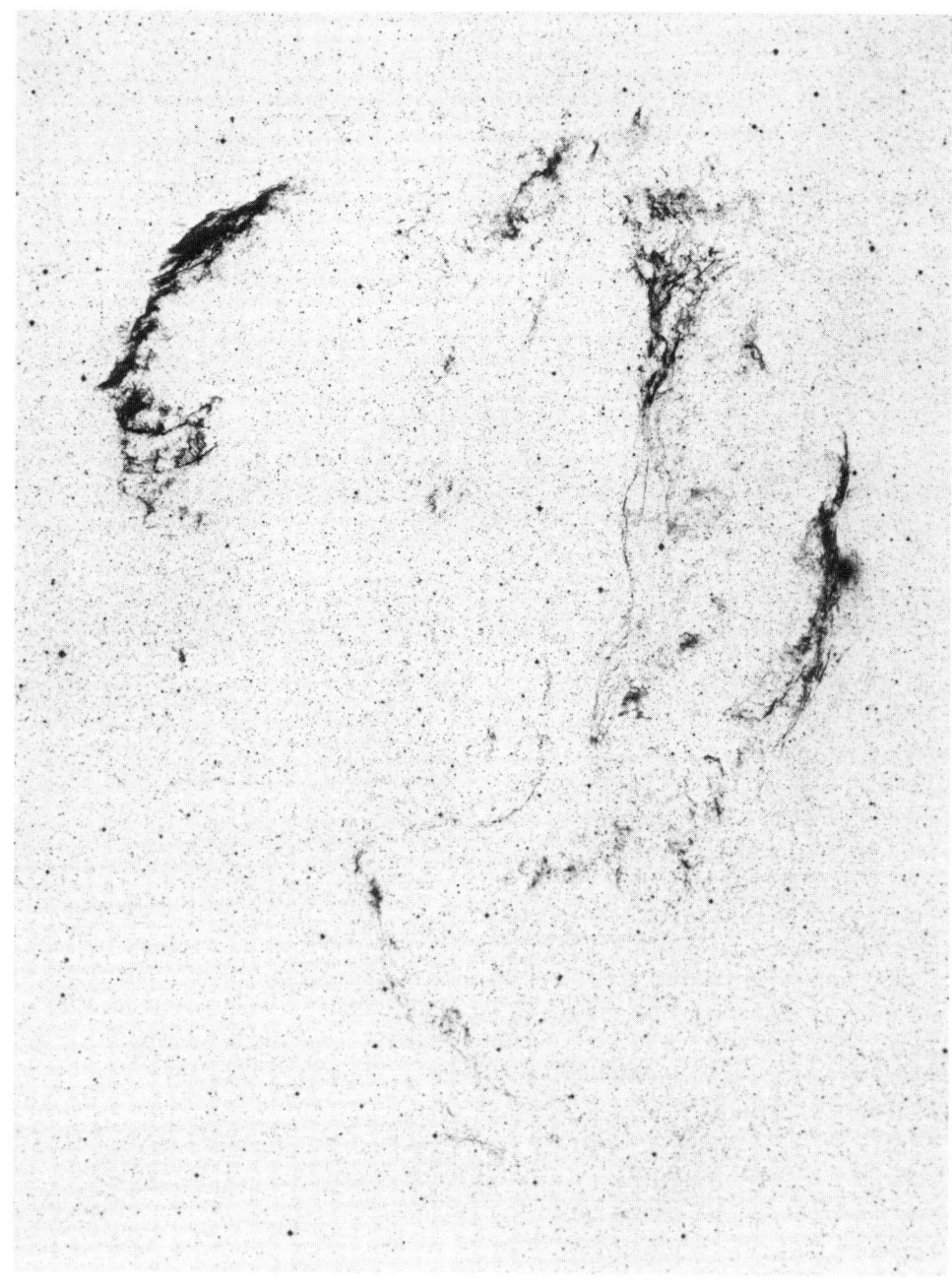

Figure 1: An optical Palomar-Schmidt photograph of the Cygnus Loop supernova remnant. Note the strong filamentary emission from radiative shocks along east and west boundaries.

2 SINGLE-FLUID SHOCKS IN DIFFUSE CLOUDS

2.1 Gas Dynamics

A shock is sometimes described as a "hydrodynamic surprise" A fluid element is "suddenly" accelerated from an initial "pre-shock" velocity to a "post-shock" velocity. Thus, shocks are supersonic disturbances driven by thermal pressure (a "piston"), radiative acceleration, or other mechanical sources. The most common sources of high velocity gas in the ISM are supernova remnants (SNRs), stellar winds, molecular outflows from pre-main-sequence stars, and infalling H I (21-cm) clouds.

At this point, we should make an important distinction between "adiabatic" and "radiative" shock waves. The sudden, discontinuous jump in density, flow velocity, and temperature is characteristic of an adiabatic shock. No energy is lost in the front (hence the term adiabatic), and the gas is heated to a large temperature subject to the constraints of mass and momentum conservation. All single-fluid shocks contain an adiabatic shock transition. If the post-shock gas can radiate away its energy in a time short compared to the flow time, the temperature drops and the post-shock gas is compressed to maintain approximately constant total pressure – this forms the radiative shock wave. A typical compression in a strong, radiative shock, limited by magnetic pressure, is about a factor of 100.

Examples of adiabatic shocks occur at the peripheries of young SNRs in the Sedov-Taylor evolution phase. The strong X-ray line emission observed toward Tycho, Kepler, and Cas A (Becker et al. 1979, 1980a,b) has been interpreted as shocked metal-rich ejecta (Shull 1982; Hamilton et al. 1985). The post-shock temperatures inferred from the He-like Si and S lines at 1.8 and 2.2 keV and from the 1-20 keV X-ray continuum (Pravdo and Smith 1979) are characteristic of non-equilibrium ionization behind a shock of several thousand km s^{-1} (Itoh 1977; Shull 1982; Gronenschild and Mewe 1982; Hamilton, Sarazin, and Chevalier 1983). The most likely source is the "reverse shock" generated by "pressure echoes" during the early stages of SNR evolution, as the ejecta interact with the ambient ISM (Gull 1973, 1974; McKee 1974). The presence of X-rays above 10 keV requires high electron temperatures if the emission is thermal. This in turn requires that ion and electron temperatures equilibrate rapidly, probably through plasma instabilities (McKee and Hollenbach 1980). Fast non-radiative shocks, which encounter H^0 more rapidly than it can be pre-ionized, are believed responsible for weak Balmer line emission (Chevalier and Raymond 1978; Chevalier, Kirshner, and Raymond 1980) and ultraviolet lines and two-photon continuum from filaments just outside the main optical filaments in the Cygnus Loop (Raymond et al. 1983) – see Fig. 1.

The structure of a "radiative shock" can approximately be divided (Fig. 2) into three regions: (1) a radiative precursor in which the ambient gas is moderately heated and partially ionized by ultraviolet photons produced in the shocked layer; (2) the "adiabatic shock front", a thin layer in which the pre-shock gas is accelerated and heated by dissipative processes; and (3) a much broader layer, in which inelastic collisions produce radiative cooling, emission, recombination, and further compression downstream from the front. The state of the gas beyond the last layer depends on boundary conditions with the driving source and the total column density of shocked gas. Magnetic fields, thermal conduction, and

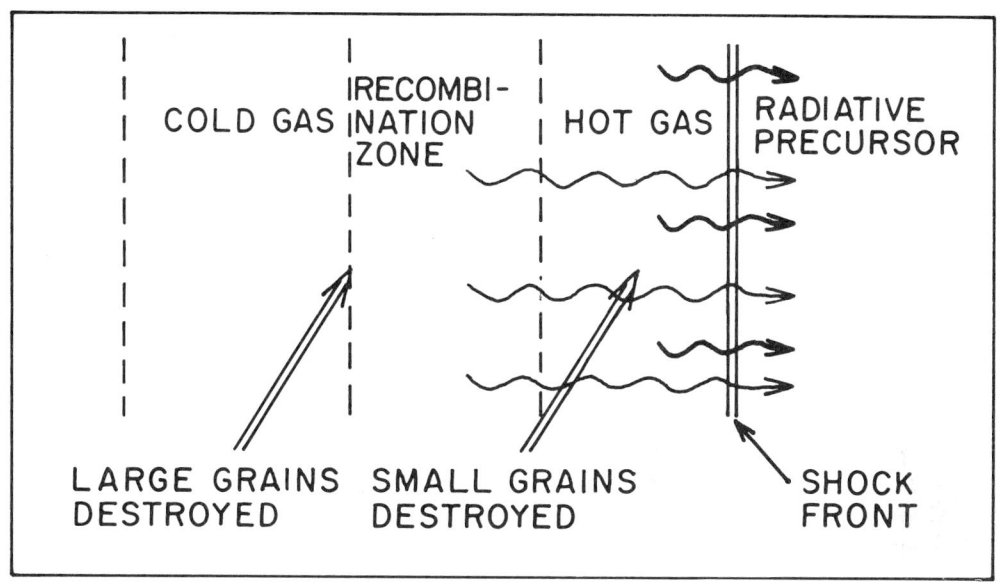

Figure 2: Schematic structure of radiative shock wave. Regions in which large and small grains are destroyed are indicated, as are sources of ionizing photons for radiative precursor.

ambient UV radiation fields often play a role in determining the density and temperature of this interface. Recent theoretical studies of radiative shocks include Raymond (1979), Shull and McKee (1979), Seab and Shull (1983, 1985), Dopita *et al.* (1984), and Cox and Raymond (1985).

The radiative precursors to fast shocks, with $V_s > 110$ km s^{-1}, produce singly ionized H and He ahead of the front (Shull and McKee 1979). This results in a shock front in which the dissipation is governed by plasma instabilities rather than collisions, and the "collisionless shock front" has a negligible thickness. For slower shocks, the gas is only partially ionized (in the absence of external sources of ionizing radiation). The ionized component still undergoes a collisionless shock, and the large H^0-H^+ charge exchange cross section $\sigma_{in} = 3 \times 10^{-15}$ cm^2 (Dalgarno and Yadav 1953, Dalgarno 1958) ensures that the ions and neutrals remain coupled. The front structure in slower ($V_s < 20$ km s^{-1}) shocks is determined by elastic H^0-H^0 collisions, with a scalelength $\lambda_{nn} = (n\sigma_{nn})^{-1} \approx (10^{15}$ cm$)n_0^{-1}$ set by the density of neutral particles and their elastic cross section.

Since the gas "jumps" discontinuously from its pre-shock to post-shock conditions, such shocks are called "J-shocks" (Draine 1980). We will discuss multi-fluid "C-shocks" in section 3. The existence of an adiabatic shock transition requires that the relative kinetic energies of pre-shock and post-shock gas be dissipated into heat over a scale length small compared to the cooling length. A radiative shock occurs when radiative cooling occurs faster than dynamical times. Thus, the shock velocity must be sufficiently slow or the gas sufficiently dense so that the cooling time, t_c, of the post-shock gas is less than the dynamical expansion time, t_d. For example, in a time $t = (10^4 \text{ yr})t_4$ after the explosion, the remnant of a supernova of energy $(10^{51} \text{ ergs})E_{51}$ in a medium of hydrogen density $n_H = (1 \text{ cm}^{-3})n_0$ has a shell radius, velocity, and temperature given by the Sedov-Taylor similarity solution,

$$R_s = (12.5 \text{ pc})(E_{51}/n_0)^{1/5}t_4^{2/5} \tag{1}$$

$$V_s = (490 \text{ km s}^{-1})(E_{51}/n_0)^{1/5}t_4^{-3/5} \tag{2}$$

$$T_s = (3.32 \times 10^6 \text{ K})(E_{51}/n_0)^{2/5}t_4^{-6/5}. \tag{3}$$

Formation of a radiative shock begins when $t_c \approx t_d$, or in terms of the radiative cooling coefficient $\Lambda(T)$, when $[3kT_s(t)/n\Lambda(T_s)] \approx R_s/V_s = (5t/2)$. Since $n = 4n_0$ behind the shell, this occurs when $[3kT_s/10n_0\Lambda(T_s)t] \approx 1$. If $\Lambda(T) \approx (1.3 \times 10^{-19} \text{ erg cm}^3 \text{ s}^{-1})T^{-1/2}$ between 10^5 and 10^7 K (Raymond, Cox, and Smith 1976; Gaetz and Salpeter 1983) we can solve for the shell formation time, radius and velocity,

$$t_{sh} = (1.91 \times 10^4 \text{ yr})E_{51}^{3/14}n_0^{-4/7} \tag{4}$$

$$R_{sh} = (16.2 \text{ pc})E_{51}^{2/7}n_0^{-3/7} \tag{5}$$

$$V_{sh} = (332 \text{ km s}^{-1})E_{51}^{1/14}n_0^{1/7}. \tag{6}$$

The resultant velocity V_{sh} is somewhat higher than usual for radiative shocks, but it is important to note that t_{sh} refers to the *onset* of a pressure drop behind the shell. A full radiative shock satisfying the steady flow condition is not set up until the velocity drops to about 200 km s^{-1}. Equations (4-6) are consistent with the blast wave velocities observed in the optical filaments of SNRs such as the Cygnus Loop and Shajn-147. These remnants are probably in transition from the Sedov to radiative phase, although the amount of radiated energy is still a small fraction of their initial energy and has not yet affected their dynamics.

The physical variables of density, velocity, pressure, and temperature behind a shock are determined by hydrodynamical equations of mass, momentum, and energy conservation for "steady flow". (Steady flow assumes a homogeneous and steady source of pre-shock gas, as well as a shock front which does not slow appreciably in the time for gas to flow through the full cooling zone.) For quantitative analyses of radiative shocks, it is convenient to consider a frame of reference in which the adiabatic front is stationary. The pre-shock gas of density ρ_1 and pressure P_1 then flows into the front at velocity $v_1 = V_s$, is compressed discontinuously to density $\rho_2 = 4\rho_1$ (for $\gamma = 5/3$), decelerated to velocity $v_2 = v_1/4$, and heated to post-shock temperature $T_2 = T_s \gg T_1$. (Physical variables ahead of the front are subscripted 1, those immediately behind the front are subscripted 2, while those at general positions downstream are unsubscripted.) If a magnetic field is present (we consider the simple case of a field $\vec{B}$ perpendicular to the flow), a sufficient level of ionization assures that B is coupled to the matter, leading to "flux freezing" in which vB and B/ρ are constant. Conservation of mass and momentum, plus the flux freezing condition, yield relations between the pre-shock variables and downstream variables:

$$\rho_1 v_1 = \rho v \tag{7}$$

$$P_1 + \rho_1 v_1^2 + B_1^2/8\pi = P + \rho v^2 + B^2/8\pi \tag{8}$$

$$B_1/\rho_1 = B/\rho. \tag{9}$$

These equations are valid both for conditions immediately behind the front and at general downstream locations. Combining eqs. (7-9) yields the downstream pressure as a function of density ρ ,

$$P(\rho) = P_1 + \rho_1 v_1^2 \left[1 - \rho_1/\rho \right] + (B_1^2/8\pi) \left[1 - \rho^2/\rho_1^2 \right]. \tag{10}$$

The temperature follows from the ideal gas law, $T = [\mu P(\rho)/\rho k]$, where μ is the mean mass per particle (ions, electrons, and neutrals). Conditions immediately behind the front are derived by combining eq. (10) with the adiabatic gas law, $P/\rho^\gamma = constant$. The density jump, ρ_2/ρ_1, at the front is then given by the solution to the quadratic equation:

$$2(2 - \gamma) \left(\frac{\rho_2}{\rho_1} \right)^2 + [(\gamma - 1)M^2 + 2\gamma(1 + \beta)] \left(\frac{\rho_2}{\rho_1} \right) - (\gamma + 1)M^2 = 0, \tag{11}$$

where we define the quantities,

$$c_{s1}^2 = P_1/\rho_1; \qquad M = v_1/c_{s1}; \qquad \beta = (B_1^2/8\pi P_1).$$

Here, c_{s1} is the isothermal sound speed in the pre-shock gas, $M = (\rho_1 v_1^2/P_1)^{1/2}$ is the isothermal Mach number, and $\beta = (v_{A1}/2c_{s1})^2$ is the ratio of magnetic pressure to gas pressure, proportional to the square of the ratio of the Alfvén speed, $v_{A1} = B_1/(4\pi \rho_1)^{1/2}$, to the isothermal sound speed. We characterize the strength of the interstellar magnetic field by the dimensionless parameter b,

$$B_1 = (10^{-6}\ G)n_1^{1/2}b; \qquad v_A = (1.84\ km\ s^{-1})b, \tag{12}$$

with n_1 measured in hydrogen nuclei cm^{-3}. In interstellar clouds the parameter b is typically of order unity (Mouschovias 1976; Troland and Heiles 1986), and diffuse-cloud interstellar shocks are "super-Alfvénic" ($V_s > v_{A1}$) as well as supersonic ($V_s > c_{s1}$). In the strong-shock limit ($M \gg 1$) the Rankine-Hugoniot jump conditions (Landau and Lifschitz 1959; but note the error in their eq. 85.9) reduce to:

$$\left(\frac{\rho_2}{\rho_1} \right) = \left(\frac{v_1}{v_2} \right) = \left(\frac{\gamma + 1}{\gamma - 1} \right) \to 4 \tag{13}$$

$$P_2 = \left(\frac{2\rho_1 v_1^2}{\gamma + 1} \right) \to \frac{3\rho_1 v_1^2}{4} \tag{14}$$

$$kT_2 = \left[\frac{2(\gamma - 1)}{(\gamma + 1)^2} \right] \mu_s v_1^2 \to \frac{3\mu_s v_1^2}{16}. \tag{15}$$

The last numbers are evaluated for $\gamma = 5/3$. For strong shocks ($M \gg 1, b \approx 1$) magnetic fields do not appreciably alter these jump conditions, and eq. (13) is a good approximation to eq. (11).

2.2 Shock Structure

Models of radiative shocks are usually parameterized by several quantities: the shock velocity V_s; the pre-shock density n_1, temperature T_1, and magnetic field B_1; and the set of elemental abundances (*e.g.*, H, He, C, N, O, Ne, Mg, Si, S, Fe). The pre-shock ionization states of these elements are also required, but in the absence of external ionizing radiation, these may be specified self-consistently by computing the structure of the radiative precursor (Shull and McKee 1979). A new ingredient to shock models (Seab and Shull 1983, 1985) is a pre-shock grain model, specifying constituents and size distributions of grains and the initial depletions of the heavy elements which compose them, primarily C, O, Si, Mg, and Fe.

We assume steady, plane-parallel flow, for which $\partial/\partial t = \partial/\partial x = 0$, so that $(d/dt) = v(d/dx) = (\rho_1 v_1/\rho)(d/dx)$ is the Lagrangian derivative following a parcel of fluid. The post-shock density in the cooling zone is derived from an energy equation,

$$\frac{d}{dx}\left[\rho v\left(\frac{v^2}{2}+U+\frac{P}{\rho}\right)+\left(\frac{B^2 v}{4\pi}\right)\right]+n^2\mathcal{L}(T)=0. \tag{16}$$

Here, the total (specific) internal energy U includes internal quantum states of excitation, with population n_i and energy E_i (*e.g.*, vibrational-rotational states of molecules, or excited levels of atoms),

$$U=\frac{3P}{2\rho}+\frac{1}{\rho}\sum_i n_i E_i. \tag{17}$$

The total loss function (cooling minus heating) is given by,

$$n^2\mathcal{L}(T) = n^2(\mathcal{L}_{rad}+\mathcal{L}_{dis})-H_{ext}$$
$$+\sum_j n_j\left[n_e(\alpha_j E_{r,j}+C_j I_j)-4\pi\int_{\nu_j}^\infty \sigma_j(\nu)(1-\nu_j/\nu)J_\nu d\nu\right], \tag{18}$$

where $\mathcal{L}_{rad}$ and $\mathcal{L}_{dis}$ are the cooling rate coefficients for collisionally excited radiative transitions and molecular dissociations, and H_{ext} is any external heating source. For species (j) of density n_j, including all ion states of all elements and molecules, α_j is the recombination rate coefficient for ion state $(j+1) \to j$. The mean energy of recombining electrons is $E_{r,j}$, the collisional ionization rate coefficient is C_j, the radiation intensity is J_ν, the ionization threshold is $I_j = h\nu_j$, and $\sigma_j(\nu)$ is the photoionization cross section. The ionization state and cooling rate behind radiative shocks are far from equilibrium, and $\mathcal{L}(T)$ differs from the radiative cooling coefficient $\Lambda(T)$. Expressing all variables in eq. (16) in terms of ρ, we solve for

$$\left(\frac{d\rho}{dt}\right)=\left[\frac{n^2\mathcal{L}(T)+\sum_i \dot{n}_i E_i}{A_1\rho^{-1}+A_2\rho-A_3\rho^{-2}}\right], \tag{19}$$

where the constants are given by:

$$A_1 = \frac{5}{2}(P_1+\rho_1 v_1^2+B_1^2/8\pi)+\sum_i n_i E_i$$

$$A_2 = \left(\frac{B_1^2}{16\pi\rho_1^2}\right); \qquad A_3 = \left(4\rho_1^2 v_1^2\right).$$

Downstream from the shock front, radiative cooling results in a large compression ($\rho \gg \rho_1$), while the total pressure ($P + \rho v^2 + B^2/8\pi$) remains constant. For no magnetic field, the thermal pressure P varies by only 33%, from its post-shock value of $3\rho_1 v_1^2/4$ to the full value of the "ram pressure" $\rho_1 v_1^2$ when $\rho \gg \rho_1$ (see eq. 10). The flow velocity relative to the pre-shock gas varies from $3V_s/4$ just behind the front to V_s far downstream. Thus, emission or absorption lines formed in the hot post-shock gas are shifted in velocity by $V_s/4$ from lines formed in the cold gas downstream and by $3V_s/4$ from the pre-shock gas. When $B = 0$, the final compression of the shock can be quite large,

$$\left(\frac{\rho_f}{\rho_1}\right) = M^2 \left(\frac{T_1}{T_f}\right) , \tag{20}$$

where ρ_f and T_f are the final density and temperature and M is the isothermal Mach number. However, the compression is limited by a realistic initial magnetic field (eq. 12), since the magnetic pressure eventually dominates the momentum flux ($B^2 \propto \rho^2$, whereas $P \propto \rho T$ and $\rho v^2 \propto \rho^{-1}$). Thus, the maximum compression in a strong magnetized shock is set by the relation $\rho_1 v_1^2 \approx B_m^2/8\pi = (B_1^2/8\pi)(\rho_m/\rho_1)^2$, or

$$\left(\frac{\rho_f}{\rho_1}\right) = \left(\frac{8\pi \rho_1 v_1^2}{B_1^2}\right)^{1/2} = 2^{1/2} \left(\frac{v_1}{v_{A1}}\right) = (77) \left(\frac{v_{s7}}{b}\right) , \tag{21}$$

where $v_{s7} = (V_s/100 \text{ km s}^{-1})$, where $b \approx 1$ is the magnetic field parameter (eq. 12), and ρ_m and B_m are maximum values of post-shock density and magnetic field. A typical compression is about a factor of 100.

2.3 Atomic and Grain Processes

The post-shock structure of radiative shocks depends on a variety of atomic processes, the most important of which are collisional ionization, photoionization, radiative and dielectronic recombination, ion charge exchange with H^0 and He^0, and radiative cooling. The emissivity in lines and continuum is dominated by electron-impact excitation of resonance, semi-forbidden, and forbidden lines of H^0, He^0, He^+ and ions of abundant elements (mostly C and O ions). The rates of these processes are temperature dependent and involve heavy element abundances, gas-grain interactions, and radiative transfer. This section describes these processes and their nonlinear interactions.

Immediately behind an adiabatic shock of $V_s = (100 \text{ km s}^{-1})v_{s7}$, the temperature (eq. 15) is $T_s = (3\mu_s V_s^2/16k) = (1.44 \times 10^5 \text{ K})v_{s7}^2$, where we have assumed that He/H = 0.1 and that H and He are singly ionized by the radiative precursor ($\mu_s = 0.636 m_H$). At these temperatures, the radiative cooling is dominated by collisional ionization of He^+ and excitation of permitted and semi-forbidden lines of $He^+(\lambda 304)$ and carbon and oxygen ions. In general, the degree of ionization of these species is lower than it would be in coronal ionization equilibrium, and the initial radiative cooling rate exceeds equilibrium values by factors of 10 to 100. The temperature dependence of the cooling rate is also steeper than in equilibrium; $\mathcal{L}(T) \propto T^\epsilon$ where ϵ ranges between 3 and 5 depending on shock velocity and distance behind the front.

The non-equilibrium ionization fractions, $f_i = n_i/n_{tot}$, of the elements are determined by integrating time-dependent differential equations of the form,

$$\left(\frac{df_i}{dt}\right) = f_{i-1}[n_e C_{i-1} + G_{i-1}] - f_i[n_e(C_i + \alpha_{i-1}) + n(H^0)Z_i + G_i]$$
$$+ f_{i+1}[n_e \alpha_i + n(H^0)Z_{i+1}], \tag{22}$$

where $C_i(T), \alpha_i(T)$, and $Z_i(T)$ are rate coefficients (cm^3 s^{-1}) for collisional ionization *from*, recombination *to*, and charge exchange *from* ionization state (i). The coefficient $C_i(T)$ is dominated by electron impact and includes both direct (valence shell) ionization as well as autoionization following inner-shell excitation. The latter is particularly important at high temperatures for ions with 1 or 2 electrons outside a closed shell. The recombination coefficients $\alpha_i(T)$ include both radiative and dielectronic recombination; dielectronic recombination dominates over radiative by a substantial factor at high temperatures ($T > 20,000$ K for most ions). Tables of $\alpha_i(T)$ and $C_i(T)$ may be found in Shull and Van Steenberg (1982), based on fits to recent calculations or measurements by many groups (*e.g.* Crandall 1981; Younger 1980a,b, 1981a,b,c; Jacobs *et al.* 1977a,b, 1978, 1980; Jacobs 1985). Photoionization cross sections may be found in a variety of sources (*e.g.*, Reilman and Manson 1979; Clark *et al.* 1985), their energy dependence conveniently fitted to the polynomial form,

$$\sigma_{ph}(E) = \sigma_T \left[\alpha \left(\frac{E_T}{E}\right)^S + (1 - \alpha) \left(\frac{E_T}{E}\right)^{S+1}\right], \tag{23}$$

where σ_T is the cross section at threshold energy E_T, and where α and S are constants (typically $S \approx 3$). To find the photoionization rate G_i(s^{-1}), one integrates $\sigma_{ph}(E)$ over the ionizing radiation spectrum in the shock layer.

Charge exchange collisions with H^0 (and sometimes He^0) are often the most effective means of reducing the ion state in shocks containing a substantial population of neutrals (Shull and McKee 1979; Butler and Raymond 1980). Dielectronic recombination rate coefficients of multiply ionized species (*e.g.* C IV or O IV) in hot post-shock gas (20,000 K to 200,000 K) typically peak at values $\alpha_i = 3 - 5 \times 10^{-11}$ cm^3 s^{-1}, whereas charge exchange coefficients with H^0 can exceed 10^{-9} cm^3 s^{-1}. Charge exchange with multiply ionized species therefore dominates recombination when the neutral fraction $n(H^0)/n(H_{tot})$ exceeds 1 to 5%. Charge exchange rate coefficients are discussed by Dalgarno and Butler (1978), McCarroll and Valiron (1976), Butler and Dalgarno 1980; Heil, Butler, and Dalgarno (1980), Butler, Heil and Dalgarno (1980), Baliunas and Butler (1980), and Dalgarno, Heil, and Butler (1981). Generally, the rates with ions of charge $z \geq +3$ are fast ($>$ 10^{-9} cm^3 s^{-1}). Rates for doubly charged ions are mixed: C III, S III, and Ne III are slow ($\sim 10^{-12}$ cm^3 s^{-1}), while N III, O III, and Si III are fast. Charge exchange of N II is slow ($\sim 10^{-12}$ cm^3 s^{-1}), but resonant charge exchange between O II and H I effectively couples the O and H ionization fractions, (O II/O I) $\approx$ (8/9)(H II/H I).

Collisional excitation rate coefficients are found from a variety of sources, and there is not always agreement on a standard set. Electron collisions dominate the excitation of the permitted, semi-forbidden, and optical forbidden lines of atoms and ions. Infrared fine structure lines are excited by collisions with electrons, H$^+$ and H^0 (Dalgarno and McCray 1972). The electron impact excitation rate coefficient C_{ij}(cm^3 s^{-1}), for a transition (i-j) of

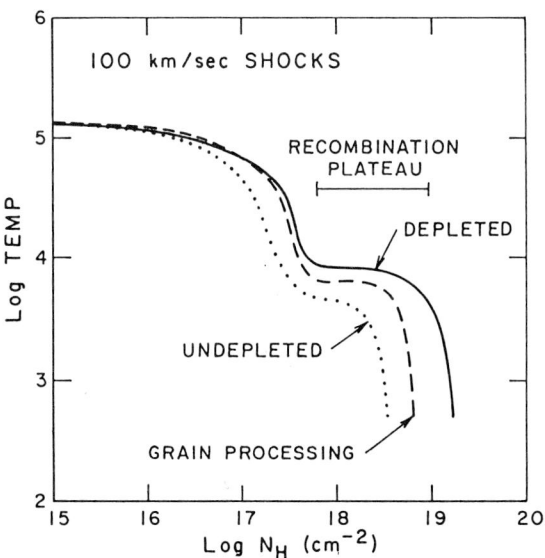

Figure 3: Post-shock temperature profiles versus column density $N_H = n_0 V_s t$ (cm^{-2}).

energy E_{ij}, is parameterized by the dimensionless "collision strength" Ω_{ij}:

$$C_{ij} = (8.616 \times 10^{-6}\ \text{cm}^3\ \text{s}^{-1}) \left(\frac{\Omega_{ij}}{g_i}\right) T^{-1/2}\exp\left(\frac{-E_{ij}}{kT}\right), \tag{24}$$

where g_i is the statistical weight of the lower state and T is the temperature (K). References for excitation of H^0, He^0, and He^+ and ions of heavy elements are found in Shull and McKee (1979). Other recent tabulations of collision strengths include: Osterbrock (1974, with revisions), Raymond and Smith (1977), Shull (1981), and Cox and Raymond (1985). Compilations of electron impact excitation data for atomic ions are available as scientific reports from Los Alamos (Merts *et al.* 1980) and JILA (Gallagher and Pradhan 1985).

Figure 3 shows the temperature profiles in three 100 km s^{-1} shock models (Seab and Shull 1985). The post-shock column density, N_H, is a convenient measure of post-shock distance or Lagrangian flow time, independent of pre-shock density n_1. By the constancy of mass flux (or nv) in one-dimensional flow, $N_H = n_1 V_s t$, where t is the flow time for a parcel of fluid to reach column N_H. The three cooling profiles in Fig. 3 represent models in which heavy elements are: (i) depleted from gas phase; (ii) initially depleted, but allowed to re-enter gas phase via grain processing; and (iii) fully in gas phase (undepleted). Evidently, the post-shock abundance of atomic coolants such as C, O, Si and Fe, can have an important effect on the total column and thus the strengths of emission lines.

Grain processing in shocks comes from grain-grain collisions and thermal and non-thermal gas-grain sputtering (more details are found in Seab 1987). Modeling is complicated by the need to specify the grain constituents and size distribution. Further complications are introduced by uncertainties in the "solid state physics": the sputtering yields in He-grain collisions (Barlow 1978; Draine and Salpeter 1979), and the rates of vaporization, partial vaporization, and shattering in grain-grain collisions (Seab and Shull 1983, 1985). In fast shocks ($V_s > 150$ km s^{-1}) grain collisions with hot post-shock ions (primarily He) dominate the sputtering of small grains, whereas grain-grain collisions and non-thermal sputtering are relatively more important for larger grains in lower velocity shocks. Crucial to these "non-thermal" processes are the large gyrovelocities of charged grains generated by "betatron acceleration" as the magnetic field is compressed with the gas in the cooling zone (Shull 1977, 1978b; Draine and Salpeter 1979). Most of the non-thermal grain destruction is produced in the strongly cooling layers around 10^4 K (Fig. 2).

As the gas recombines and cools, the radiative cooling rate falls and the temperature reaches a plateau near 6000 K. Here, ionizing photons produced in the hot post-shock layer deposit their energy by photoionizing the newly recombined H^0 and He^0. The cooling immediately behind the front is due primarily to electron-impact collisional ionization of H^0, He^0, or He$^+$, depending on shock velocity, plus electron-impact excitation of resonance and semi-forbidden lines of H$^0(Ly\alpha)$, He$^0(\lambda 584, 626)$, He$^+(\lambda 304)$, and ions of abundant heavy elements, primarily C and O. Below 20,000 K, the cooling is dominated by forbidden and semi-forbidden lines of heavy elements, such as [O III] $\lambda 5007$, [O II] $\lambda 3727$, C III] $\lambda 1909$, [S II] $\lambda 6716, 6731$ and C II] $\lambda 2326$. Because a forbidden line may be collisionally de-excited when the electron density exceeds the line's "critical density", $n_{cr} = A_{21}/C_{21}$ ranging from 10^2 to 10^6 cm^{-3}, the cooling scale may be lengthened in shocks of higher density. In the gas below 10^3 K, infrared fine structure lines dominate the cooling – for example [Si II] $34.8\mu m$, [O I] 63 and $145\mu m$, and various lines of [Fe II] (1.27, 1.6, 5.0, and $26\mu m$). If the gas contains a fraction of H$_2$, the rotational lines are also important coolants.

2.4   Molecular Processes

Thus far, we have largely ignored molecular physics, except to note that molecular cooling and dissociation play a role in the thermal balance. However, optical and Copernicus satellite observations show that many diffuse clouds contain small amounts of H$_2$, CO, CH$^+$, and other trace molecules (Morton 1975; Spitzer and Jenkins 1975). The Copernicus survey of diffuse interstellar H I and H$_2$ (Savage et al. 1977) showed that, statistically, the molecular fraction of diffuse clouds is correlated with both E(B-V) and N(H), and undergoes a transition from low values ($<1\%$) to high values ($>1\%$) when $E(B - V) \geq 0.08$ mag.

A full discussion of molecular chemistry in diffuse clouds is beyond the scope of this paper. Useful reviews on related subjects are: molecular abundances in hydrostatic cloud models (van Dishoeck and Black 1986), chemistry in molecular shocks (Hollenbach and McKee 1979; McKee and Hollenbach 1980), and a general review of interstellar molecular hydrogen (Shull and Beckwith 1982). Here we confine our discussion to the processes of H$_2$ formation and destruction.

Because radiative association of two H atoms is forbidden by dipole selection rules, interstellar H$_2$ is believed to form most rapidly on grain surfaces. When two H^0 atoms collide with a grain and stick, they migrate and eject an H$_2$ molecule with substantial kinetic, vibrational, and rotational energy (Hollenbach and Salpeter 1971). In grain-free or pre-galactic environments, H_2 may also form by slower gas-phase reactions with H^- or H_2^+ (Lepp and Shull 1984; MacLow and Shull 1986; Shapiro and Kang 1987). Dissociation of H_2 occurs either by a two-step process initiated by absorption of a UV photon in one of the Lyman ($\lambda < 1120$Å) or Werner bands ($\lambda < 1021$Å) or by collisions with H^0, H^+, or e^-. The photodissociation rate may be diminished by optical depth ("self-shielding" in the Lyman lines (Jura 1974; Shull 1978a) or by dust opacity. At low density ($n_H < 10^5$ cm^{-3}) the rate of collisional dissociation can also be reduced by "radiative stabilization" (Roberge and Dalgarno 1982; Lepp and Shull 1983), in which radiative decays decrease the populations of vibrationally and rotationally excited H_2 levels which are subject to large collisional dissociation rates in thermal (Boltzmann) populations. Rate coefficients for radiative decay of vibrational and rotational states of H_2 are given by Turner, Kirby-Docken, and Dalgarno (1977), and for H^0-H_2 collisional excitation and dissociation by Lepp and Shull (1983), revised at high temperature by MacLow and Shull (1986).

Molecular cooling in shocks arises from the excitation of rotational and vibrational states of H_2, CO, H_2O, and other abundant molecules. Since molecular shocks are often slower and the gas denser and more neutral than in diffuse clouds, the excitation comes from collisions with H^0 and H_2 as well as from electrons. In addition, magnetic fields and multi-fluid effects play an important role (see section 3). Here, we restrict our discussion to J-shocks with velocities of order 10 km s^{-1}, in which the post-shock temperatures are,

$$ T_s = \begin{cases} (2900K)(V_s/10 \text{ km s}^{-1})^2 & \text{atomic H I} \\ \\ (3900K)(V_s/10 \text{ km s}^{-1})^2 & \text{molecular } H_2, \end{cases} \qquad (25) $$

where the formulae account for the different mean masses and adiabatic indices in neutral atomic gas ($\mu_s = 1.27m_H$, $\gamma = 5/3$) and rotationally excited molecular gas ($\mu_s = 2.33m_H$, $\gamma = 7/5$). These temperatures are sufficient to excite many rotational states (J) and several ($v = 1$ and 2) vibrational states of H_2. For small J and v, the H_2 excitation temperatures are $T_e(J) = E(J)/k = (85K)J(J+1)$ and $T_e(v) = (6300K)v$.

Such J-type shocks, with $V_s = 8$ to 15 km s^{-1}, were initially thought to produce the strong 2 μm vibrational emission lines seen toward the Orion Molecular Cloud (Hollenbach and Shull 1977; London, McCray, and Chu 1977). However, it was quickly realized (Kwan 1977; Hollenbach and McKee 1980) that shocks with velocities larger than about 25 km s^{-1} would dissociate the H_2, whereas the (1-0) S(1) H_2 line exhibited wings out to 90 km s^{-1} (Nadeau and Geballe 1979). At the densities characteristic of the OMC outflow ($n > 10^5$ cm^{-3}), radiative stabilization is incapable of significantly reducing the dissociation rates. The lower temperatures in C-type shocks with magnetic precursors are now believed to be part of the solution (Draine and Roberge 1982; Chernoff, Hollenbach, and McKee 1982). The 90 km s^{-1} wings are difficult to model without postulating condensations in a bulk flow (Shull and Beckwith 1982).

2.5   Observations and Line Diagnostics

In many astrophysical situations, such as SNRs or Herbig-Haro objects, shocks are believed to dominate the line emissivity. In other environments such as emission-line galaxies, the

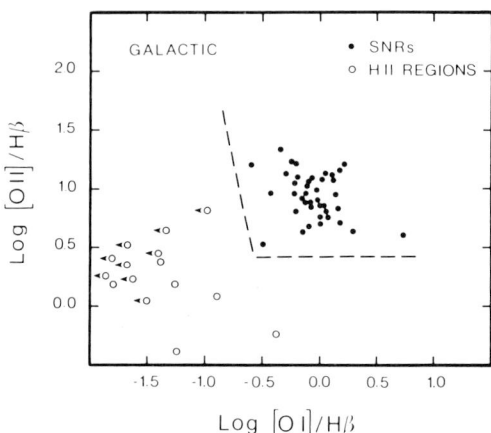

Figure 4: Spectral discrimination of SNRs in the Galaxy. The [O II] and [O I] ratios in SNRs, corrected for extinction, are clearly distinguished from H II regions.

emission source is unclear, either because of a faint image or because angular size precludes resolving the source. In both situations, theoretical models of shocks are useful.

Many authors have remarked on the spectral signatures of shock waves, as distinguished from H II regions and other photoionized regions (Baldwin, Phillips, and Terlevich 1981; Fesen, Blair, and Kirshner 1985). These distinctions are important for identifying the source of excitation in many active galaxies and "LINERs". Generally, SNRs are characterized by strong optical forbidden line emission over a wide range of ionization states. For example, [S II]/Hα is stronger in SNRs than in H II regions. The main observational features attributed to radiative shock waves in the optical are:

1. Strong emission lines, relative to Hβ , of underionized species, e.g., [O I] 6300, [N I] 5200, [O II] 3726,3729, [S II] 6716,6731.

2. A high excitation temperature ($T > 20,000$; K) measured from the intensity ratio of [O III] lines, [4363/(5007 + 4959)].

3. The presence of a range of ionization states, e.g., [O I], [O II], [O III], [Ne III], [Ne V].

4. Large ratios of [O I]/Hβ and [O II]/Hβ , relative to H II regions.

The fourth effect has been demonstrated empirically (Fig. 4) for SNRs in the Galaxy and M31/M33 (Fesen et al. 1985). New wavelength bands have opened up other shock discriminants. In the ultraviolet, shocks waves produce strong resonance and semi-forbidden lines of C II] 2326, C III] 1909, O III] 1663, N III] 1750, O IV] 1402, C IV 1549, and N V 1240.

In the infrared, the fine structure lines of [O I] 63 μm, [Si II] 34.8 μm, [Ne II] 12.8 μm, and [Fe II] (26, 1.6, 1.27, 5 μm) are strong. While these features are not unique (photoionized gas at high densities can mimic some of the optical line ratios), the combination of lines from several ion stages and varying excitation temperatures can be used to attribute the power source to shocks.

Once shocks have been identified as the source of excitation, certain line ratios can be used to constrain the shock velocity V_s, the pre-shock density n_0, and the abundances. Table I lists several line intensities, relative to Hβ, for 50, 100, and 150 km s^{-1} shocks with $n_0 = 10$ cm^{-3}, and $B_1 = 1\mu G$ (Shull, Seab, and McKee 1986). The emission lines of [O III], [Ne III], C III], C IV, and N V are the best "speedometers" since their intensities rise steeply with V_s. The temperature in the post-shock "recombination zone" may be gauged by the intensity ratios of [O III] [4363/(5007+4959)] or [Ne III] 3342/3869. Density sensitive line ratios include [O II] 3729/3726 and [S II] 6716/6731, as well as certain infrared fine structure lines. Shocks into diffuse clouds generally have recombination-zone densities less than 10^3 cm^{-3}. The reason for the absence of higher densities is clear: shocks which propagate from a lower density intercloud medium into a dense cloud are slowed by a factor equal to the square root of the density contrast (momentum flux ρv^2 is conserved). The emission from these much slower shocks ($V_s = 10$ km s^{-1}) would prove difficult to detect optically, although infrared lines of [O I], [Ne II], [Si II], and [Fe II] might be detected with a new generation of detectors.

Abundance determinations from emission lines are fraught with uncertainties. For example: (1) Changes in heavy element abundance are difficult to distinguish from differences in density, magnetic field, and velocity. (2) Intensities of strong forbidden lines "saturate" with increasing velocity or abundance, since there is only a fixed amount of energy available for these cooling lines in the recombination zone. (3) Strong UV lines such as C III] 1909 and C IV 1549 become insensitive to V_s as the post-shock temperature rises above their ionization and excitation temperature. (4) Many resonance lines (C II 1035, 1335; C IV 1549) become optically thick, and their emergent intensities are reduced by scattering in the shock layer. (5) Grain disruption by sputtering and grain-grain collisions introduces another degree of freedom: variable gas-phase abundances of C, O, Si, and Fe. To model the intensities from lines of these refractory elements one must understand the betatron acceleration of grains, and gas-grain interactions as a function of V_s, n_0, and B_1. With these uncertainties in mind, one may still make progress with carefully chosen forbidden and semi-forbidden lines of ionized species (Raymond et al. 1981; Raymond 1984; Dopita et al., 1984).

The Cygnus Loop is the best studied of the older SNRs (Fig. 1). It is generally pictured (Raymond 1984; Hester and Cox 1986) as a 400 km s^{-1} blast wave, propagating in an intercloud medium of density $n_H \approx 0.2$ cm^{-3} and driving 100 km s^{-1} shocks into clouds of density $2 - 10$ cm^{-3}. The faster, non-radiative shocks produce the observed X-rays while the slower (radiative) shocks produce the bright optical filaments. Fesen, Blair, and Kirshner (1982) obtained interference filter images of the remnant in [O III], Hα + [N II], and [S II]. Spectra of several bright filaments show line ratios which disagree radically with radiative shock models having normal abundances and full recombination zones. Some filaments show [O III]/Hβ as large as 10-25, whereas current shock models do not allow

Table 1: LINE INTENSITIES[1] IN RADIATIVE SHOCK MODELS[2]

Band	Line		50 km s⁻¹	100 km s⁻¹	150 km s⁻¹
UV:	Lyα	1216	12,000	4740	3880
	C II	1035	...	321	169
	C II	1335	1.6	70	22
	C II]	2326	23	121	49
	C III]	1909	0.13	192	64
	C IV	1549	...	113	116
	N II]	2141	0.26	22	13
	N III]	1750	...	38	34
	N V	1240	...	0.17	220
	O III]	1663	...	144	185
	O IV]	1402	...	7	109
	Si III	1206	...	43	31
	Fe II	2600	14	44	56
Opt:	Hα	6563	378	306	297
	He II	4686	...	4.5	6.6
	[O I]	6300,6364	212	127	98
	[O II]	3727	126	1070	762
	[O III]	5007,4959	...	291	276
	[O III]	4363	...	19	22
	[N I]	5200	135	72	42
	[N II]	6548,6584	11	255	314
	[S II]	6716,6731	424	475	408
	[Ne III]	3869.3967	...	70	50
IR:	[Fe II]	5 μm	72	132	140
	[Fe II]	1.27 μm	98	238	273
	[Fe II]	26 μm	47	72	73
	[C II]	158 μm	92	14	6
	[Si II]	34.8 μm	429	372	243
	[O I]	63 μm	451	185	122
	[O I]	145 μm	58	21	14
	[O III]	88 μm	...	9	7
	[Ne II]	12.8 μm	3	137	109
	[Ne III]	15.6 μm	...	43	32

[1] Intensities (erg s⁻¹ cm⁻²) relative to I(Hβ) = 100.

[2] Unpublished models, for $n_0 = 10$ cm⁻³, $B_1 = 1\mu G$, including radiative precursors and grain processing (Seab and Shull 1983, 1985; (Shull, Seab, and McKee 1986).

ratios greater than 3, owing to rapid O^{+2} - H^0 charge exchange in the recombination zone (see Table I). More generally, Fesen *et al.* (1982) showed that the distribution of line ratios, *e.g.*, [O III]/Hβ versus [O II]/Hβ or [O II]/Hβ versus [O I]/Hβ, bears little resemblance to those predicted by standard radiative shock models.

The remedy to this disagreement may be to "truncate" the shocks' recombination zones. A full radiative shock, complete with recombination zone, requires a column density $N_{rec} \approx 10^{19}$ cm^{-2}, corresponding to a flow time of $(3 \times 10^4$ yr$)(1$ cm$^{-3}/n_0)(100$ km s$^{-1}/V_s)$. If the zone with $T < 10^4$ K is missing, as a result of an inhomogeneous pre-shock medium or thermal instability, then the region which produces Hβ recombination lines will be missing and the O III charge exchange will be insignificant in the more ionized gas. A physical realization of this scenario for the optical filaments requires either many small ($< 10^{16}$ cm) cloudlets engulfed by the blast wave (Fesen, Blair, and Kirshner 1982), or variations in line-of-sight surface brightness produced by a few wavy thin sheets (Hester and Cox 1986). The latter model depends on complex interactions between the blast wave and reflection shocks or bow shocks around dense clouds. A critical question seems to be how to create a "wave train" of density fluctuations by these interactions.

Shock models have also been used to explain the optical line emission and fix the velocity and pre-shock density of Herbig-Haro objects (see the review by Schwartz 1983). Studies of bright HH-objects in the blue (Dopita, Binette, and Schwartz 1982) and ultraviolet (Brugel, Shull, and Seab 1982) demonstrated the presence of a H^0 two-photon continuum, peaking near 1500Å and containing a substantial fraction of the total luminosity. This continuum results from the collisionally excited 2s-state of H I, and its strength relative to Hβ and the Balmer continuum is anomalously high. Brugel, Shull, and Seab (1982) suggested a range of shock velocities (70 – 100 km s^{-1}) and a truncation of the recombination zone, in a manner reminiscent of the Cygnus Loop filaments. Ultraviolet variabilility in HH-1 (Brugel *et al.* 1985) is consistent with small parcels of shocked gas. However, the actual situation may involve shocks with a range of velocities due to two dimensional flow in a jet-interface or bowshock. Recent work piecing together a range of velocities in bowshocks (Hartmann and Raymond 1984; Raga and Böhm 1985; Hartigan, Raymond, and Hartmann 1986) has been successful in explaining: (i) the extraordinary line widths and double-peaked profiles of optical lines; (ii) the range of velocities needed by the models; and (iii) position-velocity diagrams.

A final observational topic concerns grain destruction by shocks. In section 2.3, we described the importance of grain processes in determining self-consistent gas-phase abundances of refractory elements in the post-shock gas. Emission lines from Si and Fe are particularly affected by the grain sputtering and grain-grain collisions in the cooling zone. But what are the global consequences of this grain processing ? Seab and Shull (1983) discussed effects on the UV selective extinction curves of shock processed gas, and many authors have looked for evidence of grain destruction in interstellar absorption studies of elemental depletions.

The *Copernicus* satellite found that significant fractions of many heavy elements are depleted from the interstellar gas, presumably locked up in dust grains. Quantitatively, we

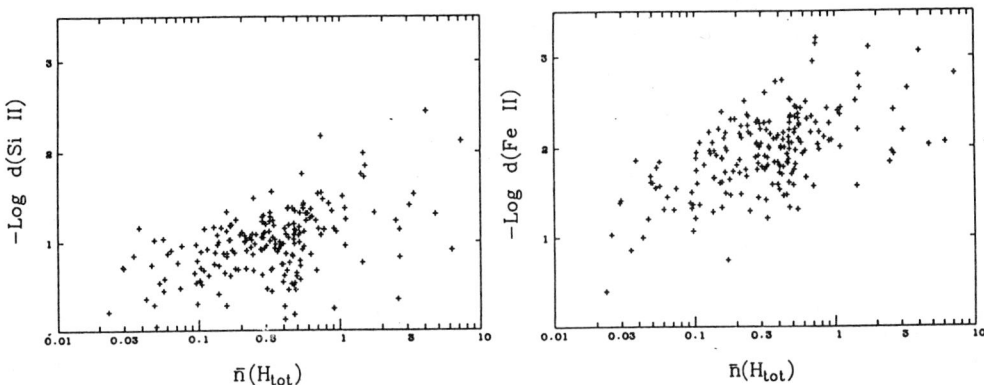

Figure 5: Depletion factors d_i for interstellar Si and Fe observed by IUE toward 225 OB stars, are correlated with mean hydrogen density $\bar{n} = N_H/r$.

define the "depletion factor" d_i of an element (i) by,

$$\log d_i = \log \left(\frac{N_i}{N_H} \right) - \log \left(\frac{N_i}{N_H} \right)_\odot \tag{26}$$

where N_i and N_H are the column densities of (i) and of hydrogen, and where $(N_i/N_H)_\odot$ is the solar or cosmic abundance (Withbroe 1971; Grevesse 1983). Generally the elements S, Ar, Cl, P, and Zn are not depleted in diffuse clouds; C, N, and O appear to be depleted by factors up to of 2 to 3 (Hobbs, York, and Oegerle 1982; York et al. 1983); and "refractory" elements such as Si, Fe, Ca, Ti, Mn, Al, and Ni are depleted by factors ranging from 10 to 1000 (Spitzer and Jenkins 1975; Phillips, Gondhalekar, and Pettini 1982).

The influence of shocks is believed to explain the correlation of refractory element depletions with cloud velocity. The optical observation (Routly and Spitzer 1952; Siluk and Silk 1974) that interstellar clouds with high velocities show systematically higher ratios of Ca II/Na I has been interpreted as evidence of selective grain destruction, which returns the highly depleted calcium back to gas phase. The same correlation of depletion with cloud velocity has been seen in UV absorption studies of Si and Fe (Shull, York, and Hobbs 1977). Both the data and theoretical models (Shull 1978b; Cowie 1978; Draine and Salpeter 1979; Seab and Shull 1983) are consistent with the general conclusion that clouds with velocities greater than about 20 km s^{-1} have larger gas-phase abundances of refractory elements than low-velocity gas.

Grain processing may also be responsible for the correlation of heavy element depletions with mean line-of-sight column density $\bar{n} = N(H)/r$ (Savage and Bohlin 1979; Phillips, Gondhalekar, and Pettini 1982; Harris, Gry, and Bromage 1984; Jenkins, Savage, and Spitzer 1986; Van Steenberg and Shull 1986). Figure 5 shows the results of an IUE survey of 225 lines of sight toward OB stars within 5 kpc (Van Steenberg and Shull 1986). The increasing depletion of Fe and Si with larger $\bar{n}$ has traditionally been interpreted as evidence that heavy elements undergo depletion onto grains in the diffuse ISM. However, the trend may also be understood with an idealized model (Spitzer 1985), based on random distributions of two types of cold clouds plus uniformly distributed warm (intercloud) H I. Each component has a different depletion pattern, with heavy elements less depleted in intercloud matter. As $\bar{n}$ increases, the relative contributions of the three components change; the low density intercloud gas dominates for $\bar{n} < 0.2$ cm^{-3} and the large clouds dominate for $\bar{n} > 3$ cm^{-3}. A possible physical mechanism for Spitzer's model is that lines of sight with low $\bar{n}$ are more likely to have had extensive shock processing of grains.

The IUE survey also suggests that certain elements may undergo selective grain destruction. For example, the Fe/Si abundance ratio is greater for halo stars: $\langle$Fe/Si$\rangle$ = 0.12 and 0.22 towards stars with $|b| < 20^{o}$ and $|b| > 20^{o}$ respectively. Perhaps fast shocks in the halo destroy relatively more Fe grains than Si grains. Studies of Mn, Zn, and S are currently underway (Van Steenberg 1986) to better understand the pattern.

The large rate of grain destruction presents us with a quandary – why do grains exist at all ? How can interstellar grains survive the large amount of shock processing predicted theoretically for standard models of shock waves into the interstellar medium (Draine and Salpeter 1979; Dwek and Scalo 1980; Seab *et al.* 1985)? The "generalized Routly-Spitzer effect" (correlation of Ca, Si, Fe abundances with cloud velocity) suggests that shocks of 20 km s^{-1} and higher destroy some grains; theoretical models suggest that 100 km s^{-1} shocks destroy between 20% and 50% of the grain population, depending on magnetic fields, pre-shock densities, and grain physics. Since the mean time between 100 km s^{-1} shocks in the diffuse ISM is about 10^{8} years (Seab and Shull 1985), while the mean time for grain astration is about 2×10^{9} years (Dwek and Scalo 1980), a typical grain population will be processed by 10 to 20 shocks if it lives long enough to be incorporated into new star formation. These numbers depend sensitively on the assumed physical model of the ISM, on the theories of shock processing, and on the processes of grain injection and growth. However, unless we have underestimated grain formation rates, it appears likely that depletion patterns, grain size distributions, and selective extinction curves are determined by the physics of gas-grain interactions in shocks and grain accretion in quiescent clouds.

3 MHD SHOCKS WITH MAGNETIC PRECURSORS

3.1 Fluid Dynamics

The fundamental concept underlying the following discussion of MHD shocks in gas of low fractional ionization is that the matter in the shocked regions may be thought of as consisting of several distinct, interpenetrating fluids. Normally one thinks of three fluids: (i) the neutral particles; (ii) the ions; and (iii) the electrons. Under some circumstances it may be useful to consider the charged dust grains to constitute a fourth fluid. The motivation for this conceptual decomposition is that under some circumstances (*e.g.*, in a shock transition) these fluids may develop appreciably different flow velocities and temperatures.

A necessary condition for a shock wave to occur in an initially quiescent medium is that a compressive disturbance be advancing into the medium at a velocity greater than the "signal speed". Otherwise "signals" will travel ahead of the shock and "inform" the quiescent medium that a compression is approaching – the element of "surprise" will be lost.

The effective signal speed here is just the velocity of propagation of long-wavelength, small amplitude compressional waves. In a fluid consisting of neutrals, ions, and electrons, a wave of sufficiently long wavelength (low frequency) must have the neutrals, ions, and electrons moving together. If no magnetic field is present, the compressional signal speed is just the sound speed $c_{s,nie} = (5P/3\rho)^{1/2}$ where $P = k(n_n T_n + n_i T_i + n_e T_e)$ is the total gas pressure, and ρ is the total density. In a magnetized fluid there are two distinct compressional modes, referred to as the "fast" and "slow" magnetosonic modes. The propagation velocity of the "fast" magnetosonic mode is (Spitzer 1962)

$$v_f = \frac{2^{1/2} v_A c_s \cos\theta}{(v_A^2 + c_s^2)^{1/2}} \left[1 - \left\{ 1 - \frac{4 v_A^2 c_s^2 \cos^2\theta}{(v_A^2 + c_s^2)^2} \right\}^{1/2} \right]^{-1/2}, \tag{27}$$

where θ is the angle between $\vec{B}$ and the direction of propagation. The "fast" magnetosonic speed for the combined neutral-ion-electron fluid, $v_{f,nie}$, may be obtained by using $v_A = v_{A,nie} = B/[4\pi(\rho_n + \rho_i)]^{1/2}$, and $c_s = c_{s,nie}$. A shock will occur if the compressive disturbance is advancing with a velocity $V_s > v_{f,nie}$.

To understand the structure of the shock, it is now useful to consider the velocity of propagation of "short" wavelength compressional waves. Because of the weak coupling between the ions and the neutrals, each fluid supports short wavelength waves which, aside from eventually suffering damping, are independent of the presence of the other fluid. Thus the short wavelength modes in the neutral gas are simply sound waves, propagating at the neutral adiabatic sound speed $c_n = (5 n_n k T_n / 3 \rho_n)^{1/2}$. The short wavelength modes in the magnetized ion-electron plasma include the "fast" magnetosonic mode, with propagation speed $v_{f,ie}$ obtained from eq. (27) using $v_A = v_{A,ie} = B/(4\pi\rho_i)^{1/2}$ and $c_s = c_{s,ie} = [5k(n_i T_i + n_e T_e)/3\rho_i]^{1/2}$.

How large are the various Alfvén velocities? Existing observations of interstellar magnetic field strengths (Troland and Heiles 1986) are consistent with an empirical "scaling law" $B = (1\mu G)(n_H/\text{cm}^{-3})^{1/2}$ for densities $10\ \text{cm}^{-3} < n_H < 10^6\ \text{cm}^{-3}$. This relation implies relatively large values of $v_{A,ie}$ in predominantly neutral clouds:

$$v_{A,ie} \approx (50\ \text{kms}^{-1}) \left[\frac{n_i/n_H}{10^{-4}} \right]^{1/2} \left[\frac{20 m_H}{\rho_i/n_i} \right]^{1/2}. \tag{28}$$

Thus, for example, a dense molecular cloud with $n_i/n_H = 10^{-7}$ would have $v_{A,ie} \approx 1500\ \text{km s}^{-1}$. Compressive disturbances with $V_s > v_{f,nie}$ will be shock waves, since long-wavelength signals cannot travel faster than the disturbance. It is now important to realize that if $V_s < v_{f,nie}$ the ion-electron plasma can send signals ahead of the disturbance to "inform" the preshock plasma that a compression is coming, so in some sense the ion-electron plasma is not totally surprised. The short wavelength ion-electron magnetosonic waves are damped, however, as the result of scattering of the ions by neutrals, so the signals only propagate a few damping lengths ahead of the shock before they die away. The region ahead of the shock which is perturbed by the magnetosonic signals is referred to as a "magnetic precursor". So far as the ion-electron plasma is concerned, the compressive wave is a sub-magnetosonic disturbance, and the ion-electron plasma is able to make a smooth, continuous transition from the quiescent preshock state to the quiescent postshock conditions; it does not undergo any sudden acceleration or deceleration.

The equations of motion for each fluid may readily be obtained by applying the laws of conservation of mass, momentum, and energy, with allowance for the fact that mass, momentum, and energy can be exchanged among the fluids by scattering processes (Draine 1986a). Alternatively, the fluid equations can be derived from the Boltzmann equation (Chernoff 1987). The simplifying assumption is made that the velocity distribution function for each species is (locally) nearly isotropic, so that viscous stresses and heat transport may be neglected. Let $\rho_f, n_f, \vec{v}_f, T_f$, and u_f be the mass density, number density, velocity, temperature, and internal energy per particle of fluid f, where the subscript f runs over n = neutrals, i = ions, and e = electrons. If the flow field is taken to be steady ($\partial/\partial t = 0$) and plane-parallel ($\partial/\partial x = \partial/\partial y = 0$), and gravity is neglected, then the following equations result for number conservation:

$$\frac{d}{dz}(n_f v_{fz}) = N_f \; ;$$

(29)

mass conservation:

$$\frac{d}{dz}(\rho_f v_{fz}) = S_f \; ;$$

(30)

momentum conservation for the neutrals:

$$\frac{d}{dz}(\rho_n \vec{v}_n v_{nz} + \hat{e}_z n_n k T_n) = \vec{F}_n \; ;$$

(31)

momentum conservation for the ions plus electrons:

$$\frac{d}{dz}\left[\rho_i \vec{v}_i v_{iz} + \hat{e}_z(n_i k T_i + n_e k T_e + \frac{B^2}{8\pi})\right] - \frac{1}{4\pi} B_z \frac{d\vec{B}}{dz} = \vec{F}_i + \vec{F}_e \; ;$$

(32)

and energy conservation for each species:

$$\frac{d}{dz}\left[n_f(\frac{3}{2}kT_f + u_f)v_{fz}\right] + n_f k T_f \frac{dv_{fz}}{dz} = G_f \; .$$

(33)

In the above equations, N_f is the rate at which the number density of fluid f increases as the result of chemical reactions, S_f is the rate per volume at which the mass of fluid f increases as the result of chemical reactions, $\vec{F}_f$ is the rate per volume at which the momentum of fluid f increases as the result of interaction with other fluids, and G_f is the rate per volume at which the thermal energy of fluid f increases as the result of interaction with other fluids, radiative cooling, endothermic or exothermic chemical reactions, etc.

In the above it has been assumed that the ions and electrons have a common flow velocity $\vec{v}_i$. This is an accurate assumption for the z-component, since otherwise the electric current would have nonzero divergence and charge separation would result – a small amount of charge separation will lead to large electric fields tending to restore charge neutrality. The transverse velocities of the electrons and ions must differ somewhat in order to provide a current consistent with the variation in the magnetic field in the shock transition, but the requisite drift velocities are quite small (Draine 1986a), and may be neglected.

The above equations for the fluid variables were obtained from conservation laws, and assumed the electromagnetic field to be known. To obtain $\vec{B}$ as a function of position one uses Maxwell's equations, plus the assumption of "perfect conductivity" (so that the electric field is given by $\vec{E} = -\vec{v}_i \times \vec{B}/c$) to obtain

$$B_z = constant \tag{34}$$

$$\frac{d}{dz}(v_{iz}B_{iz}) = B_z\frac{dv_{iz}}{dz} \tag{35}$$

$$\frac{d}{dz}(v_{ix}B_y) = B_z\frac{dv_{iy}}{dz}. \tag{36}$$

The above terms are obviously simplified if the magnetic field is exactly transverse to the direction of propagation ($i.e.$, $B_z = 0$), and most computational studies have been restricted to this special case. The more general case of the "oblique" shock has been studied recently (Wardle and Draine 1987).

If the "source terms" $N_f, S_f, \vec{F}_f, G_f$ are known functions of the local fluid variables (temperatures, velocities, compositions), then the above equations suffice to describe the variation in the fluid variables so long as the flow variables are continuous. Under some circumstances, however, the flow may contain a "jump front", or "J-front" – a thin layer, of thickness comparable to the molecular mean free path, in which the density, temperature, and velocity of one of the fluids undergo an irreversible transition to a higher density, higher temperature, and higher entropy state. Viscous stresses are important within the J-front, so that the above fluid equations – which assumed viscous stresses to be negligible – are not valid within this narrow zone. Since the J-front is so thin, however, radiative cooling and chemical reactions have no effect on its structure, and the neutral fluid variables on one side of the J-front may be related to the variables on the other side via the familiar "jump conditions", the Rankine-Hugoniot relations (eqs. 13-15).

3.2 Momentum Transfer Processes

Collisions between ions and neutral atoms or molecules result in momentum transfer at a rate

$$\vec{F}_n = -\vec{F}_i = n_n n_i \langle \sigma v \rangle \frac{m_n m_i}{m_n + m_i}(\vec{v}_i - \vec{v}_n), \tag{37}$$

where $\langle \sigma v \rangle$ is the momentum-transfer rate coefficient. If the neutral has no permanent dipole moment, then at low collision energies the interaction potential between ion and neutral is due to the induced dipole moment. The resulting r^{-4} interaction potential has the property that the scattering cross section $\sigma \propto v^{-1}$, so that $\sigma v = constant$, with the value (Osterbrock 1961),

$$\langle \sigma v \rangle = 2.41\pi \left[\frac{m_n + m_i}{m_n m_i}\right]^{1/2} e\alpha_n^{1/2}, \tag{38}$$

where α_n is the polarizability of the neutral atom or molecule. As the collision velocities are increased, due to either elevated temperature or large drift velocities, eq. (38) underestimates the rate coefficient. If the scattering cross section is taken to be a constant at high energies, then the effective momentum transfer rate coefficient is given by (Draine 1986a)

$$\langle \sigma v \rangle \approx \sigma \left[\frac{128kT_r}{9\pi m_r}\right]^{1/2} \left[1 + \frac{9\pi s^2}{64}\right]^{1/2}, \tag{39}$$

where $T_r \equiv (m_n T_i + m_i T_n)/(m_n + m_i)$ is the "reduced temperature", $m_r \equiv m_n m_i/(m_n + m_i)$ is the reduced mass of the collision partners, and $s^2 = m_r \mid v_i - v_n \mid^2 /kT_r$.

At what velocity does the transition from eq. (38) to (39) take place? Consider C^+ - H_2 collisions. The polarizability of an H_2 molecule is $\alpha = 8.04 \times 10^{-25}$ cm^3, so that the rate coefficient of eq. (38) is $\langle \sigma v \rangle = 1.93 \times 10^{-9}$ cm^3 s^{-1}. If the effective "hard sphere" cross section for C^+-H_2 elastic scattering at high energies is $\sim 1 \times 10^{-15}$ cm^2 (corresponding to a distance of closest approach of 1.8 Å), then equation (38) will underestimate the cross section for drift velocities $v_d > 20$ km s^{-1}.

Under conditions of very low ionization, dust grains can play an important role in transferring momentum between neutrals and ions. Charged dust grains are fairly well coupled to the magnetic field, and therefore have a flow velocity which differs from the neutral velocity; collisions with neutrals therefore transfer momentum to the neutrals. If the acceleration of the dust grains is neglected, then one can readily estimate the drift velocity of a charged grain, and the rate of momentum transfer to the neutral gas (Draine 1980). If the dust grains are not accelerating, then the rate of change of momentum of the magnetized ion-electron plasma is just minus the rate of change of the momentum of the neutral gas. In real shocks, the acceleration of the dust grains may not be negligible, and a more sophisticated analysis is called for.

3.3   Energy Transfer Processes

Elastic collisions between ions and neutrals will transfer energy as well as momentum between the fluids. Part of the energy transfer goes into changing the ordered kinetic energy of the fluids (for the neutral gas, for example, this appears in the fluid equations as $\vec{F}_n \cdot \vec{v}_n$, and part goes into changing the thermal kinetic energy of the fluids. For scattering with a cross section $\propto v^{-1}$ (i.e., low velocity ion-neutral scattering) the rate of heating of species α due to elastic collisions with species β is (Draine 1986a)

$$G_\alpha = \frac{n_\alpha n_\beta m_\alpha m_\beta}{(m_\alpha + m_\beta)^2} \langle \sigma v \rangle \left[3k(T_\beta - T_\alpha) + m_\beta \mid \vec{v}_\alpha - \vec{v}_\beta \mid^2 \right]. \tag{40}$$

In the case of electron-ion scattering (for which the electron-ion drift velocity is negligible) the energy transfer rate due to Coulomb scattering is (Spitzer 1962)

$$G_e = -G_i = \frac{4e^4}{m_i} \left[\frac{2\pi m_e}{k^3 T^3} \right]^{1/2} \ln(\Lambda) \, n_i n_e k(T_i - T_e), \tag{41}$$

where $\Lambda = (3/2e^3)(k^3 T_e^3/\pi n_e)^{1/2}$. When the fractional ionization is sufficiently low, elastic collisions of neutrals with streaming charged grains can be an important heating mechanism (see Section 3.4 below).

The elastic processes discussed above act to generate heat (when a velocity difference exists between fluids) and to redistribute heat (when a temperature difference exists between fluids). Inelastic processes act to reduce the thermal energy by converting it into either internal energy (by changing the quantum state of one or both collision partners) or chemical energy (*e.g.*, collisional ionization or collisional dissociation). In the case of quantum excitation, one must check whether the excited state will decay radiatively (with one or more photons removing the energy from the system) or whether collisional deexcitation (i.e., "superelastic" scattering) may occur, returning part or all of the original excitation energy to the thermal energy pool.

The dominant processes for cooling the neutral gas in C-type MHD shocks include emission from rotationally- and vibrationally-excited H_2, rotationally excited CO, rotationally excited H_2O, and the excited fine structure levels of C I, C II, and O I. When the neutral gas temperature and density are high enough (Lepp and Shull 1983), collisional dissociation of H_2 can become an important sink for thermal energy.

The electron gas can become significantly hotter than the neutrals. The electrons are cooled by elastic collisions with the neutrals, and by collisional excitation of the same atomic, molecular, and ionic excited states which are important for cooling the neutral gas. In addition, the electron gas may sometimes be hot enough to collisionally populate excited electronic states of abundant atoms and molecules.

The radiation emitted by the atoms, ions, and molecules responsible for cooling the gas may also provide a means of detecting the shock, and diagnosing the physical conditions in the shocked gas. The power per area converted into heat in a strong shock is of order $\rho_{n0}V_s^3/2$, where ρ_{n0} is the preshock mass density. For shocks in gas with an appreciable H_2 fraction, the cooling will normally be dominated by line emission from H_2 at densities $n_H < 10^3$ cm^{-3}; at higher densities the dominant coolants are O I, H_2, and H_2O (C I and CO are less important) with their relative importance varying depending upon density and shock speed (see Fig. 4 of Draine, Roberge, and Dalgarno 1983). At the present time it appears possible to detect H_2 line emission only from shocks with $n_H > 10^4$ cm^{-3}. However, some of the rotationally-excited levels which are important in cooling the gas (when they decay radiatively) may also be detectable through absorption line studies if the gas is not totally opaque to ultraviolet radiation and a suitable background star is available. Absorption line measurements of the populations of rotationally-excited levels of H_2 serve to strongly constrain shock models for the gas in the lines of sight to ς Per, o Per, χ Oph (Draine and Katz 1986b) and ς Oph (Draine 1986b).

3.4 Chemical Processes and Grain Dynamics

Chemical processes in the shocked gas can be important for four major reasons: (i) they can alter the abundances of important coolants; (ii) they can affect the fractional ionization (and hence the coupling between ionized and neutral fluids); (iii) they can directly reduce or increase the thermal energy in the fluid; (iv) they can alter the abundances of species which are important as diagnostics. In addition, chemical reactions can affect the hydrodynamics via changes in the molecular weights of the various fluids, but this is usually not a major effect.

The principal coolants which are subject to creation or destruction in the shock are H_2 (which can be destroyed if the gas temperature and density become sufficiently high), and

H_2O (which is normally not abundant in the preshock gas, but which can be produced by the endothermic reactions $O + H_2 \rightarrow OH + H$, $OH + H_2 \rightarrow H_2O + H$ if the gas temperature exceeds $\sim 500K$). Since H_2O, with its large dipole moment, can be the dominant coolant in shocks in dense clouds, it is important to accurately estimate the abundance of H_2O in such shocks.

It is obvious that for large enough shock speeds, collisional ionization will be able to raise the fractional ionization of the shocked gas sufficiently rapidly that the magnetohydrodynamic calculations must allow for the variation in fractional ionization through the shock. The fact that chemical processes could in some cases reduce the fractional ionization on a "flow" timescale was first pointed out by Flower, Pineau des Forets, and Hartquist (1985), who showed that conversion of C^+ into CH^+ (via the endothermic reaction $C^+ + H_2 + 0.4\,eV \rightarrow CH^+ + H$), followed by rapid dissociative recombination of the CH^+, could result in an appreciable decrease in the fractional ionization if most of the preshock ions were C^+ and conditions were favorable for CH^+ production. Hence, even if one were not interested in the carbon chemistry *per se*, it may be important to follow it in some detail in order to know the fractional ionization. Similar chemical effects (i.e., hydride formation, followed by dissociative recombination) may affect the contribution of other metal ions (*e.g.*, S^+, Mg^+, Fe^+) to the ionization.

Chemical processes can to some extent directly affect the thermal energy, and therefore temperature and pressure, in the gas. For example, collisional dissociation (*e.g.*, of H_2) converts thermal energy into chemical energy, while photoionization or photodissociation converts photon energy partly into thermal energy. The momentum and energy "source terms" for chemical processes which involve both neutral and charged particles (*e.g.*, radiative recombination, dissociative recombination, charge exchange, or ion-molecule reactions) are given by Draine (1986a). Finally, it may be important to follow the chemistry in the shock in order to be able to use observations of shock-produced species (*e.g.*, CH^+) to constrain shock models, or, for a given shock model, to predict abundances of various species in advance of observation.

Early studies of MHD shocks included only very rudimentary chemistry, but recent investigations have modeled the carbon chemistry in diffuse cloud shocks in considerable detail (Flower, Pineau des Forets, and Hartquist 1985; Pineau des Forets, Flower, Hartquist, and Dalgarno 1986; Draine and Katz 1986a,b; Draine 1986b). The sulfur chemistry in diffuse cloud shocks has recently been studied by Pineau des Forets, Roueff, and Flower (1986). As mentioned above, the dynamics of dust grains in MHD shocks can be of importance under conditions of low fractional ionization, since the dust grains may then play an important role in transferring momentum from the magnetized plasma to the neutral gas. It is also important to understand the motions of dust grains in these shocks in order to estimate the frequency (and violence) of grain-grain collisions, since grain-grain collisions in shocks may play a major role in the evolution of the interstellar grain population.

The coupling of a grain to the magnetic field is described by the grain gyrofrequency $\omega = (Q_{gr}B/m_{gr}c)$, where Q_{gr} and m_{gr} are the charge and mass of the grain. The coupling of the grain to the neutral gas is described by the "slowing-down" time τ such that the grain acceleration due to drag forces is $dv_{gr}/dt = -(v_{gr} - v_n)/\tau$. Then the grain is strongly coupled to the magnetic field if $\omega\tau \gg 1$, and strongly coupled to the neutral gas if

$\omega\tau \ll 1$. The motions of the dust grains in the shocks are complicated by several factors: (i) the charge on a given grain fluctuates (since charge is quantized) and the probability distribution varies with time (since the electron temperature and electron density vary as the grain travels through the shock); (ii) grains are sometimes in the regime $\omega\tau \approx 1$, so that they cannot be assumed to be strongly coupled to either the neutrals or the magnetic field; (iii) the momentum transferred by grains to the neutral gas can be appreciable, in which case the grains must exert a significant force on the magnetized plasma (*i.e.*, the current associated with the motions of the charged grains makes an appreciable contribution to the magnetic field); and (iv) the grain mass, while small compared to the total gas mass, may not be small compared with the mass in ions, so that the grain inertia may appreciably affect the motion of the plasma.

3.5   Classes of Solutions

Consider MHD shocks for which $V_s < v_{A,ie}$, so that the magnetized plasma is "submagnetosonic". There are two basic classes of solutions: (1) C-type shocks, and (2) J-type shocks with magnetic precursors. In the first class of shocks – the C-type shocks – no discontinuity is present: all the flow variables vary continuously through the shock transition, and ordinary molecular viscosity plays no role. Two distinct types of C-type shocks are possible (Roberge 1986; Chernoff 1987). The more familiar type – which we might term a "supersonic" C-type flow – has the neutral gas remaining sufficiently cold throughout the shock so that it always flows supersonically. This type of flow structure occurs when cooling is sufficiently rapid that it can keep the neutral gas cool even as it is being heated by ion-neutral collisions and by compression. [For sufficiently strong magnetic field, this type of flow structure can even occur in the absence of cooling; for the case where the shock is propagating perpendicular to the magnetic field, Chernoff (1987) has shown that C-type shocks occur, even with zero cooling, provided $V_s < 2.76 v_{A,nie}$.]

The second type – which we might term a "transonic" C-type flow – has the neutral gas (initially supersonic) flowing smoothly through a sonic point and becoming subsonic without a "jump". The gas subsequently flows smoothly through another sonic point, and becomes supersonic again far downstream. [Note that it is implicitly assumed here that the downstream postshock flow is sufficiently cool that it is supported primarily by magnetic pressure and is supersonic; the magnetic fields in interstellar clouds are generally large enough that this is true.]

The J-type shocks resemble single fluid shocks in that there is a "jump" transition in which molecular viscosity effects an irreversible change in the neutral flow variables on a length of order one molecular mean free path; for our purposes such a change is treated as a discontinuity, with the flow variables across the discontinuity related through the Rankine-Hugoniot jump conditions. However, ahead of this "J-front" the neutral gas is accelerated and heated by collisions with streaming ions – *i.e.*, the shock has a "magnetic precursor". In the frame of reference of the shock, the neutral gas is flowing supersonically upstream from the shock, and subsonically immediately downstream from the shock.

The three different classes of MHD shocks – supersonic C-type, transonic C-type, and J- type – differ significantly in their degree of computational difficulty. The supersonic

C-type shocks are relatively easy to compute, since the ordinary differential equations describing a steady, plane-parallel shock can be integrated as an initial value problem: one introduces a small (linearized) perturbation to the preshock flow variables, and integrates directly through the shock structure (Draine 1980). There are no critical points in the flow, and the integration is numerically stable as long as the neutral flow remains supersonic. There can of course be considerable "overhead" associated with providing a realistic treatment of the chemistry, heating, and cooling, but there are no numerical obstacles presented by the hydrodynamics. Transonic C-type shocks are somewhat more difficult to compute because the neutral gas flow has a subsonic portion, within which integration of the fluid equations is numerically unstable. "Shooting" techniques can be employed to determine the flow structure in this subsonic region (Roberge and Draine 1987).

The J-type shocks are most difficult from a numerical point of view. One begins the integration, as in the C-type shocks, from a perturbation added to the preshock conditions. One integrates along as before, but at some point integration of the differential equations must cease and a "jump" in the neutral fluid hydrodynamic variables (*i.e.*, a "viscous shock") must be inserted. The problem here is that one does not know, *a priori*, where (or, equivalently, at what neutral flow velocity) this jump front should be located – the position and strength of the jump is determined by the flow downstream from the shock. The neutral gas just after the jump front is subsonic, but becomes supersonic far downstream after the neutral gas has cooled to low temperatures – there must therefore be a critical point in the flow downstream from the shock where the neutral gas makes the transition from subsonic to supersonic flow. The jump front must be located at just the right point to put the flow on a postshock "trajectory" which will pass through the critical point. This problem has been discussed by Chernoff (1987), who has considered some idealized cases (adiabatic flow, isothermal flow, and various power-law cooling functions). Roberge and Draine (1987) discussed methods for treating this problem using realistic cooling laws, chemistry, etc.

3.6 Observations

The BN-KL region (in the molecular cloud OMC-1) is a site of active star formation, and exhibits high-velocity CO radio emission, intense H_2 vibrational line emission, emission from very high rotational levels of H_2 (up to $J = 19$ has been observed) and CO (up to $J = 34$ has been observed), far-infrared emission from OH, and (perhaps) [O I] $63\mu m$. Gas with conditions suitable for producing the observed H_2 line emission would cool on a timescale of ~ 1 year; the observed persistence of the emission therefore requires an ongoing source of heating. A shock wave propagating into the molecular cloud appears the most likely explanation for the observed H_2 line emission. Nonmagnetic shock models, however, appear to be unable to reproduce the observations; high velocity shocks produce very hot gas which dissociates the H_2 before it can radiate, and low velocity shocks have difficulty in reproducing the observed surface brightness and line ratios.

Magnetized, multifluid shock models appear to offer a very attractive model for the shock wave in OMC-1 (Draine and Roberge 1982; Chernoff, Hollenbach, and McKee 1982). It appears likely that a shock wave is propagating outward into OMC-1 with a shock speed $30 < V_s < 40$ km s^{-1}. The preshock gas appears to have a density in the range $2 \times 10^5 < n_H < 7 \times 10^5$ cm^{-3}, and a magnetic field strong enough to produce a preshock

Alfvén velocity $v_{A,nie} = 2 - 3$ km s^{-1}. A theoretical model of such a shock is shown in Fig. 6. If one accepts that the observed emission is due to an MHD shock propagating into the cloud, then the combination of the observed emission and the theoretical models can be considered to constitute a "measurement" of the magnetic field in the preshock gas in OMC-1, since this magnetic field strength is a very important parameter in the theoretical shock models.

The models for the OMC-1 shock put forward by the two different groups differ somewhat because of different assumptions regarding the dust extinction, and differences in adopted rate coefficients for collisional excitation of CO. Details of the theoretical models will undoubtedly evolve with time in response to further observational developments as well as improvements in rate coefficients for collisional excitation (especially of H$_2$ and CO), rate coefficients for chemical reactions, and techniques for modeling these multifluid flows.

Line emission from H$_2$ has also been detected from other "high-velocity" flow regions associated with regions of ongoing (or recent) star formation in molecular clouds (see the list in Shull and Beckwith [1982]). Once enough spectral information is obtained (primarily from lines of H$_2$ and CO), it will be of interest to see whether MHD shock models can reproduce the observations, and what preshock conditions are required.

Confrontation between observation and theory is also possible in the case of MHD shocks in diffuse clouds. Observational evidence points strongly to shock waves as the site of formation of interstellar CH$^+$ (Lambert and Danks 1986), as originally proposed by Elitzur and Watson (1978a,b). Detailed shock models have been compared to observations of well-studied lines of sight in which CH$^+$ is present, especially ς Per, o Per, χ Oph and ς Oph (Pineau des Forets, Flower, Hartquist, and Dalgarno 1986; Draine and Katz 1986b; Draine 1986b), though with differing results: Pineau des Forets et al. conclude that the observed CH$^+$ cannot be accounted for by MHD shock models whereas Draine and Katz contend that shock models can be found which are in reasonable agreement with existing observations. The differences between these two groups appear to be due in part to different assumed shock parameters, and in part to different adopted rate coefficients for C$^+$ + H$_2$ → CH$^+$ + H. One example of a possible shock model for ς Oph is shown in Fig. 7. On the basis of their shock modeling, Draine and Katz (1986b) have proposed that, in addition to being a primary source for interstellar CH$^+$, MHD shock waves may also play an important role in the overall production of CH, OH, and CO in diffuse molecular clouds.

4 PROBLEMS AND DIRECTIONS OF FUTURE RESEARCH

4.1 Atomic and Molecular Physics

In constructing theoretical models of radiative and MHD shock waves, one incorporates large amounts of atomic and molecular data. In this section we briefly mention the data needs, if the models are to be used as accurate diagnostic tools.

A primary modeling need for radiative shocks is an accurate determination of the non-equilibrium ionization state behind the shock. This requires rates for ionization, re-

Figure 6: Temperature and velocity structure of a model proposed for shock wave in OMC-1 (Draine and Roberge 1982). The shock has $V_s = 38$ km/s and the ambient gas has $n(H_2) = 3.5 \times 10^5$ cm^{-3} and $B_1 = 1.5$ mG.

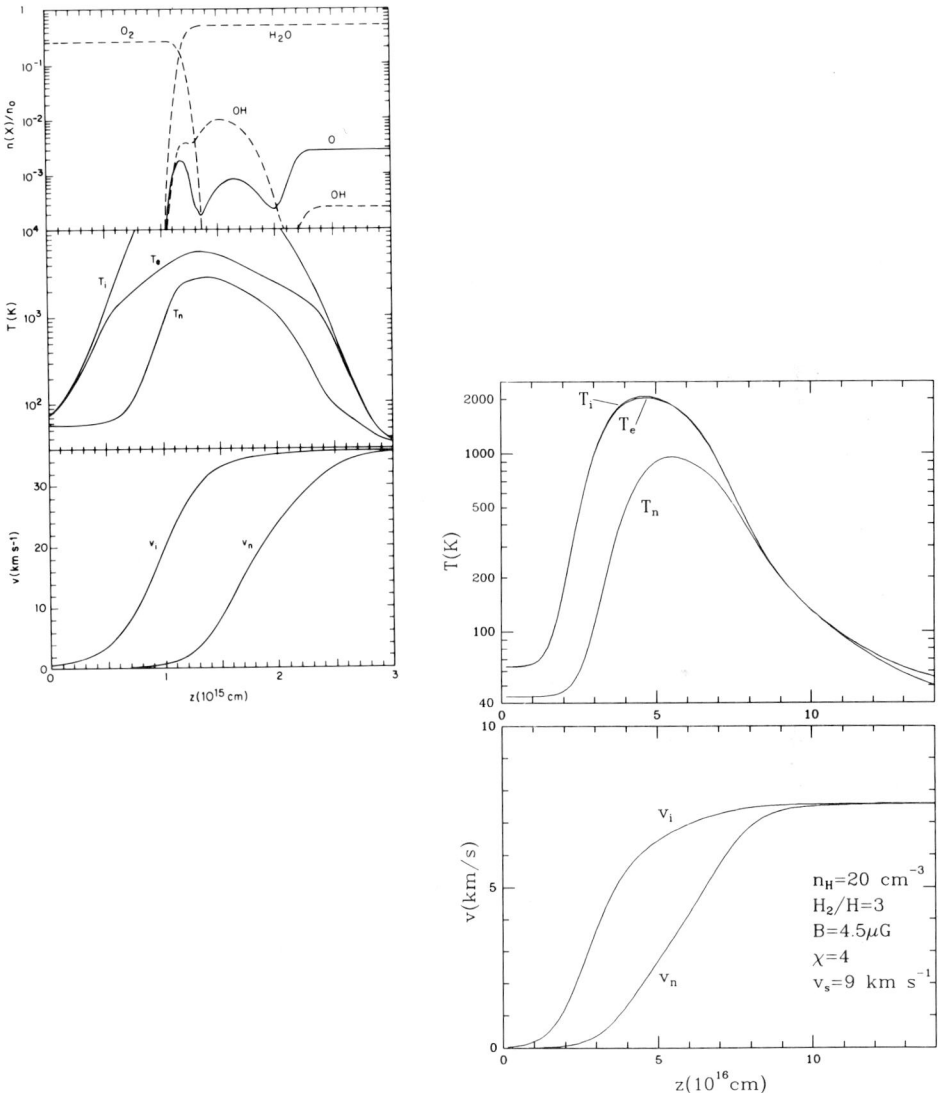

Figure 7: Temperature and velocity structure of shock model proposed to account for observed CH$^+$ in cloud toward ς Oph (Draine 1986b). Shock heating would account for much of the observed rotational excitation of H_2.

combination, and charge exchange. The rates of collisional ionization are fairly well determined, based on the agreement between theoretical and experimental cross sections (Dunn 1985). However, at high temperatures, double collisional ionization is largely unexplored in the models. Radiative recombination rates are less accurate, but will improve with the application of detailed balance to newly improved photoionization cross sections. However, dielectronic recombination dominates radiative recombination in hot gas, and several of these rates are still controversial to factors of 2 - 5. Confirmatory work based on new quantum defect methods would be welcome. Charge exchange rate coefficients exist for a number of ions, but their accuracy is also unconfirmed. [Collisions with H^0 dominate this process, but He^0 can also be important and the possibility of double charge exchange, $e.g.$, $C^{+4} + He^0 \rightarrow C^{+2} + He^{+2}$, should be explored.] Rates for the fast processes ($> 10^{-9}$ cm^3 s^{-1}) are most accurate, but because charge exchange can dominate recombination when the neutral fraction exceeds several percent, we need accurate rates for the somewhat slower processes as well.

A second category of useful atomic data is electron-impact excitation cross sections, which dominate the radiative cooling and line emission from shocks. Many lines are currently parameterized by a single number (the collision strength Ω_{ij}); one should allow for temperature dependence of Ω_{ij}, but the data is often of insufficient quality to warrant this degree of sophistication. As mentioned earlier, there is no standard set of collision strengths, although a compilation of evaluated data now exists (Gallagher and Pradhan 1985). A new generation of excitation cross sections, accurate to 10%, is currently being generated in association with the US-UK "Opacity Project" (Pradhan 1987).

For studies of low-velocity MHD shocks, the lack of accurate cross sections for a number of collisional processes may be the principal limitation. In the case of shocks in diffuse molecular clouds, the existing order-of-magnitude uncertainty in cross sections for collisional excitation of H_2 by H^0 ($cf.$ the discussion by Draine and Katz 1986b) implies a corresponding uncertainty in the cooling rate. For shocks in dense molecular clouds, the populations of the higher rotational levels of H_2 and CO, which can be observed through line emission originating from these levels, provides a wonderful diagnostic for the density and temperature in the shocked gas. Unfortunately, the inelastic cross sections involved in populating these levels remain uncertain. There has been recent progress in computing theoretical cross-sections for rotational excitation of CO in collisions with H_2 (Schinke et $al.$ 1985), but little work has been published on cross sections for rotational excitation of H_2, particularly for the high-J levels which provide useful diagnostic information at high densities.

The rates of H_2 collisional dissociation are also uncertain, owing to a lack of accurate cross sections for the states (v, J) of large rotational and vibrational excitation which dominate the total rate (Lepp and Shull 1983). To make an accurate assessment of the density dependence of the dissociation rate requires accurate excitation rates of $H_2(v, J)$. These are poorly known, even for H^0-H_2 collisions, and the likelihood of large H_2-H_2 resonant collisions involving a transfer of vibrational to rotational energy (Shull and Hollenbach 1978) makes the area of molecular excitation highly uncertain at this time.

Accurate reactive scattering cross sections are also needed. Even in the case of the important reaction $C^+ + H_2 \rightarrow CH^+ + H$, little seems to be known concerning the dependence of the reaction cross section on the initial quantum state of the reactants. The H_2 in a laboratory experiment is likely to have its rotational levels populated according to the Boltzmann distribution, but this is not likely to be the case for H_2 in a diffuse molecular cloud. Millar *et al.* (1986) have recently reported cross sections for $S^+ + H_2 \rightarrow HS^+ + H$. Also needed are measurements of reaction cross sections of H_2 with other ions (*e.g.*, Mg^+ or Fe^+).

Under some circumstances the streaming of the ions relative to the neutrals can lead to ion-molecule collisions with large center-of-mass kinetic energies. For example, a 25 km s^{-1} MHD shock may have a region in which the ion-neutral streaming velocity exceeds 15 km s^{-1} (Figs. 1–3 of Draine, Roberge, and Dalgarno 1983). A 15 km s^{-1} collision between, say, C^+ and CO has a center-of-mass kinetic energy of 10 eV. Shocks in very dense clouds tend to convert a large fraction of the gas phase oxygen into H_2O – such high energy collisions of ions with H_2O may be an important mechanism for destroying the H_2O and creating O and OH. Unfortunately, little appears to be known regarding such collisions. Finally, photoionization is a very important chemical process in diffuse molecular clouds, but many important rates, including photodissociation of CO and CH, remain uncertain by a factor of two or more (van Dishoeck 1986; van Dishoeck and Black 1986).

4.2   Grain Physics

Interactions between gas and grains have a significant effect on the gas-phase abundances and line emissivities of C, O, Si, and Fe. The infrared lines of [Si II] and [Fe II] are especially sensitive to the rates of grain sputtering and grain-grain disruption. Most uncertain are the sputtering yields near threshold, and the partial vaporization rates and shattering efficiencies during collisions between grains of different sizes.

The solid state physics of interstellar grains could improve our knowledge of grain processing in hot radiative shocks. However, it is not obvious that we even have the correct grain model. For several years, astronomers have used the "MRN model" (Mathis, Rumpl, and Nordsieck 1977), which assumes a power-law size distribution of graphite and silicate grains between $0.01\mu m$ and $0.25\mu m$. However, recent work on the UV selective extinction curve (*e.g.* Fitzpatrick and Massa 1986) challenges this model. The diffuse infrared cirrus (Low *et al.* 1984) and near-infrared emission from reflection nebulae (Sellgren 1984) now suggest the presence of a population of much smaller grains (Leger and Puget 1984; Draine and Anderson 1985). Their effect on shock models is unexplored, although they may prove to be quite hardy.

The largest question about grains is their global history in the multi-phase, violent interstellar medium. Destruction rates in shocks appear to be large, and in disagreement with grain formation rates. Our knowledge of grain physics in shocks may be faulty, the assumed global model of the ISM may be wrong, or our estimates of the rates of grain formation and growth in the diffuse ISM may be too low.

4.3   Fluid Dynamics and Shock Structure

A number of important issues remain to be investigated in the general area of fluid dynamics. It is generally assumed that interstellar shock waves can be approximated as plane-parallel, steady flows. However, Chevalier and Imamura (1982) have shown that under certain circumstances radiative shock waves may be unstable to temporal oscillations, and Vishniac (1983) has shown that spherical radiative shock waves can be unstable to nonradial perturbations. Bertschinger (1986) has given a general analysis of the stability of single-fluid radiative shock waves, but he assumed a cooling coefficient $\Lambda(T)$ characteristic of equilibrium ionization. To date, no analysis has been done of the stability of radiative multifluid shock waves to either temporal or spatial oscillations.

As remarked above, the high velocity gas flow in OMC-1 requires further study. In particular, the existing shock models (Draine and Roberge 1982; Chernoff, Hollenbach and McKee 1982), while offering an attractive explanation for the overall emission line intensities, do not provide a satisfactory explanation for the observed high-velocity wings on the H_2 emission lines (Nadeau, Geballe, and Neugebauer 1982).

New areas for single-fluid radiative shock models are: (1) line diagnostics for "truncated shocks" (see section 2.5); (2) an improvement of the "multi-shock" modeling of bow shocks and the "working surfaces" of jet interactions in star forming regions; and (3) the physics of reverse shocks in young X-ray emitting SNRs. Because each of these involves aspects of two-dimensional, non-steady flow, a standard, simplifying assumption will need to be relaxed. In the simpler, plane-parallel models, we need to explore the influence of external sources of ionizing EUV radiation, non-equilibrium radiative precursors, and radiative transfer of resonance lines. Finally, we need much more work on the state of pre-galactic shock waves, in which H, He, and H_2 dominate the cooling. These shocks may play an important role in triggering gravitational collapse of intergalactic clouds and in the early evolution of star formation in galaxies.

4.4   New Observations

Many of the needed observations have been mentioned in the text. With a new generation of space observatories, the UV and IR diagnostics of shocks will take on an important role. Especially important are the infrared fine structure lines of [Si II] $34.8\mu m$, [O I] 63, 145 μm, and [Fe II] 26, 1.27, 1.6, and $5\mu m$, which are a signature of shock processing of grains. In the UV, semi-forbidden lines of C II], C III], O III], and O IV] are useful tests of the shock model and physical parameters. The spatial and velocity resolution of the Hubble Space Telescope will enable astronomers to study shock flows in Herbig-Haro objects, filamentary structure in middle-aged SNRs such as the Cygnus Loop, and interstellar cloud depletion patterns with velocity.

At the present time, the only intensely studied dense molecular cloud shock is in OMC-1. Ground-based observations of the vibrational and rotational lines of H_2, and airborne observations of the far infrared lines of CO, OH, and O I will provide valuable information concerning physical conditions in the shocked regions of other dense molecular clouds and pre-main-sequence outflows.

High velocity shocks may be studied with lines from higher ionization states; one of the best diagnostics (O VI λ1035) will require the LYMAN spacecraft's sensitivity below 1150 Å. Spatially resolved X-ray spectroscopy of the He-like lines of Si, S, and Fe will improve our knowledge of reverse shocked supernova ejecta, nucleosynthetic abundances, and stellar progenitors to supernovae. The oxygen-rich knots in some young remnants (Cas A, N-132D, etc.) require advances in both observation and the theory of shocks in metal-rich gas.

Solving the grain-history dilemma may also require further observational input: (1) pinning down the volume filling factor of hot, low-density interstellar gas; (2) determining the true supernova rate in the galaxy; (3) measuring UV selective extinction curves and heavy element depletions in regions of shock activity; and (4) searching for evidence of grain input and growth in the ISM. The last could be done by correlating the wavelength of maximum polarization with UV indicators of heavy element abundances and selective extinction.

This work was supported by grants from NASA (NSG-7128 and NAG5-193) at the University of Colorado and from NSF (AST83-41412) at Princeton University. We thank Dr. Rob Fesen for his comments on the manuscript and Michael Van Steenberg for help with the "production".

REFERENCES

Baldwin, J.A., Phillips, M.M., and Terlevich, R. 1981, *P.A.S.P.*, **93**, 5.
Baliunas, S.L., and Butler, S.E. 1980, *Ap. J. Letters*, **235**, L45.
Barlow, M. 1978, *M.N.R.A.S.*, **183**, 367.
Becker, R., 1979, *Ap. J. Letters*, **234**, L73.
Becker, R., 1980a, *Ap. J. Letters*, **235**, L5.
———— 1980b, *Ap. J. Letters*, **237**, L77.
Bertschinger, E. 1986, *Ap. J.*, **304**, 154.
Brugel, E.W., Shull, J.M., and Seab, C.G. 1982, *Ap. J. Letters*, **262**, L35.
Brugel, E.W., Böhm, K-H., Shull, J.M., and Böhm-Vitense, E. 1985,
    Ap. J. Letters, **292**, L75.
Butler, S.E., and Dalgarno, A. 1980, *Ap. J.*, **241**, 838.
Butler, S.E., Heil, T.G., and Dalgarno, A. 1980, *Ap. J.*, **241**, 442.
Butler, S.E., and Raymond, J.C. 1980, *Ap. J.*, **240**, 680.
Chernoff, D. F. 1987, *Ap. J.*, in press.
Chernoff, D. F., Hollenbach, D. J., and McKee, C. F. 1982, *Ap. J. Letters*, **259**, L97.
Chevalier, R. A., and Imamura, J. N. 1982, *Ap. J.*, **261**, 543.
Chevalier, R.A., Kirshner, R.P., and Raymond, J.C. 1980, *Ap. J.*, **235**, 186.
Chevalier, R.A., and Raymond, J.C. 1978, *Ap. J. Letters*, **225**, L27.
Clark, R.E.H., Cowan, R.D., and Bobrowicz, F.W. 1986,
    Atomic and Nuclear Data Tables, **34**, 415.
Cowie, L.L. 1978, *Ap. J.*, **225**, 887.

Cox, D.P., and Raymond, J.C. 1985, *Ap. J.*, **298**, 651.

Cox, D.P., and Smith, B.W. 1974, *Ap. J. Letters*, **189**, L105.

Crandall, D.H. 1981, *Physica Scripta*, **23**, 153.

Dalgarno, A. 1958, *Phil. Trans. Roy. Soc., Ser. A*, **250**, 426.

Dalgarno, A., and Butler, S. 1978, *Comments Atom. Molec. Phys.*, **7**, 129.

Dalgarno, A., Heil, T.G., and Butler, S.E. 1981, *Ap. J.*, **245**, 793.

Dalgarno, A., and McCray, R. 1972, *Ann. Rev. Astr. Ap.*, **10**, 375.

Dalgarno, A., and Yadav, H.N. 1953, *Proc. Phys. Soc.*, **A66**, 173.

Dopita, M.A., Binette, L., and Schwartz, R.D. 1982, *Ap. J.*, **261**, 183.

Dopita, M., Binette, L., d'Odorico, S., and Benvenuti, P. 1984, *Ap. J.*, **276**, 653.

Draine, B. T. 1980, *Ap. J.*, **241**, 1021 [Erratum 1981, **246**, 1045].

Draine, B. T. 1986a, *M.N.R.A.S.*, **220**, 133.

Draine, B. T. 1986b, *Ap. J.*, **310**, in press.

Draine, B.T., and Anderson, N. 1985, *Ap. J.*, **292**, 494.

Draine, B. T., and Katz, N. 1986a, *Ap. J.*, **306**, 655.

Draine, B. T., and Katz, N. 1986b, *Ap. J.*, **310**, in press.

——— 1984, *Ap. J.*, **282**, 491.

Draine, B. T., and Roberge, W. G. 1982, *Ap. J. Letters*, **259**, L91.

Draine, B. T., Roberge, W. G., and Dalgarno, A. 1983, *Ap. J.*, **264**, 485.

Draine, B.T., and Salpeter, E.E. 1979, *Ap. J.*, **231**, 77.

Dunn, G.H. 1985, in *Electron Impact Ionization*, eds. T.D. Märk
    and G.H. Dunn, (Vienna: Springer-Verlag), p. 277.

Dwek, E., and Scalo, J.M. 1980, *Ap. J.*, **239**, 193.

Elitzur, M., and Watson, W.D. 1978a, *Ap. J. Letters*, **222**, L141.

——— 1978b, *Ap. J. Letters*, **226**, L157.

Elmegreen, B.G., and Lada, C.G. 1977, *Ap. J.*, **214**, 725.

Fesen, R.A., Blair, W.P., and Kirshner, R.P. 1982, *Ap. J.*, **262**, 171.

——— 1985, *Ap. J.*, **292**, 29.

Fitzpatrick, E.L., and Massa, D. 1986, *Ap. J.*, **307**, 286.

Flower, D. R., Pineau des Forets, G., and Hartquist, T. W. 1985, *M.N.R.A.S.*, **216**, 775.

Gaetz, T.J., and Salpeter, E.E. 1983, *Ap. J. Suppl.*, **52**, 155.

Gallagher, J., and Pradhan, A.K. 1985, *JILA Information Center Report*, No. 30,
    University of Colorado.

Grevesse, N. 1983, *Physica Scripta*, **T8**, 49.

Gronenschild, E.H.B.M., and Mewe, R. 1982, *Astr. Ap. Suppl.*, **48**, 305.

Gull, S.F. 1973, *M.N.R.A.S.*, **162**, 135.

——— 1974, *M.N.R.A.S.*, **171**, 237.

Hamilton, A.J.S., Sarazin, C.L., and Chevalier, R.A, 1983, *Ap. J. Suppl.*, **51**, 115.

Hamilton, A.J.S., Sarazin, C.L., Szymkowiak, A.E., and Vartanian, M.H. 1985,
    Ap. J. Letters, **297**, L5.

Harris, A.W., Gry, C., and Bromage, G.E. 1984, *Ap. J.*, **284**, 157.

Hartigan, P., Raymond, J.C., and Hartmann, L. 1986, *Ap. J.*, in press.

Hartmann, L.H., and Raymond, J.C. 1984, *Ap. J.*, **276**, 560.

Heil, T.G., Butler, S.E., and Dalgarno, A. 1980, *Phys. Rev. A*, **23**, 1100.

Hester, J.J., and Cox, D.P. 1986, *Ap. J.*, **300**, 675.

Hobbs, L.M., York, D.G., and Oegerle, W. 1982, *Ap. J. Letters*, **252**, L21.

Hollenbach, D. J., and McKee, C. F. 1979, *Ap. J. Suppl.*, **41**, 555.

———— 1980, *Ap. J. Letters.*, **241**, L47.

Hollenbach, D.J., and Salpeter, E.E. 1971, *Ap. J.*, **163**, 155.

Hollenbach, D.J., and Shull, J.M. 1977, *Ap. J.*, **216**, 419.

Itoh, H. 1977, *P.A.S.J.*, **33**, 1, and 521.

Jacobs, V.L. 1985, *Ap. J.*, **296**, 121. [Erratum **299**, 1079].

Jacobs, V.L., Davis, J., Kepple, P.C., and Blaha, M. 1977a, *Ap. J.*, **211**, 605.

———— 1977b, *Ap. J.*, **215**, 690.

Jacobs, V.L., Davis, J., Rogerson, J.E., and Blaha, M. 1978, *J. Quant. Spectrosc. Rad. Transf.*, **19**, 591.

———— 1980, *Ap. J.*, **239**, 1119.

Jenkins, E.B., Savage, B.D., and Spitzer, L. 1986, *Ap. J.*, **301**, 355.

Jura, M. 1974, *Ap. J.*, **191**, 375.

Kwan, J. 1977, *Ap. J.*, **216**, 713.

Lambert, D. L., and Danks, A. 1986, *Ap. J.*, **303**, 401.

Landau, L. D., and Lifshitz, E. M. 1959, *Fluid Mechanics* (Oxford:Pergamon).

Leger, A., and Puget, J.L. 1984, *Astr. Ap.*, **137**, L5.

Lepp, S., and Shull, J. M. 1983, *Ap. J.*, **270**, 578.

———— 1984, *Ap. J.*, **280**, 465.

London, R., McCray, R., and Chu, S.-I. 1977, *Ap. J.*, **217**, 442.

Low, F.J., 1984, *Ap. J. Letters*, **278**, L19.

McCarroll, R., and Valiron, P. 1976, *Astr. Ap.*, **53**, 83.

MacLow, M.M., and Shull, J.M. 1986, *Ap. J.*, **302**, 585.

McCray, R., and Snow, T.P. 1979, *Ann. Rev. Astr. Ap.*, **17**, 213.

McCray, R., Stein, R.F., and Kafatos, M. 1975, *Ap. J.*, **196**, 565.

McKee, C.F. 1974, *Ap. J.*, **188**, 335.

McKee, C.F., and Hollenbach, D.J. 1980, *Ann. Rev. Astr. Ap.*, **18**, 219.

McKee, C.F., and Ostriker, J.P. 1977, *Ap. J.*, **218**, 148.

Mathis, J.N., Rumpl, W., and Nordsieck, K.H. 1977, *Ap. J.*, **217**, 425.

Merts, A.L., Mann, J.B., Robb, W.D., and Magee, N.W. 1980, *Los Alamos Informal Report* No. LA-8267-MS.

Millar, T. J., Adams, N. G., Smith, D., Lindinger, W., and Villinger,H. 1986, *M.N.R.A.S.*, in press.

Mouschovias, T. Ch. 1976, *Ap. J.*, **207**, 141.

Mullan, D. J. 1971, *M.N.R.A.S.*, **153**, 145.

Morton, D.C. 1975, *Ap. J.*, **197**, 85.

Nadeau, D., and Geballe, T.R. 1979, *Ap. J. Letters*, **230**, L169.

Nadeau, D., Geballe, T. R., and Neugebauer, G. 1982, *Ap. J.*, **253**, 154.

Nussbaumer, H., and Storey, P. 1984, *Astr. Ap. Suppl.*, **56**, 293.

Osterbrock, D. E. 1961, *Ap. J.*, **134**, 270.

———— 1974, *Astrophysics of Gaseous Nebulae*, (San Francisco: Freeman).

Phillips, A.P., Gondhalekar, P.M., and Pettini, M. 1982, *M.N.R.A.S.*, **200**, 687.

Pineau des Forets, G., Flower, D. R., Hartquist, T. W., and Dalgarno, A. 1986, *M.N.R.A.S.*, in press.

Pineau des Forets, G., Roueff, E., and Flower, D. R. 1986, *M.N.R.A.S.*, submitted.

Pradhan, A.K. 1987, in Proc. Conf. on *Atomic Spectra and Oscillator Strengths in Astrophysics and Fusion Research*, to be published in *Physica Scripta* (in press).

Pravdo, S.H., and Smith, B.W. 1979, *Ap. J.*, **234**, L195.

Raga, A.C., and Böhm, K.-H. 1985, *Ap. J. Suppl.*, **58**, 201.

Raymond, J.C. 1979, *Ap. J. Suppl.*, **39**, 1.

———— 1984, *Ann. Rev. Astr. Ap.*,**22**, 75.

Raymond, J.N., Black, J.H., Dupree, A.K., Hartmann, L.H., and Wolff, R.S. 1981, *Ap. J.*, **246**, 100.

Raymond, J.N., Blair, W.P., Fesen, R.A., and Gull, T.R. 1983, *Ap. J.*, **275**, 636.

Raymond, J., Cox, D.P., and Smith, B.W. 1976, *Ap. J.*, **204**, 290.

Raymond, J.C., and Smith, B.W. 1977, *Ap. J. Suppl.*, **35**, 419.

Reilman, R.F., and Manson, S.T. 1979, *Ap. J. Suppl.*, **40**, 815.

Roberge, W. G. 1986, private communication.

Roberge, W.G., and Dalgarno, A. 1982, *Ap. J.*, **255**, 176.

Roberge, W. G., and Draine, B. T. 1987, in preparation.

Routly, P.M., and Spitzer, L. 1952, *Ap. J.*, **115**, 227.

Savage, B.D., and Bohlin, R. 1979, *Ap. J.*, **229**, 136.

Savage, B.D., Bohlin, R.C., Drake, J.F., and Budich, W. 1977, *Ap. J.*, **216**, 291.

Schwartz, R.D. 1983, *Ann. Rev. Astr. Ap.*, **21**, 209.

Schinke, R., Engel, V., Buck, U., Meyer, H. and Diercksen, G. H. F. 1985, *Ap. J.*, **299**, 939.

Seab, C.G. 1987, this volume.

Seab, C.G., and Shull, J.M. 1983, *Ap. J.*, **275**, 652.

———— 1985, "Shock Processing of Interstellar Grains", in *Interrelationships among Circumstellar, Interstellar, and Interplanetary Grains*, NASA CP-2403, p 37.

Seab, C.G., Hollenbach, D.J., McKee, C.F., and Tielens, A.G.G.M. 1985, in *Interrelationships among Circumstellar, Interstellar, and Interplanetary Grains*, NASA CP-2403, p A-28.

Sellgren, K. 1984, *Ap. J.*, **277**, 623.

Shapiro, P., and Kang, H. 1987, *Ap. J.*, in press.

Shull, J.M. 1977, *Ap. J.*, **215**, 805.

———— 1978a, *Ap. J.*, **219**, 877.

———— 1978b, *Ap. J.*, **226**, 858.

———— 1981, *Ap. J. Suppl.*, **46**, 27.

———— 1982, *Ap. J.*, **262**, 308.

Shull, J.M., and Beckwith, S. 1982, *Ann. Rev. Astr. Ap.*, **20**, 163.

Shull, J.M., and Hollenbach, D.J. 1978, *Ap. J.*, **220**, 525.

Shull, J.M., and McKee, C.F. 1979, *Ap. J.*, **227**, 131.

Shull, J.M., and Van Steenberg, M.E. 1982, *Ap. J. Suppl.*, **48**, 95 [Erratum **49**, 351].

Shull, J.M., Seab, C.G., and McKee, C.F. 1986, unpublished shock calculations.

Shull, J.M., York, D.G., and Hobbs, L.M. 1977, *Ap. J. Letters*, **211**, L139.

Siluk, R., and Silk, J. 1974, *Ap. J.*, **192**, 51.

Spitzer, L. 1962, *Physics of Fully Ionized Gases* (2d ed.; New York: Interscience)

———— 1978, *Physical Processes in the Interstellar Medium*, (New York: Wiley-Interscience).

———— 1985, *Ap. J. Letters*, **290**, L21.

Spitzer, L., and Jenkins, E.B. 1975, *Ann. Rev. Astr. Ap.*, **13**, 133.

Troland, T. H., and Heiles, C. 1986, *Ap. J.*, **301**, 339.

Turner, J., Kirby-Docken, K., and Dalgarno, A. 1977, *Ap. J. Suppl.*, **35**, 281.

van Dishoeck, E.F., and Black, J.H. 1986, *Ap. J. Suppl.*, in press.

van Dishoeck, E. F. 1986, in *I.A.U. Symposium 120 Astrochemistry*, ed. M. S. Vardya and S. P. Tarafdar.

Van Steenberg, M.E. 1986, *Ph.D. Thesis*, University of Colorado.

Van Steenberg, M.E., and Shull, J.M. 1986, *Ap. J.*, submitted.

Vishniac, E. 1983, *Ap. J.*, **274**, 152.

Wardle, M. J., and Draine, B. T. 1987, *Ap. J.*, to be submitted.

Withbroe, G. 1971, in "The Menzel Symposium", NBS SP-53, ed. K. Gebbie, (Washington: Government Printing Office).

Younger, S.M. 1980a, *Phys. Rev. A*, **22**, 111.

———— 1980b, *Phys. Rev. A*, **22**, 1425.

———— 1981a, *J. Quant. Spectrosc. Rad. Tranf.*, **26**, 329.

———— 1981b, *Phys. Rev. A*, **23**, 1138.

———— 1981c, *Phys. Rev. A*, **24**, 1272.

York, D.G., Spitzer, L., Bohlin, R.C., Hill, J., Jenkins, E.B., Savage, B.D., and Snow, T.P. 1983, *Ap. J. Letters*, **266**, L55.

Jean-Loup Puget, Louis d'Hendecourt, and Eli Dwek

THEORY OF COLLAPSE AND PROTOSTAR FORMATION

A. P. Boss
DTM, Carnegie Institution of Washington
5241 Broad Branch Road, N.W.
Washington, D.C. 20015
U.S.A.

ABSTRACT. Interstellar clouds must increase in density by a factor of more than 10^{20} in order to form stars. Because observations of the phases intermediate between dense interstellar clouds and pre-main-sequence stars are difficult, theoretical solutions presently provide the primary means for exploring the collapse phase of protostellar formation. The mathematical formulation of the protostellar collapse problem is presented, and various methods employed in solving the equations are outlined. This tutorial emphasizes the numerical approach to the study of the nonlinear, time dependent evolution of collapsing interstellar clouds, including self-gravitation, rotation, and radiative transfer. Results are summarized for the restricted cases of spherical and axisymmetric symmetry, as well as for fully three dimensional evolutions, and briefly compared to observations of star formation.

1. INTRODUCTION

The close association of young stars with interstellar clouds of gas and dust clearly demonstrates that Population I stars are formed from clouds in the interstellar medium. Because the densest interstellar clouds ($n \approx 10^5 - 10^6 \ cm^{-3}$) are roughly 10^{20} times less dense than pre-main-sequence stars, it is also clear that the formation of stars must involve contraction through an extremely large range in density. Observations are largely limited to the few orders of magnitude of density variation at either end of this range, namely to diffuse and dense interstellar clouds at one end, and to pre-main-sequence and zero-age main sequence stars at the other end. The large number of recently discovered embedded infrared objects (e.g., Baud et al. 1984; Beichman et al. 1984) could represent truly intermediate protostellar objects, or could be simply heavily obscured pre-main-sequence stars.

Theoretical models are capable of studying all phases of protostellar formation, but are limited by the mathematical difficulties inherent in the problem. Early theoretical studies of the thermodynamics of protostellar clouds (Gaustad 1963; Hayashi 1966; Larson 1969) showed that a large part of protostellar contraction must occur in the absence of significant support from thermal pressure, leading to *dynamic collapse*. During dynamic collapse, the protostar contracts on a time scale similar to that of the collapse of a pressureless cloud; large-scale subsonic and supersonic fluid motions ensue. Even after one or more quasi-equilibrium cores are formed, the residual gas must be hydrodynamically accreted. The accretion phase is responsible for adding the majority of the gas to the protostar.

D. J. Hollenbach and H. A. Thronson, Jr. (eds.), Interstellar Processes, 321–348.
© *1987 by D. Reidel Publishing Company.*

Protostellar formation is thus a highly time-dependent problem compared to the equilibrium calculations employed for main sequence stellar structure. This and other intrinsic aspects of protostellar formation, such as nonlinearity and three dimensionality, make theoretical progress difficult, but not impossible. This tutorial will present the mathematical equations
which describe protostellar collapse, outline the methods used to solve the equations, summarize the results obtained to date, and briefly compare the results to observations. The primary
emphasis will be on fully numerical solutions of the basic equations of inviscid, non-magnetic
protostellar formation. The tutorial thus updates the recent reviews of protostellar collapse
given by Bodenheimer and Black (1978), Appenzeller (1980), Tscharnuter (1980), Yorke (1981),
Bodenheimer (1981, 1983), and Tohline (1982).

2. PHYSICAL EFFECTS

A variety of physical effects can be important in star formation, such as magnetic fields, internal radiation transport and radiation from nearby OB stars, shock formation and supersonic flow, compression by externally generated shocks from HII regions or cloud collisions, self-
gravitation, tidal forces from nearby objects, thermal pressure, thermodynamics such as the dissociation of molecular hydrogen, rotational motion, compressional heating, turbulent viscosity,
as well as those effects previously noted, time-dependence, nonlinearity, and three dimensionality. Clearly these effects would produce a bewilderingly complex problem if one initially tried
to consider all possible effects. Hence there is an obvious need in the theoretical approach to
isolate a limited number of crucial physical effects, study their ramifications by solving a well-
defined problem, and then to add on other physical effects and start over again.

A pioneering phase involving the exploration of new physical effects has characterized
most of the work on protostellar collapse in the last two decades. Table I gives a concise summary of this progress, which can be compared to previous tables given by Tscharnuter (1980),
Yorke (1981), and Bodenheimer (1983). The table places special emphasis on the dimensionality of the solution, because increasing the dimensionality is generally the greatest obstacle to
progress. Whereas the previous three tables showed a fair number of neglected effects ($\approx$ 14),
it is clear from the number (8) of neglected physical effects in Table I (and the time difference
of only a few years) that the pioneering phase of protostellar collapse may soon be over. In the
absence of new physical effects to explore, theoretical attention must then turn to the more ma-
ture task of re-checking the previous work with higher precision techniques and with more thorough examinations of parameter space.

2.1 Dimensionality

Table I lists three different types of calculations, 1D, 2D, and 3D. 1D refers to calculations performed in one spatial dimension, 2D to two spatial dimensions, and 3D to three (in addition to
the temporal dimension.) Essentially all 1D calculations of protostellar collapse use the spherical coordinate radius r, and so assume that the protostar remains spherically symmetric during its collapse. Considering the slow rotation and fairly high degree of spherical symmetry
of the sun, this might seem to be an adequate approximation for single stars. The majority of
stars occur in binary or multiple stellar systems (Abt 1983), where the requirement of orbital
angular momentum breaks the spherical symmetry. Spherical symmetry may not be adequate
even for single stars, however. Our sun is accompanied by a strongly flattened planetary system
with orbital angular momentum about 100 times greater than the spin angular momentum of
the sun. Because the planets were most likely formed as a consequence of the same events that

formed the sun, spherical symmetry could not have characterized all phases of the sun's formation. Thus while spherical symmetry is an extremely useful approximation, it is certain that a complete theory of star formation must allow for rotational flattening and fragmentation to occur, which can only be rigorously studied in 2D and 3D, respectively. 2D calculations usually involve the assumption of axial symmetry, that is, the solutions do not depend on the angular variable ϕ, and so have symmetry of rotation about the rotation axis. 3D calculations relax this assumption, and allow arbitrarily shaped protostars to form.

TABLE I.

Physical effect	1D	2D	3D
Hydrodynamics & self-gravity	yes	yes	yes
Thermodynamics	yes	yes	yes
Radiative transfer (gray)	yes	yes	yes
Radiative transfer ($f(\nu)$)	yes	yes	no
Convective transport	yes	no	no
Turbulent viscosity	...	yes	yes
Molecular chemistry	yes	no	no
Magnetic fields	yes	yes	yes
Gas-dust interaction	yes	yes	no
Thermonuclear reactions	yes	no	no

2.2 Hydrodynamics & self-gravity

The most basic equations of protostellar collapse are the equations of compressible hydrodynamics, which describe the flow of the gas in interstellar clouds, and Poisson's equation, which relates the gravitational potential to the density distribution. All theoretical models of protostellar formation solve these equations at the minimum.

Interstellar clouds can be treated as compressible fluids providing that the mean free path λ for collisions between gas particles is much smaller than the length scale of interest R. This constraint is easily satisfied: $\lambda \approx 10^{-7}$ pc for an interstellar cloud with $n = 10^4$ cm^{-3}, which is considerably smaller than any length scale of gross physical importance, and this constraint lessens further as the number density increases ($\lambda \propto n^{-1}$, while $R \propto n^{-1/3}$.)

Poisson's equation is essential, because it is self-gravity that drives protostellar formation. Provided that the fluid speeds v remain small compared to the speed of light $c = 3 \times 10^{10}$ cm s^{-1}, the gravitational potential may be assumed to be instantaneously responsive to the mass distribution. Because the maximum protostellar fluid speeds are typically not much larger than the sound speed $c_s \approx 2 \times 10^4 (T/10K)^{1/2}$, where T is the gas temperature, this approximation is very good.

2.3 Thermodynamics

Cold, dense interstellar clouds are composed mainly of molecular hydrogen, along with helium and trace amounts of other gases and dust (approximately 2 % by weight). The thermodynamics of the hydrogen and helium gas thus dominate the thermodynamics of protostars, but the dust grains have a strong indirect effect through their effect on the opacity at low temperatures. The ideal gas law and the Saha equation are adequate for determining the gas pressure and state of ionization throughout most of the protostar, but dense, nearly-stellar cores may require corrections for effects such as electron degeneracy and pressure ionization.

2.4 Radiative transfer

Pre-main-sequence stars maintain hydrostatic equilibrium by balancing self-gravity and thermal pressure. Considering that the internal temperatures of stars ($\approx 10^7 \ K$) are much greater than the temperatures of star-forming interstellar clouds ($\approx 10 - 50 \ K$), it is clear that a considerable amount of heating must occur during star formation. Initially heating is caused by conversion of gravitational potential energy into thermal energy through the compression associated with the collapse. Because radiative transfer determines how much of the heat can escape from the protostar, and hence controls the thermal pressure, radiative transfer (in at least the gray, Rosseland mean approximation) is important for understanding the overall dynamics of protostellar collapse. In addition, frequency dependent ($f(\nu)$) radiative transfer is necessary for determining the spectral appearance of protostars, for purposes of comparison with observation.

2.5 Convective transport

Protostellar clouds are generally stable with respect to convection (Schwarzschild criterion) throughout the collapse phase. Convection can only occur after the final, quasi-equilibrium core has formed. Should convective instability be indicated, the convection is usually assumed to be so efficient as to keep the temperature gradient very close to adiabatic. Because nearly all 2D and 3D models have been unable to advance a protostar all the way to formation of this final core, the neglect of convective transport in 2D and 3D to date (Table I) is somewhat understandable.

2.6 Turbulent viscosity

While molecular viscosity μ is too feeble to have any effect on protostellar collapse, the situation for turbulent viscosity is not so straightforward. Turbulence in nonrotating fluids occurs when the Reynolds number Re exceeds a critical value Re_{cr} which is on the order of 10 to 10^3 for laboratory flows. Because Re for protostellar clouds is much greater than 10^3, it might be expected that protostellar collapse is turbulent. Turbulence produces a redistribution of momentum that mimics viscous stresses, leading to the concept of an effective turbulent viscosity $\mu_{turb} = (Re/Re_{cr})\mu = \rho v l / Re_{cr}$, where ρ is the density, v is a typical fluid velocity, and l a typical scale size. The size of μ_{turb} is uncertain by several orders of magnitude.

Whether or not turbulent viscosity has an effect on protostellar collapse depends largely on the assumed value of Re_{cr}: for $Re_{cr} \approx 10^3$, Tscharnuter (1978) found turbulent viscosity to be negligible, while Tscharnuter (1981) and Regev and Shaviv (1981) found turbulent viscosity to be important for $Re_{cr} \approx 10$. The correct value for Re_{cr} is unknown. The importance of μ_{turb} also depends on the formulation of the turbulent stresses; Vanajakshi and Jenkins (1984) found that with a formulation different from that of the previous workers, turbulent stresses were unlikely to dominate the collapse process.

In addition, there is some question as to whether or not the collapse phase will even be turbulent. First, the simple laboratory criterion $Re > Re_{cr}$ may be misleading in astrophysical situations because of the absence of the rigid boundaries that produce differential motion and efficiently transfer energy to large scale eddies (Cameron 1970). Second, in the presence of appreciable rotation, the appropriate criterion for turbulent instability is Rayleigh's criterion (e.g., Chandrasekhar 1961), which requires a more rapid fall-off in angular velocity as a function of cylindrical radius than typically occurs in collapsing clouds. Finally, both of these criteria apply only to incompressible fluids; when compressibility is included, Schwarzschild's criterion for instability to convection must be satisified in general, which does not happen during the collapse phase.

Even if the collapse is not laminar, protostellar collapse may occur too quickly for turbulent viscosity to be effective. However, turbulent viscosity may well be of importance after the formation of the final stellar core, when convective instability first occurs, and when quasistatic evolution allows processes with relatively long time scales to dominate. Indeed, turbulent viscosity is thought by many to be essential for transferring angular momentum between protostellar and protoplanetary material, thereby allowing a relatively massive central star to form, surrounded by a protoplanetary disk carrying the bulk of the angular momentum. Turbulent viscosity can only be self-consistently incorporated in 2D or 3D models, and with the exception of Tscharnuter's models (e.g., Morfill, Tscharnuter, and Völk 1985), no 2D or 3D models have been pushed far enough to require its inclusion.

2.7 Molecular chemistry

Observations continually detect new molecules in regions of star formation. To the extent that the formation of these molecules requires well-constrained physical conditions, the observations can be used as a diagnostic for protostellar collapse, and hence should be incorporated in theoretical models. Molecular chemistry has been included in some detail in 1D collapse models but not in 2D or 3D. This deficiency stems mainly from the belief that fine details of molecular chemistry have little effect on the gross dynamics of protostellar collapse.

2.8 Magnetic fields

While magnetic fields are known to be important for many aspects of the dynamics of relatively diffuse protostellar clouds (reviewed by Mouschovias 1981), the fact that the magnetic fields are structurally negligible in main-sequence stars implies that at some point in the star formation process, magnetic fields cease to be of importance (e.g., Nakano and Umebayashi 1986). The density at which this occurs is crucial; the subsequent evolution may be assumed to be non-magnetic, which greatly simplifies the problem. Effects which may reduce the importance of magnetic fields include: lowered fractional ionization as the cloud collapses and becomes more opaque to ionizing radiation from the cloud exterior; ambipolar diffusion, where the ions, tied to the magnetic field lines, slip past the neutral species, allowing the neutral portion to collapse unimpeded; Joule dissipation of magnetic field energy; and magnetic field line recombination.

Unfortunately, estimates of the critical density vary widely. Mouschovias (1977) found that ambipolar diffusion could decouple the magnetic field at number densities of $10^4 - 2 \times 10^6 \ cm^{-3}$, whereas Nakano (1984) prefers a critical density of at least $10^{11} \ cm^{-3}$. Thus magnetic fields could be negligible for collapse from dense interstellar cloud conditions, or could be important for much of the collapse phase. Observations often find only upper limits to the magnetic field strength in molecular clouds (e.g., Crutcher, Troland, and Heiles 1981), implying significant loss of magnetic field strength even in this relatively diffuse phase. Hence we will assume for the purposes of this tutorial that magnetic fields can be safely ignored during the collapse phase.

2.9 Gas-dust interaction

The dust component of protostellar clouds is usually assumed to be transported along with the gaseous component, and hence it has been given little special attention. This may be justifiable during the collapse phase, when both gas and dust are largely in free fall, but may not be true once the infall of the gas is arrested by formation of a quasi-equilibrium core. Subsequent sedimentation of dust grains could have an indirect effect on protostellar dynamics through the opacity, which is controlled by the dust grain population in regions with $T < 1500K$. Morfill et al. (1978) provided a preliminary treatment of dust in 2D collapse, but restricted their attention to early phases where the dust could be assumed to have little effect on the overall dynamics. While dust grain sedimentation and coagulation is of course of central importance to theories of planet formation, its importance for protostellar formation remains largely uninvestigated.

2.10 Thermonuclear reactions

The protostellar phase may be defined to terminate once nuclear burning takes place and the protostar moves onto the main sequence, marking the epoch where nuclear burning, rather than gravitational contraction, becomes the dominant energy source. By this definition, nuclear reactions are negligible for protostellar formation. However, deuterium burning can provide a temporary source of significant energy (e.g., D'Antona and Mazzitelli 1985), so some nuclear reactions may still be important for pre-main-sequence evolution. Only a few 1D collapse models have included thermonuclear reactions, and no 2D or 3D models, once again largely because the latter do not reach densities high enough to initiate thermonuclear reactions.

3. MATHEMATICAL FORMULATION

3.1 Coordinate system

For multidimensional problems, the choice of the optimum coordinate system (e.g., cartesian (x, y, z), cylindrical (R, Z, ϕ), or spherical (r, θ, ϕ) coordinates) is nontrivial, particularly when a numerical solution is attempted. The primary concern is one of providing for adequate numerical resolution of sharply varying physical quantities with a limited computational grid. Spherical objects will in general be most accurately represented in spherical coordinates, highly flattened objects with some axial symmetry will do best in cylindrical coordinates, and completely asymmetrical objects may be well represented in cartesian coordinates. A secondary concern comes from the maximum size of the time step Δt_{CFL} (the Courant-Friedrichs-Lewy constraint; Roache 1976) consistent with the numerical stability of explicit hydrodynamics techniques (see section 3.7.1). Because this time step depends directly on the grid spacing (in spherical coordinates, Δt_{CFL} is often determined by the minimum of $r sin\theta \Delta\phi/(c_s + |v|)$, where $\Delta\phi$ is the ϕ grid spacing, c_s is the sound speed, and v is the fluid velocity), some coordinate systems may require smaller time steps, and hence more computational effort, to solve a given problem. This drawback is ameliorated somewhat by the fact that the use of smaller time steps (like smaller grid spacings) formally increases the anticipated mathematical accuracy of the solution.

The spherical coordinate system has the unique advantage of being directly applicable to all three of the dimensionality assumptions used in protostellar collapse theory. That is, the cartesian coordinate system cannot be used to study problems with exact spherical or axial symmetry, and the cylindrical coordinate system cannot be used to study exactly spherical collapse. Because the spherical coordinate system can be used in all three cases (1D, 2D, and 3D), the equations of motion in this tutorial will be given in spherical coordinates. This advantage of spherical coordinates is important for numerical codes, because a 3D spherical coordinate code can then be easily converted to a 2D or 1D code with little extra effort.

3.2 Equations of Motion

The time evolution of a nonmagnetic, inviscid, 3D protostellar cloud is determined by the following set of equations. The equations will be given in Eulerian form, i.e., written with respect to a fixed frame of reference. The equations of compressible hydrodynamics include the continuity equation

$$\frac{\partial \rho}{\partial t} + \nabla \cdot (\rho v) = 0,$$

and the three momentum equations

$$\frac{\partial (\rho v_r)}{\partial t} + \nabla \cdot (\rho v_r v) = -(\rho \frac{\partial \Phi}{\partial r} + \frac{\partial p}{\partial r}) + \frac{\rho}{r}(v_\theta^2 + v_\phi^2),$$

$$\frac{\partial (\rho v_\theta)}{\partial t} + \nabla \cdot (\rho v_\theta v) = -\frac{1}{r}(\rho \frac{\partial \Phi}{\partial \theta} + \frac{\partial p}{\partial \theta}) - \frac{\rho}{r}(v_r v_\theta - v_\phi^2 cot\theta),$$

$$\frac{\partial (\rho A)}{\partial t} + \nabla \cdot (\rho A v) = -(\rho \frac{\partial \Phi}{\partial \phi} + \frac{\partial p}{\partial \phi}),$$

where t is time, ρ is the mass density, $v = (v_r, v_\theta, v_\phi)$ is the fluid velocity, $A = rsin\theta v_\phi$ is the specific angular momentum, and p is the pressure. The vector operator $\nabla \cdot \Psi$ is defined by

$$\nabla \cdot \Psi = \frac{1}{r^2}\frac{\partial}{\partial r}(r^2 \Psi_r) + \frac{1}{r \, sin\theta}\frac{\partial}{\partial \theta}(sin\theta \Psi_\theta) + \frac{1}{r \, sin\theta}\frac{\partial \Psi_\phi}{\partial \phi}.$$

The gravitational potential Φ is determined by Poisson's equation

$$\nabla^2 \Phi = 4\pi G\rho,$$

where G is the gravitational constant and ∇^2 is defined by

$$\nabla^2 \Phi = \frac{1}{r^2}\frac{\partial}{\partial r}(r^2 \frac{\partial \Phi}{\partial r}) + \frac{1}{r^2 \, sin\theta}\frac{\partial}{\partial \theta}(sin\theta \frac{\partial \Phi}{\partial \theta}) + \frac{1}{r^2 \, sin^2\theta}\frac{\partial^2 \Phi}{\partial \phi^2}.$$

Because a significant portion of protostellar collapse occurs with constant temperature (i.e., *isothermally*, see section 4.1), and because the assumption of isothermality eliminates the need to solve any further equations, more work has been done on isothermal protostellar collapse than on *nonisothermal* collapse, particularly in 2D and 3D. The isothermal approximation (or the equally simplified adiabatic approximation) has been used in 1D protostellar collapse by McNally (1964), Bodenheimer and Sweigart (1968), Hunter (1969), Buff, Gerola, and Stellingwerf (1979), Boss (1980a), Wang and Qian (1981), and Boss and Black (1982). In 2D the work includes Nakazawa, Hayashi, and Takahara (1976), Deissler (1976), Kamiya (1977), Takahara *et al.* (1977), Bodenheimer and Tscharnuter (1979), Boss (1980b,d), Tohline (1980b),

Norman, Wilson, and Barton (1980), Regev and Shaviv (1981), Qian and Wang (1981), Virgopia and Ferraioli (1981b), Villere and Black (1980, 1982), Boss and Black (1982), Boss and Haber (1982), Bastien (1983), Narita *et al.* (1983), Terebey, Shu, and Cassen (1984), Narita, Hayashi, and Miyama (1984), and Vanajakshi and Jenkins (1985). In 3D the work includes Narita and Nakazawa (1977), Norman and Wilson (1978), Cook and Harlow (1978), Boss and Bodenheimer (1979), Tohline (1980a), Bodenheimer, Tohline, and Black (1980), Różyczka *et al.* (1980), Różyczka, Tscharnuter, and Yorke (1980), Boss (1980c, 1981a,b, 1982a), Bodenheimer and Boss (1981), Wood (1981, 1982), Gingold and Monaghan (1981, 1982, 1983), Różyczka (1983), and Miyama, Hayashi, and Narita (1984).

If one wishes to follow protostars into the nonisothermal regime, one needs to calculate the temperature from the internal energy. The specific internal energy E is determined by

$$\frac{\partial(\rho E)}{\partial t} + \nabla \cdot (\rho E v) = -p\nabla \cdot v + L,$$

where L is the time rate of change of energy per unit volume due to radiative transfer.

We will consider only *gray* radiative transfer, where the frequency dependence of the radiation field is replaced by appropriately chosen mean functions (see Sandford, Whitaker, and Klein 1982 for a description of the non-gray equations). In the diffusion approximation for radiative transfer, L is given by

$$L = \nabla \cdot \left(\frac{4}{3\kappa\rho}\nabla(\sigma T^4)\right),$$

where κ is the Rosseland mean opacity, T is the temperature, σ is the Stefan-Boltzmann constant, and

$$\nabla\Psi = \frac{\partial\Psi}{\partial r}\hat{r} + \frac{1}{r}\frac{\partial\Psi}{\partial\theta}\hat{\theta} + \frac{1}{r\sin\theta}\frac{\partial\Psi}{\partial\phi}\hat{\phi}.$$

The diffusion approximation is strictly valid only for optically thick regions, whereas protostellar collapse involves optically thick, optically thin, and intermediate regions. In optically thin regions, the diffusion approximation gives the same temperature as the boundary temperature, which is approximately correct. Because of its simplicity, the diffusion approximation has been used in many studies of protostellar collapse in both 1D (Bodenheimer 1968; Larson 1969, 1972b; Appenzeller and Tscharnuter 1974; Boss 1984a) and 2D (Larson 1972a; Black and Bodenheimer 1975, 1976; Tscharnuter 1975; Bastien and Mitalas 1979).

The Eddington approximation provides a superior description of the radiation field compared to the diffusion approximation (Mihalas 1970, pp. 41-46), at the expense of solving an added equation. In the Eddington approximation,

$$L = 4\pi\kappa\rho(J - B),$$

where J is the mean intensity, and $B = \sigma T^4/\pi$ is the Planck function. The mean intensity is the normalized angular integral of the specific intensity I, and is determined by the equation

$$\frac{1}{3}\frac{1}{\kappa\rho}\nabla \cdot \left(\frac{1}{\kappa\rho}\nabla J\right) = J - B.$$

In the limit of infinite optical depth, $J \to B$, and the Eddington approximation becomes identical to the diffusion approximation. Use of the Rosseland mean opacity ensures that the radiative flux in the gray approximation is identical to that in the frequency dependent case (e.g., Mihalas 1970, p. 38). The Eddington approximation appears to be quite adequate for following the gross dynamics of protostellar collapse (Tscharnuter 1977). The Eddington approximation has been used in 1D protostellar collapse models by Kondo (1978) and Winkler and Newman (1980a,b), in 2D by Tscharnuter (1978, 1980), Tscharnuter and Winkler (1979), and Boss (1984a), and in 3D by Boss (1983, 1985, 1986).

The full set of equations for nonmagnetic protostellar collapse are a set of seven coupled, nonlinear, partial differential equations, of first and second order, and with strongly variable coefficients. The equations for 2D may be obtained by dropping all partial derivatives with respect to ϕ, while the equations for 1D may be obtained from the 2D equations by setting terms with v_θ or v_ϕ equal to zero, neglecting the equations for v_θ and $A(v_\phi)$, and by dropping terms with partial derivatives with respect to θ.

3.3 Thermodynamical relations

The seven protostellar collapse equations need to be solved for ten variables, namely $\rho, v_r, v_\theta, A,$ $\Phi, p, E, \kappa, L,$ and T. Three additional relations are thus necessary in order to close the system, which are the descriptions of how the specific internal energy, pressure, and opacity depend on the density and temperature of the gas. These three functions have been discussed in detail by Larson (1969), Black and Bodenheimer (1975), and Boss (1984a). The specific internal energy and pressure include contributions from dissociating and ionizing hydrogen, para- verus ortho-hydrogen molecules, and translational motion of all species. At the low temperatures characteristic of protostellar clouds, the opacity is controlled completely by the dust grain population, chiefly the water molecules and iron, silicate, and graphite grains. The usual stellar opacity sources such as the H^- ion and Rayleigh scattering only become dominant for temperatures in excess of $\approx 3000 K$.

3.4 Initial conditions

Because the equations of motion are time dependent, an initial value for all physical variables must be provided. Ideally the initial conditions for protostellar collapse are completely specified by observations of interstellar clouds which are on the verge of collapse. While observations do provide important global constraints, such as the total mass, average velocity dispersion, and minimum density, these estimates still fall short of the need for the precise initial conditions that are required for mathematical solutions of the partial differential equations. In order to make progress, most theoretical work has centered on simplified initial conditions where global quantities can be analytically evaluated, and where different authors can easily duplicate the initial conditions for purposes of mutual comparison. Larson's (1969) initial conditions define the standard initial model: collapse starting from a uniform density and temperature spherical cloud, initially everywhere at rest. Because of the presence of self-gravity and the complete absence of pressure gradients, the standard initial conditions are certain to produce at least a transient collapse. Larson (1972a) extended the standard initial conditions to include solid body rotation (uniform angular velocity ω_i). For isothermal, rotating clouds, there are just two parameters that specify all possible protostellar models, termed α_i and β_i (Black and Bodenheimer 1976). For the standard initial conditions,

$$\alpha_i = \frac{5 R_g R T_i}{2 G M_i \mu},$$

where R_g is the gas constant, R is the cloud radius, M_i is the cloud mass, and μ is the mean molecular weight, while

$$\beta_i = \frac{\omega_i^2}{4\pi G \rho_i}.$$

For adiabatic clouds, one must specify in addition the value of the adiabatic exponent, γ, defined by $p \propto \rho^\gamma$ ($\gamma = 1$ for isothermal clouds.) Scaling relations exist for both isothermal and adiabatic clouds (Bodenheimer and Black 1978; Boss 1980d) which allow a calculation for fixed $\alpha_i, \beta_i, \gamma$ to be scaled to a cloud of arbitrary mass, temperature, and μ. In 3D, the only addition to the standard initial model is the form of the perturbation which is to be followed, which is a purely sinusoidal perturbation to the density (Boss and Bodenheimer 1979).

For nonisothermal clouds, the possibility of heating becoming important at a certain density means that two new parameters are needed to characterize the standard model, which can be taken to be M_i and T_i. Thus for nonisothermal evolution, a four dimensional parameter space $(\alpha_i, \beta_i, M_i, T_i)$ exists.

It is generally believed that much of the important physics of protostellar collapse can be learned from collapse starting from the standard initial conditions. The four parameters which can be varied in the standard problem $(\alpha_i, \beta_i, M_i,$ and $T_i)$ represent all of the important physics of nonmagnetic star formation: thermal pressure, self-gravity, rotation, and non-isothermal heating at a density dependent on the opacity. Some workers have investigated alternatives, such as the 3D calculations of Norman and Wilson (1978), Cook and Harlow (1978), and Różyczka et al. (1980), whose models started from axisymmetric rings, but these studies involved fragmentation of previously formed rings, rather than fragmentation during collapse. Centrally concentrated initial conditions were explored by Bodenheimer and Sweigart (1968) and Boss and Black (1982). Different forms for the initial 3D density perturbations to the standard initial conditions have been studied by Tohline (1980a) and Bodenheimer, Tohline, and Black (1980), while Różyczka, Tscharnuter, and Yorke (1980) added velocity perturbations to the standard initial conditions.

For a more realistic simulation of star formation, one must study collapse starting from an equilibrium state which has been strongly perturbed by a specific physical mechanism, such as cloud compression by a shock wave (reviewed by Klein, Whitaker, and Sandford 1985) or loss of magnetic field support (see section 2.8). Equilibrium models for rotating, isothermal, axisymmetric clouds have been investigated in some detail and could provide an initial model. Simplified models based on the global properties of Maclaurin spheroids first indicated that two types of equilibrium are possible for clouds with low angular momentum (Weber 1976; Hachisu and Eriguchi 1984; Tohline 1985a): slowly rotating, nearly spherical clouds supported by external pressure (diffuse branch), and rapidly rotating, highly flattened clouds, supported by rotation (compact branch). For high angular momentum clouds, the two types merge. Because the free energy of the diffuse branch is higher than that of the compact branch, the interesting possibility exists that known physical perturbations could force a phase transition from an otherwise stable diffuse cloud to a higher density compact cloud (Tohline 1985b).

It is important to construct rigorous models of both branches and to investigate their dynamical stability, otherwise the two branches will remain mathematical abstractions rather than possible physical objects. Stahler (1983a) has calculated numerical models of the diffuse branch clouds and demonstrated their stability. The situation for the compact branch is less certain. Analytical solutions by Hayashi, Narita, and Miyama (1982) and Toomre (1982) appear to be similar to the compact branch, but have infinite central density and infinite extent. Furthermore, the analytical models are compressionally unstable when nearly spherical, dynamically unstable to ring formation when highly flattened (Hayashi et al. 1982; Schmitz 1984), and appear to be unstable to nonaxisymmetric perturbations in the intermediate range (Tohline

1985a; Boss 1986). The models closest to the compact branch are the numerical models of thin disks by Hachisu and Eriguchi (1985), which have a finite size, but still have infinite central density; their stability is unknown. Numerical collapse calculations of isothermal clouds have found the diffuse branch (Boss and Haber 1982), but have not found the compact branch. Predictions involving the existence and stability of the compact branch must still be viewed with some caution.

3.5 Boundary conditions

In addition to the initial conditions, the two second order equations require boundary conditions on Φ and either T or J, depending on whether the diffusion or Eddington approximation is used. The boundary condition for Φ is usually met by assuming that there is no mass exterior to the protostar, which results in the desired constraint on $\Phi(R)$ (see Black and Bodenheimer 1975 and Boss 1980a for details about for the 2D and 3D cases, respectively). The boundary condition on T is usually chosen to be constant $T = T_i$ (e.g., Tscharnuter 1975); in the Eddington approximation, one can choose constant J (Tscharnuter and Winkler 1979) or a boundary condition on J derivable from the constant temperature boundary condition (Boss 1984a).

The standard boundary conditions (Larson 1969) also assume that no fluid enters or leaves the cloud, and that the cloud boundary remains fixed in space (constant volume). Constant pressure boundary conditions have also been explored (e.g., Black and Bodenheimer 1976; Boss and Haber 1982).

The specification of these boundary conditions is thought to have little fundamental effect on the protostellar collapse phase (e.g., Larson 1969; Tscharnuter 1975; but see Disney 1976). This appears to be true when the cloud free fall time (defined in the next sub-section) is less than the time for a sound wave to propagate across the cloud (which is equivalent to small α_i), because in this case the cloud collapses faster than information from the boundary (sound waves) can travel inward (Bodenheimer and Sweigart 1968). A systematic study of 2D collapse found that constant volume and constant pressure boundary conditions produce easily understood differences in the collapse (Boss and Haber 1982).

3.6 Analytical and semi-analytical solution methods

No one has attempted a completely analytical solution of the full set of equations of protostellar collapse for obvious reasons. Progress has been made however in solving more idealized situations, both in purely analytical form, and in semi-analytical form. The latter are included in the same sub-section as analytical solutions, because semi-analytical solutions involve ordinary differential equations, which can usually be solved to arbitrarily high accuracy. The fully numerical solutions of the following section (particularly in 2D and 3D) normally involve considerable uncertainty as to their mathematical accuracy.

The most common simplifying assumption involves thermodynamics: the protostar is assumed to be either pressureless or isothermal. While the latter is permissible in some phases, the former can only be justified on the basis of mathematical convenience.

The most fundamental analytical work is that of Jeans (1929), who analyzed the linear stability of an initially static, uniform density, isothermal gas of infinite extent. Jeans found that gravitational instability could occur for perturbations greater than a critical size, leading to collapse for masses greater than the *Jeans mass* M_J, defined by

$$M_J = [\frac{3}{4\pi}(\frac{5R_g}{2G})^3]^{1/2}(\frac{T}{\mu})^{3/2}\rho^{-1/2}.$$

Because the initial phases of protostellar collapse occur isothermally, the Jeans mass decreases as the density increases, which Hoyle (1953) hypothesized would lead to fragmentation to progressively smaller masses.

Hunter (1962) showed that when an initially uniform density (ρ_i), static, spherical, pressureless cloud collapses, it maintains uniform density throughout the collapse, and the density becomes infinite after a time defined to be the free fall time $(t_{ff} = (3\pi/32G\rho_i)^{1/2})$. Hunter (1962) then performed a linearized analysis of the growth of spherically symmetric perturbations $(\delta\rho = \delta_i\rho_i$, with $\delta_i << 1)$ on top of the background collapse of a pressureless sphere. Because the perturbations initially have a higher density than the background sphere $(\rho_{pert} = \rho_i(1 + \delta_i))$, they also have a smaller free fall time, and so they reach infinite density first. Thus fragmentation occurs, in the sense of having $\rho_{pert}/\rho \to \infty$ in a finite time, when the background density has increased by a factor $\rho/\rho_i \propto \delta_i^{-2}$. Rapid fragmentation of this sort is referred to as *Hoyle-Hunter* fragmentation (see also McNally and Settle 1980).

This analytical approach can be extended to spheroidal (axisymmetric 2D) clouds as well. Lin, Mestel, and Shu (1965), using the work of Lynden-Bell (1962, 1964), showed that a static, uniform density, non-rotating, pressureless spheroid collapses with monotonically increasing eccentricity: oblate clouds become more oblate and collapse into disks, while prolate clouds collapse into rods. Thus purely spherical collapse was found to be singular; any small deviation from spherical symmetry grows on a dynamical time scale. Silk (1982) studied the growth of linear perturbations on top of a collapsing spheroidal cloud, and found that perturbations grew even faster than in the spherical case studied by Hunter (1962). For spheroidal perturbations, fragmentation in the sense of $\rho_{pert}/\rho \to \infty$ occurs when the background density has increased by a smaller factor $(\rho/\rho_i \propto \delta_i^{-1})$. The pressureless analyses thus imply that any initial noise can grow and lead to fragmentation prior to a time $t < t_{ff}$.

As noted above, the assumption of pressureless collapse is unrealistic. Hunter (1962, 1964) was able to approximate the effects of thermal pressure on perturbation growth, and found that for isothermal clouds, perturbations could still grow, but at an appreciably slower rate. This finding is important, because it calls into question the delicate timing required for having the perturbation reach infinite density an instant in time before the background cloud does. Tohline (1980c) used time scale arguments to show that isothermal clouds starting from $\alpha_i \approx 1$ states should not fragment prior to $t = t_{ff}$, while Boss (1983) used a 3D collapse calculation to demonstrate that Hoyle-Hunter fragmentation can indeed occur for a low $\alpha_i(= 0.01)$ cloud.

Boss (1982b) applied linear perturbation analysis to the growth of nonaxisymmetric perturbations in collapsing clouds, and developed a heuristic criterion for their initial growth, consistent with numerical calculations of the full equations. In general, however, linear stability analyses of fragmentation in 1D or 2D collapse are of limited usefulness, because protostellar fragmentation is ultimately a nonlinear process.

The collapse of spherically symmetric, isothermal clouds can be studied with *similarity methods* (Larson 1969; Penston 1969; Shu 1977; Hunter 1977; Cheng 1978), where solutions are obtained in terms of a new variable involving both the radius and time (e.g., $x = r/c_s t$). Shu (1977) showed that an isothermal sphere starting from the singular equilibrium state $\rho \propto r^{-2}$ will accrete onto the central singularity. The central regions collapse and accrete first, in free fall, with a characteristic $\rho \propto r^{-3/2}$ distribution. An expansion wave moves outward, separating the infalling regions from the static regions, and eventually the entire cloud falls onto the core. Boss and Black (1982) confirmed numerically the Shu (1977) analysis, and studied as well the accretion of rotating flows. Terebey, Shu, and Cassen (1984) used a semi-analytical technique involving matched asymptotic expansions to study the collapse of slowly rotating, isothermal spheres with $\rho_i \propto r^{-2}$. Many analytical and semi-analytical papers (too numerous to discuss here) have been written on the general problem of accretion, because of its wide astrophysical importance.

Stahler, Shu, and Taam (1980a,b, 1981) divided an accreting protostar into several spherically symmetric layers, where the physics of the different layers allowed for substantial simplification of the mathematics. In this case the partial differential equations reduce to ordinary differential equations, and by carefully matching the solutions obtained in each layer, Stahler, Shu, and Taam were able to compute the evolution of a 1D protostar to the main sequence (see section 4.1).

Tohline (1980b) and Boss (1980b) used semi-analytical and analytical methods respectively to follow the collapse of a rotating pressureless cloud in a fixed gravitational potential. Integration of the orbits of test particles showed that an outward moving density wave occurs after the innermost particles undergo centrifugal rebound. Collisions with other particles which are still collapsing inward can then account for ring formation in 2D (see section 4.2). Virgopia and Ferraioli (1981a) used analytical methods and certain simplifying assumptions to confirm the production of rings in rotating, 2D, isothermal collapse.

3.7 Numerical solution methods

3.7.1 Spherically symmetric (1D)

In 1D, there are still five equations to solve, but the retention of only one spatial variable r makes accurate numerical solutions almost trivial, compared to 2D and 3D.

The basic approach is to replace the partial differential equations with *finite difference* (FD) equations. The FD equations are generated by replacing all derivatives with approximate expressions based on Taylor series expansions on a discrete grid. For example, the time derivative of a variable X can be replaced with

$$\frac{\partial X}{\partial t} \approx \frac{X(t + \Delta t) - X(t)}{\Delta t},$$

where Δt is the time step. Analogous expressions hold for spatial derivatives, and replacements for higher order derivatives and mixed partial derivatives can also be derived. More details can be found in Roache (1976) and Tohline (1982).

The key simplification in 1D collapse involves the manner in which the time evolution proceeds. If the FD equations are set up so that the physical variables at the next time step depend solely on their values at the previous time step, the solution is termed an *explicit* solution and is relatively straightforward. However, explicit solution methods incur a restriction on the size of the time step (section 3.1) which essentially limits the solution to the dynamic collapse phase; during quasi-equilibrium phases, such as while the core is accreting the rest of the cloud, explicit methods have intolerably small time steps. This restriction does not apply to *implicit* solutions, where the physical variables at the next time step depend on one another. The penalty paid for lifting the time step restriction is this linear dependence of all the variables at all the grid points. The resulting set of linear equations must be solved simultaneously, but in 1D this is easily done using a mathematical trick called the tridiagonal method of matrix inversion (e.g., Roache 1976). Thus is it quite practical to write an implicit 1D code, capable of following the collapse of a protostar through both dynamic and quasi-equilibrium phases (e.g., Bodenheimer 1968; Larson 1969).

FD techniques require the specification of a temporal-spatial grid with finite grid spacing $(\Delta t, \Delta r)$. The FD replacements, however, are only exact replacements of the original partial differential equations in the limit of $\Delta t \to 0, \Delta r \to 0$, which formally requires an infinite number of grid points. In 1D one can afford to use a sufficiently large number of grid points $(N_r \approx 100 - 1000)$ to ensure a quite accurate solution.

3.7.2 Axially symmetric (2D)

In 2D all seven equations must be solved, but the addition of two momentum equations is not particularly bothersome. The real complication involves the addition of partial derivatives with respect to θ, which means that when an implicit solution is attempted, the set of linear equations cannot be solved by the tridiagonal technique. Coupled with the greatly increased number of grid points $(N_r \times N_\theta)$, a straightforward implicit solution would require the inversion of a very large matrix, which is computationally prohibitive. Because of this, essentially all 2D FD codes use explicit time differencing. The exception is the code developed by Tscharnuter and colleagues (Fricke, Möllenhoff, and Tscharnuter 1976; Tscharnuter and Winkler 1979), which uses a Legendre polynomial decomposition of all physical variables to remove the θ dependence from the partial differential equations, resulting in an easily solvable system dependent only on r and t. One drawback is that this scheme does not conserve mass and angular momentum (Morfill, Tscharnuter, and Völk 1985), two concerns of fundamental importance with regard to accuracy.

Kamiya (1977) developed a 2D FD code which was Lagrangian, that is, the numerical grid distorts in order to follow the fluid flow. All other 2D protostellar collapse codes are basically Eulerian, but most have a grid which roughly contracts along with the flow, giving them a mixed Eulerian-Lagrangian nature. A moving grid is necessary in order to maintain adequate spatial resolution in the increasingly dense protostellar core, while economizing on the total number of grid points.

3.7.3 Fully asymmetric (3D)

In 3D the addition of partial derivatives with respect to ϕ means that an implicit solution is even more difficult. No fully implicit 3D protostellar collapse code exists, though fully implicit 3D codes have been developed for many gas hydrodynamics problems. The 3D Eddington approximation code of Boss (1983, 1985, 1986) involves an implicit solution of the mean intensity equation and an explicit solution of the hydrodynamical equations, with the former being solved by an alternating directions implicit (ADI) method (see Roache 1976). When a fully implicit 3D code is written, it will probably involve either an ADI solution or an expansion in spherical harmonics (i.e., a generalization of Tscharnuter's technique). Różyczka et al. (1980) have developed a 3D code based on spherical harmonics, but it is fully explicit. Spherical harmonics are also used in the solution for the gravitational potential by Boss (1980a). The absence of a fully implicit 3D code means that the rigorous investigation of 3D protostellar evolution is largely limited to the dynamic collapse phase, purely for reasons of computational complexity.

Even an explicit 3D FD code requires a large amount of computer time, because of the increased demand for grid points $(N_r \times N_\theta \times N_\phi)$. Another method of solving the basic equations is considerably more economical to run, namely the *smoothed particles hydrodynamics* (SPH) method (Lucy 1977; Gingold and Monaghan 1977; Larson 1978; Miyama, Hayashi, and Narita 1984). In SPH, a number of particles (typically less than 2000) are used to represent the fluid; because the particles can move and interact while conserving momentum, SPH methods are able to globally conserve angular momentum (mass is conserved also, as long as no particles are lost). SPH clearly provides a Lagrangian description of flow.

A recent study has compared FD and SPH codes applied to the problem of the dynamic instability of a rapidly rotating polytrope (Durisen *et al.* 1986). The codes tested were the FD codes of Tohline (1980a) and Boss (1980a) and the SPH code of Gingold and Monaghan (1977). The basic conclusions with regard to numerical techniques were that SPH codes tend to be faster to write, require less computer time to run, have less intrinsic numerical dissipation, and are better suited for arbitrary deformations than FD codes. The latter advantage arises because SPH codes contain no coordinate grid (see section 3.1). The primary disadvantage of SPH codes lies in their need to smooth the discrete particles into a fluid continuum: this is no problem as long as the fluid density is fairly uniform, but in general SPH cannot be trusted when strongly varying densities are encountered. Wood (1981, 1982) has attempted to reformulate SPH in order to minimize this problem, perhaps unsuccessfully (see the comment in Durisen *et al.* 1986). While SPH codes thus have several clear advantages, FD codes appear to be necessary for treating problems with a great range of density, as occurs in protostellar collapse.

3.8 Accuracy of numerical solutions

A variety of means exist for testing numerical methods. Considering the possibilities for error at all the levels involved in obtaining an accurate numerical solution (i.e., the levels of physical effects and equations, numerical analysis, FORTRAN programming, and interpretation of results), any new 2D or 3D code must be proven accurate, rather than assumed to be accurate.

3.8.1 Conservation of mass and angular momentum

The hydrodynamical equations (section 3.2) were written in *conservation law* form. This means that if the continuity equation, for example, is integrated over the volume of the cloud, and the divergence theorem is applied, the equation relates the time derivative of the total mass to the flux of matter in or out of the volume. In the absence of such a boundary flux, the total mass is conserved (i.e., *globally* conserved). Because there are no source terms on the right hand side of the continuity equation, the same argument can be applied to any small volume in the cloud, implying that mass must be conserved *locally* as well as globally. The presence of the source terms in the momentum equations means that momentum is conserved globally, but not necessarily locally. Note that in the case of 2D, the source terms for the specific angular momentum disappear, meaning that in 2D angular momentum must be globally and locally conserved.

Writing the equations of motion in conservation law form allows FD schemes to be constructed that assure global conservation of mass and angular momentum. (SPH codes automatically conserve these quantities globally.) Local conservation of angular momentum in the absence of a physical mechanism for angular momentum transport is another concern when devising FD codes (Boss 1980a; Norman, Wilson, and Barton 1980). Future FD codes should employ higher order accuracy schemes, in order to minimize spurious numerical diffusion (Norman, Wilson, and Barton 1980; van Albada, van Leer, and Roberts 1982).

3.8.2 Comparison with analytical and semi-analytical solutions

The analytical solutions for the collapse of pressureless spheres (Hunter 1962) and spheroids (Lin, Mestel, and Shu 1965; see section 3.6) and the 1D equilibrium structure of isothermal spheres (Bonnor 1956; Ebert 1957) and 2D rotating polytropes (Bodenheimer and Ostriker 1973; Boss 1980d) provide valuable tests for any numerical code.

3.8.3 Consistency with lower dimension codes

In general, results obtained with a code of lower dimensionality will be more accurate, because they probably involve more grid points and more accurate numerical techniques. For example, radiative transfer in the Eddington approximation can be solved using a tridiagonal technique in 1D, but requires an iterative (ADI) solution in 2D or 3D, so the multidimensional results should be compared with the 1D results (Boss 1984a).

3.8.4 Consistency between different codes

Because of the many different numerical methods as well as different versions of the same basic techniques, it is important to try to achieve consistency when different numerical codes are applied to the same problem. If divergent results are obtained on a well-chosen test problem, it is certain that both codes cannot be correct (and quite possibly neither!) Bodenheimer and Tscharnuter (1979) formulated a test problem for 2D collapse (used by Boss 1980b). Boss and Bodenheimer (1979) defined a standard problem for 3D collapse, used by Gingold and Monaghan (1981, 1982; see also Bodenheimer and Boss 1981; Boss 1984b).

3.8.5 Consistency with increasing spatial resolution

Because FD approximations exactly represent the continuum partial differential equations only in the limit of an infinite number of grid points, it is important to determine how the results depend on the number of grid points. A typical test is to double the number of grid points in each direction; this can be quite painful, however, in explicit 3D codes where the computational expense thereby increases by a factor of at least 16 (2^3 more grid points, and twice as many time steps).

4. RESULTS

The result of any given calculation of protostellar collapse can be expected to depend strongly on the initial conditions (see Buff, Gerola, and Stellingwerf 1979 for an extreme point of view). For isothermal collapse, the initial conditions may be specified by just two parameters (section 3.4). The parameter α_i has been varied between ≈ 1 (necessary for Jeans instability and hence significant collapse), and 0.01, which is so low that the thermal pressure is essentially negligible. An upper limit on β_i of 1/3 for uniform density and rotation clouds is obtained by requiring that self-gravity everywhere exceeds centrifugal acceleration. Interstellar clouds which have any detectable rotation at all tend to have $\beta \sim 1/3$ (e.g., Snell 1981; Harris et al. 1983), in part because more slowly rotating clouds often have Doppler shifts smaller than the minimum observational linewidths. A lower limit on β_i of $\approx 10^{-6}$ is determined by the minimum amount of rotation a dense interstellar cloud could have, that due to its revolution about the galactic center.

For nonisothermal evolution, two more parameters must be specified. The mass M_i has varied between $0.02 M_\odot$ and $2.0 M_\odot$ in 3D, and up to $60 M_\odot$ in 1D. The temperature T_i is usually set equal to $10 K$, which is appropriate for cold, dark interstellar clouds which either have not yet collapsed or have not yet undergone massive star formation.

4.1 Spherically symmetric (1D)

This sub-section sketches the results for the 1D collapse (with radiative transfer) of solar mass sized protostars (e.g., Bodenheimer 1968; Larson 1969, 1972b; Winkler 1978; Winkler and Newman 1980a,b; Stahler, Shu, and Taam 1980a,b, 1981; see Appenzeller and Tscharnuter 1974 for a treatment of a $60M_\odot$ protostar).

Hayashi's (1966) early semi-analytical studies on protostellar collapse assumed that the contraction would occur *homologously*, that is, the shape of the density distribution would remain fixed as the density increased. Subsequent numerical work by Bodenheimer (1968) and Larson (1969) showed that initially uniform density clouds become increasingly centrally condensed as the collapse proceeds and thermal pressure resists the infall of the outer layers, leading to density distributions of the form $\rho \propto r^{-2}$. A quasi-equilibrium core forms, surrounded by an infalling envelope of gas, which is a highly *nonhomologous* situation.

Because in 1D the dynamics of collapse consist simply of radially inward infall, the thermodynamical evolution is of most interest (Figure 1). The initial phases of contraction of diffuse interstellar clouds involve a slight decrease in temperature, because the increasing density shields the cloud from exterior radiation, while dust grains efficiently cool the cloud, resulting in formation of hydrogen molecules. Because this contraction takes place without an increase in temperature, thermal pressure is unable to resist the contraction, and dynamic collapse ensues. For the next several orders of magnitude increase in density, the cloud remains nearly isothermal at $\approx 10K$, again because the dust grains are able to absorb energy from the increasingly compressed gas and to re-radiate the energy in the infrared. As long as the cloud is transparent to the infrared radiation, no heating takes place, and collapse continues.

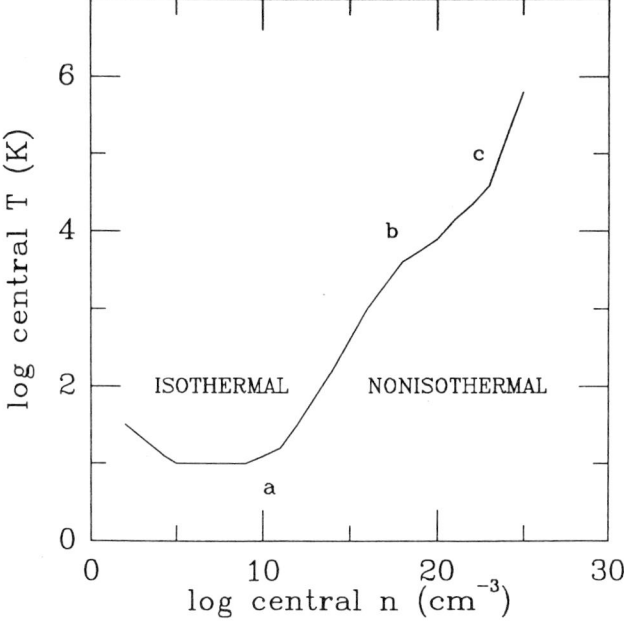

Figure 1. Schematic evolution of $T_c(\rho_c)$ during the 1D collapse of a solar type protostar. Critical events: (a) opaque to infrared radiation, (b) H_2 dissociation starts, (c) H_2 dissociation completed.

Starting at densities of about $10^{-14} g\ cm^{-3}$, the center of the protostar becomes opaque to infrared radiation, trapping the compressional energy, and initiating the heating of the protostar's central regions (nonisothermal regime). When the central region is completely opaque, it begins to evolve nearly adiabatically, and the increasing importance of thermal pressure finally stops the collapse at the center. A first core forms (at a central density $\rho_c \approx 10^{-12} - 10^{-11} g\ cm^{-3}$ and $T_c \approx 100K$). The quasi-equilibrium core is surrounded by an infalling envelope which remains nearly isothermal at $10K$. As the core accretes mass from the envelope, the envelope density distribution approaches the $\rho \propto r^{-3/2}$ form appropriate for accretion flows.

When the first core has gained mass and contracted to $T_c \approx 2000K, \rho_c \approx 10^{-7} g\ cm^{-3}$, the molecular hydrogen begins to dissociate into atomic hydrogen. Because this transformation is endothermic, the thermal pressure fails to increase as rapidly as before, and soon the central regions begin to collapse once again. This time the collapse proceeds until nearly stellar densities are obtained ($T_c \approx 10^4 - 10^5K, \rho_c \approx 10^{-1} g\ cm^{-3}$). The second, final protostellar core is supported by thermal pressure from H, H^+, e^-. The first core disappears along with its pressure support, and the final core is then surrounded by an accretion flow consisting of the remainder of the initial cloud.

Because the mass of the final core is less than one percent of the total cloud mass, the majority of the protostar mass must be accreted through the strong accretion shock that forms on the edge of quasi-equilibrium core. The large luminosity of the protostar arises primarily from this accretion shock, and once the outer envelope has become sufficiently depleted by accretion (after $\approx 10^6$ years), the protostar becomes visible as an infrared object and eventually as an optical source with temporally increasing effective temperature. The peak luminosity depends on α_i: when $\alpha_i \approx 1, L_{peak} > 10L_\odot$, while for $\alpha_i \approx 0.1, L_{peak} > 100L_\odot$. By the time that accretion has stopped, a solar mass protostar has a radius several times $R_\odot$ and a luminosity several times $L_\odot$. The subsequent evolution consists of homologous contraction along a convective track, as originally envisioned by Hayashi (1966).

4.2 Axially symmetric (2D)

The addition of a reasonable amount of angular momentum (based on observations of interstellar clouds) produces a qualitatively different collapse. When β_i is about 0.01 or greater, collapse from uniform density and rotation (see section 3.4) proceeds as in the 1D case, in that the central regions become considerably denser than the outer regions, but now the equi-density contours in the central regions become quite oblate. Flattening occurs because rotation is able to hinder collapse in the cloud, except in the direction parallel to the rotation axis. What happens next is sensitively dependent on the initial conditions (Figure 2). We begin with a discussion of the results for isothermal clouds, where most work has been done (Boss and Haber 1982).

For clouds with large values of α_i, β_i (e.g., $\alpha_i + 2\beta_i > 1$), significant collapse is prevented by the resistance of either thermal pressure or rotation (or both). In the limit of $\beta_i \to 0$, high α_i clouds collapse to form equilibrium Bonnor-Ebert spheres. For $\beta_i > 0$, the collapse forms the rotating analogue of the Bonnor-Ebert sphere, a *Bonnor-Ebert spheroid* (see also section 3.4). The standard 2D test case ($\alpha_i = 0.46, \beta_i = 0.32$) of Bodenheimer and Tscharnuter (1979) is basically consistent with this type of solution, as is the recent calculation by Narita et al. (1983). In the absence of further perturbations, this type of cloud will not collapse to form a protostar.

Clouds with lower α_i, β_i (e.g., $\alpha_i + 2\beta_i < 1$), do undergo significant collapse. Those with large β_i (greater than about 0.05) form *rings*, where the maximum density occurs well away from the rotational axis. The ring occurs as a result of the centrifugal rebound of the innermost regions away from the rotation axis, at the same time that the outer regions are still infalling, producing a density enhancement (see section 3.6). If this density enhancement is sufficiently large to become self-gravitating, it will grow and form a ring. Ring solutions were first found by Larson (1972a) and Black and Bodenheimer (1976). The theoretical attraction of the ring is that it is very likely to fragment into a multiple protostar.

Clouds with low α_i and very low β_i are likely to collapse without forming rings. Instead, they collapse through a succession of disk configurations, where the maximum density occurs on the rotational axis. The collapse continues through a sequence of disks at progressively higher densities. This behavior was first found by Norman, Wilson, and Barton (1980) and Boss and Haber (1982). The disk solution appears to be dependent on the isothermal approximation; once the collapse of a disk is halted by increasing temperatures, the disk may be unstable to the formation of rings (Narita, Hayashi, and Miyama 1984).

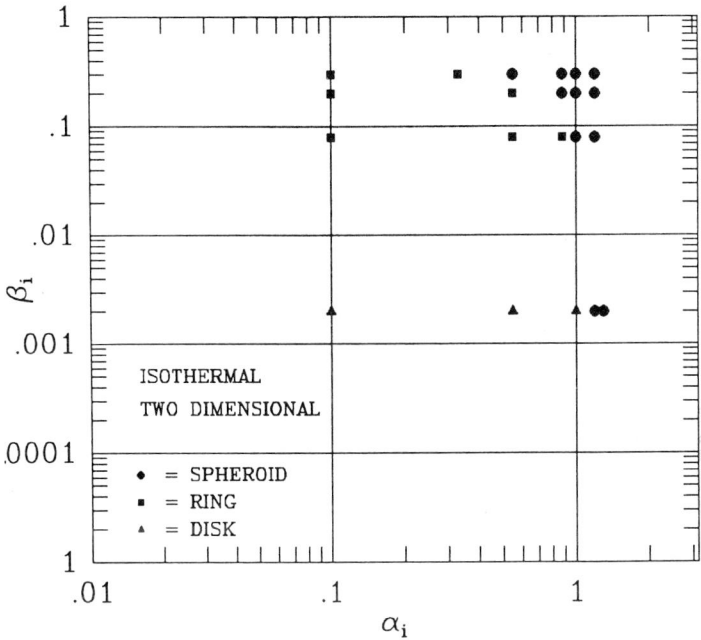

Figure 2. Results of 2D models of the collapse of rotating, isothermal clouds, as a function of α_i, β_i. Three outcomes were found: Bonnor-Ebert spheroids, rings, and collapsing disks. [Adapted from Boss and Haber (1982)].

While much controversy has surrounded the question of ring verus disk solutions (see the previous reviews), it appears that the controversy can be resolved by consideration of a number of factors, such as spurious angular momentum transport, different initial conditions (α_i, β_i), and the inability to push the disk calculations far enough to yield a ring (either in the central regions or the envelope.) Another key factor is that the standard problem of isothermal collapse from uniform density and rotation may well be singular (Norman, Wilson, and Barton 1980), in the sense that exactly uniform initial rotation may yield a collapsing disk in the innermost regions, while slightly non-uniform initial rotation produces rings. Considering that protostellar collapse is neither precisely isothermal, nor starts from precisely uniform density and rotation initial conditions, the ring versus disk controversy is evidently not of physical interest. Furthermore, 3D calculations have shown that in general fragmentation occurs without passing through a ring phase (next section).

Comparatively little work has been done on 2D nonisothermal collapse to high density. Boss (1984a) found that rings formed in $M_i \approx M_\odot$ nonisothermal clouds as long as $\beta_i > 10^{-3}$ for $\alpha_i = 0.5, T_i = 10\ K$. Lower β_i clouds formed either slightly flattened disks or nearly spherical protostars by the time the calculations were terminated by the formation of the first quasi-equilibrium core.

Tscharnuter (1978, 1981; see also Morfill, Tscharnuter, and Völk 1985) holds the record for reaching the highest densities and temperatures in a 2D nonisothermal calculation, having pushed models through to the formation of the second core. Tscharnuter's models contain substantial turbulent viscosity, which transfers angular momentum outward and mass inward, prevents the formation of a ring, and aids the cloud in collapsing to very high density, in spite of having appreciable rotation ($\beta_i \approx 10^{-2} - 10^{-4}$). Tscharnuter's models are halted by the formation of a strongly flattened central core and consequent loss of spatial resolution. At this point, the protostar is rotating so rapidly that it is subject to the rotational fission instability, which can only be studied in 3D (e.g., Durisen et al. 1986).

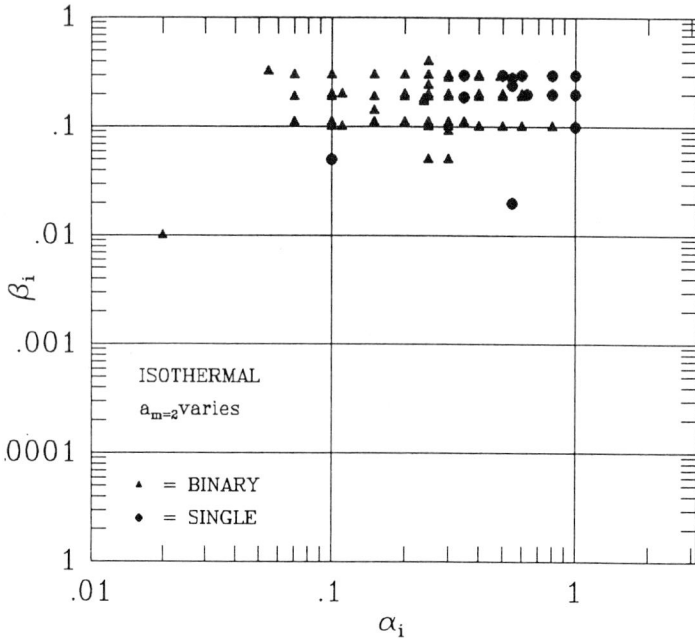

Figure 3. Results as a function of α_i, β_i for 3D isothermal collapse by all workers, classified as to whether or not fragmentation occurred. Various initial perturbations were added to the standard initial conditions.

4.3 Fully asymmetric (3D)

The production of rings in 2D collapse for high β_i implies that rapidly rotating protostellar clouds should be subject to fragmentation. This has been confirmed by 3D isothermal collapse calculations. When an initially uniform density and rotation cloud with $\beta_i > 0.1$ is given an initial sinusoidal density perturbation $(\delta\rho = a\cos(m\phi)\rho_i)$, if the cloud has low enough α_i to undergo significant collapse, it will fragment. Fragmentation generally occurs directly from growth of the initial perturbation, and not through an intermediate ring phase, unless the initial perturbation is very small, or special values of α_i, β_i are used (Bodenheimer, Tohline, and Black 1980; Boss 1981a). However, fragmentation for high α_i, β_i clouds occurs only after the formation of a highly flattened disk, well after one free fall time has elapsed. Thus this type of *rotational fragmentation* is quite different from that hypothesized by Hoyle and Hunter (section 3.6).

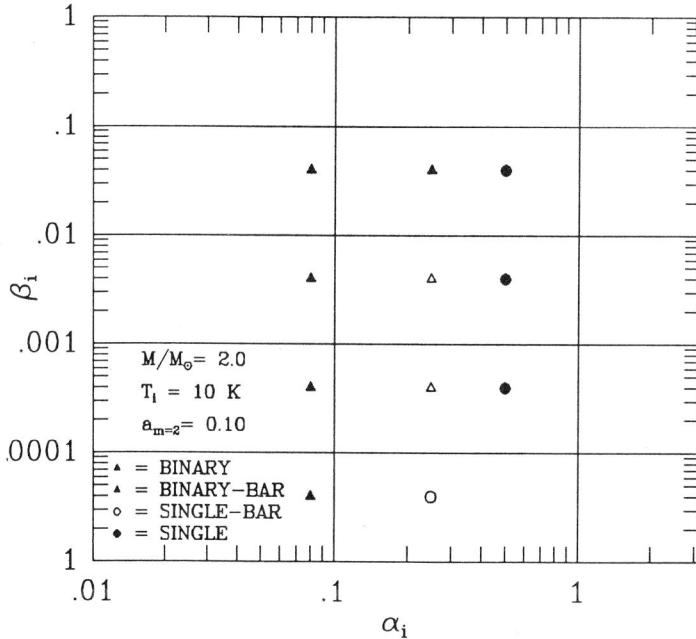

Figure 4. Results for 3D nonisothermal collapse as a function of α_i, β_i, for fixed M_i, T_i and perturbation. [Adapted from Boss (1986)].

Figure 3 shows the region of $\alpha_i - \beta_i$ space explored by 3D isothermal codes (Narita and Nakazawa 1977; Larson 1978; Boss and Bodenheimer 1979; Tohline 1980a; Boss 1980c; Bodenheimer, Tohline, and Black 1980; Różyczka, Tscharnuter, and Yorke 1980; Boss 1981b,c; Wood 1982; Gingold and Monaghan 1983; Miyama, Hayashi, and Narita 1984). Because of the restriction to isothermal thermodynamics, these studies were limited to relatively large values for α_i, β_i. Because of the limited spatial resolution possible in 3D, fragmentation can generally only involve a small number (≈ 4 or less) of fragments, and the figure lumps all fragmentation events into the binary category. In fact, binary fragmentation appears to be the preferred mode, independent of concerns about spatial resolution (Norman and Wilson 1978; Boss 1982a,b; Wood 1982). In spite of the great variety of numerical codes, the results are reasonably consistent:

analogous to the 2D Bonnor-Ebert spheroids, high α_i, β_i clouds form 3D *Bonnor-Ebert ellipsoids*, while low α_i, β_i clouds collapse and undergo binary or higher order fragmentation. The two exceptions are the single results at $\alpha_i = 0.1, \beta_i = 0.05$ by Różyczka, Tscharnuter, and Yorke 1980, and at $\alpha_i = 0.55, \beta_i = 0.02$ by Bodenheimer, Tohline, and Black 1980. Because the former model was only evolved to $1.0t_{ff}$ and the latter to $1.05t_{ff}$, these models may very well fragment when pushed farther in time, so the 3D results appear to be largely self-consistent.

The only nonisothermal models of 3D collapse are those by Boss (1983, 1985, 1986). A 3D radiative transfer code allows one to explore very low β_i, where the collapse proceeds to nonisothermal densities before non-spherically symmetric dynamics become important. Boss (1986) has performed a comprehensive survey of the available parameter space, varying α_i, β_i, M_i, and T_i (Figure 4).

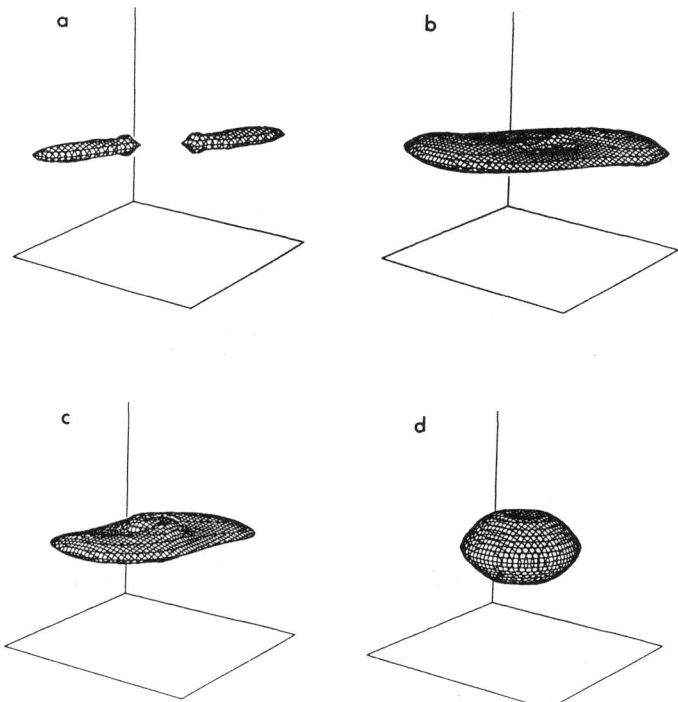

Figure 5. Surfaces of constant density for the four solutions found by Boss (1986). (a) binary, $M_i = 2.0M_\odot$, (b) binary-bar, $M_i = 0.25M_\odot$, (c) single-bar, $M_i = 0.10M_\odot$, and (d) single, $M_i = 0.020M_\odot$.

Four different types of solutions were found by Boss (1986), termed binary, binary-bar, single-bar, and single solutions, for obvious reasons (Figure 5). The four models in Figure 5 form a sequence with fixed $\alpha_i = 0.25, \beta_i = 0.04, T_i = 10 K$, but varied M_i. As the initial mass is decreased from 2.0 to 0.020 $M_\odot$, the sequence progresses from binary to binary-bar to single-bar to single, i.e., the tendency toward fragmentation into a binary system decreases. This can be understood in terms of the initial density, which increases as M_i decreases (at fixed α_i, β_i, T_i). As the density increases the initial dynamic collapse phase becomes shorter, because the density at which nonisothermal heating halts the collapse is reached more quickly. Thermal pressure is responsible for damping the growth of nonaxisymmetric structure as well as halting collapse. Dynamic fragmentation is thus smothered as the initial mass decreases. This type of fragmentation, occurring only when α_i is relatively small, is the closest realization yet achieved of Hoyle-Hunter fragmention.

Perhaps the most important result of the 3D calculations is the finding that the fragments that form in the isothermal regime (and to a more limited extent in the nonisothermal regime) have small values of α_f. This implies that the fragments will continue to collapse themselves, and fragment again and even again, leading to the possibility of *hierarchical fragmentation* (Hoyle 1953; Bodenheimer 1978; Boss 1980c). Furthermore, because the spin angular momentum of the cloud ends up primarily as the orbital angular momentum of its binary fragments, it is possible that successive stages of hierarchical fragmentation can lower the spin angular momentum enough to produce a pre-main-sequence protostar. Even in clouds with small β_i, the fragments acquire substantial angular momentum through the disordered motions of 3D collapse, implying that rotational effects must always become important for protostellar formation.

5. CONCLUSIONS: COMPARISON WITH OBSERVATIONS

Finding solutions to systems of coupled partial differential equations is principally an exercise in applied mathematics, and may seem (particularly to observational astronomers) often to be rather disjoint from astronomical reality. Of course the final worth of any theoretical effort in astrophysics can be assessed only by comparing the theoretical picture with observations.

While the general theory of star formation is not yet sufficiently advanced to permit a detailed comparison with observations (the inability to produce a 2D or 3D pre-main-sequence star is one glaring deficiency), the progress made so far allows an at least superficial comparison to be made.

The 1D models of protostellar formation are able to predict the location of optically observable subsolar mass protostars in the Hertzsprung-Russell diagram with a reasonable degree of accuracy (e.g., Stahler 1983b). The remaining discrepancies (e.g., Beichman *et al.* 1986) are probably attributable to effects such as the neglect of rotation (Mercer-Smith, Cameron, and Epstein 1984) and the termination of accretion in multidimensional flows (Shu 1986). The final resolution of these discrepancies will probably require evolving 2D and 3D protostars all the way to the pre-main sequence phase.

The multidimensional models imply that hierarchical fragmentation could take place, through rotational fragmentation in initially high thermal energy and high rotational energy clouds, and through Hoyle-Hunter fragmentation in initially low thermal energy clouds or fragments. A small cluster of protostars is likely to form from the collapse of rotating, dense interstellar clouds. Initially high thermal energy and low rotational energy clouds may collapse to form isolated single protostars.

Scalo (1985) has reviewed hierarchical structure in the interstellar medium, and finds evidence for 4 or 5 stages, though on very large scales, where effects not included in the 3D calculations discussed in this tutorial may be involved in fragmentation, such as magnetic fields and strong stellar winds. However most of Scalo's findings, such as a small number of fragments per stage, containing a small fraction of the mass at that stage, and a preponderance of prolate structures, are quite consistent with the numerical work. In addition, Schloerb and Snell (1984) have found evidence for 3 levels of a fragmentation hierarchy in Heiles Cloud 2, as well as a large scale ring reminiscent of the 2D models.

On a smaller scale, one or two stages of hierachical fragmentation appear to be necessary in order to form the plethora of binary stars and multiple systems consisting of pairs of binaries (Abt 1983; Zinnecker 1984). The finding that double-lined spectroscopic (Lucy and Ricco 1979) and solar-type contact binaries (van't Veer 1981) were formed with nearly equal mass components is consistent with the dominance of the binary fragmentation mode in the numerical work (higher order modes would produce unequal mass binaries).

Rotational fragmentation implies some alignment of physically observable quantities with the initial angular momentum vector. Observations support a certain degree of alignment. Weis (1974) found a tendency for the spin angular momentum of component stars to be aligned with the orbital angular momentum in binary systems with $10AU - 100AU$ separations. Fekel (1981) found that up to 2/3 of a sample of 20 *pairs* of visual binaries could be co-planar (some fraction of which must be associated by chance rather than birth), while Norris *et al.* (1982) and Hughes and Baines (1985) showed that strings of masers and other signs of protostellar activity are aligned in W51 and the Monoceros R2 molecular cloud, respectively. Guthrie (1982, 1984) showed that the B stars in the galactic cluster NGC 2516 tend to have their axes of rotation aligned, contrary to the situation for field stars.

Finally, the 3D nonisothermal calculations allow a rigorous estimate of the minimum protostellar mass to be made, assuming that the smallest mass stars are formed through hierarchical fragmentation (Boss 1986). The lower bound on the minimum protostellar mass of $\approx 0.01 M_\odot$ for Population I is determined by the smallest mass cloud which can fragment before rising thermal pressure prevents further fragmentation (see the discussion of Figure 5). Previous estimates of $0.003 M_\odot - 0.007 M_\odot$ were based on simple Jeans mass arguments (Low and Lynden-Bell 1976; Rees 1976; Silk 1977). The new lower bound of $\approx 0.01 M_\odot$ is consistent with recent observational evidence for *brown dwarf* stars with $M < 0.08 M_\odot$ (Kafatos, Harrington, and Maran 1986).

ACKNOWLEDGEMENTS

This research was directly supported by the National Science Foundation grants AST-8315645 and AST-8515644, and indirectly by the Innovative Research Program of the National Aeronautics and Space Administration, under grant NAGW-398.

REFERENCES

Abt, H. A. 1983, *Ann. Rev. Astr. Ap.*, **21**, 343.
Appenzeller, I. 1980, in *Star Formation*, eds. A. Maeder and L. Martinet (Sauverny: Geneva Observatory), p. 1.
Appenzeller, I., and Tscharnuter, W. 1974, *Astr. Ap.*, **30**, 423.
Bastien, P. 1983, *Astr. Ap.*, **119**, 109.
Bastien, P., and Mitalas, R. 1979, *M.N.R.A.S.*, **186**, 755.
Baud, B., *et al.* 1984, *Ap. J. (Letters)*, **278**, L53.
Beichman, C. A., *et al.* 1984, *Ap. J. (Letters)*, **278**, L45.
Beichman, C. A., *et al.* 1986, *Ap. J.*, in press.
Black, D. C., and Bodenheimer, P. 1975, *Ap. J.*, **199**, 619.
—— 1976, *Ap. J.*, **206**, 138.
Bodenheimer, P. 1968, *Ap. J.*, **153**, 483.
—— 1978, *Ap. J.*, **224**, 488.
—— 1981, in *IAU Symposium 93, Fundamental Problems in the Theory of Stellar Evolution*,

ed. D. Sugimoto, D. Q. Lamb, and D. N. Schramm (Dordrecht: Reidel), p. 5.
—— 1983, *Lectures Appl. Math.*, **20**, 141.
Bodenheimer, P., and Black, D. C. 1978, in *Protostars and Planets*, ed. T. Gehrels (Tucson: Univerity of Arizona), p. 288.
Bodenheimer, P., and Boss, A. P. 1981, *M.N.R.A.S.*, **197**, 477.
Bodenheimer, P., and Ostriker, J. P. 1973, *Ap. J.*, **180**, 159.
Bodenheimer, P., and Sweigart, A. 1968, *Ap. J.*, **152**, 515.
Bodenheimer, P., Tohline, J. E., and Black, D. C. 1980, *Ap. J.*, **242**, 209.
Bodenheimer, P., and Tscharnuter, W. 1979, *Astr. Ap.*, **74**, 288.
Bonnor, W. B. 1956, *M.N.R.A.S.*, **116**, 351.
Boss, A.P. 1980a, *Ap. J.*, **236**, 619.
—— 1980b, *Ap. J.*, **237**, 563.
—— 1980c, *Ap. J.*, **237**, 866.
—— 1980d, *Ap. J.*, **242**, 699.
—— 1981a, *Ap. J.*, **244**, 40.
—— 1981b, *Ap. J.*, **246**, 866.
—— 1981c, *Ap. J.*, **250**, 636.
—— 1982a, *Icarus*, **51**, 623.
—— 1982b, *Ap. J.*, **259**, 159.
—— 1983, *Icarus*, **55**, 181.
—— 1984a, *Ap. J.*, **277**, 768.
—— 1984b, *M.N.R.A.S.*, **209**, 543.
—— 1985, *Icarus*, **61**, 3.
—— 1986, *Ap. J. Suppl.*, in press.
Boss, A. P., and Black, D. C. 1982, *Ap. J.*, **258**, 270.
Boss, A. P., and Bodenheimer, P. 1979, *Ap. J.*, **234**, 289.
Boss, A. P., and Haber, J. G. 1982, *Ap. J.*, **255**, 240.
Buff, J., Gerola, H., and Stellingwerf, R. F. 1979, *Ap. J.*, **230**, 839.
Cameron, A. G. W. 1970, *EOS*, **51**, 628.
Chandrasekar, S. 1961, *Hydrodynamic and Hydromagnetic Stability* (Oxford: Clarendon), p. 273.
Cheng, A. F. 1978, *Ap. J.*, **221**, 320.
Cook, T. L., and Harlow, F. H. 1978, *Ap. J.*, **225**, 1005.
Crutcher, R. M., Troland, T. H., and Heiles, C. 1981, *Ap. J.*, **249**, 134.
D'Antona, F., and Mazzitelli, I. 1985, *Ap. J.*, **296**, 502.
Deissler, R. G. 1976, *Ap. J.*, **209**, 190.
Disney, M. J. 1976, *M.N.R.A.S.*, **175**, 323.
Durisen, R. H., Gingold, R. A., Tohline, J. E., and Boss, A. P. 1986, *Ap. J.*, **305**, 281.
Ebert, R. 1957, *Zs. Ap.*, **42**, 263.
Fekel, F. 1981, *Ap. J.*, **246**, 879.
Fricke, K. J., Möllenhoff, C., and Tscharnuter, W. M. 1976, *Astr. Ap.*, **47**, 407.
Gaustad, J. E. 1963, *Ap. J.*, **138**, 1050.
Gingold, R. A., and Monaghan, J. J. 1977, *M.N.R.A.S.*, **181**, 375.
—— 1981, *M.N.R.A.S.*, **197**, 461.
—— 1982, *M.N.R.A.S.*, **199**, 115.
—— 1983, *M.N.R.A.S.*, **204**, 715.
Guthrie, B. N. G. 1982, *M.N.R.A.S.*, **198**, 795.
—— 1984, *M.N.R.A.S.*, **210**, 159.
Hachisu, I., and Eriguchi, Y. 1984, *Astr. Ap.*, **140**, 259.
—— 1985, *Astr. Ap.*, **143**, 355.

Harris, A., Townes, C. H., Matsakis, D. N., and Palmer, P. 1983, *Ap. J. (Letters)*, **265**, L63.

Hayashi, C. 1966, *Ann. Rev. Astr. Ap.*, **4**, 171.

Hayashi, C., Narita, S., and Miyama, S. M. 1982, *Prog. Theor. Phys.*, **68**, 1949.

Hoyle, F. 1953, *Ap. J.*, **118**, 513.

Hughes, V. A., and Baines, J. G. N. 1985, *Ap. J.*, **289**, 238.

Hunter, C. 1962, *Ap. J.*, **136**, 594.

———— 1964, *Ap. J.*, **139**, 570.

———— 1977, *Ap. J.*, **218**, 834.

Hunter, J. H. Jr. 1969, *M.N.R.A.S.*, **142**, 473.

Jeans, J. H. 1929, *Astronomy and Cosmogony* (Cambridge: Cambridge University Press).

Kafatos, M., Harrington, R. S., and Maran S. P. (eds.) 1986, *Astrophysics of Brown Dwarfs* (Cambridge: Cambridge University Press), in press.

Kamiya, Y. 1977, *Prog. Theor. Phys.*, **58**, 802.

Klein, R. I., Whitaker, R. W., and Sandford, M. T. II 1985, in *Protostars and Planets II*, eds. D. C. Black and M. S. Matthews (Tucson: University of Arizona), p. 340.

Kondo, M. 1978, *Moon Planets*, **19**, 245.

Larson, R. B. 1969, *M.N.R.A.S.*, **145**, 271.

———— 1972a, *M.N.R.A.S.*, **156**, 437.

———— 1972b, *M.N.R.A.S.*, **157**, 121.

———— 1978, *M.N.R.A.S.*, **184**, 69.

Lin, C. C., Mestel, L., and Shu, F. H. 1965, *Ap. J.*, **142**, 1431.

Lynden-Bell, D. 1962, *Proc. Cambridge Phil. Soc.*, **58**, 709.

———— 1964, *Ap. J.*, **139**, 1195.

Low, C., and Lynden-Bell, D. 1976, *M.N.R.A.S.*, **176**, 367.

Lucy, L. B. 1977, *Astron. J.*, **82**, 1013.

Lucy, L. B., and Ricco, E. 1979, *Astron. J.*, **84**, 401.

McNally, D. 1964, *Ap. J.*, **140**, 1088.

McNally, D., and Settle, J. J. 1980, *M.N.R.A.S.*, **192**, 917.

Mercer-Smith, J. A., Cameron, A. G. W., and Epstein, R. I. 1984, *Ap. J.*, **287**, 445.

Mihalas, D. 1970, *Stellar Atmospheres* (San Francisco: Freeman).

Miyama, S. M., Hayashi, C., and Narita, S. 1984, *Ap. J.*, **279**, 621.

Morfill, G. E., Röser, S., Tscharnuter, W., and Völk, H. J. 1978, *Moon Planets*, **19**, 211.

Morfill, G. E., Tscharnuter, W., and Völk, H. J. 1985, in *Protostars and Planets II*, eds. D. C. Black and M. S. Matthews (Tucson: University of Arizona), p.493.

Mouschovias, T. Ch. 1977, *Ap. J.*, **211**, 147.

———— 1981, in *IAU Symposium 93, Fundamental Problems in the Theory of Stellar Evolution*, ed. D. Sugimoto, D. Q. Lamb, and D. N. Schramm (Dordrecht: Reidel), p. 27

Nakano, T. 1984, *Fund. Cosmic Phys.*, **9**, 139.

Nakano, T., and Umebayashi, T. 1986, *M.N.R.A.S.*, **218**, 663.

Nakazawa, K., Hayashi, C., and Takahara, M. 1976, *Prog. Theor. Phys.*, **56**, 515.

Narita, S., Hayashi, C., and Miyama, S. M. 1984, *Prog. Theor. Phys.*, **72**, 1118.

Narita, S., McNally, D., Pearce, G. L., and Sorensen, S. A. 1983, **203**, 491.

Narita, S., and Nakazawa, K. 1977, *Progr. Theor. Phys.*, **59**, 1018.

Norman, M. L., and Wilson, J. R. 1978, *Ap. J.*, **224**, 497.

Norman, M. L., Wilson, J. R., and Barton, R. T. 1980, *Ap. J.*, **239**, 968.

Norris, R. P., *et al.* 1982, *M.N.R.A.S.*, **201**, 191.

Penston, M. V. 1969, *M.N.R.A.S.*, **144**, 425.

Qian, B.-C., and Wang, R.-Y. 1981, *Ann. Shanghai Obs. Academia Sinica*, **3**, 165.

Rees, M.J. 1976, *M.N.R.A.S.*, **176**, 483.

Regev, O., and Shaviv, G. 1981, *Ap. J.*, **245**, 934.

Roache, P. J. 1976, *Computational Fluid Dynamics* (Albuquerque: Hermosa), p. 81.
Różyczka, M. 1983, *Astron. Ap.*, **125**, 45.
Różyczka, M., Tscharnuter, W. M., and Yorke, H. W. 1980, *Astron. Ap.*, **81**, 347.
Różyczka, M., Tscharnuter, W. M., Winkler, K.-H., and Yorke, H. W. 1980, *Astron. Ap.*, **83**, 118.
Sandford, M. T. II, Whitaker, R. W., and Klein, R. I. 1982, *Ap. J.*, **260**, 183
Scalo, J. M. 1985, in *Protostars and Planets II*, eds. D. C. Black and M. S. Matthews (Tucson: University of Arizona), p. 201.
Schloerb, F. P., and Snell, R. L. 1984, *Ap. J.*, **283**, 129.
Schmitz, F. 1984, *Astr. Ap.*, **131**, 309.
Shu, F. H. 1977, *Ap. J.*, **214**, 488.
—— 1986, in *Radiation Hydrodynamics in Stars and Compact Objects*, (Copenhagen, in press).
Silk, J. 1977, *Ap. J.*, **214**, 152.
—— 1982, *Ap. J.*, **256**, 514.
Snell, R. L. 1981, *Ap. J. Suppl.*, **45**, 121.
Stahler, S. W. 1983a, *Ap. J.*, **268**, 165.
—— 1983b, *Ap. J.*, **274**, 822.
Stahler, S. W., Shu, F. H., and Taam, R. E. 1980a, *Ap. J.*, **241**, 637.
—— 1980b, *Ap. J.*, **242**, 226.
—— 1981, *Ap. J.*, **248**, 727.
Takahara, M., Nakazawa, K., Narita, S., and Hayashi, C. 1977, *Prog. Theor. Phys.*, **58**, 536.
Terebey, S., Shu, F. H., and Cassen, P. 1984, *Ap. J.*, **286**, 529.
Tohline, J. E. 1980a, *Ap. J.*, **235**, 866.
—— 1980b, *Ap. J.*, **236**, 160.
—— 1980c, *Ap. J.*, **239**, 417.
—— 1982, *Fund. Cosmic Phys.*, **8**, 1.
—— 1985a, *Icarus*, **61**, 10.
—— 1985b, *Ap. J.*, **292**, 181.
Toomre, A. 1982, *Ap. J.*, **259**, 535.
Tscharnuter, W. M. 1975, *Astr. Ap.*, **39**, 207.
—— 1977, *Astr. Ap.*, **57**, 279.
—— 1978, *Moon Planets*, **19**, 229.
—— 1980, *Space Sci. Rev.*, **27**, 235.
—— 1981, in *IAU Symposium 93, Fundamental Problems in the Theory of Stellar Evolution*, ed. D. Sugimoto, D. Q. Lamb, and D. N. Schramm (Dordrecht: Reidel), p. 29.
Tscharnuter, W. M., and Winkler, K.-H. 1979, *Computer Phys. Comm.*, **18**, 171.
Vanajakshi, T. C., and Jenkins, A. W., Jr. 1985, *Ap. J.*, **294**, 502.
van Albada, G. D., van Leer, B., and Roberts, W. W. Jr. 1982, **108**, 76.
van't Veer, F. 1981, *Astr. Ap.*, **98**, 213.
Villere, K. R., and Black, D. C. 1980, *Ap. J.*, **236**, 192.
—— 1982, *Ap. J.*, **252**, 524.
Virgopia, N., and Ferraioli, F. 1981a, *Ap. Space Sci.*, **78**, 211.
—— 1981b, *Ap. Space Sci.*, **79**, 129.
Wang, R.-Y., and Qian, B.-C. 1981, *Ann. Shanghai Obs. Academia Sinica*, **3**, 170.
Weber, S. V. 1976, *Ap. J.*, **208**, 113.
Weis, E. W. 1974, *Ap. J.*, **190**, 331.
Winkler, K.-H. 1978, *Moon Planets*, **19**, 237.
Winkler, K.-H., and Newman, M. J. 1980a, *Ap. J.*, **236**, 201.
—— 1980b, *Ap. J.*, **238**, 311.
Wood, D. 1981, *M.N.R.A.S.*, **194**, 201.

—— 1982, *M.N.R.A.S.*, **199**, 331.

Yorke, H. W. 1981, in *Proceedings of the ESO Conference on "Scientific Importance of High Angular Resolution at Infrared and Optical Wavelengths"*, ed. M. H. Ulrich and K. Kjar, p. 319.

Zinnecker, H. 1984, *Ap. Space Sci.*, **99**, 41.

John Scalo

Theoretical Approaches to Interstellar Turbulence

John M. Scalo
Department of Astronomy
University of Texas

To put it more succinctly, junk your old equations and look for guidance in clouds' repeating patterns.

> \- Cvitanovic (1984, p. 4)

ABSTRACT

The properties of the interstellar medium strongly suggest that it is turbulent, in the generalized sense of nonlinear systems which exhibit unpredictable temporal behavior accompanied by self-organizing spatial fluctuations covering a wide range of size scales. This paper reviews a number of theoretical approaches to the problem. These include phenomenological models designed to explain the observed scaling relations and virialization constraints, direct numerical solutions of the hydrodynamical equations, applications of statistical turbulence closure methods, and solutions of local model equations which suggest limit cycle and chaotic behavior.

1. Introduction: The Nature of Turbulence and Its Relation to Star Formation

The word "turbulence" is difficult to define, let alone explain. Historically, and in the minds of most astronomers and physicists, turbulence refers to the tendency of incompressible fluids to undergo a transition from smooth, laminar flow to a state characterized by unpredictable fluctuations in the velocity and pressure existing within a nested vortical structure of "whirls within whirls". Turbulent flow is the norm in terrestrial environments, notably the atmosphere and ocean, and we are all experientially aware of it through encounters with local wind gusts and lulls, clear air turbulence during airflights, and the flow of our own blood. In these cases the transition to turbulence occurs when the value of a dimensionless parameter, called the Reynolds number, exceeds a critical value. Denoting characteristic length and mean velocity scales for a given system by l and u, the Reynolds number is defined as

$$Re \equiv ul/\nu \qquad (1)$$

where ν is the kinematic viscosity (dimensions cm^2 s^{-1}). The length scale is usually associated with the velocity gradient $l \sim [(du/(udx)]^{-1}$. Experimentally, turbulence results

D. J. Hollenbach and H. A. Thronson, Jr. (eds.), Interstellar Processes, 349–392.

when Re exceeds ~ 10^2-10^4, depending on the geometry of the flow. Physically, the transition to turbulence occurs when the nonlinear advective term in the momentum equation, u · ∇u, can sufficiently overpower the dissipation due to viscosity. However the transition has proven impossible to study numerically, except in a few special cases, because the number of mathematical operations required increases drastically with Re. The ultimate source of this difficulty is the extremely large range of size scales which are present and coupled, a problem which, as we shall see, also plagues all attempts to understand interstellar turbulence.

For the interstellar medium (ISM), a typical collisional cross section is roughly 10^{-15} cm^2, so the viscosity is

$$v \approx v_{th}\lambda = v_{th}/\sigma n \sim 10^6/n_3 \text{ cm}^2 \text{ s}^{-1}, \qquad (2)$$

where v_{th} is the thermal velocity (~ 10^4 cm s^{-1} for the cloud complexes of interest here), λ the mean free path, σ the collisional cross section, and n_3 the particle number density in units of 10^3 cm^{-3}. This gives

$$\text{Re} \sim 3 \times 10^7 u_5 l_{pc} n_3 \sim 10^8 l_{pc}^{-0.7} \qquad (3)$$

where u_5 is the fluid velocity in units of km s^{-1}, l_{pc} is the length scale in units of parsecs, and, in the last expression, the observed density-size and velocity dispersion-size relations (see Section 2 below) have been used: $n_3 \approx 4l_{pc}^{-1.2}$, $u_5 \approx l_{pc}^{0.5}$. So, even at the largest scales, Re is much greater for interstellar flows than the critical value found in laboratory (incompressible) flows. It was on the basis of such an argument that von Weizäcker (1951) proposed that the interstellar medium (and the universe in general) can be modelled as a turbulent fluid.

Unfortunately, very little is known about incompressible fluid turbulence; the nonlinearity and coupling of size scales makes the problem intractable both analytically and numerically. However, a key insight, due mainly to Kolmogorov (1941), is that incompressible 3-dimensional turbulence should exhibit a dissipationless cascade of energy from the largest scale, at which energy is supposed to be injected, down to the scale at which viscous dissipation becomes important. This latter scale is called the Kolmogorov microscale, l_k. When Re is large, l_k is much smaller than the overall flow scale, and so there exists a wide range of scales over which the energy cascade can occur; this dissipationless range is called the "inertial subrange" in the literature. The physical mechanism by which energy is transferred among the scales remains a largely unsolved problem, although it is generally believed that a dominant effect is the nonlinear stretching and twisting of vortex tubes. The value of the dissipation scale l_k can be derived from dimensional analysis (e.g., Tennekes and Lumley 1972) as

$$l_k = (v^3 l/u^3)^{1/4} \approx 10^{13}(l_{pc}/n_3^3 u_5^3)^{1/4} \approx 10^{12} l_{pc}^{-1} \text{ cm} \qquad (4)$$

where the observed interstellar scaling relations have again been used in the last expression. Thus l_k is very small compared to what can be observed, so one might expect a cascade down to the smallest obtainable instrumental resolution.

For the scales between the largest scale (call it L - it might be ~ 1 kpc in a galaxy) and l_k, Kolmogorov hypothesized that the power spectrum should only reflect the fact that the kinetic energy transfer rate $\varepsilon \approx u^2/(l/v) = u^3/l$ should be constant, giving

$$u \propto l^{1/3} \tag{5}$$

In terms of the energy spectrum, this relation is equivalent to $E(k) \propto k^{-5/3}$, where k is the wavenumber, the famous "Kolmogorov-Obhukov law." This phenomenologically-derived energy spectrum has been observed in several terrestrial atmospheric and oceanic flows, but departures have also been observed.

Because of the large values of Re in the ISM and the existence of an observed scaling relation only a little steeper than eq. 5, several papers have suggested that the ISM can be described in terms of a dissipationless cascade, modified in some way by compressibility (e.g., Arny 1971, Larson 1981, Fleck 1981, 1983, Ferrini *et al.* 1983), and perhaps energetically fed at the largest scales by differential galactic rotation (Fleck 1981; see also Jog and Ostriker 1986).

The assumption that the ISM can be viewed in terms of incompressible turbulence is highly questionable. First of all, the ISM exhibits *large* density fluctuations on all scales and the Mach numbers corresponding to the observed velocity fluctuations are highly supersonic (though possibly sub-Alfvenic), so compressibility is a major feature of the flow. Density fluctuations can interact with themselves and with velocity fluctuations in a nonlinear manner, considerably modifying the possible phenomena. Secondly, shock waves are expected to occur over a range of scales (although they may be modified or partially suppressed by magnetic fields), so there may be no clear separation between the dissipation scale and the other scales of interest. A model of interstellar turbulence as a field of interacting shocks was proposed by von Hoerner's group (von Hoerner 1962). Third, there is ample evidence that energy and momentum are injected into the ISM over a range of scales due to, for example, stellar H II regions, winds, and explosions, so the assumption of energy injection only at the largest scale is dubious. Finally, the behavior of the ISM is at least partially controlled by a long-range field, self-gravitation, a feature completely absent from any experimental or theoretical studies of incompressible turbulence. Since it is known from observational studies that self-gravity is at least comparable to all other forces in cold clouds of all scales, and because the magnitude of this force is sufficient to generate the observed density and velocity fluctuations, self-gravitation must be an essential ingredient in any theory of interstellar structure. A schematic illustration of the differences between incompressible turbulence and interstellar fragmentation leading to star formation is given in Figure 1.

The above paragraph suggests not that the idea of a turbulent ISM should be abandoned, but that our concept of turbulent phenomena must be enlarged. The appropriate direction for such a programme is indicated by the recognition, mostly within the last decade, that a wide variety of nonlinear dissipative systems besides fluids exhibit phenomena akin to fluid turbulence, namely unpredictable fluctuating temporal behavior rooted in extreme sensitivity to initial conditions ("chaos") and the ability to undergo self-organization into coherent spatial structures. For reviews and references from a variety of viewpoints, see Nicolas and Prigogine (1968), Lanford (1982), Eckmann (1981), Helleman (1980), Hakens (1983), Lichtenberg and Lieberman (1983), Cvitanovic (1984), and Wolffram (1983).

That the ISM should be regarded as a manifestation of this general behavior is strongly suggested by the following points:

1. The strong <u>nonlinearity</u> of the hydrodynamic equations governing the ISM and

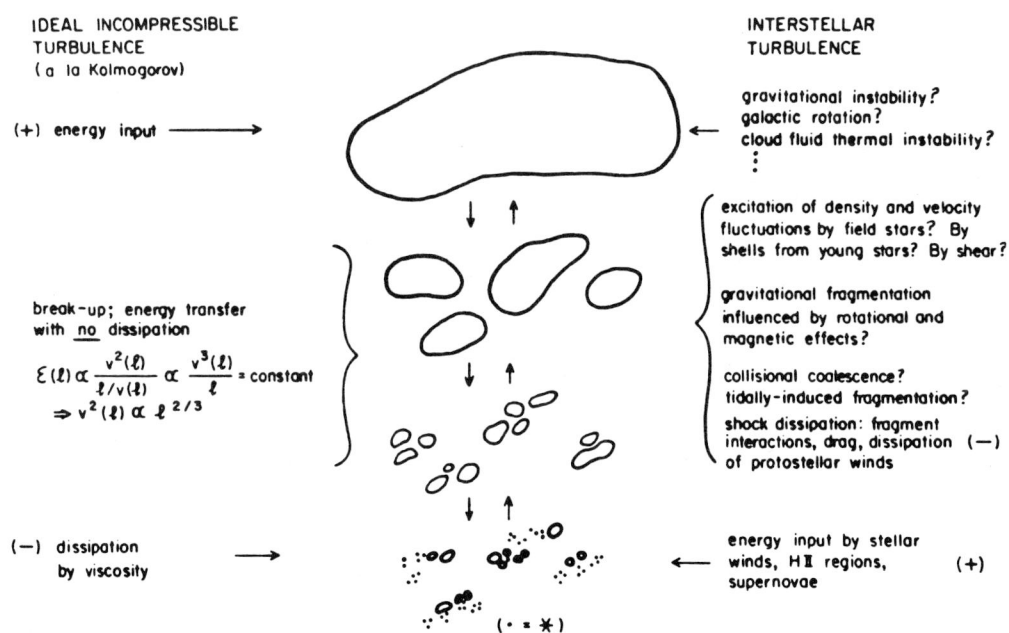

IDEAL INCOMPRESSIBLE
TURBULENCE
(a la Kolmogorov)

(+) energy input ⟶

break-up; energy transfer
with no dissipation

$$\mathcal{E}(\ell) \propto \frac{v^2(\ell)}{\ell/v(\ell)} \propto \frac{v^3(\ell)}{\ell} = \text{constant}$$
$$\Rightarrow v^2(\ell) \propto \ell^{2/3}$$

(−) dissipation
by viscosity   ⟶

INTERSTELLAR
TURBULENCE

gravitational instability?
galactic rotation?
cloud fluid thermal instability?
⋮

excitation of density and velocity
fluctuations by field stars? By
shells from young stars? By shear?

gravitational fragmentation
influenced by rotational and
magnetic effects?

collisional coalescence?
tidally-induced fragmentation?

shock dissipation: fragment
interactions, drag, dissipation (−)
of protostellar winds

energy input by stellar
winds, H II regions,   (+)
supernovae

(· = ✳)

Figure 1. Schematic illustration of the contrast between the Kolmogorov
dissipationless cascade model and the processes suspected of being important in
real interstellar turbulence. From Scalo (1985).

the nonlinear coupling between the relevant physical processes.

2. The observed prevalence of <u>fluctuations</u> in density and velocity covering a wide
range of spatial scales, a property shared with, for example, second order phase
transitions near a critical point.

3. The apparent structuring of star-forming gas complexes in a <u>hierarchical spatial
organization</u> (Scalo 1985).

4. The existence of power-law <u>scaling relations</u> between density and size, and
velocity dispersion (linewidth) and size, over several orders of magnitude in scale (see
section 2).

5. The likely importance of <u>feedback mechanisms</u>, for example, between small-
scale star formation and larger-scale structure through the action of energetic stellar
phenomena (e.g., Norman and Silk 1980, Silk 1985, Franco 1983, Franco and Cox
1983, McCray and Kafatos 1986). The coupling of local and global behavior is
characteristic of synergetic systems in general and turbulence in particular.

6. Observational evidence for possible <u>global stochastic behavior</u> in galaxies,
primarily from analyses of the star formation rates in dwarf irregular and blue compact

galaxies (see Seiden and Gerola 1982 for a review and an interpretation in terms of non-deterministic processes), and theoretical calculations indicating that at least some deterministic cloud models for the evolution of interstellar gas can undergo a transition to chaotic behavior (Struck-Marcell and Scalo 1986, see sec. 5.2).

7. Indications of <u>sensitivity to initial conditions</u> from numerical gravitational collapse calculations (Buff *et al.* 1979, Boss 1980).

Based on this list, it appears that if one had to bet on a system to exhibit self-organizing, multiscale, chaotic behavior, the ISM would be a profitable choice.

Perhaps the most important and interesting aspect of interstellar turbulence is its relation to star formation. Several papers have speculated about certain consequences of turbulence for the stellar initial mass function (e.g. Arny 1971, Fleck 1982), the time dependence of the star formation rate as a function of stellar mass (Henriksen and Turner 1984), and the formation of protostars at the scale below which turbulence becomes subsonic (Larson 1981), based on different phenomenological models. But the more fundamental problem connecting turbulence and star formation concerns the possibility of understanding the dependence of the rate of star formation on local and global physical conditions in galaxies. This question is crucial in all studies of galaxy evolution, and remains completely open. The coupling between a large range of scales in turbulent systems leads directly to the conclusion that star formation theory will have little predictive power, even in a statistically averaged sense, until the nature of interstellar turbulence is understood, no matter how well we understand the gravitational collapse of individual protostellar clouds.

The importance of multi-scale processes and structures for small scale star formation bears a remarkable resemblance to the problems encountered in terrestrial rainfall prediction. Theoretically, the local conditions required for the transition from the gas to liquid phase, as well as the kinetics of raindrop growth (in analogy with the evolution of protostars), are relatively well-understood. Yet the *rate* of rainfall at a given location cannot be reliably predicted (or even guessed, for timescales larger than a few days) because this rate depends on the coupled turbulent interactions between processes which operate from the scale of local wind fluctuations and thermal sources to global flows and climate. This analogy is, I feel, highly instructive, and suggests that future studies of interstellar turbulence should be oriented toward prediction of statistical properties of fluctuations and identification of the primary dimensionless parameters (analogous to the Reynolds number) that govern the nature of the flow, and must treat the multi-scale nature of the problem as fundamental, rather than concentrating solely on specific physical processes which operate over a narrow range of scales.

The opinions of workers in the area of the ISM and star formation concerning the nature, importance, implications, and even existence of interstellar turbulence are sharply and multiply divided. While an attempt will be made to present the flavor of this diversity, it must be recognized from the outset that these viewpoints mostly just reflect the tendency to conceptualize physical phenomena, a practice which leads to concentration on isolated aspects of the real problem. On the other hand, the very nature of nonlinear multiscale deterministic unpredictable systems does not yield easily to such dissection, but is rooted more in the couplings, interactions, and relationships between the various parts and between the parts and the whole. Worded more accurately, an understanding of turbulence will eventually necessitate a realization of the integration and vanishing of such dualities as parts and the whole, determinism and randomness, and, especially, order and chaos. Therefore, while the merits and deficiencies of certain proposed ideas will be discussed below, none of them should be either judged as correct or discarded. The aim is rather to understand how they may act in concert to yield the observed structure of the interstellar medium and the formation of stars. If this sounds like a peculiar attitude toward a scientific problem, one should consider carefully the fact that very little theoretical progress has been made in turbulence theory in nearly a century, even though turbulence, in the general sense, is a dominant feature of our direct experience.

The purpose of the present review is to briefly address the following more modest and restricted questions. 1. How can the observed scaling relations be understood phenomenologically, on the basis of various aspects of simple conceptual models? What are the major physical processes controlling these scaling relations and the observed structure of the ISM? An excess of proposed answers will emerge. 2. How can numerical solutions of the equations of hydrodynamics and the methods used in the study of the statistical equations of incompressible turbulence be used and generalized to investigate interstellar turbulence? 3. Can the rigorous equations which govern the ISM be boiled down to a set of simpler "model equations", based on conceptual models, which can be used to describe the temporal evolution of interstellar star-forming gas, and do these equations suggest the existence of chaotic behavior? If so, what are the implications for the eventual prediction of the star formation rates in galaxies?

The range of topics covered here is necessarily selective. In particular, the large and heterogeneous body of relevant observational results and detailed studies of fragmentation mechanisms are not reviewed, since the emphasis here is on theoretical approaches to the overall problem of interstellar turbulence, and these subjects were treated in a previous review (Scalo 1985). For a discussion of applications of the autocorrelation function to interstellar observational data, the reader is referred to the review by Dickman (1985), which also contains a useful general discussion of interstellar turbulence. Applications to turbulent structure in H II regions (see Roy and Joncas 1985, Roy *et al.* 1986, O'Dell 1986) and small scale fluctuations in the interstellar electron density (see Armstrong and Rickett 1981) are not discussed here; the present paper concentrates on the cool cloudy component of the ISM. For a more comprehensive look at the vast literature on incompressible fluid turbulence, the reader is referred to the monographs by, for example, Beran (1968), Tennekes and Lumley (1972), and Stanisic (1985) on statistical approaches, the reviews by Cantwell (1981) and Lumley (1981) on coherent structures in turbulent flows, and the papers in Dwoyer *et al.* (1985) which cover a broad range of topics and viewpoints.

2. Scaling Relations and Cloud Support

Larson (1981) was the first to point out that the internal densities and velocity dispersions (linewidths) of cool interstellar clouds appear to scale with size according to

$$n \propto l^p$$

$$\Delta v \propto l^q \qquad\qquad (6)$$

over a large range of cloud sizes. Larson found, using a heterogeneous compilation of data, $p \approx -1.1$, $q \approx 0.38$. That the value of q is similar to the exponent in the Kolmogorov law (equation 5) led to the suggestion that there may be a close connection with incompressible turbulence. (See Larson 1979 for work along these lines relevant to stellar groupings from star clusters to galaxies.) Larson also suggested that the fact that the interstellar value of q is somewhat larger than the Kolmogorov value might be attributed to shock dissipation. These $n(l)$ and $\Delta v(l)$ relations have been confirmed, with some modifications, by several independent studies focussing on different types of cloud structures (e.g. Leung *et al.* 1982, 16 dark globules using ^{13}CO; Myers 1983, 37 dark cloud cores using NH_3; Quiroga 1983, H I clouds; Sanders *et al.* 1985 and Dame *et al.* 1985, giant molecular clouds using ^{12}CO; Arquila and Goldsmith 1985, rotating dark clouds). All these studies indicate p = -1.1 to -1.4 (except Sanders *et al.* 1985 who obtain p = -0.75) and q = 0.3 to 0.6. Notice that even a cursory examination of large-scale photographs of the Milky Way or an examination of Lynds' (1962) dark cloud catalogue shows that p < -1, since p = -1 would imply constant column density, while it is obvious that smaller clouds are usually more opaque. In the present paper the following scaling relations are adopted as a convenient compromise: n = 4000 $l_{pc}^{-1.2}$ cm^{-3}, $\Delta v = l_{pc}^{0.5}$ km s^{-1}. It should also be noted that departures from these relations occur, especially at the smallest scales where the linewidths can be much larger than predicted by the scaling relation (e.g., Martin-Pintado *et al.* 1985).

Kleiner and Dickman (1985, 1986) have questioned the existence of any velocity dispersion-size scaling relation within the Taurus complex on the basis of the small and shallow velocity autocorrelation function for scales in the range 1-10 pc, with significant correlations occurring only at small scales ($\sim$ 0.1 pc) in the one region (Heiles' Cloud 2) studied at higher angular resolution. However, the disagreement may only reflect unavoidable preconceptions about what such a $\Delta v(l)$ relation *should* be measuring and about the nature of the turbulence itself. For example, in constructing the autocorrelation function for a finite image, it is necessary to filter out any large-scale features due to, say, rotation or contraction; but if these features are actually part of a hierarchy of structure extending to scales much larger than the observed region, then, from this point of view, the filtering is suppressing meaningful "turbulent" structure. Another problem is that a hierarchical scaling relation may only be apparent from the *joint* density-velocity correlation function if the "turbulent" structure actually conssits of a virialized hierarchy of nested high density-contrast clouds with a small volume filling factor. In that case, furthermore, the large disparity in size between the various structural components would require complete sampling over a factor of at least several hundred in spatial scale for an autocorrelation analysis to reveal a velocity scaling relation within a single complex. These points can be more easily visualized by considering the density structure. Complexes such as Taurus exhibit an obvious relation between density and size for their internal structures, but an autocorrelation analysis of a finite part of the complex with a limited dynamical range might, after filtering, reveal no correlations except at the scales corresponding to the sizes and separations of the dominant condensations in the map. Therefore, Kleiner and Dickman's conclusion that the type of scaling relation expected for incompressible space-filling turbulence does not exist in Taurus may be correct, but does not rule out the existence of a scaling relation for more general, highly compressible, turbulent hierarchies to which autocorrelation analyses are insensitive. To further confound the situation, it should be noted that the autocorrelation analysis of a complex in Perseus by Perault *et al.* (1985) *does* yield a velocity scaling relation. Further discussion is beyond the scope of this paper, and it will be assumed in what follows that the n(l) and $\Delta v(l)$ scaling relations hold not only for an ensemble of spatially unrelated clouds, but within individual complexes.

One more significant, and probably crucial, result found by Larson and confirmed by later work is that the kinetic energy associated with the linewidths approximately balances the gravitational energy: clouds covering a large range of sizes are apparently virialized:

$$\Delta v \approx (GM/l)^{1/2} \approx 0.02 \, n^{1/2} \, l_{pc} \text{ km s}^{-1}. \tag{7}$$

Something is "holding up" the collapse of clouds on all scales. The fact that clouds are supported against self-gravity is also suggested by several lines of evidence indicating that cloud lifetimes significantly exceed their free-fall times. (A review of the evidence is given in Scalo 1985). In addition, there is little evidence for dynamical collapse, at least on the larger scales. The process responsible for this virialization remains an open question; suggestions include fluid turbulence (in the incompressible sense), channeling of gravitational energy into macroscopic motions of fragments, stellar winds, and magnetic fields. Whatever the mechanism, the existence of virialization at all scales is almost certainly related to the form of the scaling relations.

The number of proposals to explain the $n(l)$ and $\Delta v(l)$ scaling relations has become numerous. On the mundane side, the $n(l)$ relation could simply be due to a selection effect in which only a small range of column density has been selected by the constraints imposed by the observing equipment. The $\Delta v(l)$ relation could be explained simply as a statistical effect due to the fact that variations in the depths of the different clouds along the line of sight would give rise to variations in the number of sub-fragments intercepted, and hence in the apparent velocity dispersion. On the more physical side, the value of $q \approx 0.5$ follows directly from virialization if $p \approx -1$, as observed; however, this is hardly an explanation, since it accounts for neither the value of p nor the mechanism of virialization. One would expect $p \approx -1$ if all clouds reflect compression in one dimension (disks), but prolate clouds seem fairly common, expecially at small scales (e.g. Schneider and Elmegreen 1979). However, most workers feel that the scaling relations reflect a deeper clue to the nature of interstellar turbulence, and so I will review a few of the most prominent suggestions. The possibility that a magnetically-dominated cloud model can account for the scaling relations is postponed to section 2.3 below.

2.1 Kolmogorov-type arguments

These proposals assume: (i) an "inertial range" of scales exists in which there is no dissipation and no energy injection; (ii) virialization occurs, although they do not specify the physical mechanism responsible; (iii) some physical variable is conserved over the range of scales of interest.

The first argument of this type, taken over directly from Kolmogorov, is an *energy cascade* to smaller scales, modified to include weak compressibility (Fleck 1983, Ferrini *et al.* 1983). If the volume rate of energy transfer

$$\varepsilon \approx n v^2/\tau \approx n v^3/l, \qquad (8)$$

where τ is a characteristic timescale taken to be l/v, is constant at all scales of the cascade, then one finds, if we take $n \propto l^p$

$$v \propto l^{(1-p)/3} \qquad (9)$$

The assumption of virialization gives

$$v \propto (M/l)^{1/2} \propto l^{(2+p)/2}. \qquad (10)$$

Equating these two expressions gives $p = -4/5$, $q = 3/5$. These values are close to those inferred from observation although, as mentioned above, p is probably smaller than -1.

Another argument (Henriksen and Turner 1984) involves an *angular momentum cascade* to *larger* scales by gravitational torques, whose importance has been emphasized by Larson (1984) and Boss (1984). In this view, the torque density

$$K = n \cdot lv \cdot (v/l) = nv^2 \qquad (11)$$

is constant on all scales. Combined with the virialization assumption, this gives p = -1, q = 0.5, again similar to the observed values. Henriksen and Turner (1984) also give a more elegant derivation of this result based on a renormalization group approach, as well as several interesting implications for star formation.

The major problems with these arguments are: (i) they do not explain why the cascade is virialized; (ii) they neglect the likely importance of energy dissipation and energy injection at intermediate scales.

2.2 Wind-driven shell fragmentation

There is abundant evidence for shell-like structures in the interstellar medium (see Scalo 1985 and Elmegreen, this volume), and several proposals have been offered to explain how such shells might arise and possibly trigger further star formation (e.g. Elmegreen and Lada 1977, Norman and Silk 1980, Elmegreen and Chiang 1982, McCray and Kafatos 1986). Particularly interesting and influential was the work of Norman and Silk who proposed that fragmentation of shells driven by T Tauri winds could lead to self-regulated star formation in molecular clouds. Some potential difficulties with the specifics of the model were discussed in Scalo (1985).

Silk (1985) has shown how such a model can be used to derive scaling relations similar to those observed. His assumptions are: 1. The pressure of the ambient medium eventually stops the motion of the swept-up shell. 2. The energy lost by the shell is radiated by the shocked layer; this brings in the density through its dependence on the cooling rate, which Silk took as $\Lambda \propto n^{0.6}$. 3. The swept-up shell fragments by means of gravitational instability; the energy criterion for the fastest-growing mode used by Silk is essentially the virial condition. Combining these gives p = -1, q = 1/2.

While Silk applied his argument only to molecular clouds containing T Tauri stars, it is true that in order to explain the observed scaling over a large range of sizes one must invoke shells from several types of sources. Luckily for this general model, there is no dearth of proposed shell sources beside T Tauri winds. These include H II regions and O star winds (see Bania and Lyon 1980 for a simulation of the expected structure), shells driven from OB associations by anisotropic radiation pressure (Elmegreen and Chiang 1982), and multiple supernova-driven supershells (McCray and Kafatos 1986). For any of these sources one might expect *roughly* the observed scaling relations, since one-dimensional compression in shells will give approximately $n \sim l^{-1}$ (except for the effects of cooling) and, assuming all these shells can gravitationally fragment, virialization will then give $\Delta v \propto l^{1/2}$. However, a detailed numerical modelling of shell fragmentation is not yet available (but see Roczyczka 1985), and it is also possible that magnetic instabilities play a role in shells (see Baierlein 1983).

One is therefore led to postulate an ensemble of expanding, fragmenting, shells of various sizes, analogous to the extragalactic bootstrapping model of Ostriker and Cowie (1981). Shell fragmentation models incorporate, in principle, gravitational forces, dissipation, and energy injections over a wide range of scales, features whose likely importance was emphasized above. Stochastic fluctuations in shell sizes and the nonlinear nature of shell interactions and gravitational fragmentation may also qualify the overall evolution and resulting structure as "turbulence" in the sense outlined in section 1.

However, it is difficult to imagine such a "bottom-up" scenario giving rise to the kind of hierarchical structure observed in interstellar cloud complexes. Although shells *are* observed at large scales (for example, H I supershells, the expanding ring of cloud complexes and young stellar groups associated with Lindblad's ring, and shells around OB associations; see Elmegreen, this volume, for references), the overall impression from large-scale high resolution column density maps of extinction-molecule complexes is one of a "top-down" hierarchy, with a tendency for roughly spherical or filamentary structure. Some illustrations for Taurus are given in Scalo 1985, and similar structure can be seen in the Sco-Oph complex, as shown in Figure 2 from Rossano (1978), the complex mapped

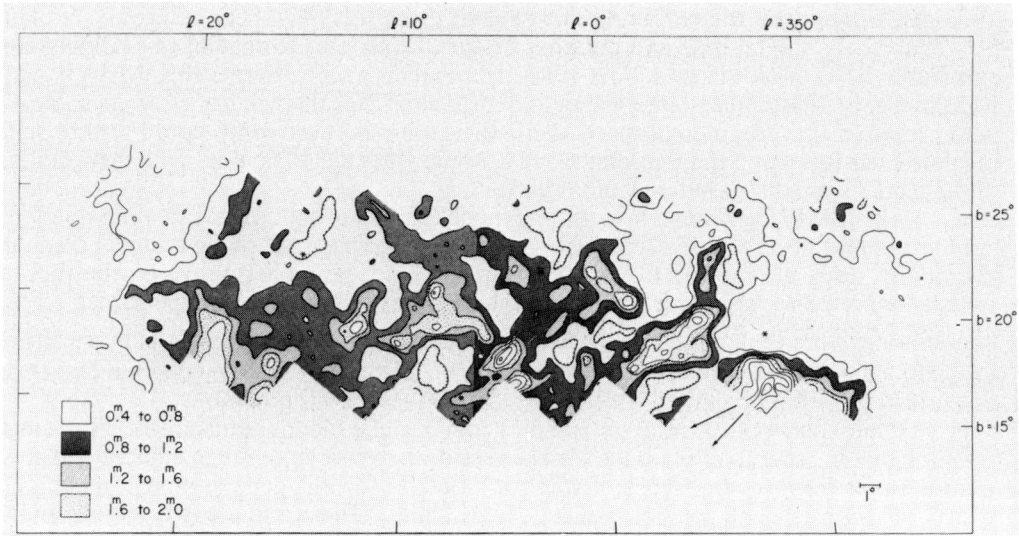

Figure 2. An extinction map of the Sco-Oph cloud complex, from Rossano (1978). The extent of the region mapped is roughly 150 pc. Contours are shown for extinctions of 0.4, 0.8, 1.2, 1.6, 2.0, 2.5, 3.0, 3.5, and 4.0 mag. Only the first four contour levels are shaded.

by Perault *et al.* (1985), and Orion (see the recent maps by Maddalena *et al.* 1986, Bally *et al.* 1986, and Schloerb *et al.* 1986), although in the case of Orion, where massive stars have formed, one can also see a few shell-like structures and holes (Schloerb *et al.* 1986). These maps suggest that wind-driven shells do not dominate the observed structure over the entire range of relevant scales. This view is reinforced by noting that the shell model, while it invokes the virial condition through the assumption that the shells fragment gravitationally, really gives no explanation as to why the observed structures on all scales appear to be supported against gravity rather than in a state of collapse.

Given these considerations, one might speculate that shells only give rise to the largest (~ 500- 1000 pc) observed structures, while a top-down process like hierarchical gravitational fragmentation within the shells controls the structure at all smaller scales. A possibly more realistic, and challenging, idea is that it is the coexistence and interaction of bottom-up shell propagation with top-down gravitational fragmentation which is responsible for interstellar turbulence. Such a model may be beyond current numerical capabilities and even eludes easy conceptualization; however, the same can be said of turbulence in general.

2.3. Energy inputs and cloud support

The necessity of some sort of energy input to maintain most cloud structures in near-virial equilibrium with self-gravity and extend their lifetimes beyond the free-fall timescale has received considerable attention in the literature, but no consensus has been reached. Only a brief list of the various proposals will be given here. Of particular importance is the fact that whatever processes are involved must operate over a large range of size scales.

2.3.1. Stellar sources

Stellar winds and explosions, as discussed above, provide an obvious candidate for cloud support, and Norman and Silk (1980) were successful in demonstrating how T Tauri winds could support moderately-massive clouds in a self-regulating manner. For an alternative protostellar wind model for driving turbulent motions, see Franco (1983, 1984). However, the doubts expressed earlier that winds can virialize clouds with a *large* range of scales, as well as the lack of observed protostellar objects in some clouds which appear virialized, argues against stellar winds as the cause of multi-scale virialization. Also, one must question whether the cloud gas will simply accumulate between the protostars in the early stage of their shell development. Lastly, the fact that the Southern Coalsack and the complex Khav 141 are highly fragmented, exhibit supersonic line widths, and show no evidence for collapse (the usual arguments for some form of supersonic turbulent support), yet contain no known protostellar sources capable of driving a wind suggests that turbulence can persist in cloud complexes without the presence of protostellar winds.

2.3.2. Collisions with small clouds

A slight generalization of a proposal by Bash *et al.* (1981) is that the energy to support clouds of a given size is supplied by random collisions with smaller clouds. These smaller clouds are supposed to interpenetrate the larger clouds, assuming the role of "turbulent eddies" whose "turbulent pressure" supports the larger clouds. The problems with this idea are: (i) the origin and assumed location of the small clouds exterior to the large clouds is unexplained, though, of course, possible if one begins with a complex filled with randomly positioned fragments covering a wide-range of sizes. However, even in this situation it is difficult to imagine the evolution of such a system into a hierarchical system, as explained elsewhere in connection with the Oort model for cloud evolution (Scalo 1985). (ii) Since, observationally, $n \propto l^p$ with $p < -1$, the smaller clouds have larger column densities, and should be capable of plowing through larger clouds without becoming lodged. (iii) The model, as formulated by Bash *et al.*, requires an extremely fine tuning, for any fluctuation in the flux of impinging small clouds lasting longer than the hydrodynamical timescale of a large cloud will increase or decrease the turbulent pressure within the large cloud, leading to its disruption or collapse.

2.3.3. Galactic rotation.

An appealing long-lasting source of turbulent energy is galactic rotation. Fleck (1980, 1981) suggested that differential galactic rotation gives rise to a shear instability in the disk gas that generates a turbulent cascade in which large scale structures are assumed to generate smaller-scale structures, in analogy with incompressible turbulence. This particular cascade picture is problematical because it offers no physical explanation of how the energy is supposed to be transferred to smaller scales (it is not at all clear that the "vorticity stretching" usually invoked for this purpose in discussions of incompressible fluid turbulence has any relevance to the compressible, supersonic, self-gravitating ISM), why virialization is observed over a range of scales, or how the smaller-scale structure survives shock dissipation. Fleck (1981) argues that the turbulence can survive dissipation and collapse for $\sim 10^{10}$ yr., but in fact the timescale he computes is the time over which the energy injection at the largest scale would slow the rotation of the entire Galaxy, demonstrating only that Galactic differential rotation is indeed a long-lived source of energy *at the largest scales*. But just how this energy is channelled to other scales and other forms (especially the motions and support of hierarchical interstellar cloud complexes), remains unknown.

A recent effort in this direction by Jog and Ostriker (1986) attempts to explain the unexpectedly (but see below) large velocity dispersion of "giant molecular cloud"-mass objects by using the energy from differential galactic rotation to drive "gravitational viscosity". Equating this "viscosity" with inelastic cloud collisional dissipation then allows a solution for the cloud velocity dispersion. Although this work overlooks the hierarchical, virialized nature of the ISM (for example, the large measured velocity dispersions which were the motivation for this work are probably artifacts if the structure is hierarchical, a point made independently by Elmegreen [this volume]), it demonstrates a physical mechanism by which galactic rotational energy can be channelled into the random motions of density fluctuations (clouds), motions which, with only a modest shift in viewpoint, can be considered akin to turbulence.

I conclude that differential Galactic rotation may very well be an important source of energy at the largest scales of any top-down model of interstellar turbulence, but probably has little to do with the nonlinear physics responsible for generating the chaotic-ordered fragmentation hierarchy which we observe to result in star formation.

2.3.4. Magnetic field support and line broadening.

The idea that interstellar clouds are supported by a magnetic field and that the source of supersonic line widths observed in these clouds is a spectrum of MHD waves is one that has persisted in spite of the extreme difficulties associated with obtaining empirical field strength measurements and with carrying out any theoretical calculations. Since the observations and theory are reviewed in detail by Heiles and Zweibel, respectively, elsewhere in this volume, I will limit myself to what I think are a few salient points and problems. Some further discussion of calculations by Falgarone and Puget (1986) is given in the next section.

(i) Despite the study of the problem by Zweibel and Josaffetson (1983), it is still not known with any confidence whether the damping time for MHD waves in interstellar clouds is large enough for survival over the estimated cloud lifetimes. One way around this potential problem is to invoke an energy source for the waves, in particular a coupling between protostellar winds and the large scale cloud field, as suggested by D. Gilden (private communication).

(ii) The linear MHD wave modes are expected to be incompressible, so the observed clumpiness of the ISM may still have to be attributed to gravity unless the nonlinear wave interactions can produce substantial density fluctuations. Numerical techniques are now available for an attack on the nonlinear aspects of this question, and will hopefully be pursued in the near future.

(iii) "Interstellar clouds" are actually observationally selected components of a hierarchical structure covering at least three orders of magnitude in spatial scale, and supersonic virialized linewidths occur over this entire range. If magnetic fields are important in the support and line broadening of interstellar clouds, it is essential to understand how a magnetic field can organize hierarchical structures over such a large range in scale. I am unaware of any published suggestions concerning this point. This is a major shortcoming of the general proposal that the evolution of star-forming interstellar structures is magnetically-dominated. Figure 4 (discussed in section 3 below) illustrates that a magnetic field *can* exhibit simultaneous self-organization and chaos, but such behavior requires that the field be coupled to *fluid* turbulence. (See Eilik and Henriksen 1984 for a discussion, in a different context, of how fluid turbulence might drive MHD waves.)

(iv) If magnetic fields support clouds and are responsible for the observed line widths, then it ought to be possible to account for the observed scaling relations. In a magnetic model we can identify the observed linewidth with the Alfven speed:

$$\Delta v = v_A = B/(4\pi\rho)^{1/2} \qquad\qquad (13)$$

where ρ is the mass density. Using the observed scaling relations for Δv and n adopted earlier, we obtain the relation between field strength and density which must be satisfied for consistency:

$$B \approx 50\ n_4^{-0.1}\ \mu G \qquad\qquad (14)$$

If we assume instead that clouds of all scales are in virial balance between gravity and magnetic forces, the virial theorem gives $B \propto nl$, and the adopted n(l) relation (notice that the Δv scaling is not needed) then gives $B \propto n^{-0.17}$, similar to the above result, and was pointed out independently by Myers (this volume) and Shu *et al.* (1986). (It should be noted that these two derivations are based on different physical assumptions, and only give the same result if p = -1 and q = 0.5.) Substituting the B(n) relation for this case into the expression for the Alfven speed gives a predicted linewidth-density scaling as $\Delta v \propto n^{-0.67}$. These results show that for the observed n(l) relation, clouds in magneto-gravitational equilibrium will also have linewidths approximately consistent with the observed $\Delta v(l)$ relation. Although neither result explains the observed n(l) scaling, the consistency is encouraging. In addition, one might expect a relation like n $\propto$ l^{-1} if gas predominantly accumulates along the field into a flattened configuration.

The field strength predicted by equation 14 is in agreement with the field strengths observed in Zeeman effect experiments in fairly low-density clouds covering a range of about two orders of magnitude in density (see the review by Heiles, this volume). Some authors (Myers, this volume, Shu *et al* . 1986) have intepreted this agreement as evidence that fragmentation and star formation are controlled by magnetic forces.

On the other hand, the truly surprising implication of equation 14 is that, if magnetic forces support clouds and cause the line broadening, the field strength must remain essentially constant over at least three or four orders of magnitude in spatial scale, the range over which the scaling relations (and the associated hierarchical structure) are reasonably well-established. If this is true, it suggests that clouds of all sizes, even if hierarchically nested, have experienced collapse along the field, attaining magneto-gravitational balance across the field and pressure-gravity balance along the field. If this were true, we would expect all clouds to be roughly disk-shaped, with the field lines parallel to the disk axis. This conclusion contradicts a large body of observational evidence. There is little evidence for flattened or disk-shaped clouds within interstellar complexes on any scale, except for the statistical study of Lynds clouds by Hopper and Disney (1974).

On the contrary, molecular clouds most often seem to be roughly globular or filamentary. For example, Schneider and Elmegreen (1979) argued that most of the elongated dark clouds that they catalogued are prolate, not rings or disks viewed edge-on, on the basis of the regular spacing of the subcondensations within these clouds. Other examples include the streamers in Ophiuchus, the strings of H II regions in Cep A (Hughes and Wouterloot 1984), and the H I structures mapped by Verschuur (1974a,b). Kleiner and Dickman (1985) suggest that the entire elongated central region of the Taurus complex is prolate based on the detection of a velocity gradient across the minor axis. Examples of more spherical-like structures are too numerous to list, but it seems highly unlikely that most of them are disks viewed face-on. Polarization maps of elongated structures are more difficult to interpret, but do not seem to support the picture of collapse along the field to a disk, and are often difficult to understand at all. For example, the Taurus filaments (almost certainly not edge-on disks) are elongated roughly perpendicular to the polarization vectors in the same area (Hsu 1984, Monetti *et al.* 1984), while the long axis of the much larger globular filament studied by McDavid (1984) does coincide with the field direction. Several additional examples, with various possible interpretations, can be found in Vrba (1977). While most of these and other studies of polarization structure do claim that the magnetic field is dynamically significant, the proposed interpretations seldom involve collapse of gas along the field into a disk.

Regardless of one's interpretation of these often ambiguous observational results, it does not seem possible that the magnetic field can maintain a roughly constant value, by means of flow along the field, within complexes whose internal cloud structures span sizes from over 100 pc down to 0.1 pc, especially when one considers that these structures do not often share a common orientation. The magnetic field may be important for the support and line broadening in some clouds, but it seems likely that such a dominance, if it occurs, is restricted to a specific range of scales.

(v) A more subtle, but perhaps more realistic model in which hydromagnetic effects stabilize clouds has been discussed in detail by Falgarone and Puget (1986). Briefly, in their model the motion of fragments ($l\sim$ few parsecs, n $\sim$ 200 cm^{-3}) orbiting in the gravitational field of a larger ($l \sim 100$ pc) cloud causes the field lines to tangle and distort, especially at the fragment boundaries. These distortions excite MHD waves which become trapped inside the fragments, eventually dissipating, primarily as hydromagnetic shocks. (Falgarone and Puget present a detailed calculation of the dissipation rates due to various processes.) However, these internal waves are found capable of feeding the internal fragment velocity dispersion for several x 10^7 yr. Furthermore, the fragments are stable against radius perturbations because the dependence of the internal velocity dispersion ($\propto v_A^2 \propto r^{-1}$) on radius is the same as that of self-gravity. Although magnetic effects are important in supporting the fragments, the motions of the fragments within their parent cloud is driven by extraction of kinetic energy from the gravitational field ("gravitational virialization" - see next section). It is therefore gravity that causes the motions which distort the field and eventually feed energy to smaller scales within the fragments. Notice that the magnetic field is not invoked to support clouds directly, through the Lorentz force, as in traditional treatments of magnetic effects, but is instead used as a conduit to transfer gravitationally-driven "turbulence" on one scale to another form of turbulence, consisting of interacting waves and shocks, on a smaller scale. If, after dissipation is complete, the fragments themselves fragment, then it is possible that the process could be heirarchical. This is an interesting and innovative scenario because it integrates the effects of self-gravity, magnetic fields, dissipation, fragment interactions, and gravitational virialization (each of which has been claimed as important by various workers), rather than the usual practice of isolating a single process and forcing it to account for the observations. On the other hand, Falgarone and Puget did not present estimates of the manner in which density, velocity dispersion, and field strength might scale with size in their model, a difficult task because of the complexity and scale dependence of the dissipation mechanisms.

2.3.5. Gravitational virialization.

Probably the only physical process which is generally accepted as a dominant influence on the evolution of interstellar structures of all sizes is self-gravity. Fragmentation of gas clouds into stars, whether it occurs in one step or in a hierarchical sequence of stages, is believed to occur during gravitational collapse, perhaps modified by magnetic fields and/or rotation. The details of the outcome of gravitational collapse and fragmentation remain uncertain because of the nonlinearity of the process, the limited spatial resolution of numerical collapse calculations imposed by computational considerations, and the sensitivity of the results of such calculations on adopted initial conditions.

Scalo and Pumphrey (1982) proposed that the change in gravitational energy of a collapsing cloud can be channelled into the random motions of internal fragments until a quasi-virialized state is reached in which the kinetic energy of the fragments supports the parent cloud against free-fall collapse. The parent cloud contracts at a rate set by the rate at which the fragments dissipate energy by mutual interactions and drag. Such a situation is naturally expected in any self-gravitating system containing subcomponents, and the physics is in essence no different than the classical quasi-static contraction of an opaque protostar, in which the collapse energy is converted into thermal motions of molecules randomized by collisions, or the virialization of a collapsing cluster of galaxies in which the collapse energy of the cluster is converted into the motions of the member galaxies, randomized by gravitational scattering. The only difference lies in the processes controlling the contraction timescale, which is the inelastic fragment collision or drag time for the interstellar case (the infrared and radio photons so generated escape freely because the clouds are optically thin to these wavelengths), the photon diffusion time for the opaque protostar case, and essentially infinite for the galaxy cluster case, since the interactions are elastic except for tidal encounters and direct collisions.

If the fragments themselves are unstable to collapse, then the process can lead to a hierarchy of virialized subsystems in which the motions of fragments within parents act both to support the parent and to broaden the spectral lines. The system can be regarded as a turbulent self-organized structure in which energy cascades to smaller scales; although nonlinear vorticity stretching may play a role in adding to the chaotic nature of the system, this "gravitational turbulence" is clearly of a different nature than ordinary incompressible turbulence.

That the observed supersonic linewidths in massive molecular clouds might be due to the motion of macroscopic subfragments was first proposed by Zuckerman and Evans (1974). The idea was met with considerable skepticism because it appeared that the dissipation timescale for these highly supersonic motions would at most equal the free-fall time of the parent cloud (e.g. Field 1978). The force of this argument has weakened considerably in recent years for two reasons. First, it was realized that most collisions will only directly involve the outer parts of the fragments, considerably reducing the dissipation efficiency (Scalo and Pumphrey 1982). Second, if the system of fragments is permeated by a magnetic field, most of the fragment interactions may involve tangling of field lines, with direct collisions being relatively infrequent (see Clifford and Elmegreen 1983, Elmegreen 1985, and Falgarone and Puget 1986), again reducing the dissipation rate by nearly an order of magnitude compared to previous estimates.

Detailed N-cloud simulations of interacting fragments within a collapsing parent cloud (Pumphrey and Scalo 1986) demonstrate quantitatively that gravitational virialization of fragment motions can occur, but only if two conditions are satisfied:

1.   The fragments must undergo significant directional deflections during interactions, in order to isotropize the fragment motions; otherwise, the fragments tend to simply fall to the center of the parent cloud where they coalesce into one or two massive slow-moving condensations. Although this isotropization was imposed artificially in some of the simulations, the work on interactions of magnetically-linked fragments cited above suggests that the required directional deflections are to be expected.

2.   The fragments must somehow be internally supported against collapse for a time much longer than their free-fall timescales. One way to accomplish such support is to invoke the likely hierarchical nature of the process, so that fragments are supported by their own subfragments. This explanation is still problematical and lacks any quantitative analysis. A much more attractive solution lies in the calculations of Falgarone and Puget (1986), as discussed in the previous section. In this model, trapped MHD waves within the fragments can feed the motions within the fragments for a time well in excess of the fragment free-fall time, and support the fragments because in this case the internal energy (motions due to Alfven waves) and gravitational energy have the same dependence on fragment radius.

An example of the results of simulations designed to examine the question of gravitational virialization are illustrated in Figure 3, which shows the evolution, in units of the system free-fall time, of the half-mass radius of the system. All the models began with 1000 fragments, all the same size and speed, and which were assumed to be internally supported. The line marked "elastic" was computed assuming hard-sphere elastic collisions; as expected, the fragment motions easily virialize with the gravitational field, with the system shrinking by a factor of two after about two free-fall times. The other models, labelled B, C, and D, include the effects of dissipative collisions and drag forces; the number of fragments remaining at various times is marked along each curve. Model B, with no collisional deflections or mass loss, suffers catastrophic collapse in a free-fall time, resulting in the runaway growth of one massive fragment at the center which will eventually accumulate all the other fragments. Assuming that 10 percent of the mass of a fragment directly involved in a collision is lost during each collision (Model C) helps the

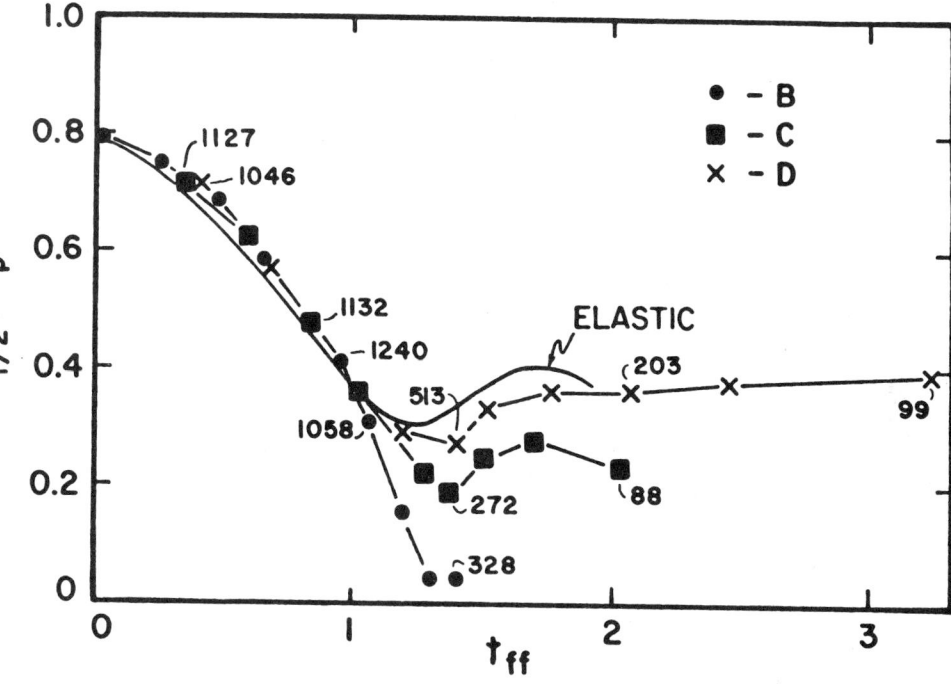

Figure 3. Evolution of the half-mass radius of self-gravitating systems of interacting fragments, according to N-body simulations of Pumphrey and Scalo (1986). Except for the case of elastic collisions, gravitational virialization is only attained in model D in which the assumed collisional deflections of fragments acts as an isotropizing agent.

situation, because the fragment sizes are reduced so much by mass loss that they can orbit through the core of the system, leading to an oscillation of the system radius; however, the model eventually results in runaway disruption due to the collisional mass loss. Model D introduces a randomizing velocity component to the fragments emerging from a collision, and still includes collisional mass loss. In this case, the system *does* evolve to a state of near-virial equilibrium. The accompanying reduction in the number of fragments increases the collision time until it greatly exceeds the cloud collapse time, and the quasi-virialized system will eventually contract on a timescale equal to the (long) collision time. This result supports the earlier suggestion (Scalo and Pumphrey 1982) that conversion of gravitational energy into random fragment motions can virialize the parent cloud. However, the assumed collisional mass loss leads to runaway cloud disruption after about 3 free-fall times. Once again, the magnetic field may save the model, since if magnetic interactions dominate over direct physical collisions, mass loss will be negligible. In addition, as discussed above, the magnetic field can provide the assumed deflections during encounters and the assumed internal support of the fragments, without drastically altering the dissipation timescale (Elmegreen 1985).

The gravitational virialization model for turbulence in cloud complexes is attractive because it automatically satisfies the observed virial relation on all scales and because it can naturally give rise to hierarchical structure. However, the observed n(*l*) relation is left unexplained, and must be due to the detailed physics of fragmentation. Of course, if we *assume* the observed n(*l*) relation, then the observed Δv (*l*) relation follows naturally from the virialization condition.

3. Approaches Based on Direct Solution of the Hydrodynamic Equations

The arguments discussed in the previous section are phenomenological, in the spirit of the Kolmogorov law for incompressible turbulence. They are all based on simple conceptual models that *assume* a self-similar process in which two physical effects are balanced, and/or in which some physical property is conserved. As such, they cannot be expected to yield much quantitative information, have very limited predictive power (unless additonal assumptions are introduced), and must be evaluated entirely on the basis of the validity (or lack of it) of their fundamental assumption(s), a very subjective judgement. Given the number of effects which various workers have claimed to be dominant in the ISM, it would be surprising if interstellar turbulence could be as simple as these models assume.

The classic example of the limitations of phenomenological theories is the Kolmogorov model, which essentially only predicts the form of the correlation function for incompressible turbulence, but which does not explain physically the observed value of the coefficient of proportionality (the "Kolmogorov constant"), the existence of large-scale coherent structures and small scale intermittency, the nature of such processes as vortex deformation, pressure fluctuations, interactions among turbulent fluctuations, or isotropization of the velocity fluctuations, all of which play a role in establishing the energy cascade, which is merely assumed to occur. Furthermore, the model cannot describe the observed deviations from the predicted power spectrum (e.g., atmospheric turbulence in situations in which local heat sources deliver buoyancy at preferred scales), why the structure of many turbulent flows appears hierarchical, or, most importantly, why the transition from laminar to turbulent flow occurs at all.

Clearly, a working theory of the nature of interstellar turbulence must be tied more closely to solutions of the hydrodynamic equations. In the present section we review the difficulties associated with direct numerical solutions of the hydrodynamic equations for turbulent flows and the clues and suggestions which have resulted from attempts in this direction, based on both interstellar studies and simulations of incompressible turbulence. The review of incompressible simulations is very incomplete (the literature is vast), and is only meant to show how such work can yield insight into effects and approaches which may carry over into the interstellar case. Discussions of statistical treatments of the hydrodynamic equations and of model equations to describe the temporal transition to turbulence are given in sections 4 and 5, respectively.

3.1. A critical survey

It is generally, though not universally, believed that fluid turbulence can be represented by solutions of the macroscopic equations of hydrodynamics. This would not be true, for example, if turbulence arises by means of the propagation of microscopic disorder to macroscopic scales (e.g. Tsuge 1974) because the Boltzmann equation from which the hydrodynamic equations are derived neglects correlations between particles. Recent experiments demonstrating that the addition of minute quantities of long chain polymers to a turbulent fluid can significantly alter the flow properties (see Gampert 1985) suggest that this idea warrants further consideration.

Ignoring this possibility, the relevant equations for *incompressible* (ρ = constant) turbulence are the Navier-Stokes equations, expressing momentum conservation

$$\frac{\partial v}{\partial t} + v \cdot \nabla v = -\frac{1}{\rho}\frac{\partial P}{\partial x} + \nu \nabla^2 v, \tag{15}$$

where v, ρ, P, and ν are the fluid velocity, density, pressure, and kinematic viscosity, and the incompressibility condition which reduces the continuity equation to

$$\nabla \cdot v = 0 \tag{16}$$

Despite their rather innocent appearance, these equations result in surprisingly complicated flows (turbulence) at large enough Reynolds numbers, and no method of numerical simulation is capable of providing solutions for Re $\gtrsim 10^3 - 10^4$. The problem is due to the presence of the nonlinear advection term $v \cdot \nabla v$ in eq. 15, which is capable of amplifying and distorting any initial flow field into an irregular and unpredictable (except, perhaps, in an statistical sense) pattern of velocity fluctuations covering an extremely large range of size scales. Computational limitations restrict the spatial resolution of all numerical techniques to such a degree that turbulent solutions are only possible for very modest Reynolds numbers. The number of required mathematical operations varies roughly as Re^6 (Orzag 1970), and exceeds 10^{30} for Re = 10^5. Recalling that the Reynolds number in interstellar complexes is $\sim 10^7$ (see eq. 3), there is no hope of *direct* solutions to the interstellar turbulence problem; turbulence, if it could occur, will be erroneously suppressed in all numerical hydrodynamic studies of the ISM. This important point can be better appreciated by noting that no one would ever have suspected the existence of turbulence in the earth's atmosphere, ocean, or anywhere else, if we were to rely on solutions of the equations of fluid mechanics. We know that turbulence exists only through direct experience. Based on this terrestrial example, the strong observational evidence for some type of turbulent behavior in interstellar cloud complexes must force us to regard numerical ISM calculations with a large grain of salt.

The ISM is not an incompressible fluid, so the Reynolds number may not be the controlling parameter; one might hope that the required range of scales will be smaller. In fact, just the opposite is more likely for two reasons. First, the observed range of structure sizes, which is itself limited by observational spatial resolution, spans at least 3 orders of magnitude, and is even much larger if one considers that we are ultimately interested in the coupling of the structure to the formation of single stars. Secondly, the hydrodynamic (or magnetohydrodynamic) equations appropriate to the ISM offer much greater opportunities for complex behavior than the comparatively simple Navier-Stokes equations. Even if we assume isothermal or adiabatic flow so that the energy equation can be omitted, the equations contain more nonlinear terms, for example those coupling the density and velocity fields, as well as at least one globally coupled field, self-gravity. When one admits that the expected flows will be supersonic (as observed), the energy equation and its attendant nonlinear terms, coupling between the temperature, density, and velocity fields, and especially the introduction of a new multiscale process, shock dissipation, suggests the throwing up of hands.

That has in fact been the posture taken by many workers in both incompressible turbulence and ISM modelling. In the case of incompressible turbulence, the most common approach has been to abandon hope of modelling the detailed flow structure and instead search for approximations which will provide some understanding of the behavior of crude statistical flow properties like the average velocity field and two-point autocorrelation function. A hybrid technique, usually called "large eddy simulation" or "subgrid modelling", directly solves the hydrodynamic equations on the large scale, but adopts some statistical closure approximation for the scales too small to be resolved. Other workers have concentrated on particular processes, like vortex interactions or intermittency. A more recent alternative approach in this area is to focus on the temporal behavior by replacing the actual hydrodynamic equations with "model equations" whose form hopefully contain the essential nonlinearities of the problem while being amenable to numerical solution. The famous Lorenz model is an example. Both the statistical and "model equation" approaches and their application to the more difficult interstellar turbulence problem will be discussed below.

In the case of interstellar-protostellar research the prevailing attitude toward direct simulation of turbulence is very different. A large number of multidimensional hydrodynamical studies of the collapse and fragmentation of self-gravitating clouds have been published (see Scalo 1985 and Boss, this volume, for detailed reviews), mostly concentrating on the effects of rotation and, more recently, initial geometry and magnetic fields. Limitations on spatial resolution, as well as numerical diffusion, eliminate any hope of revealing spatially turbulent behavior, and the fact that this limitation is rarely mentioned (see, however, Wood 1982 for an objective evaluation) has led to the impression that turbulence is unimportant in gravitational collapse and that fragmentation, at least from one scale to another, is relatively well-understood. In fact, though, turbulent behavior is *expected* in such nonlinear flows involving coupled physical processes (see the prescient remark by Peebles and Dicke 1968), as supported by the sensitivity of the calculated evolution to small changes in the initial conditions found by Buff *et al* (1979) and Boss (1980), a typical symptom of turbulent behavior. Furthermore, the nature of the fragmentation process in the published numerical calculations depends sensitively on the assumed physical initial conditions, for instance the importance of rotation, magnetic fields, geometry, or the presence of pre-existing internal density and velocity fluctuations. This clearly indicates that the modelled collapsing cloud is strongly coupled to the larger-size structures which presumably gave rise to the cloud that has been singled out for study. Unfortunately, observations offer few clues concerning the appropriate choice of initial physical conditions.

Direct numerical integration of the hydrodynamic equations may be more profitably used to study interstellar turbulence by simulations of particular processes which may play a role in controlling the overall turbulent structure. Examples include the dynamics of cloud collisions (e.g., Gilden 1984, Lattanzio *et al.* 1986), the evolution of wind-driven shells (Rozyczka 1985), the implosion or destruction of clouds by convergent shocks driven by the radiation field of nearby stars (Sanford *et al.* 1984) or by supernova blast waves (Krebs and Hillebrandt 1984), and several others. From this point of view, the studies of cloud fragmentation could be included in this category. These calculations parallel the modelling of vortex interactions used to understand the effects of the nonlinear advective term in incompressible turbulence. Unfortunately, the number of nonlinear processes potentially relevant to interstellar turbulence is extremely large.

3.2. Specific direct solutions as guides

Despite all the limitations and reservations discussed above, direct numerical solutions for both interstellar and incompressible flows *have* yielded results which may serve as guides to the types of phenomena to be expected in fluid turbulence and motivate interesting phenomenological models.

The studies of the fragmentation of collapsing clouds, though restricted to one fragmentation stage, have yielded information concerning, for example, the expected number of fragments, (usually 2-5, depending on numerical technique and initial conditions), mass efficiency of fragmentation (~ 0.2 to 0.5 in rotationally-dominated models, ~ 0.05 in a 3-dimensional magnetically-dominated model), the geometry of the fragments, the transfer of angular momentum flux to larger scales by tidal effects or nonaxisymmetric trailing perturbations ("gravitational torqueing"), and the loss of spin angular momentum by conversion to orbital motion (reduction factors are generally in the range 10-30) or by magnetic braking. Although it is not known how much weight to place on these results because they are not properly coupled to the larger and smaller scales (a coupling which is the essence of the turbulence we are attempting to model using these very calculations), such results can form the basis for a variety of phenomenological models for interstellar multiscale turbulence. For example, Bodenheimer (1978) proposed a hierarchical fragmentation model based on the numerical results concerning the efficiency of spin-orbit angular momentum loss in fragmenting clouds. Henriksen and Turner (1984), motivated by numerical studies which suggested the importance of gravitational torqueing, proposed a self-similar turbulent hierarchy controlled by the cascade of angular momentum from smaller to larger scales. The resulting $n(l)$ and $\Delta v(l)$ scaling relations were outlined in sec. 2.1. Henriksen and Turner also show how renormalization techniques can be used to derive the scaling for an assumed dissipationless self-similar cascade for other choices of conserved quantities.

In a similar spirit, the numerical studies of the other particular processes like those listed above can motivate or help quantify additional phenomenological models, e.g., the wind-driven shell hierarchy and the gravitational virialization hierarchy discussed in sec.2 above. In all these cases, however, the numerical solutions have provided no *direct* insight into the origin or evolution of interstellar turbulence: an explanation of its hierarchical appearance, its connection with star formation, virialization at all scales, and scaling relations for density and velocity dispersions, and the relation of these features to the nonlinearity, self-organization, and unpredictability expected from the nature of the hydrodynamical equations. They *have* provided fuel for a proliferation of phenomenological models which, as emphasized earlier, are of limited value.

The situation is quite different for numerical studies of incompressible turbulence.

There are only two nonlinear terms in the equations: advection $(v \cdot \nabla v)$ and the pressure gradient, which is nonlinear by virtue of its coupling to the velocity through the incompressibility condition. (Compare with the nonlinearities due to the coupling of the density, velocity, and temperature fields in the interstellar case.) The only remaining term is the viscous force, which is local, linear, and well-defined, and so presents no numerical difficulties. (Compare with the presence of self-gravity, shocks, and stellar sources in the interstellar case.) The pressure can be eliminated using the fact that the velocity field is divergence-free, so there is only one equation to solve.

The numerical solutions are time-consuming and are restricted to relatively simple flows at present. Solutions can currently be obtained for Reynolds numbers approaching $10^3 - 10^4$, so one can study the transition from laminar to turbulent flow and, with more difficulty, the nature of fully developed turbulence. Simulations at Reynolds numbers large enough for the establishment of a well-defined inertial subrange (i.e., a large separation between the scale at which energy is injected and the viscous dissipation scale) should become available in the near future. For a given Re, there are no free parameters and the solutions depend only on the initial physical conditions (e.g., a velocity shear, or a fluid at rest with density gradients), the geometry of any boundaries (e.g., pipe flow, flow between concentric cylinders, flow past a cylinder, free jets), and an assumption concerning the presence of external forcing. These choices are dictated in some cases by the type of laboratory experiment or terrestrial environment with which the simulations are to be compared (note the tremendous advantage of having controlled experimental results, compared to the interstellar case), or by numerical considerations when the interest lies more in identifying the physical interactions. The advantages and disadvantages of various numerical techniques (e.g., finite-difference, spectral techniques, particle methods) will not be discussed here.

Although many published numerical simulations of incompressible turbulence are available, I will restrict the present review to a few studies whose results suggest behavior and tactics which may be applicable to interstellar turbulence.

One of the most detailed and instructive studies is by Deissler (1984), who obtained solutions using a finite difference method on a 32^3 grid with a sinusoidal initial velocity field. The Reynolds number varied between 140 and 4400. Deissler examined, among other things, the development of turbulent fluctuations from the initial sinusoidal field, the development of shear layers, the effect of removal of shear, and the effect of a spatially periodic body force. Two results seem especially important in the context of the interstellar turbulence problem.

First, Deissler showed how the simulations could be used to interpret and compare the importance of the various terms which appear in the *statistical* turbulence equations (e.g., for the two-point correlation function), to be discussed in section 4. For example, he shows how the interaction of the pressure and velocity fluctuations can transfer the velocity fluctuations among different directional components, leading to isotropization, and identifies the processes by which transfer of energy to smaller scales and the generation of vorticity occur. His work also illustrates the limitations of the statistical approach. For example, these equations predict no effect of the pressure gradient on the two-point correlation function, while the numerical solutions do show such an effect, indicating that it must be contained in the higher-order correlation equations.

These results are important for interstellar turbulence because they suggest that examination of the statistical equations, which are much more complex in the interstellar case (see section 4), is a worthwhile exercise even if the equations cannot be closed. The interpretation of the various terms and a study of their relative importance at different scales should provide at least some understanding of the actual physical processes and couplings involved in the generation and evolution of turbulent structure (something which is only assumed in the phenomenological models) and can suggest approximations which may allow numerical solutions of particular interstellar flows which are suspected of being susceptible to turbulent behavior (e.g., gravitational collapse).

Another result of Deissler's which deserves attention is the fact that, when the temporal evolution of the velocity components at a given spatial position is projected onto a two-component phase plane, the trajectory bears a strong resemblance to the chaotic behavior found in many other nonlinear systems, with unpredictable motion consisting of a tangle of twists and loops, as illustrated in Figure 4. This suggests that studies of the purely temporal evolution of model nonlinear equations designed to represent the behavior of interstellar gas and its interaction with star formation may be useful in understanding the conditions under which the ISM is turbulent (in the sense of temporal unpredictability) and might reveal the parameters which control this behavior for different types of model equations (see section 5).

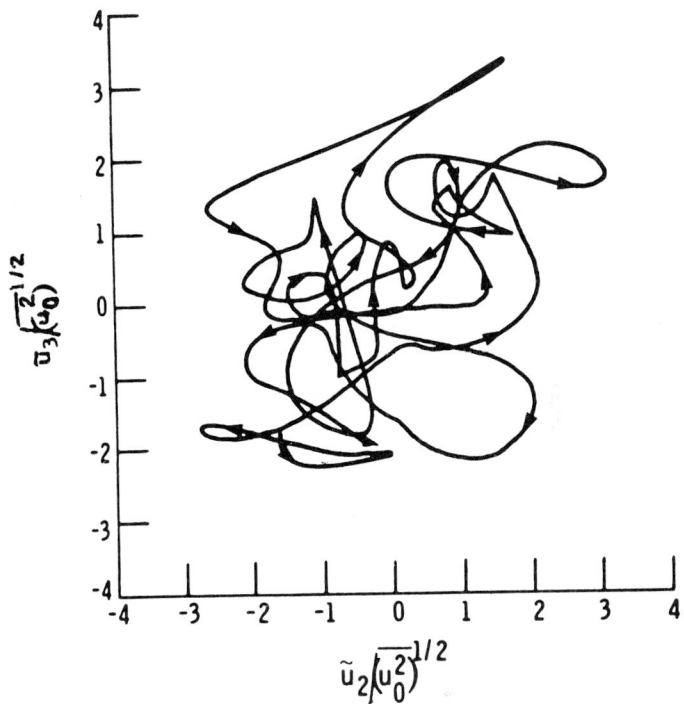

Figure 4. Trajectory of a phase point projected onto the u_2-u_3 plane in a numerical turbulence simulation by Deissler (1984). The calculation used a 32^3 grid, a Reynolds number of 138.6, and a spatially periodic body force to establish a statistically steady state. The chaotic appearance indicates that the flow lies on a strange attractor.

The general notions of "self-organization", "disorder" or "chaos", and "cascades" of quantities along a spectrum of size scales have been mentioned several times in connection with interstellar turbulence. These features are common to many nonlinear systems, and, as argued in section 1, are expected in the ISM. But the existence and nature of these properties cannot presently be studied by direct numerical solution. Nevertheless, the ISM *is* a fluid, and some useful clues may be inferred from numerical studies of incompressible turbulence.

A particularly interesting type of self-organizing behavior occurs in turbulent fluids in which more than one quantity is conserved, a situation reviewed by Hasegawa (1985). In these cases, one conserved quantity becomes spatially chaotic by means of a direct cascade from large to small scales, while the other self-organizes into large structures by undergoing an inverse cascade from small to large scales.

As quantitative examples (Hasegawa 1985), consider fluid turbulence in two dimensions. Unlike the three-dimensional case, in which only the energy is conserved, it can be shown that there is an additional conserved quantity, called the enstrophy, which is the volume integral of $(\nabla \mathrm{x} v)^2$, and so is related to the evolution of the vorticity. In this

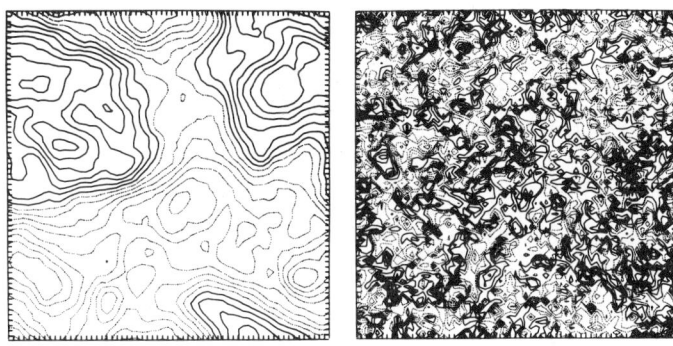

Figure 5. Contour plots of the stream function (left) and the enstrophy (right) in a numerical simulation of two-dimensional incompressible turbulence. These results show that energy cascades to larger scales (self-organizes) while enstrophy cascades to smaller scales. From Hasegawa (1985).

case, it is the enstrophy that cascades to smaller scales, establishing a chaotic-looking spatial distribution. This is illustrated in the left panel of Figure 5 which shows the distribution of vorticity at a particular time in a numerical simulation. However, the energy *inversely* cascades to larger scales, as shown in the right panel of Figure 5, which is a contour plot of the stream function ψ, defined by $V = -\nabla \mathrm{x} \psi$.

A second example occurs in three-dimensional MHD turbulence. Here a quantity called the "magnetic helicity" H_B, related to the vector potential A by $H_B = 1/2 \int A \cdot (\nabla \mathrm{x} A) dV$, is conserved in addition to the magnetic energy density $W_B = 1/2 \int B^2 dV$. The helicity inversely cascades, as shown in the contour plot of the vector potential, found in a numerical simulation, given in the top panel of Figure 6, while the field energy cascades to small scales, illustrated by the contour plot of the current density in the bottom panel of Figure 6. The occurrence of inverse cascades in MHD turbulence has been recognized for a long time (an astrophysical example using statistical turbulence equations rather than direct simulation can be found in DeYoung 1980), although the actual physical interactions which are involved remain poorly understood (just as for the energy cascade in 3-dimensional turbulence).

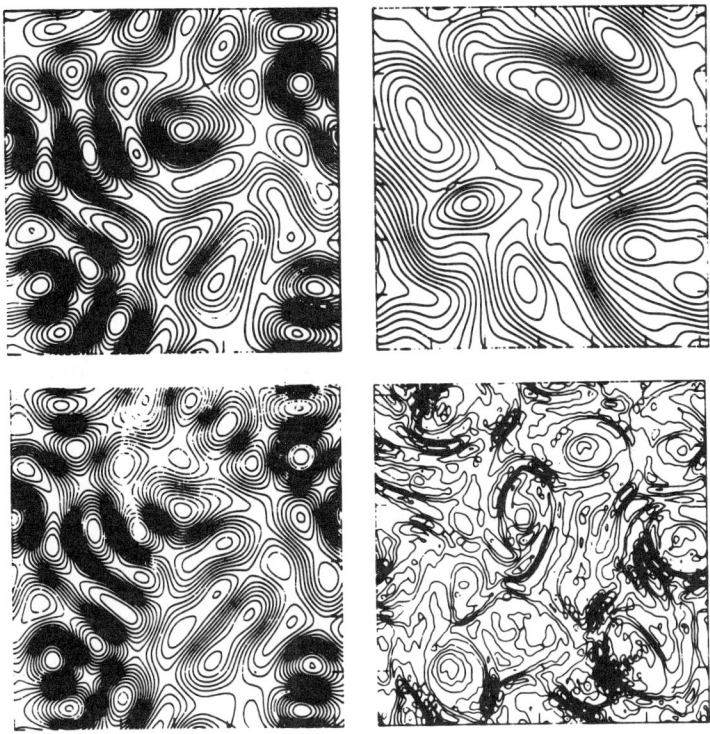

Figure 6. Contour plots of vector potential (top) and electric current density (bottom) in a numerical simulation of three-dimensional incompressible MHD turbulence for the initial state (left) and a later time (right). These results show that magnetic helicity cascades to larger scales (self-organizes) while magnetic field energy cascades to smaller scales.

The physical conditions assumed in the turbulent flows described above are certainly much different than in the highly compressible, dissipative, self-gravitating ISM, but they do point out one way in which a turbulent fluid system can develop large-scale order and small-scale disorder in two different variables. If such a situation occurs in the ISM, we should not be surprised if maps of different observed quantities exhibit very different structures.

As a very speculative example, many workers have pointed out that the magnetic field in interstellar clouds, as traced by polarization vectors, often appears fairly smooth, and argue from this that turbulence cannot be important because it would lead to a wildly tangled field configuration. Although there are various ways around this argument, the illustrations given above show that the statement need not be true; the smooth field could reflect a more sublime coexistence of order and chaos, depending on which physical quantities are being conserved. More importantly, these examples warn us not to place too much trust in simple conceptual arguments concerning nonlinear processes like turbulence.

3.3. Cellular automata solutions

One possibility for progress in direct simulations of interstellar turbulence is the development of new numerical techniques for solving the hydrodynamic equations aimed at high spatial resolution. An approach based on the use of "cellular automata" has recently sparked some optimism concerning the calculation of large Reynolds number incompressible turbulence (Frisch *et al.* 1986), so it is worthwhile to examine the possibility that this method can be extended to interstellar conditions.

A cellular automaton (see Wolffram 1983 for a comprehensive presentation) is a representation of a physical system by a regular lattice which is specified by the values of variables (e.g., density, velocity here) at each site on the lattice. The evolution of the cellular automaton, which occurs in discrete time steps, is governed by a set of rules that dictate how the variables at one site are affected by variables at another site. In general, cellular automata using a large class of interaction rules are found to possess self-organizing abilities and sensitivity to initial conditions with the development of long-range correlations and hierarchical structure, even from random initial conditions. These features are similar to the behavior of many nonlinear real and model systems, in particular turbulence. Cellular automata have been applied to a variety of problems, from structure formation in biological systems to number theory. The well-known stochastic self-propagating star formation galaxy model by Gerola and Seiden (see Seiden and Gerola 1982 for a review) is a simple cellular automaton.

It might seem, at first sight, improbable that a cellular automaton can yield solutions to the equations of hydrodynamics. However consider a lattice whose sites are occupied (or unoccupied) by particles meant to represent atoms or molecules, labelled by the direction of their velocity vectors. If we specify a set of interaction rules concerning, for example, how to alter the velocity vectors of "molecules" which collide at a lattice site, one can derive "microscopic" rate equations for the site populations. Averaging these rate equations over a statistically meaningful number of sites, one can derive the fluxes of mass and momentum through the lattice. In order to represent a fluid, these fluxes must satisfy the hydrodynamic equations.

Hardy *et al.* (1976) had attempted to construct such a lattice gas using very simple interaction rules, but found that the tensor relating the momentum flux to quadratic terms in the velocity is not isotropic, giving an unphysical stress-strain relation. Frisch *et al.*, (1986) made the remarkably simple discovery that the rotational symmetry of a *triangular*, as opposed to a square, lattice, restores isotopy to the momentum flux tensor. Using, again, very simple interaction rules for resetting the velocity vectors, the resulting macroscopic equations are the Naveir-Stokes equations, except for a higher-order nonlinear term in the momentum equation. This term, which appears because a lattice gas does not possess Galilean invariance, is negligible for Mach numbers less than about 0.3 - 0.5 according to Frisch *et al.* Furthermore, they suggest a method for extending the approach to three dimensions, and indicate that modifications of the interaction rules allow a temperature to be defined, so it may be possible to solve the energy equation using a lattice gas.

The advantage of the use of a cellular automaton is that the calculation is inexpensive compared to numerical integration on a grid, since, at each time step, one only needs to check and reset the velocity vectors, so very few arithmetic operations are required. Therefore, although a very large number of lattice sites are required, the spatial dynamic range of the computation of the macroscopic fluxes can be made much larger than in ordinary numerical hydrodynamics for the same computation time. Frisch *et al.* state that the method is already being applied to two-dimensional turbulence with satisfactory results.

The application of this method to interstellar turbulence is of obvious interest, since spatial resolution has been the major problem. However, this extension is not at all straightforward. First, the viscosity must be minimized, since it is unimportant for interstellar turbulence. Frisch *et al.* suggest a computationally efficient method for viscosity reduction, but it is not clear how far this technique can be pushed. More seriously, we are interested in supersonic flows, but in this case spurious higher-order nonlinear terms will appear in the momentum equation, as mentioned above. Finally, self-gravity is known to be important in interstellar structures, but the incorporation of such a long-range force in a lattice gas model may be problematic. One might hope to define a potential at each sight and devise clever rules for its dependence on the macroscopic density around other sites, and modifications to the particle interaction rules that would self-consistently yield the Poisson equation for the potential and the correct gravitational force in the momentum equation. This goal is possible in principle, since any differential equation can be approximated by a cellular automaton (Wolffram 1983), but this procedure would introduce a global coupling between all the lattice sites, greatly increasing the computational time. A much better procedure would be to only define the potential on a "superlattice" whose cells are much larger than those of the original lattice, computing the potential using site populations smeared over the scale of the superlattice, and then interpolating the superlattice potential onto the original lattice for use in the local interaction rules (which must produce the correct force in the momentum equation). Alternatively, one could simply keep the calculation separate from the lattice calculation by "just" solving the Poisson equation directly after each timestep of the lattice calculation (since we then know the density), again interpolating back onto the lattice for use in the next timestep's update of the automaton. In any case, the inclusion of self-gravity will greatly increase the computation time, and may at least partially offset the advantages of the entire approach.

These problems, especialy the calculation of flows with large Mach numbers, are serious, but the possibility of greatly increased spatial resolution suggests that these extensions of the cellular automaton method for solving the hydrodynamical equations should be considered in more detail. A useful starting point would be a calculation of the development and structure of *weakly* compressible (say Ma ~ 0.3) turbulence, in an imposed mean velocity field, neglecting self-gravity. Such a calculation would yield valuable information on the coupling between the density fluctuations, the velocity fluctuations, and the mean flow, a subject about which essentially nothing is known.

4. Statistical Approaches

Until very recently, the approach used in nearly all theoretical investigations of fluid turbulence was to ignore the details of the flow and instead concentrate on the solution of equations for statistically-averaged properties, particularly the 2-point velocity correlation function (or, equivalently, the power spectrum). The motivations for the statistical approach are basically the impossibility of obtaining direct numerical solutions of the Navier-Stokes equations, the strong desire to derive and quantify the Kolmogorov law for the energy spectrum using the actual flow equations rather than dimensional and phenomenological arguments, and the fact that correlation functions and power spectra can be measured in laboratory experiments and some terrestrial environments. Faced with a similar situation for interstellar turbulence, it is worthwhile to review the approximations that have been proposed to solve the statistical equations of incompressible turbulence and examine the possibility of extending these techniques to interstellar turbulence.

Two points should be recognized from the outset. First, low-order statistical properties like the correlation function or power spectrum contain only very crude information about the spatial structure of turbulence. The loss of information concerning the spatial relationships and connectivity of the structure is discussed at length in Houlahan and Scalo (1986, in preparation), who show, for example, that 2-point statistics cannot even distinguish between hierarchical structures and structures containing a random arrangement of clouds with a range of sizes. For the problem of interstellar structure, then, the primary use of the statistical approach is as a tool for identifying the relevant physical couplings and estimating their relative importance. Secondly, there is as yet no completely satisfactory method of approximation which can be used to express the statistical equations in closed form, because of their intrinsically nonlinear stochastic nature, although an astounding variety of techniques, most of them of extreme mathematical complexity, have been proposed. This famous "closure problem" will be explained below, but the only proposed closure techniques which will be discussed here are those which have actually been used in applications relevant to turbulence in cool interstellar cloud complexes. Besides, a presentation of additional techniques (e.g., expansion and diagrammatic methods) would cloud the major issues of interest in a maze of mathematical detail. Despite these reservations, it should also be pointed out that the statistical approach *has* been successfully used to examine a number of incompressible turbulence problems that are not readily accessible to direct numerical simulations (see, e.g., Poquet 1985).

The basic idea is to write each fluid variable as the sum of a mean part and a fluctuating part, since turbulence is usually thought of as consisting of fluctuations. In order to keep the equations relatively simple, we consider incompressible turbulence. The velocity and pressure fields are then $v_i = U_i + u_i$, $p = P + p'$, where u_i and p' are the fluctuations. Substituting in the Navier-Stokes equations (eqs. 15 and 16), averaging using the usual rules (e.g., $\langle u_i \rangle = \langle p' \rangle = 0$, $\langle v_i \rangle = U_i$, etc.), the mean flow momentum equation is found to be

$$\rho \frac{\partial U_i}{\partial t} = -\rho U_k \frac{\partial U_i}{\partial x_k} - \frac{\partial P}{\partial x_i} + \frac{\partial}{\partial x_k}\left[\rho v \frac{\partial U_i}{\partial x_k} - \rho \langle u_i u_k \rangle\right], \qquad (17)$$

where the cartesian summation rule applies. Because the term $\rho \langle u_i u_k \rangle$ appears to augment the viscous stress $\rho v \partial U_i/\partial x_k$, it is usually called the "turbulent stress tensor" or the "Reynolds stress". (An equation for P can also be easily derived, but we omit it for simplicity.) This equation shows that we cannot calculate the mean flow without some knowledge of the fluctuations. An equation for the fluctuating components is easily derived by subtracting the mean flow equations from the unaveraged Navier-Stokes equations; the result for the velocity fluctuations is

$$\frac{\partial u_i}{\partial t} = -\frac{\partial}{\partial x_k}\left(u_i u_k\right) - \frac{1}{\rho}\frac{\partial P}{\partial x} + v \frac{\partial^2 u_i}{\partial x_k \partial x_l} - u_k \frac{\partial U_i}{\partial x_k} - U_k \frac{\partial u_i}{\partial x_k} + \frac{\partial}{\partial x_k}\left(\langle u_i u_k \rangle\right) \qquad (18)$$

The terms on the right-hand side represent the following physical effects (Deissler 1984): 1. turbulent self-interactions which produce small-scale structure and randomization; 2.

the net force due to pressure fluctuations; 3. viscous dissipation; 4. the production of fluctuations due to the interaction of existing fluctuations with a gradient in the mean flow; 5. the convection of fluctuations by the mean flow, which generates small-scale structure by vortex stretching; 6. the mean turbulent stress. This last term only appears for inhomogenous turbulence. It should be noted that most statistical studies of turbulence assume homogeneity (invariance of all statistical properties with respect to translation) for the sake of simplification, even though a hierarchical system is *not* statistically homogeneous.

To derive the required equation for $<u_i \cdot u_k>$ at a single point (*one*-point correlations), multiply eq. 18 by u_j and the analogous equation for $\partial u_j / \partial t$ by u_i, add them, and take averages. The result is

$$\frac{\partial <u_i u_j>}{\partial t} = -\left[<u_j u_k> \frac{\partial U_i}{\partial x_k} + <u_i u_k> \frac{\partial U_j}{\partial x_k} \right] - U_k \frac{\partial}{\partial x_k} \left(<u_i u_j> \right) - \frac{\partial}{\partial x_k} \left(<u_i u_j u_k> \right)$$

$$-\frac{1}{\rho} \left[\frac{\partial}{\partial x_j} <p' u_j> + \frac{\partial}{\partial x_j} <p' u_i> \right] + \nu \frac{\partial^2 u_i u_j}{\partial x_l \partial x_l} + \frac{1}{\rho} \left[<p' \frac{\partial x_j}{\partial x_i}> + <p' \frac{\partial x_i}{\partial x_j}> \right]$$

$$= 2\nu < \frac{\partial u_i}{\partial v_i} \frac{\partial u_j'}{\partial x_i} > \tag{19}$$

The interpretation of the various terms can be found in Deissler (1984). Although this equation is quite complicated, the essential difficulty in its solution lies in the appearance of the third-order correlation $<u_i u_j u_k>$, which cannot be evaluated without deriving an even more complicated equation for the rate of change of $<u_i u_j u_k>$. This is the "closure problem".

The one-point correlation equation (19) tells us nothing about the transfer of turbulence between different scales. In order to derive an equation for the *two*-point correlation function, the equations for the fluctuations are written for two different positions, and then a similar manipulation as used in deriving equation 19 yields (e.g., Deissler 1984, Beran 1968)

$$\frac{\partial}{\partial t} \left(<u_i u_j'> \right) = - <u_k u_j'> \frac{\partial U_i}{\partial x_k} - <u_i u_k> \frac{\partial U_j'}{\partial x_k} - U_k \frac{\partial}{\partial x_k} <u_i u_j'> - U_k' \frac{\partial}{\partial x_k} \left(<u_i u_j' u_k'> \right)$$

$$-\frac{\partial}{\partial x_k} \left(<u_i u_j' u_k'> \right) - \frac{1}{\rho} \left[\frac{\partial}{\partial x_i} <p u_j'> + \frac{\partial}{\partial x_j} <u_i p'> \right] + \nu \left[\frac{\partial^2 <u_i u_j>}{\partial x_k \partial x_k} + \frac{\partial^2 <u_i u_j'>}{\partial x_k' \partial s_l'} \right] \tag{20}$$

where u and u' refer to two different spatial positions. The terms in this equation are usually interpreted by taking the Fourier transform, which gives an equation for the spectral components of $u_i u_j'$. Once again, it is the presence of the unknown triple correlations that render the equation unsolvable without some sort of assumptions.

The purpose of presenting the above equations was not only to show explicitly the appearance of the higher-order correlations, but also to illustrate the large number of couplings which arise from the nonlinear terms in the original fluid equations. It is a useful and exhausting exercise to derive the analogous equations for the compressible, self-gravitating, non-isothermal fluid equations relevant to the ISM, which suggest myriad ways in which interstellar turbulence can be generated, dissipated, convected, diffused, and transferred between scales.

For the present discussion of existing astrophysical statistical treatments of interstellar turbulence, it is sufficient to focus on the closure problem. Symbolically, we have equations of the form

$$dX/dt = XX + \text{other terms} \qquad (21)$$

where X is some fluid variable. Decomposing X into mean and fluctuating parts, $X = \langle X \rangle + X'$, gives equations for the mean variables of the form

$$d\langle X \rangle/dt = \langle X'X' \rangle + \text{other terms} \qquad (22)$$

If we derive an equation for $\langle X'X' \rangle$, we get

$$\frac{d\langle X'X' \rangle}{dt} = \langle X'X'X' \rangle + \text{other terms} \qquad (23)$$

and then

$$\frac{d\langle X'X'X' \rangle}{dt} = \langle X'X'X'X' \rangle + \text{other terms} \qquad (24)$$

and so on, leading to an infinite hierarchy of equations. Note that for a realistic treatment of interstellar turbulence, we would have correlations involving all combinations of ρ', v', and T' and their coupling to the gravitational potential.

Most of the turbulence literature consists of attempts to terminate this hierarchy using a variety of assumptions and approximations. A few of these are described below, along with astrophysical applications which might be relevant to interstellar turbulence.

4.1. Eddy viscosity closures

This approach, pioneered by Heisenberg (1948), makes use of the fact that $\langle u_i u_j \rangle$ appears with the viscous term in eq. 17, and so tries to model this correlation in terms of an "eddy viscosity" v_T. For example, one might assume

$$\langle u_i u_j \rangle = v_T \, (\partial U_i/\partial x_j + \partial U_j/\partial x_i) \qquad (25)$$

with $v_T \sim ul$, where l and u represent a characteristic length and velocity to be specified. In this form, the expression for $\langle u_i u_j \rangle$ can be used to solve the mean flow equations, as long as some parameterized form of the eddy viscosity is adopted. This is the approach used in many engineering applications.

However, the main interest here is in the calculation of the evolution of $\langle u_i u_j \rangle$, or, more specifically, the shape of the energy spectrum in wavenumber space, E(k). Heisenberg attacked this problem by noting that, if the motion of small eddies is to affect the motion of the large eddies in a similar manner as molecular motions affect laminar flow, the eddy viscosity $v_T(k)$ acting on wavenumber k must depend on the spectrum of eddies with wavenumbers in the range (k,∞). Using dimensional considerations, Heisenberg proposed a relatively simple expression for $v_T(k)$ which involves an integral from k to ∞ of $[E(k)dk/k^3]^{1/2}$, with a proportionality constant that can be determined from experiments. If the equation for the 2-point correlation function is transformed to Fourier space, the resulting spectral equation for E(k) can be solved in certain limiting cases. In particular, if one assumes that the smaller eddies are in a stationary state $(\partial E(k)/\partial t = 0)$, the predicted spectrum has the form $E(k) \propto k^{-5/3}$, in agreement with the Kolmogorov "law". Subsequent studies also showed that Heisenbeg's theory satisfactorily accounts for the observed form of the double and triple velocity correlations during the intermediate stage of turbulent decay. The main objection is a physical one, namely that the eddy viscosity idea can only be valid if the small eddies responsible for the viscosity acting at a given wavelength are much smaller than the larger eddies from which they extract energy; the spectral equations can be used to show that, in fact, energy transfer by inertial forces (which is what is being modelled by eddy viscosity) occurs mainly by interaction of eddies with slightly different wavelengths, which calls into question the basic assumption.

A number of variants based on this eddy viscosity model have been proposed. One of the most interesting of these approaches, and the only one that has been applied to interstellar turbulence, is presented by Canuto and Goldman (1985). These authors attempt to model the *large-scale* turbulence by relating the energy spectrum E(k) to the linear growth rate n(k) of the instability (whatever it may be) that generates the large scale turbulence as a function of wavenumber, using an assumed form of the eddy viscosity to deal with the nonlinear terms. Canuto and Goldman show that this yields good agreement with experimental results on thermal convection. Canuto and Battaglia (1986) have applied the model to interstellar turbulence by using the observed v(*l*) scaling relation to infer the spectrum of the turbulence-generating mechanism, n(k). The result is that n(k) must be roughly constant. As illustrations, they point out that the linear growth rate for non-magnetic Rayleigh-Taylor instability does not satisfy this constraint while the magnetic Rayleigh-Taylor instability (Parker's instability) may.

By concentrating on the large-scale generating mechanism, Canuto and Battaglia are able to minimize the importance of the nonlinear interactions which control the structure on smaller scales. However, it is not clear that this model, or any eddy viscosity model, can describe the evolution of structure over a range of scales in the interstellar medium, since this structure must be dominated by nonlinear interactions, and eddy viscosity models are purely dynamical, without any consideration of self-gravity, shock dissipation, or small-scale stellar energy inputs.

4.2 Quasi-normal closures

A large number of statistical studies of incompressible turbulence have been based on the assumption that the statistical properties of turbulence are in some sense nearly Gaussian, or "quasi-normal". One method is to assume not that the turbulent fluctuations are themselves Gaussian, but that the fourth-order moments can be related to the second-order moments in the same way as for a Gaussian stochastic process. Symbolically,

$$<X'X'X'X'> \; = \; <X'X'><X'X'>. \tag{26}$$

This closure method is called the "fourth order cumulant discard" technique in the literature.

It was later recognized that this method can yield negative energy spectra because it deprives the third-order correlations <X'X'X'> of a relaxation mechanism, which causes an unphysical overshoot. A successful, if somewhat non-physical, method to alleviate this problem is the "eddy-damped quasinormal Markovian approximation", which simulates the required relaxation essentially by writing

$$d<X'X'X'>/dt = <X'X'><X'X'> - \mu \ (<X'X'>)<X'X'X'>, \qquad (27)$$

where the relaxation function μ must be chosen on the basis of simplicity or some physical considerations. (See Poquet 1979, 1985 for reviews.) This damped quasinormal model has been used to study intergalactic MHD turbulence by DeYoung (1980), but has so far not been applied to interstellar problems, and so will not be discussed further here.

There are two astrophysical applications of the original quasinormal method which may be relevant to turbulence in self-gravitating interstellar structures. First, Olson and Sachs (1973) extended the classical study of Proudman and Reid to study the decay of vorticity in an expanding medium, meant to represent the Hubble flow. When time-reversed, their result suggests that the vorticity will grow rapidly in a contracting medium; in fact, the vorticity becomes infinite in a finite time. This suggests that vortical interactions, thought to be dominant in incompressible turbulence, may be important in contracting interstellar clouds.

The most detailed paper relevant to interstellar structure of which I am aware is that of Sasao (1973), who studied the generation of weak density fluctuations by incompressible vorticity fluctuations in a self-gravitating medium. Although his work was aimed at the fragmentation of protogalaxies, Sasao's results can be applied directly to the problem of how clumps may be generated by vortical fluctuations in interstellar clouds. The correlation equations for the spectrum of density fluctuations were solved by assuming that: 1. The initial turbulent energy spectrum $E(k)$ is specified; 2. The fourth-order cumulant discard closure method is used; however Sasao does not assume that the third-order correlations are Gaussian, which will turn out to be an important feature of his results; 3. The compressible modes are assumed to be only weakly nonlinear, so that an expansion solution can be obtained.

Without reviewing the details of Sasao's calculations and results, there are two potentially important conclusions which should be noted.

First, if the input turbulent spectrum $E(k)$ has a sharp peak at a scale above the thermal Jeans wavelength, then it forms density fluctuations whose spectrum has a sharp peak at about the same wavelength. But if $E(k)$ is a power law (e.g., a Kolmogorov spectrum), spreading to both sides of the Jeans wavelength, the result is a density fluctuation spectrum peaked at the thermal Jeans wavelength. This means that if vortical turbulence grows in a collapsing cloud, as was suggested by the Olson and Sachs (1973) analysis, then, if this turbulence has a broad energy spectrum, it will generate "fragments" of about the thermal Jeans length. The end result is therefore very similar to the ordinary conception of gravitational fragmentation, although the physical process by which the fragments are produced is very different.

Second, turbulence can assist, as well as prevent, gravitational contraction. Most discussions of the effect of turbulence on collapse view the turbulence as providing a "turbulent pressure" which resists self-gravity. This conception traces back to a derivation of the virial theorem by Chandrasekhar which gives a term $\rho \Delta v^2/2$ as the dominant effect of turbulence. However, Chandrasekhar neglected third order correlations like $<\rho'u_iu_j'>$ which Sasao shows may be important. If such terms coupling the density and velocity fluctuations are indeed important, then one could observe a cloud with a Δv^2, measured by the linewidth, larger than the gravitational energy, yet the cloud could still be bound and contracting. (An interesting alternative analysis of the Jeans criterion in a medium containing *supersonic* turbulence has recently been given by Bonazzola *et al.* [1986], who suggest that a gradient in the turbulent pressure can stabilize a cloud globally, while supersonic turbulent interactions can generate local instabilities leading to star formation; these latter interactions are related to the density-velocity correlations stressed by Sasao for *weakly* compressible turbulence.)

Obviously Sasao's treatment is restricted to a particular process for generating density fluctuations, and involves several assumptions and approximations. However, it does give some indication of how statistical turbulence theories may be used to study interstellar turbulent interactions, and suggests that there may exist major effects which would not be expected on the basis of our usual conceptions.

4.3. Probability density functional approach

A functional is a quantity which depends on *all* the values some function $y(x,t,...)$ takes in some interval of $x,t,...$, as occurs in the calculus of variations. A probability density functional, similarly, is a probability density which depends on all the values of some *stochastic* functions, which in the case of interest might be the density and velocity fields $\rho(x,t)$ and $u(x,t)$. A knowledge of the probability functional, $P[\rho(x,t),u(x,t),...]$ here, would provide a *complete* solution to any stochastic problem, since it is equivalent to the n-point correlation function with $n \rightarrow \infty$. Worded differently, the probability functional is a solution of a Liouville-type equation, rather than of some low-order moment equation, like the first exact moment equation of the BBGKY hierarchy in kinetic theory. For turbulence, the functional would give the probability of *all* possible solutions to the hydrodynamic equations subject to given initial and boundary conditions.

When the variables are constrained by the governing dynamical equations (here the hydrodynamic equations), functional differential equations can be derived for the evolution of the probability functional of the problem. The equation governing the probability functional for a weakly compressible self-gravitating turbulent fluid was presented by Sasao (1973), who used it to derive the equation for the energy spectrum. The use of probability functionals in statistical problems, especially turbulence, was pioneered by Hopf (1952), and an excellent monograph on the subject, concentrating on its relation to physical problems, is Beran (1968). An advanced treatment, applied to the incompressible MHD turbulence problem, is given in the book by Stanisic (1984).

As might be guessed, equations for the probability functional, while relatively straightforward to derive, cannot be solved except for very simple systems or with major approximations. Nevertheless, the characteristic functional (Fourer transform of the probability functional) method has been used to derive, for example, the form of the energy spectrum of turbulence for stationary, incompressible, randomly forced flows with no viscosity, the final decay stages of incompressible turbulence (no inertial or pressure terms), and incompressible MHD turbulence decay. (For references, see the texts mentioned above.) It is therefore of some interest to examine how this approach might be applied to interstellar turbulence.

The first step must be the generalization of the functional differential equation with the full continuity equation as well as the momentum equation used as constraints, so that the effects of compressibility can be explored. One paper of relevance is the study by Moiseev *et al.* (1976), who examined the statistical properties of a compressible adiabatic turbulent fluid with an imposed large amplitude external fluctuating force. The external force is required to maintain a steady state in the presence of viscous dissipation. Moiseev *et al.* use the probability functional equations to show that, if a self-similar ("inertial range") solution is assumed to exist, the energy spectrum has the form

$$E(k) \propto k^{-(5\gamma-1)/(3\gamma-1)} \tag{28}$$

where γ is the adiabatic index. This gives the Kolmogorov $k^{-5/3}$ spectrum when $\gamma \to \infty$ (incompressible) and a k^{-2} spectrum when $\gamma = 1$ (isothermal). They also give the correlation functions for the pressure, density, and vorticity fluctuations, and the compressible analogues of the Kolmogorov microscales. This result cannot be directly applied to interstellar turbulence because the adiabatic assumption suppresses entropy variations in shocks (viscosity is assumed to dominate the dissipation) and because self-gravity is not included.

A treatment of self-gravity is the second required extension of the probability functional approach. Moiseev *et al.* assumed that the external force was a prescribed process (in their case Gaussian) independent of the state of the system, but noted that the fluctuating force could also be generalized to a functional of the hydrodynamic variables subject to constraints. This coupling of the force to the flow is just what is required to include self-gravity, the constraint being Poisson's equation. Sasao (1973) included self-gravity in his formulation of the functional equations, but his assumption of weak compressibility greatly simplified the derivation.

The inclusion of shock dissipation, which may occur over a range of scales in the interstellar medium, requires that the energy equation, as well as the continuity equation, be incorporated as a constraint in deriving the equation for the probability functional. This appears quite difficult because of the nonlinear dependence of the cooling function on density and temperature.

Concerning the possibility of considering the effects of a magnetic field, the reader is referred to Stanisic's (1984) book for a treatment of the incompressible MHD case.

To summarize, there exists no convincing statistical method which can adequately describe even incompressible turbulence, although several interesting proposed closure methods have been described. None of these approaches have been generalized to treat realistic models for interstellar turbulence, a goal which should be pursued in future work. It is also important to note that these statistical methods for the most part can only give information on two-point correlation functions, which are insensitive to spatial structure (Houlahan and Scalo 1986, in preparation). For this reason, the goal of the statistical approach to interstellar turbulence should be the identification of characteristic scales at which the effects of compressibility, shocks, self-gravity, stellar momentum and energy inputs, and magnetic fields should cause *changes* in the structure, or else a demonstration that a self-similar structure can persist in the presence of these various processes and their interactions.

5. Local Model Equations and Chaotic Behavior

Fluid turbulence is a spatial *and* temporal phenomenon. The primary reason why both the direct numerical and statistical approaches discussed above have proven intractable has to do with the difficulty associated with properly treating the coupling between a large range of *spatial* scales. It is natural for studies of interstellar turbulence to concentrate on the spatial aspects of the problem, since we cannot directly observe the temporal evolution of interstellar structure. Nevertheless, from a theoretical viewpoint, an interesting question is: can we learn something about the conditions which give rise to interstellar turbulence by suppressing most or all spatial variations and concentrating on the temporal evolution? Such an approach would result in ordinary differential equations representing the evolution of properties of the interstellar medium, whose solutions may reveal unpredictable, "chaotic" behavior akin to turbulence, behavior whose dependence on physical parameters may be gleaned from a study of these equations. This approach has assumed prominence in recent years as a result of studies of ecological, dynamical, biological, fluid, chemical, and other systems. Reviews and references can be found in, for example, May (1983), Lichtenberg and Lieberman (1983), Cvitanovic (1984), and especially Spiegel (1985) for applications to fluid turbulence.

Since the ISM is a fluid, one might attack the problem by Fourier analyzing the hydrodynamic equations and then simplifying the resulting (infinite) set of equations by keeping only a small number of modes corresponding to, say, the smallest wavenumbers. This was the approach used in the pioneering paper by Lorenz (1963), who demonstrated unpredictable behavior in a deterministic third-order system of differential equations, and subsequent extensions to include more modes have shown qualitatively similar behavior. However, the ISM is enormously more complicated than the fluid systems which have been examined so far, and there is no *a priori* way of knowing which modes to retain or suppress. In addition, the multi-phase aspects of the ISM and the dominance of cloud interactions in some models suggest a similarity to the model equations used in population dynamics (e.g., predator-prey system) or chemical reaction problems. The equations which have been used to date in studies of the nonlinear behavior of the ISM are most related to these latter fields, and, for the most part, suppress all spatial variations, concentrating on the temporal behavior of certain properties.

Because such equations ignore spatial variations, they will be referred to as *local*, or one-zone, equations. In addition, these equations do not follow rigorously from the exact equations governing the system of interest, but instead are based on conceptual models or approximations for the system, so they will be called local *model* equations.

The behavior found from studies of local (*and* nonlocal) model equations depends somewhat on the system studied, but, generally, it is found that if the system is sufficiently nonlinear, and/or contains a sufficient number of interactions or feedbacks between components, the system exhibits a generic behavior which depends on the value of one or more critical parameters. As the parameter is increased, the system variables undergo a transition from a stable equilibrium state to a stable limit cycle (periodic oscillations), to multiply periodic limit cycles, to a state in which the behavior is completely unpredictable, or "chaotic", even though the governing equations are completely deterministic. The chaotic behavior results from an extreme sensitivity to initial conditions, and is thought to be related to the similar behavior of incompressible turbulent fluids, which are governed by the deterministic Navier-Stokes equations. The fascinating properties of these chaotic solutions of model equations, such as the possibility of a universal bifurcation sequence and the fractal properties of the phase space structure, are beyond the scope of the present paper. The question of interest here is: Do certain model equations describing the evolution of the ISM exhibit such behavior?

Although this approach has been adapted to the problem of the evolution of the ISM only within the last few years, a number of relevant papers have appeared in that time, not all of which will be reviewed here. For example, Shore (1981) examined the linear stability of simple model equations for the star and gas mass fractions of a galaxy and later (Shore 1983) showed how solutions and stability are modified by spatial effects represented by a diffusion term and a term meant to represent spiral arms. The present discussion will be restricted to two types of models: phase models and cloud interaction models.

5.1. Phase Models

The ISM is often thought of as consisting of a number of distinct phases with different temperatures and densities. For example, in the 3-phase model of McKee and Ostriker (1977) the ISM is divided into a hot (T $\gtrsim$ 3 x 10^5 K) rarefied phase due to supernova remnant cavities, a warm (T ~ 10^4 K) phase possibly due to evaporative interfaces between cool clouds and the hot phase, and a cool (T ~ 10-100 K) cloud component.
    Ikeuchi and Tomita (1983) investigated the linear stability and nonlinear behavior of this model. Letting X_h, X_w, and X_c denote the fractional masses of the hot, warm, and cold cloud phases, the model equations used to represent the time evolution include the following processes: (a) sweeping of warm gas into cold clouds by supernova remnants at a rate aX_w; (b) evaporation of cold clouds embedded in supernova remnant cavities at a rate $bX_cX_h^2$; and (c) radiative cooling of hot gas by interactions with the warm gas at a rate cX_wX_h. The coefficients a, b, and c contain the details of the physics, but only need to be known to order of magntiude for the present purposes. Ikeuchi and Tomita estimate a ~ 5 x 10^{-8} yr^{-8}, b ~ 10^{-7}-10^{-8} yr^{-1}, and c ~ 10^{-6}-10^{-7} yr^{-1}. The model equations are then

$$dX_h/dt = -cX_hX_w + bX_cX_h^2$$
$$dX_w/dt = -aX_w + cX_hX_w$$
$$dX_c/dt = -bX_cX_h^2 + aX_w.$$

By using the mass conservation constraint $X_h + X_w + X_c = 1$ and nondimensionalizing, these equations reduce to two equations for, say, X_c and X_h, with two parameters A = a/c and B = b/c. Ikeuchi and Tomita show that there are three types of solutions. First, if A > 1, a runaway occurs in which all the gas is converted to the hot phase. Second, if A < 1 and B > B_c $\equiv$ (1-2A)A^2, a stable stationary state results, with two or three phases, depending on the values of A and B. Third, if A < 1 and B < B_c, a stable limit cycle is produced, with the values of X_c, X_w, and X_h varying in a periodic manner. It is this last result which is of interest here, since the appearance of limit cycle behavior at a critical value of a parameter often signals the possibility of more complex behavior.
    Ikeuchi and Tomita also considered the effect of a periodic process supposed to convert warm gas into clouds, which might represent the effect of a galactic spiral shock wave. The results are quite interesting. For example, a stable equilibrium state can be altered to a limit cycle; single limit cycles become, except for special frequencies of the periodic process, multiply periodic or aperiodic, with the most complicated cases exhibiting behavior which is almost chaotic. One might expect that a stochastic, rather than a periodic, external force, as might be due to galaxy tidal interactions, would lead more easily to chaotic behavior. If it is assumed that the star formation rate depends on the fraction of mass in the cool phase, then the calculations imply that the cyclic or irregular phase changes will be accompanied by bursts of star formation.
    A detailed extension and quantification of this phase model has been given by Ikeuchi et al. (1984), who consider a six-component model and attempt to use physically realistic rate coefficients. The evolution is still controlled largely by supernova remnants, but the cool cloud phase is subdivided into three types of clouds: small diffuse clouds with M $\approx$ 40 $M_\odot$ formed by the breakup of dense shells swept up by supernova remnants, molecular clouds with M $\approx$ $10^4 M_\odot$ formed by collisions among the small clouds, and giant

molecular clouds with $M = 3 \times 10^5$ $M_\odot$ formed by gravitational instability of the ensemble of molecular clouds when the mass surface density of clouds exceeds a critical value given by the Toomre-Goldreich-Lynden-Bell condition. A discussion of their results and the dependence on parameters will not be given here because of space considerations. The important general result is that cyclic phase changes and periodic bursts of star formation were found to occur for a fairly wide range of the parameters. Unfortunately, the model is so complex that it is difficult to interpret the numerical calculations in terms of dominant physical interactions, or to guess whether the model might exhibit chaotic behavior for some values of the parameters not examined in the paper. It should also be noted that the model of Ikeuchi *et al.* is actually a hybrid between pure phase models and the cloud interaction models discussed below, since cloud collisions are supposed to convert small clouds into molecular clouds, and the cycle period is the mean cloud collision timescale in some cases.

A problem with these types of phase models is that they rest on the common notion that both small H I clouds and molecular clouds lie embedded within the hot coronal gas, while there is very little observational evidence for this picture. Instead, the cool ISM appears hierarchical, so that each cloud is embedded in a larger cloud structure.

A very different type of phase model has been discussed by Bodifee and deLoore (1983). Their model concentrates on the cloud complexes, and consists of 3 phases: a neutral component which cannot collapse to form stars, a molecular component which is assumed able to collapse because of the more efficient cooling by molecules, and an "active" component consisting of young stars and their associated H II regions. The mass leaving the system due to stellar evolution is assumed to be balanced by a reservoir of fresh material which enters the system in the atomic phase. The model equations include parameterized terms which are supposed to account schematically for molecular production, spontaneous star formation, and induced star formation. If A, M, and S are the masses of the atomic, molecular, and stellar components, the star formation rates are assumed to vary as M^{n_1}, where n_1 is a parameter, while the molecular formation rate is assumed proportional to $A^{n_2}(M + \alpha A)^{n_3} S^{-n_4}$, where n_2, n_3, n_4 and α are parameters.

A numerical exploration of the model equations for dA/dt and dM/dt (S follows from the condition $A + M + S = 1$) shows that limit cycles accompanied by bursts of star formation can occur for certain ranges of the parameters. Specifically, limit cycles occur fo $n_1 < 2$, $n_2 < 2$, $n_3 > 2$, and $\alpha < 0.2$ (i.e., molecular cooling must dominate atomic cooling in the assumed form of the term in the molecule formation rate). Whether these values, or even the forms adopted for the various terms in the model equations, are physically realistic is questionable. In particular, the assumption that gravitational collapse requires the enhanced cooling efficiency of molecules seems contrived; it is true that both star formation and molecule formation require high densities, but I know of no evidence that molecule formation is a causal agent for star formation, unless fragmentation is due to some thermal-chemical instability associated with molecular gas, which appears unlikely. However, this paper is still important as another demonstration of the fact that phase models with sufficient nonlinearity and couplings between components can give rise to limit cycles, although chaos cannot occur in this particular case because there are only two variables.

What does the nonlinear behavior found for phase models have to do with interstellar turbulence? First, it is plausible that any indications of chaotic temporal behavior in a local model will correspond to some type of irregular spatial structure. The local model equations can say nothing about the nature of the spatial structure, but they do yield the range of physical parameters under which a transition to turbulence may occur. However, the phase models studied to date do not give much indication of chaotic behavior (except possibly for externally forced models), just limit cycles. Nevertheless, as emphasized by Bodifee and deLoore (1985), many nonlinear systems are known to develop spatial self-organization ("dissipative structures") when a bifurcation to a limit cycle occurs, and such self-organization may be important in understanding interstellar structure. At this point what is needed is a systematic examination of various phase models and an attempt to generalize the model equations to include spatial structure. The addition of diffusion terms is the usual procedure in such problems; this may be a useful first step, even though diffusion may be a poor approximation to the actual hydrodynamic interactions which occur.

5.2 Cloud interaction models

By far the most common approach to the modelling of inhomogeneous interstellar structure is to idealize the ISM as a system of clouds whose interactions among themselves, with stars, and with the global gravitational field controls the temporal and spatial evolution of the system. An example is the well-known Oort cycle in which collisional coalescence of small clouds builds high-mass clouds which form stars that disrupt the massive clouds into a number of small clouds. Such models are similar in spirit to mixing length and eddy viscosity models for incompressible turbulence which, although incapable of providing details of the fluctuations and limited by the presence of unknown parameters, have proven useful in describing mean flow properties in engineering applications. In models like the Oort cycle, on the other hand, the emphasis is on dissipation and coalescence in cloud collisions and the feedback between stars and gas, rather than on phenomena associated with viscous stresses. It is these models, which I will refer to as "cloud models", that I will be discussing here.

For studies of interstellar turbulence, cloud models must be considered as an extreme idealization since they picture the clouds as discrete entities, ignore spatial correlations, and must make assumptions about the outcome of cloud interactions and the manner in which stars form and feed energy back to the clouds (for criticisms, see Scalo 1985 and Lattanzio et al. 1985). Nevertheless, several recent studies have demonstrated that cloud models can successfully account for many of the observed properties of spiral galaxies (see, e.g., Combes and Gerin 1985 and Tomisaka 1985 for collisional growth model, Hausman and Roberts 1984 for a model in which cloud collisions cause star formation) and even tidally interacting galaxies (Noguchi and Ishibashi 1986). Furthermore, the two-fluid (stars + gas) hydrodynamic simulations of Chiang and Prendergast (1985) show that small perturbations in an initially uniform medium develop behavior which is very similar to that postulated in the Oort model. In addition, the Oort cycle can be adapted to a model in which supernovae and OB wind shells drive fragmentation (McCray and Kafatos 1986). Thus, although the idealized nature and the limitations of discrete cloud models must be kept in mind, it appears that the implications of such models for interstellar turbulence deserve further study.

Scalo and Struck-Marcell (1984) tried to develop a general continuum description of cloud models by averaging a kinetic equation over the cloud mass and velocity distributions. These "cloud fluid" equations are similar to the usual hydrodynamic equations except that they contain various sources and sinks of cloud number, mass, and energy. They are also similar to the equations used by Larson (e.g., Larson 1976) in his models for protogalaxy collapse, except that they allow for more processes, like collisional growth and disruption of clouds, evaporation by a hot intercloud medium, etc., and follow the evolution of the mean cloud mass. In the language of turbulence, the cloud fluid equations are the analogues of the mean flow equations for incompressible turbulence with the fluctuation terms modelled phenomenologically. The cloud fluid equations are limited in some respects, compared with N-body simulations, but are better suited for exploratory work. Also, Tomisaka (1986) has shown that cloud fluid models agree fairly well with N-body simulations of gas in spiral galaxies.

When spatial gradients are ignored, the cloud fluid equations reduce to a set of three ordinary differential equations for the evolution of the number density of clouds n, the mean cloud mass m, and the cloud velocity dispersion c^2. If it is assumed that the mass of the system is constant, then there are only two equations, since the mass density is $\rho = nm$. The behavior of these equations for the Oort cycle, and for a modification of the model in which field stars rather than protostars are assumed to accelerate clouds, has been examined in detail by Struck-Marcell and Scalo (1986; see Scalo and Struck-Marcell 1986 for a brief summary). Although a number of parameters appear in the equations, it was found that the qualitative behavior depends only on one key parameter; which is the ratio T_d of the average time delay between the formation of a star-forming cloud and the disruption of the cloud by internal stellar activity to the equilibrium cloud collision timescale.

When T_d is small ($\lesssim 0.3 - 0.5$ in most cases), perturbations are damped and the system returns to equilibrium. However, when T_d exceeds a critical value of order unity, the system relaxes to a stable limit cycle which is driven by a coalescence overshoot, as the massive clouds can continue to grow for a finite time before they are disrupted. The cycles are accompanied by bursts of star formation and collisional dissipation. As T_d is increased further, the system develops doubly and then multiply periodic limit cycles, aperiodic behavior, and eventually, a transition to chaotic behavior. An example is given in Figure 7.

The observational implications of these results as they might apply to starburst galaxies, especially those triggered by galactic tidal interactions, are discussed in detail in Struck-Marcell and Scalo (1986). The important point for the present discussion is the appearance of chaotic behavior, the first such result in the area of ISM and star formation theory. That the evolution is truly chaotic was suggested by visual inspection of trajectories in the n-c^2 phase plane, by the divergence of models with slightly different initial conditions, and (equivalently) by an approximate estimate of Lyapunov exponents. The approach to chaos appears to be a period-doubling bifurcation cascade, and the chaos may temporarily clear in a narrow window of T_d values, with the attractor reverting to limit cycle behavior. These features are common to many systems of nonlinear ordinary differential equations which exhibit chaotic behavior.

For the model investigated in detail, chaos only occurred when $T_d \gtrsim 3$, so one may ask whether such large values of the time delay parameter are relevant to the ISM. The answer appears to be yes. A number of indirect arguments suggest that the lifetimes of massive clouds are $3 \times 10^7 - 10^8$ yr. (a summary is given in Scalo 1985), similar to or greater than common estimates for the mean cloud collision timescale in the solar neighborhood, implying $T_d \gtrsim 1$. Furthermore, if the massive cloud lifetime is independent of environmental conditions, then the ratio T_d is approximately proportional to the mean gas mass density, suggesting that T_d could become very large in regions which have suffered compression by, for example, a spiral density wave or a direct galaxy collision. The problem with this argument, and with the model in general, is that it takes no account of the hierarchical nature of interstellar cloud structure, but instead, assumes that the clouds are randomly distributed. Cloud collision timescales derived using the observed velocity dispersion and mean number of clouds per unit line of sight are ill-defined for a hierarchical structure. One way around this problem would be to assume that massive complexes are indeed formed by collisional coalescence of smaller clouds, but that their hierarchical structure is a result of internal instabilities, such as gravitational fragmentation.

In any case, it is significant that an examination of a simple cloud model yielded chaotic behavior and was able to isolate the critical physical parameter governing the evolution. The relation, if any, between this chaos and real interstellar turbulence is unclear; the same situation occurs in studies attempting to understand incompressible turbulence using model equations with chaotic solutions. Spiegel (1985, p. 304) writes that "Ordinary chaos is not turbulence, but turbulence is chaotic. The more we learn about chaos, the better we will understand turbulence." This assessment would seem to apply just as well to interstellar turbulence, and will hopefully motivate further studies of model equations based on different physical assumptions concerning cloud evolution and star formation.

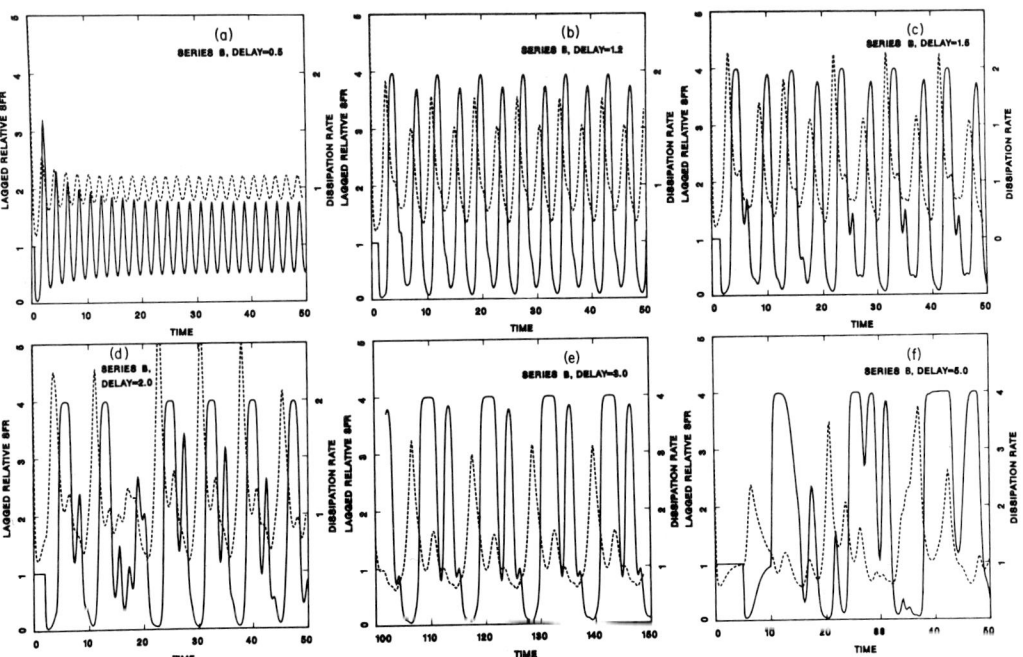

Figure 7. Time evolution of the relative star formation rate (solid curves) and collisional dissipation rate (dashed curves) for Oort cycle model equations using a number of values of the time delay parameter T_d, which is the ratio of the time between formation and disruption of the star-forming clouds to the equilibrium cloud collision timescale. The unit of time is the equilibrium cloud collision timescale. For $T_d < 0.5$ (not shown), the system damps to equilibrium. As T_d is increased, the system exhibits bifurcation to a simple limit cycle (a), doubly (b) and multiply periodic (c and d) limit cycles, with slight aperiodicity in (d), a return to a simpler periodic cycle (e), and a transition to chaotic behavior (f).

6.   Conclusion

It may seem somewhat disheartening that this review is conspicuously lacking in any firm conclusions. Instead, I have simply summarized reasons for considering the ISM as turbulent, enumerated some of the many physical interactions which may play a role in controlling the observed structure, and outlined several theoretical approaches which have been used in other turbulence problems and which might provide a useful basis for extensions to the more difficult interstellar problem. While I have indicated some weaknesses and strengths of several proposals, it would be premature to rule out any of them given the fact that the modern study of interstellar turbulence is in its infancy.

It must be remembered that astronomers recognized the importance of some form of turbulence in the 1950's, but several symposia designed to initiate contact between astronomers and turbulence theorists only succeeded in making apparent the intractability of the problem. It was apparently for this reason that the field remained dormant, except for a few scattered papers, for nearly three decades. However, mounting observational evidence, largely from millimeter line observations in the 1970's, has forced us to consider the problem anew. Fortunately, there has been substantial progress in turbulence research in the intervening time, due to advances in computer capabilities, innovations in numerical techniques, increased sophistication in experimental work, the application of new mathematical tools, and a broadening of viewpoints which brought to the field an awareness of the importance of spatial structure, self-organization, and chaotic deterministic behavior. The problem remains extremely difficult, and it is clear that any real understanding of interstellar turbulence must be a long-term goal requiring simultaneous attacks using a variety of approaches. Nevertheless, it often pays to fish in troubled waters, and the time seems ripe for the initiation of such a programme. With some luck and hard work, it should at least be possible to agree within the next few years on the relative importance of various physical processes and the scales at which they dominate, to observationally quantify the nature of interstellar structure, and perhaps even obtain a glimmer of realization concerning the apparent and ubiquitous coexistence of order and choas, so that we may begin to understand why the word "turbulence" cannot be defined.

References

Arny, T. 1971, *Ap. J.*, **169**, 289.
Armstrong, J. W., and Rickett, B. J. 1981, *M.N.R.A.S.*, **194**, 623.
Baierlein, R. 1983, *M.N.R.A.S.*, **205**, 669.
Bally, J., Dragovan, M., Langer, W. D., Stark, A. A., and Wilson, R. W. 1987, in *Summer School on Interstellar Processes*, NASA Tech. Mem., ed. D. J. Hollenbach and H. A. Thronson.
Bania, T. M., and Lyon, J. C. 1980, *Ap. J.*, **239**, 173.
Bash, F. N., Hausman, M., and Papaloizou, J. 1981, *Ap. J.*, **245**, 92.
Beran, M. J. 1968, *Statistical Continuum Theories*, (New York: Interscience).
Bodenheimer, P. 1978, *Ap. J.*, **224**, 488.
Bodifee,, G., and de Loore, C. 1985, *Astr. Ap.*, **142**, 297.
Bonazzola, S., Falgarone, E., Heyvaerts, J., Perault, M. and Puget, J. L. 1987, in *Summer School on Interstellar Processes*, NASA Tech. Mem., ed. D. J. Hollenbach and H. A. Thronson.

Boss, A. P. 1980, *Ap. J.*, **237**, 866.

———. 1984, *M.N.R.A.S.*, **209**, 543.

Buff, J., Gerola, H., and Stellingwerf, R. F. 1979, *Ap. J.*, **230**, 839.

Canuto, V. M., and Battaglia, A. 1986, preprint.

Canuto, V. M., and Goldman, I. 1985, *Phys. Rev. Letters*, **54**, 430.

Cantwell, B. J. 1981, *Ann. Rev. Fluid Mech.*, **13**, 457.

Chiang, W.-H., and Prendergast, K. H. 1985, *Ap. J.*, **297**, 507.

Clifford, P., and Elmegreen, B. G. 1983, *M.N.R.A.S.*, **202**, 629.

Combes, F., and Gerin, M. 1985, *Astr. Ap.*, **150**, 327.

Cvitanovic, P. 1984, in *Universality in Chaos*, ed. P. Cvitanovic (Bristol: Adam Hilger Ltd.), p. 3.

Deissler, R. G. 1984, *Rev. Mod. Phys.*, **56**, 223.

DeYoung, D. S. 1980, *Ap. J.*, **241**, 81.

Dickman, R. L. 1985, in *Protostars and Planets II*, ed. D. C. Black and M. S. Matthews (Tucson: Univ. of Arizona Press), p. 150.

Dwoyer, D. L., Hussaini, M. Y., and Voigt, R. G. (eds.) 1985, *Theoretical Approaches to Turbulence* (New York: Springer-Verlag).

Eckmann, J. P. 1981, *Rev. Mod. Phys.*, **53**, 643.

Eilik, J. A., and Henriksen, R. N. 1984, *Ap. J.*, **277**, 820.

Elmegreen, B. G. 1985, *Ap. J.*, **299**, 196.

Elmegreen, B. G., and Chiang, W.-H. 1982, *Ap. J.*, **253**, 666.

Elmegreen, B. G., and Lada, C. J. 1977, *Ap. J.*, **214**, 725.

Falgarone, E., and Puget, J. L. 1986, *Astr. Ap.*, in press.

Ferrini, F., Marchesoni, F., and Vulpiani, A. 1983, *Ap. Sp. Sci.*, **96**, 83.

Field, G. B. 1978, in *Protostars and Planets*, ed. T. Gehrels (Tucson: University of Arizona Press), p. 243.

Fleck, R. C. 1980, *Ap. J.*, **242**, 1019.

———. 1981, *Ap. J. (Letters)*, *246*, L151.

———. 1982, *M.N.R.A.S.*, **201**, 551.

———. 1983, *Ap. J. (Letters)*, **272**, L45.

Franco, J. 1983, *Ap. J.*, **264**, 508.

———. 1984. *Ap. J.*, **137**, 85.

Franco, J., and Cox, D. P. 1983, *Ap. J.*, **273**, 243.

Frisch, U., Hasslacher, B., and Pomeau, Y. 1986, *Phys. Rev. Lett.*, **56**, 1505.

Gampert, B. (ed.) 1985, *The Influence of Polymer Additives on Velocity and Temperature Fields* (New York: Springer-Verlag).

Gilden, D. L. 1984, *Ap. J.*, **279**, 335.

Goldsmith, P. F., and Arquila, R. 1985, in *Protostars and Planets II*, ed. D. C. Black and M. S. Matthews (Tucson: Univ. of Arizona Press), p. 137.

Hakens, H. 1983, *Synergetics* (New York: Springer-Verlag).

Hardy, J., de Pazzis, O., and Pomeau, Y. 1976, *Phys. Rev. A.,* **13**, 1949.

Hasegawa, A. 1985, *Adv. Phys.*, **34**, 1.

Hausman, M. A., and Roberts, W. W. 1984, *Ap. J.*, **282**, 106.

Heisenberg, W. 1948, *Proc. Roy. Soc.*, A**195**, 402.

Helleman, R. H. G. 1980, *Fundamental Problems in Statistical Mechanics*, **5**, 165.

Henriksen, R. N., and Turner, B. E. 1984, *Ap. J.*, **287**, 200.

Hopf, E. 1952, *J. Ratl. Mech. Anal.*, **1**, 87.

Hopper, P. B., and Disney, M. J. 1974, *M.N.R.A.S.*, **168**, 639.

Hsu, J.-C. 1984, Ph.D. dissertation, University of Texas.

Hughes, V. A., and Wouterloot, J. G. A. 1984, *Ap. J.*, **276**, 204.

Ikeuchi, S., Habe, A., and Tanaka, Y. D. 1984, *M.N.R.A.S.*, **207**, 909.

Ikeuchi, S., and Tomita, H. 1983, *P.A.S. Japan*, **35**, 56.
Jog, C. J., and Ostriker, J. P. 1987, in *Summer School on Interstellar Processes* , NASA Tech. Mem., ed. D. J. Hollenbach and H. A. Thronson.
Kleiner, S. C., and Dickman, R. L. 1985, *Ap. J.*, **286**, 255.
_____. 1985, *Ap. J.*, **295**, 466.
_____. 1986, preprint.
Kolmogorov, A. 1941, *Compt. Rend. Acad. Sci.,* URSS, **30**, 301. English translation in *Turbulence*, ed. S. Friedlander and L. Topper (New York: Interscience).
Krebs, J., and Hillebrandt, W. 1984, *Astr. Ap.*, **128**, 411.
Lanford, O. E. 1982, *Ann. Rev. Fluid Mech.*, **14**, 347.
Larson, R. B. 1979, *M.N.R.A.S.*, **186**, 479.
_____. 1981, *M.N.R.A.S.*, **194**, 809.
_____. 1984, *M.N.R.A.S.*, **206**, 197.
Larson, R. R. 1985, *M.N.R.A.S.*, **214**, 379.
Lattanzio, J. C., Monaghan, J. J., Pongracic, H., and Schwarz, M. P. 1985, *M.N.R.A.S.*, **215**, 125.
Leung, C. M., Kutner, M. L., and Mead, K. N. 1982, *Ap. J.*, **262**, 583.
Lichtenberg, A. J., and Lieberman, M. A. 1983, *Regular and Stochastic Motion* (New York: Springer-Verlag).
Lorenz, E. N. 1963, *J. Atmos. Sci.*, **20**, 130.
Lumley, J. L. 1981, in *Transition and Turbulence*, ed. R. E. Meyer (New York: Academic Press), p. 215.
Lynds, B. T. 1962, *Ap. J. Suppl.*, **7**, 1.
Maddalena, R. J., Morris, M., Moscowitz, J., and Thaddeus, P. 1986, *Ap. J.*, **303**, 375.
Martin-Pintado, J., Wilson, T. L., Johnston, K. J., and Henkel, C. 1985, *Ap. J.*, **299**, 386.
May, R. M. 1983, in *Chaotic Behavior of Deterministic Systems*, ed. G. Iooss, R. H. G. Helleman, and R. Storz (North Holland Publ. Co.), p. 515.
McCray, R., and Kafatos, M. C. 1986, preprint.
McDavid, D. 1984, *Ap. J.*, **284**, 141.
McKee, C. F., and Ostriker, J. P. 1977, *Ap. J.*, **218**, 198.
Moiseev, S. S., Tur, A. V., and Yanovskii, A. 1976, *Sov. Phys. JETP*, **44**, 556.
Monetti, A., Pipher, J. L., Helfer, H. L, McMillan, R. S., and Perry, M. L. 1984, *Ap. J.*, **282**, 508.
Myers, P. C. 1983, *Ap. J.*, **270**, 105.
Nicolas, G., and Prigogine, I. 1968, *Self-Organization in Nonequilibrium Systems* (New York: Wiley & Sons).
Noguchi, M., and Ishibashi, S. 1986, *M.N.R.A.S.*, **219**, 305.
Norman, C. A., and Silk, J. 1980, *Ap. J.*, **238**, 158.
O'Dell, C. R. 1986, *Ap. J.*, **304**, 767.
Olson, D. W., and Sachs, R. K. 1973, *Ap. J.*, **185**, 91.
Orzag, S. A. 1970, *J. Fluid Mech.*, **41**, 363.
Ostriker, J. P., and Cowie, L. L. 1981, *Ap. J. (Letters)*, **243**, L127.
Peebles, P. J. E., and Dicke, R. H. 1968, *Ap. J.*, **154**, 891.
Perault, M., Falgarone, E., and Puget, J. L. 1985, *Astr. Ap.*, **153**, 371.
Poquet, A. 1979, *Intern. J. Fusion Energy*, **2**, 39.
_____. 1985, in *Theoretical Approaches to Turbulence*, ed. D. L. Dwoyer, M. Y. Hussaini, and R. G. Voigt (New York: Springer-Verlag), p. 209.
Pumphrey, W. A., and Scalo, J. M. 1986, submitted to *Ap. J.*
Quiroga, R. J. 1983, *Ap. Sp. Sci.*, **93**, 37.

Roberts, W. W., and Hausman, M. A. 1984, *Ap. J.*, **277**, 744.
Roczyczka, M. 1985, *Astr. Ap.*, **143**, 59.
Rossano, G. S. 1978, *A. J.*, **83**, 241.
Roy, J. R., and Joncas, G. 1985, *Ap. J.*, **288**, 142.
Roy, J. R., Arsenault, R., and Joncas, G. 1986, *Ap. J.*, **300**, 624.
Sanford, M. T., Whitaker, R. W., and Klein, R. I. 1984, *Ap. J.*, **282**, 178.
Sanders, D. B., Scoville, N. Z., and Solomon, P. M. 1985, *Ap. J.*, **289**, 373.
Sasao, T. 1973, *Publ. Astr. Soc. Japan*, **25**, 1.
Scalo, J. M. 1985, in *Protostars and Planets II*, ed. D. C. Black and M. S. Matthews (Tucson: Univ. of Arizona Press), p. 201.
Scalo, J. M., and Pumphrey, W. A. 1982, *Ap. J. (Letters)*, **258** L26.
Scalo, J. M., and Struck-Marcell, C. 1984, *Ap. J.*, **276**, 60.
Schloerb, F. P., Snell, R. L., Goldsmith, P. F., and Morgan, J. A. 1986, in *Summer School on Interstellar Processes*, NASA Tech. Mem., ed. D. J. Hollenbach and H. A. Thronson.
Schneider, S., and Elmegreen, B. G. 1979, *Ap. J. Suppl.*, **41**, 87.
Seiden, P. E., and Gerola, H. 1982, *Fund. Cosmic Phys.*, **7**, 241.
Shore, S. N. 1981, *Ap. J.*, **249**, 93.
_____. 1983, *Ap. J.*, **265**, 202.
Shu, F., Lizano, S., and Adams, F. 1986, in *Star Forming Regions*, IAU Symp. No. 115, ed. M. Peimbert and J. Jugaku (Dordrecht: Reidel).
Silk, J. 1985, *Ap. J. (Letters)*. **292**, L71.
Spiegel, E. A. 1985, in *Theoretical Approaches to Turbulence*, ed. D. L. Dwoyer, M. Y. Hussaini, and R. G. Voigt (New York: Springer-Verlag), p. 303.
Stanisic, M. M. 1985, *The Mathematical Theory of Turbulence* (New York: Springer-Verlag).
Struck-Marcell, C., and Scalo, J. M. 1986, *Ap. J.*, in press.
Tennekes, H., and Lumley, J. L. 1972, *A First Course in Turbulence* (Cambridge: MIT Press).
Tomisaka, K. 1984, *Publ. Astr. Soc. Japan*, **35**, 456.
_____. 1986, *Publ. Astr. Soc. Japan*, **38**, 95.
Tsuge, S. 1974, *Phys. Fluds*, **17**, 22.
Verschuur, G. A. 1974a, *Ap. J. Suppl.*, **27**, 65.
_____. 1974b, *Ap. J. Suppl.*, **27**, 283.
von Hoerner, S. 1962, in *The Distribution and Motion of Interstellar Matter in Galaxies*, ed. L. Woltjer (New York: Benjamin), p. 193.
von Weizäcker, C. F. 1951, *Ap. J.*, **114**, 165.
Vrba, F. J. 1977, *A. J.*, **82**, 198.
Wolffram, S. 1983, *Rev. Mod. Phys.* **55**, 601.
Wood, D. 1982, *M.N.R.A.S.*, **199**, 331.
Zuckerman, B., and Evans, N. J. 1974, *Ap. J. (Letters)*, 192, L149.
Zweibel, E. G., and Josaffetson, K. 1983, *Ap. J.*, **270**, 511.

PART 2

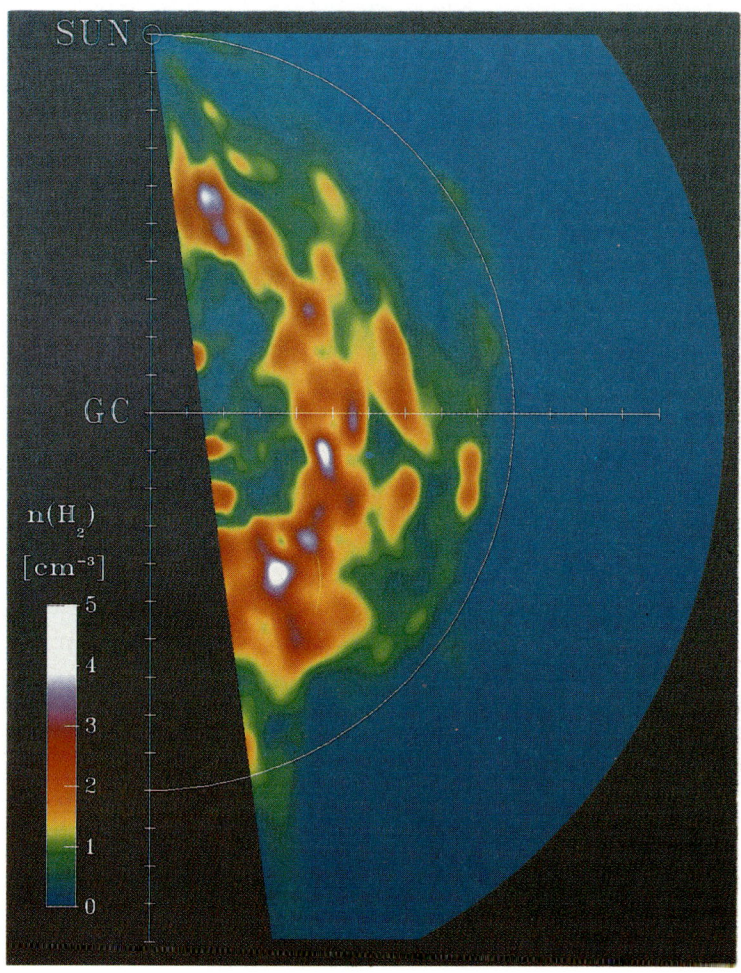

Face-on image of the peak molecular hydrogen density n(H₂) as viewed from the North Galactic Pole. This picture was produced by D. Clemens (U. Arizona), N. Scoville and D. Sanders (both Caltech) at the Digital Image Analysis Laboratory, University of Massachusetts with the assistance of D. Chesley and R. Newton. Over 40,000 individual CO spectra, obtained with the 14m telescope of the Five College Radio Astronomy Observatory, University of Massachusetts, were used to create the image.

Section I: Interstellar Dust Grains

Kris Sellgren and Diane Wooden

COMPOSITION, STRUCTURE, AND CHEMISTRY OF INTERSTELLAR DUST

A. G. G. M. Tielens[1,2] and L. J. Allamandola[1]

[1]Space Sciences Division, MS 245-6
NASA Ames Research Center
Moffett Field, California 94035, U.S.A.

[2]Space Sciences Laboratory
University of California
Berkeley, California 94720, U.S.A.

INTRODUCTION AND SUMMARY

Interstellar dust is an important component of the interstellar medium. It dominates the heating and cooling of clouds through energetic photoelectrons and gas-grain collisions (see the chapter by Black in this volume). It is also the dominant source of opacity and determines, therefore, the spectrum of dust-enshrouded objects. Unquestionably, molecular hydrogen, the most abundant gas-phase species in molecular clouds, is formed on grain surfaces. Other gas-phase molecules might result from reactions on grain surfaces as well. Grains can also influence the gas-phase composition of molecular clouds indirectly, because they may lock up some of the elements. The metals (e.g., iron) are particularly important in this respect since they tend to carry the charge inside a dense cloud and, thus, regulate the ion-molecule chemistry (see the chapter by Prasad in this volume). Such depletions will, of course, also influence the cooling balance.

Yet, despite the importance of the dust and despite 50 years of active research in this field, many of the major questions either remain unanswered or the answers that have been proposed are highly controversial. These include the two basic ones: What is the composition of the interstellar dust? and Where and how is it formed? These questions are of course interrelated and an answer to one suggests an answer to the other.

The dust components proposed to be present in the interstellar medium can be divided into two classes depending on their formation

---

This work was supported through NASA grant NCA2-1R050-405.

D. J. Hollenbach and H. A. Thronson, Jr. (eds.), Interstellar Processes, 397–469.

history. First, there is stardust, such as silicates, graphite and
amorphous carbon, which are made in the oxygen-rich or carbon-rich
outflow from late-type giants and planetary nebulae and, possibly, novae
and supernovae. Dust formation has been theoretically predicted to
occur in such outflows (Hoyle and Wickramasinghe 1962; Kameijo 1963).
This was subsequently confirmed, mainly by infrared observations, in the
early seventies. There is now abundant evidence for the formation of
dust in these high-density, high-temperature environments (cf. the
reviews by Merrill 1977 and Aitken 1981). Second, there are dust components that are formed in the interstellar medium itself. This includes
"icy" grain mantles, consisting of simple molecules (e.g., H_2O, NH_3,
CH_3OH, and CO) inside dense clouds, as well as an organic refractory
dust component, consisting of more complex molecules, in the diffuse
interstellar medium. The presence of both of these dust components in
the interstellar medium has been revealed by infrared spectroscopy of
highly obscured objects. The icy grain mantles, which form by accretion
and reaction of gas-phase species on preexisting cores, seem to be
ubiquitous inside molecular clouds (Merrill et al. 1976; Joyce and Simon
1982; Willner et al. 1982). In contrast, the organic refractory component has, at present, only been detected toward the galactic center
(Willner et al. 1979; Allen and Wickramasinghe 1981; Butchart et al.
1985). Presumably, there is a chemical, evolutionary link between these
two dust components. That is, the organic refractory dust component is
formed by energetic processing (UV photolysis or cosmic ray bombardment)
of the icy grain mantles (Greenberg 1982; Moore and Donn 1982; Strazulla
et al. 1983).

Detailed models of interstellar extinction have been constructed
based on stardust alone (e.g., Mathis et al. 1977; Draine and Lee 1984),
as well as on a combination of stardust and interstellar medium dust
(Greenberg and Hong 1974; Greenberg and Chlewicki 1984). Not too much
credence should be given to such models since the interstellar extinction curve is quite insensitive to the exact composition of the dust.
This is because other parameters (e.g., size distribution, shape, and
surface roughness) will influence the extinction curve of an ensemble of
particles as well. This non-uniqueness is compounded by the variations
in the optical properties, even within one class of materials such as
silicates, and the resulting uncertainty in the optical properties of
interstellar grain materials. At this moment, the most diagnostic
information on the dust composition results, therefore, from features in
the extinction curve, notably in the infrared, rather than from the
overall shape of the extinction curve.

In §1, a compilation of the different dust components observed to
be present in the interstellar medium and their relative importance is
presented. These include the stardust components amorphous silicates,
amorphous carbon, polycyclic aromatic hydrocarbons (PAHs), and graphite,

as well as the interstellar medium components, which are the organic
refractory grains and icy grain mantles. These data are summarized in
Table I. It is emphasized that the uncertainties in these observations
preclude a definite choice between the different proposed dust models,
although it seems that stardust may dominate the dust composition in the
solar neighborhood.

In §2, the difference between amorphous and crystalline materials
is examined. Some attention is given to the structure of truly amorphous materials (e.g., random network materials) such as glasses (e.g.,
silicates) and ices. The microcrystalline structure of soot particles
is also discussed with an emphasis on the structural differences and
similarities between graphite sheets, amorphous carbon particles, and
PAH molecules. This section is concluded with a discussion of the
structure of amorphous polymer solids, appropriate for the organic
refractory dust component.

In §3, the optical properties of amorphous and crystalline materials are contrasted. The emphasis is on silicates (amorphous), because
the properties of carbon grains are to some extent already covered in
the chapter by Allamandola, Tielens, and Barker in this volume. The
optical properties of materials divide naturally into three regions
which are discussed separately. The first is the optical and UV region,
which is dominated by the electronic properties of the material. Section 3.1 focuses on understanding the observed, enhanced absorption at
about 2 µm of silicates formed in red giant envelopes. This absorption,
which lies within the band gap of silicates, is due to localized electronic states, probably caused by the presence of impurities such as FeO
in the silicate. The second is the mid-infrared region, which for
amorphous materials is dominated by the fundamental vibrational modes of
the nearest-neighbor atoms. Special emphasis is given in §3.2 to the
large width of the 10 µm Si-O stretching vibration in amorphous silicates. It is suggested that the decreased width of the 10 µm feature in
absorption as compared to emission is due to the low temperature of the
dust in the interstellar medium. The third is the far-infrared region,
which shows the most distinctive differences between amorphous and
crystalline materials. In §3.3 the conditions for a λ^{-1} far-infrared
absorption efficiency are examined. From interstellar far-infrared
observations it is concluded that the volume of the interstellar dust is
dominated by layered materials, such as amorphous carbon or layer-
lattice silicates.

The physical principles of grain surface chemistry are discussed in
§4. This includes the surface structure of materials (§4.1), the
adsorption energy and residence time of species on a grain surface
(§4.2), the sticking probability (§4.3), the mobility of species on a
grain surface (§4.4), the reactions that are of importance for the

formation of icy grain mantles (§4.5), and the dissipation of the reaction heat (§4.6).

It is generally agreed that H_2 is formed on grain surfaces in the interstellar medium. Because of its low mass, a physically adsorbed H atom can tunnel through the activation barrier for reaction with a chemisorbed H atom and form H_2. This process is examined in some detail in §5.

In §6 the depletion of elements due to their accretion on grain surfaces in the diffuse interstellar medium is examined. It is suggested that the high depletion of elements such as Fe and Ca is due to chemisorption on grain surfaces. Other atoms (e.g., O and N) will also be chemisorbed, initially, but reaction with H will form saturated hydrides which are easily photodesorbed. Their depletion will then be fairly low.

The composition of interstellar grain mantles is discussed in §7. It results from a complex interplay of gas-phase and grain-surface reactions. Broadly speaking, three different regimes can be discerned, depending on the conditions in the gas phase. First, a reducing atmosphere in which most of the accreting gas-phase species are saturated through reactions with atomic hydrogen on the surface. Second, an oxidizing atmosphere in which surface reactions with atomic oxygen dominate. Third, an inert atmosphere in which the accreting gas-phase species essentially do not react with each other. Observations show that the reducing conditions dominate during the formation of interstellar icy grain mantles.

In §8 the formation of the refractory dust component that is observed in the diffuse interstellar medium is discussed. Essentially, the problem is to get from the simple, largely hydrogenated molecules in icy grain mantles to more complex organic refractory material, which is able to survive better in the harsh environment of the diffuse interstellar medium. Some form of energetic processing of the grain mantle is required for this transformation. The most likely sequence of events in UV photolysis producing radicals, followed by the diffusion and reaction of these radicals to produce complex molecules. In order to get a substantial yield of organic refractory dust material, it is concluded that there has to be a significant internal UV field in molecular clouds and that transient heating mechanisms such as cosmic ray heating, which promote diffusion, are important.

Finally, in §9 the contribution of grains to the gas-phase composition of molecular clouds is examined. It is emphasized that the critical temperature for grain-mantle formation is only about 30 K and that above this temperature grain-mantle formation is a slow process. Ejection mechanisms from grain surfaces are discussed in §9.1. Transient heating of photolyzed grain mantles by cosmic rays or grain-grain collisions up to about 30 K promotes diffusion and reaction of stored radicals. The liberated reaction heat will explosively evaporate an appreciable fraction of the icy grain mantle. During this process more organic refractory dust material can be produced. The signatures of grain chemistry are discussed in §9.2. It is pointed out that grain chemistry will mainly lead to simple hydrogenated molecules (e.g., H_2O, NH_3, and CH_3OH), large deuteration effects (e.g., $NH_3:NH_2D:NHD_2:ND_3 \approx 1:0.1:0.01:0.001$), saturated metal hydroxides (e.g., NaOH), and ortho/para ratios characteristic for the dust temperature.

1. COMPOSITION OF DUST IN DIFFUSE INTERSTELLAR MEDIUM

Many different dust components have been proposed to be present in the diffuse interstellar medium. These include silicates, amorphous carbon, PAHs, graphite, organic refractories, SiC, metallic oxides and many more (cf. Mathis 1986 and references therein). Most models rely on some combination of these to explain the interstellar extinction curve. Direct information on the dust components present in the diffuse inter-stellar medium comes, actually, from structure in the extinction curve; for example, the 2200 Å bump, the 10 and 20 μm features; several infra-red absorption features in the spectrum of the galactic center; a 7.6 μm absorption feature in some protostars; and the 3.3, 6.2, 7.7, and 11.3 μm emission features. A successful model should, of course, also explain the wavelength dependence of the observed extinction curve and, in particular, the amount of visual extinction per hydrogen atom A_V/N_H. These constraints can be used to identify the dust components present in the diffuse interstellar medium and to determine their rela-tive importance. In the remainder of this section we present such an analysis of these constraints. It should be emphasized, however, that not all researchers in the field of interstellar dust will agree with our interpretation of the data, in particular with the details. The reader is referred to the review by Mathis (1986) for a critical analy-sis of some of the different dust models proposed.

The spectroscopic constraints listed above can be used to identify the different dust components present and to determine their relative importance. In line with most other analyses, we attribute the ubiqui-tous 2200 Å bump and the 10 and 20 μm features to graphite and silicate, respectively. Evidence for the presence of amorphous carbon, organic refractory material, and PAHs is provided by a 7.6 μm absorption fea-ture, several other absorption features, and the IR emission features, as detailed below. In Table I we summarize the estimates of the mass and volume per H atom, M_j and V_j, of each dust component identified in the interstellar medium. These have been determined from the observa-tions using

$$M_j = \sum_i f_i \times A_i \times m_i \qquad (1)$$

Table 1: The Composition of the Dust in the Diffuse Interstellar Medium

Dust Component	Observational Evidence	Elemental Abundance[a]	Mass $[10^{-27}$ g/H atom]	Volume $[10^{-27}$ cm^{-3}/H atom]
Silicates	10(and 20)μm feature	100% Si	9.0	3.6
Amorphous Carbon	7.6μm feature in two protostars	5-10% of C	0.6	0.3
PAHs	IR emission features	1% of C	0.08	0.04
Organic Refractory	several IR absorption features in Sgr A	24% of C 6% of O	3.0	3.0
Graphite	2200Å bump (≈200Å spheres only)	25% of C	2.0	0.9
Total Accounted for			14.5	8.6
Total required	extinction curve		~16	~10

Notes: a) the fraction of an element locked up in a particular dust component relative to solar abundance.

b) "Icy" grain mantles inside dense molecular clouds contain 5-40% of the elemental oxygen and carbon, corresponding to a dust volume of up to 10^{-26} cm^{-3}/H atom.

and

$$V_j = M_j/\rho_s \ , \tag{2}$$

where f_i is the fraction of element i locked up in dust component j; A_i is the solar abundance of element i; m_i is the atomic mass of element i; ρ_s is the specific weight of the dust component; and where the summation is over all elements in a dust component. The fraction of an element locked up in a particular dust component (relative to solar abundances) is estimated from the observed strength of the absorption (or emission) features per hydrogen atom. The details of this analysis are given below.

1.1 Silicates

The 10 and 20 µm features are seen in many objects in absorption as well as emission, including oxygen-rich M giants (presumably the birth sites), HII regions, protostars, and background stars (Merrill 1977; Aitken 1981). Generally, these features are attributed to the Si-O stretching and bending vibrations in silicate materials. The width of the observed features (e.g., much broader than in terrestrial minerals) and the absence of substructure suggests that the silicates have an amorphous structure or are hydrated (Zaikowski et al. 1975; Day and Donn 1978; Day 1979).

The fraction of the silicon locked up in silicate dust grains in the diffuse interstellar medium can be estimated from the observed strength of the 10 µm silicate absorption feature per magnitude of visual extinction. For eight bright, nearby (<3 kpc) WC Wolf-Rayets and supergiants, the ratio τ (10 µm)/A_V is determined to be 18.5 + 1.5 mag^{-1} (Roche and Aitken 1984). Assuming a standard dust-to-gas ratio, N_H/A_V, appropriate for the diffuse interstellar medium of 1.9×10^{21} mag^{-1} cm^{-2} (Bohlin, Savage, and Drake 1978) and using the observed width of the 10 µm feature of 220 cm^{-1}, this translates into an integrated strength, W (10 µm), of 6.2×10^{-21} cm per hydrogen atom. Because of the disordered nature of amorphous or hydrated silicates, it is expected that all modes will be infrared-active and, thus, that the integrated strength of the Si-O stretching vibration will not vary much from one silicate material to the other (see §3.2); this is borne out by laboratory studies. Laboratory measurements of the integrated strength of the 10 µm feature in amorphous or hydrated magnesium silicates typically yield 1.2×10^{-16} cm per Si atom (Penman 1976; Day 1979; Mooney and Knacke 1986). Note that the measured peak strength and width vary by a factor of 3, but the integrated strengths agree to within 30%. Using this value requires about 150% of the solar Si abundance to be locked up in silicates. Such an elemental enrichment in the present interstellar medium is perhaps not impossible. However, larger 10 μm integrated absorption strengths have been reported for crystalline silicates (2.8 × 10^{-16} cm per Si atom; Zaikowski, Knacke, and Porco 1975). Using this value requires that at least 63% of the silicon be locked up in silicate dust grains in the interstellar medium. Quite likely, the actual fraction of Si locked up in silicates is about 100%, and that is what we have assumed in Table I. Based on cosmic abundances and condensation sequences (in thermodynamic equilibrium; thus, perhaps not completely applicable) a magnesium or iron-magnesium silicate is expected to condense out in the outflow from late-type giants (Hackwell 1971). Some Na, Ca, and Al might also be contained in interstellar silicates. A composition of $(Mg,Fe)SiO_4$ has been assumed in the calculation of the silicate mass per hydrogen atom. The specific density of terrestrial silicate materials ranges from about 2.5 to 3.5 g/cm^3. Interstellar silicate materials are probably amorphous and their specific density is on the low side of this range. In the calculation of the silicate volume per hydrogen atom (Table I) a specific density of 2.5 g/cm^3 has been assumed. Note that both these assumptions tend to maximize the contribution of silicates to the interstellar dust.

1.2 Amorphous Carbon and Polycyclic Aromatic Hydrocarbons

A broad and shallow 7.6 μm absorption feature has been observed in the spectrum of NGC 7538-IRS9 and W33A, two protostars with the largest known 10 μm optical depth (Tielens et al. 1986a). This band may be due to the CC stretching vibration in amorphous carbon dust particles, although other interpretations (in terms of molecules in icy grain mantles) are also possible. The hydrogen column density or visual extinction toward these sources is not known, but from a comparison with the 10 μm optical depth and assuming that all of the Si is in the form of silicate grains, the fraction of the carbon locked up in amorphous carbon grains is calculated to be 5-10% (Tielens et al. 1986a). If the silicon depletion in silicate grains is actually less, then this number has to be revised downward accordingly. The resulting mass and volume estimates (assuming a specific weight of 2 g/cm^3) for amorphous carbon are given in Table I.

Emission features at 3.3, 6.2, 7.7, and 11.3 μm have been observed in a variety of sources, including planetary nebulae, HII regions, reflection nebulae, and galactic nuclei (Russell, Soifer, and Willner 1977; Aitken and Roche 1984; Sellgren, Werner, and Dinerstein 1983; Cohen et al. 1986). These features have been attributed to vibration involving polycyclic aromatic hydrocarbons in various forms (Duley and Williams 1981; Leger and Puget 1984; Allamandola, Tielens, and Barker 1985). From an analysis of the emission mechanism, the fraction of carbon locked up in these molecules is estimated to be about 1%. For completeness, this component is included in Table I. It is, however, likely that this component is just the molecular domain of the size distribution of interstellar amorphous carbon grains (see §3.5).

1.3 Organic Refractory Material

Absorption features due to an organic refractory dust component have been detected along the line of sight toward the galactic center (Willner et al. 1979; Allen and Wickramasinghe 1981; Butchart et al. 1985; Tielens et al. 1986b). The dust along the line of sight toward the galactic center is presumably located in the diffuse interstellar medium (Roche and Aitken 1985). From an analysis of these spectra in terms of functional groups of simple molecules, the fraction of the carbon and oxygen locked up in this dust component has been determined to be 24% and 6%, respectively (Tielens et al. 1986b). The H column density has been determined from the estimated visual extinction, A_v = 30 mag, assuming a standard dust-to-gas ratio. The specific density of this dust component has been assumed to be 1 g/cm^3.
It should be emphasized that infrared spectra generally only yield the characteristic molecular subgroups responsible for the absorption and not the specific molecule. The intrinsic absorption strength of these molecular subgroups, and thus the calculated column density, is therefore somewhat uncertain (by a factor of 2). Furthermore, it should be noted that the silicate abundance in the galactic center is larger than in the local interstellar medium (Roche and Aitken 1985); this may also pertain to the organic refractory component. In fact, the 3 μm spectrum of the supergiant VI Cygni No. 12 (~3 kpc) does not show the 3.4 μm absorption feature (τ (3.4 μm) < 0.03; A_v = 10 mag; Gillett et al. 1975a). Assuming that the composition of the organic refractory dust in the local interstellar medium is similar to that along the line of sight to the galactic center, we arrive at an upper limit on the fraction of the carbon locked up in the dust component of about 10%. Clearly, the volume and mass values given in Table I should only be considered as order-of-magnitude estimates.

1.4 Graphite

The strong 2200 Å bump, which is a ubiquitous component of the interstellar extinction curve in the diffuse interstellar medium (cf. the review of Savage and Mathis 1979 and references therein), is generally attributed to small (~200 Å) spherical graphite grains (Gilra 1972). About 25% of the available carbon is required to explain the strength of this feature (Tielens and de Jong 1979). Note that larger carbon grains show no detectable absorption features, and their abundance can presently only be guessed at. A specific density of 2.2 g/cm^3 has been

assumed for graphite. There are some problems in attributing the 2200 Å
bump to graphite particles. The graphite transition is intrinsically
very strong and, because of the surface polarization charge, the shape
and peak position are very sensitive to shape and size (an excellent
review of the effects of the surface polarization charge on the spectrum
of small particles is given by Bohren and Huffman 1984). The observations imply 200 Å spherical graphite grains. Moreover, there may be
some problems with graphite in explaining the observed variations in the
ratio of the FUV extinction ($\lambda < 1600$) to the 2200 Å bump (Greenberg and
Chlewicki 1984). For these reasons, the 2200 Å bump has sometimes been
attributed to amorphous carbon instead of graphite grains (cf. Mathis
1986). (The structural difference between amorphous carbon and graphite
will be discussed in §3.5). Indeed, the intrinsic strength of the
amorphous carbon transition is much less (Borghesi et al. 1983), and
surface charge effects are, therefore, unimportant. The drawback of
amorphous carbon is, however, that more material is required to explain
the observed strength of the 2200 Å feature per hydrogen atom. This
does not only conflict with the amount of amorphous carbon determined
from the infrared, but actually requires more than the cosmic abundance
of carbon.

1.5 Total Extinction

A rough estimate of the total volume of dust material required to
explain the observed extinction can be made from a Kramers-Kronig analysis of the extinction curve (Purcell 1969). An extensive discussion of
this analysis is given in Spitzer (1978). For reasonable values of the
optical properties of the dust, the value in Table I results. This is
only somewhat larger than the sum of the identified components. A more
detailed analysis of the extinction curve is, however, required to
determine whether the identified components can indeed explain the full
extinction curve.

1.6 Interstellar Grain Mantles

For completeness we now mention interstellar grain mantles. Inside
dense molecular clouds, interstellar grains can acquire an icy grain
mantle, consisting of such simple molecules as H_2O, CH_3OH, CO, and NH_3
(Merrill et al. 1976; Hagen et al. 1980; Knacke et al. 1982; Lacy et al.
1984; Tielens et al. 1984, 1986a). The fraction of the carbon and
oxygen locked up in these grain mantles ranges typically from 1-40% with
a corresponding mass of 0.2-10 × 10^{-27} g per H atom and corresponding
volume of 0.2-10 × 10^{-27} cm^3 per H atom. Thus, the growth of these icy
grain mantles in molecular clouds can actually double the total dust
mass or volume. However, these icy grain mantles have never been

observed outside of molecular clouds. Presumably, these volatile materials are easily destroyed by interstellar shocks and UV photodesorption
and their abundance in the diffuse interstellar medium is, therefore,
low (Barlow 1978; Draine and Salpeter 1979); as a result, they are not
included in Table I.

1.7 Discussion

In the preceding paragraphs different dust components in the diffuse
interstellar medium have been identified (cf. Table I). These are
silicates, amorphous carbon, PAHs, organic refractory materials, and
graphite. From Table I, one may get the impression that the organic
refractory dust component and silicates dominate the composition of the
interstellar dust rather than graphite. Greenberg and his co-workers
(Greenberg and Hong 1974; Greenberg and Chlewicki 1984) have proposed
such a model for the interstellar dust. It should be borne in mind,
however, that the estimate of the total contribution of the organic
refractory component is actually quite uncertain; it may be much less
than estimated, particularly in the solar neighborhood. If that is the
case, the bulk of the dust has to be in a form without detectable IR or
UV absorption features. Large graphite particles fulfill these conditions. The electronic intraband transitions in graphite screen the
(weak) lattice vibrations of the carbon atoms and make them presently
undetectable (Draine 1984). Furthermore, if their size is large enough
(>0.1 μm) large graphite particles will have an extinction cross section
in the UV equal to the geometric cross section and, therefore, will not
show a detectable UV feature. Thus, if the volume of the organic
refractory component is much less than estimated in Table I, then large
graphite particles without detectable absorption features (a theorist's
dream!) may dominate the composition of the interstellar dust. Models
based on this premise have been developed by Mathis and his co-workers
(Mathis, Rumpl, and Nordsieck 1977; Mathis and Wallenhorst 1981) and by
Draine and Lee (1984). Choosing between these two possible models is
presently not possible. As emphasized before, the most uncertain component, besides graphite, is the organic refractory material. Determining
its composition and, hence, its volume will require studies of the
infrared absorption features along many different lines of sight, in
particular in the local diffuse interstellar medium. Nevertheless, it
should be emphasized that all the materials identified in Table I are
important components of the interstellar dust and contribute substantially to the interstellar extinction curve in at least some part of the
spectrum. In our opinion, it is quite likely that the composition of
the interstellar dust lies somewhere in between the two extremes represented by the models of Greenberg and Mathis.

2. THE DISORDERED STRUCTURE OF INTERSTELLAR DUST

Interstellar grains are generally thought to be highly disordered. This
assessment is based on two observations. First, the width of the 10 μm
silicate feature is much larger than that in crystalline silicate materials. This broadening can result from the disorder present in amorphous or hydrated silicates (Zaikowski et al. 1975; Day and Donn 1978;
Day 1979). Second, the far-infrared cross section of the interstellar
dust seems to fall off much slower with wavelength than expected for
crystalline solids. Laboratory studies suggest that this is also a
characteristic of some amorphous materials. Theoretically, disordered
structures are expected because grain condensation generally occurs
under nonequilibrium conditions. In this section we will examine the
structure of interstellar grains with emphasis on amorphous silicates.
A detailed discussion on the structure of amorphous materials and their
properties can be found in the Ehrenreich report (1972) and in the
monograph by Zallen (1983).

2.1 Structure of Amorphous Solids

Amorphous solids, sometimes referred to as glasses, are characterized by
complete chemical coordination (i.e., few dangling bonds), as are crystalline solids. Similarly, the nearest-neighbor distances (i.e., bond
lengths) are nearly constant. There is, however, a considerable spread
in bond angles. Because of this latter point, amorphous solids do not
have long-range order, in contrast to crystalline solids. This structural difference between amorphous and crystalline solids is illustrated
schematically in Figure 1. In the crystal, the nearest-neighbor equilibrium distances and the bond angles are all equal. Consequently, it
possesses long-range order and translational symmetry. The atomic
arrangement in an amorphous solid is a continuous random network that
lacks symmetry and periodicity (Zachariasen 1932). Specifically, the
nearest-neighbor distances are only almost equal. Moreover, there is
considerable spread in the bond angles. Clearly, there is a considerable amount of local order but a complete absence of long-range order in
the amorphous state (cf. Fig. 1). The presence of the short-range order
is, of course, just a consequence of the chemical bonding of the solid.
    Although not strictly applicable to interstellar solids, it may be
advantageous from a conceptual point of view to study a schematic
volume-temperature phase diagram (Fig. 2). When the temperature is
lowered, a gas will condense to the liquid phase at the boiling temperature, T_b. Essentially, this is a transition from a regime where the
intermolecular forces are unimportant to one in which they dominate.
Although the free-volume (i.e., the volume available to each molecule)
is much less in the liquid phase than in the gas phase, it is still

CRYSTALLINE   AMORPHOUS

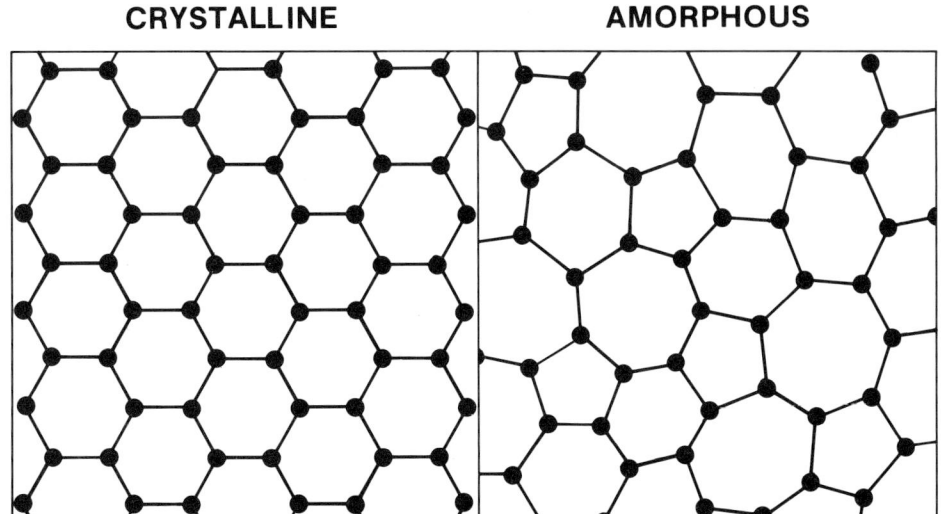

Figure 1. Atomic arrangement in a crystalline and an amorphous solid.
The dots indicate the equilibrium positions about which the atoms
vibrate, and the lines indicate the chemical bonds.

sufficiently large to allow translational motion to the molecules. A
molecule in the liquid phase is continuously making and breaking bonds
with its neighbors while it is diffusing through the liquid. This is,
of course, what gives a liquid its macroscopic fluid character. Thus,
because of the chemical bonding of the molecules, a liquid will have
considerable short-range order. But, because of the continuous rearrangement of the molecules, the bond angles and bond lengths will show a
broad distribution around the average value. Consequently, there is no
long-range order in a liquid.
     When a liquid is very slowly cooled to the freezing temperature, it
will transform to the crystal phase (route 1 in Fig. 2). That is,
crystallization centers will form and will grow at the crystal-liquid
interfaces. Like the gas-liquid transition, this process is accompanied
by an abrupt change in the free-volume available to the molecules. In
the crystal, the molecules are arranged on a translationally periodic
lattice, which exhibits long-range order. It is important to realize
that the crystallization process is essentially a kinetic one. That is,
during the nucleation and growth the molecules have to "find" each other
and arrange themselves on the crystal lattice in the energetically most
favorable orientation. This takes time. Thus, if the liquid is rapidly
cooled below its freezing temperature nucleation cannot take place and
the liquid phase will persist (i.e., it will form a supercooled liquid).

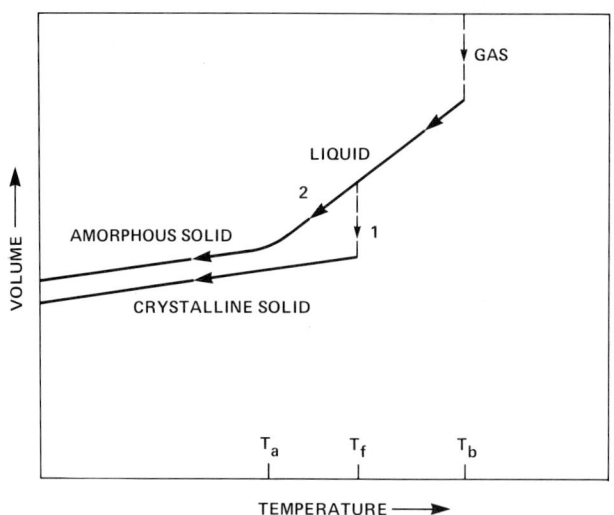

Figure 2.   Volume-temperature phase diagram.   There are two ways for the condensation of a solid to proceed: route 1 is the path to the crystalline state; route 2 is the rapid-quench path to the amorphous state.

When a fluid is cooled below its freezing temperature while nucleation is prevented (i.e., rapid cooling; route 2 in Fig. 2), the free-volume available to its molecules will shrink further and further because of the thermal contraction (i.e., the anharmonicity of the intermolecular potential).   It will finally reach a point where the available free-volume does not allow any further translationary motions of the molecules and the fluid has been transformed to an amorphous solid.   The molecules are "stuck" together in a rather random way.   The molecules will now only show oscillatory motion around a well-defined equilibrium position.   The temperature at which this transformation occurs, T_a, depends slightly on the cooling history of the solid, because the rate of kinetic processes depends on the temperature.   In contrast to the gas-liquid and liquid-crystal transformations, this transition is characterized by a smooth volume change.   Because of the decreased free-volume of the molecules the distribution of the bond lengths in the amorphous state will be much less broad than in the liquid state, resembling more the crystalline state.   The distribution of bond angles however is very broad, similar to that in the liquid state.   Consequently, although there is short-range order, there is no long-range order.

It is thus important to realize that any liquid or gas when cooled rapidly enough (i.e., nucleation prevented) will form an amorphous solid.   This amorphous state is generally metastable with respect to the crystalline state, but the latter is kinetically inaccessible. The detailed structure of a particular amorphous material will depend on the details of its formation. In particular, if the cooling is very rapid, the amorphous solid may retain the local structure of the liquid (e.g., the presence of highly strained bond angles). In contrast, when the cooling is somewhat slower, the amorphous solid may be able to rearrange itself somewhat and attain a thermodynamically slightly more favorable structure in the amorphous state. It should be emphasized that there is not just one amorphous state but many, each with a slightly different energy. In fact, there is an almost continuous distribution of amorphous states with energies up to kT_a, where T_a is the temperature for the liquid-amorphous solid transition. Thus, although amorphous solids may approach an "ideal" metastable noncrystalline state, they can be quenched into many other metastable states. The structure and properties of an amorphous solid depend, therefore, also on the preparation conditions.

Such a distribution of amorphous states will be even more prominent for the condensation conditions prevailing in interstellar space. In that case, condensation will take place directly from the vapor state to the solid state. Often the temperature of the system will be much less than the transition temperature, T_a, from the amorphous to the crystalline state (note that in this case, this transition does not go through the liquid phase and the temperature T_a differs from that defined above). Accreting molecules will, thus, have insufficient energy to arrange themselves favorably and the resulting structure will be far from that reached by cooling down from the energetically more favorable structure of an amorphous solid deposited at a temperature around T_a. Upon warm-up, an amorphous structure rapidly deposited at a low temperature can rearrange itself somewhat to an energetically more favorable amorphous structure. This annealing process will continue until the temperature at which it transforms into the crystalline phase is reached. This annealing process for the interstellar dust has been studied experimentally in considerable detail for amorphous H_2O ice (Hagen, Tielens, and Greenberg 1981) and in somewhat less detail for amorphous silicates (Day and Donn 1978; Nuth and Donn 1982).

2.2 Impurities and Defects

Up to now we have only discussed disorder caused by condensation below the amorphous-crystalline transition temperature. Disorder can, however, also be due to the presence of impurities and defects in the crystal. Impurity atoms or molecules will have slightly different sizes, polarizabilities, bond strengths, and angles than the host species. Their presence will necessitate some degree of local rearrangement of the crystal to accommodate them. This is, of course,

particularly true for an impurity ion with a valence different from the
host in an ionic crystal. Impurities will, thus, distort the surrounding host species (slightly) and thereby destroy the long-range order of
the crystal. For icy grain mantles formes in interstellar molecular
clouds this type of disorder is very important because these mantles
consist of about equal parts of H_2O and CH_3OH (or another alcohol) and
traces of other molecules such as CO and NH_3 (Tielens et al. 1986a). Of
course, such icy grain mantles are formed at temperatures (~10 K) much
below T_a and would, therefore, be highly disordered even if they had a
homogeneous composition.

Crystal lattices will also exhibit defects, owing to the presence
of vacancies in lattice positions (Schottky defects) or the promotion of
molecules or ions from substitutional to interstitial sites (Frenkel
defects). The formation of these defects in a crystalling solid has an
appreciable activation barrier. In interstellar space they might result
from cosmic ray bombardment. A concentration of point-defects may give
rise to dislocations (screw or edge dislocations). Dislocations are
generally a product of the nucleation and growth processes of crystalline solids. In fact, a crystalline solid generally grows at these
dislocations. For an amorphous solid, formed under supercooled conditions, dislocations are, however, of little importance.

2.3 Amorphous Silicates

Silicate glasses (e.g., fused quartz) are the prototype of (binary)
amorphous solids, and their structure is well represented by a continuous random network in three dimensions (the three-dimensional analog of
Fig. 1; Zachariasen 1932). In fused quartz, each silicon atom is predominantly covalently bonded to four nearest-neighbor oxygen atoms, and
each oxygen atom is bonded to two nearest-neighbor silicon atoms. Each
of the SiO_4 units consists of an almost regular tetrahedron, and most of
the structural variations are due to variations in the Si-O-Si bond
angles (Bell and Dean 1972).

The presence of electropositive elements such as Mg changes this
picture somewhat. There are now two kinds of oxygen atoms: bridging
oxygens, which link two tetrahedrons together, and non-bridging oxygens,
which belong only to one tetrahedron. These non-bridging oxygens are
bonded to the electropositive element by charge ransfer and are topologically "dead-ended." For a small number of cations (positive ions), the
resulting structure is that of a polymer with long chains cross-linked
at intervals and cations in holes in between the oxygen tetrahedrons.
This is illustrated in Figure 3a. When the number of bridging oxygen
atoms becomes less than two per tetrahedron, spatial coherence by cross-
linking is not possible. The silicate structure consists now of isolated chains (cf. Fig. 3b). If the number of cations present increases

even further, full charge separation may take place and each tetrahedron is only bonded to cations. In any case, however, the disorder in silicate glasses is due to the variable orientation of the SiO_4 tetrahedrons with respect to each other. Additional disorder can be caused by the presence of different cations in the structure.

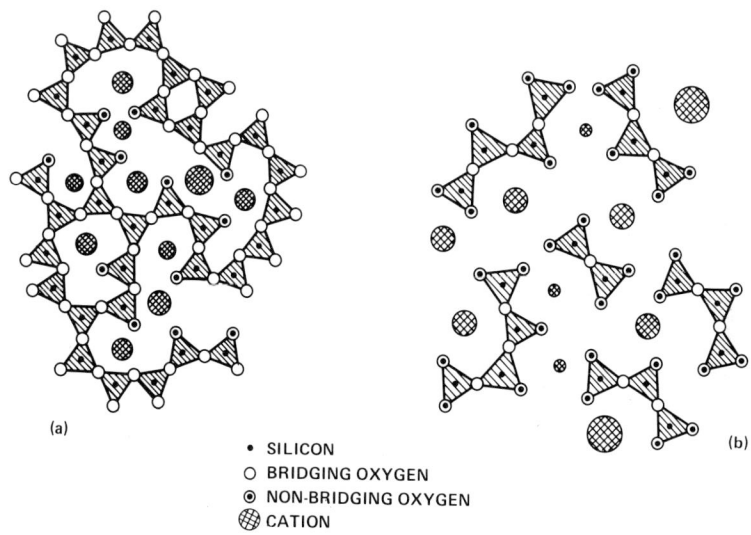

• SILICON
O BRIDGING OXYGEN
⊙ NON-BRIDGING OXYGEN
⊛ CATION

Figure 3. Two-dimensional representation of the atomic arrangement in silicate glasses. Each silicon atom is in the center of an oxygen tetrahedron. Some oxygen atoms bridge two tetrahedrons whereas others are bonded to cations. Depending on the number of cations present the SiO_4 chains will be cross-linked (a) or not (b).

2.4   Amorphous H_2O Ice

The structure of amorphous H_2O ice is very similar to that of silicates discussed above (Madden et al. 1978). Its infrared spectrum behaves very similar to that of silicates except for effects owing to hydrogen bonding. This has been discussed extensively in an astrophysical context by Hagen, Tielens, and Greenberg (1983a,b). Here we will merely examine the importance of the thermal history for the structure of interstellar amorphous ice. Consider the condensation of H_2O molecules on preexisting silicate cores in the outflow from oxygen-rich late-type giants. For the typical vapor pressures of H_2O in these outflows, the condensation temperature of H_2O is about 90 K (Nakagawa 1980; that is, at this temperature the accretion rate equals the evaporation rate). The time-scale for the amorphous-crystalline transformation is given by a Boltzman expression:

$$\tau = \nu^{-1} \exp(E_a/kT) , \tag{3}$$

where E_a is the activation energy for the transformation and ν is a characteristic frequency. Because the disorder in amorphous H_2O ice is mainly a result of a broader distribution of bond angles (Madden et al. 1978), it is appropriate to use the O-H bending frequency (5×10^{13} Hz) for ν. The time-scale for transformation observed in the laboratory, 45 min at 140 K (Hagen, Tielens, and Greenberg 1981), translates then into an activation energy of about 5500 K. This is, of course, somewhat less than the binding energy of an H_2O molecule in an H_2O ice (~8000 K) and actually very similar to the heat of formation of the crystalline state from the amorphous state. At a temperature of 90 K, appropriate for late-type giants, the transformation time-scale is then about 2×10^5 yr. This is much longer than the time-scale for the growth of the ice mantle. It is also much longer than the cooling time-scale of the grains owing to the outflow (~10^3 yr). The ice will condense in the amorphous form and will not be able to transform to the crystalline state. Thus, although the amorphous state is metastable with respect to the crystalline state, the latter is kinetically inaccessible. It is, thus, not surprising that the 3 μm H_2O ice band detected in the spectrum of the late-type giant OH0739 has a shape characteristic for amorphous H_2O ice at about 80 K (Hagen, Tielens, and Greenberg 1983b).

2.5 Polycrystalline Carbon Grains

A polycrystalline solid consists of a large number of microcrystals (10-100 Å) separated by some connective material. Within the microcrystals the molecules are positioned on a regular lattice as in a crystalline solid. The connective material at the grain boundaries is highly disordered (Fig. 4). The microcrystalline description of amorphous solids did have some popularity in the past, in particular for glasses, but has now generally been deserted for the noncrystalline description shown in Fig. 1. The two descriptions differ, of course, in the extent of the local order. It should be emphasized that fine-grained polycrystalline solids certainly exist in nature, but experiments have shown that in general amorphous solids are not microcrystalline (Zallen 1983; Phillips 1981). As will be discussed below, interstellar grains may have a polycrystalline structure because of their formation history.

Figure 5 shows schematically the structure of a carbon (soot) particle. This is essentially a polycrystalline structure. On a small scale (diameter < 50 Å) the carbon atoms are arranged on a planar honeycomb lattice, as is graphite. The carbon atoms are bonded by regular covalent sp^2 bonds to three neighbors. These bonds are localized. The fourth electron of each carbon atom is delocalized among the π bonds

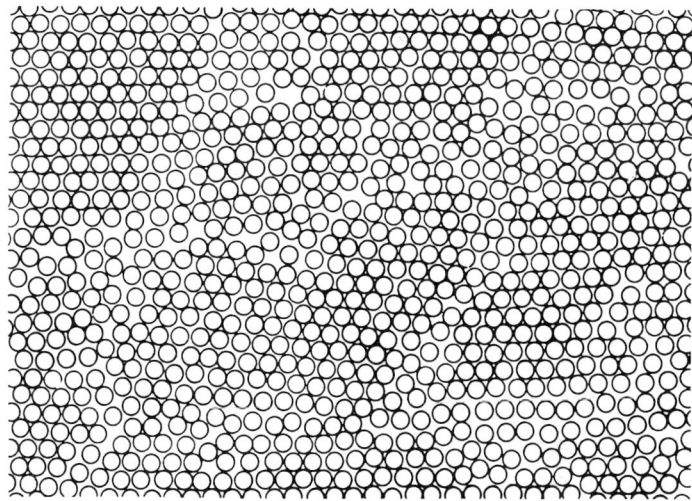

Figure 4. Structure of a polycrystalline material. Highly ordered
microcrystals are separated by highly disordered connective material.
Note that unlike the amorphous structure shown in Figure 1, there are
large regions of fairly ordered material present.

and resonates between several valence-bond structures, giving each C-C
bond a one-third double-bond character. These small planar "crystals"
are in fact just the carbon skeleton of polycyclic aromatic hydrocarbons
(PAHs). Typically, they are less than 50 Å in diameter. Several of
such PAHs are stacked together in platelets bonded by weak van der Waals
forces, as are the sheets in graphite. Often there is also a considerable content of disorganized, tetrahedrally bonded, sp^3 carbon (i.e.,
diamond structure), which distort the planar structure and can link
different layers within a platelet. Functional groups at the edge of
these platelets can also cross-link different layers within a platelet. It is these functional groups that separate different platelets
and prevent orientation of the platelets with respect to each other in a
soot particle. Although these platelets are topologically similar to
graphite, they will have quite different optical properties (cf.
Allamandola, Tielens, and Barker, elsewhere in this volume). These
platelets are assembled into a much larger, disordered unit, the soot
particle, bonded by weak van der Waals forces or by bridging aliphatic
hydrocarbon groups (i.e., tetrahedrally bound carbon skeletons rather
than aromatic skeletons). As in an amorphous solid, there is thus an
absence of long-range order, although the structural units themselves
have highly regular, planar hexagonal structures.

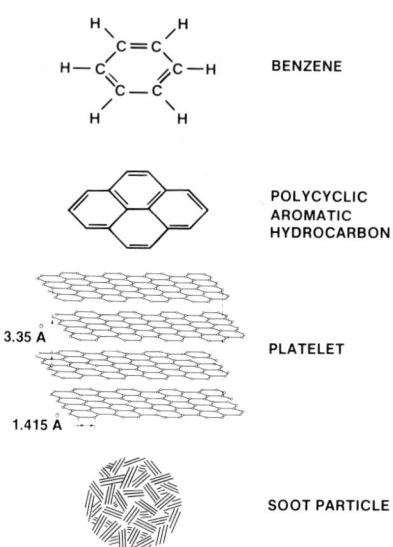

Figure 5. Structure of a soot particle. The atoms are arranged on a
planar honeycomb lattice. Each layer can be viewed as a small graphite
sheet or the carbon skeleton of a large polycyclic aromatic hydrocarbon
(PAH) molecule. The structure of benzene, the smallest PAH, is shown in
detail. Several of these layers form a platelet. These platelets are
the building blocks of soot particles.

It should be emphasized that carbon rapidly condensed on a substrate will form an amorphous solid with aromatic and tetrahedrally
bound carbon atoms "randomly" interspersed. The differences in bond
length and angle will produce a disordered structure similar to that
shown in Figure 1. The soot particles described above are from a structural point of view intermediate between this "truly" amorphous carbon
and graphite.

These two different views of interstellar amorphous carbon grains
merely reflect a difference in the condensation history. The polycrystalline particles of Figure 5 will form when many condensation nuclei
(the PAHs) are formed simultaneously in a (slowly) condensing gas. Most
of the carbon in the gas will crystallize out into these nuclei. Farther downstream in the outflowing wind these PAHs can then coagulate, or
cluster, into the much larger soot particles. In contrast, truly amorphous carbon can be formed when only a few condensation nuclei (i.e.,
PAHs) are formed in a rapidly condensing gas. The condensation of the
remainder of the carbon forms large grains with an amorphous structure. This discussion is, of course, very general, since it is likely
that the condensing species are molecules or ions rather than carbon

atoms. The high abundance of PAHs with about 20 carbon atoms in planetary nebulae suggests that the first process dominates in that case
(Allamandola, Tielens, and Barker 1986). The presence of only weak IR
emission features in the population I WC 9 Wolf-Rayets stars, suggests
that the second process may actually dominate in these outflows (Cohen
et al. 1986). In any case, the differences between amorphous carbon and
polycrystalline carbon are too subtle to be of importance at the present
state of our knowledge of interstellar carbon grains. However, as the
observations and theory progress, we anticipate that these differences
will be revealed.

2.6 Macromolecular Organic Refractory Grains

The organic refractory component of the dust in the diffuse interstellar
medium contains methyl (CH_3), methylene (CH_2), carbonyl (C=O) and probably alcohol (OH) groups. The relative abundance of these groups suggests that the molecules have a highly branched carbon skeleton or are
fairly small (Tielens et al. 1986b). The latter possibility is, however, ruled out by the requirement that they have to be able to survive
in the harsh environment of the diffuse interstellar medium. The molecular structure suggested by the IR data is that of a copolymer (cf.
Fig. 6). The side groups in this chain can be a hydrogen atom or a CH_3,
C=O, or OH group in random order. Also an occasional backbone C atom
could be in the carbonyl (C=O) form. Typically, a polymer contains
about 10^4 backbone C atoms but the polymer length is not well known in
the interstellar case.

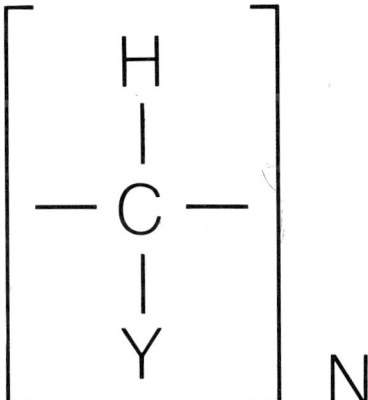

Figure 6. Monomer repeat unit within a polymer chain. The carbon atoms
in these units form a linear chain; Y indicates a side group.

The structure of such a solid polymer will be very disordered. Unlike crystalline solids or monoatomic solids there is no repeatable sub-unit (the unit cell). Figure 7 represents a perspective sketch of a small section of a polystyrene polymer to give an impression of the spatial disorder. In an isolated polymer all the bond lengths and bond angles are equal to the crystalline value. Even the third nearest neighbors have the correct crystal separation (i.e., staggered bond configuration). This set of crystal-like requirements is far stricter than for the amorphous structures discussed before (cf. Fig. 1). Yet, a polymer has a great deal of configurational freedom, which has been used here to open up the polymer as much as possible to show its molecular structure clearly. Obviously, a polymer is quite flexible and can change directions very often along its chain by rotation about a back-bone bond. In that sense, Figure 7 is somewhat misleading since it may give the incorrect impression that an amorphous polymer solid consists of aligned strands of polymer molecules. In fact, such a solid has a random-coil structure, as illustrated in Figure 8. Each polymer adopts a random-walk configuration and the structure of the organic glass consists, thus, of many intertwined random-coils. Note that the scale of Figure 8 is about 1000 times larger than that of Figure 7. The bonding between different polymer strands within an amorphous polymer solid is due to weak van der Waals forces or hydrogen bonding forces between side groups. It should be emphasized that we expect that because of their chemical formation process, the polymers in the organic refractory component of the interstellar dust are not linear, one-dimensional molecules, but rather very branched, three-dimensional molecules. That is, they are cross-linked by true chemical bonds between side groups (e.g., resins). For that reason they are more properly called macromolecules.

3. OPTICAL PROPERTIES OF AMORPHOUS MATERIALS

The amorphous nature of materials has considerable influence on their physical properties, such as specific heat and thermal conductivity, at very low temperatures (Phillips 1981). Essentially, this is due to the existence of more than one amorphous state very close in energy. Tunneling between these states can then be of importance at very low temperatures. Generally, this is not very important at interstellar temperatures (>10 K). However, the disordered structure of a solid is of considerable importance for its optical properties, even at higher temperatures. As mentioned in the introduction to §2, it is generally thought that a large fraction of the interstellar dust has an amorphous structure. This is based on the shape of the 10 μm feature and on the far-infrared absorption law ($\sim\lambda^{-1}$). The high near-infrared absorption

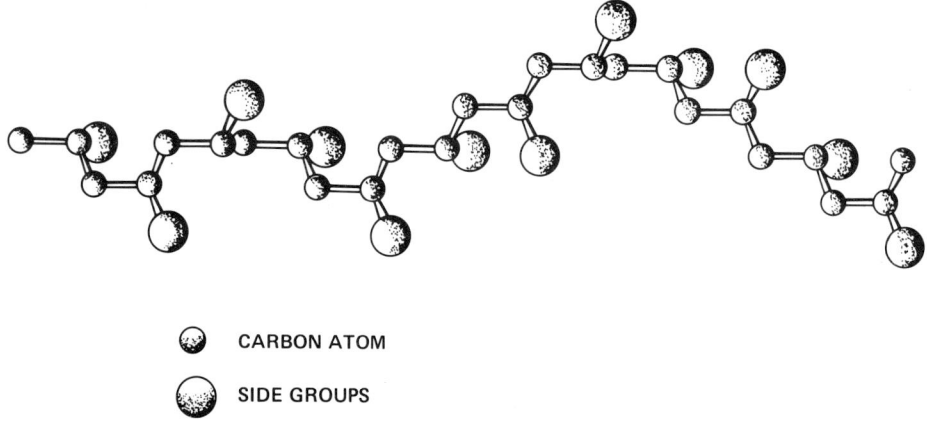

CARBON ATOM

SIDE GROUPS

Figure 7. Small section of a polystyrene polymer. The carbon atoms are tetrahedrally bonded. Alternating carbon atoms in this polymer contain a side group (a benzene ring in this case). For clarity, the hydrogen atoms have been left out.

Figure 8. Random coil network for organic glasses. One polymer chain has been drawn with a thick line for ease of visualization.

of circumstellar silicates (Jones and Merrill 1976; Bedijn 1977; Schutte and Tielens 1985) may also indicate an amorphous structure (Tielens 1983). In this section the optical properties of amorphous materials is discussed with the emphasis on silicates. The optical properties of amorphous carbon grains are discussed by Allamandola, Tielens, and Barker elsewhere in this volume. Each of these three observational

indications of the amorphous nature of interstellar grains will be
discussed in turn.

3.1 UV and Visible Absorption

Absorption in the ultraviolet and visible spectral regions corresponds
to changes in the electron structure of the material, which is dominated
by the short-range order. In general, provided the chemical bonding is
the same in crystalline and amorphous materials, the electronic density
of states will also be very similar and therefore physical characteristics such as the energy gap will be about the same. Thus, the fundamental optical absorption edge in a crystalline and amorphous material will
occur at a very similar frequency. For example, the sharp rise in the
reflection spectra of amorphous and crystalline quartz at about 10 eV
arises from the onset of the bonding→antibonding transitions of the
electrons (Tauc 1972). Often, crystalline solids have special
directions associated with the crystallographic axes, leading to anisotropy in the optical properties (i.e., dependence on crystal orientation). Amorphous materials lack such directionality, that is, they are
isotropic.

Crystalline materials can show fine structure in their UV spectrum,
particularly at low temperatures. Because each unit cell has an identical environment, such optical selectivity will arise because energy, as
well as momentum, has to be conserved in the photon absorption process
(cf. Kittell 1976). Such spectral structure is an intrinsic property of
crystalline materials only, because it is a consequence of the long-
range order (e.g., crystal symmetry). It will be smoothed out in amorphous materials, because of the absence of periodicity (e.g., the wave
vector is no longer a good quantum number). For example, crystalline
materials show a rather abrupt onset of the electronic absorption edge,
which in amorphous materials is much more gradual. Of course, spectral
substructure in the UV can also result from the presence of more electronic transitions within the molecular sub-unit. Such substructure
will be present in both crystalline and amorphous materials.

Silicates formed in the outflow from Miras show considerable
absorption around 1 μm (Jones and Merrill 1976; Bedijn 1977; Schutte and
Tielens 1985). This high, near-infrared absorption is somewhat surprising since it occurs in the energy gap region; it may indicate an amorphous structure (Tielens 1983a). For an ideal crystalline dielectric
material, there is an energy gap between the valence band and the conduction band. This energy difference is determined by the bonding-
antibonding splitting of the electronic states associated with the
chemical bonding of the nearest neighbors. This energy gap is completely devoid of electronic states, and the crystalline material will
be transparent in this region. An amorphous material will have

localized electronic states (e.g., associated with a few atoms) in the
energy gap region, and the density of electronic states is, thus, not
zero but merely low. The transition probability of these states is,
however, quite small and an amorphous material can be quite transparent
(cf. glass). As a result, an amorphous material can still be characterized by an optical band gap energy. These localized states are therefore an unlikely source of the enhanced absorption of circumstellar
silicates at about 1 μm; a more likely source is the presence of impurities (e.g., FeO) that can also produce localized electronic states in
the energy gap region. Unfortunately, there are presently no laboratory
studies that directly address this astrophysical problem.

3.2 Mid-Infrared Absorption

For low frequencies, the oscillations in the electromagnetic field are
slow enough that molecular vibrations can respond to them, leading to
absorption. Since the "force constants" between atoms and between
atomds and electrons are very similar (i.e., both follow from Coulomb
interaction), it follows from Hooke's law that while electronic transitions fall in the UV and visible, transitions involving atomic vibrations will occur in the infrared (i.e., $(m_e/M)^{1/2} \sim 10^{-3}$). Of course,
the difference in mass also implies that the absorption strength of
atomic vibrations is smaller by a factor of 10^{-3} (consider a harmonic
oscillator) than that of electronic vibrations (e.g., $f_{IR} \sim 10^{-2}-10^{-3}$
whereas $f_{UV} \sim 1-10$). Finally, it should be noted that damping in a
solid is due to collisions with neighboring atoms. The collision rate
is then approximately equal to the natural frequency of the vibration.
The damping coefficient is proportional to this collision rate, to the
anharmonicity of the interaction potential and, of course, to the number
of phonons present (i.e., temperature).
    In a crystalline solid the vibrational excitations are associated
with plane waves, and they are therefore characterized by a frequency
and a wave vector. These lattice vibrations are called phonons. Momentum conservation is more important for lattice vibrations than for
electronic excitations, and only a small number of all the possible
lattice vibrations are optically active. The infrared spectrum of a
crystalline material consists, therefore, of a few sharp lines.
    The disorder in an amorphous solid localizes the lattice vibrations
and they can no longer be viewed as plane waves. As a consequence of
the reduced symmetry, most modes will be infrared active and the infrared absorption spectrum reflects to a greater extent than it does in
crystals the density of vibrational states. It should be emphasized
that the vibrational density of states in an amorphous material is not
much different from that of the corresponding crystalline material.
Basically, it reflects the interaction between nearest neighbors, which

is very similar in both types of materials. Of course, the increased
dispersion in bond lengths and bond angles will produce a larger range
in bond strength and thus in frequency (recall Hooke's law). Because of
this and because most modes are IR active, the fundamental absorption
features in an amorphous material will be broader than in a crystalline
material.

The integrated strength of an absorption band in an amorphous
material is generally somewhat less than that of the crystalline material. That the oscillator strength is very similar is of course not
surprising, since the force constants of the oscillators are very similar. The larger disorder in an amorphous material implies, however, a
larger damping coefficient (i.e., a larger effective coupling with
phonons). As a consequence the peak strength of the absorption in an
amorphous material is smaller and the width larger (cf. a damped oscillator). The absorption strength is related to the change in dipole
moment. The high order in a crystalline material makes all dipoles
change in unison, resulting in a fairly large change in dipole moment.
In contrast, the disorder in an amorphous material will make the effective change in dipole moment less and consequently lower the integrated
absorption strength (i.e., the oscillator strength). For silicates this
difference in integrated absorption strength of amorphous and crystalline solids seems to be relatively large (a factor of 2.5; see §1.1).
This is much larger than in, for example, amorphous ice (25%; Hagen
et al. 1981). It should be noted, however, that the variation in peak
absorption strength and width are much larger than the variations in the
integrated strength. In general, it is therefore better to use the
integrated strength in interpreting the strength of absorption features
(cf. §1.1).

Finally, it is interesting to consider the effects of temperature
on the fundamental vibrations of a solid. Typically, absorption features sharpen and shift to slightly higher frequencies with decreasing
temperature, but the integrated absorption strength stays approximately
constant (Plendl 1970). These variations are a result of the thermal
shrinking of the lattice. The decreased nearest neighbor distances
imply larger force constants and thus higher frequencies (Hooke's
law). The decrease in width is due to the decreased importance of
anharmonicity and to a decrease in the number of phonons present, implying a smaller damping coefficient (cf. a harmonic oscillator). Different modes (e.g., stretching versus bending vibrations) in a material may
behave differently upon cooldown, since they may have different anharmonicity constants and they may interact with different phonons. This may
account for the experimentally measured behavior of amorphous magnesium
silicates (Day 1976).

At this point it is tempting to speculate that the decreased width
of the 10 μm feature in absorption in the interstellar medium

(presumably at about 10-20 K) as compared to the emission feature in Orion (≈250 K; Gillett et al. 1975b; Roche and Aitken 1985) is due to the effects of thermal shrinking. However, no experimental data are available on the width of the 10 μm silicate band at low (10 K) temperatures. The variation seems, however, consistent with that measured for other materials (Plendl 1970).

3.3 Far-Infrared Absorption

The far-infrared ($\lambda > 30$ μm) absorption of amorphous materials differs considerably from that of crystalline materials. In crystalline materials most of the far-infrared absorption is due to multiphonon processes (for example, one photon and one phonon in, and one phonon of higher energy out). Fundamental vibrations (i.e., one photon in, one phonon out) are optically inactive because of wave-vector conservation (cf. §3.2). Multiphonon processes are highly temperature sensitive, and at 10 K the absorptivity will be much reduced, essentially because there are few phonons around (Hadni 1970). Besides multiphonon processes, a material can also absorb in the damping wing of an infrared active fundamental vibration. Far from the natural frequency, the absorption coefficient of a damped harmonic oscillator will show a quadratic dependence with frequency, and such a dependence is expected for crystalline dielectric materials at low temperatures (Wooten 1972). In a metallic or semimetallic material (e.g., graphite), far-infrared absorption is due to interaction of the radiation with the free electrons (Wooten 1972). Again, the absorption coefficient will depend quadratically on the frequency, since the free electrons in a material are essentially "oscillators" without a restoring force. The damping is now due to scattering of the electrons by phonons (i.e., lattice vibrations).
In amorphous materials, the far infrared absorption is quite strong, often featureless, and not temperature sensitive (Hadni 1970). These differences with respect to crystalline materials are due to the loss of long-range order. The resulting breakdown of the selection rules implies that all modes are infrared active, and the far-infrared absorption spectrum reflects the density of vibrational states. Generally, the density of vibrational states is well described by the Debye law at low energies ($\sim \omega^2$; Kittell 1976). The strength of the far-infrared absorption depends on the disorder in the charge distribution in the lattice. But, since the oscillator strength is independent of the particular vibrational mode involved in the phonon regions, the frequency dependence of the far-infrared absorption reflects directly the phonon spectrum (Tielens 1986a). Thus, an amorphous material will show the same wavelength dependence of the far-infrared absorption (i.e., λ^{-2}) as a crystalline material, albeit for completely different physical reasons. The far-infrared absorption of amorphous materials

is, of course, much stronger and not temperature sensitive. The far-infrared spectrum of fused quartz shows these characteristics of amorphous materials (Hasegawa and Koike 1984).

An exception to the quadratic law for far-infrared absorption should be made for amorphous, layered materials, such as amorphous carbon, and layer-lattice silicates. The structure of such materials limits the phonons to two dimensions and, consequently, their phonon spectrum is proportional to the frequency rather than to the frequency squared (Kittell 1976). The temperature dependence of the specific heat of graphite, the classic example of a layered structure, reveals the presence of such a λ^{-1} phonon spectrum (Touloukian and Buyco 1970). Because the far-infrared absorption of this material is dominated by free electrons, the far-infrared spectrum of this crystalline, layered material shows a λ^{-2} dependence (see above). But, amorphous carbon, which has a very similar structure (cf. §2.5) does show the expected λ^{-1} far-infrared absorption law (Koike et al. 1980.). Layer-lattice silicates, which also have a layered structure, show a λ^{-n} far-infrared absorption law with n in the range 1.25 to 1.5 (Day 1976). Apparently, there is some degree of cross-linking between the layers in such materials, leading to a somewhat steeper law than expected for a layered structure (Tielens 1986a). The last case of a λ^{-1} far-infrared absorption law involves small amorphous grains. In small particles (say <30 Å) the surface modes (two-dimensional) rather than the bulk modes (three-dimensional) will dominate the far-infrared absorption (Seki and Yamamoto 1980). Again, for amorphous materials where all modes are active, the two-dimensional structure will lead to a λ^{-1} law.

Clearly, the far-infrared is a very useful tool for studying the structure of interstellar grains. Far-infrared observations of celestial objects generally show a far-infrared absorption law that is less steep than expected for crystalline materials (Campbell et al. 1976; Gatley et al. 1977; Harvey et al. 1978). This implies that most of the interstellar grains have an amorphous layered structure. Since amorphous carbon seems to be a relatively minor component of the interstellar dust (Table I), this suggests that interstellar silicates, which may make up half of the interstellar dust volume, have a layer-lattice amorphous structure. In this respect, it is worth noting that the 2.9 μm band observed in the spectrum of the galactic center can be well fitted by a spectrum of hydrated magnesium silicates, a layer-lattice silicate (Tielens et al. 1986b). Obviously, a search for this band in other sources with strong silicate bands is important for the corroboration of this conclusion.

4. PHYSICAL PRINCIPLES OF GRAIN-SURFACE CHEMISTRY

Figure 9 shows a schematic view of grain-surface chemistry. Atoms and molecules are accreted from the gas phase. They may diffuse and react on the surface, and the reaction products may remain on the surface, forming a grain mantle, or may actually be ejected from the surface by some means. In this section we will briefly discuss some of the physics of these processes. For a detailed discussion, the reader is referred to the tutorial by Watson (1976). This discussion will be geared toward H_2 formation and the formation of icy grain mantles in molecular clouds.

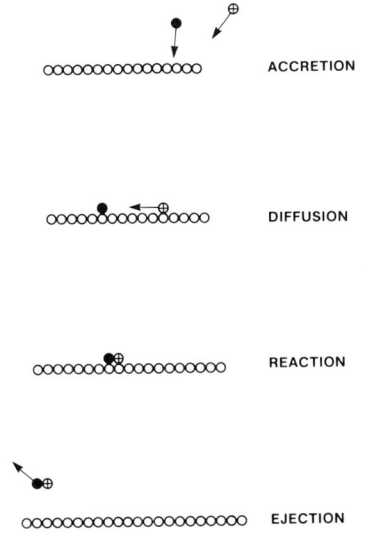

Figure 9. Grain-surface chemistry. Gas-phase species accrete, diffuse, and react on a grain surface. Possibly, the reaction product is ejected into the gas phase.

4.1 Surface Structure

The structure of the surface is of importance for many aspects of grain-surface chemistry. It should be realized that the structure of the surface will deviate from that of the bulk, in particular for a crystalline solid. In a sense, the surface is adjacent to a row of lattice vacancies, and as a result the equilibrium positions of the molecules at the surface will have been relaxed from that in the bulk. Moreover, the surface will facilitate the formation of lattice defects, which will further disturb the molecular bonds. Furthermore, the surface composition may differ from that of the bulk. Finally, a larger concentration of impurities may be present on the surface than in the bulk. Some of

these points are illustrated in Figures 10 and 11, which show some of
the structures possible at silicate and carbon surfaces.

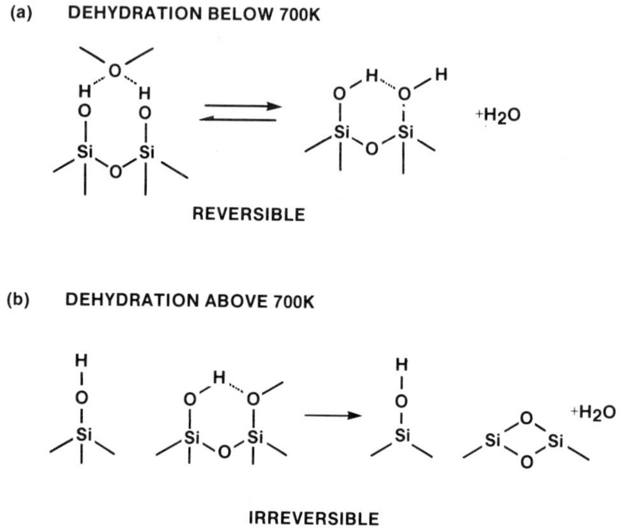

Figure 10. Some of the structures possible at silicate surfaces. Note
the presence of silanol (Si-OH) and siloxane (Si-O-Si) groups. The
former bonds H_2O by hydrogen bonding.

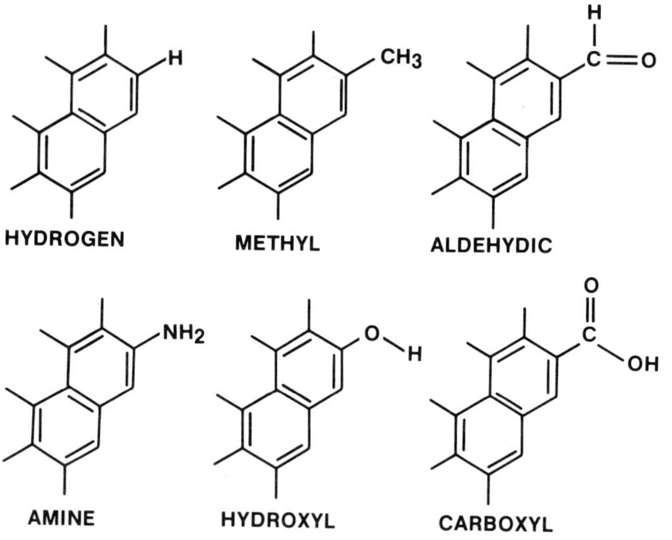

Figure 11. Some of the structures possible at the edge of the basal
plane in graphite or amorphous carbon.

Surface groups on silicates are extensively discussed in the monograph by Hair (1967) and by Snoeyink and Weber (1972). In an astrophysical context they have been briefly discussed by Duley and McCullough
(1977). Generally, a silicate surface contains two types of surface
groups, silanol groups (Si-OH) and siloxane groups (Si-O-Si). Silicates
precipitated from aqueous solutions at low temperature (<700 K) mainly
contain hydroxyl groups on the surface, a result of the tendency of the
silicon atom to complete tetrahedral coordination. Water will be
adsorbed on these surface silanol groups by van der Waals and hydrogen
bonding forces. Upon heating to 700 K, these physically adsorbed H_2O
molecules will be lost (dehydration of the silicate surface; Fig. 10).
Adjacent hydroxyl groups will also form H_2O molecules that are ejected
from the surface, leaving a strained siloxane group behind. Because of
the strain in these siloxane groups, rehydrogenation can occur, reforming the silanol groups. Warm-up above 700 K (or formation above this
temperature) enables the strained siloxane groups to anneal themselves
and, upon addition of water, no rehydrogenation will occur (e.g., the
dehydrogenation is irreversible). Above 1000 K, only isolated silanol
groups (so-called free OH or terminal OH groups) and siloxane groups
remain (Fig. 10). Such a surface will bind H_2O much more weakly than
the untreated silicates and is actually quite hydrophobic. The presence
of other atoms in the silicate surface structure, such as Al, will
increase the acidic nature of a silicate surface because of the induc-
tive effect on adjacent silanol groups. Such atoms may also bind
hydroxyl groups, which will behave slightly differently on heat
treatment.

Physical adsorption of a polar molecule, such as H_2O, on a silicate
surface will preferentially occur on the OH groups, and the heat of
adsorption will increase with increasing polarity of the adsorbed molecule, a result of hydrogen bonding. The silanol groups on a silicate
surface react as a weak acid and can, thus, react with bases (at room
temperature). Reactions with alcohols (at high temperatures) will
produce estersils (e.g., Si-O-C groups). At low temperatures, such
reactions are inhibited by activation energies. Annealed siloxane
groups are very weak bases and show, therefore, only a weak tendency to
form hydrogen bonds. They are also quite unreactive. Because interstellar silicates condense out at relatively high temperatures
(~1000 K), they are presumably initially quite hydrophobic. After
processing in a shock they will presumably have many dangling surface
bonds, which will be quickly saturated by H atoms. The resulting surface may bind H_2O more strongly. In any case, the low temperature of
interstellar dust grains will inhibit reactions with surface molecules,
and the only property of interest in interstellar grain surface chemistry is the ability of silicate surfaces to bond H_2O and other polar
molecules relatively strongly by hydrogen bonding.

Commercial carbon blacks usually have several different surface groups besides aromatic hydrogen (e.g., methyl, phenol, and carboxyl groups; Fig. 11). These groups are substituted for the aromatic hydrogen at the edge of the aromatic ring system (Puri 1976). They result from chemical reactions with the graphitic structure (at elevated temperatures). Graphite or amorphous carbon freshly injected into the interstellar medium will contain mainly aromatic hydrogen at its edges because of the overabundance of hydrogen in their birth sites. Sputtering in fast shocks will change this surface structure and may lead to dangling bonds. These will either be resaturated by atomic hydrogen or the carbon atoms in the edge may revert to a divalent s^2p^2 state. The latter structure limits, of course, the resonance of the π electrons and makes that site more reactive, but leaves no dangling bonds. In fact, vibrational spectroscopy of fluorescing interstellar PAHs, the building blocks of the carbon grains, generally shows an absence of surface groups other than aromatic hydrogen (Allamandola, Tielens, and Barker, elsewhere in this volume), and this is likely to be true for carbon grains as well.

4.2 Adsorption Energies

A species can be either physically or chemically bonded to the surface. In physical adsorption the interaction is due to van der Waals-London or electric field-dipole forces, and the interaction energy is fairly low (<0.5 eV). There is no charge exchange between the adsorbed species and the substrate in physical adsorption. The attractive force is due solely to the interaction of the (instantaneous) dipole of the adsorbed species and the neighboring surface molecules (Kittell 1976). In chemisorption, the interaction leads to actual chemical bond formation and the interaction energy is much larger (~0.5-5 eV). This interaction can entail a complete charge exchange between the surface and the adatom and the formation of an ionic bond. Usually, however, the electronic wave functions of the adatom and the surface mix into a new wave function. Essentially, the electrons responsible for the bonding move in orbitals between the surface and the adatom, and a covalent bond is formed.

The theoretical description of chemisorption is very complex and not completely understood (cf. Einstein et al. 1980). For illustration purposes consider the physical processes occurring when an adatom approaches a metal surface. A crude impression of the tendency to form a chemical bond can be obtained by comparing the adsorbate's ionization potential and electron affinity with the work function of the surface. If the ionization potential of the adatom is less than the work function of the surface, then an electron can be transferred to the surface and the adatom becomes positively charged. Charge transfer can also take

place when the work function is less than the electron affinity of the
adatom. Now the electron is transferred to the adatom. In both cases,
chemisorption is energetically feasible. However, if the work function
of the surface is larger than the electron affinity but less than the
ionization potential of the adatom, then the adatom will be physically
adsorbed. This discussion is of course very crude because the interaction between an approaching atom and a surface will lead to broadening
of the atomic states, and the ionization and affinity levels will become
bands whose widths depend on the degree of interaction (cf. Einstein
et al. 1980). These bands may partially overlap with the (Fermi) level
of the metal. When the atom is very close to the surface, strong interaction may actually lead to a complete modification of the electronic
structure of the atom and the surface.
    A simple potential energy diagram for chemisorption is shown in
Figure 12. Note that the physical adsorption forces have a longer range
than the chemical forces and, thus, an accreting species will first
encounter the physical adsorption potential. For atoms approaching a
surface there is no activation barrier associated with chemisorption.
For molecules, however, there is probably an activation barrier because
a molecular bond has to be weakened or broken. In the interstellar
medium, chemisorption of molecules is, therefore, generally unimportant
owing to the low temperatures.

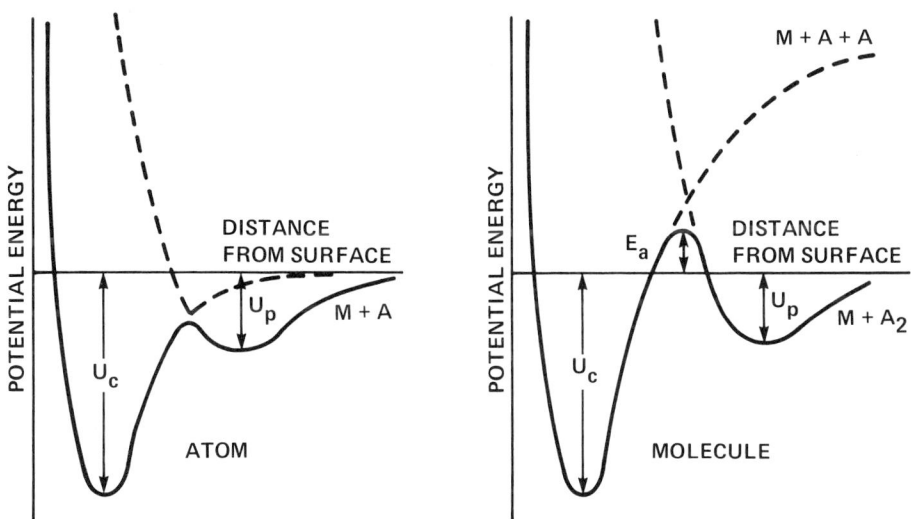

Figure 12. A potential energy diagram for the adsorption of an atom (a)
or a molecule (b) on a surface. Note that there are two adsorption
wells, the physical adsorption well with binding energy, U_p and the
chemisorption well with binding energy, U_c. For molecules, these wells
may be separated by an activation energy, E.

Hydrogen atoms on silicate surfaces have two types of binding
sites. First, a chemisorbed site with a binding energy of about 2 eV
($\sim$2.3 $\times$ 10^4 K). Second, a physical adsorbed site with a binding energy
of only about 0.09 eV ($\sim$1000 K; Gelb and Kim 1971; Wood and Wise 1962;
Hickmott 1960; de Boer and van Steenis 1952; Langmuir 1912, 1915). The
second sites should be viewed to be on top of the first one (cf.
Fig. 12) and will only be occupied when the first one is already
filled. Note that reaction between H atoms in the two layers is inhibited by an activation barrier ($\sim$1000 K) associated with breaking the
chemisorbed bond with the surface (see §5). The number of sites on a
silicate surface is about 1 $\times$ 10^{15} to 2 $\times$ 10^{15}/cm^2, corresponding to
about 7 Å^2 per site.

Although it is well known that atomic hydrogen can be chemisorbed
on the basal plane of graphite as well as on the edges (King and Wise
1963; Wood and Wise 1969; Rye 1977; Gould 1975; Balooch and Olander
1975; Robell et al. 1964), the parameters of these states are less well
known, a result of the complicated chemistry of H atoms with the edge
structure of such surfaces at elevated temperatures (300 K). At high
surface coverage, H$_2$ formation has an activation barrier of about
1200 K, suggesting an H chemisorption energy on the basal plane of
perhaps 2 eV. From diffusion experiments, McCarroll and McKee (1971)
estimated a somewhat lower binding energy of about 0.9 eV on the basal
plane. For H$_2$ formation on interstellar grains, the activation energy
is, however, more important than the binding energy (see §5), so that
this discrepancy should not worry us too much. The "aromatic" H atoms
at the edge of the basal plane are chemically bonded by about 5 eV. The
physical adsorption energy of H atoms on the basal plane of graphite is
measured to be about 800 K (Lee 1975). The number of chemisorption
sites is again of the order of 10^{15} cm^{-2}. Other atoms on silicate and
graphite surfaces are also expected to be chemisorbed on these sites.
At low temperatures, molecules will be physically adsorbed on these
surfaces.

Atoms and molecules will be physically adsorbed on icy grain.
Their binding energies can be determined from molecular beam studies and
from diffusion experiments in low-temperature matrices. Table II summarizes estimates for the binding energy of relevant atoms and molecules
on an icy surface.

Using these binding energies, the residence time of a species on a
surface can be calculated

$$\tau = \tau_0 \exp(E_b/kT) , \tag{4}$$

where τ_0 is the oscillation time of the adsorbed species perpendicular
to the surface (Frenkel 1924). Essentially, the adsorbed molecule has
replaced at least one degree of translationary freedom by a vibrational

TABLE 2

The Interaction of Atoms and Molecules with an H_2O Surface

Species	$E_b(K)$ [1]	$\tau_{ev}(sec)$ [2]		$\tau_m(sec)$ [3]			Notes
		10	30	5	10	30	
H	350	1.6×10^3	10^{-7}	10^{-12}	10^{-12}	10^{-12}	4
H_2	450	3×10^7	3×10^{-6}	4×10^{-12}	4×10^{-12}	4×10^{-12}	4
C	800		4×10^{-1}	10^8	10^{-2}	2×10^{-9}	5,6
N	800		4×10^{-1}	10^8	10^{-2}	2×10^{-9}	5,6
O	800		4×10^{-1}	10^8	10^{-2}	2×10^{-9}	5,6
S	1100		8×10^{13}		10^2	4×10^{-8}	6,7
CO	1900				6×10^{12}	2×10^{-4}	5,8
N_2	1700				10^{10}	2×10^{-5}	5,8
O_2	1600				7×10^8	9×10^{-6}	5,8
CH_4	2600					2×10^{-1}	5,8
H_2O	4000					2×10^5	9,6

Notes: 1) Estimated binding energy of a species on an H_2O surface.
2) Evaporation time-scale from an H_2O surface for grain temperatures of 10 and 30 K. No value is given when it exceeds the lifetime of a molecular cloud.
3) Migration time-scale on an H_2O surface for grain temperatures of 5, 10 and 30 K. No value is given when it exceeds the lifetime of a molecular cloud.
4) Binding energy determined from experimentally measured attractive potential corrected for zero point energy and H_2 coverage (see text).
5) Binding energy determined from measured maximum attractive potential for H_2, corrected for difference in polarizability (cf. Watson 1976).
6) Diffusion barrier determined from low temperature matrix isolation studies (Tielens and Hagen, unpublished).
7) Binding energy determined from diffusion experiments in low temperature matrices in analogy to oxygen.
8) The diffusion barrier is assumed to be 30% of the binding energy.
9) Enhanced binding energy due to hydrogen bonding with the H_2O surface.

degree of freedom of the bond with the surface. It may also have lost some degrees of rotational freedom. For a monoatomic gas which is completely mobile on the surface, we can find for τ_o from a statistical thermodynamical analysis

$$\tau_o = (h/kT)f_z , \qquad (5)$$

where f_z is the vibrational partition function of the adsorbed molecule (de Boer 1968). At high temperatures the vibration perpendicular to the surface is excited and f_z is equal to $kT/h\nu_z$, where ν_z is the vibrational frequency perpendicular to the surface and $\tau_o = \nu_z^{-1}$.

We will assume that the frequency of vibration perpendicular to the surface is equal to that parallel to the surface, which for a symmetric harmonic potential is given by (Landau and Lifshitz 1960):

$$\nu_z = [(2N_sE_b)/\pi^2m]^{1/2} , \qquad (6)$$

where N_s is the number of sites per unit surface area and m the mass of the atom. With a physical adsorption energy of 1000 K and $N_s \approx (7 \text{ Å}^2)^{-1}$, ν_z is about 5×10^{12} sec^{-1} for an H atom and about 10^{12} sec^{-1} for an O atom. For a chemisorbed site with a binding energy of 2 eV, ν_z is about 2×10^{13} and 5×10^{12} sec^{-1}, respectively. At low temperatures, the vibration perpendicular to the surface is not excited and $f_z = 1$, leading to $\tau_o = h/kT \approx 5 \times 10^{-12}$ sec, where a temperature of 10 K has been assumed. When the adsorbed species is not mobile on the surface or has lost some rotational degrees of freedom, then equation (5) has to be multiplied by the ratio of the partition functions for hindered and free translationary motion on the surface or the rotational partition functions on the surface and in the gas phase, respectively. In both these cases τ_o will be less than h/kT. In the limit that an adsorbed atom has lost all its degrees of translationary freedom, a different expression is derived:

$$\tau_o - (h/kT)f_xf_yf_z(N_sh^2/2\pi mkT) \qquad (7)$$

where the f_x, f_y are the partition functions along the surface (de Boer 1968). The last factor in this expression takes into account the number of possible ways of distributing the adsorbed species over the adsorption sites. Consider again the low-temperature limit (T = 10 K), assume a surface area per site of 7 Å^2, and a mass of 20 amu. This yields $\tau_o \approx 4 \times 10^{-12}$ sec. Thus, for the interstellar case, τ_o lies between 2×10^{-13} and 4×10^{-12} sec for an H atom physically adsorbed and chemisorbed, respectively. Using $\tau_o = 10^{-12}$ sec, we find the residence time-scales on an icy surface given in Table II.

4.3 Sticking Probabilities

A gas-phase species approaching a grain experiences an attractive force
owing to the molecules in the grain. Whether a gas-phase species will
stick or not depends on two factors: (1) the magnitude of its translational energy compared to the adsorption energy and (2) the efficiency
with which the excess kinetic energy can be transferred to the surface
atoms with which it is colliding. Some insight into the physics can be
gained by examining the classical case. Imagine an atom with mass m
colliding with a single surface atom of mass M which is attached to
the surface with a spring of a frequency ω_D (the Debye frequency).
This is the so-called "soft-cube" model. We can define a characteristic
inverse time-scale for the collision, ω_o,

$$\omega_o = v/2b = [(E_{th} + E_b)/2mb^2]^{1/2} \tag{8}$$

where v is the velocity during the approach, b is a characteristic
length scale for the repulsive interaction (≈ 0.3-0.4 Å), and E_{th} and
E_b are the kinetic and binding energies, respectively. For an H atom
at low temperatures, ω_o is typically 3×10^{14} and 6×10^{13} sec^{-1} for
chemisorption and physical adsorption, respectively. When the collision
time is short compared to the spring period (i.e., $\omega_o \gg \omega_D$) then the
energy transferred in the collision is

$$\Delta E = 4(m/M)(E_{th} + E_b) \tag{9}$$

(Hollenbach and Salpeter 1970). This transferred energy should be
compared to the initial kinetic energy of the impacting atom, E_{th}, to
determine whether sticking will occur. If the collision time-scale and
the spring period are equal (i.e., $\omega_o = \omega_D$) then the energy transferred
in the collision is one fourth that given by equation (9). If ω_o is
actually much less than the spring frequency, then other surface atoms
can couple in the collision. The energy transferred is then decreased
over that given by equation (9) by a factor of about $1/2(\omega_o/\omega_D)^3$
(Hollenbach and Salpeter 1970; Watson 1976). Consider, for example, an
H_2 molecule with a kinetic energy corresponding to 300 K colliding with
an ice surface ($E_b \approx 800$ K). The maximum energy transfer is 50% of the
total energy (cf. eq. [9]) or somewhat less than 600 K. Actually ω_D
is about equal to ω_o in this case, and the energy transferred is only
about 140 K, leaving about 160 K of kinetic energy, and sticking will
not occur in the initial collision. Because the collision is generally
not head-on, the approaching atom will, however, ricochet off and collide with other surface atoms before being finally ejected from the
surface; therefore, more energy can be transferred (Hollenbach and
Salpeter 1970). Taking this into account, the sticking coefficient of

an H_2 molecule on an H_2O surface at 300 K is calculated to be about 0.3 (Hollenbach and Salpeter 1970), close to the experimental value (Govers et al. 1980). Of course when the surface is covered with a layer of H_2 molecules, as is likely inside molecular clouds, then the sticking coefficient will increase to about unity (i.e., $\Delta E \simeq E_{th} + E_b$), in agreement with experiments (Govers et al. 1980).

In general, if the kinetic energy of the accreting atom is much less than the adsorption energy, then only a relatively small amount of energy needs to be transferred to the surface, and the sticking coefficient will be relatively large. Similarly, if the accreting atom is much heavier than the surface atoms, then the amount of kinetic energy transferred to the surface will be relatively large and the sticking coefficient will also be relatively large. Finally, for a stiffer lattice (e.g., $\omega_D \gg \omega_o$) the sticking coefficient will be lower than for a softer lattice. It should be noted that this classical description assumes that any amount of energy can be transferred in the collision and, thus, when the initial kinetic energy goes to zero, the sticking coefficient will go to unity. This is, however, not correct, and for low initial energies quantum effects have to be taken into account (cf. Leitch-Devlin and Williams 1985). The adatom has to wind up in a bound level in the interaction potential well, and, thus, a minimum amount of energy has to be transferred, which depends on the spacing of the upper levels in the potential well. Because of anharmonicity this will be of little importance for chemisorption, but it can be of importance for physical adsorption. It should also be kept in mind that the phonon spectrum of a solid decreases with energy (e.g., the phonon density is proportional to $(\omega/\omega_D)^2$, so that coupling and thus energy transfer to the lattice will be less at low impact energies. Again, this is mainly important for physical adsorption for which the interaction potential is quite small.

The sticking coefficients can be found from the theoretical and experimental results. Anticipating our discussion later on, the sticking probabilities of importance for interstellar grain chemistry are the collision of a species with a clean silicate or graphite surface on which it can be chemisorbed (e.g., in the diffuse interstellar medium) and the collision of a species with an icy grain mantle on which it will be physically adsorbed (e.g., inside dense molecular clouds). Theoretical calculations on the sticking probability of an atom on a grain surface have been evaluated for several likely gas-phase-species and grain combinations (Hollenbach and Salpeter 1970; Watson and Salpeter 1972; Burke and Hollenbach 1984; Leitch-Devlin and Williams 1985). These calculations show that an H atom colliding with a silicate or graphite surface on which it can be chemisorbed will have a sticking coefficient of unity at a gas temperature of 100 K. An H atom colliding with a clean H_2O ice surface at 10 K has a relatively low sticking coefficient (≈ 0.3). However, inside a molecular cloud the icy surface will be partially covered with H_2 molecules, and this will raise the sticking coefficient to close to unity. For heavier atoms, the larger binding energy and larger mass will insure a large sticking probability (≈ 1).

An exception to these high sticking probabilities may have to be made for ions accreting on charged grains (Watson and Salpeter 1972). Grains inside molecular clouds are negatively charged most of the time because the electrons have a greater velocity than the ions (Spitzer 1978). Charge recombination at or near the surface complicates the evaluation of the sticking coefficient considerably (Watson and Salpeter 1972). The Coulomb interaction energy is much larger than the adsorption energy of neutrals on icy grains considered above (i.e., comparable to a chemisorption energy). Thus, an accreting ion acquires a much larger approach velocity than a neutral. The Coulomb interaction may lead to the tunneling of an electron through the work function of the grain material, followed by ion-electron recombination, possibly before the ion has reached the surface. The resulting neutral experiences the much weaker, neutral-grain adsorption potential, but its kinetic energy is still that of the original ion. This is, therefore, a case in which the kinetic energy of the gas-phase species is much larger than the adsorption energy and, consequently, the sticking probability of ions will be very low (Watson and Salpeter 1972). This hinges, however, on the mobility of electrons in grains, and if electrons are actually quite localized in a grain then the ion-sticking probability will be close to unity. Of course, the Coulomb interaction will also enhance the collision cross section by about an order of magnitude (i.e., focusing).

A parameter of importance for molecule formation on grain surfaces is the time-scale at which a grain accretes species from the gas phase. This accretion time-scale is given by

$$\tau_{ac} = (n_i \sigma v S)^{-1} , \tag{10}$$

where n_i is the density of the species in the gas phase ($5 \times 10^{-4} n_o$), σ is the geometrical cross section of the grain (10^{-9} cm^2), v the average velocity (1 km/sec), and S the sticking probability (≈ 1).

Thus, inside dense molecular clouds a grain will see a very small flux of incoming atoms or molecules (e.g., a few per day at a density of 10^4 cm^{-3}). Of course, with this flux all the heavier species in the gas phase will have accreted in 10^5 yr, which is a much shorter period than the cloud's lifetime.

4.4 Surface Mobility

The binding energy of an adsorbed species varies considerably across the surface owing to the discrete nature of the lattice. For physical adsorption, the barrier against diffusion to an adjacent site, E_d, is typically about 30% of the binding energy, although sometimes much lower values have been calculated (Jaycock and Parfitt 1986). For chemisorption these variations can be comparable to the binding energy, although often it is again about half the binding energy. For a heavy species the migration time-scale, τ_m (i.e., the time to go to a neighboring site), is given by the Boltzman expression:

$$\tau = \nu_z^{-1} \exp(E_d/kT) \; . \tag{11}$$

Here ν_z is the frequency of the vibrational motion along the surface (cf. eq. [6]). For atomic hydrogen, deuterium, and helium and molecular hydrogen, migration is due to quantum mechanical tunneling. The time-scale for this process is given by

$$\tau_m = 4h/\Delta E \; , \tag{12}$$

where ΔE is the width of the lowest energy band (cf. Hollenbach and Salpeter 1970; Watson 1976). For atomic H on an H_2O surface this has been evaluated to be 30 K, yielding a surface diffusion time-scale of 10^{-12} sec. Values for the migration time-scale on an icy surface are given in Table II. It should be stressed that the actual time involved in moving from one site to a neighboring site is very small ($\approx 10^{-12}$ sec). The limiting step in the migration process is the acquisition of the energy required to overcome the diffusion barrier.

The surface diffusion process is essentially a random-walk one. The rms value of the displacement vector R, after a time Δt is then

$$\langle R^2 \rangle^{1/2} = N^{1/2} \ell \tag{13}$$

where ℓ is the length of one site and N the number of hops during this time ($N = \Delta t/\tau_m$). Thus, the average time to scan a grain surface, τ_s, is $\tau_s = A_s \tau_m/4\pi$, where A_s is the total number of sites on the surface ($\sim 10^6$ for a 1000 Å grain).

4.5 Surface Reactions

Here we will only consider the formation of icy grain mantles inside dense clouds. The reactions associated with H_2 formation are discussed in §5. The reaction network can be limited considerably by a comparison of the migration time-scale and the thermal evaporation time-scale

(Table II) with the accretion time-scale (eq. [10]; $\sim 10^5$ sec). Species with a migration time-scale much less than the accretion time-scale can diffuse on the surface until they find a co-reactant (e.g., H, H_2, C, N, O, S at 10 K). Otherwise they are trapped in their site and can only react with migrating co-reactants that visit their sites (e.g., CO, N_2, O_2, CH_4, NH_3, and H_2O at 10 K). Thus, at 10 K the grain mantle reaction network contains reactions of the migrating atoms H, C, N, and O among themselves and with the non-diffusing atoms or radicals (Tielens and Hagen 1982).

Comparison of the evaporation time-scale with the accretion time-scale shows that if atomic H does not find a reaction partner on the surface then it will evaporate before the next radical accretes. Direct recombination of two H atoms to form H_2 does not occur on icy grain surfaces in molecular clouds (cf. §5). It should be stressed that because of their low mass, H atoms can tunnel through activation barriers and react with molecules containing an even number of electrons (e.g., non-radicals), such as CO, O_2, H_2CO, H_2O_2, H_2S, and O_3 (Tielens and Hagen 1982). The time-scale for tunneling through a barrier with a height, E_a, and a width, a, is given by

$$\tau_t = \tau_o \exp[(2a/\hbar)(2mE_a)^{1/2}] \tag{14}$$

where m is the mass of the tunneling species. Assuming a barrier height of 2000 K and a width of 1 Å, this time-scale is 2×10^{-4} sec. This is much longer than the migration time-scale, and the probability for reaction is therefore very small ($\tau_m/\tau_t \approx 5 \times 10^{-9}$). Thus, generally an H atom will migrate to an adjacent site rather than react. However, over an evaporation time-scale, an H atom will visit each site many times ($\tau_{ev}/\tau_s \approx 10^{10}$ at 10 K for a 1000 Å grain). The probability of reaction p_r is therefore given by

$$p_r = (\tau_{ev}/\tau_s)(\tau_m/\tau_t) \tag{15}$$

For the numerical example considered above, p_r is about 5. Thus, H atoms can react with molecules even if there is an activation barrier as large as 2000 K. For a high concentration of co-reactants, p_r will be larger. For example, for a concentration of 1% on a grain of radius 10^3 Å, the limiting barrier an H atom can tunnel through is about 3500 K. Similarly, reactions of H_2 with an activation barrier less than about 4500 K can be important on grain surfaces (Tielens and Hagen 1982). Heavier atoms or radicals will have to thermally hop over an activation barrier and reactions with activation barriers larger than the barrier against diffusion are effectively inhibited unless the abundance of co-reactants is very large. Assuming the surface is covered with such co-reactants and considering the accretion of a new radical as the limiting step for reactions of O atoms (rather than evaporation), reactions of O atoms with an activation energy of about 450 K or less have to be taken into account at 10 K. Reactions of O atoms with negligible activation barriers include O_2 and possibly CO forming O_3 and CO_2, although the latter activation barrier is still somewhat controversial (Tielens et al. 1986a).

4.6 Dissipation of Reaction Heat

The product molecules of grain-surface reactions are generally formed in electronically and vibrationally excited states. The subsequent decay, ultimately to the ground state, stabilizes the newly formed molecule. Some of the reaction heat might be transferred into center-of-mass translational motion, possibly leading to ejection of the molecule. The excess translational energy of the ejected molecule will heat the gas. The ejected molecule may also contain excess vibrational or rotational energy, which will decay radiatively. This process might explain the relatively high excitation of $J > 4$ levels of H_2 observed in diffuse interstellar clouds, although other mechanisms are also possible (Spitzer and Cochran 1973; Dalgarno 1975).

In particular, the ejection of "hot" H_2 from grain surfaces has incited several theoretical studies. Classical mechanical trajectory calculations have been performed, mainly for H_2 formed by recombination of two physically adsorbed H atoms on a grain surface (Hollenbach and Salpeter 1970; Hunter and Watson 1978). In these calculations enough energy was transferred to the center-of-mass translational motion in the collision between the newly formed H_2 molecule and the surface potential to cause ejection. Similar calculations for heavier molecules indicate that their ejection is unlikely (Watson and Salpeter 1972). Essentially the coupling between vibrations and center-of-mass translational motion is small, because the molecular vibrational time-scale is much shorter than that for change in the repulsive interaction potential with the surface (Hunter and Watson 1978). However, these calculations suffer from two deficiencies: they ignore transfer of energy to other decay channels than center-of-mass translational motion and the quantum nature of the energy transfer process. It should be noted that molecule formation may occur along a pathway that is directly dissociative in the surface-admolecule bond (Aronowitz and Chang 1985). The general importance of this ejection mechanism is, however, unknown, and we will ignore this channel in the subsequent discussion.

The transfer of the reaction heat to other channels is very important. In fact, in the classical trajectory calculations referred to above, a somewhat arbitrary criterion had to be used to determine the fraction of H_2 molecules ejected, because in the absence of other decay channels all newly formed H_2 molecules will eventually acquire enough energy in their center-of-mass translational motion to be ejected. Among the other energy channels that can accept the excess vibrational energy of a newly formed H_2 molecule are rotations of the H_2 molecule (rapidly followed by transfer to lattice phonons), (de)localized phonons of the grain, and vibrational modes of neighboring molecules. Because of the inherent quantum nature of the transfer process, the contribution of each of these processes depends on the spacing of the energy levels involved, as well as on the nature of the interaction.

Except for energy transfer to neighboring molecules, the accepting energy channel has a much smaller energy level spacing than the initial vibrational energy channel. For example, a vibrational quantum of H_2 is 3000 to 5000 cm^{-1} (≈ 0.4-0.67 eV), which is much larger than an H_2 rotational quantum (≈ 120 cm^{-1}; ≈ 0.015 eV), a vibrational quantum of the H_2-surface bond (≈ 30 cm^{-1}; ≈ 0.004 eV), or phonons (≈ 100 cm^{-1}; ≈ 0.01 eV). Therefore, in the energy-transfer process a large number of quanta have to be created simultaneously. This process is described by an energy-gap law, and its rate decreases exponentially with the energy gap involved, that is, with the number of quanta that have to be generated (Diestler 1974; Nitzan et al. 1975; Diestler et al. 1977). Experimental studies of the energy transfer from vibrational (≈ 3000 cm^{-1}) to rotational (≈ 30 cm^{-1}) modes of small hydrides isolated in low-temperature matrices indeed show such an energy-gap law (Legay 1977).

Energy transfer to neighboring molecules can be a very important energy decay channel for vibrationally excited molecules if the energy mismatch between the donor and acceptor is not too large and if the exchange is symmetry allowed. This energy exchange is induced by (long-range) multipolar (i.e., dipole or quadrupole) interactions. Theoretically, the lifetime associated with this energy decay channel is estimated to be about 5×10^{-12} sec for resonant conditions (Brus 1981). Experimental studies of resonant energy transfer of vibrationally excited CO indicate a lifetime of about 10^{-11} sec (Legay 1977). Energy transfer in the Lorentzian wing of a transition is very inefficient and non-resonant energy transfer is, therefore, generally assisted by phonons (or other low-lying energy channels that match). This will again lead to an exponential decrease in the rate with increasing energy gap. This is confirmed by experimental studies (Legay 1977). Non-resonant energy transfer can still be very fast, if the conditions are favorable.

As an example of these principles consider the different energy-decay channels for the excess vibrational energy of a newly formed H_2 molecule one by one. Since this is a homonuclear diatomic molecule, it does not possess a dipole moment. Near-resonant quadrupole-dipole transfer to neighboring molecules of the excitation in the high vibrational levels ($\nu \sim 3000$ cm^{-1}) therefore is probably slower by a factor of 10^2 to 10^3 than the 10 psec time-scale for dipole-dipole interaction, leading to a time-scale of perhaps 10^{-9} to 10^{-8} sec. It should be noted that interstellar grain surfaces are expected to contain many functional groups with vibrational modes at about 3 μm (e.g., OH and CH groups; cf. §4.1), which can act as energy acceptors. Recall also that H_2 is very mobile on a grain surface, so that energy transfer is likely to occur between nearest neighbors. The mismatch in energy between donor and acceptor can be taken up by rotational energy of the H_2 molecule phonons or center-of-mass translational motion. For the lower levels ($\nu \sim 4000$ cm^{-1}) this decay channel is probably closed because of the large energy mismatch.

Rotations might also directly act as energy acceptors of the excess vibrational energy. The experimentally measured relaxation rate of dipolar hydrides, which couple well with the lattice, ranges from about 10^7 sec^{-1} for the lighter ones (e.g., OH and NH) to 10^3 sec^{-1} for the heavier ones (e.g., HCl; Legay 1977). The mass ratio of unity favors a high relaxation rate for this channel. However, because H_2 is nonpolar, it will not couple as well to the lattice. The transfer rate of vibration to rotation might perhaps be as fast as 10^8 sec^{-1}. Small energy mismatches (i.e., less than one rotational spacing) can be taken up by center-of-mass translational motion of the H_2 molecule or other local phonon modes. Decay of the rotational energy through transfer to local or delocalized phonon modes (one quantum at a time) will be very rapid because of the similarity in energy (for example, the Debye frequency of H_2O is 133 cm^{-1}). Some of the energy may wind up momentarily in vibrational energy of the H_2-surface bond but, again, transfer to phonons will be rapid and buildup of energy in this mode will not occur. Thus, ejection of H_2 is unlikely in this case.

Finally, direct transfer of the vibrational energy to center-of-mass translational motion has been estimated theoretically to have a rate of about 10^7 sec^{-1}, but this estimate is very uncertain (Lucas and Ewing 1981). In this case, the excess energy in this channel is sufficient to cause ejection of a translationally "hot" (≈ 0.5 eV; i.e., one vibrational quantum) H_2 molecule.

From this discussion, it seems that for the high vibrational levels of H_2, decay will occur through energy transfer to neighboring molecules. The lower levels may decay through transfer to rotational modes or to center-of-mass translational motion. If the latter process dominates, then ejection is likely. Clearly, laboratory studies of the relaxation of vibrationally excited H_2 are important to the resolution of this point. For heavier molecules, in particular for non-homonuclear molecules, energy transfer through multi-polar interaction is expected to dominate the reaction heat dissipation.

5. FORMATION OF H_2 ON GRAIN SURFACES

Ultraviolet observations have revealed that molecular hydrogen is an important constituent of diffuse interstellar clouds, such as those in front of ζ Oph (Morton 1975). It is generally assumed that H_2 is the most abundant molecule in dense molecular clouds, but direct detection of H_2 in quiescent clouds is difficult, because of the high UV opacity of such clouds and the low Einstein A values of the IR rotational tran- sitions. Ultraviolet-pumped, IR fluorescence from H_2 has, however, been observed in reflection nebulae (e.g., NGC 2023) and around HII regions (e.g., the Orion bar; Gatley et al. 1986; Sellgren 1986). It has also been observed in regions with strong molecular shocks such as the BN/KL region in Orion (Beckwith et al. 1978). This suggests that H_2 is indeed the dominant form of hydrogen in molecular clouds.

It is difficult to form H_2 through reactions in the gas phase and, therefore, it is generally presumed that it is formed on the surfaces of interstellar grains. Because of the importance of H_2 in models of interstellar chemistry, the principles of grain-surface chemistry out- lined above will be applied in this section to the formation of H_2 in the interstellar medium. First, H_2 formation on bare silicate grains in a diffuse interstellar cloud ($n_o \sim 10$ cm^{-3}, $T \sim 100$ K, and $T_d \sim 20$ K) will be considered, then H_2 formation on the basal plane of graphite and amorphous carbon will be briefly examined. This will be followed by a discussion of poisoning of sites at which H atoms can be chemisorbed. The section will close with a discussion of H_2 formation on icy grain mantles inside molecular clouds. Hydrogen formation on organic refrac- tory dust is presumably very similar to this latter case.

5.1 H_2 Formation on Silicate Surfaces

Hydrogen atoms on a silicate grain can be chemisorbed as well as physi- cally adsorbed on top of a chemisorbed layer (cf. §4.2). There are about 10^6 of each of these sites on a grain with a radius of 10^3 Å. Under the conditions in the interstellar medium, migrating physically adsorbed atoms will react with chemisorbed atoms forming H_2, which will evaporate quickly. With a sticking coefficient of unity and the gas parameters stated above, the H accretion time-scale, τ_a, on a 1000 Å grain is about 500 sec. Thermal evaporation from chemisorption sites is completely unimportant. For physical adsorption sites the evaporation time-scale is about 2×10^4 sec. The migration time-scale for physi- cally adsorbed atoms is, however, only 3×10^{-9} sec and, if there is an empty chemisorption site, the physically adsorbed atom will find it before it evaporates. Most H atoms will, thus, be bound in chemisorp- tion sites. Occasionally, a chemisorbed atom can tunnel to an adjacent site on a time-scale τ_t of about 10^7 sec, assuming a barrier with a

height of 1 eV and a width of 1 Å. If this site is empty, the H atom
will be chemisorbed again until it tunnels on. If there is another
H atom present inthis site, the entering H atom will only be physically
adsorbed. It may react with the H atom already present or migrate on.
With an activation barrier of 1000 K and a width of 1 Å, the time-scale
for reaction, τ_r, is 1.3 × 10^{-6} sec. This should be compared to the
time-scale for migration (3 × 10^{-9} sec). The probability of reaction,
p_r, is therefore only 2 × 10^{-3} and the H atom will migrate about 500
sites before it reacts (if it is not trapped in an empty chemisorbed
site before that). The probability for evaporation during this migration and reaction time is given by τ_r/τ_{ev}; for the parameters above,
this is negligible, and all accreting H atoms will form H_2.

When the dust temperature is increased, evaporation of H atoms in
the physically adsorbed layer will become important. A critical temperature, T_{cr}, can be defined for which the probability for reaction with a
chemisorbed atom is equal to the evaporation rate. This critical temperature is given by

$$T_{cr} = E_b/\ln(\tau_r \nu_z) \, , \tag{16}$$

where E_b is the binding energy and ν_z the characteristic vibration
frequency of the H atom in the physically adsorbed layer (cf.
eq. [6]). For the values given above, this critical temperature is
50 K. Note that this critical temperature is somewhat lower than that
calculated by Hollenbach and Salpeter (1971). They envisioned a situation with a (small) number of enhanced binding sites interspersed in an
otherwise physically adsorbed layer. However, they neglected the activation barrier for reaction. In the analysis above, the activation
barrier for reaction is crucial. If the activation barrier for reaction
were actually less than the binding energy of the H atom in the physically absorbed site, then H_2 formation would occur until evaporation
from the chemisorbed layer dominated the loss of H atoms. The laboratory studies measure, however, both the physical adsorption energy and
the activation energy rather directly, and the latter is indeed larger
than the former. Above the critical temperature, the reaction rate will
drop by a factor $\tau_{ev}/\tau_r \approx 3 \times 10^{-7} \exp(750/T_d)$; at 75 K, this factor is
6 × 10^{-3}. Finally when the dust temperature is very high and evapora-
tion of physically adsorbed H atoms is very rapid, only those impinging
gas-phase atoms that have enough kinetic energy to directly overcome
the activation barrier for reaction with the chemisorbed atoms will
react. The rate of H_2 formation is then reduced by a factor
$\phi_{eff} \exp(-E_a/kT) \approx 0.25 \exp(1000/T) \approx 10^{-5}$. Here ϕ_{eff}, the fraction of
a site covered by an H atom, takes into account the smaller cross sec-
tion of an H atom compared to the site area. Note that evaporation of

chemisorbed H atoms is unimportant until the grain temperature reaches
about 800 K.

5.2 H_2 Formation on Graphite Surfaces

The formation of H_2 on the basal plane of graphite in the diffuse interstellar medium is very similar to that on silicates. As on silicates,
atomic hydrogen can be both chemisorbed and physically adsorbed on
graphite with binding energies of about 2 eV and 0.07 eV, respectively
(see §4.2). The most important parameter for H_2 formation, the activation barrier for the reaction between physically adsorbed and chemisorbed H, is also very similar for the two surfaces ($\approx$1200 K; King and
Wise 1963; versus $\approx$1000 K; Wood and Wise 1962). The only difference is
the presence of so-called "active surface areas" on graphite (i.e., the
edges of the basal plane). Hydrogen atoms can attack the carbon skeleton at those areas and form saturated hydrocarbons such as methane.
There is a considerable spread in the activation barriers reported for
this set of reactions (1700-5000 K; King and Wise 1963; Wood and Wise
1969; Rye 1977; Balooch and Olander 1975), which may be due to differences in the surface structure of the samples (Veprek and Haque 1975).
In any case, the activation energy is larger than that for reaction of a
migrating H atom with a chemisorbed H atom on the basal plane; as a
result, diffusing H atoms generally form H_2 on the basal plane instead
of reacting on the edges. This is even more pronounced in the interstellar medium because H coverage on the basal plane will be high. In
this respect, it should be noted that in the one series of experiments
in which the H atom coverage on the basal plane was high, H_2 was the
dominant product at low temperatures (King and Wise 1963). Draine
(1979) has argued that a migrating H atom would abstract a chemisorbed
H atom (part of a CH_3 group) on these edge structures to form H_2 rather
than form CH_4 at low temperatures. However, this H abstraction reaction
is likely to have considerable activation energy ($\approx$8000 K; Balooch and
Olander 1975), and from the experiments we infer that it does not play a
role until the surface temperature is much higher ($\approx$800 K). At that
temperature, CH_4 becomes thermodynamically unstable, and the equilibrium
will shift toward hydrogen and (solid) carbon (Wood and Wise 1969).

5.3 Poisoning of Chemisorbed Sites

We will now consider poisoning of the chemisorbed sites by other
species. That this is important can be seen by comparing the accretion
rate with the thermal evaporation rate. For small grains ($<$100 Å) or
warm grains ($T_d > 22$ K), physically adsorbed H atoms evaporate before
the next H atom accretes, and H_2 formation through recombination of two
physically adsorbed H atoms is inefficient. In the interstellar medium,

poisoning will occur when another atom blocks the access of H to the chemisorbed site. The poisoner may be physically adsorbed as long as it is essentially immobile. Note that if there are still a few chemisorbed sites available on a grain that H_2 formation will occur by the mechanism outlined above. Poisoning is, thus, only effective if almost all of the sites are covered by poisoning atoms. The time-scale for poisoning is then the accretion time-scale per site of poisoners, which is about $5 \times 10^5(10^{-4}/X)$ yr, with X the relative gas-phase abundance of the poisoning atoms. For oxygen, nitrogen, and carbon atoms this time-scale is very short compared to the lifetime of the cloud or the H_2 photodissociation lifetime, and they might therefore be very effective poisoners; this is even more likely because they are also chemisorbed to these sites. However, H atoms will be able to tunnel through the activation barrier for reaction and form H_2O, NH_3, and CH_4, which is energetically favorable. However, in the diffuse interstellar medium, photodesorption is an effective way of cleaning a surface of these molecules, since their photodesorption time-scale in an unshielded region is only about 700 yr (see §9.1).

Poisoning by other atoms might also be important in the diffuse medium. The observed total surface area of interstellar dust is about 3×10^{-21} cm^2 per H atom. This number might be somewhat larger (50%) if the grain size distribution extends to below 100 Å. With a surface area per site of about 7 $Å^2$, this dust surface area corresponds to about 4×10^{-6} sites per H atom. Thus, species with a total gas-phase abundance in the diffuse interstellar medium comparable to this number could poison the chemisorbed sites on a grain surface. Note that we have to correct for the underlying depletion caused by the incorporation of some elements in refractory dust grains produced by stars. This implies that poisoning by atoms such as S, Si, Na, Mg, Fe, Al, Ca, and Ni can be of importance. Some of these (e.g., S and Si) will form volatile hydrides which are probably also easily photodesorbed, whereas others (i.e., Na and K) may be ejected upon electron recombination (cf. §6). The transition metals, as well as Ca, will form hydroxide groups on interstellar grain surfaces (§6). Presumably, H will also be chemisorbed on such surfaces. Moreover, although their abundance might be sufficient to completely poison the chemisorbed sites on a time-scale of only about 10^7 yr, this assumes that these species are used as efficiently as possible (i.e., one atom per chemisorbed site). This is unlikely, since these species are immobile at 20 K. It seems, therefore, that poisoning of grain surfaces cannot stop H_2 formation in the diffuse interstellar medium.

5.4 H_2 Formation on Grain Mantles

Finally, H_2 formation on icy surfaces will be considered. Inside dense molecular clouds, interstellar grains will accrete mantles consisting of such simple molecules as H_2O, NH_3, CO, and CH_3OH (see §7). Such icy grain mantles have only physical adsorption sites, with a typical binding energy of about 350 K; this is owing to the high surface coverage of H_2 (Govers et al. 1980; Tielens and Hagen 1982; see also §4.2). The thermal evaporation rate of H is then about 2 × 10^2/sec, which is lower than the accretion time-scale of H atoms on a grain (~10^5 sec for a 1000 Å grain and a typical H density of 1 cm^{-3}) and H_2 formation by direct recombination of physically adsorbed H atoms on a grain surface will not occur. Of course, by the time a grain mantle has been formed, hydrogen is already completely converted into the molecular form through reactions on bare graphite or silicate grain surfaces.

Although icy grain mantles do not have chemisorbed sites, molecules such as H_2CO, H_2S, N_2H_2, N_2H_4, and metal hydrides can act as enhanced binding sites for H_2 formation (Tielens and Hagen 1982). During its lifetime on a grain surface, an H atom may tunnel through the activation barrier associated with H abstraction from such molecules producing H_2 and leaving a reaction site (cf. §4.5). Thus, H_2 can also be formed and reformed inside dense molecular clouds. It should be stressed that because radical-radical reactions are activationless, if there are radicals on the surface, then an H atom will react with them rather than tunnel through an activation barrier and form H_2. Reaction of H with non-radicals such as O_2 and CO can also be important in this respect. Thus, when the gas-phase atomic H abundance decreases below the abundance of heavier species such as O, CO, and O_2, H_2 formation on grain surfaces will stop (Tielens 1983b). Typically, this occurs at a density of about 3 × 10^4 cm^{-3} (inert conditions; see §7). This will, however, not increase the abundance of H in the gas phase, because the H atom will now be incorporated in such molecules as H_2O, NH_3, and H_2CO.

6. DEPLETION OF THE ELEMENTS

UV absorption line studies have shown that the gas-phase abundance of many elements in the diffuse interstellar medium is less than expected from cosmic abundances. The degree of depletion varies from element to element, as well as from one line of sight to another (see chapter by Jenkins, elsewhere in this volume). The missing elemental fraction is generally thought to be locked up in grains and, thus, elemental abundance studies may yield information on the composition and physical and chemical properties of interstellar grains. Several processes can influence the gas-phase abundance of an element. Clearly, dust

condensation in the high-temperature and high-density outflow from
stellar sources (Field 1974) is important, as evidenced, for example, by
the large fraction of silicon locked up in silicates (cf. §1.1).
Destruction of dust in interstellar shocks is also important, given the
relatively high gas-phase abundance of elements such as Ca, Fe, and Si
in high-velocity clouds (i.e., recently shocked clouds) compared to the
general diffuse interstellar medium (Spitzer 1976). Finally, accretion
of atoms in the interstellar medium is also an important process (Snow
1975), since elemental abundances lower than 10^{-1} times the cosmic
abundance are difficult to achieve by stardust formation alone (see
chapter by Jura, elsewhere in this volume). The complexity of these
interplaying processes has defied a general theory of elemental depletions in the interstellar medium. Considering the scope of this book,
in this section we merely want to point out some considerations concerning the third process, the accretion of elements in the diffuse interstellar medium.

Nitrogen, oxygen, and sulfur are expected to be little depleted in
the interstellar medium, because their main condensation products around
stars are very volatile (e.g., NH_3, H_2O, and H_2S). Although these
elements are presumably chemisorbed on interstellar graphite and silicate surfaces, they are expected to react in the diffuse medium with
atomic hydrogen on grain surfaces and thereby form their volatile
hydrides. These are probably easily photodesorbed (cf. §9). Once they
are in the gas phase, photodissociation will rapidly destroy these
molecules. Sputtering in low velocity (≈ 20 km/sec) shocks can also be
of importance for the removal of these saturated molecules. Some oxygen
will be locked up in refractory silicate grains and, possibly, in
organic refractory material.

Carbon takes a special place among the more abundant heavy elements. Not only is solid carbon the thermodynamically favored phase at
high temperatures and densities (i.e., in the outflow froms stars), but
carbon can also form large molecules that can be fairly resistant to
destruction. Observations indicate that a considerable fraction of the
carbon can be locked up in grains made up of cross-linked macromolecules
(cf. §1.3). The physical processes important in the formation of the
organic refractory dust component in the interstellar medium are discussed in §8. Here we merely emphasize that carbon can be highly
depleted in the interstellar medium.

Silicon and phosphorous can form a refractory high-temperature
condensate in the outflow around stellar objects (i.e., $MgSiO_3$ and Fe_3P;
Fegley and Lewis 1980). In the diffuse interstellar medium, it is
expected, however, that both atoms when accreted on a grain surface will
form saturated, volatile hydrides (i.e., SiH_4 and PH_3) in a manner similar to that of carbon and nitrogen. Again, photodesorption will probably prevent large depletions in the diffuse medium. Silicon can

substitute for carbon in organic molecules, and its depletion in the interstellar medium may, therefore, vary through incorporation into the organic refractory dust component formed in molecular clouds (cf. §8).

The depletion pattern of the alkali and alkaline metals is somewhat puzzling in view of their closeness in the periodical system. Sodium, potassium, and magnesium are little (about a factor of two) depleted in stardust (presumably composed of silicates) and do not show large differential depletion effects in the diffuse interstellar medium. In contrast, Ca is highly depleted and shows large variations. Both Na and K have ionization potentials below the grain work function and, thus, charge effects will not complicate their sticking (i.e., $S \approx 1$; cf. §4.3). The positive ions Na^+ and K^+ have an inert-gas electronic structure, which inhibits their reaction with H upon accretion. However, once accreted an electron is expected to be able to tunnel through the work function and recombine with the alkali metal ion. If this recombination occurs in a Rydberg state (high principal quantum number), the extended electron cloud coupled with the repulsive interaction with the surface will then, presumably, lead to desorption (see the discussion on photodesorption in §9.1). Recombination may also occur directly in an antibonding valence orbital of the surface-adion bond. If this scenario is correct, then one would expect that because of their similar electronic structures and low ionization potentials, Li, Rb, and Cs are also little depleted in the interstellar medium.

Because the ionization potentials of Mg and Ca lie above the work function of interstellar grains, sticking might be complicated by recombination effects (cf. §4.3). The large depletion of Ca suggests, however, that electrons in interstellar grains are not very mobile on the short collision time-scale ($\approx 10^{-13}$ sec). Since the cations do not have an inert-gas electron structure, they probably react immediately upon accretion with atomic hydrogen on the surface, which may actually prevent their ejection upon electron recombination. Both elements will react with hydrogen to form the dihydrides. However, whereas MgH_2 has a covalent structure, CaH_2 is a salt (Mueller, Blackledge, and Libowitz 1968). Presumably, MgH_2 is therefore easily photodesorbed and CaH_2 is not. If this is indeed the essential difference, then it is expected that other elements that form covalent hydrides, such as Be, Zn, and Cd, would be little depleted (Zn is discussed below). At this point, it should be recalled that hydrogen abstraction reactions are very important on interstellar grain surfaces. The Ca hydride remaining on the surface will therefore react often with H (i.e., $CaH_2 + H \rightarrow CaH + H_2$, followed by $CaH + H \rightarrow CaH_2$). This cycle will continue until CaH is poisoned by another atom, presumably O, leading to hydroxide formation. It is expected that the $Ca(OH)_2$ will be bonded quite strongly to the surface.

The transition metals Ti, Mn, Fe, and Ni are highly depleted in the interstellar medium. In contrast, the depletion of Zn is quite small. The special feature of the electronic structure of the transition metals that presumably lead to this depletion pattern is the presence of unfilled d orbitals, which can form strong complexes with electron-rich material (Pauling 1964). The unfilled 3d orbitals available should provide strong bonds to a grain surface. These elements can also react with atomic H on the surface to form mono- and dihydrides, which could be stabilized by the formation of a coordination complex with a ligand group on the surface (e.g., highly polar or polarizable groups with unshared electron pairs such as an OH group in a silanol moiety on a silicate surface or the π electron system in PAHs, amorphous carbon particles, or graphite). This complex will still be strongly bound to a grain surface and photodesorption will be of little importance. In this respect it is important to note that Zn has all its 3d orbitals filled and does not readily form coordination complexes. Zinc hydride has a covalent character (Mueller, Blackledge, and Libowitz 1968), and is probably easily photodesorbed or sputtered. Cadmium, which has an electronic structure similar to that of Zn, also forms a covalent hydride and should, therefore, have a similar depletion pattern.

In summary, the differential elemental depletion patterns in the diffuse interstellar medium can be ascribed to the chemical properties of the elements. Of particular importance is the ability to form hydrides and coordination complexes. Photodesorption and electron recombination may play a key role in preventing depletion of some elements. In contrast to Barlow (1978), we do not think that ejection upon reaction is an important process (cf. §4.6). We also caution against the straightforward use of surface bond energies of the elements based on chemisorption theory to explain observed depletion patterns. Chemisorption is a very complex process and is far from theoretically understood (Einstein et al. 1980).

7. COMPOSITION OF ICY GRAIN MANTLES

The composition of interstellar grain mantles has been calculated, employing gas-phase as well as grain-surface reactions (Tielens and Hagen 1982; Tielens 1983b; d'Hendecourt, Allamandola, and Greenberg 1985). Figure 13 shows the composition of the gas phase and the grain mantle as a function of the gas-phase hydrogen density (Tielens and Hagen 1982). In the subsequent discussion we will concentrate on the global composition of the grain mantles and ignore to a large extent the abundances of trace species.

First, we will concentrate on the gas-phase composition of molecular clouds (Fig. 13a), since it sets the stage for the diffusion-limited

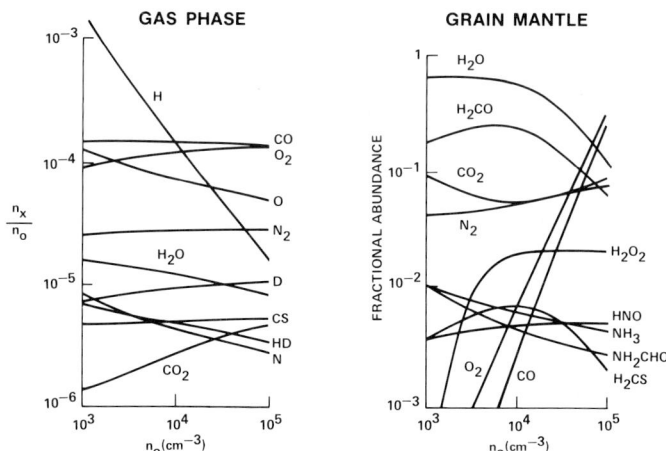

Figure 13. Molecular composition of interstellar clouds as a function of density in the gas phase (a) and grain mantle (b).

chemistry on grain surfaces. Hydrogen is mainly in molecular form (H_2). The atomic hydrogen abundance is inversely proportional to the total density, because the H_2 formation time-scale depends on the (grain) density, whereas the destruction time-scale (by cosmic rays) does not. The most abundant molecule in the gas phase after H_2 is CO, which contains most of the elemental carbon in the gas phase. Oxygen and nitrogen are either in atomic or molecular form (O_2 and N_2). Note that H_2O contains only a small fraction of the gas-phase oxygen. These results are very typical for time-independent gas-phase chemistry calculations or for time-dependent gas-phase chemistry calculations at late stages ($>10^6$ yr; Mitchell et al. 1978; Prasad and Huntress 1980; Graedel et al. 1982).

The calculated grain mantles consist mainly of the molecules H_2O, H_2CO, N_2, O_2, CO, CO_2, H_2O_2, and NH_3 and their deuterated counterparts in varying ratios. The exact composition depends strongly on the conditions in the gas phase. Depending on the gas-phase composition, three different physical regimes can be discerned, a reducing, an inert, and an oxiding "atmosphere." For low densities ($n < 5 \times 10^4$ cm^{-3}), atomic hydrogen is more abundant in the gas phase than the heavier species (e.g., O, O_2, and CO) and the gas is largely reducing. As a result, most species will be hydrogenated on the grain surface (e.g., H_2O and H_2CO). Unsaturated molecules, such as CO, are trace species in the grain mantle, whose abundance depends directly on their gas-phase abundance relative to that of atomic H and is, thus, a strong function of the total hydrogen density. An exception should be made for N_2, which does not react with atomic H. Ammonia is therefore a trace species, whose abundance depends on the gas-phase abundance of atomic N.

At higher densities ($n > 5 \times 10^4$ cm^{-3}), the atomic H density in the gas phase is less than that of the heavier species and most accreting species will not be hydrogenated. The resulting grain mantle depends then on the O to O_2 ratio in the gas phase, since O is the next most abundant, mobile radical. If this ratio is low, then only few grain surface reactions will occur (i.e., inert atmosphere). The grain mantle will then mainly reflect the gas-phase composition (e.g., CO, O_2, and N_2). If the atomic O to O_2 ratio in the gas phase is high (oxidizing atmosphere) then atomic O and CO will react; CO_2 will then be the dominant grain-mantle constituent. In both cases hydrogenated molecules such as H_2O and H_2CO are only trace species, whose abundance depends on their abundance in the gas phase and on the ratio of atomic H to heavier species.

Infrared observations show that interstellar grain mantles consist mainly of H_2O and CH_3OH; CO and NH_3 are only trace species (Tielens et al. 1986a). This implies that grain mantles are formed in a "reducing" atmosphere and, thus, at relatively low densities ($\approx 10^4$ cm^{-3}). Note that CH_3OH may result from H addition reactions to H_2CO. There is some experimental justification for the occurrence of this reaction at low temperatures (Tielens and Hagen 1982). It has, however, not been included in the calculations described above. Finally, in some cases evidence for grain mantle formation in a more inert atmosphere has also been observed (Tielens et al. 1986a).

It is interesting to consider the deuteration of species in a grain mantle. Figure 14 shows the relative abundance of a deuterated species with respect to its hydrogenated counterpart (Tielens 1983b). Generally, this ratio reflects the ratio of atomic deuterium to atomic hydrogen in the gas phase, essentially because the grain-surface reaction networks are very similar. Thus, this ratio increases strongly with increasing gas-phase density owing to the decrease in the gas-phase abundance of atomic hydrogen (cf. Fig. 13a). It is important to note that although most of the hydrogen is in molecular form, most of the deuterium is actually in atomic form in the gas phase, and the ratio of atomic deuterium to atomic hydrogen is much larger than expected from elemental abundances (Tielens 1983b). Molecular hydrogen is destroyed in the gas phase by cosmic ray ionization leading eventually to H_3^+. However, in its destruction network, HD competes with other molecules such as CO and N_2 for reaction with H_3^+ forming H_2D^+. Reaction of this molecular ion with neutral molecules such as CO and N_2 (leading to the deuterated ions DCO^+ and N_2D^+), followed by dissociative electron recombination produces atomic deuterium. On the basis of their relative abundances, one out of about every 20 cosmic ray ionizations of H_2 will lead to HD destruction; thus, the HD destruction rate is comparatively

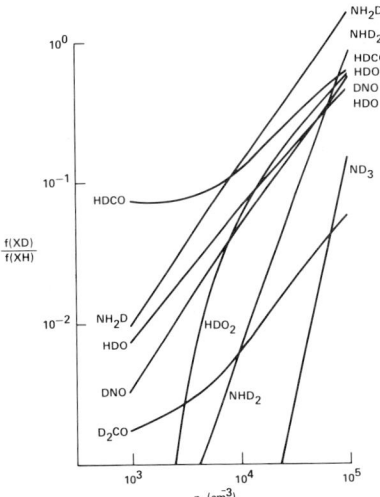

Figure 14. Concentration of the most abundant deuterated molecules relative to their hydrogenated counterparts in the grain mantle as a function of the total hydrogen density in the gas phase.

much larger than that of H_2 (Tielens 1983$\underline{b}$). In fact the high deuteration observed for molecular ions in molecular clouds implies that the D/H ratio is much larger than the elemental abundance ratio. Note that the grain-mantle abundances of some molecules (e.g., HDCO, D_2CO and to a lesser extent HDO) do not directly reflect the atomic D/H ratio in the gas phase. These result from differences in the grain-surface reaction network for these species (e.g., H abstraction reactions are of importance; Tielens 1983$\underline{b}$).

8. UV PHOTOLYSIS, TRANSIENT HEATING, AND ORGANIC REFRACTORY DUST

Greenberg and his co-workers and others (Greenberg 1963; Greenberg and Yencha 1973; Hagen et al. 1979; Greenberg 1979; d'Hendecourt et al. 1986; Donn 1960; Donn and Jackson 1970; Khare and Sagan 1973) have suggested that UV photolysis of interstellar grain mantles inside molecular clouds can produce complex molecules. These complex molecules are not as easily destroyed by interstellar shock waves as H_2O ice (Draine and Salpeter 1979) and could conceivably form an important component of the dust in the diffuse interstellar medium. Processing by cosmic rays might play a similar role (Moore and Donn 1982; Strazulla et al. 1983). Infrared spectroscopy of protostars deeply embedded inside dense molecular clouds suggests that the composition of interstellar grain mantles is more complex than can be explained by grain-surface reactions alone (Lacy et al. 1984; Tielens et al. 1984, 1986a). This may indicate the importance of energetic processing by UV photons or cosmic rays.

It should be stressed that grain-surface reactions alone only form simple, saturated hydrides (cf. §7). The formation of the organic refractory dust component, consisting of long carbon chains (i.e., polymeric carbon), in such a reducing atmosphere as the interstellar medium is difficult to understand. This section will discuss this problem and will identify the two key processes--that is, UV photolysis producing radicals and radical diffusion triggered by transient heating--that may convert interstellar grain mantles consisting of simple saturated hydrides into the complex organic refractory dust component of the interstellar medium.

In contrast to grain-surface reactions occurring during accretion of gas-phase species, UV photolysis mainly takes place in the bulk. Ultraviolet photolysis of interstellar grain mantles can produce reactive radicals, often in electronically or vibrationally excited states (Hagen 1982). These "hot" radicals can react with their neighbors or they can diffuse through the matrix and react with other species present (Hagen, Allamandola, and Greenberg 1979). Because of the large amount of excess energy available, even reactions with molecules having closed electron shells, which normally have appreciable activation barriers, might take place. Because, in general, reactions only take place when the reactants approach along well-defined reaction paths and because collisions of the hot reactant and the host material will transfer energy to the matrix (e.g., heat), reactive species can also be stored at low temperatures in the grain mantle. Upon warm-up those radicals can start diffusing through the host material and react with each other. Thus, during photolysis and warm-up complex molecules can build up, producing an organic refractory mantle on interstellar dust grains (Greenberg 1979).

The chemical processes taking place during grain mantle photolysis are complex. Often there are several photo-decomposition channels of a primary molecule. The photoproducts can also react with many of the neighboring unsaturated molecules. Moreover, the resulting secondary molecules themselves can often also be photolyzed, leading to a third generation of molecules. This complexity is compounded by the loss of volatile components owing to temperature spikes generated by grain-grain collisions (d'Hendecourt et al. 1982) or cosmic rays (Leger et al. 1985). The complexity of these processes has inhibited theoretical studies of photolyzed grain mantles. Extensive and painstaking laboratory studies are required to understand these processes. Although some laboratory studies have already been undertaken (Hagen 1982; d'Hendecourt et al. 1986; van de Bult 1986), this field is still in its infancy. Little more beyond the general outline of the chemical scheme (UV photons produce radicals which form complex molecules) can presently be said. Among the typical radicals produced by the photolysis of H_2O-rich ices are HCO, HOCO, and OH. Simple molecules that have been produced by photolysis of such mixtures are H_2CO and CO_2 (Hagen 1982; d'Hendecourt et al. 1986). A general conclusion seems to be that aliphatic rather than aromatic hydrocarbon mixtures are readily formed. Ultraviolet photolysis of oxygen-rich mixtures generally leads to the formation of residues containing large amounts of carboxylic acids. If NH_3 is also present in the initial mixture, then amino groups are also formed (Hagen, Allamandola, and Greenberg 1979; d'Hendecourt et al. 1986).

Let us consider quantitatively the conversion of interstellar grain mantles into organic refractory dust by means of UV photolysis. The absorption cross section of H_2O ice below 1600 Å is about 10^3 ℓ mole^{-1} cm^{-1}, about 4 times larger than for gas-phase H_2O. In an unattenuated interstellar FUV radiation field (10^8 photons cm^{-2} sec^{-1}; Habing 1968), the number of photons absorbed by one H_2O molecule in a grain mantle, N_{abs} ($A_v = 0$), is then about 2×10^{-9} sec^{-1}. Taking a UV dust albedo of 0.6, a mean cosine of the scattering angle of the dust of 0.7 (Lillie and Witt 1976), and a UV dust absorption cross section 3 times the visual extinction, A_v, the number of UV photons absorbed by one H_2O molecule in a grain mantle as a function of depth in a semi-infinite cloud is given by (Flannery et al. 1980) as

$$N_{abs}(A_v) = 2 \times 10^{-9} \exp(-2A_v) . \tag{17}$$

Thus, deep inside dense clouds ($A_v > 7$ mag) UV photolysis by the ambient interstellar radiation field is unimportant over the lifetime of a molecular cloud (3×10^7 yr). The UV photon field inside a dense cloud may actually be larger than estimated from this treatment because of the presence of local UV sources. For example, cosmic ray excitation of H_2 may lead to a mean intensity of the UV radiation field inside the cloud of 10^3 photons cm^{-2} sec^{-1} (Prasad and Tarafdar 1983). Thus, for a visual extinction larger than about 6 mag this field will dominate. There may also be other sources of UV photons inside dense clouds, such as shocks or newly formed stars, whose flux is difficult to estimate but which might be as high as 5×10^3 photons cm^{-2} sec^{-1} (Norman and Silk 1980). Taking 10^3 photons cm^{-2} sec^{-1} as the value for the mean intensity of the UV radiation field inside dense clouds yields, for the number of UV photons absorbed by one H_2O molecule in a grain mantle, $N_{abs} = 2 \times 10^{-14}$ sec^{-1}. Thus, over the lifetime of a molecular cloud every H_2O molecule in a grain mantle will have absorbed about 20 UV photons.

Not all of the absorbed UV photons will, however, lead to radical production. The efficiency, f_{uv}, of the photodissociation reaction (number of radicals produced per UV photon absorbed) depends on the

particular electronic state involved. In the gas phase, photodissociation is the dominant channel (Okabe 1978; Calvert and Pitts 1966); in a
solid, many other decay channels compete. These include photo-assisted
reactions with neighbors, and dissipation of the energy in the phonon
modes of the solid (i.e., heat; Bass and Broida 1966; Bondybey and Brus
1980). The accessibility of the various channels depends on the composition of the ice, the identity of the molecule that has absorbed the UV
photon, and the particular state it is excited into. Considering H_2O
under a few well-defined conditions, several pathways can be identified
and followed. The dissociated H_2O molecule may recombine immediately;
the "hot" product may recombine with a neighbor to form a new species
and it may also recombine with another radical to reform the original
molecule. In that case, the UV photon has merely promoted diffusion of
a previously stored radical. The importance of this latter process
depends on the average distance between stored radicals as compared to
the average distance a hot product will diffuse before it is trapped.
Note that this process will limit the concentration of radicals that can
be stored in a grain mantle. Laboratory studies of UV photolysis of H_2O
and CO matrices yield values for f_{uv} ranging from 0.25 to 0.005 when
the H_2O/CO ratio increases from 10^{-2} to 2 (Hagen 1982). In the former
case, the H and OH produced will react with a CO, producing HCO and
HOCO. In the latter case, there is a much greater probability of reaction of the H with an OH previously produced, thereby reforming H_2O.
The cage effect might limit f_{uv}, thus, to values less than 0.25. For
the lower CO concentrations, the efficiency is limited by the recombination of the hot product reforming H_2O.

The limiting concentration of H_2O photodissociation products in a
pure H_2O matrix is estimated to be about 1% from these experiments
(Hagen 1982). In photolysis experiments involving more complex mixtures
designed to duplicate an interstellar grain mantle, the conversion
efficiency also falls in the 1% range (d'Handecourt et al. 1986). Water
is a major component of interstellar grain mantles (Tielens and Hagen
1982; Tielens et al. 1986a). On the basis of these experiments, the UV
photolysis efficiency for interstellar grain mantles is about 0.5%.
With this efficiency, about 10% of the mantle molecules will have been
converted into photolyzed products over the lifetime of the cloud.
Thus, while photolysis will not change the composition of accreted grain
mantles drastically it can increase the complexity of the molecules in a
grain mantle. It should be emphasized, however, that most of these
photolyzed products will not be very large. In particular, photolysis
of H_2O will lead to reactions of its photolysis products, H, O, and OH
with neighboring simple molecules such as CO, thereby forming CO_2, H_2CO,
and CH_3OH. In an interstellar grain mantle, molecules containing CH_2,
CH_3, and NH_2 groups can also be photolyzed, adding to the complexity of
the mixture. However, most product molecules are expected to be

volatile and will not survive long in the harsh environment of the diffuse interstellar medium. The efficiency of the formation of very complex molecules is expected to be low. Using the values quoted above, for a molecule containing N building blocks, $f_{uv} \approx 20(0.005)^n$, where $n = {}^2\log N$. UV photolysis alone can, therefore, not account for the presence of an abundant complex organic refractory dust component in the interstellar medium.

Laboratory studies show that after warm-up of a photolyzed icy mixture, a complex molecular mixture remains. Presumably, this mixture has been partly formed by radical diffusion and reaction during the warm-up (the exploding grain mantle; see §9.1). However, the efficiency of this process has not been estimated. For the sake of argument we will assume that 1% of all the radicals present at the warm-up will form such complex molecules (i.e., 10^{-4} of all the molecules in the grain mantle). The equilibrium concentration of radicals in an H_2O-rich matrix ($\approx 1\%$) is attained on a time-scale of about 10^5 yr inside a dense cloud, which is fast compared to the accretion time-scale ($10^9/n_0$ yr). Warm-up of the grain mantles may be due to grain-grain collisions (Greenberg 1979) with a time-scale of about $10^8/(n_0)^{1/2}$ yr in molecular clouds (Draine 1984). It has also been suggested that warm-up results from transient heating by cosmic rays with a time-scale of about 3×10^4 yr (see §9.1). The limiting step seems, thus, to be the accretion on grains. Taking an accretion time-scale of 10^6 yr, appropriate for a density of 10^3 cm^{-3} and a cloud lifetime of 3×10^7 yr, implies that each atom cycles about 30 times through a grain mantle and thus has 0.3% probability of being incorporated into the organic refractory dust. Since an atom cycles about 50 times through a molecular cloud during its lifetime in the interstellar medium, there is a 15% probability that it will be part of the organic refractory dust component during some part of its lifetime.

It should be emphasized that there are two key steps in models of the UV photolysis of icy grain mantles and the production of complex organic refractory material. First, such models hinge on the presence of sources of UV radiation inside a dense cloud. Given a cloud lifetime of 3×10^7 yr, the ambient interstellar UV radiation field is unimportant for UV photolysis when A_v is larger than about 7 mag. Second, transient heating of grain mantles is required to bring the stored radicals together for reaction. The efficiency of such explosions is, however, unknown. A conversion efficiency of about 1% of the stored radicals is required to explain the presence of about 25% of the carbon in the form of organic refractory dust.

9. GRAINS AND GAS-PHASE COMPOSITION OF MOLECULAR CLOUDS

One of the most pressing problems remaining in theoretical studies of
the gas-phase composition of molecular clouds is the contribution of the
grain chemistry. Grains can act as sources and as sinks for gas-phase
species. The importance of grains can be easily illustrated by recalling the accretion time-scale for gas-phase species (eq. [10]). At a
density of 10^4 cm^{-3} this time-scale is about 10^5 yr, which is less than
the free-fall time-scale ($\tau_{ff} \approx 4 \times 10^7/[n_o]^{1/2} = 4 \times 10^5$yr; Spitzer
1978). Thus, for $n_o = 10^4$ cm^{-3}, gas-phase molecules accrete much
faster on a grain than a cloud collapses to form a star. One might
therefore expect that no molecules will be observed at such high densities in molecular clouds (Watson 1976; Greenberg 1979). This is in
stark contrast, however, with the observations. Molecules such as H_2CO
and NH_3 are routinely observed in dense cloud cores. Moreover, elemental depletions in grain mantles range up to only 40% and are typically
much less (Tielens et al. 1986a). This suggests that there is an efficient exchange between the gas phase and grain mantles. However,
although the accretion time-scale may presently be less than the free-
fall time-scale, it need not have been in the past. In particular, a
newly formed star will raise the temperature of the dust in its neighborhood above the critical temperature for grain-mantle formation
($\approx$30 K, compare the accretion time-scale of abundant species with their
evaporation time-scale; see Table II).

The high abundance of molecular hydrogen in molecular clouds also
suggests an important contribution from grain-surface chemistry to the
gas-phase composition. There is no efficient gas-phase reaction network
for the formation of H_2. For that reason molecular hydrogen is generally assumed to be formed on grain surfaces by recombination of two
hydrogen atoms or by hydrogen abstraction reactions (see §5). This
assessment is, however, based on negation of other possibilities and
there is no direct (chemical) evidence. Although the formation and
ejection of H_2 on grain surfaces does not necessarily imply that other
grain surface reactions contribute to the gas-phase composition, it is
suggestive. Certainly, there is good observational evidence for the
formation of molecules by grain-surface reactions in the composition of
grain mantles (Tielens and Hagen 1982; Tielens et al. 1984; Tielens
et al. 1986a).

As argued in §8, grain-mantle "explosions" are required to explain
the observed abundance of the organic refractory dust component in the
diffuse interstellar medium. Laboratory grain-mantle analog studies
have provided good experimental evidence for an efficient return of
grain-mantle molecules to the gas phase upon warm-up of photolyzed
mixtures, as well as for the formation of an organic refractory residue
(Hagen et al. 1979; d'Hendecourt et al. 1982; d'Hendecourt 1984;

d'Hendecourt et al. 1986; van de Bult 1986). Simple theoretical arguments, such as those in §8 and below, suggest that UV photolysis and transient heating of grains can indeed be of importance.

In this section, the physical processes that may return grain-mantle molecules back to the gas phase are discussed. Some observational signatures of grain chemistry will also be indicated. It should be emphasized that there is abundant evidence for the importance of gas-phase reactions in the formation of simple gas-phase molecules. In particular, the molecular ions, HCO^+ and DCO^+, provide key support for ion-molecule gas-phase reactions.

9.1 Ejection of Grain-Mantle Molecules

If the accretion of gas on grains is offset by ejection of the accreted material, then grains may play an important role in the composition of the gas phase. Model calculations of the gas-phase composition of molecular clouds in which grain-surface reactions were used alone or in conjunction with gas-phase reactions have been performed by several groups (Allen and Robinson 1977; Viala et al. 1979; d'Hendecourt et al. 1985). The contribution of grains to the gas-phase composition in all of these models hinges to a large extent on the identification of ejection mechanisms for molecules formed on the grain surface. Among the ejection mechanisms suggested are thermal evaporation (including single-photon heating events for very small grains), ejection upon reaction, photodesorption and release of stored chemical energy (cf. Watson 1976; Tielens and Hagen 1982; d'Hendecourt et al. 1982).

The first two processes can be of importance for H_2 because of its low binding energy and because of its small mass, which makes energy transfer to the surface relatively inefficient (cf. §4.6). For other molecules, the efficiency of these ejection processes is expected to be low.

Photodesorption is probably an efficient ejection mechanism for small molecules on grain surfaces in the diffuse interstellar medium or at the edges of molecular clouds. The absorption of a UV photon by a molecule on the surface will electronically excite this molecule. Photodesorption may then occur because of the electronic repulsion between the excited admolecule and neighboring surface molecules. Essentially, the larger orbital distribution of the electronically excited state coupled with the asymmetric force field of the surface will give the nucleus of the admolecule momentum perpendicular to the surface, and ejection may occur (Nishi et al. 1984). For small molecules the efficiency of this process can be fairly large (Greenberg 1973). Larger molecules will, however, have a relatively low photodesorption efficiency because many more energy-decay channels are available and because they may have a larger binding energy. The photodesorption yield will also be low when the excitation energy can be rapidly transferred to other molecules in the particle (e.g., near-resonant transfer with a time-scale of the order of 10^{-13}–10^{-14} sec). Thus, for an H_2O molecule on a clean substrate the photodesorption yield is much higher than for H_2O adsorbed on an H_2O surface (Greenberg 1973). Using values typical for small molecules and assuming the average interstellar UV radiation field, photodesorption can keep a grain surface clean for A_v less than about 3 mag (Tielens and Hagen 1982).

The most important mechanism for ejection of molecules in a grain mantle into the gas phase inside a dense molecular cloud is probably the release of stored chemical energy. At low temperatures, radicals produced by UV photolysis can be stored in a grain mantle up to a concentration of about 1% (cf. §8). Upon warm-up, these radicals can diffuse and react upon encounter. Laboratory studies on warm-up of photolyzed mixtures show that upon heating to about 27 K, those samples with a poor heat contact with the substrate will explosively eject most of the mixture (d'Hendecourt et al. 1982; d'Hendecourt 1984). The critical temperature of 27 K seems to be a general property of irradiated molecular ices, for it is independent of the mixture. Apparently, at this temperature rapid diffusion of most of the trapped radicals occurs simultaneously.

It should be emphasized that these explosions are not a result of a chain reaction (e.g., the reaction heat of a recombining radical driving the diffusion and reaction of other radicals). The deposition of 5 eV of reaction heat in a volume corresponding to 200 molecules (i.e., a radical concentration of 1%) raises the temperature to about 170 K (see below). The temperature relaxation time T, owing to heat diffusion, is given by (Anderson 1981) as

$$\tau_T = 4\rho C r_o^2/(\pi^2 K) \ , \tag{18}$$

where ρ is the density; C the specific heat ($\approx 2 \times 10^4 \ T^{1.3}$ erg cm^{-3} K; Leger et al. 1985); r_o the radius of the initial deposition region (≈ 12 Å); and K the thermal conductivity ($\approx 3 \times 10^4$ erg sec^{-1} cm^{-1} K^{-1}; d'Hendecourt 1984). Inserting these values yields a temperature relaxation time-scale of 3×10^{-12} sec. Within that time-scale, however, a radical can migrate at best only a few sites and, given the low concentration of radicals, it will not have found a reaction partner. Thus, the heat liberated by the reaction does not promote reaction of other stored radicals.

It is probable that the existence of a unique critical temperature is a general property of amorphous molecular matrices. At about 25 K the heat conductivity of an amorphous material changes abruptly in character (Anderson 1981). Below that temperature the energy is carried by low-frequency phonons ($\lesssim 10^{12}$ sec^{-1}) which have a long mean free path

($\approx 10^5$ Å). These phonons interact only with tunneling sites where the amorphous solid can tunnel from one amorphous state to another (cf. §2). Above that temperature the heat is carried by high-frequency phonons ($\approx 10^{12}$ sec^{-1}) with a mean free path of about 10 Å. It is these latter, localized phonons that assist in diffusion processes.

We will now estimate the amount of grain-mantle material evaporated by one cosmic ray event or a grain-grain collision. We will assume that enough energy is deposited to raise the grain temperature to the critical temperature ($\Delta E \approx 5 \times 10^{-8}$ erg for a 1000 Å grain at 27 K). For simplicity it is assumed in this calculation that the grain consists mainly of H_2O with a trace (1%; cf. §8) of radicals that will start to diffuse and react at this temperature. The reaction energy liberated raises then the temperature enough to cause appreciable evaporation of the mixture. With a 1% concentration of radicals, assuming a reaction energy of 5 eV and a specific heat of H_2O ice of about 2×10^4 $T^{1.3}$ erg cm^{-3} (T > 50 K); Leger et al. 1985), this yields a temperature of about 170 K. This is well above the evaporation temperature of H_2O ice in the interstellar medium (90 K; Nakagawa 1980), and evaporation of the mantle will dominate over other heat-loss mechanisms, such as radiation. The amount of material evaporated can then be estimated by equating the reaction heat with the heat loss through evaporation. The energy remaining in the particle at the evaporation temperature should be subtracted since this energy will be carried away by radiation. For the assumptions stated above, this leads to 0.019 eV per H_2O molecule available for evaporation, which should be compared to the binding energy of 0.05 eV. Thus, about 40% of the H_2O molecules will evaporate before the temperature will drop below the evaporation temperature. Note that this is less than estimated by Leger et al. (1985), because they assumed that grain mantles consist mainly of CO with a binding energy of only about 0.09 eV. Cosmic ray heating seems to dominate the transient grain heating process with a time-scale of about 3×10^4 yr for a grain size of 1000 Å. Grain-grain collisions are only important at high densities ($n_0 > 10^6$ cm^{-3}). With the efficiency calculated above, the ejection time-scale for a molecule is thus about 10^5 yr.

9.2 Signatures of Grain Chemistry

Molecules produced by grain-surface chemistry are typically neutral, saturated, small, and highly hydrogenated. Infrared absorption studies of protostars show that H_2O and alcohols (presumably CH_3OH) dominate the composition of interstellar grain mantles; NH_3 is also present as a trace (Tielens et al. 1986a). These molecules are the grain-mantle reaction products of the abundant gas-phase species O, O_2, CO, and N. The measured abundances of H_2O and CH_3OH in grain mantles ranges from

4×10^{-6} to 2×10^{-4} per hydrogen atom, which is much larger than their gas-phase abundances. Obviously, these molecules, which are easily made on grain surfaces, would be prime candidates for studies of the influence of grains on the gas-phase composition. In this respect, it should be noted that species produced by gas-phase chemistry are often highly unsaturated, radicals, or ions.

In general, the relative contribution of grains to the composition of the gas phase is expected to increase with density. Gas-phase chemistry is driven by ion-molecule reactions. The degree of ionization decreases, however, with increasing density, essentially because recombination increases faster with increasing density that ionization (Oppenheimer and Dalgarno 1974; Prasad and Huntress 1980). Thus, the rate of ion-molecule reactions is expected to increase slower with density than the rate of grain-surface reactions. The rate of production of species on grain surfaces is limited by the accretion rate of the appropriate reactants onto grains. At molecular cloud densities, this means that only for those molecules that are easily made on a grain surface but are slowly produced in the gas phase, grains will contribute to the gas-phase abundance.

Thus, one can test the importance of grain chemistry for the composition of the gas phase by searching for gas-phase molecules in molecular clouds that are difficult to form by ion-molecule reaction schemes. Obvious candidates are molecular species containing metal atoms (e.g., NaH or NaOH). These species are not expected to form molecular species through gas-phase reactions (cf. Oppenheimer and Dalgarno 1974; Smith et al. 1983). However, these species will react on grain surfaces.

One might expect, naively, that the overabundance of atomic hydrogen will convert these species into hydrides, as observations show has happened to oxygen, nitrogen, and carbon monoxide; however, the grain-surface chemistry of such species is somewhat more complicated. In particular, hydrogen abstraction reactions can be of importance (Barlow 1978). For example, the reaction HaH + H $\rightarrow$ Na + H_2 is exothermic by about 2.5 eV. Because a molecular bond has to be broken in order for this reaction to proceed, there will be an activation barrier that has to be overcome. Typically, such an activation barrier is about 5% of the bond energy ($\approx$1000 K), which is of little consequence on grain surfaces. Thus, in effect, Na like N_2H_4, H_2CO, and other molecules will act as chemical bonding sites for H_2 formation (cf. §5). Hydrogen addition and abstraction reactions will continue to form H_2 with Na, until it reacts with another migrating species such as O, which effectively poisons the site for H_2 formation. Since the energy barrier is expected to be low and because atomic oxygen is the next most abundant migrating radical on a grain surface, most sodium atoms are expected to wind up in NaOH molecules. Hydrogen abstraction from these molecules is

presumably inhibited by a high actuation energy. It is not yet possible
to search for the vibrational transitions of molecules with such low
abundances in interstellar grain mantles, but it is possible to search
for the rotational transitions of the gas-phase species.

Nitrogen and sulfur bearing, gas-phase molecules (e.g., NH_3 and
H_2S) seem difficult to form by ion-molecule reactions (Prasad and
Huntress 1980, 1982). This would also make them prime candidates for
studies of the importance of grains for the gas-phase composition of
molecular clouds. Both NH_3 and H_2S have tentatively been identified in
grain mantles (Knacke et al. 1982; Geballe et al. 1985). However, the
gas-phase chemistry of these compounds is still open to some debate (cf.
Graedel et al. 1982), which somewhat limits their use as grain-chemistry
indicators. Complex molecules seem also to be difficult to make by gas-
phase reactions. Grain-surface reactions will not be much help because
they generally lead to fairly simple, completely hydrogenated molecules. Ultraviolet photolysis of grain mantles will, however, convert
such molecules into more complex ones which may eventually be injected
into the gas phase. The importance of this process is difficult to
assess. The first step seems to be to identify such molecules in interstellar grain mantles.

As discussed before, grain-surface chemistry can lead to large
deuteration effects (cf. §7 and Fig. 14). Essentially, this reflects
the high gas-phase ratio of atomic D to H. This ratio is high because
of the rapid destruction of HD through reactions with H_3^+ (Tielens 1983b;
cf. §7). Generally, high density and low gas-phase temperatures lead to
high abundances of deuterated molecules with respect to their hydrogenated counterparts. The temperature dependence arises from the temperature sensitivity of the gas-phase H_2D^+ plus H_2 reactions. Prime candidates for such studies are HDO, NH_2D, DNO, HDCO, and possibly CH_2DOH.
Studies of the deuteration of ammonia may be particularly revealing,
since the ratios $NH_{3-i}D_i/NH_{2-i}D_{i+1}$ (i = 0, 1, 2) should all be equal to
the gas-phase n(H)/n(D) ratio, which at low temperatures is just
$A(D)n_o$, where A(D) is the elemental abundance of deuterium.

Finally, the ortho/para ratio can serve as a grain-chemistry indicator. Some molecules exist in two distinct states, depending on the
spin of the nucleus: the para state when the spins are anti-parallel
and the ortho state when they are parallel. Because of the symmetry of
the wave function, the para states are associated with the even rotational levels and the ortho states with the odd levels. Rotational
transitions between ortho and para states require a simultaneous transition of the nuclear spins of the molecule. The radiative lifetime for
decay from the ortho to the para state is therefore long. The energy
difference between these two states can be quite large compared to
interstellar temperatures (i.e., for H_2 about 170 K and for H_2O about
55 K), and small variations in temperature can thus have large effects.

The initial ortho-para ratio of molecules such as H_2O, NH_3, H_2CO, and CH_4, newly formed on a grain surface, is expected to be equal to the high-temperature limit (i.e., the ratio of the statistical weights of the two nuclear spin states). This is due to the large reaction heat available upon formation and the considerable rearrangement of the hydrogen nuclei involved in most reactions. Ortho-para transformation will occur on a time-scale of a day because of the interaction of the nuclear spins of neighboring molecules. An even faster ortho-para transformation time-scale may result from the presence of paramagnetic impurities or lattice defects (Tielens 1986b). The nuclear spin transition is due to the inhomogeneous magnetic dipole field generated by the nuclear spin of a neighboring molecule or the unpaired electron of a paramagnetic impurity or lattice defect. This couples to the proton magnetic moments in such a way as to cause a transition between the ortho and para states (Wigner 1933). Since the residence time of molecules produced on grain surfaces is expected to be about 10^5 yr, their ortho-para ratio will reflect the dust temperature (i.e., $n(o)/n(p) = g \exp(-E_{op}/T_d)$), where g is the ratio of the statistical weights of the nuclear spin states, E_{op} the ortho-para energy difference, and T_d the dust temperature). Because ejection into the gas phase is a rapid process, it will not influence this ratio.

It should be emphasized that once the molecules are in the gas phase, ion-molecule reactions may influence the ortho-para ratio. Probably the most important reaction in this respect is H exchange with H_3^+. Assuming a Langevin reaction rate coefficient (10^{-9} cm^3 sec^{-1}) and a relative abundance of H_3^+ of 10^{-8} (Smith and Adams 1984; Lepp et al. 1986) yields a time-scale for this process of $3 \times 10^9/n_o$ yr. This is comparable to the time-scale for accretion on grains ($10^9/n_o$ yr), which is probably the limiting step in the grain-surface reactions.

From this discussion, it is clear that in a region where the dust and gas temperature are sufficiently different relative to the ortho-para energy difference, the ortho-para ratio can be used as a discriminant for grain-gas interaction (Tielens 1986b). It should be emphasized that the ortho-para transformation process discussed above is not a chemical mechanism; strictly speaking, the ortho-para ratio will therefore not prove the importance of *grain chemistry* for the composition of the gas phase. However, the importance of grain chemistry for the composition of grain mantles is already well attested to by IR spectroscopy.

The best objects in which to study the ortho-para ratio are cold, dense, dark clouds such as Taurus. One such study (for H_2CO) has been performed for two dark clouds (and the warm gas in Orion; Kahane et al. 1984). Since the energy difference between the ortho and para states of H_2CO is small (15 K) and the lines suffer from beam dilution, the interpretation is not straightforward; nevertheless it seems that in dark clouds thermalization on grains dominates, implying an efficient gas-grain exchange mechanism.

ACKNOWLEDGMENT

We thank Dave Hollenbach for many stimulating discussions and a critical reading of an earlier version of this manuscript.

REFERENCES

Aitken, D. K.   1981, in IAU Symp. No. 96, Infrared Astronomy, eds. G. C. Wynn Williams and D. P. Cruikshank (Reidel, Dordrecht), p. 207.

Aitken, D. K., and Roche, P. F.   1984, M. N. R. A. S., 208, 751.

Allamandola, L. J., Tielens, A. G. G. M., and Barker, J. R.   1985, Ap. J. Letters, 290, L25.

Allamandola, L. J., Tielens, A. G. G. M., and Barker, J. R.   1986, in preparation.

Allen, D. A. and Wickramasinghe, D. T.   1981, Nature, 294, 239.

Allen, M., and Robinson, G. W.   1977, Ap. J., 212, 396.

Anderson, H. L.   1981, Physics Vademecum (American Institute of Physics, New York).

Aronowitz, S., and Chang, S.   1985, Ap. J., 293, 243.

Balooch, M., and Olander, D. R.   1975 J. Chem. Phys., 63, 4772.

Barlow, M. J.   1978, M. N. R. A. S., 183, 397.

Bass, A. M., and Broida, H. P.   1966, Formation and Trapping of Free Radicals (Academic Press, New York).

Beckwith, S., Persson, S. E., Neugebauer, G., and Becklin, E. E.   1978, Ap. J., 223, 464.

Bedijn, P.   1977, Ph.D. thesis, Leiden.

Bell, R. J., and Dean, P.   1972, Phil. Mag., 25, 1381.

Bohlin, R. C., Savage, B. D., and Drake, J. F.   1978, Ap. J., 224, 132.

Bohren, C. F., and Huffman, D. R.   1984, Absorption and Scattering of Light by Small Particles (Wiley and Sons, New York).

Bondybey, V. E., and Brus, L. E.   1980, Adv. Chem. Phys., 41, 269.

Borghesi, A., Bussoletti, E., and Colangeli, L.   1983, Astr. Ap., 142, 225.

Brus, L. E.   1981, J. Chem. Phys., 74, 737.

Burke, J. R., and Hollenbach, D. J.   1984, Ap. J., 265, 223.

Butchart, I., McFadzean, A. D., Whittet, D. C. B., Geballe, T., and Greenberg, J. M.   1985, Astr. Ap., 154, L5.

Calvert, J. G., and Pitts, J. N.   1966, Photochemistry (Wiley, New York).

Campbell, M. F., et al.   1976, Ap. J., 208, 396.

Cohen, M., Allamandola, L. J., Tielens, A. G. G. M., Bregman, J.,
    Simpson, J. P., Witteborn, F. C., Wooden, D., and Rank, D. 1986,
    Ap. J., **302**, 737.
d'Hendecourt, D. B., Allamandola, L. J., and Greenberg, J. M. 1985,
    Astr. Ap., **152**, 130.
d'Hendecourt, L. B. 1984, Ph. D. thesis, Leiden.
d'Hendecourt, L. B., Allamandola, L. J., Baas, F., and Greenberg,
    J. M. 1982, Astr. Ap., **109**, L12.
d'Hendecourt, L. B., Allamandola, L. J., Grim, R. J. A., and Greenberg,
    J. M. 1986, Astr. Ap., **158**, 119.
Dalgarno, A. 1975, in Atomic and Molecular Processes in Astrophysics,
    eds. M. C. E. Huber and H. Nussbaumer, (Geneva Observatory,
    Geneva), p. 1.
Day, K. L. 1976, Ap. J., **210**, 614.
Day, K. L. 1979, Ap. J., **234**, 158.
Day, K. L., and Donn, B. 1978, Ap. J. Letters, **222**, L45.
de Boer, J. H. 1968, The Dynamical Character of Absorption, (Clarendon
    Press, Oxford).
de Boer, J. H., and van Steenis, J. 1952, Proc. Kon. Ned. Akad. Wet.,
    B55, 572.
Diestler, D. J. 1974, J. Chem. Phys., **60**, 2692.
Diestler, D. J., Knapp, E-W., and Ladouceur, H. D. 1977, J. Chem,
    Phys., **68**, 4056.
Donn, B. 1960, in Formation and Trapping of Free Radicals, eds. A. M.
    Bass, and H. P. Broida (Academic Press, New York), p. 347.
Donn, D., and Jackson, W. M. 1970, B. A. A. S., **2**, 309.
Draine, B. T., and Lee, H. M. 1984, Ap. J., **285**, 89.
Draine, B. T. 1979, Ap. J., **230**, 106.
Draine, B. T. 1984, Ap. J. Letters, **277**, L71.
Draine, B. T., and Salpeter, E. E. 1979, Ap. J., **231**, 438.
Duley, W. W., and McCullough, J. D. 1977, Ap. J. Letters, **211**, L145.
Duley, W. W., and Williams, D. A. 1981, M. N. R. A. S., **196**, 269.
Ehrenreich, H. 1972, Fundamentals of Amorphous Semi Conductors,
    (National Academy of Sciences, New York).
Einstein, T. L., Hertz, J. A., and Schieffer, J. R. 1980, in Theory of
    Chemisorption, ed. J. R. Smith (Springer Verlag, Berlin), p. 183.
Fegley, B., and Lewis, J. S. 1980, Icarus, **41**, 439.
Field, G. B. 1974, Ap. J., **187**, 453.
Flannery, B. P., Roberge, W., and Rybicki, G. B. 1980, Ap. J., **236**,
    598.
Frenkel, J. 1924, Z. Physik, **26**, 117.
Gatley, I., Becklin, E. E., Werner, M. W., and Wynn Williams, C. G.
    1977, Ap. J., **216**, 277.
Gatley, I., Garden, R., Brand, P. W. J. L., Lightfoot, J., Glencross,
    W., Okuda, H., and Nagata, T. 1986, in preparation.

Geballe, T. R., Baas, F., Greenberg, J. M., and Schutte, W. 1985, Astr.
     Ap., 146, L6.
Gelb, A., and Kim, S. K. 1971, J. Chem. Phys., 55, 4935.
Gillett, F. C., Jones, T. W., Merrill, K. M., and Stein, W. A. 1975a,
     Astr. Ap., 45, 77.
Gillett, F. C., Forrest, W. J., Merrill, K. M ., Capps, R. W., and
     Soifer, B. T. 1975b, Ap. J., 200, 609.
Gilra, D. P. 1972, in The Scientific Results from the Orbiting
     Astronomical Observatory OAO-2, ed. A. D. Code, NASA SP-310,
     p. 297.
Gould, R. K. 1975, J. Chem. Phys., 63, 1825.
Govers, T. R., Mattera, L. and Scoles, G. 1980, J. Chem. Phys., 72,
     5446.
Graedel, T. E., Langer, W. D., and Frerking, M. A. 1982, Ap. J. Suppl.,
     48, 321.
Greenberg, J. M., and Hong, S. S. 1974, in I. A. U. Symp. No. 60,
     Galactic Radio Astronomy, eds. F. J. Kerr and S. C. Simonson
     (Reidel, Dordrecht), p. 155.
Greenberg, J. M. 1982, in Submillimetre Wave Astronomy, eds. J. E.
     Beckman and J. P. Phillips (Cambridge University Press, Cambridge),
     p. 261.
Greenberg, J. M. 1963, Ann. Rev. Astr. Ap., 15, 267.
Greenberg, J. M. 1979, in Stars and Stellar Systems, ed. B. Westerlund
     (Reidel, Dordrecht), p. 173.
Greenberg, J. M., and Chlewicki, G. 1984, Ap. J., 272, 563.
Greenberg, J. M., and Yencha, A. J. 1973, in I. A. U. Symp. No. 52,
     Interstellar Dust and Related Topics, eds. J. M. Greenberg and H.
     C. van de Hulst (Reidel, Dordrecht), p. 309.
Greenberg, L. T. 1973, in I. A. U. Symp. No. 52, Interstellar Dust and
     Related Topics, eds. J. M. Greenberg and H. C. van de Hulst
     (Reidel, Dordrecht), p. 413.
Habing, H. J. 1968, Bull. Astron. Inst. Neth., 19, 421.
Hackwell, J.A. 1971, Ph. D. thesis, University College, London.
Hadni, A. 1970, in Far-Infrared Properties of Solids, eds. S. S. Mitra
     and S. Nudelman (Plenum Press, New York), p. 561.
Hagen, W. 1982, Ph, D. thesis, Leiden.
Hagen, W., Tielens, A. G. G. M., and Greenberg, J. M. 1981, Chem.
     Phys., 56, 367.
Hagen, W., Tielens, A. G. G. M., and Greenberg, J. M. 1983a, Astr.
     Ap.Suppl., 51, 389.
Hagen, W., Tielens, A. G. G. M., and Greenberg, J. M. 1983b, Astr. Ap.,
     117, 132.
Hagen, W., Allamandola,L. J., and Greenberg, J. M. 1979, Astr. Space
     Sci., 65, 215.

Hagen, W., Allamandola, L. J., and Greenberg, J. M.   1980, Astr. Ap.,
     86, L3.
Hair, M. L.   1967, Infrared Spectroscopy in Surface Chemistry (Dekker,
     New York).
Harvey, P. M., Hoffmann, W. F., and Campbell, M. F.   1978, Astr. Ap.,
     70, 165.
Hasegawa, H., and Koike, C.   1984, in Laboratory and Observational
     Infrared Spectra of Interstellar Dust, eds. R. D. Wolstencroft and
     J. M. Greenberg (Oc. Rep. Royal Obs. Edinburgh), p. 137.
Hickmott, T. W.   1960, J. Appl. Phys., 31, 128.
Hollenbach, D., and Salpeter, E. E.   1970, J. Chem. Phys., 53, 79.
Hollenbach, D., and Salpeter, E. E.   1971, Ap. J., 163, 155.
Hoyle, F., and Wickramasinghe, N. C.   1962, M. N. R. A. S., 124, 417.
Hunter, D. A., and Watson, W. D.   1978, Ap. J., 226, 477.
Jaycock, M. J., and Parfitt, G. D.   1986, Chemistry of Interfaces (Wiley
     and Sons, New York).
Jones, T. W., and Merrill, K. M.   1976, Ap. J., 209, 509.
Joyce, R. R. and Simon, T.   1982, Ap. J., 260, 604.
Kahane, C., Frerking, M. A., Langer, W. D., Encrenaz, P., and Lucas,
     R.   1984, Astr. Ap., 137, 211.
Kameijo, F.   1963, Publ. Astr. Soc. Japan, 15, 440.
Khare, B. N., and Sagan, C.   1973, in Molecules in the Galactic
     Environment, eds. M. A. Gordon, and L. E. Snyder (Wiley and Sons,
     New York), p. 399.
King, A. B., and Wise, H.   1963, J. Phys. Chem., 67, 1163.
Kittell, C.   1976, Introduction to Solid State Physics (Wiley and Sons,
     New York).
Knacker, R. F., McCorkle, S., Puetter, R. C., Erickson, E. F., and
     Kratschmer, W.   1982, Ap. J., 260 141.
Koike, C., Hasegawa, H., and Manabe, A.   1980, Astr. Space Sci., 67,
     495.
Lacy, J. H., Baas, F., Allamandola, L. J., Persson, S. E., McGregor, P.
     J., Lonsdale, C. J., Geballe, T. R., and van de Bult, C. E. P.
     1984, Ap. J., 276, 533.
Landau, L. D., and Lifshitz, E. M.   1960, Course of Theoretical Physics,
     Vol. 1, Mechanics (Pergamon Press, New York).
Langmuir, I.   1912, J. Am. Chem. Soc., 34, 1310.
Langmuir, I.   1915, J. Am. Chem. Soc., 37, 417.
Lee, T. J.   1975, Astr. Space Sci., 34, 123.
Legay, F.   1977, in Chemical and Biochemical Applications of Lasers,
     vol. 2, ed. C. B. Moore (Academic Press, New York), p. 1.
Leger, A., and Puget, J. L.   1984, Astr. Ap., 137, L5.
Leger, A., Jura, M., and Omont, A.   1985, Astr. Ap., 144, 147.
Leitch-Devlin, M. A., and Williams, D. A.   1985, M. N. R. A. S., 213,
     295.

Lepp, S., Dalgarno, A., and Sternberg, A. 1986, preprint.

Lillie, C. F., and Witt, A. N. 1976, Ap. J., 208, 64.

Lucas, D., and Ewing, G. E. 1981, Chem. Phys., 58, 385.

Madden, W. G., Bergren, M.S., McGraw, W. R., and Rice, S. A. 1978, J. Chem. Phys., 69, 3497.

Mathis, J. S., and Wallenhorst, S. G. 1981, Ap. J., 244, 483.

Mathis, J. S., Rumpl, W. and Nordsieck, K. H. 1977, Ap. J., 217, 425.

Mathis, J. S. 1986, in The Interrelation of Interstellar, Circumstellar and Planetary Dust, eds. J. Nuth and R. Stencel, NASA CP 2403, p. 29.

McCarroll, B., and McKee, D. W. 1971, Carbon, 9, 301.

Merrill, K. M. 1977, in The Interaction of Variable Stars with their Environments, eds. R. Kippenhahn, J. Rahe, and W. Strohmeier (Veroff, Remeis Sternwarte, Bamberg), Bd. XI, nr 121, p. 446.

Merrill, K. M., Russell, R. W., and Soifer, B. T. 1976, Ap. J., 207, 763.

Mitchell, G. F., Ginsburg, J. L., and Kuntz, P. J. 1978, Ap. J., Suppl., 38, 39.

Mooney, T., and Knacke, R. F. 1986, preprint.

Moore, M. H., and Donn, B. 1982, Ap. J. Letters, 257, L47.

Morton, D. C. 1975, Ap. J., 197, 85.

Mueller, W. M., Blackledge, J. P., and Libowitz, G. G. 1968, Metal Hydrides (Academic Press, New York).

Nakagawa, N. 1980, in I. A. U. Symp. No. 87, Interstellar Molecules, ed. B. H. Andrew (Reidel, Dordrecht), p. 365.

Nishi, N., Shinohara, H., and Okuyama, T. 1984, J. Chem. Phys., 80, 3898.

Nitzan, A., Mukamel, S., and Jortner, J. 1975, J. Chem. Phys., 63, 200.

Norman, C. A., and Silk, J. 1980, Ap.J., 238, 158.

Nuth, J. A., and Donn, B. 1982, J. Chem. Phys., 777, 2639.

Okabe, H. 1978, Photochemistry of Small Molecules (Wiley and Sons, New York).

Oppenheimer, M., and Dalgarno, A. 1974, Ap.J., 192, 29.

Pauling, L. 1964, College Chemistry (Freeman and Co., San Francisco).

Penman, J. M. 1976, M. N. R. A. S., 175, 149.

Phillips, W. A. 1981, Amorphous Solids, Low Temperature Properties, (Springer Verlag, New York).

Plendl, J. N. 1970, in Far-Infrared Properties of Solids, eds. S. S. Mitra and S. Nudelman (Plenum Press, New York).

Prasad, S. S., and Huntress, W. T. 1980, Ap. J. Suppl., 43, 1.

Prasad, S. S., and Huntress, W. T. 1982, Ap. J., 260, 590.

Prasad, S. S., and Tarafdar, S. P. 1983, Ap. J., 267, 603.

Purcell, E. M. 1969, Ap. J., 158, 433.

Puri, B. R. 1976, in Chemistry and Physics of Carbon, 6, ed. P. L. Walker (Marcel Dekker, New York), p. 191.

Robell, A. J., Ballou, E. V., and Boudart, M. 1964, J. Phys. Chem., 68, 2748.

Roche, P. F., and Aitken, D. K. 1984, M. N. R. A. S., 208, 481.

Roche, P. F., and Aitken, D. K. 1985, M. N. R. A. S., 215, 425.

Russell, R. W., Soifer, B. T. and Willner, S. P. 1977, Ap. J. Letters, 217, L149.

Rye, R. R. 1977, Surface Sci., 69, 653.

Savage, B. D., and Mathis, J. S. 1979, Ann. Rev. Astr. Ap., 17, 73.

Schutte, W., and Tielens, A. G. G. M. 1985, in Mass Loss from Red Giants, eds. M. Morris, and B. Zuckerman (Reidel, Dordrecht), p. 87.

Seki, J., and Yamamoto, T. 1980, Astr. Space Sci., 72, 79.

Sellgren, K. 1986, Ap. J., 305, 399.

Sellgren, K., Werner, M. W., and Dinerstein, H. L. 1983, Ap. J., 271, L13.

Smith, D., and Adams, N. G. 1984, Ap. J. Letters, 284, L13.

Smith, D., Adams, N. G., Alge, E., and Herbst, E. 1983, Ap. J., 272, 365.

Snoeyink, V. L., and Weber, W. J. 1972, Prog. Surf. Membrane Sci., 5, 63.

Snow, T. P. 1975, Ap. J. Letters, 202, L87.

Spitzer, L. 1976, Comments on Astr., 6, 177.

Spitzer, L. 1978, Physical Processes in the Interstellar Medium (Wiley and Sons, New York).

Spitzer, L., and Cochran, W. D. 1973, Ap. J. Letters, 186, L23.

Strazulla, G., Calgagno, L. and Foti, G. 1983, M. N. R. A. S., 204, 59P.

Tauc, J. 1972, in Optical Properties of Solids, ed. F. Abeles (North Holland Publishing Company, Amsterdam), p. 278.

Tielens, A. G. G. M. 1983a, Ap. J., 271, 702.

Tielens, A. G. G. M. 1983b, Astr. Ap., 119, 177.

Tielens, A. G. G. M. 1986a, in preparation.

Tielens, A. G. G. M. 1986b, in preparation.

Tielens, A. G. G. M., and deJong, T. 1979, Astr. Ap., 75, 326.

Tielens, A. G. G. M., and Hagen, W. 1982, Astr. Ap., 114, 245.

Tielens, A. G. G. M., Allamandola, L. J., Bregman, J., Goebel, J., d'Hendecourt, L. B., and Witteborn, F. C. 1984, Ap. J., 287, 697.

Tielens, A. G. G. M., Allamandola, L. J., Bregman, J., Witteborn, F. C., Wooden, D., and Rank, D. 1986a, in preparation.

Tielens, A. G. G. M., Allamandola, L. J., Bregman, J., and Witteborn, F. C. 1986b, in preparation.

Touloukian, Y. S., and Buyco, E. H. 1970, Thermophysical Properties of Matter, 5 (Plenum Press, New York).

van de Bult, C. E. P. 1986, private communication.

Veprek, S., and Haque, M. R. 1975, Appl. Phys., 8, 303.

Viala, Y. P., Bel, N., and Clavel, J. 1979, Astr. Ap., **73**, 174.

Watson, W. D. 1976, in Atomic and Molecular Physics and the Interstellar Matter, eds. R. Balian, P. Encrenaz, and J. Lequeuz (Elsevier, Amsterdam), p. 177.

Watson, W. D., and Salpeter, E. E. 1972, Ap. J., **174**, 321.

Wigner, E. 1933, Z. Physik. Chem. B., **23**, 28.

Willner, S. P., et al. 1982, Ap. J., **253**, 174.

Willner, S. P., Russell, R. W., Puetter, R. C., Soifer, B. T., and Harvey, P. M. 1979, Ap. J. Letters, **229**, L65.

Wood, B. J., and Wise, H. 1962, J. Phys. Chem., **66**, 1049.

Wood, B. J., and Wise, H. 1969, J. Phys. Chem., **73**, 1368.

Wooten, B. J. 1972, Optical Properties of Solids (Academic Press, New York).

Zachariasen, W. H. 1932, J. Am. Chem. Soc., **54**, 3841.

Zaikowski, A., Knacke, R. F., and Porco, C. C. 1975, Astr. Space Sci., **35**, 97.

Zallen, R. 1983, The Physics of Amorphous Solids (Wiley and Sons, New York).

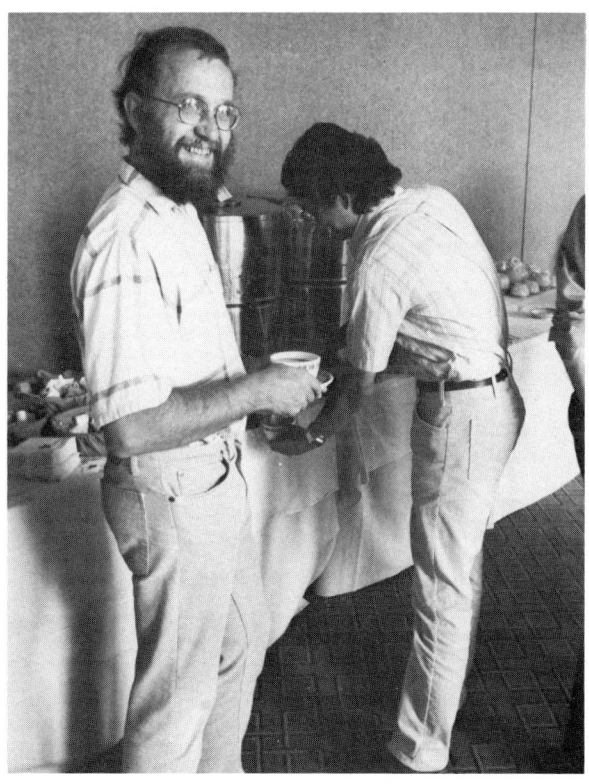

Alexander Tielens

Irene Little-Marenin, Jesse Bregman and Alexander Tielens

INFRARED ABSORPTION AND EMISSION CHARACTERISTICS OF INTERSTELLAR PAHs

L. J. Allamandola and A. G. G. M. Tielens
Space Science Division, MS 245-6
NASA/Ames Research Center
Moffett Field, CA   94035
U.S.A.

J. R. Barker
Department of Atmospheric and Oceanic Science
Space Research Building
University of Michigan
Ann Arbor, MI   48109-2143
U.S.A.

ABSTRACT.   The mid-infrared interstellar emission spectrum with features at 3050, 1610, 1300, 1150, and 885 cm^{-1} (3.28, 6.2, 7.7, 8.7 and 11.3 microns) is discussed in terms of the Polycyclic Aromatic Hydrocarbon (PAH) hypothesis. This hypothesis is based on the suggestive, but inconclusive comparison between the interstellar emission spectrum with the infrared absorption and Raman spectra of a few PAHs. The fundamental vibrations of PAHs and PAH-like species which determine the IR and Raman properties are discussed. Interstellar IR band emission is due to relaxation from highly vibrationally excited PAHs which have been excited by ultraviolet photons. The excitation/emission process is described in general and the IR fluorescence from one PAH, chrysene, is traced in detail. Generally, there is sufficient energy to populate several vibrational levels in each mode. Molecular vibrational potentials are anharmonic and emission from these higher levels will fall at lower frequencies and produce weak features to the red of the stronger fundamentals. This process is also described and can account for some spectroscopic details of the interstellar emission spectra previously unexplained. Analysis of the interstellar spectrum shows that PAHs containing between 20 and 30 carbon atoms are responsible for the emission.

1.   INTRODUCTION

Understanding the origin of the set of interstellar infrared emission bands at 3.28, 6.2, "7.7", 8.6 and 11.3 microns (3050, 1613, 1310, 1160 and 885 cm^{-1}) forms the core of an important problem in interstellar

D. J. Hollenbach and H. A. Thronson, Jr. (eds.), Interstellar Processes, 471–489.

astrophysics. Since their discovery in 1973 by Gillett, Forrest and Merrill, the bands have been found to be associated with a number of stellar objects, planetary nebulae, reflection nebulae, H-II regions and extragalactic sources. As the number and variety of objects discovered to emit these features increased, it became clear that identifying the carrier and elucidating the excitation-emission mechanism were the central issues. These observations also showed that the bands were due to a surprisingly widespread, extremely stable, interstellar constituent (see Willner, 1984 and Allamandola, 1984 for reviews of the observations and theories up to 1983).

While it has been recognized for some time that the emission was most likely pumped by ultraviolet photons, (Allamandola and Norman, 1978), identifying the carrier remained elusive. Although several models have been proposed, until recently none was very satisfactory. In brief, these models are as follows: (1) infrared fluorescence from UV-pumped, vibrationally excited, small molecules (e.g., CH_4, H_2O, NH_3, CO) frozen on 0.1 micron-sized grains at low ($\approx$ 10 K) temperatures (Allamandola and Norman 1978); (2) equilibrium thermal emission from small (0.01 micron) grains at 300 K, coated with an unspecified polymeric material (Dwek et al. 1980); (3) equilibrium thermal emission from characteristic groups on aromatic-like moieties, present at the surface of small carbon grains (Duley and Williams 1981); (4) nonequilibrium thermal emission from very small grains (0.001 micron) of unspecified composition that are temporarily heated to 1000 K by the absorption of a single UV photon (Sellgren 1984; Sellgren, Werner and Dinerstein 1983). There are drawbacks with each model. Model 1 requires a multicomponent mantle that does not change composition under a wide variety of interstellar conditions. Models 2 and 3 require values for the infrared oscillator strength (f) near unity. Typically, f varies from 10^{-5} to 10^{-3} for vibrational transitions. Model 4 relies on the uncertain assumption that a 10 Å-sized species can be treated as if it has bulk thermal and optical properties. Progress towards the hypothesis currently gaining acceptance, that aromatic hydrocarbons may be responsible for the bands, began with the suggestion by Duley and Williams (1981) that they arise from vibrations of characteristic chemical groups attached to the aromatic moities present at the surfaces of small (<0.01 microns) amorphous carbon particles. Leger and Puget (1984) and Allamandola, Tielens and Barker (1985) later proposed that vibrations in individual, molecule sized (0.001 microns) polycyclic aromatic hydrocarbons (PAHs) and PAH-like species were responsible for the bands. As a class these complex, planar, organic molecules which are made up of fused six membered rings (chicken wire-like structure) are extremely stable.

Apart from the overall resemblance of the interstellar IR band spectrum to the vibrational spectrum expected from PAH-like species (see Section 2 below), the results of recent, related observations tend to favor an aromatic hydrocarbon origin as well. Cohen et al. (1986) have shown that the fraction of total IR luminosity radiated by the 7.7 micron feature in planetary nebulae is strongly correlated with the nebular C/O ratio. Because the carriers must be produced in these nebulae under harsh conditions, they must be extremely stable and carbon rich, two characteristics completely consistent with the aromatic hydrocarbon hypothesis. Cohen et al. also show that, while there is variation among the relative IR band intensities among different objects, they are correlated, implying that a single class of chemical species is responsible.

Within the framework of the PAH hypothesis, spectroscopic analysis suggests that there are at least two classes of components which con- tribute to the total infrared emission spectrum. Free, molecule-sized PAHs produce the narrow features and amorphous carbon particles (which are primarily made up of an irregular "lattice" of interconnected PAHs, see Tielens and Allamandola, elsewhere in this volume) contribute to the broad, underlying component (Allamandola, Tielens and Barker, 1986a; Goebel, 1986.)

In view of the central role PAHs seem to play in determining the IR properties of specific objects as well as of the interstellar medium itself, this chapter will focus on the infrared characteristics of PAHs and the UV-Visible pumped IR emission mechanism.

2. THE VIBRATIONAL SPECTROSCOPY OF PAHs

Although the IR emission band spectrum resembles what one might expect from a mixture of PAHs, it does not match in details such as frequency, band profile or relative intensities predicted from the absorption spectra of any known PAH or their mixtures. In Figure 1, the emission spectrum from position 4 in Orion is compared with a schematic version of the absorption spectra of three PAHs: chrysene ($C_{18}H_{12}$), pyrene ($C_{16}H_{10}$) and coronene ($C_{24}H_{12}$). Leger (1986), shows a similar com- parison between the emission spectrum from the reflection nebula NGC 7023 and the absorption spectra of several larger PAHs suspended in KBr pellets. Similar suggestive, but inconclusive, comparisons between the interstellar emission spectra with the emission spectrum expected from the PAHs coronene and chrysene can be found in Leger and Puget (1984, hereafter LP), and Allamandola, Tielens and Barker (1985, 1986b hereafter ATBa,c). Because of the suggestive match, the assumption has been made that PAHs in some form or combination are responsible for the interstellar emission. Only when detailed laboratory spectra for the various free PAHs in their neutral and ionized forms become available can precise conclusions be drawn regarding their respective importance. Keeping this qualification in mind, the following general remarks concerning band assignments apply to virtually all PAHs.

As illustrated in Figure 1, the 3.29 micron band is highly charac- teristic of an aromatic system (Duley and Williams, 1981; Bellamy, 1958). In general, PAHs show a dominant band at about 3.28 microns, corresponding to a C-H stretch, in addition to a number of weaker bands between 3.1 and 3.6 microns which are overtone and combination bands involving lower frequency fundamentals. (Bellamy, 1958, Cyvin, et al. 1982a,b, Herzberg, 1968).

Figure 1 shows that the 6.2 micron emission band (which corresponds to a C-C stretching vibration in PAHs) is as characteristic of poly- cyclic aromatic species as is the 3.3 micron band (LP; Bellamy, 1958).

Perusal of Figure 1 also shows that the 5 to 10 micron region in
PAH spectra is richest in IR active vibrations and the largest density
of bands occurs in the 7.2 to 8.5 micron range (ATBa,c, Bellamy, 1958).
Unlike the 3.3 and 6.2 micron bands which consistently occur at nearly
the same wavelength, independent of the molecule, the precise position
of these C-C stretching bands depends on the particular molecular
structure. Thus the infrared spectrum of a mixture of PAHs could
produce a broad band, possibly with substructure, in this region. Of
course the precise peak position and profile would vary somewhat
depending on the particular PAH mixture responsible.

The small shoulder at 8.6 microns on the "7.7" micron feature which
often appears in the interstellar spectra is assigned to the in-plane
aromatic C-H bending mode in PAHs (LP; Bellamy, 1958). As shown in
Figure 1, PAHs show several bands close to this position.

The 11.3 micron feature is assigned to the out-of-plane C-H bending
vibration (Duley and Williams, 1981; Bellamy, 1958). Because this
frequency is so highly characteristic for aromatic species with edge
rings which contain only non-adjacent peripheral hydrogen atoms
(Bellamy, 1958), Duley and Williams postulated that the aromatic
containing material they believed responsible, amorphous carbon
particles, was only partially hydrogenated. Fully hydrogenated PAHs
which contain more than one H atom per edge ring possess several strong
bands in the 11-15 micron range. The discovery of the 11-13 micron
interstellar emission plateau underlying the well-known 11.3 micron band
not only relieved some of the difficulty associated with understanding
partial hydrogenation in exceedingly H rich environments, but also
showed that edge rings of PAHs responsible for the interstellar emission
can have non-adjacent as well as 2 or 3 adjacent peripheral H atoms, but
not 4 or 5. (Tielens et al. 1986, Cohen, Tielens and Allamandola 1985).

The spectra shown in Figure 1 serve to illustrate several additional points. For a free, highly symmetric PAH (e.g., coronene) with
an inversion center of symmetry, the infrared spectrum will appear
simple and the Raman and IR active vibrational modes will be mutually
exclusive due to the high molecular symmetry. However, if the vibrational force field is not so symmetric (as is the case for free, less
symmetric PAHs, or PAHs in clusters or amorphous carbon particles) the
IR and Raman spectra can be very similar. For example, although coronene has 66 C-C modes while chrysene has only 48, the IR spectrum of
chrysene is far richer in the C-C stretching region (5-10 microns)
because the molecule is less symmetric. It is for these reasons that we
legitimately could compare the Raman spectrum of soot in the 6 to 8
micron range with emission from Orion (Figure 2, ATBa), and thus point
out the striking similarity between the interstellar emission bands and
the vibrational spectra of a mixture of PAHs (amorphous carbon is made
up of PAH subunits cross linked in an irregular fashion, see, Tielens
and Allamandola, elsewhere in this volume. For comparison, Figure 3
shows the IR absorption spectrum of a different mixture of aromatic
hydrocarbons, known as a char (Mortera and Low 1983).

Note that the char spectrum also shows structure in the 3 micron
region which is similar to that shown in many of the interstellar
emission spectra (i.e., a dominant 3.3 micron band and a broader, weaker

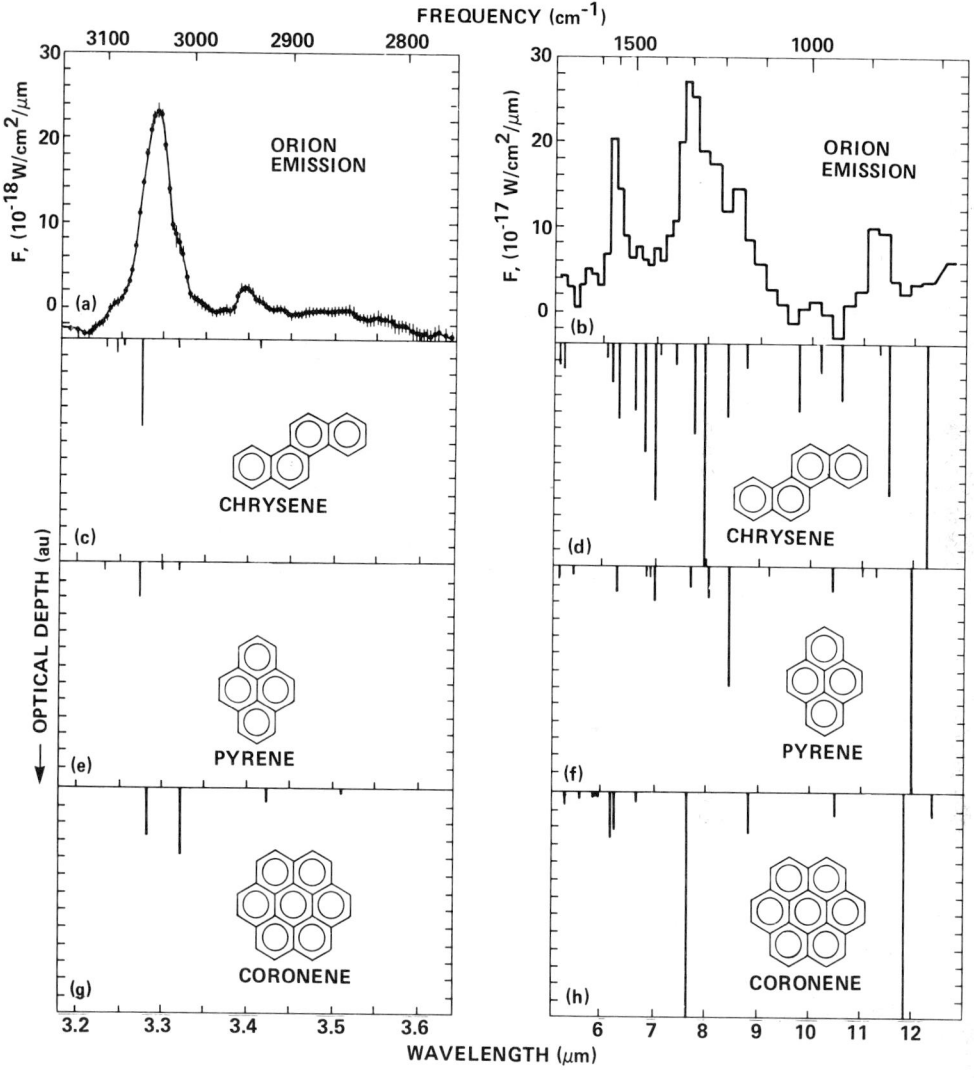

Figure 1.   The 3–13 micron emission spectrum from the Orion Bar compared
with the absorption spectra of the PAHs chrysene, pyrene and coronene
suspended in KBr pellets. (Orion, Bregman et al. 1986; Chrysene, Cyvin
et al. 1982a; Pyrene, Cyvin et al. 1979; Coronene, Bakke et al. 1979;
Cyvin et al. 1982b).   Taken from ATB,c.

component starting at about 3.15 microns, extending to about 3.65
microns, and peaking near 3.41 microns). Figure 4 shows that spectra of
individual PAHs also have similar structure. As mentioned above, the
weak absorptions by PAHs in this region are due to overtones and combi-

nations of 5–10 micron fundamentals. The spectra shown in Figures 2,3
and 4 are of PAHs in various solids where the perturbations within the
solid broaden the individual bands causing them to overlap and produce a
broad component. Free PAHs will show individual bands whose positions
and intensities are determined by the molecular structure of each PAH.
The interstellar emission component is due to the overlap of individual
emission bands which arise from different PAHs. In addition to the weak
blended contribution from overtone and combination bands, highly
vibrationally excited molecules emit from vibrational levels higher than
$V=1$. Emission from these higher levels, shifted for anharmonicity,
contributes significantly to the emission features superimposed on the
3.4 micron plateau and many of the other components of the interstellar
spectrum (see section 3.2, Barker et al. 1986 and ATBa,c). The recent
discovery of specific bands in this plateau (de Muizon, et al. 1986)
supports this explanation of the plateau in terms of overlapping
individual lines.

As shown in Figures 2 and 3, the infrared signature of amorphous
carbon soots and chars resembles the interstellar IR band spectrum
rather closely. Weak, broad features at roughly 6.2, 7.7, and 11.3
microns are also evident in the extinction curve of amorphous carbon
particles (Koike, Hasegawa and Manabe, 1980, Borghesi et al., 1983).
Sakata et al., (1984) have also suggested that the absorption spectrum
of the hydrocarbon residue deposited by a methane plasma resembles the
interstellar spectrum. The similarities and differences between the
interstellar IR emission spectrum and
spectra associated with various
forms of amorphous carbon are
discussed elsewhere (ATBb, Goebel,
1986). Since amorphous carbon is
primarily composed of randomly
oriented clusters of PAHs cross-linked
and interconnected by saturated and
unsaturated hydrocarbon chains, the
overall spectral emissivity of small
amorphous carbon particles will
resemble PAH spectra, with the
individual bands blurred out due to
the solid state effect which produce
line shifting, broadening and intensity changes (Allamandola 1984),
Thus, the infrared spectroscopic
properties of small amorphous carbon
particles will be largely determined

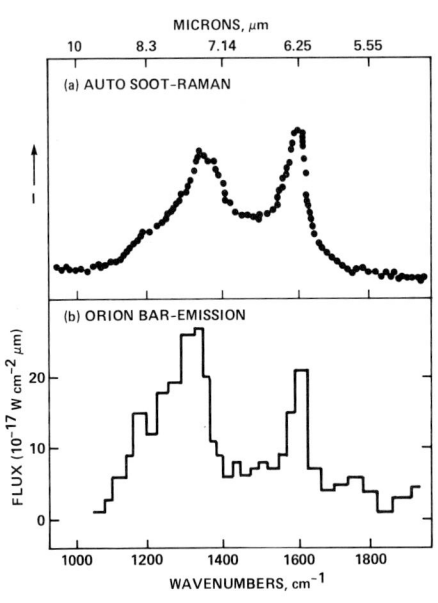

Figure 2. Comparison of the 5 to 10
micron Raman spectrum of auto soot (a
form of amorphous carbon) with the
emission from Orion (soot spectrum,
adapted from Rosen and Novakov 1978;
Orion, Bregman et al. 1984, 1986).
Taken from ATBa.

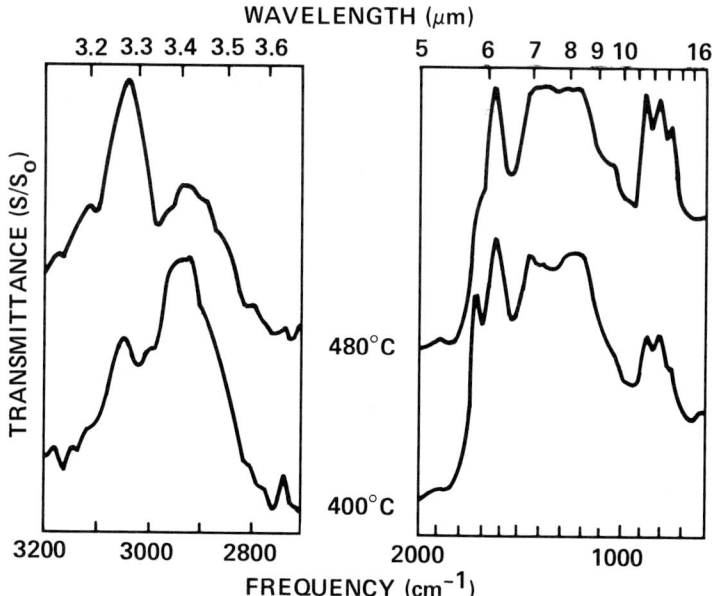

Figure 3. Infrared absorption spectra of chars at 400 and 480°C. Note the similarity between the 480°C char (a form of amorphous carbon) absorption spectrum, with the Raman spectrum of soot. (Char spectra from Mortera and Low 1983). Taken from ATBc.

by the properties of the PAHs of which they are made. As the particles get larger, bands will overlap, producing broad features that possibly retain some substructure indicative of the individual PAHs. For still larger particles, bulk properties dominate and the broad components will appear as substructure on a strong continuum which follows a $1/\lambda$ law, (see Tielens and Allamandola, this volume) producing the extinction curve reported by Koike et al. (1980) and Borghesi et al. (1983).

Summing up, spectroscopically, a collection of PAH-like species, either in solid form (amorphous carbon grains) or as "free-flying", gas phase molecules, seem to be able to account for the narrow emission features. However, as described in the next section, the intensity of the emission at 3 microns with respect to that at longer wavelengths forces us to conclude that the emission bands originate in molecule sized species.

3. THE EMISSION MECHANISM

The overall excitation-emission mechanism envisioned is as follows. A free PAH, which may be in an ionized or neutral form, absorbs an ultraviolet or visible photon and is excited into an upper electronic state. The PAH quickly converts this electronic energy into vibrational energy. For example, a PAH such as pyrene ($C_{16}H_{10}$) has 72 normal, vibrational modes, each having a specific vibrational frequency lying

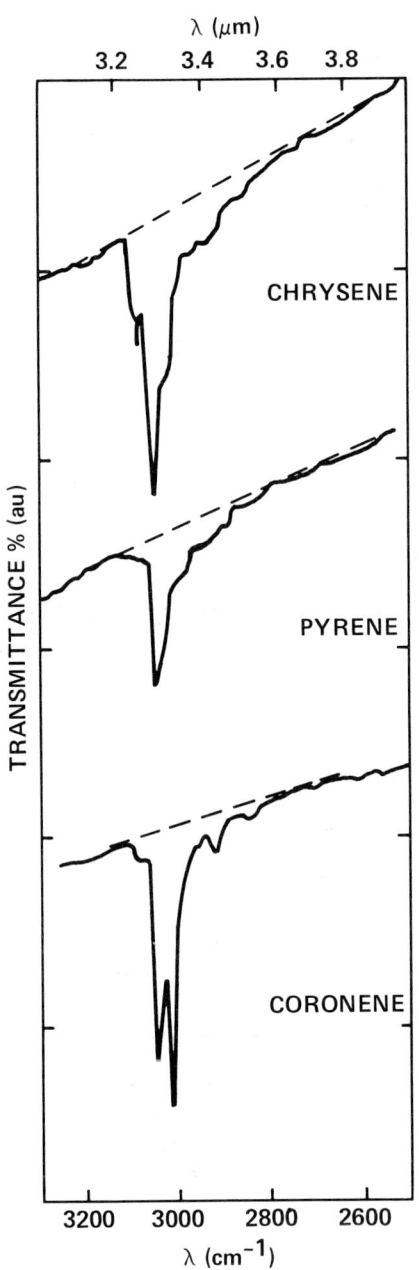

Figure 4. Infrared absorption spectra in the 3.3 micron region of the PAHs chrysene, pyrene and coronene, suspended in KBr. Spectra courtesy of Drs. Cyvin and Klaeboe, University of Trondheim, Norway.

in the 3100-200 cm^{-1} range. Within a few picoseconds of the deposition of the energy in a particular electronic state it will be spread out amongst these modes. Energy transfer from mode to mode occurs on a similar timescale. In the absence of collisions, the only routes for deactivation of the neutral, vibrationally excited molecule are electronic phosphorescence, and fluorescence and infrared vibrational fluorescence (IRF). For ions IRF is the only relaxation mechanism available. It is this relaxation process which is responsible for the interstellar infrared emission bands. The details of this process are described below.

3.1 The Molecular Emission Process

Figure 5 shows a highly schematic version of the overall excitation – emission processes for a neutral PAH (ATBa,c, see also Jortner, 1986, for a detailed discussion of these processes). Prior to the absorption of a photon, any neutral PAH is in the lowest singlet electronic state, S_o. Photons of many energies are incident on the molecule, and of those absorbed, some can ionize the molecule and others can excite it to higher singlet states ($S_1, S_2, \ldots$). The absorbed energy will quickly redistribute itself within the molecule ($<10^{-12}$s) via internal conversion (IC) ($S_f \leftarrow S_i$; i.e., transfer to a highly vibrationally excited state of the lower lying electronic singlet state) and intersystem crossing (ISC) ($T_f \leftarrow S_i$; i.e transfer to a highly vibrationally excited state of a lower lying electronic triplet state) processes. Note that the latter process leaves the molecules also with a considerable amount of electronic

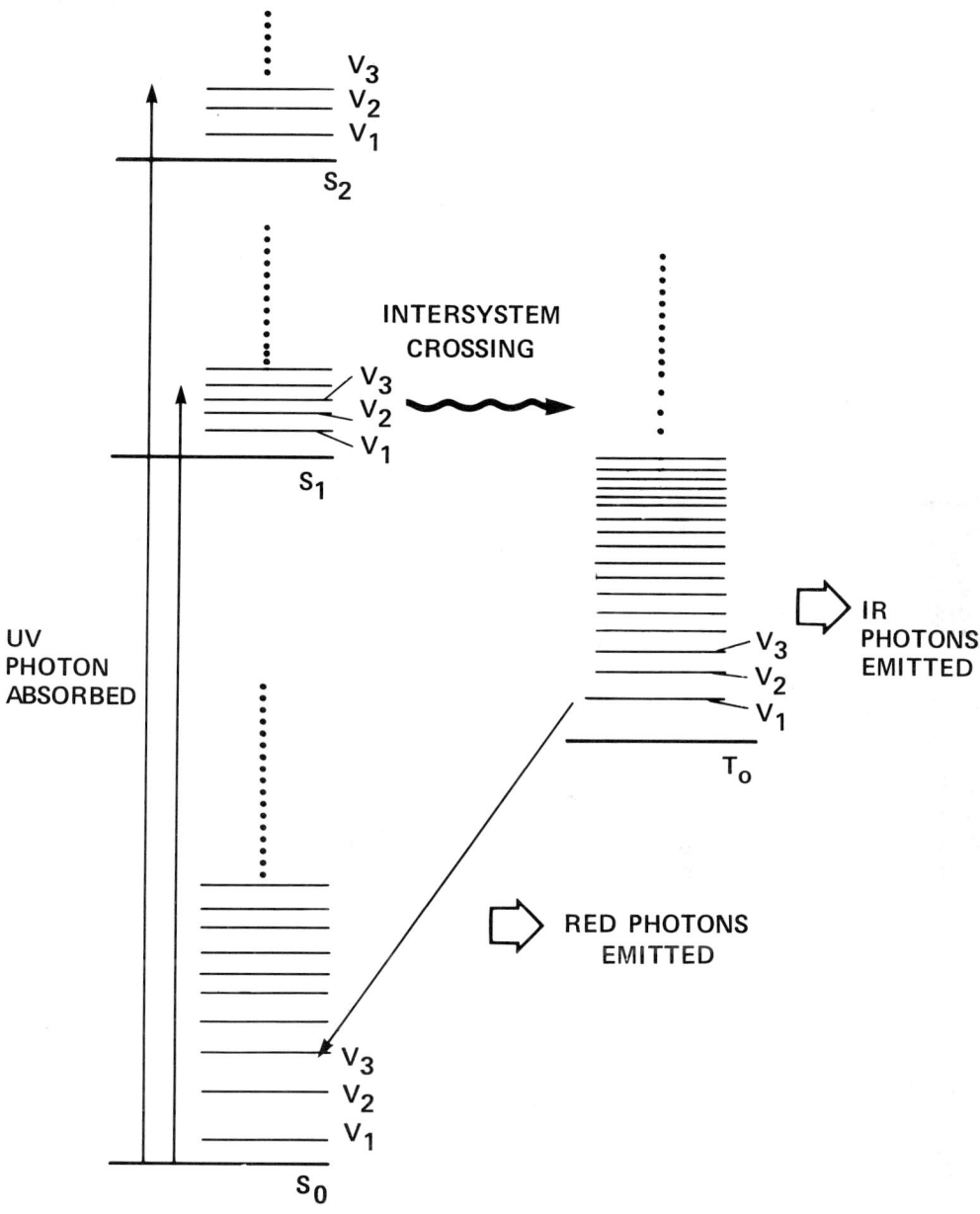

Figure 5.   Schematic energy level diagram for a neutral PAH, showing the various radiative and nonradiative excitation and relaxation channels possible.   Taken from ATBc.

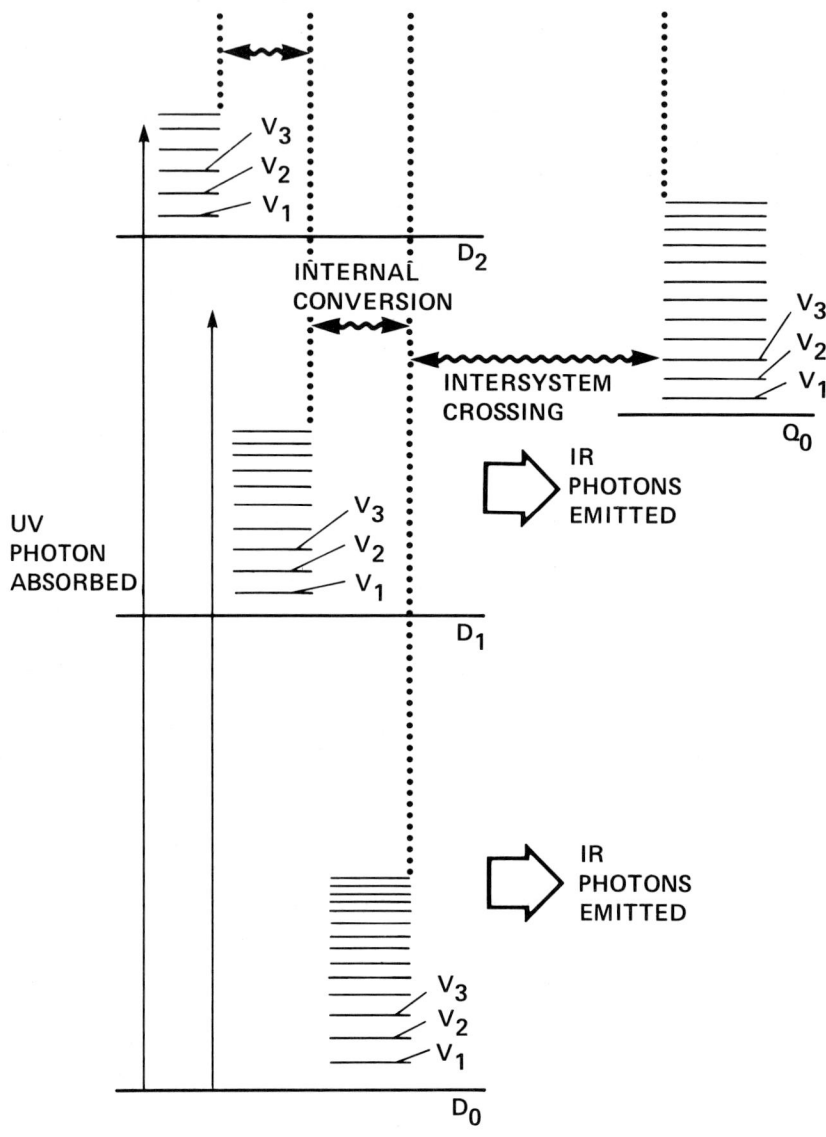

Figure 6.   Schematic energy level diagram for an ionized PAH, showing
the various radiative and non-radiative excitation and relaxation
channels possible.   Taken from ATBc.

excitation energy.   For chrysene, the strongest absorption corresponds
to the $S_3 \leftarrow S_0$ transition which peaks near 2675A.   Nearly 90% of these
excited molecules undergo rapid ISC to high vibrational levels of the
T_0 state.   Because T_0 is lower than S_1 (this is true for virtually all
neutral PAHs) and the direct electronic radiative relaxation (phospho-

rescence) from T_o to S_o occurs very slowly (on the order of several seconds) about 17,400 cm^{-1} of the 37,400 cm^{-1} photon energy is trapped in the triplet state as vibrational energy. The remainder is "stored" by the molecule which is now in the triplet electronic state. This highly vibrationally excited molecule can lose its vibrational energy only by radiating IR photons, primarily by the transitions ($V_{n-1} \leftarrow V_n$) at the fundamental vibrational frequencies of the T_o state. IRF lifetimes are on the order of a tenth of a second while the phosphorescent lifetime ($S_o \leftarrow T_o$) for chrysene is longer than three seconds.

Figure 6 shows that if the excited PAH is ionized (as we believe to be the case in the emitting zone), the energy level diagram will be quite different, involving doublet (D) and quartet (Q) states rather than singlets and triplets, with the ground electronic state being a doublet, D_o. In the case of an ion, the lowest excited doublet state, D_1, is always lower than Q_o and trapping in the Q_o state, which can only slowly radiatively relax to the ground state, D_o, is not important, (see Leach, 1986, for a complete description of this situation). Thus, for free molecular ions nearly all of the energy is quickly converted to vibrational energy in the ground electronic state, D_o, where IRF is the only deactivation route.

The calculation is described in detail elsewhere (ATBc; Barker, Allamandola and Tielens, 1986). The feature of this system which distinguishes it from a thermal system is that all molecules which absorb a photon of energy $h\nu$ have the same total energy: in a thermal system each molecule has a different total energy but the ensemble average energy equals $h\nu$. The fractional number of molecules with total vibrational energy E that have v quanta in a particular mode is a ratio equal to the number of ways of distributing the energy while that mode has v quanta excited, divided by the total number of ways to distribute the energy. For energies in excess of a few thousand wavenumbers, the statistical redistribution is completed in less than a few microseconds (Oref and Rabinovitch, 1979, Bondybey, 1984).

The infrared emission spectrum predicted for chrysene is plotted in Figure 7 as a function of vibrational energy content. From Figure 7 several conclusions may be reached. Firstly, the IRF spectrum varies dramatically as the vibrational energy is changed and it does not necessarily match the intensity distribution observed in absorption. Secondly, no significant IRF emission is observed from the 3000 cm^{-1} modes (3.3 microns), unless the molecule contains a significant amount of vibrational energy per vibrational mode. Thirdly, the intensities of the 3000 cm^{-1} modes (3.3 microns) relative to the 1000 cm^{-1} modes (10 microns) are directly related to the internal energy of the molecule.

One of the many important consequences of this treatment is that the intensity ratio of the 885 to 3030 cm^{-1} (11.3 to 3.3 micron) bands provides a measure of the vibrational energy content of the emitting species if its size is known; conversely, if the energy content of the molecules is known, the size of the emitting species can be estimated. Thus, if one knows the spectrum of the exciting UV field, the 885/3030 cm^{-1} (11.3/3.3 micron) intensity ratio indicates the number of carbon atoms in the smallest emitting species (the most intense emitters).

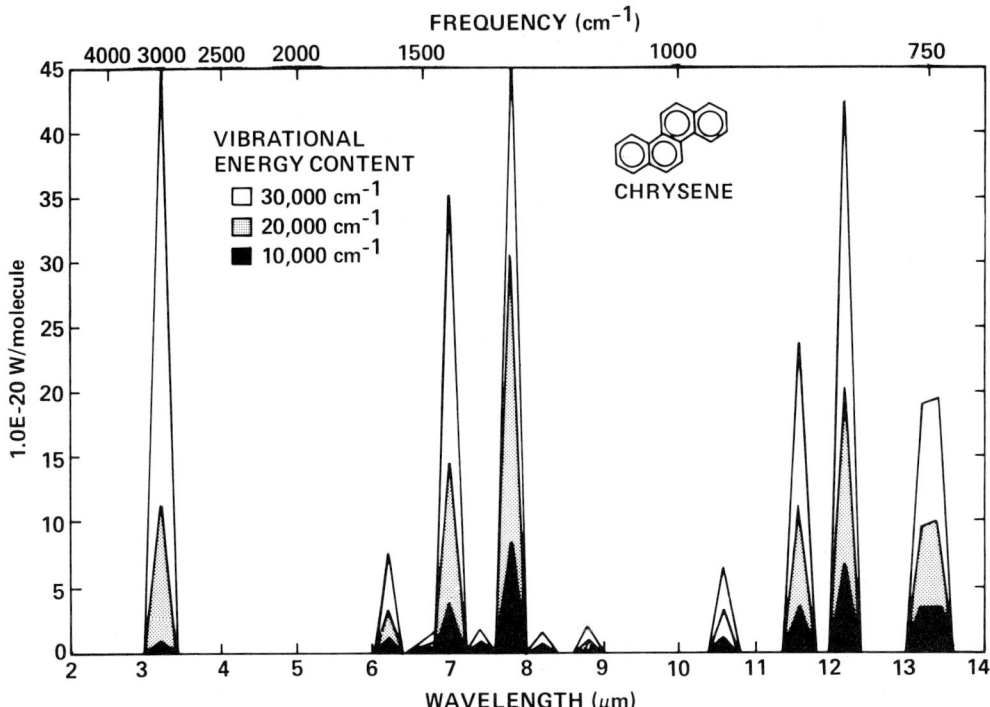

Figure 7. The IR fluorescence spectrum from chrysene as a function of vibrational energy content. Taken from ATBc.

Conversely, if one can confine the size of the most intense emitting PAH on the basis of spectroscopic constraints, one can determine the part of the incident radiation field that is most important in pumping the IR bands.

To make this estimate note that for each C-H bond one out-of-plane bending mode and one stretching mode will be present; thus, the number of modes emitting near 885 cm^{-1} (11.3 micron) equals the number of modes emitting near 3000 cm^{-1} (3.3 microns), regardless of the actual number of C-H bonds present in each molecule. Vibrational assignments for benzene (C_6H_6), azulene ($C_{10}H_8$), anthracene ($C_{14}H_{10}$), chrysene ($C_{18}H_{12}$), and perylene ($C_{20}H_{12}$) were used to predict the relative IRF emission intensities due to one C-H stretch mode and due to one C-H out-of-plane mode, assuming they have equal integrated absorption coefficients, as is approximately true in the gas for free molecular benzene (Bishop and Cheung, 1982), napthalene and anthracene (Niki, 1986).

As an initially vibrationally excited molecule relaxes by infrared emission, it must sequentially emit many infrared photons and undergo an energy cascade before it reaches equilibrium with the low temperatures of the ISM. Thus, the observed interstellar emission bands are due to emission from molecules in all stages of relaxation subsequent to

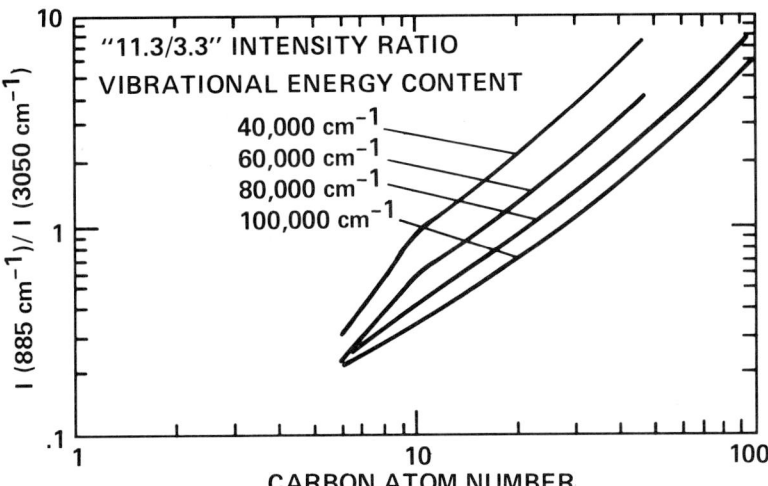

Figure 8. The 11.3 micron/3.3 micron intensity ratio plotted as a
function of carbon atom number and vibrational energy content. The
relative fluorescence intensities includes integration over the
vibrational cascade. Taken from ATBc.

excitation to some initial energy. For comparison with the emission
bands, the calculated decay rates of the IRF emission were used to
average the calculated intensities over the entire range of energy up to
the initial excitatation level. Ratios of IRF intensities (averaged
over the cascade) were calculated for several initial excitation
energies. The calculated ratios of IRF intensities are plotted in
Figure 8 as a function of the number of carbon atoms in each molecule
for several values of initial internal energy.

   The observed 11.3/3.3 intensity ratio in Orion is about three, and
the ratios observed in other objects range from 2 up to about 4 or 5
(Cohen et al. 1986). If the average emitting species was initially
excited to about 80,000 cm^{-1} (1250Å) and all of that energy was
converted to vibrational energy, it must have 20 to 80 carbon atoms to
explain the observed intensity ratios ranging from 1 to 5. At lower
excitation energies, even smaller species can explain the observed
intensity ratios, but the maximum size is limited by the maximum energy
of the photons available (about 80,000 - 100,000 cm^{-1}).

   Visual inspection of published IR absorption spectra for PAHs
suspended in KBr shows that the integrated absorption strength of the
bands in the 11-13 micron range is generally about three to five times
more intense than the bands at 3.3 microns. This differs from the near
unity values (used in the calculation) observed in the gas phase for
benzene, napthalene and anthracene (Niki, 1986). However, if the 11-13
micron modes in the larger emitting PAHs consisting of several fused
rings has an integrated absorption coefficient three times larger than
that of the 3.3 micron mode, the maximum molecular size consistent with
the observed intensity ratio range is reduced by a factor of two to
three, indicating that the band carriers are species containing about 10

to 30 carbon atoms. As discussed below, this conclusion is in agreement
with the number deduced from individual band profiles analyzed by
accounting for emission from higher vibrational levels, which are
shifted by anharmonicity.

3.2 Molecular Anharmonicity

Introductory descriptions of molecular vibrations and derivations of
vibrational frequencies are generally discussed in terms of the harmonic
oscillations of two bodies connected with a spring and described by the
quantum version of Hookes law where the frequencies are determined by
the molecular force field. The force field, which depends on the
electron distribution in the molecule, describes the shapes and depths
of the potential well which binds the atoms. For example, the frequency
range of the aromatic CH stretch is 3100-2900 cm^{-1} (3.23-3.45 microns),
while for the in-plane bend it is 1250-1000 cm^{-1} (8-10 microns) and for
the out-of-plane bend it is 900 to 555 cm^{-1} (11-18 microns) (see
Allamandola 1984, for an introductory level discussion of molecular
vibrations; Barrow 1962, for a more detailed treatment; and Herzberg
1968, for an extensive discussion). While virtually all molecular
potential wells are anharmonic, the degree of anharmonicity can vary
markedly. One of the effects of anharmonicity is to continuously
decrease the spacing between the adjacent vibrational states for higher
values of v. (In a harmonic oscillator, the spacing between all adjacent levels is constant). Thus, for a given vibration, such as a CH
stretch, $\Delta v=1$ transitions between higher v levels will occur at lower
frequencies. The extent of anharmonicity depends on the particular
vibration involved. For example, the anharmonicity for the IR allowed
aromatic CH stretch is rather large, producing a shift on the order of
100 cm^{-1}, while for the C-C stretch and C-H out-of-plane bend it is
believed to be less than 10 cm^{-1}. Although data are not available for
large PAHs, the anharmonicities for the C-H stretch of the aromatic
molecules benzene, C_6H_6; napthalene, $C_{10}H_8$; and anthracene $C_{14}H_{10}$, have
been measured (Swofford, Long and Albrecht, 1976; Reddy, Heller and
Berry, 1982). The frequencies for the $v=3\rightarrow2$, $2\rightarrow1$ and $1\rightarrow0$ transitions
in benzene are 2814 cm^{-1} (3.55 microns), 2925 cm^{-1} (3.42 microns) and
3047 cm^{-1} (3.28 microns) respectively.

If the interstellar IR emission bands originate from highly vibrationally excited molecules, emission from vibrational levels higher than
v=1 must contribute to the spectrum and anharmonicity effects are
expected. In our calculation we have also taken anharmonicity into
account.

For chrysene, the calculated emission spectrum for a single C-H
stretching vibration is shown in Figure 9a, as a function of vibrational
energy content. All other PAHs will exhibit similar spectra, with 100
to 120 cm^{-1} separation between the peaks, but the exact location of the
pair depends on the particular molecule. More realistic simulations
would include the other C-H vibrational stretch modes, which may have
slightly different frequencies, and the effects of the energy-cascade,
which we have neglected in preparing Figure 9a. Figure 9 shows there is
a remarkable resemblance to the spectra observed from NGC 7027 and S106,

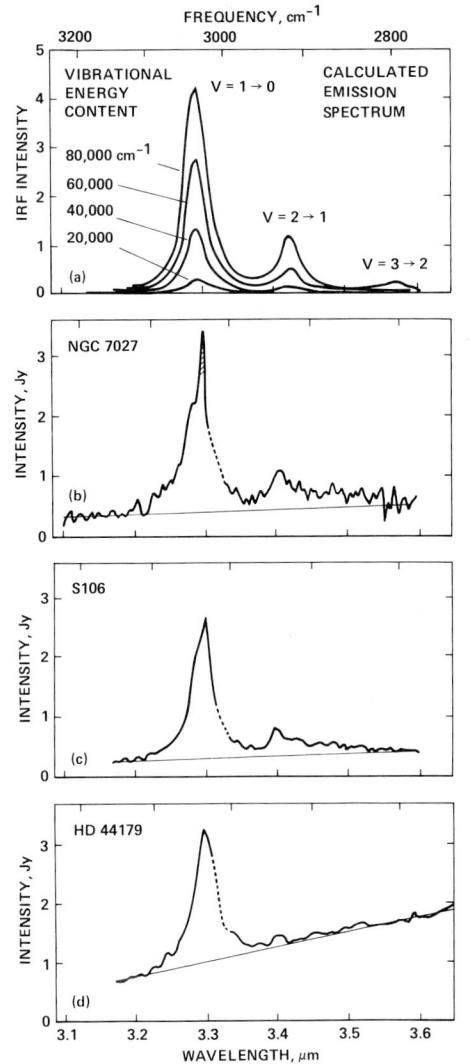

Figure 9. a) The calculated emission spectrum for chrysene in the CH stretching region as a function of vibrational energy content. Anharmonicity is assumed to be 120 cm^{-1}. b), c) and d), the observed emission spectra of NGC 7027, S106 and HD 44179 (Geballe, et al., 1985) showing how emission from higher vibrational levels depends on the availability of energetic photons. In NGC 7027, where the most energetic photons are available, emission from higher levels is important and produces a prominent v=2$\rightarrow$1 band, whereas in the relatively benign HD44179, emission from v=2 is barely discernable. S106 is intermediate energetically, and the 3.3/3.4 micron band ratio is intermediate. Taken from Barker, Allamandola and Tielens, 1986.

and possibly from HD44179. Comparison with the observed spectrum shows a distinct satellite peak at exactly the position required. Thus, the peaks observed near 2950 cm^{-1} in various objects provide additional strong support for the PAH hypothesis.

Since the relative importance of the emission from v=2 depends only on the internal energy of the excited molecule, the ratio of intensities from v=1 and v=2 provides a firmer measure of excitation energy and molecular size than is possible by comparing the intensity of the stretching (3050 cm^{-1} 3.3 micron) and out-of-plane bending (885 cm^{-1}, 11.3 micron) modes.

Predictions for the intensity of the v=1 line compared to that for the v=2 line are presented in Figure 10 as functions of molecular size and excitation energy. The effects of the energy-cascade are included in this figure. Inspection of Figure 1 shows that, in Orion the satellite band observed is about ten times less intense than the main band. In Figure 10, a ratio of ten and a maximum excitation energy of 80,000 to 100,000 cm^{-1} are consistent with molecules of about 20-30 carbon atoms, a conclusion in agreement with the range deduced above from the relative intensities of the 885 and 3050 cm^{-1} (11.3 and 3.3

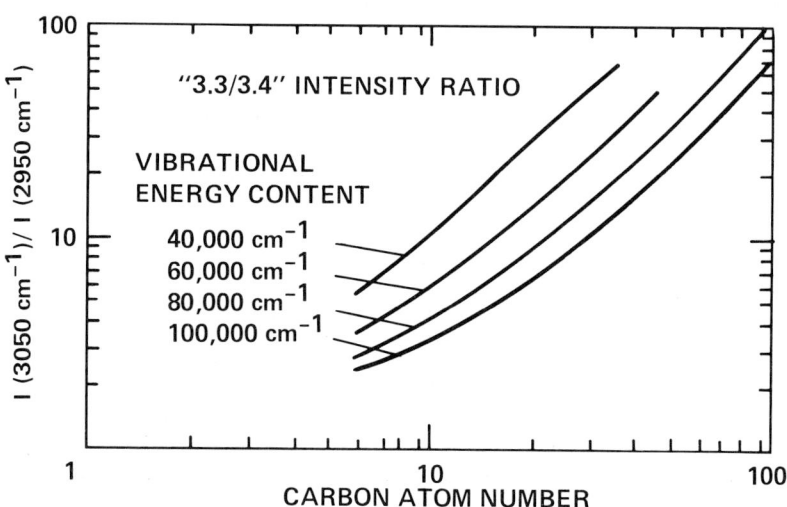

Figure 10. The 3.3 micron/3.4 micron intensity ratio as a function of
carbon atom number and vibrational energy content. The relative
fluorescence intensities includes integration over the vibrational
cascade. Taken from Barker, Allamandola and Tielens, 1986.

micron) bands.

    Experimental data are not available concerning the anharmonicities
of the other PAH modes. However, molecular force fields are understood
well enough to permit an estimation for the out-of-plane CH bend which
occurs in the 900 to 500 cm^{-1} range (11-20 microns). Anharmonicity on
the order of a few wavenumbers is expected (R. Eggers and J. Pliva,
private communication). We have calculated the effects of anharmonicity
on the 885 cm^{-1} (11.3 micron) feature assuming an anharmonicity of 5
cm^{-1} and a Lorentzian linewidth of 10 cm^{-1}. The vibrational frequency
is much lower than that for the CH stretch (800 compared to 3050 cm^{-1}).
Consequently, because the energy is assumed to be distributed statistically among all the modes, emission from considerably higher vibrational
states (v levels) is expected to contribute to the band. Although the
anharmonicity is only a few wavenumbers it will effectively broaden the
band by skewing it to lower frequencies. Thus, as shown in Figure 11,
for highly vibrationally excited molecules the band will be rather
assymetric. It will also be considerably broader and peaked at slightly
lower frequencies than that from molecules containing less vibrational
energy.

    This variation in behavior has been observed. Aitken and Roche
(1983) have measured the interstellar 11.3 micron band at several
positions in NGC 7027. In the central, ionized region where the most
energetic photons are available, the band is substantially broader and
markedly more assymetric than the band originating in the outer neutral

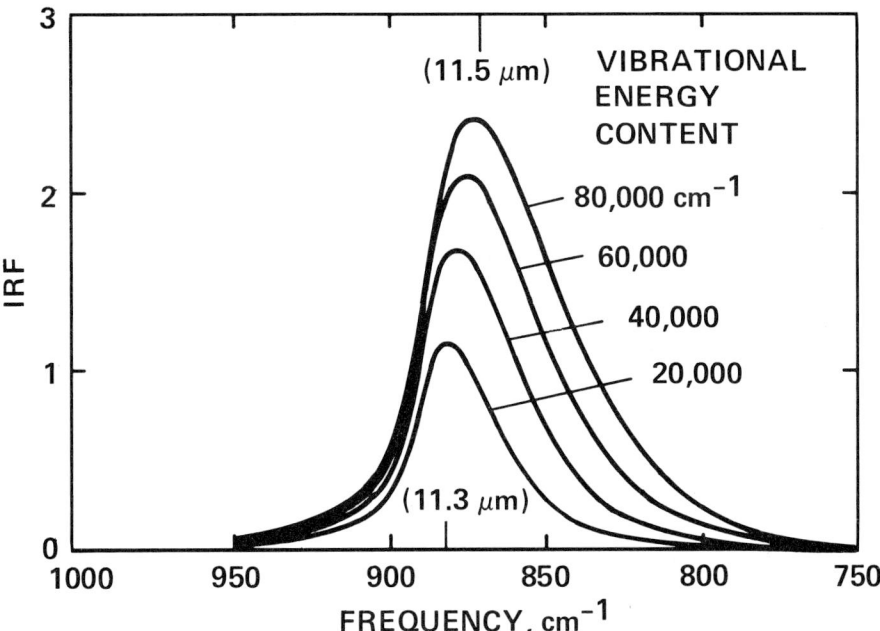

Figure 11. The calculated emission spectrum in the CH out-of-plane bending region as a function of vibrational energy content. Anharmonicity is assumed to be 5 cm^{-1}, natural linewidth 10 cm^{-1} and v=1 0 at 880 cm^{-1}. Taken from Barker, Allamandola and Tielens, 1986.

zone, where far less energetic photons are available. Until a more quantitative measure of the anharmonicity for these lower frequency modes is available, one cannot determine the size of the emitting species responsible as precisely as the analysis of the 3050 cm^{-1} .2930 cm^{-1} (3.3 micron/3.4 micron) band intensity ratio allows. Within the framework of the assumptions made here (anharmonicity 5 = cm^{-1}, Lorentzian width 10 = cm^{-1}), the size of the emitting species, assuming 100,000 cm^{-1} of vibrational energy, is again in the 20-30 carbon atom range.

In concluding this section on the emission process, we wish to point out that while one can model the emission phenomenon using a thermal model approximation, such approximations are of severely limited validity (ATB,c; Barker, Allamandola and Tielens, 1986). We stress that in addition to being able to explain the observed variations in emission ratio and account for band profile behavior, the molecular approach permits one to address questions regarding photostability, hydrogen coverage, reactivity, photoisomerization and deuterium fractionation in a completely self-consistent, more rigorous, general manner.

ACKNOWLEDGEMENTS: We are indebted to Professor Cyvin of Trondheim University, Norway for kindly sending us original spectra and pointing out some important spectroscopic properties of condensed aromatics. J.

R. Barker acknowledges partial support from the U. S. Department of
Energy, Office of Basic Energy Sciences.

4. REFERENCES

Aitken, D. K. and Roche, P. F. (1983), Mon. Not. R. Ast., Soc. 202,
    1233.
Allamandola, L. J. (1984), "Absorption and Emission Characteristics of
    Interstellar Dust" in Galactic and Extragalactic Infrared
    Spectroscopy, eds. Kessler, M. F. and Philips, J. P. (D. Reidel
    Publishing Co., Dordrecht), 5.
Allamandola, L. J. and Norman, C. A. (1978), Astro. Ap. 66, 129.
Allamandola, L. J., Tielens, A. G. G. M. and Barker, J. R. (1985), Ap.
    J., Letters, 290, L25 (ATBa).
Allamandola, L. J., Tielens, A. G. G. M., and Barker, J. R. (1968a,
    ATBb), Chapter in Polycyclic Aromatic Hydrocarbons and
    Astrophysics, eds. Leger, A. and d'Hendecourt, L. B. (D. Reidel
    Publishing Co., Dordrecht).
Allamandola, L. J., Tielens, A. G. G. M. and Barker, J. R. (1986b) Ap.
    J., submitted, (ATBc).
Bakke, A., Cyvin, B. N., Whitmer, J. C., Cyvin, S. J., Gustavsen, J. E.,
    and Klaeboe, P. (1979), Z. Naturforsch, 34a, 579.
Barker, J. R., Allamandola, L. J., and Tielens, A. G. G. M. (1986), Ap.
    J. (Letters) submitted.
Barrow, G. M., 1962, Introduction to Molecular Spectroscopy. (McGraw
    Hill, New York).
Bellamy, L. J., 1958, The Infrared Spectra of Complex Organic Molecules,
    (John Wiley and Sons, 2nd ed. New York).
Bishop, D. M. and Cheung, L. M. (1982), J. Phys. Chem. Ref. Data, 11,
    119.
Bondybey, V. (1984), Ann. Rev. Phys. Chem. 35, 591.
Borghesi, A., Bussoletti, E., Colangeli, L., Minafra, A. and Rubini, F.,
    (1983), Infrared Physics, 23, 321.
Bregman, J., Allamandola, L. J., Simpson, J. Tielens, A. and Witteborn,
    F. (1984) NASA/ASP Symposium, Airborne Astronomy, NASA/Ames
    Research Center (NASA CP 2353).
Bregman, J. et al., (1986) in preparation.
Cohen, M., Tielens, A. G. G. M., and Allamandola, L. J. (1985), Ap. J.
    (Letters), 299, L93.
Cohen, M., Allamandola, L. J., Tielens, A. G. G. M., Bregman, J.,
    Simpson, J. P., Witteborn, F. C., Wooden, D. and Rank, D. (1986)
    Ap. J., 302, 737.
Cyvin, S. J., Cyvin, B. N., Brunvoll, J., Whitmer, J. C., Klaeboe, P.,
    and Gustavsen, J. E., (1979), Z. Naturforsch, 34a, 876.
Cyvin, S. J., Cyvin, B. N., Brunvoll J., Whitmer, J. C., and Klaeboe, P.
    (1982b), Z. Naturforsch, 379, 1359.
Cyvin, B. N., Klaeboe, P., Whitmer, J. C. and Cyvin, S. J. (1982a), Z.
    Naturforsch, 37a, 251.
de Muizon, M., Geballe, T. R., d'Hendecourt, L. B., and Bass, F., (1986)
    Ap. J. Lett., 306, L105.

Duley, W. W. and Williams, D. A. (1981), Mon. Not. R. Astro. Soc., 196, 269.

Dwek, E., Sellgren, K., Soifer, B. T. and Werner, M. W., 1980, Ap. J. 238, 140.

Geballe, T. R., Lacy, J. H., Persson, S. E., McGregor, P. J., and Soifer, B. T. (1985), Ap. J. 292, 500.

Gillett, F. C., Forrest, W. J., and Merrill, K. M., 1973, Apt. J. 183, 87.

Goebel, J. (1986) Chapter in Polycyclic Aromatic Hydrocarbons and Astrophysics, eds. Leger, A. and d'Hendecourt, L. B. (D. Reidel Publishing Co., Dordrecht).

Herzberg, G. H., (1968), Infrared and Raman Spectra of Polyatomic Molecules, (D. van Nostrand Co., Princeton).

Jortner, J. (1986), Chapter in Polycyclic Aromatic Hydrocarbons and Astrophysics, eds. Leger, A. and d'Hendecourt, L. B. (D. Reidel Publishing Co., Dordrecht).

Koike, C., Hasegawa, H., and Manabe, A., (1980), Astrophys. Space Sci. 67, 495.

Leach, S. (1986), Chapter in Polycyclic Aromatic Hydrocarbons and Astrophysics, eds. Leger, A. and d'Hendecourt, L. B.)D. Reidel Publishing Co., Dordrecht).

Leger, A., (1986), Chapter in Polycyclic Aromatic Hydrocarbons and Astrophysics, eds. Leger, A. and d'Hendecourt, L. B. (D. Reidel Publishing Co., Dordrecht).

Leger, A., and Puget, J. L. (1984), Astro. Ap. 137, L5, (LP).

Mortera, C., and Low, M. J. D. (1983), Carbon, 21, 283.

Niki, H. (1986), (private communication)

Oref, I. and Rabinovitch, B. S. (1979), Acc. Chem. Res. 12, 166.

Reddy, K. V., Heller, D. F. and Berry, M. J. (1982), J. Chem. Phys. 76, 2814.

Rosen, H., and Novakov, T., (1978), Atmospheric Environment 12, 923.

Sakata, A., Wada, S., Tanabe, T., and Onaka, T. (1984), Ap. J. (Letters), 287, L51.

Sellgren, K. (1984), Ap. J., 277, 623.

Sellgren, K., Werner, M. W., and Dinerstein, H. L. (1983), Ap. J. (Letters), 217, L149.

Swofford, R. L., Long, M. E., and Albrecht, A. C. (1976), J. Chem. Phys. 65, 179.

Tielens, A. G. G. M., Allamandola, L. J., Barker, J. R. and Cohen, M. (1986), Chapter in Polycyclic Aromatic Hydrocarbons and Astrophysics, eds. Leger, A. and d'Hendecourt, L. B. (D. Reidel Publishing Co., Dordrecht).

Willner, S. P., (1984), "Observed Spectral Features of Dust" in Galactic and Extragalactic Infrared Spectroscopy, eds. Kessler, M. F. and Phillips, J. P. (D. Reidel Publishing Co., Dordrecht), 37.

Constance Walker and Chris Henkel

GRAIN DESTRUCTION, FORMATION, AND EVOLUTION

C. Gregory Seab
Virginia Institute for Theoretical Astronomy
Univeristy of Virginia
P.O. Box 3818, University Station
Charlottesville, VA   22903

ABSTRACT.   Grains do not exist statically in the interstellar medium
(ISM), but undergo processing depending on their environment. Timescales
are used to estimate the relative importance of various grain processes.
Destruction in supernova shock waves is shown to occur on a shorter
timescale than any other process, including the injection of new grains
from red giants and supernova ejecta. This finding, if it holds true,
implies that most of the grain mass in the diffuse ISM was formed by
accretion in dark clouds, rather than by thermal condensation in cool
star atmospheres and supernova ejecta.   The physics of grain destruction
in shocks is briefly outlined, as well as accretion and coagulation
processes in dark clouds.   A general scenario for the life cycle of
grains in the ISM is presented.

1.   INTRODUCTION

1.1.   Overview

Grains participate in the life of the violent ISM.   They effect star
formation, the propagation of shocks, and dark cloud chemistry, light,
and energy balance.   The grains themselves are affected by their
environment in the ISM.   They will grow volatile ice mantles and
possibly add to their refractory cores by accretion in dark clouds; they
may also coagulate to make larger grains.   Outside of dark clouds, they
will be stripped of the volatile part of their mantles.   Processing
through shocks will peel off some of the refractory core material, and
may vaporize all or part of the grain.   If a grain survives destruction
in shocks, it will ultimately be incorporated into new-formed stars.

Because grains are continually being destroyed, it is necessary to
inject new grains and grain material into the interstellar medium.   It
is known that new grains form in the outer atmospheres of cool
supergiants.   It is also probable that grains form out of the enriched
ejecta from supernovae.   Grains formed in these environments will
characteristically be high temperature condensates such as enstatite,

D. J. Hollenbach and H. A. Thronson, Jr. (eds.), Interstellar Processes, 491–512.
© *1987 by D. Reidel Publishing Company.*

olivine, and graphite. A fundamental open question about grains is
whether or not these sources of new grains are sufficient to maintain
the population of interstellar grains against the various destructive
processes.

The key to answering this question lies in the evaluation of the
grain destruction rates in supernova shock waves. If these shock waves
are as effective in destroying grain material as some calculations
indicate (Seab and Shull 1983; Draine and Salpeter 1979b; Dwek and Scalo
1980), then the classical sources of new grains are inadequate to
maintain the grain population. In this case, new sources of grain
material must be invoked for most of the mass of grains. If, on the
other hand, grain destruction in shocks is inefficient, then there is
little problem with supplying the observed amount of grain material.

If the grain destruction rates are significantly larger than the
classical formation rates, then the most likely new source for grain
material is accretion in dark clouds. Grains are known to form volatile
icy mantles in such clouds. These mantles have binding energies less
than 1 ev and are easily lost outside of the sheltered environment of
the cloud. The problem here is to form refractory grain material with
binding energies of at least several ev. Greenberg (1982, 1984 and
references therein) offers a possible mechanism for forming organic
refractory material, sometimes called 'yellow stuff' or Greenberg's
glue, in dark clouds. Other mechanisms are possible and may be necessary
in order to understand the pattern of depletions of elements in the ISM
(Jenkins, this volume). Grain material formed in dark clouds is likely
to differ significantly from the high-temperature condensates formed in
cool stars and supernova ejecta.

Because grain destruction fuels the whole question of how much
grain formation is necessary, we will start with an examination of shock
destruction mechanisms. From there, we will look briefly at ideas about
accretion and coagulation in dark clouds. In the last section, we will
comment on observational constraints on grain processing theories, and
present a rather general scenario outlining a grain's evolutionary life
cycle in the ISM.

1.2. Timescales

We can get a idea of the relative importance of various grain processes
in the ISM by looking at timescales. Since supernova play a major role
in establishing the structure of the interstellar medium (McKee and
Ostriker 1977; Cox and Smith 1974; McCray and Snow 1979), we can guess
that they are important in determining the life cycle of interstellar
grains. Therefore we will first look at the timescale for destruction
of grains in a supernova generated shock wave.

Consider a supernova blast wave propagating through the hot phase
of a cloudy ISM. Sedov-Taylor expansion of the remnant relates the
radius R to the blast wave velocity $v_7 \equiv v/10^7$ cm s^{-1} by

$$R = 36.1 \ v_7^{-2/3} \ (E_{51}/n_a)^{1/3} \quad pc \tag{1}$$

where E_{51} is the supernova energy in units of 10^{51} ergs and n_a is the ambient hydrogen density. When the blast wave hits a cloud, the shock velocity drops by approximately the square root of the density contrast, i.e., $v_c \sim v(n_a/n_c)^{1/2}$, where v_c is the cloud shock velocity in a cloud with initial density n_c. The total mass of shocked gas is given by

$$M_s = \mu n_c V_s f_c = 6.7 \times 10^3 \; v_{c,7}^{-2} E_{51} f_c \qquad M_\odot \qquad (2)$$

where V_s is the volume of the supernova remnant, μ is the mass per H atom, and f_c is the filling factor of the clouds of density n_c. This shocked mass is independent of the density of the clouds, but is linearly dependent on the filling factor. Most of the grain destruction will occur in the phases with the largest volumes (with the exception of the hot phase, where the density is too low to effectively destroy grains). In the McKee and Ostriker (1977) three-phase model, destruction in the warm medium (n~0.25 cm^{-3}, T~10^4 K, f~0.25) will dominate. Models with larger filling factors of warm phase gas will have correspondingly larger grain destruction rates.

To translate the shocked mass of equation (2) into a timescale, we multiply it by an efficiency factor ε for destruction of grains in shocks and compare the result to the total mass of the ISM. Seab and Shull (1983) show that 100 km sec^{-1} shocks can destroy about 50% of the silicate grain material. We therefore take $\varepsilon = 0.5$ for the efficiency and a filling factor $f_c = 0.25$ corresponding to the warm phase in the McKee-Ostriker model. The timescale for grain destruction in shocks is

$$t_{shock} = t_{SN} (M_{ISM}/\varepsilon M_s) = 1.7 \times 10^8 \text{ yrs} \qquad (3)$$

for $M_{ISM} = 5 \times 10^9$ M$_\odot$ and a supernova time $t_{SN} = 30$ yrs. More recent work by McKee et al. (1986) and Tielens et al. (1987) suggest that the Seab and Shull results overestimate the destructiveness of the shocks by a factor of several, so that a more accurate grain destruction timescale would be near 5×10^8 yrs.

Grains in the hot coronal phase of the ISM will be subject to thermal sputtering. The 10^6 K hot phase of the McKee and Ostriker (1977) model has a very low density, in the neighborhood of 10^{-3} cm^{-3}. The sputtering rate for refractory grains in this medium is only 10^{-6} Å yr^{-1} (Tielens et al. 1987; Draine and Salpeter 1979a) for refractory grains. Thus a 1000 Å grain will survive for 10^9 yr in this phase. Since the hot phase only contains about 0.1% of the mass of the ISM, then the galactic average lifetime of a grain against thermal sputtering is on the order of 10^{12} yrs.

Thermal sputtering of refractory grains requires a temperatures above a few times 10^5 K, so that these grains survive well in the cooler phases of the ISM. We will see, however, that thermal sputtering behind very fast shocks (v≥200 km s^{-1}) in warm clouds can be important since both the densities and temperatures can be high.

The destruction of icy grains and volatile grains mantles is easily accomplished in the diffuse phases of the ISM (cf. Draine and Salpeter 1979) by thermal sputtering and by destruction in low velocity shocks.

These volatile materials are confined to dark clouds (Harris, Woolf, and
Rieke 1978).

Cloud-cloud collisions also produce shocks in the ISM. However,
the typical cloud dispersion velocities are only on the order of
10 km s^{-1}, which is not enough to produce significant destruction of
refractory grains. Such low velocities phenomena may play a role in
grain growth in the ISM (Meyers et al. 1985).

Whereas there have been arguments against shock destruction of
grains, the star formation process provides an unequivocal end to some
grains by incorporating them into new stars. The star formation rate for
the galaxy is usually taken to be about 3 M_Θ yr^{-1}. Comparing this to a
total ISM mass of 5 × 10^9 M_Θ gives a timescale of about 1.6 × 10^9 yrs
for grain destruction by astration. This is a reasonably hard upper
limit for grain lifetimes.

These destruction timescales are to be compared with the time for
injection of new grains from red giant winds and supernova ejecta. Dwek
and Scalo (1980) show that dust injection from supernovae dominates
other sources, if indeed grains can form and survive in the ejecta.
Assume that a typical supernova ejects 4 M_Θ of heavy elements into the
ISM. If oxygen dominates the ejecta material, then only that portion of
it that can bond with refractory elements will go into non-volatile
grains. In the general ISM, only about a quarter of the available
oxygen is depleted. This is just the amount needed to form silicate
materials such as olivine and enstatite using essentially all of the
available Si, Fe, Mg, and other refractories. The remaining oxygen can
only form volatile ices such as CO and H_2O, which are not our concern
here. We therefore assume that at most 25% of the mass of the supernova
ejecta can be formed into refractory grains. If we further estimate
that the actual grain formation process is optimistically 50% efficient,
then each supernova can produce about 0.5 M_Θ of new grains. A supernova
time of one every 30 yrs will therefore replenish the interstellar grain
population at a rate of 1.67 M_Θ per century, or a timescale of

$$t_{inject} = 5 \times 10^7 \; M_\Theta \; / \; 0.0167 \; M_\Theta \; yr^{-1} = 3 \times 10^9 \; yr, \qquad (4)$$

assuming a gas-to-dust ratio of 100 in the ISM. Red giants winds
contribute more total mass to the ISM, but less dust since they are not
so enriched in heavy elements. If 0.5 M_Θ yr^{-1} is injected (Jura, this
volume) with 1% forming grains, then the timescale for replenishing the
ISM from this source is 10^{10} yrs, or about three times slower than for
supernovae.

Comparison of these timescales shows that shock processing of
grains is at least potentially the dominant destruction mechanism. The
comparison also shows the root of the problem of understanding the life
cycle of interstellar grains: the destruction timescales are shorter
than the injection times for producing new grains. This difference is
hard to reconcile with the observed large depletions of refractory
elements in the ISM without invoking grain formation in dark clouds.
One would expect that the destruction timescales would be much longer
than the formation timescale in order to produce depletions of 90% or
more for the typical refractory elements. The problem would not be so

severe if the shock destruction rates were grossly overestimated. The
next step will be to examine how grains are destroyed in shocks.

2. GRAIN DESTRUCTION IN SHOCKS

2.1. Radiative and Non-Radiative Shocks

Grains are destroyed in shocks by three principal mechanisms: thermal
sputtering, non-thermal sputtering, and grain-grain collisions. The
last two mechanisms depend on the betatron acceleration of grains in the
cooling layers of radiative shocks. In shocks slow enough to be
radiative, these mechanisms strongly dominate thermal sputtering.
Thermal sputtering only turns on at the high temperatures produced by
fast non-radiative shocks. Therefore the problem breaks conveniently
into two parts depending on whether the shock is radiative or non-
radiative.

The division between radiative and non-radiative shocks in this
context depends on a comparison between the cooling time of the post-
shock gas and the expansion time of the supernova remnant. Hollenbach
and McKee (1979) show that the cooling time for a shock is given
approximately by

$$t_{cool} = 2 \times 10^{10} \ v_s^{3.2} / n_0 \tag{5}$$

where v_s is the shock velocity and n_0 is the preshock density. For
Sedov-Taylor expansion in a uniform medium of density n_a, the age of the
supernova remnant is

$$t_{exp} = 4.46 \times 10^{12} \ v_7^{-5/3} \ (E_{51}/n_a)^{1/3}. \tag{6}$$

The expansion time of the supernova remnant must take into account the
cloudy nature of the ISM: a blast wave propagating through a thin
intercloud medium suffers a drop in velocity when it hits a cloud.
Incorporating this factor into equations (5) and (6) shows that the
cloud shock will be radiative if

$$v_7 \leq 2.95 \ (n_c \ n_a^3 \ E_{51}^2)^{1/30}. \tag{7}$$

For an ambient density of 10^{-3} and a cloud density of 1 cm^{-3}, this gives
a critical velocity of about 160 km s^{-1}.

2.2. Grains in Radiative Shocks

Steady-state radiative shocks have been relatively well studied by
various authors including Shull and McKee (1979), Cox and Raymond
(1985), and many others. A typical temperature-density structure for a
shock is shown in Figure 1. The abscissa in the figure is given as the
total hydrogen column density $n_0 v_s t$ behind the shock front in order to
scale approximately for shocks propagating into a range of different
density clouds in the ISM. By the term "shock" in this paper we will

normally mean to include the full post-shock cooling region shown in Figure 1 as well as the (thin) shock front itself.

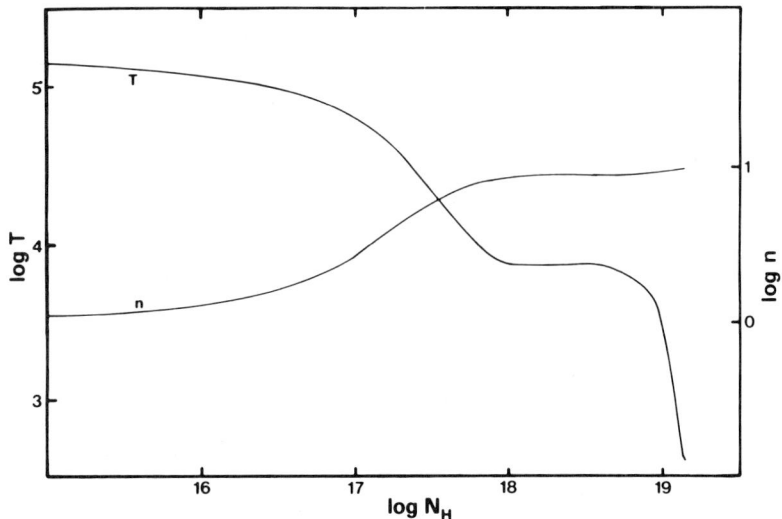

Figure 1. The post-shock temperature and density curves for a typical 100 km s^{-1} shock in a cloud of preshock density $n_0 = 0.25$ cm^{-3} and B = 1μG.

2.2.1. <u>Grain Velocities</u>. In the shock frame, grains enter the front of the shock with a velocity of about $0.75v_s$ for strong shocks. Since the grains are charged, they will gyrate in the magnetic field component parallel to the front. The magnetic field increases along with the density as the gas cools, and the grains attempt to conserve the magnetic moment

$$\mu = \frac{1}{2}mv_{gr}^2/B,$$ (8

leading to a spin-up or betatron acceleration of the grain (Spitzer 1976). Betatron acceleration can also be expressed as

$$\frac{d \ln(v_{gr})}{dt} = \frac{1}{2}\frac{d \ln\chi}{dt,}$$ ($

where χ is the density enhancement over the preshock density. We see that the relative enhancement in the grain velocity is limited to the square root of the density enhancement. Furthermore, betatron acceleration drops to zero in the "recombination plateau" region that occurs at $T \leq 10^4$ K, where the density and temperature curves in Figure 1 level off until hydrogen is fully recombined. This allows the drag forces to bring the grains to a halt relative to the gas. Very large

grains of mm to cm size are not betatron accelerated, since they have
Larmor radii larger than the cooling length of the shock. These grains
will barrel through the cooling region and drag to a halt in the dense
region behind the shock without being very much affected. However, the
typical interstellar grains of radius ≤ 0.25 μm are small enough to be
effectively betatron accelerated and destroyed in radiative shocks.

Betatron acceleration is only half the story. Drag forces on the
grains will halt the gyromotion early in the post-shock layers for small
grains, thereby limiting their destruction. Drag forces come in two
flavors: collisional drag and plasma drag. Collisional drag results
from the moving grain impacting on atoms in the gas, giving a
deceleration

$$\left(\frac{dv_{gr}}{dt}\right)_{coll} = - \frac{3\rho v_{gr}^2}{4a\rho_{gr}} (1+\frac{128kT}{9\pi\mu_H})^{1/2} \tag{10}$$

where ρ is the gas density, ρ_{gr} is the solid density for a grain of
radius a, and the second term in the parentheses is a small correction
for the thermal motion of the gas. The plasma drag results from coulomb
interactions between the charged grain and the ionized gas, and may be
expressed as

$$\left(\frac{dv_{gr}}{dt}\right)_{pl} = - \frac{3nkT}{a\rho_{gr}} \phi^2 \ln(\lambda) \sum_i x_i z_i^2 G(s_i), \tag{11}$$

where $\ln(\lambda)$ is the plasma parameter, $\phi=Z_{gr}e^2/akT$ is the grain charge
parameter that has a classical value of 2.51. The summation is over all
gas phase species with charge z_i and abundance fraction x_i relative to
hydrogen, and

$$G(s_i) \equiv G(m_i v_{gr}^2/2kT) \sim s_i (1.5\pi^{1/2}+2s_i^3)^{-1} \tag{12}$$

is an approximation due to Draine and Salpeter (1979a).

Seab and Shull (1983) show that the velocities and destruction
rates of the grains in a shock peak just before the recombination
plateau. In fact, the relatively steep density gradient at this point
adds an extra boost of acceleration to the grains. It therefore becomes
important to know whether or not a grain of a given size and density
will maintain its velocity through to this part of the shock. We can
get an approximate answer by comparing the cooling time (which is also
the acceleration time) with the stopping time evaluated just behind the
shock front. The result is that the grain stops early in the shock if

$$a_{-5} \leq \frac{(1+1.21g\phi^2)^2}{5.8} \frac{v_{s,7}^{4.2}}{\rho_{gr}}, \tag{13}$$

where g is a normalized version of the function in equation (12), and
$a_{-5}\equiv a/10^{-5}$ cm is the grain radius. The value of $g\phi^2$ varies strongly
with shock velocity and time, with typical values between 6.0 and 0.1.
The smaller values occur in faster shocks. In a 100 km s^{-1} shock,

silicate grains ($\rho=3.2$ g cm^{-3}) smaller than about 0.5μm are stopped.
Figure 2 shows the velocities of three sizes of silicate grains through
the shock of Figure 1. The smallest grains stop almost immediately,
while the largest grains are accelerated to about twice their original
velocity. On the basis of this analysis of grain motion, we therefore
expect small grains to survive radiative shocks relatively unscathed.
However, we shall see that they fare less well in adiabatic fast shocks.

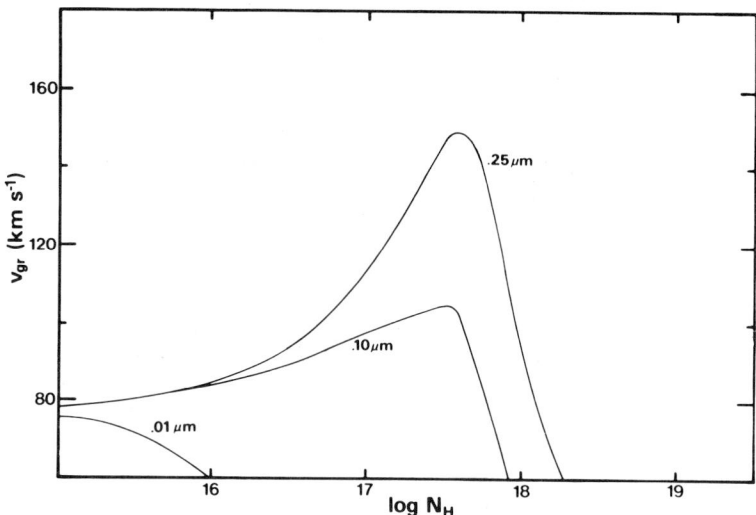

Figure 2. Grain velocities for three sizes of silicate grains
as a function of column density in the shock of Figure 1.

2.2.2. <u>Non-thermal Sputtering</u> The high velocity of the grains through
the gas drives the destructive processes in radiative shocks. In
adiabatic shocks, the temperature of the gas produces thermal sputtering
of the grain. The difference between thermal and non-thermal sputtering
is that in the former, sputtering occurs due to the thermal motion of
particles in the gas, while in the latter, the grain motion causes
impacts with the gas. The sputtering process is the same, but the sum
over the gaseous species is done differently. We will look first at the
sputtering equations, and then consider briefly the results of grain-
grain collisions.

    Sputtering is the ejection of atoms from a solid by the impact of
high-energy ions. Unfortunately for astrophysics, most of the
experimental studies of sputtering have been done for Kev to Mev
energies of heavy ions impacting single-element solids. The latest and
probably best formula for low-energy sputtering yield in ejected atoms
per impact is probably that of Andersen and Bay (1981):

$$Y(E/E_{th}) = 0.017 \ a \ M_T \ \gamma^{5/3}(E/E_{th})^{1/4}(1-E_{th}/E)^{7/2}, \tag{14}$$

where M_T is the atomic mass of the target material, E_{th} is the threshold impact energy for sputtering, $\gamma \equiv 4M_p M_t / (M_p + M_t)^2$ is the energy transfer factor, and M_p is the mass of the gas ion. The threshold energy is found to be approximately 4 times the binding energy of atoms in the solid for most situations not involving extreme values of the target/projectile mass ratio. The formula gives an excellent fit to the data for a considerable variety of energies and target/projectile combinations, provided that the constant in front of the equation and the threshold energy are adjusted according to the data. The problem, inevitably, is that there is very limited data for energies and projectile/target combinations of astrophysical interest.

Betz and Wehner (1983) show that sputtering in compounds and in materials with impurities can be either higher or lower than measured rates in pure materials. Coincidentally, sputtering of SiO and SiO$_2$ compounds seem to be identical to sputtering rates for pure silicon. We can therefore have better confidence is guessing the interstellar silicate grain sputtering rates from the available laboratory data. However, we expect even the best fits to the data to be factor-of-two approximations to the actual sputtering rates for the complex and disordered interstellar grain material.

The best-fit values for silicate and graphite sputtering are given in Table I and in Figure 3, based on collected data in Andersen and Bay (1981) and Betz and Wehner (1983). The last two columns give the

TABLE I

Material	E_{th}(ev)	a	E_a(ev)	λ
Silicate sputtering				
He	40.	1.33	2670	0.32
H	30.	0.57	1160	0.026
CNO	30.	0.75	11460	1.71
Graphite sputtering				
He	40.	0.87	4000	0.22
H	35.	0.36	414	0.036
CNO	63.	0.75	7800	1.17

parameters for a fit to the high energy sputtering formula given by Draine and Salpeter (1979a):

$$Y(E) = 2\lambda s_n (E/E_A) \tag{15}$$

where the "reduced nuclear stopping power" can be approximated by

$$\log[s_n(x)] = -0.466 - 0.230\log(x) - 0.176[\log(x)]^2 \quad \text{for } x \leq 10 \quad (16)$$

$$s_n(x) = (2x)^{-1} \ln(1.294x) \quad \text{for } x > 10 \quad (17)$$

which is accurate to 5% in the range $0.01 < E/E_A < 10.0$. The last formula is from Lindhard et al. (1968). Both these sputtering formulae include a rough correction for angle-of-incidence effects given variously as a factor of 1.5 (Barlow 1978) or 2.0 (Draine and Salpeter 1979a); the latter value is used here. Figure 3 plots the silicate sputtering rate from Draine and Salpeter (1979a) for comparison with the present curve. The threshold energies are found to be somewhat higher than they assumed, while the low energy sputtering rates are lower.

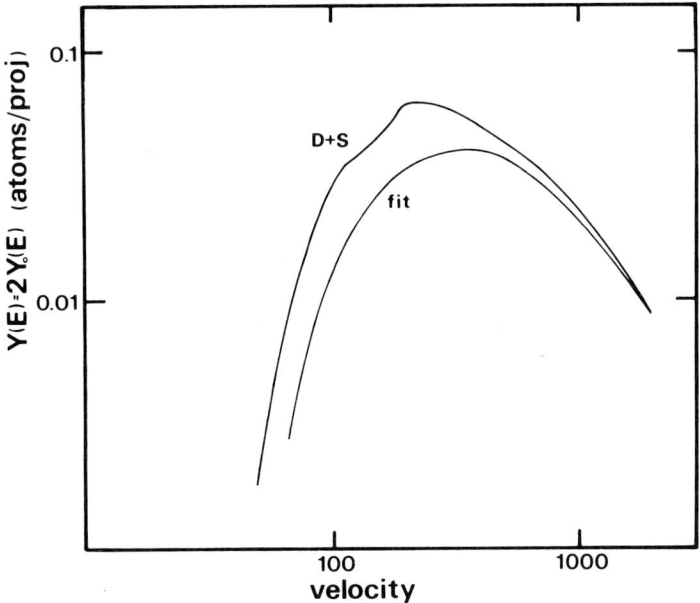

Figure 3. Non-thermal sputtering rates for silicates as a function of grain velocity. Rates from Draine and Salpeter (1979a) are given for comparison. Y_0 refers to sputtering rates at zero angle of incidence.

2.2.3. <u>Grain-grain Collisions</u>. The sputtering process is reasonably well studied, even though the actual numbers chosen are uncertain by about a factor of two. In contrast, the processes involved in grain-grain collisions are much more complicated and less understood. Grains gyrating in the magnetic field with random phase velocities will collide very energetically. A head-on collision between two grains each traveling at 100 km s^{-1} gives an energy of about 1000 ev per atom (with average atomic mass of 20). This compares to a binding energy between 5 and 10 ev for refractory grains. If only 1% of the collision energy goes into breaking chemical bonds, then both the grains would be vaporized. It is therefore very important to try to evaluate the results of grain-grain collisions despite the inherent uncertainties.

Seab and Shull (1983) included grain-grain collisions in their shock models, using a Mathis, Rumpl, and Nordsieck (1977) type of power law grain size distribution (dex -3.5 with sizes from 0.01 to 0.25μm for both silicate and graphite grains). However, they also used a very crude scheme for deciding the fate of colliding grains. They considered a grain totally vaporized if there was sufficient energy in the collision to break every bond in the grain; otherwise the grain suffered no damage at all. The two grains involved in the collision were treated separately, with the lighter grain being preferentially destroyed. It was thought that this scheme would overestimate the number of grains totally destroyed but underestimate the amount of material vaporized from grains only partially destroyed. With this scheme, they found that grain-grain collisions actually dominated the destruction of the larger grain sizes, confirming the potential importance of the process.

Unfortunately, very little real data is available on what happens in high speed collisions between submicron sized grains. Experiments have been done for impacts of mm and cm sized grains on semi-infinite target slabs at velocities of a few tens of km s^{-1}. However, it is very dangerous to try to extrapolate this sort of data over 10 orders of magnitude in mass; even very small uncertainties in the scaling laws will be greatly magnified in such a large extrapolation. Work is currently being pursued by Tielens et al. (1987) to get a more realistic model of what happens in grain-grain collisions, including both partial vaporization and shattering effects. Preliminary results are that the per unit volume thresholds for grain vaporization are several times higher than the minimum requirement used by Seab and Shull (1983), while shattering thresholds are lower.

The inclusion of the shattering process in the grain-grain processing scheme will have a significant impact on the destruction results. Previous results suggested that small grains would survive radiative shocks while large grains are destroyed. With shattering, radiative shocks would actually produce more small grains than originally present, but at the expense of an even greater destruction of the large grains. These small grains will in turn be vulnerable to thermal sputtering in adiabatic shocks.

2.2.4. Model results. The results of radiative shock processing of grains has been modeled by Seab and Shull (1983) and more recently by McKee et al. (1986). Figure 4 presents some of the results of the latter paper. Grain-grain collision results are plotted separately from the sputtering results. Thermal sputtering is included, but is not significant. The old Seab and Shull (1983) grain processing scheme is used in this run; shattering has not yet been included. The grain-grain collisional destruction is larger than the sputtering result for grains larger than about 1000 Å. In total, 37% of the silicate and 4% of the

graphite grain material was vaporized or sputtered away for this case.
The difference in results for the two types of grains is due to the
larger binding energy and lower density of graphite in the Seab and
Shull processing scheme. This scheme will be updated by Tielens et al.
(1987).

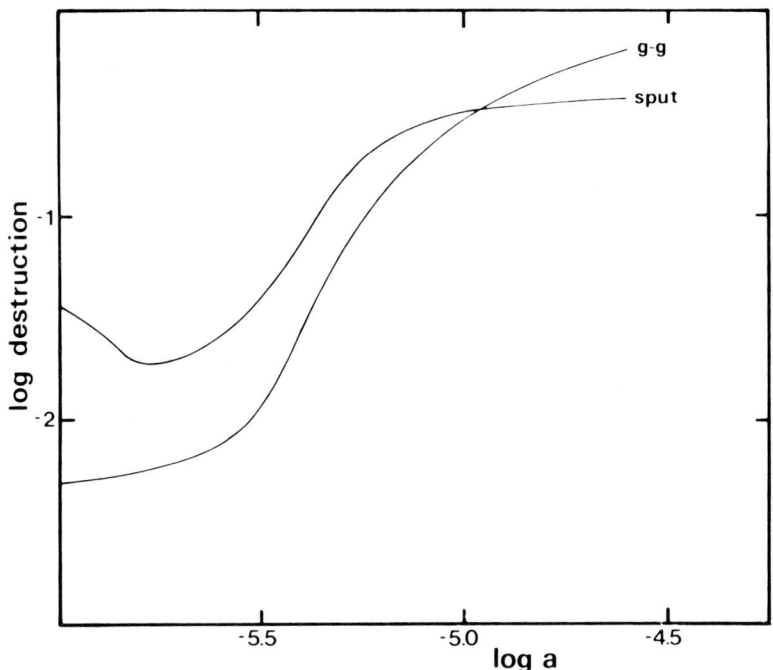

Figure 4. Results of grain processing through the shock of
Figures 1 and 2, plotted as a function of grain size for a
power law grain model. The ordinate gives the fraction of
each size of grain destroyed by either grain-grain collision
or by non-thermal sputtering.

        Some of the features of this plot can be understood in terms of the
grain motion equations given in section 2.2.1. The drag terms in
equations (10) and (11) are inversely proportional to the grain size,
while the betatron acceleration in equation (9) is size-independent. We
therefore expect that the larger grains will suffer generally more
destruction that the small grains, as is evident in the figure. (The
upturn in the sputtering curve at the lowest sizes is due to grain
charge effects.) The largest change in destruction rates occurs around
the break radius of about 500 Å calculated from equation (13). This
shows the difference between grains that are dragged to a stop relative
to the gas early in the shock, and those that maintain their velocity
into the hydrogen recombination zone where grain destruction rates peak.

2.3.   Thermal Sputtering in Non-radiative Shocks

Thermal sputtering rates are obtained by a suitable integration of the sputtering equations (14) - (17) over a maxwellian velocity distribution and summed over H, He, and CNO group projectiles. The results of this process are shown in Figure 5, along with the Draine and Salpeter (1979a) results for comparison. Temperatures in shocks of less than

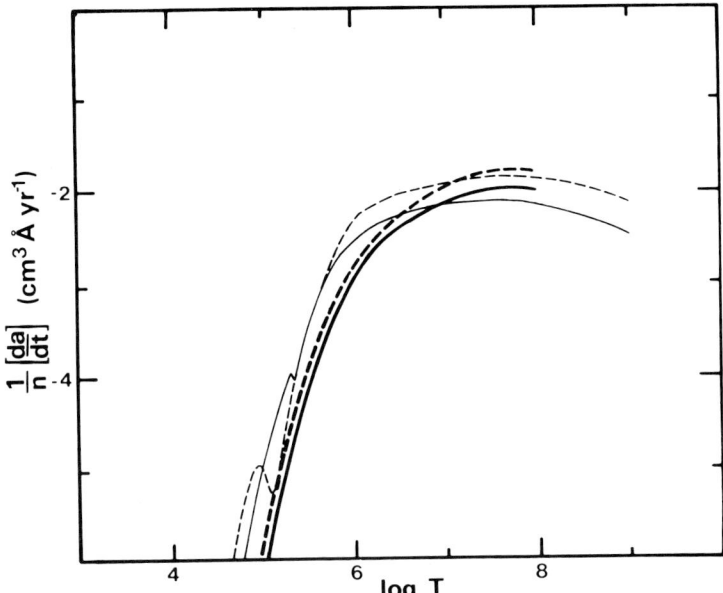

Figure 5. Thermal sputtering rates for silicates (dashed lines) and graphite (solid lines). Thick lines are results from Tielens et al. (1987), while the thin lines are from Draine and Salpeter (1979a) for comparison.

about 200 km s^{-1} are not high enough to produce significant thermal sputtering in comparison to the rates of non-thermal sputtering and grain-grain collisional destruction. However, in faster shocks where the temperature reaches near 10^6 K and post-shock cooling is not sufficient to drive betatron acceleration of the grains, then thermal sputtering becomes the dominant destruction mechanism. Figure 6 shows the destructive results of thermal sputtering as a function of shock velocity in a cloud of density 0.25 cm^{-3}. The ordinate in this figure represents the thickness of grain material removed from each grain. Grains smaller than Δa are totally vaporized.

The preferential destruction of small sized grains is a significant aspect of thermal sputtering in adiabatic shocks. The sputtering rate

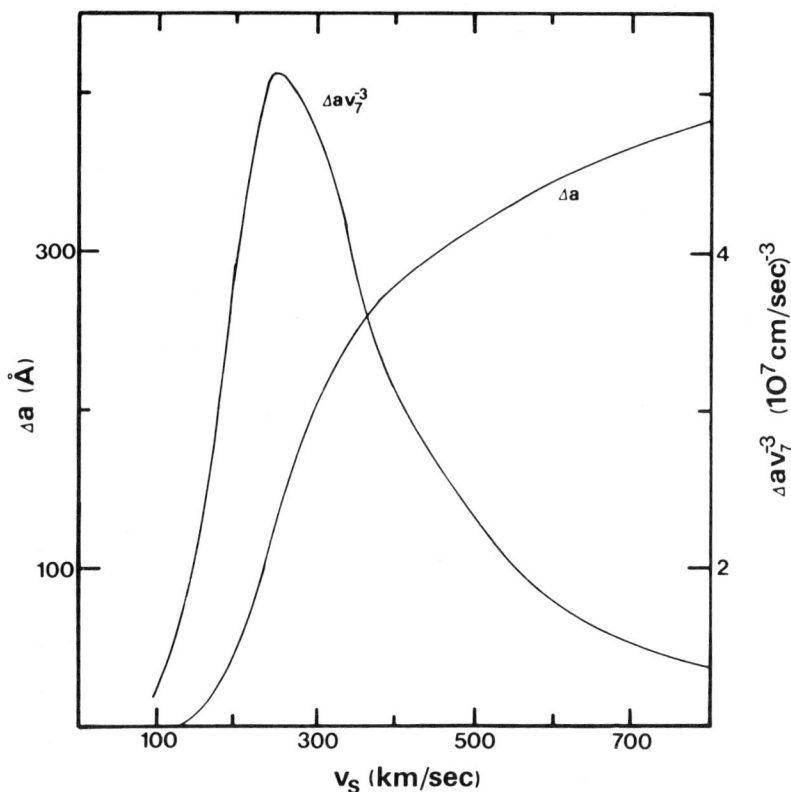

Figure 6. Thermal sputtering rates for adiabatic shocks in $n_0 = 0.25$ cm s^{-1} clouds. Here Δa is the change in radius of the grain, while Δav^{-3} roughly factors in the relative shock frequency.

depends on the exposed surface area of a grain, and therefore the smaller grains are more vulnerable. This is particularly interesting because small grains tend to survive the radiative shocks discussed previously; drag forces are more effective on these grains. Moreover, the shatter products of large grains processed by radiative shocks would likely be small enough to be vaporized by an adiabatic shock. It has sometimes been argued that shattering might preserve the grain material from shock destruction by turning big grains into little ones. We now see that the shatter products will be more vulnerable to destruction in very fast shocks.

The importance of thermal vaporization of small grains depends on the timescale for grains to be hit by very fast shocks. If we return to the formalism of equations (1) - (3), we see that the blast wave radius

$R \propto v^{-2/3}$, so the volume of the remnant goes as v^{-2}. Therefore a 300 km s^{-1} shock occurs about a ninth as often as a 100 km s^{-1} shock. But from Figure 6, this shock destroys 100% of grains smaller than 200 Å, so the efficiency for these grains is 1.0 instead of the estimated 0.5 used in equation (3). Scaling from this equation gives

$$t_{a<200Å} \sim (1.7\times10^8) \times 9/2 \sim 7.7\times10^8 \text{ yrs} \qquad (18)$$

which is still less than the injection time for grains of all sizes.

The discrepancy between the timescales for grain destruction and injection remains a problem. For the grains affected by thermal sputtering, the problem is less severe than for large grains, since shattering can be invoked to resupply the ISM with small grains. Incorporating shattering into the grain-grain collision schemes will help with the small grains, but it will shorten the timescale for large grain destruction. Coagulation can then be invoked to recreate large grains from small ones. But coagulation does not do anything about the material vaporized or sputtered from the grains. Is seems necessary in view of the long timescales for the injection of new grains that this vaporized material be reaccreted onto existing grain cores by some process in the ISM itself.

3. GRAIN GROWTH IN THE ISM

If the grain destruction rates presented in the preceding section are anywhere near correct, then it seems necessary that grain growth rates must be far higher than estimated from the injection of new grains from red giants and supernova ejecta. The most natural explanation is that grains grow in the ISM via accretion and coagulation in dense clouds.

3.1. Accretion

3.1.1. Accretion timescales. Assuming that refractory grain material accretes onto existing grains, we ask what the timescale for this process is in the various phases of the ISM. The accretion time for any element can be written

$$t_{accr} = \frac{1.}{\alpha n_H \Sigma v} \sim 2.19\times10^9 \, (\frac{\mu}{m_H T})^{1/2} \, \frac{1}{\alpha n_H \Sigma_{21}} \text{ yrs}, \qquad (19)$$

where α is the sticking efficiency of the atom, μ is its atomic weight, and $\Sigma_{21} \equiv \Sigma/10^{-21}$ cm^{-2} H^{-1} is the total cross section of grains per H atom. Accretion times for oxygen as representative of heavy elements are given in Table II below, based on the three phase ISM model of McKee and Ostriker (1977). The last column gives approximate shock times for grains being hit by 100 km s^{-1} or faster shocks in that particular phase of the ISM [the times in equations (3) and (18) are galactic averages]. Shocks are assumed not to penetrate significantly into dark clouds, while shocks in the hot phase don't do enough damage to the grains to be important.

TABLE II

Phase	T	n	t_{accr}	t_{shock}
Hot	10^6	10^{-3}	8.7(9) yr	--
Warm	10^4	0.25	3.5(8)	6.9(6) yr
Dif Cld	80.	20.	4.9(7)	2.6(9)
Dark Cld	10.	10^3	2.8(6)	--

Apparently grain accretion in dark clouds is easily fast enough to explain the observed depletions of refractory elements. In fact, accretion is so fast that the opposite problem occurs: how to explain the observed presence of any heavy elements at all in the gas phase of dark clouds (Greenberg 1982, 1984). The controlling timescale for accretion becomes the mixing time for the accreted grains to reenter the more diffuse phases of the ISM. This time is not well known, but is generally taken to be on the order of a galactic rotation time, or about 10^8 yrs. The mixing time must be significantly shorter than the grain destruction time (about 5×10^8 yrs) to make grain growth in dark clouds effective in replenishing the interstellar grains.

There is a problem with accretion as a mechanism for growing grains. As Spitzer (1978) points out, each grain will accrete atoms in proportion to its surface area. But if its volume grows according to the surface area, then

$$\delta V = 4\pi a^2 \delta a \propto \pi a^2 / \Sigma \tag{20}$$

and the a^2 terms cancel out so that the change in radius δa is independent of the grain radius. Since the small grains have larger surface area, then most of the accreted mass will be added to these grains rather than the larger grains. It is very difficult, then, to produce large grains through purely accretion processes.

The radius increase of the grains due to accretion depends on the amount of accretable material and on the size distribution of the grains. With a power law grain size distribution as in Mathis, Rumpl, and Nordsieck (1977), Draine (1984) calculates that the accretion of icy mantles will add 175 Å to the radius of each grain. A comparable amount of refractory material could be added if the reaccretion of refractory material occurs immediately behind a grain destroying shock. This much accretion onto small grains would seriously skew the grain size distribution, giving an unreasonable UV extinction curve. It is more likely that accretion of refractory materials will occur in small increments, since most of the refractory elements are already locked up in existing grains. This type of accretion will produce smaller changes in the ultraviolet extinction curve, and can be useful in explaining some of the observed variations in extinction (Joseph et al. 1986).

However, it is evidently very difficult to grow large grains through the accretion process since all grains add the same increment in radius. If large grains are destroyed at all in shocks, then coagulation is needed to recreate them from the smaller grains that accrete new refractory material.

3.1.2. <u>Accretion mechanisms</u>. If grain destruction is to be balanced by the reaccretion of elements in the ISM, then it is necessary that refractory grain material be formed in the accretion process. It is known that ice mantles form in dark clouds because the $3\mu m$ ice feature is observed at cloud extinctions as low as $A_v = 3$ mag (Goebel 1984; Harris, Woolf, and Rieke 1978; see also review by Savage and Mathis 1979). However, these ice mantles are not seen in the diffuse ISM. They are expected to come off from photodesorption, thermal sputtering, or non-thermal sputtering in frequent low-velocity shocks once the grains move out of the dark clouds. Either the accretion of refractory materials occurs in diffuse clouds where ice mantling is not a problem, or some mechanism operates to convert ice mantles into more refractory substances.

A number of authors have proposed mechanisms for making refractory material from volatile mantles, including Greenberg (1982), Sagan and Khare (1979), Salpeter (1977), and others. The mechanism described by Greenberg (1982) consists of photolysis of icy mantles by low intensity UV photons in dark clouds to form free radicals. At low grain temperatures, these radicals are isolated by the ice matrix. An event such as a very low velocity collision between grains heats the mantle enough to allow the radicals to migrate through the matrix to chemically combine with other radicals. The chemical reactions can produce enough additional heat to spark a chain reaction that explosively removes the ice mantle, leaving behind a small amount of residue variously called "yellow stuff", "glue", or "organic refractory material". This process has been observed in the laboratory, but its efficiency in an astrophysical context is unknown. Further processes relevant to accretion in dark clouds are reviewed by Draine (1984).

In contrast to this dark cloud process, Barlow (1978) proposes a chemical mechanism for selectively depleting elements in diffuse clouds. Burke and Hollenbach (1983) show that the sticking coefficient of heavy elements onto grain surfaces is very nearly unity. Barlow examines the heats of formation of various metallic hydrides to determine if the metal atom will remain adsorbed on the surface or will evaporate off. If the chemical energy liberated by adding a single H atom to the metal or to a lower stage metal hydride is larger than the binding energy of the adsorbed atom, then it is assumed to be ejected from the surface in analogy to H_2 formation on grain surfaces. If the chemical energy is not sufficient, then the metal atom remains on the surface and may be incorporated into the refractory grain material. This scheme explains why elements such as Na and K are very little depleted in the ISM, in contrast to other elements such as Fe and Ti that are heavily depleted. A problem with the scheme is that it is not clear that the timescale for accretion in diffuse clouds is short enough to replenish the grain material vaporized by shocks. Further, the adsorbed hydride still must

be converted to refractory material by breaking and reforming chemical
bonds.

The two mechanisms briefly discussed here are illustrative of the
kinds of accretion processes that might produce refractory grain
material in the ISM. Neither one is entirely satisfactory, but together
they indicate that accretion in the ISM is at least conceptually
possible.

3.2. Coagulation

The natural counterpart of grain shattering in interstellar shocks is
grain coagulation in dark clouds. Apart from theoretical requirements,
there is observational evidence for coagulation of grains. Jura (1980)
notes that the large λ_{max} polarization wavelength and high
$R \equiv A_V/(A_B-A_V)$ ratio in the ρ Oph cloud, combined with the fact that
the dust-to-gas ratio is lower, strongly implies that the average grain
size is larger due to coagulation. We also note that comet dust, as
represented by Brownlee particles collected at the top of the
atmosphere, frequently show an aggregate structure of micron and sub-
micron sized grains coagulated into a heterogeneous mass (Fraundorf,
Brownlee, and Walker 1982).

Coagulation of grains in dark clouds requires a turbulent velocity
field, since thermal or brownian motions of the grains is much too low
velocity to bring them together frequently enough (Draine 1984).
However, the turbulent velocities cannot be too high, because the grain
might shatter if the collision velocity is over about 0.1 km s^{-1}. While
the grains will probably adhere if they collide at velocities below
about 0.001 km s^{-1} = 1 cm s^{-1}, the results of collisions in the large
range between this velocity and 0.1 km s^{-1} are very uncertain (Draine
1984). Since the turbulent velocity fields are not known either, it is
difficult to calculate the rate of grain coagulation with any
confidence. Furthermore, it is necessary that the grains adhere
strongly enough to survive in the diffuse ISM. If the coagulation
occurs in dark clouds, then the ice mantles will be bonded. The grains
will likely come apart again as the ice mantles erode away in the
diffuse medium, unless chemical or photoprocessing manages to convert
the icy material into something more tightly bound.

Despite the problems, it still seems necessary to invoke grain
coagulation in order to balance the processes of shattering and
vaporization of large grains in interstellar shocks. As with accretion,
there is some evidence that the process does take place in at least some
clouds in the ISM.

3.3. Nucleation

The formation of brand new dust grains out of free atoms, ions, and
molecules via nucleation requires a very special environment. The
density needs to be on the order of 10^9 cm^{-3} or higher, while the
temperature is less than the typical condensation temperatures around
1000 - 2000 K. These conditions exist primarily in outflows from red
giant stars, and possibly in condensations or knots in supernova ejecta.

General reviews of grain nucleation theory are given by Salpeter (1977) and by Draine (1979). Since nucleation only occurs in these special environments that are only nominally part of the ISM, we will not go further into the subject here.

4. LIFE CYCLE OF GRAINS

4.1. Observations

The depletion of refractory elements (Si in particular) in the ISM is observed to be on the order of 90% or higher. One would therefore expect that the timescales for grain destruction should be significantly longer than the grain formation times. The fact that theoretical results on grain destruction give relatively short destruction times poses a fundamental problem in the understanding of interstellar grains. This problem has been recognized previously by such authors as Draine and Salpeter (1979b) and Dwek and Scalo (1980), and is being pursued at present by McKee et al. (1986) and coworkers. Clearly, either the grain destruction rates are far too high, or the grain growth rates are far too low. At this point it is difficult to see how the destruction rates can be pushed to be an order of magnitude longer than the injection rates.

Direct observation of grain destruction in shocks is very difficult. The best evidence at present is correlation of depletion of certain elements with cloud velocity. For example, Shull, York, and Hobbs (1977) show that Si and Fe depletions are significantly less in high velocity clouds, with gas phase abundances approaching cosmic for cloud velocities of 100 km s^{-1}. Routly and Spitzer (1952) and Siluk and Silk (1974) show that high velocity clouds also tend to have higher Ca/Na ratios. These results have been interpreted (Jura 1976; Spitzer 1976; Shull 1977; Barlow and Silk 1977; Cowie 1978) as evidence for shock processing of grains. The cloud velocity in this interpretation is a lower limit to the shock velocity because of projection and deceleration effects.

On the other hand, Gondhalekar (1985b) argues that relatively high depletions (90% depletion of Si, e.g.) can be observed at cloud velocities up to 200 km s^{-1}. It is not clear how serious this objection is, since improved models of grain processing in shocks demonstrate that in some cases, destruction rates in individual shocks can be much reduced (McKee et al. 1986).

Galactic destruction rates for grains are further influenced by the choice of model of the ISM. Very preliminary calculations done by Seab et al. (1985) based on the McKee and Ostriker (1977) ISM model and a Mathis, Rumpl, and Nordsieck (1977) grain model (cf. Seab and Shull 1983) produced grain lifetimes on the order of a few times 10^8 yrs for all sizes of grains. In this model, most of the grain destruction occurred in the warm intercloud medium characterized by a density of 0.25 cm^{-3} and containing 5% of the total ISM mass. This may be a low estimate of the mass fraction of this phase. Increasing the mass fraction in the warm medium would have the effect of increasing the galactic grain destruction rates, shortening grain lifetimes and creating more of a discrepancy with grain injection times. On the other hand, putting more mass into dark clouds, where the grains are protected from shocks, would increase the grain lifetimes and reduce the need for grain growth in the ISM (Gondhalekar 1985a).

The choice of grain model can have a major influence on the grain-grain destruction results. The results in Figure 4 were based on the Mathis, Rumpl, and Nordsieck (1977) power law type silicate grain plus graphite grain model, chosen because it matches the observed interstellar extinction curve from infrared to UV wavelengths. The Hong and Greenberg (1980) bimodal grain model would give less destruction from grain-grain collisions because it lacks the intermediate-sized grains that are most effective in destroying grain material in the Seab and Shull (1983) scheme. Grain sputtering results, however, would not be substantially changed. Witt (1985) has resently presented data suggesting that neither of these models agrees with scattering phase function data from reflection nebula. Eventually it may be possible to construct a consistent grain model based on grain evolution considerations.

4.2. Life Cycle of Grains

The results of grain processing calculations suggests a preliminary galactic model of the life cycle and evolution of interstellar grains.

1. Refractory grain cores nucleate in cool stars and supernova ejecta, then mix into the general ISM.
2. Diffuse media (n $\leq$ 20 cm^{-3}) shocks erode, vaporize, and shatter grains.
3. Dark clouds reaccrete grain materials into a refractory form probably distinct from the high temperature condensates from step 1; small grains are coagulated into larger ones.
4. Ultimately, star formation eats everything.
5. Go back to step 1.

This is a very general statement of a grain history scenario, and is similar to that presented by Greenberg (1984). The differences between these grain scenarios are determined by the assumed or calculated values for the galaxy averaged grain destruction rates. If the grain destruction rates are high, then most of the material making up grains will have been grown in dark clouds in the ISM. If the averaged vaporization rates are low, then most of the refractory grain material will be in the form of high temperature condensates from red giants and possibly supernova ejecta. Shattering and coagulation might still be allowed to alter the size distribution of the grains without radically affecting their substance. At present it seems likely that the grain destruction rates are high, so that the first possibility applies. Further evaluation of grain destruction models is in progress to attempt to clarify what really happens to grains in shocks and to grains in the galaxy (McKee et at. 1986; Tielens et al. 1987, and succeeding papers in the series).

ACKNOWLEDGEMENTS

I am indebted to D. J. Hollenbach, C. F. McKee, and A. G. G. M.
Tielens, my coauthors on a forthcoming series of papers on grain
evolution, for many hours of discussion, insight, and hard work. In
addition, I wish to thank T. P. Snow, J. M. Shull, and C. L. Joseph
for many helpful conversations about interstellar dust grains.

REFERENCES

Andersen, H. H., and Bay, H. L. 1981, in *Sputtering by Particle
    Bombardment*, **I**, ed. R. Behrisch (Berlin: Springer-Verlag), p. 145.
Barlow, M. J. 1978, *M.N.R.A.S.*, **183**, 417.
Barlow, M., and Silk, J. 1977, *Ap. J. (Letters)*, **211**, L83.
Betz, G., and Wehner, G. K. 1983, in *Sputtering by Particle
    Bombardment*, **II**, ed. R. Behrisch (Berlin: Springer-Verlag), p.11.
Burke, J. R., and Hollenbach, D. J. 1983, *Ap. J.*, **265**, 223.
Cowie, E. E. 1978, *Ap. J.*, **225**, 887.
Cox, D. P., and Raymond, J. C. 1985, *Ap. J.*, **298**, 651.
Cox, D. P., and Smith, B. W. 1974, *Ap. J. (Letters)*, **189**, L105.
Draine, B. T. 1979, *Ap. Sp. Sci.*, **65**, 313.
_____. 1984, in *Protostars and Planets*, **II**, eds. D. C. Black and
    M. S. Matthews (Tucson: University of Chicago Press).
Draine, B. T., and Salpeter, E. E. 1979a, *Ap. J.*, **231**, 77.
_____. 1979b, *Ap. J.*, **231**, 438.
Dwek, E., and Scalo, J. M. 1980, *Ap. J.*, **239**, 193.
Fraundorf, P., Brownlee, D. E., and Walker, R. M. 1982, in *Comets*, ed.
    L. L. Wilkening (Tucson: University of Arizona Press), p. 383.
Goebel, J. H. 1983, *Ap. J. (Letters)*, **268**, L41.
Gondhalekar, P. M. 1985a, *Ap. J.*, **293**, 230.
_____. 1985b, *M.N.R.A.S.*, **216**, 57P.
Greenberg, J. M. 1982, in *Submillimeter Wave Astronomy*, ed.
    J. E. Beckman and J. P. Phillips (Cambridge: Cambridge University
    Press), p. 261.
_____. 1984, in *Laboratory and Observational Infrared Spectra of
    Interstellar Dust*, ed. R. D. Wolstencroft and J. M. Greenberg
    (Edinburgh: Royal Observatory), p. 1.
Harris, D. H., Woolf, N. J., and Rieke, G. H. 1978, *Ap. Sp. Sci.*, **65**,
    69.
Hollenbach, D., and McKee, C. F. 1979, *Ap. J. Suppl.*, **41**, 555.
Hong, S. S., and Greenberg, J. M. 1980, *Astr. Ap.*, **88**, 194.
Joseph, C. L., Snow, T. P., Seab, C. G., and Crutcher, R. M. 1986,
    Ap. J., submitted.
Jura, M. 1976, *Ap. J.*, **206**, 691.
Jura, M. 1980, *Ap. J.*, **253**, 63.
Lindhard, J., Nielsen, V., and Scharff, M. 1968, *Kgl. Danske
    Vid. Selsk. Mat.-Fys. Medd.*, **36**, No. 10.

Mathis, J. S., Rumpl, W., and Nordsieck, K. H.   1977, *Ap. J.*, **217**, 425.
McCray, R., and Snow, T. P.   1979, *Ann. Rev. Astr. Ap.*, **17**, 213.
McKee, C. F., and Ostriker, J. P.   1977, *Ap. J.*, **218**, 148.
McKee, C. F., Seab, C. G., Tielens, A. G. G. M., and Hollenbach, D. J.
      1986, preprint.
Meyers, K. A., Snow, T. P., Federman, S. R., and Breger, M.   1985,
      Ap. J., **288**, 148.
Routly, P. M., and Spitzer, L. 1952, *Ap. J.*, **115**, 227.
Sagan, C., and Khare, B. N. 1979, *Nature*, **227**, 102.
Salpeter, E. E. 1977, *Ann. Rev. Astr. Ap.*, **15**, 267.
Savage, B. D., and Mathis, J. S.   1979, *Ann. Rev. Astr. Ap.*, **17**, 73.
Seab, C. G., and Shull, J. M. 1983, *Ap. J.*, **275**, 652.
Seab, C. G., Hollenbach, D. J., McKee, C. F., and Tielens, A. G. G. M.
      1985, abstract in *Interrelationships Among Circumstellar,
      Interstellar, and Interplanetary Dust* (NASA Conference Publication
      No. 2403), p. A-28.
Siluk, R. S., and Silk, J.   1974, *Ap. J.*, **192**, 51.
Shull, J. M. 1977, *Ap. J.*, **215**, 805.
Shull, J. M., and McKee, C. F. 1979, *Ap. J.*, **227**, 131.
Shull, J. M., York, D. G., and Hobbs, L. M.   1977, *Ap. J. (Letters)*,
      211, L139.
Spitzer, L.   1976, *Comments Ap.*, **6**, 157.
_____. 1978, *Physical Processes in the Interstellar Medium* (New York:
      Wiley).
Tielens, A. G. G. M., McKee, C. F., Seab, C. G., and Hollenbach, D. J.
      1987, in preparation.
Witt, A. N.   1985, *B.A.A.S.*, **17**, 571.

Edwine Van Dishoeck

INTERSTELLAR GRAINS IN THE SOLAR SYSTEM

D. E. Brownlee
University of Washington
Dept. of Astronomy
Seattle, Washington 98195
U.S.A.

ABSTRACT. It is likely that presolar interstellar grains are preserved inside comets and asteroids. If these particles can be collected, identified and properly analyzed in the laboratory, they could provide a vital link between the study of the early solar system and the interstellar materials that preceded it. Abundant isotopic evidence from meteorites indicates that interstellar components have survived in asteroids although it is not evident that unaltered individual grains survived. Continued work on meteorites, interplanetary dust and samples directly retrieved from comets will likely lead to new insight into interstellar and circumstellar grain properties and processes and provide constraints on grain models.

1. INTRODUCTION

The solar system formed by the gravitational collapse of an interstellar gas cloud that contained more than 10^{46} interstellar dust grains. During the isothermal collapse phase that preceded the formation of the solar nebula disk, probably all condensable compounds and nearly all elements heavier than oxygen were originally in the solid grains. Some of these particles probably survived the formation of the solar system without being melted or vaporized and are preserved today inside minor bodies in the solar system. Analysis of meteorites has shown that the isotopic signature of chemically separable presolar components has been preserved, although to date no single individual grain has been shown to be preserved presolar material.

If presolar grains can be found and analyzed in the laboratory, their properties could provide important insights into the nature of contemporary interstellar and circumstellar grains that can only be studied by spectral features. The detailed study of preserved presolar grains could in a sense provide some level of "ground truth" for interstellar grain models. For example, in a set of presolar grains it might be possible to identify the actual phases that produce spectral effects like the 9.7μm silicate feature and 2175A hump. Using the powerful techniques of modern analytical electron microscopy it is possible to determine structure and elemental composition on 100A grains. If the UV feature is due to

D. J. Hollenbach and H. A. Thronson, Jr. (eds.), Interstellar Processes, 513–530.
© *1987 by D. Reidel Publishing Company.*

graphite it would be possible to directly image the individual grains and their lattice structure. An example of the synergism that can occur between astronomical results and laboratory sample studies is the recent work by Sandford (1986) on IR features measured on collected samples of interplanetary dust. Some of the particles he examined contained a 6.8μm feature that is comparable to the similar spectral feature seen in W33A (figure 1). By electron microscopy and chemical etching, Sandford proved that the 6.8μm feature in the dust is caused by carbonates. Electron microscopy showed that carbonate is a constituent of the particles. After acid dissolution, the 6.8μm feature feature and carbonates were absent but other spectral features and components remained unchanged. While this endeavor does not by any means prove that the 6.8μm feature in any astronomical source is due to carbonates it does provide new insight derived from direct analysis of samples that are probably close interstellar grain analogs.

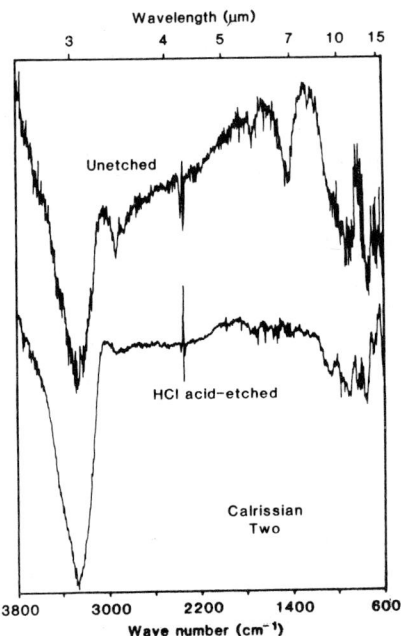

Figure 1. The IR absorption spectrum of a hydrated interplanetary dust particle collected in the stratosphere. The 6.8μm feature in the particle is absent after removal of carbonates by acid etching. The 3.1μm and 10μm features are respectively due to water and SiO stretch mode vibrations.(Figure from Sandford 1986).

2. SURVIVAL OF PRE-SOLAR GRAINS

For presolar grains to exist in the present solar system they must have survived the initial accretion into the solar nebula, processing in the nebula and "planetary" processing in the host body in which they were stored for most of the age of the solar system. Particles that accreted into minor bodies in the outer regions of the nebula are the most likely to have survived all three of these potentially destructive phases. In the outer solar nebula the original free-fall velocity onto

the planetary disk was only a few kms^{-1}. In the inner regions the accretion velocities onto the disk must have been tens of kms^{-1} over most of the formation
time of the planetary system and the resulting drag heating may have melted
or vaporized presolar solids (Wood, 1984). Evidence from meteorites suggests
that the thermal environment alone was sufficient to vaporize refractory grains
within three AU of the nebula center (Boynton 1985). Beyond a critical radial
distance that was probably beyond the asteroid belt, the midplane temperatures
were probably never high enough to destroy refractory components of presolar
grains. Volatile grain components may have been altered throughout the planetary region. In addition to steady state heating, other nebular processes such as
chemical reaction with nebular gas, collisions, sputtering and transient thermal
events may also have altered grains. Even if grains survive up to the point of accretion of planetesimals, they must still survive residence inside a carrier body.
Grains inside even minor solar system bodies can be destroyed by heating, shock
effects or chemical effects such as aqueous alteration. Individual grains cannot
survive for aeons outside of a larger parentbody because their orbits decay by
Poynting-Roberson drag.

   All of the presolar grains in the solar nebula were destroyed except those
that were incorporated into asteroids, comets and perhaps some of the small
satellites. Of these bodies the comets appear to be the best carriers for preserving grains. The asteroids were formed in the transition zone between the terrestrial and Jovian planets and it is likely that most of their components were totally reprocessed by nebular and parentbody processes. Evidence from meteorites
indicates that only trace components of the meteorite-producing asteroids could
contain preserved presolar material. Less is known about comets because confirmed cometary samples have never been examined in the laboratory. However,
the volatile contents and presumed history of comets suggest that these bodies
are the most favorable carriers of presolar materials. It is generally believed that
comets accreted near Uranus and Neptune and have never been heated internally. At the outer fringes of the planetary nebula it is likely that the intrinsic
nebular temperature was not sufficient to vaporize presolar grains. Even if grains
were not altered by accretion onto the nebular disk or by the local thermal environment, it is still possible that they were altered by collisional processes near
Uranus and Neptune. In order to be ejected beyond the orbits of the planets, the
comets must have been perturbed by a planetary mass sufficient to also gravitationally stir up relative planetesimal velocities to high values. The inevitable
collisions should have shock processed some proportion of cometary materials to
levels commonly seen in meteorites. The presence of volatiles in comets suggests
that shock processing was not so severe as to destroy all primordial materials.
If comets formed beyond the planetary region of the nebula they could have assembled gently and never been exposed to post-accretion high velocity collisions
(Weissman 1986). Comets that formed beyond Neptune could be pure assemblages of well preserved presolar grains and their volatile coatings, while other
comets may have been extensively altered.

3. SAMPLES

The extraterrestrial samples currently available for searches for presolar materials are meteorites and interplanetary dust. In the future it is hoped that samples will be directly retrieved from a comet and that some cometary and contemporary extrasolar dust grains will be recovered with Earth satellites.

Meteorites appear to be samples of asteroids. A few meteorites have been shown to be lunar in origin and some appear to be fragments from Mars but no evidence indicates that any of the meteorites are cometary. While a cometary origin cannot be ruled out, the thermal history of meteorites inferred by observed thermal metamorphism effects, seems incompatible with long term storage in a comet. Typical meteorites are reasonably well preserved samples of materials that accreted to form small planetesimals between Mars and Jupiter.

Interplanetary dust particles are collected in the stratosphere above the level where contamination by large terrestrial particles is a serious problem. The collected particles range in size from 2μm to over 50μm although the typical particle that is examined in detail is only 10μm in size. All particles are heated during atmospheric entry but the heat pulse lasts only for a few seconds and does not typically exceed 600C for particles smaller than 10μm. The density of cosmic ray tracks in the particles indicates that they were exposed in space as small particles for 10,000 years. Prior to this exposure they must have resided in either comets or asteroids, the only parental materials than can replenish the solar system dust cloud and balance the loss effects of collisions and light pressure effects. It is generally believed that comets are the major source of interplanetary dust but dust from the asteroid belt may also provide a significant contribution (Zook and McKay 1986).

The critical importance of dust and the reason why it is intensively studied in the laboratory, is that some of the dust particles must be cometary. The major fraction of dust with earth-crossing orbits is cometary and the dust that survives atmospheric entry should be a relatively unbiased sample of the interplanetary population. Small particles decelerate at altitudes near 100km and never experience the high dynamic ram pressures that produce fragmentation in the larger conventional meteorites. Ten micron particles that are not heated to their melting point, are not exposed to ram pressures as high as the pressure that is observed to fragment meteors in cometary meteor streams. While it is possible that no fragment of cometary material has ever survived atmospheric entry in a piece large enough to be found on the ground as a conventional meteorite, there is no known mechanism that could prevent cometary matter from surviving as dust.

In the future it is likely that there will be primitive samples directly returned from space. The European Space Agency (ESA) has chosen the earth return of a pristine comet sample as one of the four cornerstone missions of its Horizon 2000 program of future space experiments. The mission will rendezvous with a short period comet and recover several kilograms of cometary dust and ice. This type of mission is the only possible means of recovering substantial quantities of bona-fide cometary material. If the collected sample contained abundant and well preserved presolar materials, it would provide a unique and probably revolutionary means of studying the nature, origin and evolutionary history of interstellar grains.

Smaller scale experiments have also been proposed for collection of cometary samples. They can provide valuable information, but they cannot collect pristine material or substantial masses of sample. A flyby mission through a comet coma with a free earth-return trajectory can collect thousands of cometary particles in the 1-100μm size range. The particles would be altered due to the high velocity of collection but their analysis can provide direct measurement of elemental and isotopic composition. It is also possible to collect a few cometary particles with devices in earth orbit. With accurate electronic measurement of the impact velocity and direction it is possible to determine the orbital parameters of the collected particle and in some cases identify its origin. As is the case with the comet coma flyby, the samples are collected as residue in craters, as debris in porous media or as material coating the interior walls of a "capture cell", an enclosed volume that a particle enters by penetration of a thin membrane.

The same techniques used for collection of cometary grains from spacecraft in earth orbit could also in principle be used to collect contemporary interstellar grains passing through the solar system. The signature of an interstellar grain would be its hyperbolic orbit about the sun. The Pioneer 8 and 9 spacecraft launched into solar orbits in the 1967 and 1968 measured orbital parameters for several dust particles that had incoming hyperbolic paths (Berg and Grun 1973). At a bare minumum, analysis of residue in craters formed by such objects could provide direct information on elemental composition of micron size interstellar grains.

4. METEORITES

4.1 Alteration in the early solar system

Most stony meteorites are primitive materials whose elemental compositions are close to that of the Sun for elements that are not highly volatile. Anders and Ebihara (1982) have made a strong case that the relative elemental abundances in CI (type 1) carbonaceous chondrites represent those of the primordial solar system to a typical accuracy of better than 10% for most elements. Although many chondrites are primitive materials they are not simply collections of presolar grains. The detailed elemental, mineralogical and isotopic study of meteorites has shown that they are composed almost entirely of materials that were extensively processed by nebular and parent body processes.

A major fraction of the mass of average chondrites consists of chondrules, recrystallized silicate spheroids that were at one time molten droplets suspended in nebular gas. In some regions of the nebula, at least 80% of the solid grains were melted by mysterious and yet unknown processes to form these millimeter-sized droplets. In regions where chondrules formed it is likely that most presolar grains were destroyed.

The bulk elemental composition of the various chondrite groups suggests that nebular condensation played a major role in forming solid grains in the asteroidal region of the nebula. The implication is that vaporization of preexisting solids was widespread in the terrestrial planet region of the nebula. The chondrite groups have bulk elemental fractionation effects that are probably related to secondary condensation in the nebula. The major effects are for volatile elements, refractory elements and the siderophile "iron loving" elements. None of these effects would be expected for rocks that formed by simple accretion of presolar grains. All meteorite groups are depleted in volatile elements relative to CI chondrites, the meteorite class that most closely represents solar relative abundances. In the ordinary chondrites, highly volatile elements such as Tl, Bi and In show correlated depletions of over two orders of magnitude below the CI norm. The depletion of volatiles is believed to be the result of incomplete condensation from nebular gas (Larimer and Anders, 1967).

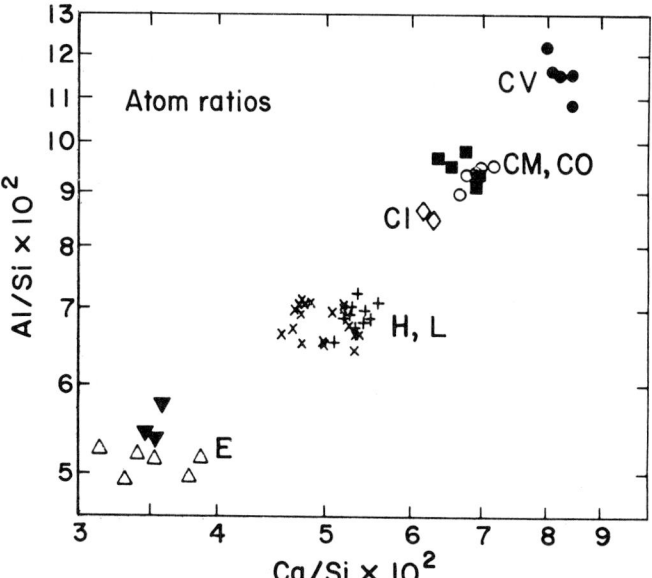

Figure 2. Bulk Ca and Al abundances for the various chondrite classes. The correlated depletions and excesses of these two elements relative to CI chondrites are evidence for large scale fractionation of refractory elements in the solar nebula. Fractionation due to condensation or evaporation of these elements occurs near 1400K under nebular conditions.(Figure adapted from Wasson 1985).

The chondrite groups also have significant correlated fractionation effects for elements that condense from nebular gas at temperatures above 1400K. The abundances of high temperature elements such as Ca, Al. Ti.Sc when normalized to Si, range by a factor of two among the chondrite groups. As can be seen in figure 2 the meteorites cluster into well defined groups indicative of large scale chemical fractionation of solids in the nebula. Depletion of these elements can occur in a region where an early condensate, rich in these elements, was removed by accretion into bodies or by transport out of the region. The overabundance of the refractory elements in the CV meteorites is an indication that these objects acquired more than their share of the early condensate. The CV meteorites contain a high abundance of white inclusions that are rich in the high temperature elements.

The final chemical fractionation pattern in chondrites that provides clues for large scale nebular effects is for Fe and the siderophile elements. Both the Fe/Si ratio and the oxidation state of Fe vary significantly among meteorite classes (figure 3). Both Fe and Si condense at similar temperatures, under equilibrium conditions, and the Fe/Si fractionation may be the result of the differential accretion efficiencies of metallic and silicate grains. Anders (1971) has estimated the temperature of this accretion to be near the curie point of FeNi alloy, on the basis of depletion of volatile siderophiles. The oxidation state of Fe is surely due to gas-solid reactions. If equilibrium exists between solids and gas, all Fe should exist as oxide for temperatures of 400K or less, while Fe at its condensation temperature is only stable as metal. In the carbonaceous chondrites most of the iron is oxidized, while in the E chondrites, essentially all of the iron is reduced. A similar but less extreme range is seen in the oxidation state of terrestrial planets that is implied by their bulk densities.

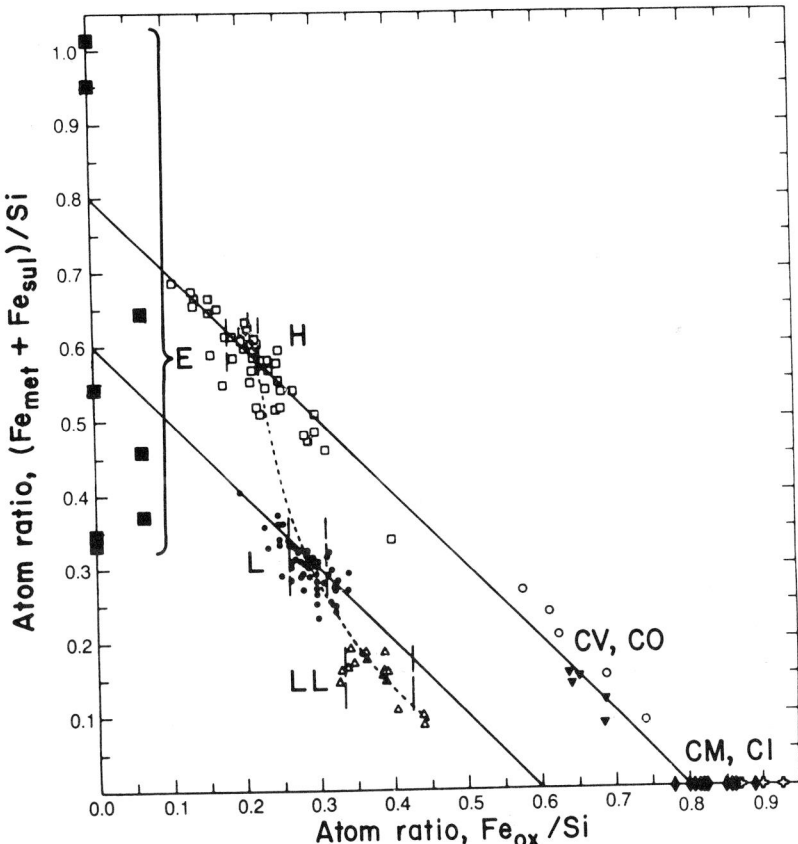

Figure 3. The range of oxidized and reduced iron in marked chondrite groups. The solid lines are lines of constant total Fe/Si. (Figure from Wasson 1985).

An intriguing clue to the complexity of condensation and possible evaporation effects in the nebula is seen in the rare earth abundances of Ca,Al rich inclusions (CAI) in the Allende meteorite. The rare earth elements (REE) are chemically very similar and strong fractionation does not occur by igneous processes. The condensation temperatures however differ for these elements and uniquely large deviations from solar REE patterns are seen in some of the CAI's. Order of magnitude depletion of Eu and Yb, the two most volatile REE, are interpreted as incomplete condensation at high temperature. Some patterns are seen (figure 4) that show depletions in both the volatile REE and the most refractory REE. Boynton (1985) has suggested that this type of REE pattern is due to loss of an early ultra-refractory condensate and incomplete condensation of the low temperature REE.

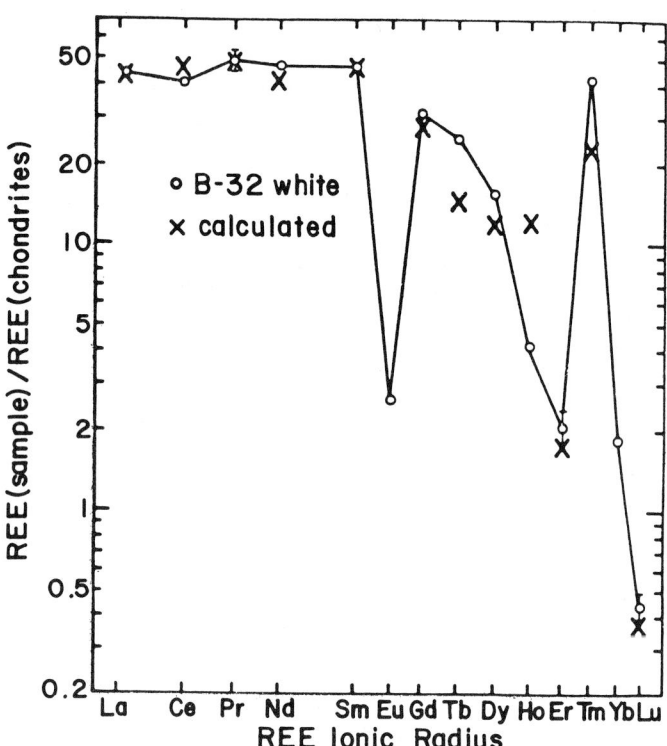

Figure 4. Normalized rare earth element abundances in a Ca,Al inclusion from the Allende meteorite. The large depletions of the most volatile REE, Eu and Yb, and the most refractory REE, Lu and Er, indicate a complex history of high temperature condensation. (Figure from Boynton 1985).

Once grains accreted into planetesimal bodies, they were subjected to collisional shocks and internal heating from a variety of sources. The parent bodies of all known meteorite classes experienced heating that resulted in some level of alteration. Diffusion, chemical equilibration of mineral grains and even recrystallization are common metamorphic effects seen in chondrites. Mild heating even occurred in the parentbody of the CI carbonaceous chondrites where warm temperatures allowed liquid water to alter silicates and form actual veins of water soluble phases such as sulfates. In extreme cases, heating in the parent body led to melting as is the case for the basaltic achondrite meteorites.

The evidence is fairly compelling that the asteroids and presumably other inner solar system bodies were formed predominantly by assembly of second generation solids initially created by condensation in the nebula. The condensation and accretion processes were apparently complex and produced fractionation patterns that would not be expected for simple assemblages of interstellar grains.

4.2 Evidence for pre-solar components

While the bulk of material in meteorites appears to have formed in the solar system there is strong evidence for minor components that retain presolar information. Meteorites contain a variety of isotopic effects that are tracers of pre-solar events. Most of these effects are thought to be the result of nuclear processes at the site of element formation although one effect (hydrogen)is apparently due to chemical fractionation.

Isotopic effects in meteorites indicate the former presence of the following extinct radioactive isotopes.

TABLE I

Extinct parent	Half-life	Product
^{244}Pu	83 my	fission spectrum of Xe isotopes
^{129}I	17my	^{129}Xe
^{107}Pd	6.5 my	^{107}Ag
^{26}Al	0.7my	^{26}Mg
^{22}Na	2.6y	^{22}Ne

Except for ^{22}Na the majority of effects seen for these isotopes are from decay of atoms that were still in existence when the solar system formed and their presence did not require survival of presolar solids. The ^{22}Na decay product and possibly some of the ^{26}Al effects are due to fossil effects carried in presolar grains. Initially it was thought that the ^{26}Al effect in meteorites might have been related to a supernova just prior to the formation of the nebula, in which case ^{26}Al would most likely still have been alive when the nebula formed. The necessity of a local origin of the ^{26}Al was eased by the findings of the HEAOS and Solar Maximum Mission spacecraft (Share *et al.* 1985) that indicate that ^{26}Al is apparently widespread and potentially uniformly distributed along the Galactic plane. Since ^{26}Al is a common component of the ISM it is likely that some meteoritic materials could carry fossil ^{26}Mg effects frozen into presolar grains as suggested by Clayton(1982, 1984). Well documented analyses of meteoritic materials have shown however that probably most of the excess ^{26}Mg effects that have been studied in meteorites are in fact due to in-situ decay of live ^{26}Al in the solar system. An illustration of this is the excellent correlation of excess ^{26}Mg with the Al/Mg ratio in minerals from the inclusion EGG-3 from the Allende meteorite(figure 5). EGG-3 is a 2cm spheroidal inclusion with igneous textures of large intergrown crystals. It is a solar system product and certainly not a presolar interstellar grain. The implied ^{26}Al/^{27}Al ratio of 5X10^{-5} for EGG-3 is fairly typical for inclusions that show a ^{26}Mg anomaly, although smaller values are seen.

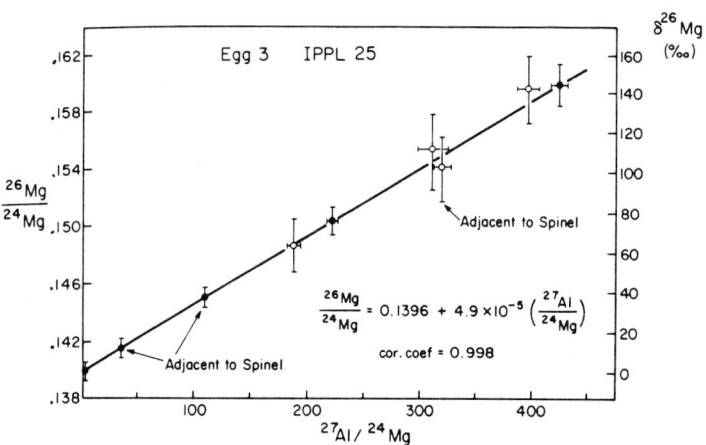

Figure 5. Excess ^{26}Mg correlated with the Al/Mg ratio of minerals in the coarse grained Allende inclusion Egg 3. The well correlated excess in this large clearly igneous inclusion is strong evidence for the decay of live ^{26}Al in the solar system. (Figure from Wasserburg 1985).

^{22}Na is the inferred parent for a nearly pure ^{22}Ne component found in some carbonaceous chondrites. It was isolated in its purest form by step-wise pyrolysis of a silicate-free residue of the Orgueil carbonaceous chondrite(figure 6). At temperatures above 600C. nearly pure ^{22}Ne was released from the low density residue sample. The enrichment of ^{22}Ne is a three order of magnitude enhancement from the isotopic composition of neon normally found in meteorites and terrestrial samples. This exotic component is called "neon E" and it appears to be gas that was trapped in grains following the decay of its parent ^{22}Na. There is no evidence that the ^{22}Na production occurred in the solar system and the neon E is almost certainly a presolar component. The trapping of neon E could have occurred by condensation of nova ejecta within a few days of an outburst that produced ^{22}Na.

Neon E is contained in a high temperature, carbon rich material that is a trace component of meteorites. This generic class of material also contains isotopically anomalous Xe, Kr, nitrogen, carbon and hydrogen. It is possible that some of the effects in these other elements are truly correlated although it has not been demonstrated that they really occur in exactly the same carrier phase.

The carbon rich residue contains a Xe component that matches the calculated production pattern for the s process and is quite unlike the solar isotopic pattern (figure 7). Probably correlated with the s process Xe is a heavy carbon component that has a $^{12}C/^{13}C$ ratio of 42 (Swart *et al.* 1983). This heavy carbon is similar to many estimates for the interstellar medium but is half of the 88 to 93 range for terrestrial carbon. The heavy carbon fraction is only 5ppm of the mass of the Allende and Murchison meteorites where it was found. The heavy carbon is found in acid resistant residue that is combusted at temperatures in the 600-1200C range. Even heavier carbon with ratios of 20 has been found by ion microprobe analysis of micron-size grains of spinel and hibonite (Niederer *et al.* 1985). These are aluminum rich oxide crystals found in acid resistant residue. Hydrogen in the acid resistant residues from some chondrites has been found to have extremely large enhancements in deuterium (Yang and Epstein 1983; Kerridge 1983). The observed enhancements of 35 over the probable nebula composition cannot plausibly occur by isotopic exchange in the nebula (Lewis and Anders 1983). The extreme isotopic fractionation seen in hydrogen is most likely a remnant of presolar events. Ion molecule reactions in cold molecular clouds can produce large fractionation of deuterium and are commonly interpreted as the source of high D/H ratios seen in objects such as Orion A.

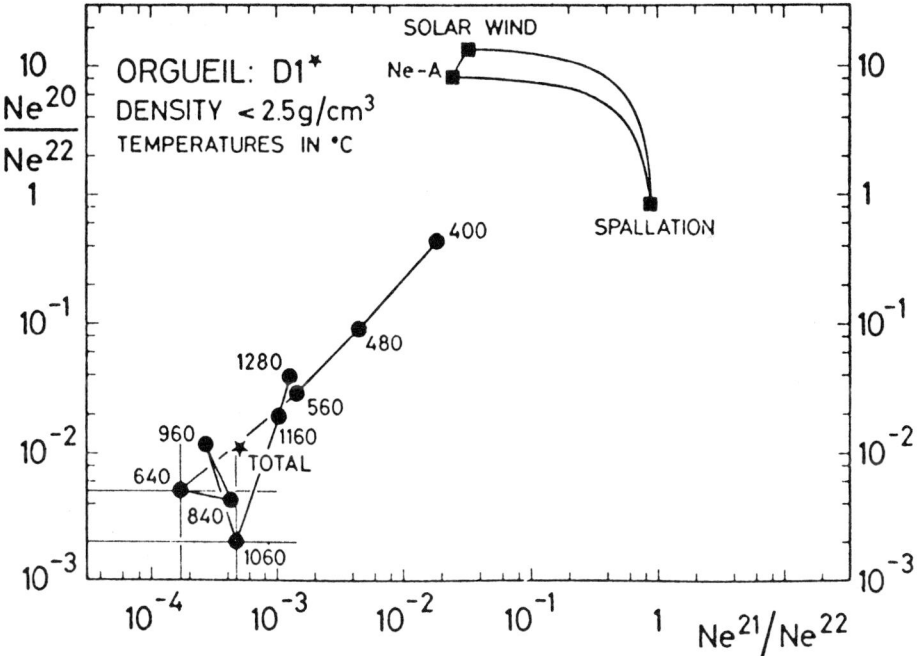

Figure 6. The isotopic composition of neon E released by stepwise heating of a low density silicate-free fraction of the Orgueil CI chondrite. Normal meteoritic values lie in the area bounded by solar wind, neon A and cosmic ray spallation. The neon released above 600C is nearly pure ^{22}Ne. (Figure from Eberhardt *et al.* 1981).

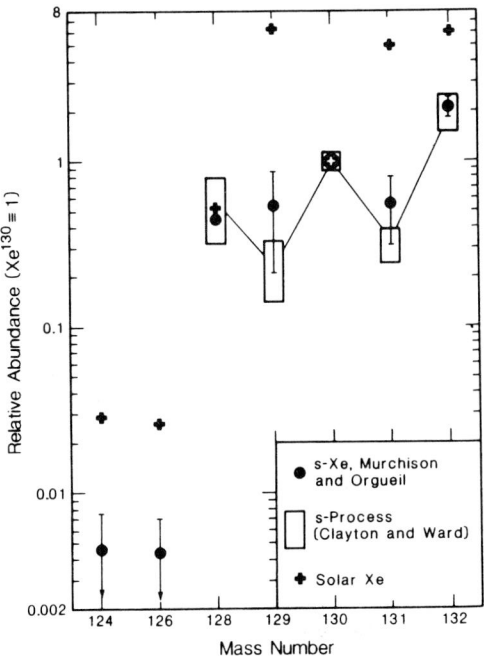

Figure 7. The relative abundance of xenon isotopes released from acid-resistant residue of the Murchison and Orgueil carbonaceous chondrites. The abundance pattern closely matches the calculated values for s-process nucleosynthesis in red giants. The isotopes 124 and 126 are shielded from s-process production. (Figure from Alaerts *et al.* 1982).

Isotopic effects attributable to presolar materials are also found in some of the Ca, Al rich inclusions from Allende and some other carbonaceous and even ordinary chondrites. The most exotic effects are found in the very rare "FUN" inclusions which show large mass dependent isotopic fractionation (F) and additional anomalies that are often due to unknown nuclear (UN) effects. The FUN inclusions contain presolar material that did not totally mix and homogenize with other average solar system matter when the solar system formed. It is not at all clear however that any presolar grains were actually preserved in any of the CAI. They probably contain presolar matter that was diluted with materials that were molten or were condensed in the early solar system at relatively high temperature. One of the major goals of work on isotopic anomalies in meteorites is to find correlated effects that can be attributed to a particular site of nucleosynthesis. Recent advance in this area has been made by Papanastassiou (1986) who found correlated excesses in Cr, Ca and Ti in Allende inclusions. These are all neutron rich isotopes and their excesses are consistent with predictions for the products of neutron rich equilibrium burning that occurs in the core of a massive evolved star prior to detonation of a supernova. Although the individual grains may not have survived inside the meteorite the preservation of the isotopic effect requires that the supernova products were carried to the solar nebula and became incorporated into the CAI as discrete grains. Enhancements in r process isotopes seen in the FUN inclusions are apparently also the result of incorporation of supernova condensates (Wasserburg *et al.* 1979).

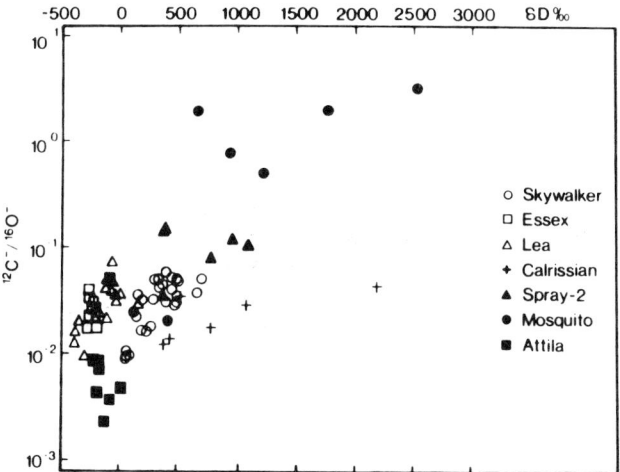

Figure 8. Correlation of excess deuterium with the C/O ratio for individual micronsized fragments of different interplanetary dust particles. The delta D values refer to the D/H ratio relative to a terrestrial standard in parts per thousand. (Figure from McKeegan *et al.* 1985).

One of the most remarkable isotopic effects is for oxygen, the most abundant element in chondrites and an element that is divided between grains and gas in the ISM. Meteorites show linear fractionation in oxygen due to exchange reactions but in addition they have a non-linear, apparently nuclear effect, that is evidence for widespread heterogeneity in the nebula. All meteorites have bulk excesses or depletions of ^{16}O relative to the terrestrial mass fractionation relationship for the three oxygen isotopes. Single crystals show effects consistent with addition of up to five percent of pure ^{16}O relative to terrestrial oxygen. This effect is generally believed to be the result of mixing of presolar components with different nucleosynthetic histories although there is some possibility that the observed effects could be due to non-nuclear effects in the solar nebula (Clayton *et al.* 1985). The presolar components could be the IS gas and at least two grain components with different isotopic compositions.

4. INTERPLANETARY DUST

Interplanetary dust particles (IDP's) that survive atmospheric entry without melting are of particular interest in the search for presolar grains because it is likely that a substantial fraction of the collected particles are cometary. All of the collected particles are small and cannot be studied with all of the sophisticated techniques used on the much larger conventional meteorites. In particular most of the isotopic analyses done on meteorites are not currently possible for the typical nanogram interplanetary particles. As an example, it is not feasible to identify some trace components such as neon E, both because of the limited number of atoms of this species in a single particle and because the small sample mass would not allow the chemical separation and stepwise heating used to isolate the pure component. Most of the isotopic work on individual dust samples will have to be done on fairly abundant elements such as C, H, Mg, Ca, and Si. Using the ion microprobe it is possible to do some of these analyses on grains of only micron size. If any of these particles really do contain preserved grains and not just presolar components diluted with ordinary solar system material then it is possible that some isotopic effects might be very large at the individual grain level.

The principle isotopic effect that has been observed in interplanetary dust is bulk enrichment in deuterium. The ion probe work of (McKeegan *et al.* 1985) indicates that a large fraction of the particles examined have deuterium enrichments. The largest effect seen was a 260% increase above the terrestrial D/H standard used for normalization. The effect was variable on the micron scale and so the highest D/H ratio in the sample could be much higher. As can be seen in figure 8 the magnitude of the effect correlates with the C/O ratio and so it appears to be carried in a carbon rich phase. While many of the dust particles have net D/H effects, only one carbonaceous chondrite and two ordinary chondrites have been shown to have bulk effects. Most meteorites may contain minor heavy hydrogen components that can be chemically isolated, but their bulk compositions are essentially normal.

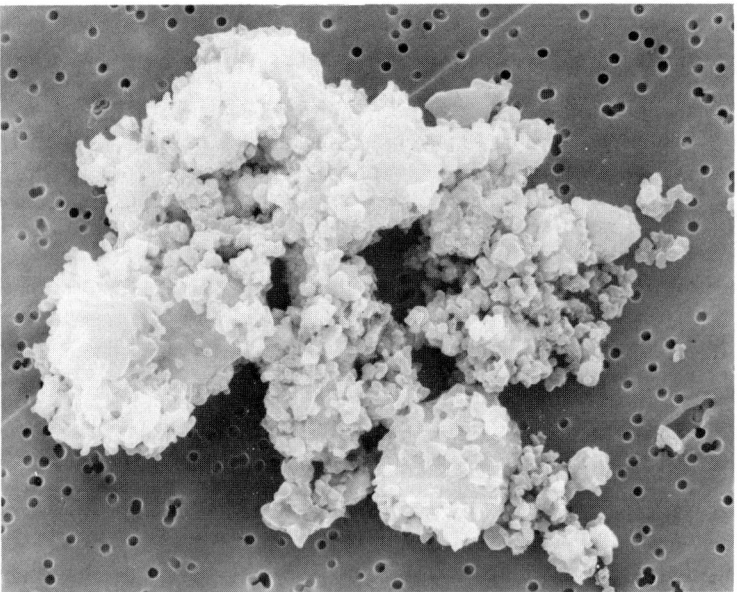

Figure 9. Scanning electron micrograph of a 18 μm long anhydrous particle of interplanetary dust. The stratospheric particle is composed of hundreds of thousands of mineral grains and it has a bulk chondritic elemental composition. Optically the particle is black.

A substantial effort has gone into the elemental, mineralogical and morphological study of interplanetary dust. Most of the collected IDP's are chondritic in the sense that they contain essentially undifferentiated elemental compositions similar to the carbon rich carbonaceous chondrites. Some particles are found that do not have this cosmic abundance pattern but most of these are single mineral grains or clusters of a few minerals such as FeS, olivine and pyroxene. In most cases the non-chondritic particles appear to be materials that were previously imbedded in a matrix of fine grained material identical to the chondritic composition particles. The source bodies for most interplanetary dust particles are composed of fine-grained materials that have chondritic abundances for typical micron-size clumps. They are composed of mineral grains and amorphous material that are commonly of sub-micron size.

Because all minor bodies produce interplanetary dust there are many types of particles. Most of the chondritic composition particles can however be grouped into two major divisions. One group contains hydrated silicates and the other is composed entirely of anhydrous minerals. The hydrated particles tend to be compact and are in many ways mineralogically similar to the CI/CM carbonaceous chondrites which are also largely composed of hydrated silicates. Although the materials are similar, they are not identical and in some cases they have significant mineralogical differences (Buseck and Tomeoka 1986). Like the chondrites, the hydrated particles might be samples of asteroids.

The anhydrous particles have chondritic compositions but they have a morphology and mineral composition that is unlike any known chondrite group. Among other properties, they contain FeNi carbide (Christoffersen and Buseck 1983; Bradley et al. 1984) and unique whisker forms of $MgSiO_3$ (Bradley et al. 1983) that have been shown to be vapor phase products. The anhydrous particles are the most porous meteoritic materials (figure 9) and their fragile structure is consistent with models commonly envisioned for cometary meteors. Although the link has not been proven, it is likely that the porous, anhydrous particles are cometary. The pore spaces in the particles may have originally been filled with ice or other volatiles. The anhydrous particles are open aggregates of grains that range in size from 100A to several microns. The typical component grain is a few 1000A across. Some of the components are single mineral grains, some are amorphous carbon and some of the small "grains" are themselves micro-aggregates of sub-femto gram mineral grains imbedded in carbonaceous material. These last components are called "tar balls" and they dominate some particles (Figure 10). They perhaps are the only materials in interplanetary dust that resemble the core-mantle IS grain models. The tar balls no not have single cores with a single mantle but they are rather like tapioca pudding with many cores permeating a carbonaceous matrix.

The interplanetary dust particles are complex materials that have many interesting properties. If it is ever possible to provide direct links between any of the collected particles and IS grains, their analysis will provide valuable constraints on interstellar processes. Possible links to IS material might be made by IR measurements such those made by Sandford et al. (1985), by isotopic effects, by irradiation effects or by unique chemical or mineralogical properties. One possible link would be to find condensates such as SiC or MgS that can form around

carbon stars but could not easily form gas with a solar C/O ratio. If common
anhydrous IDP's were largely composed of presolar grains, it could be concluded
that high D/H was common, that well crystallized grains up to micron-size were
common and that both Fe metal and true graphite were exceedingly rare. Also
rare would be SiC, core-mantle grains and hydrated silicates.

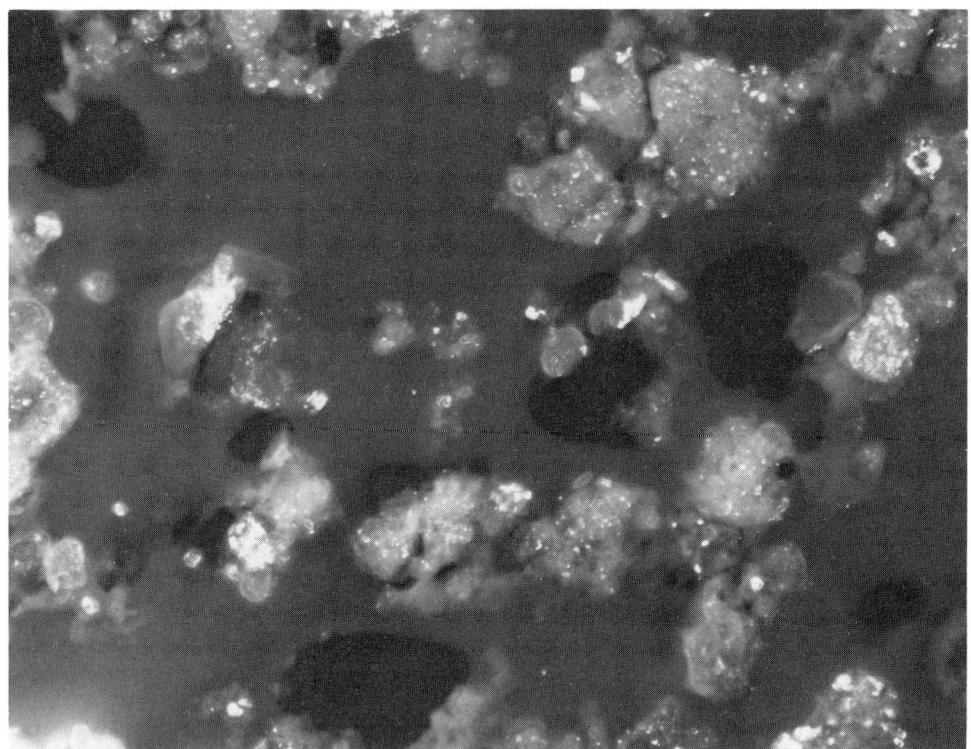

Figure 10. A transmission electron microscope image of a 900A thick microtome
section of a IDP similar to that in figure 9. The uniform dark grey areas are
epoxy that fills the original void spaces of this highly porous particle. Some of
the individual component grains are minerals such as silicates and FeS, some are
structureless carbon and most are the tar balls, mixtures of mineral grains and
low z material. The width of the image is 2.2   um.

5.DISCUSSION

The interdisciplinary field of research on primitive solar materials and comparison
with extrasolar grains is an expanding field with considerable promise. Even if
the search for unaltered presolar grains fails, further study of the many presolar
isotopic records in meteorites will provide a rich base of data for linking interstellar and circumstellar processes with samples that can be studied in the laboratory with ever more sophisticated and powerful tchniques.

    At the present time one of the most interesting and fundamental results
from the sample work is the demonstration that chemically distinct phases (min-

erals) formed around stars where they incorporated isotopic effects particular to their source, and they then survived transportation through the interstellar medium to the nebular disk that surrounded the forming Sun. It is not yet possible to quantitatively determine what fraction of the heavy elements in the solar system were in such grains, but the fact that these grains form and do survive in the ISM should be factored into models of grain formation and destruction.

REFERENCES

Anders, E. 1971, Meteorites and the early solar system. *Ann. Rev. Astron. Astrophys.*, **9**:1-34.

Anders, E., and Ebihara, M. 1982, Solar-system abundances of the elements. *Geochim. Cosmochim. Acta* **46**:2363-2380.

Alaerts, L., Lewis, R. S., Masuda, J. I. and Anders, E. 1980, Isotopic anomalies of noble gases in meteorites and their origins. *Geochim. Cosmochim. Acta* **44**:189-209.

Berg, O. E., and Grun, E. 1973, Evidence of hyberbolic cosmic dust particles, in *Space Research* **XIII**:1047.

Boynton, W. V. 1985, Meteoritic evidence concerning conditions in the solar nebula, in *Protostars and planets II* eds. D. C. Black and M. S. Mathews (Tucson: Univ. of Arizona Press),pp 772-787.

Bradley, J. P., Brownlee, D. E., and Veblen, D. R. 1983, Pyroxene whiskers and platelets in interplanetary dust: evidence for vapor phase growth.*Nature* **301**:473-477.

Bradley, J. P., Brownlee, D. E., and Fraundorf, P. 1984, Carbon compounds in interplanetary dust: evidence for formation by heterogeneous catalysis. *Science* **223**:56-58.

Buseck, P. R., and Tomeoka, K. 1986, A carbonate-rich hydrated interplanetary dust particle: possible residue from protostellar clouds. *Science* **231**:1544-1546.

Christoffersen, R., and Buseck, P. R. 1983, Epsilon carbide: A low temperature component of interplanetary dust particles. *Science* **222**:1327-1329.

Clayton, D. D. 1982, Cosmic chemical memory: a new astronomy. *Quart. Jour. Roy. Astron. Soc.* **23**:174:212.

Clayton, D. D. 1984, [26]Al in the interstellar medium. *Astrophys. J.*, bf 280:144-149.

Clayton, R. N., Mayeda, T. K., and Molini-Velsko, C. A. 1985, Isotopic variations in solar system material: evaporation and condensation of silicates, in *Protostars and Planets II* eds. D. C. Black and M. S. Mathews (Tucson: Univ. of Arizona Press) pp755-771.

Eberhardt, P., Jungck, M. H., Meier, F. O., and Niederer, F. R. 1981, A neon-E rich phase in Orgueil: Results obtained on density seperates. *Geochim. Cosmochim. Acta* **45**:1515-1528.

Kerridge, J. F. 1983, Isotopic composition of carbonaceous chondrite-kerogen: evidence for an interstellar origin of organic matter in meteorites. *Earth Planet. Sci. Lett.* **64**:186-200.

Larimer, J.W., and Anders, E. 1967, Chemical fractionations in meteorites-I. condensation of the elements. *Geochim. Cosmochim. Acta* **31**:1215-1238.

McKeegan, K. D., Walker, R. M., and Zinner, E. 1985, Ion microprobe isotopic measurements of individual interplanetary dust particles. *Geochim. Cos-*

        mochim. Acta **49**:1971-1987.

Niederer, F. R., Eberhardt, P., Geiss, J., and Lewis, R. L. 1985, Carbon isotope
        abundances in Murchison residue 2C10c. *Meteoritics* **20**:716-718.

Papanastassiou, D.A. 1986, Chromium isotopic anomalies in the Allende mete-
        orite. *Astrophys. J. Lett.* (in press).

Sandford, S. A. 1986, Acid dissolution experiments: carbonates and the 6.8um
        bands in interplanetary dust particles. *Science* **231**:1450-1541.

Sandford, S. A., and Walker, R. M. 1985, Laboratory infrared transmission spec-
        tra of individual interplanetary dust particles from 2.5-25um. *Astrophys. J.*
        291:838-851.

Share, G. H., Kinzer, R. L., Kurfess, J. D., Forrest, D. J., Chupp, E. L., and
        Rieger, E. 1985, Detection of galactic ^{26}Al gamma radiation by the SMM
        spectrometer. *Astrophys. J.* **292**:L 61-L 65.

Srinivasan, B. and Anders, E. 1978, Noble gases in the Murchison meteorite: pos-
        sible relics of the s-process nucleosynthesis. *Science* **201**:51-56.

Swart, P. K., Grady, M. M., Pillinger, C. T., Lewis, R. S. and Anders, E. 1983,
        Interstellar carbon in meteorites. *Science* **220**:406-410.

Wasserburg, G. J., Papanastassiou, D. A., and Lee, T. 1980, Isotopic hetero-
        geneities in the solar system. in *Early Solar System Processes and the Present*
        Solar System(Bologna:Soc.Italiana di Fisica), pp. 144-191.

Wasserburg, G. J. 1985, Short-lived nuclei in the early solar system, in *Protostars*
        and Planets II eds. D. C. Black and M. S. Mathews (Tucson: Univ. of Ari-
        zona Press), pp 703-737.

Wasson, J. T. 1985, *Meteorites* (New York:Springer-Verlag).

Wasson, J. T. 1978, Maximum temperatures during the formation of the solar
        nebula. in *Protostars and Planets* ed. T. Gehrels (Tucson: Univ. of Arizona
        Press), pp 488-501.

Weissman, P. R. 1986, Are cometary nuclei primordial rubble piles? *Nature* **320**:242-
        244.

Wood, J. A. 1984, On the formation of meteoritic chondrules by aerodynamic
        drag heating in the solar nebula. *Earth Planet. Sci. Lett.* **70**:11-26.

Yang, J., and Epstein, S.,1983, Interstellar organic matter in meteorites. *Geochim.*
        Cosmochim. Acta **47**:2199-2216.

Zook, H. A, and McKay, D. S. 1986, On the asteroidal component of cosmic dust.
        Lunar Planet. Sci. **XVII**:977-978.

Section II: Interstellar Chemical Processes

Robert Haskell (Security) and Shane Burns

ELEMENT ABUNDANCES
IN THE INTERSTELLAR ATOMIC MATERIAL

Edward B. Jenkins
Princeton University Observatory
Princeton, NJ 08540
USA

ABSTRACT. In the interstellar medium, a significant fraction of the free atoms for elements heavier than helium must have condensed into solids, in the form of small, interstellar dust grains. Observations of optical and ultraviolet absorption lines in the spectra of hot stars indicate that the depletions below the cosmic abundances of various elements range from almost none at all to a factor 10^{-4} and seem to be enhanced along lines of sight with higher mean density. Other methods of measuring abundances confirm the absorption line results. The available evidence supports the notion that the grains form in both the atmospheres of cool stars and in dense interstellar clouds, and they are destroyed by shocks which frequently pass through the low density gas.

1. INTRODUCTION

The interstellar medium within the disk of our galaxy is constantly being replenished by material from the outer layers of stars. This transfer of mass results from a variety of phenomena, ranging all the way from the gentle, outward breeze of gases from late-type giants, through the robust winds from early-type stars, and finally up to the violent explosions from supernovae. Out of material within very dense accumulations of interstellar gas, new stars are created -- a reverse process which completes the cycle of matter between the stars and the vast volumes of space between them. Since much of the exchange of mass occurs on a timescale short compared to that for the chemical evolution of our galaxy, it is reasonable to expect that the interstellar medium will have relative abundances for various elements virtually identical to those of the Population I stars. Such an abundance pattern, the so-called "cosmic abundances", are deduced from observations of the sun (both the photosphere and the corona), other stars, and meteorites. If we accept the principle that the interstellar medium as a whole indeed has this cosmic composition[1], the main issue to explore is the distribution of the elements in various

---

[1] A very recent observational confirmation of this notion for the case of O *vs.* H + He has been discussed by Schattenburg and Canizares (1986). They measured x-ray absorptions on either side of the oxygen K-edge toward the Crab Nebula and found a relative abundance of O very close to the cosmic value. The absorption of x-rays is not influenced by the chemical state of the atoms, except for atoms in large ($> 0.5\mu$ diameter) grains where self-shielding can be important. See also an earlier discussion by Ryter, Cesarsky and Audouze 1975.

D. J. Hollenbach and H. A. Thronson, Jr. (eds.), Interstellar Processes, 533–559.

physical forms, such as free atoms or ions, gaseous molecules, and solid particles.

In the previous century, Struve (1847) noted from star counts that there
must be a haze of intervening substances which absorbed starlight. Even
before it was fashionable to regard space as being filled with a low density,
atomic gas, Barnard (1910, 1919) observed irregular dark patches in the sky
where stars were blotted out. As observations and interpretations of this interstellar attenuation became more refined, we attained a better grasp of the
nature of the material responsible for this effect, i.e., small, solid particles
("dust grains") having different sizes and compositions. This paper will not
recount the different types of extinction observations and the attendant,
rather specific arguments which constrain our insights on the properties of
this dust, except to point out one straightforward concept which defines an
order of magnitude for the total distributed mass of solid material present
(see chapter 7 of Spitzer 1978 for details).

Consider the simplest case where a line of sight of length d is evenly filled
with small spherical grains, all with a uniform radius a and internal density ρ_s.
If the projected surface number density of these grains is N_g, the overall average density $<\rho_g>$ is given by

$$<\rho_g> = \frac{N_g}{d} \times \frac{4\pi a^3}{3} \times \rho_s .$$ (1)

Now if optical extinction could be described as just a simple geometrical block-
age of light by opaque spheres, we'd have the optical depth along the path
given by

$$\tau = N_g \pi a^2 .$$ (2)

This elementary concept of shadowing does not fit reality, however. At any
given wavelength λ, one must explicitly calculate the extinction efficiency Q_e
for the interaction of light with a substance having a complex index of refraction and include this factor on the right hand side of eq. (2). It turns out,
nevertheless, that this added term greatly simplifies our argument. The Kra-
mers - Kronig relationship

$$\int_0^\infty Q_e \, d\lambda = 4\pi^2 a \left(\frac{m^2 - 1}{m^2 + 2} \right) ,$$ (3)

where m is the index of refraction at low frequencies, adds an extra factor of a
to eq. (2) and makes $\int \tau d\lambda$ proportional to a^3. Thus when we use eq. (2) to
solve for N_g in eq. (1), we obtain a result which is independent of grain radius
a,

$$<\rho_g> = \frac{\rho_s \int_0^\infty \tau d\lambda}{3\pi^2 d} \left[\frac{m^2 + 2}{m^2 - 1} \right] ,$$ (4)

which in turn allows to deal with realistic grain populations with any distribution of different radii. In this equation, the parenthetical term containing the

m's is probably around 2 for many common substances. The only practical problem with this concept is that we are insensitive to large particles because we can not measure τ out to extraordinarily long wavelengths.

At visible wavelengths, the mean extinction by dust grains in our part of the galaxy is $A_V = 1.8$ mag kpc^{-1}. If we combine this value with relative extinctions over wavelengths from the far ultraviolet (1000Å) to the far infrared ($\gtrsim 20\mu$) and assume $\rho_s = 3$, we find

$$<\rho_g> = 1.8\times10^{-26} \text{ g cm}^{-3}, \tag{5}$$

which is about 0.006 times the overall density of hydrogen in the interstellar medium. If the grains are composed primarily of H_2O, both ρ_s and m decrease, leaving little change in the estimated $<\rho_g>$. Also, if the grains are not spherical and have random orientations, the result is not changed appreciably.

The significance of the mass fraction of grains derived above is that some, but by no means all, of the carbon, oxygen and nitrogen (the three commonest elements other than H and He) should be depleted out of the gas phase of the interstellar medium, since their cosmic abundances constitute 0.014 of the total by mass. Elements beyond Ne in the periodic table, comprising 0.003 of the total, could in principle be severely depleted as they are incorporated into grains.

For H II regions and diffuse H I regions, relatively few atoms are consumed by the formation of molecules. H I clouds having an extinction $A_V \lesssim 0.3$ usually have less than 10^{-3} of the H atoms tied up into H_2, and for $0.3 \lesssim A_V \lesssim 1.0$ the fraction is of order 5-50% (Savage, et al. 1977). It is not surprising that absorption line studies in the visible and ultraviolet indicate that the relative concentration of other molecules, those containing heavier elements, is also very low for these diffuse accumulations of gas. Most of these molecules are presumably created by complicated networks of reactions, most of which ultimately depend on ion exchanges with H_2 (van Dishoeck and Black 1986 and references cited).

Observations of radio line emissions show that the interior portions of large, dense clouds have significant concentrations of various types of molecules; the high densities promote rapid molecular production and the regions are well shielded from dissociating radiation. Under these conditions one might expect that there is little to prevent the almost complete consumption of free atoms, i.e., those not already locked into grains, into some bound form by a broad variety of possible reaction routes. However the observations of infrared line emission from the fine-structure levels of neutral carbon atoms in some dense clouds by Phillips and Huggins (1981) seem to contradict this simple picture. They found unexpectedly large column densities of C I coexisting with the CO molecules (the principal molecule which would deplete the free carbon atoms), and they conjecture that the lifetimes of the clouds may be shorter than the chemical equilibrium times ($\sim 10^6$ yr). As an alternative, material from the center may be dredged toward the surface in this short a time scale by churning motions, and some molecules may be destroyed by exposure to dissociating radiation.

Generally speaking, there have been relatively few measurements of various atomic concentrations in the dense, molecular clouds. By far the greatest body of information applies to regions of space which are subject to scrutiny by observing ultraviolet and visible absorption lines. Such regions, by necessity, must not contain so much dust that the attenuation of light is severe. Thus one is restricted to regions of moderate density, in the approximate range 0.01 to 100 atoms cm^{-3}. While most of the discussion which follows will concentrate on phases of the interstellar gas which are accessible to the optical absorption line measurements, we will touch briefly on the results of radio recombination line observations in a few dense, H I regions which are exposed to large ionizing fluxes, and also we will explore some of the abundance determinations for dense H II regions based on measurements of optical emission line strengths.

Our principal objective will be to understand how the atomic abundances vary from place to place and element to element. This information should help to elucidate the physical processes which govern the formation and destruction of interstellar grains. A knowledge of the concentrations of heavier atoms in space is also useful to other studies on the interstellar medium. For instance, the primary source of free electrons in H I regions is from starlight ionization of atoms which have their first ionization potential below that of neutral hydrogen (at an energy of 13.6 eV), since the more energetic photons are blocked by the hydrogen continuum opacity. Carbon is by far the most important contributor here, and an estimate its concentration has a direct bearing on our estimate for the electron fraction in H I regions. For H I regions, the only means for cooling the gas radiatively, outside of emissions from molecules, is from collisional excitation and the spontaneous decay of the excited fine-structure levels of such atoms and ions as C I, C II, 0 I, Si II, and Fe II (Dalgarno and McCray 1972). Below 10^4K, 0 II, 0 III and N II are the principal coolants for fully ionized (H II) regions (Spitzer 1978, chapter 6).

2. ABSORPTION LINE STUDIES OF H I REGIONS

2.1. SOME HISTORY AND BASIC CONCEPTS

Because densities are low, nearly all atoms in the interstellar medium are in the ground electronic state. Thus, surveys of absorption lines must be restricted to only those transitions which arise from ground levels. Unfortunately, most spectrum lines of this sort are found at ultraviolet wavelengths inaccessible to ground-based observatories -- the visible spectrum offers only a few such lines, mostly from minor constituents in unfavorable stages of ionization.

While it is true that the first indication of a gaseous component of the interstellar medium came from systematic studies of stationary, narrow Ca II *H* and *K* lines in the spectra of binary stars (Hartmann 1904, Slipher 1909), followed later by observations of Na I (Heger 1919, 1921) Ti II (Dunham and Adams 1937, Dunham 1937a), Ca I (Dunham 1937b), K I (Dunham 1937b), and Fe I (Adams 1943), most of the early work was qualitative. Definitive determinations of abundances were difficult to achieve because corrections for line saturation and ionization to higher stages were required. Later, in spite of the uncertainities in electron densities and ionizing radiation fluxes, it was correctly perceived that atomic calcium was depleted relative to its cosmic abundance with respect to sodium (Strömgren 1948, Spitzer 1954). Furthermore, the interstellar ratio of calcium to sodium seemed to increase for clouds which had a higher than normal velocity with respect to the local medium (Routly and Spitzer 1952; see Siluk and Silk 1974 for more recent data), although initially there was some question whether Na was selectively ionized in the high velocity clouds. The advent of reliable quantitative data from the ground on total abundances for a heavy element came from a survey of absorption lines by Ti II by Wallerstein and Goldsmith (1974); this element is in a favored ionization state and thus there is no question that its interstellar abundance is well below the cosmic value. Stokes (1978) has surveyed extensively the abundance of this ion.

The fact that nearly all of the interesting absorption lines are in a domain accessible only to instruments launched into space explains abrupt genesis of widespread atomic abundance studies in 1972, with the launch of the *Copernicus* satellite which contained an instrument specifically designed for this type of research. Following some initial studies of abundances of a number of principal elements along various lines of sight (Morton, *et al.* 1973, Rogerson, York, *et al.* 1973), Morton (1975) compiled a comprehensive list of ultraviolet interstellar lines toward the star ζ Oph, which led to depletions (or limits) of some 20 elements in a dense foreground cloud. Even today, Morton's study of the interstellar absorptions toward ζ Oph is a valuable atlas of lines which one might expect to see from a moderately dense cloud (including many unidentified lines), and serves as a useful template for future studies.

Conforming to traditions established long ago with the study of visible absorption lines (Unsöld, Struve, and Elvey 1930, Beals 1936, Wilson and Merrill 1937, Wilson 1939), most of the abundances have been derived using lines of different strength and assuming they saturate in a manner similar to those arising from a single volume of gas having a Maxwellian velocity distribution. For a discussion of the mathematical formalism which describes this behavior (the standard "curve of growth"), see Spitzer 1968, chapter 2 or Münch 1968. Often, this is the only alternative one can employ when the lines are not so weak that they are sure to be unsaturated, but if one is careless (see Pitfall 1 described below) the outcome can have considerable error (Nachmann and Hobbs 1973).

Following the research on ζ Oph, intensive studies of other lines of sight were performed, thus broadening the base of interstellar conditions sampled. Table 1 summarizes the more prominent investigations available in the literature, arranged chronologically. Many studies were done with enough care to evaluate the abundances in separate velocity components, even though they were partially blended by the instrumental profile of the *Copernicus* U1 detector (15 km s^{-1} FWHM).

A different research tactic is embodied in studies which survey the abundance of one or just a few species over many lines of sight. Here, instead of a detailed model of a particular line of sight, the emphasis is on learning how the abundances change over widely varying environmental conditions. Table 2 summarizes some references where this approach has been taken.

TABLE 1. ABUNDANCE STUDIES TOWARD SPECIFIC TARGETS

Star	Reference
ζ Oph	Morton 1975, Snow and Meyers 1979
ξ Per	Gómez-Gonzàlez and Lequeux 1975
γ Ara	Morton and Hu 1975
o Per	Snow 1976
μ Col	Shull and York 1977
HD 28497	" "
HD 50896	Shull 1977
ζ Per	Snow 1977
ζ Pup	Morton 1978
α Vir	York and Kinahan 1979
γ Vel	Morton and Bhavsar 1979
ε and π^5 Ori	Shull 1979
15 Mon	Shull 1980
χ Oph	Frisch 1980
ε Per	Martin and York 1982, Vidal-Madjar, *et al.* 1982
δ Per	Martin and York 1982
μ Oph	Cardelli and Böhm-Vitense 1982
λ Sco	York 1975, 1983

TABLE 2. SURVEYS OF ABUNDANCES TOWARD MANY TARGETS
(10 or more stars)

Species	Reference
Cl I, Cl II, P II	Jura and York 1978
Na I, K I, Ca II	Hobbs 1974, 1978
Ti II	Stokes 1978
Zn II	York and Jura 1982, Harris, Bromage and Blades 1983 Harris and Mas Hesse 1986a
Fe II	Savage and Bohlin 1979
C I	Jenkins and Shaya 1979, Jenkins, Jura and Loewenstein 1983
K I	Chaffee and White 1982
Fe II	Shull, Van Steenberg and Seab 1983
Cl I, Cl II	Harris and Bromage 1984
Fe II, P II Cl I, Cl II Mn II, Mg II	Jenkins, Savage and Spitzer 1986
Mg I, Mg II Cr II, Mn II Fe II, Zn II	de Boer, *et al.* 1986
Li I	White 1986
S II	Harris and Mas Hesse 1986b

2.2. PITFALLS

Almost any field of research has its share of classic difficulties or misleading interpretations, and a critical scholar of absorption line studies soon becomes aware of recurrent problems. Listed below are ones which this author can recall, in approximately decreasing order of importance:

(1) *Unjustified trust in the standard (Gaussian) curve of growth for strongly saturated lines.* This is probably the most widespread form of abuse. Even though it is well known that interstellar lines have complicated velocity structures (Hobbs 1969)/ which can result in curves of growth which differ from the standard one [e.g., Fig. 4 of Snow (1977) and Fig. 5 of Morton and Bhavsar (1979)], investigators sometimes derive column densities assuming the slope of a limited portion of the standard curve can properly unravel one to three orders of magnitude saturation in the equivalent widths! Often, one has no alternative but to employ standard curve of growth analysis techniques, but many authors[2] have understated the magnitude of the resulting errors in column density. It is difficult to generalize how far up on the curve of growth one can safely venture, since signal-to-noise ratios, the reliability of relative f-values, and the probable disparity of velocity dispersions b of different components are all important determinants. Generally speaking, one should be suspicious of results where the curve of growth indicates that the weakest line has a central optical depth τ_0 greater than 2. In terms of typical numbers for familiar quantities, for $b = 3$ km s^{-1} and $N(H_{tot}) = 10^{21}$ cm^{-2} we should require for a transition with an f-value f at a wavelength λ (in Å) that the quantity A(element) $+ \log(f\lambda)$ should be less than -5.4, where A(element) is the log of the element's abundance relative to hydrogen in the interstellar gas. On a more optimistic note, Jenkins (1986) has shown that if the equivalent widths are each a composite of a large number of components drawn from a population of τ_0 and b whose distribution is very broad but not bizarre, errors in the saturation corrections seem to converge to a small number. His calculations indicated that only mild systematic underestimates in column densities are obtained (0-20%), even when the typical contribution to the *weakest* line has $\tau_0 = 4$.

(2) *Errors in transition probabilities.* There is a broad range in the relative accuracies of transition f-values. Some of the values in the literature are accurate to within about 10% or better. However, there have been many instances in the recent past where older values have been revised by as much as a factor of 10 (Zeippen, Seaton, and Morton 1977, Lugger, *et al.* 1982, Hibbert, *et al.* 1983). If, as is often the case, all of the f-values used in the study of a particular atom or ion are off by some consistent factor (i.e., when a single multiplet is used), abundances will be wrong by this same factor. However, sometimes the *ratios* of f-values may be wrong and then incorrect b values and saturation corrections may result. It is unfortunate that some of the most useful lines for measuring abundances, the semiforbidden intersystem lines which produce unsaturated absorption features, seem to have the worst uncertainties. For a perspective on the experimental and theoretical methods plus sources of information, see Wynne 1984.

[2] Including myself, on some occasions.

(3) *Uncertainties in cosmic abundances (for deriving depletions).* For most elements, accurate and consistent results for cosmic abundances have been obtained. There are a few exceptions, however: different determinations for O, Ne, and Ar disagree by as much as 0.4 dex (Pottasch 1985), and extremes in the abundance measurements of Cl span almost one order of magnitude (Jura and York 1978).

(4) *Misapplication of curves of growth across different species.* Sometimes, one can (or is forced to) use the equivalent widths of two or more lines from one atom or ion to deduce the saturation correction for the single line of another. This procedure is acceptable only if one has good reason to believe the two cases should have nearly the same velocity profiles. Usually, the correspondence is not valid when one element is in its favored ionization state for an H I region (e.g., O I, N II, Si II, etc.), while another is in a state below the favored one (as with most of the elements which can be seen at visible wavelengths, such as Na and K). For the latter, the local density scales with $n(H)^2$ instead of just $n(H)$, thus emphasizing compact concentrations of gas which generally have low b -- see the curve of growth plots in Spitzer and Jenkins 1975 and Gómez-Gonzàlez and Lequeux 1975. Some ion stages may be abundant in either H I or H II regions (e.g., Si II), while others are dominant in only one or the other (e.g., N I, N II -- see Sarazin 1977). One might also suspect that highly depleted species would have larger b values than less depleted ones, reflecting the Routly-Spitzer effect mentioned earlier. (For a counter example, see the plot of b for Fe *vs.* b for P in Jenkins, Savage and Spitzer 1986. There seems to be no systematic difference between the two, despite their large differences in depletion.)

(5) *Ignoring contributions from H II regions.* With some exceptions[3], one can say that in an H I region virtually all atoms of a particular element are in the lowest stage of ionization having an ionization potential greater than that of hydrogen. Thus, the conventional method of deriving the abundance of an element, relative to that of hydrogen, is to compare the column density $N(element)$ of the favored stage of ionization to measurements of $N(H_{tot}) \equiv N(H\ I) + 2N(H_2)$ along the same line of sight. However contributions from H II regions may increase $N(element)$ with no change in $N(H_{tot})$, thus tricking one into overestimating the relative logarithmic abundance $A(element) \equiv \log[N(element)/N(H_{tot})]$ in the H I region. This effect is probably important only for relatively nearby stars. For a concrete example, we can refer to York's (1983) detailed analysis of the line of sight to λ Sco. He found column densities of C II, S II, Si II, Fe II and P II in a velocity component identified with an H II region to be approximately the same as those associated with the foreground H I component. York and Kinahan (1979) came to a similar conclusion for α Vir.

---

[3]In dense clouds ($n(H) \sim 100$ cm^{-3})appreciable fractions of Ca may be singly, rather than doubly ionized (Spitzer 1968, chapter 4; Snow 1984). If much H_2 is present, a good fraction of Cl II may be converted into Cl I via an ion-molecule reaction involving the production of HCl followed by dissociation (Jura 1974).

(6) *Confusion with circumstellar components.* Some Be stars, but probably only those identified as shell stars from visible spectra, produce their own absorption components which mimic those from the interstellar medium (Oegerle and Polidan 1984). Such stars are also undesirable to use in surveys which attempt to correlate element depletions with various measures of interstellar dust absorption because measures of color excess are unreliable.

(7) *Failure to recognize the influence of correlated errors in scatter diagrams.* Many studies of element depletions make use of correlations (or lack thereof) with other observables. For instance, a common relationship to plot is $\log[N(\text{element})/N(H_{tot})]$ vs. $\log[N(H_{tot})/d]$, i.e., $A(\text{element})$ against the logarithm of the average line-of-sight density $\bar{n}(H_{tot})$. When statistically significant, the correlation coefficient of these two variables (on log-log scales) is a negative number (see Fig. 1). Experimental errors in $N(H_{tot})$, however, tend to enhance this negative correlation since they displace the points in the scatter diagram along a diagonal with slope -1. For datasets spanning the large differences in $\log \bar{n}(H_{tot})$, as depicted in Fig. 1, this effect is small. However comparisons of depletions with some other variable, such as the ratio of some type of extinction to gas, may not cover a large dynamic range. In such a case, errors in variables appearing in either addition or subtraction in both the x and y axes can be important. Jenkins, Savage and Spitzer (1986) discuss a method of calculating what the correlation coefficients would have been in the absence of such errors, provided one knows their magnitudes.

(8) *Not handling upper or lower limits properly.* In many investigations, a significant fraction of the column densities are either upper limits (no line seen) or lower limits (where lines may be saturated by an arbitrarily large amount). If one consistently ignores these limits, the conclusions may be biased. There are several ways to avoid this problem. First, in cases where lines are often undetectable, one should report literally the negative determinations (i.e., downward noise excursions) in an evenhanded manner along with the marginal or definitely positive measurements. It may seem silly to report a determination -1 ± 2 when one could just as well say the 2σ upper limit is 3, but when many such measurements are combined with the positive results and the errors cancel, one can obtain a meaningful and fair global result. Now suppose we have no control over the situation because we are working with someone else's results, and that person only reported limits! One alternative is to use some advanced mathematical treatments which account for the presence of these limits without imposing a bias, such as the methods described by Schmitt (1985). Another trick, which sometimes works, is to evaluate the median value of something, rather than its mean, with the provisional assumption that the limits are actual values. If most of the upper limits are below this median and the lower limits above, then the median is not changed appreciably if the true values are somewhat displaced (in the right direction) from the limiting values. We are then left with only having to justify that the distribution of the variable is well enough behaved (i.e., not terribly skewed) that the median is a good measure of what we want.

(9) *The ζ Oph bandwagon.* To some this may appear frivolous, but I regard it as an interesting sociological problem in interstellar matter research. Well before the advent of ultraviolet observations, the line of sight toward ζ Oph was recognized to contain an unusually rich and interesting assortment of atomic and molecular lines (Herbig 1968). As a consequence, interstellar lines in the spectrum of this star have received a disproportionately large share of astronomers' attention, both observationally and theoretically. While the results for ζ Oph present an intriguing example of interstellar physics and chemistry, we sometimes are enticed by the completeness of information to use this line of sight to exemplify it as a "typical dense cloud". This cloud, however, may be far from typical.

2.3. DENSITY DEPENDENCE OF DEPLETIONS IN H I REGIONS

An important finding from the general abundance surveys is that most elements have greater depletions when the average density $\bar{n}(H_{tot})$ along a line of sight is large (Savage and Bohlin 1979, Phillips, Gondhalekar and Pettini 1982, Murray, *et al.* 1984, Harris and Bromage 1984, Phillips, Pettini and Gondhalekar 1984, Jenkins, Savage and Spitzer 1986, Harris and Mas Hesse 1986). This effect is clearly attributable to differences in the amount of material locked into solid form, since the cloud densities are generally far short of those where free molecules other than H_2 have an appreciable abundance. The anticorrelations of relative abundances with total column densities of hydrogen $N(H_{tot})$ do not seem to be as good as with $\bar{n}(H_{tot})$ (Harris, Gry and Bromage 1984, Gondhalekar 1985). One might have expected a better correlation with $N(H_{tot})$ if the largest, densest cloud nearly always dominated over others along any particular line. Since we know that the interstellar medium contains marked density inhomogeneities, a plausible picture is that $\bar{n}(H_{tot})$ indicates the relative proportion of gas in dense clouds and a lower density intercloud medium.

Spitzer (1985) modeled the depletion variations, together with the fluctuations caused by small number statistics, for a three-component model of the interstellar medium consisting of (1) a uniform, low-density gas, (2) diffuse clouds, and (3) large, dense clouds. The basic properties of the latter two phases (i.e., internal densities, filling factors, granularity, etc.) have been suggested by earlier data on the average values and fluctuations in the interstellar extinction data (Spitzer 1978, chapter 7). The quality of existing depletion measurements do not do full justice to all of the definable parameters in Spitzer's model. To aid in the interpretation of information which will be presented below, we may crudely characterize cases where the uniform intercloud medium dominates as those where $\bar{n}(H_{tot}) < 0.2$ cm^{-3}, with diffuse clouds taking over in the range $0.4 < \bar{n}(H_{tot}) < 1.0$ cm^{-3}, and large clouds being most influential when $\bar{n}(H_{tot}) > 3$ cm^{-3}.

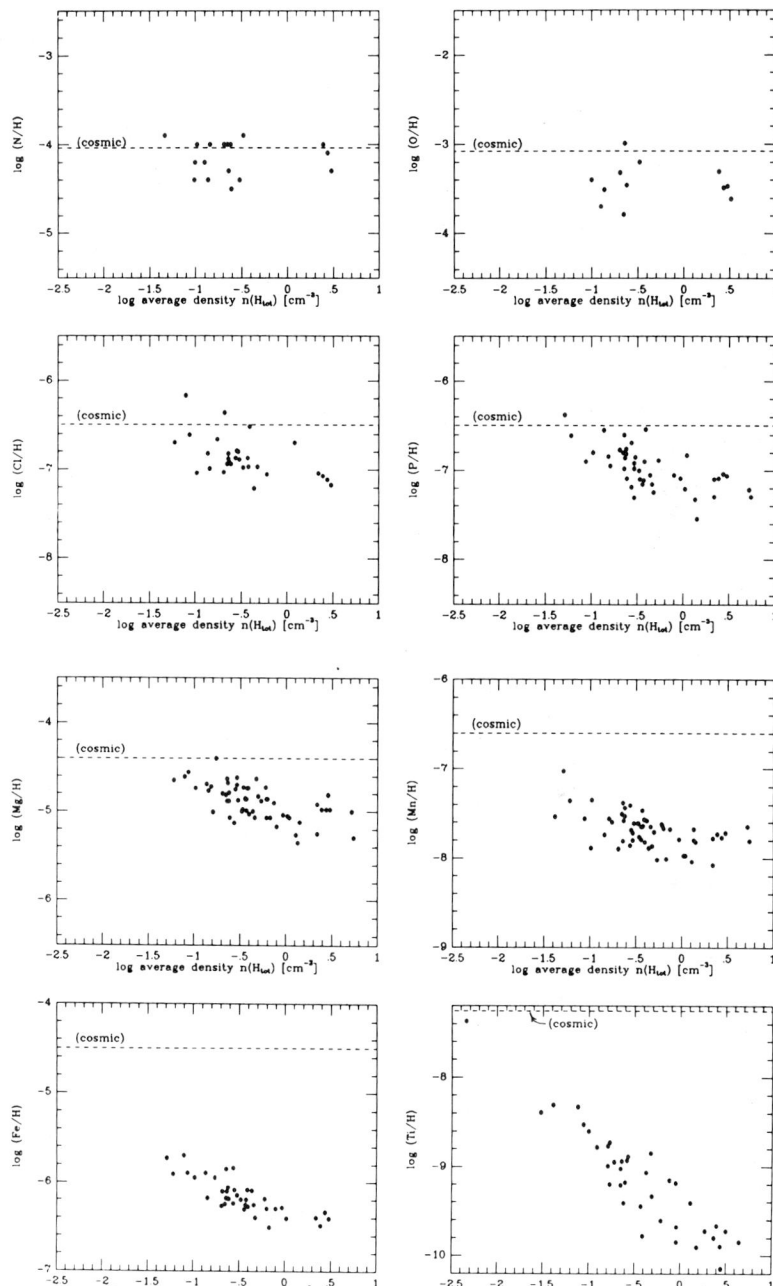

Figure 1 -- Logarithmic abundances A relative to hydrogen in the interstellar medium for the elements N, O, Cl, P, Mg, Mn, Fe, and Ti plotted against the logarithms of the average line-of-sight densities $\bar{n}(H_{tot})$. Cosmic abundances are shown by the horizontal dashed lines.

TABLE 3. BEST FITS TO EQUATION (6)

Element	Adopted $A(\text{elem.})_{cosmic}$	$A(\text{elem.})_{min}$ $+ \log(f\lambda)_{min}$	d_0	m	Data Source
N	-4.04	-6.8	-0.12±0.05	-0.01±0.09	1
O	-3.08	-6.3	-0.36±0.07	-0.01±0.11	2
Cl	-6.50	-5.8	-0.36±0.04	-0.27±0.08	3
P	-6.50	-5.9	-0.43±0.03	-0.32±0.06	3
Mg	-4.40	-6.0	-0.44±0.02	-0.28±0.05	3
Mn	-6.60	-5.9	-1.06±0.02	-0.22±0.02	3
Fe	-4.50	-5.7	-1.65±0.02	-0.38±0.05	3
Ti	-7.26	-6.8	-1.86±0.04	-0.84±0.06	4
Cr	-6.30	-6.4	-1.58±0.09	-0.50±0.11	5
Zn	-7.40	-5.4	-0.38±0.07	-0.11±0.09	6
Si	-4.45	-5.2	-1.09±0.11	-0.49±0.15	7

Key to data sources:

1: Hibbert, Dufton and Keenan 1985
2: Keenan, Hibbert and Dufton 1985
3: Jenkins, Savage and Spitzer 1986
4: Stokes 1978
5: de Boer, et al. 1986
6: Harris and Mas Hesse 1986a
7: Barker, et al. 1984

Figure 1 shows some selected logarithmic element abundances plotted against $\log \bar{n}(\text{H}_{tot})$. Least squares linear fits of these points, plus those for three additional elements, are listed in Table 3. The coefficients d_0 and m are defined by the following equation for logarithmic depletions $D(\text{element})$,

$$D(\text{element}) \equiv A(\text{element})_{meas.} - A(\text{element})_{cosmic}$$
$$= d_0 + m\left[\log n(\text{H}_{tot}) + 0.5\right] \tag{6}$$

The intercept d_0 is calculated about an approximate mean value of $\log \bar{n}(\text{H}_{tot})$ = -0.5 for most of the datasets, so that errors in the intermediate depletion values will not be strongly dependent on errors in slope m. For deriving depletions, the adopted cosmic abundances (listed in the 2nd column of the table and shown as dashed lines in the figure) were those chosen by the investigators of the respective surveys. Standard errors for the coefficients listed in the table are only the formal errors for an unweighted fit. No allowances were made for errors in f-values or cosmic abundances (Pitfalls 2 and 3). Omitted from the diagrams and the regression evaluations are measurements which are merely limits or those which have inordinately large errors.[4] Thus, it is important to emphasize that the results shown here are in clear violation of the admonition given under Pitfall (8) discussed above.[5]

---

[4] Except for Zn, where many points which had upper error bounds reaching to ∞ had to be admitted.
[5] A more elaborate analysis which properly handled lower and upper limits was

Concerning some of the other pitfalls, Pitfall (7) is not important here; assuming the probable errors in $\bar{n}(H_{tot})$ are only of order 20% (Bohlin, Savage and Drake 1978), the true slopes differ from those listed in Table 3 by only about 0.01. Pitfall (5) could be of some consequence, since stars with low $\bar{n}(H_{tot})$ could more easily have relatively influential contributions from their H II regions, thus spuriously enhancing the perceived element abundances on the left ends of the diagrams. Pitfall (1) could introduce systematic errors for some of the element relationships, i.e., those predicated on measurements of strong lines only. Except for Ti II, all of the plots in Fig. 1 were generated from a special survey of *weak lines* recorded by the *Copernicus* satellite (Bohlin, *et al.* 1983). The fact that these results probably do not suffer from significant saturation errors is shown by the numbers in the third column of the table which gives A(element) + $\log(f \lambda)$ for the weakest line used and at log $\bar{n}(H_{tot}) \sim 0.5$. For results from other sources, all but the last two numbers in this column are less than the threshold of -5.4 discussed under Pitfall (1). The weakest lines of Zn II (Harris and Mas Hesse 1986a) and Si II (Barker, *et al.* 1984) have A(elem.)$_{min}$ + $\log(f \lambda)_{min} \gtrsim$ -5.4, and thus are only marginally safe. For these two cases we should be aware that saturation errors will probably alter our conclusions in a systematic manner since the dense clouds may produce narrow profiles which are lost within the cores of lines coming from the intercloud medium. Thus, as Nachmann and Hobbs (1973) have demonstrated, significant underestimates of the total column densities may ensue in the right-hand part of the diagrams. Ti II column densities were derived from the high resolution visible data of Stokes (1978), where optical depths are integrated over velocities -- a much safer procedure than the use of curves of growth. Unfortunately, we must refrain from studying the density relationships of a number of interesting elements, such as S (Harris and Mas Hesse 1986b), Al (Barker, *et al.* 1984) and Ar (Duley 1980) since their A(elem.)$_{min}$ + $\log(f \lambda)_{min}$ are of order -4.7, -4.5 and -4.4, respectively.

It seems remarkable that points within some of the panels in Figure 1, such as those for Mg and Fe, show very little scatter about the linear relationships defined by eq. (6) and the coefficients in table 3. (Lines representing these best-fit determinations were purposely omitted from the scatter plots, to allow the reader's eye to obtain an unbiased appraisal of the points' layout.) Quantitative determinations of the surprisingly small spreading of points from the general trends have been made: Jenkins, Savage and Spitzer (1986) found that the real *rms* vertical dispersion of points about a best fit in their diagrams of this type was about 0.1 dex, after compensating for added scatter from known experimental errors. In addition to telling us something about the predictability of depletions in different circumstances, the well defined behavior indicates that variations in $\bar{n}(H_{tot})$ are really caused by a swapping

---

not carried out here for lack of time. If one examines the original data in the literature with proper error bars or limit arrows, one has the impression that the omission of less definitive data does not appreciably alter the appearance of the relationships, except possibly for N/H and O/H where there is a preponderance of points at large $\bar{n}(H_{tot})$ which are lower limits rather close to the cosmic abundance ratio.

between the principal components of H I gas at different density. In turn, we can surmise that either this H I material occupies a large fraction of the volume of space, or if not, the overall distribution of some other, unseen phase of matter with a large filling factor must be quite uniform.

In several cases in Figure 1, there is a hint that the slopes of the trends become shallower for $\log \bar{n}(H_{tot}) > 0$. This qualitative change is consistent with the expected behavior of Spitzer's (1985) model; see the lines drawn in the figures in Jenkins, Savage and Spitzer (1986). Unfortunately, there are too few points above $\log \bar{n}(H_{tot}) > 0.5$ to ascertain any effect from Spitzer's large clouds.

From element to element, there seems to be no simple, coherent scheme which ties the overall magnitude of the depletion D to the slope of the D vs. $\log \bar{n}(H_{tot})$ relationship. Figure 2 plots $|m|$, the steepness of the change of with $\log \bar{n}(H_{tot})$, against what we could define as a "base depletion", namely the value of D at $\log \bar{n}(H_{tot}) = -1.5$ (i.e., simply $d_0 - m$). The incoherence between these two variables is probably a result of there being various physical processes which are responsible for the growth and destruction of grains, and these mechanisms operate on different elements in an uneven manner. We will discuss some possibilities in §3.

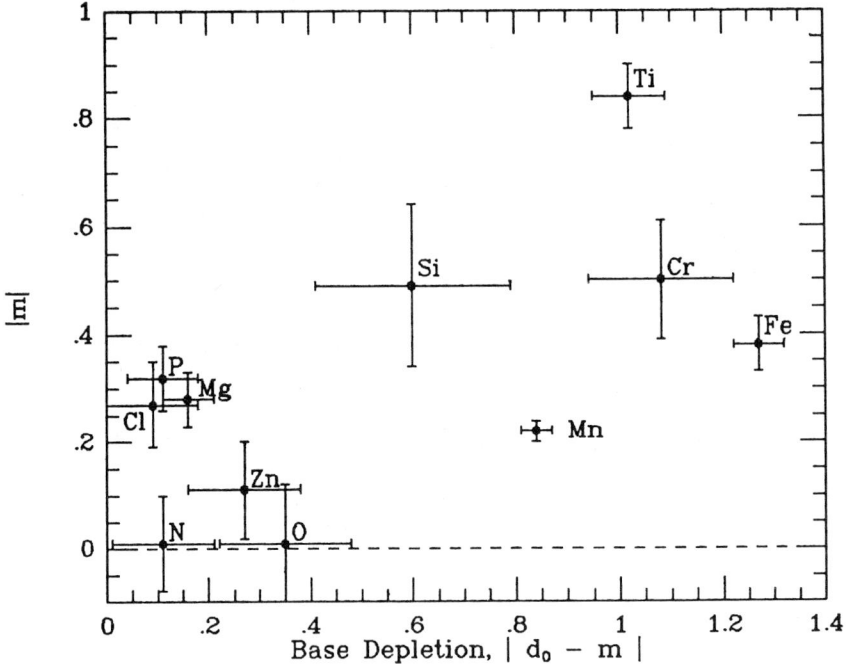

Figure 2 -- A plot comparing the sensitivity of logarithmic depletions D to the logarithm of the average density $\bar{n}(H_{tot})$ against the values of D in the intercloud medium ($\bar{n}(H_{tot}) \approx 0.03$ cm^{-3}).

The use of $\bar{n}(H_{tot})$ as an indicator of local interstellar conditions is, of course, a very crude index. With the successes we have seen so far, it would seem worthwhile to pursue other methods of differentiating densities. Gondhalekar (1985) attempted to measure depletions of Ti and Cl against local density as indicated by the fine-structure excitation of C I. His correlations, however, seem pretty marginal. Considerable promise in identifying especially dense clouds may come from the work of Joseph, *et al.* 1986, who used the presence of CN to single out clouds whose internal densities are higher than normal. They found that the depletions of various elements were markedly higher for CN-rich clouds in front of 5 stars. The most striking effect was with Mn, where the depletions approached those of Fe (compare with Fig. 1).

Beyond the 11 elements covered so far, we can ascertain the depletions (or their limits) for a number of other elements which have accessible transitions from their favored stages of ionization in H I regions. A principal problem, however, is that at one or another $\bar{n}(H_{tot})$ their lines are either too strong (giving us saturation problems) or too weak to be seen.

2.4. NEUTRAL ATOMS WITH IONIZATION POTENTIALS BELOW 13.6 eV

If we neglect the few cases where charge exchange (Solomon and Klemperer 1972) or certain ion-molecule reactions (Jura 1974) have some influence, the abundance of an atom which can be ionized by starlight photons longward of the Lyman limit is governed by the ionization equilibrium

$$n^0 = \frac{\alpha(T) n_e}{\Gamma} n^+ , \qquad (7)$$

where $\alpha(T)$ is the temperature dependent recombination coefficient with free electrons, n_e is the local electron density, and Γ is the ionization rate for the neutrals. (For elements like Ca which are normally doubly ionized in H I regions, one can, to a good approximation, still use eq. (7) with different α and Γ and substitute n^+ for n^0 and n^{++} for n^+.) The value of Γ is governed by the integral over frequency of the local starlight photon density times the ionization cross section. For most elements and interstellar conditions n^0/n^+ is very small.

As discussed in §2.1, good quantitative work on abundances using visual lines was hampered by uncertainties in the parameters appearing in eq. (7). One way to circumvent this difficulty was employed by White (1974). He measured lines of Ca II and Ca I and used the ratio of column densities to deduce n_e and the radiation density. With this information, he concluded that Ca was strongly underabundant and Na was also mildly depleted, since he could solve for the amount of Na II and Ca III present along the line of sight (also see Pottasch 1972). One drawback of this method is that it assumes the conditions are homogeneous for the velocity component under consideration. (To see why this is so, one can picture a cloud with a compact, dense core which produces nearly all of the Ca I and Ca II absorption, surrounded by an extended, low density envelope of gas composed almost entirely of Na II and Ca III.)

Another way to restrict the range of possibilities for the range of n_e (but not Γ) is to make use of the fine-structure level populations of an atom to determine the local hydrogen density and then to assume that the free electrons all come from the ionization of carbon. Jenkins, Jura and Loewenstein (1983) used this method to obtain the abundance of C in H I regions. Once again, heterogeneity of the physical conditions along the line of sight can spoil things.

An elegant way to cope with the uncertainties in the ionization equilibria is not to attempt to derive the absolute abundance of a single element, but instead to compare the column densities of the neutral forms of two (or more) elements. To first order, the ratios of column densities are influenced only by the element abundance ratios and some simple constants related to the atomic physics of each element's recombination or ionization cross section. Absolute values of n_e or starlight intensity, or even substantial variations thereof, should have no effect since they cancel out when a ratio is evaluated. Also, changes in temperature from region to region will not be important, since $\alpha(T)$ scales approximately as $T^{-0.7}$ for different elements (except at very high temperatures, above the energy threshold for dielectronic recombinations).

In the interpretation of the ratios of neutral species (York 1980, Chaffee and White 1982, White 1986), the only problem to be wary of is uncertainties or changes in the *energy spectrum* of the ionizing photons. Jenkins and Shaya (1979) speculated that this could be an important effect which explained the progressive increase in the ratio of C I to Na I and K I for lines of sight penetrating clouds with ever greater extinctions. The window for carbon ionizations is a narrow one between the Lyman limit and about 1110Å, where dust absorption is much stronger than at longer wavelengths and a reasonable fraction of photons can be absorbed by H_2 Lyman and Werner bands. Snow (1984) noted, however, that for a broader assortment of elements there seems to be no correlation between an atom's ionization potential and changes of depletions within the denser parts of clouds. In fact, he found the ratios of depletions indicated by neutrals to be virtually the same as their ionized counterparts. This, he concluded, suggests that either the clouds are homogeneous (i.e., the ions coexist with the neutrals rather evenly) or, alternatively, that depletions increase toward the center of a cloud for all elements by the same amount. The latter possibility is in accord with more direct findings for the ρ Oph cloud by Snow and Jenkins (1980).

3. INTERPRETATION

3.1. BASIC CONCEPTS

A principal application of the observational results discussed so far is to check their consistency with theories of grain formation and destruction. We will probably find it difficult to synthesize a complete description of the creation and destruction of grains and confirm it in detail, since in all likelihood there are several processes at work at both ends of the cycle, and some of the theoretical reconstructions are approximate or controversial. Nevertheless there are some aspects of the depletion patterns which give us encouragement that we are on the right track in many of the general concepts.

The nuclei of grains are probably formed in the atmospheres of cool giant stars or stellar nebulae where the densities, time scales and temperatures are about right for the condensation of certain elements into refractory compounds (Grossman and Larimer 1974, Draine 1981). We can imagine that at the highest densities and temperatures, certain solids condense out and full chemical equilibrium with the surrounding gas results in very low partial pressures for some atomic species. Radiation pressure on the grains and their drag on the gas then drives an outflow of the outer atmosphere (Gehrz and Woolf 1971, Tielens 1983). As material drifts away and cools, the pressures and time scales become too small to achieve equilibrium, and thus elements which condense at lower temperatures are only partly depleted. There is ample observational evidence in the infrared (Toombs, et al. 1972, Forrest, et al. 1978, Sutton, et al. 1978, Werner, et al. 1980) and radio (Morris, et al. 1979) that cool stars are surrounded by dust shells containing silicates (or for carbon stars, amorphous carbon [Jura 1986]).

Shortly after the first results on atomic abundances were available from the *Copernicus* satellite, Field (1974) proposed that the picture just described was the principal mechanism for the formation and growth of grains found in the interstellar medium. He produced a persuasive exhibit to substantiate this notion, namely, a plot of logarithmic element depletions D vs. their respective condensation temperatures T_C in an oxygen-rich environment, where T_C is defined to be the temperature where half of the atoms are depleted at a given equilibrium pressure (see Figs. 4-6). While we have noted there is ample independent evidence to support Field's claim that refractory compounds and eventually grains are formed in circumstellar environments and subsequently cast out, the proposal that this is the *dominant* method of depleting atoms runs into some difficulty. The principal problem is that some elements with depletions approaching 10^{-3} to 10^{-4} in dense clouds, such as Ca and Ti, are created in supernova explosions. We are then forced to accept the doubtful premise that less than one part in 10^3 of the gas now in the medium has never been cycled through a cool stellar atmosphere or nebula. Also, each time any gas is blown off from the atmosphere of an early-type star the grains probably can not reform, since the rate of mass loss (and hence pressure decrease through the important temperatures for forming compounds) is too rapid. Still another difficulty with the concept that grains are formed only near stars is the dependence of depletions on $\bar{n}(\mathrm{H_{tot}})$. This is not a devastating argument however, since grains are probably preferentially destroyed by shocks in regions of low density.

Not long after the publication of Field's paper, Snow (1975) presented plots of D vs. each element's first ionization potential, and except for the alkali elements Li, Na and K, these plots looked about as good as Field's. Snow used these diagrams to support the idea that grains grew in the dense interstellar clouds. While it is not completely clear why ionization potentials should be important, he proposed that ions sticking to grains would soon recombine with an electron and that the probability an atom would be jarred loose would be proportional to the energy liberated by the recombination. The low depletions of alkali metals could be explained by the formation and instant ejection of their saturated hydrides from the grains (Watson and Salpeter 1972).

A fundamental problem in comparing various, specific chemical properties to depletions for testing competing theories is that, except for a few elements, these properties are highly correlated. For the two examples discussed above, it is no accident that the free energy for the formation of compounds is related to the ionization potentials of the constituents. In simplest terms it has not been easy (and it still isn't) to use the relative depletions from element to element to favor one process over another.

The fact that grains are destroyed in shocks seems clear from various studies of abundances in high velocity gas. In all cases depletions were noticeably curtailed[6] (Shull and York 1977, Shull, York and Hobbs 1977, Shull 1977, 1979, 1980, Cowie, Songaila and York 1979, Frisch 1980, Walborn and Hesser 1982, Laurent, Paul and Pettini 1982, Jenkins, Wallerstein and Silk 1984) or, for some very high velocity shocks which could be directly related to a known supernova explosion, there was virtually no depletion at all (Jenkins, Silk and Wallerstein 1976). These findings substantiated the original results of Routly and Spitzer (1952) that shocks were an important means for returning atoms within the grains to the gas phase. They also give us an important alternative to the explanation that enhanced condensation rates in dense clouds are responsible for the correspondences of D with $\bar{n}(H_{tot})$. Strong shocks should propagate frequently through the intercloud medium (McKee and Ostriker 1977, McCray and Snow 1979), but as they penetrate the clouds, they should soften and be much less effective in eroding or destroying grains.

Following the initial proposals which related depletions to the history of grains by Field and Snow, there followed more refined theoretical descriptions of grain formation (Barlow 1978c, Duley and Millar 1978) and destruction (Jura 1976, Barlow and Silk 1977, Barlow 1978a,b, Spitzer 1976, Draine and Salpeter 1979a,b, Seab and Shull 1983). Once enough depletion data had been acquired to give more complete information on the variations from element to element and some convincing evidence for the density dependence, slightly more meaningful tests of theory could be applied. One such test was a plot by Phillips, Pettini and Gondhalekar (1984) showing the dependence of m, as defined in eq. 6, with the quantity $E_{ads}/A^{\frac{1}{2}}$, where E_{ads} is the adsorption energy of an atom on a graphite grain calculated by Barlow (1978c), and A is the atomic mass. The supposition here is that a quasi-equilibrium will be established in the balance between atoms sticking to grains (at a rate proportional to their velocity and hence $A^{-\frac{1}{2}}$) and being removed at a rate inversely related to E_{ads}. The value of using m instead of the depletion at some particular density regime is based on the premise that differentials between the dense clouds and the intercloud medium indicate shifts in the balance between growth and destruction.

3.2. UPDATE

In this section, we will re-examine some of the plots discussed above. These diagrams differ somewhat from those which have appeared in the literature in that a) they are based on more recent and comprehensive information bases, b) data on some elements which, in the opinion of this author, are of inferior reliability have been omitted, and c) for the condensation temperature

---

[6]In reviewing the literature on this subject, the reader should note that many authors compared their depletions to that of ζ Oph rather than the warm, neutral, intercloud medium, not recognizing that the latter is a more appropriate standard for recognizing anomalously low depletions.

diagrams a distinction will be made on the basis of the density regime (see also Savage and Mathis 1979). Consideration c) not only permits us to examine different interpretations, but also offers slight modifications in the selections of elements (since some can be measured reliably only at high or low $\bar{n}(H_{tot})$).[7]

We shall first focus on the relationship between m and $E_{ads}/A^{1/2}$. Figure 3 presents this information based on the numbers in Table 3 with the addition of slopes for Ca and Na provided by Phillips, Pettini and Gondhalekar (1984), who used measurements of Mg I/Mg II and the methods described in §2.4 to correct for ionization. Their determinations probably have some error beyond what they recognized, since they thought they could determine $N(H_{tot})$ from $N(S\ II)$ and $N(P\ II)$ and assume that P and S were always undepleted.

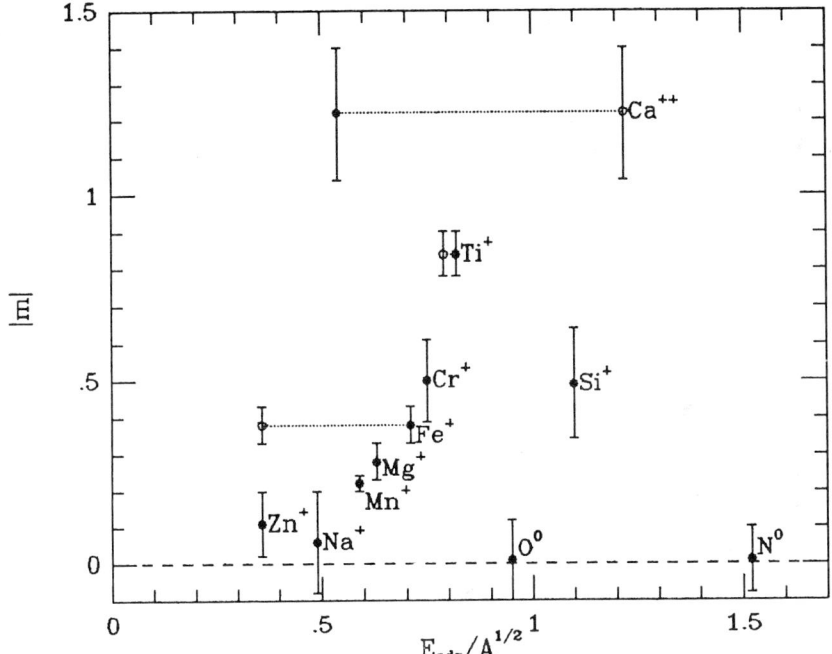

Figure 3 -- A plot comparing the sensitivity of logarithmic depletions D to the logarithm of the average density $\bar{n}(H_{tot})$ against Barlow's (1978c) adsorption energy E_{ads} divided by the square root of the atomic mass. Solid points are for atoms sticking to graphite grains, while open points are for bonds to silicate grains. The charge designations with each element show the dominant stage of ionization in H I regions.

The points in Fig. 3 do not exhibit the neat trend of Fig. 4b in Phillips, Pettini and Gondhalekar (1984), primarily because different elements are shown.

[7]For a recent compilation of depletions containing a broader selection of elements, see Cowie and Songaila (1986).

For capture on graphite grains, nitrogen and oxygen are probably at a disadvantage because they are neutral in H I regions while the other elements are ionized and thus attracted to the negatively charged graphite grains. According to Barlow (1978c) Ca, Ti and Fe will stick to silicate grains: with the extra (open) points added one can see that there is no serious inconsistency for these elements. Si seems to show a departure from trend which is not easily reconciled.

Figures 4, 5 and 6 show plots of D $vs.$ T_C under different circumstances (again, we must be watchful for differences in appearance due to the inclusion of different elements). In principle, a comparison of Figs. 4 and 5, representing the two extremes of $\bar{n}(H_{tot})$, might give us some insight on whether the density effect on depletions is governed primarily by grain growth in dense clouds or by protection from shocks. If the former alternative were the most important, we'd expect the base depletions to show a better relationship with T_C than the depletions at high density, while the converse would hold if grain destruction in the intercloud medium was the dominant effect. Unfortunately, the two diagrams seem to have about equal degrees of raggedness in their relationships. As pointed out by Jura and York (1978) phosphorus is a troublesome element; very recently this element has been joined by Mg, after a downward revision of the f-values for the Mg II λ1240 lines by 0.56 dex by Hibbert, et $al.$ (1983).

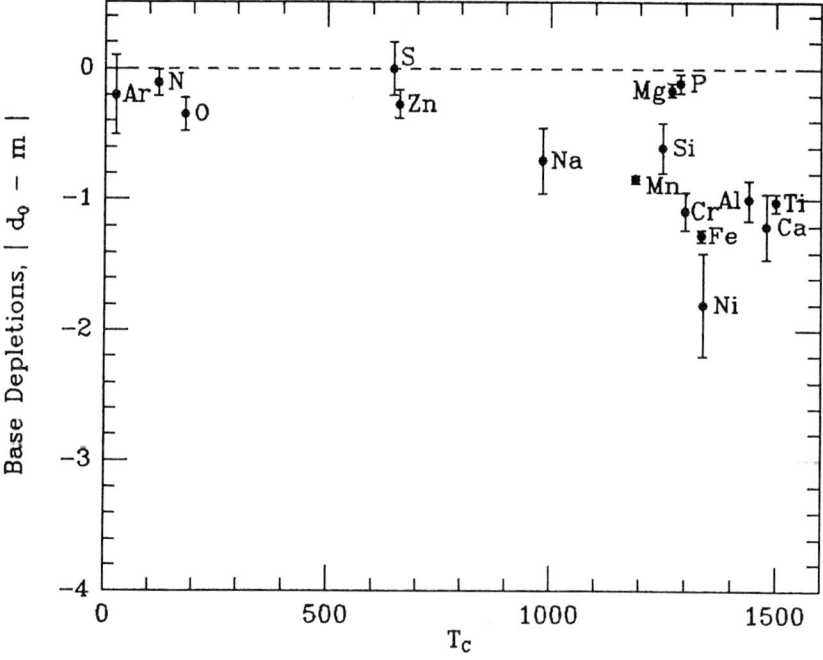

Figure 4 -- Logarithmic depletions for log $\bar{n}(H_{tot}) = -1.5$ $vs.$ the condensation temperature T_C.

Sources of information other than the coefficients in Table 3:

Al: Barker, *et al.* 1984.
S: Harris and Mas Hesse 1986b.
Ar: Shull 1977, Morton 1978, Morton and Bhavsar 1979, Vidal-Madjar, *et al.* 1982, Ferlet, *et al.* 1980.
Ni: Morton 1978.
Ca and Na: Phillips, Pettini and Gondhalekar 1984.

Values for T_C in Figs. 4-6 are from Field (1974), Grossman and Olsen (1974), and Wai and Wasson (1977).

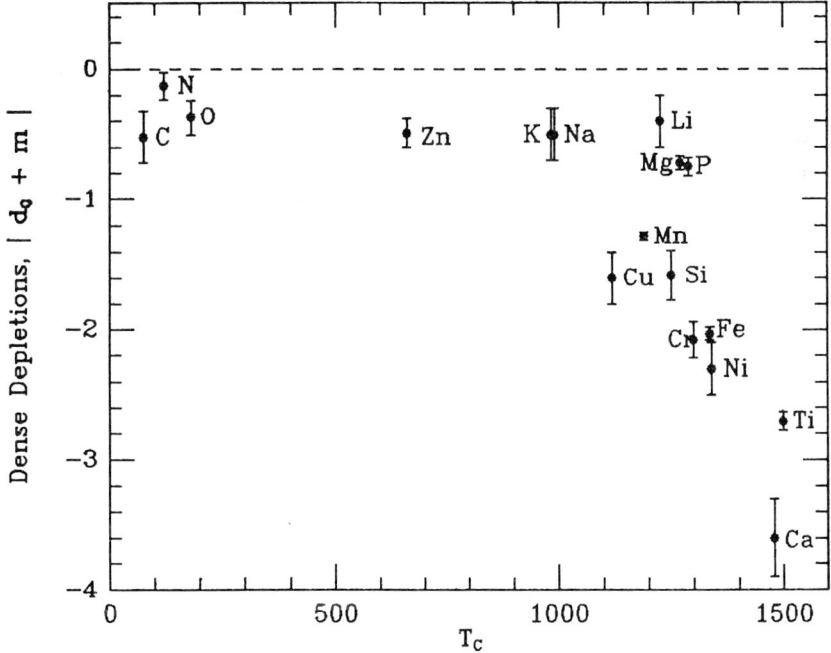

Figure 5 -- Logarithmic depletions for log $\bar{n}(H_{tot}) = 0.5$ *vs.* the condensation temperature T_C.

Sources of information other than the coefficients in Table 3:

Ni: Morton 1975, Jenkins, Savage and Spitzer 1986.
C: Jenkins, Jura and Loewenstein 1983, Hobbs, York and Oegerle 1982.
Ca: Phillips, Pettini and Gondhalekar 1984.
Na: Chaffee and White 1982, Phillips, Pettini and Gondhalekar 1984.
Cu: Jenkins, Savage and Spitzer 1986.
K: Chaffee and White 1982.
Li: White 1986.

Cardelli (1984) analyzed *IUE* spectra of B component (spectral type B2.5 V) binary system α Sco. This star shines through the circumstellar envelope of the M1.5 Iab primary. This measurement, in principle, represents an ideal way to learn how grains forming in a star's atmosphere will create the initial depletions for gases injected into space. Unfortunately, most of the elements reported in this study seriously violate our condition that their weakest lines should have $\tau_0 \lesssim 2$. Also, one might question his assumption that absorption from interstellar material is insignificant. Nevertheless, the neatness of the points in Fig. 6 (aside from Mg, which was revised upward in this presentation to account for the new f-values of the weak lines) entices us to speculate that the observations may be not far from correct.

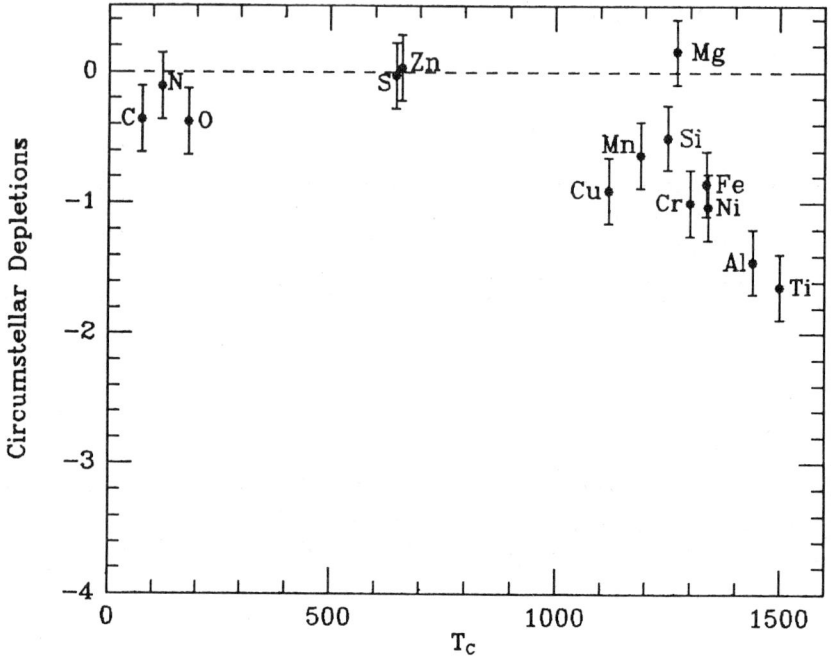

Figure 6 -- Logarithmic depletions for the envelope of α Sco A *vs.* the condensation temperature T_c. [From the results of Cardelli (1984) except for Mg, whose abundance was raised by 0.56 dex -- see text for details.]

4. H II REGIONS

Over the lifetime of a very dense H II region, volatile compounds should erode away from grains by sputtering (Draine and Salpeter 1979a). Refractory substances which contain Fe and Si are resistant to sputtering at temperatures below 10^5K however. Superficially, it is probably no surprise that data for the Orion Nebula, perhaps the densest H II region analyzed (using emission line strengths), show relative abundances consistent with those discussed in §3. For instance, the depletions of Fe and Cl are virtually identical to those we find for the dense H I clouds (see a review by Pottasch 1985). A similar conclusion follows from York's (1983) analysis of the H II region in front of λ Sco -- a very favorable case where the H II region's velocity is separated from that of the intervening H I. Relative to the abundance of S, York found the depletions of P and Fe to be close to those reported here for log $\bar{n}(H_{tot})$ = -1.5, that of Si II close to H I at log $\bar{n}(H_{tot})$ = -0.5 and Cl II corresponding to log $\bar{n}(H_{tot})$ = 0.5.

The interpretation presented above is probably overly simplistic, since it is based on the simple picture of a static, photoionized volume of gas. In reality, H II regions which surround stars with strong stellar winds will have their interior portions subjected to shocks which can heat the gas to $\sim 10^6$K (Castor, McCray and Weaver 1975, Weaver, et al. 1977) which in turn should rapidly destroy the grains. At the same time however, an H II region will sweep up new H I material as it evolves, and there should also be photoevaporation from embedded, dense H I clouds (McKee, Van Buren and Lazareff 1984, McKee 1986). These last two processes should drive the gaseous mixture in the direction of higher depletions.

5. RADIO RECOMBINATION LINES

As a final note, we should touch upon an observational tool which gives us some insight on the depletions in regions which are too dense to probe with ultraviolet or visible optical observations. If such clouds envelop O or B-type stars or border H II regions which contain associations of these very hot stars, there will be enough ultraviolet flux to dissociate molecules and support the ionization of atoms within an appreciably large parcel of gas (Pankonin 1980, Brown 1980). As a result of recombinations and collisional re-excitations, a sufficient number of the atoms have their outermost electrons in high n-levels which can decay radiatively in the radio spectrum. Line emissions from different elements are differentiated from each other by the small frequency shifts caused by changes in the electron's reduced mass.

In spite of some complicated modeling for the expected emissions of different elements from the clouds and the response of radio telescopes at different frequencies where beam dilution effects are important, the results seem to indicate that the ratio of carbon to sulfur in the dense cloud associated with ρ Oph is about 1/4 the solar ratio (Falgarone 1980) and the group of heavier elements (Mg, Fe and Si) are more heavily depleted (by factors 5-10)(Brown 1980). It is interesting to note from this example that the carbon depletion in a region with $n(H_{tot})$ = 3000 − 3×10^4 cm^{-3} is not an order of magnitude more than indications from the uv data for less dense clouds. This could be an anomalous case, however, since there seems to be a general consensus that there are fewer regions emitting detectable carbon recombination fluxes than expected.

The preparation of this paper was supported by grant NAG5-616 from the U. S. National Aeronautics and Space Administration.

REFERENCES

Adams, W. S. 1943, *Ap. J.*, **97**, 105.
Barker, E. S., Lugger, P. M., Weiler, E. J. and York, D. G. 1984, *Ap. J.*, **280**, 600.
Barlow, M. J. 1978a, *M. N. R. A. S.*, **183**, 367.
Barlow, M. J. 1978b, *M. N. R. A. S.*, **183**, 397.
Barlow, M. J. 1978c, *M. N. R. A. S.*, **183**, 417.
Barlow, M. J. and Silk, 1977, *Ap. J. (Letters)*, **211**, L83.
Barnard, E. E. 1910, *Ap. J.*, **31**, 8.
Barnard, E. E. 1919, *Ap. J.*, **49**, 1.
Beals, C. S. 1936, *M. N. R. A. S.*, **96**, 661.
Bohlin, R. C., Hill, J. K., Jenkins, E. B., Savage, B. D., Snow, T. P., Spitzer, L., and York, D. G. 1983, *Ap. J. (Suppl.)*, **51**, 277.
Bohlin, R. C., Savage, B. D., and Drake, J. F. 1978, *Ap. J.*, **224**, 132.
Brown, R. L. 1980, in *Radio Recombination Lines*, P. A. Shaver, ed., (Dordrecht: Reidel), p. 127.
Cardelli, J. A. 1984, *A. J.*, **89**, 1825.
Cardelli, J. and Böhm-Vitense 1982, *Ap. J.*, **262**, 213.
Castor, J., McCray, R., and Weaver, R. 1975, *Ap. J. (Letters)*, **200**, L107.
Chaffee, F. H. and White, R. E. 1982, *Ap. J. (Suppl.)*, **50**, 169.
Cowie, L. L. and Songaila, A. 1986, *Ann. Rev. Astr. Ap.*, **24**, in press.
Cowie, L. L., Songaila, A. and York, D. G. 1979, *Ap. J.*, **230**, 469.
Dalgarno, A. and McCray, R. A. 1972, *Ann. Rev. Astr. Ap.*, **10**, 375.
de Boer, K. S., Lenhart, H., van der Hucht, K. A., Kamperman, T. M., Kondo, Y., and Bruhweiler, F. C. 1986, *Astr. Ap.*, in press.
Draine, B. T. 1981, in *Physical Processes in Red Giants*, ed. I. Iben and A. Renzini, (Dordrecht: Reidel), p. 317.
Draine, B. T. and Salpeter, E. E. 1979a, *Ap. J.*, **231**, 77.
Draine, B. T. and Salpeter, E. E. 1979b, *Ap. J.*, **231**, 438.
Duley, W. W. 1980, *M. N. R. A. S.*, **190**, 683.
Duley, W. W. and Millar, T. J. 1978, *Ap. J.*, **220**, 124.
Dunham, T. 1937a, *Nature*, **139**, 246.
Dunham, T. 1937b, *Pub. A. S. P.*, **49**, 26.
Dunham, T. and Adams, W. S. 1937, *Pub. A. A. S.*, **9**, 5.
Falgarone, E. 1980, in *Radio Recombination Lines*, P. A. Shaver, ed., (Dordrecht: Reidel), p. 141.
Ferlet, R., Vidal-Madjar, A., Laurent, C., and York, D. C. 1980, *Ap. J.*, **242**, 576.
Field, G. B. 1974, *Ap. J.*, **187**, 453.
Forrest, W. J., Gillett, F. C., Houck, J. R., McCarthy, J. F., Merrill, K. M., Pipher, J. L., Puetter, R. C., Russell, R. W., Soifer, B. T., and Willner, S. P. 1978, *Ap. J.*, **219**, 114.
Frisch, P. C. 1980, *Ap. J.*, **241**, 697.
Gehrz, R. D. and Woolf, N. J. 1971, *Ap. J.*, **165**, 285.
Gómez-Gonzàlez, J. and Lequeux, J. 1975, *Astr. and Ap.*, **38**, 29.
Gondhalekar, P. M. 1985, *Ap. J.*, **293**, 230.
Grossman, L. and Larimer, J. W. 1974, *Rev. Geophys Space Phys.* **12**, 71.
Grossman, L. and Olsen, E. 1974 *Geochim. Cosm. Acta*, **38**, 173.
Harris, A. W. and Bromage, G. E. 1984, *M. N. R. A. S.*, **208**, 941.
Harris, A. W., Bromage, G. E., and Blades, J. C. 1983, *M. N. R. A. S.*, **203**, 1225.

Harris, A. W., Gry, C. and Bromage, G. E. 1984, *Ap. J.*, **284**, 157.
Harris, A. W. and Mas Hesse, J. M. 1986a, *M. N. R. A. S.*, **220**, 271.
Harris, A. W. and Mas Hesse, J. M. 1986b, preprint (RAL-86-017).
Hartmann, J. 1904, *Ap. J.*, **19**, 268.
Heger, M. L. 1919, *Lick Obs. Bull.*, **10**, 59.
Heger, M. L. 1921, *Lick Obs. Bull.*, **10**, 141.
Herbig, G. H. 1968, *Zs. f. Ap.*, **68**, 243.
Hibbert, A., Dufton, P. L., and Keenan, F. P. 1985, *M. N. R. A. S.*, **213**, 721.
Hibbert, A., Dufton, P. L., Murray, M. J., and York, D. G. 1983, *M. N. R. A. S.*, **205**, 535.
Hobbs, L. M. 1969, *Ap. J.*, **157**, 135.
Hobbs, L. M. 1974, *Ap. J.*, **191**, 381.
Hobbs, L. M. 1978, *Ap. J. Suppl.*, **38**, 129.
Hobbs, L. M., York, D. G. and Oegerle, W. 1982, *Ap. J. (Letters)*, **252**, L21.
Jenkins, E. B. 1986, *Ap. J.*, **304**, 739.
Jenkins, E. B., Jura, M., and Loewenstein, M. 1983, *Ap. J.*, **270**, 88.
Jenkins, E. B., Savage, B. D. and Spitzer, L. 1986, *Ap. J.*, **301**, 355.
Jenkins, E. B. and Shaya, E. J. 1979, *Ap. J.*, **231**, 55.
Jenkins, E. B., Silk, J., and Wallerstein, G. 1976, *Ap. J. (Suppl).*, **32**, 681.
Jenkins, E. B., Wallerstein, G., and Silk, J. 1984, *Ap. J.*, **278**, 649.
Joseph, C. L., Snow, T. P., Seab, C. G. and Crutcher, R. M. 1986, *Ap. J.*, in press.
Jura, M. 1974, *Ap. J. (Letters)*, **190**, L33.
Jura, M. 1976, *Ap. J.*, **206**, 691.
Jura, M. 1986, *Ap.J.*, **303**, 327.
Jura, M. and York, D. G. 1978, *Ap. J.*, **219**, 861.
Keenan, F. P., Hibbert, A. and Dufton, P. L. 1985, *Astr. Ap.*, **147**, 89.
Laurent, C., Paul, J. A., and Pettini, M. 1982, *Ap. J.*, **260**, 163.
Lugger, P., Barker, E., York, D. G., and Oegerle, W. 1982, *Ap. J.*, **259**, 67.
Martin, E. R. and York, D. G. 1982, *Ap. J.*, **257**, 135.
McCray, R. and Snow, T. P. 1979, *Ann. Rev. Astr. Ap.*, **17**, 213.
McKee, C. F. 1986, *Ap. Space Sci.*, **118**, 383.
McKee, C. F. and Ostriker, J. P. 1977, *Ap. J.*, **218**, 148.
McKee, C. F., Van Buren, D., and Lazareff, B. 1984, *Ap. J. (Letters)*, **278**, L115.
Morris, M., Redman, R., Reid, M. J., and Dickinson, D. F. 1979, *Ap. J.*, **229**, 257.
Morton, D. C. 1975, *Ap. J.*, **197**, 85.
Morton, D. C. 1978, *Ap. J.*, **222**, 863.
Morton, D. C. and Bhavsar, S. P. 1979, *Ap. J.*, **228**, 147.
Morton, D. C., Drake, J. F., Jenkins, E. B., Rogerson, J. B., Spitzer, L., and York, D. G. 1973, *Ap. J. (Letters)*, **181**, L103.
Morton, D. C. and Hu, E. M 1975, *Ap. J.*, **202**, 638.
Münch, G. 1968, in *Stars and Stellar Systems*, vol. **7** , G. P. Kuiper and B. M. Middlehurst, ed. (U. Chicago Press: Chicago), p. 365.
Murray, M. J., Dufton, P. L., Hibbert, A., and York, D. G. 1984, *Ap. J.*, **282**, 481.
Nachman, P. and Hobbs, L. M. 1973, *Ap. J.*, **182**, 481.
Oegerle, W. R. and Polidan, R. S. 1984, *Ap. J.*, **285**, 648.
Pankonin, V. 1980, in *Radio Recombination Lines*, P. A. Shaver, ed., (Dordrecht: Reidel), p. 111.
Phillips, A. P., Gondhalekar, P. M., and Pettini, M. 1982, *M. N. R. A. S.*, **200**, 687.
Phillips, A. P., Pettini, M., and Gondhalekar, P. M. 1984, *M. N. R. A. S.*, **206**, 337.

Phillips, T. G. and Huggins, P. J. 1981, *Ap. J.*, **251**, 533.

Pottasch, S. R. 1972, *Astr. Ap.*, **17**, 128.

Pottasch, S. R. 1985, in *The Milky Way Galaxy*, (IAU Symp. 106), H. van Woerden, R. J. Allen, and W. B. Burton, eds. p. 575.

Rogerson, J. B., York, D. G., Drake, J. F., Jenkins, E. B., Morton, D. C., and Spitzer, L. 1973, *Ap. J. (Letters)*, **181**, L110.

Routly, P. M. and Spitzer, L. 1952, *Ap. J.*, **115**, 227.

Ryter, C., Cesarsky, C. J., and Audouze, J. 1975, *Ap. J.*, **198**, 103.

Sarazin, C. L. 1977, *Ap. J.*, **211**, 772.

Savage, B. D. and Bohlin, R. C. 1979, *Ap. J.*, **229**, 136.

Savage, B. D., Bohlin, R. C., Drake, J. F., and Budich, W. 1977, *Ap. J.*, **216**, 291.

Savage, B. D. and Mathis, J. S. 1979, *Ann. Rev. Astr. Ap.*, **17**, 73.

Schattenburg, M. L. and Canizares, C. R. 1986, preprint.

Schmitt, J. H. M. M. 1985, *Ap. J.*, **293**, 178.

Seab, C. G. and Shull, J. M. 1983, *Ap. J.*, **275**, 652.

Shull, J. M. 1977, *Ap. J.*, **212**, 102.

Shull, J. M. 1979, *Ap. J.*, **233**, 182.

Shull, J. M. 1980, *Ap. J.*, **238**, 560.

Shull, J. M., Van Steenberg, M. and Seab, C. G. 1983, *Ap. J.*, **271**, 408.

Shull, J. M. and York, D. G. 1977, *Ap. J.*, **211**, 803.

Shull, J. M., York, D. G., and Hobbs, L. M. 1977, *Ap. J. (Letters)*, **211**, L139.

Siluk, R. S. and Silk, J. 1974, *Ap. J.*, **192**, 51.

Slipher, V. M. 1909, *Lowell Obs. Bull.*, no. 51.

Snow, T. P. 1975, *Ap. J. (Letters)*, **202**, L87.

Snow, T. P. 1976, *Ap. J.*, **204**, 759.

Snow, T. P. 1977, *Ap. J.*, **216**, 724.

Snow, T. P. 1984, *Ap. J.*, **287**, 238.

Snow, T. P. and Jenkins, E. B. 1980, *Ap. J.*, **241**, 161.

Snow, T. P. and Meyers, K. A. 1979, *Ap. J.*, **229**, 545.

Solomon, P. M. and Klemperer, W. 1972, *Ap. J.*, **178**, 389.

Spitzer, L. 1954, *Ap. J.*, **120**, 1.

Spitzer, L. 1968, *Diffuse Matter in Space*, (New York: Wiley Interscience).

Spitzer, L. 1976, *Comments Astr.*, **6**, 177.

Spitzer, L. 1978, *Physical Processes in the Interstellar Medium*, (New York: Wiley Interscience).

Spitzer, L. 1985, *Ap. J. (Letters)*, **290**, L21.

Spitzer, L. and Jenkins, E. B. 1975, *Ann. Rev. Astr. Ap.*, **13**, 133.

Stokes, G. M. 1978, *Ap. J. (Suppl.)*, **36**, 115.

Strömgren, B. 1948, *Ap. J.*, **108**, 242.

Struve, F. G. W. 1847, *Etudes d'Astronomie Stellaire*, Académie Impériale des Sciences, St. Petersburg.

Sutton, E. C., Storey, J. W. V., and Townes, C. H. 1978, *Ap. J. (Letters)*, **224**, L123.

Tielens, A.G.G.M. 1983, *Ap.J.*, **271**, 702.

Toombs, R. I., Becklin, E. E., Frogel, J. A., Law, S. K., Porter, F. C., Westphal, J. A. 1972, *Ap. J. (Letters)*, **173**, L71.

Unsöld, A., Struve, O. and Elvey, C. T. 1930, *Zs. f. Ap.*, **1**, 314.

van Dishoeck, E. F. and Black, J. H. 1986, *Ap. J. (Suppl.)*, in press.

Vidal-Madjar, A., Ferlet, R., Laurent, C., and York, D. G. 1982, *Ap. J.*, **260**, 128.

Wai, C. M and Wasson, J. T. 1977 *Earth Planet. Sci Letters*, **36**, 1.
Walborn, N. R. and Hesser, J. E. 1982, *Ap. J.*, **252**, 156.
Wallerstein, G. and Goldsmith, D. 1974, *Ap. J.*, **187**, 237.
Watson, W. D. and Salpeter, E. E. 1972, *Ap. J.*, **174**, 321.
Weaver, R., McCray, R., Castor, J., Shapiro, P., and Moore, R. 1977, *Ap. J.*, **218**, 377.
Werner, M. W., Beckwith, S., Gatley, I., Sellgren, K., Berriman, G., and Whiting, D. L. 1980, *Ap.J.*, **239**, 540.
White, R. E. 1974, *Ap. J.*, **187**, 449.
White, R. E. 1986, *Ap. J.*, in press.
Wilson, O. C. 1939, *Ap. J.*, **90**, 244.
Wilson, O. C. and Merrill, P. W. 1937, *Ap. J.*, **86**, 44.
Wynne, J. J. 1984, *Current Trends in Atomic Spectroscopy*, (Report of an NRC Committee Workshop on Line Spectra of the Elements -- Atomic Spectroscopy), (Washington: National Academy Press).
York, D. G. 1975, *Ap. J. (Letters)*, **196**, L103.
York, D. G. 1980, in *Astrophysics from Spacelab*, P. L. Bernacca and R. Ruffini, eds. (Dordrecht: Reidel), p. 609.
York, D. G. 1983, *Ap. J.*, **264**, 172.
York, D. G. and Jura, M. 1982, *Ap. J.*, **254**, 88.
York, D. G. and Kinahan, B. F. 1979, *Ap. J.*, **228**, 127.
Zeippen, C. J., Seaton, M. J., and Morton, D. C. 1977, *M. N. R. A. S.*, **181**, 527.

Michael Haas, Harriet Dinerstein and Lawrence Aller

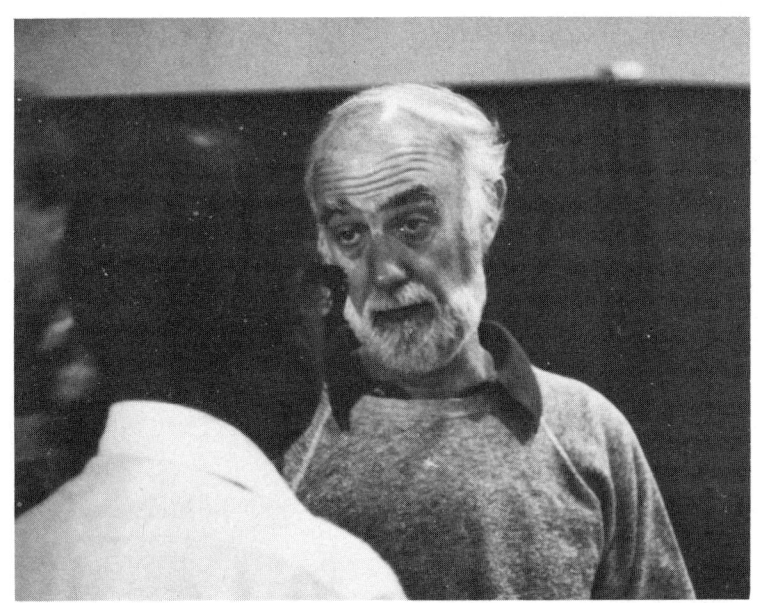

Alexander Dalgarno

CHEMICAL ABUNDANCES IN MOLECULAR CLOUDS

W. M. Irvine, P. F. Goldsmith
Five College Radio Astronomy Observatory
Department of Physics and Astronomy
University of Massachusetts, Amherst, MA   01003

Å. Hjalmarson
Onsala Space Observatory, Chalmers
University of Technology, S-43900 Onsala, Sweden

ABSTRACT.   At present approximately 70 interstellar molecules are known. We discuss methods for determining chemical abundances in interstellar clouds and present results for the best studied regions, which include the "spiral arm" clouds seen towards distant continuum sources, quiescent dark and giant clouds, and the gas in regions of active star formation. For many simple molecules abundances are rather uniform over a range of densities and temperatures in quiescent clouds, in accord with gas phase, ion-molecule chemical models. Some striking chemical differences do exist both within and among clouds, however, particularly in star-forming regions. This chapter is organized as follows:

D. J. Hollenbach and H. A. Thronson, Jr. (eds.), Interstellar Processes, 561–609.
© 1987 by D. Reidel Publishing Company.

1. INTRODUCTION

    The growing field of astrochemistry includes the measurement of the
chemical composition of the molecular component of the ISM; the determination of how this composition varies as a function of such physical
parameters as temperature and density, of cloud age and/or past history,
and of initial conditions or other relevant factors; the determination
of those processes and reactions which produce the observed abundances;
and the use of astronomical observations of interstellar material to
study the intrinsic properties of individual molecular species or reactions among molecules. In regard to the last point, note that the low
temperature, low density, and lack of containing walls in the ISM
provide a "laboratory" which has been successfully used to identify and
characterize a number of molecules, radicals, and ions, both prior to
and in more detail then has been possible in the terrestrial laboratory
(e.g., Thaddeus, 1981; Thaddeus et al., 1985a).
    The fundamental importance of astrochemistry becomes clearly
evident when we realize that molecular clouds are the most massive
objects in the universe whose composition is subject to the chemical
bond - their chemical nature and the processes producing that composition thus become problems fundamental to our understanding of the
universe around us. The unexpected characteristics of this chemistry
are typified, for example, by the realization that it has been less
than twenty years since the very existence of polyatomic molecules in
the ISM was discovered, and by the perhaps still surprising result
that, in a universe in which hydrogen is orders of magnitude more abundant than heavier reactive atoms, many of the organic constituents of
dense clouds have highly unsaturated chemical bonds.
    Moreover, astrochemistry is a subject important for more than its
own sake or to a few specialists. The trace molecules not only demonstrated the existence of a previously unknown component of the ISM,
dense molecular clouds, but they provide the major probes of the interior physics of these regions in which the processes of star formation
are otherwise hidden by optical obscuration due to dust. These molecules are indeed more than probes; they provide cooling and heating
pathways which govern cloud evolution, including the enhancement or
suppression of instabilities leading toward clump formation, fragmentation, and ultimately the formation of protostars. Thus, the physical
and chemical evolution of a cloud are intimately coupled and must be
analyzed together.
    On a more speculative note, it is still uncertain whether inter-

stellar molecular material was incorporated into objects forming in the
solar nebula, and if so, the subsequent role played by such material.
Such incorporation is suggested, however, by the deuterium/hydrogen
ratio found in certain carbonaceous components in some meteorites
(Kerridge and Chang, 1985; McKeegan, Walker, and Zinner, 1985) and by
inferences from the observed radicals and ions detected in comets
(e.g., Biermann et al., 1982). Critical information on this question
will surely be provided by the several probes to Halley's Comet, which
data is only just being analyzed at this writing. If a link can be
established between the solar nebula and interstellar material, it may
ultimately be possible to specify in some detail the type of interstellar cloud from which the Sun and planets formed. Moreover, if
those models in which the bulk of the Earth's volatiles were provided
by a late bombardment of cometary type material prove correct (cf.
Lazcano-Araujo and Oro, 1981), there may indeed be a fascinating
pathway leading from astrochemistry to the origin of life on Earth.

This chapter concentrates on the observational astrochemistry of
the gas phase component of denser interstellar clouds. The composition
of the particulate material (grains) in the interstellar medium is discussed in chapters by Seab, Tielens, and Allamandola, while one view of
the role of grains in the overall chemistry of molecular clouds is
provided in the second chapter by Tielens. Other recent discussions of
the nature and role of grains are provided in the volumes by Nuth and
Stencel (1986) and by Black and Matthews (1985). Astrochemical theory
is discussed in the chapter by Herbst, at recent symposia (Black and
Matthews, 1985; Tarafdar and Vardya, 1986) and (e.g.) by Dalgarno
(1986). The chemistry of the so-called diffuse clouds, which can be
observed by way of the absorption features produced in the spectra of
background stars at optical and ultraviolet wavelengths, presents problems different from but related to those discussed here. In particular,
the lower densities and higher ionizing fluxes produce a somewhat different chemical environment (e.g., Black, 1986).

Reviews of observational astrochemistry have been presented recently
by Irvine et al., (1985), Hjalmarson (1985), Guelin (1985) and in
several papers included in Tarafdar and Vardya (1986). We have endeav-
ored to update overlapping material from these earlier reviews, and
perhaps to present a somewhat more pedagogical approach. Following
this Introduction, Section 2 deals with the observational approach to
determining chemical abundances in interstellar sources, Section 3
presents results for some of the now standard sources, and Section 4
discusses these results. We have avoided multitudinous references to
original papers insofar as such research is cited in the above reviews
or more recent publications.

2. MEASUREMENT OF THE CHEMICAL COMPOSITION OF INTERSTELLAR CLOUDS

2.1. Introduction

The goal of determining the chemical composition as a function of
position within a molecular cloud is an ambitious one, and is reason-

ably enough broken down into various steps, only some fraction of which
are generally completed. Molecular observations at millimeter and
centimeter wavelengths have the complication that the population can be
distributed over a very large number of energy levels. The degree of
excitation is often ill-determined, meaning that the fractional population associated with the transition(s) being observed is not accurately
known. This problem can be overcome by observations of several transitions, and this technique is being increasingly utilized as the
frequency range available for high sensitivity measurements is broadened. A problem common to many wavelength ranges is that of line
saturation, which can be addressed by study of hyperfine ratios and
comparison with isotopically substituted variants, although neither
method is guaranteed to yield highly accurate results. Finally,
progressing from telescope beam-averaged column densities to local
distributions is a major challenge in view of the often complex but not
well understood structure of molecular clouds. Given these obstacles,
it is not surprising that determination of the chemical composition of
molecular clouds is fraught with uncertainties, and that even the
errors are difficult to determine with accuracy.

2.2. Optically Thin Emission

This is clearly the most favorable situation for column density
determinations. If we consider a homogeneous cloud and observe a
transition at frequency $v = (E_u - E_\ell)/h$, the antenna temperature of
the line above the background (observed by a lossless antenna above
the atmosphere and perfectly coupled to the source) is given by (cf.
Penzias, 1975)

$$\Delta T_A = \{T_* [\exp(T_*/T_{EX}) - 1]^{-1} - T_{BG'}\} \{1 - \exp(-\tau)\} \qquad (1)$$

where $T_* = hv/k$, T_{EX} is the excitation temperature of the transition
being observed, and τ is its optical depth. $T_{BG'}$ is the antenna temperature produced by the background; in the case of an optically thick
blackbody at T_{BG}, it is just $T_*/(\exp(T_*/T_{BG}) - 1)$ which is equal to T_{BG}
in the Rayleigh-Jeans limit ($T_*/T_{BG} \ll 1$). In addition to the cosmic
background radiation, the background at long wavelengths can include
free-free emission and synchrotron radiation, and at short millimeter
wavelengths dust emission can be significant. If T_{EX} is large compared to T_{BG} we can neglect the background term; this is the case for
typical interstellar clouds, although caution is necessary in dealing
with large dipole moment molecules in cold clouds. Of course, neglect
of T_{BG} must be justified in terms of some other method of determining
T_{EX}. If optically thin emission makes a major contribution to the
background, it is more difficult to show analytically when its effect
might be appreciable. This is rarely a major consideration except for
maser amplification by an inverted population distribution (cf. Broten
et al. 1976), but column density determination in this situation is in
any case of doubtful accuracy.

For optically thin lines, ignoring the background, we have

$$T_A = \tau T_* / [\exp(T_*/T_{EX})-1] \quad . \tag{2}$$

The optical depth is

$$\tau = h\nu\phi_\nu [B_{\ell u} N_\ell - B_{u\ell} N_u]/c \tag{3}$$

where ϕ_ν is the line profile function ($\int\phi_\nu d\nu = 1$), $B_{\ell u}$ and $B_{u\ell}$ are the stimulated emission coefficients for lower to upper and upper to lower state transitions, respectively, and N_ℓ and N_u are the column densities in the two states. With the relations (g_ℓ and g_u are the statistical weights)

$$N_\ell/N_u = (g_\ell/g_u) \exp(T_*/T_{EX}), \tag{4a}$$

$$B_{\ell u}/B_{u\ell} = g_u/g_\ell, \tag{4b}$$

we obtain

$$T_A = h^2\nu^2\phi_\nu B_{u\ell}N_u/ck \quad . \tag{5}$$

From the relation between stimulated and spontaneous rates, $A_{u\ell} = B_{u\ell}8\pi h\nu^3/c^3$, we can write

$$T_A = (hc^2\phi_\nu/8\pi k\nu) A_{u\ell}N_u \quad . \tag{6}$$

This expression gives the antenna temperature as a function of frequency, in terms of the line profile function. Although discarding considerable kinematic information as well as possible variations with velocity along the line of sight, integrating over the line does at least improve the signal-to-noise ratio. If we express this in velocity units, we find

$$\int T_A dv \text{ [K km s}^{-1}] = 10^{-5} (hc^3/8\pi k\nu^2) A_{u\ell}N_u. \tag{7}$$

Rewriting this in terms of convenient units we find

$$N_u[cm^{-2}] = 1.94\times10^{+3} \nu^2[GHz] \int T_A dv[K \text{ km s}^{-1}]/A_{u\ell}[s^{-1}]. \tag{8}$$

We see that the column density in the upper level is directly proportional to the antenna temperature observed. Since in the optically thin limit we are simply counting photons produced by spontaneous emission, the preceeding expression represents the upper level column density integrated along the line of sight irrespective of variations in the molecular excitation. If the optically thin limit is established (discussed further in Section 2.6 below), the next step is to obtain the total column density of the species in question.

2.3. Determination of Total Molecular Column Density

The simplest model for obtaining the total column density of a particular molecular species from observations of a single transition

is to assume that all levels are populated according to a Boltzmann
distribution at a single temperature T. Then the total population is
related to that in the upper level of the observed transition by

$$N_{TOT} = F(u)N_u \quad , \tag{9a}$$

$$F(u) = f_u{}^{-1} = Z\exp(E_u/kT)/g_u \quad , \tag{9b}$$

$$Z = \sum_i g_i \exp(-E_i/kT) \quad . \tag{9c}$$

Here, f_u is the fractional population of the upper level of the
transition observed, and Z is the partition function. The use of this
equilibrium population distribution can be motivated by a number of
arguments.

For a single optically thin transition with spontaneous decay rate
A, a critical density may be defined by $n^* = A/\langle\sigma v\rangle$ where $\langle\sigma v\rangle$ is a
characteristic rate coefficient for collisional de-excitation of the
transition. The excitation temperature of a single transition will
approach the gas kinetic temperature when the density exceeds the
critical density.

If the collisional de-excitation rate exceeds the spontaneous decay rate (i.e., $n > n^*$) for all transitions with significant population
at temperature T, then the assumption of thermalized level populations
is essentially correct and the overall molecular excitation will be
characterized by the gas kinetic temperature. For a linear molecule,
the level which has half the fractional population of the level with
maximum fractional population is

$$J' = (2T/B_0)^{1/2}, \tag{10}$$

where B_0 is the molecular rotational constant expressed in Kelvins.
For ^{13}CO B_0 is 2.65 K and J' is 5 for T = 35 K. The corresponding
spontaneous decay rates are given by

$$A_{JJ-1}(s^{-1}) = 8.4 \times 10^{-7}(\mu_0[D])^2(B_0[K])^3 J^4/(2J+1), \tag{11}$$

where μ_0 is the permanent dipole moment in Debyes (= .112 for carbon
monoxide). Thus A_{54} is 1.3×10^{-5} s^{-1} for ^{13}CO. Taking a typical do-
excitation rate coefficient for CO, $\langle\sigma v\rangle = 10^{-10}$ cm^3 s^{-1}, a density
$n(H_2) = 1.3 \times 10^5$ cm^{-3} is required to assure a thermalized population
distribution at the 35 K temperature considered above. Most molecules
have much larger permanent dipole moments than CO and hence require
higher densities for thermalized population ladders.

Except for the case of thermalization at the kinetic temperature,
there is little reason to expect a single temperature to characterize
the level populations, since the spontaneous decay rates are a rapidly
increasing function of transition frequency. The situation can be
examined by use of a rotation diagram obtained by plotting upper level
population versus upper state energy. From eq. 8, we can write this
relation as (in cgs units except as indicated)

$$\ln\{(10^5 8\pi k\nu^2/g_u A_{u\ell} hc^3)\int T_A dv[K \text{ km s}^{-1}]\} = \ln(N_{TOT}/Z) - (E_u/kT) \quad (12)$$

If several transitions have been observed and the relation follows a linear distribution on a semi-logarithmic plot as a function of E_u, then, as indicated by the above equation, we can determine the single temperature which characterizes the population distribution. Note that this technique is not restricted to linear molecules, but for other structures the appropriate spontaneous decay rates rather than those given by eq. 11 must be used. The partition function for a linear molecule in equilibrium is particularly simple, with $Z=T/B_0$ for $T \gg B_0$ (expressing B_0 in kelvins). The partition functions for symmetric top molecules and for asymmetric tops are discussed by Townes and Schawlow (1955; p. 75 and 101). For a number of other molecules of astrophysical interest, Poynter and Pickett (1980) give the partition function, but only at 300K. The accuracy of the approximate expressions, with $Z \propto T^{3/2}$ for symmetric and asymmetric top molecules, will not be very good for T<50K, and a numerical summation over the energy levels is necessary for high accuracy.

Some examples of this technique taken from observations of the giant molecular cloud Sgr B2 by Cummins, Linke, and Thaddeus (1986) are shown in Figure 1. The quality of the fits is variable. Since the structure of the Sgr B2 cloud core, discussed in Sections 3.2 and 4.3, is complex, it is hardly surprising that a distribution at a single temperature fails to reproduce the relative population of levels having a wide range of E_u. Nonetheless, one can assess from such a plot whether a single temperature gives a sufficiently accurate method of calculating the partition function and hence of determining the total column density. This can certainly be the case even if details of the excitation are not properly described by such a simple model.

2.4. Multi-Transition Studies

More detailed models of molecular excitation can be developed which are particularly relevant if data from several transitions are available to constrain the critical parameters of H_2 density, kinetic temperature, and total molecular column density. If at least two transitions are available, one can attempt to reproduce the observed intensities by fitting the results of a statistical equilibrium model. Of course accurate collisional rates enhance the value of such modeling. For two transitions, a solution can generally be found, but not all parameters will be determined. For example, if one has data for the J=1-0 and J=2-1 lines of a rare isotope of CO for which the emission is optically thin, the hydrogen density can be determined if the kinetic temperature is assumed. In Figure 2 we show the implications for determining the total column density in terms of F (eq. 9a). The figure indicates that the J=2-1 transition is subject to less uncertainty in terms of model parameters than the J=1-0 line, for determining the total column density, if the hydrogen density is greater than 10^3 cm^{-3}. This is because the J=2 level (E_u = 16.5 K) is closer to the center of the population distribution, even for moderately subthermal excitation.

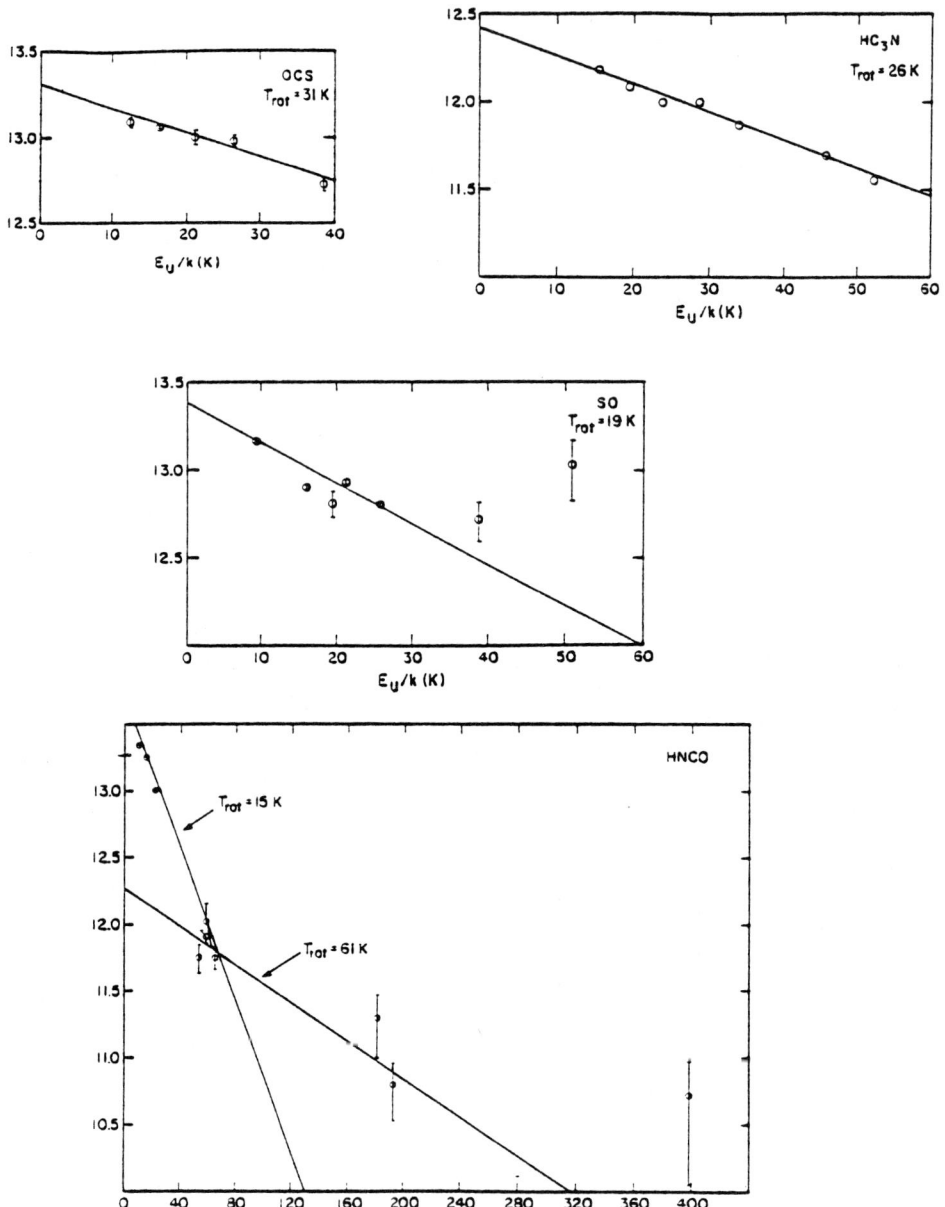

Figure 1.   Rotation diagrams for different molecular species observed towards Sgr B2 by Cummins, Linke, and Thaddeus (1986).   The ordinate is the left hand side of eq. 12 expressed as a base 10 logarithm.

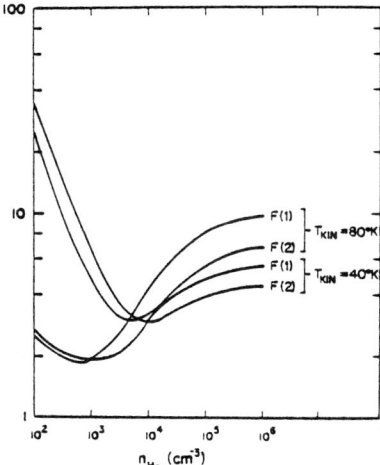

Figure 2. F(u) function (eq. 9) for CO, relating the total column
density to that in the upper level of an observed transition, for J=1-0
and J=2-1. Values of F(1) and F(2) are plotted for kinetic temperatures
40 K and 80 K as a function of hydrogen density.

     Multi-transition studies become increasingly powerful as the number
of lines observed increases. For determination of column densities, it
is particularly important that the upper state energies cover a range
comparable to the width of the distribution of populated levels. A
study of three molecular cloud cores using four transitions of the CS
molecule was carried out by Snell et al. (1984). A similar study using
$C^{34}S$ by Mundy et al. (1986) yielded essentially consistent results.
In both cases, the intensities of the observed transitions were calculated using a set of models varying n (the hydrogen density) and N_{CS}
(the CS column density), with the kinetic temperature taken from ^{12}CO
data. The "best fit" model was determined by minimizing the χ^2 of
the differences between model and data. The results for two positions
in each cloud are shown in Figure 3, where in the upper row representative positions having good fits and in the lower row having poor fits
are included. The best fit n and N_{CS} are also indicated. The best
fit densities do not give excitation temperatures which are independent
of J, since the transitions are subthermally excited. Most of the CS
transitions in these cloud cores also have opacities $\simeq$ 1, so that
radiation trapping affects the excitation temperatures as well as
emergent line intensities. For the 5-4 transition the critical density
$n^* = A/<\sigma v>$ is $7x10^6$ cm^{-3}, a factor of 5-15 greater than the best fit
densities. Although the results of this study have important implications for the density distribution and structure of the cloud cores,
the key point for the determination of abundances is that the column
density of CS is very well determined. In Figure 4 we show the variation in n and N_{CS} produced by changing the kinetic temperature of the
model cloud. Although the hydrogen density is appreciably affected,
the molecular column density is essentially unchanged, with the varia-

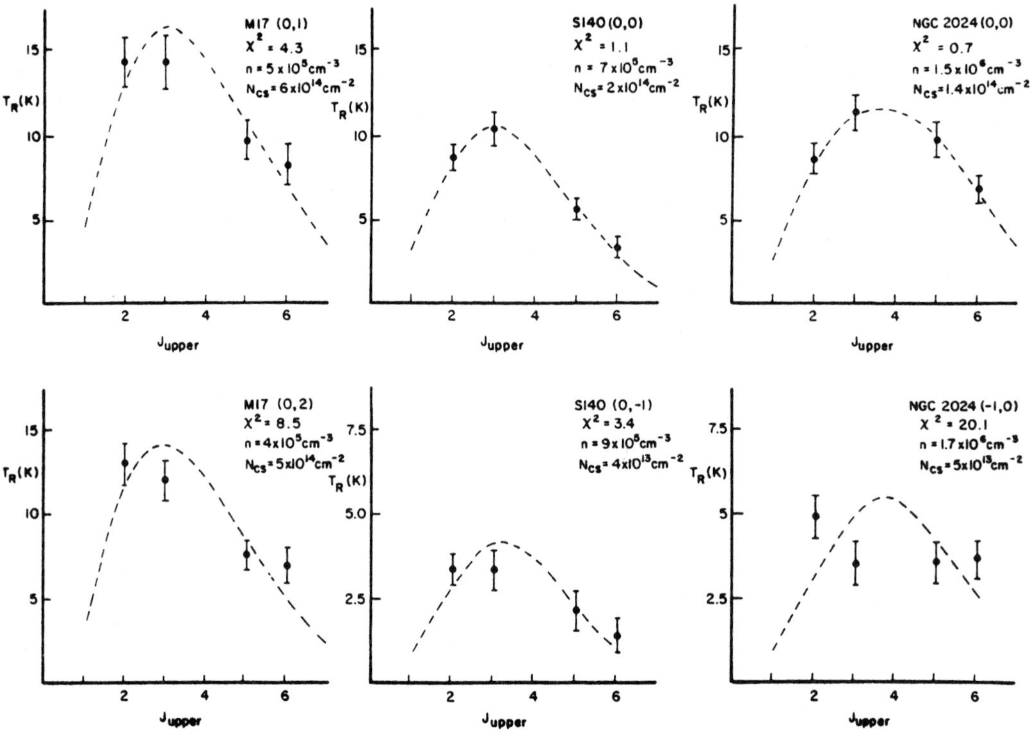

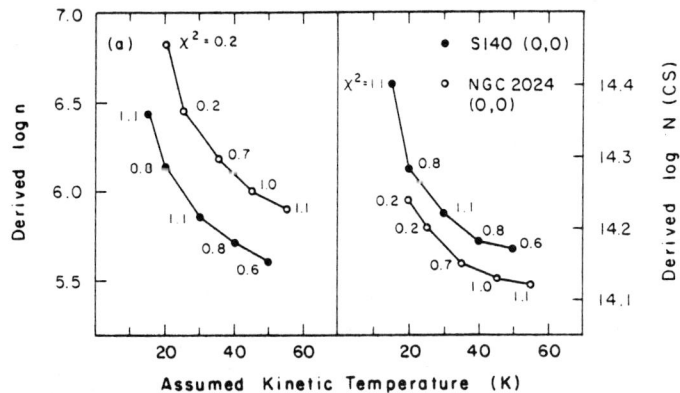

Figure 3. Observed and calculated (best fit) intensities for four CS transitions observed in 3 sources by Snell et al. (1984).

Figure 4. Variation of hydrogen density (n) and CS column density (N(CS)) as functions of assumed kinetic temperature, for central positions of two clouds studied by Snell et al. (1984).

tion in log N_{CS} < 0.2. This is due to the model itself being fairly well constrained by the data. It is also a consequence of the four upper levels for which the column densities have been determined containing a large fraction of the total molecular column density under any plausible set of excitation conditions. A rotation diagram analysis of the 4 transitions observed, ignoring optical depth effects, gives column densities (for the central positions of the clouds) which agree with the modelling results within a factor of two. This might be taken as one indication of the uncertainties produced by the rotation diagram method for relatively uniform clouds; clumping and/or major variations in physical conditions along the line of sight will certainly decrease the accuracy of column density determinations. The multitransition methods are less susceptible to these complications (cf. discussion by Snell et al. 1984 and Mundy et al. 1986).

2.5. Optically Thick Emission

Optically thick lines present a challenge for determination of column densities since the velocity field and hence the radiative transfer in molecular clouds is not well understood. The most often-used model (LVG ≡ large velocity gradient) assumes that large scale motions dominate the velocity field, although there is virtually no detailed evidence that this is the case. The issue of mass motions is particularly critical for clouds with embedded heating sources where large temperature gradients are present. The gas temperature at the depth in the cloud where lines become moderately thick plays a critical role in determining the line intensity, and, as the temperature distribution is not accurately known, large uncertainties in the column density result (cf. Goldsmith and Mao 1983).

The availability of multi-transition data mitigates the errors introduced by optical depth effects, as indicated by the agreement in hydrogen density determined by the studies of CS by Snell et al. (1984) and $C^{34}S$ by Mundy et al. (1986). In addition, simulations of two component models (high and low density material) fit to the data showed that the effect of quite substantial opacity in the foreground gas (τ=2) produced only a factor of 3 change in the derived CS column density. This is a result of the partially compensating effects of line saturation on high- and low-J CS transitions in terms of the total derived molecular column density. Thus, multi-transition data appears of particular value in column density determinations of species with optically thick emission. It would be of great value to include the lowest transition(s) to ensure that low excitation material has not been excluded from the total column density.

In the limit of weak emission lines, the emitted intensity can be proportional to the column density of the molecular species in question, irrespective of the optical depth of the transition being observed. This is due to the fact that every upwards collision ultimately results in a photon escaping from the cloud if the downwards collision rate is much slower than the effective spontaneous decay rate (Penzias 1975). This condition can be written

$$\langle \sigma v \rangle \; n_{H2} < \beta A,$$

where β is the photon escape probability. This converts to the limit on the observed antenna temperature

$$T_A < (T_\star/4) \; \exp(-T_\star/T_{KIN}),$$

(Linke et al. 1977), where T_{KIN} is the gas kinetic temperature and the effect of any background radiation has been ignored. For a typical millimeter transition with $T_\star = 4$ K arising in a medium with $T_{KIN} = 20$ K, the above indicates that lines with $T_A < 0.8$ K will be in the limit where antenna temperature is proportional to column density whether they are optically thin or thick. This method is susceptible to problems if there are major temperature variations along the line of sight, but when applied with appropriate caution can be an important aid in analyzing emission when little is known about the opacity of the transition observed.

2.6. Optical Depth Determination

The two most widely used techniques for determination of optical depths are hyperfine component and isotopic variant intensity ratios. Both of these techniques give moderate accuracies, but neither can be convincingly demonstrated to give a definitive determination of the optical depth.

Hyperfine intensity ratios are in principle a very useful tool; the splitting is often larger than the linewidths, but generally sufficiently small that the different components can be observed simultaneously, thus reducing calibration problems. This technique has been used extensively with NH_3 (cf. Ho and Townes 1983), CN (Allen and Knapp 1978; Churchwell 1980) and HCN (Wannier et al. 1984; Gottlieb et al. 1975), as well as other molecules. In LTE the observed ratio of the hyperfine components and the relative line strengths together can be used to determine the opacity. However, when examined in detail, this procedure has serious possibilities for erroneous results. The causes include different excitation temperatures for the different hyperfine components due to saturation of other transitions (Kwan and Scoville 1975) and due to line overlaps (Guilloteau and Baudry 1981). It is also possible that collisional discrimination among hyperfine levels (Stutzki and Winnewisser 1985a) will significantly perturb the populations. At least for NH_3, far IR excitation can also upset the relative hyperfine populations (Stutzki and Winnewisser 1985b). The situation at present is not entirely clear, but most observers do not feel that HCN hyperfine ratios are of any use in deriving opacity (since, for example, the ratios often cannot be matched by any homogeneous cloud model, irrespective of optical depth). Ammonia hyperfine component ratios are taken as indicative of the optical depth of transitions in dark clouds (where there are negligible effects from IR radiation). In some giant molecular clouds the use for determining column density is compromised by the result that the distribution of molecular material appears to be highly inhomogeneous (cf. Ho and

Townes 1983). In dark clouds this does not appear to be a major effect, and hyperfine ratios can be used with moderate accuracy (~30%; Stutzki and Winnewisser 1985c) to correct for opacity effects and thus improve the accuracy of column density determinations.

The relative intensity of lines from different isotopic species can be an indicator of optical depth if the isotopic abundance ratio is known. This is rarely the case (at least to high accuracy) for molecular clouds, where both variations in elemental composition (cf. Penzias 1980) and chemical isotopic fractionation can change the isotope ratio for a particular molecular species in a temperature-and position-dependent manner (see, e.g., Langer et al. 1984). When combined with the possibility that for the same transition different isotopic species may have different excitation temperatures as a result of radiative trapping, it is clear that accurate determinations of opacity cannot be obtained with this technique. Nevertheless, it is widely used to determine that a particular spectral line is not highly saturated. For example, if $T_A(^{12}CO)/T_A(^{13}CO)=30$, then it is reasonable to conclude that $30<N(^{12}CO)/N(^{13}CO)<100$ (a reasonable upper limit to the ratio), from which it follows that $\tau(^{12}CO)<3.2$, assuming that the two species have equal excitation temperatures (cf. eq. 1). This can be used to make an approximate correction for the column density of ^{12}CO in the upper level of the observed transition, but the problem of the partition function remains. For two isotopic species whose relative abundance has been established by observations of a variety of molecular species, and where the observed ratio in the species in question is close to the standard value, it is plausible that the lines being observed are optically thin. This technique is of particular use with isotopic pairs for which the expected abundance ratio is not too large, e.g. $^{32}S/^{34}S$, for which the solar system ratio is 22.5.

2.7. Cloud Structure and Molecular Abundances

Converting molecular column densities to abundances is necessarily a difficult task, given the complex structure of molecular clouds. The uncertainties, in general, increase as one progresses from more global quantities, such as beam-averaged column density, to local quantities such as number density.

The most directly determined quantity from spectral line observations (after some assumptions about molecular excitation) is, as discussed in preceeding sections, the column density of a particular species. This is a telescope beam-averaged quantity in that it refers to the column density (cm^{-2}) spread over the entire beam area projected through the cloud that is required to produce the observed line intensity (or intensities). [The actual size of the region to which this column density refers is a function of the coupling efficiency used to correct the observed antenna temperature to a brightness temperature. Coupling efficiency to a uniform disk source is often used, in which case column density refers to that region. If coupling to a source having a Gaussian distribution is used, the column density derived refers to peak values (cf. Ulich and Haas 1976)]. The data given in Table IV are an example of this relatively unprocessed type of measurement. Even determination of the beam-averaged column density is subject to errors due to variations in excitation along the line of sight (arising from regions with differing gas temperature or density), as well as optical depth effects; these have been touched on in the preceeding sections. With availability of multiple transitions and avoidance of very strong and hence likely saturated lines, these problems can be contained to the extent that beam-averaged column density can be measured to within a factor of two except in unusually complex sources.

The beam-averaged fractional abundance is generally defined as the ratio of the beam-averaged column density of a given species compared to that of H_2; the latter quantity is itself determined indirectly. Some of the methods are (1) using CO (or one of its isotopic variants) as a H_2 column density tracer (cf. Frerking, Langer, and Wilson 1982), (2) using continuum emission from dust as a tracer of total column density (cf. Hildebrand 1983), and (3) using visual extinction as a tracer of hydrogen column density (cf. Bohlin, Savage, and Drake 1978). Each method has limits to its validity and particular uncertainties. There is also the possibility of source to source variations; these do not seem to be very large. The abundances given in Table III (see Section 3.3) are referenced to H_2 column densities obtained from CO column densities and with the assumption of a CO abundance ratio of 8×10^{-5} with respect to H_2 (cf. Irvine et al., 1985 and Guelin, 1985).

It is of course true that if significant clumpiness is present, the molecules are concentrated in a fraction of the cloud volume. Consequently, the actual density of a particular species will certainly be higher than derived from the beam-averaged column density. If there is a way to estimate the total clump volume and the hydrogen density there, actual fractional abundances can be obtained to compare, for example, with chemical reaction networks.

The beam-averaged abundance estimates with respect to a reference species derived when the cloud is highly inhomogeneous can be misleading if the reference column density tracer has significantly different dependence on density than the molecular species of interest. For example, simultaneous interpretation of ^{13}CO J=2-1 and 1-0 line intensities and CS line ratios led Plambeck and Williams (1977) to suggest clumps of density $n = 10^6$ cm^{-3} immersed in $n = 2 \times 10^3$ cm^{-3} interclump medium. Mundy et al. (1986) conclude that in the presence of this degree of clumpiness, multitransition CS data and modelling significantly underestimate the beam-averaged column density and thus the abundance, if the area filling factor is <0.5. In this situation, the CS emission is primarily from the clumps, while CO emission arises from both high and low density components. Consequently, the CS to CO ratio derived could be seriously in error. Based on the increasing evidence for clumping in molecular cloud cores (e.g. Snell et al. 1984; Wilson 1985a,b; Mundy et al. 1986) this may well be a real issue, as well as likely the most difficult obstacle to be overcome in the determination of molecular fractional abundances in interstellar clouds.

3. RESULTS

3.1. Identified Interstellar Molecules

New molecular constituents of the ISM continue to be regularly
identified by radio astronomical techniques at short-centimeter, millimeter, and submillimeter wavelengths. Since a review of interstellar
chemistry two years ago (Irvine et al., 1985; see also Guelin, 1985),
several species have been added (in some cases tentatively) to the list
of known interstellar molecules, as shown in Table I. The ions $HCNH^+$
(Ziurys and Turner, 1986), H_2D^+ (Phillips et al., 1985), and SO^+
(Churchwell et al., 1986) are expected constituents of molecular clouds,
as is the radical C_5H (Cernicharo et al., 1986) and the closed-shell
molecule CH_3C_5N (Snyder et al., 1984); nonetheless, their detection
provides important new information on the chemical state of the ISM.
In contrast, the identification of the first interstellar hydrocarbon
ring, C_3H_2 (Thaddeus et al., 1985b; Matthews and Irvine, 1985), was
something of a surprise, and reopens questions about the relative
abundance of cyclic species in the interstellar gas; and the detection
of HCl (Blake et al., 1985) provides the first chlorine-containing
molecule in the ISM and gives information on depletion of chlorine in
dense regions. It is interesting that three of these possible identifications involved observations in the still largely unexplored
sub-millimeter region ($\nu > 300$ GHz), while five were made in the more
familiar portion of the radio spectrum.
The degree of confidence ascribed to the new identifications varies
(as it does for some of the previously "known" interstellar molecules).
In the case of H_2D^+, for example, only one transition has been
observed, and confirmation by the detection of additional lines is
clearly required. The cases for HCl and SO^+ are somewhat more secure,
because two-line hyperfine or Λ-doubling patterns were observed. In
contrast, the identifications of C_3H_2 and C_5H are now completely
definitive, resting on a multitude of astronomical and laboratory lines
(Bogey and Destombes, 1986; Gottlieb et al., 1986; Vrtilek et al.,
1986) for both the main isotopic species and (for C_3H_2) ^{13}C variants.
Both CH_3C_5N and $HCNH^+$ fall somewhat in between, since multiple
transitions have been observed, but the possibility of "interlopers"
produced by other molecular species is not entirely ruled out.
The detection of two molecules listed in previously published lists
of interstellar species has been questioned, based on recent data. The
detection of HNO has not been supported by searches for additional
transitions (Guelin, 1985), while the evidence for NaOH seems too weak
to include it here (the two transitions observed have very different
line shapes and occur at different source velocities, and it has been
suggested that they may be assignable to other species such as isotopes
of H_2CS). In addition, a line corresponding in frequency to a transition of H_3O^+ has been observed (Wootten et al., 1986), but may in
fact result from an unassigned line of CH_3OH (Herbst, private communication). Another interesting case is HOC^+. The sole detected line,
in Sgr B2, agrees within measurement uncertainty with laboratory data.
There has been considerable experimental and theoretical research on

TABLE I.

INTERSTELLAR MOLECULES

<u>SIMPLE HYDRIDES, OXIDES, SULFIDES, AND RELATED MOLECULES:</u>

H_2	CO	NH_3	CS
HCℓ	SiO	SiH_4*	SiS
	H_2O	CH_4*	OCS
	SO_2		H_2S
	CC		HNO ?

<u>NITRILES, ACETYLENE DERIVATIVES, AND RELATED MOLECULES:</u>

HCN	$HC{\equiv}C-CN$	$H_3C-C{\equiv}C-CN$	H_3C-CH_2-CN
H_3CCN	$H(C{\equiv}C)_2-CN$	$H_3C-C{\equiv}CH$	$H_2C=CH-CN$
CCCO	$H(C{\equiv}C)_3-CN$	$H_3C-(C{\equiv}C)_2-H$	$HN=C$
$HC{\equiv}CH$*	$H(C{\equiv}C)_4-CN$	$H_3C-(C{\equiv}C)_2-CN$	$HN=C=O$
$H_2C=CH_2$*	$H(C{\equiv}C)_5-CN$		$HN=C=S$

<u>ALDEHYDES, ALCOHOLS, ETHERS, KETONES, AMIDES, AND RELATED MOLECULES:</u>

$H_2C=O$	H_3COH	$HO-CH=O$	H_2CNH
$H_2C=S$	H_3CCH_2OH	$H_3C-O-CH=O$	H_3CNH_2
$H_3C-CH=O$	H_3CSH	$H_3C-O-CH_3$	H_2NCN
$NH_2-CH=O$		$H_2C=C=O$	

<u>CYCLIC MOLECULES:</u>   <u>IONS:</u>

C_3H_2	CH^+	HCS^+
SiC_2	H_2D^+ ?	$HCNH^+$
	HN_2+	SO^+
	$HOCO^+$	HOC^+ ?
		HCO^+

<u>RADICALS:</u>

CH	C_2H	CN	HCO
OH	C_3H	C_3N	NO
	C_4H	NS	SO
	C_5H		

* Detected only in the envelope around the evolved star IRC+10216.
? Claimed but not yet confirmed.

the HOC^+/HCO^+ system, which seems to suggest that detectable HOC^+ should only exist in the ISM under rather special circumstances (e.g., Jarrold, et al., 1986; cf. also McMahon and Keberle, 1985). A sensitive search for a confirming line (which practically speaking has to be the J=3-2 transition at 268 GHz) has not yet been carried out, in part because of the expected weakness of the line.

Finally, a series of doublets in the spectrum of IRC+10216 may correspond to the radical HSiCC or to HSCC (Guelin et al., 1986).

Two aspects of the list of molecular species in Table I provide immediate evidence of the very non-equilibrium character of interstellar chemistry. The most obvious of these is the number of highly reactive radicals and ions that have been identified. The radicals in particular, containing by definition one or two unpaired electrons, are very short-lived under most laboratory conditions and in several cases were observed astronomically before being studied on Earth. The identification of, e.g., C_3H_2, C_3H, or C_3N present fascinating stories of deductive logic applied to a pattern of unidentified interstellar lines, with subsequent confirmation in the laboratory (Thaddeus et al., 1985b; Thaddeus et al., 1985a; Johansson et al., 1984; Guelin and Thaddeus, 1977; Friberg et al., 1980).

The second striking characteristic of the molecular inventory, given that the cosmic abundance of hydrogen is 3 to 4 orders of magnitude greater than that of C, N, and O, is the degree of unsaturation of many of the species in Table I. This reflects the dominance of kinetics over thermodynamics in the interstellar chemical reaction schemes, and supports the current view that much of the relevant chemistry is carried out via binary gas phase reactions, which ultimately derive their energy from cosmic rays and ultraviolet photons. That chemically saturated species (e.g., CH_3CH_2CN rather than HC_3N) are present in some circumstances (the Orion "hot core" source; see below) thus suggests that different processes may become dominant under higher temperature conditions.

3.2. Individual Molecular Clouds

It might naively be expected that the chemical composition of interstellar clouds would vary significantly as a function of such physical conditions as kinetic temperature and density, as well as potentially of other variables including cloud age and past history, elemental depletions, flux of energetic particles and photons, influx of material from stellar outflows, etc. It is therefore essential to try to estimate such conditions as accurately as possible for the regions being investigated (of course, this information will come primarily from observations of the molecular line spectrum itself, by way of the techniques discussed elsewhere in this volume). In fact, the detailed quantitative chemical analysis of molecular clouds is currently limited to very few sources; a major area for future studies will surely be the expansion of the data base to other regions. Fortunately, those sources that have been studied include a variety of cloud types, which may be at least partly representative of rather general conditions: the Orion KL core region, which contains at least

four identifiable subsources of differing chemistry; the Galactic
Center Cloud Sgr B2, the most massive known giant molecular cloud; the
nearby cold, dark clouds TMC-1 and L134N; and, to a lesser extent,
the "spiral arm" clouds seen in absorption (usually) against bright,
background continuum sources such as HII regions or the supernova
remnant Cas A. We shall now briefly discuss these clouds, referring
the reader to previous reviews for more detailed descriptions (a comparison with the composition of the envelope expelled by the evolved
carbon star IRC+10216 = CW Leo is given in Irvine et al., 1985).

3.2.1. <u>Orion KL</u>. The giant molecular cloud in Orion is about 500 pc
from the sun, making it the nearest region of massive star formation
(see references in Blake, 1985; Irvine et al., 1985; and Hjalmarson,
1985). A complex of molecular clouds extends over nearly 30 square
degrees (Kutner et al., 1977), the inner two square degrees of which
are shown in a figure presented at this meeting by Schloerb et al.
This region is one of the most extensively studied in the sky, by
virtually all techniques available to astronomers (cf. Glassgold,
Huggins, and Shucking, 1982). A smaller, dense core of molecular
material lies approximately behind the well known HII region M42, and
includes young and massive embedded stars which are completely obscured
at visible wavelengths by gas and dust. This central core is known as
OMC-1 and includes the KL nebula (Kleinmann and Low, 1967). The most
intrinsically luminous of the embedded sources is IRc 2, which is
believed to power the outflows described below.
     The molecular material towards Orion KL includes several sub-
regions which have distinctive spatial and spectral distributions. The
more quiescent gas is distributed in an <u>extended ridge</u> of material
extending from northeast to southwest and characterized by emission
lines with velocity halfwidths $\Delta V \sim 2.5\text{-}4$ km s^{-1}, kinetic temperatures $T_{KIN} \sim 60K$ and molecular hydrogen densities reaching at least
$n \sim 10^5$ cm^{-3} (Figures 5 and 6). There is a sharp velocity gradient
near the position of IRc 2, suggesting the division of this ridge into
northern ($V_{LSR} \sim 10$ km s^{-1}) and southern ($V_{LSR} \sim 8$ km s^{-1}) components,
although there have been suggestions that the velocity signature may
actually indicate rotation of this core region (Hasegawa et al., 1984).
Near the interface of the two cloud components there is ample evidence
of violent activity which seems to be powered by IRc 2. Because of the
characteristic spectral shape given to emission lines, this outflow has
been referred to as the <u>plateau</u> source. Data from both millimeter
interferometers and high resolution single dish maps suggest that this
turbulent gas comprises both a lower velocity "<u>disk</u>" or doughnut
($V_{LSR} \simeq 9$ km s^{-1}, $\Delta V \simeq 20$ km s^{-1}) in which the flow is perhaps constrained by the surrounding ridge clouds, and a roughly orthogonal
<u>bipolar outflow</u> of higher velocity ($\Delta V > 30$ km s^{-1}). The structure
of this region is shown schematically in Figure 7. The outflow may in
turn produce the more spatially extended vibrationally excited H$_2$
emission and high rotational quantum number CO emission where this gas
drives shocks into surrounding quiescent material. Spectrally distinct
but difficult to spatially resolve from the outflow is the <u>hot core</u>,
which may be a large clump of material left over from the formation of

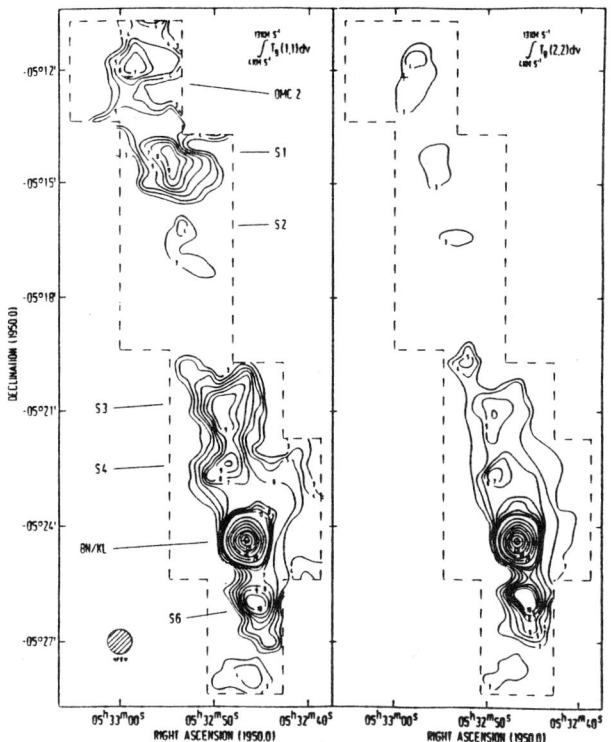

Figure 5. The ridge of dense gas forming the central portion of the
Orion Molecular Cloud, showing both OMC-1 (which contains BN/KL) and
the northern condensation known as OMC-2. Contour plots of integrated
emission from NH_3 (1,1) transition (left) and (2,2) transition (right).
Data from MPIfRA 100 m antenna by Batrla et al. (1983).

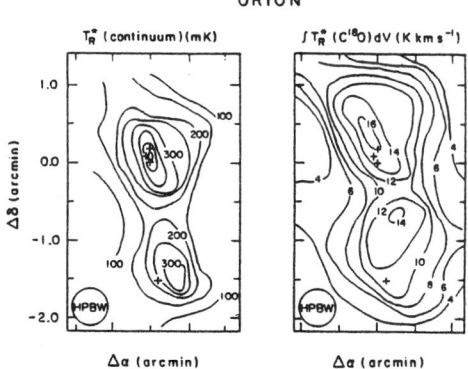

Figure 6. OMC-1 as seen in continuum emission from dust and integrated
emission over $C^{18}O$ J=2-1 line. Coordinates are offset from α(1950) =
$05^h32^m46^s.8$, δ(1950) = $-5°24'28"$. Data from FCRAO 14 m antenna
(Schloerb et al., 1986).

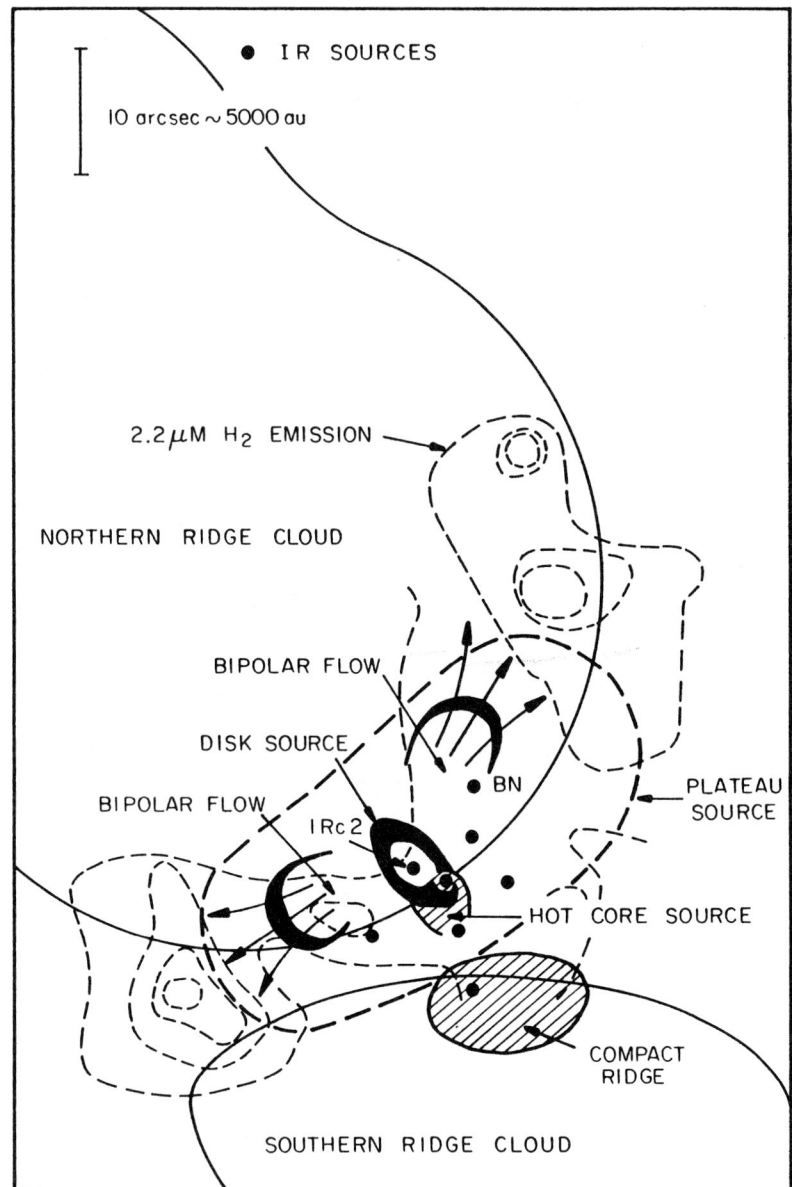

Figure 7. Schematic of the Orion KL region showing the source regions discussed in the text. Adapted from Irvine et al. (1985) and Vogel et al. (1984).

IRc 2 which has been heated by that object but not yet dispersed, prob-
ably because of its density and mass ($V_{LSR} \sim 5$ km s^{-1}, $\Delta V \sim 5$-10 km s^{-1}).
An additional recognizable source region is the <u>southern condensation</u>
or <u>compact ridge cloud</u> which shares the V_{LSR} and ΔV of the extended
southern ridge cloud but is noticeably warmer (T $\sim$ 150K) and has some
unique chemical aspects. The compact ridge may well represent an area
where outflow from IRc 2 interacts with the ambient molecular cloud
material.

3.2.2. <u>Sgr B2</u>. The Sgr B2 molecular cloud represents a relatively
extreme case of high luminosity star formation taking place in a very
massive giant molecular cloud. Direct measurements of the radio
continuum emission imply a total excitation equivalent to a stellar
luminosity $\sim 3 \times 10^7$ L$_\odot$, about one third of which arises from compact
sources studied with the VLA (Benson and Johnston 1984). Infrared
photometry indicates a luminosity of 7×10^6 L$_\odot$, but in a recent paper
Thronson and Harper (1986) suggest that the true luminosity may be a
factor of 10 higher due to spatially extended emission at relatively
long wavelengths. Molecular observations reveal an extended envelope
40 pc in diameter, with a mean density 5×10^3 hydrogen molecules per
cubic centimeter (Scoville, Solomon, and Penzias 1975), traced out by
CO emission. A core of 5-10 pc diameter contains an extremely rich
variety of molecular species. Part of the reason for this is the very
large molecular column density along the line of sight through the Sgr
B2 core, 2-10×10^{23} cm^{-2}, with higher values present within small
regions. The core region has a relatively high mean density 3-30×10^4
cm^{-3} extending over a long path. Together with the elevated temperatures in the central portion of the cloud, the large space densities
produce a high degree of molecular excitation. Temperatures deduced
from molecules whose emission comes from the central region of the
cloud range from 90K (CH_3CN; Linke <u>et al.</u> 1982) to 175K (NH_3;
Wilson <u>et al.</u> 1982), with other molecules giving intermediate values.
Churchwell et al. (1986) have pointed out the importance of far
infrared excitation of HNCO in environments such as this. Infrared
pumping can also populate vibrationally excited levels of HC_3N (cf.
Goldsmith <u>et al.</u> 1982, 1986c). It thus appears plausible that different
molecular species have quite different degrees of excitation, depending
on their coupling to the effective background temperature via radiative
transitions, and to the gas kinetic temperature via collisions. Since
the radiation intensity and gas density as well as grain and gas temperature will be strong functions of position, single component models
for these quantities must be viewed with caution. Different actual
distributions of various species together with varying excitation makes
comparison of column densities in other than a beam-averaged sense
extremely difficult. The large distance of the Sgr B2 cloud -- approximately 8.5 kpc (cf. discussion in Rohlfs <u>et al.</u> 1986) -- makes the
likelihood of blending together of emission from quite different
regions a problem, and interferometers and large millimeter single
antennas have revealed structure on the smallest observable angular
scales (cf. Table II). Further discussion of the small scale structure
of particular molecular species in Sgr B2 is given in Section 4.3.2.

TABLE II.

SPATIAL RESOLUTION FOR MM OBSERVATIONS OF TYPICAL MOLECULAR CLOUDS

Source	Distance(pc)	Resolution(pc) for HPBW[a]=		
		135"	40"	10"
Dark Clouds	~ 150	0.1	0.03	0.007
Orion KL	~ 500	0.3	0.1	0.02
Sgr B2	~ 8500	6	2	0.4

[a]Half-Power Beam Width. Values are characteristic of surveys by Cummins et al. (1986) at BTL 7m (180"-90"); Johansson et al. (1984) at OSO 20m (50"-40") and Sutton et al. (1985) at OVRO 10m (~30"); and interferometer observations by Vogel et al. (1984) at Hat Creek (~10").

3.2.3. TMC-1/L134N. These are well studied examples of dark clouds, low mass molecular clouds which lack embedded high-luminosity sources and thus appear dark against the stellar background. Portions of such clouds are extremely quiescent, having linewidths close to that expected from thermal broadening alone (e.g., Rydbeck et al., 1977). As the nearest relatively dense molecular clouds, such objects may be studied with high angular resolution (Table II). Although the kinetic temperature ($T_{KIN} \simeq 10K$) and density ($n \sim 10^4 cm^{-3}$) appear very similar for TMC-1 and L134N, they have interesting chemical differences, as discussed below. Each has, moreover, a definite sub-structure, with individual "clumps" having masses of a few $M_\odot$. Such regions are presumably formation sites for solar type stars, and TMC-1 itself appears to be part of a ring of material such as is predicted by models of the collapse of a rotating molecular cloud (Schloerb and Snell, 1984). To what extent the denser cores of such clouds are surrounded by lower density halos remains uncertain (e.g., Schloerb et al., 1983).

3.2.4. "Spiral Arm" Clouds. At sufficiently low densities the ISM is predominately atomic rather than molecular, as described in earlier chapters of this volume. At densities intermediate between these HI regions and the giant molecular clouds and dark clouds discussed above, radiative decay will dominate collisional excitation for most molecules that survive the UV radiation field, and the rotational temperature will come into equilibrium with the cosmic background (apart from any nearby luminous sources). Under these conditions observations of molecules at millimeter wavelengths become very difficult, since back-ground continuum sources are much weaker than at longer wavelengths. The material of this nature which has traditionally been observed by radio astronomers at centimeter wavelengths often seems to define the

Galactic spiral arms which are also visible in 21 centimeter HI radiation (e.g., studies of OH, H_2CO, and CH; Weinreb et al., 1963, Rydbeck et al., 1976). The advent of more sensitive millimeter systems has, however, now made it possible to see these "spiral arm" clouds at millimeter wavelengths, and hence to identify many of the molecular species known in denser regions. We shall operationally define spiral arm clouds as those regions in which high dipole moment molecules like CS, HCO^+ and HCN are seen in absorption against background continuum sources, while CO and ^{13}CO are seen in emission. Density estimates of such clouds, for example toward the distant HII region W49 or the supernova remnant Cas A, are $\sim 10^3$ cm $^{-3}$. Some controversy exists as to whether much of this material may not actually consist of denser dark clouds which are too distant to be seen optically (Troland, Crutcher and Heiles, 1985).

3.3. Chemical Abundances

The most accurate relative abundances for a particular astronomical object will be obtained through uniformly calibrated observations with a single telescope, over a frequency range throughout which the beam size does not vary greatly. To date only three systematic spectral surveys at radio frequencies have been published: of the Orion KL region (and IRC+10216) from 70 to 90 GHz with the Onsala Space Observatory 20 meter telescope (Johansson et al., 1984, 1985); of Orion KL from 216 to 265 GHz at the Owens Valley Radio Observatory with a 10 m telescope (Sutton et al., 1985, Blake et al., 1986); and of Sgr B2 from 70 to 150 GHz (with some gaps) with the AT&T Bell Telephone Laboratories 7 m system (Cummins et al., 1986). By a happy coincidence, the OSO beam size at 3 millimeters is similar to that at OVRO at the higher frequencies observed, and the abundances deduced from these two surveys are very consistent. These data bases provide multiple transition and/or multiple isotope observations for almost all species listed in Table III.

No spectral surveys yet exist for cold, dark clouds, so that the data for TMC-1 and L134N are necessarily more heterogeneous. There are both technical and physical reasons for this. Dark clouds are very cold and quiescent, and the resulting narrow line profiles require high spectral resolution; the combination of high resolution and broad total bandwidth (necessary for covering a large frequency range in a survey) is simply not available at most radio observatories. Moreover, at the low kinetic temperatures in these regions most molecules have only the very lowest energy states populated; the corresponding transitions cover a very wide frequency range if we consider both light polyatomic species such as HCN and the heavier species now found in dark clouds, making the use of more than one telescope system a necessity. We have tried in the data incorporated into Table III to select observations made with approximately the same beam size and to insure against saturation by requiring observations of rare isotopic species or multiple hyperfine components (recent observations of the weak hyperfine components of C_2H in TMC-1, for example, have shown this molecule to be much more abundant then previously realized; Friberg, 1986). The abundance ratio L134N/TMC-1 has, when possible, been obtained from results by the same investigator and telescope. Unless noted, the dark cloud data refer to the positions (1950): $\alpha = 04^h38^m38.6^s$, $\delta = 25° 35'45"$ (TMC-1) and $\alpha = 15^h51^m30^s$, $\delta = -02°43'31"$ (L134N). For a discussion of possible abundance gradients within these clouds, see Section 4.3.

Our resulting chemical abundances with respect to H_2 are given in Table III and Figures 8 and 9. The fractional abundances ($\equiv f(X)$ for species X) have been calculated primarily by ratioing the observed total column density of each species to that of CO, and then adopting $[CO]/[H_2]=8\times10^{-5}$, as deduced for dark clouds and warmer Orion gas (cf. Section III.B. in Irvine et al., 1985; there is some evidence that the CO abundance may increase in warmer regions, Scoville et al., 1983). For the spiral arm clouds Nyman (1983, 1984) has also used H_2CO as a tracer of H_2, via measurements of H_2CO and visual extinction. Note that the values given are antenna-beam-averages and probably may not be relied upon to better than an order of magnitude. However, abundance ratios other than to H_2 should be accurate to within a factor of a few, or even better, if the molecules in question are observed with the same telescope.

4. INTERPRETATION

4.1. General Uniformities in Abundance

In spite of the considerable uncertainties in the molecular abundance data (including the effect of beam-averaging over widely different cloud sizes -- cf. Table II), an abundance uniformity over a rather wide range of density and temperature for quiescent clouds is apparent from Table III and Figure 8 (see also Guelin, 1985). Since molecules in the gas phase would be expected to stick onto grains upon collision, a strong inverse dependence of gas phase abundances on density might have been expected. Such an inverse dependence had indeed been claimed in some earlier observations of CO, H_2CO, C_2H, HCN and HCO^+ in a large sample of clouds (cf. Wootten et al. 1982). However, it now seems clear that these early, pioneering studies suffered from serious systematic errors (assumptions of unsaturated emission lines, homogeneous source regions). It has been shown by Stenholm (1983) that simplifications in the radiative transfer analysis paired with the inverse correlation between errors in density and abundance may artificially produce such results. In support of our present conclusion, we note that Frerking et al. (1982) in their extensive dark cloud studies find no evidence for CO depletion in regions of visual extinction up to $A_v \sim 20$ mag (cf. also recent IRAM observations summarized by Guelin, 1986). Likewise, Schloerb et al. (1986) compare column densities of $C^{18}O$ and dust in three clouds and obtain results consistent with a fixed CO/dust ratio. Moreover, the H_2CO abundance does not seem to decrease appreciably with increasing density in dark and molecular clouds according to Wilson (1985a). It appears from this work that the H_2CO concentration estimated by LVG (Large Velocity Gradient) methods only agrees with

TABLE III

MOLECULAR ABUNDANCES FOR SEVERAL REGIONS

Species	Name	Abundance vs. H_2 ($\times 10^8$)			Abundance ratio		Remarks and References
		Orion Ridge	TMC-1	Sgr B2	Plateau Ridge	L134N / TMC-1	
Assumed H_2 Column Density ($\times 10^{-22}$):		10	1	20	1	1/1	(29)
CO	Carbon monoxide	8000	8000	8000	1	1	Adopted values, see text
CH_4	Methane	<80					(1)
C	Atomic carbon	>1000			<0.5		L134N >400
C_2	Carbon dimer		5				20 arcmin from std. position
OH	Hydroxyl		30			0.25	beam size >> that for heavier species
CH	Methylidyne		2			0.5	beam size >> that for heavier species
C_2H	Ethynyl	1	5-10	>0.5	<0.2	<1	(2) (3)
C_3H	Propynylidyne		0.05				(4)
C_4H	Butadiynyl	<0.03	2			0.05	
C_3H_2	Cyclopropenylidene		2	0.1		<0.2	(5) (6) (31)
CH_3C_2H	Methyl acetylene	0.5	0.6	0.4		<0.2	(propyne)
CN	Cyanogen	0.5	3	2	<0.2	<0.1	
HCl	Hydrogen chloride	~1					(7)
HCN	Hydrogen cyanide	2	2	2	25	0.2	(8) (9)
HNC	Hydrogen isocyanide	0.04	2	0.3	2	0.3	(10) (11) *
$HCNH^+$				0.1			(13)
CH_3CN	Methyl cyanide	0.04/0.08*	0.1	0.07		<1	(acetonitrile) (12)
HC_3N	Cyanoacetylene	0.04	0.6	0.2	20	0.03	
HC_5N	Cyanodiacetylene	0.006	0.3	0.04		0.03	
CH_2CHCN	Vinylcyanide		0.02	0.02		<0.5	(acrylonitrile)
C_3N	Cyanoethynyl	<0.006	0.1	<0.02		<0.2	
C_3O	Tricarbon Monoxide	<0.003	0.01	<0.002		<0.5	(28)

TABLE III (Continued)

CH$_3$C$_3$N	Methylcyanoacetylene		0.05				
CH$_3$C$_4$H	Methyldiacetylene		0.2				
HNCO	Isocyanic acid	<0.03	0.02	0.9			
N$_2$H$^+$		0.02	0.05				(14)
NH$_3$	Ammonia	20	2	1–10	10?	1	Apparently position dependent, (15) (16)
HCO$^+$	Formyl ion	0.3	0.8	1	5	10	
HCO$_2^+$						1	"Plateau" emission extended
HDO	Deuterated water	<0.02		0.3	10		
		<0.04/0.4*					
CH$_3$OH	Methanol	/40*	0.4	20		1	(17)
H$_2$CO	Formaldehyde	3/30*	2	2–10	1	1	(18)
H$_2$C$_2$O	Ketene	/0.2*	0.01	0.08			(19)
(CH$_3$)$_2$O	Dimethyl ether	/2*		0.25			
HCOOCH$_3$	Methyl formate	/2*	<0.1	0.2			
CS	Carbon monosulfide	0.4	1	1	10	~0.1	(20) (21)
HCS$^+$	Thioformyl ion	0.02	0.06	0.02	<1	~0.1	(22)
H$_2$CS	Thioformaldehyde	/0.2*		0.3			
H$_2$S	Hydrogen sulfide	<0.1			100		
OCS	Carbonyl sulfide	/0.9*	0.2	1	8	1	(23)
SO	Sulfur monoxide	0.2/0.5*	0.5	0.2	300	4	
SO$_2$	Sulfur dioxide	/0.4	<0.1	0.3–2	300	>4	Orion position 1.'5 S of KL, (24)
SiO	Silicon monoxide	<0.1	<0.0005	0.04	>100		From isotopes for Sgr B2, (25)
SiS	Silicon sulfide	<0.001		0.03	>100		
HC$_7$N	Cyanohexatriyne		0.1	0.1		<0.02	(30)
HC$_9$N	Cyano-octatetra-yne		0.03				
HC$_{11}$N	Cyano-decapenta-yne		0.01				
HOC$^+$		<0.001	<0.002	0.003			(26)
HCO	Formyl	<0.02					
CH$_3$CHO	Acetaldehyde	<0.02	0.06	0.1		1	
HC$_2$CHO	Propiolaldehyde		<0.06				(propynal)

TABLE III (Continued)

CH$_2$CHCHO	Acrolein	<0.02			
CH$_3$CH$_2$OH	Ethanol	<0.05		0.3	
HCOOH	Formic acid	/.03*		~0.2	(32)
CH$_3$COOH	Acetic acid	<0.5			
CH$_3$NC	Methyl isocyanide	<0.005	<0.01		Tentative detection in TMC-1, (27)
CH$_3$CH$_2$CN	Ethyl cyanide	<0.03	<0.1	0.03	
NH$_2$CN	Cyanamide	<0.02	<0.01	0.01	
NH$_2$CHO	Formamide	<0.03	<0.2	0.04	
CH$_3$SH	Methyl mercaptan	<0.06		0.08	
HNCS	Isothiocyanic Acid			0.01	
(NH$_2$)$_2$CO	Urea	<0.07	<0.04		
NH$_2$CH$_2$COOH	Glycine II	<0.05			
C$_4$H$_4$O	Furan	<0.07			beam size >> that for most species
C$_4$H$_5$N	Pyrrole	<0.03	<0.04		beam size >> that for most species
C$_3$N$_2$H$_4$	Imidazole	<0.1	<0.03		beam size >> that for most species
CH$_3$NH$_2$	Methylamine	<0.1			
PN	Phosphorus nitride	<0.003			
PO	Phosphorus monoxide	<0.1			
NO	Nitric Oxide	<5		~10	Detected in Orion

For Orion ridge, values given as x/y* refer to the extended and the compact ridge, respectively. All values are beam-averages. Compiled from Irvine et al. (1985), Blake (1985), Cummins et al. (1986), and Guelin (1985), plus references in last column: (1) Knacke, et al. (1985); (2) TMC-1 from Friberg, P. (unpublished FCRAO data); (3) Sgr B2 from Ziurys, L. (unpublished; calculated from Cummins et al. (1986); (4) Thaddeus et al. (1985a); (5) Matthews and Irvine (1985); (6) Irvine et al (1986); (7) Blake et al. (1985); (8) L134N from Swade (1987); (9) Sgr B2 from Ziurys and Turner (1986); (10) cf. Goldsmith et al (1986a); (11) Sgr B2 discussed by Ziurys and Turner (1986); (12) Orion from Andersson (1985); (13) Ziurys and Turner (1986); (14) Estimates include unpublished FCRAO dark cloud data by D. Swade and S. Madden; (15) Sgr B2 from Winnewisser et al (1979) and Gusten et al (1981); (16) for comparison, Zeng et al (1984) find 3 in ρ OphB; (17) Dark clouds from Friberg et al (1986); (18) cf. Bastien et al (1985) and Wilson et al (1980); (19) TMC-1 from Matthews and Sears (1986); (20) TMC-1 from Suzuki, H. (unpublished NRO data); (21) Sgr B2, see Frerking et al (1980); (22) L134N from Swade (1987); (23) Dark clouds from Matthews et al (1986); (24) Irvine et al (1983); (25) TMC-1 from Ziurys et al (1985); (26) Bell and Matthews (1985); (27) Irvine and Schloerb (1984); (28) Brown et al (1985); (29) see also Goldsmith et al (1985); (30) L. Ziurys (unpublished FCRAO data); (31) Vrtilek et al 1986 and unpublished FCRAO data on ^{13}C variant (S. Madden); (32) Sgr B2, S. Cummins, private communication.

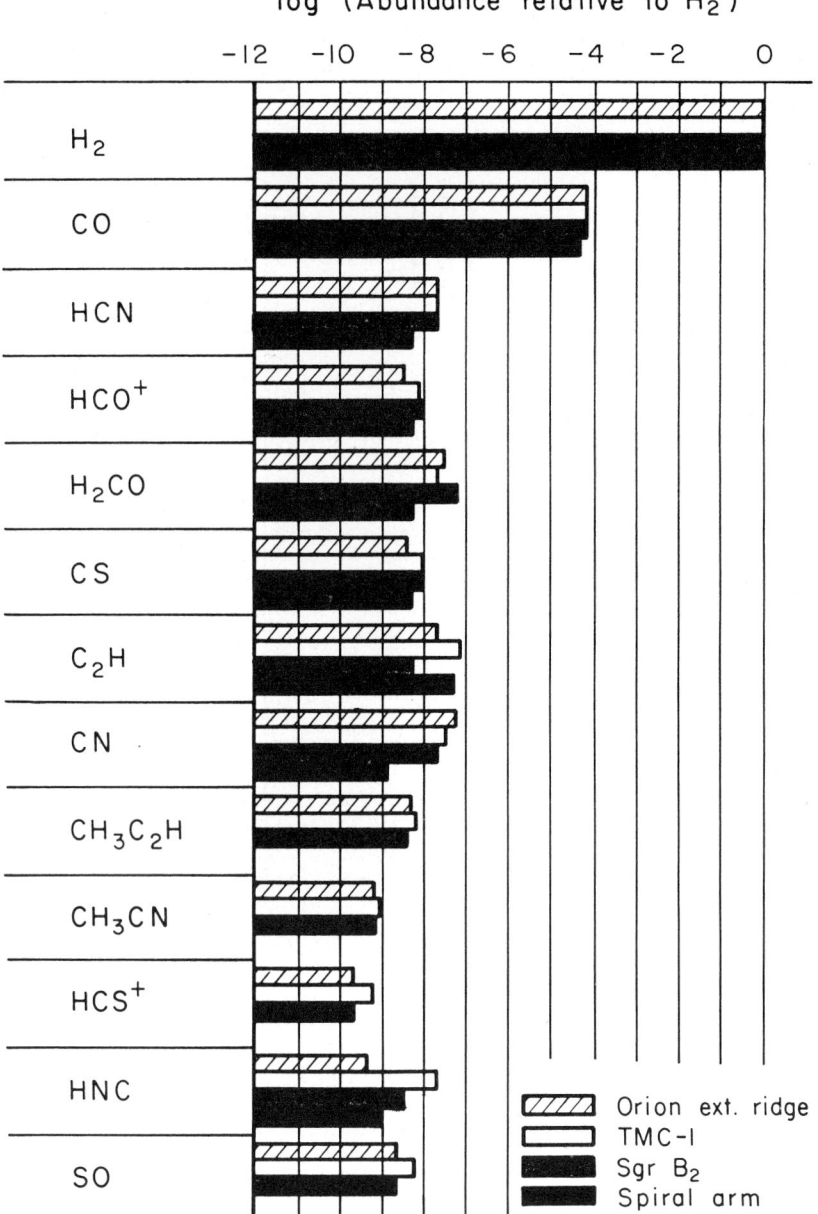

Figure 8.   Abundances relative to H_2 on a logarithmic scale, for the
Orion extended ridge cloud, TMC-1, Sgr B2, and the "spiral arm" clouds
towards W49N.   From Table III and (for spiral arm clouds) from Irvine
et al. (1985) and recent Onsala measurement by L.Å. Nyman.   CN abundance may be higher toward some other "spiral arm type" clouds (Irvine
et al., 1985, Table XI).

that directly derived (from column density, resolved clump size, and H_2 density) if very large velocity gradients are used. The average value of the velocity gradients estimated from the ratio of observed line width to clump size are 10-15 km s^{-1} pc^{-1}, which should be compared to the typical value of 1-2.5 km s^{-1} pc^{-1} used in the early LVG studies (cf. Wootten et al. 1982). Finally, since one would not expect "freeze out" on grains to be equally efficient for all molecules, it seems significant that Wootten et al. (1984) find a great uniformity of the $H^{13}CO^+/^{13}CO$ abundance ratio among a large number of clouds including Orion KL and L134N. Likewise, Huggins et al. (1984) conclude that the $C_2H/^{13}CO$ ratio does not vary much across and among GMC's.

The multi-transition mapping of CS and $C^{34}S$ across cloud cores (Snell et al. 1984, Mundy et al. 1986; discussed in Section 2) are also interesting in this regard. The fact that only small H_2 density variations were found across the cloud cores, while the CS column density varied by an order of magnitude, seems to be best explained if the CS emission arises in numerous dense ($n=10^5-10^6 cm^{-3}$) clumps, with little emission from the interclump medium. The beam filling factor (of clumps) then is close to unity at the center of a cloud core but decreases by an order of magnitude at the core edges. The CS fractional abundances estimated toward two cloud cores from CS and $C^{18}O$ (J=2-1) column densities (the latter from Goldsmith et al. 1985) fall in the range $(2-7)\times10^{-9}$, which is close to our Table III concentrations.

Our conclusion of abundance uniformity over a large range of cloud densities (and temperatures) poses a problem, since the time scale for molecular absorption on to dust grains is expected to be short compared to estimated cloud ages (cf. Draine 1985; Leitch-Devlin and Williams 1984, 1985). Efficient desorption mechanisms are needed. Such processes have been considered by Boland and deJong (1982), Greenberg (1983), Williams and Hartquist (1984), Draine (1985), and Leger et al. (1985). On the other hand, most abundance values are in reasonable agreement with purely gas phase ion-molecule chemical models, such as that of Herbst and Leung (1986).

4.2. Chemical Differences Among Clouds

4.2.1. Dark clouds. The cyanopolyyne region in TMC-1 stands out relative to L134N and the GMC's due to its considerably enhanced abundances of carbon-rich species (C_3N, HC_3N, C_4H, HC_5N, etc.; Table III). The abundances of C_3H_2, CH_3C_2H, CN, CS and HCS^+ also appear to be enhanced in TMC-1. It seems that NH_3 is depleted in TMC-1 relative to L134N. Likewise, the abundances of SO and SO_2 are higher in L134N. There may, however, be variations with positions in the clouds (see Section 4.3).

Since TMC-1 and L134N have similar temperature and density, other factors may determine the chemical differences (relative elemental abundances, UV field, cloud age or history; cf. Millar and Freeman, 1984). Differential depletion onto grains, leading to variations in the gas phase C/O ratio, may be a likely cause.

4.2.2. <u>Giant molecular clouds</u>. There is very little reliable,
comprehensive data on chemical abundances in GMC's apart from the
spectral surveys of Orion and Sgr B2 mentioned previously. Earlier
studies in a number of sources (<u>e.g.</u>, Wootten <u>et al.</u>, 1982) tended to
rely on single transitions for each molecule and are now known to be
affected by saturation and inaccurate assumptions about source homogeneity. Careful research on ratios of the isomers HCN and HNC has
found differences of more than an order of magnitude among clouds
(Goldsmith <u>et al.</u>, 1981). However, the nearest and best studied GMC is
that in Orion, within which there are striking chemical heterogeneities,
as discussed in Section 4.3.3. This suggests that differences among
GMC's may at least in part reflect different mixes of the type of
heterogeneities that are apparent in Orion, but which would be totally
unresolved at larger distances.

        This problem of spatial resolution is particularly troubling for
the Galactic center source Sgr B2 (almost 20 times more distant than
Orion). We shall, nonetheless, discuss differences in the survey
results for Orion and Sgr B2, selecting for Orion the abundances of the
extended plus compact ridge clouds. With the repeated precaution that
the Sgr B2 abundances will be averages over an area $\sim 20^2 = 400$ times
larger than that for Orion (the situation is actually worse, because of
the larger beam size used by Cummins <u>et al.</u> (1986) for the Sgr B2
survey; cf. Table II), we note that for many ubiquitous molecules Table
III nonetheless indicates abundance <u>uniformity</u> (see also Figure 8).

        We may then proceed to discuss <u>possible abundance differences</u>
apparent in Table III. The Sgr B2 abundances have been estimated from
ratioing with total column densities of $C^{18}O$ calculated for a terrestrial $^{16}O/^{18}O$ value and a population distribution ("rotation
temperature") of 20 K – the average ^{12}CO brightness temperature.
Since the emission from a number of species seem to mainly originate
in denser, warmer regions (30–120K, cf. Goldsmith <u>et al.</u> 1986b), this
could lead to an underestimate of the relevant CO column density and
hence an overestimate of other beam averaged abundances by a factor of
a few. On the other hand, the regions of enhanced emission probably
only cover a small part of the Bell Laboratories antenna beam (cf.
Section 4.3.2. on structure and possible abundance variations within
this source), which would have the opposite effect.

        From Table III we note that the CH_3OH abundances may be similar
in warm localized clouds within Orion A and Sgr B2. Comparing
abundance <u>ratios</u> with CH_3OH, we find that in Orion A (compared to Sgr
B2) the more complex species $HCOOCH_3$ (methyl formate) and $(CH_3)_2O$
(dimethyl ether) are clearly enchanced, while CH_3CH_2OH (ethanol, an
<u>isomer</u> of dimethyl ether), CH_3CHO (acetaldehyde) and $HCO_2{}^+$ are
depleted. The abundance of H_2C_2O (ketene) seems similar in the two
sources. (On the other hand, if all Sgr B2 abundances were too high by
a factor of 4, which is certainly possible, we would conclude that the
warm Orion clump exhibits enhanced abundances of H_2CO, CH_3OH, H_2C_2O,
$HCOOCH_3$ and $(CH_3)_2O$, while those of CH_3CHO and CH_3CH_2OH may be similar
in the two sources.)

        The pronounced <u>chemical selectivity</u> for complex species may be
quantified by the following examples comparing Sgr B2 and Orion,

respectively: $f[(CH_3)_2O]/f[CH_3CH_2OH] = 1$ and > 40, $f(CH_3OH)/f(HCOOCH_3)$ $= 100$ and 20, $f(HCOOCH_3)/f(HCOOH) = 1$ and 70, and $f(CH_3OH)/f(CH_3CH_2OH)$ $= 70$ and > 800. It is also interesting to note that $HCOOCH_3$ is rather abundant in Orion while its isomer CH_3COOH (acetic acid) has not been detected. However, the limit is only $f(CH_3COOH)<f(HCOOCH_3)/4$.

These differences in abundances of oxygen-containing molecules seem to be primarily due to the characteristics of the Orion compact ridge cloud, and may be generally understood in terms of known processes and reaction rates (Section 4.3.3 below). Table III also indicates an abundance enhancement in Sgr B2 for HNCO, HC_3N, SO_2, and SiS compared to the Orion Ridge. This may mainly result from localized, warmer regions in Sgr B2 that are similar to the Orion plateau and hot core (cf. Figure 9).

4.2.3. <u>Influences of cloud temperature</u>. For a few species we are able to point out clear abundance variations which may be due to temperature (and density?). While $f(CH_3OH)>10^{-7}$ in warm, dense regions of Orion A and Sgr B2, our recent detections in TMC-1 and L134N yield $f(CH_3OH)=4\times10^{-9}$ (Friberg <u>et al.</u>, 1986). Menten <u>et al.</u> (1985) estimate $f(CH_3OH)\sim10^{-5}$ in a methanol maser cloud close to the W3(OH) compact HII region – a considerable over-abundance. Since CH_3OH probably is formed from H_2O and $CH_3{}^+$, this may indicate that evaporation of icy grain mantles produces H_2O and hence CH_3OH in warm clouds. Consistent with this scenario is the case of HDO, which so far has only been detected in warm, dense regions of Orion A and W51M (Olofsson 1984; Moore <u>et al.</u>, 1986). The very high abundance of NH_3 ($\sim10^{-5}$) estimated in the Orion hot core and other similar regions may also reflect a release of NH_3 from grain mantles (see Section 4.3.3.).

Variations in the HCN/HNC isomer abundance ratio are clearly correlated with cloud temperature (Figure 10). While this ratio is close to unity in cold, dark clouds as predicted by the simplest chemical schemes, it increases to 2 - 50 in GMC's (Goldsmith <u>et al.</u> 1981). Recently we have shown that the HCN/HNC abundance ratio varies from $\sim$ 6 in the northern, colder parts of the Orion A ridge to $\sim$ 60 in the warmer ridge gas towards KL, and increases to $\sim$ 250 in the plateau and hot core sources (Goldsmith <u>et al.</u>, 1986a). For comparison we note that the result of hot stellar atmosphere chemistry in the IRC+10216 envelope is about 280. The Sgr B2 discovery by Ziurys and Turner (1986) of the HCN and HNC "precursor" ion $HCNH^+$, at orders of magnitude higher abundance than expected by ion-molecule schemes, will be important in sorting out the relevant reaction schemes.

4.3. Chemical Differences Within Clouds

4.3.1. <u>L134N (L183)</u>. The dark cloud L134N, also known as L183, provides perhaps the most suggestive evidence for abundance gradients within quiescent cloud material. In an effort to disentangle possible gradients in physical characteristics such as temperature and density from true chemical gradients, D. Swade at the University of Massachusetts has carried out an extensive series of observations with the FCRAO 14 m radio telescope, supplemented by some lower frequency

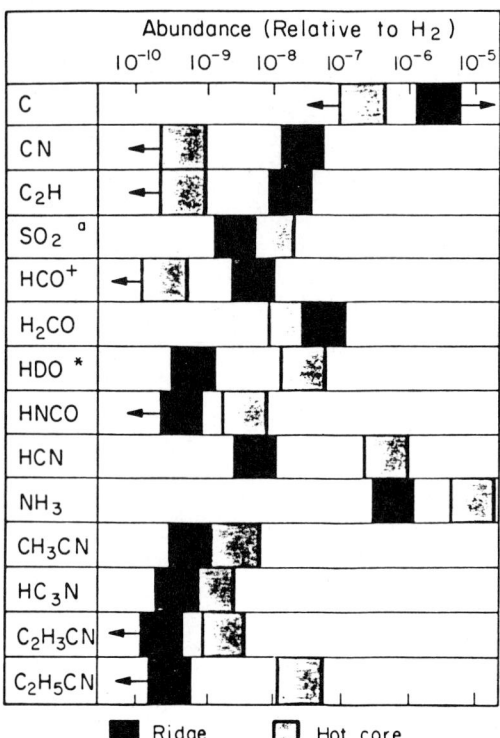

Figure 9. Abundances relative to H$_2$ for the Orion ridge and hot core. Ridge values for HDO and SO$_2$ refer to the compact ridge source and to a position 1.5' south of KL, respectively. Adapted from Blake (1985).

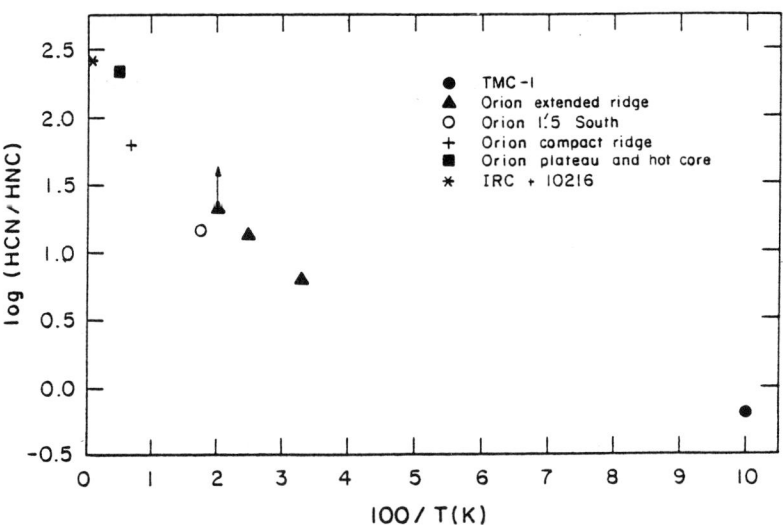

Figure 10. Variation of the abundance ratio [HCN]/[HNC] versus the inverse of the temperature. Adapted from Goldsmith et al. (1986a).

data from the Haystack Observatory and NRAO Green Bank (cf. Swade et al., 1985). The typical spatial resolution achieved was ~ 0.05 pc, which is about the radius of the Oort cloud at the distance of L134N.

From a map of the ^{13}CO emission covering approximately one square degree, Swade estimates an overall cloud mass of $200M_\odot$, a spatial extent of about 2.4 pc, and a mean H_2 density of 10^3 cm^{-3}. The cloud shows indications of sub-structure comprising several components, and the following discussion refers to the most massive of these sub-clouds, which is also the one which has been most thoroughly studied in the past. Observations of ^{12}CO indicate that the kinetic temperature is a rather constant 10 K in this core region. Although this is the type of dense core often believed to be the site of low mass star formation, there are no identified IRAS point sources in this portion of L134N. Let us now consider the molecular maps in more detail.

The distribution of ^{13}CO velocity-integrated intensity is rather uniform across the dense core region, indicating that this species is optically thick. In contrast, the map of $C^{18}O$ J=1-0 emission (Figure 11a) shows considerable structure, including a general east-west extension with peaks near the extremities in right ascension, and with the western peak being resolved into two parts. The total mass of the core region is estimated to be $15M_\odot$, indicating that the individual peaks are on the order of a few solar masses. Although even $C^{18}O$ may not be optically thin in some portions of the core (Guelin, Langer, and Wilson 1982), the low dipole moment for carbon monoxide suggests that this map should provide an approximate representation of the distribution of total molecular column density in this region (See Section 2).

The distribution of $H^{13}CO^+$ emission, also given in Figure 11a, presents a striking contrast to that of $C^{18}O$. The emission is extended in a north-south direction with at least three peaks which do not coincide with any feature in the carbon monoxide map (a fourth peak, immediately adjacent to an emission minimum, does correspond to a secondary peak in the $C^{18}O$ structure). Data for $HC^{18}O^+$ (Guelin et al., 1982) indicates that the $H^{13}CO^+$ emission is not strongly saturated over the area of the map. Because of its large dipole moment, HCO^+ is generally believed to be a good tracer of the dense portions of molecular clouds. This interpretation of the $H^{13}CO^+$ distribution is supported by its similarity to the ammonia map given in Swade et al. (1985); ammonia is expected to trace gas of moderate density (Ho and Townes, 1983).

In spite of the considerable intensity variations, the above data may be consistent with uniform chemical abundances for the species considered, although they do not prove that such uniformity exists. In contrast, the distribution of SO emission (Figure 11b) appears to imply that abundance differences exist between the small sub-condensations within the dense core region of L134N. Of course, before one may draw this conclusion from comparison of the morphology of these maps, it is necessary to consider whether the SO emission may be saturated. Swade has addressed this question by observing not only the 3_2-2_1 line illustrated, but also at selected points the corresponding line of ^{34}SO, as well as the 2_2-1_1 line of SO. A preliminary examination of the results suggests that, although the 3_2-2_1 line is moderately

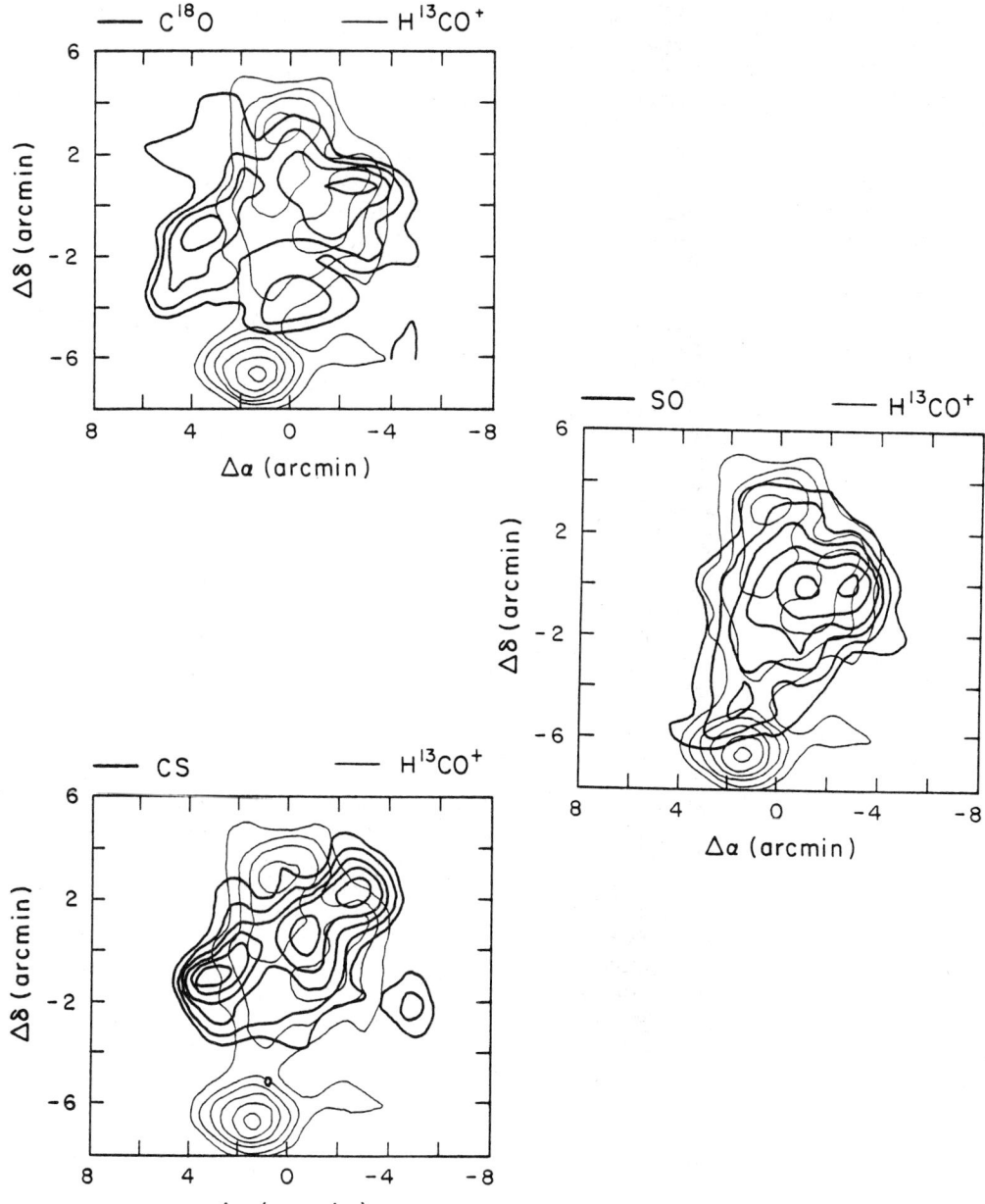

Figure 11. Contour diagrams of the velocity-integrated emission from several molecular transitions in the core of the dark cloud L134N (= L183). Superimposed on the contours (light) for $H^{13}CO^+$ (J=1-0) are a.) $C^{18}O$ (J=1-0); b.) SO (3_2-2_1); c.) CS (J=2-1). From Swade (1987) and Swade et al. (1985).

optically thick at the map peaks, its intensity is well correlated with the weaker lines, indicating that Figure 11b probably does represent the distribution of SO reasonably well. Since the Einstein A coefficients for $H^{13}CO^+$ (1-0) and SO (3_2-2_1) differ by less than a factor of two (2.8×10^{-5} s^{-1} and 1.6×10^{-5} s^{-1}, respectively), it seems unlikely that the excitation of these two transitions is very different. Thus, the SO abundance appears to vary on a scale of ~ 0.25 pc.

The situation is made even more complex by examining the emission from CS (Figure 11c). Although in general this map has a similar morphology to that of $C^{18}O$, the Einstein A coefficient for CS (1.7×10^{-5} s^{-1}) is much more similar to that of $H^{13}CO^+$ or SO than to the corresponding coefficient for CO (7.5×10^{-8} s^{-1}). The CS lines do seem to be rather strongly saturated, however. Data taken for $C^{34}S$ suggests that $\tau(CS)$ ~ 5-10 in much of the region surveyed. Moreover, these same $C^{34}S$ data do indicate the presence of significant CS at the eastern peak in Figure 11, suggesting that the CS distribution is unlike that characterizing the dense gas structure indicated by $H^{13}CO^+$ and NH_3, and also unlike that of SO. We can therefore not simply ascribe the SO distribution to some anomaly characterizing all sulfur-containing molecules.

These data refer to integrated intensities; examination of line profiles and maps at particular velocities reinforces the case for differences among subregions within L134N. It will be interesting to see whether detailed chemical models can explain such differences with reasonable assumptions.

4.3.2. <u>Sgr B2</u>. The large distance of the Sgr B2 molecular cloud has until recently prevented study of the internal structure of this region on a reasonably fine linear scale. Only H_2O (maser) and OH (in emission and absorption) have been observed interferometrically; the OH and H_2O masers are grouped into three condensations (Elmegreen <u>et al.</u> 1980): north (N), middle (M), and south (S), separated by approximately 45 arcseconds in declination (1.85 pc at a distance of 8.5 kpc). A study of the distribution of a number of species was recently completed by Goldsmith <u>et al.</u> (1986b) using the Nobeyama Radio Observatory (NRO) 45m telescope with a beamwidth of 15 arcseconds (0.6 pc) at a frequency of 109 GHz. The distributions of emission from four different molecular species are shown in Figure 12. The (0,0) map position corresponds to Sgr B2 (M) at RA(1950) = $17^h44^m10^s.5$; decl.(1950) = $-28°22'05''$. It is clear that the emission in these transitions shows very different distributions. It is difficult, however, to disentangle the effects of abundance variations and excitation differences. The relatively weak intensities of these lines together with their likely thermalized excitation (see below) suggest low optical depth; this is consistent with observations of isotopic variants carried out with larger beamwidths. The HC_3N transition has the highest spontaneous emission rate of the ground vibrational state transitions observed (A = 1.0×10^{-4} s^{-1}) and a critical density $n^*(H_2)$ = A/$\langle\sigma v\rangle$ = 1×10^6 cm^{-3}. Its extensive, uniform emission suggests that it has a fairly uniform abundance and that the hydrogen density is greater than 10^5 cm^{-3} throughout the ~ 3.5 pc diameter region. This density should be adequate to thermalize

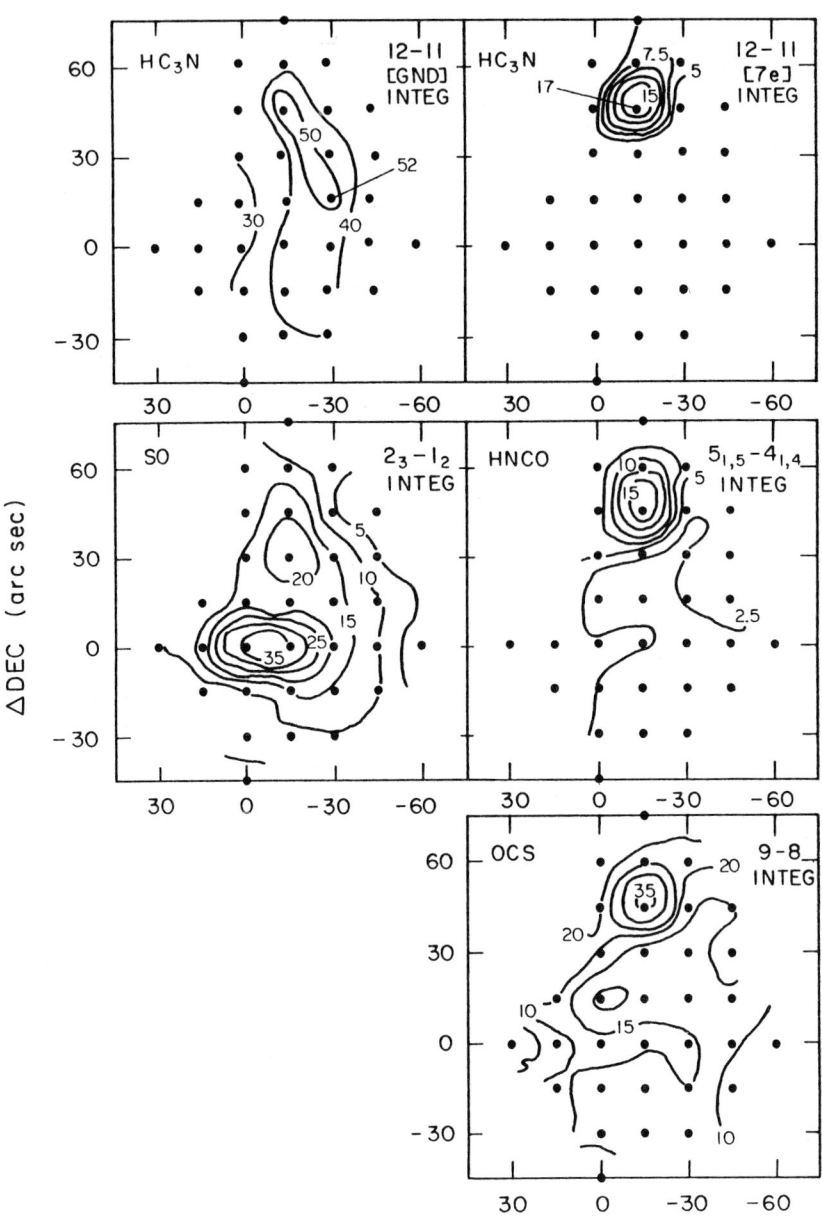

Figure 12.   Distribution of HC3N, SO, HNCO, and OCS integrated
intensity from the central region of Sgr B2; all the transitions
observed are near 109 GHz.   The (0,0) position is RA(1950) =
17^{h}44^{m}10^s.5, decl.(1950) = -28°22'05".   From Goldsmith et al. (1986b).

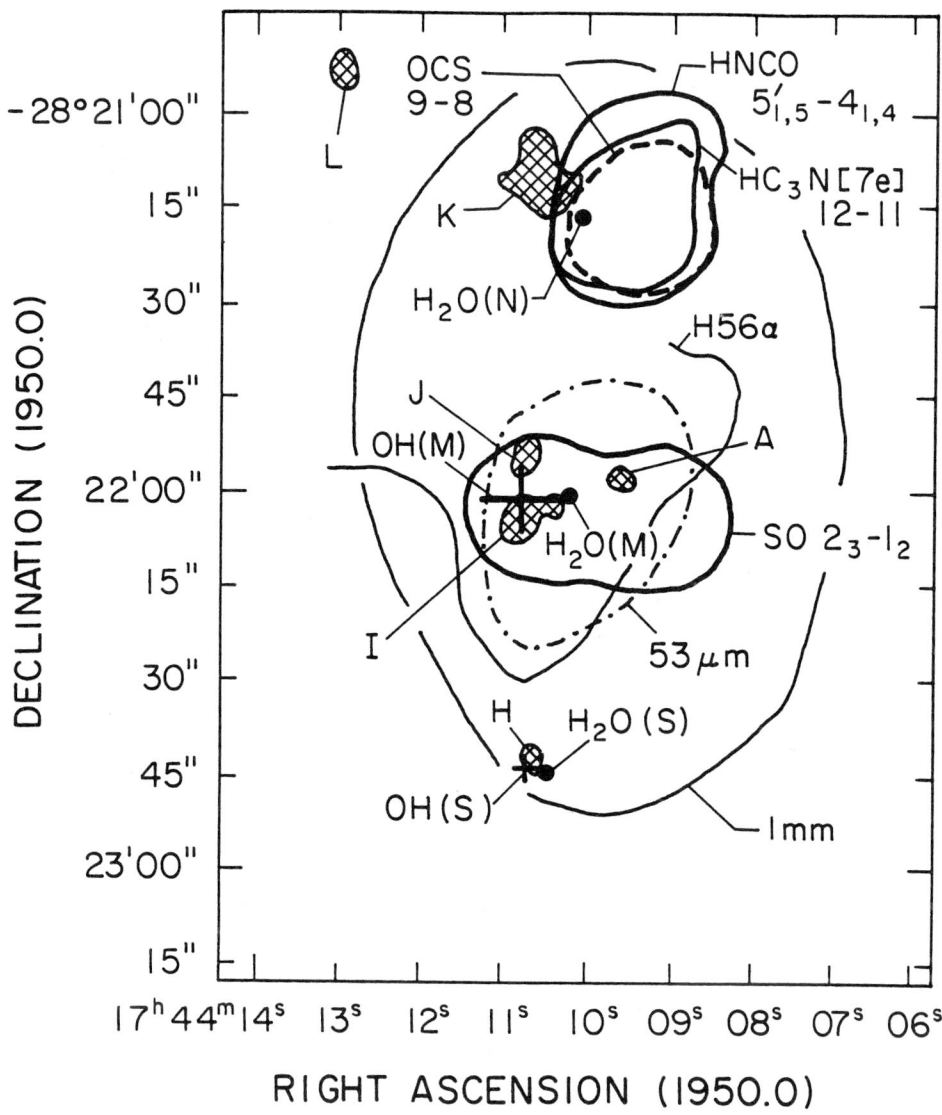

Figure 13. Comparison of molecular emission (from Fig. 12) and compact radio continuum sources (hatched regions; from Benson and Johnston 1984), 1mm continuum (Westbrook et al. 1976), 53 micron emission (Harvey, Campbell, and Hoffmann 1977), H56α emission (Morimoto et al. 1984), and H_2O (Elmegreen et al. 1980) and OH (Raimond and Eliasson 1969) masers. From Goldsmith et al. (1986b).

the SO and OCS transitions, which have A-coefficients smaller by
factors of 9 and 26, respectively. These two species show remarkably
different distributions, however, with SO strongly peaked near Sgr B2
(M) and OCS concentrated around Sgr B2 (N). The HC_3N vibrationally
excited emission and the HNCO are concentrated around Sgr B2(N). The
HC_3N excitation may well be by infrared pumping (Goldsmith et al.
1982, 1986c). The lower level of the observed HNCO transition lies 52
K above the molecular ground state and can spontaneously decay to the
$K_a=0$ ladder at a rate $\sim 6 \times 10^{-3}$ s^{-1}. A density of $\sim 10^{+5}$ cm^{-3} will
not readily populate this level, even allowing for moderate opacity in
the $\lambda \sim 280 \mu m$ transition. Churchwell et al. (1986) have suggested
that far-infrared pumping may be responsible for the population of the
$K_a > 0$ levels of HNCO. Thus, the enhancement of the intensity we observe
near Sgr B2 (N) could be a reflection of increased radiation there.
However, the relatively long wavelength of the b-type transitions connecting the $K_a=0$ and $K_a=1$ ladders (compared, for example, to the $45 \mu m$
wavelength for excitation of v=1 HC_3N) suggests that the enhancement
should not be highly localized; we may thus also be seeing an enhancement in the abundance of HNCO in the vicinity of Sgr B2 (N). In Table
IV we give LTE, optically thin column densities obtained using a rotational temperature of 80 K for the Sgr B2 north and middle positions.
The 80 K rotational temperature reflects a reasonable estimate of the
gas kinetic temperature in the molecular cloud core, but is certainly
an underestimate for the immediate vicinity of Sgr B2 (N), where the
ratio of the HC_3N J=12-11 [v_7=1] to [Gnd] transitions yields a vibrational temperature $\sim$ 290 K.

It is also striking that there is not stronger vibrationally
excited HC_3N emission from Sgr B2 (M), since there is HC_3N in the
vicinity and comparable heating from embedded stars to that powering
Sgr B2 (N) , judging from radio continuum emission studies (Benson and
Johnston 1984). The overall situation is summarized in Figure 13,
showing radio continuum regions (hatched), 1mm and $53 \mu m$ emission
(from dust), and contours of emission from different molecular transitions taken from Figure 12. We see that the molecular emission in
Sgr B2(N) peaks just adjacent to the radio continuum source (K), but
that there is no detectable 53 micron emission from this direction.
One explanation is that the Sgr B2 (N) region is located toward the
back of the Sgr B2 core, and there is a very large foreground column
density of cold dust which absorbs the radiation from the hot material
immediately surrounding the HII region. The minimum molecular hydrogen
column density required is 10^{24} cm^{-2}. The picture of this region
shows how different types of transitions of various molecular species
can contribute to our understanding of complex regions within molecular
clouds -- and also reminds us that it is very difficult to know a
priori what molecules or lines will convey the most information.

The column densities of the different species given in Table IV
clearly show major variations between the two locations of local maxima
-- Sgr B2 (N) and (M). Given the aforementioned uncertainties in
excitation, it is difficult to be certain about the degree of chemical
differentiation that is present; factors of two variation can certainly
be ascribed to excitation differences. However, given the small

TABLE IV.

COLUMN DENSITIES OF MOLECULAR SPECIES IN SGR B2 NORTH AND MIDDLE SOURCES[a]

	HC_3N	SO	OCS	HNCO
Sgr B2 (N)	1.1	4.2	27	5.0
Sgr B2 (M)	0.6	8.9	8.7	0.6
Ratio (North/Middle)	2	0.5	3	8
CLT	0.4	0.4	2.1	0.4-1.4

[a]From Goldsmith et al. (1986b). Column densities in units of 10^{15} cm^{-2}, coupling efficiency of 0.25 assumed for both sources, and a rotational temperature of 80 K adopted for all species. CLT refers to column densities given by Cummins, Linke, and Thaddeus (1986).

(factor of 4) difference in decay rates between OCS and SO, it appears that there is a significant composition difference between Sgr B2 (N) and (M). The HNCO column density also shows appreciable variation. We have also included in Table IV the column densities obtained by Cummins, Linke, and Thaddeus (1986, the source of most Sgr B2 entries in Table III) with a beamwidth of 2.9-1.5 arcminutes, pointed at a position approximately 30 arcseconds south of the (0,0) position of Figures 12 and 13. This clearly will produce an unequal effect for species having different spatial distributions. Cummins, Linke, and Thaddeus used rotational temperatures determined from their multi-transition data (as discussed in Section 2), and these are quite low, in the 15-30 K range. However, it should be noted that the fits for SO and HNCO (see also Churchwell et al. (1986) are particularly poor and strongly suggest a much higher rotational temperature T_{ROT} for higher-lying transitions ($E_u > 30$ K). There is thus clear evidence for a temperature distribution more complex than that specified by a single value of T_{ROT}. The OCS and HC_3N emission is more widespread than that from the other species, and hence the rotation diagram is reason-ably well fit by a single modest temperature, as long as relatively low energy levels exclusively are considered. The vibrationally excited HC_3N emission discussed above clearly shows gross deviations in excitation from the values inferred for low-lying transitions.

     It is particularly difficult to disentangle the emission produced by the Sgr B2 (N) and (M) "clumps" from that arising in the larger cloud core. If we use the observed NRO linewidths (which are generally

narrower than those found in larger-beamwidth studies) and the virial
theorem to estimate the mass of the north and middle condensations, we
find $M \sim 2 \times 10^4$ $M_\odot$, a density of 3×10^6 hydrogen molecules per cubic
centimeter, and a H_2 column density of 3×10^{24} cm^{-2}. This last is
a factor of 15 larger than the column density used in Table III for the
overall core. The resulting fractional abundances for OCS and HNCO are
approximately the same in Sgr B2 (N) as those given in Table III. The
abundance of SO is about the same in both clumps compared to the more
extended region; the abundance of HC_3N is clearly lower in both of
the clumps; and OCS and HNCO are less abundant in Sgr B2 (M). The diversity of the variations between lower and higher density regions, and
between the immediate surroundings of different regions of star formation for various molecular species, is a clear challenge for chemical
modelling of active regions of the interstellar medium.

4.3.3. <u>Orion KL</u>. This nearest of the giant molecular clouds provides
a fascinating example of heterogeneity in physical conditions and
chemical abundances within a relatively small spatial region. Many of
the chemical differences among the sub-sources in the Orion KL region
seem to be directly related to processes of star formation. The
following discussion draws heavily upon Blake (1985) and Irvine <u>et al.</u>
(1985).

The chemical similarity between the <u>extended ridge</u> and that of
other quiescent molecular clouds, even with somewhat different temperatures and densities, has been mentioned above (Section 4.1.) and is
in general agreement with calculations in which the dominant processes
are ion-molecule gas phase reactions (e.g., Herbst and Leung, 1986). A
major question in such comparisons arises from the time dependence of
the theoretical models. Calculations which assume an elemental abundance ratio C/O < 1 (the cosmic value is about 0.4) find that steady
state values for the abundance of many carbon-rich species are vastly
underestimated. Instead, the abundance of such molecules peaks at
relatively short times ($\sim 10^6$ years) and then declines sharply as
most of the available carbon is converted to CO. This problem is in
addition to the apparent lack of "freeze out" onto grains for most
molecules which should occur on time scales much less then the expected
age of the clouds. Although various scenarios have been suggested for
liberating atomic carbon from CO or from molecules frozen onto grains
(e.g. Keene <u>et al.</u>, 1985), an alternative suggestion is the possibility
that the gas phase C/O ratio may be greater than unity (Langer <u>et al.</u>,
1984). This may not be so difficult as it might at first appear, since
atomic C, N, and O colliding with grains should react with atomic
and/or molecular hydrogen on the grain mantle to produce fully hydrogenated species such as CH_4, NH_3, and H_2O. Since CH_4 is nonpolar,
it will most easily evaporate from the cold grain surfaces, and thus
would preferentially return carbon to the gas phase (Blake, 1985).

The chemistry in the <u>plateau</u> source shows clearly the effects of
the energetic outflow driven by the deeply embedded but highly luminous
source IRc 2 (Table III). The abundances of several species are
strongly enhanced relative to the quiescent ridge cloud, including such
sulfur and silicon-containing species as SiO, H_2S, SO_2, SO, and to

a lesser extent OCS and CS. The abundances of HDO, HCN, and HC_3N are also considerably greater in the plateau than in the ridge cloud, suggesting that H_2O in the plateau may be very abundant indeed (cf. Olofsson, 1984). These values refer primarily to the disk-like outflow. Relatively little is known about abundances in the very high velocity wings (bipolar-type outflow), although emission is clearly visible here in CO, SiO, SO, SO_2, HCN, HCO^+, and HDO. The $f(HCO^+)/f(CO)$ ratio increases still further (by about an order of magnitude) in the spatially extended high velocity flow which coincides with the region of vibrationally excited H_2 emission (Olofsson et al. 1982; Vogel et al., 1984). Although anomalously high abundances of sulfur-containing species have sometimes been suggested as diagnostic of shock chemistry (Hartquist et al. 1980), the abundances observed in the plateau for SO, SO_2, SiO, and SiS can be produced in gas phase ion-molecule models if S and Si are not strongly depleted (Prasad and Huntress, 1982). In any case, the chemical composition of the plateau would seem to imply a gas phase C/O ratio which is solar or even oxygen enriched. Relative to the extended ridge cloud, such a composition could arise either from outflow from young, oxygen-rich stars (IRc 2?) or from evaporation of oxygen-rich grain mantles. A large oxygen abundance is supported by the observations of Werner et al. (1984) of neutral atomic oxygen at 63.2 μm.

Nonetheless, shocks are clearly present in this region, as is evident from the vibrationally excited H_2 emission, and are a likely cause for the enhanced temperature observed. The temperature then plays an important role in determining the chemical composition, since a number of endothermic reactions which are not possible in the ambient cloud material can now take place. An example is $H_2 + CN \rightarrow HCN + H$, which not only increases the HCN abundance but is probably responsible for the diminished CN abundance relative to the extended ridge cloud.

There are, moreover, interesting anomalies in the plateau chemistry which illustrate the complexity of conditions in this source. Thus, the abundance of H_2CO is very similar to that in the extended ridge, while that of HDO is significantly enhanced. These results are surprising, both because these molecules are relatively fragile and should not survive the passage of a strong shock front, and because the degree of deuterium fractionation at high temperature should be very small. The observations may perhaps be understood as resulting from the presence of a significant fraction of the outflow in high density clumps whose composition has not yet been significantly altered from the ambient material (the enhancement of HDO in the gas phase may reflect evaporation of grain mantles within which water has been fractionated either in situ (Tielens, 1983) or in the gas phase prior to incorporation into the grain mantle).

Perhaps the most massive clump (or clumps) leftover from the formation of IRc 2 is represented by the hot core. As with the plateau source, the abundances of HCN and HDO are strongly enhanced relative to the extended ridge cloud, while reactive species such as CN, C_2H, and atomic carbon are sharply reduced (Figure 9). In addition, a number of species which are not detected elsewhere in Orion are found in this region, including HNCO and the heavily hydrogenated molecules CH_2CHCN

and CH_3CH_2CN. Likewise, the ammonia abundance appears to be significantly higher than in the ambient cloud. There do appear to be
differences with respect to the plateau, however, since the sulfur
and silicon-containing species such as SO_2 and SiO are not clearly
detectable (and SO appears to be differently distributed then the other
hot core species; Friberg, 1984), while CS seems clearly underabundant.
Another interesting feature of the chemistry of the hot core is the
predominance of nitrogen-containing species in the observed spectrum.

There is considerable evidence that this region is physically close
to a luminous infrared source (e.g., Goldsmith et al. 1983). Blake
(1985) suggests that the hot core may provide a good example of gas-
grain interactions for the chemistry of dense molecular clouds with
such embedded possible protostars or YSO's. A viable model for the
chemistry may consist of a dense clump that is radiatively heated to
temperatures high enough to release volatile material from grain
mantles (H_2O and NH_3), while not destroying the more refractory
grain cores which may contain silicon and sulfur. The enhanced abundance of chemically saturated molecules like ethyl cyanide would then
reflect the presumed high efficiency of hydrogenation for molecules
adsorbed onto grain surfaces.

A further example of the variety of chemical regimes which may
exist within molecular clouds is provided by the compact ridge source.
This warm, spatially confined condensation may well be a pre-protostellar
candidate, since multitransition CH_3OH observations show that linewidths and temperature increase with level energy above the ground
state -- a signpost of contraction (or expansion). The effects of
cloud evolution may therefore be important in the chemical modeling of
this region (Hjalmarson, 1986). The emission from a number of large,
oxygen-rich species in Orion seems to come predominately from the
compact ridge. In particular, the abundances of HDO, CH_3OH, and H_2CO
are strongly enhanced with respect to the more extended ridge, while
H_2C_2O, $(CH_3)_2O$, $HCOOCH_3$, H_2CS, OCS, and HCOOH have been detected
in the compact ridge but are not obvious in the more extended cloud
(more careful mapping should be carried out to try to quantify the
apparent enhancements in the compact ridge). On the other hand, CH_3CN
has recently been found by Andersson (1985) to have a rather similar
beam-averaged abundance in the compact and the extended ridge cloud.
Although sulfur dioxide has been seen south of the KL region, this
emission probably originates from the hot, dense core also observed in
400 μm continuum emission and in the SiO J=2-1 line (Irvine et al.,
1983; Ziurys et al., 1985).

The presence of a number of oxygen-rich molecules, including HDO,
suggests that the compact ridge source may result from an interaction
between the outflow producing the plateau source (where clearly O/C>1)
and the quiescent southern ridge cloud. The quite selective nature of
the observed species also provides important clues to the processes
taking place. For example, both dimethyl ether (CH_3OCH_3) and methyl
formate ($HCOOCH_3$) have abundances strongly enhanced relative to both
Sgr B2 (Table III) and theoretical gas phase models (Herbst and Leung,
1986), while their isomers ethanol (CH_3CH_2OH) and acetic acid (CH_3COOH)
have not been detected in Orion. Methanol is likewise significantly

enhanced in the compact ridge. In contrast, formic acid (HCOOH) and acetaldehyde (CH_3CHO) have an abundance similar to or less than that found in Sgr B2, and more in accordance with theoretical gas phase predictions.

It is, we hope, indicative of an increasing sophistication in astrochemical models that just this kind of chemical specificity can be explained in terms of differences in rates of radiative association leading to precursors of the complex molecules (Blake, 1985). For example, the reaction $CH_3{}^+ + H_2O \rightarrow CH_3OH_2{}^+ + h\nu$ is expected to be about two orders of magnitude faster than $HCO^+ + H_2O \rightarrow H_2COOH^+ + h\nu$, so that the abundance of CH_3OH should be enhanced to a much greater extent than that of HCOOH, as is observed. Moreover, the reaction of $CH_3OH_2{}^+$ with H_2CO then leads directly to the ionic precursor of methyl formate ($HCOOCH_3$), <u>not</u> to that of acetic acid (CH_3COOH). Since radiative association reactions are not favored at higher temperatures or small fractional ionizations, the failure to detect these large complex molecules in the plateau and hot core may also be understood.

4.3.4. <u>TMC-1</u>. There have been repeated suggestions in the literature that there may be abundance gradients along the ridge of this dark cloud. It is certainly clear that the emission from NH_3 and that from the cyanopolyynes peak at different locations, at least if one compares low J transitions for the latter. Likewise, the carbon chain species (including C_2H and CH_3C_2H) have a more limited distribution than $C^{18}O$ and CH_3OH (Friberg <u>et al.</u>, 1986). There is disagreement as to how much of these apparent differences may be due to varying physical conditions which produce gradients in excitation, rather than to abundance gradients (cf. Bujarrabal <u>et al.</u> 1981; Schloerb <u>et al.</u> 1983; Avery <u>et al.</u> 1982; Walmsley and Wilson, 1985). There is clearly small scale clumping and intricate velocity structure. There is also the problem that some transitions that have been studied, particularly for the cyanopolyynes, may not be optically thin.

5. CONCLUSIONS

We may conclude on a note of increasing optimism regarding the understanding of chemical abundances and processes in dense inter-stellar clouds. Although there are still problems of uniqueness, there is a general agreement between recent theoretical calculations and observational abundances for many molecules observed in quiescent clouds. Moreover, some of the striking differences observed between regions, including some of the very evident selectivity in chemical processes, are beginning to be understood as a natural result of the kinetic chemistry that follows from the physical conditions found in regions of active star formation (e.g., the prevalence of chemically saturated species in the Orion hot core and of only certain complex oxygen-containing molecules in the compact ridge; Section 4.3.).

Nonetheless, the field remains in its infancy and certain very fundamental questions remain to be answered:

1) Why is depletion of molecules onto cold grains in dense cloud

cores not more apparent, and what desorption processes return molecules
to the gas phase?

2) Why is the abundance of carbon-chain molecules so strikingly
different from one dark cloud to another, in spite of apparently very
similar physical conditions?

3) Are there real tracers of "shock chemistry", as opposed to high
temperature chemistry produced (for example) by radiative heating from
nearby embedded sources?

4) What is the true nature of the interstellar grains, and what
chemical processes occur on them and contribute to observed gas phase
abundances?

5) What additions to the list of known interstellar molecules will
occur as more systematic spectral studies take place for cold dark
clouds, particularly at short centimeter wavelengths, and for all types
of regions as the sub-millimeter spectrum is opened up? We note that
abundance upper limits for potential interstellar complex molecules are
often not very low because of their large partition functions -- here
absence of evidence is not evidence of absence.

6) Will there be surprises as more sophisticated chemical models
are developed which include simultaneously the chemical and physical
evolution of a collapsing cloud?

7) What is the chemical form in which a number of as yet
undetected elements exist in dense clouds, including metals (such as
Fe, Mg, Ca, etc.) and other moderately abundant elements such as
phosphorus?

ACKNOWLEDGEMENTS

This research was supported in part by grants from NSF, NASA, and
the Swedish Natural Science Research Council (NFR). The FCRAO is
operated with permission from the Metropolitan District Commission of
the Commonwealth of Massachusetts.

REFERENCES

Allen, M., and Knapp, G.R. (1978), Ap.J., 225, 843.
Andersson, M. (1985), in (Sub)Millimeter Astronomy, ed. P.A. Shaver and
    K. Kjar, ESO Conf. Workshop Proc. No. 22, pp. 353-364.
Avery, L.W., MacLeod, J.M. and Broten, N.W. (1982), Ap.J. 254, 116.
Bastien, P., Batrla, W., Henkel, C., Pauls, T, Walmsley, C.M. and
    Wilson, T.L. (1985), Ast. Ap., 146, 86.
Batrla, W., Wilson, T.L., Bastien, P., and Ruf, K. (1983), Ast. Ap.
    128, 279.
Bell, M.B. and Matthews, H.E., (1985), Ap.J. (Lett.) 291, L63 .
Benson, J.M., and Johnston, K.J. (1984), Ap.J. 277, 181.
Biermann, L., Giguere, P.T., and Huebner, W.F. (1982), Ast. Ap., 108,
    221.
Black, D.C. and Matthews, M.S., eds. (1985), Protostars and Planets. II.
    (Tucson: U. Arizona Press).

Black, J. (1986), in Astrochemistry (IAU Symp. 120), ed. S.P. Tarafdar
    and M.S. Vardya (Dordrecht: Reidel).
Blake, G.A., Keene, J. and Phillips, T.G. (1985), Ap.J. 295, 501.
Blake, G.A., Sutton, E.C., Masson, C.R., and Phillips, T.G. (1986),
    Ap.J. Suppl., in press.
Blake, G.A. (1985), Ph.D. Dissertation, California Institute of
    Technology.
Bogey, M. and Destombes, J.L. (1986), Ast. Ap. 159, L8.
Bohlin, R.C., Savage, B.D., and Drake, J.F. (1978), Ap.J. 224, 132.
Boland, W., and deJong, T. (1982), Ap.J. 261, 110.
Broten, N.W., MacLeod, J.M., Oka, T., Avery, L.W., Brooks, J.W., McGee,
    R.X., and Newton, L.M. (1976), Ap.J. (Lett.) 209, L143.
Brown, R.D., Godfrey, P.D., Cragg, D.M., Rice, E.H.N., Irvine, W.M.,
    Friberg, P., Suzuki, H., Ohishi, M., Kaifu, N., and Morimoto, M.
    (1985), Ap.J. 297, 302.
Bujarrabal, V., Guelin, M., Morris, M. and Thaddeus, P. (1981), Ast. Ap.
    99, 239.
Cernicharo, J., Kahane, C., Gomez-Gonzalez, J., and Guelin, M. (1986),
    Ast. Ap., in press.
Churchwell, E. (1980), Ap.J., 240, 811.
Churchwell, E., Wood, D., Myers, P., and Myers, R.V. (1986), Ap.J. 305,
    405.
Churchwell, E., Woods, R.C., Dickman, R.L., and Irvine, W.M. (1986),
    data from NRAO and FCRAO (in preparation).
Cummins, S.E., Linke, R.A., and Thaddeus, P. (1986), Ap.J. Suppl., 60,
    819.
Dalgarno, A. (1986), QJRAS 27, 83.
Draine, B.T. (1985), in Protostars and Planets. II., ed. D.C. Black and
    M.S. Matthews (Tucson: U. Arizona Press), pp. 621-640.
Elmegreen, B.G., Genzel, R., Moran, J.M., Reid, M.J., and Walker, R.C.
    (1980), Ap.J. 241, 1007.
Frerking, M.A., Langer, W.D. and Wilson, R.W. (1982), Ap.J. 262, 590.
Frerking, M.A., Wilson, R.W., Linke, R.A., and Wannier, P.G. (1980),
    Ap. J. 240, 65.
Friberg, P. (1986), private communications.
Friberg, P. (1984), Ast. Ap. 132, 265.
Friberg, P., Hjalmarson, Å., Irvine, W.M., and Guelin, M. (1980),
    Ap.J. (Lett.) 241, L99.
Friberg, P., Irvine, W.M., Madden, S.C., and Hjalmarson, Å. (1986), in
    Astrochemistry (IAU Symp. 120), ed. S.P. Tarafdar and M.S. Vardya
    (Dordrecht: Reidel).
Glassgold, A.E., Huggins, P.J., and Shucking, E.L., eds. (1982), Symp.
    on the Orion Nebula, Ann. NY Acad. Sci. 395.
Goldsmith, P.F., and Mao, X.-J. (1983), Ap.J. 265, 791.
Goldsmith, P.F., Irvine, W.M., Hjalmarson, Å., and Ellder, J. (1986a),
    Ap.J., in press.
Goldsmith, P.F., Krotkov, R., and Snell, R.L. (1986c), Ap.J., 299, 405.
Goldsmith, P.F., Krotkov, R., Snell, R.L., Brown, R.D., and Godfrey, P.
    (1983), Ap.J. 274, 184.
Goldsmith, P.F., Langer, W.D., Ellder, J., Irvine, W.M. and Kollberg, E.
    (1981), Ap.J. 249, 524.

Goldsmith, P.F., Snell, R.L., Deguchi, S., Krotkov, R., and Linke, R.A. (1982) Ap.J. 260, 147.

Goldsmith, P.F., Snell, R.L., Erickson, N.R., Dickman, R.L., Schloerb, F.P., and Irvine, W.I. (1985), Ap.J. 289, 613.

Goldsmith, P.F., Snell, R.L., Hasegawa, T., and Ukita, N. (1986b), submitted to Ap.J.

Gottlieb, C.A., Gottlieb, E.W., and Thaddeus, P. (1986), Ast. Ap., submitted.

Gottlieb, C.A., Lada, C.J., Gottlieb, E.W., Lilley, A.E., and Litvak, M.M. (1975), Ap.J., 202, 655.

Greenberg, J.M. (1983), in Cosmochemistry and the Origin of Life, ed. C. Ponnamperuma (Dordrecht: Reidel), pp. 71-112.

Guelin, M. and Thaddeus, P. (1977), Ap.J. (Lett.). 212, L81.

Guelin, M., Cernicharo, J., Kahane, C., and Gomez-Gonzalez, J. (1986), Ast. Ap. 157, L17.

Guelin, M., Langer, W.D. and Wilson, R.W. (1982), Ast. Ap. 107, 107.

Guelin, M. (1985), in Molecular Astrophysics – State of the Art and Future Directions, ed. G.H.F. Dierckson, W.F. Huebner, and P.W. Langhoff (Dordrecht: Reidel), pp. 23-44.

Guelin, M. (1986), in Astrochemistry (IAU Symp. 120), ed. S.P. Tarafdar and M.S. Vardya (Dordrecht: Reidel).

Guilloteau, S., and Baudry, A. (1981), Ast. Ap., 97, 213.

Gusten, R., Walmsley, C.M. and Pauls, T. (1981), Ast. Ap. 103, 197.

Hartquist, T.W., Oppenheimer, M., and Dalgarno, A. (1980), Ap.J. 236, 182.

Harvey, P.M., Campbell, M.F., and Hoffmann, W.F. (1977), Ap.J. 211, 786.

Hasegawa, T., Kaifu, N., Inatani, J., Morimoto, M., Chikada, Y., Hirabayashi, H., Iwashita, H., Morita, K., Tojo, A., and Akabane, K. (1984), Ap.J. 283, 117.

Herbst, E. and Leung, C.M. (1986), MNRAS, in press.

Hildebrand, R.H. (1983), QJRAS, 24, 267.

Hjalmarson, Å. (1985), in (Sub)-Millimeter Astronomy, ed. P.A. Shaver and K. Kjar, ESO Conf. Workshop Proc. No. 22, pp. 285-326.

Hjalmarson, Å. (1986), Phys. Scripta 33, in press.

Ho, P.T.P. and Townes, C.H. (1983), Ann. Rev. Ast. Ap. 21, 239.

Huggins, P.J., Carlson, W.J., and Kinney, A.L. (1984), Ast. Ap. 133, 347.

Irvine, W.M., Good, J.C., and Schloerb, F.P. (1983), Ast. Ap. 127, L10.

Irvine, W.M., Matthews, H.E., Friberg, P., Madden, S.C., Swade, D.A., and Ziurys, L.M. 1986, in preparation.

Irvine, W.M. and Schloerb, F.P. (1984), Ap.J., 282, 516.

Irvine, W.M., Schloerb, F.P., Hjalmarson, Å., and Herbst, E. (1985), in Protostars and Planets II., ed. D. Black and M. Matthews (Tucson: U. Arizona Press), pp. 579-620.

Jarrold, M.F., Bowers, M.T., DeFrees, D.J., McLean, A.D. and Herbst, E. (1986), Ap.J. 303, 392.

Johansson, L.E.B., Andersson, C., Ellder, J., Friberg, P., Hjalmarson, Å., Hoglund, B., Irvine, W.M., Olofsson, H. and Rydbeck, G. (1984), Ast. Ap. 130, 227.

Johansson, L.E.B., Andersson, C., Ellder, J., Friberg, P., Hjalmarson, Å., Hoglund, B., Irvine, W.M., Olofsson, H. and Rydbeck, G.

(1985), Ast. Ap. Suppl. 60, 135.

Keene, J., Blake, G.A., Phillips, T.G., Huggins, P.J., and Beichman, C.A. (1985), Ap.J. 299, 967.

Kerridge, J.F. and Chang, S. (1985), in Protostars and Planets II., ed. D. Black and M. Matthews (Tucson: U. Arizona Press), pp. 738-754.

Kleinmann, D.W. and Low, F.J. (1967), Ap.J. (Lett.) 149, L1.

Knacke, R.F., Geballe, T.R., Noll, K.S. and Tokunaga, A.T. (1985) Ap.J. (Lett.) 298, L67.

Knacke, R.F., McCorkle, S. Puetter, R.C., Erickson, E.F., and Kratschmer, W. (1982), Ap.J. 260, 141.

Kutner, M.L., Tucker, K.D., Chin, G. and Thaddeus, P. (1977), Ap.J. 215, 521.

Kwan, J., and Scoville, N.Z. (1975), Ap.J., 195, L85.

Langer, W.D., Graedel, T.E., Frerking, M.A. and Armentrout, P.B. (1984), Ap.J. 277, 581.

Lazcano-Araujo, A. and Oro, J. (1981), in Comets and the Origin of Life, ed. C. Ponnamperuma (Dordrecht: Reidel), pp. 191-226.

Leger, A., Jura, M. and Omont, A. (1985), Ast. Ap. 144, 147.

Leitch-Devlin, M.A. and Williams, D.A. (1984), MNRAS 210, 577.

Leitch-Devlin, M.A. and Williams, D.A. (1985), MNRAS 213, 295.

Linke, R.A., Cummins, S.E., Green, S., and Thaddeus, P. (1982), in Regions of Recent Star Formation, ed. R.S. Roger and P.E. Dewdney, (Boston: Reidel), 391.

Linke, R.A., Goldsmith, P.F., Wannier, P.G., Wilson, R.W., and Penzias, A.A. (1977), Ap.J. 214, 50.

Madden, S.C., (1986), private communication (FCRAO data).

Matthews, H.E. and Irvine, W.M. (1985), Ap.J. (Lett.). 298, L61.

Matthews, H.E. and Sears, T.J. (1986), Ap.J. 300, 766.

Matthews, H.E., MacLeod, J.M., Broten, N.W., Friberg, P., and Madden, S.C. (1986), Ap.J., in press.

McKeegan, K.D., Walker, R.M. and Zinner, E. (1985), Geochim. Cosmochim. Acta 49, 1971.

McMahon, T.B. and Kebarle, P. (1985), J. Chem. Phys. 83, 3919.

Menten, K.M., Johnston, K.J., Wilson, T.L., Walmsley, C.M., Mauersberger, R., and Henkel, C., (1985), Ap.J. (Lett.) 293, L83.

Millar, T.J. and Freeman, A. (1984), MNRAS 207, 425.

Moore, E.L., Langer, W.D., and Huguenin, G.R. (1986), Ap.J. 306, 682.

Morimoto, M., Ohishi, M., and Kanzawa, T. (1984), Ap.J. 288, L11.

Mundy, L.G., Snell, R.L., Evans, N.J., Goldsmith, P.F. and Bally, J. (1986), Ap.J., in press.

Nuth, J.A. and Stencel, R.E. (1986), Interrelationships Among Circumstellar, Interstellar, and Interplanetary Dust, NASA Conf. Publ. 2403.

Nyman, L.Å. (1983), Ast. Ap. 120, 307.

Nyman, L.Å. (1984), Ast. Ap. 141, 323.

Olofsson, H. (1984), Ast. Ap. 134, 36.

Olofsson, H., Johansson, L.E.B., Hjalmarson, Å, and Nguyen-Quang-Rieu (1982), Ast. Ap. 107, 128.

Penzias, A.A. (1975), in Atomic and Molecular Physics and the Interstellar Matter (Les Houches Session 26), ed. R. Baliau et al. (Amsterdam: North-Holland), pp. 375-408.

Penzias, A.A. (1980), Science, 208, 663.

Phillips, T.G., Blake, G.A., Keene, J., Woods, R.C., and Churchwell, E. (1985), Ap.J. (Lett.) 294, L45.

Plambeck, R.L., and Williams, D.R.W. (1977), Ap.J., 227, L43.

Poynter, R.L., and Pickett, H.M. (1980), Submillimeter, Millimeter, and Microwave Spectral Line Catalogue, JPL Publication 80-23.

Prasad, S.S. and Huntress, W.T., Jr. (1982), Ap.J. 260, 590.

Raimond, E., and Eliasson, B. (1969), Ap.J. 155, 817.

Rohlfs, K., Chini, R., Wink, J.E., and Bohme, R. (1986), Ast. Ap., 158, 181.

Rydbeck, O.E.H., Kollberg, E., Hjalmarson, Å., Sume, A., Ellder, J., and Irvine, W.M. (1976), Ap.J. Suppl. 31, 333.

Rydbeck, O.E.H., Sume, A., Hjalmarson, Å., Ellder, J., Ronnang, B.O., and Kollberg, E. (1977), Ap.J. (Lett.) 215, L35.

Schloerb, F.P. and Snell, R.L. (1984), Ap.J., 283, 129.

Schloerb, F.P., Snell, R.L., and Schwartz, P.R. (1986), Ap.J., submitted.

Schloerb, F.P., Snell, R.L., and Young, J.S. (1983), Ap.J. 267, 163.

Scoville, N.Z., Kleinmann, S.G., Hall, D.N.B., and Ridgway, S.T. (1983), Ap.J. 275, 201.

Scoville, N.Z., Solomon, P.M., and Penzias, A.A., (1975), Ap.J., 201, 352.

Snell, R.L., Mundy, L.G., Goldsmith, P.F., Evans, N.J., and Erickson, N.R. (1984), Ap.J. 276, 625.

Snyder, L.E., Dykstra, C.E., and Bernholdt, D. (1985), in Masers, Molecules and Mass Outflows in Star Forming Regions, ed. A. Haschick (NEROC Haystack Obs.), pp. 9-22.

Snyder, L.E., Wilson, T.L., Henkel, C., Jewell, P.R., and Walmsley, C.M. (1984), Bull. AAS 16, 959.

Stenholm, L.G. (1983), Ast. Ap. 117, 41.

Stutzki, J., and Winnewisser, G. (1985a), Ast. Ap. 144, 1.

Stutzki, J., and Winnewisser, G. (1985b), Ast. Ap. 144, 13.

Stutzki, J., and Winnewisser, G. (1985c), Ast. Ap. 148. 254.

Sutton, E.C., Blake, G.A., Masson, C.R., and Phillips, T.G. (1985), Ap.J. Suppl. 58, 341.

Suzuki, H. (1986), private communication (NRO data).

Swade, D.A. (1987), Ph.D. Dissertation, University of Massachusetts.

Swade, D.A., Schloerb, F.P., Irvine, W.M. and Snell, R.L. (1985), in Masers, Molecules, and Mass Outflows in Star Forming Regions, ed. A. Haschick (NEROC Haystack Observ.), pp. 73-87.

Tarafdar, S.P. and Vardya, M.S. (1986), Astrochemistry (IAU Symp. 120), (Dordrecht: Reidel).

Thaddeus, P. (1981), Phil. Trans. Roy. Soc. London, A 303, 469.

Thaddeus, P., Gottlieb, C.A., Hjalmarson, Å., Johansson, L.E.B., Irvine, W.M., Friberg, P., and Linke, R.A. (1985a), Ap.J. (Lett.), 294, L49.

Thaddeus, P., Vrtilek, J.M. and Gottlieb, C.A. (1985b), Ap.J. (Lett.). 299, L63.

Thronson, H.A., and Harper, D.A. (1986), Ap.J. 300, 396.

Tielens, A.G.G.M. (1983), Ast. Ap. 119, 177.

Townes, C.H., and Schawlow, A.L. (1955), Microwave Spectroscopy (New

York: McGraw-Hill).

Troland, T.H., Crutcher, R.M., and Heiles, C. (1985), Ap.J. 298, 808.

Ulich, B.L., and Haas, R.W. (1976), Ap.J. Suppl. 30, 247.

Vogel, S.N., Wright, M.C.H., Plambeck, R.L., and Welch, W.J. (1984), Ap.J. 283, 655.

Vrtilek, J.M., Gottlieb, C.A., and Thaddeus, P. (1986), Ap.J., in press.

Walmsley, C.M. and Wilson, T.L. (1985), in Nearby Molecular Clouds (IAU. Region. Symp. 80).

Wannier, P.G., Encrenaz, P.J., Wilson, R.W., and Penzias, A.A. (1984), Ap.J., 190, L77.

Watt, G.D., Millar, T.J., White, G.J., and Harten, R.H. (1986), Ast. Ap. 155, 339.

Weinreb, S., Barrett, A.H., Meeks, M.L., and Henry, J.C. (1963), Nature 200, 829.

Werner, M.W., Crawford, M.K., Genzel, R., Hollenbach, D.J., Townes, C.H., and Watson, D.M. (1984), Ap.J. (Lett.) 282, L81.

Westbrook, W.E., Werner, M.W., Elias, J.H., Gezari, D.Y., Hauser, M.G., Lo, K.Y., and Neugebauer, G. (1976), Ap.J. 209, 94.

White, G.J., Avery, L.W., Richardson, K.J., and Lesurf, J.C.G. (1986), Ap.J. 302, 701.

Williams, D.A. and Hartquist, T.W. (1984), MNRAS 210, 141.

Wilson, T.L. (1985a), Comm. Astrophys. 11, 83.

Wilson, T.L. (1985b), in (Sub)-Millimeter Astronomy, ed. P.A. Shaver and K. Kjar, ESO Conf. Workshop Proc. No. 22, pp. 401-420.

Wilson, T.L., Ruf, K., Walmsley, C.M., Martin, R.N., Pauls, T.A., and Batrla, W. (1982), Ast. Ap. 115, 185.

Wilson, T.L., Walmsley, C.M., Henkel, C., Pauls, T., and Mattes, H. (1980), Ast. Ap. 91, 36.

Winnewisser, G., Churchwell, E. and Walmsley, C.M. (1979), Ast. Ap. 72, 215.

Wootten, A., Boulanger, F., Ziurys, L.M., Bogey, M., Combes, F., Encrenaz, P., and Gerin, M. (1986), preprint.

Wootten, A., Loren, R.B., Sandquist, A., Friberg, P., and Hjalmarson, Å. (1984), Ap.J. 279, 633.

Wootten, A., Loren, R.L. and Snell, R.L. (1982), Ap.J. 255, 160.

Zeng, Q., Batrla, W. and Wilson, T.L. (1984), Ast. Ap. 141, 127.

Ziurys, L.M. (1986), private communication (FCRAO data).

Ziurys, L.M. and Turner, B.E. (1986), Ap.J. (Lett.) 302, L31.

Ziurys, L.M., Friberg, P. and Irvine, W.M. (1985), Bull. AAS 17, 563.

John Bally and Laura Danly

GAS PHASE CHEMICAL PROCESSES IN MOLECULAR CLOUDS

Eric Herbst
Department of Physics
Duke University
Durham, NC 27706
USA

ABSTRACT. The important gas phase rate processes governing the abundances of molecules in dense interstellar (molecular) clouds are thoroughly reviewed. Recent contributions to our understanding of gas phase rate processes are explored. The use of these processes in chemical models is discussed with an emphasis on the basic details of model construction. Isotopic fractionation and the formation of complex molecules are considered.

1. INTRODUCTION

Dense interstellar (molecular) clouds are complex assemblies of gas and dust in which considerable turbulence often occurs, shock waves propagate, bipolar flows exist, and stars are forming. Nevertheless, the physical conditions present in most of the volume of a typical cloud are reasonably homogeneous and severely constrain the types of chemical processes that can occur in the gas phase. These physical conditions include very low gas densities by terrestrial standards and temperatures far below those associated with most chemical processes. Let us determine how and why these conditions affect the cloud chemistry.

The gas densities n are sufficiently low except in the immediate vicinity of protostars to preclude the possibility of ternary reactions, in which three atoms or molecules are involved. Such reactions occur when two species A and B collide and form a complex labelled AB* which, before dissociating, collides with a third body C. The complex can last for times between $1\times10(-14)$ sec and $1\times10(-3)$ sec depending on its energy, size, structure, and potential well (Herbst 1980; 1985a). The time interval between collisions in an interstellar cloud is far longer than a millisecond so that the probability of the complex lasting until a third body collides with it is quite small. Thus it is safe to say that the gas phase chemistry of interstellar clouds is comprised solely of binary reactions.

The low kinetic temperatures of molecular clouds also affect the chemistry strongly. Chemical reactions are either exothermic (give off energy) or endothermic (require energy). Endothermic reactions cannot occur appreciably under normal interstellar conditions. To see this, consider a reaction between species A and B to form products C and D:

$$A + B \;\text{---}\!\!> C + D. \tag{1}$$

D. J. Hollenbach and H. A. Thronson, Jr. (eds.), Interstellar Processes, 611–629.

Chemists normally write the rate of a reaction in terms of the concentrations of reactants and a rate coefficient k. For process (1), the rate can be written as

$$d[C]/dt = -d[A]/dt = k[A][B] \qquad (2)$$

where the symbol [ ] stands for concentration and the rate coefficient k is given by the expression

$$k = <\sigma v> \qquad (3)$$

where σ is the total cross section, v is the relative velocity between reactants, and the averaging is performed over the thermal distribution. For endothermic reactions the cross section is zero unless the energy of the reactants exceeds a certain value E. Under these conditions, the rate coefficient can be derived to be

$$k = A(T) \exp(-E/k_B T) \qquad (4)$$

where k_B is the Boltzmann constant, and the so-called pre-exponential factor A(T), a weak function of the temperature, depends upon details of the collision, such as the attractive long-range potential (Levine and Bernstein 1974) . Thus, if $E >> k_B T$, as is true for strongly endothermic reactions under typical molecular cloud conditions ($E \approx 1$ eV, $k_B T <$ 0.01 eV), the rate coefficient k is negligible. Endothermic reactions can occur in shock waves if the collisional energy is sufficiently high. (See the chapter by Draine and Shull in this volume.) The formation of the molecule CH^+ via the well-studied endothermic reaction ($E \approx 0.4$ eV)

$$C^+ + H_2 \dashrightarrow CH^+ + H \qquad (5)$$

is critical to all shock models although the current view is that the energy to power the reaction is due to streaming between ions and neutrals rather than a high temperature.

Even most exothermic reactions have rate coefficients that can be expressed via equation (4). The reason for this is that chemical reactions normally involve the breaking of chemical bonds before the formation of new ones. In reactions between neutral species, the potential energy during a reaction increases by an amount called the activation energy (here labelled E) as the old bonds are broken before declining to a value below that of the reactants. (Endothermic reactions may also possess activation energy in excess of their endothermicity.) As in endothermic reactions, the cross section is zero unless the system possesses energy $\geq$ E, and the rate coefficient assumes the form in eq. (4) where E is now the activation energy. Since activation energies are also on the order of 1 eV, one can easily see that most exothermic reactions between neutral species do not occur appreciably under molecular cloud conditions.

Some neutral reactions are known from laboratory measurements to occur without activation energy. These reactions normally involve either atoms or reactive species called "radicals" which are most often in non-singlet ground electronic states. For these systems A(T) is approximately 1×10^{-11} cc/sec at room temperature and varies perhaps as $\sqrt{T}$ (Gardiner 1972). As has been discussed by a wide variety of authors, however, starting with Herbst and Klemperer (1973), exothermic ion-molecule reactions constitute a more important exception to the existence of activation energy barriers. These reactions have been studied by many different investigators using a variety of techniques; an excellent summary of this work has just been provided by Anicich and Huntress (1986). At room temperature, the rate coefficients of ion-molecule reactions are normally (but not always) quite large, ranging from 1 x 10(-09) cc/sec up to 1 x 10(-08) cc/sec, and do not exhibit activation energy barriers. Unfortunately, few low temperature studies have been reported and rate coefficients at interstellar temperatures must usually be estimated or calculated (see Section 2.2 for a discussion). A simple reason for the lack of existence of activation energy barriers is the strong long-range attractive force, given by the potential

$$V(R) = -e^2/2\alpha R^4 \tag{6}$$

where e is the electronic charge in e.s.u., α (cc) is the polarizability of the neutral species, and R is the distance between reactants. This potential can be used (Levine and Bernstein 1974) in a simple treatment of ion-molecule collisions to obtain a rate coefficient given by the expression

$$k_L = 2\pi e (\alpha/\mu)^{1/2} \tag{7}$$

where μ is the reduced mass of the reactants. In this treatment, first formulated by Langevin, $k \approx 10(-9)$ cc/sec and possesses no temperature dependence. In general, the prediction is born out by laboratory studies. However, if the neutral species has a permanent dipole moment, the Langevin model is inadequate (see Section 2.2).

Ions are produced in molecular clouds by cosmic ray bombardment and by UV photoionization. The primary cosmic ray ionization processes involve the dominant species molecular hydrogen (formed on interstellar dust grains) and helium:

$$H_2 + Cosmic\ Ray \longrightarrow H_2^+ + e + Cosmic\ Ray \tag{8a}$$

$$H_2 + Cosmic\ Ray \longrightarrow H^+ + H + e + Cosmic\ Ray \tag{8b}$$

$$He + Cosmic\ Ray \longrightarrow He^+ + e + Cosmic\ Ray \tag{8c}$$

Rates for these processes and the penetration ability of cosmic rays into molecular clouds are reasonably well understood (Herbst and Klemperer 1973). Until recently, little attention was paid to UV photoionization except in diffuse clouds. The reason is that grains attenuate the photon field to such an extent that the process would appear to be a negligible one (Herbst and Leung 1986a). However, it has become recently understood (Prasad and Tarafdar 1983) that cosmic ray bombardment of molecular hydrogen results in UV photon production internal to molecular clouds and calculations of the importance of this effect as regards photodissociation have recently been undertaken (Lepp, Dalgarno, and Sternberg 1986). The role of internally produced photons in causing ionization should also be pursued.

2. ION-MOLECULE REACTIONS

2.1 Synthetic Processes

How can ion-molecule reactions form complex gas phase molecules? Indeed, how can these processes even form simple polyatomic species? These questions have been addressed by a wide variety of authors in the last decade. For a thorough recent review, see Winnewisser and Herbst (1986). As an example of the synthesis of a simple polyatomic molecule, consider the synthesis of water starting from molecular hydrogen, its cosmic ray-produced ion, and atomic oxygen:

$$H_2^+ + H_2 \longrightarrow H_3^+ + H \tag{9}$$

$$O + H_3^+ \longrightarrow OH^+ + H_2 \tag{10}$$

$$OH^+ + H_2 \longrightarrow H_2O^+ + H \tag{11}$$

$$H_2O^+ + H_2 \longrightarrow H_3O^+ + H \tag{12}$$

$$H_3O^+ + e \longrightarrow H_2O + H \tag{13}.$$

Reactions (9) - (13) have been studied in the laboratory, but the neutral products of reaction (13) are unknown. Note that reactions with molecular hydrogen as a reactant occur on a short time scale because of the large abundance of this species (it is essentially equal to the gas density in molecular clouds). Note also that the protonated water ion does not react with molecular hydrogen.

Process (13) is one example of a type of reaction known as dissociative recombination; these reactions occur with large rate coefficients ($k \approx 10(-6)$ cc/sec) that possess a weak inverse temperature dependence (Mul and McGowan 1980; Adams, Smith, and Alge 1984). Unfortunately, the branching ratios for formation of neutral products have not been measured and a variety of theories have been proposed for these ratios (Herbst 1978; Green and Herbst 1979; Bates 1986a). The theories of Herbst (1978) and Green and Herbst (1979) predict that the radical OH is produced in addition to water in reaction (13); the theory of Bates (1986a) suggests that water is the only product. If the theory of Bates (1986a) is correct, then OH in molecular clouds must be produced via a different mechanism. Lepp, Dalgarno, and Sternberg (1986) have suggested that photons produced via cosmic ray bombardment photodissociate water at a sufficiently rapid rate to produce the observed amount of interstellar OH. The topic of dissociative recombination is discussed more fully in Section 4.

Similar schemes to reactions (9)-(13) can be invoked to produce molecules such as methane (CH_4) and ammonia (NH_3). (Radiative association is involved in the synthesis of methane - see Section 3). Once a simple hydrocarbon such as methane is produced, it can be utilized to produce larger species. Consider the following synthesis of the interstellar radical C_4H, one of several pathways leading to this species (Mitchell and Huntress 1979):

$$C^+ + CH_4 \longrightarrow C_2H_2^+ + H_2 \tag{14a}$$

$$C^+ + CH_4 \longrightarrow C_2H_3^+ + H \tag{14b}$$

$$C_2H_3^+ + e \longrightarrow C_2H_2 + H \tag{15}$$

$$C_2H_2^+ + C_2H_2 \longrightarrow C_4H_2^+ + H_2 \tag{16}$$

$$C_4H_2^+ + e \longrightarrow C_4H + H \qquad (17).$$

Once again, the ion-molecule reactions utilized have been measured in the laboratory but the ion-electron branching ratios have not been measured. In this instance, all theories agree that the neutral products shown above are dominant ones.

For many of the complex molecules observed in molecular clouds, ion-molecule syntheses based on normal ion-molecule reactions and ion-electron dissociative recombination reactions are not possible because of roadblocks in the form of endothermic reactions. Such a roadblock is met in the synthesis of methane; the reaction

$$CH_3^+ + H_2 \longrightarrow CH_4^+ + H \qquad (18)$$

is endothermic and terminates a chain of reactions ending in the precursor ion CH_5^+. To avoid this roadblock, it is necessary to consider radiative association reactions. These exothermic processes, in which two smaller species coalesce to form a larger one, are considered in Section 3.

2.2 Rate Coefficients at Temperatures Under 50 K

Until several years ago, most investigators assumed that the Langevin model was correct for ion-molecule reactions and that rate coefficients measured at room temperature pertained at temperatures as low as 10 K. In general, this view is probably accurate although it is now recognized that ion-molecule reactions in which the neutral reactant has a permanent dipole moment behave differently. In addition, some reactions not in this category and studied at temperatures down to 10 K in the laboratory are known to possess a peculiar temperature dependence. Let us first consider ion-polar neutral reactions.

Based on experimental evidence in the range 205-540 K and a new theory by Clary (1985), Adams, Smith, and Clary (1985) have suggested that the rate coefficients of ion-polar reactions increase significantly as the temperature is lowered. Earlier theoretical work on this same topic was undertaken by the Japanese group (Sakimoto and Takayanagi 1980) and other investigators. It had been known from experimental measurements for some time that at room temperature ion-polar neutral reactions are typically more rapid than the Langevin model predicts by up to a factor of five. With the latest spate of effort, predictions that rate coefficients could increase another order of magnitude and approach 10(-7) cc/sec at temperatures near 10 K have been made. Limited confirmation has recently come from as yet unpublished laboratory work from the Meudon group (Rowe 1985) at 28K. Herbst and Leung (1986b) have shown that in their models of molecular clouds, use of the large rate coefficients does not make a significant difference in calculated abundances for many species, presumably because both formation and depletion rates are changed equally. However, there are species, typically protonated ions of polar neutrals (HCS^+, $HCNH^+$), for which the large ion-polar neutral rate coefficients lead to a dramatic increase in calculated abundance, in good agreement with observation (Millar $et\ al.$ 1985). It is not clear that these new theories are correct in their prediction that all ion-polar neutral rate coefficients become large at low temperatures because of the possibility of short-range repulsive effects (Herbst 1986a). More experiments are needed!

As more ion-molecule reactions are studied in the laboratory at very low temperature by groups such as those at Meudon (e.g. Marquette $et\ al$. 1985) and Boulder, CO (e.g., Luine and Dunn 1985), one can expect some surprises in measured rate coefficients.

Consider the reaction

$$NH_3^+ + H_2 \; -----> \; NH_4^+ + H \qquad\qquad (19)$$

which is slower than normal at room temperature and decreases in rate as the temperature is lowered, behaving like a system with activation energy. However, at a temperature of 50 K, the rate coefficient starts to increase and reaches the significant value of $\approx 10(-12)$ cc/sec at 10 K (Luine and Dunn 1985). Theoretical studies are underway to understand the cause of this effect.

3. ASSOCIATION REACTIONS

Association reactions are those in which two species, here labelled A^+ and B, coalesce to form a larger molecule AB^+ via formation and subsequent stabilization of a short-lived complex AB^{+*}. Although this process in the laboratory is almost always ternary in nature - a third body C collisionally stabilizes the complex - under interstellar conditions the dominant process is radiative. Although several radiative association reactions have been studied in the laboratory (e.g. Barlow *et al.* 1984), the rate coefficients of most important interstellar processes must be calculated or inferred from ternary measurements, typically at higher temperatures. Calculations have been performed by three major groups, including myself (Herbst 1985a), Bates (1986b,c), and Bowers and coworkers (Bass *et al.* 1981), where only the most recent references have been cited. The theoretical techniques can be checked partially by utilizing them for ternary reactions, where laboratory measurements are routinely available, and it has been found that theory can reproduce measured rate coefficients to perhaps one order of magnitude. The general importance of radiative association reactions in molecular clouds has been recognized by several authors, most notably Smith and Adams (1978) and Huntress and Mitchell (1979).

In all the theories utilized, radiative association proceeds via a two-step process in which the complex formed in the collision of A^+ and B either redissociates into reactants or is stabilized. This process can be written as follows:

$$A^+ + B \; -----> \; AB^{+*} \qquad\qquad (20)$$

$$AB^{+*} \; -----> \; A^+ + B \qquad\qquad (21)$$

$$AB^{+*} \; -----> \; AB^+ + h\nu \qquad\qquad (22)$$

where an ion-molecule process is considered because in most neutral-neutral systems activation energy barriers prevent complex formation. At steady state

$$d[AB^{+*}]/dt = k_{20}\,[A^+]\,[B] - (k_{21} + k_{22})\,[AB^{+*}] = 0$$

and the overall rate law becomes

$$d[AB^+]/dt = k_{22}\,[AB^{+*}] = k_{ra}\,[A^+][B] \qquad\qquad (23)$$

where

$$k_{ra} = k_{20} k_{22} / (k_{21} + k_{22}) \tag{24}$$

and the numerical subscripts refer to the specific rate processes. Note that k_{21} and k_{22} are not normal rate coefficients given by eq. (3) but are so-called "unimolecular" rates with units of s^{-1}. Normally, the complex redissociates into reactants more rapidly than it stabilizes via radiation $(k_{21} >> k_{22})$, and

$$k_{ra} \approx (k_{20} / k_{21}) k_{22} \tag{25}.$$

The ratio k_{20}/k_{21} is calculated in the thermal model (Herbst 1980), the simplest of the theoretical approaches, by using the relevant value at thermal equilibrium (the so-called "equilibrium coefficient"), which is a function of the partition functions of the reactants and complex (Hill 1960). This function possesses a severe inverse temperature dependence, which is moderated slightly in more complex versions of the theory which include conservation of angular momentum to varying degrees of rigor (see, e.g., Bates 1983; Herbst 1981). Nevertheless, the qualitative predictions of the thermal model are correct. These predictions are that the rate of radiative association increases as temperature is decreased, as molecular complexity is increased, and as the dissociation energy of the compound molecule AB^+ increases. To complete the calculation, the rate of radiative stabilization k_{22} is required. Up to quite recently, the dominant view was that radiative stabilization occurs via the emission of an infra-red photon between vibrational levels of the ground electronic state of the complex at a rate of perhaps 100- 3,000 s^{-1} depending on a variety of factors (Herbst 1985a). However, Bates (1986b,c) has recently shown that in order for theory to reproduce the measured value (Barlow et al. 1984) of the rate coefficient for the reaction

$$CH_3^+ + H_2 \; ----> \; CH_5^+ + h\nu \tag{26}$$

a larger radiative stabilization rate of $\approx 3.5 \times 10(4) \; s^{-1}$ is required. To achieve such a large rate, he hypothesizes that an electronic transition is involved and that such transitions are often important in radiative association processes. His view of the radiative association process is that the reactants form a complex which undergoes a radiationless transition into an excited electronic state, if one is energetically available, and then relaxes to the ground electronic state radiatively. If Bates is correct, then many of the radiative association rate coefficients tabulated in Herbst (1985a) will have to be revised upwards by an order of magnitude or so. These currently tabulated rate coefficients show that calculated values range downward from values equal to the collision rate coefficient ($\approx 10(-9)$ cc/sec) to values many orders of magnitude smaller. Thus, radiative association cannot be invoked without careful attention to the particular system. It should be noted, however, that because molecular hydrogen is so abundant a species in molecular clouds, radiative association processes involving this species can be important even if the rate coefficient is quite small. An example is the reaction

$$C^+ + H_2 \; ----> \; CH_2^+ + h\nu \tag{27}$$

which is calculated to possess a rate coefficient of only $\approx 10(-15)$ cc/sec at 10 K but which is primarily responsible for the formation of CH in diffuse clouds (Black and Dalgarno 1977).

In addition to being important in the synthesis of methane - reaction (26) circumnavigates the road block of reaction (18) - radiative association reactions are needed in the syntheses of a wide variety of interstellar species (Leung *et al.* 1984). Let us first consider here the case of the well-known interstellar ring molecule C_3H_2. The dominant synthesis is most probably (Herbst *et al.* 1984)

$$C^+ + C_2H_2 \ ----> \ C_3H^+ + H \tag{28}$$

$$C_3H^+ + H_2 \ ----> \ C_3H_3^+ + h\nu \tag{29}$$

$$C_3H_3^+ + e \ ----> \ C_3H_2 + H \tag{30}.$$

The rate of reaction (29) has been estimated both theoretically and from ternary measurements at 80 K to be $\approx 10(-11)$ cc/sec at 10 K. It is needed because the normal ion-molecule reaction between the reactants is slightly endothermic. The difficulty in this mechanism is that it is unclear what the structure of the $C_3H_3^+$ ion is; two structures exist for this species - one of which has the ring shape. If process (29) leads to a high percentage of the ring isomer, one can expect that process (30) will produce the ring version of C_3H_2. Smith and Adams (Birmingham, UK) are currently investigating the products of reaction (29).

A series of important radiative association reactions in molecular clouds is thought to occur between the CH_3^+ ion and assorted neutrals (Smith and Adams 1978). This ion possesses a reasonably large abundance because it is depleted only slowly by radiative association (see reaction 26). It is known to associate, at least from ternary studies in the laboratory, with a variety of species (Smith and Adams 1978). As an example of an important radiative association, consider the synthesis of methanol:

$$CH_3^+ + H_2O \ ----> \ CH_3OH_2^+ + h\nu \tag{31}$$

$$CH_3OH_2^+ + e \ ----> \ CH_3OH + H \tag{32}.$$

Some of the ternary association reactions between CH_3^+ and neutral species as measured in the laboratory occur in competition with normal binary channels (Smith and Adams 1978). Herbst (1985b; 1986b) has considered the possibility of radiative association when the complex can dissociate to products C^+ and D in addition to the original reactants; this leads to a more complex rate law than eq. (25) and also to a diminution of the calculated radiative association rate, which can be negligible or important depending on details of the process. There are important interstellar radiative association reactions in this category including the one between CH_3^+ and NH_3 (Herbst 1985h).

The fact that radiative association reactions are calculated to become more rapid as molecular complexity increases until they occur on every collision has led over the years to the view that these processes might be responsible for the formation of molecules as complex as interstellar grains, or to rephrase the argument in more modern terms, PAH's. There are a number of reasons for caution, however. First, there is insufficient laboratory evidence to compare with theoretical treatments of radiative association. Some reactions previously predicted by theoretical treatments to be rapid do not occur, presumably because of activation energy barriers (Herbst *et al* . 1983). Secondly, it is not clear that interstellar clouds live long enough for very complex species to be formed via gas phase reactions at such low gas densities. Preliminary work on this subject by myself (Herbst 1985c) suggests that a time of $\approx 10(7)$ yr. is required to produce hydrocarbons with twenty carbon atoms. More work on this subject is clearly needed, as is an experimental study of which reactions - both normal and associative - result in ring closure to form PAH-like structures. Even if there exist gas phase routes to interstellar PAH's, one must remember that these molecules will also be depleted by gas phase reactions (Duley and Williams 1986) and their high abundance is by no means assured.

4. ION-ELECTRON DISSOCIATIVE RECOMBINATION

From the examples discussed above, it should be apparent that ion-molecule syntheses of neutral molecules A are thought to proceed most often through a protonated ion precursor AH^+ which dissociatively recombines with electrons to form the desired neutral and a hydrogen atom. Unfortunately, there exists little experimental evidence that ion-electron dissociative reactions proceed preferentially in this manner. Indeed, other than one study of the reaction between $H_3^+ + e$ (Mitchell et al. 1983) in a merged beam apparatus, in which the H_3^+ ion was vibrationally excited, there are no published laboratory studies dealing with the neutral products of these reactions. Ever since the theoretical work of Bates (1950), it has been realized that ion-electron recombination occurs rapidly only via a dissociative process in which the stable ion potential curve is crossed by a repulsive curve of the parent neutral. Once a jump from the ion to the neutral potential curve is made, the system rapidly dissociates. If a suitable neutral potential curve does not intersect with the ion potential curve, no reaction occurs. An example is the reaction $H_3^+ + e$, where recent experimental work by Smith and Adams (1984) has corroborated a theoretical prediction of a negligible reaction rate coefficient unless the ion is vibrationally excited (Michels and Hobbs 1984). However, our current understanding of the mechanism of the process does not really disclose what the eventual neutral fragments are and authoritative ab initio theoretical treatments of the possible neutral fragments are unavailable.

In 1978, Herbst published a statistical theory of ion-electron neutral branching ratios, in which the probability of forming a particular set of products is related to the density of quantum states of the products subject to energy and angular momentum constraints. In this approach, ion-electron dissociative recombination reactions can lead to a large number of possible products but breaking of weak chemical bonds is preferred to the breaking of strong bonds. For example, the linear ion $HCNH^+$ recombines with electrons to form the species

$$HCNH^+ + e ----> HCN + H; \quad HNC + H; \quad CN + H_2 \qquad (33)$$

which are produced by breaking what chemists would term "single" bonds. Products such as CH and NH, which could only be formed by breaking the "multiple" strong CN bond, are not computed to be important. In addition, the neutrals HCN and HNC are formed at approximately the same rate, an important result. The model's prediction that multiple bonds are not broken when the ion and electron recombine means that the synthetic power of ion-molecule reactions in producing complex ionic structures is not vitiated by ion-electron processes in which the complex ions are broken apart into tiny fragments. Green and Herbst (1979) modified the conclusions of Herbst (1978) somewhat and viewed the ion-electron dissociative recombination process as one in which the neutral fragments would have to depart quite rapidly, favoring pathways in which one or possibly more hydrogen atoms would be ejected but not including pathways unlikely kinematically,

such as the CN + H_2 channel in (33) or those involving large structural changes. These
authors also showed that their model leads to a prediction of significant abundances for a
variety of metastable isomers of stable species analogous to HNC in molecular clouds.
The physical basis of this prediction is the unlikelihood of large structural changes in the
skeleton of the ion undergoing dissociative recombination. Based on this approach, a
detailed calculation of the interstellar abundance ratio between the metastable isomer
CH_3NC and the stable isomer CH_3CN has been performed recently by De Frees *et al.*
(1985). These authors calculated the abudance ratio between the ions CH_3NCH^+ and
CH_3CNH^+ in molecular clouds and, assuming hydrogen atom fragmentation to be the
dominant mode of bond rupture upon electron recombination, then used this abundance
ratio to calculate the one between the two neutral species.

An interesting semi-quantitative way of looking at ion-electron dissociative
recombination processes has just been undertaken by Bates (1986a) based on his original
work (Bates 1950) on diatomic molecules. According to this treatment, which arrives at
somewhat different conclusions from earlier theories, the favored neutral products are
those involving the minimum rearrangment of chemical bonds. More specifically, after the
electron attaches itself to the molecular ion, one bond where the positive charge was
located is preferentially disrupted. As an example, consider the protonated water ion
H_3O^+, which consists of a positively charged oxygen atom surrounded by three
hydrogens. According to Bates' view, the dissociation products upon recombination are
primarily H_2O and H. It is unclear to this author, however, whether one can exclude the
possibility that the water produced can decompose secondarily to produce OH and H.
After all, there is a large amount of available energy in such reactions. As a more unusual
but equally important example, consider the ion CH_5^+, which is pictured by Bates to be a
loose cluster between CH_3^+ and H_2. In Bates' view, the electron attaches itself to the
carbon ion in CH_3^+ and preferentially breaks one of the three C-H bonds to produce CH_2
and H. Meanwhile, the molecular hydrogen gently breaks away. Thus, the overall
process looks like

$$CH_5^+ \;+\; e \;\text{-----}\!> \; CH_2 \;+\; H \;+\; H_2 \tag{34}$$

which is quite distinct from the previous view that the dominant neutral products would be
CH_4 and possibly CH_3. (The Bates model also leads to the formation of CH_3 and H_2 if a
molecular orbital connecting these two species can be broken.) If Bates is correct, then
interstellar methane is not formed via the dissociative recombination of CH_5^+ to a
significant extent. However, an alternative ion-molecule pathway

$$CH_5^+ \;+\; CO \;\text{-----}\!> \; CH_4 \;+\; HCO^+ \tag{35}$$

is sufficiently rapid to produce copious amounts of methane.

Now let us consider an ion such as $C_2H_3^+$ with a strong carbon-carbon bond.
According to the views of Herbst (1978) and Green and Herbst (1979), likely products
upon dissociative recombination are CCH, CCH_2 (vinylidene) and HCCH (acetylene).
According to Bates, if one writes the structure of $C_2H_3^+$ as $HC(+)CH_2$, then the favored
neutral products are obtained by breaking the bonds of C(+):

$$C_2H_3^+ \;+\; e \;\text{-----}\!> \; CCH_2 \;+\; H; \; CH \;+\; CH_2 \tag{36}.$$

Current quantum chemical calculations show that vinylidene (CCH_2) is an unstable isomer of acetylene and quickly reverts to its more stable form. In addition, the acetylene may have sufficient internal energy upon formation to dissociate to CCH and H. If secondary decomposition does occur, the major difference between the model of Bates and earlier approaches in this instance is the prediction that breaking of the multiple CC bond can occur. A further complication in this and other systems is that molecular ions often have loosely bound structures and undergo large amplitude motions. Quantum chemists have long suspected that $C_2H_3^+$ easily isomerizes between the "classical" form mentioned above and a "non-classical" form in which one hydrogen atom lies in between the two carbons and the positive charge is spread over much of the molecule.

In general, the predictions of Bates for the neutral products of dissociative recombination reactions are that single hydrogen atom removal is a major channel if the atom lies near the center of positive charge and that, immediately upon H atom removal, a neutral molecule is formed with a geometrical shape similar to the parent ion. Isomerization of the newly produced neutral will occur if it is not stable or metastable (viz., CCH_2 --> HCCH). Secondary decomposition of the newly produced neutral may or may not occur. Finally, there is the possibility at least that multiple bonds can be broken with more facility that either Herbst (1978) or Green and Herbst (1979) would countenance. The somewhat divergent theoretical views will only be resolved by experimental studies. Meanwhile, it would be an interesting project to run a model of interstellar cloud chemistry using branching ratios for ion-electron dissociative recombination reactions based on the ideas of Bates (1986c) once they are quantified.

What is the outlook for experimental studies of the neutral products? In the one published experiment undertaken successfully to date, Mitchell et al. (1983) found that the favored products in the reaction between H_3^+ and electrons are three hydrogen atoms rather than a hydrogen atom and a molecule. Since the H_3^+ in this experiment is vibrationally hot, it is unclear whether the results represent a serious disagreement with the Bates model. The Meudon group has recently built a different type of apparatus to study recombination reactions; preliminary results on the system H_2O^+ + electrons were presented in a poster session at the Goa meeting on astrochemistry and showed oxygen atoms to be a product. One awaits definitive results from this study with anticipation.

5. PHOTODESTRUCTION

The question of how shielded molecules are from UV radiation in molecular clouds has been discussed for some time ever since the pioneering studies of Stief et al. (1972). These authors showed that the lifetimes against photodissociation for a representative sample of small interstellar molecules increase dramatically as a function of visual extinction. This result led to the realization that interstellar clouds could be divided into diffuse regions of low visual extinction where photons penetrate appreciably and photodestructive events affect the chemistry and dense (molecular) clouds where the visual extinction is sufficiently large that photodestructive events need not be considered. Pioneering models of diffuse interstellar clouds such as that of Black and Dalgarno (1977) showed how strongly stellar UV radiation incident upon a cloud affects the molecular composition of the cloud; indeed diffuse cloud chemistry is primarily the chemistry of atoms and diatomic molecules. Important new work on photodissociation rates of small molecules and their importance has been undertaken by van Dishoeck and collaborators (see, for example, van Dishoeck 1986; van Dishoeck and Black 1986).

The complete neglect of photodissociation and photoionization in molecular clouds has always been a contentious subject with several authors' arguing that preferential forward scattering by the grains could result in more UV radiation inside clouds than customarily assumed (see Sandell 1978). Nevertheless, use of the customary grain parameters (Mathis *et al.* 1983) shows that even at a rather small visual extinction of $A_V = 5$, most photodestructive processes are not rapid compared with competing gas phase processes. This conclusion is based on a compendium of photodissociation and ionization rates recently compiled by Herbst and Leung (1986a), who utilized the available literature regarding cross sections and branching ratios for photodestructive processes involving neutral polyatomic species. Following Black and Dalgarno (1977), these authors fitted their calculated results for photodestruction processes as a function of A_V to the form

$$k \ (s^{-1}) \ = \ A exp(-CA_V) \eqno(37).$$

As an example of their results, consider the fitted parameters for the dissociation of methanol - $A = 7 \times 10(-10) \ s^{-1}$ and $C = 1.72$. For a cloud with $A_V = 5$, a minimum value thought to pertain to the center of the small molecular cloud TMC-1 (Millar and Freeman 1984), $k \approx 10(-13) \ s^{-1}$. This number is comparable to the depletion rate of methanol via ion-molecule reactions. To appreciate this point, consider a cloud of density $10(4) \ cm^{-3}$ and a total fractional abundance of reacting ions of $5 \times 10(-9)$. With an estimated rate coefficient of $5 \times 10(-9)$ cc/sec (methanol is polar), the depletion rate due to ion-molecule reactions is $2.5 \times 10(-13) \ s^{-1}$. In their model calculations, Herbst and Leung (1986a) showed that for $A_V = 5$, calculated abundances are affected by photodissociation and ionization reactions but the effect is not a dramatic one for most species. An exception is C I which has a significantly increased abundance due to photodissociation of CO. For giant molecular clouds, estimated values of A_V are much greater than 5 and, unless the scattering of photons by grains is strongly forward in direction, photodestrucion via external UV photons appears to be unimportant except at cloud edges.

As mentioned previously, Prasad and Tarafdar (1983) have recently studied the effects of internal UV fields in molecular clouds generated by cosmic ray excitation of molecular hydrogen. These authors concluded that photodestruction rates of up to $\approx 10(-14) \ s^{-1}$ could be expected. Their particular interest lay in attempting to determine a reason for the large abundance of atomic carbon observed in molecular clouds (C I is not calculated to have a large steady-state abundance) and they concluded that photodissociation of CO via internally produced photons was a viable mechanism. Internally generated UV photons have also been considered by Lepp, Dalgarno, and Sternberg (1986) to solve the problem of OH formation in molecular clouds if, as according to Bates (1986c), OH cannot be formed from the precursor ion H_3O^+. The effect of internally generated photons on other species remains to be worked out in detail.

6. A BRIEF LOOK AT GAS PHASE MODELS AND ABUNDANCE CALCULATIONS

How does one utilize processes of the types discussed in Sections 1-5 in gas phase models of the chemistry of interstellar clouds? The first step is to write kinetic equations for each of the species in the model. Consider the hypothetical molecule X^+ which is formed via the reaction

$$Q^+ \ + \ M \ ----> \ X^+ \ + \ Y \eqno(38)$$

and destroyed via the reaction

$$X^+ \ + \ Z \ ----> \ Products \tag{39}.$$

The rate law (kinetic equation) for the change of concentration of X^+ with time is given by the equation

$$d[X^+]/dt \ = \ k_{38} \ [Q^+] \ [M] \ - \ k_{39} \ [\ X^+] \ [\ Z] \tag{40}.$$

A chemical model consists of simultaneous differential equations of type (40) which can be solved with varying degrees of complexity depending upon the assumed physical conditions and changes in them. The simplest approach is to assume that steady state pertains. In this approach, one sets the time derivatives of concentrations equal to zero and obtains coupled algebraic equations, which are normally relatively easy to solve. In the above example, eq. (40) reduces to

$$[X^+]_{ss} \ = \ k_{38} \ [Q^+]_{ss}[M]_{ss} \ / \ k_{39} \ [Z]_{ss} \tag{41}.$$

Solution of coupled equations such as these leads to concentrations which can be compared with observation by either converting them into column densities or by dividing them by the total gas density and obtaining so-called fractional abundances. The steady-state approach for molecular clouds was taken by Herbst and Klemperer (1973) who solved coupled algebraic equations as a function of the total gas density with the cosmic abundances of the elements as constraints. Assuming molecular hydrogen to be formed on the grains (a feature of all "gas phase" models), they calculated abundances for a variety of small polyatomic species and obtained reasonable agreement with observation. Their approach has been since improved and expanded by several authors, including Glassgold and Langer (1976), Mitchell *et al* . (1978), Herbst *et al* . (1984), and Millar and Freeman (1984). It is far from clear, however, that steady-state conditions can ever be achieved considering that interstellar clouds are collapsing to form stars and that gas phase species are both being adsorbed onto and being desorbed from the grains.

The next degree of complexity, pioneered by Prasad and Huntress (1980 a,b), is to solve the time-dependent kinetic equations given a set of initial concentrations and fixed physical conditions (temperature, density). This approach shows that steady-state abundances in the gas can be reached in perhaps $10(7)$ yrs. Unfortunately, the time necessary for gas phase species to strike grains is much smaller than this ($t \approx 3 \times 10(5)$ yrs; Winnewisser and Herbst 1986) and unless the sticking probability is small or there is an efficient desorption mechanism (see, e.g.Leger *et al.* 1985), it is difficult to understand the relative success of calculated steady-state abundances. The approach of Prasad and Huntress (1980a,b) has been utilized and modified by Langer and co-workers (Graedel *et al.* 1982; Langer *et al.* 1984) to investigate the dependence of the results on assumed gas phase elemental abundances and to study isotopic fractionation of heavy isotopes; and extended by Leung *et al* . (1984), Herbst and Leung (1986a,b), and Millar and Nejad (1985) to include complex molecules. The calculations involving complex molecules are discussed below.

The most ambitious type of model calculation, discussed by Prasad in this volume, is the evolutionary approach, in which time-varying physical conditions are included. This approach, pioneered by Tarafdar *et al.* (1985), is conceptually the most satisfying but is quite computationally intensive.

6.1 Deuterium Fractionation

In addition to complete models, a large body of papers in the literature has appeared in which a limited number of ion-molecule reactions is utilized to explain an observation or to demonstrate a chemical effect. Steady-state calculations are normally undertaken in such approaches and are indeed appropriate if the reaction time scales are short. (In this context, it should be remembered that the long time interval required for steady-state to be achieved in complete models is due to only certain reactions.) Most important among this class of papers are those involving isotopic fractionation, principally deuteration. Pioneered by Watson (1977) and Guelin et $al.$ (1977), the study of large deuterium isotopic fractionation in trace constituents of molecular clouds remains an active field of current research (see, e.g., Adams and Smith 1986; Croswell and Dalgarno 1985; Herbst et $al.$ 1986). The basic principle behind deuterium fractionation is the necessity for rapid ion-molecule reactions under interstellar conditions to be exothermic. Consider the important reactions

$$H_3^+ + HD \Longleftrightarrow H_2D^+ + H_2 \tag{42}$$

studied in the laboratory down to 80 K by Adams and Smith (1981) and theoretically by Herbst (1982). The left-to-right process is exothermic by a small amount whereas the right-to-left process is endothermic by this amount, approximately $E/k_B \approx 230$ K. The result is that at temperatures considerably lower than 230 K, the endothermic reaction is repressed (see eq. 4) and the abundance of H_2D^+ relative to H_3^+ is enhanced markedly compared with the HD/H_2 abundance ratio of approximately $4 \times 10(-5)$. The analysis is of course complicated by the fact that H_2D^+ can be depleted by other channels such as reaction with CO. Note that it cannot be depleted by reaction with electrons (Smith and Adams 1984), as once thought (Guelin et $al.$ 1982). Still, given normal interstellar abundances of species such as CO, theory can predict the H_2D^+/H_3^+ ratio as a function of temperature, or alternately one can utilize theory and observation of this ratio to determine the cloud temperature and upper limits to the abundances of species such as CO which deplete H_2D^+. Unfortunately, this ratio cannot be measured in molecular clouds, because H_3^+ has no permanent dipole moment. However, the large H_2D^+ enhancement, especially pronounced in the lower temperature molecular clouds, translates into a large DCO^+ abundance via the reaction

$$H_2D^+ + CO \longrightarrow DCO^+ + H_2 \tag{43}$$

and careful measurements of the DCO^+/HCO^+ abundance ratio (Guelin et $al.$ 1982) can be compared with theory (Herbst 1982) in exact analogy with the H_2D^+/H_3^+ case. Thus, a value for this ratio of ≈ 0.1 in TMC -1 is consistent with a temperature of 10 K and a CO fractional abundance of $\approx 10(-4)$. Interestingly enough, the more direct exothermic process to form DCO^+:

$$HCO^+ + HD \longrightarrow DCO^+ + H_2 \tag{44}$$

is known not to occur from laboratory studies. What about isotopic fractionation processes involving other deuterated interstellar molecules? An early and still useful review by Watson (1976) contains a discussion of the various possible processes as does a more recent paper by Herbst *et al.* (1986). A recent comparison of ion-molecule calculations of abundance ratios between deuterated and normal species with observation has been given by Wootten (1986).

Within the last year, observations of the deuterated species CCD and C_3HD have been reported (Combes *et al.* 1985; Vrtilek *et al.* 1985; Friberg 1986; Bell *et al.* 1986) that are of special interest. The reported CCD/CCH abundance ratio appears to be 0.05 in Orion (T ≈ 50 K) and only 0.01 in TMC-1 (T ≈ 10 K). Normally deuterium fractionation is more pronounced at lower temperatures because endothermic reactions decrease in rate as temperature is lowered. In the case of CCD/CCH, Herbst *et al.* (1986) argue that although the precursor to CCD, C_2HD^+, is formed primarily in the reaction system

$$C_2H_2^+ + HD \iff C_2HD^+ + H_2 \qquad (45)$$

where the right-to-left reaction is endothermic by $E/k_B \approx 550$ K, the C_2HD^+ ion is depleted principally via a radiative association reaction with H_2, which *increases* in rate as the temperature is lowered, and probably leads to a lower CCD/CCH abundance ratio at the lower temperatures. The analysis is complicated by other fractionation mechanisms, however. The C_3HD (deuterated cyclopropenylidene) fractionation has been analyzed by Bell *et al.* (1986) and utilized to obtain an estimate of the abundance of the optically thick C_3H_2 in TMC-1.

The HD molecule is not the only species that provides a source of deuterium for fractionation. Dalgarno and co-workers (Croswell and Dalgarno 1985; Dalgarno and Lepp 1984) have shown how deuterium atoms, formed in the dissociative recombination of DCO^+, can also function in this regard and, in selective instances, be more important than HD. Finally, the deuterated species H_2D^+ and DCO^+ can themselves function as deuterators of neutral species via reactions of the sort (Watson 1976)

$$AH + H_2D^+ \longrightarrow AHD^+ + H_2 \qquad (46)$$

$$AHD^+ + e \longrightarrow AD + H \qquad (47).$$

6.2 Chemistry in Star Formation Regions

The above discussion has been limited to models of chemistry in ambient regions of molecular clouds where gas densities are in the range 10(3) < n < 10(6) and kinetic temperatures are under 100 K; shock models are discussed in another chapter of this volume. Specialized models of small non-shocked regions in Orion associated with star formation such as the compact ridge and the hot core also exist (Blake *et al.* 1986; Blake 1985). According to Blake (1985), the unique chemistry of the compact ridge source in Orion, in which complex oxygen-containing organic molecules are especially abundant, is caused by a large abundance of water in the outflow from the shock driven plateau source. The copious water abundance is processed by radiative association reactions to form large amounts of more complex species such as methanol and methyl formate. On the other hand, the chemistry of the hot core region surrounding the luminous source IRc 2 can be interpreted in terms of a high density and temperature leading to the release of adsorbed species from the grains, the occurrence of gas phase neutral-neutral reactions, and significant grain processing of adsorbed molecules including hydrogenation reactions to form more saturated species (Williams 1986). The study of the chemistries prevalent in star formation regions such as these and their differences from the ion-molecule dominated chemistry promises to become a major new area of investigation as observations with higher and higher resolution become more commonplace.

6.3   Complex Molecule Formation

Polyatomic molecules as complex as $HC_{11}N$ have been seen unambiguously in molecular clouds, far larger species such as PAH's may also exist there. A discussion of the ion-molecule pathways to complex molecules is given in a review by Winnewisser and Herbst (1986). What are the results of gas phase models containing the more complex interstellar molecules? The answer to this question depends upon which models one investigates. In the pseudo-time dependent (non-evolutionary) models of Leung and Herbst (Leung $et\ al$. 1984; Herbst and Leung 1986a,b), abundances of molecules as complex as C_4H, HC_5N, and CH_3OCH_3 (dimethyl ether) are calculated to have significant abundances only at times well before steady state conditions are achieved. At these times (t $\approx$ 3 x 10(5) yrs) , the atomic carbon abundance is also predicted to be high, in agreement with observation, unlike the situation at steady-state when little C I is predicted. Indeed, these authors show how atomic carbon aids in the syntheses of complex carbon-containing molecules. However, in a similar but smaller model by Millar and Nejad (1985), the time dependence of the calculated abundances of the other workers is preserved but the quantitative abundances of most polyatomic species are much higher and in reasonable agreement with observation at steady state. The discrepancies between the two models range up to three or more orders of magnitude for species such as HC_3N. The causes of these discrepancies are currently being investigated by Millar $et\ al$. (1987) and appear to be primarily: (i) the inclusion of unmeasured neutral-neutral and ion-molecule reactions by Herbst and Leung involving oxygen atoms, and (ii) the choice of dissociative recombination branching ratios. The apparent extreme sensitivity of the calculated abundances of complex species to different choices of whether and how to include unmeasured reactions and branching ratios militates strongly for additional laboratory measurements!

Despite the discrepancy among results of different workers, Herbst and Leung (1987) are extending their reaction set to include predictions of the abundances of molecules as complex as C_9H, CH_3C_6H, HC_9N, CH_3C_7N, etc. The complex hydrocarbons are produced via both condensation reactions (Mitchell and Huntress 1979) and reactions involving C I and C II (Winnewisser and Herbst 1986). The complex nitrogen-containing molecules are formed primarily via reactions between N atoms and hydrocarbon ions and selected radiative association processes analogous to those measured in the laboratory and involved in the synthesis and depletion of HC_3N (Federer $et\ al$. 1986; Knight $et\ al$. 1986; Raksit and Bohme 1985). It is hoped that the strong observational evidence on the abundance ratios between members of the cyanoacetylene series will constrain the chemistry and help determine which group of workers is closer to the truth.

7. SUMMARY

The various gas phase rate processes that can occur in interstellar clouds have been discussed. These processes include ion-molecule and neutral-neutral reactions, radiative association reactions, ion-electron dissociative recombinations, and photodissociation and ionization. The topics of why ion-molecule reactions dominate in importance for regions removed from star formation and how these reactions synthesize molecules have been discussed, while the limited present understanding of the products of ion-electron dissociative recombination reactions has been demonstrated. The key but not exclusive role of radiative association reactions in synthesizing complex molecules has been illuminated. Finally, the role of photodestruction events in molecular clouds caused both by externally and internally produced photons has been explored.

After the discussion of the various gas phase rate processes, the topic of gas phase chemical models of molecular clouds has been introduced with an emphasis on the basic details of constructing such models. Isotopic fractionation and complex molecule formation have been given attention. Some mention has been made of the chemistry occurring in star forming regions.

ACKNOWLEDGEMENTS

This work has been supported by the National Science Foundation (U.S.) via grant AST - 8513151.

REFERENCES

Adams, N. G. and Smith, D. 1981, *Ap. J.* , **248**, 373.
Adams, N. G. and Smith, D. 1986, in IAU Symposium No. 120 on *Astrochemistry* , eds. M. S. Vardya and S. P.. Tarafdar (Dordrecht: Reidel).
Adams, N. G., Smith, D., and Alge, E. 1984, *J. Chem. Phys.* , **81**, 1778.
Adams, N. G., Smith, D., and Clary, D. C. 1985, *Ap. J. (Letters)* , **296**, L31.
Anicich, V. G. and Huntress, W. T., Jr. 1986, *A Survey of Bimolecular Ion-Molecule Reactions For Use in Modeling the Chemistry of Planetary Atmospheres, Cometary Comae, and Interstellar Clouds* (Jet Propulsion Laboratory pre-print).
Barlow, S. E., Dunn, G. H., and Schauer, K. 1984, *Phys. Rev. Letters* , **52**, 902.
Bass, L. M., Kemper, P. R., Anicich, V. G., and Bowers, M. T. 1981, *J. Am. Chem. Soc.* , **103**, 5283.
Bates, D. R. 1950, *Phys. Rev.* , **78**, 492.
Bates, D. R. 1983, *Ap. J.* , **270** , 564.
Bates, D. R. 1986a, *Ap. J. (Letters)* , **306**, L45.
Bates, D. R. 1986b, *Ap. J.* , in press.
Bates, D. R. 1986c, *J. Chem. Phys.,* in press.
Bell, M. B., Feldman, P. A., Matthew, H. E., and Avery, L. 1986, pre-print.
Black, J. H. and Dalgarno, A. 1977, *Ap. J. Suppl.* , **34**, 405.
Blake, G. A. 1985, Ph. D. Thesis, California Institute of Technology.
Blake, G. A., Sutton, E. C., Masson, C. R., and Phillips, T. G.1986, pre-print.
Clary, D. C. 1985, *Molec. Phys.* , **54**, 605.
Combes, F. *et al.* 1985, *Astr. Ap.* , **147**, L25.
Croswell, K. and Dalgarno, A. 1985, *Ap. J.* , **289**, 618.

Dalgarno, A. and Lepp. S. 1984, *Ap. J. (Letters)* , **287**, L47.

DeFrees, D. J., McLean, A. D., and Herbst, E. 1985, *Ap. J.* , **293**, 236.

Duley, W. W. and Williams, D. A. 1986, *M. N. R. A. S.* , **219**, 859.

Federer, W., Villinger, H., Lindinger, W., Richter, R., and Ferguson, E. E. 1986, *Chem. Phys. Letters* , in press.

Friberg, P. 1986, private communication.

Gardiner, W. C. 1972, *Rates and Mechanisms of Chemical Reactions* (Menlo Park, CA: Benjamin).

Glassgold, A. E. and Langer , W. D. 1976, *Ap. J.* , **206**, 85.

Graedel, T. E., Langer, W. D., and Frerking, M. A. 1982, *Ap. J. Suppl.* , **48**, 321.

Green, S. and Herbst, E. 1979, *Ap. J.* , **229**, 121.

Guelin, M., Langer, W. D., Snell, R. L., and Wootten, H. A. 1977, *Ap. J. (Letters)* , **217**, L165.

Guelin, M., Langer, W. D., and Wilson, R. W. 1982, *Astr. Ap.* , **107**, 107.

Herbst, E. 1978, *Ap. J.* , **222**, 508.

Herbst, E. 1980, *Ap. J.* , **237**, 462.

Herbst, E. 1981, *J. Chem. Phys.* , **75**, 4413.

Herbst, E. 1982, *Astr. Ap.* , **111**, 76.

Herbst, E. 1985a, *Ap. J.* , **291**, 226.

Herbst, E. 1985b, *Ap. J.* , **292**, 484.

Herbst, E. 1985c, *Origins of Life* , **16**, 3.

Herbst, E. 1986a, *Ap. J.* , **306**, 667.

Herbst, E. 1986b, *Ap. J.* , in press.

Herbst, E., Adams, N. G., and Smith, D. 1983, *Ap. J.* , **269**, 329.

Herbst, E., Adams, N. G., and Smith, D. 1984, *Ap. J.* , **285**, 618.

Herbst, E., Adams, N. G., Smith, D., and DeFrees, D. J. 1986, *Ap. J.* , in press.

Herbst, E. and Klemperer, W. 1973, *Ap. J.* , **185**, 505.

Herbst, E. and Leung, C. M. 1986a, *M. N. R. A. S.* , in press.

Herbst, E. and Leung, C. M. 1986b, *Ap. J.* , in press.

Herbst, E. and Leung, C. M. 1987, in preparation.

Huntress, W. T., Jr. and Mitchell, G. F. 1979, *Ap. J.* , **231**, 456.

Hill, T. L. 1960, *An Introduction to Statistical Thermodynamics* (Reading, Mass : Addison-Wesley).

Knight, J. S., Freeman, C. G., McEwan, M. J., Smith, S. C., Adams, N. G., and Smith, D. 1986, *M. N. R. A. S.* , **219**, 89.

Langer, W. D., Graedel, T. E., Frerking, M. A., and Armentrout, P. B. 1984, *Ap. J.* , **277**, 581.

Leger, A., Jura, M., and Omont, A. 1985, *Astr. Ap.* , **144**, 147.

Lepp, S., Dalgarno, A., and Sternberg, A. 1986, *Ap. J.* , submitted.

Leung, C. M., Herbst, E., and Huebner, W. F. 1984, *Ap. J. Suppl.* , **56**, 231.

Levine, R. D. and Bernstein, R. B. 1974, *Molecular Reaction Dynamics* (New York: Oxford University Press).

Luine, J. A. and Dunn, G. H. 1985, *Ap. J. (Letters)* , **299**, L67.

Marquette, J. B., Rowe, B. R., Dupeyrat, G., and Roueff, E. 1985, *Astr. Ap.* , **147**, 115.

Mathis, J. S., Mezger, P. G., and Panagia, N. 1983, *Astr. Ap.* , **128**, 212.

Michels, H. H. and Hobbs, R. H. 1984, *Ap. J. (Letters)* , **286**, L27.

Millar, T. J., Adams, N. G., Smith, D., and Clary, D. C. 1985, *M. N. R. A. S.* , **216**, 1025.

Millar, T. J. and Freeman, A. 1984, *M. N. R. A. S.*, **207**, 405.
Millar, T. J. and Nejad, L. A. M. 1985, *M. N. R. A. S.*, **217**, 507.
Millar, T. J., Leung, C. M., and Herbst, E. 1987, in preparation.
Mitchell, G. F., Ginsburg, J. L., and Kunz, P. J. 1978, *Ap. J. Suppl*, **38**, 39.
Mitchell, G. F. and Huntress, W. T., Jr. 1979, *Nature*, **278**, 722.
Mitchell, J. B. A. *et al. 1983*, *Phys. Rev. Letters*, **51**, 885.
Mul, P. M. and McGowan, J. Wm. 1980, *Ap. J.*, **237**, 749.
Prasad, S. S. and Huntress, W. T., Jr. 1980a, *Ap. J. Suppl.*, **43**, 1.
Prasad, S. S. and Huntress, W. T., Jr. 1980b, *Ap. J.*, **239**, 151.
Prasad, S. S. and Tarafdar, S. P. 1983, *Ap. J.*, **267**, 603
Raksit, A. B. and Bohme, D. K. 1985, *Can. J. Chem.*, **63**, 854.
Rowe, B. R. 1985, private communication.
Sakimoto, K. and Takayanagi, K. 1980, *J. Phys. Soc. Japan*, **48**, 2076.
Sandell, G. 1978, *Astr. Ap.*, **69**, 85.
Smith, D. and Adams, N. G. 1978, *Ap. J. (Letters)*, **220**, L87.
Smith, D. and Adams, N. G. 1984, *Ap. J. (Letters)*, **284**, L13.
Stief, L. J., Donn, B., Glicker, S., Gentieu, E. P., and Mentall, J. E. 1972, *Ap. J.*, **171**, 21.
Tarafdar, S. P., Prasad, S., Huntress, W. T., Jr., Villere, K. R., and Black, D. C. 1985, *Ap. J.*, **289**, 220.
van Dishoeck, E. F. 1986, in IAU Symposium 120 on *Astrochemistry*, eds. M. S. Vardya and S. P. Tarafdar (Dordrecht: Reidel).
van Dishoeck, E. F. and Black, J. H. 1986, *Ap. J. Suppl.*, in press.
Vrtilek, J. M., Gottlieb, C. A., Langer, W. D., Thaddeus, P., and Wilson, R. W. 1985, *Ap. J. (Letters)*, **296**, L35.
Watson, W. D. 1976, *Rev. Mod. Phys.*, **48**, 513.
Watson, W. D. 1977, in *CNO Processes in Astrophysics*, ed. J. Androuze (Dordrecht: Reidel), p. 105.
Williams, D. A. 1986, in IAU Symposium 120 on *Astrochemistry*, eds. M. S. Vardya and S. P. Tarafdar (Dordrecht: Reidel).
Winnewisser, G. and Herbst, E. 1986, *Organic Molecules in Space* in *Topics in Current Chemistry* (Berlin: Springer Verlag), in press.
Wootten, H. A. 1986, in IAU Symposium 120 on *Astrochemistry*, eds. M. S. Vardya and S. P. Tarafdar (Dordrecht: Reidel).

John Skilling

CHEMICAL EVOLUTION OF MOLECULAR CLOUDS

Sheo S. Prasad[1,2], Sankar P. Tarafdar[3], Karen R. Villere[4], and
Wesley T. Huntress,Jr.[1]

[1]Jet Propulsion Laboratory (Mail Stop 183/601)
California Institute of Technology
4800 Oak Grove Drive, Pasadena, California 91109, USA

[2]Aerojet Electrosystems Company (Dept. 8701)
P. O. Box 296, Azusa, California 91702, USA

[3]Astrophysics Theory Group
Tata Institute of Fundamental Research
Homi Bhabha Road, Colaba, Bombay 400 005, INDIA

[4]Department of Physics and Astronomy
San Francisco State University
San Francisco, CA 94132, USA.

ABSTRACT.   Chemical evolution of molecular clouds seems to be governed
by both chemistry and dynamics. We briefly describe the basics of the
coupled chemical-dynamical evolution. Current problems involving the
simplest species, such as C, CO, O_2, H_2O, in quiescent clouds are
then discussed. We emphasize the simplest molecules, because their
chemistry is much less uncertain than the chemistry of the larger, more
complex molecules. This allows the role of dynamics to be elucidated
with a little bit more confidence. A particularly interesting issue in
this area is the possibility that a molecule or a suite of molecules may
serve as signpost of quiescent dynamical evolution. It is recognized
that complex molecules may be more useful for this purpose, once their
chemistry is at least fairly understood. Energetic dynamical phenomenon,
such as shocks, produce dramatic chemical signatures, such as the
dramatically enhanced CH^+ abundance, due to the opening of otherwise
forbidden chemical reaction paths. As a corollary, chemistry may have
the potential to improve our understanding of the impulsive dynamical
processes in a significant manner. Equally impressive are the footprints
of star formation on the chemical evolution of the parent dense cloud.
This has been discussed with respect to the star forming region in the
Orion. It seems possible that chemistry might act as a tracer of
embedded protostellar or stellar objects which manage to elude direct
observations.

D. J. Hollenbach and H. A. Thronson, Jr. (eds.), Interstellar Processes, 631–666.
© *1987 by D. Reidel Publishing Company.*

1. INTRODUCTION

Chemical evolution of molecular clouds continues to attract considerable research effort. Figure 1 illustrates the various molecules that have been detected in molecular clouds (see also Irvine et al. 1987). Many of these molecules are quite complex indeed, considering the chemically unproductive low density (n $\sim$ 10^3 - 10^6 cm^{-3}), low temperature (T $\sim$ 15 - 20K) and often dark environment in molecular clouds. More

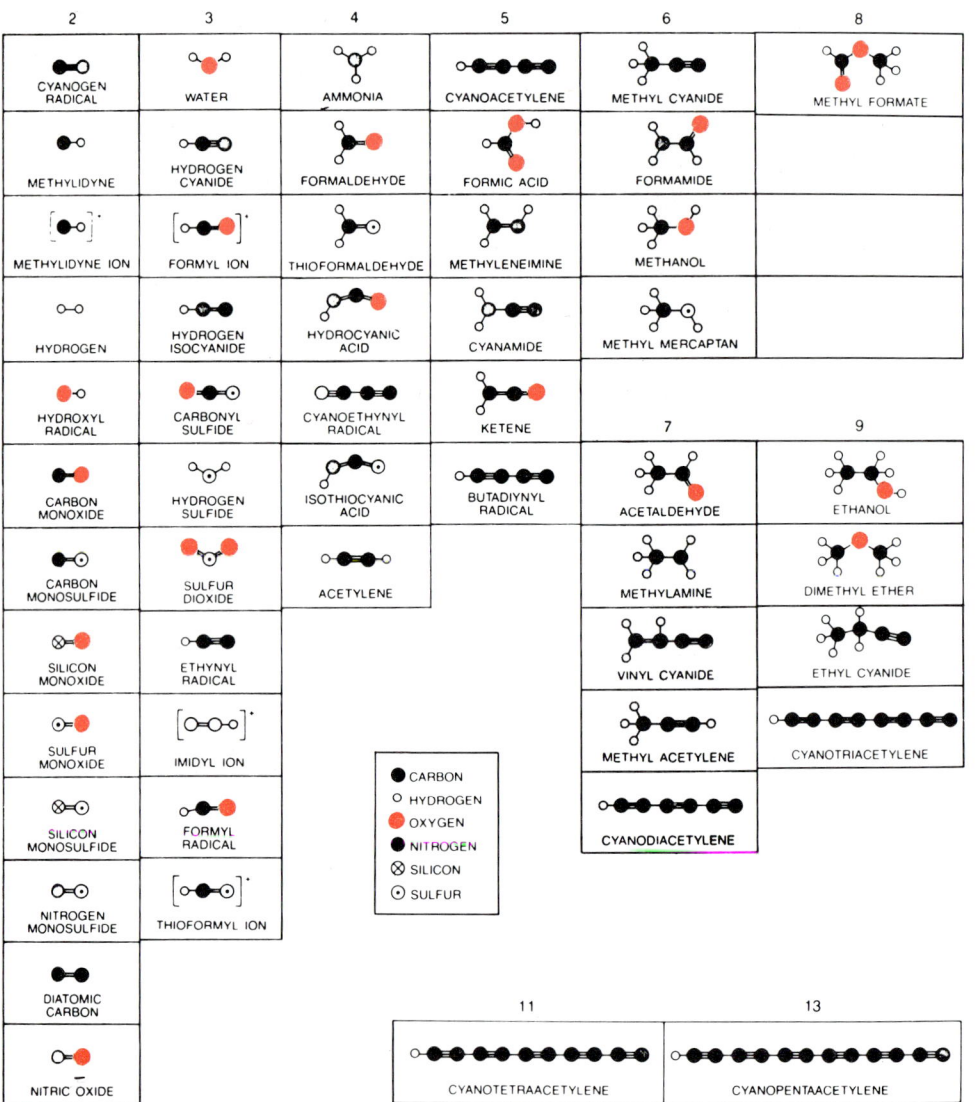

Figure 1. Examples of molecules found in molecular clouds (From Blitz, 1984. © 1984 by Scientific American, Inc.).

recently, a cyclic molecule (C_3H_2), and large clusters of molecules,
such as the polycyclic aromatic hydrocarbons (PAHs) shown in Figure 2
have been added to the list of interstellar species. The discovery of
cyclopropenylidene (C_3H_2) is a culmination of the efforts of
Thaddeus et al. (1981), Matthews et al. (1985) and Vrtilek et al.
(1985). This discovery removes the long standing absence of ring
molecules among interstellar species. Thirty years ago Platt (1956) had
suggested that PAHs may be present in the interstellar medium. Since
then they fell into oblivion, and were revived recently by a number of
astronomers (Sellgren 1984, Leger and Puget 1984, Allamandola et al.
1985, Draine and Anderson 1985, Omont 1986). Understanding the chemical
processes that lead to the observed abundances of these molecules and
clusters of molecules is a challenging task in its own right.

Molecular clouds are the dominant mass phase of of the interstellar
medium, and are of utmost importance in astronomy because they are the
sites of star formation. Chemical processes affect the evolution of
molecular clouds up to the threshold of star formation. For example,
together with C and C^+, abundant molecules such as CO and H_2O serve
as coolants (Goldsmith and Langer 1978, Takahasi et al. 1983), and are
thus of fundamental importance in the formation of dense clouds by the
gravitational collapse of diffuse clouds. The state of ionization, which
plays a dominant role in the cloud dynamics and evolution (Black and
Scott 1982, Paleologou and Mouschovias 1983, Shu 1984), is determined

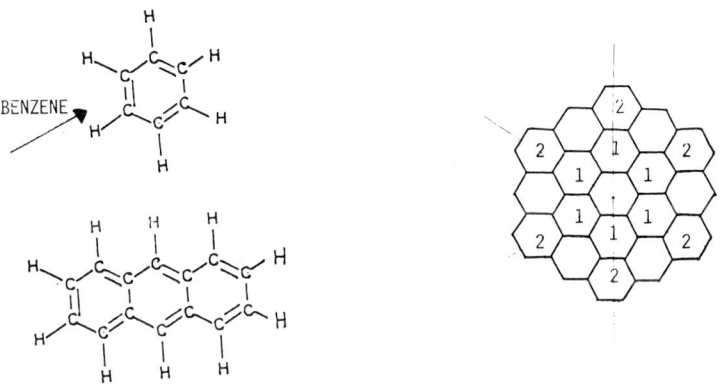

Figure 2. Representative Polycyclic Aromatic Hydrocarbons (PAHs).
The right hand side bottom sketch is the structure of anthracene
($C_{14}H_{10}$) which is a member of n-cenes ($C_{4n+2}H_{2n+4}$) with n
linearly annealed benzene rings. Other members of n-cenes include
napthalene, tetracene, pentacene etc. The right hand side sketch is
the sketch of circum-coronene ($C_{54}H_{18}$) which is a member of the
more stable catacondensed quasi-circular variety of PAHs with the
general formula $C_{6p*p} H_{6p}$ where p-1 is the number of crowns
around the central cycle and $3p^2-3p+1$ is the number of hexagonal
cycles.

almost exclusively by the chemistry. Once a star is formed, then it
affects the chemical evolution of the surrounding dense cloud in
spectacular ways. For example, the bipolar flows associated with new
born stars leave an impressive footprint in terms of dramatically
altered chemical composition in the shocked gas. All taken together, by
studying the chemical evolution of molecular clouds we may hope to gain
a better understanding of the star formation process -- better than that
gained by molecular line shape studies alone.

2. THE PLAN OF PRESENTATION

Our paper focuses on the chemical evolution of molecular clouds. To be
specific, we mean the evolution of the aggregate of the molecules which
characterize the chemical composition or makeup of the clouds. Note that
we have purposely avoided mentioning either diffuse or dense clouds.
This is in anticipation of some link between them via dynamical
evolution. We also make a subtle distinction between the molecular
evolution and the chemical evolution. Molecular evolution is supposed to
be concerned with the growth in the complexity of molecules, e.g., from
diatomic to triatomic to polyatomic and so on. For our purpose,
understanding the molecular evolution is a means - not the goal itself.
The goal is to understand the chemical evolution. The plan of the paper,
therefore, is as follows. We first describe the diffuse and the dense
states of interstellar clouds. Then comes a brief discussion of the
coupled chemistry of molecular families and an outline of the schemes
proposed for the growth of molecular complexity. This section is limited
to only those ideas that are necessary to understand the subsequent
discussions of the chemical evolution. Throughout the paper, our
emphasis is on simpler molecules and current issues. Chemistry of the
simple molecules is less uncertain than the chemistry of the larger,
more complex molecules. This enables us to better appreciate the role of
dynamics. Finally, we present some suggestions for future research.

3. THE DIFFUSE AND THE DENSE STATES OF MOLECULAR CLOUDS

Molecular clouds consist of gas phase atoms, molecules, electrons and
ions, and micron and sub-micron sized solid particles. These solid
particles are also referred to as dust or grains. Hydrogen and helium
are by far the overwhelming gas phase species. All other elements, such
as C, N, O, S, Si, Na, Mg, Fe, are very much less abundant. For example,
O and C are only about 0.0007 and 0.0003 of the total hydrogen density,
while the remaining elements are even less abundant. The actual values
of the abundances of these elements in a cloud appear to depend upon the
local hydrogen density (Jenkins et al. 1986, Keenan et al. 1986).
According to Tarafdar et al. (1983) there may also be a dependence on
the hydrogen column density. Roughly, only 1% of the cloud mass is in
the form of dust. In contrast to the well known nature of the gas phase
constituents, the nature of the grains remains uncertain. However, it is
generally believed that the grains are composed of the heavy elements,

such as C, Si, Mg. Properties of grains have been deduced mostly from
the extinction and the polarization of the ultraviolet and visible
starlight passing through the dust laden clouds (see, for example,
Spitzer 1978 as a general reference on this subject matter). In addition
to extinction, the dust grains also present a non-negligible surface on
which gas phase atoms and molecules can condense and be altered by
surface reactions.

Molecular clouds have been broadly classified as diffuse and dense
clouds. Representative parameters of diffuse clouds are as follows: (i)
density, n, of all particles ~ a few hundred cm^{-3}, (ii) kinetic
temperature, T_k, ~ 25 - 100 K, total continuum extinction due to
dust grains at the visual wavelength, $A_v \leqslant 1$. Except for hydrogen, all
gas phase chemical elements (He, C, N, O, S, Si, Na, Mg, Fe etc.) are in
atomic or ionic form, with the most abundant molecule CO being only 0.01
of the total gas phase carbon. In the exceptional case of hydrogen, the
atomic and molecular forms may be comparable in abundance. The
abundances of the heavier elements relative to hydrogen in these clouds
are generally less than the cosmic (or solar) abundances (Morton 1974,
Jenkins et al. 1985, Jenkins 1987). With the observed variation of the
dust extinction with the wavelength, the extinction in the uv (where
most of the photodissociations occur) is typically about three times
that at the visible. Thus, the photorates at the center of diffuse
clouds may be reduced by no more than a factor of about ten from those
at the cloud edges. At certain frequencies, however, absorption by
spectral lines (particularly of H_2) can be very much greater. The
radiative flux incident upon diffuse clouds can usually be approximated
by the galactic background flux [see, for example, Draine (1978), Mathis
et al. (1983) and van Dishoeck (1987) for the intensity - wavelength
relations]. Penetration of this radiation inside the diffuse clouds
provides the energy needed to keep these clouds warm, and all low
ionization potential elements (C, S, Si, Na etc.) fully ionized. Cosmic
ray ionization is a relatively smaller source of energy in these clouds.

Dense cloud conditions are quite different. Figure 3 gives a schematic
representation of these clouds. Densities in the core of dense clouds
approach 10^6 - 10^7 cm^{-3}, particularly in clouds at the threshold
of star formation. These cloud cores are also quite cold, with T_k on
the order of 10 - 20 K, because the the external galactic radiation
cannot penetrate the core and the high energy cosmic ray ionization
cannot provide enough energy. There is also a significant density
gradient between the core and the edge of a dense clouds, so that the
density at the edge may be on the order of 10 - 100 cm^{-3}. Complex
polyatomic molecules are found mostly in the core where they are
protected from destruction by photodissociation. Low ionization
potential elements are predominantly in the ionic form at the edge, but
are mostly neutral in the interior. Except for some turbulence, the
clouds are quiescent till a star forms. After star formation, these
clouds become quite perturbed by the ultraviolet and the mass outflows
from the star. As we shall see later, the chemical evolution in the star
forming regions is quite different from that in the quiescent regions.

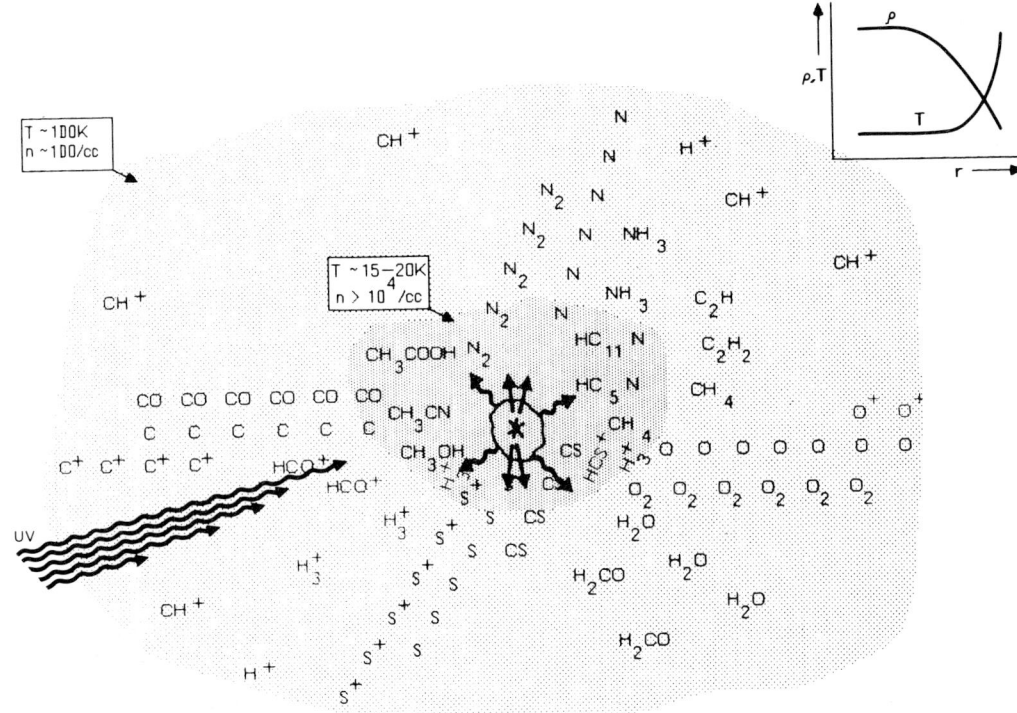

Figure 3. Schematic representation of the core-halo structure of
dense molecular clouds. The density and temperature variations with
distance is shown schematically by the insert in the top right hand
corner. Ultraviolet radiation (represented by the wavy lines) from
outside can not reach the cloud core. Distributions of the atoms,
simpler molecules and complex polyatomic molecules are also
illustrated in the figure. Star formation occurs in the dense core.
This is symbolized in the figure by the encircled * . A nascent
star generates a copious amount of ultraviolet radiation and drives
outflows of matter (shown by solid arrows emanating from the star).

4. MOLECULAR EVOLUTION IN INTERSTELLAR CLOUDS

There is a vast amount of literature on the evolution of molecules in
interstellar clouds (see citations in the sections 5 and 6). Proceedings
of the IAU Symposium # 120 on Astrochemistry (Vardya and Tarafdar 1987),
proceedings of the NATO Advanced Research Workshop on Molecular
Astrophysics: State of the Art and Future Directions (Dercksen et al.
1985), and this proceeding should be few of the most convenient source
books on the current state of interstellar chemistry. This section
briefly describes the evolution of molecular complexity for the sake of
the completeness of this paper, and tries to highlight some of the
currently unresolved issues.

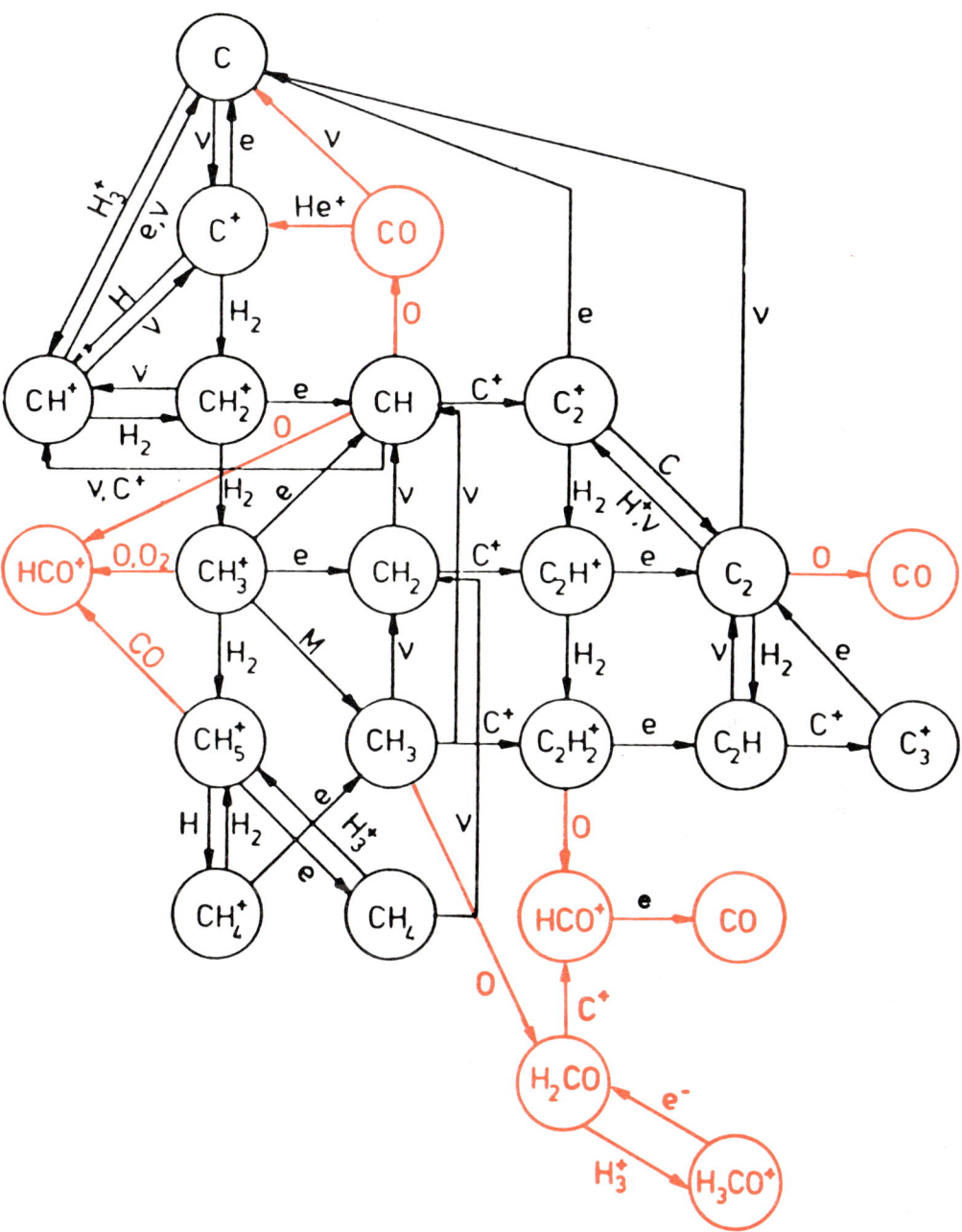

Figure 4. A simplified version of carbon chemistry and its coupling with oxygen chemistry.

4.1 Formation of simple molecules:

Formation of the H_2 molecule is the prerequisite of any significant molecular evolution in interstellar clouds by gas phase chemistry. The hydrogen molecule forms very efficiently on dust grains (van de Hulst 1949, Hollenbach and Salpeter 1970, 1971 and references therein). Once H_2 is formed, the carbon chemistry (i.e., the formation of carbon bearing molecules) in diffuse clouds is started by the radiative association of the abundant C^+ ion with H_2:

$$C^+ + H_2 \quad \text{-->} \quad CH_2^+ + h\nu \tag{1}$$

Chemical reactions leading to the formation of simple hydrocarbons, such as CH, CH_2, CH_3, CH_4, C_2, C_2H_2, can be easily seen from the Figure 4. Cosmic ray ionization of both H and H_2 produces H^+ ions. Slightly endothermic charge transfer reaction between H^+ and O:

$$H^+ + O \text{ --> } H + O^+ \tag{2}$$

leads to the formation of the O^+ ions needed to start the oxygen chemistry in diffuse clouds via the reaction chain:

$$O^+ \xrightarrow{\quad H_2 \quad} OH^+ \xrightarrow{\quad H_2 \quad} OH_2^+ \xrightarrow{\quad H_2 \quad} OH_3^+ \tag{3}$$

followed by the dissociative recombinations and photodissociation:

$$OH_2^+ + e^- \quad \text{-->} \quad OH + H \tag{4}$$

$$OH_3^+ + e^- \quad \begin{array}{l} \text{-->} \quad H_2O + H \\ \text{-->} \quad OH + H_2 \end{array} \tag{5} \tag{6}$$

$$H_2O + h\nu \text{ --> } OH + H \tag{7}$$

Considering the energy of the available uv photons, O^+ ions cannot be produced by the photoionizations of O in interstellar clouds. Their production by the reaction (2) is quite sensitive to the cloud temperature due to the endothermicity of the reaction. Consequently, the OH and H_2O concentrations in diffuse clouds are also sensitive to the cloud temperature. Figure 5 gives a simplified view of the reactions involved in oxygen chemistry and its coupling with the carbon chemistry.

Synthesis of the H_3^+ ions via the ion - molecule reaction:

$$H_2^+ + H_2 \text{ --> } H_3^+ + H \tag{8}$$

(also shown in the Figure 5) is one of the most crucial steps in the interstellar chemistry, because H_3^+ ion plays innumerable pivotal roles in the molecular evolution. For example, in the cold dark clouds, where the endothermic reaction (2) can not occur and C^+ abundance is very small, the oxygen and carbon chemistry are sustained by the H_3^+ ions via the reactions:

$$O + H_3^+ \; --> \; OH^+ + H_2 \tag{9}$$

and

$$C + H_3^+ \; --> \; CH^+ + H_2 \tag{10}$$

Figures 4 and 5 both show the extensive coupling between the oxygen and carbon chemistry. The production of ubiquitous CO molecule via the reactions CH + O --> CO + H and OH + C --> CO + H, for example, is an extremely important outcome of this coupling. The color coding in Figures 4 and 5 is meant to help in the appreciation of this coupling.

Essentials of nitrogen and sulfur chemistry are shown in the Figures 6 and 7. Ammonia (NH_3) and hydrogen cyanide (HCN), found mostly in the dense regions shielded from uv, are two quite important molecules, because they serve as useful temperature and density probes. The formation of NH_3 is initiated by the following reactions:

$$N^+ + H_2 \; --> \; NH^+ + H \tag{11}$$

and

$$N + H_3^+ \; --> \; NH_2^+ + H \tag{12}$$

The formation of HCN, on the other hand, involves the coupling between nitrogen and carbon chemistry as illustrated in Figure 6. As with the nitrogen, significant processing of sulfur occurs only in the dense and shielded regions where SO, SO_2, CS, HCS^+, OCS and H_2S have been observed in a number of clouds. Important reactions involved in their production and loss are shown in the figure 7. These sulfur molecules are the simplest yet quite interesting. For example, CS is one of the most widely used density probes. Also, as will be seen a little later, H_2S and the CS and HCS^+ pair highlight some of the difficult problems and interesting developments of the current interstellar chemistry.

Reactions (11) and (12) involved in the initiation of the nitrogen, particularly ammonia, chemistry have problems. Consequently, the current understanding of the ammonia chemistry is incomplete. Reaction (11) is endothermic and does not occur in the cold dense regions where ammonia is mostly found. There is, however, a mitigating factor. In dense clouds N^+ is formed mainly by the dissociative charge transfer reaction:

$$N_2 + He^+ \; --> \; N^+ + N + He \tag{13}$$

which produces translationally hot N^+ ions. According to a measurement reported by Adams et al. (1985), the translational excitation in the N^+ ion from reaction (13) is sufficient to overcome the endothermicity of the reaction (11). Reaction (12) is somewhat unusual, since the H_3^+ usually transfers a proton to the reacting neutral partner. Even

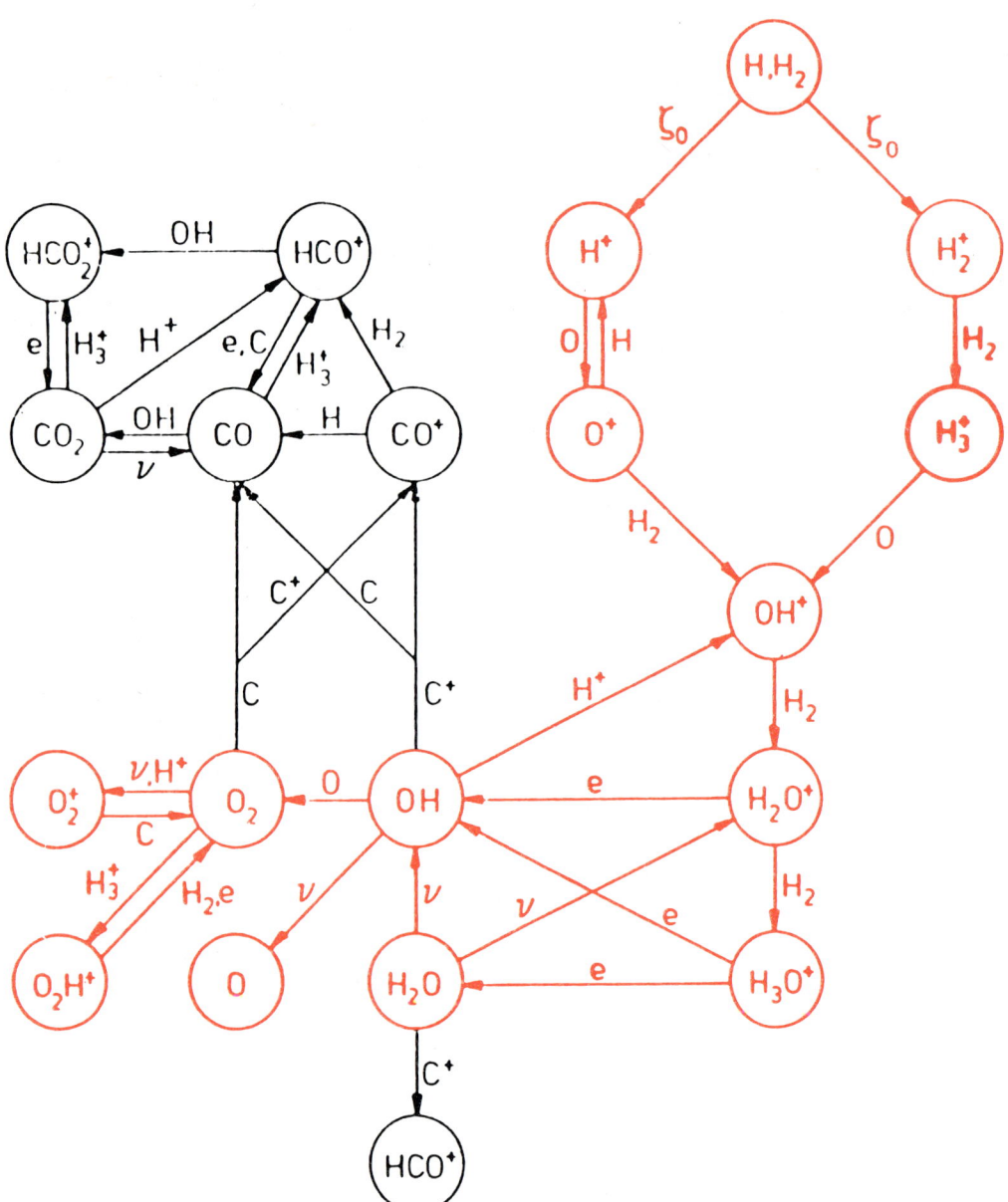

Figure 5. A simplified version of oxygen chemistry and its coupling
with carbon chemistry.

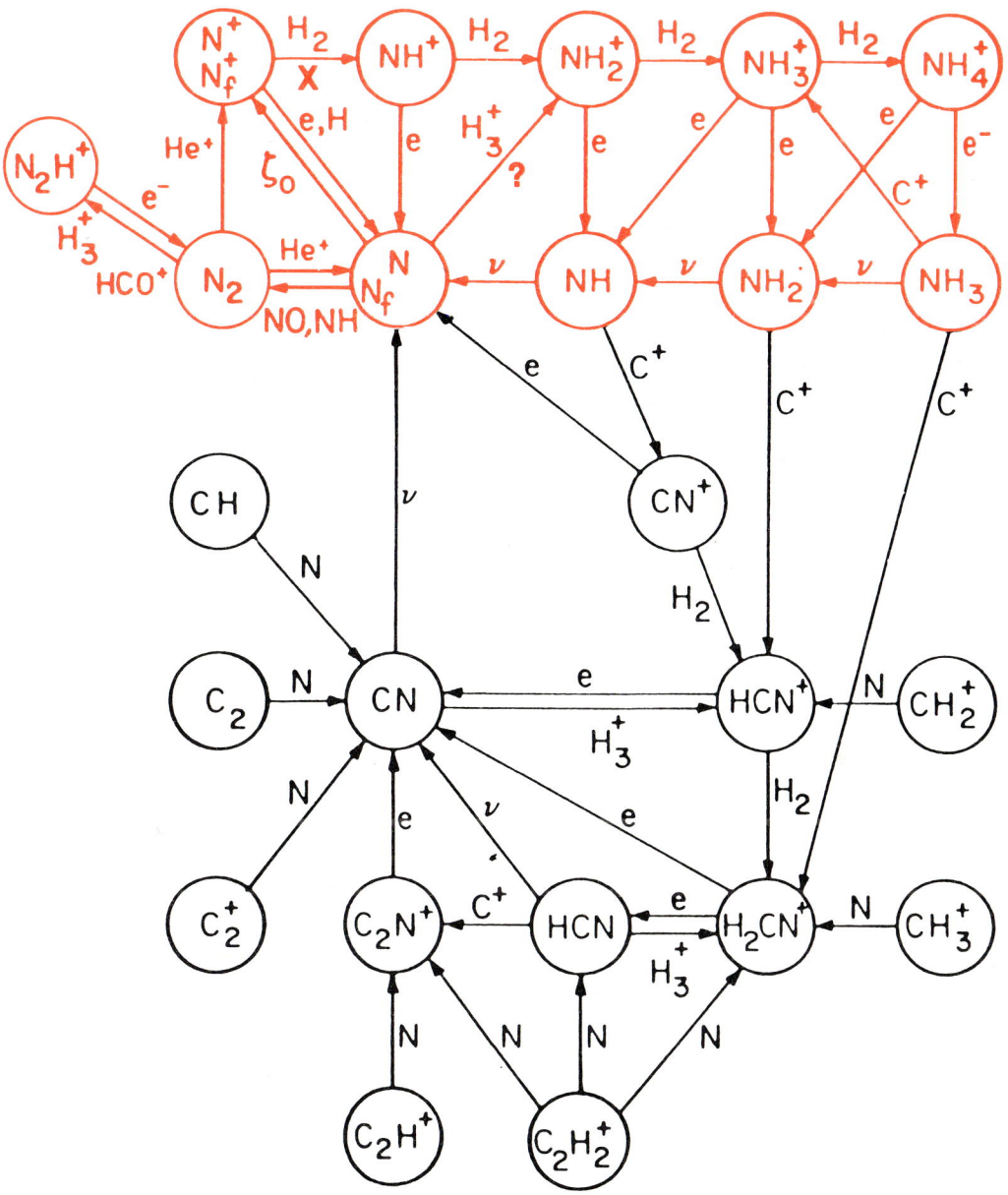

Figure 6. A simplified version of nitrogen chemistry and its coupling with carbon chemistry.

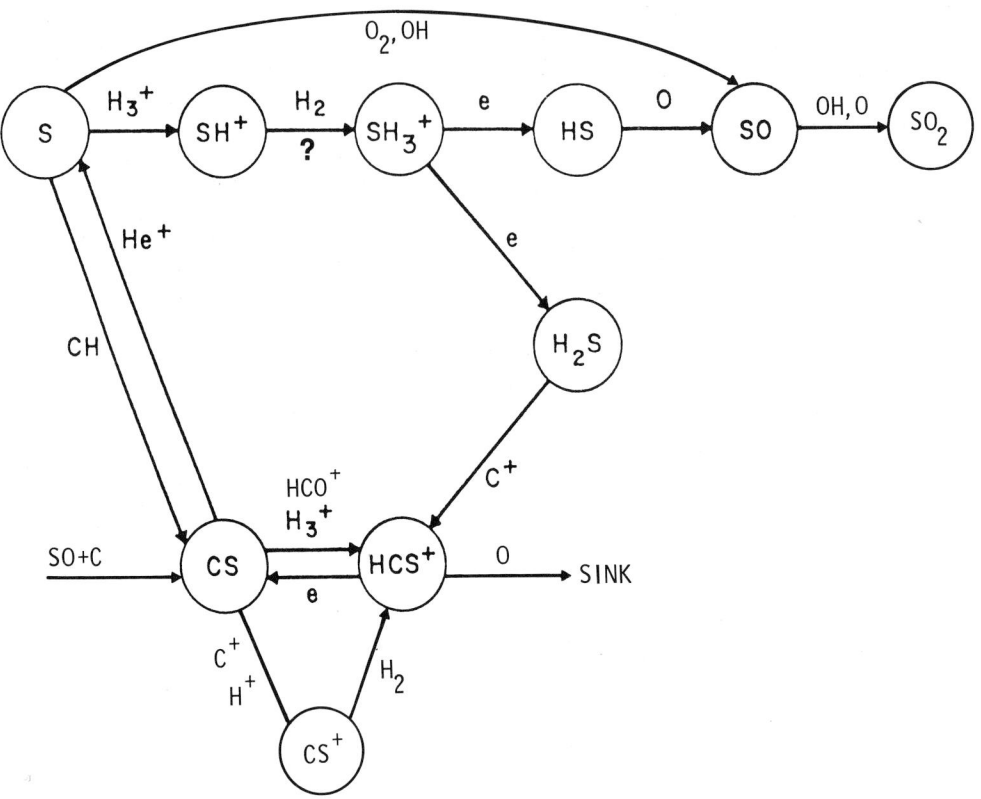

Figure 7. A simplified version of sulfur chemistry.

so, it has been retained in the interstellar chemistry because N appears to react with H_3^+ ion and because the proton transfer channel is endothermic. Definitive measurements of the reaction rate coefficient is not available at this time. It is, however, hoped that the now certain high abundance of H_3^+ will ensure sufficient NH_2^+ formation even if the rate coefficient of the reaction (12) turned out to be small due to the unusual nature of the reaction. High abundance of the H_3^+ ion is certain so long both the photodissociation and dissociative recombination of the ion remains as slow as it is currently thought to be on the basis of the studies by van Dishoeck (1986) and Smith and Adams (1984).

The observed abundance of H_2S in quiescent dense clouds has eluded a satisfactory explanation. The problem arises because the reaction

$$HS^+ + H_2 \quad --> \quad H_2S^+ + H \tag{14}$$

does not occur in interstellar clouds under normal conditions. There is, therefore, no obvious efficient route to the production of the H_3S^+ ion needed to form H_2S. It has been speculated that the radiative association of the HS^+ ion with H_2 might lead to the formation of the precursor H_3S^+ ion (Prasad and Huntress 1982). Duley et al. (1980) have invoked surface reactions to synthesize H_2S. However, this process produces too much H_2S. (Surface reactions as a means of molecule formation are further discussed later.) Until recently, the high $HCS^+/$ CS abundance ratio on the order of 0.1 observed in cold dark clouds (Irvine et al. 1983) had also defied explanation. It now appears that the problem is probably explicable in terms of the rapid increase of the rate coefficient of ion-molecule reactions involving molecules of very large dipole moments at very low temperatures on the order of 10 - 15 K or less (Clary 1985, Millar et al. 1985).

4.2 Evolution towards molecular complexity:

Formation of larger and relatively more complex molecules by gas phase reactions is now beginning to be understood. Radiative ion-molecule association reaction is invariably the key reaction leading to the formation of the precursor ion whose dissociative recombination yields a given complex molecule. Figure 8 illustrates the point with respect to

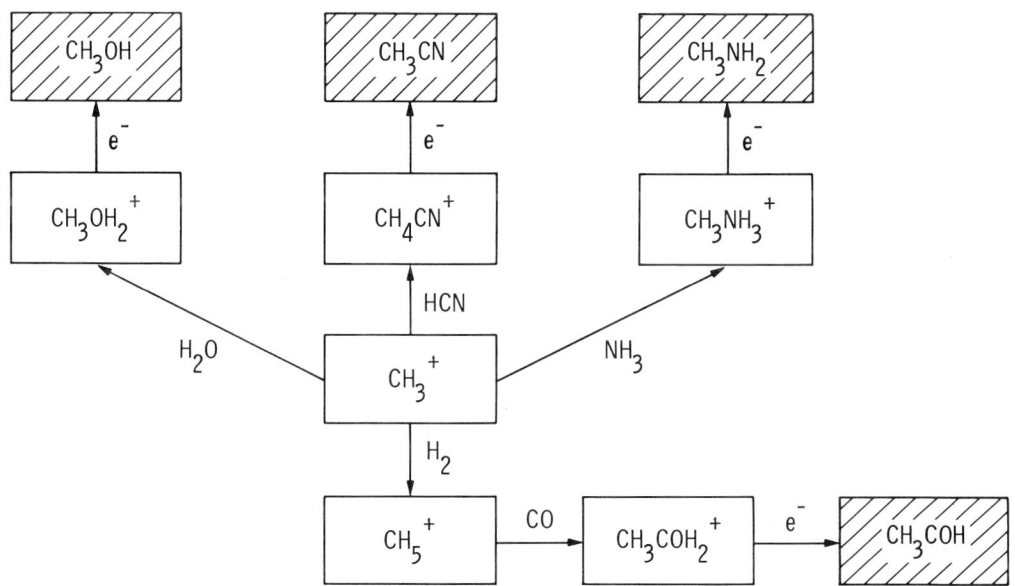

Figure 8. Examples of the role of ion-molecule radiative association reactions in the formation of complex molecules.

the formation of methanol (CH$_3$OH), methylcyanide (CH$_3$CN), and
acetaldehyde (CH$_3$COH). Working with the CH$_3^+$ ion, Barlow et al.
(1984) and McEwan et al. (1980) have experimentally demonstrated that
the radiative association reaction could indeed be very fast at low
temperatures of the dark clouds. Experimental data for almost all other
interstellar ions are currently unavailable. Theoretical treatments are,
therefore, needed for almost all of the radiative association reactions
relevant to interstellar chemistry. Bates (1979, 1983a,b), and Herbst
(1982, 1983, 1985, 1987) have provided the foundation for the
theoretical treatment. The formation of other complex hydrocarbons and
molecules containing long carbon chain takes place via a combination of
the condensation, fixation and radiative association reactions, as
discussed by Huntress (1977), Mitchell et al. (1979), Herbst (1983),
Leung et al. (1984) and Bohme (1986). Carbon atoms and ions play a
crucial role in these schemes. This is illustrated in Figure 9.

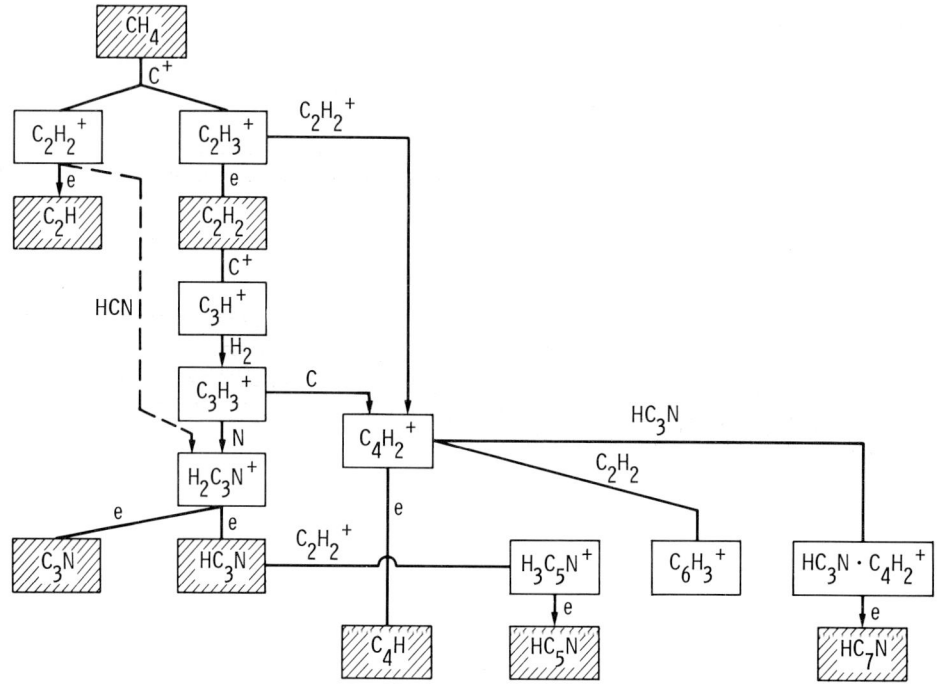

Figure 9. Further examples of the formation of complex molecules,
particularly of molecules with long carbon chain.

4.3 <u>Grain surface chemistry and its implication for the molecular
evolution</u>:

Grain surface reactions can also play a role in the formation of gas
phase interstellar molecules. In contrast to the chemical schemes of
forming interstellar molecules by gas phase reactions, there are a

number of serious uncertainties in reaction schemes involving grain
surface reactions. The latter schemes are, therefore, still in the
formative stages. Briefly speaking, interstellar molecule formation by
grain surface reactions involves accretion of gas phase species on the
grains, their migration on the surface, exothermic reactions with other
species accreted on the surface and, finally, the ejection of the
molecules from the grain surface back into the gas phase. Watson and
Salpeter (1972) reviewed these processes and concluded that the ejection
of molecules from the grain surfaces is an extremely inefficient process
in the cold and dark environment of dense clouds. Thus, the grains would
act as sinks for for the gas phase molecules in dark clouds. If the
probability that an atom or molecule will stick onto the grain during a
collision is assumed unity, as is indicated by theoretical models
(Watson and Salpeter 1978, Burke and Hollenbach 1983), then the
molecules would be removed from the gas phase on a time scale of

$$t_s = 2 \times 10^9 (m/m_H)^{1/2} / (n T^{1/2}) \text{ yr} \tag{15}$$

where m is mass of the sticking atom or molecule, m_H is the mass of
the hydrogen atom, and n and T are the hydrogen density and temperature
of the clouds (Glassgold 1985). Even for moderate densities ($n > 10^5$
cm^{-3}), the sticking time scale becomes very short (e.g., $< 10^4$ yr).
As a corollary, there should be a marked decrease in the abundances of
molecules relative to hydrogen with increasing hydrogen density. Until
recently, that appeared to be the case (Wootten et al. 1978). However,
it is now believed that the inverse correlation between the relative
abundances and density was an artifact of inaccurate data reduction
procedure (see Irvine et al. 1985 for a discussion). The absence of the
expected marked inverse correlation implies the existence of efficient
desorption mechanisms, given the near unity sticking probability and the
certainty of the gas - grain collisions. Although many desorption
mechanisms have been suggested and shown to work in a qualitative
manner, many uncertainties still remain. Following are some of the
suggested mechanisms.

According to Allen and Robinson (1975, 1977), the exothermicity of a
surface reaction would be able to heat the smaller grains (r < 0.005
microns) to a temperature at which thermal ejection is efficient. In
fact, they constructed a purely grain surface chemistry model which had
a mixed success in predicting the observed abundances. The abundance of
small grains in dense clouds is, however, uncertain because in these
clouds small grains coagulate to form larger grains. Recently, Leger et
al. (1985) pointed out that the spot heating by the heavier component of
the cosmic rays should produce the temperature spike needed to desorb
molecules from the grains. Smallness of grains is not a requirement in
this mechanism. d'Hendecourt et al. (1982, 1985) have argued for
explosive release of molecules triggered by grain - grain collisions.
Grain - grain collisions are assumed to heat the grains to a temperature
of about 27 K needed to induce fast diffusion of accreted radicals. It
is thought that the onset of fast diffusion leads to a chain of
exothermic radical - atom or radical - radical reactions culminating in

explosive mantle evaporation and the release of all accreted molecules from the grain surface back into the gas phase.

Dynamical processes have also been considered for the desorption of molecules from the grain surface. Boland and de Jong (1982) suggested the possibility that dense clouds may have circulation currents which bring depleted gas and grain from the dark and cold interior to the warm and lighted exterior on a time scale shorter than the time needed for the complete accretion of molecules onto the grains. As the grains approach the cloud edges, the intense uv radiation would photodesorb the accreted molecules into the gas phase. Williams and Hartquist (1984) prefer to desorb molecules by sputtering due to shocks. Dense clouds, in their opinion, are exposed to slow shocks at the average time interval of a few million years. Molecules accreted onto the grains during this time interval are desorbed when the grains encounter a shock wave. These ideas have important implications for the chemical evolution of dense clouds. To the extent these ideas are correct, chemical equilibrium may never be reached in dense clouds. We shall return to this point later.

5.   CHEMICAL EVOLUTION OF DIFFUSE CLOUDS

5.1 Traditional Approach:

Traditionally both the chemical composition and the physical conditions in diffuse clouds have been thought to be in steady state. The assumption of steady state in the physical conditions is an inheritance from the isothermal collapse studies of the early days (the sixties), which indicated that warm diffuse clouds, particularly those of low mass, are stable against gravitational contraction (Hunter 1969, Penston 1969, Disney et al. 1969). Chemical abundances in diffuse clouds were assumed to be in equilibrium, because the abundances of the important species are controlled by either photodissociation or rapid reactions for which time constants are short, typically 10^4 yr or less. Diffuse cloud chemistry models of Glassgold and Langer (1974, 1976), Barsuhn and Walmsley (1977), Black and Dalgarno (1977), Federman and Glassgold (1980), Oppenheimer and Dalgarno (1975), Prasad and Huntress (1980a,b), van Dishoeck and Black (1986) belong to this class of models. While most of the cited models have considered density and temperature to be uniform through out the cloud, some of them have allowed for their spatial variations. Except for the H_2 formation on grains, these models use only gas phase chemistry. Models using both gas phase and grain surface chemistry have been published by Pickles and Williams (1977a,b, 1980), and Mann and Williams (1984, 1985).

5.2 Results from the Traditional Approach and Their Discussion:

Table 1 shows a comparison between the theoretical and the observed column densities of the most studied molecular species in the diffuse clouds towards ζ Oph and ζ Per. Theoretical values are from van Dishoeck and Black's (1986) model. The spread in the theoretical values

Table 1. Comparison of the Observed and the Theoretical Model
Column Densities of Molecules in Diffuse Clouds.

Spec- ies	ζ Per		ζ Oph	
	Model	Observed	Model	Observed
OH	(4.3 - 5.3) 13	(4.2 ± 0.5) 13	(4.3 - 5.0) 13	(4.8 ± 0.5) 13
CO	(0.2 - 2.0) 14	(5.4 ± 2.6) 14	(0.9 - 2.0) 14	(2.0 ± 0.3) 15
C_2	(0.7 - 2.9) 13	(1.1 ± 0.2) 13	(0.7 - 1.8) 13	(1.5 ± 0.2) 13
CH	(1.6 - 2.7) 13	(2.0 ± 0.1) 13	(1.3 + 2.1) 13	(2.5 ± 0.1) 13
CH^+	(2.8 - 5.3) 10	(3.5 ± 0.4) 12	(2.8 - 4.7) 10	(2.9 ± 0.1) 13
CN	(1.0 - 2.3) 12	(3.0 ± 0.1) 12	(0.9 - 1.6) 12	(2.5 ± 0.1) 12
N	1.1 17	>2.5 17	(8.0 - 8.3) 16	(5.2 ± 3.3) 16
NH	(3.3 - 6.8) 11	<6.3 11	(3.8 - 4.9) 11	<7.5 12

Note that entries like 1.1 17 mean 1.1 x 10^{17}.

are due to the ranges in their model parameters. The observed values are
from various sources as listed in van Dishoeck and Black's (1986) paper.
Disregarding CH^+ for a moment, there seems to a reasonably good accord
between the observed and the equilibrium model abundances of molecules
in diffuse clouds. Quantitatively, however, there are differences due to
possible deficiencies in the theoretical model. By analyzing these
differences we can improve our understanding of the physical conditions
and the chemistry of diffuse clouds. van Dishoeck and Black's (1986)
paper gives several interesting examples of chemistry problems. In the
following paragraphs, therefore, we shall discuss the differences
between the theory and observations from a perspective highlighting the
usefulness of chemistry in elucidating time dependent evolution and
impulsive processes.

5.2.1 The CO Problem and the Possible Inadequacy of Equilibrium Models:

According to van Dishoeck and Black (1986), the best theoretically
predicted CO column density (1 - 2) X 10^{14} is two to five times lower
than the best observational value. The principal source of CO in diffuse
clouds is the reaction between C^+ and OH, and it is difficult to
visualize any other source in the gas phase models. This suggests a
possibility that the photodissociation cross sections of CO may be less
than their theoretically obtained values. This might happen if the CO

photodissociation is limited by the E - X (0,0) line absorption with small predissociation probability, and if the shorter wavelength transitions do not predissociate. In this case the photodissociation of CO would be decreased by an order of magnitude and the self shielding of CO would occur at even shallower depths. However, available laboratory data do not support the above conjecture, and more likely suggest even larger CO photodissociation rates. The possibility that the production of CO in the shock heated gas might make up the difference is also unworkable, as the inferred velocity dispersion of CO (Wannier et al. 1982, Liszt 1979) argue against significant contribution from the shocked region. This dilemma entices an inquisitive mind to explore beyond the steady state models.

5.2.2 The Evolutionary Approach and its Implications:

Could it be possible that steady state models are not totally adequate for diffuse clouds, so that the their physical-chemical evolution should be considered in predicting the molecular composition? Tarafdar, Prasad, Huntress, Villere and Black (1985 , hereinafter called TPHVB) have studied this issue. It is helpful to recall here that the common notion that diffuse clouds resist gravitational contraction is based on the assumption of an isothermal temperature structure. Using the more realistic temperature structures, TPHVB have constructed evolutionary models of interstellar clouds in which initially diffuse and warm interstellar gas gravitationally evolves into a dense cloud at the threshold of star formation. The key to their success lies in the fact that the steep outward increase of the temperature provides a pressure gradient force that assists gravity in the initial stages of the collapse (see Figure 10). The variation of CO column density with the visual extinction (A_v) in dynamically evolving clouds in their diffuse phase is shown in Figure 11. Theoretical curves are for clouds with different masses and initial densities. The sources of the observational data is also given in the insert. The agreement between the theoretical and the observational data presented in the Figure 11 suggests that evolutionary models have considerable potential. To date there have been very limited number of studies with the evolutionary models. One urgently needed refinement is the inclusion of a better treatment of H_2 and CO photodissociation and self-shielding effects in the evolutionary models. This improvement will enable us to make a better assessment of the evolutionary models.

An important implication of the evolutionary models discussed by TPHVB is that depending upon the stage of its evolution, a given mass of interstellar gas would manifest the properties of both diffuse and dense clouds. Gravitational collapse, therefore, appears to provide a common thread through the diversity of interstellar clouds. In other words, diffuse and dense clouds may be related in a dynamical and evolutionary sense. Previously, this relation was thought to be very weak, in the sense that the diffuse clouds were assumed to exist as individual entities until disturbed sporadically by violent shocks or spiral density waves.

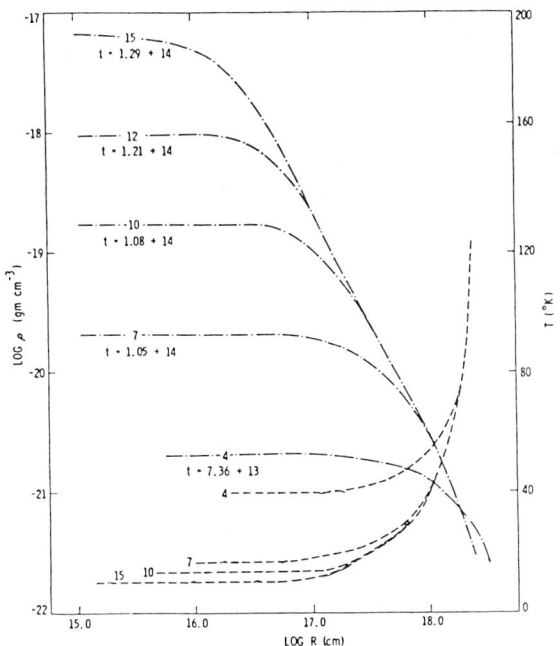

Figure 10. Evolution of the density (-.-.-) and temperature (---) structure for a cloud of 40 $M_\odot$. Increasing numbers on the curves denote progress of evolution with time, t, as indicated.

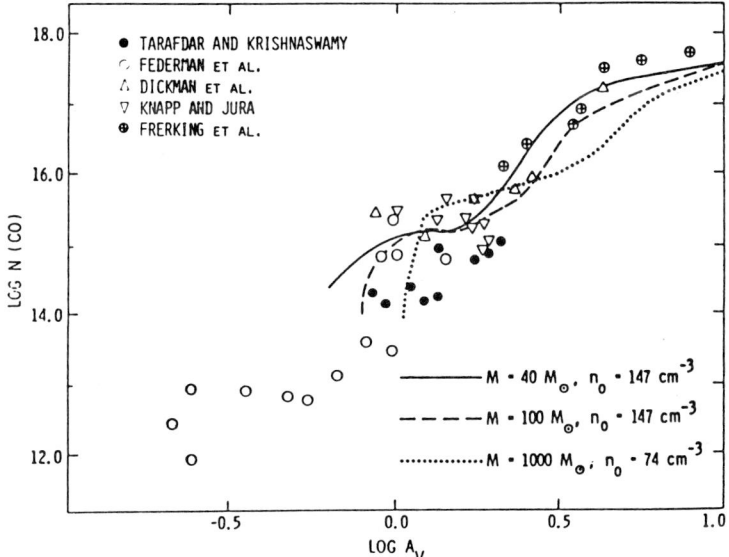

Figure 11. The observed and the predicted abundances of CO.

5.2.3 <u>Chemistry</u> <u>and</u> <u>Impulsive</u> <u>Dynamical</u> <u>Events</u>:

Chemistry has the potential to serve as an impressive signposts of impulsive dynamical events, such as shocks. Although the existence of shocks can be inferred from line profile studies, their unique chemical consequences dramatize their occurrences. The outstanding case of the CH^+ ions in Table 1 illustrates this point. The predicted column densities of this ion are two to three orders of magnitude smaller than the observed densities. There is hardly any relief even if all uncertain rate coefficients (e.g., the dissociative recombination of CH^+) are adjusted in the favorable direction. Time dependent evolutionary models of TPHVB do no better either. Given the 0.4 ev endothermicity of the reaction

$$C^+ + H_2 \longrightarrow CH^+ + H_2 - 0.4 \text{ ev} \qquad (16)$$

there is no efficient source of CH^+ in diffuse molecular clouds under normal conditions. If any region of the molecular cloud is dynamically perturbed, then the situation changes dramatically. The high kinetic temperature in the shocked region can easily overcome the endothermicity of the above reaction, so that there is a large production of the CH^+ ions in the shock heated regions of these clouds. Shocked cloud models of Elitzur and Watson (1980) and Mitchell and Deveau (1983) appear to account for the observed CH^+ densities, although improved non-magnetic or single fluid shock models tend to produce too little CH^+ by factors of 2 to 5 (Graff and Dalgarno 1986, Mitchell and Watt 1985, Pineau des Forets et al. 1986). These models, however, produce too much OH and CH. Magnetohydrodynamic models, in which the mean velocities of the ions and neutrals differ, are able to reproduce the observed CH^+ without producing too much OH and CH (Draine and Katz 1986, Draine 1986).

5.2.4 <u>Possible</u> <u>Unimportance</u> <u>of</u> <u>Surface</u> <u>Chemistry</u> <u>in</u> <u>Diffuse</u> <u>Clouds</u>:

Finally, the debate over the importance of the grain surface reactions in diffuse clouds continues. According to Crutcher and Watson (1976), the ratio of NH and OH colum abundance ($\leq$ 1/100) observed towards the the star O Per suggests that CH and OH are not the results of surface reactions. The latest study of the possible importance of grain surface chemistry in diffuse clouds is by Mann and Williams (1985, 1986). They discuss CO, H_2O, OH, CH, CN, and NH observations towards a number of diffuse clouds, and compare the obsevational data with model calculations with and without grain surface chemistry. They find that uncertainties in the observational data are currently so great that model chemistries with and without a contribution from grain surface reactions provide a satisfactory fit to the observations. Thus, at present there is no compelling evidence for the importance of surface reactions in diffuse clouds -- excepting the case of H_2 formation.

In summary, gas phase equilibrium models of diffuse interstellar clouds are good first order approximations. There are, however, some basic issues which require evolutionary models.

6. CHEMICAL EVOLUTION IN DENSE CLOUDS

Dense clouds are sites of star formation. This divides the studies of
chemical evolution into two broad topics: (i) in quiescent dense clouds,
and (ii) in dense clouds containing nascent stars. The second topic is a
class in itself, because the impact of star formation on the chemical
evolution of the surrounding dense cloud is dramatic.

6.1 Chemical evolution in quiescent dense clouds;

There are significant differences between the chemical evolution of
diffuse clouds and quiescent dense clouds, let alone the dense clouds
impacted by star formation. While static equilibrium models were a good
first approximation in the case of diffuse clouds, that may not be the
case for quiescent dense clouds, even for the very simplest atoms and
molecules such as C and CO. With the advances in infrared technologies,
studies of molecules condensed on the grain surfaces as mantles is
rapidly becoming a productive research field. This is in sharp contrast
to the situation in diffuse clouds where grains probably do not have any
significant mantle. Last, but not the least in importance, the continual
discovery of exotic molecules in dense clouds is keeping these objects
ever so more fascinating than the diffuse clouds. Let us now see where
we stand in our understanding of the chemical evolution in dense clouds.

6.1.1 Approach:

Closely paralleling the case of diffuse clouds, the earliest studies of
the dense cloud chemistry investigated the equilibrium solutions of the
coupled chemical rate equations assuming a spatially and temporally
fixed density and temperature (Herbst and Klemperer 1973, Mitchell et
al. 1978, Millar and Freeman 1984, Tielens and Hollenbach 1985a,b). A
variant of this approach uses spatially varying, but temporally fixed
background density and temperature (de Jong et al. 1980, Boland and de
Jong 1984). In the pseudo time-dependent approach, these solutions of
the rate equations are obtained as functions of time (Iglesias 1977,
Prasad and Huntress 1980a,b, Henning 1981, Graedel et al. 1982, Leung et
al. 1984, Millar and Nejad 1985). Another approach has been to use
background density and temperature in accordance with the isothermal
free-fall or some arbitrary modification of the free-fall model (Kiguchi
et al. 1974, Suzuki et al. 1976, Leung and Herbst 1986). In the more
advanced version of this approach, which was mentioned earlier also, the
density and temperature are allowed to vary in both space and time in
accordance with the equations of motion and heat balance (Gerola and
Glassgold 1978, TPHVB).

Certain features of Gerola and Glassgold and TPHVB approach, not shared
by the other approaches, should be mentioned before proceeding to other
aspects of our discussions. As shown in Figure 10, with the passage of
time dynamically evolving clouds develop a well defined core-envelope
structure (see also Figure 3). The density is highest and practically
uniform in the core. It falls off rapidly in the envelope as r^{-p} where

r is the radius and p lies between 1.5 and 2 when averaged over the entire envelope. The size of the core is usually 1/10 or less of the entire cloud. The mass of the core is also small (30% or less) than the total cloud mass. An important parameter of the evolutionary approach is the dynamical time scale of the cloud. This time scale is defined as the time needed to change the core density by a given factor, typically a factor of 10 or a factor of e (= 2.7). The dynamical time scale is large when the cloud is diffuse, but deceases very significantly when the core density becomes 10^4 or more. As a result, clouds spend most of their lifetime as diffuse clouds and pass through the high density states rather quickly. The actual values of the time scales are quite model dependent. For example, time scales are larger if random magnetic fields are assumed to pervade the clouds, because the magnetic pressure opposes and, therefore, retards the gravitational collapse. Table 2 shows this effect for a 40 M_0 cloud.

Table 2. Time (yr) Required to Change the Central Density From a Given Initial Density to an Order of Magnitude Higher Density. (Magnetic Field Strengths Are in Micro-Gauss)

Initial Density	Final Density	Non-Magnetic	B_0=2.0 k=1/2	B_0=4.0 k=1/3
1.0 +2	1.0 +3	2.0 +6	2.2 +6	2.3 +6
1.0 +3	1.0 +4	5.5 +5	6.8 +5	9.0 +5
1.0 +4	1.0 +5	1.9 +5	2.7 +5	4.4 +5
1.0 +5	1.0 +6	5.7 +4	1.0 +5	1.6 +5

$(1.0 +2 = 1.0 \times 10^2)$

The calculations for the magnetic case were done using the flux-freezing assumption, so that the magnetic field scaled as:

$$B / B_0 = (\rho / \rho_0)^k \tag{17}$$

where B and ρ are respectively the magnetic field and the density, the subscript 0 denotes initial values, and k may range between 1/3 and 1/2. The temperature distribution in the cloud was governed by the empirical formulae of TPHVB. Introduction of turbulence, rather than the magnetic field would produce similar results. In any case, the clouds spend most of their life time as diffuse clouds, and the higher density states are relatively short-lived. Thus, steady state models are good first order approximations for diffuse clouds but may not be so for dense clouds.

Almost all of the cited dense cloud models use only gas phase chemistry. Models of dense clouds incorporating the formation and desorption of

molecules by grain surface related processes are due to Allen and
Robinson (1975), Duley et al. (1980) and d'Hendencourt et al. (1985).
Generally, gas phase molecule formation by grain surface reactions in
cool dark clouds have been viewed with considerable skepticism, due
mostly to uncertainties in the desorption processes and the nature of
the desorbed molecules. In contrast, pure gas phase models have been
quite popular, because their basic foundations have been experimentally
verifiable and because they have been successful in reproducing the
observed abundances in a large number of cases. Their success in
reproducing the observed abundances must be considered <u>fortuitous</u>, due
to reasons that are at present not well understood. One obvious
possibility is a near cancellation between the condensation of the gas
phase molecules onto the grains and their re-evaporation by the
currently not well quantifiable mechanisms mentioned in section 4.3.

6.1.2 <u>Some</u> <u>Current</u> <u>Issues</u> <u>and</u> <u>their</u> <u>Implications</u>:

Considerable understanding of the chemical evolution in dense clouds has
been gained by the theoretical models working in concert with the
observational data. To appreciate the current issues in a simple and
effective manner, let us consider the recent dense cloud observations of
atomic carbon (Phillips and Huggins 1981, Keene et al. 1985), and the
non-detection of O_2 in attempts at its observation (Lizst and Vanden
Bout 1985, Lizst 1985). These two species are amongst the simplest
possible interstellar chemical species. They have been chosen due to
their extreme simplicity and due to their chemistry being free from the
uncertainties that surround the formation of complex molecules by the
radiative association reactions.

(i) <u>Atomic</u> <u>Carbon</u> <u>Abundance</u>: The observed atomic carbon abundance
relative to the CO abundance (also denoted as the C/CO ratio) in dense
clouds ranges between 0.5 and 0.01. In sharp contrast, all constant
density, constant temperature pseudo time-dependent models predict a
steep decline in C/CO ratio at about 10^5 yr after the uv shut off or
the onset of the dark condition. By about 10^6 yr, the C/CO ratio
becomes very small (Graedel et al. 1982, Leung et al. 1984), so that the
calculated ratios are three to four orders of magnitude smaller than
their observed values. In models allowing for the spatial variation of
the background density (e.g., the hydrostatic models), it is possible to
maintain a high C/CO abundance at all times. In this case, however, the
bulk of C resides in the tenuous envelope while most of the CO is
located in the dense core. This is contrary to the observations, which
indicate that C and CO coexist even in the dense core. Lately, Langer et
al. (1984) and Leung and Herbst (1985) have tried the possibility of a
significant depletion of available oxygen at high densities, so that the
gas phase C/O elemental abundance ratio may exceed one in dense clouds.
Using C/O ratio = 1.28, they are able to reproduce the observed C/CO
ratios. However, the fractional abundance of other carbon compounds,
such as C_2H, C_3H, HC_3N, were predicted to be much higher than
their observed values. For example, the predicted abundance of HC_3N
relative to H_2 (Leung and Herbst 1985) is 9×10^{-8} whereas

the observed values are only $(0.4 - 6) \times 10^{-9}$ (Irvine et al. 1985). Three types of solutions for these difficulties with constant density models have been proposed.

One type of solutions envisages specialized physical conditions. Thus, Tielen and Hollenbach (1985a,b) have argued that the observed atomic carbon emissions can be produced from the surfaces of molecular clouds exposed to far-ultraviolet fluxes 10^3 to 10^6 times more intense than the ambient interstellar field. From their observational data Keene et al. (1985) suggest that this theory may be inadequate.

The second type of solutions argues that chemical evolution in time invariant or constant density clouds may be interrupted by dynamical processes at the interval of about 10^6 yr, so that chemical equilibrium is not attained and the complete conversion of C to CO is prevented. According to Boland and de Jong (1982), this might happen if dense clouds have turbulent transport currents which brings the core material to the edges or the other way around in a time interval shorter than that needed for complete C --> CO conversion. As the material from the cloud center arrives at the edge, the photodissociation of CO would recover the atomic carbon. The turbulent transport in Boland and de Jong model is assumed to occur on time scales of R_C/V_T, where R_C and V_T are respectively the cloud radius and the turbulent broadening velocity of the spectral lines. As emphasized by Williams and Hartquist (1984), the turbulent transport is a diffusive process. Hence, a more realistic estimate of the time needed to transport matter from the cloud core to the edge would be $(R_C/L_T)^2 L_T/V_T$, where L_T is the size of the largest eddies. Unless L_T is nearly equal to R_C, Boland and de Jong may have severely underestimated the transport time.

The third type of solution eliminates the concept of constant density. It is a by-product of our enquiry into the basic question: How did the dense clouds become dense in the first place? One mechanism is the gravitational contraction where by an initially diffuse mass of interstellar gas contract and form dense clouds. In this scheme the lifetime of a cloud in the high density state is quite small (see Table 2), smaller than or on the order of the time needed for complete C to CO conversion. Consequently, a fairly large amount of C can coexist with CO almost throughout a dense cloud, including the regions where $n \sim 10^4$ - 10^5 cm^{-3} and $A_v \sim 10 - 20$, but excluding the very dense core where n may approach 10^6 cm^{-3} (see Figures 6 and 7 of TPHVB). Cosmic ray excitations of the H_2 Lyman and Werner bands maintains a weak uv radiation field capable of dissociating CO at a significant rate deep inside the dense clouds where the external galactic uv can not penetrate (Prasad and Tarafdar 1983). Consequently, C and CO may coexist even in the very dense core region provided the O_2 density is kept low as implied by its non-detection. If due to any reason the high density phase lasts longer than implied by our present evolutionary models, then atomic carbon will, of course, avoid the densest core region. These theoretical model results have been discussed in a little greater detail elsewhere (Prasad 1987).

What happens to the clouds when the core density approaches 10^6 cm^{-3} is an important question in the proposed solution of the C/CO problem on the basis of the evolutionary models. In TPHVB it was assumed that they all proceed to form stars. There is a possibility that this assumption may lead to star formation rates in excess of the observed rates. By including either the randomly oriented magnetic field or the ab-initio calculation of the temperature in our model, we now find that the clouds may bounce back after the core attains such high densities and become diffuse clouds again. If this new result survives further scrutiny, then it will significantly reduce the star formation efficiency of gravitationally contracting clouds and will bring our evolutionary models in better accord with the observed star formation rates.

(ii) The Non-Detection of O_2: Another important development of the recent years is the non-detection of O_2 in observational searches cited earlier. These searches suggest that O_2/ CO abundance ratio is probably $\ll 0.1$ in dense clouds. In contrast, constant density - constant temperature chemical equilibrium models of dense clouds suggest that O_2 should be almost as abundant as CO, i.e., O_2/ CO abundance ratios of 0.3 - 0.5 with the possibility of some variation due to variation in the metal abundances. The observed low upper limits on the abundance of O_2 coupled with the abundance of H_2O inferred from HDO observations in dense clouds (Moore et al. 1986) suggests that oxygen left over from the C --> CO conversion remains mostly atomic with f_o (the abundance of O relative to H_2) approaching 10^{-4}. As a second possibility, the left over oxygen may be on the grains as ice. High f_O poses some difficult problems. Given the observed f_{OH}, a high f_O and the non-detection of O_2 jointly requires that the reaction

$$O + OH \rightarrow O_2 + H \tag{18}$$

has an activation temperature of 80 - 90K. Otherwise, there must be some efficient mechanisms to dissociate O_2 in dense clouds. Both conditions are extremely difficult to fulfill. For example, most recent critical evaluation of rate coefficients pertinent to the reaction (18) suggest a negative activation temperature (DeMore et al. 1985). Furthermore, high f_O has potential problems with the observed HCS$^+$/ CS abundance ratios of 0.1 or more in cool dark clouds (Irvine et al. 1983), even with the high rate coefficient for the H_3^+ + CS --> HCS$^+$ + H_2 reaction now predicted by certain theories (e.g., Clary 1985) and partly verified by experiments (Millar et al. 1985). The potential for conflict arises if rate coefficient of the reaction HCS$^+$ + O --> products increases at the low temperatures (15 - 20K) and makes the loss of HCS$^+$ ions by reactions with O more important than their destruction by the dissociative recombinations. Stated differently: either atomic oxygen reacts only slowly (rate constant 10^{-10} or less) even at the low dark cloud temperatures, or the abundance of atomic oxygen is significantly smaller than 10^{-4} where HCS$^+$/ CS ratio is 0.1 or more. The second possibility that the left over oxygen is mostly on the grains as solid water can be realized if atomic oxygen is converted into H_2O more rapidly than into O_2, and if the water is preferentially retained by the grains during the desorption process. Rapid conversion of O into H_2O would be possible if the dissociative recombinations of the H_3O^+ gives H_2O almost exclusively (Bates 1986). Preferential retention of water by the grains would be possible because water has the highest freezing point relative to all other interstellar species. Clearly, we have an interesting unsolved problem for future research. Perhaps, the difficult task of integrating the gas phase and grain surface chemistry cannot be postponed any longer.

(iii) <u>Chemical</u> <u>Differences</u> <u>Between</u> <u>Clouds</u>: An important observational result is that the chemical composition of clouds with strikingly similar physical properties may be strikingly different. Furthermore, abundances of a given species may also very significantly with position in the cloud (Irvine et al. 1985). For example, relative to both L134N and the Orion ridge, the abundances of TMC-1 are enhanced for the acetylenic species such as C_3N, HC_3N, C_4H, HC_5N. In contrast, abundances among these clouds are similar for CS and C_2H. Ammonia in TMC-1 is depleted relative to the Orion and L134N. Equilibrium chemical models have tried to explain this in terms of different elemental abundances (Millar and Freeman 1984a,b). In evolutionary models the differences between the chemical composition of clouds with quite similar physical properties are attributed to differences in the evolutionary tracks due to differences in the cloud masses and initial densities. The possibility that contracting clouds often rebound after attaining high core density has the potential to be another important cause of variability in the chemical composition of clouds with similar physical properties. This is easily understood. The chemical composition of a cloud at a given density in the contracting phase will certainly be different from the chemical composition at the very same density in the expanding phase, due simply to the greater time and wider spectrum of density available for chemical processing in arriving at that density in the expansion phase.

(iv) <u>Understanding</u> the <u>Icy</u> <u>Mantles</u> on the <u>Grains</u>: Recent strides in the observational tools for the middle infrared (2 - 20 microns) have resulted in numerous spectra of infrared sources, many still embedded in dense clouds. These spectra provide a powerful means of identifying both gas phase and grain mantle molecules in dense interstellar clouds (Kessler and Phillips 1984). For example, Whittet et al. (1983) have have reported 3 micron ice bands along six lines of sight towards both background stars and dust embedded sources in the Taurus Molecular Cloud (TMC). These observations suggest that ice mantles may form on grains at A_v as low as 4. Furthermore, the band carrier has been shown to be widely distributed in the cloud and not limited to the circumstellar environment. More recently, Lacy et al. (1984) have found rather positive identification of frozen CO in the direction of W33 A and other protostellar objects. Models of gas - grain interaction in molecular clouds are needed to interpret these advances in the observational studies of grain mantles. These models would also help in resolving the difficulties in simultaneously modeling O_2 and H_2O abundances in the gas phase as discussed earlier.

Unfortunately, it is very difficult to construct satisfactory models of interstellar chemistry incorporating gas - grain interactions, given the current uncertainties in the desorption processes. d'Hendecourt et al. (1985) have, nevertheless, developed dense cloud models which examines the continuous interaction between a limited set of grain surface reactions and and gas phase chemistry schemes. Explosive release of molecules triggered by grain - grain collisions was used as the desorption mechanism by which molecules from the grain mantles are returned to the gas phase. Furthermore, molecules in the solid phase were kept the same as the molecules in the gas phase. In other words, their limited surface reactions did not make any molecule other than those made by the gas phase reactions. This restriction was imposed to insure computational stability, although its validity is limited. Table 3 shows their results

Table 3. Gas phase and solid phase abundances of molecules for a dense cloud with $n = 10^5$ cm^{-3} and $A_v = 8$ from d'Hendecourt et al. (1985)

Molecules	Time (yr)			
	10^6		10^8	
	Gas	Solid	Gas	Solid
H_2O	1.6 (-4)	2.2 (-4)	2.4 (-6)	5.3 (-6)
CH_4	3.5 (-5)	5.0 (-5)	4.0 (-7)	6.6 (-7)
NH_3	1.9 (-5)	2.9 (-5)	3.0 (-7)	6.0 (-7)
O_2	6.6 (-6)	6.5 (-6)	7.4 (-5)	8.0 (-5)
CO	2.6 (-5)	2.3 (-5)	3.5 (-5)	5.6 (-5)
C	2.0 (-11)	--	2.0 (-11)	--

for a dense cloud. Their gas phase results for both C/CO and O_2/CO do not agree with the observations either at the early or at the late times. They also predict probably too much water vapor at the early times. This is an outcome of the deficiencies in their gas phase model. Even so, their study is significant because it shows that a plausible desorption mechanism can prevent a complete accretion of gas phase molecules onto the grains. Williams and Hartquist (1984) have presented a quite different version of gas - grain interaction in dense clouds. Instead of a continuous interaction, they have proposed an intermittent desorption of molecules from grains at the interval of about 10^6 yr due to shocks. The model gives encouraging results for C and CO. Results for other molecules were not presented in the currently available version of the model.

6.2 IMPACT OF STAR FORMATION ON THE CHEMICAL EVOLUTION IN DENSE CLOUDS:

Stars lose mass in the form of supersonic outflows of matter. They also
generate intense fluxes of electromagnetic radiation. These stellar
activities profoundly influence the dynamic conditions and the chemical
evolution of the surrounding dense cloud. Figure 12 sketches the
perturbed dynamical conditions in a star forming region which is
nearest to us. Due to its proximity, this region in Orion is perhaps the
most thoroughly studied example of a dense cloud core affected by
intense star formation activity. The mass, density, and temperature of
the ridge cloud are respectively 100 - 1000 $M_\odot$, 10^4 - 10^6 cm^{-3}
and 50 - 100 K (Bastien et al. 1981, Schloerb and Loren 1982, Johansson
et al. 1982). The ridge cloud has been nicknamed "spike" due to the
narrowness of its molecular spectral lines. The ambient ridge (spike) is
most easily visible in the CS or the H_2CO 2_{12} --> 2_{11} transition
and lies on both northern and southern sides of the infrared sources.
The southern ridge contains a localized warmer clump often referred to
as the southern condensation. The southern condensation is characterized
by the relatively large abundances of certain complex molecules

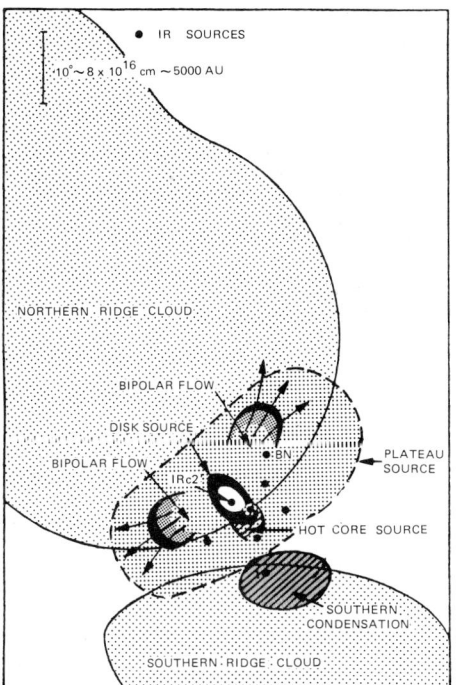

Figure 12. Schematic of the core of the Orion Molecular Cloud (OMC-1)
              perturbed by recent star formations

(Johansson et al. 1984, Olofsson 1984). Excepting this clump, the chemical composition of the ridge is similar to that of a warm dense cloud. The mass, density and temperature of the plateau source are, respectively, 3 - 30 $M_\odot$, 10^6- 10^7 cm^{-3} (Johansson et al. 1984, Frieberg 1984, Plambeck et al. 1982). The "plateau" source derives its name from the characteristic very broad shape of its molecular lines (and not for any such spatial structure). The plateau contains several infrared sources, such as BN and IRc 2. Particularly noteworthy is the disk like low velocity (i.e., Δv ~ 18 km/s) plateau centered on the IRc 2. The low velocity plateau is coincident with the low velocity maser centroid and the Hat Creek interferometric SO maps (Plambeck et al. 1982, Genzel et al. 1981).

Orthogonal to the low velocity plateau (or disk) are the high velocity (Δv ~ 20 - 25 km/s) bipolar flows seen in the CO, H_2O and a number of other species. In addition to the high and low velocity sources, there is also a hot core source containing about 10 $M_\odot$ of interstellar matter at a high density (10^6 - 10^7 cm^{-3}) and a high temperature (100 - 200 K). The heating of the hot core is probably radiative due to the luminous IRc 2 (Genzel et al. 1982, Goldsmith et al. 1983). All of these sources are related, in one way or other, with the mass loss from one or more of the pre-main sequence objects formed in the plateau.

The chemistry of the star forming region responds dramatically, to the dramatic changes in the physical conditions described above. Figure 13

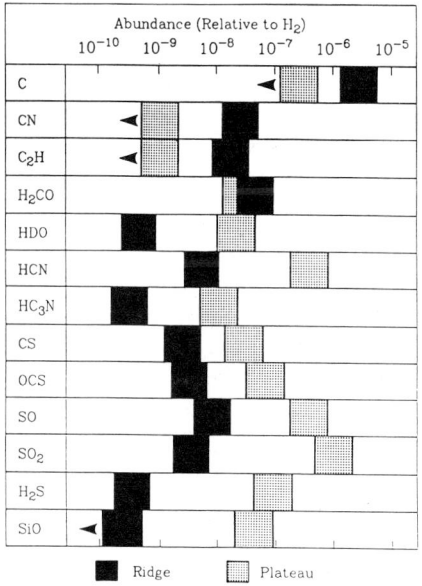

Figure 13. A schematic comparison of abundances in the Orion ridge and plateau (Blake 1986)

shows the impressive chemical differences between the ambient ridge and the star forming plateau in a schematic way. Differences in the physical conditions were quantified from studies of the molecular lines shapes and infrared emissions. As we discuss the chemistry of this region, it will become self evident that chemistry has the potential to complement and/or supplement the information gained by line shape and infrared studies. Relative to the ridge, SO, SO_2, H_2S, SiO, HDO, HCN are two orders of magnitude or more abundant in the plateau. Species like H_2CO have only barely enhanced abundances, while C, C_2H, CN have actually decreased abundances in the plateau. Abundances of sulfur compounds in quiescent clouds imply a factor of 100 - 1000 depletion of the elemental sulfur (Prasad and Huntress 1982). Considering this and the difficulties in the H_2S formation, Blake (1986) has suggested that a factor of 10 - 100 enhancements of SO, SO_2 and H_2S in the high velocity plateau requires release of sulfur from the grains and also a high temperature to efficiently process S into H_2S. According to him, shocks would probably satisfy both requirements. Sputtering, for example, may release the needed sulfur from at least the smaller grains. The chemistry in the hot post shock gas can then synthesize the observed amount of SO, SO_2 and H_2S, given the elevated amount of elemental sulfur (Hartquist et al. 1980, Dalgarno 1982). Enhancement of HCN and H_2O, and the decrease of C, CN and C_2H can also occur in the shocked chemistry (Iglesias and Silk 1978, Mitchell 1984). The fact that shocks exist in the high velocity flow region is quite evident from vibrationally excited H_2 and far infrared high-J CO emissions. It should, however, be noted that the need to release sulfur from the grains may not be compelling. Prasad and Huntress' conclusion about the depletion of sulfur is critically dependent upon the assumption that there is no activation energy in the reaction S + O_2 --> SO + O. If this reaction has a small activation energy or if the O_2 abundance is too small, then sulfur may remain mostly as neutral atoms in cold dense environments. In this scenario, one would need only shock heating to explain the amounts of SO, SO_2, and H_2S observed in the plateau.

The low velocity (18 km/s) flow should also drive strong shocks in the surrounding molecular cloud. The apparent absence of the signatures of hot gas chemistry in the disk source (i.e., the low velocity outflow) may simply be due to observational and sensitivity effects. The high density in this source will rapidly cool the post shock gas resulting in a very thin shock front and thereby unobservable column density.

The chemistry in the hot core appears to be substantially different from that in the ridge and the plateau. This is most easily appreciated from the comparative chemical compositional data presented in Figure 14. Relative to the ridge, the abundances of fully hydrogenated molecules, such as H_2O, HCN and NH_3, are quite enhanced in the hot core. The abundances of reactive carbon species like C, CN, CS and C_2H are also smaller in the hot core relative to the ridge. These are similar to the situation in the plateau and suggest that a high temperature chemistry is operating in the hot core also. On the other hand, the abundances of SiO, SO, SO_2 in the hot core is very much less than those in the

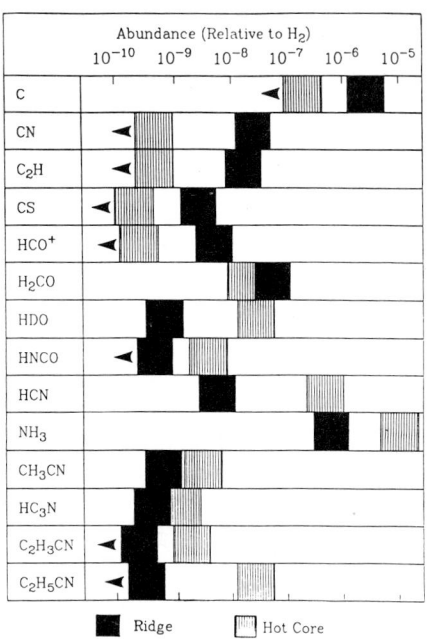

Figure 14. Comparison of the molecular abundances in the Orion
ridge and hot core (Blake 1986)

plateau (Irvine et al. 1985, Blake 1986). Clearly, there must be a basic
difference between the heating mechanism in the plateau and the hot
core. Since the abundances of SO and SO_2 are not elevated in the hot
core, the heating mechanism in the hot core seems to be incapable of
releasing the sulfur from the grains. Radiative heating from IRc 2 seems
to meet this condition. Such a heating would passively increase the gas
temperature to 150 - 300 K needed to enhance the abundances of HCN (for
example) at the expense of CN via the reaction

$$CN + H_2 \longrightarrow HCN + H \tag{18}$$

or to decrease the abundance of CS (for example) via the reaction:

$$CS + O \longrightarrow CO + S \tag{19}$$

Radiative heating would not break any grains. Consequently, there would
not be any enhancement in the abundances of sulfur compounds like SO and
SO_2 in the hot core. Radiative heating would also warm up the grains
and, thereby, induce the thermal evaporation of their icy mantles. This
might help in understanding the remarkable increase in the abundances of
fully hydrogenated molecules, such as H_2O and NH_3 (Switzer 1978).

From the above brief and qualitative discussions, it is obvious that dynamical and radiative processes following star formation profoundly affects the chemical evolution of the surrounding dense cloud. Our discussions have been qualitative and brief, because the observational data is still maturing especially with respect to the abundances in the hot core. Nevertheless, it appears that chemistry has the potential to complement and supplement the knowledge gained from molecular line shapes. This is exemplified by the role of chemistry in elucidating the subtle difference in the heating mechanisms of the plateau and the hot core.

7.   SOME REMARKS ON FUTURE DIRECTIONS

Coupled chemical-dynamical modeling of interstellar clouds are still in their infancy and present a fertile field of research. The present models must be improved by incorporating magnetic fields, turbulence, and a better treatment of the self-shielding effects. The improved models should initially focus on the problems involving the simpler species, such as C, CO, O_2. Once these problems are understood, bigger and more complex molecules should be stressed. The emphasis at this stage should be on testing the potentials of chemistry to serve as signposts of dynamical evolution. At present there is a considerable uncertainty in grain surface reactions and ejection mechanisms. Nevertheless, attempts at integrating the gas phase and grain surface chemistry should be undertaken with a sense of urgency. This is suggested by the difficult problems that we encountered in budgeting the gas phase oxygen. Advances in infrared technology allowing direct observations of grain mantles would also make that attempt highly rewarding. Special conditions prevailing in the hot core and the southern condensations present an uncharted realm for exploring new dimensions in interstellar chemistry at very high density, temperature and radiation field intensity. Hopefully, this might lead to chemical diagnostics of hidden stellar or protostellar objects.

8.   ACKNOWLEDGMENTS

The work done at the Jet Propulsion Laboratory, California Institute of Technology, was supported by a contract with the National Aeronautics and Space Administration. SPT and KRV were consultant at JPL during a part of this work. Contributions of SPT were also provided under the terms of a cooperative research program with the Tata Institute of Fundamental Research sponsored by the Smithsonian Institution.

9.   REFERENCES

Adams, N. G., Smith, D., and Millar, T. J. 1984, Mon. Not. R. Astr. Soc., 211, 857.
Allen, M., and Robinson, G. W. 1975, Ap. J., 195, 91.

Allen, M., and Robinson, G. W. 1977, Ap. J. , 212, 396.
Allamandola, L. J., Tielens, A. G. G. M., and Barker, J. R. 1985, Ap. J. (Letters), 290, L25.
Barlow, S. E., Dunn, G. H., and Schauer, M. 1984, Phys. Rev. Lett., 52, 902.
Barsuhn, J., and Walmsley, C. M. 1977, Astr. Ap., 54, 345.
Bastien, P., Bieging, J., Henkel, C., Martin, R. N., Pauls, T., Wamsley, C. M., and Ziurys, L. M. 1981, Astr. Ap. 98, L94.
Bates, D. R. 1979, J. Phys. B, 12, 4135.
------. 1983a, Ap. J. (Letters), 267, L121.
------. 1983b. Ap. J., 270, 564.
Black, J. H., and Dalgarno, A. 1977, Ap. J. Suppl., 34, 405.
Black, D. C., and Scott, E. H., 1982, Ap. J., 263, 697.
Blake, G. A. 1986, Ph. D. Thesis, California Institute of Technology.
Blitz, L. 1984 (April), Scientific American, page 88.
Bohme, D. K. 1986, Nature ,319, 473.
Boland, W., and de Jong, T. 1982, Ap. J., 261, 110.
------. 1984, Astr. Ap., 134, 87.
Burke, J. R., and Hollenbach, D. J. 1983, Ap. J., 265, 223.
Clary, D. C. 1985, Mol. Phys., 54, 605.
Dalgarno, A. 1982, Phil. Trans. R. Soc. Lond., A303, 513.
de Jong, T., Dalgarno, A., and Boland W. 1980, Astr. Ap., 91, 68.
DeMore, W. B. et al., NASA Panel for Data Evaluation 1985, JPL Publication 85-37, Jet Propulsion Laboratory, California Institute for Technology, Pasadena, California, USA.
Dickman, R. L., Somerville, W. B., Whittet, D. C. B., McNally, D., and Blades, J. C. 1983, Ap. J. Suppl., 53, 55.
Disney, M. J., McNally, D., and Wright, A. E. 1969, Mon. Not. R. Astr. Soc., 146, 123.
Draine, B. T. 1978, Ap. J. Suppl., 36, 595.
------. 1986, Submitted to Ap. J.
------, and Anderson, N. 1985, Ap. J., 292, 491.
------, and Katz, N. 1986, Ap. J., 306, 655.
Duley, W. W., Millar, J. J., and Williams, D. A. 1980, Mon. Not. R. Astr. Soc., 192, 945.
Elitzur, M., and Watson, W. D. 1980, Ap. J., 236, 172.
Federman, S. R., and Glassgold, A. E. 1980, Astr. Ap., 89, 113.
------, Jenkins, E. B., and Shaya, E. J. 1980, Ap. J., 242, 545.
Frerking, M. A., Langer, W. D., and Wilson, R. W. 1982, Ap. J., 262, 590.
Friberg, P. 1984, Astr. Ap., 132, 265.
Genzel, R., Downes, D., Ho, P. T. P., Bieging, J. 1982, Ap. J. (Letters), 259, L103.
Gerola, H., and Glassgold, A. E. 1978, Ap, J. Suppl., 37, 1.
Glassgold, A. E. 1985 , in Protostars and Planets. II., Ed. D. C. Black and M. Matthews, (Tucson: University of Arizona Press)
------, and Langer, W. D. 1974, Ap. J., 193, 73.
------. 1976, Ap. J., 206, 85.
Goldsmith, P. F., and Langer, W. D. 1978, Ap. J., 222, 881.
Graff, M., and Dalgarno, A. 1986, submitted to Ap. J.

Graedel, T. E., Langer, W. D., and Frerking, M. A. 1982, Ap. J. Suppl., 48, 321.

Goldsmith, P. F., Krotkov, R., Snell, R. L., Brown, R. D., and Godfrey, P. 1983, Ap. J.,274, 184.

Hartquist, T. W., Oppenheimer, M., and Dalgarno, A. 1980, Ap. J., 236, 180.

d'Hendecourt, L. B., Allamandola, L. J., Bass, F., and Greenberg, J. M. 982, Astr. Ap., 109, L12.

-----,Allamandola, L. J., and Greenberg, J. M. 1985, Astr. Ap., 152, 130.

Henning, K. 1981, Astr. Ap. Suppl., 44, 405.

Herbst, E. 1982, Ap. J., 252, 810.

-----. 1983a, Ap. J., 270, 564.

-----. 1983b, Ap. J. (Letters), 267, L121.

-----. 1985, Ap. J., 291, 226.

-----. 1987. This volume.

-----, and Klemperer, W. 1973, Ap. J., 185, 505.

Hollenbach, D. J., and Salpeter, E. E. 1970, J. Chem. Phys., 82, 3152.

-----. 1971, Ap. J., 163, 155.

-----, Werner, M. W., and Salpeter, E. E. 1971, Ap. J., 163, 165.

Hunter, J. H. 1969, Mon. Not. R. Astr. Soc., 142, 473.

Huntress, W. T., Jr. 1977, Ap. J. Suppl., 33, 495.

Iglesias, E. R. 1977, Ap. J., 218, 697.

-----, and Silk, J. 1978, Ap. J., 226, 851.

Irvine, W. M., Good, J. C., and Schloerb, F. P. 1983, Astr. Ap., 127, L10.

Irvine, W. M., Schloerb, F. P., Hajlmarson, A., and Herbst, E. 1985, in Protostars and Planets. II., Ed. D. C. Black and M. Matthews (Tucson: University of Arizona Press)

Irvine, W. M., Goldsmith, P. F., and Hjalmarson, A. 1987. This volume.

Jenkins, E. B. 1987. This volume.

-----, Savage, B. D., and Spitzer, L. 1986, Ap. J., 301, 355.

Johansson, L. E. B., Andersson, C., Ellder, J., Friberg, P., Hjalmarson, A., Hoglund, B., Irvine, W. M., Olofsson, H., and Rydbeck, G. 1984, Astr. Ap., 130, 227.

Keenan, F. P., Dufton, P. L., Hilbert, A., and Murray, M. J. 1986. Mon. Not. R. Astr. Soc., 222, 143.

Keene, J., Blake, G. A., Phillips, T. G., Huggins, P. J., and Beichman, C. A. 1985, Ap. J., 299, 967.

Kessler, M. F., and Phillips, J. P. 1984, Galactic and Extra Galactic Infrared Spectroscopy, Ed. M. F. Kessler and J. P. Phillips (Reidel: Dordrecht)

Kiguchi, M., Suzuki, H., Sata, K., Miki, S., Tominatsu, A., and Nakagawa, Y. 1974, Pub. Astr. Soc. Japan, 26, 499

Knapp G., and Jura, M. 1976, Ap. J., 209, 782.

Lacy, J. H., Bass, F., Allamandola, L. J., Persson, A. C., McGregor, P. J., Landsdale, C. J., Gaballe, T. R., and van de Bult, C. E. P. M. 1984, Astr. Ap., 276, 533.

Langer, W. D., Graedel, T. E., Frerking, M. A., and Armentrout, P. B. 1985, Ap. J., 277, 581.

Leger, A., and Puget, J. L. 1984, Astr. Ap., 137, L5.

Leger, A., Jura, M., and Omont, A. 1985, Astr. Ap., 144, 147.
Leung, C. M., Herbst, E., and Huebner, W. F. 1984, Ap. J. Suppl., 56, 231.
Leung, C. M., and Herbst, E., 1986. In press with Mon. Not. R. Astr. Soc.
Lizst, H. S. 1979, Ap. J. (Letters), 233, L147.
Liszt, H. S. 1985, Ap. J., 298, 281.
-----, and Vanden Bout, P. A. 1985, Ap. J., 291, 178.
Mann, A. P. C., and Williams, D. A. 1984, Mon. Not. R. Astr. Soc., 209, 33.
-----. 1985,Mon. Not. R. Astr. Soc., 214, 279.
Mathis, J. A., Merzer, P. C., and Panagia, N. 1983, Astr. Ap., 128, 212.
Matthews, H. E., Friberg, P., and Irvine, W. M. 1985, Ap. J., 290, 609.
McEwan, M. J., Anicich, V. G., Huntress, W. T., Jr., Kemper, P. R., and Bowers, M. T. 1980, Chem. Phys Letters., 75, 278.
Millar, T. J., and Freeman, A. 1984a, Mon. Not. R. Astr. Soc., 207, 405.
-----. 1984b, Mon. Not. R. Astr. Soc., 204, 425.
Millar, T. J., Adams, N. G., Smith, D., and Clary, D. C. 1985, Mon. Not. R. Astr. Soc., 216, 1025
-----, T. J., and Nejad, L. A. M. 1985, Mon. Not. R. Astr. Soc., 217, 507.
Mitchell, G. M. 1984, Ap. J. Suppl., 54, 81.
-----, and Deveau, T. J. 1983, Ap. J., 266, 646.
-----, and Watts, G. D. 1985, Astr. Ap., 151, 121.
-----, Ginzberg, J. L., and Kunz, P. J. 1978, Ap. J. Suppl., 38, 39.
Moore, E. L., Langer, W. D., and Huguenin, G. R. 1986, Ap. J., 306, 682.
Morton, D. C. 1975, Ap. J., 197, 85.
Olofsson, H. 1984, Astr. Ap., 134, 36.
-----, Ellder, J., Hjalmarson, A., and Rydbeck, G. 1982, Astr. Ap., 113, L18.
Omont, A. 1986, Astr. Ap., 164, 159.
Oppenheimer, M., and Dalgarno, A. 1975, Ap. J., 200, 419.
Paleogolou, E. V., and Mouschovias, T. Ch. 1983, Ap. J., 275, 838.
Penston, M. V. 1969, Mon. Not. R. Astr. Soc., 145, 457.
Phillips, T. G., and Huggins, 1981, Ap. J., 251, 533.
Pickles, J. B., and Williams, D. A. 1977a, Astr. Space. Sci., 52, 443.
-----. 1977b, Astr. Space Sci., 52, 453.
-----.1981, Mon. Not. R. Astr. Soc., 197, 429.
Pineau des Fortes, G., Flower, D. R., Hartquist, T. W., and Dalgarno, A. 1986, Mon. Not. R. Astr. Soc., 220, 801.
Plambeck, R. L., Wright, M. C. H., Welch, W. J., Bieging, J. H., Baud, B., Ho, P. T. H., and Vogel, S. N. 1982, Ap. J., 259, 617.
Platt, J. R. 1956, Ap. J., 123, 486.
Prasad, S. S. 1987, to appear in IAU Symposium 120, Asrochemistry, eds. M. S. Vardya and S. P. Tarafdar (Dordrecht:Reidel)
-----. 1985, in Molecular Astrophysics: State of the Art and Future Directions, eds. G. H. G. Dercksen, W. F. Huebner, and P. W.

Langhoff (Dordrecht:Reidel)

Prasad, S. S., and Huntress, W. T., Jr. 1982, Ap. J., **260**, 590.

------. 1980a, Ap. J. Suppl., **43**, 1.

------. 1980b, Ap. J., **239**, 151.

Prasad, S. S., and Tarafdar, S. P. 1983, Ap. J., **267**, 603.

Schloerb, F. P., Loren, R. B. 1982, in Symposium on the Orion Nebula to Honor Henry Draper, eds A. E. Glassgold, P. J. Huggins, and E. L. Shuckling (Ann N. Y. Acad. Sci., New York)

Sellgren, K. 1984, Ap. J., **277**, 623.

Shu, F. 1983, Ap. J., **273**, 202.

Spitzer, L. 1978, Physical Processes in the Interstellar Medium (Interscience:New York)

Smith, D., and Adams, N. G. 1984, Ap. J. (Letters), **284**, L13.

Suzuki, H., Miki, S., Sata, K., Kiguchi, M., and Nakagawa, Y. 1976, Progr. Theoret. Phys. (Japan), **56**, 1111.

Sweitzer, J. S. 1978, Ap. J., **225**, 116.

Takahasi, T., Hollenbach, D. J., and Silk, J. 1983, Ap. J., **275**, 145.

Tarafdar, S. P., and Krishnaswamy, K. S. 1982, Mon. Not. R. Astr. Soc., **200**, 431.

Tarafdar, S. P., Prasad, S. S., and Huntress, W. T., Jr. 1983, Ap. J., **267**, 156.

Tarafdar, S. P., Prasad, S. S., Huntress, W. T. Jr., Villere, K. R., and Black, D. C. 1985, Ap. J., **289**, 220 (TPHVB).

Tielens, A. G. G. M., and Hollenbach, D. J. 1985a, Ap. J., **291**, 722.

------. 1985b, Ap. J., **291**, 747.

Thaddeus, P., Guelin, M., and Linke, R. A. 1981, Ap. J. (Letters), **246**, L41.

van de Hulst, H. C. 1949, Rech. Astr. Obs. Utrecht, **11**, part 2.

van Dishoeck, E. F. 1987, to appear in IAU Symposium 120, Astrochemistry, eds. M. S. Vardya and S. P. Tarafdar (Reidel: Dordrecht)

------, and Black, J. H. 1986, Ap. J. Suppl. in press.

Vardya, M. S., and Tarafdar, S. P. 1987, eds. Astrochemistry, IAU Symposium 120 (Reidel:Dordrecht).

Vrtileck, J. M., Thaddeus, P., and Gottlieb, C. A. 1985, Bull. Am. Astr. Soc., **17**, 568.

Wannier, P. G., Penzias, A. S., and Jenkins, E. B. 1982, Ap. J., **254**, 100.

Watson, W. D., and Salpeter, E. F. 1972, Ap. J., **174**, 321.

Williams, D. A., and Hartquist, T. W. 1984, Mon. Not. R. Soc., **210**, 41.

Whittet, D. C. B., Bode, M. F., Longmore, A. J., Baines, D. W. T., and Evans, A. 1983, Nature, **303**, 218.

Wootten, A., Evans, N. J., Snell, R., and Vanden Bout, P. 1978, Ap. J., (Letters), **225**, L143.

Wynn-Williams, C. G., Genzel, R., Becklin, E. E., and Downes, D. 1984, Ap. J., **281**, 172.

CHEMICAL EVOLUTION OF GALAXIES

S. Torres-Peimbert and M. Peimbert
Instituto de Astronomía
Universidad Nacional Autónoma de México, México.

ABSTRACT. We review some of the present ideas on chemical evolution of irregular galaxies and disks of spiral galaxies. Based on the O abundances in stars and H II regions we present a discussion of the yield; we also discuss the effects of large scale mass flows and of cold dark matter to explain variations in the observed yield. The C, N, O observed abundances are compared to models of galactic chemical evolution.

I. INTRODUCTION.

It has been increasingly clear that it is necessary to construct models of galactic chemical evolution to fit into a consistent picture the observed abundances in the interstellar medium of our galaxy and other galaxies. Such models predict the time variation of chemical composition depending on the assumed: a) initial mass function, b) time dependence of the stellar birth rate, c) chemical composition of the ejecta of stars during their evolution, and d) large scale mass flows, like infall from the halo, outflow to the intergalactic medium or radial flows within the galaxy. The observed abundances are used as test for the input assumptions, since these physical parameters are not generally known in galaxies.

Dwarf galaxies, with giant H II regions, are very well suited to study the early chemical evolution of galaxies. These objects are usually gas rich and have not been very efficient in their star formation processes, thus they present a relatively low level of heavy element contamination of the interstellar medium. The O/H ratio gives an indication of the level of contamination and provides us with a good indication of the heavy element abundance by mass, Z. By comparing the C/H and N/H values in metal rich and metal poor objects, it is possible to constrain the models of galactic chemical evolution. Previous reviews of some of these topics are present in the literature, (e.g., Peimbert, Serrano and Torres-Peimbert 1984; Pagel 1985; Peimbert 1985a; and references therein).

There is also a wealth of information on the heavy element abundance and its gradients in the Milky Way and in external galaxies; this

D. J. Hollenbach and H. A. Thronson, Jr. (eds.), Interstellar Processes, 667–678.

information is not only reliable but the number of observed objects is
relatively large (e.g., Peimbert 1979; Pagel and Edmunds 1981; Shaver *et
al*. 1983; Mc Call *et al*. 1985). Several authors have computed models of
galactic chemical evolution based on the observed data both for external
galaxies and the Milky Way (e.g., Twarog 1980; Talbot 1980; Tosi 1982;
Serrano and Peimbert 1983; Tosi and Diaz 1985). It is therefore possible
to compare the main chemical evolution parameters (initial mass function,
star formation rate, infall of gas, etc.) in the Milky Way and their corresponding values in other galaxies. A recent study by Tosi (1986)
evaluates the uniqueness of the models of chemical evolution of the Galaxy; the conclusion is that there is not a unique solution to the problem
of chemical evolution of galaxies, but that strong limits can be assigned
to the range of variability of model parameters.

II. OXYGEN ABUNDANCE.
a) Heavy Element Yield.

Detailed description of various aspects related to the behaviour of the
yield have been presented by Edmunds and Pagel (1984a) and Mould (1984).
    The heavy element yield is defined by

$$p = M(Z)/M_*$$
                                                                       (1)

where $M(Z)$ is the mass that a generation of stars ejects as newly formed
heavy elements to the interstellar medium and M_* is the mass of the generation of stars that remains locked into stellar remnants and long-
lived stars, where we include the low mass end of the initial mass function, IMF, which might comprise objects that do not become stars.
    An effective yield of primary elements can be obtained from

$$p(SM) = Z/ \{\ell n (M_{total}/M_{gas}) \}$$
                                                                       (2)

which is based on the 'simple model', that is, it is based on the following assumptions: a) galaxies evolve as closed well mixed systems, b)
the instantaneous recycling approximation, IRA, applies, and c) a constant yield prevails over the history of the system. From this relation
in the simple model it can be shown that in a region where the gas has
almost been exhausted $p = Z$, where z is the mean stellar   abundance.
    In Table 1 we present typical values of $p(SM)$, as defined in equa-
tion (2) for different objects. It is clear from this table that either
the yield is not constant, or the simple model does not apply, or both.

b) Non-baryonic Matter.

Under the assumption that $\Omega = 1$ it has been suggested that most of the
mass in the Universe is in the form of cold dark matter. It has been
argued that the increase in the M/L ratio from the center to the outer
regions of spiral galaxies is due to the increase of the ratio of cold
dark matter to baryonic matter; the increase in the M/L

TABLE 1. EFFECTIVE YIELDS

Objects	p(SM)	References
Irregular and blue compact galaxies	0.002 - 0.004	(1)
Solar neighborhood	0.006 - 0.008	(1)
M51, M83 and NGC 5055	0.012 - 0.02	(1,2,3)
Galactic center	0.04 - 0.06	(4)

(1) Peimbert and Serrano 1982; (2) Mc Call *et al*. 1985; (3) Edmunds and
    Pagel (1984<u>b</u>); (4) Whitford and Rich 1983.

ratio with decreasing mass of elliptical galaxies could be explained in
the same way (Frenk 1986). If the cold dark matter does not participate
in the star formation process then the M_{total} term in equation (2) should
include only baryonic matter and if a large fraction of the total mass is
non baryonic then p(SM) should increase. The effect goes in the right
direction but is not strong enough to explain the full difference of a
factor of 30 present in Table 1. At most a change of a factor of 2 or 3
in the yield when going from dwarf irregular galaxies to the nucleus of
our galaxy could be accounted for; this would be the case if most of the
matter in the nucleus of our galaxy is baryonic and only about 10% of the
matter in irregular galaxies is in baryonic form.

c) Large Scale Gas Motions.

Inflow, outflow and radial gas flow models of galactic chemical evolution have been proposed in the literature to explain the large differences in p(SM), other than allowing for strong variations in the heavy
element yield, (Larson 1972; Twarog 1980; Tinsley 1981; Peimbert and
Serrano 1982; Tosi 1982; Matteucci and Chiosi 1983; Mould 1984; Lacy and
Fall 1985; Matteuci and Tosi 1985).

   The observations that support the possibility of infall are: a)
collision of an infalling cloud of high velocity with the disk of the
Galaxy, (Mirabel 1981; Mirabel and Morras 1984; Hulbosch 1985), b) 21-cm
absorption redshifted with respect to 21-cm emission in 6 out of 7 objects (Dickey 1982; Mirabel 1983). c) Neutral H I haloes in 38% of the
irregular and blue compact galaxies, observed by Huchtmeier *et al*. (1981),
d) extended H I haloes 10-20% of well observed nearby galaxies
Huchtmeier and Seiradakis 1985).

   Inflow models, where unprocessed material is continuously falling
into the star forming regions of the galaxies, have been characterized
by γ the ratio of the accretion rate to the star formation rate. With
this representation for a closed model, $\gamma = 0$; and for a case where infall (or accretion) exactly balances star formation, $\gamma = 1$, which leads
to the limit that when $M_{gas}/M_{total} \rightarrow 0$, $p \rightarrow Z$; the case of $0 < \gamma < 1$ requires galaxies to diminish their gaseous mass as a result of star for-
mation then, $p > p(SM)$, and $p < Z$ for low M_{gas}/M_{total} ratios.

   Since the average Z value for irregular and dwarf compact galaxies
is about 0.005 it is not possible to reconcile a value of the yield
higher than p(SM) by more than a factor of 2 for moderate amount of ac-

cretion ($\gamma < 1$).

The assumption of the same value of the yield for objects as different as irregular and dwarf compact galaxies and M51, M83 and the galactic center, is very extreme; it would require the infall rate to be very high ($\gamma > 1$) in the former objects, these values would imply that M_{gas} increases with time in irregular and blue compact galaxies.

Models by Twarog (1980), Tinsley (1981), Tosi (1982) and Serrano and Peimbert (1983) require that $\gamma \sim 1/2$ in the solar neighborhood to explain: the age metallicity relationship, a self consistent model of disk formation and nucleochronology, the isotopic CNO ratios as well as the N/O and O/H ratios.

For outflow models it can be shown that

$$p = Z (1 + \lambda/\alpha), \qquad (3)$$

where λ is the ratio of the outflow rate of processed material to the star formation rate and α is the fraction of matter that remains in stellar remnants and long lived stars, for the solar vicinity IMF it has been found that $\alpha \sim 0.7$ (Sarmiento and Peimbert 1985). From equations (2) and (3) it can be obtained

$$p(SM) = p/(1 + \lambda/\alpha). \qquad (4)$$

The situation is similar to the case of infall, it is very difficult to explain with the same yield all types of objects. For p = 0.05, values of λ/α of the order of 20 would be necessary to explain the Z values in dwarf galaxies. This solution would require that more than 90% of matter to have been ejected from these objects in the form of gas during its lifetime; this large amount of matter would be observable.

In disks of galaxies it has been proposed that in addition to inflow of unprocessed gas there are also radial flows of partially processed material that have to be included in the equations (e.g., Lacey and Fall 1985).

d) Variable Yield.

It is possible that inflow (and/or outflow) fixes the run of surface density, gas fraction, yields, and abundances; alternatively it is also possible that the IMF changes or that stellar evolution properties change with metallicity or density.

It has been suggested that the yield varies with position within a galaxy or for each galaxy. To explain values similar to those in Table 1, Peimbert and Serrano (1982) proposed that

$$p = 0.002 + 0.6 Z, \qquad (5)$$

while Edmunds and Pagel (1984b) proposed

$$p = 0.003 + 7 \times 10^{-6} \sigma_d, \qquad (6)$$

where σ_d is the total surface density in $M_\odot pc^{-2}$; both relations explain reasonably well the data and imply that there is a tight correlation between surface density and heavy element abundances. Also Güsten and Mezger (1983) proposed that a bimodal star formation can help explain the observed properties of the galaxy. They describe separately the IMF of the spiral arms (preponderance of medium and high mass stars) and interarm region, this introduces a variable yield which increases toward the inner part of the disks.

A yield increasing with metallicity possibly implies that galaxies with low O/H are more efficient in the formation of stars with masses smaller than 1 $M_\odot$ than galaxies with high O/H. Bruzual et al. (1982) have fitted the continuum spectrum of the nucleus of M81, which is metal rich, with an old stellar population; they find that the number of stars with M < 0.3 $M_\odot$ is negligible in agreement with the idea that at least part of the variation of the yield is due to a change in the lower mass end of the IMF.

Observations of the IMF in the Large and the Small Magellanic Clouds (Butcher 1977; Dennefeld and Tammann 1980) and of integral properties of irregular galaxies (Lequeux et al. 1981; Gallagher et al. 1984) seem to indicate that there are no substantial changes in the upper mass end of the IMF. Alternatively, other authors (e.g., Burki 1977; Viallefond 1983; Terlevich and Melnick 1984) find that the smaller the O/H ratio of an H II region the higher the effective temperature of the hottest stars; this trend would increase the yield for low Z, contrary to what is required to explain the differences in effective yields.

From the IMF of the solar vicinity, Peimbert (1985a) obtained that p ∿ 0.03 which is high according to the values of Table 1 and might imply that above a certain mass the stars become black holes, or that about 2/3 of the stars more massive than 10 $M_\odot$ become black holes. This result seems to imply that there is no simple relationship between the upper mass end of the IMF and the heavy element yield.

e) Metallicity Gradients in Disk Galaxies.

The oxygen abundance gradients present in spiral galaxies have been measured for the Galaxy and several disk galaxies (Peimbert 1979; Shaver et al. 1983; Dennefeld and Kunth 1981; Blair et al. 1982; Kwitter and Aller 1981; Mc Call 1982; Dufour et al. 1983; Shields and Searle 1978; Rayo et al. 1982). Recently, Garnett and Shields (1986) found strong radial gradients in the oxygen abundance of the disk of M81, the earliest type galaxy for which extensive observations have been made. M81 has an abundance gradient and an overall metallicity similar to those of late-type galaxies of similar mass, luminosity and gas fraction. Fierro et al. (1986) from observations of bright H II regions in NGC 2403 were able to estimate the gradients in the mass and in the effective temperature of the most massive stars at different galactocentric distances. They found for the disk of NGC 2403 that the mass of the most massive star is very similar in all the observed H II regions and therefore, it is independent of the local value of the heavy element abundance.

Various approaches have been followed by different authors to explain the abundance gradients observed in spiral galaxies: Tinsley and

Larson (1978) argue in favor of gas infall as residual of the gravitational collapse of the protogalactic cloud. Diaz and Tosi (1984) and
Tosi and Diaz (1985) models assume an infall rate which is constant and
uniform after the disk formation, as due to gas in the intergalactic
medium trapped in the gravitational field of the galaxy. Mayor and
Vigroux (1981) and Lacey and Fall (1985) consider radial flows in the
disks; Peimbert and Serrano (1982) and Güsten and Mezger (1983) invoke
yields varying with metallicity and bimodal star formation, respectively.

Diaz and Tosi (1984) analyzed the oxygen abundance gradients for
five nearby spirals, including our own. They found that although the
abundance distributions in the galactic disks are different from galaxy
to galaxy and the uncertainties on gas and total mass data are rather
high, their variation with galactocentric distance can be described relatively well with the same class of models with a suitable choice of
only two free parameters: the infall flux and the star formation rate.
The values for the infall rate are in the range $0 < F < 5 \times 10^{-3}$ $M_\odot$
kpc^{-2} yr^{-1}. For the galaxy, in fact, the infalling gas mass as derived
from high and very high velocity clouds is 1-2 $M_\odot$ yr^{-1} (e.g., van Woerden et $al.$ 1985), consistent with the value of 1.8 $M_\odot$ yr^{-1} corresponding to the value of Tosi and Diaz, $F = 4 \times 10^{-1}$ $M_\odot$ yr^{-1}.

Tosi and Diaz (1985) infer that small late type spirals evolve with
rapidly decreasing star formation rates (they assumed exponentially decreasing rate with $\tau = 5$ Gyr) and no infall, while in large late-type
spirals the SFR is also decreasing ($\tau = 5-10$ Gyr) but the infall rate is
larger than about 2×10^{-3} $M_\odot$ kpc^{-2} yr^{-1}. For earlier type galaxies, the
star formation is more constant ($\tau = 10$ Gyr).

Lacey and Fall (1985) have considered radial inflows combined with
infall to explain the radial gradients in the galaxy, present star formation rate, metallicity of H II regions, and of cepheids, K giants and
open clusters and real variations of the surface density of stars and
gas. They conclude that in the absence of spatial temperature variations
in the IMF, radial gas flows are probably required to explain the metallicity gradient. And that infall is required to fit the observations,
inflows are included in their models. They require at the solar neighborhood an infall rate of 0.1 to 1 $M_\odot$ yr^{-1}.

On the other hand, by comparing the observed chemical abundance
with models of galactic chemical evolution by Serrano and Peimbert
(1983) Fierro et $al.$ (1986) found that infall to the disk is not important for NGC 2403.

f) The Mass-Metallicity Relation.

Lequeux et $al.$ (1979) found that the metallicity of irregular and blue
compact galaxies follows a very strong correlation with mass, in agreement with the trend presented by elliptical galaxies.

With available observations of irregular and spiral galaxies, as
summarized by Garnett and Shields (1986) with the addition of the data
for NGC 2403 by Fierro et $al.$ (1986), and the data for elliptical galaxies taken from Mould (1984), we present in Figure 1 the diagram of
O/H versus total mass. As has been pointed out by Mould, the transformation of colors to metallicities in elliptical galaxies is very uncertain.

Garnett and Shields also find a very good correlation between the 'mean' metallicity of a galaxy and its mass. They conclude that abundances in galaxies correlate better with galaxy mass than with morphological type.

Galactic chemical evolution models should be able to account for this trend. This correlation is related to the apparent increase of the effective yield with metallicity.

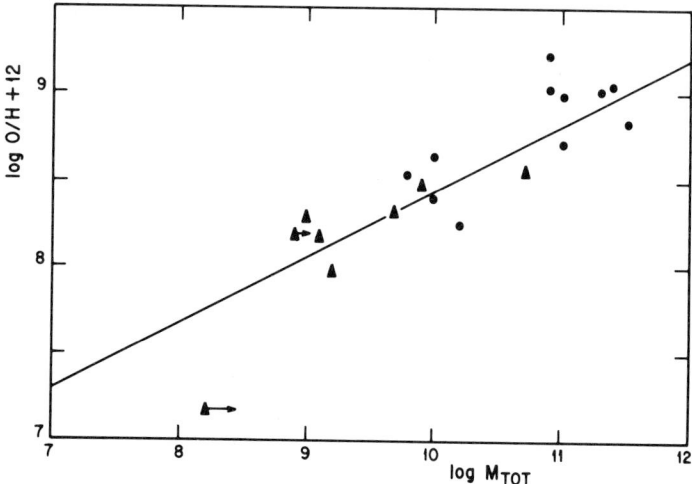

Fig. 1. Mass-metallicity relation. Where we have assumed that [O/H] ∝ [M/H] and that 12 + log (O/H)⊙ = 8.92. The solid line was derived by Mould (1984) from a sample of elliptical galaxies. The data for spiral and irregular galaxies come from Garnett and Shields (1986); we have also added the O/H value of Fierro et al. (1986) for NGC 2403 and the mass derived by Rogstad and Shostak (1972). Triangles represent irregular galaxies and filled circles spiral galaxies.

III. CARBON ENRICHMENT OF INTERSTELLAR MEDIUM

The C enrichment of the interstellar medium is due to massive stars, MS, that explode like supernovae with $M_i \geq 10$ $M_\odot$ and to intermediate mass stars, IMS, that produce planetary nebulae before becoming white dwarfs with 10 $M_\odot \geq M_i \geq 1$ $M_\odot$, where M_i is the initial mass in the main sequence.

It has been proposed that the previously estimated $^{12}C(\alpha,\gamma)^{16}O$ reaction rate should be increased (e.g., Trautvetter et al. 1984; Arnett and Thielemann 1984). This increase in the reaction rate reduces the C/O ratio predicted in models of SN ejecta prior to 1983. Arnett and Thielmann (1984) find that C/O = 0.376 (C/O)⊙, for model of a single star with an 8 $M_\odot$ He-core, corresponding to a 20-25 $M_\odot$ star, with the $^{12}C(\alpha,\gamma)^{16}O$ rate increased by a factor of five. Since it is assumed that practically all the O is produced in massive stars, and if the 20-25 $M_\odot$ stellar model is representative of massive objects it follows that the fraction of carbon produced by massive stars, ε is < 0.38. The fraction could be smaller than 0.38 due to time delays in the C ejection by intermediate mass stars in the solar vicinity.

According to Matteucci (1986) including new yields from massive stars computed with a revised rate for the $^{12}C(\alpha,\gamma)^{16}O$ reaction give the best agreement with the observations; moreover, carbon must have been mostly produced by low and intermediate mass stars (70%).

From the solar and Orion nebula C/H values (Lambert 1978; Torres-Peimbert *et al.* 1980; Perinotto and Patriarchi 1980) it is possible to obtain the C yield which together with the IMF by Serrano (1978) and the C production by intermediate mass stars (Renzini and Voli 1981), for $\alpha = 1.5$ and $\eta = 1/3$, give $\varepsilon = 0.36$ (Sarmiento and Peimbert 1985).

The number of objects for which C has been determined is very limited. Thus we depend on the determination of the C abundances in irregular and blue compact galaxies to study the C enrichment of the ISM (Dufour *et al.* 1982; Gondhalekar 1983; Dufour *et al.* 1985; Dufour 1985; Bergvall 1985; Sarmiento and Peimbert 1985; Matteucci 1986).

In Figure 2 it is also shown a chemical evolution model by Tinsley (1979) with the following characteristics: a) a star formation rate of 0.05 $M_{galaxy}/10^9$ years, b) O is primary and formed by stars more massive than 10 $M_\odot$, and c) C is also primary but formed by stars in the 4.5-6.5 $M_\odot$ range. In this model each star ejects 0.42 $M_\odot$ of C; this value is very high but it can be reduced considering the C production by massive stars and by stars in the 1 - 4.5 $M_\odot$ range. The increase in the C/O ratio as O/H increases is just due to the delay in the C enrichment between IMS and MS.

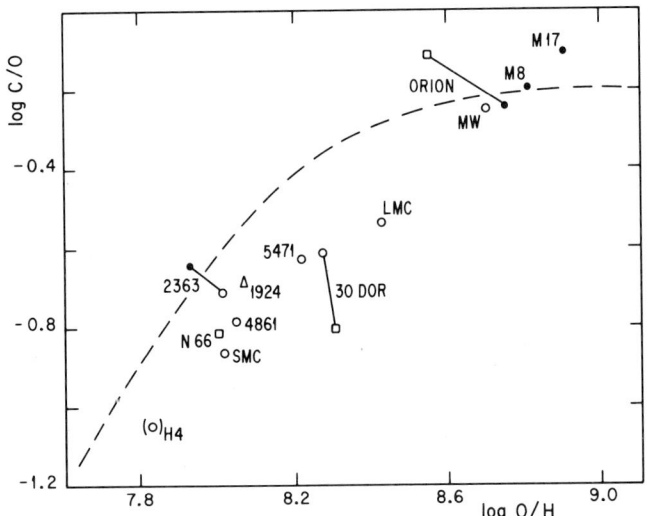

Fig. 2. log C/O vs. log O/H diagram based on observations of H II regions in our galaxy and in irregular and compact galaxies. The open circles correspond to data from Dufour *et al.* (1985), filled circles from Peimbert *et al.* (1986), Torres-Peimbert *et al.* (1980), Peimbert (1979), NGC 1924 from Bergvall (1985), and the squares come from Mathis *et al.* (1985). The dotted lines correspond to a model of galactic chemical evolution (Tinsley 1979).

IV. NITROGEN ENRICHMENT OF THE INTERSTELLAR MEDIUM

From the observed N/O vs. O/H relation for different objects and adopting the simple model, it has been suggested that a significant fraction of N is of primary origin (Smith 1975; Edmunds and Pagel 1978; Peimbert 1979; Alloin *et al.* 1979; Lequeux *et al.* 1979). The predictions differ when the hypotheses of instant recycling approximation, IRA, and closed system (no gas flows) break down. For example, the N/O vs. O/H diagram for irregular and spiral galaxies can be explained under the assumption that most of the N is of secondary origin, that most of the N is produced by stars in the $1 < M/M_\odot < 5$ mass range, that accretion plays an important role in the chemical evolution of galaxies and that the yield increases with metallicity (Serrano and Peimbert 1983). In Figure 3 we show schematically the observed locus in the N/O vs. O/H diagram of extragalactic H II regions, irregular galaxies and the Milky Way.

The models by Serrano and Peimbert (1983) can explain the variations of N/O vs. O/H in the disk of spiral galaxies in terms of a variation of the parameters of infall rate, γ and age. Most of the spiral galaxies follow a nearly constant N/O relation and it can be seen that this result is compatible with disk formation at the same age but with different rate of infall.

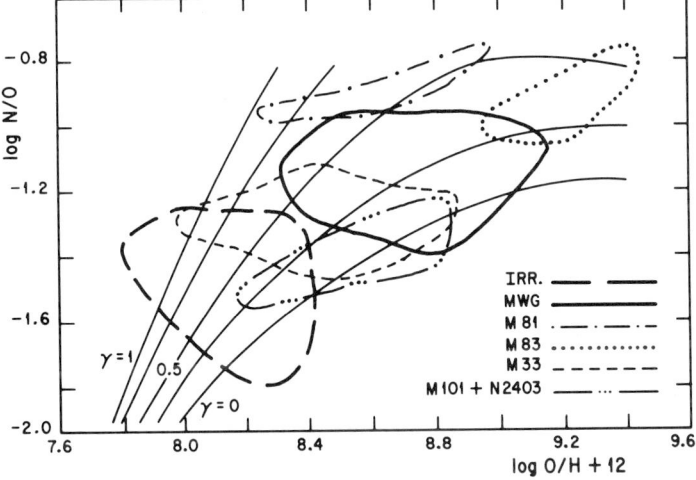

Fig. 3 log N/O vs. log O/H for extragalactic H II regions, irregular galaxies and the Milky Way. The data have been taken mostly from the summary by Pagel (1985), for M81 from Garnett and Shields (1986), and for NGC 2403 from Fierro *et al.* (1986). Superimposed are the models by Serrano and Peimbert (1983) for different values of accretion star formation rate, γ.

Matteucci and Tosi (1986) explain the N/O vs. O/H diagram for dwarf irregular galaxies based on: a) recent models of stellar evolution, b) a Salpeter IMF and galactic winds with a mass loss rate ranging from 0 to 20 times the star formation rate, their models imply that from 30 to 50% of the N is of primary origin. Since the value of the N/C ratio is virtually uniform in nearly all the galactic and extragalactic H II re-

gions, spanning about two orders of magnitude in C/H (log N/C $\sim$ -0.8±
0.15). The value of the H II regions is in agreement with a constant
N/C value derived from field dwarf stars (Laird 1985). The uniformity of
the N/C ratio has led several authors to conclude that most of the N is
of primary origin (Gondhalekar 1983; Dufour 1985a); from the C/N ratio
in dwarf stars of the solar neighborhood, Laird (1985) also concludes
that most of the N is of primary origin. Again, this conclusion is based
on the IRA, or on the hypothesis that C and N have been ejected by the
same type of stars.

Since the IRA does not apply to the enrichment of C and N and since
the stellar mass fractions ejected as newly formed ^{12}C and ^{14}N do not
have the same behavior for stars of different masses, it follows that N
is not necessarily of primary origin. Based on observations of blue
compact galaxies, Pagel (1986) also finds that a considerable fraction
of N is of secondary origin.

C has a primary origin because it is produced from He, alternatively
according to models N is produced in stars of different masses by primary and secondary mechanisms. From observations of planetary nebulae it
has been argued that most of the N produced by them is of secondary
origin (see Serrano and Peimbert 1983 and references therein); moreover
even in the case of planetary nebulae of Type I which show N excesses
larger than the initial C abundance it has been argued (e.g., Peimbert
1985b and references therein) that N is of secondary origin but coming
from O instead of C. An extreme example of this situation is provided
by observations of η Car where it has been found that knots of N and H,
without C and O, are present but where the N/H ratio is equal to the
[(C+N+O)/H]☉ value implying that most of the C and O have been transformed into N (Dufour 1985b). Novae produce N by a primary mechanism
and it has been suggested that they are the main producers of the N enrichment of the ISM (Williams 1982) nevertheless a more recent determination of the N enrichment due to novae indicates that N enrichment of
the ISM is negligible in comparison with that provided by planetary nebulae (Peimbert and Sarmiento 1985).

REFERENCES

Alloin, D., Collin-Souffrin, S., Joly, M., and Vigroux, L. 1979, *Astron.
    Astrophys.,* 78, 200.
Arnett, W.D. and Thielemann, F.K. 1984, in *Stellar Nucleosynthesis*, eds.
    C. Chiosi and A. Renzini (Dordrecht, Reidel), p. 145.
Blair, W.P., Kirschner, R.P., and Chevalier, R.A. 1982, *Ap. J.*, 254, 50.
Bergvall, N. 1985, *Astron. Astrophys.*, 146, 269.
Bruzual, A.G., Peimbert, M., and Torres-Peimbert, S. 1982, *Ap. J.*, 260,
    495.
Burki, G. 1977, *Astron. Astroph.*, 57, 135.
Butcher, H. 1977, *Ap. J.*, 216, 372.
Dennefeld, M. and Kunth, D. 1981, *Astron. Astroph.*, 83, 275.
Diaz, A.I. and Tosi, M. 1984, *M.N.R.A.S.*, 208, 365.
Diaz, A.I. and Tosi, M. 1986, *Astron. Astroph.*, 60, 66.

Dickey, J.M. 1982, *Ap. J.*, 263, 87.

Dufour, R.J., Shields, G.A., and Talbot, R.J. 1982, *Ap. J.*, 252, 461.

Dufour, R.J. 1985a, in *Future of UV Astronomy based on six years of IUE Research*, eds. J.M. Mead, R.D. Chapman, and Y. Kondo (NASA Conf. Publ. 2349), p. 107.

Dufour, R.J. 1985b, private communication.

Dufour, R.J., Shiffer III, F.H., and Shields, G.A. 1985, in *Future of UV Astronomy based on six years of IUE Research*, eds. J.M. Mead, R.D. Chapman, and Y. Kondo (NASA Conf. Publ. 2349), p. 111.

Edmunds, M.G. and Pagel, B.E.J. 1978, *M.N.R.A.S.*, 185, 77.

Edmunds, M.G. and Pagel, B.E.J. 1984a, in *Stellar Nucleosynthesis*, ed. C. Chiosi (Dordrecht, Reidel), p. 341.

Edmunds, M.G. and Pagel, B.E.J. 1984b, *M.N.R.A.S.*, 211, 507.

Fierro, J., Torres-Peimbert, S., and Peimbert, M. 1986, *P.A.S.P.*, in press.

Frenk, C. 1986, private communication.

Gallagher, J.S., Hunter, D.A., and Tutukov, A. 1984, *Ap. J.*, 284, 544.

Garnett, D.R. and Shields, G.A. 1986, *Ap. J.*, submitted.

Gondhalekar, P.M. 1983, *Adv. Space Res.*, 2, 163.

Güsten, R. and Mezger, P. 1983, *Vistas in Astronomy*, 26, 159.

Hulbosch, A.N.M. 1985, in *The Milky Way, IAU Symposium No. 106*, eds. H. van Woerden, R.J. Allen, and W.B. Burton (Dordrecht, Reidel) p. 409.

Huchtmeier, W.K., Seiradakis, J.H., and Materne, J. 1981, *Astron. Astrophys.*, 102, 134.

Huchtmeier, W.K. and Seiradakis, H.H. 1985, *Astron. Astrophys.* 143, 216.

Kwitter, K.B. and Aller, L.H. 1981, *M.N.R.A.S.*, 195, 939.

Lacy, C.G. and Fall, S.M. 1985, *Ap. J.*, 290, 154.

Laird, J.B. 1985, *Ap. J.*, 289, 556.

Lambert, D.L. 1978, *M.N.R.A.S.*, 182, 249.

Larson, R.B. 1972, *Nature, Phys. Sci.*, 236, 21.

Larson, R.B. 1986, *M.N.R.A.S.*, 218, 409.

Lequeux, J., Maucherat-Joubert, M., Deharveng, J.M., and Kunth, D. 1981, *Astron. Astrophys.*, 103, 305.

Lequeux, J., Peimbert, M., Rayo, J.F., Serrano, A., and Torres-Peimbert, S. 1979, *Astron. Astrophys.*, 80, 155.

Mallik, D.C.V. and Mallik, S.V. 1985, *J. Astrophys. Astron.*, in press.

Mathis, J.S., Chu, Y.H., Peterson, D.F. 1985, *Ap. J.*, 292, 155.

Matteucci, F. 1986, *P.A.S.P.*, in press.

Matteucci, F. and Chiosi, C. 1983, *Astron. Astroph.*, 123, 121.

Matteucci, F. and Tosi, M. 1985, *M.N.R.A.S.*, 217, 391.

Matteucci, F. and Tosi, M. 1986, preprint.

Mayor, M. and Vigroux, L. 1981, *Astron. Astrophys.*, 98, 1.

Mc Call, M.L. 1982, Doctoral dissertation, University of Texas.

Mc Call, M.L., Rybski, P.M., and Shields, G.A. 1985, *Ap. J. Suppl.*, 57, 1.

Mirabel, I.F. 1981, *Ap. J.*, 247, 97.

Mirabel, I.F. 1983, *Ap. J.*, 270, L35.

Mirabel, I.F. and Morras, R. 1984, *Ap. J.*, 279, 86.

Mould, J.R. 1984, *P.A.S.P.*, 96, 773.

Pagel, B.E.J. 1985, in *Production and Distribution of C, N, O, Elements,* eds. I.J. Danziger, F. Matteucci, and K. Kjar (ESO, Garching), p. 155.

Pagel, B.E.J. 1986, preprint.

Pagel, B.E.J. and Edmunds, M. 1981, *Ann. Rev. Astron. Astrophys.*, 19, 77.

Peimbert, M. 1979, in *Les Elements et leurs Isotopes dans L'Universe*, (Université de Liege), p. 451.

Peimbert, M. 1985a, in *Star Forming Dwarf Galaxies and Related Objects*, eds. D. Kunth and T.X. Thuan (Editions Frontiere, Paris), p. 403.

Peimbert, M. 1985b, *Rev. Mexicana Astron. Astrofis.*, 10, 125.

Peimbert, M., Peña, M., and Torres-Peimbert, S. 1986, *Astron. Astrophys.*, 158, 266.

Peimbert, M. and Serrano, A. 1980, *Rev. Mexicana Astron. Astrofis.*, 5, 9.

Peimbert, M. and Serrano, A. 1982, *M.N.R.A.S.*, 198, 593.

Peimbert, M. and Sarmiento, A. 1985, *Astronomy Express*, 1, 97.

Peimbert, M., Serrano, A., and Torres-Peimbert, S. 1984, *Science*, 224, 345.

Perinotto, M. and Patriarchi, P. 1980, *Ap. J.*, 235, L13.

Rayo, J.F., Peimbert, M., and Torres-Peimbert, S. 1982, *Ap. J.*, 255, 1.

Renzini, A. and Voli, M. 1981, *Astron. Astrophys.*, 94, 175.

Rogstad, D.H. and Shostak, G.S. 1972, *Ap. J.*, 176, 315.

Sarmiento, A. and Peimbert, M. 1985, *Rev. Mexicana Astron. Astrofis.*, 11, 73.

Serrano, A. 1978, Doctoral dissertation, University of Sussex, U.K.

Serrano, A. and Peimbert, M. 1981, *Rev. Mexicana Astron. Astrofis.*, 5, 109.

Serrano, A. and Peimbert, M. 1983, *Rev. Mexicana Astron. Astrofis.*, 8, 117.

Shaver, P.A., Mc Gee, R.X., Newton, L.M., Danks, A.C., and Pottasch, S. R. 1983, *M.N.R.A.S.*, 204, 54.

Shields, G.A. and Searle, L. 1978, *Ap. J.*, 222, 821.

Smith, H.E. 1975, *Ap. J.*, 199, 591.

Talbot Jr., R.J. 1980, *Ap. J.*, 235, 821.

Terlevich, R.T. and Melnick, J. 1984, private communication.

Tinsley, B.M. 1979, *Ap. J.*, 229, 1048.

Tinsley, B.M. 1981, *Ap. J.*, 250, 758.

Tinsley, B.M. and Larson, R.B. 1978, *Ap. J.*, 221, 554.

Torres-Peimbert, S., Peimbert, M., and Daltabuit, E. 1980, *Ap. J.*, 238, 133.

Tosi, M. 1982, *Ap. J.*, 254, 699.

Tosi, M. 1986, preprint.

Tosi, M. and Diaz, A.I. 1985, *M.N.R.A.S.*, 217, 571.

Trautvetter, H.P., Gorres, J., Kettner, K.U., and Rolfs, C., 1984, in *Stellar Nucleosynthesis*, eds. C. Chiosi and A. Renzini (Dordrecht, Reidel), p. 145.

Twarog, B.A. 1980, *Ap. J.*, 242, 242.

van Woerden, H., Schwarz, U.J. and Hulbosch, A.N.M. 1985, in *The Milky. IAU Symposium No. 106*, eds. H. van Woerden, R.J. Allen, and W.B. Burton (Dordrecht, Reidel), p. 387.

Viallefond, F. 1983, private communication.

THE SOLAR SYSTEM/INTERSTELLAR MEDIUM CONNECTION: GAS PHASE ABUNDANCES

Barry L. Lutz
Planetary Research Center
Lowell Observatory
Mars Hill Road, 1400 West
Flagstaff, Arizona 86001, U.S.A.

ABSTRACT. Gas-phase abundances in the outer solar system are presented as diagnostics of the interstellar medium at the time of the solar system formation, some 4.55 billion years ago. Possible influences of the thermal and chemical histories of the primitive solar nebula and of the processes which led to the formation and evolution of the outer planets and comets on the elemental and molecular composition of the primordial matter are outlined. The major components of the atmospheres of the outer planets and of the comae of comets are identified, and the cosmogonical and cosmological implications are discussed.

1. INTRODUCTION

The solar system is imprinted with the chemical signatures of primordial material 4.55 billion years old. Neither the events which initiated the collapse of the interstellar medium into the protosolar nebula, nor the processes which led to the subsequent formation of the solar system, are well understood; yet the Sun, a cache of planetary bodies, and a swarm of cometary nuclei have evolved out of this primordial material, shielding it from the alchemy of subsequent stellar nucleosynthesis.

These signatures are encoded in the present-day composition of the solar system, but they have been costumed by the varied evolutionary scenerios that led to the formation of each of the different members of the solar family. A significant goal of solar system research is to separate these disguises from the primeval character and discern our physical and chemical origins. The composition of the atmospheres of the outer planets, of their satellites, and of comets contain clues to the search for these origins, and in this review, the current state of our knowledge is summarized.

D. J. Hollenbach and H. A. Thronson, Jr. (eds.), Interstellar Processes, 679–704.

2. ORIGIN AND EVOLUTION OF THE PRIMITIVE SOLAR NEBULA

2.1. Thermal History of the Primitive Solar Nebula

The thermal history of the evolving solar nebula has influenced the chemical distribution of the elements and compounds that were fossilized in the solar system bodies. Thus, the degree to which these bodies have retained the pristine chemical composition of primordial matter is largely uncertain. Heating of the infalling gas as it impinges on the standing shock that bounds the solar core and the exterior nebula can pyrolize complex molecules, while thermochemical reactions driven by the increased temperatures and pressures can alter the equilibrium molecular species that are dominant in the nebula. Similarly, frictional heating of the dust component can vaporize or pyrolize any volatiles that may have accreted as mantles, leaving only the more refractory residues while yielding a source of simple gases.

Spatial and temporal variations in the temperatures and pressures within the accreting nebula can strongly impact these physical and chemical processes. Radial gradients in both are produced within the nebula by the infall process and by differential rotation which provides a channel for the viscous dissipation of energy. These gradients further change as the protosun and its nebula evolve, so a time-dependent analysis is required to formulate a detailed understanding of any alterations to the primordial composition.

In the case of the outer solar system, these effects were minimized. Cameron's (1978) pioneering work, though exploratory rather than definitive, characterizes the physics which is believed to have driven the evolution of the primitive solar nebula. His analysis yielded thermal histories which indicate that beyond the orbit of Jupiter the nebular grains and gas may have never reached temperatures much in excess of 200 K or densities greater than 4×10^{-10} g/cm^3.

Cameron's scenerio which leads to this conclusion assumes the basic physical theory of viscous accretion disks developed by Lynden-Bell and Pringle (1974): the primitive solar nebula was formed by the accretion of highly turbulent, interstellar material. Angular momentum intrinsic to the collapsing interstellar cloud was transferred to the accreting nebula. What triggered the collapse of this cloud is highly speculative; but once initiated, interstellar gas and grains continued to fall into the protosun at nearly free-fall velocities, adding mass to the nebula and increasing its temperature. The grains aggregated during this period and settled into a dust layer at the midplane of a rotating, disk-shaped, solar nebula with the protosun forming at the center.

The period of infall lasted about 4.7×10^4 years during which time the temperatures in the disk rose while its total mass, spread over a radius of several hundred astronomical units, increased to somewhat in excess of one solar mass. At that time a transition from accretion to mass loss occurred, bringing with it similar decreases in radius and temperatures. At first the outer boundaries accounted for most of the loss as the disk continued to spin mass outwards, but as the total disk mass decreased, the inner areas assumed responsibility and the disk radius contracted to the regime of the outer planets.

In the infancy of evolution little mass was at the center of the disk. Bodies with specific angular momenta of the planets today would orbit the central core at radial distances much larger than in the current gravitational configuration. As the central mass of the protosun increased during accretion and into the early phases of mass loss, these areas of specific angular momentum would gradually spiral inwards, asymptotically approaching the orbits of the modern planets. Figure 1, taken from Cameron (1978), summarizes this temporal variation of orbital radii of specific angular orbital momentum for regions which are associated with regions of planet formation.

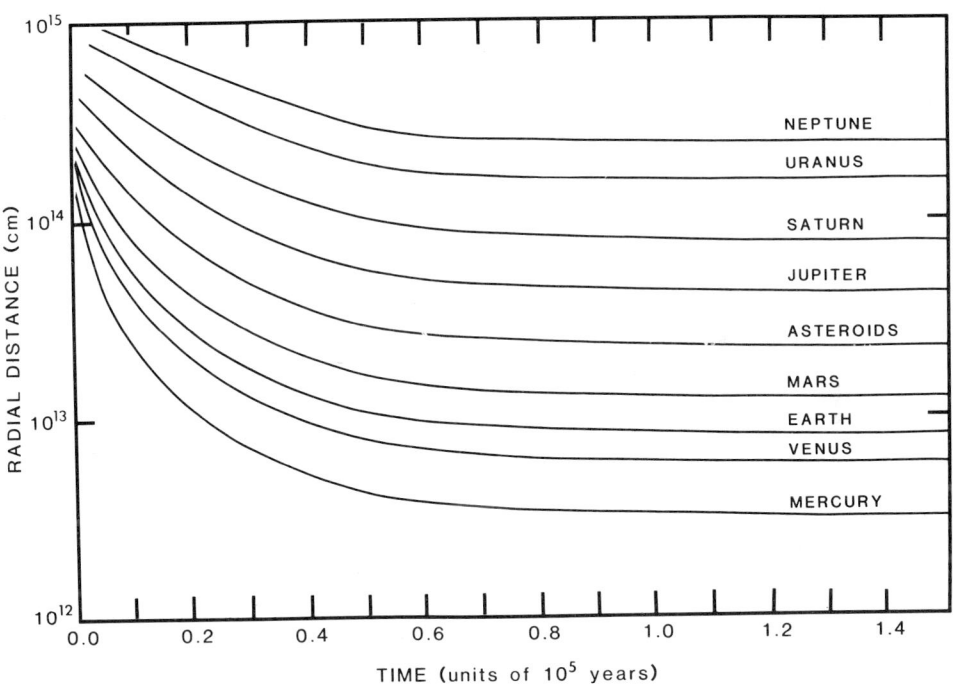

Figure 1. Radial distances of the planet formation regions as a function of time, centered on the specific angular momenta of the present planets (adapted from Cameron 1978).

Figure 2, also extracted from Cameron (1978), takes this analysis one step further, following the thermal history of these regions as they converge to existing planetary orbits. Each of the planets began its birth in the very cold, outer reaches of the primordial nebula, the temperature of which could have been as low as that of the material in the original dense interstellar cloud from which it accreted. As the nebula heated up and the matter in it gravitated towards the protosolar core, the temperatures of these planet-forming regions increased, until mass loss began to cool off the nebular dust and gases and any primitive planetary bodies that had formed from them.

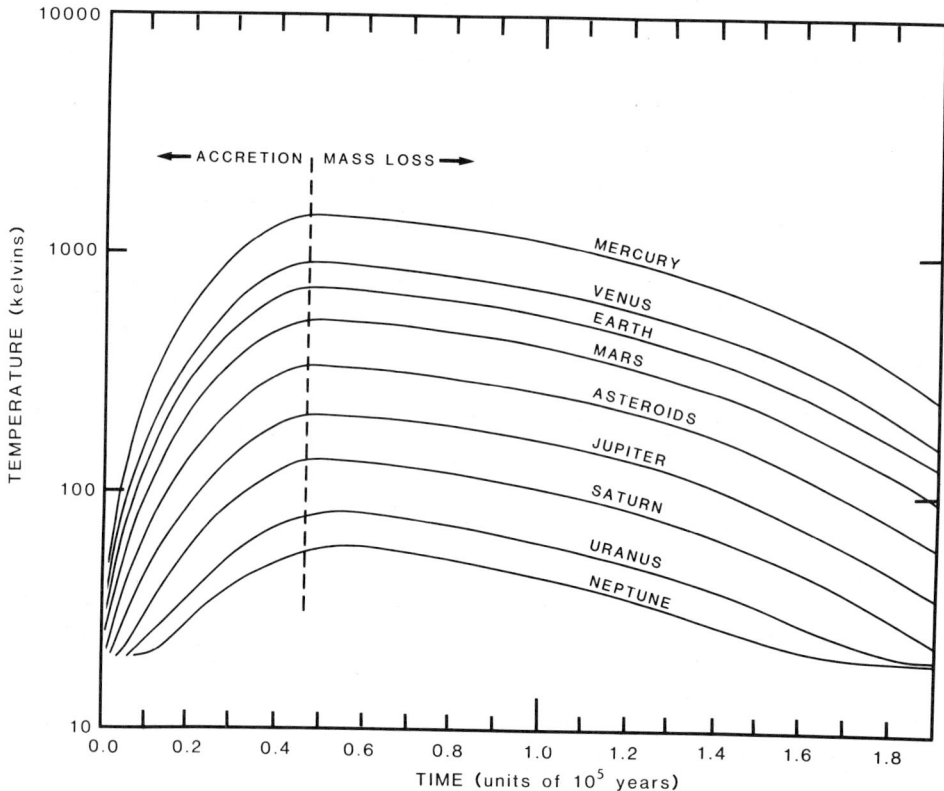

Figure 2. Temperatures in the regions of planetary formation as a function of time for each of the present planets (adapted from Cameron 1978).

Cameron's standard model suggests that the highest nebular temperatures were attained at about the time of the switch from net accretion to net mass loss. At this time, temperatures in the region of Mercury's formation may have reached as high as 1500 K, while temperatures in the regions farther out were progressively lower, topping no more than about 50 K for Neptune's progenitor. Using these parameters as guidelines, we can investigate to what degree physical and chemical processes may have altered the composition of the primordial matter and what influences may be present in the atmospheres of the outer planets, of their satellites and of comets.

2.2. Chemical Evolution of the Primitive Solar Nebula

The degree of physical and chemical processing of the primitive interstellar material which occurred throughout the evolution of the protosolar nebula depends on the radial distance from the protosolar core
and the extent of radial mixing that convection induced. In the hot
inner regions of the nebular disk the mantles on interstellar grains
would have completely evaporated or pyrolized. Even the grains
themselves may have evaporated, although the degree of any such
destruction is determined by the efficiency of transport of the outer,
cooler material into the inner, warm regions and by the time scales for
accumulation into large aggregates which were no longer coupled with
the gas.

As the net nebular heating shifted to net nebular cooling after
completion of the accretion phase, evaporation and pyrolization were
replaced by condensation, either onto any unvaporized refractory
grains, forming new mantles, or directly into liquids or solids, and
aggregation proceeded more vigorously. These complexities in the
thermal and chemical histories and the uncertainties in the extent of
evaporation and re-equilibration make interpretation of observed gas
phase abundances in terms of their original character very difficult in
the hot, inner regions.

In contrast, the ambient temperatures of the material in the
extreme, outer regions never exceeded a few tens of Kelvins. The
interstellar dust and gas suffered neither pyrolization nor significant
evaporation, but rather were preserved in their primordial
compositions. It is in these regions that ices and clathrates, crystal
lattices which serve as cages that trap various atoms and molecules,
would have survived, and, with the grains, would have accreted to form
volatile-rich nuclei of comets throughout the thermal history of
primitive solar nebula.

Between these two extremes lie the giant outer planets, Jupiter,
Saturn, Uranus and Neptune, and their satellites. For regions in which
Uranus and Neptune formed, temperatures probably never rose above about
80 K, and, with the exceptions of those components with high degrees of
volatility such as the noble gases, the primitive composition of the
interstellar grains would have remained unchanged. Thus, in the more
distant, outer regions, volatile-rich planetesimals, representative of
the composition of the interstellar medium of the time, would have
accreted to form the cores of the nether planets. The major species of
any gravitationally trapped gas would also have retained the memory of
their interstellar origins. At low temperatures, gas-phase reactions
involving the abundant molecules in the nebular material are slow
compared to any accretion time scales, so that little change from their
primordial abundances would be expected.

The regions of Jupiter and Saturn formation may have experienced a
more complex chemical history. The volatile components of the mantles

on the grains would have undergone some degrees of evaporation and
pyrolysis, injecting additional gases into the nebular matter, and
leaving the dust more refractory. If there were no transport of this
gas and dust into warmer, inner regions of the nebula where
temperatures exceeded 500 K, few chemical transformations would be
expected because of the slow reaction rates in the gas phase, although
the presence of ions or free radicals could have activated such
changes, including condensation. Convective transport would have
significantly altered this picture.

Large-scale movement of the relatively cool, solid material from
the regions of Jovian planet formation to the warmer inner reaches of
the nebular disk would have furthered the evaporation and pyrolysis of
the mantles, and contributed to the evaporation of the grain material
itself. The reverse processes would have occurred as the warm nebular
matter which replaced the convected cooler material condensed onto
available, unvaporized refractory grains or directly into icy crystals.
Complete evaporation and recondensation of the interstellar material
may have been possible, depending upon the efficiency and extent of the
convective transport, and multiple cycles seem plausible.

Changes in the elemental and molecular compositions of the gas
also would be likely, as the higher temperatures and pressures of the
inner regions would have promoted chemical reactions and determined the
equilibrium densities. As an example, Figure 3, taken from the work of
Lewis and Prinn (1980), shows the dominant carbon, nitrogen, and oxygen
gases that would result from strict thermodynamical equilibrium in a
solar-nebular environment. At temperatures above about 680 K, the
equilibrium kinetics would produce a CO-rich nebula at the expense of
CH_4; above about 330 K, N_2 would dominate NH_3. However, chemical
equilibrium may not have been maintained throughout the nebula, and
below these transition temperatures, conversion to CH_4 and NH_3
dominance may not have occurred.

If the temperatures in the innermost regions of the fledgling
solar nebula were sufficiently high, the chemical time scales would
have been shorter than the cooling time scale and chemical equilibrium
would have been established, yielding CO and N_2 as the prevailing
carbon and nitrogen species. However, chemical time scales generally
increase more rapidly with decreasing pressure than cooling time
scales, so that at some distance farther out in the nebula these two
characteristic times would have been equal, forming what is called a
"quench surface." Beyond this quench surface, chemical equilibrium
could not be established, and the disequilibrium concentrations were
governed by the continuity equations. Furthermore, if the "chemical
length" of a species, that is the e-folding distance for conversion of
one species to another, was less than the mixing length of convection,
then the mixing ratios of the species at the quench surface were
preserved in all regions outside the surface.

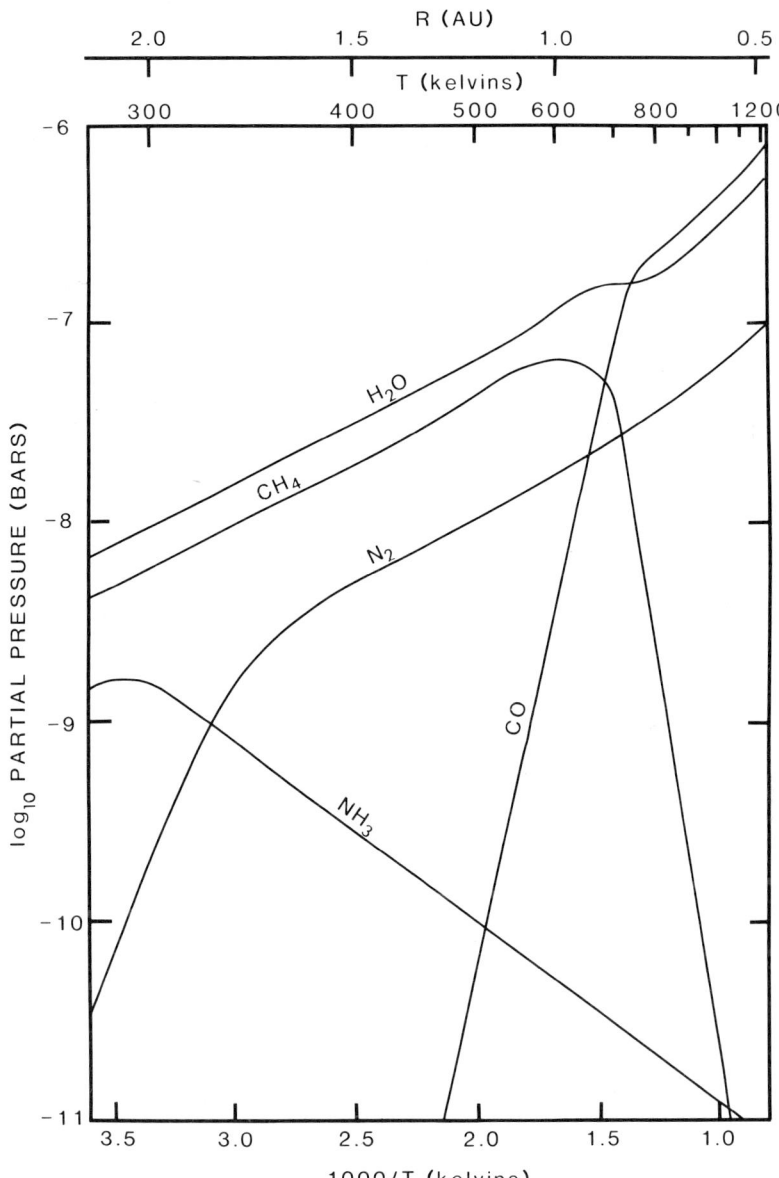

Figure 3. Partial pressures of the major C-, N-, and O-containing gases as function of temperature and distance from the Sun, along a hypothetical solar nebula adiabat (adapted from Lewis and Prinn 1980).

The positions of the chemical time scales and chemical lengths are functions of numerous variables, depending upon the species involved and the chemical reaction paths available, so the position of the quench surface and equilibrium/disequilibrium balance varied with the specific nature of the chemical transformations. But for the conditions in the primitive solar nebula, the quench surface is at about 1000 K for CO to CH_4 transformations and at about 2000 K for N_2 to NH_3. Since the regions for thermochemical transformation lie well outside these surfaces, conversion could not have taken place and only small amounts of CH_4 and NH_3 would have been present in the nebula. Thus, N_2 and CO would have been the dominant nitrogen and carbon gases throughout.

Chemical conversions in the solar nebula are not limited to gas phase channels, but also various reactions can be catalytically promoted on the grain surfaces, depending upon the details of the grain size and composition and upon the specific nebular environment. In the case of the example of the carbon-nitrogen-oxygen chemistry, Lewis and Prinn (1980) also have shown that the presence of stable metallic iron grains (which could exist in the temperature regime of about 680-1340 K) would promote the reduction of N_2 to NH_3 and CO to CH_4, but current (and very model-dependent) estimates of the efficiencies of these processes suggest they are too low to affect the chemical equilibrium in any substantial way.

Similarly, adsorption of the H_2 and CO, and of any residual NH_3 on various grain surfaces could lead to the spontaneous synthesis of numerous compounds including H_2O and CO_2 through Fischer-Tropsch type processes proposed to have occurred in the solar nebula (Anders, Hayatsu and Studier 1973). Such compounds may have been retained on the grains, rather than ejected back into the gas phase, and accreted with the grains as they formed the planetesimals during the cooling of the nebular matter. Like catalysis of gas phase material on grain surfaces, theoretical modelling of these processes is extremely difficult, and laboratory studies fall short in approximating a nebular environment, so that any conclusive evidence that these processes played a role in the chemical evolution of the the primitive solar nebula has been elusive.

Other physical and chemical processes could have altered the chemical composition of the solar nebula throughout its formative years, but our present knowledge of ambient conditions during this primitive period is too limited to assess the roles of these processes with any confidence. Photochemical reactions driven by ultraviolet radiation from the young Sun could have promoted disequilibration in the inner regions, but the rates of dissociation would have depended upon the solar uv output at the time, a rather controversial topic. Independent of the solar uv output, thermochemical reactions involving the nonequilibrium species in the regions of higher densities and temperatures (T > 1000 K) would have been so rapid as to minimize any net yield. Products from photochemical dissociations may have been able to survive the thermochemistry of the tenuous and cool, outer regions where the photochemistry would have been driven by the light from stars surrounding the nebular disk. However, dissociation rates resulting from the interstellar uv flux would have been orders of magnitude smaller than those resulting from solar uv. Where or how the transition from the solar-uv-dominated zone to the interstellar-uv-dominated regime was effected may have determined the overall influence of photochemical reactions on the chemical evolution of the solar nebula. That transition was strongly coupled to time-dependent composition of the gas, dust and ices in the solar disk, the influence that this composition had on the radiative transfer characteristics as the nebula evolved, especially the uv opacity, and the roles that convection and turbulence played in the mixing of the hot and cold regions.

Quantitative assessments of the influences of ionizing radiation and of electrical discharges and their associated acoustic shocks on the gas and grain material are similarly lacking. The high opacity of the protosolar nebula would have severely limited the effectiveness of cosmic rays in all but the outer layers of the nebula, but evidence of ^{26}Al in the early solar system opens the door for the presence of possible internal sources of high-energy positrons and gamma rays which could have initiated ion-molecule reactions. Lightning and thunder shocks resulting from charge separation induced by turbulent mass transport could have potentially released enormous amounts of energy which could have gone into chemical transformations, but quantification of the problem and of the consequences has yet to be done.

3. ORIGIN AND EVOLUTION OF THE OUTER PLANETS AND COMETS

3.1 Planetary Formation in the Primitive Solar Nebula

The uncertainty in our knowledge of the effects of the thermal and chemical histories of the primitive solar nebula increase rapidly for those regions of planetesimal formation approaching the inner disk. In the outer extremes where these planetesimals aggregate to form cometary nuclei, the record of our primitive origins is best preserved. Uranus and Neptune probably contain similarly pristine primordial material, while the composition of Jupiter and Saturn may or may not have undergone some degree of chemical transformation, although the primordial elemental abundances were probably retained. However, this primeval information is inexorably interwoven with the scenerios that led to the formation of these outer solar system objects, and our ability to extrapolate observations of a few species in the outer layers of inhomogeneous bodies to general bulk properties is limited.

If indeed comets formed in the dark outskirts of the primitive solar nebula, say 100 to 1000 AU from the protosun, it was by cold accretion of the icy grains dragged in from the interstellar medium by the collapsing solar nebula. Such nuclei never would have accreted sufficient mass to gravitationally capture any of the nebular gases to form an atmosphere. As resources of pristine primitive matter, they would consist of the ices and refractory materials that made up the cold, solid matter of the interstellar medium, deficient in hydrogen and helium, and in other highly volatile gases which did not condense out at the ambient temperatures.

An alternate scenerio suggests that comets could have evolved from a direct condensation of the cooling gas and dust within 50 AU of the protosun, perhaps as close as 5 AU. In this case, the nuclei also would have been too small to have retained any solar composition atmosphere, but inner solar system thermal and chemical histories could have changed the elemental and molecular abundances resident in the primeval material, similar to the possible effects of these histories on the makeup of the outer planets.

Until the era of spacecraft encounters, our knowledge of this cometary material and tests of these formation options were largely based on circumstantial inferences from remote observations of transient phenomena induced by varying solar insolation as the icy conglomerates approached and then receded from the Sun. The nucleus of a comet and its physical and chemical natures were hidden from view by the dust and gases released, the chemical structures of which were altered by photodissociation, thermochemical reactions, and interactions with the solar wind. Even with **in situ** data such as retrieved by the spacecrafts **ICE, Giotto, Suisei, Sakigake,** and **Vegas 1** and 2, the actual composition of the nucleus must be reconstructed from the atomic and molecular fragments left in the wake of these processes, making the interpretation of these data a challenge to our skills.

Two similar formation scenerios can applied to the outer planets. In the first, the giant planets are assumed to have formed by gravitational condensation of a local instability in the primordial nebula. The planets underwent hydrodynamical collapse, and then settled into the slow contraction of hydrostatic equilibrium which led to the state they are in today. In such a scenerio, the main components of the atmospheres would have retained protosolar composition, since the planets would have remained homogeneous during the gravitational collapse and hydrostatic contraction.

In the second scenerio, called the nucleation model, the region of planet formation occurred at low temperatures and began with the accretion of the ices, clathrates, and refractory grains of the environment to form a core. When this solid core of rocks and ices had reached a sufficent mass, it would have captured an atmosphere of nearly nebular composition, including H_2, N_2, CO and the noble gases as well as any trace components which had not condensed. Accretional heating would have revaporized some of the ices and clathrates, enriching this envelope in less volatile gases such as CH_4, NH_3 and H_2O. Recent models of the interiors of the outer planets favor large cores, and for Uranus and Neptune large fractions of ices are required, in congruence with this nucleation scenerio. Saturn's satellite Titan more certainly fits this theory, with its atmosphere entirely the result of the revaporization of the core ices. However, it was probably formed in the proto-Saturnian nebula, under conditions which could have been significantly different from those in the solar nebula.

3.2. Chemical Evolution of Planetary Atmospheres

Both the homogeneous collapse and nucleation models predict atmospheres of nearly nebular composition for the outer planets, with temperature and pressure gradients that insure further chemical processing. The atmospheres of Jupiter and Saturn are believed to be convective to great depths (and temperatures) where thermochemical equilibrium is established, so that the assumption of adiabatic structure becomes the basis for determining their bulk chemical compositions.

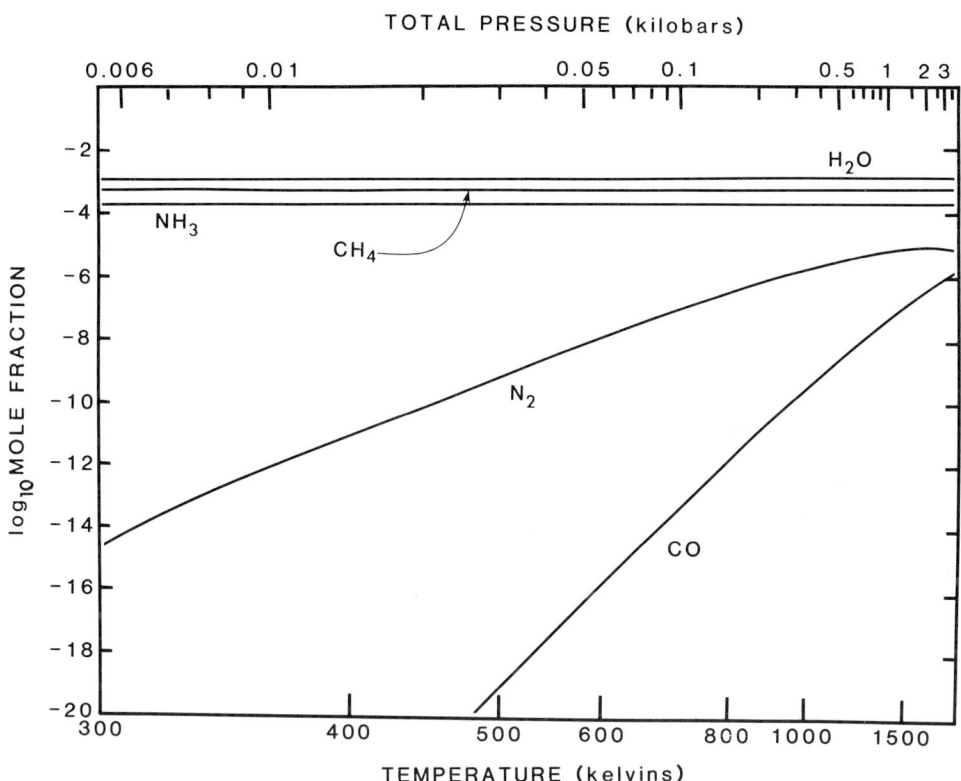

Figure 4. Equilibrium abundances of the major C-, N-, and O-containing gases along a Jupiter adiabat (adapted from Barshay and Lewis 1978).

Barshay and Lewis (1978), assuming a solar composition for the Jovian nebula, calculated the equilibrium abundances for the abundant elements along a typical Jovian adiabat, and demonstrated that the atmosphere would have undergone reduction, converting nebular N_2 to NH_3 and nebular CO to CH_4. Figure 4, which was extracted from Barshay and Lewis's Figures 2 and 3, characterizes the equilibrium abundances of

the major carbon-, nitrogen- and oxygen-containing gases along their
adopted adiabat, summarizes this result. Although these calculations
have been carried out only for levels between 300 K (approximately the
6 bar pressure level) and 2000 K (approximately 4 kbar), strong
vertical convection provides a uniform mixing throughout the
troposphere.

No such calculations specific to any of the other outer planets
have yet been published, but Prinn **et al.** (1984) have scaled the Jovian
results for application to Saturn. Equilibrium thermochemistry should
have also resulted in qualitatively similar, reduced atmospheres of
Uranus and Neptune, but the quantitative details are uncertain because
of their different thermal structures.

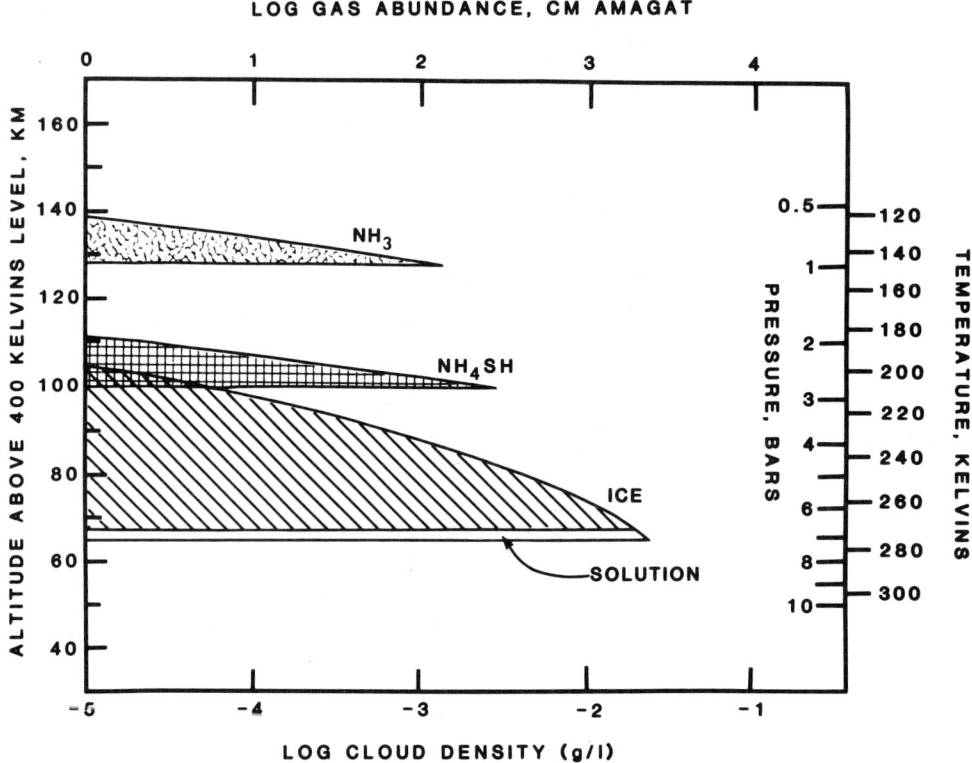

Figure 5. Nominal atmospheric profile for a solar composition Jupiter
showing the computed cloud densities (adapted from Weidenschilling and
Lewis 1973).

In all these cases, a complication arises as the warmer material
from the deep atmosphere cools in its upward travels and some of the
gases condense out as clouds. These condensibles deviate from the uniform mixing assumption, making the interpretation of their gas-phase
abundances in terms of the primordial elements more difficult, although
the total gas-cloud composition adheres to the primordial

concentrations. Figure 5, based on the work of Weidenschilling and Lewis (1973), illustrates the principle cloud layers expected for a solar composition Jupiter with their nominal atmospheric pressure-temperature profile. NH_3, the principal nitrogen-bearing compound in the atmosphere, and H_2O, the principal oxygen-bearing compound, constitute these layers in various chemical and aqueous solution forms, thus depleting their contributions to the atmospheric gases. Weidenschilling and Lewis have also computed such wet adiabatic models for Saturn, Uranus, and Neptune. For both Uranus and Neptune, their models predict methane clouds in their upper tropospheres if CH_4 is greatly enriched with respect to the solar carbon abundance.

Titan is a special case. Its atmosphere is almost entirely secondary, derived from the outgassing of the volatiles contained in the rocks and ices of the nebula out of which the satellite formed. But which nebula: solar or Saturnian? In the solar nebula the dominant carbon and nitrogen forms were CO and N_2, and Strobel (1981) argued that a plausible case can be made for a N_2-dominated atmosphere, derived from a primordial N_2-clathrate hydrate trapped in Titan's core.

On the other hand, Prinn and Fegley (1981) showed that radial mixing and cooling in the Saturnian nebula may have been sufficiently rapid to convert these nebular gases to CH_4 and NH_3, suggesting these gases would be the dominant ones contained in Titan's ices. However, Atreya, Donahue, and Kulin (1978) demonstrated that if NH_3 were the dominant nitrogen-bearing component trapped in the nebular ices, then photochemical processes initiated by solar ultraviolet radiation would have irreversibly converted any of it that was outgassed to N_2, if the surface temperature of the satellite were in excess of 150 K at some time during its evolution. The present thermal structure of Titan's atmosphere would inhibit this conversion channel, but if Titan did have a warmer period during its early phase a N_2-dominated atmosphere would also result.

Whether or not photochemical processes led to N_2 as the dominant nitrogen-bearing form in the atmosphere of Titan, Strobel pointed out that they are important to the hydrocarbon chemistry in it. A more complete investigation of these processes by Yung, Allen and Pinto (1984) yields the conclusion that over geologic time, Titan's atmosphere lost significant amounts of CH_4, N_2 and CO, making the interpretation of its present atmospheric composition in terms of primordial abundances difficult. For the outer planets, such disequilibrating processes also operate (cf. Prinn and Owen 1976 and Prinn **et al.** 1984), albeit only in the upper atmosphere, and they must be considered when interpreting the abundances of trace species.

These uncertainties in the routes of formation of the outer planets, their satellites and comets adds to the mystery of interpreting modern-day observations of the solar system. One insight into unraveling this mystery that derives from the exercise of this and the previous sections is that, for the most part (some care must be taken with Titan), the primordial elemental abundance ratios that existed in the primitive solar nebula have been unchanged by the thermal and chemical evolution of the nebula and by the formation of the planets; they have been only disguised in chemical transformations that have shifted their distributions into the present-day atomic and molecular compositions. And within this framework of disguises lie clues, both to the primitive matter from which we evolved and to the formational processes that led to our present existence. In the following section, the gas-phase abundances of the outer solar system are discussed along with the constraints that these abundances pose to cosmogonic and cosmologic theories.

4. GAS PHASE ABUNDANCES IN THE OUTER SOLAR SYSTEM

4.1. The Atmospheres of the Outer Planets

The outer planets Jupiter through Neptune all possess atmospheres whose compositions are overwhelmingly dominated by H_2, He, and CH_4 (Prinn and Owen 1976; Trafton 1981; Prinn et al. 1984). NH_3 is a significant minor constituent on Jupiter and Saturn but its contributions to compositions of Uranus and Neptune are undetected.

These observations are qualitatively consistent with our expectations, based on the evolutionary scenarios outlined in the previous sections. The bulk of the atmospheres should consist of primordial H (in the form of H_2) and He, the primary components of the interstellar medium that accreted to form the protosolar nebula. The less abundant elements, carbon and nitrogen, would have been chemically processed by the primitive solar nebula and then again in the planetary atmospheres to the fully reduced states of CH_4 and NH_3. Clouds of ammonia-based compositions would have depleted the upper tropospheres (from which most of our observations are derived) of the more condensible NH_3 gas, with higher degrees of completion as the atmospheres become progressively colder, farther out in the solar system. CH_4 should be free of the effects of condensation, although thick methane clouds could form on Uranus and Neptune if CH_4 is greatly enriched. The observations do not appear to support such clouds.

On the basis of these arguments, H_2, He, and CH_4 should possess uniform mixing ratios throughout each of the outer planet atmospheres, and provide reliable estimates of the elemental abundance ratios He/H and C/H in them. Figure 6 summarizes the helium mass fractions for Jupiter, Saturn and Uranus, as determined by the **Voyager** encounters (Gautier **et al.** 1981; Conrath **et al.** 1984; Hanel **et al.** 1986) and compares these values with the most recent estimate of the solar abundance derived from helium emission lines (Heasley and Milkey 1978). There is considerable controversy over the solar value, and a large scatter exists between measurements by different techniques (cf. Gautier and Owen 1983a and Conrath **et al.** 1984), but estimates based on solar evolutionary models and on measurements of solar oscillations appear to support this emission line result.

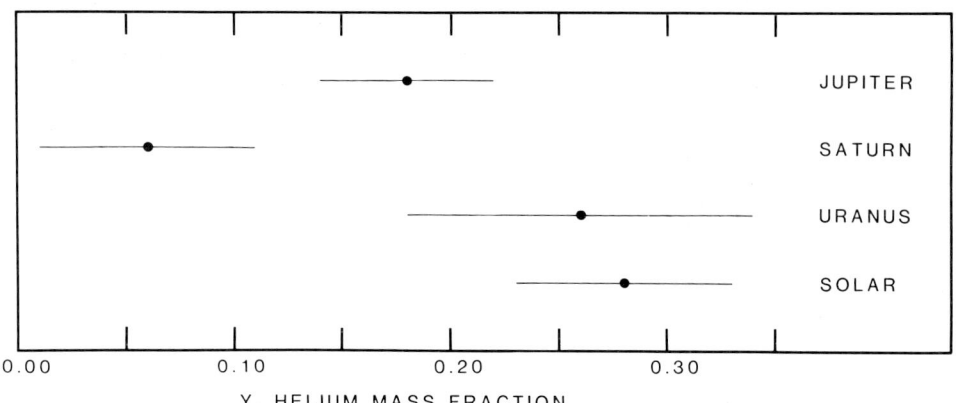

Figure 6. Helium mass fractions measured in the atmospheres of Jupiter, Saturn, and Uranus by **Voyager** compared to the solar value derived from helium line emission. See text for references.

Figure 7 provides a similar comparison of the methane-to-hydrogen mixing ratios, CH_4/H_2, for Jupiter (Gautier and Owen 1983b), Saturn (Courtin **et al.** 1984), Uranus (Baines and Bergstralh 1986), and Neptune (Bergstralh and Neff 1983). The "solar" value is twice the solar C/H ratio (Lambert 1978). The Jovian and Saturnian ratios were obtained from **Voyager** measurements, while those for Uranus and Neptune were derived from ground-based observations.

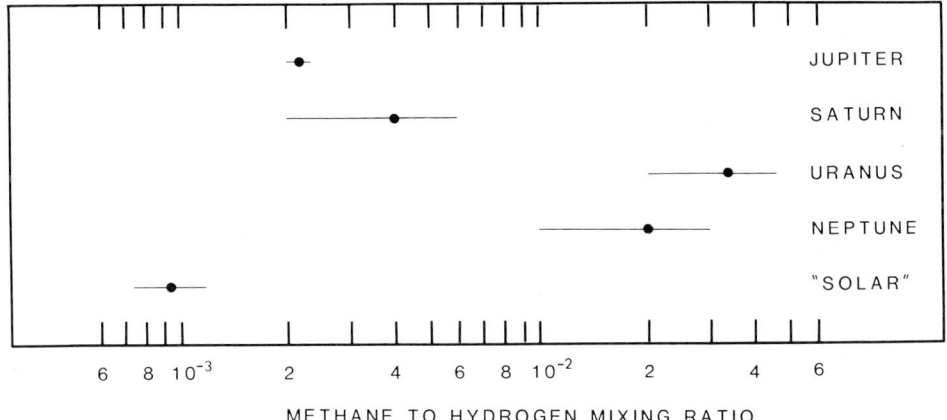

Figure 7. Methane-to-hydrogen mixing ratios observed in the atmospheres of the outer planets compared to the value corresponding to the solar carbon abundance. See text for references.

Titan's atmosphere is composed primarily of N_2 (65-98%) and CH_4 (2-10%), with a substantial fraction (up to 25%) of a heavier gas possibly required by the measured tropospheric mean molecular weight of 28.6 amu (Hunten **et al.** 1984). Argon is the most likely candidate for this constituent. The atmosphere contains only about 0.2% H_2. Since Titan's atmosphere is secondary and its constituents are believed to have suffered differential losses over geologic time, these mixing ratios are not directly comparable to those observed in the outer planet atmospheres.

Deuterium has been detected on Jupiter, Saturn, Uranus, Neptune and Titan. Except on Titan, the most abundant form is HD, but it has been observed also in the singly substituted form of methane, CH_3D, in all but Neptune (de Bergh **et al.** 1986). Owen, Lutz, and de Bergh (1986) have suggested that because of the difficulties inherent in the analysis of HD observations, the currently most reliable measures of D/H in the outer solar system may be those derived from CH_3D/CH_4. For Titan, only CH_3D provides a D/H ratio.

Figure 8, taken from Owen, Lutz, and de Bergh's (1986) paper, compares the D/H ratio in methane $[D/H = (1/4)(CH_3D/CH_4)]$ in the atmospheres of the outer planets and Titan with an estimate of the protosolar value derived from studies of ^3He in meteorites and in the solar wind (Geiss and Reeves 1981) and with a value for the interstellar medium (Vidal-Madjar **et al.** 1983). Planetary D/H ratios deduced from the HD/H_2 mixing ratios appear to be nearly all the same, about 1×10^{-4} (Cochran and Smith 1983; Smith and Macy 1986).

4.2. Cometary Abundances

Our knowledge of the chemical composition of comets is still in its infancy. We have no direct information on the composition or structure of a cometary nucleus, and it is only the recent armada of spacecraft to Comet P/Halley that has given us the first pictures of one. Even with this addition of **in situ** measurements of a cometary environment, our data bases for compositional studies are still limited to observations of secondary products released from the surface by interaction of solar radiation with the nuclear ices and grains. We are still left with the formidable task of working backwards from these daughter products (sometimes granddaughter species) to identify their parents and establish the physical and chemical makeup of the primordial icy conglomerate. This difficult analysis is further complicated by the fact that cometary atmospheres are transient phenomena, with time scales as short as minutes and perhaps as long as months, depending upon the particular comet studied.

In Table I, the observed species in cometary spectra are summarized. For no single comet have the production rates of all these species been measured, nor is there any comet for which a complete inventory has been published for a single date. The results of the International Halley Watch, whose efforts were to obtain such a comprehensive data base during the 1985/86 apparition of Comet P/Halley, are not yet in, but at most it will detail the physics and chemistry of only one comet which may or may not be representative of the comet population.

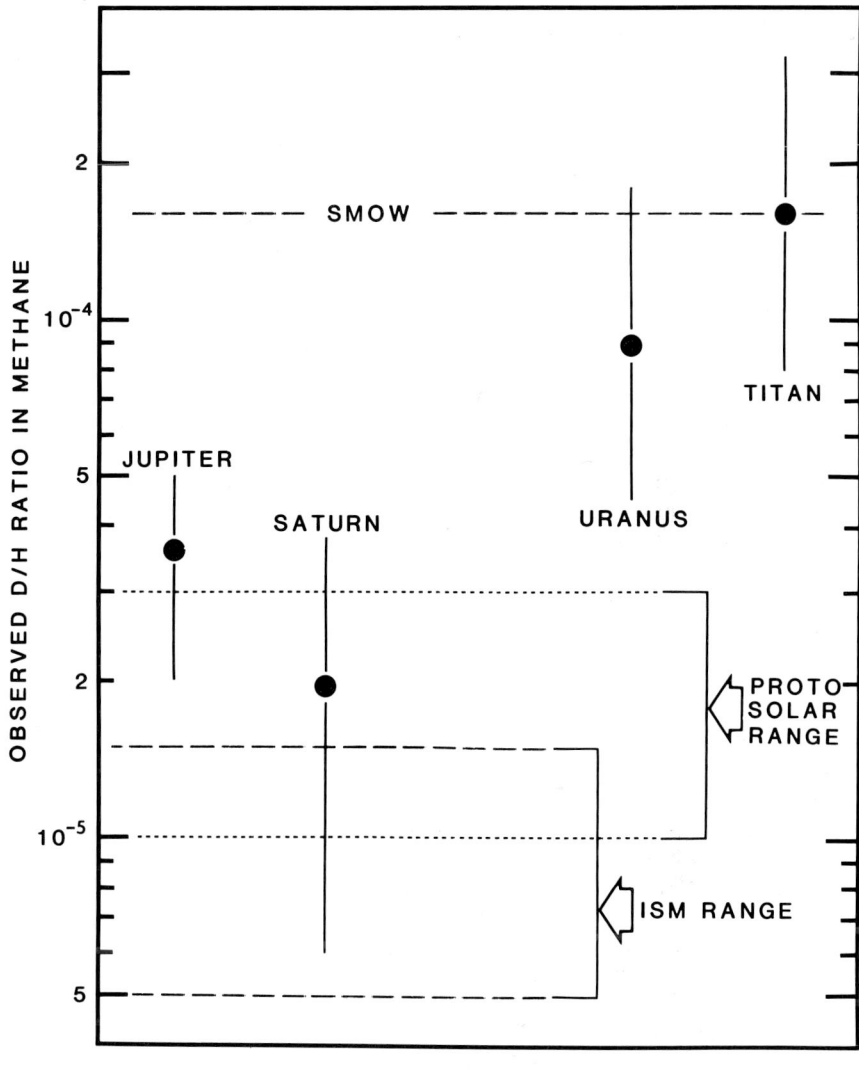

Figure 8. Stoichiometric D/H ratios in the methane component of Jupiter, Saturn, Uranus and Titan compared to the range of values estimated for the protosolar nebula and the interstellar medium. SMOW refers to the Standard Mean Ocean Water value on Earth (adapted from Owen, Lutz, and de Bergh 1986).

An interim approach to characterizing the chemical composition of cometary nuclei is to build a heuristic model which represents the average of a group of comets. Delsemme (1982) developed this approach using recent bright comets as his representative group, and his estimates of the mean abundances of the light elements in cometary

TABLE I. Observed Species in Cometary Spectra

Organic	Inorganic	Metals	Ions	Dust
C	H	Na	C^+	Silicates
C_2	NH	K	CO^+	
C_3	NH_2	Ca	CO_2^+	
CH	O	V	CH^+	
CN	OH	Mn	H_2O^+	
CO	H_2O	Fe	OH^+	
CS	S	Co	Ca^+	
HCN	S_2	Ni	N_2^+	
CH_3CN		Cu	CN^+	

volatiles are given in Table II and compared with the cosmic (solar system) abundances proposed by Cameron (1982). Assuming a C I chondritic elemental composition for cometary dust and a dust-to-gas mass ratio of 0.8, Delsemme also computed total elemental abundance ratios for the total nuclear conglomerate. These too are provided in Table II.

TABLE II. Elemental Abundances in Comets

Element	Cometary gas	Gas and dust	Cosmic
H	1.50	1.09	1445.7
C	0.20	0.17	0.64
N	0.10	0.07	0.13
O	1.00	1.00	1.00

Although cometary material represents our best chance to explore the primitive matter that was contained in the protosolar nebula, matter that has been least affected by the processes of formation and evolution of the solar system, its study is an area least developed and understood. New and continuing efforts, dually focused on establishing statistically significant data bases of ground-based and earth-orbiting observations, and on carrying out more spacecraft encounters, especially an orbiting rendezvous with surface probes, are essential to fundamental progess in this area. Without them this chance may be lost.

4.3. Cosmogonic Implications

The elemental abundance ratios He/H, C/H and D/H, derived from the observations of He, CH_4 and CH_3D discussed in section 4.1 are given in Table III. Although these ratios are qualitatively consistent with the formation and evolutionary scenerios outlines in sections 2 and 3, there are surprising, large quantitative differences among the various solar system members. In many cases the differences are considerably larger than the uncertainties of measurement.

Values for the He/H ratio span a factor of three, ranging from a nominal low of 0.034 for Saturn to a nominal high of 0.100 for the Sun, with probable errors that support the difference as real. Similarly, the He/H ratio derived for Jupiter is substantially lower than the solar value, although the difference is not as extreme as for Saturn. Uranus is in agreement with the solar value.

TABLE III. Solar System Abundance Ratios

Object	He/H	D/H*	C/H
Jupiter	0.056 ± 0.012	$(2.6\pm1.0)\times10^{-5}$	$(1.09\pm0.09)\times10^{-3}$
Saturn	0.034 ± 0.027	$(1.5^{+1.5}_{-1.0})\times10^{-5}$	$(2\pm1))\times10^{-3}$
Titan	.....	$\mathbf{(1.65^{+1.65}_{-0.8})\times10^{-4}}$	.....
Uranus	0.088 ± 0.030	$(9^{+9}_{-4.5})\times10^{-5}$	$(1.6\pm0.6)\times10^{-2}$
Neptune	.....	.....	$(1.0\pm0.5)\times10^{-2}$
Solar	0.100 ± 0.018	$(2\pm1)\times10^{-5}$	$(4.67^{+1.21}_{-0.96})\times10^{-4}$

*Values in boldface refer to D/H in the methane component of the atmosphere. See text.

In any scenerio for the formation of these planets, both hydrogen and helium are assumed to exist purely in gaseous form in the primitive solar nebula because of their high vapor pressures at low temperatures, and no fractionation of He is expected (Hubbard and MacFarlane 1980). However, Smoluchowski (1967) suggested that, during subsequent evolution, differentiation of He from H could have occurred since over the range of temperatures and pressures appropriate to the interiors of Jupiter and Saturn the gases could be immiscible. The low values of He/H for these planets can be taken as evidence that this differentiation has begun on both. No such fractionation of He is expected during the evolution of Uranus, and the apparent agreement between the solar He/H ratio and the He/H ratio in its atmosphere seems to support the contention that the He content of the Uranian atmosphere corresponds to the helium abundance in the protosolar nebula.

C/H ratios derived from the methane-to-hydrogen mixing ratios, $[C/H = (1/2)(CH_4/H_2)]$, are enhanced over the solar value by a nominal factor of 2.3 on Jupiter and by about twice as much on Saturn. The estimated uncertainties in these factors indicate that enrichments of carbon are certain for these atmospheres, thus providing a strong argument for the nucleation model of planetary formation. The much larger enhancements of C/H in the atmospheres of Uranus and Neptune, by an order of magnitude or more over the solar ratio, further strengthen this argument. Dilution of the enriching gases emanating from the large icy cores by their thinner hydrogen envelopes would be much less than on Jupiter or Saturn.

Although CH_4 in the atmosphere of Titan provides no comparable insight to the formation of the planets, it does add an additional constraint on the composition of the nebular material in its environment

at the time of its formation. It is most likely derived from methane
ices and clathrates that accreted to form the core of this Saturnian
satellite, suggesting that CH_4, not CO, was the dominant carbon-bearing
molecule in the forming nebula. This conclusion could be interpreted as
support of Prinn and Fegley's (1981) contention that the
proto-Saturnian nebula had converted the protosolar N_2 and CO to NH_3
and CH_4, but an alternative interpretation is that the radial mixing in
the protosolar nebula was insufficient to have effected the conversion
of the infalling carbon- and nitrogen-bearing gases to N_2 and CO
throughout, and that the observed CH_4 is of interstellar origin. The
dominance of N_2 in Titan's atmosphere cannot distinguish between these
two options, since it can be obtained from either N_2-clathrate hydrates
or NH_3 ices in Titan's core.

With the exception of Titan, D/H ratios in methane, shown in
Figure 8, are not direct measures of the bulk D/H ratios in the outer
planet atmospheres. They sample only a small fraction of the total
atmospheric deuterium, and a correction must be applied to account for
deuterium exchange between methane and the hydrogen where most of the
deuterium resides. The sense of this fractionation is to enhance the
concentration of deuterium in methane at the expense of that in
hydrogen, and the degree of enrichment over the stoichiometric value,
$[(1/4)(CH_3D/CH_4)]$, is a strong function of temperature. At high
temperatures, the stoichiometric value is asymptotically approached,
while at low temperatures D/H in methane can exceed that in hydrogen by
orders of magnitude, if there has been sufficient time for the exchange
reactions to have reached equilibrium.

Beer and Taylor (1973) considered this problem for Jupiter and
calculated that there should be an enhancement of D/H in methane by a
factor of 1.37 over the D/H in hydrogen. Homologous arguments suggest a
similar amount of enrichment of deuterium in methane is expected for
Saturn. Application of this correction factor to the CH_3D measurements
for these two planets yields the bulk D/H ratios for each given in
Table III. These ratios are in excellent agreement with the value
deduced from the ^3He measurements, listed as solar, and they appear to
be representative of the deuterium content in the gaseous component of
the protosolar nebula. The D/H ratios of 10^{-4} derived from observations
of HD and H_2 in the atmospheres of Jupiter and Saturn do not appear
consistent with either. This discrepancy between the CH_3D and HD
results is possibly due to an underestimation of the effects of
scattering processes on the very weak HD absorptions and to
contamination of these absorptions by very weak methane features.

No correction factors have been applied to the D/H ratios for
Uranus and Titan in Table III. Methane is the major source of deuterium
in Titan's atmosphere and its stoichiometrically derived D/H ratio is
the bulk atmospheric value. This measured ratio, 1.65×10^{-4}, is enhanced
over the protosolar (i.e., the Jovian/Saturnian/^3He) value by a factor
of 8. Deuterium enrichment in Titan's atmosphere could be effected by
the dissociation of methane and subsequent differential escape of H

over D, but Pinto **et al.** (1986) showed that over geologic time, the maximum enhancement of D/H due to this process would be a factor of 2. Owen, Lutz, and de Bergh (1986) concluded that, within this factor of 2, the D/H ratio in Titan's methane corresponds to the methane ices trapped in its core which had been enriched in deuterium already in the protosolar nebula. They further argued that this enrichment in the primitive methane ices was evidence of two distinct reservoirs of deuterium in the interstellar medium, set before the formation of the solar system: one, the gaseous hydrogen component, characterized by the atmospheres of Jupiter and Saturn, the other, the volatiles frozen in the cold, solid component.

Uranus appears to lack the interior heat sources to drive convection sufficient to establish deuterium-fractionation equilibrium between hydrogen and any methane that emanated from its core. Since D/H in methane may be uncoupled from that in hydrogen, the D/H listed for Uranus in Table III is also derived stoichiometrically from CH_3D/CH_4. The D/H ratio in Uranus's atmospheric methane appears to be enhanced by a factor of 4.5. The only plausible process which could be responsible for this enhancement is the outgassing from deuterium-enriched methane ices in the core of Uranus, again indicating that deuterium enrichment had been established in the protosolar nebula before the formation of the solar system. Within the uncertainties, the D/H enhancement on Uranus is indistinguishable from that observed on Titan, but the the nominal values are suggestive of nearly a factor of 2 less enhancement on Uranus. Dissociation of methane and the differential escape of hydrogen over deuterium do not play a significant role in Uranus's atmospheric chemistry and could perhaps account for any such difference.

The conclusion that the CH_3D in Uranus's atmosphere derives from primordial ices also implies that methane existed in the protosolar nebula in the region of Uranus's formation. The presence of CH_4 requires that nebular CO was converted to CH_4 in the vicinity of Uranus's formation, or that radial mixing in the protosolar nebula was not sufficient to convert interstellar CH_4 to CO. Since the first of these options seems unlikely, it is possible that the methane ices in the cores of both Uranus and Titan are of interstellar origin.

4.4. Cosmologic Implications

Gas-phase abundances in the present-day solar system serve not only as unique probes of the protosolar nebula and the interstellar medium at the time of formation of the solar system, but also provide effective diagnostics of the evolution of our galaxy and of the primordial abundances at the time of its creation. Both Kunde **et al.** (1982) and Gautier and Owen (1983a) used the D/H ratio derived for Jupiter as a fiducial point to discriminate between the viabilities of two classes of chemical evolutionary models for the galaxy. They concluded that infall models in which extragalactic gas of primordial big-bang composition is assumed to enter the galaxy during evolution seem to be excluded by the Jovian observations.

Figure 9 is adapted from Figure 2 of Gautier and Owen (1983a), but the evolutionary abundance curves for Audouze and Tinsley's (1974) infall and no-infall models have been replaced by evolutionary abundance envelopes for a range of D/H ratios at the time of solar system formation consistent with the Jovian and Saturnian values listed in Table III. For comparison, the D/H ratio deduced from ^{3}He is also shown, as is the interstellar medium value shown in Figure 8, which corresponds to D/H at the current epoch.

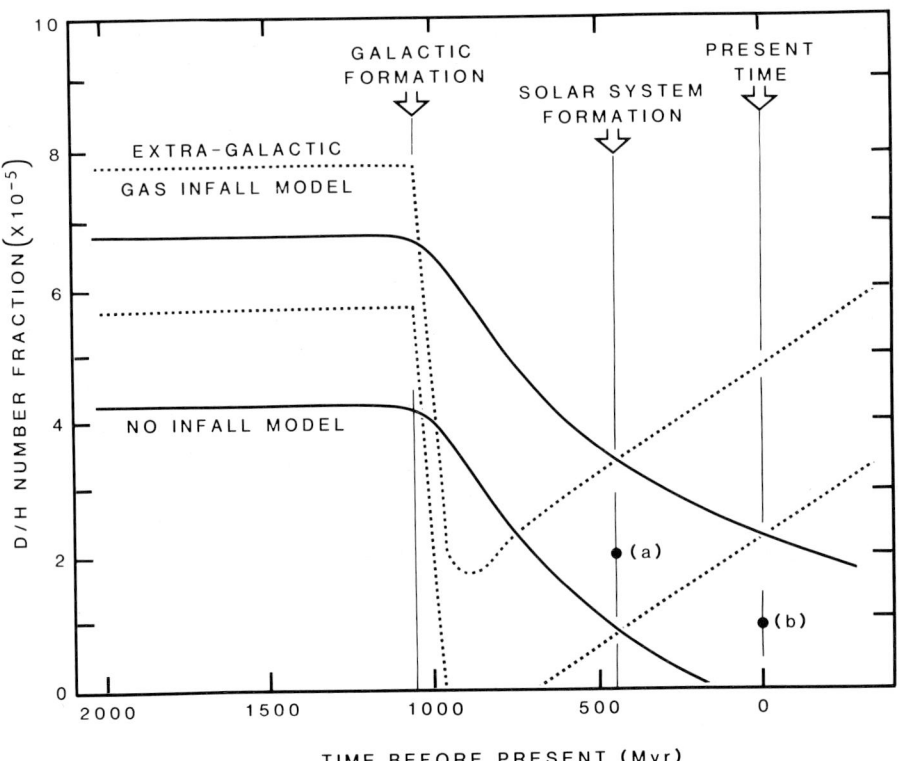

Figure 9. Evolutionary abundance curve envelopes for D/H for two contrasting models of galactic chemical evolution, normalized to the range of D/H ratios consistent with a protosolar D/H ratio determined from the atmospheres of Jupiter and Saturn. The data points shown are measurements for the protosolar nebula (a) and the interstellar medium (b) taken from Figure 8 (adapted from Gautier and Owen 1983a).

Galactic D/H in the no-infall model monotonically decreases with time following the formation of the galaxy, while in the infall model it reaches a local minimum shortly after formation, then increases subsequently. Specifically, the infall model predicts a rise in the galactic D/H ratio from the time of the solar system formation to the present. The comparison of the solar system value and the interstellar medium value is consistent with the no-infall model, and seems to exclude the infall model.

These models can also be used to extrapolate back in time to estimate the primordial D/H ratio which resulted from synthesis in the big bang. For the no-infall model, the solar system range adopted in Figure 9, $0.9 \times 10^{-5} < D/H < 3.3 \times 10^{-5}$, corresponds to a primordial value in the range $4.2 \times 10^{-5} < [D/H]^P < 6.8 \times 10^{-5}$. Kunde **et al.** (1982) have used this type of extrapolation to estimate the present-day baryon density and concluded that the Universe is open.

The production of primordial deuterium in the standard big bang model is tied to the production of primordial helium, and the binding rope is the baryon-to-photon number density ratio at the time of creation. Since both elements are produced at the same time, their production must correspond to the same baryon-to-photon number density ratio, so the abundance of one determines the abundance of the other. The production of each also depends on the number of neutrino families, which may be greater than 2. The production curves for fractional mass abundances of D and He, X_D^P and Y^P, respectively, as a function of baryon-to-photon number density ratio are shown in Figure 10 for 2, 3, and 4 neutrino families. This figure is adapted from the work of Rana (1982) and is based on the calculations of Yang **et al.** (1979).

The primordial deuterium mass fraction corresponding to the nominal primordial D/H ratio obtained from Figure 9 is 8.3×10^{-5}. This value is shown in Figure 10, along with the primordial helium mass fractions implied by a single baryon-to-photon number density ratio, for each assumed number of neutrino families shown. Table IV summarizes the range of primordial helium mass fractions permitted by the range of primordial D/H ratios in each case and gives the corresponding He/H ratios.

TABLE IV. Primordial Helium Abundance Implied by Solar System
          D/H for Various Numbers of Neutrino Families, N_ν

N_ν	Y^P	He/H
2	0.234-0.238	0.077-0.079
3	0.249-0.253	0.083-0.085
4	0.263-0.267	0.090-0.092

*$0.9 \times 10^{-5} < D/H < 3.3 \times 10^{-5}$
$4.2 \times 10^{-5} < [D/H]^P < 6.8 \times 10^{-5}$

The number of neutrino families is not known unambiguously, but Olive **et al.** (1981) conclude that there could be at most 4. Independent estimates of the primordial helium abundance could help constrain this ambiguity and test the galactic chemical evolution theory employed to extrapolate the solar system D/H ratio back into time. In reviewing the

observations Conrath **et al.** (1984) found the most stringent measurement, based on the analysis of Rayo, Peimbert and Torres-Peimbert's
(1982) observations of NGC 5471 suggested $Y^p < 0.23$, while an analysis
of 25 galaxies by Kunth (1981) favored $Y^p < 0.25$. In either case, the
viability of 4 neutrino families seems ruled out. The contest between 2
and 3 families is much less clear, particulary considering the
uncertainties inherent in this speculative analysis of the D/H ratio in
the solar system, but if more definitive measures of the primordial
helium abundance can be made, the distinction might be possible, and
refinements in our understanding in the evolution of our galaxy may
result.

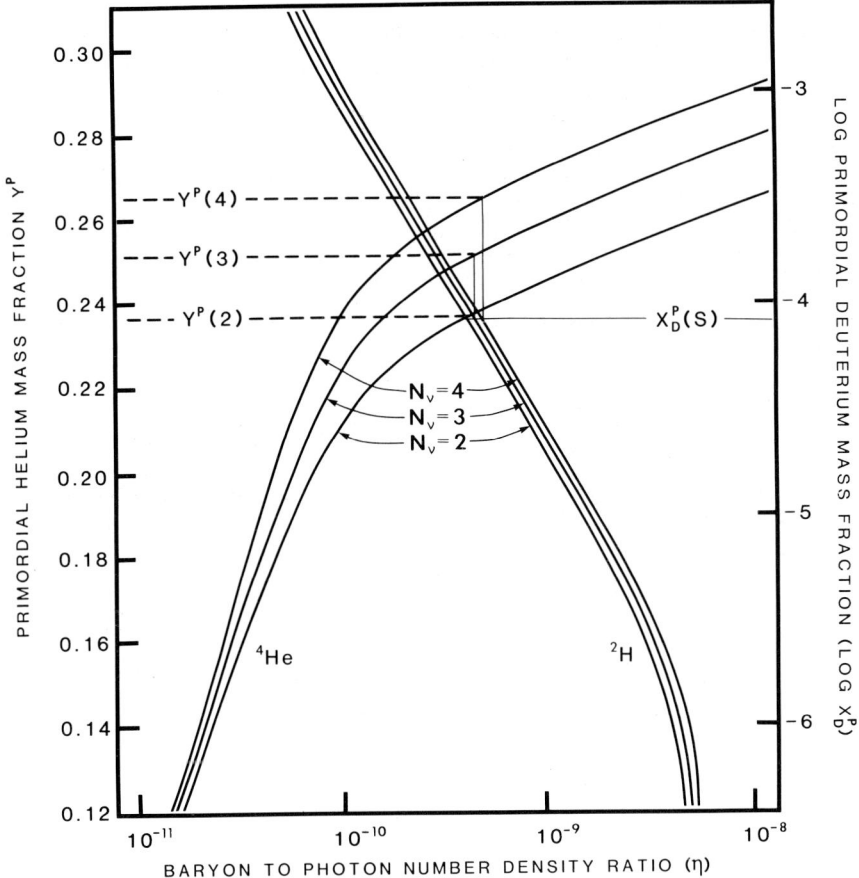

Figure 10. Calculated primordial abundances of ^{4}He and D as a function
of the baryon-to-photon number density ratio (adapted from Rana 1982).

5. EPILOGUE

The purpose of this paper is to highlight the complex and important
relationships between the solar system and the interstellar medium and
to illustrate the synergism that can result if the techniques and

expertises developed in each area are brought to focus on a common problem. I hope it is successful. And speaking of success, this effort would not have been possible without the National Aeronautics and Space Administration and the National Science Foundation who have supported my participation in the work described in it.

REFERENCES

Anders, E., Hayatsu, R., and Studier, M. H. 1973, **Science,** 182, 781.
Atreya, S. K., Donahue, T. M., and Kulin, W. R. 1978, **Science,** 20, 611.
Audouze, J., and Tinsley, B. M. 1974, **Ap. J.,** 192, 487.
Baines, K. H., and Bergstralh, J. T. 1986, **Icarus,** 65, 406.
Barshay, S. S., and Lewis, J. S. 1978, **Icarus,** 33, 593.
Beer, R., and Taylor, F. W. 1973, **Ap. J.,** 179, 309.
Bergstralh, J. T., and Neff, J. H. 1983, **Icarus,** 55, 40.
Cameron, A. G. W. 1978, **Moon and Planets,** 18, 5.
Cameron, A. G. W. 1982, in **Essays in Nuculear Astrophysics** (C. A. Barnes, D. D. Clayton, and D. N. Schramm, eds.), p. 23. Cambridge University Press, London.
Courtin, R., Gautier, D., Marten, A., Bezard, B., and Hanel, R. 1984, **Ap. J.,** 287, 287, 899.
Cochran, W. D., and Smith, W. H. 1983, **Ap. J.,** 271, 859.
Conrath, B. J., Gautier, D., Hanel, R. A., and Hornstein, J. 1984, **Ap. J.,** 282, 807.
de Bergh, C., Lutz, B. L., Owen, T., Brault, J., and Chauville, J. 1986, **Ap. J.** (in press).
Delsemme, A. H. 1982, in **Comets** (L. L. Wilkening, ed.), p. 85. University of Arizona Press, Tucson.
Gautier, D., Conrath, B., Flasar, M., Hanel, R., Kunde, V., Chedin, A., and Scott, N. 1981, **J. Geophys. Res.,** 86, 8713.
Gautier, D., and Owen, T. 1983a, **Nature,** 302, 215.
Gautier, D., and Owen, T. 1983b, **Nature,** 304, 691.
Geiss, J., and Reeves, H. 1981, **Astron. Ap.,** 93, 189.
Hanel, R., Conrath, B., Flasar, F., Kunde, V., Maguire, W., Pearl, J., Pirraglia, J., Samuelson, R., Cruikshank, D., Gautier, D., Gierasch, P., Horn, L., and Schulte, P. 1986, **Science,** 233, 70.
Heasley, J. N., and Milkey, R. W. 1978, **Ap. J.,** 221, 677.
Hubbard, W. B., and MacFarlane, J. J. 1980, **Icarus,** 44, 676.
Hunten, D. M., Tomasko, M. G., Flasar, F. M., Samuelson, R. E., Strobel, D. F., and Stevenson, D. J. 1984, in **Saturn** (T. Gehrels and M. S. Matthews, eds.), p. 671. University of Arizona Press, Tucson.
Kunde, V., Hanel, R., Maguire, W., Gautier, D., Baluteau, J. P., Marten, A., Chedin, A., Husson, N., and Scott, N. 1982, **Ap. J.,** 263, 443.
Kunth, D. 1981, Ph.D. Thesis, Paris VII.
Lambert, D. L. 1978, **M.N.R.A.S.,** 182, 249.
Lewis, J. S., and Prinn, R. G. 1980, **Ap. J.,** 238, 357.
Lynden-Bell, D., and Pringle, J. E. 1974, **M.N.R.A.S.,** 168, 603.
Olive, K. A., Schramm, D. N., Steigman, G., Turner, M. S., and Yang, J. 1981, **Ap. J.,** 246, 557.
Owen, T., Lutz, B. L., and de Bergh, C. 1986, **Nature,** 320, 244.

Pinto, J. P., Lunine, J. I., Kim, S. J., and Yung, Y. L. 1986, **Nature,** 319, 388.

Prinn, R. G., and Fegley, B., Jr. 1981, **Ap. J.,** 249, 308.

Prinn, R. G., Larson, H. P., Caldwell, J. J., and Gautier, D. 1984, in **Saturn** (T. Gehrels and M. S. Matthews, eds.), p. 88. University of Arizona Press, Tucson.

Prinn, R. G., and Owen, T. 1976, in **Jupiter,** (T. Gehrels, ed.), p. 319. University of Arizona Press, Tucson.

Rana, N. C. 1982, **Phys. Rev. Letters,** 48, 209.

Rayo, J. F., Piembert, M., and Torres-Piembert, S. 1982, **Ap. J.,** 255, 1.

Smith, W. H., and Macy, W., Jr. 1986, **Icarus** (in press).

Smoluchowski, R. 1967, **Nature,** 215, 691.

Strobel, D. F. 1981, **Planet. Space Sci.,** 30, 839.

Trafton, L. 1981, **Rev. Geophys. Space Phys.,** 19, 43.

Vidal-Madjar, A., Laurent, C., Gry, C., Bruston, P., Ferlet, R., and York, D. G. 1983, **Astron. Ap.,** 120, 58.

Weidenschilling, S. J., and Lewis, J. S. 1973, **Icarus,** 20, 465.

Yang, J., Schramm, D. N., Steigman, G., and Rood, R. T. 1979, **Ap. J.,** 227, 697.

Yung, Y. L., Allen, M., and Pinto, J. P. 1984, **Ap. J. Suppl.,** 55, 465.

Section III: Heating, Cooling and Radiative Processes

Christopher McKee

OBSERVATIONS OF THE COOLING OF THE INTERSTELLAR GAS

T. G. Phillips
California Institute of Technology 320-47
Pasadena, California
USA

ABSTRACT. Cooling of the gas of the dense interstellar clouds takes place primarily at submillimeter wavelengths. At about 1mm cooling is due to the many rotational lines of heavy moleules, but at shorter wavelengths light molecules (hydrides) become the dominant species, with some assistance from the fine structure transitions of atomic carbon. Photodissociation regions at the surfaces of molecular clouds are cooled by fine structure transitions of atoms and atomic ions. A presentation is given of some aspects of the observed emission due to heavy molecules, light molecules, atoms and ions.

1. INTRODUCTION

This paper deals with the observational aspects of interstellar gas cooling processes. Theoretical aspects are dealt with in various talks in this Summer School, but particularly in those of Black and Flower. Since there is very little observational information on the cooling lines of diffuse gas, but a great deal on the dense gas, I will concentrate on describing the observed spectra of the dense star forming molecular clouds of the Galaxy and nearby galaxies. The discussion will be from the point of view of a spectroscopist so that an impression will be given of the range of transitions that have been observed rather than of the differing molecular cloud objects.

The typical region of strong emission is a dense cloud ($n_{H2} \gtrsim 10^4 cm^{-3}$) which has already formed stars. These stars may be ionizing, dissociating and heating the gas from outside the cloud or from inside, depending on the time that has passed since their formation and the nature of the evolution of the cloud. A specific case is that of the Orion cloud in which a group of stars, lying outside the current molecular cloud, provides an ionization and photodissociation region (Zuckerman 1973; Tielens and Hollenbach, 1985) and on the same line of sight is a group of imbedded protostellar or stellar sources which heat the molecular gas and probably cause dynamic effects such as outflow. Such active regions are the easiest to observe because they provide the strongest lines and the most varied chemistry. We observe emission from the photodissociation region, the hot core region near the internal heating sources, the outflow (or plateau) region, the extended high density (ridge) region and the general quiescent molecular cloud. The various regions have differing chemical signatures (Blake et al. 1986). The quiescent cloud is carbon rich in the gas, whereas the hot core is nitrogen rich, the compact ridge is oxygen rich and the plateau seems oxygen, sulphur and silicon rich. This means that the cooling of a specific region of gas cloud depends somewhat on its chemical as well as physical circumstances. Also, it allows us to observe a wide variety of species. These species range from atoms and atomic ions to light molecules (those whose moment of inertia is dominated by hydrogen atoms) and heavy molecules The atomic and atomic ion emission is mainly from the photodissociation regions, although CI emission is apparently largely seen from the bulk of the molecular cloud. Most molecules are detected in the dense cores of clouds, but CO being highly abundant is detected throughout the molecular regions.

D. J. Hollenbach and H. A. Thronson, Jr. (eds.), Interstellar Processes, 707–730.
© *1987 by D. Reidel Publishing Company.*

An overview of the situation can be obtained by picturing the approximate emission spectrum of a gas cloud. Figure 1 shows a simplified spectrum of a possible 30K gas cloud. The gas emission is dominated by rotation of heavy molecules at millimeter wavelengths, hydride molecules at submillimeter wavelengths and fine structure transitions of atomic atoms and ions in the far-infrared. Underlying the line emission is the dust continuum emission which is optically thin and very weak at long wavelengths, but becomes optically thick towards $100\mu m$.

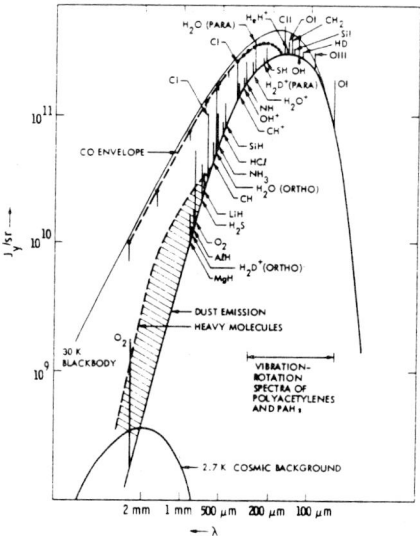

Figure 1. — A spectrum showing the extraordinary extent of spectral information contained in the anticipated emission profile of a typical interstellar cloud. At long wavelengths the spectrum is dominated by rotational transitions of heavy molecules and at short wavelengths by light molecules, atoms and dust particles. Only a few of the most fundamental of the host of spectral features have been included.

The CO molecule emission is particularly important for cooling since the lines are optically thick at long wavelengths and only lose emission strength at short wavelengths where the high J levels become deexcited. Of the light molecules, probably the most important is H_2O, although there is little observational information on this as yet. Vibrational modes of heavy molecules could also play a role in cooling, but these have not been observed directly.

Let us look now at the way that molecular excitation and deexcitation takes place in the millimeter and submillimeter regimes. The likely processes are:

- Hydrogen molecule collisions, which are largely independent of the trace molecule dipole moment, μ_D, and are usually allowed for all values of ΔJ, but with even values preferred (Green, 1974).
- By radiation, which is proportional to μ_D^2, has $\Delta J = \pm 1$ allowed transitions, and the Einstein A coefficient increases with ν^3.
- Electron collisions, which are proportional to μ_D^2 and have $\Delta J = \pm 1$ allowed transitions.

Typically, the H_2 collision cross sections are of the order of $10\overset{\circ}{A}^2$. The electron cross sections are larger, but have a reduced effect due to the paucity of electrons relative to H_2. Also the $\Delta J = \pm 1$ rule makes it hard to distinguish the effects of electrons and photons, but it can be shown that the low lying populations can be modified somewhat when the $n(e)/n(H_2)$ ratio is greater than $\sim 10^{-5}$ (Dickinson *et al.* 1977).

In order to examine some of the cases of interest, let us take two types of molecules as examples. Figure 2 shows the energy level diagrams for linear molecule rotation and for an asymmetric rotor. For the low lying (millimeter transitions) levels of the heavy linear (e.g. CO, CS, HCN...) molecules the collisions with H_2 are dominant. If the column density is high, there will be photon trapping, which helps to thermalize the level populations and therefore raise the population of higher levels if the H_2 collisions were not adequate to overpower the deexcitation from the spontaneous emission (Einstein A). For cases of low optical depth, τ, at the line frequency, the density required to balance collisional excitation with spontaneous emission is:

$$n(H_2) \sim A/< \sigma v >,\qquad(1)$$

and under the escape probability formalism for higher optical depths

$$n(H_2) \sim A/\tau< \sigma v >,\qquad(2)$$

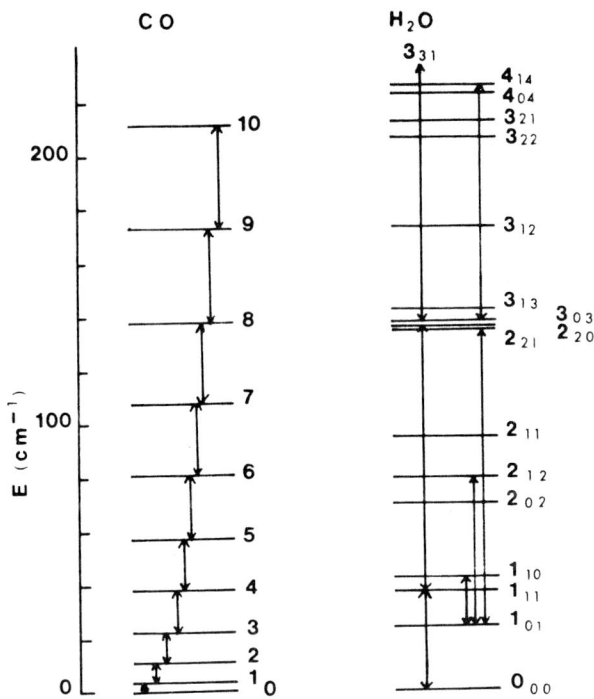

Figure 2. — Typical excitation ladder for linear (CO) molecules and asymmetric (H_2O) molecules.

In all of these cases it is important to consider the effect of emission from the dust. For the low lying levels of the heavy linear rotor the dust is ineffective because the dust emission is optically thin and is well below the value expected from a black body at the gas temperature. In fact, it is dropping with wavelengths as $\sim 1/\lambda^3$ or $1/\lambda^4$ (Figure 1). The dust temperature may be slightly above or below the gas temperature depending on the heating and cooling processes, but at this point we can consider the two temperatures as comparable (say $\gtrsim$ 10 K) and considerably greater than $h\nu/k$ for the lines. The density required from equation (1) is the order of 10^4 or 10^5cm^{-3} for $\mu_D \sim 1$ Debye.

In order to excite the submillimeter lines of the linear rotors, much higher temperatures and densities are required, since the A coefficient is increasing with ν^3. As the photon processes operate with $\Delta J = \pm 1$ only, the dust emission and trapping are not able to play a role unless the gas temperature and density are adequate to populate the high levels.

A major point emerges, which is that the submillimeter transitions of linear molecules provide information on high temperature, high density regions in a way that millimeter lines cannot. Examples of the observations are given below.

For the light rotors in warm regions the situation is very different. From Figure 2 it can be seen that collisions are less likely to be useful for the fundamental transitions and are easily overpowered by dust emission. At the shorter wavelengths the dust emission is likely to approach thermal values and can drive the transitions into equilibrium, providing large populations in the upper states which can then be redistributed among the nearby levels by collisions. For short wavelengths the transitions are typically seen in absorption because the molecule is cooling the dust, but the opposite is true at long wavelengths. There is then a close relationship between the dust emission and that of light rotors such as water. It is clear that the submillimeter lines will provide a more direct probe than millimeter lines since the small splitting populations are influenced by many factors and may even provide masing action.

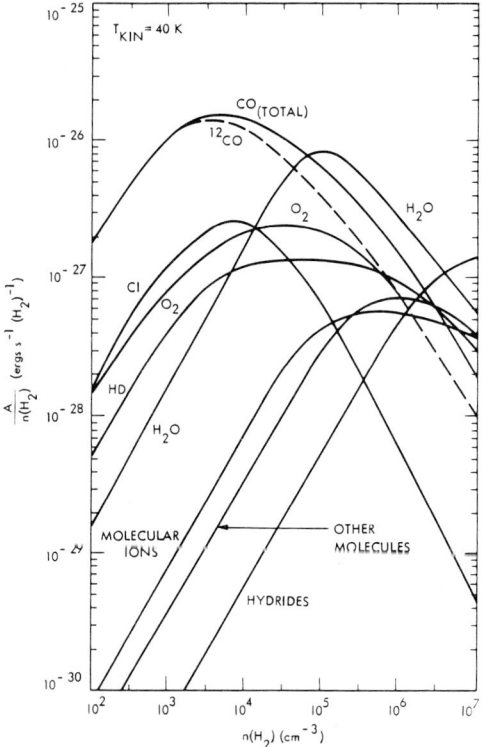

Figure 3. — Cooling rates per H_2 molecule, showing the overall dominance of H_2O and hydrides for the high densities of star–forming regions (Goldsmith and Langer 1978).

Although we are not concerned with theory in this talk it is important to understand qualitatively the relative importance of the various species to the gas cooling. A calculation of Goldsmith and Langer (1978) for a 40 K cloud is shown in Figure 3. In this model CO is dominant at low gas densities but H_2O and finally other light hydride molecules take over at higher densities. The submillimeter observations of H_2O and other hydride molecules are of great importance.

2. OBSERVATIONAL TECHNIQUES

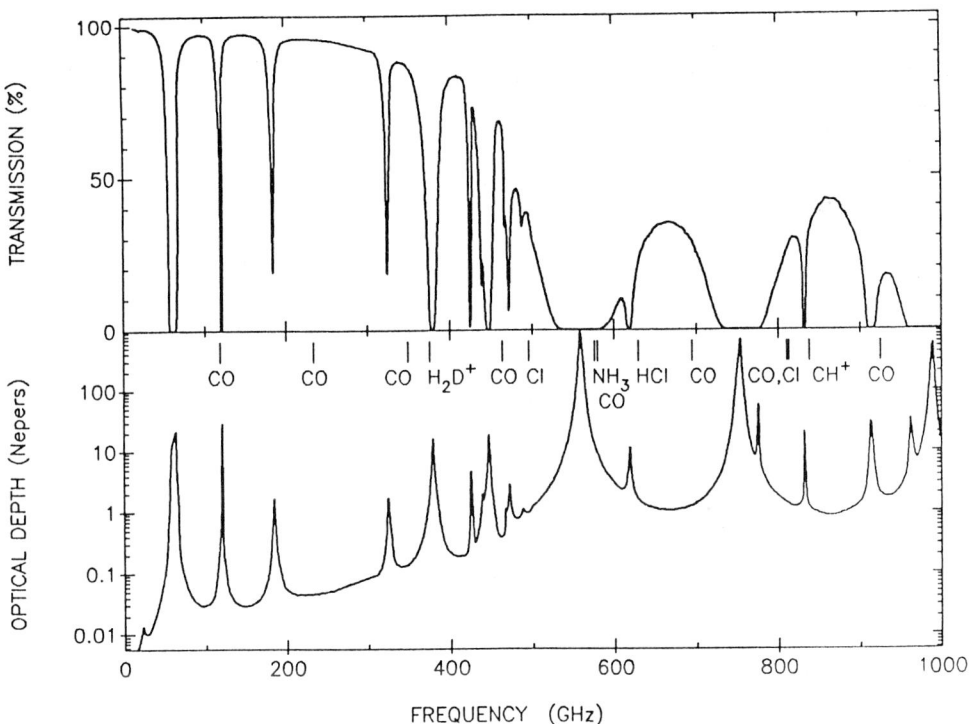

Figure 4. — Transmission of the atmosphere at Mauna Kea for 1.5 mm precipitable H_2O. Line frequencies are shown for CI, CO and four other representative molecules.

Spectroscopy in the millimeter, submillimeter and far–infrared bands is still a relatively immature field. This is because the techniques needed to build accurate surfaces for radio telescopes and to make detection devices whose performance approaches theoretical limits have only recently become available. A further reason is that the Earth's atmosphere is largely opaque at wavelengths shortwards of about 2mm, from sea level, so that high mountain sites or space or airborne platforms are required. The reason that so little is known about interstellar H_2O, for instance, is that even from airborne altitudes the water vapor in the atmosphere blacks out the interstellar emission wavelengths. Figure 4 shows the atmospheric transmission from a 14,000' mountain site such as Mauna Kea, where for instance the 10.4m diameter Caltech Submillimeter Telescope is placed. A subset of the line frequencies of Figure 1 is drawn on the figure, to show that some species fall into the transparent windows. Unfortunately the atmosphere is essentially black above 1,000 GHz (300 μm), so that no ground based work is possible in that part of the submillimeter or far-infrared. For this reason the Kuiper Airborne Observatory has been very important, even though it has only a 91.5 cm telescope.

The observations to be discussed below are really of two types. Those involving heterodyne or radio style detection and those using direct detection with a monochromator such as a cooled grating or Fabry–Perot. Heterodyne detection for the millimeter band and the submillimeter band has been reviewed by Penzias and Burrus (1973) and by Phillips and Woody (1982). Direct detection techniques with Fabry–Perots are described by Storey (1985). On the whole the longer wavelength lines are studied with heterodyne receivers and the shorter wavelength lines with direct detectors. This is demonstrated by a calculation (Phillips and Watson, 1984) of signal to noise ratio achieved in a specific observation by the two techniques as a function of frequency of the line and spectral resolution required (Figure 5).

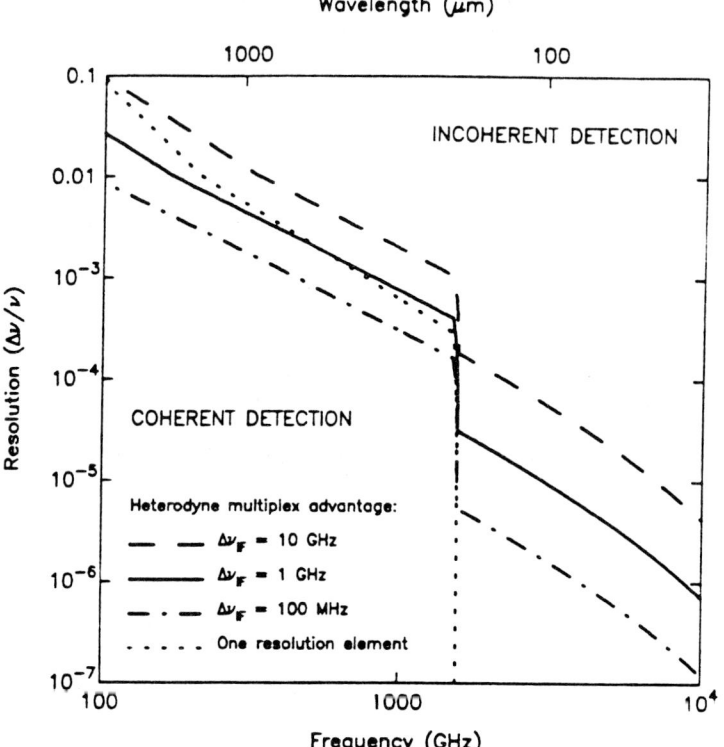

Figure 5. — Relative sensitivity of heterodyne and direct detection as a function of frequency and spectral resolution, for single detectors of each type. The domains in which either detection method is more sensitive than the other are separated by the locus of points for which $(S/N)_{heterodyne} = (S/N)_{direct}$. The discontinuity at λ = 200 μm arises from the use in the direct–detection instruments of bolometers at longer wavelengths and photoconductors at shorter wavelengths, the latter having much lower detector/preamplifier noise.

The calculation is for a warm telescope at 200 K with a 5% emissivity and for heterodyne detectors with 10% quantum efficiency and bolometer direct detectors with noise equivalent powers of 10^{-16} WHz$^{-1/2}$. At wavelengths shorter than 200 μm photoconductors can be used instead of bolometers in which case the noise is reduced. The result is that heterodyne detectors are preferred for frequencies up to about 2,000 GHz for spectral line studies, but direct detectors are preferable for line studies at very short wavelengths and for continuum detection at all wavelengths.

3. HEAVY MOLECULES

Over the past few years great progress has been made in the study of heavy molecule rotation emission from dense clouds. Most of the emphasis of this work has been in the area of interstellar chemistry or in spatial mapping or in dynamical studies of particular clouds. However, in new work, very wide band spectral scans can be used to make estimates of the total emission from spectral features and compare these with continuum emission.

The scan which provides the best overall display of the molecular features is probably that of OMC-1 made with one of the Caltech 10m telescopes in the 207–263 GHz band. A detailed discussion of this scan is given by Sutton *et al.* (1985); Blake *et al.* (1986, 1987). It is reproduced here in Figure 6. Almost 900 spectral features can be identified in this scan!

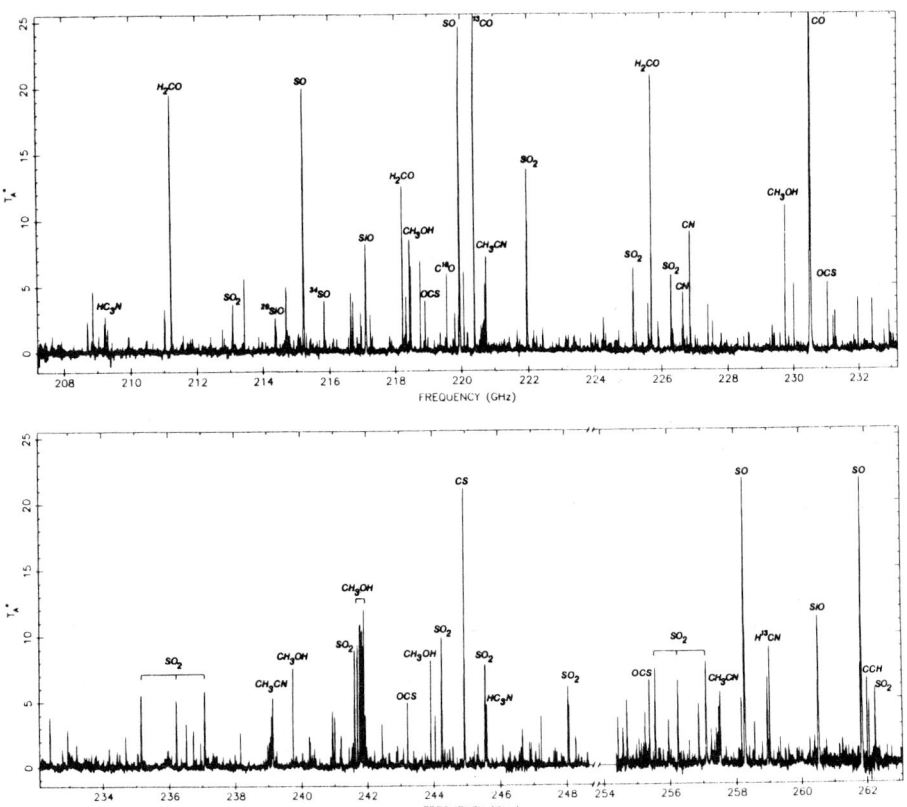

Figure 6. — A spectral scan of OMC-1 containing about 900 lines due to heavy molecules.

Even more remarkable is that, when combined with the longer wavelength spectral surveys (Onsala - Johansson *et al.* 1984; Columbia/Bell - Cummins, Linke, and Thaddeus 1986), with careful laboratory studies (De Lucia 1976; Pickett *et al.* 1981; Blake *et al.* 1984) and previously developed catalogs (Lovas 1983; Poynter and Pickett 1981) nearly all the lines can be identified as the spectra of a set of 28 molecules! The results of this analysis are summarized in Table 1.

Table 1

Detected Species and Line Survey Parameters

CO	HCN	H_2S	HC_3N
CN	HNC	HNCO	CH_3CH
NO	HCO^+	H_2CO	CH_3CCH
CS	CHS^+	H_2CS	C_2H_3CH
SiO	OCS	H_2CCO	C_2H_5CN
SO	SO_2	HCOOH	$HCOOCH_3$
C_2H	HDO	CH_3OH	CH_3OCH_3

Source: OMC-1   Coverage: 208-263 GHz
Sensitivity: $\lesssim 0.2$-0.3 K   Resolution: 1 MHz
Lines identified: $\gtrsim 825$ Unidentified Lines: 32
Isotopes Detected: $H, D, {}^{12}C, {}^{13}C, {}^{14}N, {}^{15}N, {}^{16}O, {}^{17}O, {}^{18}O$
$^{28}Si, ^{29}Si, ^{30}Si, ^{32}S, ^{33}S, ^{34}S, ^{35}Cl$

A cursory inspection of Figure 6 suggests that there is a considerable integral power emitted by the heavy molecules, which can be estimated by summing the effect of the individual lines. The one $^{12}C^{16}O$ line in the spectrum runs off scale to about $T_A^* \approx 100K$ and the other lines are of various strengths down to the minimum detectable of about 0.1 K. The number of lines as a function of their strength is plotted in Figure 7 for a fraction of the band (Sutton *et al.* 1984). This allows us to estimate the total flux and to verify that the contribution from the weaker lines is converging.

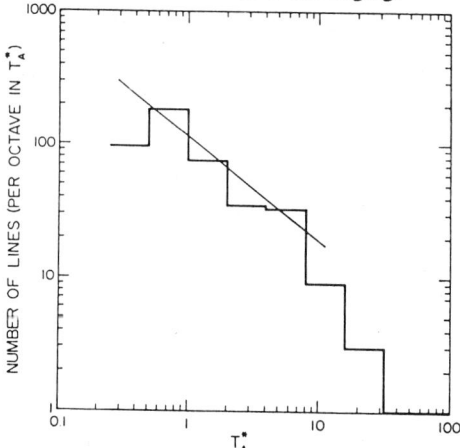

Figure 7. — Distribution of peak corrected antenna temperatures for lines in the 215-247 GHz band.

The measured integrated line flux is about 25 Jy at 1.3 mm and 50 Jy at 1.15mm wavelength. These values can be compared with the bolometric continuum flux which includes the effect of both lines and dust continuum. Figure 8 includes both the line and bolometric measurements and shows that for OMC-1 the line contribution is apparently about one half of the total in the vicinity of 1mm wavelength.

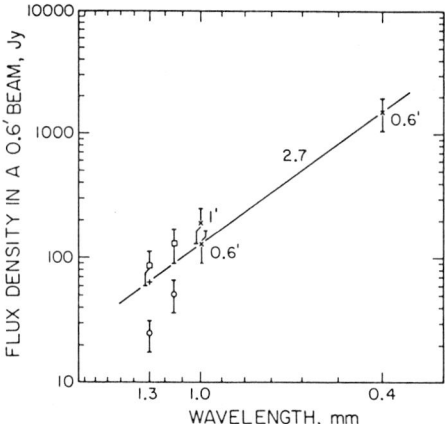

Figure 8. — Broad-band and spectral line flux densities in a 0'.6 beam centered on BN. Open circles show the line fluxes measured by Sutton *et al.* (1984) and Blake *et al.* (1986). Open boxes indicate the total broad-band fluxes measured simultaneously. Crosses represent the broad-band fluxes of Keene, Hildebrand, and Whitcomb (1982) and Elias *et al.* (1978), the latter both for a 1' beam and corrected for a 0'.6 beam. Plus sign represents the extrapolation of these short wavelength measurements to 1.3 mm.

This result is interesting in the sense that it shows that heavy molecules are significant coolants of the whole cloud not just the gas. Also of interest is the fact that the CO molecule accounts for about one third of the molecular emission in the band. These results are only for OMC-1, of course, and this is a special source in that the lines are often wide due to various dynamical effects. This tends to emphasize the molecular emission because it enhances the effect due to the optically thick transitions. Further measurements are required to determine the generality of the line contribution to the bolometric continuum measurements, but it should be noted that wide spectral features are typical of large active molecular clouds.

Spectral surveys have been carried out in the millimeter band, but they will be even more interesting in the submillimeter where the molecular emission strength should be even greater. For instance we would expect the flux to vary with ν^3 for the optically thick transitions of a linear molecule, or ν^5 for the optically thin case. The variation with ν should be even more pronounced for symmetric and asymmetric tops. At high frequencies the flux will drop due to lack of molecular excitation.

4. LIGHT MOLECULES

From Figure 1 we can see that light molecules have their fundamental rotational transitions in the submillimeter band, as opposed to more complex molecules which pervade the millimeter spectrum. The CO molecule is a special case, because, although it is heavy in the sense that the first rotational transition is at 2.6mm, it has a very small dipole moment (~ 0.1 Debye) and the A coefficient is small. This means that it is easily excited (see equations 1 and 2) and therefore strong emission persists well into the submillimeter band, for hot dense sources. For this reason CO is often included with the 'light molecules' and is a major feature of the submillimeter spectrum and a major contributor to gas cooling.

Table 2 provides a list of light molecule lines detected in the submillimeter and far-infrared. Most of the detections are from the Kuiper Airborne Observatory, but some of the CO studies are from mountain based telescopes. Several of these observations will be discussed individually, below.

Table 2

Observed Submillimeter and Far-Infrared Lines of Light Molecules

Species	Transition	Wavelength	Reference
H_2O	$3_{13} - 2_{20}$	$1{,}640\mu m$	Waters *et al.* (1980)
	$4_{14} - 3_{21}$	$789\mu m$	Phillips, Kwan, and Huggins (1980)
H_2D^+	$1_{10} - 1_{11}$	$806\mu m$	Phillips *et al.* (1985)
H_3O^+	$1_1 - 2_1$	$974\mu m$	Wootten *et al.* (1986)
OH	$^2\Pi_{3/2}(J=5/2\text{-}3/2)$	$119\mu m$	Storey, Watson, and Townes (1981)
	$^2\Pi_{3/2}(J=7/2\text{-}5/2)$	$85\mu m$	Watson *et al.* (1984)
	$^2\Pi_{1/2}(J=3/2\text{-}1/2)$	$163\mu m$	Genzel *et al.* (1985)
CH	$^2\Pi_{1/2}(J=3/2\text{-}1/2)$	$149\mu m$	Stacey, Lugten and Genzel (1986)
NH_3	$J_K = 1_0 - 0_0$	$524\mu m$	Keene, Blake, and Phillips (1983)
	$J_K = 4_3 - 3_3$	$125\mu m$	Townes *et al.* (1983)
HCl	J=1-0	$479\mu m$	Blake, Keene, and Phillips (1985)
CO	$J = 3 - 2$	$870\mu m$	Phillips *et al.* (1977)
	$J = 4 - 3$	$652\mu m$	Phillips, Kwan and Huggins (1980)
	$J = 6 - 5$	$434\mu m$	Goldsmith *et al.* (1981)
	$J = 7 - 6$	$372\mu m$	Harris *et al.* (1985)
	$J = 16 - 15$	$163\mu m$	Stacey *et al.* (1983)a
	$J = 17 - 16$	$153\mu m$	Stacey *et al.* (1982)
	$J = 21 - 20$	$124\mu m$	Watson *et al.* (1980)
	$J = 22 - 21$	$119\mu m$	Watson *et al.* (1980)
	$J = 26 - 25$	$100\mu m$	Watson *et al.* (1985)
	$J = 27 - 26$	$97\mu m$	Storey *et al.* (1981)
	$J = 30 - 29$	$87\mu m$	Storey *et al.* (1981)
	$J = 31 - 30$	$84\mu m$	Watson *et al.* (1984)
	$J = 34 - 33$	$77\mu m$	Watson *et al.* (1984)

4.1 H_2O

Water is probably the most important molecule for the thermal properties of the dense interstellar gas, but because of the great opacity of water vapor in the earth's atmosphere no really effective observations can be made except from a space platform. Also water demonstrates a particular aspect of excitation, that is the strong interaction between the dust and the line emission.

Of course, we know about the presence of interstellar water from the famous masing transition at 22 GHz in the ortho branch ($6_{16} - 5_{23}$), but this involves highly excited states (>400 cm^{-1}) and does not provide much information about the general abundance of water or its cooling properties. Those lower lying transitions which have been detected from the airborne telescope are relatively weak lines which were observed by making use of interstellar velocity shifts to avoid the atmospheric line cores (Figure 9).

The relationship of the detectable transitions to the overall level scheme is shown in Figure 10. The really important cooling transitions for the dense cool gas lie in the lower part of the diagram. The $1_{10} - 1_{01}$ ortho transition at 557 GHz and the $1_{11} - 0_{00}$ para transition at 1,113 GHz are the fundamental transitions, which would be valuable to observe, but are so badly blocked by the atmosphere, even from airborne altitudes.

Because of the large dipole moment and the large splitting between the lowest levels, radiative effects tend to dominate collisions and the populations are largely determined by dust radiation pumping.

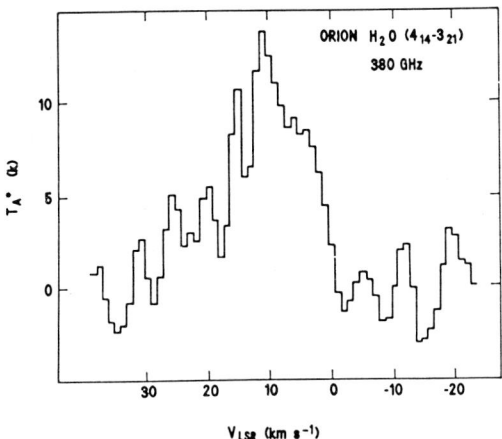

Figure 9. — A spectrum of water emission from OMC-1 taken from airborne altitudes. The poor signal to noise ratio is due to the strong atmospheric absorption. The line is wide indicating that the bulk of the water is confined to the active regions (hot core and plateau). Unfortunately, there are no general studies of water emission which would permit an estimate of the relative abundances in shocked or active regions as opposed to dynamically quiescent gas.

Collisions with the gas can redistribute the populations. Takahashi, Hollenbach and Silk (1985) have developed a model to predict the spectral output of a cloud consisting of H_2 gas, dust, water, CO and neutral carbon atoms. The cloud is assumed to be spherical with constant temperature dust and constant density gas throughout. A velocity gradient ($V \propto r$) is assumed and the calculation of emergent flux is carried out by the escape probability formalism. Figure 11 shows the spectral output of the cloud. It is most interesting to note that for the ortho-water lines treated, those longer than about $100\mu m$ appear in emission because of the lower dust opacity in this region. Those shortward of $100\mu m$ appear in absorption. What is happening is that the water molecules are effective at absorbing the shorter wavelength dust photons, thereby heating themselves and the rest of the gas and reradiating at longer wavelengths where the cloud is optically thin.

Water is an essential component of the heating-cooling balance for the cloud and must be studied over a wide frequency range to fully comprehend its role.

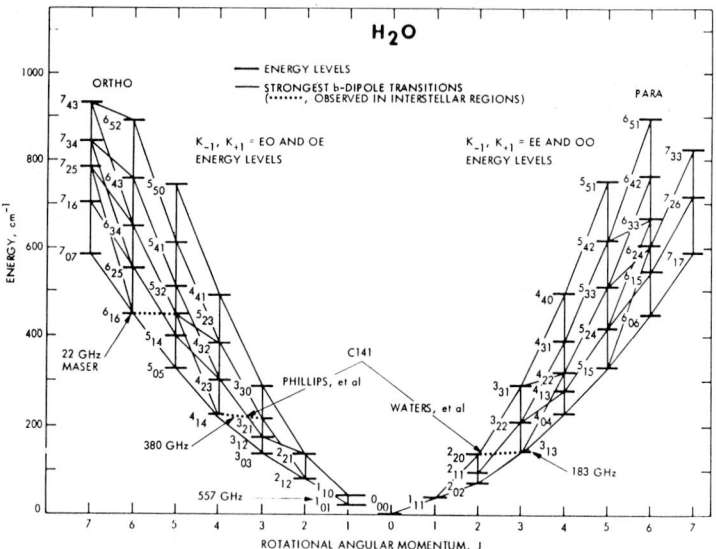

Figure 10. — The ground rotational energy level diagram for H_2O.

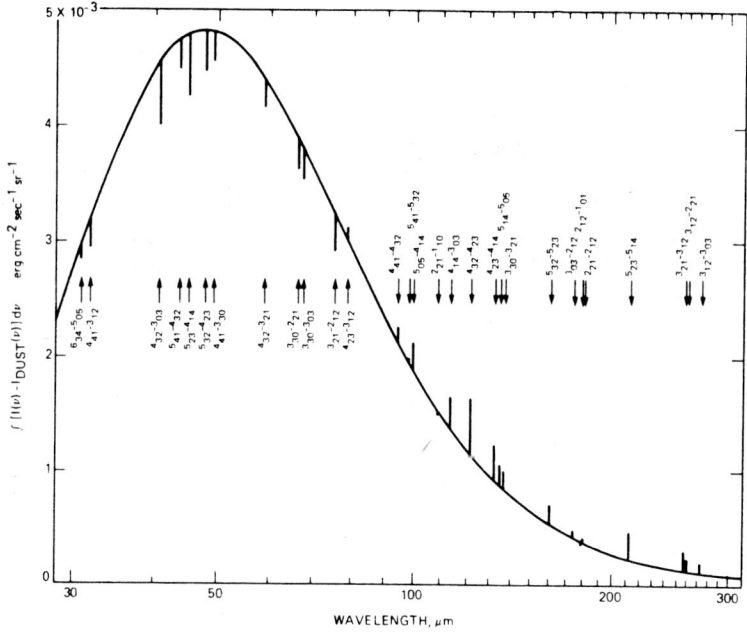

Figure 11. — The spherical cloud model of Takahashi, Hollenbach and Silk (1985), showing emission and absorption features for some of the lines of ortho H_2O.

4.2 OTHER HYDRIDES

As is clear from Figure 3, the hydride molecules are collectively useful in cooling the hot dense regions of gas. If we could make a wide spectral scan equivalent to that of Figure 6, but in the submillimeter band, we could begin to make serious observational estimates of the hydride cooling rates. However, the best that has been managed so far is a set of detections of individual lines of a few species. A restricted spectral scan will be possible with the new high mountain submillimeter telescopes, but a clear unobstructed view of the overall cooling spectrum will best be obtained by a space mission.

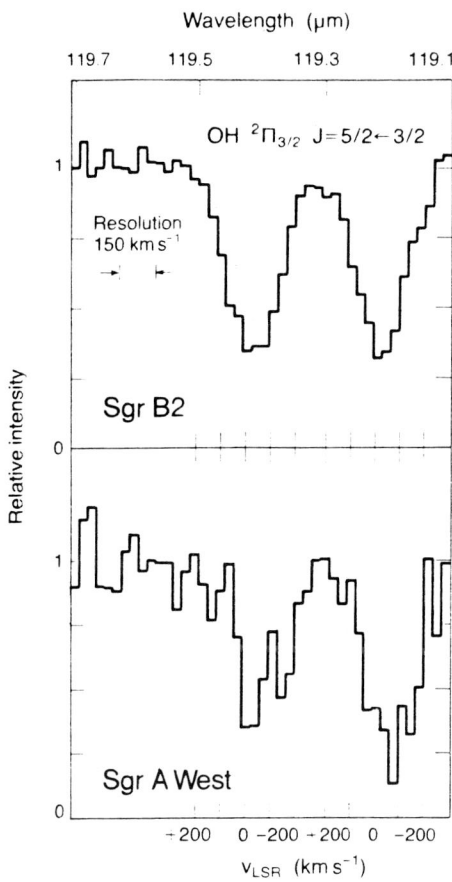

Figure 12. — The lowest rotational transitions of OH, seen in absorption toward SgrA West and Sgr B2.

OH is easily observed by means of its Λ doubling radio lines, but the fundamental rotational transitions occur at 119μm for the $^2\Pi_{3/2}$ ladder and at 163μm for the $^2\Pi_{1/2}$ ladder. These have been detected from the KAO. Figure 12 shows the spectrum of OH $^2\Pi_{3/2}$ J = (5/2-3/2) observed in absorption towards Sgr B2 (Storey, Watson and Townes, 1981) and SgrA West by Genzel *et al* 1985.

More recently (Figure 13 - left side) the 163μm line has been seen in emission from the OMC-1 cloud (Crawford *et al*. 1986). These lines have been observed with the Berkeley tandem Fabry-Perot spectrometer which does not have the very high spectral resolution of the heterodyne receivers. Nevertheless, for regions displaying great activity, such as the Galactic Center or high velocity outflow sources, dynamic information is obtained. When heterodyne receivers are available in this wavelength range it will be most interesting to study the interplay between absorption and emission profiles for the typical quiescent clouds.

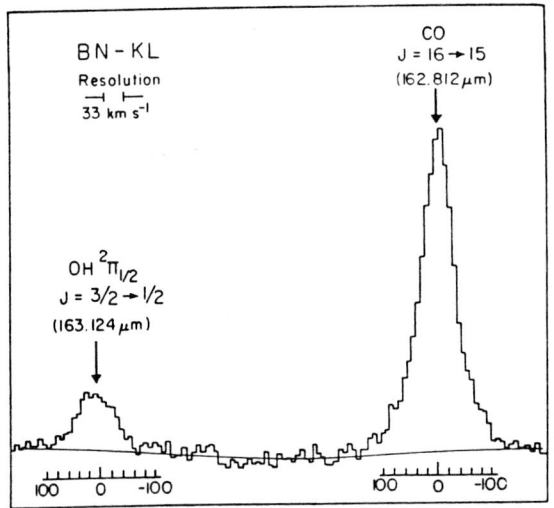

Figure 13. — OH $^2\Pi_{1/2}$J = 3/2 $\rightarrow$ 1/2 and the CO J = 16 $\rightarrow$ 15 emission lines in the shocked molecular region in the core of the Orion moleular cloud.

Ammonia is also easily observed in the radio by means of its inversion doubling lines, but again the fundamental rotation transitions lie in the submillimeter and far-infrared. Two of the rotation lines have recently been observed, the $J_K = 1_0 - 0_0$ transition was observed from the KAO with an InSb heterodyne bolometer (Keene, Blake, and Phillips 1983) and shows the necessary spectral resolution to determine the line shape and central velocity. The observations were of the core of the Orion Molecular Cloud and indicated that the NH_3 emission originated mostly from the dense quiescent cloud at V_{LSR} = 9 km/sec, but with a large optical depth. The spectrum is shown in Figure 14. The $4_3 - 3_3$ emission detected by Townes *et al*. (1983) is thought to be emanating from the 'hot core', which is a dense hot region near the cloud center with a V_{LSR} of about 5 km/sec. It is most encouraging that information on the ground states of one molecule can be obtained with such disparate techniques and wavelengths.

A goal of submillimeterwave spectroscopy in the interstellar medium has been to establish the viability of the use of metal hydride rotational spectra for astronomy. Many of the simple hydrides presumably can only be observed in the submillimeter, at least in dense clouds, so the technique is vital

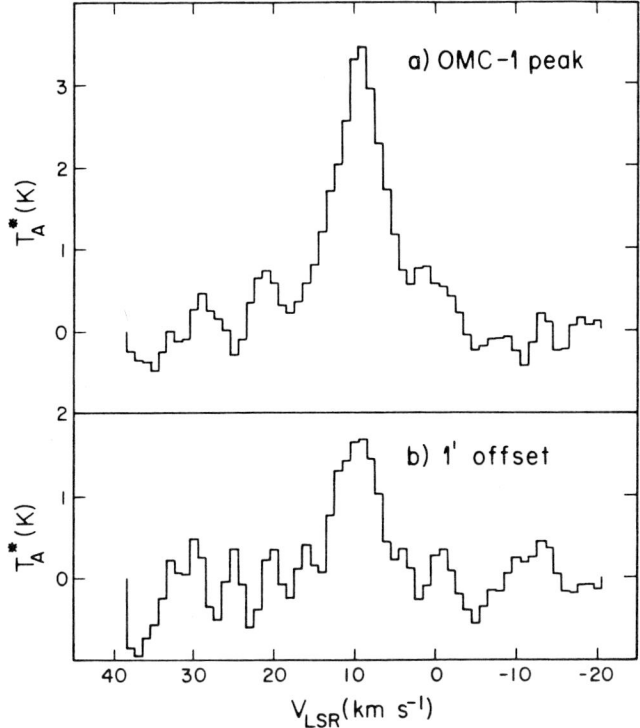

Figure 14. — NH$_3$ J$_K$ = 1$_0$ − 0$_0$

for monitoring the abundances of the metal hydrides, and therefore of the metals themselves, in the gas of the galaxy. It is exciting to report the detection of HCl (Blake, Keene and Phillips, 1985) as the first of the 'metal' hydrides to be detected in this way, and incidentally representing the first detection of chlorine in the molecular gas. HCl is expected to be a prime coolant molecule (Dalgarno *et al.* 1974; Jura, 1976). The molecule possesses a nuclear spin and concomitant hyperfine structure, so that there are three lines which can be resolved in heterodyne spectroscopy in the ground J = 1-0 transition. This is a vital aid to identification, for although the line frequencies are accurately determined by laboratory spectroscopy, the interstellar medium possesses many weak lines due to excited states of complex molecules. Figure 15 shows the KAO observed spectrum at 479μm, again in the direction of the core of the Orion Molecular Cloud. The expected line positions and relative strengths are indicated by the vertical lines. From these data it has been tentatively deduced that the order of 10% of the interstellar chlorine is in the HCl molecules.

Many other hydride species will have detectable lines. Although not necessarily important for cooling, the abundances of ion molecules such as H$_2$D$^+$ and H$_3$O$^+$ are of great interest to interstellar chemistry. H$_2$D$^+$ is a tracer for H$_3^+$, the fundamental ion of the ion–molecule gas phase chemistry scheme and H$_3$O$^+$ is in the chain for production of H$_2$O. The possible detections lead to estimates of $\sim 10^{-9}$ for X(H$_3^+$) and 5×10^{-10} for X(H$_3$O$^+$).

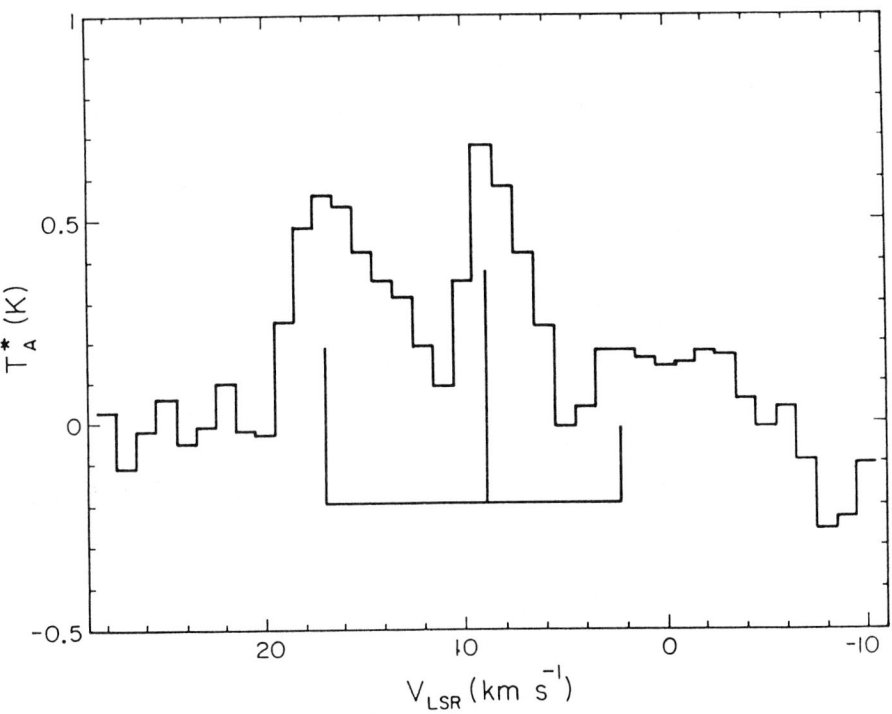

Figure 15. — HCl J = 1 − 0.

4.3 CO

From studies at millimeter wavelengths of the low J rotational transitions of CO it is expected that this molecule is the most abundant (apart from H_2) in the interstellar medium and that it is responsible for about 1/3 of the flux in the millimeter band from molecules. It is most important to have observations of as many of the CO transitions as possible, to obtain the best possible picture of the physics of the clouds and to compute the molecular cooling.

Table 1 gives the list of submillimeter CO lines observed. Up to J = 7 the detections are with heterodyne techniques, the higher J's are all detected with direct detection systems. The spectra for Orion are similar, except that the heterodyne data, e.g., J = (4 - 3) of Figure 16, reveal the separation between the 'spike' feature due to quiescent gas and the underlying 'plateau' feature due to the shocked gas.

Unfortuately, the real interstellar medium is physically complex. One of the major powers of submillimeter observations will be in the use of the ladder of lines from the highly abundant CO molecule to help unravel the problems. Suppose that we wanted to test a nonuniform cloud model of a centrally heated object. Since the μ_D for CO is $\sim$ 0.1, it is easily excited and the lower lines will be optically thick. As is well known observationally they have similar antenna temperatures, revealing the gas kinetic temperature. An example of this is the DR 21 data of White *et al.*, 1986 (Figure 17). As we proceed up the ladder, absorption dips may occur in certain parts of the line profile which are due to the optical depth

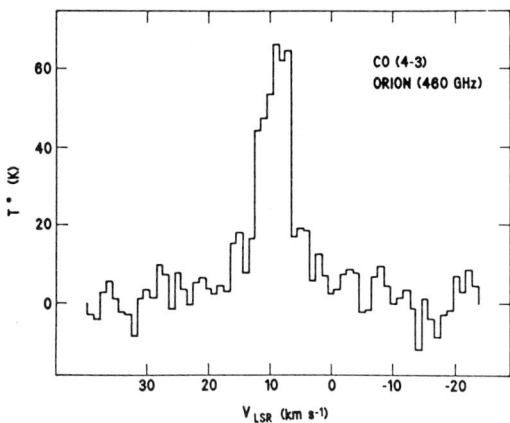

Figure 16. — A submillimeter CO spectrum of the OMC-1 region. The quiescent cloud (spike) narrow line is strong compared to the outflow region (plateau) broad line, mainly because of the beam dilution effect on the small plateau source (∼ 20") as viewed by the large KAO telescope beam (∼ 150").

increasing in diffuse outer parts of the cloud or foreground gas which may be deexcited or cool. As we go higher up the ladder yet, the temperatures and densities required for excitation go above those of the general molecular cloud regions, so that transitions such as $J = 7 - 6$ may probe the high density hot cores (Jaffe *et al.* 1986). The probe is too high in energy for the dense cool foreground gas to be effective as an absorber. This is a major use for the submillimeter lines of CO, and provides an excellent indication of the cooling mechanisms available to the central star formation regions.

Continuing to yet higher J values, where the atmosphere is truly opaque from the ground, we find that yet hotter regions are observed. CO has been detected from the KAO at various J values from 16 to 34 in the direction of the Orion region (Figure 18., Watson *et al.* 1985). The gas observed is thought to be shock excited. The lines are optically thin and fit a model in which the CO is subthermally excited at a temperature of about 750 K with an H_2 density of $2.7 \times 10^6 cm^{-3}$. The delineation of shocked regions with high spectral and spatial resolution is well performed by CO, because of course it is hard to dissociate and survives shocks which would destroy fragile molecules.

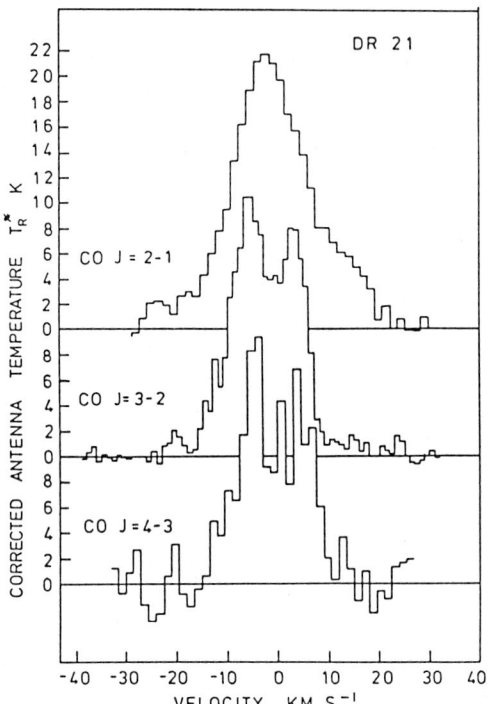

Figure 17. — CO J = 2 − 1, 3 − 2, and 4 − 3 spectra of DR21 (after White *et al.* 1986).

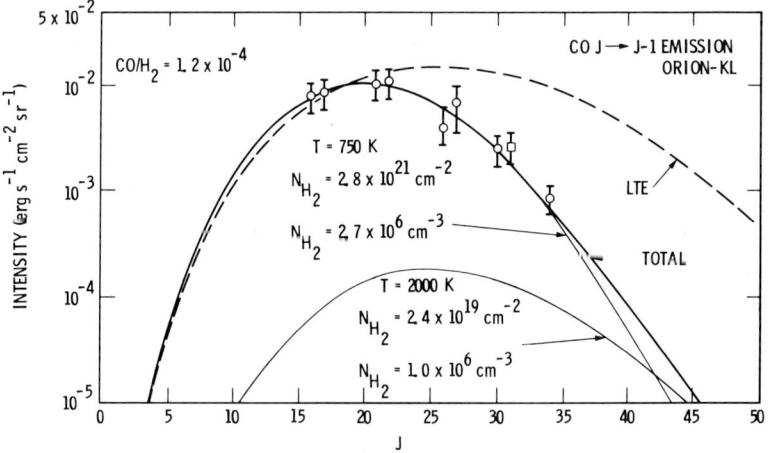

Figure 18. — Observed CO line intensities, compared to a thermal equilibrium model for T = 750 K, N(H$_2$) = 2.8 × 10^{21}cm^{-2} and a two component model incorporating a 2000 K region (after Watson *et al.* 1985).

5. ATOMS AND ATOMIC IONS

Atomic fine structure lines are expected to be major coolants for the diffuse interstellar gas. Now that we understand the basically molecular nature of the dense gas we might not expect either atoms or atomic ions to be very active in dense cloud cooling. However, it appears that neutral carbon, at least, is prevalent within many molecular cloud regions. Also ionized carbon and neutral oxygen can be observed at the edges of moleuclar clouds in the photodissociation regions. Table 3 gives a list of species and transitions which have been detected either within or on the periphery of molecular clouds.

Table 3
Submillimeter and Far-Infrared Atomic Fine Structure Lines

Species	Transition	Wavelength	Reference
CI	$^3P_1 - {}^3P_0$	$609 \mu m$	Phillips et al. (1980)
	$^3P_2 - {}^3P_1$	$370 \mu m$	Jaffe et al. (1985)
CII	$^2P_{3/2} - {}^2P_{1/2}$	$158 \mu m$	Russell et al. (1980)
OI	$^3P_1 - {}^3P_2$	$63 \mu m$	Melnick, Gull, and Harwit (1979)
	$^3P_0 - {}^3P_1$	$146 \mu m$	Stacey et al. (1983)b

5.1 CI

A current view of the nature of interstellar clouds is that they consist of large, relatively low density or diffuse regimes, permeated by the stellar ultraviolet flux, within which exist regions of higher density ($> 10^3$ H atoms or H_2 molecules per cm^3). Information on the properties of the diffuse medium, from UV studies with the Copernicus satellite (Morton et al. 1973) shows that most ($\sim 90\%$) of the gas-phase carbon is ionized. However, a detailed Copernicus study of C I abundances in the diffuse medium by Jenkins and Shaya (1979), shows that the CI column densities can approach 10^{16} cm^{-2} in the thickest clouds observable by their techniques. Chemical theories of a few years ago for the dense medium (e.g. Langer 1976a; Iglesias 1977; Prasad and Huntress 1980) predicted that near the surface of dense clouds, within the UV penetration depth, there would be a large C I content, but these same models predicted little or no C I in the deep interiors of the clouds, where the gas-phase carbon should be mostly in the form of CO. The prevailing approximate picture of the state of carbon in the interstellar medium (ISM) was then: C II in the diffuse medium, C I in dense cloud edges and CO in the bulk of the dense clouds. A newly forming cloud was expected to have a large abundance of C I for the first 10^6 years or so (e.g. Langer 1976b; Iglesias 1977).

The ground-state fine-structure transitions of C I are useful for investigations of carbon throughout the Galaxy. Due to spin orbit interactions, the ground ^3P state of C I is split into three levels, whose energies have been determined by a series of laboratory investigations. UV spectroscopy (e.g. Herzberg 1958) led to values of the splittings which were refined to the accuracy needed for radio spectroscopy techniques by Saykally and Evenson (1980). By chance, the magnetic dipole C I $^3P_1 - {}^3P_0$ transition has a comparable Einstein A coefficient to that of the weak electric dipole CO J = 1 - 0 transition, i.e. about 7.9×10^{-8} s^{-1} (Nussbaumer 1971). Since the collision cross sections of CO (1 - 0) and C I $^3P_1 - {}^3P_0$ with H or H_2 are also comparable (Yau and Dalgarno 1976; Launay and Roueff 1977; Flower, Launay, and Roueff 1978), it seems probable that CO (1 - 0) and C I ($^3P_1 - {}^3P_0$) should be observed in similar states of excitation throughout the interstellar medium, provided that both are present and that the temperature is high enough to populate the CI $^3P_1$ state, for which $E_1 \sim 24K$. Because of this and because of the interest in carbon chemistry and the relative abundances of CI and CO, it seems natural to compare the spectra, spatial distributions and column densities determined from the lines of CO (1 - 0) and C I ($^3P_1 - {}^3P_0$). The availability of the $^3P_2 - {}^3P_1$ line of C I should make a considerable improvement in the accuracy of C I column density measurements (Jaffe et al. 1985).

Figure 19 shows the OMC-1 spectrum of CI ($^3P_1 - {}^3P_0$). The line velocity and shape is very similar to that of CO (1 - 0). Subsequent studies (Phillips and Huggins, 1981) showed that this was the case for many clouds in the galaxy and more precisely the CI and ^{13}CO line shapes and widths are often almost indistinguishable.

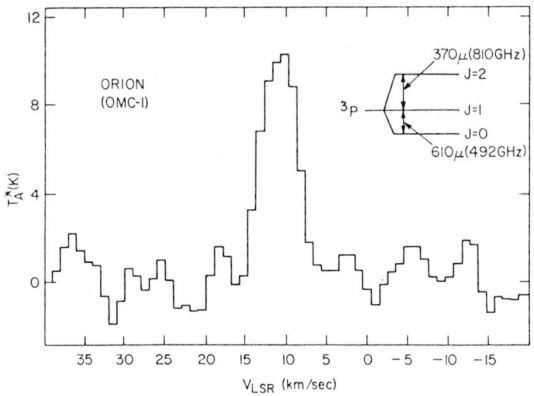

Figure 19. — A spectrum of neutral carbon emission, $^3P_1 - {}^3P_0$, at 492 GHz, taken in the direction of OMC-1. Insert: the ground-state fine-structure splitting of CI; higher levels are in the optical and UV.

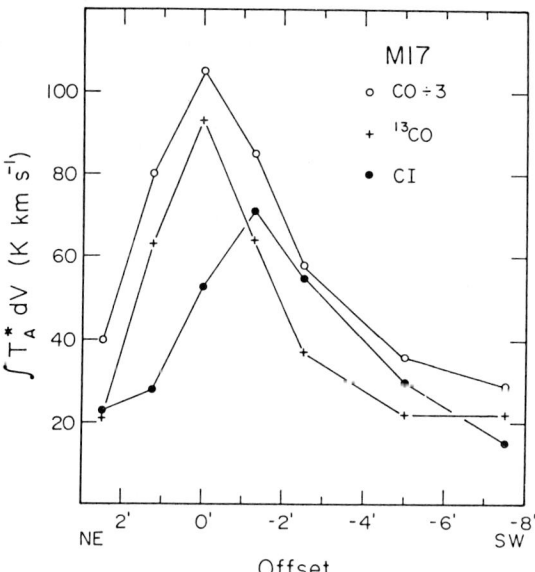

Figure 20. — Comparison of CI and CO antenna temperatures in M17 integrated over velocity. The ionization front is to the left. The CI and ^{13}CO are on the same scale; the ^{12}CO temperatures have been divided by 3.

Also it was shown that the CI column density is usually within a factor of 10 of that of ^{12}CO. This is also seen from the CI ($^3P_2 - ^3P_1$) work of Zmuidzinas *et al.* (1986) who find optical depths approaching unity for several cloud core regions. Recent KAO studies by Frerking *et al.* (1987) and Keene *et al.* (1985) have shown that C I extends throughout molecular clouds and is not simply contained in a skin the thickness of the UV penetration depth. Further, it is found by Keene *et al.* (1985) that in cases of strong optical illumination at the cloud surface the C I distribution peaks into the cloud further from the radiation source than does CO (Figure 20). The high absolute abundance of C I, its high abundance relative to CO and its distribution throughout the ISM (which is quite similar to that of ^{13}CO) are the subjects of several recent theoretical studies. Possible explanations involve higher-than-usual external UV flux (Tielens and Hollenbach 1985), internal UV flux (Prasad and Tarafdar 1983), redistribution of surface clumps (Boland and de Jong 1982), high metal atom abundances (Graedel, Langer, and Frerking 1982), high [C/O] ratios in the dense gas phase (Langer *et al.* 1984) and time dependent effects (Tarafdar *et al.* 1985; Gerola and Glassgold 1978).

While not the dominant cooling agent, CI nevertheless provides a fascinating probe of the chemical structure of molecular clouds.

5.2 CII AND OI

The 158μm line of CII and the 63 and 146μm lines of OI are powerful radiators. They are the brightest lines in the far-infrared spectrum of galaxies and together carry about 1% of the bolometric luminosity of gas rich galaxies (Genzel and Stacey; 1986; Watson 1986). A great deal of work has been carried out by the Townes group at Berkeley, using the tandem Fabry-Perot spectrometer on the KAO to determine the spatial distribution of the CII and OI emission. These emission lines emanate largely from the photodissociation regions and correlate strongly with UV illuminated surfaces of molecular clouds.

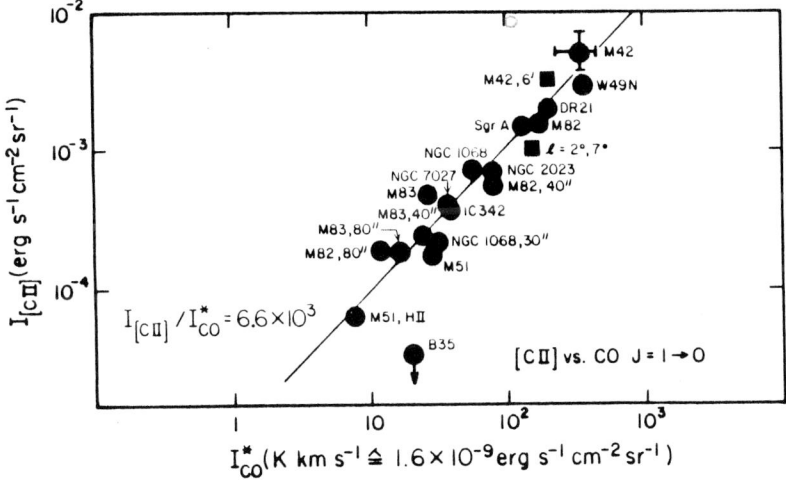

Figure 21. — CII 157.7 μm and CO J = 1 − 0 observations of galaxies with high star formation rates and of Galactic regions of star formation.

In the direction of the Orion OMC-1 region the 63μm OI line emits about 600L$_\odot$ or 2×10^{-3} of the total far-infrared luminosity of the central 5' of the source (Melnick, Gull and Harwit 1979). The velocity and width of the line (Lugten, Crawford and Genzel 1985) correlates well with that of the C 109α radio recombination line (Jaffe and Pankonin 1978) indicating that the OI and CII emission regions are the same.

Spatial maps of OI and CII emission from the OMC-1 region are well correlated (Genzel and Stacey 1985; Ellis and Werner 1985) with each other and with CO (1 - 0) emission. Since this region is viewed face-on, that is consistent with the concept of the photodissociation region being the source for CII and OI lines with the warm cloud behind it producing the CO emission.

A further confirmation of the close relationship between CO and CII emission is found in the work of Crawford *et al.* (1985) where the CII intensity is measured for a range of galactic and extragalactic sources and compared with that of CO (Figure 21). A remarkable correlation is found for most sources. This is in agreement with the model of Tielens and Hollenbach (1985), which proposes that the CII emission is correlated with the energy density of the UV field, provided that the CO emission also can be related to the UV energy density.

REFERENCES

Blake, G. A., Keene, J., and Phillips, T. G. 1985, *Ap.J.*, **295**, 501.
Blake, G. A., Sutton, E. C., Masson, C. R. and Phillips, T. G. 1986, *Ap. J. Suppl.*, **60**, 357.
Blake, G. A., Sutton, E. C., Masson, C. R. and Phillips, T. G., 1987, *Ap. J.*, submitted.
Boland, W. and de Jong, T. 1982, *Ap.J.*, **261**, 110.
Crawford, M. K., Genzel, R., Townes, C. H. and Watson, D. M. 1985, *Ap. J.*, **291**, 755.
Crawford *et al.* 1986, *Ap.J.*, in press.
Cummins, S. E., Linke, R. A., and Thaddeus, P. 1986, *Ap.J. Suppl.*, **60**, 819.
Dalgarno, A., de Jong, T., Oppenheimer, M. and Black, J. H. 1974, *Ap. J. (Letters)*, **192**, L37.
DeLucia, F. C. 1976, *Molecular Spectroscopy: Modern Research*, **2**, 69.
Dickinson, A. S., Phillips, T. G., Goldsmith, P. F., Percival, I. C. and Richards, D. 1977, *Astr.Ap.*, **54**, 645.
Elias, J. H. *et al.* 1978, *Ap.J.*, **220**, 25.
Ellis, B. and Werner, M. 1985, referred to in Genzel and Stacey (1986).
Flower, D. R., Launay, J. M., and Roueff, E. 1978, in *Spectra of Simple Molecules in the Laboratory and Astrophysics*, 137.
Frerking, M. A., Keene, J. B., Blake, G. A., Phillips, T. G. and Beichman, C. A. 1987, *Ap.J.*, submitted.
Genzel R., Watson, D. M., Crawford, M. K. and Townes, C. H. 1985, *Ap.J.*, **298**, 316.
Genzel, R. and Stacey, G. J. 1986 *Mittg. Der. Astr.*, Gesellschaft.
Gerola, M. and Glassgold, A. E. 1978, *Ap.J.Suppl.*, **37**, 1.
Goldsmith, P. F., *et al.* 1981, *Ap.J.(Letters)*, **243**, L79.
Goldsmith, P. F. and Langer, W. D. 1978, *Ap.J.*, **222**, 881.
Graedel, T. E., Langer, W. D. and Frerking, M. A. 1982, *Ap.J.Suppl.*, **48**, 321.
Green, S., 1974, *Physique Atomique et Moleculaire et Matiere Interstellaire,* Ed. Balian, Encrenaz, Lequeux. North Holland, Les Houches XXVI, p. 83..
Harris, A. E., Jaffe, D. T., Silber, M. and Genzel, R. 1985, *Ap. J. (Letters)*, **294**, L93.
Herzberg, G. 1958, *Proc. Roy. Soc. London, A*, **248**, 309.
Iglesias, E. 1977, *Ap.J.*, **218**, 697.
Jaffe, D. T. and Pankonin, V. 1978, *Ap.J.*, **226**, 869.
Jaffe, D. T. Harris, A. I., Silber, M., Genzel, R. and Betz, A. L. 1985, *Ap.J.(Letters)*, **290**, L59.
Jenkins, E. B and Shaya, E. J. 1979, *Ap.J.*, **231**, 55.
Johansson, L. E. B., *et al.* 1984, *Astr.Ap.*, **130**, 227.
Jura, M. 1974, *Ap.J. (Letters)*, **190**, L33.
Keene, J., Hildebrand, R. H. and Whitcomb, S. E. 1982, *Ap.J. (Letters)*, **252**, L11.
Keene, J. B., Blake, G. A. and Phillips, T. G. 1983, *Ap. J. (Letters)*, **271**, L27.

Keene, J. B., Blake, G. A., Phillips, T. G., Huggins, P. J. and Beichman, C.A. 1985, *Ap.J.*, **299**, 967.

Langer, W. D. 1976a, *Ap.J.*, **206**, 699.

Langer, W. D. 1976b, *Ap.J.*, **210**, 328.

Langer, W. D., Graedel, T. E., Frerking, M. A. and Armentrout, P. B. 1984, *Ap.J.*, **227**, 581.

Launay, J. M. and Roueff, E. 1977, *Astr.Ap.*, **56**, 289.

Lovas, F. J. 1983, private communication.

Lugten, J. B., Crawford, M. K. and Genzel, R. 1985, referred to in Genzel and Stacey (1986).

Melnick, G., Gull, G. E. and Harwit, M. 1979, *A. J. (Letters)*, **227**, L29.

Morton, D. C., Drake, J. F., Jenkins, E. B., Roberson, J. B. Spitzer, L. and York, D. G. 1973, *Ap.J.(Letters)*, **181**, L103.

Nussbaumer, H. 1971, *Ap.J.*, **166**, 411.

Penzias, A. A. and Burrus, C. A. 1973, *Ann.Rev.Astr.Ap.*, **11**, 51.

Phillips, T. G., Huggins, P. J., Neugebauer, G. and Werner, M. W. 1977, *Ap. J. (Letters)*, **217**, L161.

Phillips, T. G., Huggins, P. J. Kuiper, T. B. H., and Miller, R.E. 1980, *Ap. J. (Letters)*, **238**, L103.

Phillips. T. G., Kwan, J. and Huggins, P. J., 1980 I.A.U. Symposium #87, P21, (Interstellar Molecules, Reidel).

Phillips, T. G. and Huggins, P. J. 1981, *Ap.J.*, **251**, 533.

Phillips, T. G. and Woody, D. P. 1982, *Ann.Rev.Astr.Ap.*, **20**, 285.

Phillips, T. G. and Watson, D. M., 1984, The Large Deployable Reflector: Instruments and Technology (JPL report D-2214).

Phillips, T. G., Blake, G. A., Keene, J., Woods, R. C. and Churchwell, E. 1985, *Ap. J. (Letters)*, **294**, L45.

Pickett, H. M., Cohen, E. A., Brinza, E. D. and Schaefer, M. M. 1981, *J.Mol.Spec.*, **89**, 542.

Poynter, R. L. and Pickett, H. M. 1981, JPL publication 80-23.

Prasad, S. S. and Huntress, W. T. 1980, *Ap.J.*, **239**, 151.

Prasad, S. S. and Tarafdar, S. P. 1983, *Ap.J.*, **267**, 603.

Russell, R. W., Melnick, G., Gull, G. E. and Harwit, M. 1980, *Ap. J. (Letters)*, **240**, L99.

Saykally, R. J. and Evenson, K. 1980, *Ap.J.(Letters)*, **238**, L107.

Stacey, G. J. Smyers, S. D., Kurtz, N. T., Harwit, M., Russell, R. W. and Melnick, G. 1982, *Ap. J. (Letters)*, **257**, L37.

Stacey, G. J., Smyers, S. D., Kurtz, N. T. and Harwit, M. 1983a, *M.N.R.A.S.*, **202**, 25P.

Stacey, G. J., Smyers, S. D., Kurtz, N. T. and Harwit, M. 1983b, *Ap. J. (Letters)*, **265**, L7.

Stacey, G. J., Lugten, J. B. and Genzel, R. 1986, preprint.

Storey, J. W. V., Watson, D. M. and Townes, C. H. 1981, *Ap. J. (Letters)*, **244**, L27.

Storey, J. W. V., Watson, D. M., Townes, C. H., Haller, E. E., and Hansen,W. L. 1981, *Ap. J.*, **247**, 136.

Storey, J. W. V., 1985, *Infrared Phys.*, **25**, 583.

Sutton, E. C., Blake, G. A., Masson, C. R. and Phillips, T. G., 1984, *Ap. J. (Letters)*, **283**, L41.

Sutton, E. C., Blake, G. A., Masson, C. R. and Phillips, T. G. 1985, *Ap. J. Suppl.*, **58**, 341.

Takahashi, T., Hollenback, D. J. and Silk, J. 1985, *Ap.J.*, **292**, 192.

Tarafdar, S. P., Prasad, S. S., Huntress, W. T., Villare, K. R. and Black, D. C. 1985, *Ap.J.*, **289**, 220.

Tielens, A. G. G. M. and Hollenbach, D. 1985, *Ap.J.* **291** 722 and 747.

Townes, C. H., Genzel, R., Watson, D. M. and Storey, J. W. V. 1983, *Ap. J. (Letters)*, **269**, L11.

Waters, J. W., Gustincic, J. J., Kakar, R. K., Kuiper, T. B. H., Roscoe, H. K., Swanson, P. N., Rodriquez Kuiper, E. N., Kerr, A. R., and Thaddeus, P. 1980, *Ap. J.*, **235**, 57.

Watson, D. M., Storey, J. W. V., Townes, C. H., Haller, E. E. and Hansen, W. L. 1980, *Ap. J. (Letters)*, **239**, L129.

Watson, D. M., Genzel, R., Townes, C. H. and Storey, J. W. V. 1985, *Ap.J.*, **298**, 316.

Watson, D. M. 1986, *Physica Scripta*, in press.

White, G. J., Phillips, J. P., Richardson, K. J. and Harten, R. H. 1986, *Astr. Ap.*, **159**, 309.

Wootten, A., Boulanger, F., Bogey, M., Combes, F., Encrenaz, P. and Gerin, M. 1986, preprint.

Yau, A. W. and Dalgarno, A. 1976, *Ap.J.*, **206**, 652.

Zmuidzinas, J., Betz, A. L. and Goldhaber, D. M. 1986, *Ap. J. (Letters)*, **307**, L75.

Zuckerman, B. 1973, *Ap.J.*, **183**, 863.

Constance Walker and John Bally

HEATING AND COOLING OF THE INTERSTELLAR GAS

John H. Black
Steward Observatory
University of Arizona
Tucson, AZ 85721 USA

ABSTRACT. Basic considerations of the global heating and cooling of the interstellar gas are summarized. The various energy sources are reviewed. Expressions for the rates of a number of typical heating and cooling processes are given. General comments are made about the conditions in the several phases of the interstellar medium in thermal balance.

1. INTRODUCTION

The subject of heating and cooling of the interstellar gas is a very broad one. The rates of heating and cooling govern the temperatures of the various components of the interstellar medium. These rates depend in turn upon the composition and density of the interstellar matter as well as the underlying sources of energy. This subject is related to the ways in which interstellar matter participates in the birth of new stars in the Galaxy and in which it reacts to the deaths of old stars. In a short lecture it is impossible to do justice to the entire subject. Some aspects of this subject are discussed elsewhere in this volume (Flower 1986, Phillips 1986). The reader is directed to the excellent book of Spitzer (1978) and earlier reviews (e.g. Dalgarno and McCray 1972) for more extensive background information.

A discussion of global aspects of the thermal balance of the interstellar gas cannot be divorced from a treatment of the microscopic processes by which energy is exchanged between any particular volume of gas and its larger environment. The interstellar medium is far from a state of thermodynamic equilibrium: its physical state can be characterized not by a single temperature, but by the complicated balance among a potentially large number of processes. An early discussion of the energy balance of the interstellar medium by Eddington (1926) appeared at a time when the first evidence for the existence of diffuse interstellar matter was arising. Even so, Eddington was able to make useful estimates of the state of ionization and steady-state temperature of the dilute gas. Heating and cooling in ionized regions around hot stars (H II regions) is a well-developed subject in its own right (see, e.g., Osterbrock 1974) that can be generalized to apply to a broader range of photoionized nebulae such as QSO emission line regions, the "atmospheres" of stellar X-ray sources, etc. Global descriptions of the interstellar medium include two (Field, Goldsmith, and Habing 1969) or more (McKee and Ostriker 1977) distinct phases of the gas that are thermally stable. Observations at X-ray and

731

D. J. Hollenbach and H. A. Thronson, Jr. (eds.), Interstellar Processes, 731–744.

ultraviolet wavelengths have revealed evidence for hot $(T \geq 10^5$ K) components of the interstellar medium, while observations at millimeter wavelengths have led to the identification of giant molecular clouds at low $(T \leq 100$ K) temperatures.

It is conventional to evaluate the rates of heating and cooling by many processes according to the assumption that the microscopic motions of the participating atoms, electrons, and molecules can be described by a Maxwellian velocity distribution at a well-defined kinetic temperature T. Often, this is a very good assumption because the cross sections and rates for elastic collisions that tend to thermalize gas motions tend to be much larger than the cross sections and rates for inelastic processes. The heating of gas by suprathermal particles (cosmic rays) and by extremely energetic photons may require a more complicated description because much of the initial energy deposited may be lost to radiation almost immediately before it can be converted efficiently into kinetic energy in the gas.

2. SOURCES OF ENERGY AND HEATING RATES

A typical point in our Galaxy, for example the Sun's location, is exposed to external energy sources of various kinds: background radiation, energetic particles, mechanical inputs, and magnetic fields. These sources and the ways in which they interact with the interstellar gas are considered in turn.

2.1 Background radiation

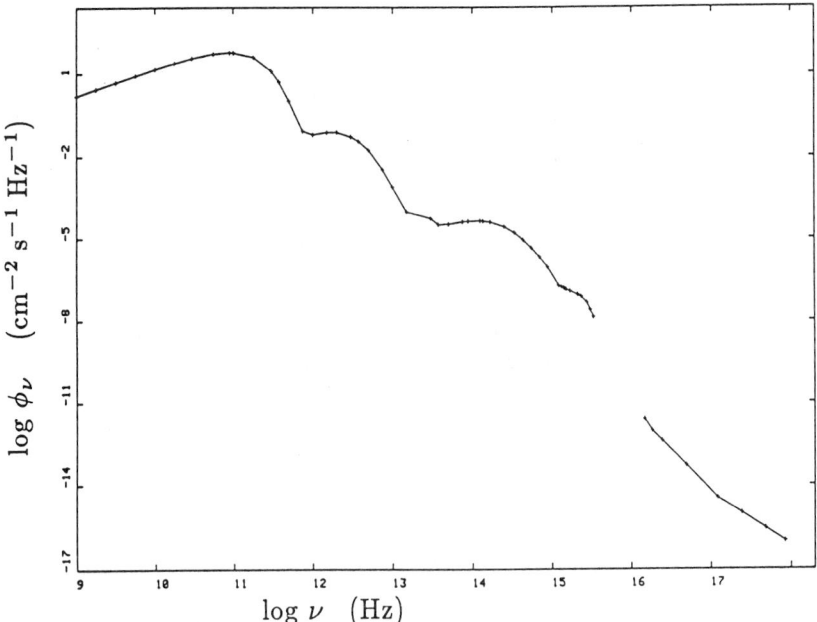

Figure 1 shows the frequency dependence of a composite interstellar radiation field that extends from microwave to X-rays. This information is displayed in the form of a photon flux ϕ_ν in cm^{-2} s^{-1} Hz^{-1}, which is related to the flux in energy units by $\phi_\nu = f_\nu/h\nu$. This spectrum is the sum of a $T = 2.7$ K blackbody; the background radiation in the infrared, visible and ultraviolet tabulated by Mathis, Mezger, and Panagia (1983) for the solar vicinity; and the soft X-ray background reviewed by Bregman and Harrington (1985).

A corresponding graph of νf_ν versus ν (Figure 2) indicates that comparable

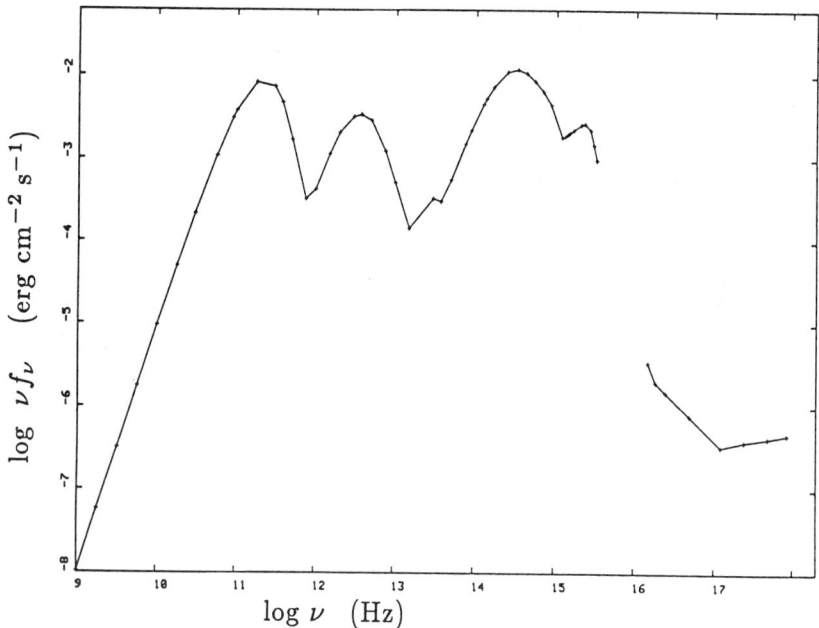

amounts of energy per decade are present in the submillimeter region (peaking near $log\ \nu = 11.3$), the far-infrared ($log\ \nu = 12.55$), the visible ($log\ \nu = 14.5$), and the ultraviolet ($log\ \nu = 15.3$). These components of the radiation can be attributed to the $T = 2.7$ K cosmic background radiation, the thermal emission of cool interstellar dust, average starlight, and the light of the O- and B-type stars including diffuse scattering by interstellar dust, respectively. The far-infrared component arises in dust that has been heated by starlight and is thus not a primary energy source itself. Photons with energies $13.6 < h\nu < 54$ eV are presumed to be absent from the general interstellar radiation owing to the effects of absorption by interstellar H and He atoms that confine these photons to the immediate neighborhoods of the hot stars that produce them. The X-ray component ($log\ \nu > 16.1$ Hz) consists of a contribution from extragalactic sources such as QSOs and nearby active galaxies, and a local contribution from hot interstellar gas that has been heated by expanding supernova remnants, stellar wind bubbles, etc. In detail, the local X-radiation is actually primarily in spectral lines rather than continuous emission because it arises in a very dilute plasma, while the extragalactic component may be continuous. Gamma rays of even higher energy are also observed. These arise from pion decay following interactions of cosmic ray protons with ambient protons, from the inverse Compton effect involving cosmic ray electrons, and from discrete sources like Cyg

X-3. Although the diffuse gamma radiation is useful for tracing the distributions of interstellar matter and of cosmic rays in the Galaxy (see, e.g., Bloemen *et al.* 1984), it is completely inconsequential as a heating source for the interstellar gas because the photons are rare and their interactions are improbable.

The integrated flux in the above radiation field is $F = \int f_\nu d\nu = 0.08$ erg s^{-1} cm^{-2}, which corresponds to an energy density of $U_r = F/c = 0.83$ eV cm^{-3}, where c is the speed of light. Note that the thermal energy density of interstellar gas of hydrogen number density n_H cm^{-3} and temperature T K is

$$U_{gas} = \frac{3}{2}n_H kT = 1.29 \left(\frac{n_H}{100}\right)\left(\frac{T}{100}\right) \text{ eV cm}^{-3}. \tag{1}$$

The rate at which radiation heats the gas depends upon both the spectrum of radiation (Figure 1) and the details of the processes by which the radiation interacts with the gas. It is useful to distinguish between continuous and discrete absorption processes. For a continuous process such as photoionization, with threshold ν_0 and absorption cross section σ_ν, the product species (e.g. photoelectrons and ions) carry off an amount of energy $h(\nu - \nu_0)$ that is typically added to the kinetic energy budget of the gas through subsequent elastic collisions until the motions of the energetic product species have been thermalized. In some cases, ν_0 may be an "effective" threshold for heating rather than the threshold for absorption in order to compensate for excited-state products that decay directly by the production of photons that escape without adding kinetic energy to the gas. The heating rate is then

$$\Gamma = n_i \int h(\nu - \nu_0)\sigma_\nu \phi_\nu d\nu \quad \text{erg s}^{-1} \text{ cm}^{-3} \tag{2}$$

for absorption by species i of density n_i. Although the interstellar dust does not maintain close thermal contact with the gas through collisions in many interstellar regions, dust particles are thought to be important sources of heat through emission of photoelectrons (Watson 1972, de Jong 1977, Draine 1978). The rate of grain photoelectric heating depends upon the photoelectron yield, the photoelectric threshold of the grain material, the mean electrical charge on the grains, and the grain abundance. Because the absorption process attenuates the radiation, the flux ϕ_ν must, of course, be evaluated as a function of depth through an interstellar region. For typical grain properties, a grain photoelectric heating rate in the unattenuated radiation field of the order of

$$\Gamma_{ge} \approx 4 \times 10^{-26} n_H \quad \text{erg s}^{-1} \text{ cm}^{-3} \tag{3}$$

is expected (cf. equation 15 of de Jong, Dalgarno, and Boland 1980). The variation of this heating rate with depth through a relatively thin interstellar cloud has been illustrated by Roberge, Dalgarno, and Flannery (1981) on the basis of an accurate treatment of the radiative transfer.

In neutral regions, atomic carbon, for example, can be photoionized at $11.2 < h\nu < 13.6$ eV with an average yield of approximately 1 eV per photoelectron, for a heating rate of

$$\Gamma_{pi} \approx 2 \times 10^{-22} n(\text{C}) \quad \text{erg s}^{-1} \text{ cm}^{-3} \tag{4}$$

where $n(C)$ is the number density of neutral carbon atoms. In highly ionized regions where C^{+3} may be the most abundant form of carbon, the corresponding yield is approximately 21 eV per X-ray absorption and the heating rate is

$$\Gamma_{pi} \approx 1.6 \times 10^{-25} n(C^{+3}) \quad \text{erg s}^{-1} \text{ cm}^{-3}. \tag{5}$$

In the case of X-ray photoionization of C^{+3}, absorption of a photon with $0.0645 < h\nu < 0.4$ KeV removes a single 2s electron, while absorption of a photon with $h\nu > 0.4$ KeV removes a K-shell electron directly and leads to the ejection of a second electron through the Auger effect (cf. Weisheit 1974).

The above examples illustrate cases in which the gas is heated directly as a result of absorption of radiation. An important heating source in some neutral regions of high density arises from collisions between gas molecules and dust grains that have been heated by the near-infrared radiation field. An atom or molecule that strikes the surface of a grain at temperature T_{gr}, that comes into thermal equilibrium with that surface, and then re-enters the gas phase, may gain kinetic energy from the grain if $T_{gr} > T$, or lose kinetic energy to it if $T > T_{gr}$. The rate of heating or cooling depends upon the temperature difference, the frequency of collisions, the composition of the gas, the properties of the particular kind of solid, and the propensity of atoms to stick to the surface (Burke and Hollenbach 1983). If it is understood that a negative heating rate signifies cooling, then

$$\Gamma_{g-gr} = n_{gr} n_H \sigma_{gr} \left(\frac{8kT}{\pi m_H} \right)^{1/2} \alpha_T 2k(T_{gr} - T) \quad \text{erg s}^{-1} \text{ cm}^{-3} \tag{6}$$

where the average thermal accomodation coefficient $\alpha_T \approx 0.35$ for various types of cold grains ($T_{gr} < 100$ K) in molecular gas at $T \leq 100$ K (Burke and Hollenbach 1983). Interstellar extinction measurements suggest that the typical grain abundance and geometrical cross section are

$$n_{gr} \sigma_{gr} \approx 1.5 \times 10^{-21} n_H \quad \text{cm}^{-1} \tag{7}$$

so that the heating/cooling rate becomes

$$\Gamma_{g-gr} \approx 2.1 \times 10^{-33} n_H^2 T^{1/2} (T_{gr} - T) \quad \text{erg s}^{-1} \text{ cm}^{-3} \tag{8}$$

which is 2.8 times larger than the expression of Boland and de Jong (1984), owing primarily to differences in the adopted grain cross sectional areas.

The grain temperature itself is governed by the balance between absorption of visible and near-infrared starlight and emission of infrared radiation by the grains (see, e.g., Mathis et al. 1983; Cox, Krügel, and Mezger 1986). It is interesting to consider briefly the energy balance of the dust in a uniform interstellar cloud exposed to the radiation field of Figure 1. Assume that all of the starlight in the wavelength range $0.1 - 8.0$ μm incident on the cloud is absorbed by the grains: this provides an integrated flux of $F_{in} = 0.0225$ erg s^{-1} cm^{-2}. The integrated flux radiated by the dust cloud of optical radius τ_ν is

$$F_{out} = \pi \int_0^\infty \left(1 - e^{-\tau_\nu} \right) B_\nu(T_{gr}) d\nu \tag{9}$$

where B_ν is the Planck function. In the special case of an opacity law of the form

$$\tau_\nu = \alpha \left(\frac{\nu}{c}\right)^\beta, \tag{10}$$

which is a good approximation for cold interstellar grains in the far-infrared and sub-millimeter region, the integral can be evaluated directly

$$F_{out} = \sigma_{sb} T_{gr}^4 \Psi \tag{11}$$

where

$$\Psi = \frac{15}{\pi^4} \sum_{m=1}^{\infty} (-1)^{m+1} \frac{\gamma^m}{m!} \, \varsigma(4 + \beta m) \, \Gamma(4 + \beta m), \tag{12}$$

and σ_{sb} is the Stefan-Boltzmann constant. The variable $\gamma = \alpha(kT/hc)^\beta$, while ς is the Riemann zeta function and Γ is the gamma function. When $\gamma \ll 1$ (often the case of interest) the series in equation (12) converges rapidly. For example, if $\tau_\nu = 0.01$ at $\nu/c = 100$ cm^{-1} (i.e., $\lambda = 100$ μm) and if $F_{in} = F_{out} = 0.0225$, then $T_{gr} = 15$ K for opacity laws (equation 10) with $1 \leq \beta \leq 2$. Now, if we compare the heating rate for gas-grain collisions at $T = 10$ K, for example, with the $unattenuated$ rate for grain photoelectric heating (equation 3), we find that $\Gamma_{g-gr}/\Gamma_{ge} \approx 2.5 \times 10^{-6} n_H$. This ratio increases with increasing depth into a cloud as the ultraviolet radiation is attenuated more rapidly than the near-infrared radiation.

At this point it is worth mentioning that the total amount of radiant energy involved in heating and cooling the grains tends to be much larger than that involved in heating and cooling the gas, because the efficiency for heating the grains directly is larger than the efficiency for converting radiant energy into kinetic energy in the gas and because the visible and near-infrared photons effective in heating the grains have a larger integrated energy flux than the ultraviolet photons and X-ray photons that are effective in heating the gas directly. This means that the luminosity of cooling radiation from the grain component of the interstellar medium will usually surpass that of the gaseous component.

There are also line absorption processes which deposit kinetic energy into the gas. One example is the spontaneous radiative dissociation of interstellar H$_2$ following absorption in the Lyman and Werner system lines (Stecher and Williams 1967), which yields an average of 0.4 eV per dissociated atom pair in a radiation field like the adopted one (Milgrom, Panagia, and Salpeter 1973; Stephens and Dalgarno 1973). Neglecting effects of radiative transfer on the flux of radiation, the heating rate for such a line absorption process can be written

$$\Gamma_{line} = \varepsilon \frac{\pi e^2}{mc} f_{line} \phi_\nu \quad \text{erg s}^{-1} \text{ cm}^{-3} \tag{13}$$

where ε is an average yield of kinetic energy, f_{line} is the absorption oscillator strength of the line, e and m are the charge and mass of the electron, respectively. The H$_2$ dissociative heating rate is $\Gamma_d(\text{H}_2) \approx 6.4 \times 10^{-13} k_d n(\text{H}_2)$ erg s^{-1} cm^{-3}, where k_d is the severely depth-dependent dissociation rate (Black and Dalgarno 1977; van Dishoeck and Black 1986). The same Lyman and Werner line absorptions that lead to dissociation are even more likely to be followed by fluorescence to bound, vibrationally excited levels of the ground electronic state. In molecular regions exposed to intense ultraviolet fluxes, this fluorescent excitation followed by superelastic collisions can be a significant heating source (Tielens and Hollenbach 1985).

A second example of heating via line absorption involves submillimeter transitions of the H_2O molecule in warm dust clouds as discussed recently by Takahashi, Hollenbach, and Silk (1983) and by Cheng-Yue et al. (1985). This is actually an indirect way of heating the gas by radiation through intermediary grains: near infrared light (either from the general background or from a local source) heats the dust grains, which in turn radiate photons at longer wavelengths; H_2O then absorbs some of this radiation in its rotational lines; and those molecules that are collisionally de-excited before they have time to radiate themselves contribute the excess energy of excitation to the kinetic energy budget of the gas. Under some circumstances—typically conditions of high dust temperatures and large H_2O abundances—this heating mechanism is more efficient than direct collisional interactions between gas and grains (equation 8). The heating efficiency depends in a very complicated way on details of the structure, velocity fields, abundances, and densities in interstellar clouds and can be evaluated reliably only within the context of a specific detailed model through a careful treatment of H_2O energy level populations and radiative transfer (Takahashi, Hollenbach, and Silk 1983).

2.2 Cosmic Rays

Galactic cosmic rays (principally protons) represent a potentially large source of energy input into the interstellar medium. Their total energy density is estimated to be 0.8 eV cm^{-3} (Spitzer 1978). Cosmic ray protons of relatively low energy, say 2 - 10 MeV, are most effective in ionizing and heating the gas. However, low-energy cosmic rays are also most severely affected by interstellar magnetic fields; therefore, although they may be produced in supernova shocks, they seem not to propagate widely through the interstellar medium. Cosmic ray fluxes in the galactic plane can be inferred from direct measurements within the solar system, but at low energies these measurements require large corrections for local modulation by the solar wind. Cosmic ray fluxes can also be deduced from the intensity of diffuse gamma radiation (see § 2.1) and from the analysis of the ionization and chemistry of neutral clouds (Black and Dalgarno 1973, O'Donnell and Watson 1974). The most recent analysis of ionization and molecular abundances in diffuse clouds suggests a typical hydrogen ionization rate $\varsigma_0 = 7 \times 10^{-17}$ s^{-1} (van Dishoeck and Black 1986). This corresponds to a primary ionization rate $\varsigma_p = \varsigma_0/(1 + \phi) = 4 \times 10^{-17}$ s^{-1}, where $\phi \approx 0.7$ is the correction for secondary ionizations (Dalgarno and McCray 1972, Cravens and Dalgarno 1978). The determination of the heating rate due to cosmic rays is very complicated owing to the myriad processes by which the cosmic rays and their secondary ionization products lose energy. The heating efficiency is sensitive to the composition and density of the gas and to its degree of ionization. In a neutral, molecular gas, the mean heating input per primary ionization (including secondary processes) probably lies in the range, $\epsilon_h = 5.7$ to 7.3 eV, the higher value applying at densities $n_H > 10^4$ cm^{-3} (Cravens and Dalgarno 1978). Thus the cosmic ray heating rate in a molecular cloud is

$$\Gamma_{cr} \approx 4 \times 10^{-28} \left(\frac{\zeta_p}{4 \times 10^{-17}} \right) \left(\frac{\epsilon_h}{6 \text{ eV}} \right) n(\text{H}_2) \quad \text{erg s}^{-1} \text{ cm}^{-3} \qquad (14)$$

while that in an atomic cloud of low ionization is

$$\Gamma_{cr} \approx 3.5 \times 10^{-28} \left(\frac{\zeta_p}{4 \times 10^{-17}} \right) n(\text{H}) \quad \text{erg s}^{-1} \text{ cm}^{-3} \qquad (15)$$

(Dalgarno and McCray 1972). In a highly ionized gas a much larger fraction of the secondary electron energy goes into heating, because elastic encounters with other electrons are much more probable than other processes. In this case, an approximate heating rate can be written

$$\Gamma_{cr} \approx \left(2 \times 10^{-27}(1 - x) + 2 \times 10^{-26}x \right) n_H \left(\frac{\zeta_p}{4 \times 10^{-17}} \right) \quad \text{erg s}^{-1} \text{ cm}^{-3} \qquad (16)$$

where the first term represents heating following ionization of neutral H, the second term describes direct heating of the electron gas, $x = n(e)/n_H$ is the fractional ionization, and $n_H = n(\text{H}) + 2n(\text{H}_2) + n(\text{H}^+)$ is the density of hydrogen in all forms.

2.3 Hydrodynamical Heating

The heating sources discussed above derive from microscopic processes. Various kinds of evidence suggest that macroscopic gas motions also provide a significant amount of heating to the interstellar gas. The effects of ordered motions in shock waves can be estimated from the velocities, sizes, momenta, and energies of expanding supernova remnants, stellar wind bubbles, spiral density waves, etc. More elusive is the influence of turbulent motions. Except in the very most quiescent dark clouds, widths of interstellar spectral lines tend to be measurably larger than the thermal Doppler widths expected for the measured gas temperatures. Other phenomena ranging from pulsar scintillation to the possible acceleration of ultra-high-energy cosmic rays in a galactic wind hint that interstellar turbulence is pervasive and present on almost all scales.

The injection of energy into the gas by supernovae and their remnants has been discussed many times, notably by Cox and Smith (1974), McKee and Ostriker (1977), and Cox (1979, 1981). For example, Cox (1979) has estimated that a typical interstellar cloud of modest density receives a time-averaged input of mechanical heating of the order of

$$\Gamma_{mech} \approx 10^{-25} \quad \text{erg s}^{-1} \text{ cm}^{-3}. \qquad (17)$$

Such estimates are based upon an average supernova rate and energy release in the context of a global description of the interstellar medium. If we are concerned with the input of mechanical energy into a particular cloud, we must apply a model for the dissipation of this energy. In general, if there exists turbulence of velocity v_t on a scale R_c equal to the radius of an interstellar region whose mass density

$\rho = 2.17 \times 10^{-24} n_H$ grams cm^{-3} (including normal abundances of helium and heavier elements), then turbulent heating enters at a rate

$$\Gamma_{turb} \approx 3.5 \times 10^{-28} v_t^3 n_H \left(\frac{1 \text{pc}}{R_c} \right) \quad \text{erg s}^{-1} \text{ cm}^{-3} \tag{18}$$

where v_t is in km s^{-1}. Recent discussions of turbulence and its dissipation in the interstellar medium include the work of Fleck (1981), Scalo and Pumphrey (1982), and Higdon (1984).

In the specific context of turbulence generated by gravitationally bound condensations of radius R, mass density ρ and volume filling factor f_v that move through a more dilute medium of density ρ_0 with characteristic velocity v, Falgarone and Puget (1985) estimate a heating rate due to dissipation of turbulence in the dilute medium of

$$\Gamma_{turb,0} \approx 2 \times 10^{-27} \left(\frac{n_H}{100} \right) \left(\frac{v}{1 \text{km s}^{-1}} \right)^3 \left(\frac{f_v}{0.03} \right) \left(\frac{1 \text{ pc}}{R} \right) \quad \text{erg s}^{-1} \text{ cm}^{-3}. \tag{19}$$

Within the boundary layer of the condensation, the heating rate is expected to scale as

$$\Gamma_{turb} = \Gamma_{turb,0} \left(\frac{\rho_0}{\rho} \right)^2. \tag{20}$$

An alternative formulation of turbulent heating of molecular clouds (Boland and de Jong 1984) yields a heating rate

$$\Gamma_{turb} \approx 2.7 \times 10^{-30} \left(\frac{\alpha^3}{1 + \alpha^2} \right) n_H T^{1.5} \quad \text{erg s}^{-1} \text{ cm}^{-3} \tag{21}$$

where $\alpha = v_t / v_{thermal}$ is a parameter that characterizes both the turbulent velocity and length scale.

A collapsing cloud can be heated by compression. As an example, consider gravitational collapse at the free-fall rate of a fully molecular cloud of density n_H, temperature T, and pressure p,

$$\Gamma_{collapse} = \rho p \frac{d}{dt} \left(\frac{1}{\rho} \right) \approx 2.6 \times 10^{-31} n_H^{1.5} T \quad \text{erg s}^{-1} \text{ cm}^{-3}. \tag{22}$$

Mention should be made of the suggestion that young stellar objects with strong stellar winds are sufficiently common inside molecular clouds to provide a large source of mechanical energy for heating the gas and driving the turbulence (Norman and Silk 1979). Although the total amount of energy supplied by such sources and the time scales over which it will be available are uncertain, the presence of energetic bipolar molecular outflows in many interstellar clouds lends some credence to this idea (Lada 1985).

2.4 Other Heating Sources

There are various ways in which the interstellar magnetic field might couple thermal gas motions to motions on large scales. The dissipation of Alfvén waves produced by the twisting of field lines within a differentially rotating cloud can heat the gas at a rate

$$\Gamma_{Alfvén} \approx 2 \times 10^{-27} f n_H \quad \text{erg s}^{-1} \text{ cm}^{-3} \tag{23}$$

where f is the energy stored in Alfvén waves as a fraction of the gravitational potential and where it is assumed that the magnetic field strength scales as $B \propto n_H^{0.5}$ (Hartquist 1977).

More generally, the relative motions of neutral and ionized components of a magnetized plasma (magnetic ion slip or ambipolar diffusion) lead to frictional heating of the gas as discussed by Scalo (1977) and others.

In molecular regions, there are possibly significant chemical sources of heating (Dalgarno and Oppenheimer 1974). Only exoergic reactions are likely to occur in cold interstellar clouds and much of the energy released in these reactions goes into kinetic energy of the product species. Ultimately the energy involved derives from the ultraviolet radiation or cosmic rays that generated the formation of reactive ions in the first place. Thus chemical heating processes can be viewed as a means of increasing the heating efficiencies of ultraviolet photons and cosmic rays.

3. COOLANTS AND COOLING RATES

The full description of the cooling of interstellar gas can be at least as complicated as the discussion of heating processes. Aside from macroscopic mechanisms like adiabatic expansion, most of the cooling of the interstellar gas is accomplished by microscopic collision processes which convert some of the kinetic energy of impact into radiation that escapes the volume of gas under consideration.

From an observational point of view, it is possible to measure (or possible to imagine measuring) all of the radiation emerging from a nebula. This may sometimes be a good first approximation to the total cooling rate of the region; however, not all of the observable radiant energy necessarily comes at the expense of the reservoir of gas kinetic energy. Moreover, in situations where the region is optically thick to some of its cooling radiation, the measured luminosity cannot provide directly any information on the local cooling rate at a particular interior location.

3.1 Ionized Regions

Ionized regions are cooled predominantly by collisional excitation of atoms and ions that subsequently decay via line emission that escapes. The most effective coolants tend to be ions of the relatively abundant elements carbon, nitrogen, oxygen, and neon, that have low-lying energy levels which can be excited in thermal collisions with electrons or protons at the relevant temperatures. Hydrogen and helium also contribute to the cooling, but since their most prominent emission lines are formed via recombination, much of the radiated energy comes from the binding energy of the atom rather than the thermal energy of the captured electron. The cooling rate for collisionally-excited line radiation in a transition between upper level u and lower level l is

$$\Lambda_{ul} = (E_u - E_l) \sum_i n(i) \left(n_l \gamma_{lu}^i - n_u \gamma_{ul}^i \right) \quad \text{erg s}^{-1} \text{ cm}^{-3} \tag{24}$$

where E_j and n_j are the energy of and number density in level j, respectively, and γ^i_{jk} is the rate coefficient for collisional transitions from level j to level k induced by collision partner i whose number density is $n(i)$. Equation (24) applies strictly only when every line photon created escapes. When the optical depth τ_ν at the line frequency $\nu = (E_u - E_l)/h$ is large, $\tau_\nu \geq 1$, then equation (24) must be modified to account for the trapping of radiation. The optical depth at line center can be expressed

$$\tau_\nu \approx \frac{A_{ul}}{8\pi(\nu/c)^3} \frac{N(j)}{\Delta V} (x_l g_u/g_l - x_u) \tag{25}$$

where A_{ul} is the spontaneous transition probability in s^{-1}; $N(j)$ is the total column density, cm^{-2}, of the radiating species; ΔV is a characteristic linewidth (usually the full width at half-maximum) in velocity units; x_m is the fraction of the radiating species in level m; and g_m is the statistical weight of level m.

Common ions like N^+, O^+, O^{+2}, etc. have relatively low-lying excited terms of the same parity as the ground term, which can be readily excited by electron impact and can undergo electric-dipole-forbidden transitions in the visible and ultraviolet. These transitions tend to have sufficiently small A_{ul} that re-absorption is unlikely. On the other hand, if the electron density is fairly high, $n(e) \geq A_{ul}/\gamma_{ul}$, then the excited ion can be de-excited by collision before it radiates.

Species such as O, O^{+2}, C, C^+, etc. also have closely spaced fine-structure levels in the ground term that give rise to forbidden transitions. Because of the small energy separations involved, these transitions can contribute effectively to the cooling over a very wide range of temperatures.

In a few cases, notably the C IV 1550 Å doublet, collisionally-excited resonance lines may contribute to the cooling. Here, the optical depth is more likely to be large and the cooling may be diminished by self-absorption.

The atomic data that are needed to evaluate cooling rates, such as collision rate coefficients and transition probabilities, have been summarized for major atomic coolants by Mendoza (1983).

As mentioned above, radiative recombination contributes to the cooling. In normal H II regions, recombination in H and He tends not to add much to the total cooling, but in coronal plasmas at high temperatures, dielectronic recombination in He and heavier elements may be important owing to the large excitation energies involved in the inelastic collisions.

There are also continuous radiation processes that provide cooling. For example, thermal bremsstrahlung cools a hydrogen/helium plasma at a rate of approximately

$$\Lambda_b \approx 1.42 \times 10^{-27} T^{1/2} \left(n(H^+) + n(He^+) + n(He^{+2}) \right) n(e) \quad \text{erg s}^{-1} \text{ cm}^{-3}. \tag{26}$$

Finally, in high-temperature gas, collisional ionization may add significantly to the cooling.

3.2 Neutral Regions

The cooling processes in neutral regions were reviewed by Dalgarno and McCray (1972). Since that time, much additional information has become available about cooling by molecules and the cooling of neutral, shocked regions. It is fairly common for molecular clouds to be at least partly opaque to their own cooling radiation; therefore, the treatment of cooling requires a lot of attention to radiative transfer. Another feature of molecular clouds is that the gas, the ambient background radiation of brightness temperature T_b, and the cooling photons all have similar energies: $kT \approx kT_b \approx h\nu$. This means that the radiative transfer can be complicated by the coupling of lines and continuum and in particular by radiative coupling between the gas and the local dust.

In diffuse neutral clouds, or in the outer layers of molecular clouds, collisionally excited fine-structure line emission in C, C$^+$, and O will often dominate the cooling. References to the relevant atomic data can be found in van Dishoeck and Black (1986) and in a recent paper by Keenan et al. (1986). Inside molecular clouds, collisionally-excited rotational transitions in abundant molecules tend to control the cooling. At high densities, gas-grain interactions may act as cooling processes (see equation 8). Because of its abundance, its favorable rotation constant, and the ease with which it is excited, CO is usually considered to be the premier molecular coolant. For a discussion of CO cooling rates, see Gilden (1984) and references therein. Molecular hydrogen may be important, particularly in hotter clouds, and even species like HD (Dalgarno and Wright 1972) and HCl (Dalgarno, de Jong, Oppenheimer, and Black 1974) have been suggested as non-negligible minor coolants under some circumstances.

In order to study the depth dependence of temperature and its relation to cloud structure and stability, it is necessary to determine local cooling rates. In order to evaluate local cooling rates, it is necessary to solve simultaneously the equations of radiative transfer, of statistical equilibrium, and of chemical balance. The radiative transfer calculations can be simplified in various ways. For example, special assumptions can be made about the character of the velocity field—systematic, uniform collapse or expansion—that permit cooling rates to be evaluated analytically (see, e.g., de Jong, Chu, and Dalgarno 1975). Another useful approach is to treat the line formation with an escape probability formalism (see, e.g., de Jong, Dalgarno, and Boland 1980).

3.3 Shocked Gas

An understanding of radiative cooling of shock-heated gas is essential to the description of the structure and propagation of shock waves through the interstellar medium. For high-velocity shocks, see the review of McCray and Snow (1979). Cooling processes in molecular shocks have been discussed by Hollenbach and McKee (1979), McKee et al. (1982), and Draine, Roberge, and Dalgarno (1983).

4. CONCLUDING REMARKS

In the best tradition of summer school lectures, the most important conclusions will be left as exercises to the reader. One possibly instructive example is suggested here.

Consider the ionization and thermal balance in a pure hydrogen plasma exposed to a power-law radiation field of flux

$$f_\nu = 10^{-20}(\nu_0/\nu) \quad \text{erg cm}^{-2}\text{s}^{-1}\text{Hz}^{-1} \tag{27}$$

where ν_0 is the frequency corresponding to the ionization potential of H. Solve the coupled equations of ionization

$$n(\mathrm{H})\left(\gamma_p + n(e)\gamma_c\right) = n(e)n(\mathrm{H}^+)\alpha_H \tag{28}$$

and thermal balance

$$\Gamma_{pi} = \sum_i \Lambda_i \tag{29}$$

to find T at low densities, $n_H \approx 10^{-5}$ to 10 cm^{-3}, subject to $n(\mathrm{H}^+) = n(e)$, where γ_p and γ_c are the rate coefficients for photoionization and electron-impact collisional ionization, respectively, and α_H is the recombination rate coefficient. Convenient analytic representations of the relevant rates can be found in Black (1981). The resulting equilibrium temperature can be shown to lie in the range $T = 9000-46000$ K. One interesting feature of this physical system is that the equation of state

$$P \propto n_H^{13/15} \tag{30}$$

is accurately barytropic for $n_H > 10^{-4}$ cm^{-3} and, indeed, is the equation of state of a polytrope of index $N = -15/2$. Properties of negative-index polytropes are discussed by Viala and Horedt (1974a,b).

Several aspects of interstellar heating and cooling require much more work. It remains difficult to account for the moderately high temperatures ($T \approx 50-100$ K) observed in many diffuse clouds. Evidently a major heating source has been overlooked or underestimated, if these regions are actually in thermal steady state. Despite recent attention, heating by dissipation of turbulence is not yet well understood. There is a continuing need for more and better atomic and molecular data on the microscopic collision processes that participate in heating and cooling. Finally, more elaborate theoretical descriptions of the internal structure and radiative transfer in molecular clouds will be instructive.

Acknowledgements. I am grateful to Dr. E. F. van Dishoeck for helpful comments and to NASA for support through grant NAGW-763.

REFERENCES

Black, J. H. 1981, *M. N. R. A. S.*, **197**, 553.

Black, J. H. and Dalgarno, A. 1977, *Ap. J. Suppl.*, **34**, 405.

Black, J. H. and Dalgarno, A. 1973, *Ap. J. (Letters)*, **184**, L101.

Bloemen, J. B. G. M., Blitz, L., and Hermsen, W. 1984, *Ap. J.*, **279**, 136.

Boland, W. and de Jong, T. 1984, *Astr. Ap.*, **134**, 87.

Bregman, J. N. and Harrington, J. P. 1985, preprint.

Burke, J. R. and Hollenbach, D. J. 1983, *Ap. J.*, **265**, 223.

Cheng-Yue, Z., Yuan, T., and De-Lin, X. 1985, *Acta Astronomica Sinica*, **26**, 224.

Cox, D. P. and Smith, B. W. 1974, *Ap. J. (Letters)*, **189**, L105.

Cox, D. P. 1979, *Ap. J.*, **234**, 863.

Cox, D. P. 1981, *Ap. J.*, **245**, 534.

Cox, P., Krügel, E., and Mezger, P. G. 1986, *Astr. Ap.*, **155**, 380.

Cravens, T. E. and Dalgarno, A. 1978, *Ap. J.*, **219**, 750.

Dalgarno, A., de Jong, T., Oppenheimer, M., and Black, J. H. 1974, *Ap. J. (Letters)*, **192**, L37.

Dalgarno, A. and McCray, R. A. 1972, *Ann. Rev. Astr. Ap.*, **10**, 375.
Dalgarno, A. and Oppenheimer, M. 1974, *Ap. J.*, **192**, 597.
Dalgarno, A. and Wright, E. L. 1972, *Ap. J. (Letters)*, **174**, L49.
de Jong, T. 1977, *Astr. Ap.*, **55**, 137.
de Jong, T., Chu, S.-I, and Dalgarno, A. 1975, *Ap. J.*, **199**, 69.
de Jong, T., Dalgarno, A., and Boland, W. 1980, *Astr. Ap.*, **91**, 68.
Draine, B. T. 1978, *Ap. J. Suppl.*, **36**, 595.
Draine, B. T., Roberge, W. G., and Dalgarno, A. 1983, *Ap. J.*, **264**, 485.
Eddington, A. S. 1926, *Proc. Roy. Soc. (London)*, **A111**, 424.
Falgarone, E. and Puget, J. L. 1985, *Astr. Ap.*, **142**, 157.
Field, G. B., Goldsmith, D., and Habing, H. J. 1969, *Ap. J. (Letters)*, **155**, L49.
Fleck, R. C. 1981, *Ap. J. (Letters)*, **246**, L151.
Flower, D. R. 1986, this volume.
Gilden, D. L. 1984, *Ap. J.*, **283**, 679.
Hartquist, T. W. 1977, *Ap. J. (Letters)*, **217**, L45.
Higdon, J. C. 1984, *Ap. J.*, **285**, 109.
Keenan, F. P., Lennon, D. J., Johnson, C. T., and Kingston, A. E. 1986, *M. N. R. A. S.*, **220**, 571.
Lada, C. J. 1985, *Ann. Rev. Astr. Ap.*, **23**, 267.
Mathis, J. S., Mezger, P. G., and Panagia, N. 1983, *Astr. Ap.*, **128**, 212.
McCray, R. A. and Snow, T. P. 1979, *Ann. Rev. Astr. Ap.*, **17**, 213.
McKee, C. F. and Ostriker, J. P. 1977, *Ap. J.*, **218**, 148.
McKee, C. F., Storey, J. W. V., Watson, D. M., and Green, S. 1982, *Ap. J.*, **259**, 647.
Mendoza, C. 1985, in *Planetary Nebulae*, D. R. Flower, ed., (Dordrecht: D. Reidel), p. 143.
Milgrom, M., Panagia, N., and Salpeter, E. E. 1973, *Ap. J. (Letters)*, **14**, 73.
Norman, C. A. and Silk, J. 1979, *Ap. J.*, **228**, 197.
O'Donnell, E. J. and Watson, W. D. 1974, *Ap. J.*, **191**, 89.
Osterbrock, D. E. 1974, *Astrophysics of Gaseous Nebulae*, (San Francisco: W. H. Freeman).
Phillips, T. G. 1986, this volume.
Roberge, W. G., Dalgarno, A., and Flannery, B. P. 1981, *Ap. J.*, **243**, 817.
Scalo, J. M. 1977, *Ap. J.*, **213**, 705.
Scalo, J. M. and Pumphrey, W. A. 1982, *Ap. J. (Letters)*, **258**, L29.
Spitzer, L., Jr. 1978, *Physical Processes in the Interstellar Medium*, (New York: Wiley-Interscience).
Stecher, T. P. and Williams, D. A. 1967, *Ap. J. (Letters)*, **149**, L29.
Stephens, T. L. and Dalgarno, A. 1973, *Ap. J.*, **186**, 165.
Strömgren, B. 1939, *Ap. J.*, **89**, 526.
Takahashi, T., Hollenbach, D. J., and Silk, J. 1983, *Ap. J.*, **275**, 145.
Tielens, A. G. G. M. and Hollenbach, D. 1985, *Ap. J.*, **291**, 722.
van Dishoeck, E. F. and Black, J. H. 1986, *Ap. J. Suppl.*, **62**, in press.
Viala, Y. and Horedt, G. P. 1974a, *Astr. Ap.*, **33**, 195.
Viala, Y. and Horedt, G. P. 1974b, *Astr. Ap. Suppl.*, **16**, 173.
Watson, W. D. 1972, *Ap. J.*, **176**, 103; addendum, **176**, 271.
Weisheit, J. C. 1974, *Ap. J.*, **190**, 735.

ATOMIC AND MOLECULAR PHYSICS OF INTERSTELLAR HEATING AND COOLING

D. R. Flower
Physics Department
The University
Durham DH1 3LE (U.K.)

ABSTRACT. This review will be concerned with microscopic processes
occurring in the interstellar medium which can lead to a change in the
mean kinetic energy of the particles composing the gas. More
precisely, the review will be concerned with some of these processes.
Their number and diversity are large and form a suitable topic for a
Summer School in their own right. The emphasis will be placed on
non-reactive collisional phenomena. Consideration will be given to
the cold (T $\lesssim$ 100 K) gas and to regions which have been heated by the
passage of a shock wave (T $\simeq$ 1000 K).

1. INTRODUCTION

The interstellar clouds gain energy from and lose energy to their
environment - the Galaxy. Input of energy occurs through the action
of cosmic rays and stellar ultraviolet radiation which are incident on
the clouds. The energy may be deposited directly in the gas, as in
cosmic ray ionisation or photoionisation and photodissociation, or
indirectly owing to the presence of dust grains, as in photoelectron
heating. The energy which is gained in this way is subsequently
shared with the particles of the gas, leading ultimately to a
Maxwellian distribution and its characteristic temperature.
    At the microscopic level, energy is lost from a cloud through
radiative processes following collisional excitation of the gas
particles. Collisional energy transfer may involve the electronic,
rotational or vibrational degrees of freedom of these particles.
Excitation will occur principally in collisions with the more abundant
"light" particles composing the gas - H, H_2, He, and electrons. If
the fractional ionisation is less than about 10^{-3}, excitation by
electrons will be relatively unimportant for the overall thermal balance
of the gas, although the electron temperature may be significantly
affected by such processes in regions, such as MHD shocks, where the
electron, ion and neutral temperatures can differ.
    Collisional mechanisms involving H, H_2 and He may be divided into
those where a single adiabatic potential energy curve or surface

745

D. J. Hollenbach and H. A. Thronson, Jr. (eds.), Interstellar Processes, 745–761.
© 1987 by D. Reidel Publishing Company.

intervenes, and those involving more than one such curve or surface. An example of the former category would be the rotational excitation of CO by He at low energies, and of the latter the excitation of the C^+ $^2P^o_{1/2}$ - $^2P^o_{3/2}$ fine structure transition by H_2. Multi-surface collisional phenomena involve coupling between the motions of the nuclei and of the electrons and, as such, represent violations of the Born-Oppenheimer approximation. These phenomena, often qualified by the double negative "non-adiabatic" (= diabatic), constitute a very important category of interstellar processes.

In the following Section 2, we shall briefly discuss the formulation of single- and multi-surface collision problems. Section 3 will be concerned with new results relating to a selection of collision processes which are directly relevant to studies of the interstellar medium.

2.   THEORY

2.1 Collisions on a single adiabatic potential energy surface

Let us consider the collision between a closed-shell atom A and a closed-shell diatomic molecule M. As noted in the Introduction, an example would be A = $He(1s^2$ $^1S)$ and M = $CO(X$ $^1\Sigma^+)$. At the energies encountered in interstellar clouds (E << 1eV, that is, T << 10^4K), excited electronic states of A and M cannot be attained, and the collision will proceed along the lowest $^1\Sigma$ potential surface of the system AM. The word "surface" is used advisedly here, as the strength of the interaction depends on more than one coordinate, in this case, R,r and θ, as shown in Fig. 1. To facilitate the collision calculation,

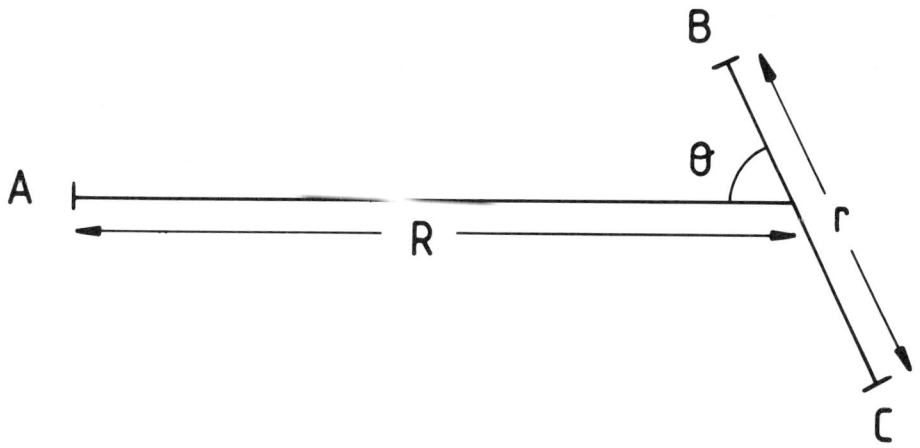

Figure 1. Coordinate system appropriate for studying the interaction between a closed-shell atom A and a closed-shell diatomic molecule BC.

the potential may be expanded in terms of a complete set of orthogonal functions of the angle θ at each pair of values of R and r, that is,

$$V(R,r,\theta) = \sum_{\lambda=0}^{\infty} V_\lambda(R,r)P_\lambda(\cos\theta) \tag{1}$$

where P_λ denotes a Legendre polynomial. The coefficients of the expansion $V_\lambda(R,r)$ may be determined from the calculated values of $V(R,r,\theta)$. In practice, V is evaluated on a finite grid of values of the coordinates, and the $V_\lambda(R,r)$ are often determined by a least squares fitting procedure. As the grid of geometries is finite, only a finite number of terms, $\lambda \leq \lambda_{max}$, in the expansion (1) is determined.

The computation of the adiabatic potential energy surface, $V(R,r,\theta)$, is a major task. In the case of collisions between neutral particles, as in the example considered above, the interaction potential is only weakly attractive, the depth of the well being typically 10^{-2} eV (100 K). The well results from the long-range van der Waals attraction and the short-range repulsion of the electron clouds. Its accurate calculation generally involves large basis set SCF-CI calculations. In practice, the configuration interaction (CI) computations are too expensive to perform for the requisite number of geometries when many-electron molecules such as CO are involved. Techniques based upon perturbation theory are often employed instead.

At collision energies well below the spacing of the vibrational energy levels, the molecule vibrates many times during a collision and $V(R,r,\theta)$ may be evaluated at the mean internuclear separation, $r=\bar{r}$. The dimensionality of the problem is correspondingly reduced. Many of the studies of interstellar molecular interactions have been directed towards the cold gas where this approximation is appropriate. On the other hand, shocks can heat the gas to temperatures such that vibrational excitation can occur, in which case the dependence of V on r needs to be known. The extension of the calculations to a grid of values of r is far from trivial and, to date, very few reliable calculations of vibrational excitation cross-sections have been performed. In this area, laboratory measurements are ahead of theory.

Classical, semi-classical and quantal techniques have been applied to solving the collision problem, with varying degrees of success. Where very low energies are involved, as in the cold gas, quantum mechanics should be used. In this case, Schroedinger's equation may be reduced to a set of coupled second order ordinary differential equations which are expressible in the matrix form

$$\left[\underset{\sim}{1}\frac{d^2}{dR^2} + \underset{\sim}{W}(R)\right]\underset{\sim}{F}(R) = 0 , \tag{2}$$

where $\underset{\sim}{1}$ is the unit matrix, $\underset{\sim}{W}$ is the interaction matrix, and $\underset{\sim}{F}$ denotes the matrix of solutions. Applying the boundary conditions appropriate to the collision problem leads to the scattering matrix $\underset{\sim}{S}$ from which the cross-sections are readily derived.

A great deal of effort has been expended on developing numerical techniques for the efficient solution of the coupled equations (2).

Amongst the more successful have been the method introduced by Gordon
(1969) and the R-matrix propagator algorithm of Stechel, Walker and
Light (1978). Several approximations have been advanced which lead to
partial decoupling of the equations and greatly facilitate their
solution. We mention here the coupled states (CS) approximation,
introduced by McGuire and Kouri (1974), and the infinite order sudden
(IOS) approximation of R.T. Pack and coworkers (see Secrest 1975).

2.2. Multi-surface Collisions

Let us consider the collision between an open-shell atom or ion A
comprising one active electron and a closed-shell diatomic molecule M.
The example mentioned in the Introduction is $A = C^+(2p\ ^2P^O)$ and
$M = H_2(X\ ^1\Sigma_g^+)$. In this case, the system may be viewed as shown in
Fig. 2, and the potential expanded as

$$V(R,\underset{\sim}{r},\underset{\sim}{\rho}) = \underset{\lambda_1\lambda_2\mu}{\Sigma}\ V_{\lambda_1\lambda_2\mu}(R,r,\rho)\,Y_{\lambda_1\mu}(\underset{\sim}{\hat{r}})\,Y_{\lambda_2-\mu}(\underset{\sim}{\hat{\rho}}) \qquad (3)$$

where Y denotes a spherical harmonic function, and $\underset{\sim}{\hat{\rho}},\underset{\sim}{\hat{r}}$ denote the polar
coordinates of the electron and the intramolecular axis, respectively,
in a coordinate system in which the line joining the nucleus of the ion
to the centre of mass of the molecule is taken to be the z-axis. In

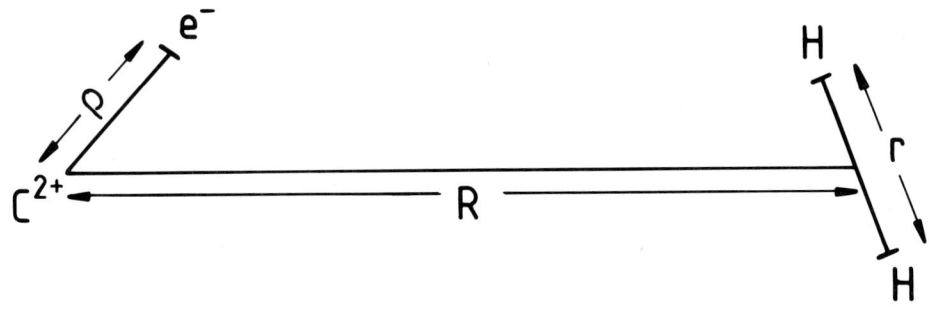

Figure 2. Coordinate system appropriate for studying the interaction
between the C^+ ion, consisting of a C^{2+} core and a 2p valence electron,
and H_2.

low-energy, non-reactive collisions, we may take $r = \bar{r}$ and $\rho = \bar{\rho}$, where $\bar{r}$ is the mean intramolecular distance and $\bar{\rho}$ the mean distance of the valence electron from the ion core.

In the presence of the H_2 molecule, group theoretical arguments show that normalised electronic orbitals of appropriate symmetry are

$$\psi(^2A_1) = Y_{10}(\hat{\rho})$$
$$\psi(^2B_1) = [Y_{11}(\hat{\rho}) - Y_{1-1}(\hat{\rho})]/2^{1/2}$$
$$\psi(^2B_2) = [Y_{11}(\tilde{\rho}) + Y_{1-1}(\hat{\rho})]/2^{1/2}$$

when the intramolecular axis is perpendicular to the z-direction and where $^2A_1$, $^2B_1$ and $^2B_2$ label the molecular states. The corresponding adiabatic potential energy curves for this geometry may be obtained as the expectation values of the potential with respect to each of these states. Thus,

$$V(^2A_1) = \langle Y_{10}(\hat{\rho}) | V(R,\bar{r},\bar{\rho}) | Y_{10}(\hat{\rho}) \rangle$$

where the braces indicate integration over the polar coordinates $\hat{\rho}$. On the other hand, if the intramolecular axis lies along the z-direction, a collinear geometry obtains and $V(^2B_1) = V(^2B_2) \equiv V(^2\Pi)$, that is, the $^2B_1$ and $^2B_2$ states are degenerate and are denoted $^2\Pi$. The $^2A_1$ state is denoted $^2\Sigma$ in this geometry.

From calculations of the five adiabatic potential energy curves mentioned above, it is possible to determine five coefficients in the potential energy expansion (3), namely, V_{000}, V_{200}, V_{020}, V_{220}, and V_{222}. These coefficients are expressible as linear combinations of the adiabatic potential energies and are, of course, responsible for inducing the $^2P^o_{1/2} - ^2P^o_{3/2}$ transition in the ion. The cross-section for the process is determined by solving coupled differential equations of the form (2) above.

In reality, the $^2A_1$, $^2B_1$, $^2B_2$, $^2\Sigma$ and $^2\Pi$ potential energy curves are not known to high accuracy, and the computed cross-sections are subject to considerable uncertainty at low energies as a consequence (Flower and Launay 1977). Indeed, it is characteristic of much work in heavy particle dynamics that the accuracy of the final results (cross-sections) is limited by the precision of the potentials which intervene in the dynamical calculations.

3.   RECENT RESULTS

It would not be possible to review all recent work in the field of heavy particle dynamics which is relevant to studies of the inter-stellar medium. I shall restrict myself to a few examples of recent studies which are pertinent and which also serve to illustrate the applications of the theory outlined in Section 2.

3.1   Rotational excitation in H_2-H_2 collisions

Observations with the Copernicus satellite showed H_2 to be present in diffuse clouds in excited rotational states (Spitzer et al. 1973; Spitzer and Cochran 1973). If shock excitation proves to be the explanation (Aannestad and Field 1973; Shull and Hollenbach 1978), cross-sections for the excitation of H_2 by other H_2 molecules are required to interpret the observations. This rotational excitation process is also a major source of cooling in regions of the diffuse interstellar gas which have undergone shock heating (Elitzur and Watson 1980; Draine 1980; Flower et al. 1986).

Cross-sections and rates for the rotational excitation of H_2 by H_2 have been computed by Green (1975), Ramaswamy, Rabitz and Green (1977), Green, Ramaswamy and Rabitz (1978), and by Monchick and Schaefer (1980). The calculations of Monchick and Schaefer are the most accurate of those published, particularly from the viewpoint of the interaction potential which was employed. However, Monchick and Schaefer (1980) tabulate rate coefficients involving only a handful of rotational levels, for $60 \leq T \leq 150$ K. In studies of postshock gas, results for both higher rotational states and higher temperatures ($T \approx 1000$ K) are required.

Accordingly, Schaefer (1986) and Danby and Flower (1986) have carried out independent calculations which extend the earlier study of Monchick and Schaefer (1980). As is well known, H_2 exists in the forms of para-H_2, in which the resultant nuclear spin $I=0$ and the rotational quantum number $J=0,2,4,\ldots$, and ortho-H_2, in which $I=1$ and $J=1,3,5,\ldots$ Collisions between two para-H_2 molecules or two ortho-H_2 molecules involve identical species, and care must be taken when dealing with the statistics of the collision process. Thus, Green (1975) gives an expression for the cross-section σ for the rotational transition $J_1 J_2 \rightarrow J_1' J_2'$, where 1 and 2 denote the H_2 molecules, such that $\sigma(J_1 J_2 \rightarrow J_1' J_2') \propto (1+\delta_{J_1 J_2})(1+\delta_{J_1' J_2'})$. On the other hand, the corresponding expression of Monchick and Schaefer (1980) has $\sigma(J_1 J_2 \rightarrow J_1' J_2') \propto (1+\delta_{J_1 J_2} \delta_{J_1' J_2'})$. Thus, the cross-sections are identical when $J_1 \neq J_2$ and $J_1' \neq J_2'$ (in which case the molecules are distinguishable in both the initial and final channels) but differ by a factor 2 when $J_1 = J_2$ and/or $J_1' = J_2'$ (in which case the molecules are indistinguishable in the initial and/or final channels).

This difference may be interpreted by means of the following example. When identical molecules are involved in the initial and/or the final channels, as in the transitions $00 \rightarrow 02$ or $00 \rightarrow 20$, the cross-section defined by Green (1975) provides a measure of the probability that either of the two identical molecules undergoes the specified change in its rotational state. In this case, the number of H_2 molecules (per unit volume and time) which are raised to the rotational state $J = 2$ is given by $\langle\sigma v\rangle n^2(J=0)/2$, when $\langle\sigma v\rangle$ denotes the Maxwellian average of the cross-section and $n(J=0)$ the number density of H_2 molecules in the initial state $J=0$. The factor 2 arises because independent pairs of (initially) identical molecules must be considered. On the other hand, the cross-section as defined by Monchick and Schaefer (1980) is a measure of the probability that one of the H_2 molecules changes its rotational state. In this case, the molecules are effectively being considered to be distinguishable, and the rate of the rotational transition is simply $\langle\sigma v\rangle n^2(J=0)$. Thus, the rate of production of H_2 molecules in the $J=2$ state, per unit

volume and time, is the same whether the cross-section as defined by
Green (1975) or by Monchick and Schaefer (1980) is employed. Indeed,
the definitions of the rates must be the same, as the rates are (in
principle) measurable quantities.

In Fig. 3, we compare rate coefficients, $<\sigma v>$, for the process
$00 \leftarrow 02$, as calculated by Green (1975), Monchick and Schaefer (1980),
and Danby and Flower (1986). The results which are plotted adhere to
the definition of the cross-section given by Green (1975). All these
calculations used the close-coupling method, consisting of the exact
solution of the coupled equations (2) for a given set of rotational
basis functions. Also shown are the results of calculations which used
the centrifugal decoupling or CS approximation, to which reference was
made in Section 2 above. The agreement with the close-coupling results
may be seen to be good.

The differences between the various calculations arise principally
from the potentials which were employed. From this standpoint, the

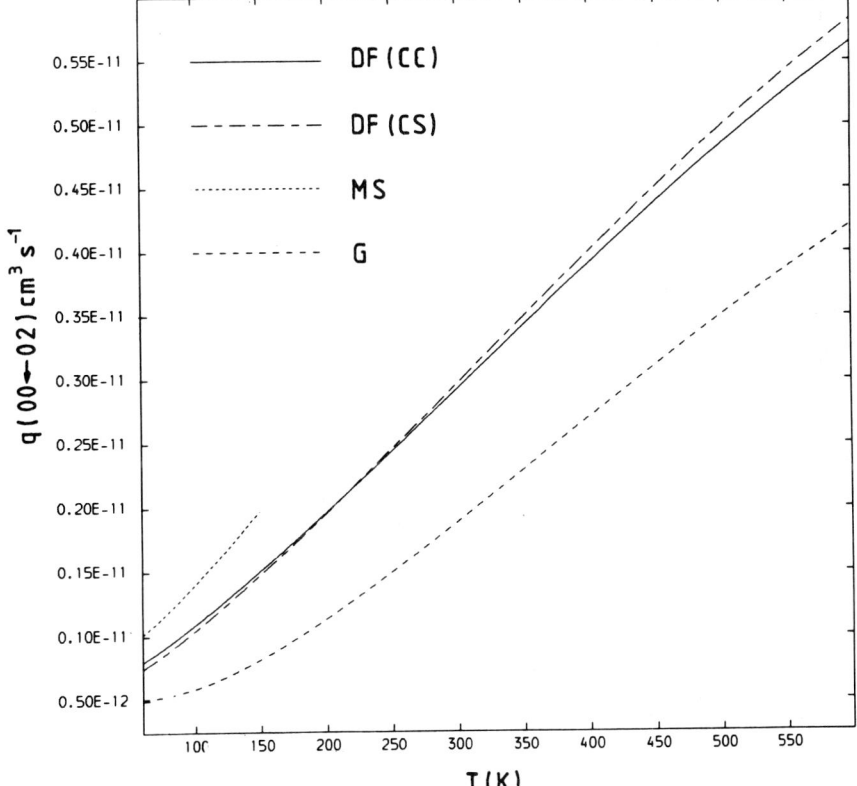

Figure 3.   Rate coefficients (cm^3 s^{-1}) for the rotational deactivation
of H_2(J=2) by H_2(J=0). MS: Monchick and Schaefer (1980);   G: Green
(1975);   DF: Danby and Flower (1986). Both the close-coupling (CC) and
coupled states (CS) results of Danby and Flower (1986) are shown.   CC
calculations by Schaefer (1986) practically coincide with those of
Danby and Flower (1986).

calculations of Danby and Flower (1986), which incorporated the 1980 potential of W. Meyer and J. Schaefer, should be the most accurate of those plotted. More work is required on this important system and is proceeding (Danby and Flower 1986; Schaefer 1986).

3.2 Rotational excitation of NH_3 by para-H_2

Rotation-inversion transitions of the ammonia molecule have been observed from interstellar clouds for a number of years (see Ho and Townes 1983). Laboratory data consist of both double-resonance and line broadening measurements (Daly and Oka 1970; Klaassen, ter Meulen and Dymanus 1983; Broquier and Picard-Bersellini 1985). There have been several theoretical studies of the rotational excitation of NH_3 by He, using both quantum mechanical and semi-classical methods (Green 1981; Davis 1985; Billing, Poulsen and Diercksen 1985).

In the interstellar clouds, excitation by H_2 is likely to be the dominant process, and there have been recent studies of this process by Billing and Diercksen (1985, 1986) and Danby et al. (1986a,b). The work of Billing and Diercksen uses semi-classical scattering techniques, whereas Danby et al. use the quantum mechanical close-coupling method. The comparisons made by Billing and Poulsen (1984) suggest that the semi-classical approximation may give satisfactory results for this system, at least at relatively high collision energies (E = 250 cm^{-1}). The studies of Billing and Diercksen (1985, 1986) and of Danby et al. (1986a,b) also differ in the potential surfaces which have been used in the dynamical calculations. Whilst the same SCF data are adopted by both groups, the correlation energy derives from different forms of perturbation theory. In view of these differences, it is hardly surprising that there are significant discrepancies between the computed rotational excitation cross-sections.

Danby et al. (1986b) have attempted to isolate the main reasons for these discrepancies. In Fig. 4 is shown a comparison of the NH_3-H_2 interaction potentials, averaged over the orientations of the H_2 intramolecular axis, at a centre of mass separation R=6 a.u. (1 a.u. = 0.5292 x 10^{-10}m), which is in the vicinity of the minimum in the interaction potential. The potentials differ substantially, notably in the local maximum which is present in the potential of Billing and Diercksen but absent in that of Danby et al.

As noted in Section 2, the coefficients of potential energy expansions are often determined by means of least squares fitting procedures. Danby et al. (1986b) show that the major difference between the potentials plotted in Fig. 4 arises from the quality of the fits to the same SCF data. At R=6 a.u., the rms deviation from the ab initio points of the potential used by Billing and Diercksen (1985, 1986) is about 20 times larger than the rms deviation of the fit used by Danby et al. (1986a,b). The local potential maximum seen in Fig. 4a is attributable to this discrepancy and is, therefore, an artefact of the fit of Billing and Diercksen to the SCF data.

The number of geometries at which the potential is evaluated may also have a substantial effect on the quality of the fit, by prematurely truncating the potential energy expansion. Thus, the

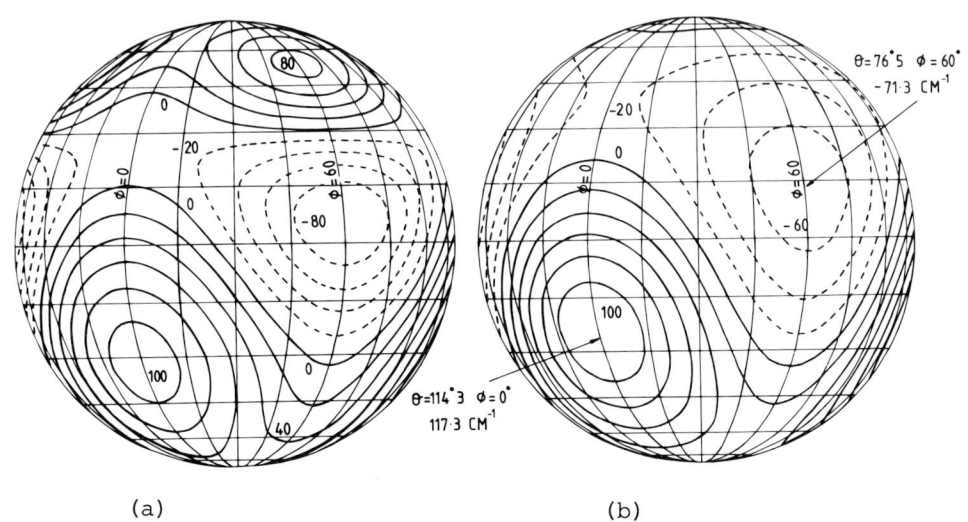

(a)   (b)

Figure 4. Contours of the NH_3-H_2 interaction potential (cm^{-1}) at an intermolecular distance R=6 a.u., averaged over the orientations of the H_2 intramolecular axis. (a) Billing and Diercksen (1986); (b) Danby et al. (1986b). Note the additional local maximum in (a) towards the side of the N-atom.

correlation energy used by Billing and Diercksen derives from many-body perturbation theory (MBPT) and should be more accurate than the second-order perturbation theory (SOPT) results used by Danby et al. However, the angular grid on which the MBPT results were obtained is much coarser than that adopted in the SOPT calculations and insufficient to reliably map the anisotropy (angular dependence) of the interaction (Danby et al. 1986b).

A question which is often raised is whether He and para-H_2 in its rotational ground state (J=0) behave similarly as perturbers. The question is important because interactions with He are easier to study than those with H_2, and rate coefficients for the excitation of inter-stellar molecules (M) by He are more widely available than for excitation by H_2.

The relevant differences are in the M-He and M-H_2 interaction potentials (the latter averaged over the orientations of the H_2 intra-molecular axis), and in the reduced mass, μ, of the perturber. If the mass of M is much greater than that of He or H_2, then $\mu(He) \approx 2\mu(H_2)$. Danby et al. (1986b) show that the differences in the interaction potentials and the reduced masses tend to compensate in the calculations of the <u>cross-sections</u>. The rate coefficients, $\langle \sigma v \rangle$, for excitation by para-H_2 are then larger by roughly a factor $2^{1/2}$ than those relating to excitation by He, owing to the higher velocity of H_2 molecules at a given temperature. This conclusion is in accord with the suggestion of Green (1981). The rates of excitation by ortho-H_2, or by para-H_2 which is not artificially constrained to its rotational ground state, may be expected to be larger still at low temperatures owing to the presence of additional long-range interactions involving the non-vanishing multipole moments of the H_2 molecule.

3.3 Rotational excitation of OH by para-H_2

Collisions involving the excitation of the Λ-doubled rotational states of OH by para-H_2 (J=0) involve two potential energy surfaces, denoted A' and A" for a general interaction geometry. As such, these collisions fall in the category discussed in Section 2.2 above.

The initial study by Bertojo, Cheung and Townes (1976) of the rotational excitation of OH by H_2 (also by H and He) indicated that the ground state ($^2\Pi_{3/2}$ J=$^3/_2$) Λ-doublet would be anti-inverted. However, it has since been recognised that this prediction was based upon two compensating errors, one in the symmetry assignments of the components of the doublets (Alexander and Dagdigian 1984; Dewangan and Flower 1985) and the other in the treatment of the collision dynamics. Bertojo et al. (1976) assumed that <u>either</u> of the adiabatic potential energy surfaces would be followed during a given collision event, whereas the work of Green and Zare (1975) shows that the interaction may be expressed as a diabatic combination of <u>both</u> potential surfaces, A' and A".

Subsequent work by Dewangan and Flower (1981, 1983) corrected the error in the treatment of the collision dynamics but failed to correct the assignment error (see Dewangan and Flower 1985). Earlier studies by Dixon and Field (1979a,b) had made the same predictions as Dewangan and Flower (1981, 1983), namely that the ground state Λ-doublet should be inverted by collisions with H_2, but were also found to be in error owing to an inconsistent treatment of phases in the interaction potential (Dixon, Field and Zare 1985). It follows that all these authors agree that H_2 collisions anti-invert the rotational ground state of OH, which accords with the crossed-beam measurements of Andresen, Häusler and Lülf (1984).

Schinke and Andresen (1984) have also computed cross-sections for the rotational excitation of OH by para-H_2(J=0), using the CS approximation. These calculations refer to a centre of mass collision energy E = 83 meV, which corresponds to the crossed-beam measurements of Andresen et al. (1984). Schinke and Andresen (1984) found reasonable agreement with the measured values of the Λ-doublet summed and averaged cross-sections,

$$\sigma(J=^3/_2,\ \bar{\Omega}=^3/_2 \to J',\bar{\Omega}') = \frac{1}{2}\sum_{\epsilon\epsilon'}\sigma(J=^3/_2,\ \bar{\Omega}=^3/_2,\ \epsilon \to J',\ \bar{\Omega}',\ \epsilon')$$

(4)

where J denotes the rotational state, $\bar{\Omega}$ is the magnitude of the projection of J on the internuclear axis, and ε is related to the parity, p, of a component of a Λ-doublet by

$$p = (-1)^{J-\frac{1}{2}}\varepsilon .$$ (5)

The calculations of Schinke and Andresen also showed that one component of each Λ-doublet is preferentially excited in such collisions. Subsequent spontaneous radiative decay then results in anti-inversion of the populations of the components of the ground state Λ-doublet, as noted above. On the other hand, the degree of preferential excitation, particularly in the $^2\Pi_{3/2}$ rotational ladder, as measured by the ratio $\sigma^+(J', \bar{\Omega}'=^3/_2)/\sigma^-(J', \bar{\Omega}'=^3/_2)$, where

$$\sigma^{\pm}(J', \bar{\Omega}') = \frac{1}{2\varepsilon}\Sigma\sigma(J=^3/_2, \bar{\Omega}=^3/_2, \varepsilon \to J', \bar{\Omega}', \varepsilon' = \pm 1) ,$$ (6)

was much in excess of the measurements.

Dewangan and Flower (1986) have recently extended their earlier close-coupling calculations (Dewangan and Flower 1983) to energies which encompass the value of E = 83 meV at which the measurements were carried out. Their results are compared with the experimental values in Figs. 5 and 6. The agreement is good for transitions within the

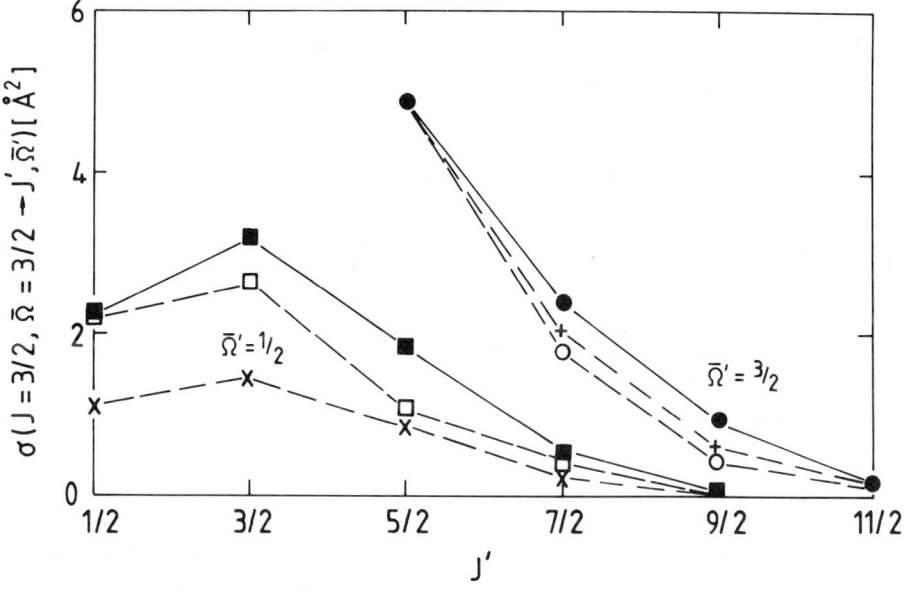

Figure 5. A comparison of computed and measured values of the Λ-doublet summed and averaged cross-sections ($\AA^2$) σ (J=$^3/_2$, $\bar{\Omega}=^3/_2 \to$ J', $\bar{\Omega}'$); open circles and squares: measurements of Andresen et al. (1984) at E=680 cm^{-1}; full circles and squares: calculations of Schinke and Andresen (1984) at E=680 cm^{-1}; crosses: averages of the results at E=556 cm^{-1} and E=834 cm^{-1}, from Dewangan and Flower (1986). All theoretical and experimental results are normalised to the value of σ(J=$^3/_2$, $\bar{\Omega}=^3/_2 \to$ J'=$^5/_2$, $\bar{\Omega}'=^3/_2$) computed by Schinke and Andresen (1984).

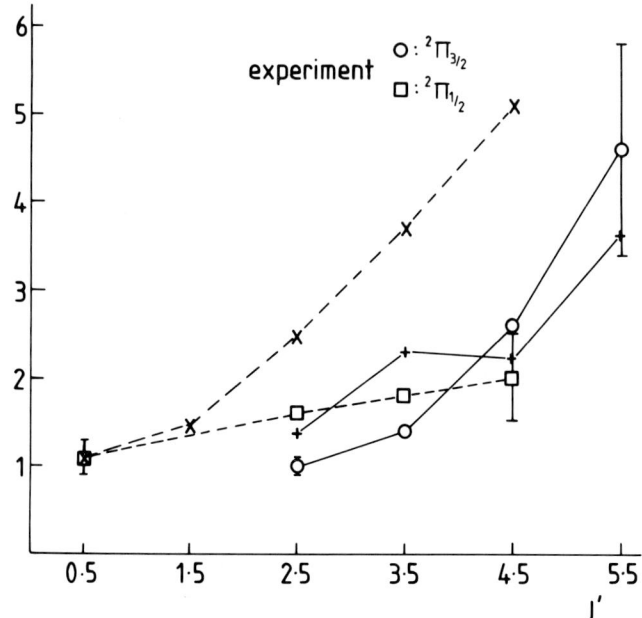

Figure 6. A comparison of computed (crosses) and measured values (with error bars) of the ratios $\sigma^+(J',\bar{\Omega}')/\sigma^-(J',\bar{\Omega}')$ for transitions within the $^2\Pi_{3/2}$ ladder and of σ^-/σ^+ for transitions to the $^2\Pi_{1/2}$ ladder. The measurements were at $E=680$ cm^{-1} and the computed values are averages of results at $E=556$ cm^{-1} and $E=834$ cm^{-1}, from Dewangan and Flower (1986).

$^2\Pi_{3/2}$ ladder, whereas the calculations appear to overestimate the ratio σ^-/σ^+ for transitions to the $^2\Pi_{1/2}$ ladder. The results of Dewangan and Flower (1986) are in qualitative but not quantitative agreement with the computations of Schinke and Andresen (1984). The reasons for these discrepancies are currently being investigated. Work by Alexander and Corey (1986) suggests that the CS approximation, employed by Schinke and Andresen, is not responsible.

3.4   Rotational excitation of CO by H_2

The rotational excitation of CO by H_2 has been the subject of a number of theoretical studies, largely in view of its astrophysical importance. The first detailed quantum mechanical calculations were performed by Green and Thaddeus (1976) and related to excitation by a simulated para-H_2(J=0) molecule. More recently, rate coefficients have been published by Schinke et al. (1985), again for excitation by para-H_2 (J=0), and by Flower and Launay (1985), for excitation by both para-H_2 and ortho-H_2. All of these calculations used different interaction potentials, of which that of Schinke et al. (1985) should be the most reliable.

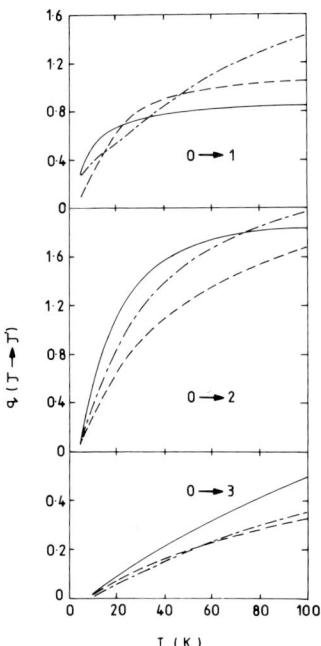

Figure 7. A comparison of computed rate coefficients, q (in units of
10^{-10} cm^3 s^{-1}), for rotational transitions J→J' in CO induced by
para-H$_2$, as a function of kinetic temperature, T. Continuous line:
Schinke et al. (1985); dashed line: Green and Thaddeus (1976); dash-dot
line: Flower and Launay (1985).

    Fig. 7 shows a comparison of the computed rate coefficients for
the first three transitions out of the rotational ground state of CO,
induced by para-H$_2$(J=0). The calculations of Schinke et al. (1985)
were based upon the CS approximation, which was shown to agree to
within about 20 per cent with close-coupling results for this system.
The computed rate coefficients agree remarkably well at low temperatures
(T ≲ 100 K) in spite of the differences in the approach to the
calculation of the interaction potential.
    Rotational transitions involving large values of the CO rotational
quantum number (J > 20) are observed in the Orion molecular cloud
(Watson et al. 1980; Storey et al. 1981), and the corresponding rate
coefficients are required for shock modeling of these observations
(Draine and Roberge 1984). McKee et al. (1982) used the IOS approximation to calculate rate coefficients for excitation by He of CO
rotational states J ≤ 32 and 100 ≤ T ≤ 2000 K. These authors
suggest that the corresponding results for excitation by para-H$_2$(J=0)
may be obtained by scaling with a factor 1.37 (≈ $2^{1/2}$), the ratio of the
velocities of H$_2$ and He at a given collision energy.
    Schinke et al. (1985) have shown that, whilst this scaling
procedure is valid for the lower ΔJ transitions at a given collision

energy, it fails dramatically for large values of ΔJ. This failure is attributable to the breakdown of the IOS approximation as ΔJ increases. The IOS approximation is valid when the collision time is much smaller than the rotational period of the molecule, that is, for collision energies E much larger than the rotational excitation energy ($\propto J^2$). As the collision time varies inversely with $v = (2E/\mu)^{1/2}$, where μ is the reduced mass, the IOS approximation fails for smaller values of ΔJ when He is the perturber than when H_2 is the perturber. Thus, if the He-CO potential energy surface is used but the H_2 mass is substituted in the dynamical calculations, much smaller cross-sections are obtained for large values of ΔJ (see Fig. 8). Clearly, the He-CO and H_2-CO cross-sections are not related by a simple scaling factor when ΔJ is large.

3.5 Excitation of hyperfine transitions in molecules

Cross-sections between hyperfine levels of molecules such as HCN, NH_3, and OH are required for a proper analysis of astronomical observations of these molecules. Stutzki and Winnewisser (1985) used the CS and IOS

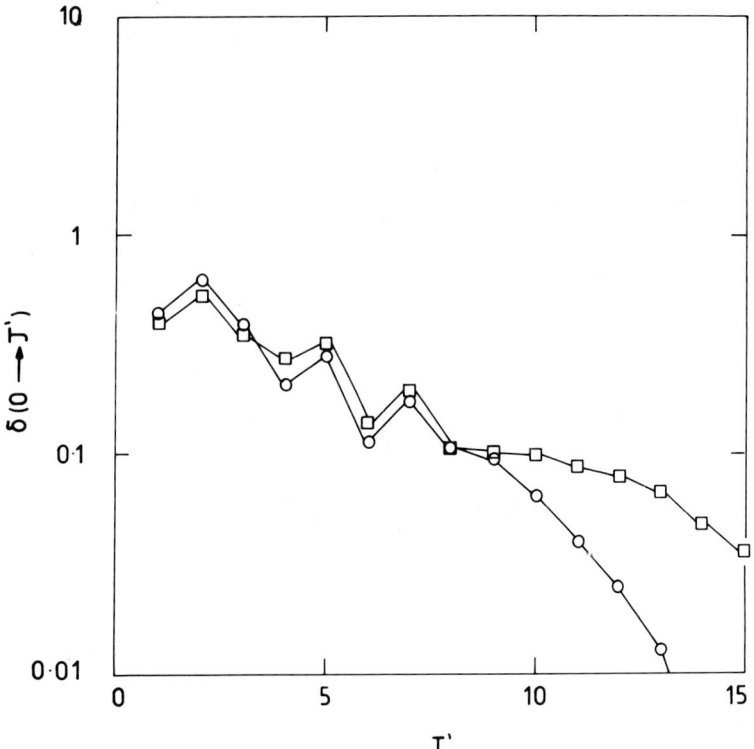

Figure 8. A comparison of cross-sections, σ (in units of 10^{-16} cm^2), for rotational transitions 0 → J' in CO induced by para-H_2 at a collision energy E=75 meV. Squares: perturber has mass of He; circles: perturber has mass of H_2. The same interaction potential was used in both calculations (from Schinke et al. 1985).

approximations in a study of hyperfine transitions in NH_3 and HCN, excited by He. Their results for HCN were in poor agreement with earlier work of Monteiro (1984), where the CS approximation was also employed. However, as recently shown by Monteiro and Stutzki (1986), great care must be exercised when applying centrifugal decoupling approximations to problems involving hyperfine coupling, that is, coupling between the rotational and nuclear spin angular momenta of a molecule. The further assumption, inherent in the IOS approximation, that the molecular rotational period is long compared with the collision time, is also of dubious reliability for temperatures $T \lesssim 100$ K, particularly for a relatively light molecule such as NH_3.

Monteiro and Stutzki (1986) avoid these uncertainties by using the full close-coupling method to compute the cross-sections for hyperfine transitions in HCN, induced by He. Their results are compared, in Table 1, with the earlier calculations of Monteiro (1984) and Stutzki

J	F	J'	F'	MS	M	SW
1	1	2	1	21	17	23
		2	2	60	61	36
		2	3	19	23	41
1	1	3	2	30	29	29
		3	3	61	62	42
		3	4	9	9	30
2	2	3	2	17	17	27
		3	3	70	65	37
		3	4	13	18	38

Table 1. Relative values of rate coefficients, q, for hyperfine transitions JF → J'F' in HCN, induced by collisions with He. The values given are percentages, 100 x q (JF → J'F')/$\sum_{F'}$ q(JF → J'F'), calculated at a kinetic temperature T=20 K. MS: Monteiro and Stutzki (1986); M: Monteiro (1984); SW: Stutzki and Winnewisser (1985).

and Winnewisser (1985). The former are evidently more accurate than the latter.

Cross-sections for hyperfine transitions in OH are required for the analysis of OH maser transitions. Results for excitation by para-H_2 (J=0) should soon be available (Corey and Alexander 1986). It is also our intention to extend the close-coupling study of NH_3-para-H_2 collisions (Danby et al. 1986a,b) to incorporate the hyperfine structure of the ammonia molecule.

4. REFERENCES

Aannestad, P.A., and Field, G.B., 1973, Ap. J., 186, L29.

Alexander, M.H., and Corey, G., 1986, personal communication.

Alexander, M.H., and Dagdigian, P.J., 1984, J. Chem. Phys., 80, 4325.

Andresen, P., Haüsler, D. and Lülf, H.W., 1984, J. Chem.Phys., 81, 571.

Bertojo, M., Cheung, A.C., and Townes, C.H., 1976, Ap. J., 208, 914.

Billing, G.D., and Diercksen, G.H.F., 1985, Chem. Phys. Lett., 121, 94.

Billing, G.D., and Diercksen, G.H.F., 1986, Chem. Phys., 105, 145.

Billing, G.D., and Poulsen, L.L., 1984, J. Chem. Phys., 81, 3866.

Billing, G.D., Poulsen, L.L., and Diercksen, G.H.F., 1985, Chem. Phys.,
    98, 397.

Broquier, M., and Picard-Bersellini, A., 1985, Chem. Phys. Lett., 121,
    437.

Corey, G., and Alexander, M.H., 1986, work in progress.

Daly, P.W., and Oka, T., 1970, J. Chem. Phys., 53, 3272.

Danby, G., and Flower, D.R., 1986, work in progress.

Danby, G., Flower, D.R., Kochanski, E., Kurdi, L., Valiron, P., and
    Diercksen, G.H.F., 1986a, J. Phys. B: At. Mol. Phys., in press.

Danby, G., Flower, D.R., Valiron, P., Kochanski, E., Kurdi, L., and
    Diercksen, G.H.F., 1986b, J. Phys. B: At. Mol. Phys., submitted.

Davis, S.L., 1985, Chem. Phys., 95, 411.

Dewangan, D.P., and Flower, D.R., 1981, J. Phys. B: At. Mol. Phys., 14,
    2179.

Dewangan, D.P., and Flower, D.R., 1983, J. Phys. B: At. Mol. Phys., 16,
    2157.

Dewangan, D.P., and Flower, D.R., 1985, J. Phys. B: At. Mol. Phys., 18,
    L137.

Dewangan, D.P., and Flower, D.R., 1986, to be published.

Dixon, R.N., and Field, D., 1979a, Proc. R. Soc. A, 368, 99.

Dixon, R.N., and Field, D., 1979b, M.N.R.A.S., 189, 583.

Dixon, R.N., Field, D., and Zare, R.N., 1985, Chem. Phys. Lett., 122,
    310.

Draine, B.T., 1980, Ap. J., 241, 1021.

Draine, B.T., and Roberge, W.G., 1984, Ap. J., 282, 491.

Elitzur, M., and Watson, W.D., 1980, Ap. J., 236, 172.

Flower, D.R., and Launay, J.M., 1977, J. Phys. B: At. Mol. Phys., 10,
    3673.

Flower, D.R., and Launay, J.M., 1985, M.N.R.A.S., 214, 271.

Flower, D.R., Pineau des Forets, G., and Hartquist, T.W., 1986,
    M.N.R.A.S., 218, 729.

Gordon, R.G., 1969, J. Chem. Phys., 51, 14.

Green, S., 1975, J. Chem. Phys., 62, 2271.

Green, S., 1981, NASA Technical Memorandum 83869.

Green, S., Ramaswamy, R., and Rabitz, H., 1978, Ap. J. Suppl., 36, 483.

Green, S., and Thaddeus, P., 1976, Ap. J., 205, 766.

Green, S. and Zare, R.N., 1975, Chem. Phys., 7, 62.

Ho, P.T.P., and Townes, C.H., 1983, Ann. Rev. Astr. Ap., 21, 239.

Klaassen, D.B.M., ter Meulen, J.J., and Dymanus, A., 1983, J. Chem.
    Phys., 78, 767.

McGuire, P., and Kouri, D.J., 1974, J. Chem. Phys., 60, 2488.

McKee, C.F., Storey, J.W.V., Watson, D.M., and Green, S., 1982, Ap. J., 259, 647.

Monchick, L., and Schaefer, J., 1980, J. Chem. Phys., 73, 6153.

Monteiro, T.S., 1984, M.N.R.A.S., 211, 257.

Monteiro, T.S., and Stutzki, J., 1986, M.N.R.A.S., in press.

Ramaswamy, R., Rabitz, H., and Green, S., 1977, J. Chem. Phys., 66, 3021.

Schaefer, J., 1986, personal communication.

Schinke, R., and Andresen, P., 1984, J. Chem. Phys., 81, 5644.

Schinke, R., Engel, V., Buck, U., Meyer, H., and Diercksen, G.H.F., 1985, Ap. J., 299, 939.

Secrest, D., 1975, J. Chem. Phys., 62, 710.

Shull, J.M., and Hollenbach, D.J., 1978, Ap. J., 220, 525.

Spitzer, L., and Cochran, W.D., 1973, Ap. J., 186, L23.

Spitzer, L., Drake, J.F., Jenkins, E.B., Morton, D.C., Rogerson, J.B., and York, D.G., 1973, Ap. J., 181, L116.

Stechel, E.B., Walker, R.B., and Light, J.C., 1978, J. Chem. Phys., 69, 3518.

Storey, J.W.V., Watson, D.M., Townes, C.H., Haller, E.E., and Hansen, W.L., 1981, Ap. J., 247, 136.

Stutzki, J., and Winnewisser, G., 1985, Astr. Ap., 144, 1.

Watson, D.M., Storey, J.W.V., Townes, C.H., Haller, E.E., and Hansen, W.L., 1980, Ap. J., 239, L129.

Chris Henkel

MASERS IN THE INTERSTELLAR MEDIUM

Moshe Elitzur
Department of Physics and Astronomy
University of Kentucky
Lexington, KY 40506
U.S.A.

ABSTRACT.   The basic concepts that govern the production and propagation of maser radiation in interstellar space are introduced and discussed. Specific interstellar masers, in particular those in star forming regions, are described in some detail.

1. INTRODUCTION

The radio radiation of interstellar molecules often shows emission patterns with marked deviations from thermal equilibrium. This became evident shortly after the first radio detection of an interstellar molecule, the hydroxyl radical (OH), by Weinreb et al. (1963). The rotation states of this molecule are split into four levels by Λ-doubling and hyperfine interactions (see fig. 1), resulting in four allowed (radio) transitions within each rotation state. The wavelengths of the ground state transitions, where the original discovery was made, are approximately 18 cm. The simultaneous detection of all four lines showed that line ratios were significantly different from the values expected if the population was distributed according to thermal equilibrium.

With the discovery of more radio lines in interstellar space it became evident that thermal equilibrium for molecular level populations is the exception rather than the rule. One of the most extreme examples of deviation from equilibrium occurs when for certain transitions, the upper state ends up with a higher population (per magnetic sub-level) than the lower one. The transition is then "inverted" and the rate for stimulated emission per unit volume in the line exceeds that for absorption. The net effect of the interaction of the line radiation with matter is that of amplification instead of attenuation. The familiar absorption factor $\exp(-\tau)$ becomes an amplification factor since τ is now negative (and hence called a "gain"). The radiation intensity of the inverted line can become very large, leading to spectacular brightness temperatures.

Strong maser activity turned out to be one of the signposts of the star formation process. Masers gave the first indication for the high

763

D. J. Hollenbach and H. A. Thronson, Jr. (eds.), Interstellar Processes, 763–780.

velocity flows in star forming regions. This review discusses the basic
concepts associated with the theory of masers in the interstellar
medium, how this intense radiation is produced in the relevant sources,
and what we can learn about the emission regions. Recent reviews that
can supplement the present discussion and provide more details on
various aspects are: Reid and Moran (1981) - observations; Elitzur
(1982) - theory; Elitzur (1986) - pumping schemes for masers in star
forming regions; Genzel (1986) - maser observations in star forming
regions.

2.   MASER THEORY

2.1. Basic Concepts

The first question regarding astronomical masers is "How do we know that
the radiation from a particular source is indeed due to maser action?
Could it not be incoherent spontaneous emission?" To answer this basic
question it is useful to introduce the concept of "brightness
temperature", defined in the following way:

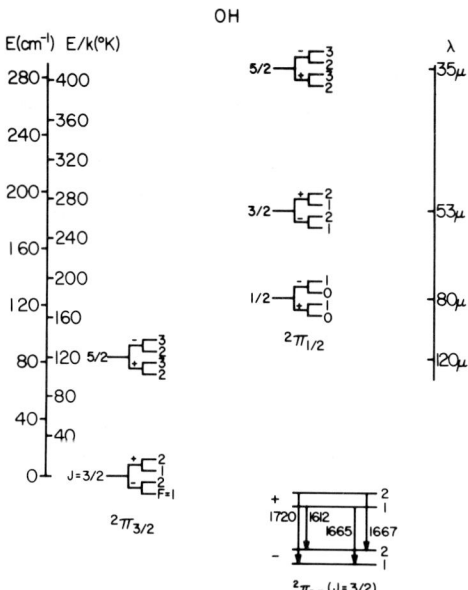

Figure 1.   The rotation levels of OH that couple radiatively to the
ground state $^2\Pi_{3/2}$(J=3/2). F is the total angular momentum, including
nuclear spin. Level splitting is not to scale. The strong maser transitions are detected in the ground state, which is plotted separately on
an expanded scale at the right-hand corner with frequencies (in MHz)
marked on the arrows.

The intensity of the radiation field of a blackbody at temperature
T is given by the familiar Planck function

$$B_\nu(T) = \frac{2h\nu^3}{c^2} \cdot \frac{1}{\exp(h\nu/kT) - 1} \tag{1}$$

For radiation intensity I_ν in a given source, the brightness temperature
T_b is defined as the temperature of an equivalent blackbody that will
emit the same amount of radiation at the given frequency ν, namely

$$I_\nu = B_\nu(T_b) \tag{2}$$

For most lines of interest in radio astronomy the Rayleigh-Jeans limit
($h\nu < kT$) is applicable and it follows from eqs. (1) and (2) that the
brightness temperature is given by

$$kT_b = \frac{c^2}{2\nu^2} I_\nu \tag{3}$$

Obviously, the brightness temperature is a formal quantity, depending on
the wavelength in general. It becomes wavelength independent only for a
source that emits according to the blackbody law. However, the significance of the brightness temperature becomes evident when we recall that
according to the laws of thermodynamics, a blackbody is the perfect
emitter since it is also the perfect absorber. Therefore, for a given
temperature nothing can emit more than a blackbody. The observed
brightness temperature therefore provides a lower limit for the
temperature of the source.
    To formulate this important conclusion more rigorously, let us
start with the equation of radiative transfer

$$\frac{dI_\nu}{d\ell} = -\kappa_\nu I_\nu + \varepsilon_\nu \tag{4}$$

where ℓ measures distances along the ray propagation path, and κ_ν and ε_ν
are the absorption and emission coefficients, respectively. For line
radiation they are given by

$$\varepsilon_\nu = N_2 A_{21} h\nu\Phi(\nu)/4\pi \tag{5.a}$$

$$\kappa_\nu = (N_1 B_{12} - N_2 B_{21}) h\nu\Phi(\nu)/4\pi \tag{5.b}$$

where N_1 (N_2) is the number density of the population of the lower
(upper) level; $h\nu$ is the energy separation of the levels; $\Phi(\nu)$ is the
normalized line profile ($\int\Phi(\nu)d\nu = 1$); and A and B are the appropriate
Einstein coefficients for spontaneous emission and for induced
processes, respectively.
    The absorption coefficient includes the correction for stimulated

emission, which is accounted for as negative absorption. This term can
be neglected most of the time since the population of the upper level is
usually much smaller than that of the lower one. However, under certain
circumstances the stimulated emission term can become important, even to
the point of dominating the absorptive one. To see how that can happen,
recall the Einstein relation

$$g_1 B_{12} = g_2 B_{21}$$

and introduce the populations per sub-level

$$n_i = N_i / g_i \qquad i = 1,2$$

where g_i are the statistical weights. The absorption coefficient
becomes

$$\kappa_\nu = g_2 B_{21} (n_1 - n_2) h\nu\Phi(\nu)/4\pi \qquad (6)$$

The stimulated emission will therefore dominate, resulting in a negative
absorption coefficient, when $n_2 > n_1$. The levels are inverted, leading
to maser action, when the population per sub-level of the upper state
exceeds that of the lower state.

We are frequently interested only in the total line emission, and
not in its detailed spectral shape. It is then advantageous to
introduce the profile-averaged line intensity

$$I = \int I_\nu \Phi(\nu) d\nu \qquad (7)$$

and integrate the equation of radiative transfer over frequency. The
result is

$$\frac{dI}{d\ell} = -\kappa I + \varepsilon \qquad (8)$$

where κ and ε are given by expressions similar to eqs. (5.a,b) with $\Phi(\nu)$
replaced by the inverse of the linewidth $\Delta\nu$, defined via

$$\Delta\nu = \frac{1}{I} \int I_\nu d\nu . \qquad (9)$$

We have seemingly managed to eliminate altogether the complicated
issue of the frequency dependence of the line radiation. However, this
function is in fact hidden in the innocent looking linewidth (eq. 9),
which we cannot evaluate without a detailed knowledge of the shape of
the intensity profile. This, of course, requires a solution of the
full, frequency dependent radiative transfer problem. In practice we
simply assume that the linewidth is known, and can be estimated from the

profile function $\Phi(\nu)$.

The equation of radiative transfer (eq. 8) is usually re-written as

$$\frac{dI}{d\tau} = -I + S \tag{10}$$

where we have introduced the optical depth element $d\tau = \kappa d\ell$ and the source function $S = \varepsilon/\kappa$. Using the second Einstein relation

$$A_{21} = B_{21} \frac{2h\nu^3}{c^2}$$

and the explicit expressions for ε and κ (eqs. 5.a and 6), we find

$$S = \frac{2h\nu^3}{c^2} (n_1/n_2 - 1)^{-1} \tag{11}$$

If we now introduce the transition excitation temperature, T_x, via

$$n_2/n_1 = \exp(-h\nu/kT_x) \tag{12}$$

then the source function for the line finally becomes

$$S = B_\nu(T_x), \tag{13}$$

namely – it is the Planck function at the line excitation temperature. With this result and the definition of the brightness temperature (eq. 2) the equation of radiative transfer can be written as

$$\frac{dB_\nu(T_b)}{d\tau} = -B_\nu(T_b) + B_\nu(T_x) \tag{14}$$

The Planck function is monotonically increasing with temperature at any fixed frequency. From eq. 14 we therefore see that radiation generated inside the source will be self absorbed unless

$$T_b \leqslant T_x, \tag{15}$$

namely – the brightness temperature cannot exceed the line excitation temperature.

Equation 14 is often written as an equation among the appropriate temperatures, assuming they all obey the Rayleigh–Jeans limit,

$$\frac{dT_b}{d\tau} = -T_b + T_x \tag{16}$$

The solution, assuming a constant excitation temperature, is

$$T_b = T_x(1 - e^{-\tau}) + T_c e^{-\tau} \qquad (17)$$

where T_c is the brightness temperature of an external continuum source that may illuminate the backside of the cloud, and the explicit expression for the line optical depth τ is

$$\tau = h\nu g_2 B_{21}(n_1 - n_2)\ell/(4\pi\Delta\nu) \qquad (18)$$

We again see that in the absence of a background source, the brightness temperature will always be smaller than the line excitation temperature and will approach it only in the limit of large optical depth ($\tau \gg 1$).

The significance of the brightness temperature is that it provides a lower limit estimate for the line excitation temperature at the source. While T_x need not coincide with the source kinetic temperature, it usually has the same order of magnitude. However, the brightness temperatures observed in interstellar OH masers are frequently as high as 10^{12}K, and those in H_2O masers even reach $10^{15} - 10^{16}$K (see e.g. Genzel 1986). Obviously, these temperatures bear no resemblance to the kinetic temperatures at the sources since molecules dissociate at a few 10^3K. In addition, the linewidths indicate that the temperatures are no more than $\sim$ 100K.

The resolution of this puzzle is that the levels are indeed inverted. Both the optical depth and the excitation temperature are negative in this case. Equation 16 becomes

$$\frac{dT_b}{d|\tau|} = T_b + |T_x|$$

and it is obvious that there is no bound on T_b. In fact, eq. 17 shows that T_b will now increase exponentially with τ, which is therefore called the maser gain. A gain of 20 will introduce an amplification factor of 10^8.

It is instructive to ask whether a direct determination of the excitation temperature is possible, showing that T_x indeed becomes negative in maser sources. The difficulty is that eq. 17, which could be used to determine T_x, contains the additional unknown τ. A clever radio-astronomy technique around this difficulty is to select clouds with an extragalactic (i.e. point) continuum source in the background, and measure the brightness temperature in directions that contain and avoid the background source. These "on-" and "off-source" measurements provide two independent brightness temperatures that can be used to solve for both τ and T_x. Rieu et al. (1976) used this technique to measure the excitation temperatures in the cloud in front of 3C123 and found $T_x \approx -10$K for the 1720 MHz line. The cloud was acting as an amplifier for the background source at this frequency!

2.2. Phenomenological Maser Model

The maser behavior is described by the statistical rate equations for the level populations, coupled with the equation of radiative transfer for the maser radiation. Because of the high intensity of maser radiation it is not obvious *a-priori* that the familiar linear equations should apply here too. Fortunately, this <u>does</u> turn out to be the case for astronomical masers.

The phenomenological theory of astronomical masers was first worked out by Litvak (1970) and later extended by Goldreich, Keely and Kwan (1973) to include polarization effects. It is based on the formalism developed by Lamb (1964) for laboratory laser radiation. In this semi-classical approach the radiation field is treated as a classical wave while the maser levels are described as a quantum mechanical system. The coupling between matter and radiation results from the interaction of the wave's electric field with the dipole moment of the molecules. The derivation of the equations will not be reproduced here. A fairly detailed presentation can be found in Elitzur (1982).

The resulting equation of radiative transfer for astronomical masers is identical to eq. (8). The level populations are governed by the familiar equation

$$\frac{dN_2}{dt} = P_2 - \Gamma_2 N_2 - A_{21} N_2 - J(N_2 B_{21} - N_1 B_{12}) - N_2 C_{21} + N_1 C_{12} \quad (19)$$

where P_2 and Γ_2 are, respectively, the rates for population gain and loss due to interaction with states other than the maser levels; C_{ij} are the rate coefficients for collisional exchange between the maser levels; and J is the angle averaged intensity

$$J = \int I \frac{d\Omega}{4\pi} \quad (20)$$

The rate equation for level 1 is similar, with the exchange terms changing sign.

We see that the populations of the maser levels, as well as the maser intensity, are governed by equations that are identical to the case of non-maser radiation. This is a non-trivial result that may be in error in the case of the strong H_2O maser sources. This point is discussed below, and will be ignored for the time being.

The steady-state solution of the level population equations is a straightforward exercise in algebra. The expressions are greatly simplified when the statistical weights (g) and loss rates (Γ) are the same for both levels, so we will assume that this is the case. The spontaneous decay rate is usually smaller than all other rates, and can therefore be neglected. To simplify matters further we also neglect the effect of collisions. After all of these simplifications, the population difference, $\Delta n = n_2 - n_1$, is given by

$$\Delta n = \frac{\Delta p}{\Gamma + 2BJ} = \frac{\Delta n_0}{1 + J/J_s} = \frac{\Delta n_0}{1 + I/I_s} \qquad (21)$$

where

$$\Delta n_0 = \Delta p/\Gamma; \quad J_s = \Gamma/2B; \quad I_s = J_s 4\pi/\Omega_m \qquad (22)$$

and $\Delta p = p_2 - p_1$ is the difference in pumping rates per sub-level ($p_i = P_i/g_i$) for the two maser states. The levels will be inverted when p_2, the pump rate into the upper level, exceeds p_1. The relation between I_s and J_s, the "saturation" intensities, follows from eq. (20) and the fact that maser radiation is usually beamed into a narrow solid angle, Ω_m. The beaming occurs because the maser photons are seeking paths that maintain good coherence in the line-of-sight velocity to obtain a large amplification. Also, Alcock and Ross (1985) have recently shown that small asymmetries in the cloud shapes will produce strong beaming of the maser radiation. Subsequently, astronomical masers have the shape of elongated tubes, a geometry that was first studied by Goldreich and Keely (1972).

The absorption coefficient, likewise, takes the form

$$\kappa = \frac{\kappa_0}{1 + J/J_s} = \frac{\kappa_0}{1 + I/I_s} \qquad (23)$$

where we omit the absolute value sign, which should be understood for all inverted quantities.

The expressions for the population inversion and the absorption coefficient show that, depending on the magnitude of J relative to J_s, the maser can display two different types of behavior. The first one is called "unsaturated" (in contrast with the "saturated" behavior described shortly.) In this regime $J \ll J_s$ and the population difference is $\Delta n = \Delta n_0$, independent of the maser radiation field; the level populations are determined exclusively by the pump and loss processes. The inverted medium provides an amplifying background, unaffected by the propagating maser photons which multiply by stimulated emission. The brightness temperature is described by the relation derived above (eq. 17), which for maser radiation becomes

$$T_b = (T_x + T_c)\exp(\kappa_0 \ell) - T_x , \qquad (24)$$

displaying an exponential amplification. This is due to the shower of induced emission photons which can be generated by one seed photon. The power series expansion of the exponential function describes the successive generations of induced photons.

The exponential growth cannot proceed indefinitely. Eventually, the radiation becomes so intense that the induced processes begin to compete with the loss rate Γ. This happens, obviously, when $J > J_s$. Because the level population is inverted, the number of stimulated emissions exceeds the number of absorption events and the interaction with the radiation *decreases* the inversion. The combined effect of increased intensity and decreased inversion is that the product $J\Delta n$

approaches the limiting value $J_s \Delta n_0$, as evident from eq. (21), and the maser "saturates". The volume production rate of a saturated maser, the difference between stimulated emission and absorption events, is

$$\Phi_m = BJ\Delta N = \Delta P/2 = \eta P \qquad (25)$$

where $P = (P_1 + P_2)/2$ is the average pump rate into the maser levels and $\eta = \Delta P/2P$ is an efficiency factor. We therefore see that *each pumping event leads to the production of a maser photon with an efficiency η (<1) which is determined by the details of the pumping scheme.* The saturated maser is really a linear converter rather than an amplifier, converting the pumping events to maser photons. This is the most efficient mode of producing maser photons. The strong interstellar masers are almost certainly saturated.

The effect of collisions, which we neglected earlier, does not alter the basic results derived here, although it modifies the specific expressions. The details can be found in Elitzur (1982).

2.3. Pumping Considerations

Almost all inversion schemes proposed for astronomical masers are based on either radiative or collisional pumping. Chemical pumps, where the inversion results from the molecular production process, can usually be dismissed because in steady state they cannot compete with collisions (Litvak 1969).

2.3.1 <u>Radiative Pumps:</u> In most of these pumping schemes the molecules are radiatively excited to higher levels and the inversion occurs as a result of the subsequent cascades. The rate of pumping events is then equal to the number of photons absorbed in the pump lines, per unit time. If N_m is the number of maser photons emitted, N_p the number of pump photons absorbed, and η the pump efficiency then

$$N_m = \eta N_p . \qquad (26)$$

This leads to the following stringent constraint on radiative pump models: *the number of photons emitted in the pump lines must exceed the number of observed maser photons.* This can sometimes eliminate radiation as the potential pump at the outset.

Equation (26) can also be used to derive the brightness temperature of a fully saturated maser in the case of radiative pumping. The pumping radiation is frequently due to a nearby source occupying a solid angle $\Delta\Omega_p$ and emitting as a blackbody at temperature T_p. If the surface area of the maser is ΔS then the number of maser photons absorbed per second in a pump line at frequency ν_p is

$$N_p = B_{\nu_p}(T_p)\Delta\Omega_p \Delta S \Delta\nu_p / h\nu_p \qquad (27)$$

where $\Delta\nu_p$ is the width of the pump line (we assume here that the maser is optically thick in the pump line and hence absorbs all the pump photons emitted toward it. If this is not the case, N_p will be reduced by τ_p, the optical depth in the pump line.) The number of maser photons emitted per unit time, N_m, obeys a similar relation where the index m replaces p in an obvious manner. Equation (26) therefore becomes

$$B_{\nu_m}(T_m) = \eta B_{\nu_p}(T_p)\left(\frac{\Delta\Omega_p}{\Delta\Omega_m}\right)\cdot\left(\frac{\Delta\nu_p/\nu_p}{\Delta\nu_m/\nu_m}\right) \tag{28}$$

Figure 2 displays the brightness temperature that will be obtained for an OH maser pumped by radiation at T_p = 500K through its 35μ rotation lines, assuming that the overall conversion efficiency to maser radiation (including the ratios of solid angles and relative bandwidths) is 50%. An analytic expression for T_m, which also explains the results displayed in the figure, can be obtained in the Rayleigh-Jeans limit. Eq. (28) then becomes

$$T_m = \eta T_p\left(\frac{\Delta\Omega_p}{\Delta\Omega_m}\right)\cdot\left(\frac{\nu_p}{\nu_m}\right)^2 \tag{29}$$

where we have also assumed the same relative linewidth for the maser and pump radiation. The high brightness temperatures of the maser radiation are the result of the factor $(\nu_p/\nu_m)^2$, which reflects the difference in

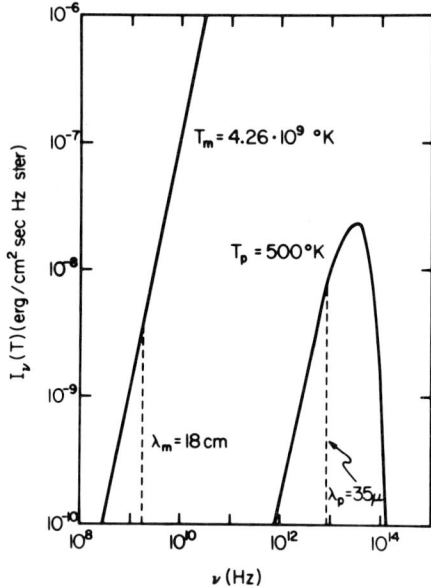

Figure 2.   The Planck distribution function $B_\nu(T)$ at temperatures T_p and T_m, as marked on the curves. The temperature T_m was determined from the condition $B_{\nu_m}(T_m) = B_{\nu_p}(T_p)/2$.

photon phase space density between the pump and maser frequencies. The
inverted population enables the maser to efficiently shift high
frequency pump photons to the low maser frequencies, where their number
density is much higher than that allowed in thermal equilibrium. It
should also be noticed that the factor responsible for these spectacular
brightness temperatures ensures at the same time that the energy flux
carried by the maser radiation will never become significant for the
energy budget of the source, because the photons are degraded to lower
energies by the inversion process.

2.3.2 Collisional Pumps: These can be broadly divided to two
categories. The first one is similar to the radiative pumps just
mentioned and can be further divided to two sub-classes. In both of
these the molecules are collisionally excited to higher levels and then
cascade. Inversion can occur in one of two ways: first, as a result of
the decay process, a situation essentially identical to radiative
pumping, with collisions replacing radiation as the excitation agent.
Alternately, certain excited levels may be selectively pumped because of
the collision rules, and the preferential excitation can translate to
inversion upon decay. Schemes that involve either type of inversion
mechanism have been proposed.

Collisional pumps of this first category are usually constrained in
a similar manner to radiative pumps, although the arguments are not as
straightforward and precise as in that case. At its root lies the
detailed-balance relation for collision rates between a lower level l
and an upper level u, separated by an energy gap ΔE:

$$\frac{C_{lu}}{C_{ul}} = \frac{g_u}{g_l} \exp(-\Delta E/kT) \qquad (30)$$

Hence, when the transitions in both directions are collision dominated,
the population follows the thermal, Boltzmann distribution and maser
action is impossible. Downward transitions therefore must be radiative
and this sets an upper limit on the density of the maser region.
Because the pump rate is proportional to the density of the colliding
particles, this limits also the maser output -- to luminosities that are
usually similar to what can be obtained by radiative pumping in the same
source. For details, see Elitzur (1982).

The second type of pumping schemes involves collisions that violate
the detailed balance relation (eq. 30). This can happen, for example,
when the colliding particles do not follow the Maxwell-Boltzmann
distribution (as, for instance, in the case of a stream) or when they
are comprised of two or more populations at different temperatures.
There is no limit on the densities in this case, since the collisions do
not lead to thermalization, and the maser output can become arbitrarily
high.

Specific examples of some of the various types of pumps are
discussed in the next section.

3. MODELS OF INTERSTELLAR MASERS

The molecules displaying strong maser action in interstellar space are
OH, H_2O and SiO. They will now be discussed in this order, which is
also the order of their original discoveries. All three appear in star
forming regions, and the OH and H_2O are used as signposts of the star
formation process. There is a close association between these masers
and very compact HII regions, compact IR sources and thermal mass
outflow (Genzel, 1986).

3.1. OH

3.1.1 Satellite Lines in Clouds: Of the three strong interstellar
maser molecules, OH is the only one also detected in cool, extended
clouds. Its emission pattern almost never follows thermal equilibrium.
While the main line ratios (1665 and 1667 MHz) are close to equilibrium,
the satellite lines (at 1612 and 1720 MHz) are almost always anomalous.
One of them usually appears in emission and the other one in absorption.
The emissivity deviates from the value expected from the main lines
ratio, even for the line whose behavior resembles that of the main
lines.

From the level diagram (fig. 1) it is evident that this anomalous
pattern can be explained quite simply with population transfer between
the ground state levels with different total angular momentum F. The
transfer of molecules from, say, F=1 to F=2 within the same Λ-doublet
component (same parity), would lead to an inversion of the 1720-MHz line
and "anti-inversion" of the 1612 line, with the main lines remaining in
equilibrium. The opposite transfer, from F=2 to F=1, would reverse the
roles of the two satellite lines, without affecting the main lines.

The required transfer occurs because the rotational transitions to
the ground state are usually optically thick in sources that allow OH
detection. The radiative decay rate is proportional to A/τ in this
case, and is independent of the line-strength (see e.g. Elitzur 1982).
The rate of cascade to each ground state level is therefore determined
by the number of transitions that land in it. Consider first the
cascade from the $^2\Pi_{3/2}$ (J=5/2) state. Its F=2 level can decay to either
ground state level and will therefore share its population with both of
them. The F=3 level, on the other hand, decays exclusively to F=2
because decays to F=1 would involve a ΔF=2 change, which is forbidden by
the dipole selection rules. The cascade from $^2\Pi_{3/2}$ (J=5/2) therefore
over-populates the ground state F=2 at the expense of F=1, leading to
inversion of the 1720 line and "anti-inversion" of the 1612 line.
Precisely the opposite happens when the cascade is from the $^2\Pi_{1/2}$ (J=1/2)
state, because its F=1 level can decay to both ground state levels,
while F=0 cannot decay to F=2.

When the two cascade modes compete, the 1612 inversion wins. In
the case of radiative pumping, the $^2\Pi_{3/2}$ (J=1/2) state is always
populated and the 1612 inversion is activated. This explains the OH
maser emission from late-type stars. To invert the 1720 line, the
$^2\Pi_{3/2}$ (J=5/2) state must be preferentially excited. This can happen when
the excitation is collisional at temperatures that are not too high or

when the optical depths are in a narrow range enabling the operation of the 1720 inversion mechanism, but not that of the 1612 line. The clouds that show a preponderance of 1720 emission should therefore be the weak sources, and should also have higher densities. This has been verified by observations (Haynes and Caswell, 1977).

3.1.2 HII/OH Regions: The strongest OH maser sources appear in regions of star formation and are associated with compact HII regions. However, the exact relation between the maser spots and the underlying HII region is not entirely clear yet. Different observational studies arrived at conflicting conclusions regarding the precise placement of the maser spots in relation to the HII region associated with them (see Norris et al. 1982; Ho et al. 1983; Garay et al. 1985).

A controversy surrounds also the motion of the maser spots. The only detailed theoretical model for HII/OH regions, by Elitzur and de Jong (1978), places the maser spots at the edge of the HII region where they partake in its expansion. This proposal is supported by the recent MERLIN observations of Baart et al. (1986). Reid et al. (1980), on the other hand, suggest that the masers are falling onto the HII region, because the centroid of their velocities is red-shifted with respect to it. This proposal was subsequently supported by Garay et al. (1985) but disputed by Norris and Booth (1981). Berulis and Ershov (1983) were then able to show that optical depth effects can produce an apparent infall velocity for the maser spots, even when they are located at the edge of an HII region and participate in its expansion.

These are important issues. If the infall idea is correct it would imply that newly born stars are accreting matter even at the late stages of their evolution at which the HII regions have already formed and are expanding.

One of the main reasons for the placement of the masers in the compressed shell between the shock and ionization fronts in the Elitzur and de Jong model, is that the shock chemistry enhances the OH abundance in this region. It should be emphasized that the same conclusion would apply also to the accreting fragments proposed by Reid et al. (1980), if a shock developed at their edge.

The HII/OH masers are almost certainly pumped by collisions, since the number of photons observed in the various potential pump wavelengths is smaller than the number of maser photons. A possible exception is the strong far-IR emission observed by Thronson and Harper (1979) in the $40 - 150\mu$ range. However, this radiation most likely comes from a cool dust shell located outside the maser region and probably cannot provide adequate pumping.

Collisional pumping schemes based on circulation of the molecules through excited rotation states run to the problem of thermalization at high densities, and have difficulties in producing the observed maser intensities. It appears that the only plausible inversion mechanism is collisions that violate the detailed balance relation (eq. 30). A possible mechanism was proposed by Johnston (1967) who showed that for collisions with electron streams, the rate for excitations is larger than for de-excitations for $\Delta m = 0$ transitions. Such collisions can therefore produce arbitrarily strong maser radiation, which is also

highly polarized -- as observed. The problem is that the electrons,
because of their low mass, would require large ordered velocities (~ 40
km/sec) to create a "stream". The electrons could perhaps be
substituted by ions, which "stream" at only ~ 1 km/sec (Elitzur 1979).
Although the viability of this proposal is not yet clear, because the
cross sections are not known with sufficient accuracy, recent magnetic
shock calculations make it particularly attractive. These calculations
show that large streaming velocities for the charged particles may be
common behind shocks that propagate perpendicular to a magnetic field
(e.g. Draine 1980). Such shocks fit rather well in the HII/OH scheme of
Elitzur and de Jong (1978) and Elitzur (1979). In addition, if charged
particle streams could indeed invert the OH levels then the observed
maser emissivities would be easily explained (Evans et al. 1979).

3.2. H_2O

The water molecule is an asymmetric rotor with a rather complex energy
level diagram (fig. 3). Maser radiation is detected in the transition
between the two excited states 6_{16} and 5_{23}, which are accidentally close
in energy (the transition frequency is 22 GHz or wavelength of 1.35cm).
    Unlike the case of HII/OH masers, it is not clear what is the
underlying object associated with H_2O masers. Although they usually
appear close to compact HII regions, they are not coincident with them.

H_2O

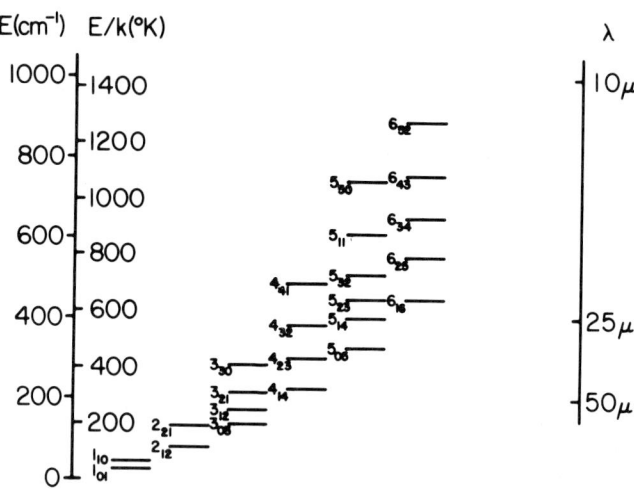

Figure 3.   Excited rotation levels of ortho H_2O with energies below
$1000 cm^{-1}$ and $J \leqslant 6$. The notation is $J_{K_- K_+}$, where J is the total angular
momentum and K_- and K_+ its projections on two molecular axes.

All attempts to detect either radio continuum or IR emission from the maser region have failed so far. The first detection of emission in any wavelength other then the H_2O 1.35cm line came recently, when Turner and Welch (1984) discovered enhanced HCN emission at the center of the H_2O maser region in W3(OH). This implies the existence of a high density ($\sim$ $10^6 - 10^7 cm^{-3}$), warm (T > 130K) clump, with a total mass of $\sim$ 2 - 3 $M_\odot$.

While the exact nature of the underlying source is not clear, there is good evidence that H_2O masers mark the phase of mass outflow from newly formed stars in the last stages of their pre-main-sequence evolution (Genzel and Downs, 1983). The maser features, whose line-widths are only $\sim$ 0.5 to 1.5 km/sec, cover a velocity range of a few hundreds km/sec. Other manifestations of this outflow phase are high velocity thermal molecular emission, IR emission in high excitation lines, optical "jets" and Herbig-Haro objects.

A couple of models that link the high velocity outflow and the maser effect have been proposed. The first one is the "interstellar bullets" of Norman and Silk (1979). In this model the maser knots are ram-pressure confined bullets, generated by instabilities in high velocity winds. The general scheme appears promising, although no detailed calculations of the inversion were performed. Tarter and Welch (1986) proposed recently that the masers occur at the interface of colliding clouds. The energy generated at the collision surface diffuses radiatively through the clouds and pumps the maser. The inversion scheme belongs to the "hot dust - cold gas" category of Goldreich and Kwan (1974) where the pumping is by the internal near-IR radiation of the dust. While Tarter and Welch manage to obtain the high power emitted from the maser sources, they are able to do this only by making the masers bigger than observed. Their model therefore fails to produce the high brightness temperatures.

This failure is not surprising and is shared by all radiative pumping schemes as well as collisions that obey the detailed balance relation (eq. 30). The problem can be traced directly to eq. (29) which shows that in such schemes, the temperature characterizing the pump can be enhanced during the maser process by $(\lambda_m/\lambda_p)^2$, at most. Pumping through the vibration transitions at $\lambda_p = 6.3\mu$ can boost the temperature by a factor of only 5×10^6, considerably less than the amount required to explain the observations. Tarter and Welch actually managed to do a bit better, because in their model the pump energy diffuses away from the collision plane. This way one effectively gets a series of maser sources, increasing the over all brightness temperature by ~10 above the strict thermodynamic limit. However, this is still not enough.

It appears that the only way to explain the extreme brightness temperatures of H_2O masers is by collisions violating the detailed balance relation. Such a scheme was proposed by Strelnitskij (1980, 1984), wherein the H_2O molecules collide simultaneously with neutral and charged particles held at different temperatures. Kylafis and Norman (1986) have recently performed detailed calculations utilizing this idea and were able to obtain high brightness temperatures, albeit at rather high densities ($\sim 10^{12} cm^{-3}$). They suggest that the pump conditions may be realized behind magnetic shock fronts. It is worth noting that in this environment, the inversion by charged particle streams, advocated

above for OH masers, might be applicable also for the H_2O masers.   If
that is the case, lower densities will probably suffice since the
inversion would involve direct collisions across the maser levels rather
than indirect circulation through other rotation states.

It is possible that the standard maser theory is in error in the
case of the strong H_2O masers.   The derivation of the linear expressions
in the standard phenomenological model (sec. 2.2.) assumes that the
bandwidth of the maser radiation exceeds the rates of all microscopic
processes in the problem.   This assumption eventually fails for the
induced processes, when the maser radiation becomes very intense, and
the relation for the variation of the gain with intensity becomes more
involved (Litvak 1970).   This may happen in the strong H_2O masers and it
is conceivable that their theoretical modeling will require some
modifications (Elitzur 1986).

3.3.   SiO

The SiO maser is unique in that its transitions occur in the excited
vibration states (figure 4), unlike the case of OH and H_2O.   This maser
therefore requires rather high temperatures for its excitation.   Maser
emission has been discovered in the Orion molecular cloud (Snyder and
Buhl 1974) and very recently in two more star forming regions (Hasegawa

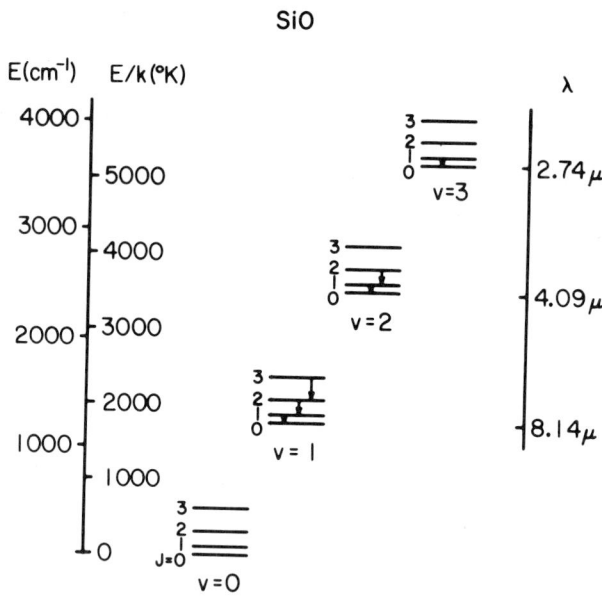

Figure 4.   The vibration levels of SiO with v and $J \leqslant 3$.   The rotation
energy separations were increased by factor 50.   Maser emission has been
detected in all the transitions marked by arrows, and also in the v = 2,
J = 3 → 2 and some of the higher rotation transitions.

et al. 1986). The latter two are less well studied.

The Orion SiO maser is located in the innermost part of the flow from IRc2, the center of activity in that well studied source. The maser is pumped by collisions with the neutral particles, and appears to probe the region where the motions acquire a high degree of order (Elitzur 1982b). It therefore provides a unique opportunity for studying the formation of the wind in a high velocity flow. Barvainis (1984) modeled the maser profile and polarization and concluded that it is located in an expanding-rotating disk. These results agree with other evidence for a disk structure in the immediate vicinity of IRc2 (Genzel 1986).

4. SUMMARY

All three strong masers are providing excellent probes of small scale structure in star forming regions. The difficulties in explaining these masers can be directly attributed to this fact -- so far there were not many other tools for high spatial resolution studies. The great progress in interferometric techniques may finally supplement maser observations with other small-scale-structure data. This should enable us to better understand the clues provided by maser radiation, ultimately leading to a better understanding of the emission regions and of the star formation process.

The partial support of NSF grant AST-8304895 is gratefully acknowledged.

REFERENCES

Alcock, C. and Ross, R.R. 1985, *Ap. J.* **290**, 433.
Baart, E.E., Cohen, R.J., Davies, R.D., Norris, R.P. and Rowland, P.R. 1986, *M.N.R.A.S.*, to be published.
Barvainis, R. 1984, *Ap. J.* **279**, 358.
Berulis, I.I. and Ershov, A.A. 1983, *Sov. Astr. Lett.*, 9, 341.
Draine, B.T. 1980, *Ap. J.* **241**, 1021.
Elitzur, M. 1979, *Astr. Ap.* 73, 322.
Elitzur, M. 1982a, *Rev. Mod. Phys.* **54**, 1225.
Elitzur, M. 1982b, *Ap. J.* **262**, 189.
Elitzur, M. 1986, in *Masers, Molecules and Mass Outflows in Star Forming Regions*, ed. A.D. Haschick (Haystack Observatory), p. 299.
Elitzur, M. and de Jong, T. 1978, *Astr. Ap.* **67**, 323.
Evans, N.J., Beckwith, S., Brown, R.L. and Gilmore, W. 1979, *Ap. J.* **227**, 450.
Garay, G., Reid, M.J. and Moran, J.M. 1985, *Ap. J.* **289**, 681.
Genzel, R. 1986, in *Masers, Molecules and Mass Outflows in Star Forming Regions*, ed. A.D. Haschick (Haystack Observatory), p. 233.
Genzel, R. and Downs, D. 1983, in *Highlights of Astronomy*, ed. R. West, vol. 6, 689.
Goldreich, P. and Keely, D.A. 1972, *Ap. J.* **174**, 517.

Goldreich, P., Keely, D.A. and Kwan, J. 1973, *Ap. J.* **179**, 111.

Goldreich, P. and Kwan, J. 1974, *Ap. J.* **191**, 93.

Haynes, R.F. and Caswell, J.L. 1977, *M.N.R.A.S.* **178**, 219.

Hasegawa, T. et al. 1986, in *Masers, Molecules and Mass Outflows in Star Forming Regions*, ed. A.D. Haschick (Haystack Observatory), p. 275.

Ho, P., Haschick, A.D., Vogel, S.N. and Wright, M. 1983, *Ap. J.* **265**, 295.

Johnston, I.D. 1967, *Ap. J.* **150**, 33.

Kylafis, N.D. and Norman, C. 1986, *Ap. J.* **300**, L73.

Lamb, W.E. 1964, *Phys. Rev.* **134**, 1492A.

Litvak, M.M. 1969, *Science* **165**, 855.

Litvak, M.M. 1970, *Phys. Rev.* **A 2**, 2107.

Norman, C. and Silk, J. 1979, *Ap. J.* **228**, 197

Norris, R.P. and Booth, R.S. 1981, *M.N.R.A.S.* **195**, 213.

Norris, R.P., Booth, R.S. and Diamond, P.J. 1982, *M.N.R.A.S.* **201**, 191.

Reid, M.J., Haschick, A.D., Burke, B.F., Moran, J.M., Johnston, K.J. and Swenson, G.W. 1980, *Ap. J.* **239**, 89.

Reid, M.J. and Moran, J.M. 1981, *Ann. Rev. Astr. Ap.* **19**, 231.

Rieu, N.Q., Winnberg, A., Guibert, J., Lepine, J.R.D., Johansson, L.E.B., and Goss, W.M. 1976, *Astr. Ap.* **46**, 413.

Snyder, L.E. and Buhl, D. 1974, *Ap. J.* **189**, L31.

Strelnitskij, V.S. 1980, *Pisma v. Astron. Zh.* **6**, 354.

Strelnitskij, V.S. 1984, *M.N.R.A.S.* **207**, 339.

Tarter, J.C. and Welch, W.J. 1986, *Ap. J.* (in press)

Thronson, A.H. and Harper, D.A. 1979, *Ap. J.* **230**, 133.

Turner, J.L. and Welch, W.J. 1984, *Ap. J.* **287**, L81.

Weinreb, S.A., Barrett, A.H., Meeks, M.L. and Henry, J.C. 1963, *Nature* **200**, 829.

Subject Index

David Hollenbach

Harley Thronson